The Mantle and Core

Citations
Please use the following example for citations:
Palme H. and O'Neill H.St.C. (2003) Cosmochemical estimates of mantle composition, pp. 1–38. In *The Mantle and Core* (ed. R.W. Carlson) Vol. 2 *Treatise on Geochemistry* (eds. H.D. Holland and K.K. Turekian), Elsevier–Pergamon, Oxford.

Cover photo: Thin section photograph of a garnet peridotite xenolith from the Monastery kimberlite in South Africa. Polarizers are only partially crossed to allow the large garnet in the upper left corner to remain transparent. (Photograph provided by Richard W. Carlson)

The Mantle and Core

Edited by

R. W. Carlson
Carnegie Institution of Washington, DC, USA

TREATISE ON GEOCHEMISTRY
Volume 2

Executive Editors

H. D. Holland
Harvard University, Cambridge, MA, USA

and

K. K. Turekian
Yale University, New Haven, CT, USA

ELSEVIER

2005

AMSTERDAM – BOSTON – HEIDELBERG – LONDON – NEW YORK – OXFORD
PARIS – SAN DIEGO – SAN FRANCISCO – SINGAPORE – SYDNEY – TOKYO

ELSEVIER B.V.
Radarweg 29
P.O. Box 211, 1000 AE Amsterdam
The Netherlands

ELSEVIER Inc.
525 B Street, Suite 1900
San Diego, CA 92101-4495
USA

ELSEVIER Ltd
The Boulevard, Langford Lane
Kidlington, Oxford OX5 1GB
UK

ELSEVIER Ltd
84 Theobalds Road
London WC1X 8RR
UK

First edition 2005

Library of Congress Cataloging in Publication Data
A catalog record is available from the Library of Congress.

British Library Cataloguing in Publication Data
A catalogue record is available from the British Library.

ISBN: 0-08-044848-8 (Paperback)

Printed and bound in the United Kingdom

Transferred to Digital Print 2011

DEDICATED TO

TED RINGWOOD (1930–1993)

Photograph provided by Roberta L. Rudnick

Contents

Executive Editors' Foreword

H. D. Holland

Harvard University, Cambridge, MA, USA

and

K. K. Turekian

Yale University, New Haven, CT, USA

Geochemistry has deep roots. Its beginnings can be traced back to antiquity, but many of the discoveries that are basic to the science were made between 1800 and 1910. The periodic table of elements was assembled, radioactivity was discovered, and the thermodynamics of heterogeneous systems was developed. The solar spectrum was used to determine the composition of the Sun. This information, together with chemical analyses of meteorites, provided an entry to a larger view of the universe.

During the first half of the twentieth century, a large number of scientists used a variety of methods to determine the major-element composition of the Earth's crust, and the geochemistries of many of the minor elements were defined by V. M. Goldschmidt and his associates using the then new technique of emission spectrography. V. I. Vernadsky founded biogeochemistry. The crystal structures of most minerals were determined by X-ray diffraction techniques. Isotope geochemistry was born, and age determinations based on radiometric techniques began to define the absolute geologic timescale. The intense scientific efforts during World War II yielded new analytical tools and a group of people who trained a new generation of geochemists at a number of universities. But the field grew slowly. In the 1950s, a few journals were able to report all of the important developments in trace-element geochemistry, isotopic geochronometry, the exploration of paleoclimatology and biogeochemistry with light stable isotopes, and studies of phase equilibria. At the meetings of the American Geophysical Union, geochemical sessions were few, none were concurrent, and they all ranged across the entire field.

Since then the developments in instrumentation and the increases in computing power have been spectacular. The education of geochemists has been broadened beyond the old, rather narrowly defined areas. Atmospheric and marine geochemistry have become integrated into solid Earth geochemistry; cosmochemistry and biogeochemistry have contributed greatly to our understanding of the history of our planet. The study of Earth has evolved into "Earth System Science," whose progress since the 1940s has been truly dramatic.

Major ocean expeditions have shown how and how fast the oceans mix; they have demonstrated the connections between the biologic pump, marine biology, physical oceanography, and marine sedimentation. The discovery of hydrothermal vents has shown how oceanography is related to economic geology. It has revealed formerly unknown oceanic biotas, and has clarified the factors that today control, and in the past have controlled the composition of seawater.

Seafloor spreading, continental drift and plate tectonics have permeated geochemistry. We finally understand the fate of sediments and oceanic crust in subduction zones, their burial and their exhumation. New experimental techniques at temperatures and pressures of the deep Earth interior have clarified the three-dimensional structure of the mantle and the generation of magmas.

Moon rocks, the treasure trove of photographs of the planets and their moons, and the successful search for planets in other solar systems have all revolutionized our understanding of Earth and the universe in which we are embedded.

Geochemistry has also been propelled into the arena of local, regional, and global anthropogenic problems. The discovery of the ozone hole came as a great, unpleasant surprise, an object lesson for optimists and a source of major new insights into the photochemistry and dynamics of the atmosphere. The rise of the CO_2 content of the atmosphere due to the burning of fossil fuels and deforestation has been and will continue to be at the center of the global change controversy, and will yield new insights into the coupling of atmospheric chemistry to the biosphere, the crust, and the oceans.

The rush of scientific progress in geochemistry since World War II has been matched by organizational innovations. The first issue of *Geochimica et Cosmochimica Acta* appeared in June 1950. The Geochemical Society was founded in 1955 and adopted *Geochimica et Cosmochimica Acta* as its official publication in 1957. The International Association of Geochemistry and Cosmochemistry was founded in 1966, and its journal, *Applied Geochemistry*, began publication in 1986. *Chemical Geology* became the journal of the European Association for Geochemistry.

The Goldschmidt Conferences were inaugurated in 1991 and have become large international meetings. Geochemistry has become a major force in the Geological Society of America and in the American Geophysical Union. Needless to say, medals and other awards now recognize outstanding achievements in geochemistry in a number of scientific societies.

During the phenomenal growth of the science since the end of World War II an admirable number of books on various aspects of geochemistry were published. Of these only three attempted to cover the whole field. The excellent *Geochemistry* by K. Rankama and Th.G. Sahama was published in 1950. V. M. Goldschmidt's book with the same title was started by the author in the 1940s. Sadly, his health suffered during the German occupation of his native Norway, and he died in England before the book was completed. Alex Muir and several of Goldschmidt's friends wrote the missing chapters of this classic volume, which was finally published in 1954.

Between 1969 and 1978 K. H. Wedepohl together with a board of editors (C. W. Correns, D. M. Shaw, K. K. Turekian and J. Zeman) and a large number of individual authors assembled the *Handbook of Geochemistry*. This and the other two major works on geochemistry begin with integrating chapters followed by chapters devoted to the geochemistry of one or a small group of elements. All three are now out of date, because major innovations in instrumentation and the expansion of the number of practitioners in the field have produced valuable sets of high-quality data, which have led to many new insights into fundamental geochemical problems.

At the Goldschmidt Conference at Harvard in 1999, Elsevier proposed to the Executive Editors that it was time to prepare a new, reasonably comprehensive, integrated summary of geochemistry. We decided to approach our task somewhat differently from our predecessors. We divided geochemistry into nine parts. As shown below, each part was assigned a volume, and a distinguished editor was chosen for each volume. A tenth volume was reserved for a comprehensive index:

(i) *Meteorites, Comets, and Planets*: Andrew M. Davis

(ii) *Geochemistry of the Mantle and Core*: Richard Carlson

(iii) *The Earth's Crust*: Roberta L. Rudnick

(iv) *Atmospheric Geochemistry*: Ralph F. Keeling

(v) *Freshwater Geochemistry, Weathering, and Soils*: James I. Drever

(vi) *The Oceans and Marine Geochemistry*: Harry Elderfield

(vii) *Sediments, Diagenesis, and Sedimentary Rocks*: Fred T. Mackenzie

(viii) *Biogeochemistry*: William H. Schlesinger

(ix) *Environmental Geochemistry*: Barbara Sherwood Lollar

(x) *Indexes*

The editor of each volume was asked to assemble a group of authors to write a series of chapters that together summarize the part of the field covered by the volume. The volume editors and chapter authors joined the team enthusiastically. Altogether there are 155 chapters and 9 introductory essays in the Treatise. Naming the work proved to be somewhat problematic. It is clearly not meant to be an encyclopedia. The titles *Comprehensive Geochemistry* and *Handbook of Geochemistry* were finally abandoned in favor of *Treatise on Geochemistry*.

The major features of the Treatise were shaped at a meeting in Edinburgh during a conference on Earth System Processes sponsored by the Geological Society of America and the Geological Society of London in June 2001. The fact that the Treatise is being published in 2003 is due to a great deal of hard work on the part of the editors, the authors, Mabel Peterson (the Managing Editor), Angela Greenwell (the former Head of Major Reference Works), Diana Calvert (Developmental Editor, Major Reference Works),

Bob Donaldson (Developmental Manager), Jerome Michalczyk and Rob Webb (Production Editors), and Friso Veenstra (Senior Publishing Editor). We extend our warm thanks to all of them. May their efforts be rewarded by a distinguished journey for the Treatise.

Finally, we would like to express our thanks to J. Laurence Kulp, our advisor as graduate students at Columbia University. He introduced us to the excitement of doing science and convinced us that all of the sciences are really subdivisions of geochemistry.

Contributors to Volume 2

C. J. Ballentine
University of Manchester, UK

V. C. Bennett
The Australian National University, Canberra, Australia

C. R. Bina
Northwestern University, Evanston, IL, USA

J. D. Blundy
University of Bristol, UK

J.-L. Bodinier
CNRS and Université de Montpellier 2, France

D. Canil
University of Victoria, BC, Canada

M. J. Drake
University of Arizona, Tucson, AZ, USA

Y. Fei
Carnegie Institution of Washington, DC, USA

M. Godard
CNRS and Université de Montpellier 2, France

E. H. Hauri
Carnegie Institution of Washington, DC, USA

D. R. Hilton
University of California San Diego, La Jolla, CA, USA

A. W. Hofmann
Max-Plank-Institut für Chemie, Mainz, Germany and
Lamont-Doherty Earth Observatory, Palisades, NY, USA

J. Li
University of Illinois at Urbana Champaign, IL, USA

R. W. Luth
University of Alberta, Edmonton, Canada

W. F. McDonough
University of Maryland, College Park, USA

J. D. Morris
Washington University, St. Louis, MO, USA

Hugh St. C. O'Neill
Australian National University, Canberra, ACT, Australia

H. Palme
Universität zu Köln, Germany

D. G. Pearson
Durham University, UK

D. Porcelli
University of Oxford, UK

K. Righter
NASA Johnson Space Center, Houston, TX, USA

J. G. Ryan
University of South Florida, Tampa, FL, USA

S. B. Shirey
Carnegie Institution of Washington, DC, USA

P. E. Van Keken
University of Michigan, Ann Arbor, MI, USA

M. J. Walter
Okayama University, Misasa, Japan

B. J. Wood
University of Bristol, UK

Volume Editor's Introduction

R. W. Carlson

Carnegie Institution of Washington, Washington, DC, USA

1 INTRODUCTION

Almost all of the Earth lies inaccessible beneath our feet. We have no hope in the foreseeable future of being able to drill deep enough to directly sample the mantle, let alone the core. Nevertheless, as this Volume attempts to demonstrate, we know much about the composition, and compositional variability, of both the mantle and core. Of course, this bold conclusion risks future discoveries that will render it more a statement of our arrogance than our understanding. Were Earth the static, rigid sphere envisioned only 50 years ago, we would have little hope of defining the composition of Earth's interior beyond the broad limits allowed by "remote sensing" techniques such as seismic imaging. With the recognition that new crust is being continually created at ocean ridges, and continually destroyed by subduction into the mantle at convergent margins, plate tectonic theory led to the realization that Earth's interior is in constant motion. This motion drives a slow two-way exchange of material between Earth's surface and interior. As a result, we have some direct, and many indirect, samples of Earth's mantle available at the surface for detailed study. Perhaps more important, the convective motions associated with plate tectonics appear to effectively stir if not the whole mantle, then at least a significant portion of the mantle. As a result, Earth's interior seems to have avoided the natural tendency of planetary bodies to undergo stable density stratification through chemical differentiation, with two important exceptions, the crust and the core. The effectiveness of mantle mixing allows us to use samples of the shallow mantle to deduce the compositional characteristics of the whole, or at least most of the mantle. With estimates of the average composition of Earth derived from analyses of the planetary objects from which Earth formed (Volume 1), coupled with our understanding of crustal compositions (Volume 3), knowledge of the bulk composition of the mantle then allows us to calculate the composition of the core by difference.

This overview offers a tour, with commentary, through the detailed discussions contained in the individual chapters of Volume 2. Our understanding of the composition of the mantle and core relies greatly on contributions from a range of fields beyond simple analytical chemistry applied to natural materials (Chapters 2.03–2.06, 2.11, and 2.13). Because Earth's interior cannot be directly placed in a mass spectrometer, many of its chemical properties must be inferred by less direct means. Astrophysics, cosmochemistry, and comparative planetology provide information on the allowable range of materials and processes involved in forming Earth from what was once a diffuse molecular cloud (Volume 1, and Chapters 2.01 and 2.15). Seismology provides a direct measure of the density and elastic properties of Earth's interior, which are sensitive to mineralogy and composition (Chapter 2.02). Experimental petrology and mineral physics determine the physical properties of candidate compositions subject to the high pressures and temperatures of Earth's interior and allow erupted magmas to serve as probes of their mantle source compositions (Chapters 2.02, 2.07, 2.08, and 2.14). Theoretical studies, in both geodynamics and element partitioning, allow understanding of the physical and chemical processes that determine how the composition of Earth's interior can be modified by partial melting and convective stirring (Chapters 2.09, 2.10, and 2.12).

2 SOLAR INHERITANCE

The inclusion of the subjects covered in Volume 1 of this Treatise illustrates the recognition that one critical avenue to understanding "geo" chemistry is to understand the solar environment in which Earth formed. Chapter 2.01 of this volume compares the composition of Earth with that of various primitive meteorite classes and with the spectroscopically determined composition of the Sun. Chemical variability in these meteorites reflects primarily two processes: (i) volatility and (ii) affinity for metal (the so-called siderophile elements) over silicate (lithophile elements). Perhaps the most surprising outcome of this comparison is that Earth's mantle has a bulk composition that is close to solar, at least for refractory lithophile elements. As detailed in Chapter 2.01, the mantle's most obvious departures from solar composition are its deficiencies in volatile and siderophile elements. The latter is easily understood in that Earth has a large metallic core that extracted the missing siderophile elements from the mantle (Chapter 2.15).

3 WHOLE-MANTLE CHARACTERISTICS

Because the part of the mantle that we can access (Chapters 2.04 and 2.05) shares compositional similarities with primitive meteorites and the Sun, the approach pursued in Chapter 2.01 has proved to be very powerful in predicting the composition expected for the bulk Earth, knowledge of which allows estimates of the degree to which mantle composition has been affected by core (Chapters 2.14 and 2.15) and crust (Chapter 2.03) formation. One must ask, however, whether the mantle that we can sample from the surface is representative of the mantle as a whole. To address this question, we turn to two quite different approaches: examination of Earth's interior with seismic waves (Chapter 2.02) and mass-balance modeling relating crust, mantle, and core compositions (Chapters 2.03 and 2.15).

3.1 Seismic Geochemistry

The speed at which seismic waves propagate through the Earth is a function of composition, mineralogy, and temperature (Chapter 2.02). As a geochemical tool, seismology is a blunt instrument. Only significant changes in major-element composition have a detectable effect on seismic wave velocities in the mantle, and the temperature dependence makes it difficult to distinguish unambiguously compositional from thermal effects. Also, given the wavelengths of seismic waves, whatever velocity, and hence compositional, information is returned is averaged over tens to thousands of kilometers, depending on the details of seismometer and earthquake spacing. Seismology thus provides only a low-resolution picture of the compositional variation in Earth's interior, but it offers one huge advantage over geochemical approaches; the depth of a particular seismic observation can be well determined. Geochemistry, with a few exceptions (Chapters 2.04 and 2.05), offers only vague constraints on the ultimate depth of origin of a material sampled at the surface.

Chapter 2.02 concludes that within broad bounds, an isochemical mantle with composition similar to that inferred from upper mantle rocks is consistent with the seismic properties of the whole mantle. This includes not only wave speeds, but also the appearance of seismic discontinuities at 410 km and 660 km depth that mark the expected phase transformations as upper mantle olivine changes first into spinel and then into perovskite structured high-pressure polymorphs. As discussed in Chapter 2.02, because the phase diagrams for these mineralogical transformations can be determined in the laboratory, the exact depth of the corresponding seismic discontinuities can be used as sensitive temperature and compositional probes in areas of expected thermal anomalies, e.g., near cold subducting plates and hot rising plumes.

3.2 Crust–Mantle Mass Balance

Beyond the broad major-element constraints afforded by seismic imaging, the abundance of many trace elements in the mantle clearly records the extraction of core (Chapters 2.01 and 2.15) and continental crust (Chapter 2.03). Estimates of the bulk composition of continental crust (Volume 3) show it to be tremendously enriched compared to any estimate of the bulk Earth in certain elements that are incompatible in the minerals that make up the mantle. Because the crust contains more than its share of these elements, there must be complementary regions in the mantle depleted of these elements—and there are. The most voluminous magmatic system on Earth, the mid-ocean ridges, almost invariably erupt basalts that are depleted in the elements that are enriched in the continental crust (Chapter 2.03). Many attempts have been made to calculate the amount of mantle depleted by continent formation, but the result depends on which group of elements is used and the assumed composition of both the crust and the depleted mantle. If one uses the more enriched estimates of bulk-continent composition, the less depleted estimates for average depleted mantle, and the most incompatible elements, then the mass-balance calculations allow the whole mantle to have been depleted by continent formation. If one uses elements that are not so severely enriched in the continental crust, for example, samarium and neodymium, then smaller volumes of depleted mantle are required in order to satisfy simultaneously the abundance of these elements in the continental crust and the quite significant fractionation of these elements in the depleted mantle as indicated by neodymium isotope systematics.

3.3 A Chemically Layered Mantle?

This latter result led to the suggestion that the depleted mantle was contained in the upper mantle, above the 660 km seismic discontinuity, and that the lower mantle was "primitive" in composition. Models of a chemically layered mantle derived from this approach have had a deep and long-lasting impact on solid Earth science, well beyond the level of certainty afforded by such mass balance modeling. As seismology began to convincingly resolve subducting oceanic plates penetrating through the 660 km discontinuity, and as the mid-mantle phase diagram was refined to show that the 660 km discontinuity need be no more than an expected phase change in an isochemical mantle (Chapter 2.02), support for the idea that this discontinuity represents a chemical boundary has faded, but not vanished. More recent approaches to this issue push the chemical boundary layer deeper in the mantle, to the point where seismology loses resolution on the fate of subducted slabs. As discussed in Chapter 2.06, a remaining observation that provides perhaps the strongest argument in favor of some minimally differentiated mantle remaining in Earth's interior comes from observations of the noble gases. Noble gases emitted from volcanoes in different tectonic settings show a wide range of isotopic compositions. The most relevant to the question of whether undegassed portions of the mantle remain are the high $^3He/^4He$ ratios found in a number of ocean islands and in a few continental hot-spot volcanoes. Unlike other radiometric systems where igneous differentiation results in melt and residue with complementary parent/daughter elemental ratios, the noble gases are assumed to be largely outgassed during eruption. Consequently, if helium is more incompatible than uranium during mantle melting, the residue will have a low He/U ratio and hence eventually a low $^3He/^4He$ ratio; but because of outgassing on eruption, the melt also will have a low He/U ratio and, if returned to the mantle by subduction, will also eventually have low $^3He/^4He$. As a result, the high $^3He/^4He$ observed in some hot-spot volcanoes is interpreted as a sign of the contribution from an undegassed, and by inference, undifferentiated, portion of the mantle.

4 THE IMPORTANCE OF RECYCLING

Crust–mantle chemical mass-balance models offer important constraints on compositional variations in the mantle, but their constraints on the size of the various reservoirs involved depend critically on uncertainties in the estimates of the bulk composition of the continental crust, the degree of depletion of the complementary depleted mantle, and the existence of "enriched" reservoirs in Earth's interior, for example, possibly significant volumes of subducted oceanic crust. This last item was left out of the mass-balance models that suggested that the upper and lower mantle are chemically distinct. Chapter 2.03 makes it clear that much of the chemical and isotopic heterogeneity observed in oceanic volcanic rocks reflects various mixtures of depleted mantle with different types of recycled subducted crust. With this realization, and excepting the noble gas evidence for undegassed mantle, some of the characteristics of what was once labeled "primitive" mantle are now recognized simply as points on mixing arrays between depleted mantle and enriched, recycled, crust.

4.1 The Fate of Crust Subducted into the Mantle

The fate of subducted crust is a critical issue in any consideration of the composition of the mantle. Over the age of the Earth, assuming a constant rate of production equal to the current rate, the volume of oceanic crust produced is over 10 times the current volume of the continental crust. If this amount of basaltic crust were permanently removed from the mantle, the remaining mantle would be much more depleted in incompatible elements than is observed even for the source of mid-ocean ridge basalts. Earth has a mechanism for erasing the effect of oceanic crust generation, and that mechanism is subduction of oceanic crust followed by convective stirring of the crust into the mantle. Chapter 2.12 presents calculations indicating that convective stirring is very efficient in the mantle, almost too efficient, in that the models have difficulty preserving heterogeneities for more than a billion years, yet isotopic model ages of oceanic basalts suggest preservation of heterogeneities for close to two billion years. One aspect of this discrepancy is that subducting slabs of oceanic crust are chemically processed by metamorphism, fluid release, and perhaps partial melting on their way into the mantle (Chapter 2.11). This processing results in convergent margin volcanism, through which material is transferred, sometimes permanently, from the oceanic to the continental crust. As a result, the crust that survives passage through the convergent margin and that is remixed into the mantle does not have the same composition as new crust produced at an ocean ridge. Studies of convergent margin volcanism and subduction-zone metamorphic rocks show a sequence of element modifications related to the degree of fluid mobility of different elements (Chapter 2.11). Some of the most fluid-mobile elements, such as boron, arsenic, and antimony may be completely returned to the overlying crust and hence not returned to the mantle at all. Other, less mobile elements, for example, uranium and thorium, may survive transit through convergent margin volcanism and may be returned in large part to the mantle.

4.2 Oceans out of, or into, the Mantle?

Perhaps the most important element to enter a subduction zone in abundance is hydrogen (as H_2O), but the fate of this element remains elusive in spite of concerted research efforts (Chapter 2.07). This is unfortunate, because water exerts great influence on the melting temperature and strength of the mantle. The return of water to the mantle through subduction may explain why Earth has been able to maintain a more or less continuous plate tectonic cycle through most of its history, while our planetary neighbors either died long ago (Mars, Mercury, and the Moon) or have found other, more episodic, ways in which to release their interior heat (Venus). Not too many years ago, there was a dearth of known mineral phases that could transport water past the upper 150 km of the mantle, with the consequent deduction that water was quantitatively stripped from a subducting plate. Chapter 2.07 describes a number of water-bearing minerals and shows that the pressure–temperature stability of many of these do lead to breakdown and dewatering in the shallow mantle. Whether or not all of this water returns to the surface, however, is a matter of some debate. Recent discoveries of a number of mineral phases containing structural water that are stable to high pressure and temperature, along with the recognition that normally anhydrous mantle minerals, such as olivine and garnet, can dissolve large amounts of hydrogen, potentially allow for the transport of substantial quantities of water into the deep mantle (Chapter 2.07). Similar arguments can be made for CO_2, and the presence of carbon isotopic compositions in some diamonds that overlap those of biogenic sediments (Chapter 2.05) indicates that carbon, like water, can find mineral carriers that allow its transport into the deep mantle.

5 THE CHEMICAL CONSEQUENCES OF PARTIAL MELTING

To this point, the discussion has focused on the broad geochemical properties of the mantle. As with many earth-science problems, the question of the degree of compositional variability in the mantle is a function of the size scale being considered. Although the mantle may well be moderately homogeneous at the tens of km^3 volumes sampled by melting along mid-ocean ridges, the heterogeneous nature of the mantle becomes more obvious at the meter to kilometer scale shown in exposed mantle sections (Chapter 2.04) and particularly at the grain-size scales at which mantle xenoliths (Chapter 2.05) are studied. The most obvious conclusion to be derived from examination of exposed mantle sections is that, over centimeter to meter length scales, the mantle shows obvious effects of chemical differentiation through the extraction and migration of partial melts. Veins and pods of various clinopyroxene-rich lithologies testify to the ubiquity of "basaltic" components in the mantle, and their melt-depleted complements, dunites and harzburgites, are abundant. Chapter 2.08 documents that the compositional variability shown by modern abyssal peridotites is consistent with the notion that they are the residues of partial melt extraction at mid-ocean ridges under physical (temperature, pressure, degree of melting) conditions consistent with those deduced from a variety of direct geophysical measurements and indirectly from the results of experimental petrology for the generation of mid-ocean ridge basalts.

5.1 Controls on Element Partitioning

A factor that complicates this issue is that trace-element studies of mid-ocean ridge basalts suggest that the initiation of melting takes place at depths sufficient to have garnet present in the mantle source, whereas the experimental petrology of the erupted basalts is better fit by shallower melting. The argument about the presence of garnet depends critically on the precise values used for the melt–solid distribution coefficients of key elements in garnet and clinopyroxene. Chapter 2.09 points out that because of the changing composition of clinopyroxene with pressure, the partitioning characteristics of clinopyroxene may approach those of garnet at moderate mantle depths, at least to the degree necessary to explain the "garnet" signature in mid-ocean ridge basalts. Whatever the eventual resolution of this question, future advances in the modeling of trace-element behavior during partial melting and fractional crystallization will have to include changes in the partitioning behavior of minerals as the pressure, temperature, and composition of the melting region or magma chamber changes during progressive melting or crystallization. The theoretical framework presented in Chapter 2.09 offers a means to make systematized predictions of element partitioning behavior that potentially can be incorporated into thermodynamics-based computer models of melting and crystallization and thus allow prediction not only of major-element trends, but also trace-element behavior during the melt evolution.

6 IS WHAT WE SEE TODAY THE WAY IT HAS ALWAYS BEEN?

6.1 The Effect of Secular Changes in Mantle Temperature

Geochemistry and geophysics are providing quite consistent pictures of the conditions of melting in Earth's most voluminous active volcanic system, the mid-ocean ridges. Determining whether these conditions have been maintained over Earth history is a job for geochemistry. Most xenoliths from the continental mantle are residues of partial melt extraction (Chapter 2.05). The Re–Os isotopic system tracks this type of differentiation event, dating the melt depletion events experienced by subcontinental mantle. Chapter 2.05 documents the longevity of mantle beneath continents, providing examples of ca. 3 Ga mantle underlying Archean crustal sections of southern Africa, Siberia, and North America. Because the degree of melt depletion experienced by old lithospheric mantle imparts compositional buoyancy, these thick continental lithospheric keels play an important role in the survivability of old continental crust. Mantle xenoliths sample melting events that occurred at various times in Earth history; hence their compositions can be used to examine changes in the conditions of melting with time. Both Chapters 2.05 and 2.08 show that mantle xenoliths indicate a secular decrease in degree of melt extraction with time. Modern melt regions, for example, as sampled in abyssal peridotites, have compositions suggestive of between 5% to 25% partial melt extraction at pressures between 1 GPa and 2 GPa (Chapter 2.08). In contrast, samples of Archean mantle indicate much higher average degrees (ca. 45%) and greater pressures (4.5 GPa) of melt extraction. If explained purely as the consequence of a cooling mantle rather than a dehydrating one, this change in average extent of melting suggests that the mantle has cooled by ~350 °C since the Archean (Chapter 2.08). Though the difference in average extent of melting between Archean and modern mantle samples is significant, both data sets indicate that the partial melt extraction affecting most mantle samples has occurred at depths of less than 150 km throughout at least the last 3.5 Gyr of Earth history. Coupled with the evidence for recycling of crustal materials to serve as "enriched" sources for some types of mantle-derived volcanism (Chapter 2.03), this suggests that the main process causing chemical variation in the Earth's interior is the shallow melting and crustal recycling typical of the plate tectonic cycle.

6.2 Magma Oceanography

Where this uniformitarian model is least likely to apply is early in Earth history. The mere existence of the core, separated from the mantle within 30 Myr of Earth formation (Chapters 2.14 and 2.15), along with evidence in atmospheric xenon for the decay products of short-lived ^{129}I and ^{244}Pu, testify to the importance of early differentiation events in the history of both core–mantle and solid-earth–atmosphere separation.

Our study of neighboring planetary objects, including the Moon, Mars, and the source of the differentiated meteorites, indicates the importance of early events in determining their compositions. A number of lines of evidence suggest that a significant fraction of the Moon experienced an event that segregated a thick crust from a compositionally layered lunar mantle by 4.50 Ga, and a similar story is inferred for meteorites possibly derived from Mars and the large parent asteroid of the eucrites and diogenites. Our planetary neighbors thus suggest that Earth experienced significant chemical differentiation early in its history, but the chemical evidence of this event in the current mantle is subdued at best. Perhaps the most convincing evidence for a terrestrial magma ocean comes from an analysis of the abundance of moderately siderophile elements in the mantle (Chapter 2.10). The Ni/Co ratio in the mantle is well established to be close to chondritic, yet nickel is much more siderophile than cobalt at upper mantle temperatures and pressures. Consequently, core formation should have left the mantle with low Ni/Co. One explanation for the higher than expected Ni/Co of the mantle is that the partition coefficients for nickel and cobalt between metal and silicate become similar at elevated pressure and temperature. Chapter 2.10 discusses the experimental results that show this to be the case and arrives at the conclusion that the mantle Ni/Co was established during a magma ocean event early in Earth history that resulted in metal–silicate equilibrium at 40–50 GPa and 3,000 °C.

Why other expected features of magma ocean differentiation are missing from the Earth is unclear, though the effective role of mantle convection (Chapter 2.12) in mixing the mantle is a suspect. Mantle mixing is not an instantaneous process; consequently, rocks formed early in Earth history may retain a better record of the chemical effects of early differentiation events than would modern mantle samples. Examining the ancient rock record for the signature of early differentiation events is the approach followed in Chapter 2.13. The difficulty with this approach is that rocks that have survived for more than three billion years at the Earth's surface have not had an easy life. The continuing dynamic activity of our planet has ensured that most of these old rocks have experienced at least one, if not many, episodes of alteration, metamorphism and/or metasomatism. As a result, discerning the composition of these rocks when they were emplaced at Earth's surface in the Archean is not straightforward. With these difficulties in mind, Chapter 2.13 shows that the degree of chemical heterogeneity in the mantle between 3 Ga and 4 Ga may well have been greater than at present. If so, this is further testimony to the efficiency of mantle convection in erasing the evidence for chemical differentiation of the solid Earth, but it is also tantalizing evidence that at least some fraction of Earth's chemical characteristics may have been influenced by events that occurred within the first few hundred million years of Earth history.

7 PROBING THE CORE

One clear consequence of early Earth differentiation is the separation of the core from the mantle. Much of what we know about the composition of the core comes from comparison with meteoritic compositional trends, from the seismologically determined density of the core, and from the recent experimental determinations of possible core materials at pressures and temperatures appropriate to the Earth's core (Chapter 2.14). Iron metal is the clear top candidate for being the major element of the core based on solar abundance considerations, meteoritic analogues, and a nearly appropriate density. One of the key issues that is just beyond the grasp of modern experimental techniques is determination of the melting temperature of iron alloys at pressures corresponding to the inner–outer core boundary. The coexistence of liquid and solid at this boundary will provide a unique pinning point for the temperature deep in Earth's interior if an appropriate composition can be found. The ~1,800 °C range in estimates of the melting point of iron at the inner core boundary in table 1 of Chapter 2.14 suggests the difficulty of this measurement with current techniques.

Iron alone has been known for decades to be too dense compared to seismically estimated outer-core densities, instigating speculation regarding the "light element" that must be in the core. This question can be addressed from both experimental (Chapter 2.14) and cosmochemical (Chapter 2.15) approaches. Answers to this question will provide not only a better estimate of core composition that can be used in experiments studying the properties of core material at appropriate pressure and temperature conditions, but also could provide important information on the conditions of core formation. For example, if silicon is the light element, this may suggest that the early Earth was much more reducing than the modern mantle. Conversely, if oxygen is the light element in the core, then an oxidizing mantle was present essentially from the beginning of Earth history. A further consideration is whether radioactive species such as potassium and uranium are present in the core, since these could provide the energy necessary to maintain a liquid outer core and to drive the geodynamo that creates Earth's magnetic field.

Neither element is soluble in iron metal at low pressures, but experimental evidence suggests that at high-pressure potassium might be soluble in sulfide melts, which could partition into the core (Chapter 2.14). Conversely, the mass-balance arguments presented in Chapter 2.15 suggest that only an insignificant portion of Earth's potassium can be sequestered in the core. This difference of opinion, along with many others that appear throughout the volume, are harbingers of future research that will expand our understanding of the geochemistry of Earth's interior.

8 CLOSING THOUGHTS

In closing, I would like to thank all the authors in this volume for finding the time in their busy schedules to put together the comprehensive summaries contained herein, and the reviewers of these lengthy tomes for working to ensure the quality and completeness of the coverage. Their efforts provide an in-depth look at the current state of knowledge concerning the geochemistry of the mantle and core. I hope that the reader finds these discussions as interesting and as educational as I have found them in my role as editor.

2.01
Cosmochemical Estimates of Mantle Composition

H. Palme
Universität zu Köln, Germany
and
Hugh St. C. O'Neill
Australian National University, Canberra, ACT, Australia

2.01.1 INTRODUCTION AND HISTORICAL REMARKS

In 1794 the German physicist Chladni published a small book in which he suggested the extraterrestrial origin of meteorites. The response was skepticism and disbelief. Only after additional witnessed falls of meteorites did scientists begin to consider Chladni's hypothesis seriously. The first chemical analyses of meteorites were published by the English chemist Howard in 1802, and shortly afterwards by Klaproth, a professor of chemistry in Berlin. These early investigations led to the important conclusion that meteorites contained the same elements that were known from analyses of terrestrial rocks. By the year 1850, 18 elements had been identified in meteorites: carbon, oxygen, sodium, magnesium, aluminum, silicon, phosphorous, sulfur, potassium, calcium, titanium, chromium, manganese, iron, cobalt, nickel, copper, and tin (Burke, 1986). A popular hypothesis, which arose after the discovery of the first asteroid Ceres on January 1, 1801 by Piazzi, held that meteorites came from a single disrupted planet between Mars and Jupiter. In 1847 the French geologist Boisse (1810–1896) proposed an elaborate model that attempted to account for all known types of meteorites from a single planet. He envisioned a planet with layers in sequence of decreasing densities from the center to the surface. The core of the planet consisted of metallic iron surrounded by a mixed iron–olivine zone. The region overlying the core contained material similar to stony meteorites with ferromagnesian silicates and disseminated grains of metal gradually extending into shallower layers with aluminous silicates and less iron. The uppermost layer consisted of metal-free stony meteorites, i.e., eucrites or meteoritic basalts. About 20 years later, Daubrée (1814–1896) carried out experiments by melting and cooling meteorites. On the basis of his results, he came to similar conclusions as Boisse, namely that meteorites come from a single, differentiated planet with a metal core, a silicate mantle, and a crust. Both Daubrée and Boisse also expected that the Earth was composed of a similar sequence of concentric layers (see Burke, 1986; Marvin, 1996).

At the beginning of the twentieth century Harkins at the University of Chicago thought that meteorites would provide a better estimate for the bulk composition of the Earth than the terrestrial rocks collected at the surface as we have only access to the "mere skin" of the Earth. Harkins made an attempt to reconstruct the composition of the hypothetical meteorite planet by compiling compositional data for 125 stony and 318 iron meteorites, and mixing the two components in ratios based on the observed falls of stones and irons. The results confirmed his prediction that elements with even atomic numbers are more abundant and therefore more stable than those with odd atomic numbers and he concluded that the elemental abundances in the bulk meteorite planet are determined by nucleosynthetic processes. For his meteorite planet Harkins calculated Mg/Si, Al/Si, and Fe/Si atomic ratios of 0.86, 0.079, and 0.83, very closely resembling corresponding ratios of the average solar system based on presently known element abundances in the Sun and in CI-meteorites (see Burke, 1986).

If the Earth were similar compositionally to the meteorite planet, it should have a similarly high iron content, which requires that the major fraction of iron is concentrated in the interior of the Earth. The presence of a central metallic core to the Earth was suggested by Wiechert in 1897. The existence of the core was firmly established using the study of seismic wave propagation by Oldham in 1906 with the outer boundary of the core accurately located at a depth of 2,900 km by Beno Gutenberg in 1913. In 1926 the fluidity of the outer core was finally accepted. The high density of the core and the high abundance of iron and nickel in meteorites led very early to the suggestion that iron and nickel are the dominant elements in the Earth's core (Brush, 1980; see Chapter 2.15).

Goldschmidt (1922) introduced his zoned Earth model. Seven years later he published details (Goldschmidt, 1929). Goldschmidt thought that the Earth was initially completely molten and separated on cooling into three immiscible liquids, leading on solidification to the final configuration of a core of FeNi which was overlain by a sulfide liquid, covered by an outer shell of silicates. Outgassing during melting and crystallization produced the atmosphere. During differentiation elements would partition into the various layers according to their geochemical character. Goldschmidt distinguished four groups of elements: siderophile elements preferring the metal phase, chalcophile elements preferentially partitioning into sulfide, lithophile elements remaining in the silicate shell, and atmophile elements concentrating into the atmosphere. The geochemical character of each element was derived from its abundance in the corresponding phases of meteorites.

At about the same time astronomers began to extract compositional data from absorption line spectroscopy of the solar photosphere, and in a review article, Russell (1941) concluded: "The average composition of meteorites differs from that of the earth's crust significantly, but not very greatly. Iron and magnesium are more abundant and nickel and sulfur rise from subordinate positions to places in the list of the first ten. Silicon, aluminum, and the alkali metals, especially potassium, lose what the others gain." And Russell continued: "The composition of the earth as a whole is probably much more similar to the meteorites than that of its 'crust'." Russell concludes this paragraph by a statement on the composition of the core: "The known properties of the central core are entirely consistent with the assumption that it is composed of molten iron—though not enough to prove it. The generally accepted belief that it is composed of nickel–iron is based on the ubiquitous appearance of this alloy in metallic meteorites," and, we should add, also on the abundances of iron and nickel in the Sun.

Despite the vast amount of additional chemical data on terrestrial and meteoritic samples and despite significant improvements in the accuracy of solar abundances, the basic picture as outlined by Russell has not changed. In the following sections we will demonstrate the validity of Russell's assumption and describe some refinements in the estimate of the composition of the Earth and the relationship to meteorites and the Sun.

2.01.2 THE COMPOSITION OF THE EARTH'S MANTLE AS DERIVED FROM THE COMPOSITION OF THE SUN

The rocky planets of the inner solar system and the gas-rich giant planets with their icy satellites of the outer solar system constitute the gross structure of the solar system: material poor in volatile components occurs near the Sun, while the outer parts are rich in water and other volatile components. The objects in the asteroid belt, between Mars at 1.52 AU and Jupiter at 5.2 AU (1 AU is the average Earth–Sun distance, ca. 1.5×10^8 km), mark the transition between the two regimes. Reflectance spectroscopy of asteroids shows bright silicate-rich, metal-containing objects in the inner belt and a prevalence of dark icy asteroids in the outer parts (Bell *et al.*, 1989). Apart from the structure of the asteroid belt, there is little evidence for compositional gradients within the inner solar system as represented by the terrestrial planets and the asteroids. There are no systematic variations with distance from the Sun, either in the chemistry of the inner planets, Mercury, Venus, Earth, Moon, Mars, and including the fourth largest asteroid Vesta, or in any other property (Palme, 2000). One reason is the substantial radial mixing that must have occurred when the terrestrial planets formed. In current models of planet formation, the Earth is made by collisions of a hundred or more Moon- to Mars-sized embryos, small planets that formed within a million years from local feeding zones. The growth of the Earth and the other inner planets takes tens of millions of years, and material from various heliocentric distances contributes to the growth of the planets (e.g., Wetherill, 1994; Canup and Agnor, 2000; Chambers, 2001). Hence, no large and systematic changes in chemistry or isotopic composition of planets are to be expected with increasing heliocentric distance. The composition of the Earth will therefore allow conclusions regarding the formation and the origin of a large fraction of matter in the inner solar system, in particular when considering that the Earth has more than 50% of the mass of the inner solar system. One should, however, keep in mind that the inner solar system comprises only a very small fraction of the total solar system which extends beyond the orbit of Pluto (39.5 AU) out to some 500 AU. This region is now called Kuiper belt, a disk of comets orbiting the Sun in the plane of the ecliptic. Lack of a chemical gradient in the inner solar system (2 AU) therefore, does, not allow any conclusions regarding the entire solar system.

The abundances of all major and many minor and trace elements in the Sun are known from absorption line spectrometry of the solar photosphere. Accuracy and precision of these data have continuously improved since the early 1930s. In a compilation of solar abundances, Grevesse and Sauval (1998) list the abundances of 41 elements with estimated uncertainties below 30%. A first approximation to the composition of the bulk Earth is to assume that the Earth has the average solar system composition for rock-forming elements, i.e., excluding extremely volatile elements such as hydrogen, nitrogen, carbon, oxygen, and rare gases.

The six most abundant, nonvolatile rock-forming elements in the Sun are: Si (100), Mg (104), Fe (86), S (43), Al (8.4), and Ca (6.2). The numbers in parentheses are atoms relative to 100 Si atoms. They are derived from element abundances in CI-meteorites which are identical to those in the Sun except that CI-abundances are better known (see Chapter 1.03). From geophysical measurements it is known that the Earth's core accounts for 32.5% of the mass of the Earth. Assuming that the core contains only iron, nickel, and sulfur allows us to calculate the composition of the silicate fraction of the Earth by mass balance. This is the composition of the bulk silicate earth (BSE) or the primitive earth mantle (PM). The term primitive implies the composition of the Earth's mantle before crust and after core formation.

As the mantle of the Earth contains neither metal nor significant amounts of Fe^{3+}, the sum of all oxides (by weight) must add up to 100%:

$$MgO + SiO_2 + Al_2O_3 + CaO + FeO = 100 \quad (1)$$

By inserting into (1) average solar system abundance ratios, e.g., Si/Mg, Ca/Mg, and Al/Mg (see Chapter 1.03), one obtains

$$2.62 \times MgO + FeO = 100 \quad (2)$$

Considering that iron is distributed between mantle and core, the mass balance for iron can be written as

$$Fe_{core} \times 0.325 + Fe_{mantle} \times 0.675 = Fe_{total} \quad (3)$$

and similarly for magnesium, assuming a magnesium-free core,

$$Mg_{mantle} \times 0.675 = Mg_{total} \quad (4)$$

By assuming that sulfur is quantitatively contained in the core and accounting for nickel in the core ($Fe_{total}/Ni = 17$), the amount of iron

Table 1 Composition of the mantle of the Earth assuming average solar system element ratios for the whole Earth.

	Earth's mantle solar model	*Earth's mantle based on composition of upper mantle rocks*[a]
MgO	35.8	36.77
SiO_2	51.2	45.40
FeO	6.3	8.10
Al_2O_3	3.7	4.49
CaO	3.0	3.65

[a] See Table 2.

in the core, Fe_{core}, is calculated to be 75%. From Equations (2) to (4) and by using the solar abundance ratio for Fe_{total}/Mg_{total}, the hypothetical composition of the Earth's mantle is obtained as given in Table 1. The Earth's mantle composition derived from the analyses of actual upper mantle rocks, as described in the next section, is listed for comparison.

The remarkable result of this exercise is that by assuming solar element abundances of rock-forming elements in the bulk Earth leads to a mantle composition that is in basic agreement with the mantle composition derived from upper mantle rocks. On a finer scale there are differences between the calculated mantle composition of the solar model and the mantle composition derived from the analyses of mantle rocks that makes a solar model of the Earth unlikely: (i) the Ca/Mg and Al/Mg ratios of mantle rocks are significantly higher than in the Sun, reflecting a general enhancement of refractory elements in the Earth; (ii) the Mg/Si ratio of the Earth's mantle is different from the solar ratio; and (iii) minor and trace element data indicate that the Earth is significantly depleted in volatile elements when compared to solar abundances. Hence, the sulfur content of the core assuming the solar composition model for the Earth is grossly overestimated. In order to assess the significance of these differences properly, some understanding of the variability of the chemical composition of chondritic meteorites is required.

2.01.3 THE CHEMICAL COMPOSITION OF CHONDRITIC METEORITES AND THE COSMOCHEMICAL CLASSIFICATION OF ELEMENTS

The thermal history of planetesimals accreted in the early history of the solar system depends on the timescale of accretion and the amount of incorporated short-lived radioactive nuclei, primarily ^{26}Al with a half-life of 7.1×10^5 years. In some cases there was sufficient heat to completely melt and differentiate the planetesimals into a metal core and a silicate mantle, while in other cases even the signatures of a modest temperature increase is absent. *Undifferentiated meteorites* derived from unmelted parent bodies are called *chondritic meteorites*. Their chemical composition derives from the solar composition more or less altered by processes in the early solar nebula, but essentially unaffected by subsequent planetary processes. Their textures, however, may have been modified by thermal metamorphism or aqueous alteration since their formation from the nebula. Chemically, chondritic meteorites are characterized by approximately similar relative abundances of silicon, magnesium, and iron indicating the absence both of metal separation (which would lead to low Fe/Si in residual silicates) and of partial melting, which would produce low Mg/Si melts and high Mg/Si residues. The roughly solar composition is, however, only a first-order observation. Nebular processes, primarily fractionation during condensation and/or aggregation, have produced a rather wide range of variations in the chemistry of chondritic meteorites. The extent to which individual elements are affected by these processes depends mainly on their volatility under nebular conditions. Nebular volatilities are quantified by condensation temperatures formally calculated for a gas of solar composition (see Wasson, 1985). In Table 2, elements are grouped according to their condensation temperatures. In addition, the geochemical character of each element is indicated, i.e., whether it is lithophile or siderophile and/or chalcophile. Based on condensation temperatures the elements may be grouped into five categories that account for variations in the bulk chemistry of chondritic meteorites (Larimer, 1988).

(1) The refractory component comprises the elements with the highest condensation temperatures. There are two groups of refractory elements: the *refractory lithophile* elements (RLEs)—aluminum, calcium, titanium, beryllium, scandium, vanadium, strontium, yttrium, zirconium, niobium, barium, REE, hafnium, tantalum, thorium, uranium, plutonium—and the *refractory siderophile* elements (RSEs)—molybdenum, ruthenium, rhodium, tungsten, rhenium, iridium, platinum, osmium. The refractory component accounts for ~5% of the total condensible matter. Variations in refractory element abundances of bulk meteorites reflect the incorporation of variable fractions of a refractory aluminum, calcium-rich component. Ratios among refractory lithophile elements are constant in all types of chondritic meteorites, at least to within ~5%.

(2) The *common lithophile* elements, magnesium and silicon, condense as magnesium silicates with chromium and lithium in solid solution. Together with metallic iron, magnesium

Table 2 Cosmochemical and geochemical classification of the elements.

	Elements	
	Lithophile (silicate)	*Siderophile + chalcophile (sulfide + metal)*
Refractory	T_c = *1,850–1,400* K Al, Ca, Ti, Be, Ba, Sc, V, Sr, Y, Zr, Nb, Ba, REE, Hf, Ta, Th, U, Pu	Mo, Ru, W, Re, Os, Ir, Pt
Main component	T_c = *1,350–1,250* K Mg, Si, Cr, Li	Fe, Ni, Co, Pd
Moderately volatile	T_c = *1,230–640* K Mn, P, Na, B, Rb, K, F, Zn	Au, As, Cu, Ag, Ga, Sb, Ge, Sn, Se, Te, S
Highly volatile	T_c < *640* K Cl, Br, I, Cs, Tl, H, C, N, O, He, Ne, Ar, Kr, Xe	In, Bi, Pb, Hg

T_c—condensation temperatures at a pressure of 10^{-4} bar (Wasson, 1985; for B, Lauretta and Lodders, 1997).

silicates account for more than 90% of the mass of the objects of the inner solar system. Variations in Mg/Si ratios among meteorites may be ascribed to separation of early condensed magnesium-rich olivine (forsterite with an atom ratio of Mg/Si = 2), either by preferred accretion of forsterite (high Mg/Si reservoir) or loss of an early forsterite component (low Mg/Si reservoir).

(3) In a gas of solar composition all iron condenses as an *FeNi alloy*, which includes cobalt and palladium with similar volatilities. Parent bodies of chondritic meteorites have acquired variable amounts of metallic FeNi. Some separation of metal and silicate must have occurred in the solar nebula, before aggregation to planetesimals had occurred.

(4) The *moderately volatile elements* are in order of decreasing condensation temperature: gold, manganese, arsenic, phosphorus, rubidium, copper, potassium, sodium, silver, gallium, antimony, boron, germanium, fluor, tin, selenium, tellurium, zinc, and sulfur (50% condensation temperatures; Wasson, 1985). The condensation temperatures range between those of magnesium silicates and FeS. All elements of this group are trace elements except the minor element, sulfur, which condenses by the reaction of gaseous S_2 with solid iron metal. The trace elements are condensed by dissolution into already condensed major phases such as silicates, metal, and sulfides. In general, abundances of moderately volatile elements normalized to average solar abundances decrease in most groups of chondritic meteorites with decreasing condensation temperatures, i.e., the more volatile an element is, the lower is its normalized abundance (Palme *et al.*, 1988).

(5) The group of *highly volatile elements* with condensation temperatures below that of FeS includes the lithophile elements—chlorine, bromine, iodine, caesium, thallium—the chalcophile elements—indium, bismuth, lead, mercury—and the atmophile elements—hydrogen, carbon, oxygen, neon, argon, krypton, xenon. The lithophile and siderophile elements of this group are fully condensed in CI-chondrites, whereas the atmophile elements are depleted in all groups of meteorites, even in CI-meteorites relative to solar abundances (see Chapter 1.03).

Figure 1 plots the abundances of several elements in the various groups of chondritic meteorites and in the Sun. All abundances are normalized to silicon. One group of meteorites, the CI-chondrites, with its most prominent member the Orgueil meteorite, has, for most elements shown here, the same abundance ratios as the Sun (Figure 1). The agreement between the composition of the Sun and CI-meteorites holds for most elements heavier than oxygen and with the exception of rare gases (see Chapter 1.03). Differences among the various types of chondritic meteorites include differences in chemical composition (Figure 1): mineralogy, texture, and degree of oxidation. More recently, new chondrite groups have been discovered (see Chapter 1.05). Only the most important groups are shown here. The compositional variations observed among chondritic meteorites may be explained in terms of the five components mentioned above.

In Figure 1 aluminum is representative of the refractory component. All types of carbonaceous chondrites are enriched in refractory elements, whereas ordinary and enstatite chondrites are depleted. Variations in Mg/Si (Figure 1) are smaller and may be the result of preferred accumulation or loss of olivine as discussed

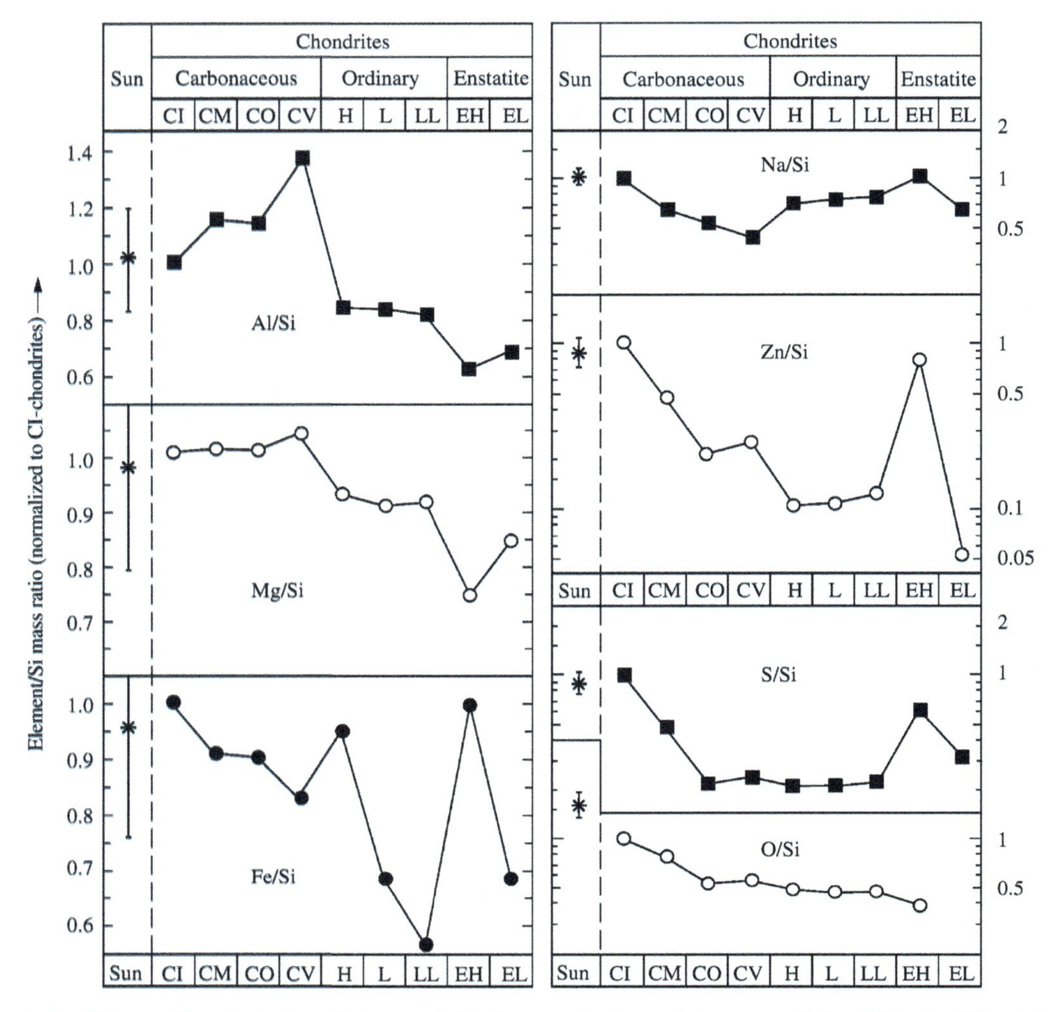

Figure 1 Element/Si mass ratios of characteristic elements in the major groups of chondritic (undifferentiated) meteorites. Meteorite groups are arranged according to decreasing oxygen content. The best match between solar abundances and meteoritic abundances is with CI-meteorites. For classification of meteorites, see Chapter 1.05.

above. The variations in Fe/Si ratios reflect variable incorporation of metallic iron into a chondrite parent body. Variations in the moderately volatile elements—sodium, zinc, and sulfur—are large and demonstrate the uniqueness of CI-chondrites as best representing solar abundances. The decrease in O/Si ratios (Figure 1) indicates average decreasing oxygen fugacities from carbonaceous through ordinary to enstatite chondrites. The average fayalite content of olivine (or ferrosilite content of orthopyroxene) decreases in the same sequence.

In the following section we will derive the composition of the mantle of the Earth from the chemical analyses of upper mantle rocks. The resulting mantle composition will then be compared with the composition of chondritic meteorites. In order to avoid circular arguments, we will use as few assumptions based on the above cosmochemical observations as possible. However, some assumptions are essential, and these will be clearly indicated.

2.01.4 THE COMPOSITION OF THE PRIMITIVE MANTLE BASED ON THE ANALYSIS OF UPPER MANTLE ROCKS

2.01.4.1 Rocks from the Mantle of the Earth

The principal division of the Earth into core, mantle, and crust is the result of two fundamental processes. (i) The formation of a metal core very early in the history of the Earth. Core formation ended at ~30 million years after the beginning of the solar system (Kleine *et al.*, 2002). (ii) The formation of the continental crust by partial melting of the silicate mantle. This process has occurred with variable intensity throughout the history of the Earth.

The least fractionated rocks of the Earth are those that have only suffered core formation but have not been affected by the extraction of partial melts during crust formation. These rocks should have the composition of the PM, i.e., the mantle before the onset of crust formation. Such rocks are typically high in MgO and low in Al_2O_3, CaO, TiO_2, and other elements incompatible with mantle minerals. Fortunately, it is possible to collect samples on the surface of the Earth with compositions that closely resemble the composition of the primitive mantle. Such samples are not known from the surfaces of Moon, Mars, and the asteroid Vesta. It is, therefore, much more difficult to reconstruct the bulk composition of Moon, Mars, and Vesta based on the analyses of samples available from these bodies.

Rocks and rock fragments from the Earth's mantle occur in a variety of geologic settings (e.g., see, O'Neill and Palme, 1998, Chapters 2.04 and 2.05): (i) as mantle sections in ophiolites representing suboceanic lithosphere; (ii) as massive peridotites, variously known as Alpine, orogenic, or simply high-temperature peridotites; (iii) as abyssal peridotites, dredged from the ocean floor—the residue from melt extraction of the oceanic crust; (iv) as spinel- (rarely garnet-) peridotite xenoliths from alkali basalts, mostly from the subcontinental lithosphere but with almost identical samples from suboceanic lithosphere; (v) as garnet peridotite xenoliths from kimberlites and lamproites—these fragments sample deeper levels in the subcontinental lithosphere and are restricted to ancient cratonic regions.

Typical mantle peridotites contain more than 50% olivine, variable amounts (depending on their history of melt extraction and refertilization) of orthopyroxene, clinopyroxene, plus an aluminous phase, whose identity depends on the pressure (i.e., depth) at which the peridotite equilibrated—plagioclase at low pressures, spinel at intermediate pressures, and garnet at high pressures. Peridotites have physical properties such as density and seismic velocity propagation characteristics that match the geophysical constraints required of mantle material. Another obvious reason for believing that these rocks come from the mantle is that the constituent minerals of the xenoliths have chemical compositions which show that the rocks have equilibrated at upper mantle pressures and temperatures. In the case of the xenoliths their ascent to the surface of the Earth was so fast that minerals had no time to adjust to the lower pressure and temperature on the Earth's surface. Most of the information used for estimating the chemical composition of the mantle is derived from spinel-lherzolite xenoliths originating from a depth of 40 km to 60 km. Garnet lherzolites, which sample the mantle down to a depth of ~200 km and a temperature of 1,400 °C, are much rarer (see Chapter 2.05).

The early compositional data on peridotites have been summarized by Maaløe and Aoki (1977). A comprehensive review of more recent data on mantle rocks is given by O'Neill and Palme (1998).

2.01.4.2 The Chemical Composition of Mantle Rocks

The chemistry of the mantle peridotites is characterized by contents of MgO in the range of 35–46 wt.% and SiO_2 from 43 wt.% to 46 wt.%, remarkably constant abundances of FeO (8 ± 1 wt.%), Cr_2O_3 (0.4 ± 0.1 wt.%), and Co (100 ± 10 ppm). Bulk rock magnesium numbers (100 × Mg/Mg + Fe, in atoms) are generally ≥0.89. The comparatively high contents of nickel (2,200 ± 500 ppm), a siderophile element, and of iridium (3.2 ± 0.3 ppb), a compatible highly siderophile element, are diagnostic of mantle rocks. Figure 2 plots the bulk rock concentrations of SiO_2, Al_2O_3, and CaO versus MgO for peridotites from the Central Dinaric Ophiolite Belt (CDOB) in Yugoslavia, a typical occurrence of mantle rocks associated with ophiolites. The samples are recovered from an area comprising a large fraction of the former Yugoslavia (Lugovic, 1991). The abundances of the refractory elements CaO, Al_2O_3 and also of Ti, Sc, and the heavy REE are negatively correlated with MgO. Both CaO and Al_2O_3 decrease by a factor of 2 with MgO increasing by only 15%, from 36% to 43% (Figure 2). There is a comparatively small decrease of 3.5% in SiO_2 over the same range. Figure 3 plots Na, Cr, and Ni versus MgO for the same suite of mantle rocks. The decrease of sodium is about twice that of calcium and aluminum. The chromium concentrations are constant, independent of MgO, while nickel concentrations are positively correlated with MgO.

Samples of massive peridotites and of spinel and garnet lherzolite xenoliths from worldwide localities plot on the same or on very similar correlations as those shown here for the CDOB samples (McDonough, 1990; BVSP, 1981; McDonough and Sun, 1995; O'Neill and Palme, 1998). It is particularly noteworthy that trends for xenoliths and massive peridotites are statistically indistinguishable (McDonough and Sun, 1995). An example is given in Figure 4, where FeO versus MgO plots for samples from two massive peridotites, the CDOB and Zabargad island in the Red Sea, are compared with xenolith data from two localities, Vitim (Baikal region, Russia) and the Hessian Depression (Germany). Two important

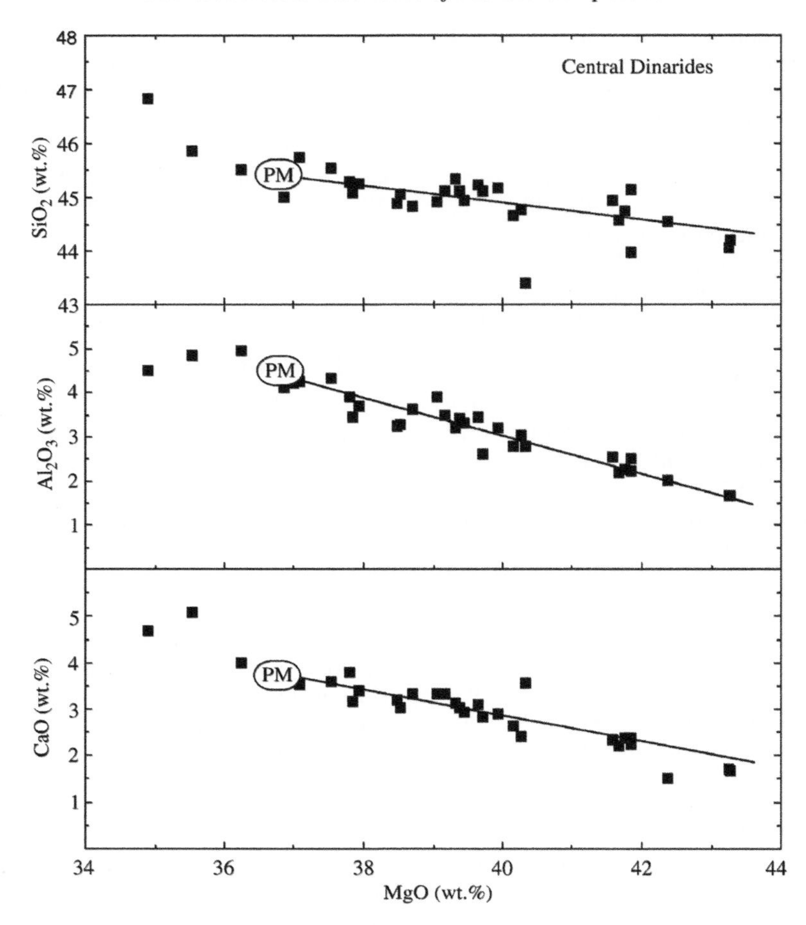

Figure 2 Correlations of SiO_2, Al_2O_3, and CaO with MgO for a suite of mantle samples from the CDOB in Yugoslavia, a typical occurrence of mantle rocks associated with ophiolites. The samples are recovered from an area comprising a large fraction of the former Yugoslavia (Lugovic, 1991). The abundances of the refractory elements CaO and Al_2O_3 are negatively correlated with MgO, indicating preferred partitioning of these elements into the partial melt. The effect for SiO_2 is comparatively small. These correlations are characteristic of a large number of occurrences of mantle xenoliths and massive peridotites (see McDonough, 1990; BVSP, 1981; McDonough and Sun, 1995; O'Neill and Palme, 1998). PM indicates the composition of the primitive mantle derived in this chapter.

conclusions can be drawn from this figure: (i) the FeO contents are independent of MgO in all occurrences of mantle rocks and (ii) the average FeO content is the same in all four localities. The two statements can be generalized. On the left side of Figure 5 average FeO contents of samples from 11 suites of xenoliths are compared with those of 10 suites of massif peridotites. Each of the points in Figure 5 represents the average of at least six, in most cases more than 10, samples (see O'Neill and Palme (1998) for details). The errors assigned to the data points in Figure 5 are calculated from the variations in a given suite. The range in average FeO contents from the various localities is surprisingly small and reflects the independence of FeO from MgO. Also, the average FeO contents are in most cases indistinguishable from each other. The average SiO_2 contents of the same suites plotted on the right side show a somewhat larger scatter, reflecting the slight dependence of SiO_2 on MgO (see Figure 2).

Peridotites with the lowest MgO contents have, in general, the highest concentrations of Al_2O_3, CaO, and other incompatible elements that preferentially partition into the liquid phase during partial melting. Such peridotites are often termed "fertile," emphasizing their ability to produce basalts on melting. Most peridotites are, however, depleted to various extents in incompatible elements, i.e., they have lower contents of CaO, Al_2O_3, Na_2O, etc., than a fertile mantle would have, as shown in Figures 2 and 3. By contrast, an element compatible with olivine, such as nickel, increases with increasing MgO contents (Figure 3). Thus, the trends in Figures 2 and 3 have been interpreted as reflecting various degrees of melt extraction (Nickel and Green, 1984; Frey *et al.*, 1985). Following this reasoning, the least-depleted

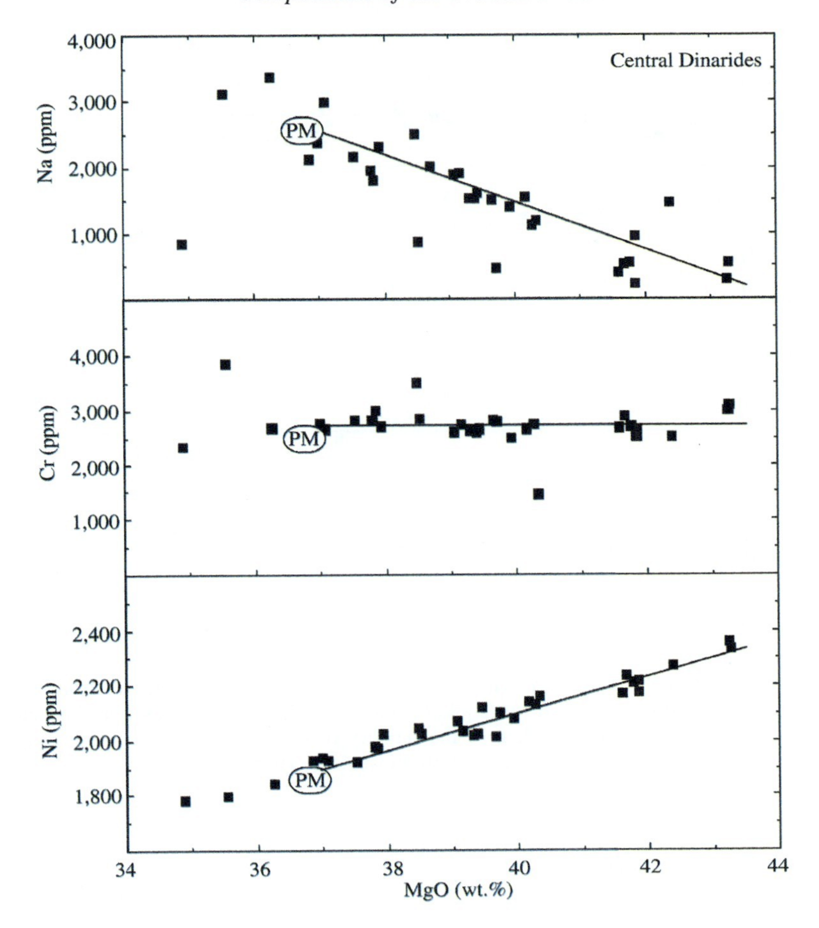

Figure 3 Na, Cr, and Ni for the same suite of rocks as in Figure 2. Na is incompatible with mantle minerals, Ni is compatible, and Cr partitions equally between melt and residue.

peridotites, i.e., highest in CaO and Al_2O_3, should be the closest in composition to the primitive mantle.

In detail, however, the picture is not so simple. All mantle peridotites (whether massive peridotites or xenoliths) are metamorphic rocks that have had a complex subsolidus history after melt extraction ceased. As well as subsolidus recrystallization, peridotites have undergone enormous amounts of strain during their emplacement in the lithosphere. Massive peridotites show modal heterogeneity on the scale of centimeters to meters, caused by segregation of the chromium-diopside suite of dikes, which are then folded back into the peridotite as deformation continues. The net result is more or less diffuse layers or bands in the peridotite, which may be either enriched or depleted in the material of the chromium-diopside suite, i.e., in climopyroxene and orthopyroxene in various proportions, ±minor spinel, and ±sulfide. This process should cause approximately linear correlations of elements versus MgO, broadly similar to, but not identical with, those caused by melt extraction. Indeed, there is no reason for melt extraction to cause a linear relationship. Partial melting experiments and corresponding calculations show that for many elements, such trends should not be linear (see O'Neill and Palme, 1998, figure 2.12). Although the banding may not be so obvious in xenoliths as in the massive peridotites, since it operates over lengthscales greater than those of the xenoliths themselves (but it is nevertheless often seen, e.g., Irving (1980)), the similarity in the compositional trends between xenoliths and massive peridotites shows that it must be as ubiquitous in the former as it may be observed to be in the latter.

In addition, many peridotites bear the obvious signatures of metasomatism, which re-enriches the rock in incompatible components subsequent to depletion by melt extraction. Where this is obvious (e.g., in reaction zones adjacent to later dikes) it may be avoided easily; but often the metasomatism is cryptic, in that it has enriched the peridotite in incompatible trace elements without significantly affecting major-element chemistry (Frey and Green, 1974). Peridotites thus have very variable contents of highly incompatible trace

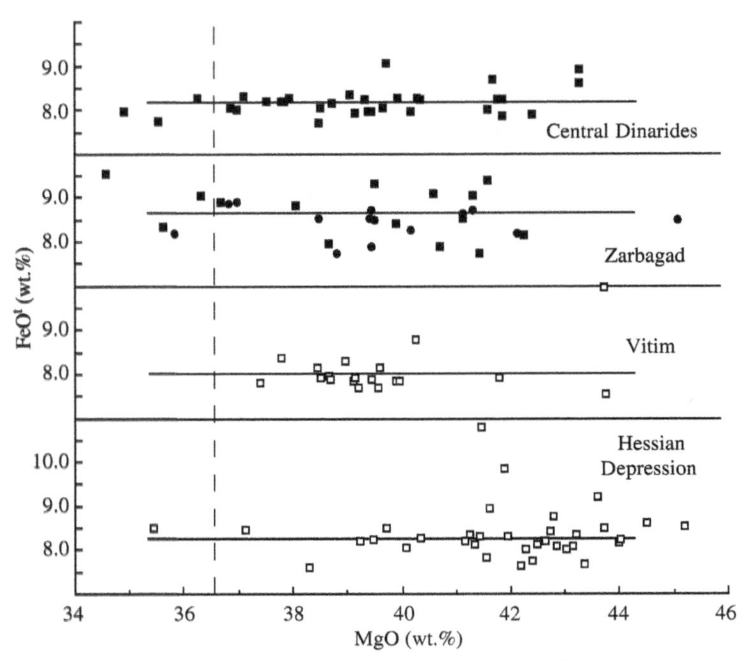

Figure 4 The FeO contents of samples from the same suites as in Figures 2 and 3, samples from Zabargad island and samples from two xenoliths suites from Vitim (Baikal region) and Hessian Depression (Germany). The FeO contents are, similar to Cr in Figure 3, independent of the fertility of the mantle rocks as reflected in their MgO contents (source O'Neill and Palme, 1998).

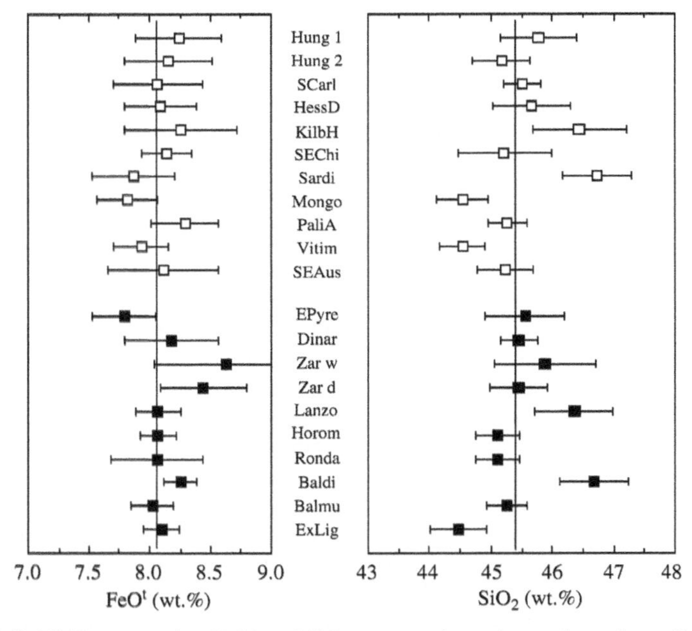

Figure 5 Average FeO (all Fe assumed as FeO) and SiO_2 contents in various suites of xenoliths (open symbols) and massive peridotites (full symbols). All suites give the same FeO value within 1σ SD. An average PM FeO content of 8.07 ± 0.06% is calculated from these data (see O'Neill and Palme, 1998). The SiO_2 concentrations are somewhat more variable as SiO_2 depends on MgO (see Figure 2) (source O'Neill and Palme, 1998).

elements like thorium, uranium, niobium, and light REEs, and these elements cannot be used to distinguish melt extraction trends from those caused by modal banding.

In summary, then, the trends of CaO, Al_2O_3, and Na (Figures 2 and 3) and also TiO_2 and many other moderately incompatible trace elements (Sc, HREE) with MgO that appear approximately linear in suites of massive peridotites or xenoliths, are due to a combination of two processes: melt extraction, followed by metamorphic modal banding; and possibly by metasomatism. This explains a number of features of these trends: the scatter in the data; the tendency to superchondritic Ca/Al ratios (Palme and Nickel, 1985; O'Neill and Palme, 1998, figure 1.13), and, importantly, the existence of peridotites with higher CaO or Al_2O_3 than that inferred for the PM. Thus, it is not possible simply to say that those peridotites richest in incompatible elements are closest to the PM, and instead some other method of reconstructing the PM composition must be used. Here we describe a new procedure for calculating the composition of PM that is independent of the apparent linear correlations discussed above. The method was first introduced by O'Neill and Palme (1998).

2.01.4.2.1 *Major element composition of the Earth's primitive mantle*

The two elements calcium and aluminum are RLEs. The assumption is usually made that all RLEs are present in the primitive mantle of the Earth in chondritic proportions. Chondritic (undifferentiated) meteorites show significant variations in the absolute abundances of refractory elements but have, with few exceptions discussed below, the same relative abundances of lithophile and siderophile refractory elements. By analogy, the Earth's mantle abundances of refractory lithophile elements are assumed to occur in chondritic relative proportions in the primitive mantle, which is thus characterized by a single RLE/Mg ratio. This ratio is often normalized to the CI-chondrite ratio and the resulting ratio, written as $(RLE/Mg)_N$, is a measure of the concentration level of the refractory component in the Earth. A single factor of $(RLE/Mg)_N$ valid for all RLEs is a basic assumption in this procedure and will be calculated from mass balance considerations.

The FeO^t (total iron as FeO, which is a very good approximation for mantle rocks) and MgO contents are estimated from two empirical constraints.

(i) As discussed above, FeO^t is remarkably constant worldwide, in massive peridotites and in alkali-basalt-derived lherzolite xenoliths (Figures 4 and 5) and moreover it does not depend on MgO contents for reasonably fertile samples ($MgO < 42$ wt.%). This is expected, as FeO^t is not affected by melt extraction at low degrees of partial melting ($D_{Fe}^{liq/sol} \cong 1$), and is also insensitive to differing olivine/pyroxene ratios caused by modal segregation and remixing (see Chapter 2.08). As discussed above, O'Neill and Palme (1998) considered data from 21 individual suites of spinel lherzolites. These individual suites usually show standard deviations of the order of ± 0.5 wt.% FeO^t for the least-depleted samples, but the mean values are remarkably coherent (Figure 5). By averaging O'Neill and Palme (1998) obtained $FeO^t = 8.10$ wt.%, with a standard error of the mean of 0.05 wt.%.

Next, the MgO content is constrained from the observation that molar $Mg/(Mg + Fe^t)$ (i.e., bulk-rock Mg#) ought not to be affected by modal segregation, since mineral Mg#s are nearly the same in olivine and in pyroxenes (as shown by experimental phase equilibrium studies and empirical studies of coexisting phases in peridotites). Hence, $(Mg\#)_{ol} = (Mg\#)_{bulk}$, to a very good approximation, in spinel lherzolites. Extraction of partial melt causes an increase in Mg#, so that the most primitive samples should have the lowest Mg#. Histograms of olivine compositions in spinel lherzolites from a wide variety of environments show that the most fertile samples that have not been grossly disturbed by metasomatism have Mg#s from 0.888 to ~0.896. Since the uncertainty with which olivine Mg#s are determined in individual samples is typically ~0.001, we interpret this distribution as implying an Mg# for the primitive depleted mantle of 0.890, which value we adopt. Hence,

$$MgO = (FeO^t \times 0.5610 \times Mg\#)/(1 - Mg\#) \quad (5)$$

with 8.10 ± 0.05 wt.% FeO^t, and $Mg\# = 0.890 \pm 0.001$, this gives $MgO = 36.77 \pm 0.44$ wt.%.

(ii) The second step uses the cosmochemical observation that only seven elements make up nearly 99% of any solar or chondritic composition, if the light elements (hydrogen, carbon, nitrogen, and associated oxygen) are removed. These elements are oxygen, magnesium, aluminum, silicon, sulfur, calcium, iron, and nickel. In the PM, nickel and sulfur are not important, and in any case their abundances are well known. The amount of oxygen is fixed by the stoichiometry of the relevant oxide components; hence,

$$MgO + Al_2O_3 + SiO_2 + CaO + FeO^t = 98.41 \text{ wt.\%} \quad (6)$$

The uncertainty is probably only $\pm 0.04\%$. The robustness of this constraint arises because the abundances of minor components which constitute 1.34% of the missing 1.59% (i.e., Na_2O, TiO_2, Cr_2O_3, MnO, and NiO, plus the excess oxygen in Fe_2O_3) vary little among reasonably fertile peridotites, and all other components (H_2O, CO_2, S, K, O, and the entire remaining trace element inventory) sum to <0.15%.

Although not quite as invariant as the FeO^t content, the concentration of SiO_2 is also relatively insensitive to melt extraction and modal segregation processes. The SiO_2 concentration from SiO_2 versus MgO trends at 36.77% MgO is 45.40 ± 0.3 wt.% (see, e.g., Figure 2) conforming to fertile spinel-lherzolite suites worldwide (Figure 5).

Since calcium and aluminum are both RLEs, the CaO/Al_2O_3 ratio is constrained at the chondritic value of 0.813. Unfortunately, calcium is rather more variable than other RLEs in chondrites, probably due to some mobility of calcium in the most primitive varieties (like Orgueil); hence, this ratio is not known as well as one might like. A reasonable uncertainty is ±0.003 (see Chapter 1.03). Then,

$$Al_2O_3 = (98.41 - MgO - SiO_2 - FeO^t)/(1 + 0.813) \quad (7)$$

With the concentrations of FeO, MgO, and SiO_2 given above, $Al_2O_3 = 4.49$ wt.%; hence, CaO = 3.65 wt.%. The RLEs/Mg ratio, normalized to this ratio in CI-chondrites, is then 1.21 ± 0.10, corresponding to an unnormalized ratio of RLE in the Earth to RLE in CI of 2.80. Hence, under the assumption of constant RLE ratios in the Earth, this constrains the abundances of all other RLEs.

Putting the algorithm in the form of a simple equation makes the effects of the uncertainties transparent. The main weakness of the method is that the final result for CaO, Al_2O_3, and the $(RLE/Mg)_N$ ratio is particularly sensitive to the chosen value of Mg#. However, the values of CaO and Al_2O_3 look very reasonable when compared to actual analyses of least-depleted peridotite samples, and we calculate through propagation of errors that uncertainties in $\sigma(SiO_2) = \pm 0.3$ wt.%, $\sigma(FeO^t) = \pm 0.05$ wt.%, and $\sigma(Mg\#) = \pm 0.001$ give a realistic uncertainty in $(RLE/Mg)_N$ of ±0.10, or ±8%. The algorithm also gives Al/Si = 0.112 ± 0.008 and Mg/Si = 1.045 ± 0.014. The resulting major element composition of PM is given in Table 3.

While all spinel-lherzolite facies suites show remarkably similar compositional trends as a function of depletion, some garnet peridotite xenoliths in kimberlites and lamproites from ancient cratonic lithospheric keels show significantly different trends (e.g., see Boyd, 1989; Chapters 2.05 and 2.08). Most of these xenoliths are extremely depleted; extrapolation of the trends back to the PM MgO of 36.7% gives similar concentrations of SiO_2, FeO^t, Al_2O_3, and CaO to the spinel lherzolites (O'Neill and Palme, 1998); the difference in their chemistry is due to a different style of melt extraction, and not a difference in original mantle composition.

Table 3 Major element composition of the PM, and comparison with estimates from the literature.

	This work	*Ringwood (1979)*	*Jagoutz et al. (1979)*	*Wänke et al. (1984)*	*Palme and Nickel (1985)*	*Hart and Zindler (1986)*	*McDonough and Sun (1995)*	*Allègre et al. (1995)*
MgO	36.77 ± 0.44	38.1	38.3	36.8	35.5	37.8	37.8	37.77
Al_2O_3	4.49 ± 0.37	3.3	4.0	4.1	4.8	4.06	4.4	4.09
SiO_2	45.40 ± 0.30	45.1	45.1	45.6	46.2	46.0	45.0	46.12
CaO	3.65 ± 0.31	3.1	3.5	3.5	4.4	3.27	3.5	3.23
FeO^t	8.10 ± 0.05	8.0	7.8	7.5	7.7		8.1	7.49
Total	98.41 ± 0.10	97.6	98.7	97.5	98.6		98.8	98.7
$(RLE/Mg)_N$	1.21 ± 0.10	1.02	1.03–1.14	1.1–1.4	1.3–1.5	1.06–1.07	1.17	1.05–1.08
Mg#	0.890 ± 0.001	0.895	0.897	0.897	0.891		0.893	0.900

Mg#—molar Mg/(Mg + Fe); FeO^t—all Fe as FeO; $(RLE/Mg)_N$—refractory lithophile elements normalized to Mg- and CI-chondrites.

2.01.4.2.2 Comparison with other estimates of PM compositions

Ringwood (1979) reconstructed the primitive upper mantle composition by mixing appropriate fractions of basalts (i.e., partial melts from the mantle) and peridotites (the presumed residues from the partial melts). He termed this mixture *pyrolite*. In more recent attempts the PM composition has generally been calculated from trends in the chemistry of depleted mantle rocks. Jagoutz *et al.* (1979) used the average composition of six fertile spinel lherzolites as representing the primitive mantle. This leads to an MgO content of 38.3% and a corresponding Al_2O_3 content of 3.97% for PM (Table 3).

In RLE versus MgO correlations, the slopes depend on the compatibility of the RLE with the residual mantle minerals. The increase of the incompatible TiO_2 with decreasing MgO is much stronger than that of less incompatible Al_2O_3 which in turn has a steeper slope than the fairly compatible scandium. The ratio of these elements is thus variable in residual mantle rocks, depending on the fraction of partial melt that had been extracted. As the PM should have chondritic ratios, Palme and Nickel (1985) estimated the PM MgO content from variations in Al/Ti and Sc/Yb ratios with MgO in upper mantle rocks. Both ratios increase with decreasing MgO and at 36.8% MgO they become chondritic, which was chosen as the MgO content of the PM. Palme and Nickel (1985), however, found a nonchondritc Ca/Al ratio in their suite of mantle rocks and concluded that aluminum was removed from the mantle source in an earlier melting event. O'Neill and Palme (1998) suggested that the apparent high Ca/Al ratios in some spinel lherzolites are the result of a combination of melt extraction, producing residues with higher Ca/Al ratios, and modal heterogeneities, as discussed above.

Hart and Zindler (1986) also based their estimate on chondritic ratios of RLE. They plotted Mg/Al versus Nd/Ca for peridotites and chondritic meteorites. The two refractory elements, neodymium and calcium, approach chondritic ratios with increasing degree of fertility. From the intersection of the chondritic Nd/Ca ratio with observed peridotite ratios, Hart and Zindler (1986) obtained an Mg/Al ratio of 10.6 (Table 2).

McDonough and Sun (1995) assumed 37.7% as upper mantle MgO content and calculated RLE abundances from Al_2O_3 and CaO versus MgO correlations.

Allègre *et al.* (1995) took the same Mg/Al ratio as Hart and Zindler (1986) and assumed an Mg/Si ratio of 0.945 which they considered to be representative of the least differentiated sample from the Earth's mantle to calculate the bulk chemical composition of the upper mantle.

All these estimates lead to roughly similar PM compositions. The advantage of the method presented here is that it is not directly dependent on any of the element versus MgO correlations and that it permits calculation of realistic uncertainties for the PM composition. With error bars for MgO, SiO_2, and FeO of only ~1% (rel), most other estimates for MgO and FeO in Table 3 fall outside the ranges defined here, whereas the higher uncertainties of Al_2O_3 and CaO of ~10% encompass all other estimates.

2.01.4.2.3 Abundance table of the PM

In Table 4 we have listed the element abundances in the Earth's primitive mantle. The CI-chondrite abundances are given for comparison (see Chapter 1.03). Major elements and associated errors are from the previous section (Table 3). Trace element abundances are in many cases from O'Neill and Palme (1998). In some instances new estimates were made. The abundances of refractory lithophile elements (denoted RLE in Table 4) are calculated by multiplying the CI-abundances by 2.80, as explained above. The two exceptions are vanadium and niobium. Both elements may, in part, be partitioned the core. The errors of the other RLEs are at least 8%, reflecting the combined uncertainties of the scaling element and the aluminum content plus the CI-error. This leads to uncertainties of 10% or more for many of the RLEs. The relative abundances of the RLEs are, however, better known, in particular when CI-chondritic ratios among RLEs are assumed. In this case the uncertainty of a given RLE ratio may be calculated by using the uncertainties of the CI-ratios given in Table 4.

In estimating trace element abundances of the BSE, it is useful to divide elements into compatible and incompatible elements with additional qualifiers such as moderately compatible or highly incompatible. A measure of compatibility is the extent to which an element partitions into the melt during mantle melting, quantified by the solid/melt partition coefficient $D^{(\mathrm{solid/melt})}$. As Earth's crust formed by partial melting of the mantle, the crust/mantle concentration ratio of an element is an approximate measure for its compatibility (see O'Neill and Palme, 1998). Crust/mantle ratios for 64 elements are listed in Table 5. The abundances of elements in the continental crust were taken from Chapter 3.01), while the PM abundances are from Table 4. The RLEs in Table 5 (in bold face) comprise a large range of compatibilities, from highly incompatible elements (thorium, uranium) to only slightly incompatible elements (scandium). Abundances of RLEs in the Earth's

Table 4 Composition of the PM of the Earth ($Z < 42$ in ppm, $Z \geq 42$ in ppb, unless otherwise noted).

Z	*Element*	*CI*	*SD*	*Earth's mantle*	*SD*	*Comment*	*References*
1	H (%)	2.02	10	0.012	20	Mass balance	ON98, com
3	Li	1.49	10	1.6	20	Data on mantle rocks	Ja79, Ry87, Se00, com
4	Be	0.0249	10	0.070	10	RLE	
5	B	0.69	13	0.26	40	B/K = $(1.0 \pm 0.3) \times 10^{-3}$	Ch94
6	C	32,200	10	100	u	Mass balance	Zh93
7	N	3,180	10	2	u	Mass balance	Zh93
8	O (%)	46.5	10	44.33	2	Stoichiometry, with $Fe^{3+}/\sum Fe = 0.03$	Ca94
9	F	58.2	15	25	40	F/K = 0.09 ± 0.03, F/P = 0.3 ± 0.1	com
11	Na	4,982	5	2,590	5	versus MgO	ON98
12	Mg (%)	9.61	3	22.17	1	Major element	th.w.
13	Al	8,490	3	23,800	8	Major element, RLE	th.w.
14	Si (%)	10.68	3	21.22	1	Major element	ON98, th.w.
15	P	926	7	86	15	P/Nd = 65 ± 10	McD85, La92
16	S	54,100	5	200	40	versus MgO, komatiites	ON91
17	Cl	698	15	30	40	Mass balance	Ja95, com
19	K	544	5	260	15	Mean of K/U and K/La	ON98
20	Ca	9,320	3	26,100	8	Major element, RLE	th.w.
21	Sc	5.90	3	16.5	10	RLE	
22	Ti	458	4	1,280	10	RLE	
23	V	54.3	5	86	5	versus MgO	ON98
24	Cr	2,646	3	2,520	10	versus MgO	ON98
25	Mn	1,933	3	1,050	10	versus MgO	ON98
26	Fe (%)	18.43	3	6.30	1	Major element	ON98, th.w.
27	Co	506	3	102	5	versus MgO	ON98
28	Ni	10,770	3	1,860	5	versus MgO	ON98
29	Cu	131	10	20	50		ON91, com
30	Zn	323	10	53.5	5	versus MgO	ON98
31	Ga	9.71	5	4.4	5	versus MgO	ON98
32	Ge	32.6	10	1.2	20	versus SiO_2	ON98
33	As	1.81	5	0.066	70	As/Ce = 0.037 ± 0.025	Si90
34	Se	21.4	5	0.079	u	Se/S = 2,528, chondritic	Pa03
35	Br	3.5	10	0.075	50	Cl/Br = 400 ± 50	Ja95, com

37	Rb	2.32	5	0.605	10	Rb/Sr = 0.029 ± .002 (Sr isotopes), Rb/Ba = 0.09 ± 0.02	Ho83
38	Sr	7.26	5	20.3	10	RLE	
39	Y	1.56	3	4.37	10	RLE	
40	Zr	3.86	2	10.81	10	RLE	
41	Nb (ppb)	247	3	588	20	RLE	com
42	Mo	928	5	39	40	Mo/Ce = 0.027 ± 012, Mo/Nb = 0.05 ± .015	Si90, Fi95
44	Ru	683	3	4.55	4	HSE, Ru/Ir = CI-chondrite	com
45	Rh	140	3	0.93	5	HSE, Rh/Ir = CI-chondrite	com
46	Pd	556	10	3.27	15	HSE, Pd/Ir = 1.022 ± 0.097, H-chondrite	Mo85, com
47	Ag	197	10	4	u	Ag/Na = 1.6 ± 1.0 × 10^{-6} in MORB, lherzolites, crust	th.w., com
48	Cd	680	10	64	u	Cd/Zn = 1.2 ± 0.5 × 10^{-3}	th.w., com
49	In	78	10	13	40	In/Y = 3 ± 1 × 10^{-3}, spinel lherzolites	Yi02, BV81
50	Sn	1.680	10	138	30	Sn/Sm = 0.32 ± 0.06	Jo93
51	Sb	133	10	12	u	Sb/Pb = 0.074 ± 0.031, Sb/Pr = 0.02 in mantle, 0.05 in crust	Jo97
52	Te	2,270	10	8	u	Te/S = CI	th.w., com
53	I	433	20	7	u	Mass balance	ON98, com
55	Cs	188	5	18	50	Cs/Ba = 1.1 × 10^{-3} in mantle, 3.6 × 10^{-3} in crust	McD92
56	Ba	2,410	10	6,750	15	RLE	
57	La	245	5	686	10	RLE	
58	Ce	638	5	1,786	10	RLE	
59	Pr	96.4	10	270	15	RLE	
60	Nd	474	5	1,327	10	RLE	
62	Sm	154	5	431	10	RLE	
63	Eu	58.0	5	162	10	RLE	
64	Gd	204	5	571	10	RLE	
65	Tb	37.5	10	105	15	RLE	
66	Dy	254	5	711	10	RLE	

(continued)

Table 4 (continued).

Z	*Element*	*CI*	*SD*	*Earth's mantle*	*SD*	*Comment*	*References*
67	Ho	56.7	10	159	15	RLE	
68	Er	166	5	465	10	RLE	
69	Tm	25.6	10	71.7	15	RLE	
70	Yb	165	5	462	10	RLE	
71	Lu	25.4	10	71.1	15	RLE	
72	Hf	107	5	300	10	RLE	
73	Ta	14.2	6	40	10	RLE	
74	W	90.3	4	16	30	W/Th = 0.19 ± 0.03	Ne96
75	Re	39.5	4	0.32	10	Re/Os = 0.0874 ± 0.0027, H-chondrite	Wa02, com
76	Os	506	5	3.4	10	HSE, Os/Ir = 1.07 ± 0.014, H-chondrite	Ka89, com
77	Ir	480	4	3.2	10	av. PM analyses	Mo01, com
78	Pt	982	4	6.6	12	HSE, Pt/Ir, CI-chondrite	com
79	Au	148	4	0.88	11	HSE, Ir/Au = 3.63 ± 0.13, H-chondrite	Ka89, com
80	Hg	310	20	6	u	Hg/Se = 0.075	th.w., com
82	Tl	143	10	3	u	Tl/Rb = 0.005 in crust	He80, com
82	Pb	2,530	10	185	10	$^{238}U/^{204}Pb$ = 8.5 ± 0.5, 206/204 = 18, 207/204 = 15.5, 208/204 = 38	Ga96
83	Bi	111	15	5	u	Bi/La = 8×10^{-3} in crust	th.w., com
90	Th	29.8	10	83.4	15	RLE	
92	U	7.8	10	21.8	15	RLE	

SD—standard deviation in %; RLE—refractory lithophile element; HSE—highly siderophile element; th.w.—this work; com—comments in text; u—uncertain, error exceeding 50%; CI—values, Chapter 1.03; References: BV81—BVSP (1981); Ca94—Canil *et al.* (1994); Ch94—Chaussidon and Jambon (1994); Fi95—Fitton (1995); Ga96—Galer and Goldstein (1996); He80—Hertogen *et al.* (1980); Ho83—Hofmann and White (1983); Ja79—Jagoutz *et al.* (1979); Ja95—Jambon *et al.* (1995); Jo93—Jochum *et al.* (1993); Jo97—Jochum and Hofmann (1997); Ka89—Kallemeyn *et al.* (1989); La92—Langmuir *et al.* (1992); McD85—McDonough *et al.* (1985); McD92—McDonough *et al.* (1992); Mo01—Morgan *et al.* (2001); Mo85—Morgan *et al.* (1985); Ne96—Newsom *et al.* (1996); ON91—O'Neill (1991); ON98—O'Neill and Palme (1998); Pa03—Palme and Jones (2003); Ry87—Ryan and Langmuir (1987); Se00—Seitz and Woodland (2000); Si90—Sims *et al.* (1990); Wa02—Walker *et al.* (2002); Zh93—Zhang and Zindler (1993); Yi02—Yi *et al.* (2000).

Table 5 Enrichment of elements in the bulk continental crust over the PM (abundances in ppb, except when otherwise noted, refractory lithophile elements are in bold face).

	Element	*Mantle (PM)*	*Crust*	*Crust/Mantle*		*Element*	*Mantle (PM)*	*Crust*	*Crust/Mantle*
1	Tl	3	360	120.0	**31**	**Dy**	**711**	**3,700**	**5.20**
2	W	16	1,000	62.5	**32**	**Ho**	**159**	**780**	**4.91**
3	Cs	18	1,000	55.6	**33**	**Yb**	**462**	**2,200**	**4.76**
4	Rb (ppm)	0.605	32	52.9	**34**	**Er**	**465**	**2,200**	**4.73**
5	Pb	185	8,000	43.2	**35**	**Y**	**4.37**	**20**	**4.58**
6	**Th**	**83.4**	**3,500**	**42.0**	**36**	**Tm**	**72**	**320**	**4.44**
7	**U**	**21.8**	**910**	**41.7**	**37**	**Lu**	**71.7**	**300**	**4.18**
8	**Ba** (ppm)	**6.75**	**250**	**37.1**	**38**	**Ti** (ppm)	**1,282**	**5,400**	**4.21**
9	K (ppm)	260	9,100	35.0	39	Ga (ppm)	4.4	18	4.09
10	Mo (ppm)	39	1,000	25.6	40	In	13	50	3.85
11	**Ta**	**40**	**1,000**	**25.0**	41	Cu (ppm)	20	75	3.75
12	**La**	**686**	**16,000**	23.3	**42**	**Al** (%)	**2.37**	**8.41**	**3.55**
13	**Be** (ppm)	**0.07**	**1.5**	**21.4**	43	Au	0.88	3	3.41
14	Ag	4	80	20.0	**44**	**V (ppm)**	**86**	**230**	**2.67**
15	**Ce**	1,785	33,000	18.5	**45**	**Ca** (%)	**2.61**	**5.29**	**2.03**
16	**Nb** (ppm)	**0.6**	**11**	**18.3**	**46**	**Sc (ppm)**	**16.5**	**30**	**1.82**
17	Sn	138	2,500	18.1	47	Re	0.32	0.5	1.56
18	Sb	12	200	16.7	48	Cd	64	98	1.53
19	As	66	1,000	15.2	49	Zn (ppm)	53.5	80	1.50
20	**Pr**	**270**	**3,900**	**14.4**	50	Ge (ppm)	1.2	1.6	1.33
21	**Sr** (ppm)	**20.3**	**260**	**12.8**	51	Mn (ppm)	1,050	1,400	1.33
22	**Nd**	**1,327**	**16,000**	**12.1**	52	Si (%)	21.22	26.77	1.26
24	**Hf**	**300**	**3,000**	**10.0**	53	Fe (%)	6.3	7.07	1.12
25	**Zr** (ppm)	**10.8**	**100**	**9.26**	54	Se	79	50	0.63
26	Na (ppm)	2,590	23,000	8.88	55	Li (ppm)	2.2	1.3	0.59
25	**Sm**	**431**	**3,500**	**8.12**	56	Co (ppm)	102	29	0.28
27	**Eu**	**162**	**1,100**	**6.79**	57	Mg (%)	22.17	3.2	0.14
28	**Gd**	**571**	**3,300**	**5.78**	58	Cr (ppm)	2,520	185	0.073
29	B (ppm)	0.26	1.5	5.77	59	Ni (ppm)	1,860	105	0.056
30	**Tb**	**105**	**600**	**5.71**	60	Ir	3.2	0.1	0.031

Sources of data: mantle—Table 4; continental crust—Chapter 3.01.

mantle are calculated by scaling RLEs to aluminum and calcium, as discussed above.

Abundances of nonrefractory incompatible lithophile elements (potassium, rubidium, caesium, etc.) or partly siderophile/chalcophile elements (tungsten, antimony, tin, etc.) are calculated from correlations with RLE of similar compatibility. This approach was first used by Wänke *et al.* (1973) to estimate abundances of volatile and siderophile elements such as potassium or tungsten in the moon. The potassium abundance was used to calculate the depletion of volatile elements in the bulk moon, whereas the conditions of core formation and the size of the lunar core may be estimated from the tungsten abundance, as described by Rammensee and Wänke (1977). This powerful method has been subsequently applied to Earth, Mars, Vesta, and the parent body of HED meteorites. The procedure is, however, only applicable if an incompatible refractory element and a volatile or siderophile element have the same degree of incompatibility, i.e., do not fractionate from each other during igneous processes. In other words, a good correlation of the two elements over a wide concentration range and over all the important differentiation processes that have occurred in the planet is required. In estimating abundances of trace elements from correlations, care must be taken to ensure that all important reservoirs are included in the data set (see O'Neill and Palme (1998) for details).

A particularly good example is the samarium versus tin correlation displayed in Figure 6 modified from figure 8 of Jochum *et al.* (1993). Both basalts, partial melts from the mantle, and mantle samples representing the depleted mantle after removal of melt plot on the same correlation line. From the Sm/Sn ratio of 0.32 obtained by Jochum *et al.* (1993), a mantle abundance of 144 ppb is calculated, reflecting a 35-fold depletion of tin relative to CI-chondrites in the Earth's mantle. The origin of this depletion is twofold: (i) a general depletion of moderately volatile elements in the Earth (see below) and (ii) some partitioning of the siderophile element tin into the core of the Earth. Many of the abundances of moderately volatile lithophile, siderophile, and chalcophile elements listed in Table 4 were calculated from such correlations with refractory lithophile elements of

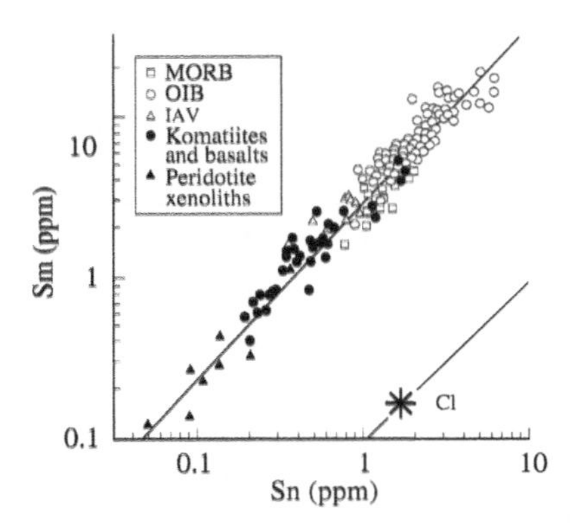

Figure 6 Correlation of Sm versus Sn in partial melts and residues of mantle melting. The constant Sm/Sn ratio suggests that this ratio is representative of the PM. Sn is depleted relative to CI-chondrites by a factor of 35. The knowledge of the abundance of the refractory element Sm in PM allows the calculation of the Sn PM abundance. Many of the PM abundances of siderophile and chalcophile elements are calculated from similar correlations (after Jochum *et al.*, 1993).

similar compatibility. Well-known examples are K/U, K/La, W/Th, P/Nd, Rb/Ba, etc.

The abundances of elements marked "com" in Table 4 are more difficult to estimate and require a more sophisticated approach. Detailed explanations on how the abundances are estimated are given below.

Hydrogen. The mass of the atmosphere is 5.1×10^{18} kg, of which 1.3% by mass is ^{40}Ar, derived from ^{40}K decay. The potassium content of the PM is 260 ppm; hence, the amount of ^{40}Ar produced from ^{40}K over the age of the Earth is calculated to 1.34×10^{17} kg. Thus, the amount of ^{40}Ar in the atmosphere corresponds to 0.5 of the amount produced in the PM, which is, therefore, taken to be the fraction of the mantle that has degassed. The amount of H_2O in the hydrosphere is 1.7×10^{21} kg (oceans, pore water in sediments, and ice). If this corresponds to the water originally contained in the degassed mantle, it would imply a PM abundance of 850 ppm. However, H_2O, unlike argon, is recycled into the mantle by subduction of the oceanic crust, such that MORB, products of the degassed mantle, contain ~0.2 wt.% H_2O. If MORB is produced by 10% partial melt and melting completely extracts H_2O, this corresponds to 200 ppm in the degassed mantle. Allowing 0.5 wt.% for the chemically bound water in the continental crust gives a total PM abundance of 1,100 ppm H_2O, or 120 ppm H. Since the exospheric inventory for H_2O is quite well constrained, the uncertainty of this estimate is only about ±20%.

Lithium. Ryan and Langmuir (1987) estimate 1.9 ± 0.2 ppm Li for the PM based on the analyses of peridotites. As Li is sited in the major minerals of upper mantle rocks and behaves as moderately incompatible, we have calculated bulk contents of spinel and garnet lherzolites from the mineral data of Seitz and Woodland (2000) which suggests an upper limit of 1.3 ppm for fertile mantle rocks. The range in unmetamorphosed fertile peridotites analyzed by Jagoutz *et al.* (1979) is somewhat higher, from 1.2 ppm to 2.07 ppm with an average of 1.52 ppm. We take here the average of all three estimates as 1.6 ppm Li.

Halides (fluorine, chlorine, bromine, and iodine). Fluorine as F^- substitutes readily for OH^- in hydroxy minerals, implying that it probably occurs in all "nominally anhydrous minerals" in the same way as OH^-. Fluorine is not significantly soluble in seawater. Both these properties make its geochemical behavior quite different from the other halides. F/K and F/P ratios in basalts are reasonably constant (Smith *et al.*, 1981; Sigvaldason and Oskarsson, 1986) with ratios of 0.09 ± 0.04 and 0.29 ± 0.1, respectively. Both ratios yield the same value of 25 ppm for the PM abundance which is listed in Table 4. However, the F/K ratio in the continental crust appears distinctly higher and the F/P ratio lower (Gao *et al.*, 1998), indicating that the incompatibility of these elements increases in the order $P < F < K$.

Jambon *et al.* (1995) estimated the amount of chlorine in the "exosphere" to be 3.8×10^{19} kg, made up of 2.66×10^{19} kg in seawater, the rest in evaporites. If this is derived by depletion of 50% of the mantle (^{40}Ar argument), it corresponds to a contribution of 19 ppm from the depleted mantle. The average amount of chlorine in the rocks of the continental crust is only a few hundred ppm (Wedepohl, 1995; Gao *et al.*, 1998) and can therefore be neglected. The amount of chlorine recycled back into the depleted mantle is more difficult to estimate, as the chlorine in primitive, uncontaminated MORB is highly variable and correlates poorly with other incompatible elements (Jambon *et al.*, 1995). Adopting a mean value of 100 ppm and assuming 10% melting adds another 10 ppm (the assumption here is that chlorine is so incompatible that all the chlorine in the MORB source mantle is from recycling and that MORB is assumed to be a 10% partial melt fraction). This gives a total chlorine content of 30 ppm for PM.

The Cl/Br ratio of seawater is 290, but that of evaporites is considerably higher (>3,000), such that the exospheric ratio is ~400 ± 50. The ratio in MORB and other basalts is the same (Jambon *et al.*, 1995). Iodine in the Earth is concentrated in the organic matter of marine sediments; this reservoir contains 1.2×10^{16} kg I (O'Neill and Palme, 1998), corresponding to 6 ppb if this iodine comes from 50% of the mantle. MORBs have ~8 ppb I (Déruelle *et al.*, 1992) implying ~1 ppb in the depleted (degassed) mantle, for a PM abundance of 7 ppb.

Copper. The abundance of copper in the depleted mantle raises a particular problem. Unlike other moderately compatible elements, there is a difference in the copper abundances of massive peridotites compared to many, but not all, of the xenolith suites from alkali basalts. The copper versus MgO correlations in massive peridotites consistently extrapolate to values of ~30 ppm at 36% MgO, whereas those for the xenoliths usually extrapolate to <20 ppm, albeit with much scatter. A value of 30 ppm is a relatively high value when chondrite normalized ($(Cu/Mg)_N = 0.11$), and would imply Cu/Ni and Cu/Co ratios greater than chondritic, difficult to explain, if true. However, the copper abundances in massive peridotites are correlated with sulfur, and may have been affected by the sulfur mobility postulated by Lorand (1991). Copper in xenoliths is not correlated with sulfur, and its abundance in the xenoliths and also inferred from correlations in basalts and komatiites points to a substantially lower abundance of ~20 ppm (O'Neill, 1991). We have adopted this latter value.

Niobium. Wade and Wood (2001) have suggested that niobium, potentially the most siderophile of the RLEs, is depleted in the PM by some extraction into the core. The depletion of niobium estimated from Nb/Ta would be ~15 ± 15% (Kamber and Collerson, 2000). As an RLE, niobium would have a mantle abundance of 690 ppb. Considering that 15% is in the core reduces this number to 588 ppm, which is listed in Table 2.

Silver. The common oxidation state is Ag^{1+}, which has an ionic radius between Na^{1+} and K^{1+}, and nearer to the former. Rather gratifyingly, the ratio Ag/Na is quite constant between peridotites (BVSP), MORB and other basalts (Laul *et al.*, 1972; Hertogen *et al.*, 1980), and continental crust (Gao *et al.*, 1998) at $(1.6 \pm 1.0) \times 10^{-6}$, giving 4 ppb in the PM. Silver reported by Garuti *et al.* (1984) in massive peridotites of the Ivrea zone is much higher and does not appear realistic.

Cadmium. Cadmium appears to be compatible or very mildly incompatible, similar to zinc. Almost nothing is known about which minerals it prefers. From a crystal-chemical view, cadmium has similar ionic radius and charge to calcium, but a tendency to prefer lower coordination due to its more covalent bonding with oxygen (similar to zinc and indium). Cadmium in spinel lherzolites varies from 30 ppb to 60 ppb (BVSP) and varies in basalts from about 90 ppb to 150 ppb (Hertogen *et al.*, 1980; Yi *et al.*, 2000). Cd/Zn is $\sim 10^{-3}$ in peridotites (BVSP) and the continental crust (Gao *et al.*, 1998), and $\sim 1.5 \times 10^{-3}$ in basalts (Yi *et al.*, 2000). We adopt the mean of these ratios (1.2×10^{-3}).

Tellurium. Tellurium is chalcophile (Hattori *et al.*, 2002), but is not correlated with other chalcophile elements, or anything else. Empirically, tellurium appears to be quite compatible. Morgan (1986) found 12.4 ± 3 ppb for fertile spinel-lherzolite xenoliths (previously published in BVSP), mainly from Kilbourne Hole, with somewhat lower values for more depleted samples. Yi *et al.* (2000) found 1–7 ppb for MORB, most OIB and submarine IAB, but with samples from Loihi extending to 29 ppb. These MORB data confirm the earlier data of Hertogen *et al.* (1980), who analyzed five MORBs with tellurium from 1 ppb to 5 ppb, and two outliers with 17 ppb, which also had elevated selenium. The important point is that tellurium in peridotites is often higher than in basalts, plausibly explained by retention in a residual sulfide phase. Thus, future work on tellurium needs to address the composition of mantle sulfides. If Te/S were chondritic, the PM with 200 ppm S would have 8 ppb Te. We have adopted this as the default value, as to use anything else would have interesting but unwarranted cosmochemical implications.

Mercury. Very few pertinent data exist, and the high-temperature geochemical properties of mercury are very uncertain. Flanagan *et al.* (1982) measured mercury in a variety of standard rocks, including basalts. Mercury in basalts is variable (3–35 ppb) and does not correlate with other elements. Garuti *et al.* (1984) report higher levels in massive peridotites (to 150 ppb, but mostly 20–50 ppb), of doubtful reliability (cf. silver). Wedepohl (1995) suggests 40 ppb in the average continental crust, but largely from unpublished sources. Gao *et al.* (1998) report ~9 ppb for the continental crust of East China. On the assumption that mercury is completely chalcophile in its geochemical properties, we obtain a crustal ratio of Hg/Se = 0.075 from Gao *et al.* (1998); on the further assumption that this ratio is conserved during mantle melting, this gives Hg = 6 ppb.

Thallium. Tl^{1+} has an ionic radius similar to Rb^{1+}. The crustal ratio Tl/Rb is quite constant at 0.005 (Hertogen *et al.*, 1980). Since >60% of the Earth's rubidium is probably in the continental crust, and rubidium is reasonably well known from Rb/Sr isotope systematics, this ratio should supply a reasonable estimate for the PM.

Bismuth. Bi^{3+} has an ionic radius similar to La^{3+}, indicating that it may behave geochemically like this lightest of the REEs, i.e., highly incompatibly. Bismuth in spinel lherzolites is 1–2 ppb, and is much less variable than lanthanum, with which it shows no correlation at all (BVSP). Nor does bismuth correlate with any other lithophile element. It is, however, noticeably higher in metasomatized garnet peridotites with elevated lanthanum. Bismuth in MORB is also quite constant at 6–9 ppb (Hertogen *et al.*, 1980), but much higher in the incompatible-element enriched BCR-1 (46 ppb). The crustal abundance is distinctly elevated, and also unusually variable, with 90–1,400 ppb in various crustal units from East China units tabulated by Gao *et al.* (1998). Their mean is 260 ppb, with a Bi/La ratio of 0.008. This would imply PM bismuth of 5 ppb.

Highly siderophile elements. The six platinum-group elements (PGEs: osmium, iridium, platinum, ruthenium, rhodium, palladium) as well as rhenium and gold are collectively termed highly siderophile elements (HSEs). The basis for estimating HSE abundances in the Earth's mantle is the rather uniform distribution of iridium in mantle peridotites. Morgan *et al.* (2001) estimate 3.2 ± 0.2 ppb Ir. Data for other HSE show considerably larger scatter. As the $^{187}Os/^{186}Os$ ratios of upper mantle rocks are similar to H-chondrites but different from carbonaceous chondrites (Walker *et al.*, 2002), mantle abundances are estimated by assuming H-chondrite ratios among the HSEs. Kallemeyn and Wasson (1981) determined an Os/Ir ratio of 1.062 for CI-chondrites and a mean of 1.07 ± 0.04 for 19 carbonaceous chondrites, which is identical to the Os/Ir ratio of 1.072 ± 0.014 for 22 H-chondrites analyzed by Kallemeyn *et al.* (1989). For the calculation of the osmium mantle abundance we have used an Os/Ir ratio of 1.07, resulting in an osmium mantle content of 3.4 ppb. For platinum, rhodium, and ruthenium, CI-ratios with iridium were used (Table 4) as they are more accurately known than H-chondrite ratios. The Pd/Ir and Pd/Au ratios are different between CI-chondrites and H-chondrites. Morgan *et al.* (1985) determined an average Pd/Ir ratio of 1.02 ± 0.097 for 10 ordinary chondrites, leading to a PM concentration of 3.3 ppb. Kallemeyn *et al.* (1989) obtained a mean Ir/Au ratio of 3.63 ± 0.13 for H-chondrites. This value is near the CI-ratio of 3.24 calculated from the abundances in Table 4. The H-chondrite ratio leads to an upper mantle gold content of 0.88 ppb which is listed in Table 4.

2.01.4.3 Is the Upper Mantle Composition Representative of the Bulk Earth Mantle?

It has been proposed that there is a substantial difference in major element chemistry, particularly in Mg/Si and Mg/Fe ratios, between the upper mantle above the 660 km seismic discontinuity and the lower mantle below this discontinuity. A chemical layering may have occurred either as a direct result of inhomogenous accretion without subsequent mixing, which has not to our knowledge been seriously suggested in recent times; or by some process akin to crystal fractionation from an early magma ocean, which has. The latter process would also imply gross layering of trace elements, and would invalidate the conclusions drawn here concerning PM volatile and siderophile element abundances. For example, the nearly chondritic Ni/Co ratio observed in the upper mantle would be a fortuitous consequence of olivine flotation into the upper mantle (Murthy, 1991; see Chapter 2.10).

Three kinds of evidence have been put forward in support of a lower mantle with a different composition from the upper mantle. The first was the apparent lack of a match between the seismic and other geophysical properties observed for the lower mantle, and the laboratory-measured properties of lower mantle minerals ($MgSiO_3$-rich perovskite and magnesiowüstite) in an assemblage with the upper mantle composition (meaning, effectively, with the upper mantle's Mg/Si and Mg/Fe ratios). Jackson and Rigden (1998) reinvestigated these issues and conclude that there is no such mismatch (see Chapter 2.02).

The second line of evidence is that crystal fractionation from an early magma ocean would inevitably lead to layering. The fluid-dynamical reasons why this is not inevitable (and is in fact unlikely) are discussed by Tonks and Melosh (1990) and Solomatov and Stevenson (1993).

The third kind of evidence is that the upper mantle composition violates the cosmochemical constraints on PM compositions that are obtained from the meteoritic record. A detailed comparison of PM compositions with primitive meteorite compositions is given below and it is shown that the PM composition shows chemical fractionations that are similar to the fractionations seen in carbonaceous chondrites.

The geochemical evidence against gross compositional layering of the mantle seems to us conclusive. This evidence is the observation that the upper mantle composition as deduced from the depleted mantle/continental crust model conforms, within uncertainty, to the cosmochemical requirement of having strictly chondritic ratios of the RLEs. These ratios would not have survived large-scale differentiation by fractional crystallization. Experimental data (e.g., Kato *et al.*, 1988) show that Sm/Nd and Lu/Hf would not remain unfractionated if there had been any significant $MgSiO_3$-perovskite or majoritic garnet fractionation. The most precise line of argument comes from the observed secular evolution of ε_{Nd} and ε_{Hf} towards their present values in the depleted mantle reservoir. With chondritic ratios among hafnium, lutetium, neodymium, and samarium, the observed mantle array can be explained by assuming the depleted mantle to represent ancient garnet-bearing residues from normal mantle melting. Although perovskite fractionation during cooling of a magma ocean in the early history of the Earth could explain the Hf–Lu systematics, it would not account for the neodymium isotopic signature of the oceanic basalts (Blicher-Toft and Albarede, 1997).

Thus, there is neither evidence for nonchondritic bulk Earth RLE ratios nor for nonchondritic ratios in the upper or lower mantle.

Finally, a well-mixed compositionally uniform mantle is also supported by geophysical evidence (i.e., tomography), showing that slabs penetrate into the lower mantle, which would support whole mantle convection (Van der Hilst *et al.*, 1997; see Chapter 2.02).

2.01.5 COMPARISON OF THE PM COMPOSITION WITH METEORITES

As discussed earlier in the chapter, the existing models of the accretion of the Earth assume collisional growth by successive impacts of Moon- to Mars-sized planetary embryos, a process lasting for tens of millions of years and implying significant radial mixing (e.g., Chambers, 2001). The growth of the embryos from micrometer-sized dust grains to kilometer-sized planetesimals is completed in less than one million years, with material derived from local feeding zones. Meteorites are fragments of disrupted planetesimals that did not accumulate to larger bodies, perhaps impeded by the strong gravitational force of Jupiter. Thus meteorites may be considered to represent early stages in the evolution of planets, i.e., building blocks of planets.

As pointed out before, many of the variations in chemical composition and oxidation state observed in chondritic meteorites must have been established at an early stage, before nebular components accreted to small planetesimals. As nebular fractionations have produced a variety of chondritic meteorites, it is necessary to study the whole spectrum of nebular fractionations in order to see if the proto-earth material has been subjected to the same processes as those recorded in the chondritic meteorites. As the Earth makes up more than 50% of the inner solar system, any nebular fractionation that affected the Earth's composition must have been a major process in the inner solar system. We will begin the discussion of nebular fractionations in meteorites and in the Earth with refractory elements and continue with increasingly more volatile components as outlined in Table 2.

2.01.5.1 Refractory Lithophile Elements

Figure 7 shows the abundances of the four refractory lithophile elements—aluminum, calcium, scandium, and vanadium—in several groups of undifferentiated meteorites, the Earth's upper mantle and the Sun. The RLE abundances are divided by magnesium and this ratio is then normalized to the same ratio in CI-chondrites. These $(RLE/Mg)_N$ ratios are plotted in Figure 7 (see also Figure 1). The level of refractory element abundances in bulk chondritic meteorites varies by less than a factor of 2. Carbonaceous chondrites have either CI-chondritic or higher Al/Mg ratios (and other RLE/Mg ratios), while rumurutiites (highly oxidized chondritic meteorites), ordinary chondrites, acapulcoites, and enstatite chondrites are depleted in refractory elements. The $(RLE/Mg)_N$ ratio in the mantle of the Earth is within the range of carbonaceous chondrites.

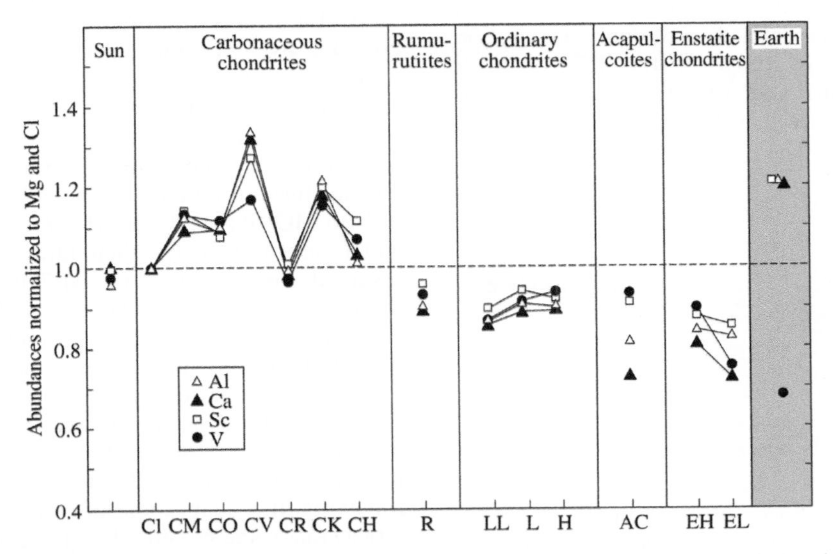

Figure 7 Element/Mg ratios normalized to CI-chondrites of RLEs in various groups of chondritic meteorites. The carbonaceous chondrites are enriched in refractory elements; other groups of chondritic meteorites are depleted. The PM has enrichments in the range of carbonaceous chondrites. The low V reflects removal of V into the core (source O'Neill and Palme, 1998).

The apparent depletion of vanadium in the Earth's mantle has two possible causes. (i) Vanadium is one of the least refractory elements. It is, e.g., significantly less enriched in CV-chondrites compared to other RLEs (Figure 7). Thus, vanadium is an exception to the general rule of constant ratios among RLEs. (ii) Vanadium is slightly siderophile. Consequently a certain fraction of vanadium is sequestered in the Earth's core (see below). Since neither aluminum, calcium, scandium, or magnesium are expected to partition into metallic iron, even at very reducing conditions, the RLE/Mg ratios should be representative of the bulk Earth. Thus, relative to CI-chondrites bulk Earth is enriched in refractory elements by a factor of 1.21 when normalized to magnesium, and by a factor of 1.41 when normalized to silicon, within the range of carbonaceous chondrites but very different from ordinary and enstatite chondrites. These chondrite groups are depleted in refractory elements (Figure 7). The small variations among RLE ratios in chondritic meteorites evident in Figure 7 probably reflect, with the exception of vanadium, the limited accuracy of analyses and inhomogeneities in analyzed samples. Whether there are minor variations (<5%) in RLE ratios among chondritic meteorites is unclear. It cannot, however, be excluded based on available evidence.

The constancy of refractory element ratios in the Earth's mantle, discussed before, is documented in the most primitive samples from the Earth's mantle. Figure 8 plots (modified from Jochum *et al.*, 1989) the PM-normalized abundances of 21 refractory elements from four fertile spinel lherzolites. These four samples closely approach, in their bulk chemical composition, the primitive upper mantle as defined in the previous section. The patterns of most of the REEs (up to praseodymium) and of titanium, zirconium, and yttrium are essentially flat. The three

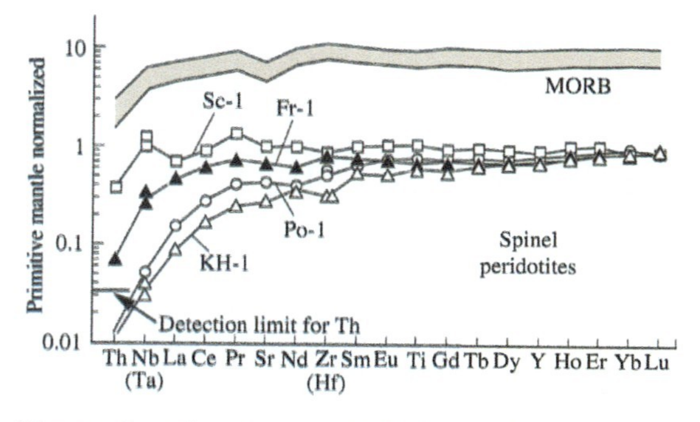

Figure 8 Abundances of RLEs in fertile spinel-lherzolite xenoliths from various occurrences. Compatible elements have constant enrichment factors. Abundances decrease with increasing degree of incompatibility, reflecting removal of very small degrees of partial melts (after Jochum *et al.*, 1989).

elements—calcium, aluminum, and scandium (not shown in Figure 8)—have the same PM-normalized ratios, i.e., they would plot at the level of ytterbium and lutetium in Figure 8. Only strongly incompatible elements—such as thorium, niobium, lanthanum, and caesium—are depleted. A very small degree of partial melting is sufficient to extract these elements from the mantle and leave the major element composition of the residual mantle essentially unaffected.

Chondritic relative abundances of strongly incompatible RLEs (lanthanum, niobium, tantalum, uranium, thorium) and their ratios to compatible RLEs in the Earth's mantle are more difficult to test. The smooth and complementary patterns of REEs in the continental crust and the residual depleted mantle are consistent with a bulk REE pattern that is flat, i.e., unfractionated when normalized to chondritic abundances. As mentioned earlier, the isotopic compositions of neodymium and hafnium are consistent with chondritic Sm/Nd and Lu/Hf ratios for bulk Earth. Most authors, however, *assume* that RLEs occur in chondritic relative abundances in the Earth's mantle. However, the uncertainties of RLE ratios in CI-meteorites do exceed 10% in some cases (see Table 4) and the uncertainties of the corresponding ratios in the Earth are in same range (Jochum *et al.*, 1989; Weyer *et al.*, 2002). Minor differences (even in the percent range) in RLE ratios between the Earth and chondritic meteorites cannot be excluded, with the apparent exception of Sm/Nd and Lu/Hf ratios (Blichert-Toft and Albarede, 1997).

That the assumption of chondritic relative abundances of refractory elements in the Earth is not without problems is demonstrated by the U/Nb ratio. Hofmann *et al.* (1986) showed that Nb/U is constant over several orders of magnitude of niobium or uranium concentrations in both MORBs and OIBs, with an Nb/U ratio of $\sim 47 \pm 10$, implying that niobium and uranium have similar incompatibilities during mantle melting, and that this ratio is also the ratio in the source of both MORBs and OIBs. However, because Nb and U are both RLEs, their ratio in the PM is assumed to be chondritic, i.e., 31.7. It turns out that the high ratio in the MORB and OIB source is compensated by a low ratio in the continental crust of ~ 10, indicating that niobium and uranium did not behave congruently during the processes responsible for crust formation. Table 4 lists a niobium concentration for PM that is 15% below the average normalized RLE abundance of the PM, following the suggestion of Wade and Wood (2001) that some niobium has partitioned into the core. Although this would reduce the PM Nb/U ratio from 31.7 to 27.6, it would not affect conclusions regarding the complementary behavior of uranium and niobium during MORB melting and crust formation. The lower than chondritic Nb/LRE ratio proposed for the Earth's mantle demonstrates how difficult it is to establish that an RLE ratio is indeed chondritic in the mantle. The 15% depletion of niobium relative to other RLE was only noticed from high accuracy studies of Nb/Ta abundances in mantle and crustal reservoirs (Münker *et al.*, 2003).

The Th/U ratio is of great importance in geochemistry, as uranium and thorium are major heat-producing elements in the interior of the Earth. In addition, the three lead isotopes—^{206}Pb, ^{207}Pb, and ^{208}Pb—the decay products of the two uranium isotopes and of ^{232}Th are used for dating. The chondritic ratio of the two refractory elements, thorium and uranium, is not as well established as one would generally assume. Although a formal uncertainty of 15% is calculated for the Th/U ratio in CI-chondrites obtained from data in Table 4, the ratio is better known. Rocholl and Jochum (1993) estimate a chondritic Th/U ratio of 3.9 with an error of 5%. Based on terrestrial lead-isotope systematics, Allègre *et al.* (1986) concluded that the average Th/U ratio of the Earth's mantle is 4.2, significantly above the CI-chondritic ratio of 3.82 listed here (Table 4) or the chondritic ratio of 3.9 given by Rocholl and Jochum (1993). There is little evidence for such a high Th/U ratio, as, e.g., shown by Campbell (2002) who used a correlation of Th/U versus Sm/Nd to establish a Th/U ratio of ~3.9 at a chondritic Sm/Nd ratio.

In summary, it is generally assumed that all refractory lithophile elements are enriched in the Earth's mantle by a common factor of 2.80 × CI. This factor can be directly verified for the less incompatible refractory elements by studying the most fertile upper mantle rocks, and there is indirect evidence that this factor is also valid for the highly incompatible refractory elements, although deviations from chondritic relative abundances of the order of 5–10% cannot be excluded for some elements. Nonchondritic refractory element ratios in various mantle and crustal reservoirs are interpreted in terms fractionation processes within the Earth. The absolute level of refractory elements, i.e., the refractory to nonrefractory element ratios in the Earth's mantle, is above that of CI-chondrites, somewhere between type 2 and type 3 carbonaceous chondrites, depending on magnesium or silicon normalization, but very different from ordinary chondrites.

2.01.5.2 Refractory Siderophile Elements

RSEs comprise two groups of metals: the HSEs—osmium, rhenium, ruthenium, iridium, platinum, and rhodium with metal/silicate partition coefficients $>10^4$—and the two moderately siderophile elements—molybdenum and tungsten (Table 2). As the major fractions of these elements are in the core of the Earth, it is not possible to establish independently whether the bulk Earth has chondritic ratios of RLE to RSE, i.e., whether ratios such as Ir/Sc or W/Hf are chondritic in the bulk Earth. Support for the similar behavior of RLE and RSE in chondritic meteorites is provided by Figure 9. The ratio of the RSE, Ir, to the nonrefractory siderophile element, Au, is plotted against the ratio of the RLE, Al, to the nonrefractory lithophile element, Si. Figure 9 demonstrates that RLEs and RSEs are correlated

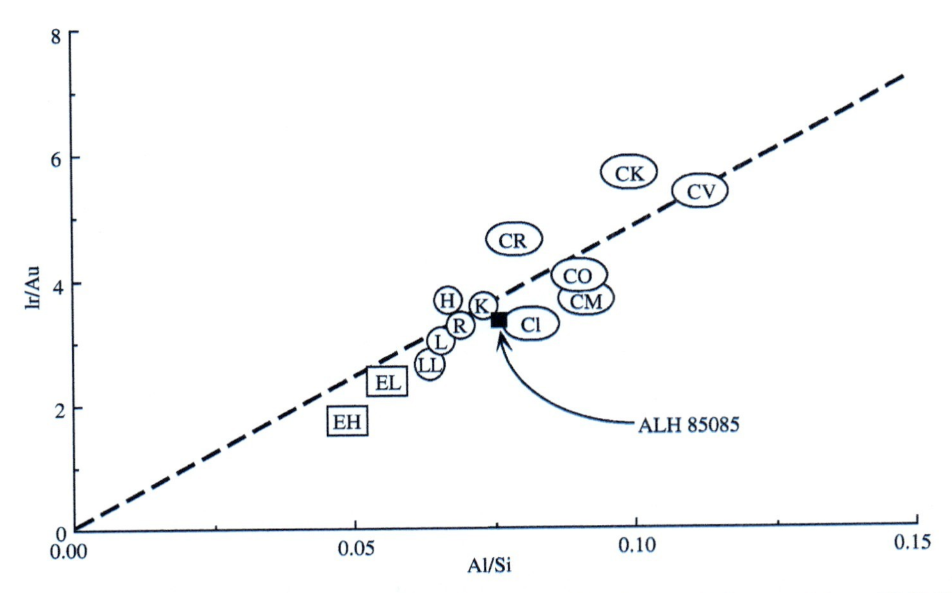

Figure 9 Ir/Au versus Al/Si in various types of chondritic meteorites. Ir is an RSE and Al is an RLE. The figure shows the parallel behavior of RSE and RLE. Enstatite chondrites (EH and EL) are low in RSE and RLE and carbonaceous chondrites (CM, CO, CR, CK, CV) are high in both (source O'Neill and Palme, 1998).

in chondritic meteorites, and one may therefore confidently assume that the bulk Earth RSEs behave similarly to the bulk Earth RLEs. This is important as it allows us to calculate bulk Earth abundances of tungsten, molybdenum, iridium, etc., from the bulk Earth aluminum or scandium contents. However, as shown in a later section, the Earth has excess iron metal, which provides an additional reservoir of siderophile elements. Thus, the RSE abundances calculated from RLE abundances provide only a lower limit for the RSE content of the bulk earth.

The refractory HSEs presently observed in the Earth's mantle are probably of a different origin than the bulk of the refractory HSEs which are in the core. This inventory of HSEs in the PM will be discussed in a later section.

2.01.5.3 Magnesium and Silicon

The four elements—magnesium, silicon, iron, and oxygen—contribute more than 90% by mass to the bulk Earth. As stated above, magnesium, silicon, and iron have approximately similar relative abundances (in atoms) in the Sun, in chondritic meteorites, and probably also in the whole Earth. On a finer scale there are, however, small but distinct differences in the relative abundances of these elements in chondritic meteorites. Figure 10 shows the various groups of chondritic meteorites in an Mg/Si versus Al/Si plot. As discussed before, the Earth's mantle has Al/Si and Al/Mg ratios within the range of carbonaceous chondrites, but the Mg/Si ratio is slightly higher than any of the chondrite ratios. The "nonmeteoritic" Mg/Si ratio of the Earth's mantle has been extensively discussed in the literature. One school of thought maintains that the bulk Earth has to have a CI-chondritic Mg/Si ratio and that the observed nonchondritic upper mantle ratio is balanced by a lower than chondritic ratio in the lower mantle (e.g., Anderson, 1989). However, layered Earth models have become increasingly unlikely, as discussed before, and there is no reason to postulate a bulk Earth CI–Mg/Si ratio in view of the apparent variations of this ratio in various types of chondritic meteorites. Ringwood (1989) suggested that the Mg/Si ratio of Earth's mantle is the same as that of the Sun implying that CI-chondrites have a nonsolar Mg/Si ratio. This is very unlikely in view of the excellent agreement between elemental abundances of CI-chondrites and the Sun (Chapter 1.03). Other attempts to explain the high Mg/Si ratio of the Earth's mantle include volatilization of silicon from the inner solar system by high-temperature processing connected with the early evolution of the proto-sun and the solar nebula (Ringwood, 1979, 1991). Another possibility is loss of some silicon into the metal core of the Earth (e.g., Hillgren *et al.* (2000) and references therein). The latter proposition has the advantage that it would, in addition, reduce the density of the outer core, which is ~10% too low for an FeNi alloy (Poirier, 1994; see Chapters 2.14 and 2.15). That the Earth's core contains some silicon is now

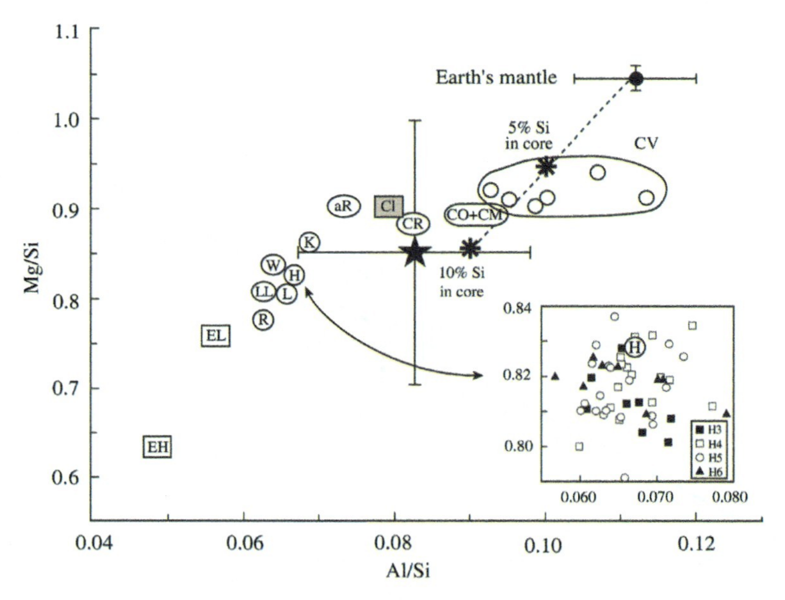

Figure 10 Mg/Si versus Al/Si for chondritic meteorites and the Earth's mantle. Carbonaceous chondrites are on the right side of the solar ratio (full star), ordinary and enstatite chondrites are on the left side, reflecting depletion of refractory elements. The Earth's mantle plots above CI-chondrites. Putting 5% Si into the core of the Earth leads to a PM composition compatible with CV-chondrites (after O'Neill and Palme, 1998).

assumed by most authors, but the amount is unclear. About 18% silicon is required to produce the 10% density reduction. This number is far too high as argued by O'Neill and Palme (1998), who concluded that other light elements must be involved in reducing the density of the core. The problem may be somewhat less severe in light of new density estimates of liquid FeNi alloys at high pressures and temperatures by Anderson and Isaak (2002) leading to a density deficit of only 5%. Based on these data McDonough (see Chapter 2.15) estimated a core composition with 6% silicon and 1.9% sulfur as light elements.

The effect of silicon partitioning into the core is shown in Figure 10. If the core has 5% silicon, the bulk Earth Mg/Si and Al/Si ratios would match with CV-chondrites which define a fairly uniform Mg/Si ratio of 0.90 ± 0.03, by weight (Wolf and Palme, 2001). Such a composition is not unreasonable in view of the similarity in moderately volatile element trends of carbonaceous chondrites and the Earth, as shown below.

2.01.5.4 The Iron Content of the Earth

Figure 11 plots the Mg/Fe ratios of various chondrites against their Si/Fe ratios. Carbonaceous chondrites fall along a single correlation line, with the Earth plotting at the extension of the carbonaceous chondrite line in the direction of higher iron contents. The bulk Earth iron content used here was calculated by assuming a core content of 85%Fe + 5%Si. The rest, comprising 10%, is made up of 5% Ni and some additional light element(s). Ordinary chondrites (H) and enstatite chondrites (EH and EL) do not plot on the same correlation line. If the upper mantle composition as derived here is representative of the bulk Earth's mantle, then the bulk Earth must have excess iron, at least 10% and more likely 20% above the iron content of CI-chondrites. Earlier McDonough and Sun (1995) have emphasized that the Earth has higher Fe/Al ratios than chondritic meteorites.

The growth of the Earth by accumulation of Moon- to Mars-sized embryos could provide an explanation for the excess iron in the Earth. Large impacts will remove some material from the Earth, preferentially silicates, if the Earth had a core at the time when the impact occurred. The amount of dispersed material is uncertain. It may be several percent of the total mass involved in a collision (Canup and Agnor, 2000). A model of collisional erosion of mantle has been proposed to explain the large metal fraction of Mercury by Benz *et al.* (1988). Because of the large size of the Earth, the colliding object must be very large or the collisional erosion occurred at a time when the Earth was not fully accreted, but nevertheless had an FeNi core. Some collisional erosion may also have occurred on the Moon- to Mars-sized embryos. The present excess of iron is then the integral effect over a long period of impact growth. Alternatively the high bulk Earth iron content would reflect the average composition of the "building blocks of the earth," i.e., the

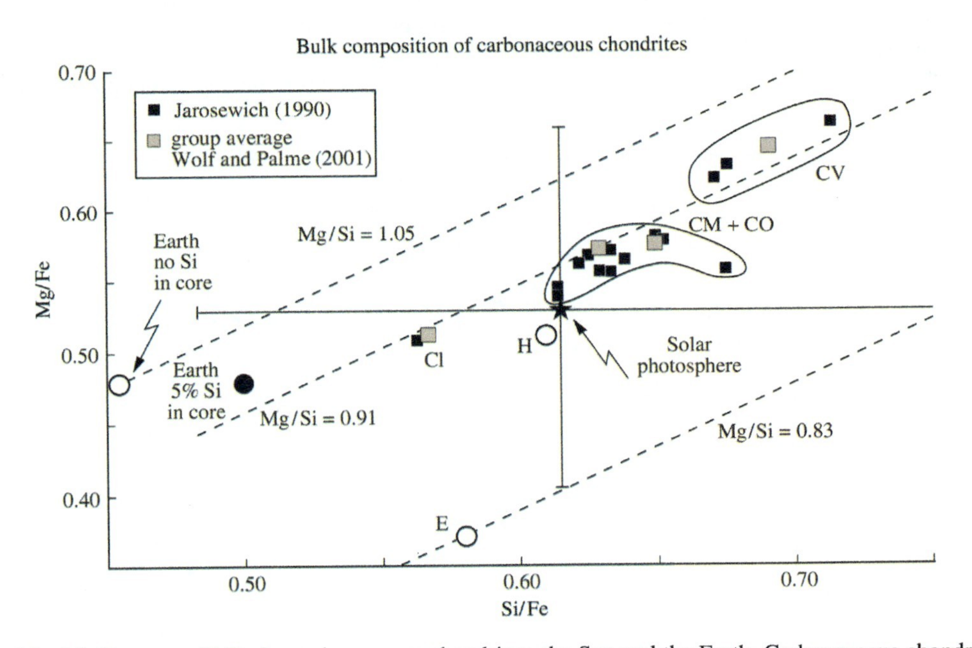

Figure 11 Mg/Fe versus Si/Fe for carbonaceous chondrites, the Sun and the Earth. Carbonaceous chondrites have constant Mg/Si ratios, but variable Fe contents. With 5% Si in the core the bulk Earth has the Mg/Si ratio of carbonaceous chondrites (see Figure 10). The bulk Earth has ~10% excess Fe compared to CI-chondrites (filled square—Jarosewich (1990) and gray square—group average Wolf and Palme (2001)).

Moon-sized embryos that produce the Earth through mutual collisions. Either the Fe/Mg ratios of these embryos were, on average, higher than the solar ratio or there were comparatively large variations in the Fe/Mg ratios of individual embryos, and a population with high Fe/Mg ratios was accidentally accumulated by the growing Earth. On the scale of meteorites, large variations in bulk iron contents are found, e.g., in the recently discovered group of CH-chondrites (not shown in Figure 1). These meteorites are compositionally and mineralogically related to carbonaceous chondrites (Bischoff *et al.*, 1993) and there is no doubt that the enrichment of iron is the result of nebular and not of planetary fractionations. It is, however, not clear if the large variations in metal/silicate ratios that occur on the scale of small parent bodies of chondritic meteorites would still be visible in the much larger Moon-sized embryos that made the Earth. The collisional erosion of silicates appears to us a more plausible possibility (Palme *et al.*, 2003).

2.01.5.5 Moderately Volatile Elements

The concentrations of four typical moderately volatile elements—manganese, sodium, selenium, and zinc—in the various classes of chondritic meteorites are shown in Figure 12, where elements are normalized to magnesium and CI-chondrites. Again there is excellent agreement between solar abundances and CI-meteorites. A characteristic feature of the chemistry of carbonaceous chondrites is the simultaneous depletion of sodium and manganese in all types of carbonaceous chondrites, except CI. Ordinary and enstatite chondrites are not or only slightly depleted in both elements, but their zinc contents are significantly lower than those of the carbonaceous chondrites. The Earth fits qualitatively with the carbonaceous chondrites in having depletions of manganese, sodium, and zinc. Selenium has a similar condensation temperature to zinc and sulfur. The selenium content of the Earth's mantle is extremely low, because a large fraction of the terrestrial selenium is, together with sulfur, in the core. In addition, the initial endowment of sulfur and selenium is very low based on the low abundances of other elements of comparable volatility (Dreibus and Palme, 1996).

Figure 13 plots the abundances of aluminum, calcium, magnesium, silicon, chromium, and of the three moderately volatile elements—manganese, sodium, and zinc—in the various groups of carbonaceous chondrites and in the Earth. All elements are normalized to the refractory element titanium and to CI-abundances. By normalizing to titanium the increasing enrichment of the refractory component from CI to CV is transformed into a depletion of the nonrefractory elements. There is a single depletion trend for magnesium, silicon, and the moderately volatile elements. The sequence of elements, their absolute depletion from aluminum to zinc, is basically in the order of decreasing condensation temperatures or increasing nebular volatility; the higher the volatility, the lower the abundance. It thus appears that this trend is the result of processes that occurred in the solar nebula and not by igneous or metamorphic activities on a parent body (Palme *et al.*, 1988; Humayun and Cassen, 2000).

Figure 13 plots only the lithophile moderately volatile elements. However, in carbonaceous chondrites siderophile and chalcophile moderately

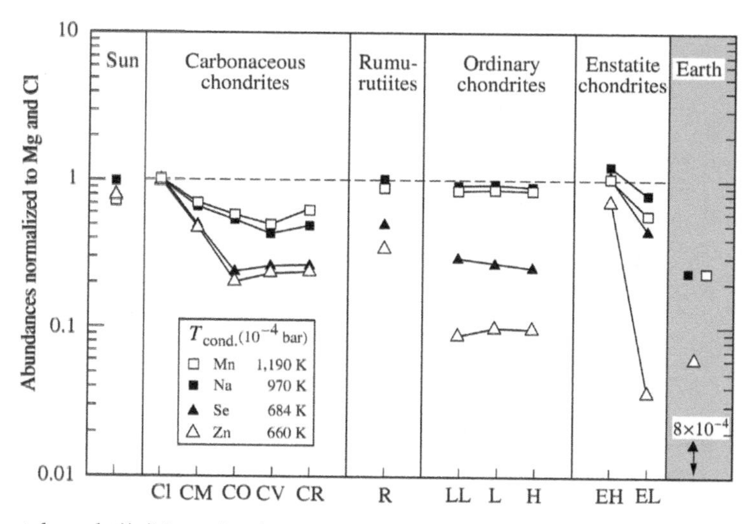

Figure 12 Moderately volatile/Mg ratios in various types of chondritic meteorites. All groups of chondritic meteorites are depleted in moderately volatile elements, none is enriched. The two elements Mn and Na are depleted in carbonaceous chondrites and in the Earth but not in ordinary and enstatite chondrites. The Earth is also depleted (source O'Neill and Palme, 1998).

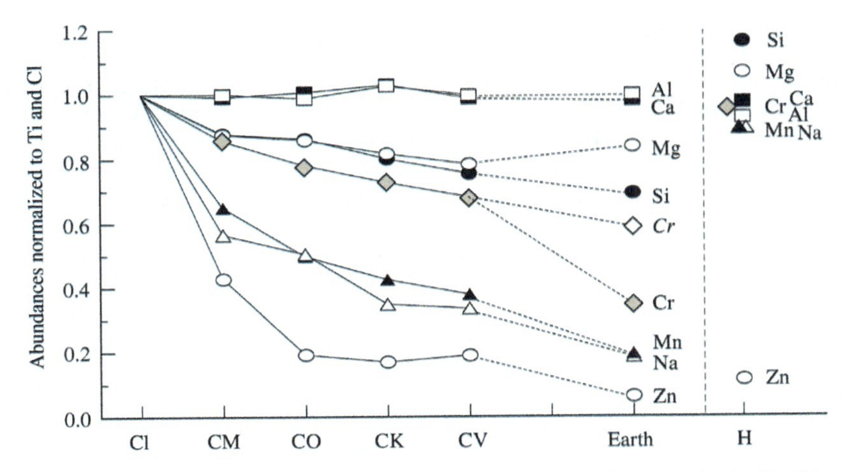

Figure 13 Major and moderately volatile elements in carbonaceous chondrites and in the Earth's mantle. All data are normalized to the RLE Ti. There is a single trend for RLE, Mg–Si, and moderately volatile elements. The Earth may be viewed as an extension of the carbonaceous chondrite trend. The low Cr content in the present mantle (full symbol) is the result of Cr partitioning into the core. The open symbol is plotted at the extension of the carbonaceous chondrite trend. Data for ordinary chondrites are plotted for comparison. Similar chemical trends in carbonaceous chondrites and the Earth are evident. H-chondrites are very different (sources Wolf and Palme, 2001; Wasson and Kallemeyn, 1988).

volatile elements show exactly the same behavior as lithophile elements as shown in Figure 14, where the ratio of moderately volatile elements in CV-chondrites to those in CI-chondrites is plotted against condensation temperature. The depletions increase with decreasing condensation temperatures independently of the geochemical character of the elements. For elucidating volatile element behavior in the Earth's composition, most of the elements plotted in Figure 14 are not useful, as siderophile and chalcophile moderately volatile elements may be depleted by core formation too. For the purpose of comparing moderately volatile elements between meteorites and the Earth, only lithophile elements can be used.

In Figure 13 the Earth's mantle seems to extend the trend of the moderately volatile elements to lower abundances, at least for sodium, manganese, and zinc (zinc behaves as a lithophile element in the Earth's mantle (see Dreibus and Palme, 1996)). The elements lithium, potassium, and rubidium which are not plotted here, show similar trends. The carbonaceous chondrite trend of iron is not extended to the Earth, as most of the iron of the Earth is in the core. The magnesium abundance of the Earth shows a slightly different trend. If the core had 5% silicon (previous section) and if that would be added to the bulk Earth silicon, then the bulk Mg/Si ratio of the Earth would be the same as that of carbonaceous chondrites (Figure 10) and the silicon abundance of the Earth's mantle in Figure 13 would coincide with the magnesium abundance.

The strong depletion of chromium (filled symbol in Figure 13) observed in the Earth's

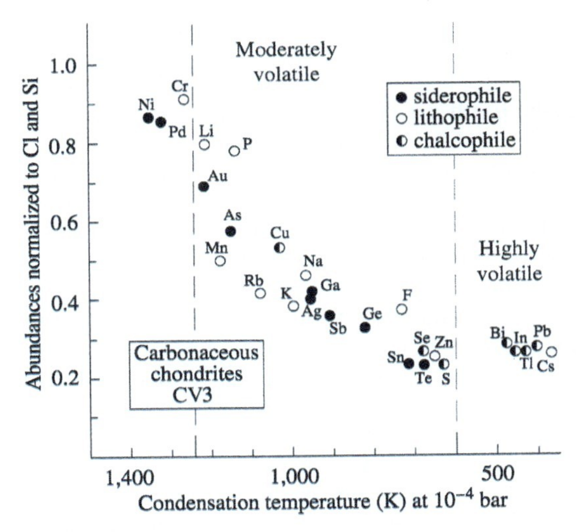

Figure 14 Abundances of volatile elements in CV3 chondrites (e.g., Allende) normalized to CI-chondrites and Si. There is a continuous decrease of abundances with increasing volatility as measured by the condensation temperature. The sequence is independent of the geochemical characters of the elements indicating that volatility is the only relevant parameter in establishing this pattern (source Palme, 2000).

mantle indicates that a significant fraction of chromium is now in the core of the Earth. Extending the carbonaceous chondrite volatility trend to the Earth leads to an estimated bulk Earth chromium content that is represented by an open symbol in Figure 13.

On the right-hand side of Figure 13 we have plotted the abundances of the same elements for average H-chondrites according to the compilation of meteorite data by Wasson and Kallemeyn (1988) and using the same normalization as for the carbonaceous chondrites in Figure 13. Ordinary and enstatite chondrites have apparently very different chemical compositions when compared to carbonaceous chondrites: (i) the depletion of refractory elements in ordinary chondrites leads to the high abundances of silicon and magnesium in Figure 13; (ii) magnesium and silicon are fractionated; and (iii) magnesium and sodium are only slightly depleted (see Figure 12). Enstatite chondrites show similar, though more enhanced, compositional trends for ordinary chondrites (Figures 9, 11, and 12). In summary, Figure 13 demonstrates quite clearly that the Earth's mantle composition resembles trends in the chemistry of carbonaceous chondrites and that these trends are not compatible with the chemical composition of ordinary or enstatite chondrites.

Figure 15(a) plots CI and magnesium-normalized abundances of lithophile moderately volatile elements against their condensation temperatures. The trend of decreasing abundance with increasing volatility is clearly visible. As mentioned above only few elements can be used to define this trend; most of the moderately volatile elements are siderophile or chalcophile and their abundance in the Earth's mantle is affected by core formation.

The additional depletions of siderophile elements are shown in Figure 15(b). Gallium, the least siderophile of the elements plotted here falls on the general depletion trend. All other elements are more or less strongly affected by core formation. The abundances of these elements in the Earth's mantle provide important clues to the mechanism of core formation. For example, the similar CI-normalized abundances of cobalt and nickel (Figure 15(b)) are surprising in view of the large differences in metal/silicate partition coefficients (e.g., Holzheid and Palme, 1996). Core formation should remove the more strongly siderophile element nickel much more effectively than cobalt leaving a mantle with low Ni/Co behind. The stronger decrease in metal/silicate partition coefficients of nickel compared to cobalt with increasing temperature and pressure provides a ready explanation for the relatively high abundance of nickel and the chondritic Ni/Co ratio in the Earth's mantle and allows us to estimate the $P-T$ conditions of the last removal of metal to the core (see Walter *et al.* (2000) and references therein; Chapter 2.10). The HSEs show the strongest depletion with a basically chondritic pattern. The significance of the HSEs will be discussed in more detail below.

Figure 15(c) shows chalcophile element abundances which are compared with the general depletion trend. The distinction between siderophile and chalcophile elements is not very clear. Analyses of separate phases in meteorites have shown that the only element apart from sulfur that is exclusively concentrated, i.e., truly chalcophile, in sulfides is selenium. All other elements have higher concentrations in metal than in coexisting sulfide, with the possible exceptions of indium and cadmium, for which the data are not known. This is particularly true for molybdenum, silver, tellurium, and copper elements that are traditionally considered to be chalcophile (Allen and Mason, 1973). In the absence of metallic iron, however, some siderophile elements will become chalcophile and partition into sulfide, such as the elements plotted in Figure 15(c). The significance of the trends shown in Figure 15(c) is twofold: (i) the depletion of the three elements—sulfur, selenium, and tellurium—is stronger than that of the highly siderophile elements in Figure 15(b) and (ii) elements more volatile than sulfur and selenium are less depleted than are these elements (see below for more details).

2.01.5.5.1 *Origin of depletion of moderately volatile elements*

The depletion of moderately volatile elements is a characteristic feature of every known body in the inner solar system (with the exception of the CI parent bodies, if these are inner solar system objects). The degree of depletion of an element is a function of nebular volatility but is independent of its geochemical character, i.e., whether an element is lithophile, chalcophile or siderophile, or compatible or incompatible (see Figure 14). Thus, volatile element depletions are not related to geochemical processes such as partial melting or separation of sulfide or metal. Also evaporation, e.g., by impact heating, is unlikely to have produced the observed depletion of moderately volatile element. (i) Volatilization on a local scale would produce enrichments through recondensation, but only depletions are observed (Palme *et al.*, 1988). (ii) Volatilization would lead to isotopic fractionations. Such fractionations are not observed even in rocks that are very low in volatile elements (Humayun and Cassen, 2000). (iii) The sequence of volatility-related losses of elements depends on oxygen fugacity. Volatilization will produce oxidizing conditions while very reducing conditions prevail during nebular condensation. A striking example for the different behavior during condensation and evaporation are the two moderately volatile elements—manganese and sodium. Both elements have similar condensation temperatures and their ratios are chondritic in all undifferentiated meteorites despite significant variations in absolute concentrations of manganese and sodium in carbonaceous chondrites

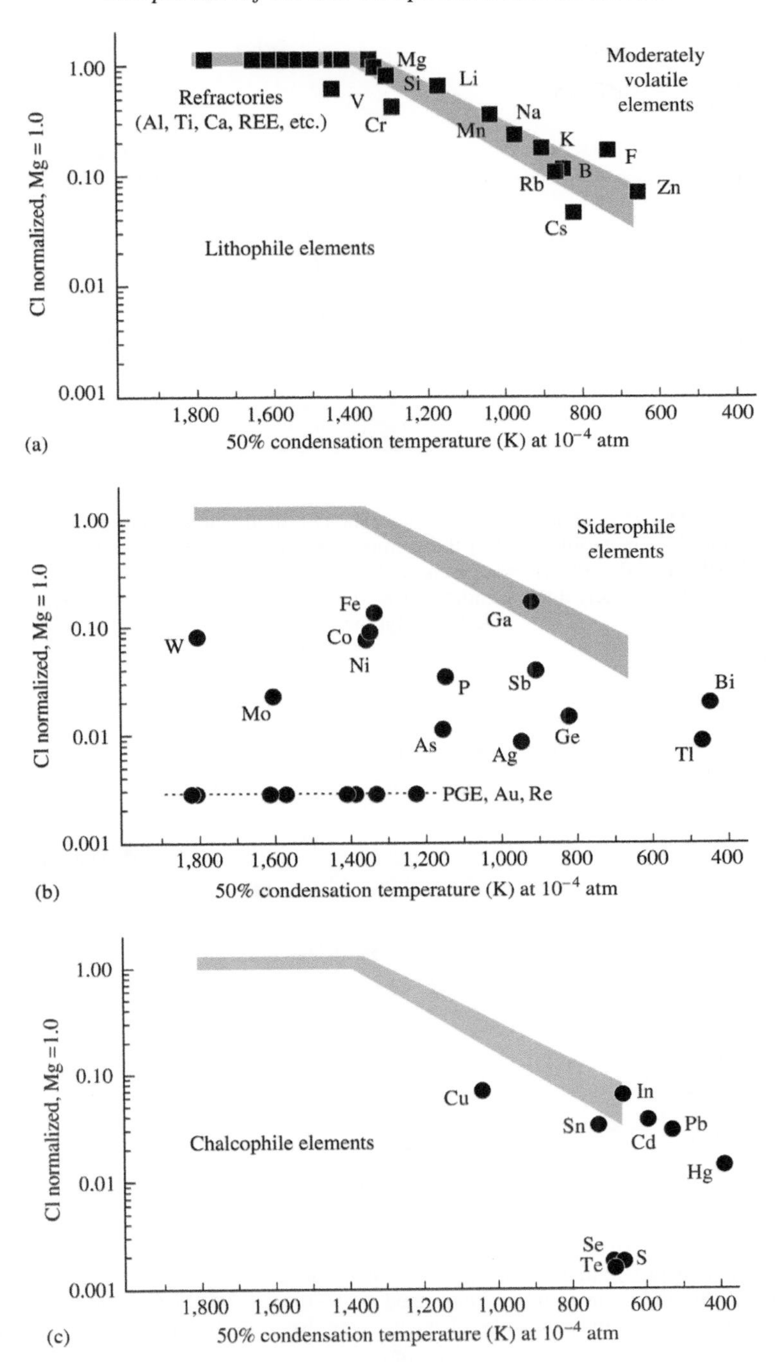

Figure 15 Abundances of moderately volatile elements in the Earth's mantle versus condensation temperatures: (a) lithophile elements define the volatility trend; (b) siderophile elements have variable depletions reflecting the process of core formation; and (c) chalcophile elements. The difference between siderophile and chalcophile elements is not well defined, except for S and Se. The large depletions of S, Se, and Te are noteworthy (see text) (after McDonough, 2001).

(Figures 12 and 16). The Mn/Na ratio of the Earth's mantle is also chondritic (Figures 12 and 16). Heating of meteorite samples to temperatures above 1,000 °C for a period of days will inevitably lead to significant losses of sodium and potassium but will not affect manganese abundances (e.g., Wulf *et al.*, 1995). The chondritic Mn/Na ratio of the Earth must, therefore, be attributed to nebular fractionations of proto-earth material (O'Neill and Palme, 1998). If some manganese had partitioned into the core, as is often assumed, the chondritic Mn/Na ratio

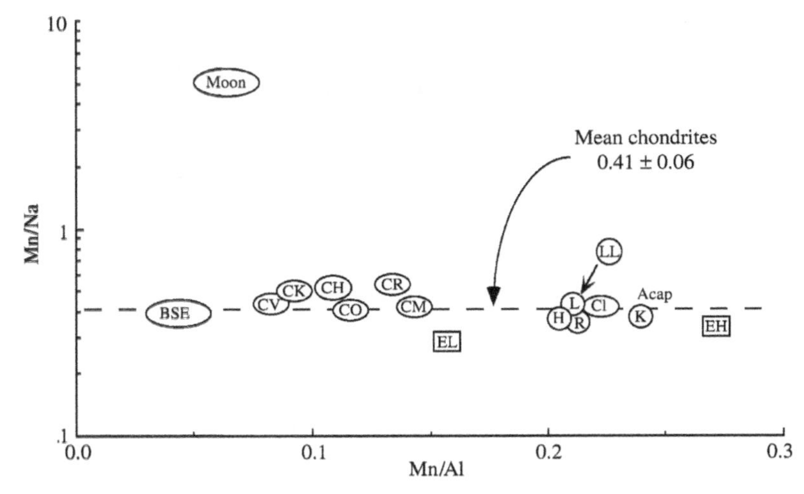

Figure 16 Mn/Na versus Mn/Al in chondritic meteorites. The two moderately volatile elements Na and Mn have the same ratio in all chondritic meteorites and in the primitive Earth's mantle, here designated as BSE. The low Mn/Al content of the Earth's mantle reflects enrichment of Al and depletion of Mn. Because of the chondritic Mn/Na ratio of the Earth's mantle, it is unlikely that a significant fraction of the Earth's inventory of Mn is in the core (source O'Neill and Palme, 1998).

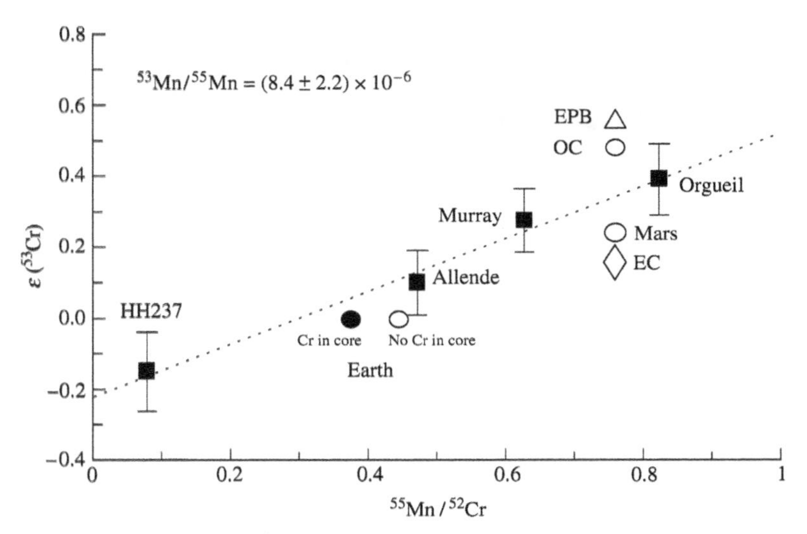

Figure 17 ^{53}Cr excess in bulk meteorites and bulk planets versus Mn/Cr ratios. The line through Orgueil, Allende, and Earth and the CH meteorite HH237 defines a $^{53}Mn/^{55}Mn$ ratio of $(8.40 \pm 2.2) \times 10^{-6}$. This is three million years earlier than the time of differentiation of Vesta (Lugmair and Shukolyukov, 1998). The $^{53}Mn/^{55}Mn$ ratio of carbonaceous chondrites and the Earth could be interpreted as the time of Mn/Cr fractionation in the solar system. The fractionation of other moderately volatile elements may have occurred at the same time and by the same process.

of the Earth's mantle would be the fortuitous result of the reduction of an originally higher Mn/Na ratio to the chondritic ratio. The similarity of the terrestrial pattern of moderately volatile elements, including manganese and sodium, with that of carbonaceous chondrites implies that the proto-earth material underwent similar fractionation processes for carbonaceous chondrites.

The depletion of moderately volatile elements such as manganese, sodium, rubidium in the Earth's mantle must have occurred very early in the history of the solar system. Figure 17 plots $^{53}Cr/^{52}Cr$ versus $^{55}Mn/^{52}Cr$ ratios of several chondritic meteorites and samples of differentiated planets. The data are taken from the work of Lugmair and Shukolyukov (1998) and Shukolyukov and Lugmair (2000, 2001). Samples from the metal-rich CH-chondrite HH237, the Earth, Allende (CV), Murray (CM), and Orgueil (CI) define an approximately straight line in Figure 17. If this line is interpreted as an isochron, the time of Mn–Cr fractionation can be calculated from the slope. The slope in Figure 17 corresponds to a $^{53}Mn/^{55}Mn$ ratio of $(7.47 \pm 1.3) \times 10^{-6}$. Nyquist *et al.* (2001) have determined chromium isotopes of individual chondrules with variable Mn/Cr ratios from the Bishunpur and Chainpur meteorites and they obtained "isochrons" in both cases with $^{53}Mn/^{55}Mn$ ratios of $(9.5 \pm 3.1) \times 10^{-6}$ for Bishunpur and $(9.4 \pm 1.7) \times 10^{-6}$ for Chainpur, respectively. These ratios are, within error limits, the same as indicated in Figure 17. Nyquist *et al.* (2001) believe that the time defined by the $^{53}Mn/^{55}Mn$ ratio of the chondrules date the Cr–Mn fractionation in the chondrule precursors. The fact that chondrules, bulk carbonaceous chondrites, and the Earth have, within error limits, the same $^{53}Mn/^{55}Mn$ ratio suggests that the Mn/Cr fractionation in these objects occurred at about the same time, at the beginning of the solar system.

In Figure 17, bulk ordinary chondrites (OCs) lie somewhat above the "isochron." This may reflect a reservoir with a slightly different initial $^{53}Cr/^{52}Cr$ ratio. The Mn/Cr ratios of Mars and of the eucrite parent body in Figure 17 are assumed to be the same as those of ordinary chondrites. This is, however, quite uncertain, as the chromium contents estimated for Mars (based on SNC meteorites) and on Vesta (based on EPB meteorites) are model dependent. The position of the Earth in Figure 17 is critically dependent on the assumed Mn/Cr ratio of the bulk Earth. As mentioned above, we take the view here that the parallel behavior of sodium and manganese in the carbonaceous chondrites and in the Earth's mantle follows from their empirically observed similar nebular volatilities in all groups of chondritic meteorites (Figure 16) and we assume that the Earth's core is, therefore, free of manganese (O'Neill and Palme, 1998). In contrast, the strong decrease of chromium in the Earth's mantle (see Figure 13) compared to CV-chondrites indicates that a certain fraction of chromium is in the Earth's core. For calculating the bulk Earth chromium content, we have extrapolated the CV-trend for chromium in Figure 13 (open symbol). With this procedure the bulk Earth chromium point of Figure 17 lies within the error of the carbonaceous chondrite isochron.

In summary, it appears from Figure 17 that the bulk composition of the Earth is related to the carbonaceous chondrites, suggesting the same event of manganese depletion for carbonaceous chondrites and the Earth. This supports the hypothesis that the depletion of volatile elements in the various solar system materials is a nebular event at the beginning of formation of the solar system as suggested by Palme *et al.* (1988), Humayun and Cassen (2000), and Nyquist *et al.* (2001).

2.01.5.6 HSEs in the Earth's Mantle

As discussed above, the HSEs—osmium, iridium, platinum, ruthenium, rhodium, palladium, rhenium, and gold—have a strong preference for the metal phase as reflected in their high metal/silicate partition coefficients of $>10^4$. Ratios among refractory HSEs are generally constant in chondritic meteorites to within at least 5%. There is, however, a notable exception. Walker *et al.* (2002) found that the average Re/Os ratio in carbonaceous chondrites is 7–8% lower than that in ordinary and enstatite chondrites. Both elements, osmium and rhenium, are refractory metals. They are, together with tungsten, the first elements to condense in a cooling gas of solar composition (Palme and Wlotzka, 1976). There is no apparent reason for the variations in Re/Os ratios. This example demonstrates that the assumption of constant relative abundances of refractory elements in chondritic meteorites may not be justified in all cases, although it is generally taken as axiomatic.

The HSE abundances in the Earth's mantle are extremely low (Figure 15(b)), ~0.2% of those estimated for the core, the major reservoir of the HSEs in the Earth. Most of the HSE data on Earth's mantle samples come from the analyses of iridium, which is sufficiently high in mantle rocks to allow the use of instrumental neutron activation analysis (e.g., Spettel *et al.*, 1990). The iridium contents in spinel lherzolites from worldwide occurrences of xenoliths and massive peridotites are, on average, surprisingly uniform. In particular, HSE abundances do not depend on the fertility of the mantle peridotite, except for rhenium (Pattou *et al.*, 1996; Schmidt *et al.*, 2000; Morgan *et al.*, 2001). The latter authors estimated an average PM iridium content of 3.2 ± 0.2 ppb, corresponding to a CI-normalized abundance of 6.67×10^{-3}. Other HSEs have similarly low CI-normalized abundances, as indicated by the approximately flat CI-normalized pattern of these elements that is generally observed, at least at the logarithmic scale at which these elements are usually displayed.

Some HSE ratios in upper mantle rocks often show significant deviations from chondritic ratios. For example, Schmidt *et al.* (2000) reported a 20–40% enhancement of ruthenium relative to iridium and CI-chondrites in spinel lherzolites from the Zabargad island. Data by Pattou *et al.* (1996) on Pyrenean peridotites, analyses of abyssal peridotites by Snow and Schmidt (1998), and data by Rehkämper *et al.* (1997) on various mantle rocks suggest that higher than chondritic Ru/Ir ratios are widespread and may be characteristic of a larger fraction, if not of the whole of the upper mantle. A parallel enrichment is found for rhodium in Zabargard rocks (Schmidt *et al.*, 2000). There are, however, few data on rhodium and it remains to be seen if this is a general signature of upper mantle. Further clarification is needed to see if these nonchondritic ratios are a primary signature of the Earth's mantle or are the result of later alterations (e.g., Rehkämper *et al.*, 1999). This question has important consequences for the understanding of the origin of the highly siderophile elements in the Earth's mantle.

Two of the HSEs, palladium and gold, are moderately volatile elements. Their abundances in chondrites are not constant. To some extent, they follow the general trend of moderately volatile lithophile elements (see Figures 9 and 14). For example, Morgan *et al.* (1985) found an average Pd/Ir ratio of 1.02 ± 0.097 for 10 ordinary chondrites, whereas E-chondrites have significantly higher ratios as shown by Hertogen *et al.* (1983) who determined an average ratio of 1.32 ± 0.2.

The concentrations of palladium in upper mantle rocks are also much more variable than those of the refractory siderophiles, with anomalously high Pd/Ir ratios with up to twice the CI-chondritic ratio (Pattou *et al.*, 1996; Schmidt *et al.*, 2000). These variations exceed those in chondritic meteorites considerably.

The high palladium content of some upper mantle rocks has led to a number of speculations attempting to explain the excess palladium. McDonough (1995) has suggested addition of outer core metal, high in nonrefractory palladium and low in refractory iridium. Other possibilities are discussed by Rehkämper *et al.* (1999), Schmidt *et al.* (2000), and Morgan *et al.* (2001). As the abundance of palladium in upper mantle rocks is so variable, we have chosen to calculate the average upper mantle palladium content from the H-chondrite Pd/Ir ratio of 1.022 ± 0.097 (Morgan *et al.*, 1985), as H-chondrite fit with osmium isotopes of Earth's mantle rocks as explained above (see Table 4 and explanations).

The element gold is even more variable than palladium in upper mantle rocks, presumably because gold is more mobile. It appears that variations in gold are regional. Antarctic xenoliths analyzed by Spettel *et al.* (1990) have an average gold content of 2.01 ± 0.17 ppb, while seven xenoliths from Mongolia analyzed by the same authors have less than 1 ppb Au. The average upper mantle abundance of gold is unclear. Morgan *et al.* (2001) suggested a very high value of 2.7 ± 0.7 ppb, based on extrapolation of trends in xenoliths. We take the more conservative view that the Ir/Au ratio in the upper mantle of the Earth is the H-chondrite ratio, which we take from Kallemeyn *et al.* (1989) as 3.63 ± 0.13. This value is near the CI-ratio of 3.24 calculated from the abundances in Table 4. This leads to an upper mantle gold content of 0.88 ppb which is listed in Table 4.

2.01.5.7 Late Veneer Hypothesis

Although HSE concentrations are low in the Earth's mantle, they are not as low as one would expect from equilibrium partitioning between core forming metal and residual mantle silicate, as emphasized by new data on metal/silicate partition coefficients for these elements (Borisov and Palme, 1997; Borisov *et al.*, 1994). Murthy (1991) suggested that partition coefficients are dependent on temperature and pressure in such a way that at the high $P-T$ conditions where core formation may have occurred, the observed mantle concentrations of HSEs would be obtained by metal/silicate equilibration. This hypothesis has been rejected on various grounds (O'Neill, 1992), and high $P-T$ experiments have not provided support for the drastic decrease of metal/silicate partition coefficients of HSE required by the Murthy model (Holzheid *et al.*, 1998).

Thus core–mantle equilibration can be excluded as the source of the HSEs in the Earth's mantle. It is more likely that a late accretionary component has delivered the HSEs to the Earth's mantle, either as single Moon-sized body which impacted the Earth after the end of core formation or several late arriving planetesimals. The impactors must have been free of metallic iron, or the metallic iron of the projectiles must have been oxidized after the collision(s) to prevent the formation of liquid metal or sulfide that would extract HSEs into the core of the Earth. The relative abundances of the HSEs in the Earth's mantle are thus the same as in the accretionary component, but may be different from those in the bulk Earth. The late addition of PGE with chondritic matter is often designated as the late veneer hypothesis (Kimura *et al.*, 1974; Chou, 1978; Jagoutz *et al.*, 1979; Morgan *et al.*, 1981; O'Neill, 1991). This model requires that the mantle was free of PGE before the late bombardment established the present level of HSEs in the Earth's mantle.

The late veneer hypothesis has gained additional support from the analyses of the osmium isotopic composition of mantle rocks. Meisel *et al.* (1996) determined the $^{187}Os/^{188}Os$ ratios of a suite of mantle xenoliths. Since rhenium is more incompatible during mantle partial melting than osmium, the Re/Os ratio in the mantle residue is lower and in the melt higher than the PM ratio. By extrapolating observed trends of $^{187}Os/^{188}Os$ versus Al_2O_3 and lutetium, two proxies for rhenium, Meisel *et al.* (1996) determined a $^{187}Os/^{188}Os$ ratio of 0.1296 ± 0.0008 for the primitive mantle. This ratio is 2.7% above that of carbonaceous chondrites, but within the range of ordinary and enstatite chondrites (Walker *et al.*, 2002). Thus, the time-integrated Re/Os ratio of the Earth's primtive mantle is similar to that of H- and E-chondrites, but 7.6% higher than that of carbonaceous chondrites. The ordinary chondrite signature of the late veneer does not contradict the general similarity of Earth's mantle chemistry with the chemistry of carbonaceous chondrites, because the fraction of material added as late veneer material is less than 1% of the mantle, and the origin of this material may, therefore, be quite different from the source of the main mass of the Earth, according to popular models of the accretion of the inner planets (Wetherill, 1994; Chambers, 2001; Canup and Agnor, 2000). In any case, the precisely determined chondritic Re/Os ratio in the primitive mantle of the Earth is a very strong argument in favor of the late veneer hypothesis.

The small amount of the late veneer (<1% chondritic material) would not have had a measurable effect on the abundances of other elements besides HSEs, except for some chalcophile elements, most importantly sulfur, selenium and tellurium (Figure 15(c)). The amount of sulfur presently in the Earth's mantle (200 ppm, Table 4) corresponds to only 0.37% of a nominal CI-component, while the iridium content suggests a CI-component of 0.67%. O'Neill (1991) has, therefore, suggested that the late veneer was compositionally similar to H-chondrites which contain only 2% S (Wasson and Kallemeyn, 1988). Because H-chondrites have higher iridium (780 ppb) than CI-chondrites, the required H-chondrite fraction would only be 0.41% based on iridium. This would correspond to 82 ppm S delivered by the late veneer to the mantle. In this case the Earth's mantle should have combined ~120 ppm S before the advent of the late veneer. If core formation in the Earth (or in differentiated planetesimals that accreted to form the Earth) occurred while the silicate portion was molten or partially molten, some sulfur must have been retained in this melt (O'Neill, 1991).

2.01.6 THE ISOTOPIC COMPOSITION OF THE EARTH

The most abundant element in the Earth is oxygen. In a diagram of $\delta^{17}O$ versus $\delta^{18}O$ the oxygen isotopic compositions of terrestrial rocks plot along a line with a slope of 0.5, designated as the terrestrial fractionation line in Figure 18. Most carbonaceous chondrites plot below and ordinary chondrites above the terrestrial mass fractionation line. Enstatite chondrites, extremely reduced meteorites with low oxygen content, and the most oxidized meteorites, CI-chondrites with magnetite and several percent of water, both plot on the terrestrial fractionation line (Figure 18). Other groups of carbonaceous chondrites have different oxygen isotopic compositions.

Variation in the oxygen isotopic composition of solar system materials has long been regarded as demonstrating the nonuniformity in isotopic composition of the solar system (e.g., Begemann, 1980). However, oxygen must be considered a

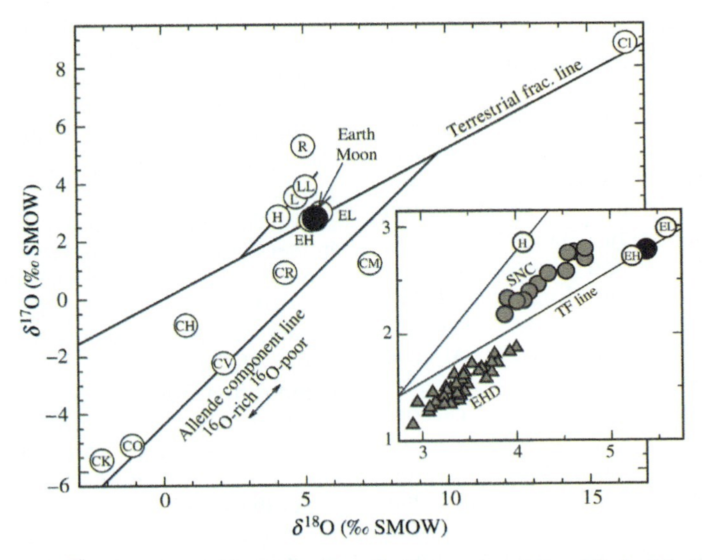

Figure 18 $\delta^{17}O$ versus $\delta^{18}O$ for chondritic meteorites. Ordinary chondrites (H, L, LL, and R) plot above the terrestrial fractionation line, and carbonaceous chondrites below. The most oxidized (CI) and the most reduced (E) chondrites also plot on terrestrial fractionation line. The larger bodies of the solar system, for which oxygen isotopes have been determined, Earth, Moon, Mars (SNC-meteorites), Vesta (EHD-meteorites), plot on or close to TFL (source Lodders and Fegley, 1997).

special case, because the major fraction of oxygen in the inner solar system was gaseous at the time of formation of solid objects, as oxygen is not fully condensed in any inner solar system material (see Figure 1). Gas–solid exchange reactions between individual meteoritic components with gases of different oxygen isotopic composition played an important role in establishing the variations (Clayton, 1993). The largest variations are found in the smallest meteorite inclusions. Individual components of carbonaceous chondrites span an extremely wide range in oxygen isotopes, such that the bulk meteorite value is of little significance (e.g., Clayton and Mayeda, 1984). The larger the object the smaller the variations in oxygen isotopes. The Earth and the Moon, the largest and the third largest body in the inner solar system for which oxygen isotopes are known, and which comprise more than 50% of the mass of the inner solar system, have exactly the same oxygen isotopic composition (Wiechert *et al.*, 2001). These two bodies may well represent the average oxygen isotopic composition of the bulk solar system, i.e., the Sun. The oxygen isotopic composition of Mars and Vesta, two other large inner solar system bodies, are not very different from that of Earth and Moon (Figure 18).

For most other elements there is no difference between the isotopic composition of carbonaceous chondrites and the Earth. As of early 2000s, only two exceptions, chromium and titanium, are known; for these two elements very small differences in the isotopic composition between carbonaceous chondrites and the Earth were found. Bulk carbonaceous chondrites have isotope anomalies in chromium and titanium. Isotopically unusual material may have been mixed to the CC-source after proto-earth material has accumulated to larger objects.

For chromium, the anomaly is only in ^{54}Cr and this effect is limited to carbonaceous chondrites (Rotaru *et al.*, 1992; Podosek *et al.*, 1997; Shukolykov and Lugmair, 2000). Titanium appears to be anomalous in ^{50}Ti and again the effect has only been found in carbonaceous chondrites (Niemeyer and Lugmair, 1984; Niederer *et al.*, 1985). In both cases the anomalies are larger in Ca, Al-inclusions of the Allende meteorite than in bulk meteorites.

The atmophile elements hydrogen, carbon, nitrogen, and the rare gases are strongly depleted in the Earth compared to chondritic meteorites. Pepin (1989) concluded that it appears that "simple 'veneer' scenarios in which volatiles are supplied from sources resembling contemporaneous meteorite classes" cannot explain the observed isotopic compositions. It is, therefore, often assumed that the isotopic compositions of these elements were affected by the process that led to their depletion (e.g., hydrodynamic escape) (Chapter 2.06).

2.01.7 SUMMARY

As regards the rock-forming elements, the bulk composition of the Earth is basically chondritic (i.e., solar) with approximately equal abundances of magnesium, silicon, and iron atoms. In detail, however, there are some variations in chemistry among chondritic meteorites, and from a detailed comparison with meteorites it is concluded that the bulk Earth composition has similarities with the chemical composition group of carbonaceous chondrites.

(i) The Earth and most groups of carbonaceous chondrites are enriched in refractory elements, other types of chondrites are depleted.

(ii) If the Earth's metal core contains 5% Si, then the Earth and carbonaceous chondrites have the same CI Mg/Si ratios. Ordinary and enstatite chondrites have significantly lower ratios.

(iii) Although the depletions of moderately volatile elements in the Earth are larger than in any group of carbonaceous chondrites, the Earth and carbonaceous chondrites show similar patterns of depletion of the moderately volatile elements: in particular, both are depleted in the alkali elements and in manganese. Enstatite and ordinary chondrites are also depleted in volatile elements, but their depletion patterns are different and sodium and manganese are not depleted relative to silicon.

(iv) The Earth and carbonaceous chondrites lie on the same ^{53}Cr/^{52}Cr versus ^{53}Cr/^{55}Mn isochron, indicating that the depletion of manganese and probably of all other moderately volatile elements in the Earth and in carbonaceous chondrites occurred shortly after the first solids had formed in the solar nebula.

There are also some differences between the chemistry of carbonaceous chondrites and the Earth.

(i) The bulk Earth has excess iron, reflected in higher Fe/Mg ratios of bulk Earth than the common groups of chondritic meteorites. It is suggested that silicates were lost during the collisional growth of the Earth involving giant impacts.

(ii) The Earth has a different oxygen isotopic composition from most carbonaceous chondrites.

(iii) Carbonaceous chondrites have isotope anomalies in chromium and titanium that are not observed in the Earth.

As the Earth makes up more than 50% of the inner solar system, we conclude that carbonaceous chondrites and the Earth reflect the major fractionation processes experienced by the material of the inner solar system. Other types of meteorites reflect fractionation processes on a more local scale.

REFERENCES

Allègre C. J., Dupre B., and Lewin E. (1986) Thorium/Uranium ratio of the Earth. *Chem. Geol.* **56**, 219–227.

Allègre C. J., Poirier J.-P., Humler E., and Hofmann A. W. (1995) The chemical composition of the Earth. *Earth Planet. Sci. Lett.* **134**, 515–526.

Allen O. R. and Mason B. (1973) Minor and trace elements in some meteoritic minerals. *Geochim. Cosmochim. Acta* **37**, 1435–1456.

Anderson D. (1989) *Theory of the Earth.* Blackwell, Boston, 366pp.

Anderson O. L. and Isaak D. G. (2002) Another look at the core density deficit of Earth's outer core. *Phys. Earth Planet. Int.* **131**, 10–27.

Begemann F. (1980) Isotope anomalies in meteorites. *Rep. Prog. Phys.* **43**, 1309–1356.

Bell J. F., Davies D. R., Hartmann W. K., and Gaffey M. J. (1989) Asteroids: the big picture. In *Asteroids II* (eds. R. P. Binzel, T. Gehrels, and M. S. Matthews). University of Arizona Press, Tucson, pp. 921–945.

Benz W., Slattery W. L., and Cameron A. G. W. (1988) Collisional stripping of mercury's mantle. *Icarus* **74**, 516–528.

Bischoff A., Palme H., Schultz L., Weber D., Weber H. W., and Spettel B. (1993) Acfer 182 and paired samples, an iron-rich carbonaceous chondrite: similarities with ALH85085 and its relationship to CR chondrites. *Geochim. Cosmochim. Acta* **57**, 2631–2648.

Blicher-Toft J. and Albarede F. (1997) The Lu–Hf isotope geochemistry of chondrites and the evolution of the mantle-crust system. *Earth Planet. Sci. Lett.* **148**, 243–258.

Borisov A. and Palme H. (1997) Experimental determination of the solubility of platinum in silicate melts. *Geochim. Cosmochim. Acta* **61**, 4349–4357.

Borisov A., Palme H., and Spettel B. (1994) Solubility of palladium in silicate melts: implications for core formation in the earth. *Geochim. Cosmochim. Acta* **58**, 705–716.

Boyd F. R. (1989) Compositional distinction between oceanic and cratonic lithosphere. *Earth Planet. Sci. Lett.* **96**, 15–26.

Brush S. G. (1980) Discovery of the Earth's core. *Am. J. Phys.* **48**, 705–723.

Burke J. G. (1986) *Cosmic Debris: Meteorites in History.* University of California Press, Berkely, CA.

BVSP (1981) *Basaltic Volcanism on the Terrestrial Planets.* Pergamon, New York, 1286pp.

Campbell I. H. (2002) Implications of Nb/U, Th/U and Sm/Nd in plume magmas for the relationship between continental and oceanic crust formation and the development of the depleted mantle. *Geochim. Cosmochim. Acta* **66**, 1651–1662.

Canil D., O'Neill H. St. C., Pearson D. G., Rudnick R. L., McDonough W. F., and Carswell D. A. (1994) Ferric iron in peridotites and mantle oxidation states. *Earth Planet. Sci. Lett.* **123**, 205–220.

Canup R. M. and Agnor C. B. (2000) Accretion of the terrestrial planets and the Earth–Moon system. In *Origin of the Earth and Moon* (eds. R. M. Canup and K. Righter). University of Arizona Press, Tucson, pp. 101–112.

Chambers B. (2001) Making more terrestrial planets. *Icarus* **152**, 205–224.

Chaussidon M. and Jambon A. (1994) Boron content and isotopic composition of oceanic basalts: geochemical and cosmochemical implications. *Earth Planet. Sci. Lett.* **121**, 277–291.

Chou C. L. (1978) Fractionation of siderophile elements in the Earth's upper mantle. *Proc. Lunar Planet. Sci. Conf.* **9**, 219–230.

Clayton R. N. (1993) Oxygen isotopes in meteorites. *Ann. Rev. Earth Planet. Sci.* **21**, 115–149.

Clayton R. N. and Mayeda T. K. (1984) The oxygen isotope record in the Murchison and other carbonaceous chondrites. *Earth Planet. Sci. Lett.* **67**, 151–161.

Déruelle B., Dreibus G., and Jambon A. (1992) Iodine abundances in oceanic basalts: implications for earth dynamics. *Earth Planet. Sci. Lett.* **108**, 217–227.

Dreibus G. and Palme H. (1996) Cosmochemical constraints on the sulfur content in the Earth's core. *Geochim. Cosmochim. Acta* **60**, 1125–1130.

Fitton J. G. (1995) Coupled molybdenum and niobium depletion in continental basalts. *Earth Planet. Sci. Lett.* **136**, 715–721.

Flanagan F. J., Moore R., and Aruscavage P. J. (1982) Mercury in geologic reference samples. *Geostand. Newslett.* **6**, 25–46.

Frey F. A. and Green D. H. (1974) The mineralogy, geochemistry, and origin of lherzolite inclusions in Victorian basalts. *Geochim. Cosmochim. Acta* **38**, 1023–1059.

Frey F. A., Suen C. J., and Stockman H. W. (1985) The Ronda high temperature peridotite: geochemistry and petrogenesis. *Geochim. Cosmochim. Acta* **49**, 2469–2491.

Galer S. J. G. and Goldstein S. L. (1996) Influence of accretion on lead in the Earth. In *Earth Processes: Reading the Isotopic Code,* Geophysical Monograph 95 (eds. A. Basu and S. Hart). American Geophysical Union, pp. 75–98.

Gao S., Luo T. C., Zhang B. R., Zhang H. F., Han Y. W., Zhao Z. D., and Hu Y. K. (1998) Chemical composition of the continental crust as revealed by studies in East China. *Geochim. Cosmochim. Acta* **62**, 1959–1975.

Garuti G., Gorgoni C., and Sighinolfi G. P. (1984) Sulfide mineralogy and chalcophile and siderophile element abundances in the Ivrea–Verbano zone mantle peridotites (Western Italian Alps). *Earth Planet. Sci. Lett.* **70**, 69–87.

Goldschmidt V. M. (1922) Über die Massenverteilung im Erdinnern, verglichen mit der Struktur gewisser Meteoriten. *Naturwissenschaften* **10**(Heft 42), 918–920.

Goldschmidt V. M. (1929) The distribution of the chemical elements. *Proc. Roy. Inst. Great Britain* **26**, 73–86.

Grevesse N. and Sauval A. J. (1998) Standard solar composition. *Space Sci. Rev.* **85**, 161–174.

Hart S. R. and Zindler A. (1986) In search of a bulk-earth composition. *Chem. Geol.* **57**, 247–267.

Hattori K. H., Arai S., and Clarke D. B. (2002) Selenium, tellurium, arsenic, and antimony contents of primary mantle sulfides. *Can. Mineral.* **40**, 637–650.

Hertogen J., Janssens M.-J., and Palme H. (1980) Trace elements in ocean ridge basalt glasses: implications for fractionations during mantle evolution and petrogenesis. *Geochim. Cosmochim. Acta* **44**, 2125–2143.

Hertogen J., Janssens M.-J., Takahashi H., Morgan J. W., and Anders E. (1983) Enstatite chondrites: trace element clues to their origin. *Geochim. Cosmochim. Acta* **47**, 2241–2255.

Hillgren V. J., Gessmann C. K., and Li J. (2000) An experimental perspective on the light element in the Earth's core. In *Origin of the Earth and Moon* (eds. R. M. Canup and K. Righter). University of Arizona Press, Tucson, pp. 245–263.

Hofmann A. W. and White W. M. (1983) Ba, Rb, and Cs in the Earth's mantle. *Z. Naturforsch.* **38a**, 256–266.

Hofmann A. W., Jochum K. P., Seufert M., and White W. M. (1986) Nb and Pb in oceanic basalts: new constraints on mantle evolution. *Earth Planet. Sci. Lett.* **79**, 33–45.

Holzheid A. and Palme H. (1996) The influence of FeO on the solubility of Co and Ni in silicate melts. *Geochim. Cosmochim. Acta* **60**, 1181–1193.

Holzheid A., Sylvester P., O'Neill H., St C., Rubie D. C., and Palme H. (1998) Late chondritic veneer as source of the highly siderophile elements in the Earth's mantle: insights from high pressure–high temperature metal-silicate partition behavior of Pd. *Nature* **406**, 396–399.

Humayun M. and Cassen P. (2000) Processes determining the volatile abundances of the meteorites and the terrestrial planets. In *Origin of the Earth and Moon* (eds. R. M. Canup and K. Righter). University of Arizona Press, Tucson, pp. 3–23.

Irving A. J. (1980) Petrology and geochemistry of composite ultramafic xenoliths in alkalic basalts and implications for magmatic processes within the mantle. *Am. J. Sci.* **280A**, 389–426.

Jackson I. and Rigden S. M. (1998) Composition and temperature of the Earth's mantle: seismological models interpreted through experimental studies of Earth materials. In *The Earth's Mantle: Structure, Composition, and Evolution—The Ringwood Volume* (ed. I. Jackson). Cambridge University Press, Cambridge, pp. 405–460.

Jagoutz E., Palme H., Baddenhausen H., Blum K., Cendales M., Dreibus G., Spettel B., Lorenz V., and Wänke H. (1979) The abundances of major, minor, and trace elements in the Earth's mantle as derived from primitive ultramafic nodules. In *Proc. 10th Lunar Planetary Science*. Pergamon Press, New York, pp. 1141–1175.

Jambon A., Déruelle B., Dreibus G., and Pineau F. (1995) Chlorine and bromine abundance in MORB: the contrasting behaviour of the Mid-Atlantic Ridge and East Pacific Rise and implications for chlorine geodynamic cycle. *Chem. Geol.* **126**, 101–117.

Jarosewich E. (1990) Chemical analyses of meteorites: a compilation of stony and iron meteorite analyses. *Meteoritics* **25**, 323–337.

Jochum K. P. and Hofmann A. W. (1997) Constraints on earth evolution from antimony in mantle-derived rocks. *Chem. Geol.* **139**, 39–49.

Jochum K. P., McDonough W. F., Palme H., and Spettel B. (1989) Compositional constraints on the continental lithospheric mantle from trace elements in spinel peridotite xenoliths. *Nature* **340**, 548–550.

Jochum K. P., Hofmann A. W., and Seufert H. M. (1993) Tin in mantle-derived rocks: constraints on earth evolution. *Geochim. Cosmochim. Acta* **57**, 3585–3595.

Kallemeyn G. W. and Wasson J. T. (1981) The compositional classification of chondrites: I. The carbonaceous chondrite groups. *Geochim. Cosmochim. Acta* **45**, 1217–1230.

Kallemeyn G. W., Rubin A. E., Wang D., and Wasson J. T. (1989) Ordinary chondrites: bulk composition, classification, lithophile-element fractionations, and composition-petrographic type relationships. *Geochim. Cosmochim. Acta* **53**, 2747–2767.

Kamber B. S. and Collerson K. D. (2000) Role of hidden deeply subducted slabs in mantle depletion. *Chem. Geol.* **166**, 241–254.

Kato T., Ringwood A. E., and Irifune T. (1988) Experimental determination of element partitioning between silicate perovskites, garnets and liquids: constraints on early differentiation of the mantle. *Earth Planet. Sci. Lett.* **89**, 123–145.

Kimura K., Lewis R. S., and Anders E. (1974) Distribution of gold and rhenium between nickel-iron and silicate melts: implications for abundance of siderophile elements on the earth and moon. *Geochim. Cosmochim. Acta* **38**, 683–701.

Kleine T., Münker C., Mezger K., and Palme H. (2002) Rapid accretion and early core formation on asteroids and the terrestrial planets from Hf-W chronometry. *Nature* **418**, 952–955.

Langmuir C. H., Klein E. M., and Plank T. (1992) Petrological systematics of mid-ocean ridge basalts: constraints on melt generation beneath ocean ridges. In *Mantle Flow and Melt Generation at Mid-Ocean Ridges,* Geophysical Monograph 71 (eds. J. P. Morgan, D. K. Blackman, and J. M. Sinton). American Geophysical Union, Washington, DC, pp. 183–280.

Larimer J. W. (1988) The cosmochemical classification of the elements. In *Meteorites and the Early Solar System* (eds. J. F. Kerridge and M. S. Matthews). University of Arizona Press, Tucson, pp. 375–393.

Laul J. C., Keays R. R., Ganapathy R., Anders E., and Morgan J. W. (1972) Chemical fractionations in meteorites: V. Volatile and siderophile elements in achondrites and ocean ridge basalts. *Geochim. Cosmochim. Acta* **36**, 329–345.

Lauretta D. S., and Lodders K. (1997) The cosmochemical behavior of beryllium and boron. *Earth Planet. Sci. Lett.* **146**, 315–327.

Lodders K. and Fegley B., Jr. (1997) An oxygen isotope model for the composition of mars. *Icarus* **126**, 373–394.

Lorand J. P. (1991) Sulphide petrology and sulphur geochemistry of orogenic lherzolites: A comparative study between Pyrenean bodies (France) and the Lanzo massif (Italy). In *Orogenic Lherzolites and Mantle Processes* (ed. M. A. Menzies) *J. Petrol.* **32**, pp. 77–95.

Lugmair G. W. and Shukolyukov A. (1998) Early solar system timescales according to ^{53}Mn-^{53}Cr systematics. *Geochim. Cosmochim. Acta* **62**, 2863–2886.

Lugovic B., Altherr R., Raczek I., Hofmann A. W., and Majer V. (1991) Geochemistry of peridotites and mafic igneous rocks from the Central Dinaric Ophiolite Belt, Yugoslavia. *Contrib. Mineral. Petrol.* **106**, 201–216.

Maaløe S. and Aoki K.-I. (1977) The major element composition of the upper mantle estimated from the composition of lherzolites. *Contrib. Mineral. Petrol.* **63**, 161–173.

Marvin U. B. (1996) Ernst Florens Chladni (1756–1827) and the origins of modern meteorite research. *Meteorit. Planet. Sci.* **31**, 545–588.

McDonough W. F. *et al.* (1985) Isotopic and geochemical systematics in Tertiary–Recent basalts from southeastern Australia and implications for the evolution of the sub-continental lithosphere. *Geochim. Cosmochim. Acta* **49**, 2051–2067.

McDonough W. F. (1990) Constraints on the composition of the continental lithospheric mantle. *Earth Planet. Sci. Lett.* **101**, 1–18.

McDonough W. F. (1995) An explanation for the abundance enigma of the highly siderophile elements in the Earth's mantle. *Lunar Planet. Sci.* **XXVI**, 927–928.

McDonough W. F. (2001) The Composition of the Earth. In *Earthquake Thermodynamics and Phase Transformations the Earth's Interior* (eds. R. Teisseyre and E. Majewski). Academic Press, pp. 3–23.

McDonough W. F. and Sun S.-S. (1995) The composition of the Earth. *Chem. Geol.* **120**, 223–253.

McDonough W. F., Sun S.-S., Ringwood A. E., Jagoutz E., and Hofmann A. W. (1992) Potassium, rubidium, and cesium in the Earth and Moon and the evolution of the mantle of the Earth. *Geochim. Cosmochim. Acta* **56**, 1001–1012.

Meisel T., Walker R. J., and Morgan J. W. (1996) The osmium isotopic content of the Earth's primitive upper mantle. *Nature* **383**, 517–520.

Morgan J. W. (1986) Ultramafic xenoliths: clues to Earth's late accretionary history. *J. Geophys. Res.* **91**(B12), 12375–12387.

Morgan J. W., Wandless G. A., Petrie R. K., and Irving A. J. (1981) Composition of the Earth's upper mantle: I. Siderophile trace elements in ultramafic nodules. *Tectonophysics* **75**, 47–67.

Morgan J. W., Janssens M. J., Takahashi H., Hertogen J., and Anders E. (1985) H-chondrites: trace elements clues to heir origin. *Geochim. Cosmochim. Acta* **49**, 247–259.

Morgan J. W., Walker R. J., Brandon A. D., and Horan M. F. (2001) Siderophile elements in Earth's upper mantle and lunar breccias: data synthesis suggests manifestations of the same late influx. *Meteorit. Planet. Sci.* **36**, 1257–1275.

Münker C., Pfänder J. A., Weyer S., Büchl A., Kleine T., and Mezger K. (2003) Evolution of planetary cores and the earth–moon system from Nb/Ta systematics. *Science* **30**, 84–87.

Murthy V. R. (1991) Early differentiation of the Earth and the problem of mantle siderophile elements: a new approach. *Science* **253**, 303–306.

Newsom H. E., Sims K. W. W., Noll P. D., Jr., Jaeger W. L., Maehr S. A., and Beserra T. B. (1996) The depletion of tungsten in the bulk silicate Earth: constraints on core formation. *Geochim. Cosmochim. Acta* **60**, 1155–1169.

Nickel K. G. and Green D. H. (1984) The nature of the uppermost mantle beneath Victoria, Australia as deduced from Ultramafic Xenoliths. In *Kimberlites, Proc. 3rd Inter. Kimberlite Conf.* (ed. J. Kornprobst). Elsevier, Amsterdam, pp. 161–178.

Niederer F. R., Papanastassiou D. A., and Wasserburg G. J. (1985) Absolute isotopic abundances of Ti in meteorites. *Geochim. Cosmochim. Acta* **49**, 835–851.

Niemeyer S. and Lugmair G. (1984) Ti isotopic anomalies in meteorites. *Geochim. Cosmochim. Acta* **48**, 1401–1416.

Nyquist L., Lindstrom D., Mittlefehldt D., Shih C.-Y., Wiesmann H., Wentworth S., and Martinez R. (2001) Manganese-chromium formation intervals for chondrules from the Bishunpur and Chainpur meteorites. *Meteorit. Planet. Sci.* **36**, 911–938.

O'Neill H. St. C. (1991) The origin of the moon and the early history of the Earth—a chemical model: Part 2. The Earth. *Geochim. Cosmochim. Acta* **55**, 1159–1172.

O'Neill H. St. C. (1992) Siderophile elements and the Earth's formation. *Science* **257**, 1282–1285.

O'Neill H. St. C. and Palme H. (1998) Composition of the silicate Earth: implications for accretion and core formation. In *The Earth's Mantle: Structure, Composition, and Evolution—the Ringwood Volume* (ed. I. Jackson). Cambridge University Press, Cambridge, pp. 3–126.

Palme H. (2000) Are there chemical gradients in the inner solar system. *Space Sci. Rev.* **92**, 237–262.

Palme H. and Nickel K. G. (1985) Ca/Al ratio and composition of the Earth's upper mantle. *Geochim. Cosmochim. Acta* **49**, 2123–2132.

Palme H. and Wlotzka F. (1976) A metal particle from a Ca, Al-rich inclusion from the meteorite Allende, and the condensation of refractory siderophile elements. *Earth Planet. Sci. Lett.* **33**, 45–60.

Palme H., Larimer J. W., and Lipschutz M. E. (1988) Moderately volatile elements. In *Meteorites and the Early Solar System* (eds. J. F. Kerridge and M. S. Matthews). University of Arizona Press, Tucson, pp. 436–461.

Palme H., O'Neill H. St. C., and Benz W. (2003) Evidence for collisional erosion of the Earth. *Lunar Planet. Sci.* **XXXIV**, 1741.

Pattou L., Lorand J. P., and Gros M. (1996) Non-chondritic platinum group element ratios in the Earth's mantle. *Nature* **379**, 712–715.

Pepin R. O. (1989) Atmospheric compositions: key similarities and differences. In *Origin and Evolution of Planetary and Satellite Atmospheres* (eds. S. K. Atreya, J. B. Pollack, and M. S. Matthews). University of Arizona Press, Tucson, pp. 291–305.

Podosek F. A., Ott U., Brannon J. C., Neal R. C., Bernatowicz T. J., Swan P., and Mahan S. E. (1997) Thoroughly anomalous chromium in Orgueil. *Meteorit. Planet. Sci.* **32**, 617–627.

Poirier J.-P. (1994) Light elements in the Earth's outer core: a critical review. *Phys. Earth Planet. Inter.* **85**, 319–337.

Rammensee W. and Wänke H. (1977) On the partition coefficient of tungsten between metal and silicate and its bearing on the origin of the moon. *Proc. 8th Lunar Sci. Conf.* pp. 399–409.

Rehkämper M., Halliday A. N., Barfod D., Fitton J. G., and Dawson J. B. (1997) Platinum-group element abundance patterns in different mantle environments. *Science* **278**, 1595–1598.

Rehkämper M., Halliday A. N., Alt J., Fitton J. G., Zipfel J., and Takzawa E. (1999) Non-chondritic platinum-group element ratios in oceanic mantle lithosphere: petrogenetic signature of melt percolation? *Earth Planet. Sci. Lett.* **65**, 65–81.

Ringwood A. E. (1979) *Origin of the Earth and Moon*. Springer, New York, 295pp .

Ringwood A. E. (1989) Significance of the terrestrial Mg/Si ratio. *Earth Planet. Sci. Lett.* **95**, 1–7.

Ringwood A. E. (1991) Phase transformations and their bearing on the constitution and dynamics of the mantle. *Geochim. Cosmochim. Acta* **55**, 2083–2110.

Rocholl A. and Jochum K. P. (1993) Th, U and other trace elements in carbonaceous chondrites: implications for the terrestrial and solar-system Th/U ratios. *Earth Planet. Sci. Lett.* **117**, 265–278.

Rotaru M., Birck J. L., and Allegre C. J. (1992) Clues to early solar system history from chromium isotopes in carbonaceous chondrites. *Nature* **358**, 465–470.

Russell H. N. (1941) The cosmical abundance of the elements. *Science* **94**, 375–381.

Ryan J. G. and Langmuir C. H. (1987) The systematics of lithium abundances in young volcanic rocks. *Geochim. Cosmochim. Acta* **51**, 1727–1741.

Schmidt G., Palme H., Kratz K.-L., and Kurat G. (2000) Are highly siderophile elements (PGE, Re and Au) fractionated in the upper mantle? New results on peridotites from Zabargad. *Chem. Geol.* **163**, 167–188.

Seitz H.-M. and Woodland A. B. (2000) The distribution of lithium in peridotitic and pyroxenitic mantle litholoies—an indicator of magmatic and metsomatic processes. *Chem. Geol.* **166**, 47–64.

Shukolyukov A. and Lugmair G. W. (2000) Cr isotope anomalies in the carbonaceous chondrites Allende and Orgueil and a potential connection between 54-Cr and oxygen isotopes. *Meteorit. Planet. Sci.* **35**, A146.

Shukolyukov A. and Lugmair G. W. (2001) Mn–Cr isotope systematics in bulk samples of the carbonaceous chondrites (abstr.). *Meteorit. Planet. Sci.* **36**, A188.

Sigvaldason G. E. and Oskarsson N. (1986) Fluorine in basalts from Iceland. *Contrib. Mineral. Petrol.* **94**, 263–271.

Sims K. W. W., Newsom H. E., and Gladney E. S. (1990) Chemical fractionation during formation of the Earth's core and continental crust: clues from As, Sb, W and Mo. In *Origin of the Earth* (eds. H. E. Newsom and J. H. Jones). Oxford University Press, Oxford, pp. 291–317.

Smith J. V., Delaney J. S., Hervig R. I., and Dawson J. B. (1981) Storage of F and Cl in the upper mantle: geochemical implications. *Lithos* **41**, 133–147.

Snow E. J. and Schmidt G. (1998) Constraints on the Earth accretion deduced from noble metals in the oceanic mantle. *Nature* **391**, 166–169.

Solomatov V. S. and Stevenson D. J. (1993) Suspension in convecting layers and style of differentiation of a terrestrial magma ocean. *J. Geophys. Res.* **98**, 5375–5390.

Spettel B., Palme H., and Wänke H. (1990) Siderophile elements in the primitive upper mantle. *Lunar Planet. Sci.* **XXI**. Lunar and Planetary Institute, Houston, pp. 1184–1185.

Taylor S. R. and McLennan S. M. (1985) *The Continental Crust: Its Composition and Evolution*. Blackwell, Oxford, 312pp.

Tonks W. B. and Melosh H. J. (1990) The physics of crystal settling and suspension in a turbulent magma ocean. In *Origin of the Earth* (eds. H. E. Newsom and J. H. Jones). Oxford University Press, pp. 151–174.

Van der Hilst R., Widiyantoro S., and Engdhal E. R. (1997) Evidence for deep mantle circulation from global tomography. *Nature* **386**, 578–584.

Wade J. and Wood B. J. (2001) The Earth's missing niobium may be in the core. *Nature* **409**, 75–78.

Walker R. J., Horan M. F., Morgan J. W., Becker H., Grossman J. N., and Rubin A. E. (2002) Comparative ^{187}Re-^{187}Os systematics of chondrites: implications regarding early solar system processes. *Geochim. Cosmochim. Acta* **66**, 4187–4201.

Walter M. J., Newsom H. E., Ertel W., and Holzheid A. (2000) Siderophile elements in the Earth and Moon: metal/silicate partitioning and implications for core formation. In *Origin of the Earth and Moon* (eds. R. M. Canup and K. Righter). University of Arizona Press, Tucson, pp. 265–289.

Wänke H., Baddenhausen H., Dreibus G., Jagoutz E., Kruse H., Palme H., Spettel B., and Teschke F. (1973) Multielement analyses of Apollo 15, 16, and 17 samples and the bulk composition of the moon. *Proc. Lunar Sci. Conf. 4th*, 1461–1481.

Wänke H., Dreibus G., and Jagoutz E. (1984) Mantle chemistry and accretion history of the Earth. In *Archean Geochemistry* (ed. A. Kröner). Springer, Berlin, pp. 1–24.

Wasson J. T. (1985) *Meteorites: Their Record of Early Solar-System History*. W. H. Freeman, New York.

Wasson J. T. and Kallemeyn G. W. (1988) Compositions of chondrites. *Phil. Trans. Roy. Soc. London* **A325**, 535–544.

Wedepohl K. H. (1995) The composition of the continental crust. *Geochim. Cosmochim. Acta* **59**, 1217–1232.

Wetherill G. W. (1994) Provenance of the terrestrial planets. *Geochim. Cosmochim. Acta* **58**, 4513–4520.

Weyer S., Münker C., Rehkämper M., and Mezger K. (2002) Determination of ultra-low Nb, Ta, Zr, and Hf concentrations and the chondritic Zr/Hf and Nb/Ta ratios by isotope dilution analyses with multiple collector ICP-MS. *Chem. Geol.* **187**, 295–313.

Wiechert U., Halliday A. N., Lee D.-C., Snyder G. A., Taylor L. A., and Rumble D. (2001) Oxygen isotopes and the moon-forming giant impact. *Science* **294**, 345–348.

Wolf D. and Palme H. (2001) The solar system abundances of P and Ti and the nebular volatility of P. *Meteorit. Planet. Sci.* **36**, 559–572.

Wulf A. V., Palme H., and Jochum K. P. (1995) Fractionation of volatile elements in the early solar system: evidence from heating experiments on primitive meteorites. *Planet. Space Sci.* **43**, 451–486.

Yi W., Halliday A. N., Alt J. C., Lee D. C., Rehkämper M., Garcia M. O., and Su Y. J. (2000) Cadmium, indium, tin, tellurium, and sulfur in oceanic basalts: implications for chalcophile element fractionation in the Earth. *J. Geophys. Res.* **105**, 18927–18948.

Zhang Y. and Zindler A. (1993) Distribution and evolution of carbon and nitrogen in earth. *Earth Planet. Sci. Lett.* **117**, 331–345.

2.02 Seismological Constraints upon Mantle Composition

C. R. Bina

Northwestern University, Evanston, IL, USA

2.02.1 INTRODUCTION

2.02.1.1 General Considerations

Direct sampling of mantle rocks and minerals is limited to tectonic slices emplaced at the surface (see Chapter 2.04), smaller xenoliths transported upwards by magmatic processes (see Chapter 2.05), and still smaller inclusions in such far-traveled natural sample chambers as diamonds (see Chapter 2.05). Because of such limited direct access to mantle materials, knowledge of mantle structure, composition, and processes must be augmented by geophysical remote sensing. What can various seismological observations tell us about the major-element composition of the upper mantle? How can they constrain possible differences in chemical composition between the upper and lower mantle? What light can they shed upon the nature of velocity heterogeneities in both the upper and lower mantle? It is these questions that we shall seek to address in this chapter.

Most seismological constraints on mantle composition are derived by comparison of values of seismic wave velocities inferred for particular regions within the Earth to the values measured in the laboratory for particular minerals or mineral assemblages, with such comparisons being made under comparable regimes of pressure (P) and temperature (T). The primary parameters of interest, then, are the compressional (or P-) wave velocities (V_P) and the shear (or S-) wave velocities (V_S). These wave velocities are simply related to the density (ρ) and to the two isotropic elastic moduli, the adiabatic bulk modulus (K_S)

and the shear (or "rigidity") modulus (G), via $V_P^2 = [K_S + (4/3)G]/\rho$ and $V_S^2 = G/\rho$, respectively.

2.02.1.2 Bulk Sound Velocity

Straightforward measurements of elastic properties of materials can be made via high-pressure static compression experiments, in which X-ray diffraction (XRD) is used to measure the molar volume (V), or equivalently the density (ρ), of a material as a function of pressure (P). The pressure dependence of volume is expressed by the "incompressibility" or isothermal bulk modulus (K_T), where $K_T = -V(\partial P/\partial V)_T$.

This isothermal bulk modulus (K_T) measured by static compression differs slightly from the aforementioned adiabatic bulk modulus (K_S) defining seismic velocities in that the former (K_T) describes resistance to compression at constant temperature, such as is the case in a laboratory device in which a sample is slowly compressed in contact with a large thermal reservoir such as the atmosphere. The latter (K_S), alternatively describes resistance to compression under adiabatic conditions, such as those pertaining when passage of a seismic wave causes compression (and relaxation) on a time-scale that is short compared to that of thermal conduction. Thus, the adiabatic bulk modulus generally exceeds the isothermal value (usually by a few percent), because it is more difficult to compress a material whose temperature rises upon compression than one which is allowed to conduct away any such excess heat, as described by a simple multiplicative factor $K_S = K_T(1 + T\alpha\gamma)$, where α is the volumetric coefficient of thermal expansion and γ is the thermodynamic Grüneisen parameter.

Experimentally, the bulk modulus is the simplest parameter to measure, but the seismological parameters of primary interest, V_P and V_S, both involve the shear modulus as well. It is convenient, therefore, to define a new parameter, the "bulk sound velocity" (V_ϕ), which eliminates all dependence upon the shear modulus (G) through a judicious linear combination of the squares of the two seismic wave velocities: $V_\phi^2 = K_S/\rho = V_P^2 - (4/3)V_S^2$. This new parameter (sometimes thought of as the P-wave velocity of an "equivalent" fluid, for which $G = 0$) can be determined directly from static compression data: $V_\phi^2 = K_S/\rho = (1 + T\alpha\gamma)(\partial P/\partial\rho)_T$. The bulk sound velocity possesses another desirable feature, in that it can also be constrained indirectly through chemical equilibrium experiments. Chemical equilibria describe free energy minima; the pressure dependence of free energy is described by the molar volume, and the pressure dependence of volume (or density) is described by K_T and hence V_ϕ. Thus, experimental determinations of equilibrium phase boundaries can provide independent constraints upon V_ϕ (Bina and Helffrich, 1992).

Again, chemical composition in regions of Earth's interior is primarily constrained by mapping values of seismic velocities in those regions and comparing the values to those determined for various candidate mineral assemblages. The primary observables in the seismological studies consist of measured travel times of various P- and S-waves, some of which can in principle constrain density contrasts (Shearer and Flanagan, 1999) from a large enough set of which values of V_P and V_S can be determined through mathematical inversion. Secondary observables include the measured amplitudes of various arriving P- and S-waves, some of which can in principle constrain density contrasts (Shearer and Flanagan, 1999) but these are much more sensitive to complex properties such as anelasticity than are simple travel times. Seismological observables, then, directly yield V_P and V_S, but static compression experiments directly yield V_ϕ. It is common, therefore, to seek to combine seismological V_P and V_S models to generate a V_ϕ model for a region, for comparison to mineralogically constrained V_ϕ values. Such combinatoric procedures can introduce additional errors, however, in that the V_P and V_S profiles used may often reflect seismic ray paths or frequency bands that differ from one another. Indeed, special source–receiver geometries (such as those whose epicentral distances are so small as to ensure nearly overlapping ray paths) may be necessary to obtain robust estimates of V_ϕ values (Bina and Silver, 1997). Nonetheless, comparisons to V_ϕ, rather than to V_P and V_S, are commonly used to constrain compositions in the deep interior, because (as noted below) mineralogical values for G (and hence for V_S) become more problematic with increasing depth.

2.02.1.3 Acoustic Methods

Direct laboratory measurements of V_P and V_S for materials can be made through acoustic methods, such as Brillouin spectroscopy and ultrasonic techniques. Brillouin spectroscopy (or "Brillouin scattering") determines acoustic velocities in a single crystal through measurements of the Doppler shifts experienced by visible light scattering off the faces of a single crystal in which thermally induced acoustic waves are propagating. Because the technique employs single crystals, full elasticity tensors (rather than just isotropic V_P and V_S) can be determined, which are useful in studies of velocity anisotropy. The technique requires only small samples, and because visible light is employed, the samples may be placed in an optically transparent diamond-anvil cell (DAC) in order to make measurements at elevated pressures and temperatures. DAC studies yielding V_P and V_S may be combined with XRD analyses yielding ρ, to allow for direct determination of K_S and G (and hence V_ϕ) for single crystals (Zha *et al.*, 1998a).

The second group of acoustic methods, the ultrasonic techniques, require larger samples but can be performed on either single crystals or polycrystalline aggregates. They require experimental measurement (via interferometry) of the travel times of two consecutive ultrasonic echoes from the ends of a shaped sample, combined with measurement of the length of the sample. These measurements directly yield V_P and V_S, rather than full anisotropic elasticity tensors, but this is usually sufficient (unless patterns of seismic velocity anisotropy are to be used to map strain fields in the mantle). Samples may be placed in a multi-anvil cell (MAC) to make measurements at elevated pressures and temperatures, and simultaneous (usually synchrotron) XRD analysis can then be used to keep track of associated changes in sample length. Again, MAC studies yielding V_P and V_S may be combined with ρ measurements from the XRD analyses to directly yield K_S and G (and hence V_ϕ) for samples (Li *et al.*, 2001).

The appealing flexibility of acoustic techniques is leading to rapid expansion of the regime of pressures and temperatures in which experimental measurements can be made. However, their potential utility continues to suffer from one poorly understood factor. While Brillouin and ultrasonic methods determine V_P and V_S at frequencies in the MHz–GHz range, seismological observations constrain V_P and V_S at frequencies in the mHz–Hz range. There is ample room for the poorly understood frequency dependence of these velocities ("dispersion") potentially to confound petrological interpretations over these many orders of magnitude. Unhappily, V_S is much more subject to dispersion than is V_P, and the magnitude of such effects upon V_S (and hence G) only grows with increasing temperature (and hence depth). Thus, until V_S values can be both measured at simultaneous high P and T and extrapolated from GHz to mHz with confidence (and quantifiable error bounds), there remain important roles for the parameter V_ϕ. Not only does V_ϕ remain free of the dispersive and thermal complications of G, if determined in part by static compression methods it also benefits from being constrained by measurements near 0 Hz, closer to the seismic frequency band than acoustic techniques permit.

2.02.2 UPPER-MANTLE BULK COMPOSITION

2.02.2.1 Overview

Based upon the compositions of mantle rocks emplaced at the surface in large slices (see Chapter 2.04) or as small xenoliths (see Chapter 2.05), the compositions of mantle-derived melts (see Chapter 2.08), various cosmochemical arguments (see Chapters 2.01 and 2.15), and simple geophysical considerations, the mineralogy of the upper mantle is commonly concluded to resemble that of some sort of peridotite (McDonough and Rudnick, 1998). Perhaps the most frequently invoked model composition is that of "pyrolite" (Ringwood, 1975, 1989), which contains ~60% olivine by volume, the depth-varying properties of which are dominated by progressive high-pressure phase transitions from olivine (α) to wadsleyite (β, also called "modified spinel") to ringwoodite (γ, also called "silicate spinel") and thence to a mixture of silicate perovskite (pv) and magnesiowüstite (mw, also called "ferropericlase"). The remaining, nonolivine, components are orthopyroxene (opx), clinopyroxene (cpx), and garnet (gt), and these undergo more gradual high-pressure transitions as the pyroxenes dissolve into the garnet, with the resulting "garnet–majorite" solid solution (gt–mj) eventually transforming to silicate perovskite as well (Figure 1). While this model of a peridotitic upper mantle has regularly been challenged, the proposed alternatives have

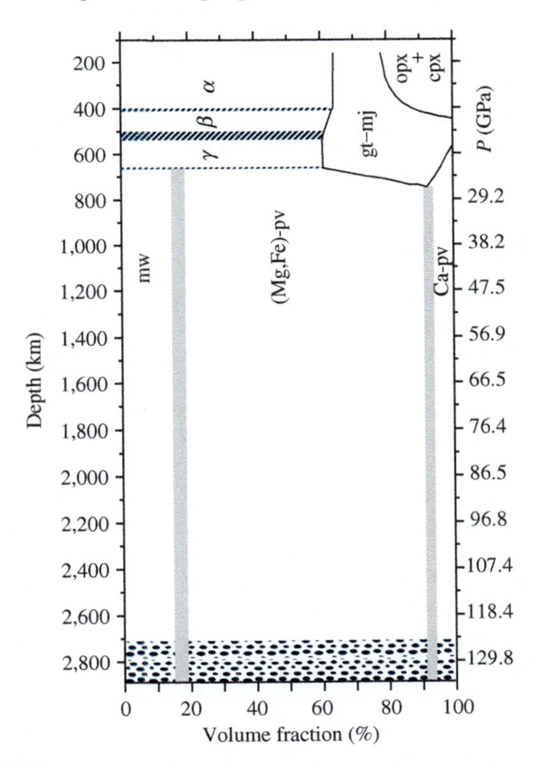

Figure 1 Depth-varying phase proportions in a pyrolite model mantle after the manner of Ringwood (1989), Ita and Stixrude (1992), and Bina (1998b). Phases are: (α) olivine, (β) wadsleyite, (γ) ringwoodite, (opx) orthopyroxene, (cpx) clinopyroxene, (gt–mj) garnet–majorite, (mw) magnesiowüstite, ((Mg,Fe)-pv) ferromagnesian silicate perovskite, and (Ca-pv) calcium silicate perovskite. Patterned region at base denotes likely heterogeneity near core–mantle boundary.

evolved over time to accommodate increasing quantities of olivine: from an eclogite containing little or no olivine (Anderson, 1979, 1982, 1984), through various "piclogite" models containing 16% (Bass and Anderson, 1984), 22% (Anderson and Bass, 1984), 30% (Anderson and Bass, 1986), or 40% (Duffy and Anderson, 1989; Duffy *et al.*, 1995) olivine. Thus, "pyrolite" and "piclogite" represent broad families of mantle compositions that are distinguished primarily by the former having >50% olivine by volume and the latter having <50% olivine.

Estimates of a suitable peridotite composition have also varied: from 40–70% olivine (Weidner, 1986), to 66–74% (Bina and Wood, 1987), to ~52% (Bina, 1993), for example. Indeed, for several years these two end-member models seemed to be converging towards 50% olivine (Ita and Stixrude, 1992; Jeanloz, 1995; Agee, 1998; Shearer and Flanagan, 1999). As the arguments hinge upon comparisons of seismic wave velocities in the upper mantle with velocity profiles computed for candidate mineral assemblages, firmer constraints upon this number require not only better experimental measurements of the simultaneous dependence of the elastic properties of mineral assemblages upon both temperature and pressure (Sinogeikin *et al.*, 1998; Zha *et al.*, 1998b) but also increased seismological resolution of the laterally varying velocity contrasts at depth within the upper mantle (Melbourne and Helmberger, 1998).

2.02.2.2 Velocity Contrasts

Seismology constrains these contrasts through observations of *P*- and *S*-wave travel times. Mathematical inversion of large numbers of travel times, observed at a variety of distances between source (earthquake) and receiver (seismometer), results in velocity profiles that represent the variation of V_P and V_S (or V_ϕ) as functions of depth. The increase in velocities across seismic "discontinuities" in these models can then be compared to the velocity changes across phase changes in olivine, as calculated from laboratory data, in order to estimate mantle olivine content. Unfortunately, seismic velocity profiles determined from such inversions generally are not very sensitive to, and therefore do not well constrain, the magnitudes of velocity discontinuities, and this is especially true of globally averaged seismic velocity models. Local or regional studies that include travel times of special seismic arrivals that have interacted directly with (reflected by or undergone *P–S* conversions at) the seismic discontinuities are best able to provide such constraints.

Measurements of elastic wave velocities in olivine and wadsleyite at high pressures and temperatures (Li *et al.*, 2001) generate "remarkable consistency" between a standard pyrolite model containing ~60% olivine and high-resolution seismic velocity profiles of the transition zone. Arguments about ±5% olivine aside, we can perhaps safely refer to the upper mantle as a peridotite, as the IUGS classification of ultramafic rocks defines a peridotite as containing 40% or more olivine.

2.02.2.3 Discontinuity Topography

Perhaps one of the most important consequences of a peridotite composition for the upper mantle is that the phase transitions in olivine that are manifested as seismic discontinuities should exhibit thermally controlled variations in their depth of occurrence that are consistent with the measured Clapeyron slopes (Bina and Helffrich, 1994) of the transitions. In particular, the olivine–wadsleyite transition at 410 km should be deflected upwards in the cold environment of subduction zones while the disproportionation of ringwoodite to silicate perovskite and magnesiowüstite at 660 km should be deflected downwards, thereby locally thickening the transition zone. In anomalously warm regions (such as the environs of mantle plumes as described below), the opposite deflections at 410 and 660 should locally thin the transition zone. The seismically observed topography of 20–60 km on each of the 410 and 660 is consistent with lateral thermal anomalies of 700 K or less (Helffrich, 2000; Helffrich and Wood, 2001).

Other consequences of thermally perturbed phase relations in mantle peridotite (Figure 2) are also supported by seismological observations. These include anticorrelation of transition zone thickness and transition zone delay times, whereby positive delays in travel times, which imply slow velocities (and therefore high temperatures), are observed to correlate with negative changes in (thinning of) transition zone thickness (Gu and Dziewonski, 2002; Lebedev *et al.*, 2002).

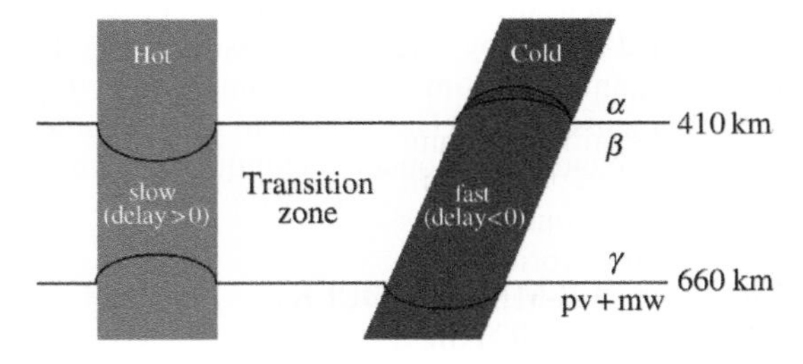

Figure 2 Schematic illustration of thermal control of olivine phase transformations in the transition zone, showing perturbations to transition zone thickness, transition zone seismic velocities (delay times), and depths of individual phase transformations.

Similarly, observations of anticorrelation of transition zone thickness and the depth to the 410 km discontinuity are consistent with thermal deflection of phase boundaries, as a positive increase in transition zone thickness (implying low temperatures) correlates with a negative change in depth to (implying uplift of) the 410 (Kind *et al.*, 2002). Furthermore, seismological observations suggest the presence of a thicker (colder) transition zone under continents than under (warmer) oceans (Gu and Dziewonski, 2002), as well as a thicker transition zone beneath (colder) subduction zones than beneath the (warmer) mid-Pacific (Gu and Dziewonski, 2002; Shearer *et al.*, 2002). Although evidence for the expected anticorrelation of topography on the 410 and the 660, whereby one may be expected to deflect upwards in any locale where the other deflects downwards, has been less robust than these other observations (Helffrich, 2000; Shearer *et al.*, 2002), such anticorrelation may be obscured by the dependence of absolute depth estimates upon assumptions about shallower velocity structures, as discussed later in the context of the depths of origin of mantle plumes. Clear anticorrelation of 410 and 660 topography may also be confounded by frequency-dependent effects (Helffrich and Bina, 1994; Helffrich, 2000): while the 660 may remain sharp in both cold and warm environments, the 410 should grow sharper in warm regions and more diffuse in cold regions, so that these two discontinuities may respond differently to seismic waves of different wavelengths at different temperatures. Overall, however, the bulk of the observational evidence indicates that topography on seismic discontinuities in the transition zone is caused by thermal perturbations of equilibrium phase transformations in a mantle of peridotite composition (Helffrich and Wood, 2001).

2.02.2.4 Sharpness

Other arguments about the composition of the transition zone have focused specifically upon the observed seismic "sharpness" or depth extent of the 410 km discontinuity, which sometimes appears to occur over a narrower depth interval than might be expected for the olivine–wadsleyite phase transition. A number of phenomena have been invoked to explain apparent variations in transition sharpness. These include kinetic effects on phase transformations (Solomatov and Stevenson, 1994), whereby low-pressure phases persist metastably for a finite extent before abruptly transforming to the stable high-pressure phases, thus eliminating what might otherwise be a finite mixed-phase regime. The probable nonlinearity of multivariant phase changes is another factor (Helffrich and Bina, 1994; Stixrude, 1997), whereby a gradual transition appears seismically to be sharper because a large fraction of the associated velocity change is concentrated within a particular portion of the mixed-phase regime. Differential solubility of water within minerals across phase changes (Wood, 1995; Helffrich and Wood, 1996; Smyth and Frost, 2002) can also affect transition sharpness, in that small amounts of dissolved H_2O should broaden the $\alpha \rightarrow \beta$ transition at 410, while an excess of H_2O resulting in a free fluid phase may be expected to sharpen the same transition. Because the sharpness of the 410 may be particularly sensitive to water, some studies have begun to attempt to map water contents in the transition zone by examining the manner in which its sharpness appears to vary as a function of the frequency (and hence wavelength) of the interacting seismic waves (van der Meijde *et al.*, 2002).

Studies of multiphase Mg–Fe partitioning between coexisting olivine, wadsleyite, pyroxene, and garnet have also suggested that such partitioning can act to sharpen the $\alpha \rightarrow \beta$ transition at 410 (Irifune and Isshiki, 1998; Bina, 1998a). It is somewhat ironic that the nonolivine phases that exhibit a very broad pyroxene–garnet transition can, nonetheless, induce the already sharp olivine–wadsleyite transtition to grow yet sharper simply by slightly shifting the effective Mg/Fe ratio in olivine through cation exchange.

2.02.2.5 Broadening and Bifurcation

As noted above, low temperatures alone can serve to broaden the 410 by expanding the depth extent of $\alpha + \beta$ mixed-phase stability field (Katsura and Ito, 1989; Bina and Helffrich, 1994). Even more confusingly, however, low temperatures can give rise to bifurcation of the $\alpha \rightarrow \beta$ transition (Figure 3), resulting in a strongly uplifted $\alpha \rightarrow \alpha + \gamma$ transition, which is seismically diffuse, overlying a less strongly uplifted $\alpha + \gamma \rightarrow \alpha + \beta$ or $\alpha + \gamma \rightarrow \beta + \gamma$ transition, which is seismically sharp (Green and Houston, 1995; Vacher *et al.*, 1999; Bina, 2002). As a result, α still transforms to β (and eventually to γ), but it does so by a two-step process, and whether a strongly uplifted broadened transition or a weakly uplifted sharpened transition is observed may depend upon the frequency of the interacting seismic waves. While understanding of such sharpening and broadening processes may be important for resolving fine details of the thermal structure of the transition zone, they would seem to have less bearing upon the overall bulk chemistry of the upper mantle, with the possible exception of constraints on local volatile contents (Wood *et al.*, 1996).

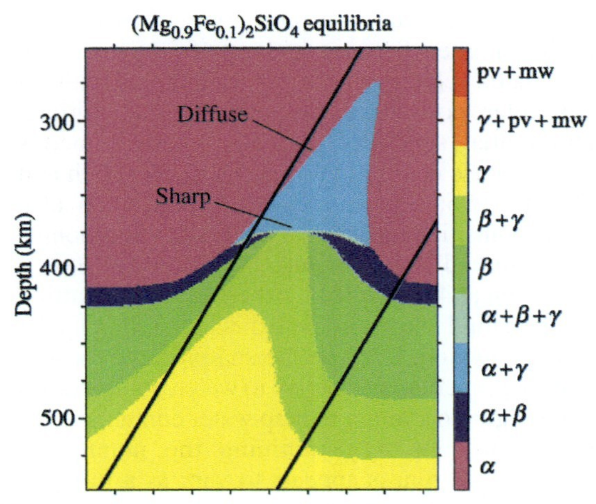

Figure 3 Effects upon olivine phase equilibria near 410 km depth of low temperatures in subduction zones, for mineral thermodynamic parameters of Fei *et al.* (1991). Dark lines denote boundaries of subducting slab. Phases are: (α) olivine, (β) wadsleyite, (γ) ringwoodite, (mw) magnesiowüstite, and (pv) ferromagnesian silicate perovskite. Note that the $\alpha \rightarrow \beta$ transition near 410 km is first uplifted and then bifurcates into a strongly uplifted diffuse $\alpha \rightarrow \alpha + \gamma$ transition overlying a weakly uplifted sharp boundary (after Bina, 2002) (vertical resolution is 1 km).

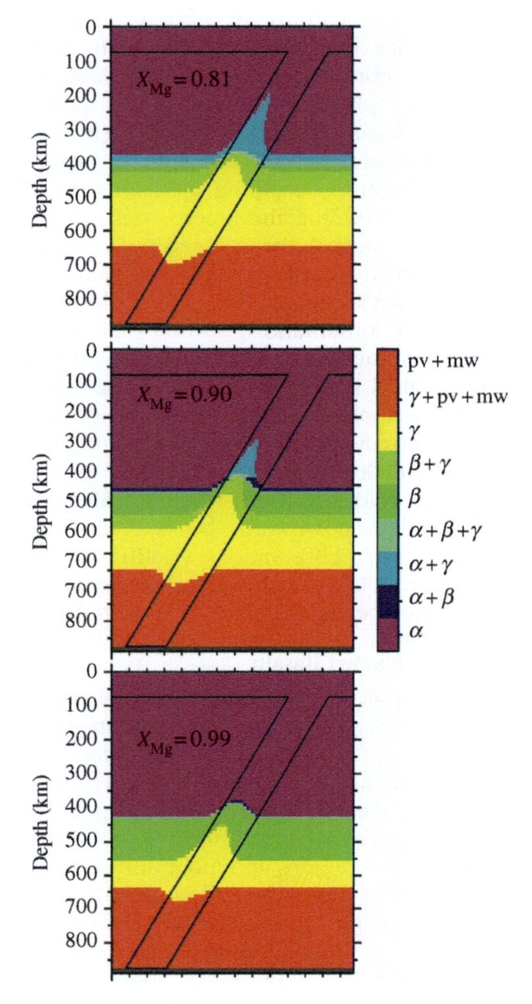

Figure 4 Effects upon olivine phase equilibria of variations in Mg/(Mg + Fe) for pyrolite (center), Fe enrichment (top), and Mg enrichment (bottom) (vertical resolution is 10 km).

However, within the cold environment of subduction zones, the temperature dependence of phase relations in olivine near depths of 410 km does exhibit particular sensitivity to bulk Mg/(Mg+Fe) ratios. Indeed, the effect on olivine phase relations of iron enrichment is largely analogous to that of lower temperatures noted above. For pyrolitic values (Mg/(Mg+Fe) = 0.90), equilibrium phase relations (Fei *et al.*, 1991) predict uplift and broadening of the sharp $\alpha \rightarrow \beta$ transition in the cold slab, replacement of the sharp $\alpha \rightarrow \beta$ transition by a more diffuse $\alpha \rightarrow \alpha+\gamma$ transition overlying a sharper $\alpha+\gamma \rightarrow \alpha+\beta$ or $\alpha + \gamma \rightarrow \beta+\gamma$ transition within the colder interior of the slab, and uplift of the broad $\beta \rightarrow \beta+\gamma \rightarrow \gamma$ transition (Figures 3 and 4); this is equivalent to the bifurcated scenario discussed above for low temperatures. Further magnesium enrichment (Mg/(Mg+Fe) = 0.99) would result in smaller uplift of a sharper $\alpha \rightarrow \beta$ transition, little or no replacement of the sharp $\alpha \rightarrow \beta$ transition by a more diffuse $\alpha \rightarrow \alpha+\gamma$ transition within the cold interior of the slab, and an uplifted $\beta \rightarrow \gamma$ transition which is much sharper within both the slab and the ambient mantle; such sharpening is analogous to the effect of slightly warmer temperatures. This latter scenario would also correspond to a globally sharp 520 km discontinuity, as well as a globally sharp 410 km discontinuity which remains sharp when uplifted (Figure 4). However, iron enrichment (Mg/(Mg+Fe) = 0.81) would result in complete replacement of the sharp $\alpha \rightarrow \beta$ transition within both the slab and the ambient mantle by a more diffuse and strongly uplifted $\alpha \rightarrow \alpha+\gamma$ transition overlying a sharper $\alpha+\gamma \rightarrow \alpha+\beta \rightarrow \beta$ or $\alpha+\gamma \rightarrow \beta+\gamma$ transition, as well as further broadening of the uplifted $\beta \rightarrow \beta+\gamma \rightarrow \gamma$ transition; this would be equivalent to the effect of very low temperatures. Such a scenario would correspond to a globally very diffuse 520 km discontinuity and a 410 km discontinuity consisting of a broad velocity gradient overlying a sharp velocity jump (Figure 4).

Other complexities of transition zone seismic structure may also indicate thermal or compositional effects. Apparent bifurcation of the 520 into superposed 500 km and 550 km discontinuities (Deuss and Woodhouse, 2001) may reflect distinct signatures of the $\beta \rightarrow \beta+\gamma \rightarrow \gamma$ transition and the exsolution of calcium silicate perovskite from majoritic garnet (Figure 1), and the offset in depth between these two features may also be expected to change with temperature. Furthermore, in colder regions majoritic garnet may transform to silicate ilmenite (ilm) within the 550–600 km depth range, prior to eventual disproportionation to silicate perovskite and magnesiowüstite, but this interposed gt → ilm transition is unlikely to express a significant seismic velocity signature, at least not within subducting slabs (Vacher *et al.*, 1999). Overall, seismological observations of a sharp 410 km discontinuity—which is occasionally broader in places, exhibits some topography, and is accompanied by a sporadic 520 km discontinuity—are primarily consistent with pyrolitic Mg/(Mg + Fe) values of 0.90, but local variations certainly cannot be excluded.

2.02.3 UPPER-MANTLE HETEROGENEITY

2.02.3.1 Subducted Basalts

The fate of the basalts and gabbros in the oceanic crust as they are subducted into a peridotite upper mantle (Chapter 2.03) can also be studied using seismological methods. Anhydrous metabasalts may be expected to undergo equilibrium transformation to an eclogite assemblage (with growth of garnet at the expense of plagioclase) around depths of 20–50 km (Wood, 1987; Peacock, 1993; Hacker, 1996). Such eclogites should be ~2% faster than surrounding pyrolite at depths of ~280 km (Helffrich *et al.*, 1989; Helffrich and Stein, 1993). However, most of this velocity contrast arises solely from the temperature contrast between cold slab (1,000 °C at 9.6 GPa) and warmer mantle, with the composition difference alone giving rise to a contrast of only ~0.5% (Helffrich *et al.*, 1989). Consideration of silica-oversaturated basalt compositions can expand this range of velocity contrasts somewhat, with anhydrous coesite eclogites being ~2–4% fast and, subsequent to the transition from coesite to stishovite at ~220–240 km depth, stishovite eclogites being ~4–6% fast (Connolly and Kerrick, 2002). Again, however, it is important to bear in mind that much of this contrast arises due to low slab temperatures alone, so that thermally equilibrated eclogites lingering long in the upper mantle would appear only slightly (if at all) fast relative to ambient mantle.

In large part, this ability of anhydrous basaltic eclogites to seismically blend into an ultramafic mantle arises from the behavior of elastic moduli in pyroxenes. In the shallowest upper mantle, orthopyroxene is ~6% slow in V_P relative to olivine and about equal in V_S. However, the bulk modulus of orthopyroxene exhibits a strong and nonlinear increase with pressure, so that there is little significant difference in either V_P or V_S between orthopyroxene and olivine by ~200 km depth (Flesch *et al.*, 1998; James *et al.*, 2001). This absence of effective velocity contrasts between anhydrous eclogites and mantle peridotites within most of the upper mantle (Helffrich *et al.*, 1989; Helffrich, 1996) is also evident in the observation that model upper mantles of both pyrolite (~60% olivine) and piclogite (~40% olivine) composition exhibit similar velocities over the 100–400 km depth range (Vacher *et al.*, 1998).

If anhydrous metabasalts in an eclogite assemblage can generate only small fast velocity anomalies or no anomalies at all, then a puzzle emerges in understanding subduction zone structures. Seismological observations in Japan, Tonga, Alaska, and other active subduction zones demonstrate the presence of 4–10% slow velocities in a layer 2–10 km thick along the upper surfaces of subduction zones in the depth range 100–250 km (Helffrich, 1996; Connolly and Kerrick, 2002). One explanation that has been advanced to explain the presence of such low-velocity layers involves kinetic hindrance in cold slabs. Rather than equilibrium transformation of slow anhydrous gabbro to fast eclogite at depths of 20–50 km, a model of metastable persistence of gabbro into the blueschist and eclogite stability fields, perhaps below 100 km depth, has been invoked (Hacker, 1996; Connolly and Kerrick, 2002). However, it appears that this model of metastable anhydrous gabbro may not be appropriate (Helffrich, 1996), not only because oceanic basalts commonly are found to be hydrothermally altered but also because metastable gabbro appears to be too slow seismically (Connolly and Kerrick, 2002).

Indeed, hydrothermal alteration of basalts may be the key to understanding the low-velocity layers in subducting slabs. In hydrous metabasalts under subduction zone conditions, lawsonite blueschist is expected to be the initially dominant facies (Peacock, 1993). At 65 km depth in subducting basaltic crust, lawsonite blueschist would be ~7% slower than the overlying peridotite mantle and ~8% slower than the underlying garnet harzburgite layer of the subducting lithosphere, suggesting that hydrous metabasalts may be the cause of the seismologically detected low-velocity layers (Helffrich, 1996). As hydrous gabbroic crust subducts beyond the blueschist regime, the resulting equilibrium lawsonite eclogites should still be 3–7% slow (Connolly and Kerrick, 2002). With continued subduction, garnet increases and lawsonite decreases in abundance, until the coesite–stishovite transition near 220–240 km further destabilizes lawsonite, giving rise to stishovite eclogites which should be 4–6% fast (Connolly and Kerrick, 2002). Such a change from slow lawsonite eclogites or coesite eclogites to fast stishovite eclogites is consistent with the observed termination of low-velocity layers and, indeed, with the occasional presence of high-velocity layers, below 250 km depth.

Interestingly, a dominant role for lawsonite eclogites in subducting oceanic crustal material may be echoed in the occurrence of lawsonite (or pseudomorphs thereafter) among the glaucophane eclogites of the Alpine Sesia zone, a complex of continental provenance presumably exhumed after subduction to at least 60 km depth (Reinsch, 1979; Pognante, 1989). Furthermore, the compositions of majoritic and sodium-rich garnets occasionally found as inclusions in diamonds (Chapter 2.05) are also consistent with equilibrium phase relations for basaltic crust which has penetrated into the transition zone (Ono and Yasuda, 1996). Within the transition zone, evidence from seismic tomography suggests that some subducting slab material penetrates directly into the lower mantle while, in other subduction zones, some is deflected horizontally at depths shallower than 1,000 km (Takenaka *et al.*, 1999; Fukao *et al.*, 2001). By the time subducted basaltic material enters the lower mantle, it should be largely dehydrated and should adopt a simpler, high-pressure perovskitite mineralogy. The expected seismic signatures of such basaltic material in the lower mantle will be examined in a later section.

2.02.3.2 Plume Origins

While much attention has been focused upon the seismological properties of subduction zones, in part because of their significant spatial extent and associated mass flux (Chapter 2.11), seismological studies of hotspot areas are also illuminating, particularly with regard to ascertaining the depth of origin of plume structures. There are two primary types of seismic evidence for constraining the depth of origin of mantle plumes. The first of these consists of seismic tomographic imaging, in which one might expect the achievement of sufficiently fine spatial resolution for imaging narrow plume conduits to be a significant challenge. Indeed, this challenge is further aggravated by the fact that the resolution of such methods tends to decay within the crucial region of the mantle transition zone.

Nevertheless, seismic tomography has been employed in efforts to image the roots of mantle plumes (VanDecar *et al.*, 1995) and to determine whether or not they arise from deep-seated sources. An instructive example is the case of the Iceland hotspot, which tomographic images suggest may be connected to a deep-seated plume source (Wolfe *et al.*, 1997). Such a conclusion is consistent with the fact that low seismic velocity anomalies appear to extend downward into the transition zone beneath the hotspot while they do not extend below 150 km beneath the rest of the mid-Atlantic ridge (Montagner and Ritsema, 2001), an observation supported by both global and regional tomography (Ritsema *et al.*, 1999; Allen *et al.*, 2002). However, this interpretation has been challenged. Foulger *et al.* (2001), claiming tomographic resolution to 450 km depth, conclude that the shape of the imaged low-velocity anomaly changes from cylindrical to tabular near the top of the transition zone, and the investigators argue on the basis of this apparent change in morphology that the plume does not extend to deeper levels. Alternatively, Allen *et al.* (2002), claiming resolution to 400 km depth, report a simple cylindrical morphology at depth. Detailed numerical tests of the spatial resolution of tomographic imaging under Iceland (Keller *et al.*, 2000) suggest that a deep-seated plume may not be required to explain the observed seismic delay times. In a similar vein, Christiansen *et al.* (2002) argue that there is a dearth of convincing seismic evidence for a plume extending to depths below 240 km beneath the Yellowstone hotspot. All of this serves to highlight some of the persistent ambiguities present in interpretation of seismic tomographic images beneath hotspots, images which remain nonunique results of the application of a variety of optimization functions to different observational data sets.

The second type of seismic evidence used to constrain the depth of origin of mantle plumes consists of analyses of boundary-interaction phases. Such phases consist of seismic waves which, by interacting with the boundaries generally known as seismic "discontinuities," have undergone conversion (repartitioning of energy between longitudinal (P) and transverse (S) waves) and/or reflection (repartitioning of energy between upgoing and downgoing waves). For the purposes of studying mantle plumes, the crucial measurements are differential travel times between those phases which interact with the 410 discontinuity and those which interact with the 660 discontinuity. Such differential times translate (via a reference velocity model) into measures of the thickness (depth extent) of the transition zone lying between these two discontinuities.

Given the opposing signs of the Clapeyron slopes of the primary phase transitions associated with these seismic discontinuities, any elevated mantle temperatures associated with thermal plumes may be expected to yield thinning of the transition zone (Figure 2), via depression of the 410 and uplift of 660 (Shen *et al.*, 1998; Bina, 1998c; Lebedev *et al.*, 2002). Some global and broad regional studies (Vinnik *et al.*, 1997; Chevrot *et al.*, 1999) have detected no clear correlation between such estimates of transition zone thickness and locations of hotspots (Keller *et al.*, 2000). Since the late 1990s, however, a number of more localized studies have measured transition zone thinning of several tens of kilometers, suggesting hot thermal anomalies of a few hundred degrees, over regions with diameters of hundreds of kilometers, beneath such presumed thermal plume features as the Snake River plain (Dueker and Sheehan, 1997), Iceland (Shen *et al.*, 1998), Yellowstone (Humphreys *et al.*, 2000), Hawaii (Li *et al.*, 2000), and the Society hotspot (Niu *et al.*, 2002). By contrast, no apparent thinning has been found beneath tectonically inactive areas such as the northern North Sea (Helffrich *et al.*, 2003). A straightforward interpretation of these results is that the transition zone beneath plumes is hotter than "normal" mantle, with thermal plumes originating either deep in the lower mantle below the transition zone or (at the shallowest) in a hot thermal boundary layer at the base of the transition zone.

The picture grows less simple, however, if we attempt to inquire into how the causes of this transition zone thinning are distributed between 410 and 660. Such inquiry involves estimating the actual absolute depths of these two seismic discontinuities beneath plumes. While the use of differential times to estimate thickness requires a reference velocity model within the transition zone, the use of absolute times to estimate individual depths further requires a (laterally varying) reference velocity model from the transition zone all the way up to the surface. A simple model of a deep-seated thermal plume suggests that we should observe a depressed 410 with an uplifted 660, which is what seems to be imaged beneath Iceland (Shen *et al.*, 1998). However, we can also find a flat 410 with an uplifted 660 beneath Hawaii (Li *et al.*, 2000), a weakly uplifted 410 with a strongly uplifted 660 beneath the Snake River plain (Dueker and Sheehan, 1997) and Yellowstone (Humphreys *et al.*, 2000; Christiansen *et al.*, 2002), or a depressed 410 with a flat 660 beneath the Society hotspot (Niu *et al.*, 2002). Taken at face value, a depressed (hot) 410 with an uplifted (hot) 660 suggests a plume origin in the lower mantle (Shen *et al.*, 1998). A flat (normal) 410 with an uplifted (hot) 660 suggests either an origin in the lower mantle (Li *et al.*, 2000) or the presence of a thermal boundary layer within the transition zone, as does an uplifted (cold) 410 with an uplifted (hot) 660. A depressed (hot) 410 with a flat (normal) 660 suggests an origin within the transition zone (Shen *et al.*, 1998). Interestingly, the one combination which would strongly suggest an origin within the shallow upper-mantle, a depressed (hot) 410 with a depressed (cold) 660, is not observed. It is complexities such as these that have led several investigators to argue against the idea of deep-seated plumes in favor of the dominance of upper-mantle processes in the origins of hotspots (Anderson, 1994, 2001; Saltzer and Humphreys, 1997; Christiansen *et al.*, 2002), including such detailed proposals as "propagating convective rolls organized by the sense of shear across the aesthenosphere" (Humphreys *et al.*, 2000). This latter proposal would explain a cold 410 overlying a hot 660 (and underlying another hot region near 200 km depth), e.g., through localized convection at depths shallower than 400 km (Humphreys *et al.*, 2000).

There are a range of other possible explanations for these seeming complexities, however, a primary factor being the aforementioned reliance of absolute depth estimates for seismic discontinuities upon accurate models of shallow velocity structures (Walck, 1984; Helffrich, 2000; Niu *et al.*, 2002). If absolute depth estimates are so sensitive to assumptions about shallower structures, then these various and seemingly paradoxical combinations of apparent deflections may arise simply from inaccurate representations of structure outside of the regions of study. Conclusions drawn from differential times, which are free of such dependence upon assumptions about distal regions, may be judged more robust. Such a stance also renders more tractable an understanding of the magnitudes of the implied thermal anomalies. If all of the observed thinning of the transition zone were caused by deflection of either the 410 or 660 alone, then hot temperature anomalies of ~400 K would be required, but the size of the requirement falls to ~200 K if the thinning is shared between anticorrelated 410 and 660 deflections (Helffrich, 2000; Niu *et al.*, 2002).

Another factor to consider is an inherent seismological bias toward underestimating topography. For example, the "Fresnel zones" that describe the regions of the discontinuities that are sampled by boundary-interaction phases can be both large in extent and irregular in shape (Helffrich, 2000; Niu *et al.*, 2002), so that the measured travel times incorporate entwined interactions with both deflected and undeflected portions of a given discontinuity. Moreover, such seismological biases can yield greater underestimates of topography at 410 than at 660 (Neele *et al.*, 1997; Helffrich, 2000), which is consistent with the apparent "cold 410" puzzle noted above. Indeed, some topography simply may not be clearly visible. While the $\alpha \rightarrow \alpha + \beta \rightarrow \beta$ transition should grow sharper at high temperatures as well as being depressed, any small-scale topography on or "roughening" of discontinuity surfaces can render undetectable the very *P*-to-*S* conversions which indicate the presence of topography (van der Lee *et al.*, 1994; Helffrich *et al.*, 2003). Again, such effects may be more significant near 410, where the magnitudes of Clapeyron slopes and hence of topography may be larger (Bina and Helffrich, 1994), but they may also be significant near 660 (van der Lee *et al.*, 1994). Furthermore, the apparent magnitude of discontinuity topography will vary with the frequency of the seismic waves used to probe it (Helffrich, 2000).

Finally, there is yet another possible contributor to the apparent "cold 660" puzzle noted above. It is the negative Clapeyron slope of the $\gamma \rightarrow$ pv + mw transition that predicts uplift of the 660 in hot plumes. The $\gamma \rightarrow$ pv + mw transition, however, may be replaced by a $\beta \rightarrow$ pv + mw transition at high temperatures, the latter exhibiting a positive Clapeyron slope and so allowing 660 depression instead of uplift (Liu, 1994). However, $\beta \rightarrow$ pv + mw appears to succeed $\gamma \rightarrow$ pv + mw only in pure Mg_2SiO_4 compositions, so this particular mechanism is unlikely to operate in real multicomponent mantle compositions (Bina and Liu, 1995; Niu *et al.*, 2002).

The primary point of this discussion, then, is that estimates of lateral variations in transition zone thicknesses from differential seismic travel times are more robust than estimates of lateral variations in the absolute depths of discontinuities from absolute travel times. There are many factors, largely unrelated to plumes, that may cause some underestimation of the former but which induce serious complications in the latter. Certainly, it is possible that the overall tectonic system, including some hotspot-designated volcanic chains, may be controlled to a significant extent from above via the lithosphere rather than from below via the deep mantle (Anderson, 2001). However, the simplest interpretations consistent with observations of transition zone thinning, despite some apparent inconsistencies in estimates of absolute depths of discontinuities, strongly suggest that some hotspots are associated with plume-like thermal anomalies that penetrate the transition zone. Given the great disparities in heat and mass flux among different hotspots, however, there is certainly room for diversity in the family of thermal plumes, and it would not be surprising ultimately to discover various classes of plumes associated with different depths of origin (Kerr, 2003).

2.02.4 LOWER-MANTLE BULK COMPOSITION

2.02.4.1 Bulk Fitting

Given that the properties of the 660 km seismic discontinuity are in excellent agreement with the predicted behavior of an isochemical phase transformation, it might seem reasonable to assume that the lower mantle below this depth possesses largely the same bulk composition as the upper mantle above. This assumption has been regularly challenged, however, based largely upon cosmochemical concerns (Anderson, 1989; Bina, 1998b), upon estimates of mass fluxes between geochemical reservoirs (Helffrich and Wood, 2001), or upon driving forces for chemical differentiation across phase transitions (Garlick, 1969; Kumazawa *et al.*, 1974; Walker and Agee, 1989; Bina and Kumazawa, 1993; Liu and Ågren, 1995). Concern has focused primarily upon whether the lower mantle might be enriched in iron and/or silicon relative to the upper mantle.

A simple way to address this question is to compute profiles of density and bulk sound velocity (thereby avoiding the large uncertainities associated with extrapolating shear moduli), along plausible lower-mantle adiabats, for a variety of candidate lower-mantle compositions and to compare these model profiles to a reference seismological model such as *ak135* (Kennett *et al.*, 1995). Examination of the r.m.s. misfit between such models over the entire lower mantle quickly reveals several important principles. Density (Figure 5) is primarily sensitive to Mg/(Mg + Fe) but not to silica content. Velocity (Figure 5) sensitivity, alternatively, exhibits a trade-off between Mg/(Mg + Fe) and Si/(Mg + Fe). This suborthogonal nature of density and velocity sensitivities allows the two together (Figures 5 and 6) to constrain lower-mantle composition via the intersection of their respective misfit minima. The resulting family of allowable lower-mantle bulk compositions (Figure 6) includes a pyrolite mantle composition. While some uncertainty in silicon content remains, there is no evidence for iron enrichment of the lower mantle (Figure 6). The extent to which silicon enrichment of the lower mantle can be accommodated by the seismological constraints increases as the assumed temperature (at the root of the adiabat) of the lower mantle is increased (Figure 7). These schematic results are for a simple lower mantle mineralogy limited to ferromagnesian silicate perovskite (Fiquet *et al.*, 2000), magnesiowüstite (Fei *et al.*, 1992), and stishovite (Li *et al.*, 1996). Effects of other components (e.g., calcium, aluminum, sodium) have been neglected, and a depth-varying Mg–Fe partitioning coefficient between perovskite and magnesiowüstite (Mao *et al.*, 1997) has been employed. However, repetition of these analyses with the inclusion of calcium-silicate perovskite (Wang *et al.*, 1996) or the use of a depth-invariant partitioning coefficient (Kesson *et al.*, 1998) results in only very minor perturbations.

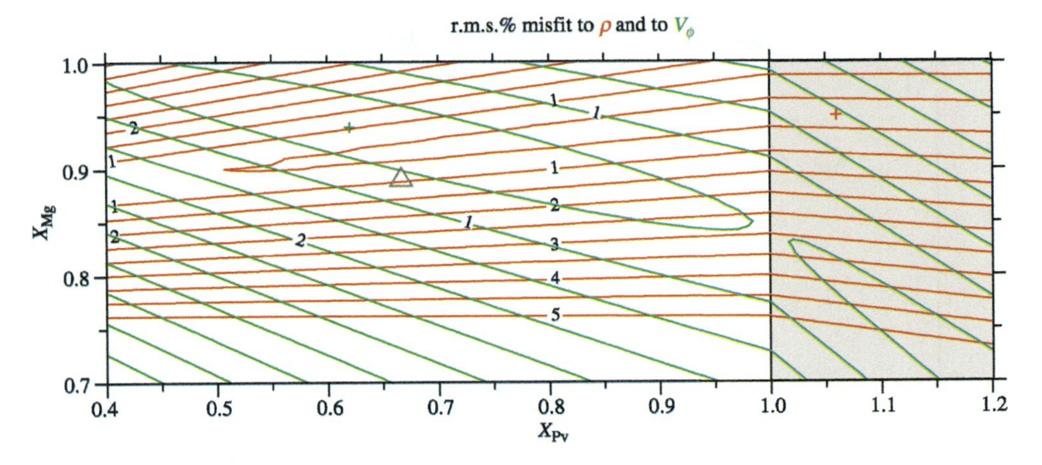

Figure 5 Contours of r.m.s misfit (%) to seismological reference model *ak135* of density (red) and bulk sound velocity (green) for candidate lower-mantle compositions, parametrized in terms of Mg/(Mg + Fe) ($=X_{Mg}$) and Si/(Mg + Fe)($=X_{Pv}$), over the entirety of the lower mantle. Shaded region at $X_{Pv}>1$ indicates free silica. Triangle denotes pyrolite. Plus signs denote minima of r.m.s. misfit. Root of lower-mantle adiabat is 2,000 K at 660 km depth.

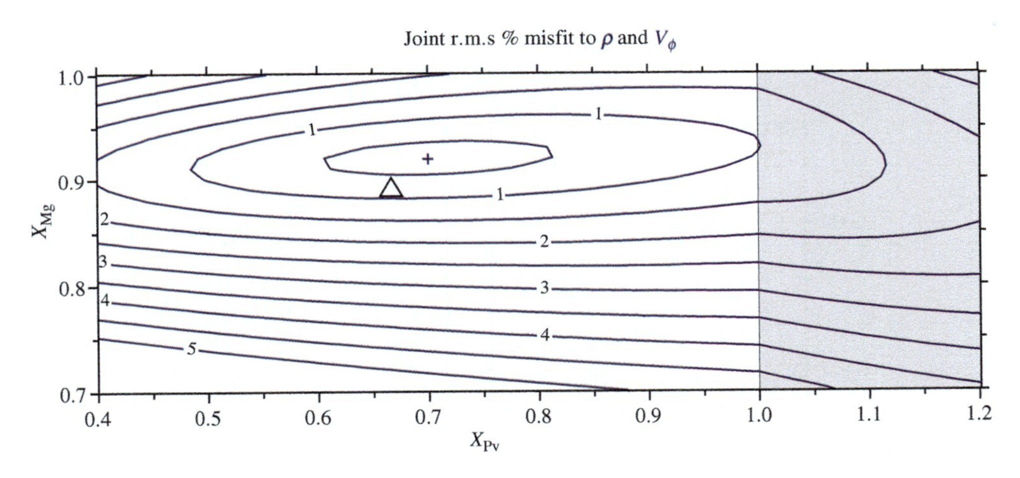

Figure 6 Contours of joint (blue) r.m.s. misfit (%) to seismological reference model *ak135* of density and bulk sound velocity for candidate lower-mantle compositions, parametrized in terms of Mg/(Mg + Fe)($=X_{mg}$) and Si/(Mg + Fe)($=X_{Pv}$), over the entirety of the lower mantle. Shaded region at $X_{Pv}>1$ indicates free silica. Triangle denotes pyrolite. Plus sign denotes minimum of r.m.s. misfit. Root of lower-mantle adiabat is 2,000 K at 660 km depth.

2.02.4.2 Depthwise Fitting

A different way of examining these relationships is to plot the best-fitting lower-mantle compositions within 10 km thick depth slices. Again, it is apparent (Figure 8) that density constrains only Mg/(Mg + Fe) while both density and velocity together are required to constrain Si/(Mg + Fe). Bulk sound velocity alone, as shown by the unstable oscillations in best-fitting compositions (Figure 8), does not effectively constrain either compositional parameter within such small depth slices. Throughout the bulk of the lower mantle, there is no evidence for iron enrichment, and the deviation from a pyrolite composition, in terms of both Mg/(Mg + Fe) and Si/(Mg + Fe), falls within the overall 1% r.m.s. misfit contour (Figure 7). Indeed, the only statistically significant deviations of the best-fitting composition from pyrolite occur in the top ~300 km of the lower mantle and in the bottom ~200 km. For the former region, this deviation is not surprising, as we probably have not fully incorporated the appropriate mineralogy. While the $\gamma \rightarrow$ pv + mw transition in the olivine component occurs at 660 km depth, the attendant gt $\rightarrow$ pv transition in the majorite component may not achieve completion until 100 km deeper or more, due to the solubility of aluminum and ferric iron in both garnet–majorite and silicate perovskite (Wood and Rubie, 1996; McCammon, 1997). Thus, the anomalous best-fitting compositions in

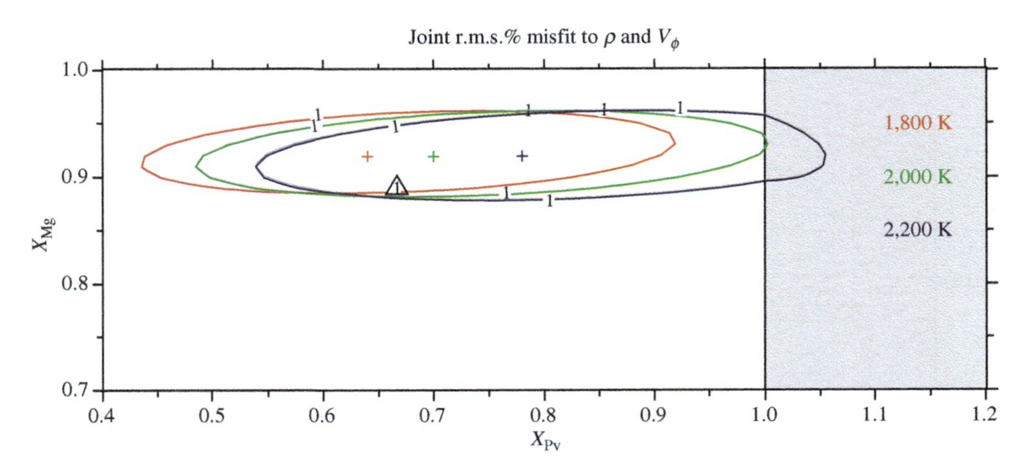

Figure 7 Contours of joint r.m.s. misfit (1%) to seismological reference model *ak135* of density and bulk sound velocity for candidate lower-mantle compositions, parametrized in terms of Mg/(Mg + Fe)($=X_{Mg}$) and Si/(Mg + Fe)($=X_{Pv}$), over the entirety of the lower mantle. Shaded region at $X_{Pv} > 1$ indicates free silica. Triangle denotes pyrolite. Plus signs denote minima of r.m.s. misfit. Roots of lower-mantle adiabats are 1,800 K (red), 2,000 K (green), and 2,200 K (blue) at 660 km depth.

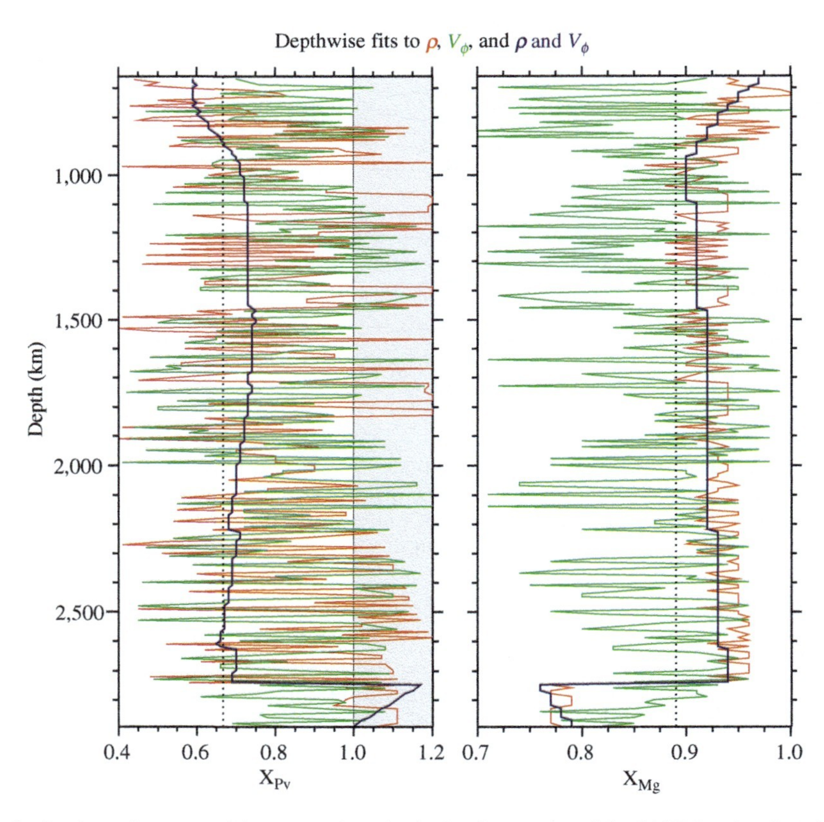

Figure 8 Depthwise best-fit compositions to seismological reference model *ak135* for density alone (red), bulk sound velocity alone (green), and both density and bulk sound velocity jointly (blue), with compositions parametrized in terms of Mg/(Mg + Fe) (X_{Mg}) and Si/(Mg + Fe) (X_{Pv}), in 10 km depth slices through the lower mantle. Shaded region at $X_{Pv} > 1$ indicates free silica. Dotted lines (at $X_{Pv} = 0.67$ and $X_{Mg} = 0.89$) denote pyrolite. Root of lower-mantle adiabat is 2,000 K at 660 km depth.

the top 200–300 km of the lower mantle probably arise simply from our omission of garnet from the model mineralogy. The situation in the bottom ~200 km of the lower mantle is more intriguing. Certainly, extrapolation of mineral properties are most uncertain in this region, and a globally averaged seismological model such as *ak135* may not accurately reflect details of structure near the core–mantle boundary. However, it is interesting to note that the implied iron enrichment and presence of free silica (Figures 8 and 9) are not inconsistent with what one might expect from interactions between silicate lower mantle and metallic core or from accumulation of subducted basaltic material at the core–mantle boundary.

Aside from the core–mantle boundary region, a pyrolite lower-mantle composition appears to be consistent with seismological constraints. Silica enrichment of the lower mantle can be accommodated if the lower mantle is hotter than expected for a simple adiabat rooted at the 660 km $\gamma \rightarrow$ pv + mw transition (Figure 9). Because any chemical boundary layer between the upper and lower mantle would be accompanied by a corresponding thermal boundary layer, such a model of a chemically distinct and hot lower mantle is also internally consistent. This trade-off has been evident for decades (Birch, 1952; Jackson, 1983, 1998; Jeanloz and Knittle, 1989; Bina and Silver, 1990, 1997; Stixrude *et al.*, 1992; Zhao and Anderson, 1994). However, the seismological evidence (discussed above) that the transition zone capping the lower mantle behaves like a set of thermally governed isochemical phase transformations, coupled with the absence of seismic evidence (e.g., a globally sharp seismic reflector displaying hundreds of kilometers of dynamically induced topographic undulations) for a chemical boundary in the lower mantle, lends considerable support to the minimalist assumption that the bulk composition of the lower mantle greatly resembles that of upper-mantle peridotite.

2.02.5 LOWER-MANTLE HETEROGENEITY

2.02.5.1 Overview

Seismic velocity heterogeneity in the mantle, e.g., as revealed by seismic tomography, is often

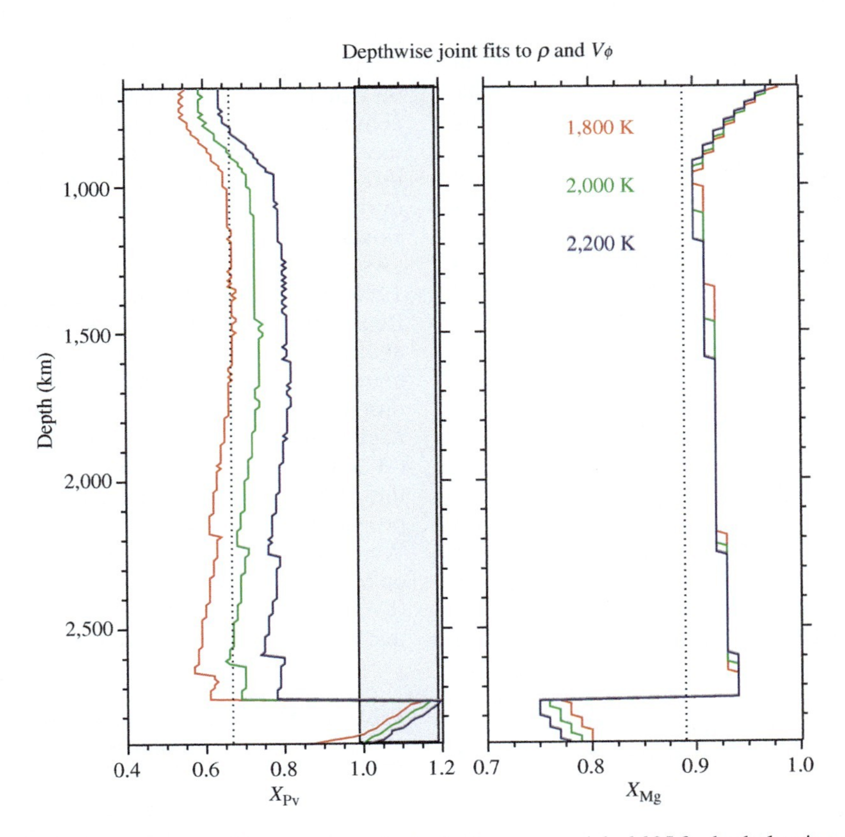

Figure 9 Depthwise best-fit compositions to seismological reference model *ak135* for both density and bulk sound velocity jointly, with compositions parametrized in terms of Mg/(Mg + Fe) (X_{Mg}) and Si/(Mg + Fe) (X_{Pv}), in 10 km depth slices through the lower mantle. Shaded region at $X_{Pv} > 1$ indicates free silica. Dotted lines (at $X_{Pv} = 0.67$ and $X_{Mg} = 0.89$) denote pyrolite. Roots of lower-mantle adiabats are 1,800 K (red), 2,000 K (green), and 2,200 K (blue) at 660 km depth.

interpreted in terms of strictly thermal origins. However, lateral variations in seismic velocity within the lower mantle (Hedlin *et al.*, 1997; Niu and Kawakatsu, 1997; Castle and Creager, 1999; Deuss and Woodhouse, 2002) may arise from a number of sources, including temperature anomalies (δT), local phase changes (δX_{ϕ}), and chemical heterogeneity (δX_i), and they should be accompanied by associated density anomalies. Thus, lower-mantle seismic velocity anomalies may reflect local heterogeneity in chemicall composition, such as variations in iron–magnesium ratio and silica content, rather than simply variations in temperature. Together, thermal and chemical variations may jointly explain the manner in which the r.m.s. amplitudes of seismic velocity anomalies appear to vary with depth (Bina and Wood, 2000). The distinction is important because velocity anomalies arising from compositional differences are not subject to the same temporal decay as those due to thermal perturbations, due to different timescales for chemical diffusion and thermal conduction.

Temperature anomalies may arise due to low temperatures within cold subducted slab material, but such thermal anomalies will decay with time (and hence depth of penetration into the mantle) as the slab is thermally assimilated into the warmer mantle. Local phase changes are unlikely to occur on significant scales below transition zone depths, except perhaps in exotic compositions containing free oxides (Bina, 1998b). Chemical heterogeneity thus seems a reasonable candidate as a source of seismic velocity heterogeneity in the lower mantle, and an obvious source of major-element chemical heterogeneity is subducted slab material that retains the chemical differentiation acquired during its formation at spreading ridges.

2.02.5.2 Subducted Oceanic Crust

Formation of oceanic lithosphere involves chemical differentiation (by partial melting) of mantle lherzolite parent material (see Chapter 2.03 and Chapter 2.08). The complex structure of oceanic lithosphere closely approximates a simple model of a basaltic–gabbroic crustal layer overlying a depleted harzburgite layer, which in turn overlies lherzolitic peridotite mantle material. During subduction, these layers undergo phase transformations to denser phase assemblages with increasing depth of penetration into the mantle. For example, the basalt and gabbro components progressively transform to eclogite in the upper mantle, to garnetite in the transition zone, and to perovskitite in the lower mantle (Vacher *et al.*, 1998; Kesson *et al.*, 1998; Hirose *et al.*, 1999). Upon deep subduction, all of the components of a petrologically layered slab should transform to a lower-mantle mineralogy, consisting of some subset of the phases (Mg, Fe, Al)SiO_3 ferromagnesian silicate perovskite, (Fe, Mg)O magnesiowüstite, $CaSiO_3$ calcium silicate perovskite, and SiO_2 stishovite, along with minor amounts of sodium-bearing and other phases. The seismic velocities and densities of these layers will differ due, e.g., to the coupled effects of silicon enrichment and magnesium depletion of the basaltic melt relative to the parent mantle.

Let us represent a subducting slab using a simple three-layered slab model (Helffrich and Stein, 1993), consisting of a basaltic crustal layer overlying a harzburgite layer overlying lherzolite mantle material. To represent the range of possible compositions of the basaltic crustal layer, we test both the "eclogite" of Helffrich *et al.* (1989) and the "gabbro" of Helffrich and Stein (1993), the latter being significantly richer in silicon and magnesium (Table 1). For the harzburgite layer, we adopt the "harzburgite" composition of Helffrich *et al.* (1989). For the mantle layer, we test both the "lherzolite" of Helffrich *et al.* (1989) and the "peridotite" of Helffrich and Stein (1993), the latter being slightly richer in silicon and iron (Table 1). We transform each of these bulk compositions into its stoichiometrically equivalent lower-mantle mineralogy (Table 2).

To isolate compositional and thermal effects, we calculate elastic properties for the various bulk compositions along adiabats rooted at a temperature of 2,000 K at 665 km depth (Bina, 1998b), using equations of state for (Mg, Fe)SiO_3 perovskite (Fiquet *et al.*, 2000), (Fe, Mg)O magnesiowüstite (Fei *et al.*, 1992), $CaSiO_3$ perovskite (Wang *et al.*, 1996), and SiO_2 stishovite (Li *et al.*, 1996; Liu *et al.*, 1996). We calculate Voigt–Reuss–Hill-averaged bulk sound velocities, to avoid the large uncertainties associated with extrapolating shear moduli to lower-mantle conditions. In these simple calculations, we have neglected the role of Al_2O_3 in silicate perovskite (Wood and Rubie, 1996; Wood, 2000), but the effects of this component upon the elastic properties—depressed bulk moduli (Zhang and Weidner, 1999)—are expected to be important only in the shallowest part of the lower mantle (Brodholt, 2000). Furthermore, we have assumed that any free silica is present as stishovite, rather

Table 1 Molar bulk compositions for model slab components.

	Mg/(Mg + Fe)	*Si/(Mg + Fe)*
Eclogite	0.57	1.65
Gabbro	0.64	2.58
Harzburgite	0.92	0.66
Lherzolite	0.90	0.68
Peridotite	0.89	0.69

Table 2 Molar quantities at 665 km depth for model slab components.

	pv-MgSiO$_3$	*pv-FeSiO*$_3$	*mw-MgO*	*mw-FeO*	*st-SiO*$_2$
Eclogite	0.57	0.43	0.00	0.00	0.65
Gabbro	0.64	0.36	0.00	0.00	1.58
Harzburgite	0.63	0.03	0.29	0.05	0.00
Lherzolite	0.64	0.04	0.26	0.06	0.00
Peridotite	0.65	0.04	0.24	0.07	0.00

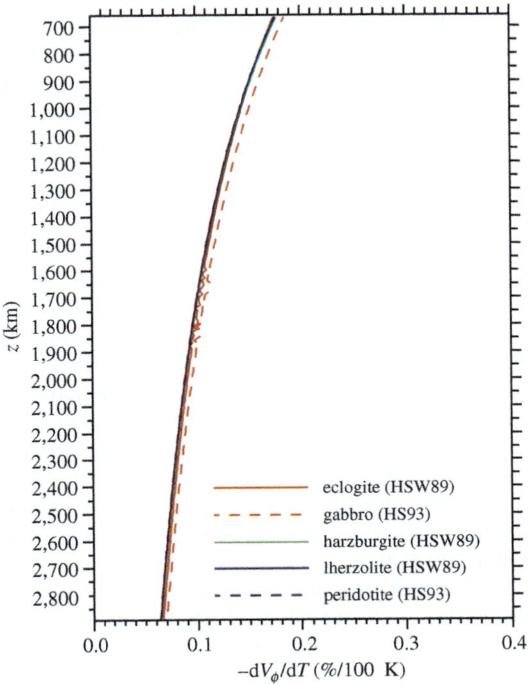

Figure 10 Temperature dependence of the bulk sound velocity versus depth in the lower mantle, for candidate bulk compositions for the basaltic layer (red), the underlying harzburgite layer (green), and the basal lherzolite (blue). Note that temperature sensitivity falls with increasing depth.

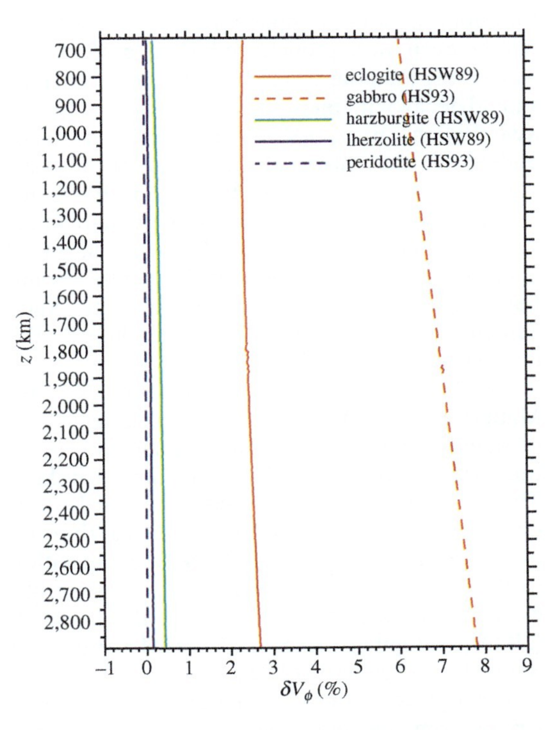

Figure 11 Bulk sound velocity anomalies relative to peridotite mantle versus depth in the lower mantle (blue); for candidate bulk compositions for the basaltic layer (red); and the underlying harzburgite layer (green). Note that the basaltic layer is fast and grows faster with increasing depth.

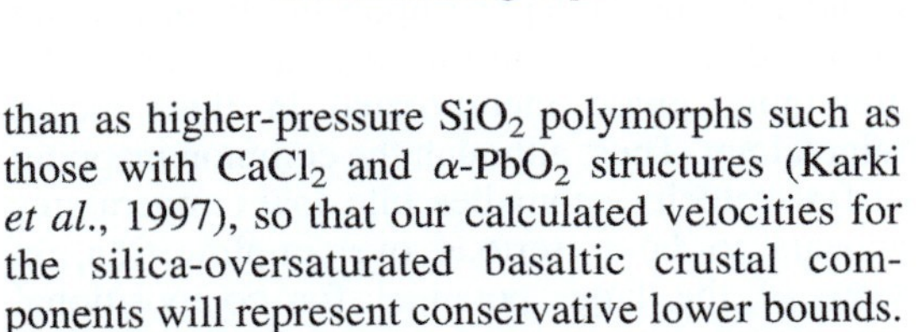

than as higher-pressure SiO_2 polymorphs such as those with $CaCl_2$ and α-PbO_2 structures (Karki *et al.*, 1997), so that our calculated velocities for the silica-oversaturated basaltic crustal components will represent conservative lower bounds.

The temperature dependence of the bulk sound velocity ($\partial V/\partial T$) for all of the compositions in the layered slab falls from below 0.2% per 100 K at the top of the lower mantle to below 0.1% per 100 K at the base of the lower mantle (Figure 10). As the magnitude of thermal anomalies associated with cold slab material will also fall with increasing depth as slabs thermally assimilate, this behavior suggests that it will be difficult to explain any large velocity anomalies at depth in terms of temperature alone. Indeed, even if thermal anomalies ~100 K were to somehow survive down to the base of the lower mantle, they would only give rise to velocity anomalies ~0.1%.

Alternatively, the composition dependence of bulk sound velocity ($\partial V/\partial X_i$) is significantly greater (Figure 11), and bulk chemical anomalies should not change significantly with depth. The velocities for the two representative models of the underlying mantle layer, the "lherzolite" (Helffrich *et al.*, 1989) and "peridotite" (Helffrich and Stein, 1993) compositions, are virtually identical. Relative to the "peridotite" composition, the "harzburgite" (Helffrich *et al.*, 1989) layer is less than 0.5% fast. However, the basaltic crustal compositions yield significantly fast velocity anomalies, and anomaly magnitudes increase with depth. The "eclogite" (Helffrich *et al.*, 1989) rises from below 2.4% fast near 660 km depth to above 2.6% fast near 2,890 km depth, and the more silica-rich "gabbro" (Helffrich and Stein, 1993) rises from ~7.0% fast to ~7.8% fast. Hence, it should be easier to generate large velocity anomalies through compositional variations, such as those associated with subducted slabs.

Thus, for temperature effects only, velocity anomalies should be small and decay with increasing depth. This effect would occur even for depth-invariant temperature anomalies, but it will be even more pronounced in the case of cold subducting material whose temperature anomalies should decay with increasing depth due to thermal assimilation. Even for large temperature anomalies, in the range of hundreds of degrees, velocity anomalies remain below 1%. For compositional effects, however, velocity anomalies due to chemical contrasts between basaltic/gabbroic material and peridotitic mantle should be large and grow with increasing depth. Such velocity anomalies should fall in the range 2–8% or greater.

Similar results have been reported by Mattern *et al.* (2002), using more recent equations of state for lower-mantle minerals and incorporating the solubility of alumina in silicate perovskite. They also used a three-layered slab model (mid-ocean ridge basalt (MORB) over harzburgite over pyrolite), but with a MORB composition ($Si/(Mg + Fe) = 2.29$) intermediate between our extreme end-members of the Helffrich *et al.* (1989) "eclogite" (1.65) and the Helffrich and Stein (1993) "gabbro" (2.58).

The high velocities in the basaltic material arise largely from the presence of free silica. As we have assumed that all free silica is in the form of stishovite, rather than as higher-pressure SiO_2 polymorphs, actual lower-mantle velocity perturbations arising from basaltic crustal material may be even larger than the values calculated here. Free silica in the former crustal material, however, should react with magnesiowüstite in the surrounding lower-mantle material to form silicate perovskite. Thus, survival of free silica in the lower mantle may require formation of perovskite rinds to preserve the free silica from reaction with magnesiowüstite, just as porphyroblasts can protect inclusions (which would otherwise become reactants) to form "armored relics" in more familiar metamorphic rocks.

Damping and smearing arising from regularization in seismic tomography typically cause narrow, intense anomalies to be imaged as broader, more subdued anomalies (Garnero, 2000). Thus, layers of basaltic/gabbroic material ~10 km thick that are ~5% fast, e.g., might perhaps appear as ~0.5% velocity anomalies distributed over 100 km thick slabs. If this is the case, then one might seek frequency-dependent effects in seismological observations, in which the apparent magnitude of velocity anomalies rises with the spatial resolving power of the seismic probe. Studies of lower-mantle seismic scatterers have suggested that bodies of 100 km or less in size exhibit velocity anomalies of several percent (Kaneshima and Helffrich, 1998; Garnero, 2000), especially near subduction zones. Some studies find evidence of 0.8–2.0% lower-mantle *P*-wave velocity heterogeneities at length scales of less than 8 km (Hedlin *et al.*, 1997), as well as lower-mantle *S*-wave velocity structures that are either >4% slow and ~8 km thick or >8% fast (Kaneshima and Helffrich, 1999). Such narrow but fast velocity anomalies are consistent with our model of lower-mantle velocity anomalies in subducted oceanic lithosphere.

Of course, we have calculated only bulk sound velocities here. Detailed comparison with seismological observations will require computation of *P*- and *S*-wave velocities, which in turn must await better constraints on the pressure and temperature dependence of the shear moduli of lower-mantle mineral phases. Furthermore, the velocity perturbations that we have calculated to arise from thermal and compositional anomalies will be accompanied by associated density anomalies. While the harzburgite is ~1% less dense than mantle peridotite, basaltic crustal material should be 1–7% denser than peridotite under lower-mantle conditions (Figure 12). The magnitude of this latter density anomaly should fall with increasing depth (Figure 12), as should the density consequences of thermal anomalies (Figure 13). These extrapolations suggest that metagabbros should remain denser than periodite throughout the lower mantle, as do others which additionally incorporate the affects of calcium and aluminum (Mattern *et al.*, 2002). (A contrary prediction by Ono *et al.* (2001), that metagabbro should attain neutral buoyancy deep in the lower mantle, seems to depend upon the large amount of low-density, calcium-ferritic-structured, sodic aluminous phase incorporated in their model mineralogy. Thus, although the common mapping of fast-velocity anomalies into low temperatures appears to be a serious oversimplification, in view of the large potential for compositional sources, the usual mapping of fast anomalies into positive density (i.e., negative buoyancy) anomalies does appear to survive in the presence of chemical differentiation.

2.02.6 SUMMARY

The properties of upper-mantle seismic reflectors, especially the observed lateral variations in the seismological properties of the mantle transition zone, indicate that the upper mantle possesses a peridotite composition, approaching

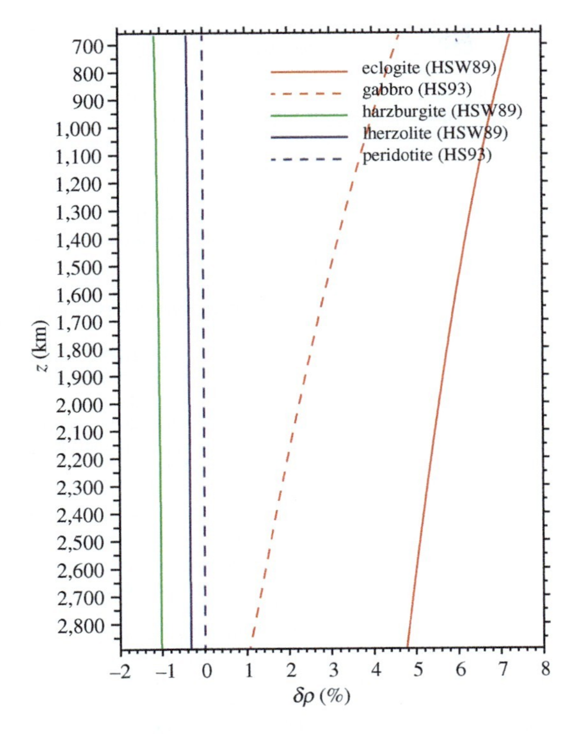

Figure 12 Density anomalies relative to peridotite mantle versus depth: in the lower mantle (blue); for candidate bulk compositions for the basaltic layer (red); and the underlying harzburgite layer (green). Note that the basaltic layer is denser but grows less so with increasing depth.

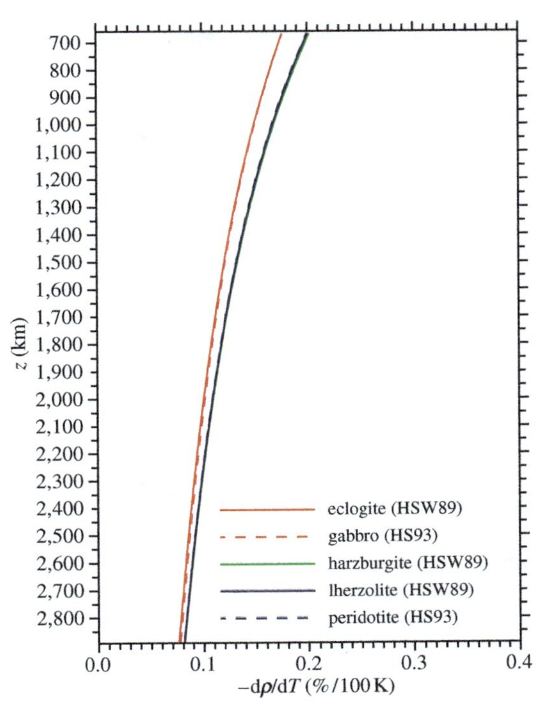

Figure 13 Temperature dependence of density versus depth in the lower mantle: for candidate bulk compositions for the basaltic layer (red); the underlying harzburgite layer (green); and the basal lherzolite (blue). Note that temperature sensitivity falls with increasing depth.

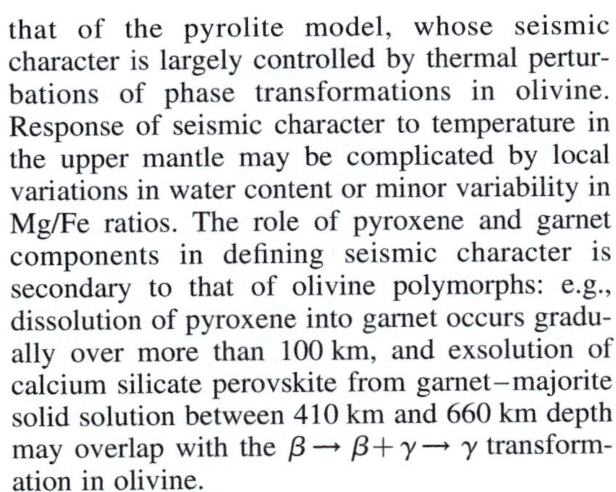
that of the pyrolite model, whose seismic character is largely controlled by thermal perturbations of phase transformations in olivine. Response of seismic character to temperature in the upper mantle may be complicated by local variations in water content or minor variability in Mg/Fe ratios. The role of pyroxene and garnet components in defining seismic character is secondary to that of olivine polymorphs: e.g., dissolution of pyroxene into garnet occurs gradually over more than 100 km, and exsolution of calcium silicate perovskite from garnet–majorite solid solution between 410 km and 660 km depth may overlap with the $\beta \rightarrow \beta + \gamma \rightarrow \gamma$ transformation in olivine.

Subducted basaltic crust is chemically distinct from its surrounding peridotite upper mantle. However, aside from a small signature arising solely from temperature gradients, anhydrous basalt should rapidly become seismically invisible as it sinks below 100–200 km depth. Features exhibiting low seismic velocities in association with the upper surfaces of subducting slabs in the 100–250 km depth range probably arise from hydrous (lawsonite-bearing) metabasalts in the blueschist and eclogite metamorphic regimes. The disappearance of such slow features, and the occasional appearance of fast velocity anomalies, below 250 km may arise from the passage of silica-saturated eclogites into the stishovite stability field.

The use of seismic tomography to determine the depth of origin of mantle plumes yields conflicting results. Analyses of travel times of seismic boundary-interaction phases, in terms of lateral changes in the absolute depths to the 410 km and 660 km seismic discontinuities, also give rise to seeming inconsistencies, perhaps in part because of the dependence of such methods upon laterally heterogenous models of shallow mantle structure. The use of (more robust) differential travel times, however, to assess lateral changes in the vertical distance between the 410 km and 660 km seismic discontinuities (i.e., the thickness of the transition zone), suggests that at least some hotspots are associated with thermal plumes which penetrate the entire transition zone.

Densities and bulk sound velocities in the lower mantle are also consistent with the high-pressure mineralogy of a bulk composition approximating an upper-mantle peridotite, such as pyrolite. Seismic constraints provide no support for iron enrichment of the lower mantle relative to such an upper mantle. Silica enrichment of the lower mantle can be accommodated if the lower mantle is anomalously hot, as would be consistent with a thermal boundary layer of several hundred degrees across a chemical boundary layer. However, silica enrichment alone is unlikely to generate an intrinsic density contrast sufficient to stabilize a chemical boundary layer against convective rehomogenization. Failure to observe the expected seismic signature of such a chemical boundary, in the form of a globally sharp seismic reflector exhibiting significant dynamic topography (>100 km) in the lower mantle, confirms that a peridotite whole mantle that is largely homogenous in its major element chemistry is most consistent with seismological observations.

Seismic scatterers within the lower mantle are more likely to represent chemical than thermal heterogeneities, with subducted slab material (especially the basaltic crustal component thereof) constituting a likely candidate. The very base of the mantle, nearest the core–mantle boundary, may also be characterized by significant major-element chemical heterogeneity.

REFERENCES

Agee C. B. (1998) Phase transformations and seismic structure in the upper mantle and transition zone. *Rev. Mineral.* **37**, 165–203.

Allen R. M., Nolet G., Morgan W. J., Vogfjörd K., Bergsson B. H., Erlendsson P., Foulger G. R., Jokobsdóttir S., Julian B. R., Pritchard M., Ragnarsson S., and Stefánsson R. M. (2002) Imaging the mantle beneath Iceland using integrated seismological techniques. *J. Geophys. Res.* **107**(B12), doi: 10.1029/2001JB000595.

Anderson D. L. (1979) The upper mantle transition region: Eclogite? *Geophys. Res. Lett.* **6**, 433–436.

Anderson D. L. (1982) Chemical composition and evolution of the mantle. In *High Pressure Research in Geophysics* (eds. S. Akimoto and M. H. Manghnani). Center for Academic Publication, Tokyo, pp. 301–318.

Anderson D. L. (1984) The Earth as a planet: paradigms and paradoxes. *Science* **223**, 347–355.

Anderson D. L. (1989) Composition of the Earth. *Science* **243**, 367–370.

Anderson D. L. (1994) The sublithospheric mantle as the source of continental flood basalts: the case against the continental lithosphere and plume head reservoirs. *Earth Planet. Sci. Lett.* **123**, 269–280.

Anderson D. L. (2001) Top-down tectonics? *Science* **293**, 2016–2018.

Anderson D. L. and Bass J. D. (1984) Mineralogy and composition of the upper mantle. *Geophys. Res. Lett.* **11**, 637–640.

Anderson D. L. and Bass J. D. (1986) Transition region of the Earth's upper mantle. *Nature* **320**, 321–328.

Bass J. D. and Anderson D. L. (1984) Composition of the upper mantle: geophysical tests of two petrological models. *Geophys. Res. Lett.* **11**, 229–232.

Bina C. R. (1993) Mutually consistent estimates of upper mantle composition from seismic velocity contrasts at 400 km depth. *Pure Appl. Geophys.* **141**, 101–109.

Bina C. R. (1998a) Olivine emerges from isolation. *Nature* **392**, 650–653.

Bina C. R. (1998b) Lower mantle mineralogy and the geophysical perspective. *Rev. Mineral.* **37**, 205–239.

Bina C. R. (1998c) Free energy minimization by simulated annealing with applications to lithospheric slabs and mantle plumes. *Pure Appl. Geophys.* **151**, 605–618.

Bina C. R. (2002) Phase transition complexity and multiple seismic reflectors in subduction zones. *Eos, Trans., AGU* **83**(47), Fall Meet. Suppl., Abstract S52C-05.

Bina C. R. and Helffrich G. R. (1992) Calculation of elastic properties from thermodynamic equation of state principles. *Ann. Rev. Earth Planet. Sci.* **20**, 527–552.

Bina C. R. and Helffrich G. (1994) Phase transition Clapeyron slopes and transition zone seismic discontinuity topography. *J. Geophys. Res.* **99**, 15853–15860.

Bina C. R. and Kumazawa M. (1993) Thermodynamic coupling of phase and chemical boundaries in planetary interiors. *Phys. Earth Planet. Inter.* **76**, 329–341.

Bina C. R. and Liu M. (1995) A note on the sensitivity of mantle convection models to composition-dependent phase relations. *Geophys. Res. Lett.* **22**, 2565–2568.

Bina C. R. and Silver P. G. (1990) Constraints on lower mantle composition and temperature from density and bulk sound velocity profiles. *Geophys. Res. Lett.* **17**, 1153–1156.

Bina C. R. and Silver P. G. (1997) Bulk sound travel times and implications for mantle composition and outer core heterogeneity. *Geophys. Res. Lett.* **24**, 499–502.

Bina C. R. and Wood B. J. (1987) The olivine–spinel transitions: experimental and thermodynamic constraints and implications for the nature of the 400 km seismic discontinuity. *J. Geophys. Res.* **92**, 4853–4866.

Bina C. R. and Wood B. J. (2000) Thermal and compositional implications of seismic velocity anomalies in the lower mantle. *Eos, Trans., AGU* **81**(22), Western Pacific Suppl., Abstract T32B-03, pp. WP193–WP194.

Birch F. (1952) Elasticity and constitution of the Earth's interior. *J. Geophys. Res.* **57**, 227–286.

Brodholt J. P. (2000) Pressure-induced changes in the compression mechanism of aluminous perovskite in the Earth's mantle. *Nature* **407**, 620–622.

Castle J. C. and Creager K. C. (1999) A steeply dipping discontinuity in the lower mantle beneath Izu-Bonin. *J. Geophys. Res.* **104**, 7279–7292.

Chevrot S., Vinnik L., and Montagner J.-P. (1999) Global-scale analysis of the mantle Pds phases. *J. Geophys. Res.* **104**, 20203–20220.

Christiansen R. L., Foulger G. R., and Evans J. R. (2002) Upper-mantle origin of the Yellowstone hotspot. *Geol. Soc. Am. Bull.* **114**, 1245–1256.

Connolly J. A. D. and Kerrick D. M. (2002) Metamorphic controls on seismic velocity of subducted oceanic crust at 100–250 km depth. *Earth Planet. Sci. Lett.* **204**, 61–74.

Deuss A. and Woodhouse J. H. (2001) Seismic observations of splitting of the mid-mantle transition zone discontinuity in Earth's mantle. *Science* **294**, 354–357.

Deuss A. and Woodhouse J. H. (2002) A systematic search for mantle discontinuities using SS-precursors. *Geophys. Res. Lett.* **29**(8), doi: 10.1029/2002GL014768.

Dueker K. G. and Sheehan A. F. (1997) Mantle discontinuity structure from midpoint stacks of converted P and S waves across the Yellowstone hotspot track. *J. Geophys. Res.* **102**, 8313–8328.

Duffy T. S. and Anderson D. L. (1989) Seismic velocities in mantle minerals and the mineralogy of the upper mantle. *J. Geophys. Res.* **94**, 1895–1912.

Duffy T. S., Zha C. S., Downs R. T., Mao H. K., and Hemley R. J. (1995) Elasticity of forsterite to 16 GPa and the composition of the upper mantle. *Nature* **378**, 170–173.

Fei Y., Mao H.-K., and Mysen B. O. (1991) Experimental determination of element partitioning and calculation of phase relations in the MgO–FeO–SiO_2 system at high pressure and high temperature. *J. Geophys. Res.* **96**, 2157–2169.

Fei Y., Mao H.-K., Shu J., and Hu J. (1992) $P-V-T$ equation of state of magnesiowüstite $(Mg_{0.6}Fe_{0.4})O$. *Phys. Chem. Min.* **18**, 416–422.

Fiquet G., Dewaele A., Andrault D., Kunz M., and LeBihan T. (2000) Thermoelastic properties and crystal structure of $MgSiO_3$ perovskite at lower mantle pressure and temperature conditions. *Geophys. Res. Lett.* **27**, 21–24.

Flesch L. M., Li B., and Liebermann R. C. (1998) Sound velocities of polycrystalline $MgSiO_3$-orthopyroxene to 10 GPa at room temperature. *Am. Mineral.* **83**, 444–450.

Foulger G. R., Pritchard M. J., Julian B. R., Evans J. R., Allen R. M., Nolet G., Morgan W. J., Bergsson B. H., Erlendsson P., Jakobsdottir S., Ragnarsson S., Stefansson R., and Vogfjörd K. (2001) Seismic tomography shows that upwelling beneath Iceland is confined to the upper mantle. *Geophys. J. Int.* **146**, 504–530.

Fukao Y., Widiyantoro S., and Obayashi M. (2001) Stagnant slabs in the upper and lower mantle transition region. *Rev. Geophys.* **39**, 291–323.

Garlick G. D. (1969) Consequences of chemical equilibrium across phase changes in the mantle. *Lithos* **2**, 325–331.

Garnero E. J. (2000) Heterogeneity of the lowermost mantle. *Ann. Rev. Earth Planet. Sci.* **28**, 509–537.

Green H. W., II and Houston H. (1995) The mechanics of deep earthquakes. *Ann. Rev. Earth Planet. Sci.* **23**, 169–213.

Gu Y. J. and Dziewonski A. M. (2002) Global variability of transition zone thickness. *J. Geophys. Res.* **107**(B7), doi: 10.1029/2001JB000489.

Hacker B. R. (1996) Eclogite formation and the rheology, buoyancy, seismicity, and H_2O content of oceanic crust. In *Subduction: Top to Bottom*, Geophys. Monogr. 96 (eds. G. Bebout, D. Scholl, S. Kirby, and J. Platt). AGU, Washington, DC, pp. 337–346.

Hedlin M. A. H., Shearer P. M., and Earle P. S. (1997) Seismic evidence for small-scale heterogeneity throughout the Earth's mantle. *Nature* **387**, 145–150.

Helffrich G. (1996) Subducted lithospheric slab velocity structure: observations and mineralogical inferences. In *Subduction: Top to Bottom*, Geophys. Monogr. 96 (eds. G. Bebout, D. Scholl, S. Kirby, and J. Platt). AGU, Washington, DC, pp. 215–222.

Helffrich G. and Bina C. R. (1994) Frequency dependence of the visibility and depths of mantle seismic discontinuities. *Geophys. Res. Lett.* **21**, 2613–2616.

Helffrich G. and Stein S. (1993) Study of the structure of the slab–mantle interface using reflected and converted seismic waves. *Geophys. J. Int.* **115**, 14–40.

Helffrich G., Stein S., and Wood B. J. (1989) Subduction zone thermal structure and mineralogy and their relationship to seismic wave reflections and conversions at the slab/mantle interface. *J. Geophys. Res.* **94**, 753–763.

Helffrich G., Asencio E., Knapp J., and Owens T. (2003) Transition zone structure under the northern North Sea. *Geophys. J. Int.* (in press).

Helffrich G. R. (2000) Topography of the transition zone seismic discontinuities. *Rev. Geophys.* **38**, 141–158.

Helffrich G. R. and Wood B. J. (1996) 410-km discontinuity sharpness and the form of the olivine $\alpha-\beta$ phase diagram: resolution of apparent seismic contradictions. *Geophys. J. Int.* **126**, F7–F12.

Helffrich G. R. and Wood B. J. (2001) The Earth's mantle. *Nature* **412**, 501–507.

Hirose K., Fei Y., Ma Y., and Mao H.-K. (1999) The fate of subducted basaltic crust in the Earth's lower mantle. *Nature* **397**, 53–56.

Humphreys E. D., Dueker K. G., Schutt D. L., and Smith R. B. (2000) Beneath Yellowstone: evaluating plume and non-plume models using teleseismic images of the upper mantle. *GSA Today* **10**(12), 1–7.

Irifune T. and Isshiki M. (1998) Iron partitioning in a pyrolite mantle and the nature of the 410-km seismic discontinuity. *Nature* **392**, 702–705.

Ita J. J. and Stixrude L. (1992) Petrology, elasticity and composition of the transition zone. *J. Geophys. Res.* **97**, 6849–6866.

Jackson I. (1983) Some geophysical constraints on the chemical composition of the Earth's lower mantle. *Earth Planet. Sci. Lett.* **62**, 91–103.

Jackson I. (1998) Elasticity, composition and temperature of the Earth's lower mantle: a reappraisal. *Geophys. J. Int.* **134**, 291–311.

James D. E., Boyd F. R., Bell D., and Carlson R. (2001) Xenolith constraints on seismic velocities in the upper mantle beneath southern Africa. Abstracts of the Slave-Kaapvaal Workshop "A Tale of Two Cratons," Merrickville, Ontario, Canada, 2pp.

Jeanloz J. (1995) Earth dons a different mantle. *Nature* **378**, 130–131.

Jeanloz J. and Knittle E. (1989) Density and composition of the lower mantle. *Phil. Trans. Roy. Soc. London A* **328**, 377–389.

Kaneshima S. and Helffrich G. (1998) Detection of lower mantle scatterers northeast of the Mariana subduction zone using short-period array data. *J. Geophys. Res.* **103**, 4825–4838.

Kaneshima S. and Helffrich G. (1999) Dipping low-velocity layer in the mid-lower mantle: evidence for geochemical heterogeneity. *Science* **283**, 1888–1892.

Karki B. B., Stixrude L., and Crain J. (1997) *Ab initio* elasticity of three high-pressure polymorphs of silica. *Geophys. Res. Lett.* **24**, 3269–3272.

Katsura T. and Ito E. (1989) The system $Mg_2SiO_4-Fe_2SiO_4$ at high pressures and temperatures: precise determination of stabilities of olivine, modified spinel, and spinel. *J. Geophys. Res.* **94**, 15663–15670.

Keller W. R., Anderson D. L., and Clayton R. W. (2000) Resolution of tomographic models of the mantle beneath Iceland. *Geophys. Res. Lett.* **27**, 3993–3996.

Kennett B. L. N., Engdahl E. R., and Buland R. (1995) Constraints on seismic velocities in the Earth from traveltimes. *Geophys. J. Int.* **122**, 108–124.

Kerr R. A. (2003) Plumes from the core lost and found. *Science* **299**, 35–36.

Kesson S. E., Fitzgerald J. D., and Shelley J. M. G. (1998) Mineralogy and dynamics of a pyrolite lower mantle. *Nature* **393**, 252–255.

Kind R., Li X., Yuan X., Sobolev S. V., Hanka W., Ramesh S. D., Gu Y. G., and Dziewonski A. M. (2002) Global comparison of SS precursor and Ps conversion data from the upper mantle seismic discontinuities. *Eos, Trans., AGU* **83**(47), Fall Meet. Suppl., Abstract S52C-07.

Kumazawa M., Sawamoto H., Ohtani E., and Masaki K. (1974) Post-spinel phase of forsterite and evolution of the Earth's mantle. *Nature* **247**, 356–358.

Lebedev S., Chevrot S., and van der Hilst R. D. (2002) Seismic evidence for olivine phase changes at the 410- and 660-kilometer discontinuities. *Science* **296**, 1300–1302.

Li B., Rigden S. M., and Liebermann R. C. (1996) Elasticity of stishovite at high pressure. *Phys. Earth Planet. Inter.* **96**, 113–127.

Li B., Liebermann R. C., and Weidner D. J. (2001) $P-V-V_P-V_S-T$ measurements on wadsleyite to 7 GPa and 873 K: implications for the 410-km seismic discontinuity. *J. Geophys. Res.* **106**, 30575–30591.

Li X., Kind R., Priestley K., Sobolev S. V., Tilmann F., Yuan X., and Weber M. (2000) Mapping the Hawaiian plume conduit with converted seismic waves. *Nature* **405**, 938–941.

Liu J., Topor L., Zhang J., Navrotsky A., and Liebermann R. C. (1996) Calorimetric study of the coesite-stishovite transformation and calculation of the phase boundary. *Phys. Chem. Min.* **23**, 11–16.

Liu M. (1994) Asymmetric phase effects and mantle convection patterns. *Science* **264**, 1904–1907.

Liu Z.-K. and Ågren J. (1995) Thermodynamics of constrained and unconstrained equilibrium systems and their phase rules. *J. Phase Equil.* **16**, 30–35.

Mao H. K., Shen G., and Hemley R. J. (1997) Multivariable dependence of Fe/Mg partitioning in the Earth's lower mantle. *Science* **278**, 2098–2100.

Mattern E., Matas J., and Ricard Y. (2002) Computing density and seismic velocity of subducting slabs. *Eos, Trans., AGU* **83**(47), Fall Meet. Suppl., Abstract S51A-1011.

McCammon C. A. (1997) Perovskite as a possible sink for ferric iron in the lower mantle. *Nature* **387**, 694–696.

McDonough W. F. and Rudnick R. L. (1998) Mineralogy and composition of the upper mantle. *Rev. Mineral.* **37**, 139–164.

Melbourne T. and Helmberger D. (1998) Fine structure of the 410-km discontinuity. *J. Geophys. Res.* **103**, 10091–10102.

Montagner J.-P. and Ritsema J. (2001) Interactions between ridges and plumes. *Science* **294**, 1472–1473.

Neele F., de Regt H., and VanDecar J. (1997) Gross errors in upper-mantle discontinuity topography from underside reflection data. *Geophys. J. Int.* **129**, 194–204.

Niu F. and Kawakatsu H. (1997) Depth variation of the mid-mantle seismic discontinuity. *Geophys. Res. Lett.* **24**, 429–432.

Niu F., Solomon S. C., Silver P. G., Suetsugu D., and Inoue H. (2002) Mantle transition-zone structure beneath the South Pacific Superswell and evidence for a mantle plume underlying the Society hotspot. *Earth Planet. Sci. Lett.* **198**, 371–380.

Ono S. and Yasuda A. (1996) Compositional change of majoritic garnet in a MORB composition from 7 to 17 GPa and 1,400 to 1,600 °C. *Phys. Earth Planet. Inter.* **96**, 171–179.

Ono S., Ito E., and Katsura T. (2001) Mineralogy of subducted basaltic crust (MORB) from 25 to 37 GPa and chemical heterogeneity of the lower mantle. *Earth Planet. Sci. Lett.* **190**, 57–63.

Peacock S. M. (1993) The importance of blueschist → eclogite dehydration reactions in subducting oceanic crust. *Geol. Soc. Am. Bull.* **105**, 684–694.

Pognante U. (1989) Tectonic implications of lawsonite formation in the Sesia zone (western Alps). *Tectonophysics* **162**, 219–227.

Reinsch D. (1979) Glaucophanites and eclogites from Val Chiusella, Sesia-Lanzo zone (Italian Alps). *Contrib. Mineral. Petrol.* **70**, 257–266.

Ringwood A. E. (1975) *Composition and Petrology of the Earth's Mantle*. McGraw-Hill, New York.

Ringwood A. E. (1989) Constitution and evolution of the mantle. *Spec. Publ. Geol. Soc. Australia* **14**, 457–485.

Ritsema J., van Heijst H. J., and Woodhouse J. H. (1999) Complex shear wave velocity structure imaged beneath Africa and Iceland. *Science* **286**, 1925–1928.

Saltzer R. L. and Humphreys E. D. (1997) Upper mantle *P* wave velocity structure of the eastern Snake River Plain and its relationship to geodynamic models of the region. *J. Geophys. Res.* **102**, 11829–11841.

Shearer P. M. and Flanagan M. P. (1999) Seismic velocity and density jumps across the 410- and 660-kilometer discontinuities. *Science* **285**, 1545–1548.

Shearer P. M., Flanagan M. P., Reif C., and Masters G. (2002) Seismic constraints on mantle discontinuities: what we know and what we don't know. *Eos, Trans., AGU* **83**(47), Fall Meet. Suppl., Abstract S52C-08.

Shen Y., Solomon S. C., Bjarnason I. T., and Wolfe C. J. (1998) Seismic evidence for a lower mantle origin of the Iceland plume. *Nature* **395**, 62–65.

Sinogeikin S. V., Katsura T., and Bass J. D. (1998) Sound velocities and elastic properties of Fe-bearing wadsleyite and ringwoodite. *J. Geophys. Res.* **103**, 20819–20825.

Smyth J. R. and Frost D. J. (2002) The effect of water on the 410-km discontinuity: an experimental study. *Geophys. Res. Lett.* **29**(10), doi: 10.1029/2001GL014418.

Solomatov V. S. and Stevenson D. J. (1994) Can sharp seismic discontinuity be caused by non-equilibrium phase transformation? *Earth Planet. Sci. Lett.* **125**, 267–279.

Stixrude L. (1997) Structure and sharpness of phase transitions and mantle discontinuities. *J. Geophys. Res.* **102**, 14835–14852.

Stixrude L., Hemley R. J., Fei Y., and Mao H. K. (1992) Thermoelasticity of silicate perovskite and magnesiowüstite stratification of the Earth's mantle. *Science* **257**, 1099–1101.

Takenaka S., Sanshadokoro H., and Yoshioka S. (1999) Velocity anomalies and spatial distributions of physical properties in horizontally lying slabs beneath the Northwestern Pacific region. *Phys. Earth Planet. Inter.* **112**, 137–157.

Vacher P., Mocquet A., and Sotin C. (1998) Computation of seismic profiles from mineral physics: the importance of the non-olivine components for explaining the 660 km depth discontinuity. *Phys. Earth Planet. Inter.* **106**, 275–298.

Vacher P., Spakman W., and Wortel M. J. R. (1999) Numerical tests on the seismic visibility of metastable minerals in subduction zones. *Earth Planet. Sci. Lett.* **170**, 335–349.

VanDecar J. C., James D. E., and Assumpção M. (1995) Seismic evidence for a fossil mantle plume beneath South America and implications for plate driving forces. *Nature* **378**, 25–31.

van der Lee S., Paulssen H., and Nolet G. (1994) Variability of P660s phases as a consequence of topography on the 660 km discontinuity. *Phys. Earth Planet. Inter.* **86**, 147–164.

van der Meijde M., Marone F., van der Lee S., and Giardini D. (2002) Seismological evidence for the presence of water near the 410 km discontinuity. *Eos, Trans., AGU* **83**(47), Fall Meet. Suppl., Abstract S52C-03.

Vinnik L., Chevrot S., and Montagner J.-P. (1997) Evidence for a stagnant plume in the transition zone? *Geophys. Res. Lett.* **24**, 1007–1010.

Walck M. C. (1984) The *P*-wave upper mantle structure beneath an active spreading center: the Gulf of California. *Geophys. J. Roy. Astro. Soc.* **76**, 697–723.

Walker D. and Agee C. (1989) Partitioning "equilibrium," temperature gradients, and constraints on Earth differentiation. *Earth Planet. Sci. Lett.* **96**, 49–60.

Wang Y., Weidner D. J., and Guyot F. (1996) Thermal equation of state of $CaSiO_3$ perovskite. *J. Geophys. Res.* **101**, 661–672.

Weidner D. J. (1986) Mantle model based on measured physical properties of minerals. In *Chemistry and Physics of Terrestrial Planets* (ed. S. K. Saxena). Springer, New York, pp. 251–274.

Wolfe C. J., Bjarnason I. T., VanDecar J. C., and Solomon S. C. (1997) Seismic structure of the Iceland mantle plume. *Nature* **385**, 245–247.

Wood B. J. (1987) Theromodynamics of multicomponent systems containing several solid solutions. *Rev. Mineral.* **17**, 71–95.

Wood B. J. (1995) The effect of H_2O on the 410-kilometer seismic discontinuity. *Science* **268**, 74–76.

Wood B. J. (2000) Phase transformations and partitioning relations in peridotite under lower mantle conditions. *Earth Planet. Sci. Lett.* **174**, 341–354.

Wood B. J. and Rubie D. C. (1996) The effect of alumina on phase transformations at the 660-kilometer discontinuity from Fe–Mg partitioning experiments. *Science* **273**, 1522–1524.

Wood B. J., Pawley A., and Frost D. R. (1996) Water and carbon in the Earth's mantle. *Phil. Trans. Roy. Soc. London A* **354**, 1495–1511.

Zha C.-S., Duffy T. S., Downs R. T., Mao H.-K., and Hemley R. J. (1998a) Brillouin scattering and X-ray diffraction of San Carlos olivine: direct pressure determination to 32 GPa. *Earth Planet. Sci. Lett.* **159**, 25–33.

Zha C.-S., Duffy T. S., Downs R. T., Mao H.-K., Hemley R. J., and Weidner D. J. (1988b) Single crystal elasticity of the α and β of Mg_2SiO_4 polymorphs at high pressure. In *Properties of Earth and Planetary Materials at High Pressure and Temperature*, Geophys. Monogr. 101 (eds. M. H. Manghnani and T. Yagi). AGU, Washington, DC, pp. 9–16.

Zhang J. and Weidner D. J. (1999) Thermal equation of state of aluminum-enriched silicate perovskite. *Science* **284**, 782–784.

Zhao Y. and Anderson D. L. (1994) Mineral physics constraints on the chemical composition of the Earth's lower mantle. *Phys. Earth Planet. Inter.* **85**, 273–292.

2.03
Sampling Mantle Heterogeneity through Oceanic Basalts: Isotopes and Trace Elements

A. W. Hofmann

Max-Plank-Institut für Chemie, Mainz, Germany and Lamont-Doherty Earth Observatory, Palisades, NY, USA

2.03.1 INTRODUCTION

2.03.1.1 Early History of Mantle Geochemistry

Until the arrival of the theories of plate tectonics and seafloor spreading in the 1960s, the Earth's mantle was generally believed to consist of peridotites of uniform composition. This view was shared by geophysicists, petrologists, and geochemists alike, and it served to characterize the compositions and physical properties of mantle and crust as "Sial" (silica-alumina) of low density and "Sima" (silica-magnesia) of greater density. Thus, Hurley and his collaborators were able to distinguish crustal magma sources from those located in the mantle on the basis of their initial strontium-isotopic compositions (Hurley *et al.*, 1962; and Hurley's lectures and popular articles not recorded in the formal scientific literature). In a general way, as of early 2000s, this view is still considered valid, but literally thousands of papers have since been published on the isotopic and trace-elemental composition of oceanic basalts because they come from the mantle and they are rich sources of information about the composition of the mantle, its differentiation history and its internal structure. Through the study of oceanic basalts, it was found that the mantle is compositionally just as heterogeneous as the crust. Thus, geochemistry became a major tool to decipher the geology of the mantle, a term that seems more appropriate than the more popular "chemical geodynamics."

The pioneers of this effort were Gast, Tilton, Hedge, Tatsumoto, and Hart (Hedge and Walthall, 1963; Gast *et al.*, 1964; Tatsumoto, *et al.*, 1965; Hart, 1971). They discovered from isotope analyses of strontium and lead in young (effectively zero age) ocean island basalts (OIBs) and mid-ocean ridge basalts (MORBs) that these basalts are isotopically not uniform. The isotope ratios $^{87}Sr/^{86}Sr$, $^{206}Pb/^{204}Pb$, $^{207}Pb/^{204}Pb$, and $^{208}Pb/^{204}Pb$ increase as a function of time and the respective radioactive-parent/nonradiogenic daughter ratios, $^{87}Rb/^{86}Sr$, $^{238}U/^{204}Pb$, $^{235}U/^{204}Pb$, and $^{232}Th/^{204}Pb$, in the sources of the magmas. This means that the mantle must contain geologically old reservoirs with different Rb/Sr, U/Pb, and Th/Pb ratios. The isotope story was complemented by trace-element geochemists, led primarily by Schilling and Winchester (1967, 1969) and Gast (1968) on chemical trace-element fractionation during igneous processes, and by Tatsumoto *et al.* (1965) and Hart (1971). From the trace-element abundances, particularly rare-earth element (REE) abundances, it became clear that not only some particular parent–daughter element abundance ratios but also the light-to-heavy REE ratios of the Earth's mantle are quite heterogeneous. The interpretation of these heterogeneities has occupied mantle geochemists since the 1960s.

This chapter is in part an update of a previous, more abbreviated review (Hofmann, 1997). It covers the subject in greater depth, and it reflects some significant changes in the author's views since the writing of the earlier paper. In particular, the spatial range of equilibrium attained during partial melting may be much smaller than previously thought, because of new experimental diffusion data and new results from natural settings. Also, the question of "layered" versus "whole-mantle" convection, including the depth of subduction and of the origin of plumes, has to be reassessed in light of the recent breakthroughs achieved by seismic mantle tomography. As the spatial resolution of seismic tomography and the pressure range, accuracy, and precision of experimental data on melting relations, phase transformations, and kinetics continue to improve, the interaction between these disciplines and geochemistry *sensu stricto* will continue to improve our understanding of what is actually going on in the mantle. The established views of the mantle being engaged in simple two-layer, or simple single-layer, convection are becoming obsolete. In many ways, we are just at the beginning of this new phase of mantle geology, geophysics, and geochemistry.

2.03.1.2 The Basics

2.03.1.2.1 Major and trace elements: incompatible and compatible behavior

Mantle geochemists distinguish between major and trace elements. At first sight, this nomenclature seems rather trivial, because which particular elements should be called "major" and which "trace" depends on the composition of the system. However, this distinction actually has a deeper meaning, because it signifies fundamental differences in geochemical behavior. We define elements as "major" if they are essential constituents of the minerals making up a rock, namely, in the sense of the phase rule. Thus, on the one hand, silicon, aluminum, chromium, magnesium, iron, calcium, sodium, and oxygen are major elements because they are essential constituents of the upper-mantle minerals—olivine, pyroxene, garnet, spinel, and plagioclase. Adding or subtracting such elements can change the phase assemblage. Trace elements, on the other hand just replace a few atoms of the major elements in the crystal structures without affecting the phase assemblage significantly. They are essentially blind passengers in many mantle processes, and they are therefore immensely useful as tracers of such processes. During solid-phase transformations, they will redistribute themselves locally between the newly formed mineral phases but, during melting, they are partitioned to a greater or lesser degree into the melt. When such a melt is transported to the Earth's surface, where it can be sampled, its trace elements carry a wealth of information about the composition of the source rock and the nature of the melting processes at depth.

For convenience, the partitioning of trace elements between crystalline and liquid phases is usually described by a coefficient D, which is just a simple ratio of two concentrations at chemical equilibrium:

$$D^i = \frac{C_s^i}{C_l^i} \tag{1}$$

where D^i is the called the partition coefficient of trace element i, C_s^i and C_l^i are the concentrations (by weight) of this element in the solid and liquid phases, respectively.

Goldschmidt (1937, 1954) first recognized that the distribution of trace elements in minerals is strongly controlled by ionic radius and charge. The partition coefficient of a given trace element between solid and melt can be quantitatively described by the elastic strain this element causes by its presence in the crystal lattice. When this strain is large because of the magnitude of the misfit, the partition coefficient becomes small, and the element is partitioned into the liquid. This subject is treated in detail in Chapter 2.09.

Most trace elements have values of $D \ll 1$, simply because they differ substantially either in ionic radius or ionic charge, or both, from the atoms of the major elements they replace in the crystal lattice. Because of this, they are called *incompatible*. Exceptions are trace elements such as strontium in plagioclase, ytterbium, lutetium, and scandium in garnet, nickel in olivine, and scandium in clinopyroxene. These latter elements actually fit into their host crystal structures slightly better than the major elements they replace, and they are therefore called *compatible*. Thus, most chemical elements of the periodic table are trace elements, and most of them are incompatible; only a handful are compatible.

Major elements in melts formed from mantle rocks are by definition compatible, and most of them are well buffered by the residual minerals, so that their concentrations usually vary by factors of less than two in the melts. In contrast, trace elements, particularly those having very low partition coefficients, may vary by as many as three orders of magnitude in the melt, depending on the degree of melting. This is easily seen from the mass-balance-derived equation for the equilibrium concentration of a trace element in the melt, C_l, given by (Shaw, 1970)

$$C_l = \frac{C_0}{F + D(1 - F)} \tag{2}$$

where the superscript i has been dropped for clarity, C_0 is the concentration in the bulk system, and F is the melt fraction by mass. For highly incompatible elements, which are characterized by very low partition coefficients, so that $D \ll F$, this equation reduces to

$$C_l \approx \frac{C_0}{F} \tag{3}$$

This means that the trace-element concentration is then inversely proportional to the melt fraction F, because the melt contains essentially all of the budget of this trace element. An additional consequence of highly incompatible behavior of trace elements is that their concentration ratios in the melt become constant, independent of melt fraction, and identical to the respective ratio in the mantle source. This follows directly when Equation (3) is written for two highly incompatible elements:

$$\frac{C_l^1}{C_l^2} \approx \frac{C_0^1}{F} \times \frac{F}{C_0^2} = \frac{C_0^1}{C_0^2} \tag{4}$$

In this respect, incompatible trace-element ratios resemble isotope ratios. They are therefore very useful in complementing the information obtained from isotopes.

2.03.1.2.2 Radiogenic isotopes

The decay of long-lived radioactive isotopes was initially used by geochemists exclusively for the measurement of geologic time. As noted in the introduction, their use as tracers of mantle processes was pioneered by Hurley and co-workers in the early 1960s. The decay

$$^{87}\text{Rb} \rightarrow {}^{87}\text{Sr} \qquad (\lambda = 1.42 \times 10^{-11}\ \text{yr}) \tag{5}$$

serves as example. The solution of the decay equation is

$$^{87}\text{Sr} = {}^{87}\text{Rb} \times (e^{\lambda t} - 1) \tag{6}$$

Dividing both sides by one of the nonradiogenic isotopes, by convention ^{86}Sr, we obtain

$$\frac{^{87}\text{Sr}}{^{86}\text{Sr}} = \frac{^{87}\text{Rb}}{^{86}\text{Sr}} \times (e^{\lambda t} - 1) \approx \frac{^{87}\text{Rb}}{^{86}\text{Sr}} \times \lambda t \tag{7}$$

The approximation in Equation (7) holds only for decay systems with sufficiently long half-lives, such as the Rb–Sr and the Sm–Nd systems, so that $\lambda t \ll 1$ and $e^{\lambda t} - 1 \approx \lambda t$. Therefore, the isotope ratio ^{87}Sr/^{86}Sr in a system, such as some volume of mantle rock, is a linear function of the parent/daughter chemical ratio Rb/Sr and a nearly linear function of time or geological age of the system. When this mantle volume undergoes equilibrium partial melting, the melt inherits the ^{87}Sr/^{86}Sr ratio of the entire system. Consequently, radiogenic isotope ratios such as ^{87}Sr/^{86}Sr are powerful tracers of the parent–daughter ratios of mantle sources of igneous rocks. If isotope data from several decay systems are combined, a correspondingly richer picture of the source chemistry can be constructed.

Table 1 shows a list of long-lived radionuclides, their half-lives, daughter isotopes, and radiogenic-to-nonradiogenic isotope rates commonly used as tracers in mantle geochemistry. Noble-gas isotopes are not included here, because a separate chapter of this Treatise is devoted to them

Table 1 Long-lived radionuclides.

Parent nuclide	*Daughter nuclide*	*Half life* (yr)	*Tracer ratio* (*radiogenic/nonradiogenic*)
^{147}Sm	^{143}Nd	106×10^9	^{143}Nd/^{144}Nd
^{87}Rb	^{87}Sr	48.8×10^9	^{87}Sr/^{86}Sr
^{176}Lu	^{176}Hf	35.7×10^9	^{176}Hf/^{177}Hf
^{187}Re	^{187}Os	45.6×10^9	^{187}Os/^{188}Os
^{40}K	^{40}Ar	1.25×10^9	^{40}Ar/^{36}Ar
^{232}Th	^{208}Pb	14.01×10^9	^{208}Pb/^{204}Pb
^{238}U	^{206}Pb	4.468×10^9	^{206}Pb/^{204}Pb
^{235}U	^{207}Pb	0.738×10^9	^{207}Pb/^{204}Pb

(see Chapter 2.06). Taken together, they cover a wide range of geochemical properties including incompatible and compatible behavior. These ratios will be used, together with some incompatible trace-element ratios, as tracers of mantle reservoirs, crust–mantle differentiation processes, and mantle melting processes in later sections of this chapter.

2.03.2 LOCAL AND REGIONAL EQUILIBRIUM REVISITED

How do we translate geochemical data from basalts into a geological model of the present-day mantle and its evolution? The question of chemical and isotopic equilibrium, and particularly its spatial dimension, has always played a fundamental role in this effort of interpretation. The basic, simple tenet of isotope geochemists and petrologists alike has generally been that partial melting at mantle temperatures, pressures, and timescales achieves essentially complete chemical equilibrium between melt and solid residue. For isotope data in particular, this means that at magmatic temperatures, the isotope ratio of the melt is identical to that of the source, and this is what made isotope ratios of volcanic rocks apparently ideal tracers of mantle composition. The question of spatial scale seemed less important, because heterogeneities in the mantle were thought to be important primarily on the 10^2–10^4 km scale (Hart *et al.*, 1973; Schilling, 1973; White and Schilling, 1978; Dupré and Allègre, 1983). To be sure, this simple view was never universal. Some authors invoked special isotopic effects during melting, so that the isotopic composition of the melt could in some way be "fractionated" during melting, in spite of the high temperatures prevailing, so that the isotope ratios observed in the melts would *not* reflect those of the melt sources (e.g., O'Hara, 1973; Morse, 1983). These opinions were invariably raised by authors not directly familiar with the analytical methods of isotope geochemistry, so that they did not realize that isotopic fractionation occurs in every mass spectrometer and is routinely corrected in the reported results.

2.03.2.1 Mineral Grain Scale

Some authors invoked mineral-scale isotopic (and therefore also chemical) disequilibrium and preferential melting of phases, such as phlogopite, which have higher Rb/Sr, and therefore also higher ^{87}Sr/^{86}Sr ratios than the bulk rock, in order to explain unusually high ^{87}Sr/^{86}Sr ratios in OIBs (e.g., O'Nions and Pankhurst, 1974; Vollmer, 1976). Hofmann and Hart (1978) reviewed this subject in light of the available diffusion data in solid and molten silicates. They concluded that mineral-scale isotopic and chemical disequilibrium is exceedingly unlikely, if melting timescales are on the order of thousands of years or more. More recently, Van Orman *et al.* (2001) have measured REE diffusion coefficients in clinopyroxene and found that REE mobility in this mineral is so low at magmatic temperatures that chemical disequilibrium between grain centers and margins will persist during melting. Consequently, the melt will not be in equilibrium with the bulk residue for geologically reasonable melting times, if the equilibration occurs by volume diffusion alone. This means that the conclusions of Hofmann and Hart (1978) must be revised significantly: The slowest possible path of chemical reaction no longer guarantees attainment of equilibrium. However, it is not known whether other mechanisms such as recrystallization during partial melting might not lead to much more rapid equilibration. One possible test of this would be the examination of mantle clinopyroxenes from oceanic and ophiolitic peridotites. These rocks have undergone various extents of partial melting (Johnson *et al.*, 1990; Hellebrand *et al.*, 2001), and the residual clinopyroxenes should show compositional zoning if they had not reached equilibrium with the melt via volume diffusion. Although the above-cited studies were not specifically conducted to test this question, the clinopyroxenes were analyzed by ion microprobe, and these analyses showed no significant signs of internal compositional gradients. It is, of course, possible in principle that the internal equilibration occurred after extraction of the melt, so this evidence is not conclusive at present. Nevertheless, these results certainly leave open the possibility that the crystals re-equilibrated continuously with the melt during melt production and extraction. There is at present no definitive case from "natural laboratories" deciding the case one way or the other, at least with respect to incompatible lithophile elements such as the REE.

Osmium isotopes currently provide the strongest case for mineral-to-mineral disequilibrium, and for mineral–melt disequilibrium available from observations on natural rocks. Thus, both osmium alloys and sulfides from ophiolites and mantle xenoliths have yielded strongly heterogeneous osmium isotope ratios (Alard *et al.*, 2002; Meibom *et al.*, 2002). The most remarkable aspect of these results is that these ophiolites were emplaced in Phanerozoic times, yet they contain osmium-bearing phases that have retained model ages in excess of 2 Ga in some cases. The melts that were extracted from these ophiolitic peridotites contained almost certainly much more radiogenic osmium and could, in any case, not have been in osmium-isotopic equilibrium with all of these isotopically diverse residual phases.

Another strong indication that melts extracted from the mantle are not in osmium-isotopic equilibrium with their source is given by the fact that osmium isotopes in MORBs are, on average, significantly more radiogenic than osmium isotopes from oceanic peridotites (see also Figure 9 further below). Although it may be argued that there is no one-to-one correspondence between basalts and source peridotites, and further, that the total number of worldwide MORB and peridotite samples analyzed is still small, these results strongly suggest that, at least with regard to osmium, MORBs are generally not in isotopic equilibrium with their sources or residues. However, osmium-isotopic disequilibrium does not automatically mean strontium, neodymium, lead, or oxygen-isotopic disequilibrium or incompatible-trace-element disequilibrium. This is because osmium is probably incompatible in all silicate phases (Snow and Reisberg, 1995; Schiano *et al.*, 1997; Burton *et al.*, 2000) but very highly compatible with nonsilicate phases such as sulfides and, possibly, metal alloys such as osmiridium "nuggets," which may form inclusions within silicate minerals and might therefore be protected from reaction with a partial silicate melt. At the time of writing, no clear-cut answers are available, and for the time being, we will simply note that the geochemistry of osmium and rhenium is considerably less well understood than that of silicate-hosted major and trace elements such as strontium, neodymium, lead, and their isotopic abundances.

2.03.2.2 Mesoscale Heterogeneities

By "mesoscale" I mean scales larger than about a centimeter but less than a kilometer. This intermediate scale was addressed only briefly by Hofmann and Hart, who called it a "lumpy mantle" structure. It was specifically invoked by Hanson (1977) and Wood (1979), and others subsequently, who invoked veining in the mantle to provide sources for chemically and isotopically heterogeneous melts. Other versions of mesoscale heterogeneities were invoked by Sleep (1984), who suggested that preferential melting of ubiquitous heterogeneities may explain ocean island-type volcanism, and by Allègre and Turcotte (1986), who discussed a "marble cake" structure of the mantle generated by incomplete homogenization of subducted heterogeneous lithosphere. These ideas have recently been revived in several publications discussing a mantle containing pyroxenite or eclogite layers, which may melt preferentially (Phipps Morgan *et al.*, 1995; Hirschmann and Stolper, 1996; Phipps Morgan and Morgan, 1998; Yaxley and Green, 1998; Phipps Morgan, 1999).

One of the main difficulties with these mesoscale models is that they have been difficult to test by direct geochemical and petrological field observations. Recently, however, several studies have been published which appear to support the idea of selective melting of mesoscale heterogeneities. Most important of these are probably the studies of melt inclusions showing that single basalt samples, and even single olivine grains, contain chemically and isotopically extremely heterogeneous melt inclusions. Extreme heterogeneities in REE abundances from melt inclusions had previously been explained by progressive fractional melting processes of uniform sources (Sobolev and Shimizu, 1993; Gurenko and Chaussidon, 1995). In contrast, the more recent studies have demonstrated that source heterogeneities must (also) be involved to explain the extreme variations in isotopic and chemical compositions observed (Saal *et al.*, 1998; Sobolev *et al.*, 2000). While the spatial scale of these source heterogeneities cannot be directly inferred from these melt inclusion data, it seems highly plausible that it is in the range of what is here called "mesoscale."

Other, more circumstantial, evidence for preferential melting of mesoscale heterogeneities has been described by Regelous and Hofmann (2002/3), who found that the Hawaiian plume delivered MORB-like magmas ~80 Ma ago, when the plume was located close to the Pacific spreading ridge. Unless this is a fortuitous coincidence, this implies that the same plume produces "typical" OIB-like, incompatible-element-enriched melts with elevated $^{87}Sr/^{86}Sr$ and low $^{143}Nd/^{144}Nd$ ratios when the degree of melting is relatively low under a thick lithosphere, and typically MORB-like, incompatible-element-depleted melts when the degree of melting is high because of the shallow melting level near a spreading ridge. Such a dependence on the extent of melting is consistent with a marble-cake mantle containing incompatible-element-rich pyroxenite or eclogite layers having a lower melting temperature than the surrounding peridotite matrix. This melting model is further corroborated by the observation that at least three other plumes located at or near spreading ridges have produced MORB-like lavas, namely, the Iceland, the Galapagos, and the Kerguelen plume. The overall evidence is far from clear-cut, however, because the Iceland and Galapagos plumes have also delivered OIB-like tholeiites and alkali basalts more or less in parallel with the depleted MORB-like tholeiites or picrites.

To sum up, the question of grain-scale equilibration with partial melts, which had apparently been settled definitively by Hofmann and Hart (1978), has been reopened by the experimental work of Van Orman *et al.* (2001) and by recent osmium isotope data. The mesoscale equilibrium involving a veined or marble-cake mantle consisting of a mixture of lherzolite (or harzburgite) and pyroxenite (or eclogite) has also received substantial support in the recent literature. In either case, the isotopic composition of the melt is likely to change as a function of the bulk extent of melting, and the melts do not provide quantitative estimates of the isotopic composition of the bulk sources at scales of kilometers or more. It will be seen in subsequent sections that this has ramifications particularly with respect to quantitative estimates of the sizes and spatial distributions of the reservoirs hosting the geochemical mantle heterogeneities observed in basalts. While this defeats one of the important goals of mantle geochemistry, it will be seen in the course of this chapter that the geochemical data can still be used to map large-scale geochemical provinces of the mantle and to reveal much about the smaller-scale structure of the mantle heterogeneities. In addition, they remain powerful tracers of recycling and mixing processes and their history in the mantle.

2.03.3 CRUST–MANTLE DIFFERENTIATION

Before discussing the internal chemical structure of the mantle, it is necessary to have a general understanding of crust–mantle differentiation, because this has affected the incompatible trace-element and isotope budget of the mantle rather drastically. This topic has been covered by Hofmann (1988), but the most important points will be summarized here again. The treatment here differs in detail because more recent estimates have been used for the bulk composition of the continental crust and of the bulk silicate earth (BSE), also called "primitive mantle."

2.03.3.1 Enrichment and Depletion Patterns

The growth of the continental crust has removed major proportions of the highly incompatible elements from the mantle, and this depletion is the chief (but not the sole) cause of the specific isotope and trace-element characteristics of MORBs. The effects of ionic radius and charge, described in Section 2.03.1.2.1, on this enrichment–depletion process can be readily seen in a diagram (Figure 1) introduced by Taylor and McLennan (1985). It is obvious from this that those trace elements that have ionic properties similar to the major silicate-structure-forming elements, namely, nickel, cobalt, manganese, scandium, and chromium are not enriched in the continental crust but remain in the mantle. In contrast, elements with deviating ionic properties are more or less strongly enriched in the crust, depending on the magnitude of the deviation. Two main transfer mechanisms are available for this differentiation, both of them are ultimately driven by mantle convection. The first is partial melting and ascent of the melt to the surface or into the already existing crust. The second involves dehydration (and decarbonation) reactions during subduction, metasomatic transfer of soluble elements via hydrothermal fluid from the subducted crust-plus-sediment into the overlying

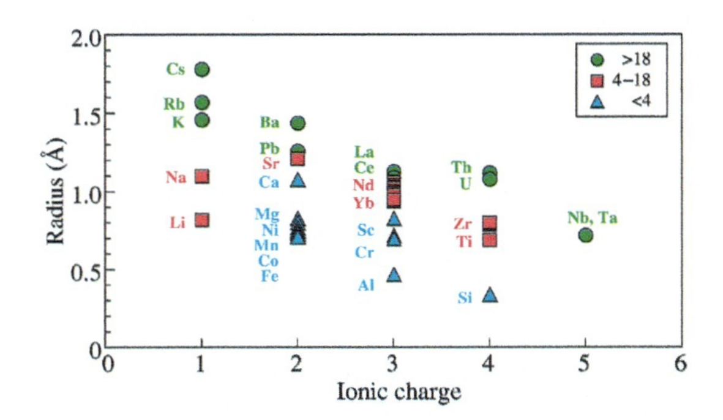

Figure 1 Ionic radius (in angstrom) versus ionic charge for lithophile major and trace elements in mantle silicates. Also shown are ranges of enrichment factors in average continental crust, using the estimate of (Rudnick and Fountain, 1995), relative to the concentrations in the primitive mantle (or "bulk silicate Earth") (source McDonough and Sun, 1995).

mantle "wedge," and partial melting of the metasomatized (or "fertilized") region. This partial melt ascends and is added to the crust, carrying the geochemical signature caused by mantle metasomatic transfer into the crust. Both mechanisms may operate during subduction, and a large body of geochemical literature has been devoted to the distinction between the two (Elliott *et al.*, 1997; Class *et al.*, 2000; Johnson and Plank, 2000; see Chapter 2.11). Continental crust may also be formed by mantle plume heads, which are thought to produce large volumes of basaltic oceanic plateaus. These may be accreted to existing continental crust, or continental flood basalts, which similarly add to the total volume of crust (e.g., Abouchami *et al.*, 1990; Stein and Hofmann, 1994; Puchtel *et al.*, 1998). The quantitative importance of this latter mechanism remains a matter of some debate (Kimura *et al.*, 1993; Calvert and Ludden, 1999).

Hofmann (1988) showed that crust formation by extraction of partial melt from the mantle could well explain much of the trace-element chemistry of crust–mantle differentiation. However, a few elements, notably niobium, tantalum, and lead, do not fit into the simple pattern of enrichment and depletion due to simple partial melting (Hofmann *et al.*, 1986). The fundamentally different behavior of these elements in the MORB–OIB environment on the one hand, and in the subduction environment on the other, requires the second, more complex, transfer mechanism via fluids (Miller *et al.*, 1994; Peucker-Ehrenbrink *et al.*, 1994; Chauvel *et al.*, 1995). Thus, local fluid transport is essential in preparing the mantle sources for production of continental crust, but the gross transport of incompatible elements from mantle to crust is still carried overwhelmingly by melting and melt ascent.

The simplest case discussed above, namely, crust–mantle differentiation by partial melting alone is illustrated in Figure 2. This shows the abundances of a large number of chemical elements in the continental crust, as estimated by (Rudnick and Fountain, 1995) and divided by their respective abundances in the primitive mantle or bulk silicate earth as estimated by (McDonough and Sun, 1995). Each element is assigned a nominal partition coefficient D as defined in Equation (1), calculated by rearranging Equation (2) and using a nominal "melt fraction" $F = 0.01$. In this highly simplified view, the continental crust is assumed to have the composition of an equilibrium partial melt derived from primitive mantle material. Also shown is the hypothetical solid mantle residue of such a partial melt and a second-stage partial melt of this depleted residue. This second-stage melt curve may then be compared with the actual element abundances of "average" ocean crust. Although this "model" of the overall crust–mantle differentiation is grossly oversimplified, it can account for the salient

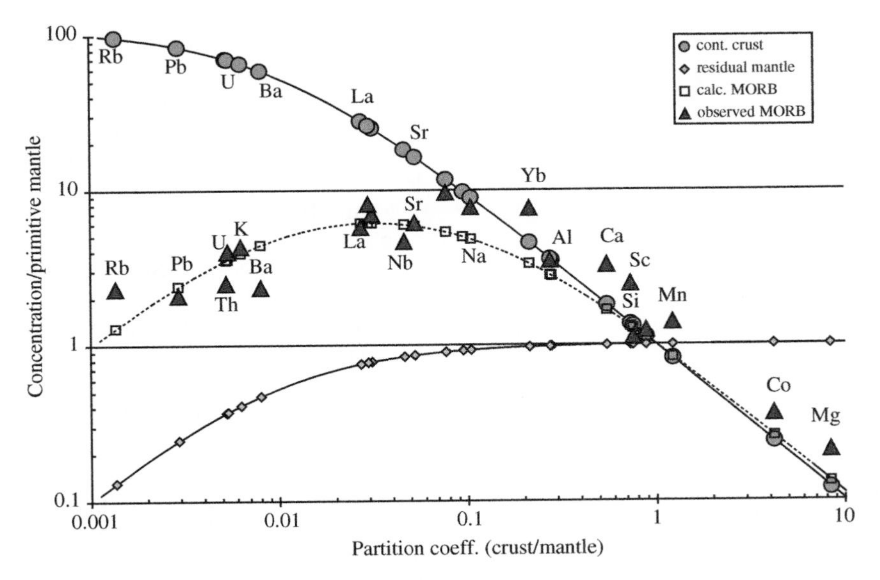

Figure 2 Comparison of the abundances of trace and (some) major elements in average continental crust and average MORB. Abundances are normalized to the primitive-mantle values (McDonough and Sun, 1995). The "partition coefficient" of each element is calculated by solving Equation (2) for D, using a melt fraction $F = 0.009$ and its abundance value in the continental crust (Rudnick and Fountain, 1995). The respective abundances in average MORB are plotted using the same value of D and using the average ("normal") MORB values of (Su, 2002), where "normal" refers to ridge segments distant from obvious hotspots.

features of the relationship between primitive mantle, continental crust, depleted mantle, and oceanic crust quite well. This representation is remarkably successful because Equation (2) is essentially a mass-balance relationship and because these major reservoirs are in fact genetically related by enrichment and depletion processes in which partial melting plays a dominant role.

The above model of extracting continental crust and remelting the depleted residue also accounts approximately for the isotopic relationships between continental crust and residual mantle, where the isotopic composition is directly represented by MORB, using the assumption of complete local and mesoscale equilibrium discussed in Section 2.03.2. This is illustrated by Figure 3, which is analogous to Figure 2, but shows only the commonly used radioactive decay systems Rb–Sr, Sm–Nd, Lu–Hf, and Re–Os. Thus, the continental crust has a high parent–daughter ratios Rb/Sr and Re/Os, but low Sm/Nd and Lu/Hf, whereas the mantle residue has complementary opposite ratios. With time, these parent–daughter ratios will generate higher than primitive $^{87}Sr/^{86}Sr$ and $^{187}Os/^{188}Os$, and lower than primitive $^{143}Nd/^{144}Nd$ and $^{176}Hf/^{177}Hf$, ratios in the crust and complementary, opposite ratios in the mantle, and this is indeed observed for strontium, neodymium, and hafnium, as will be seen in the review of the isotope data.

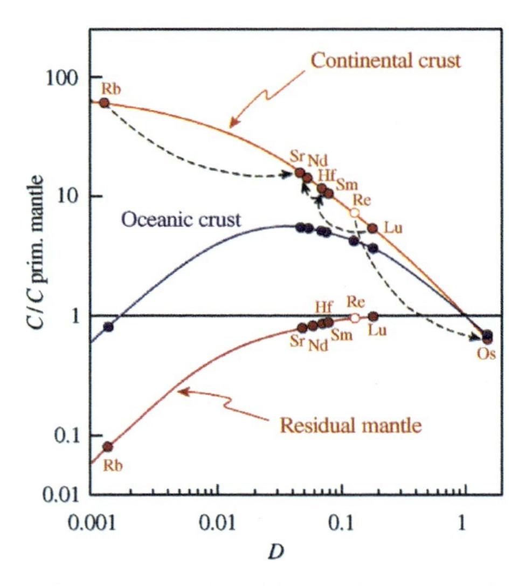

Figure 3 Crust–mantle differentiation patterns for the decay systems Rb–Sr, Sm–Nd, Lu–Hf, and Re–Os. The diagram illustrates the depletion-enrichment relationships of the parent–daughter pairs, which lead to the isotopic differences between continental crust and the residual mantle. For example, the Sm/Nd ratio is increased, whereas the Rb/Sr ratio is decreased in the residual mantle. This leads to the isotopic correlation in mantle-derived rocks plotted in Figure 4(a). The construction is similar to that used in Figure 2, but D values have been adjusted slightly for greater clarity.

The case of lead isotopes is more complicated, because the estimates for the mean parent–daughter ratios of mantle and crust are similar. This similarity is not consistent with purely magmatic production of the crust, because the bulk partition coefficient of lead during partial mantle melting is expected to be only slightly lower than that of strontium (see Figure 1), but significantly higher than the coefficients for the highly incompatible elements uranium and thorium. In reality, however, the enrichment of lead in the continental crust shown in Figure 2 is slightly higher than the enrichments for thorium and uranium, and the $^{206}Pb/^{204}Pb$ and $^{208}Pb/^{204}Pb$ ratios of continental rocks are similar to those of MORB. This famous–infamous "lead paradox," first pointed out by Allègre (1969), will be discussed in a separate section below.

How do we know the parent–daughter ratios in crust and mantle? When both parent and daughter nuclides have refractory lithophile character, and are reasonably resistant to weathering and other forms of low-temperature alteration, as is the case for the pairs Sm–Nd and Lu–Hf, we can obtain reasonable estimates from measuring and averaging the element ratios in representative rocks of crustal or mantle heritage. But when one of the elements was volatile during terrestrial accretion, and/or is easily mobilized by low-temperature or hydrothermal processes, such as rubidium, uranium, or lead, the isotopes of the daughter elements yield more reliable information about the parent–daughter ratios of primitive mantle, depleted mantle, and crust, because the isotope ratios are not affected by recent loss (or addition) of such elements. Thus, the U/Pb and Th/Pb ratios of bulk silicate earth, depleted mantle and continental crust are essentially derived from lead isotope ratios.

Similarly, the primitive mantle Rb/Sr ratio was originally derived from the well-known negative correlation between $^{87}Sr/^{86}Sr$ and $^{143}Nd/^{144}Nd$ ratios in mantle-derived and crustal rocks, the so-called "mantle array" (DePaolo and Wasserburg, 1976; Richard *et al.*, 1976; O'Nions *et al.*, 1977; see also Figure 4(a)). To be sure, there is no guarantee that this correlation will automatically go through the bulk silicate earth value. However, in this case, the primitive mantle (or "bulk silicate earth") Rb/Sr value has been approximately confirmed using element abundance ratios of barium, rubidium, and strontium. Hofmann and White (1983) found that Ba/Rb ratios in mantle-derived basalts and continental crust are sufficiently similar, so that the terrestrial value of Ba/Rb can be estimated within narrow limits. The terrestrial Ba/Sr ratio (comprising two refractory, lithophile elements) can be assumed to be identical to the ratio in chondritic meteorites, so that the terrestrial Rb/Sr ratio can be estimated as

$$\left(\frac{\mathrm{Rb}}{\mathrm{Sr}}\right)_{\mathrm{terr.}} = \left(\frac{\mathrm{Rb}}{\mathrm{Ba}}\right)_{\mathrm{terr.}} \times \left(\frac{\mathrm{Ba}}{\mathrm{Sr}}\right)_{\mathrm{chondr.}} \quad (8)$$

The terrestrial Rb/Sr ratio estimated in this way turned out to be indistinguishable from the ratio estimated by isotope correlations, and therefore the consistency between isotope and element abundance data is not circular. This example of internal consistency has been disturbed by the more recent crustal estimate of $\mathrm{Ba/Rb} = 6.7$ (Rudnick and Fountain, 1995), which is significantly lower than the above mantle estimate $\mathrm{Ba/Rb} = 11.0$ (Hofmann and White, 1983), based mostly on MORB and OIB data. This shows that, for many elements, there are greater uncertainties about the composition of the continental crust than about the mantle. The reason for this is that the continental crust has become much more heterogeneous than the mantle because of internal differentiation processes including intracrustal melting, transport of metamorphic fluids, hydrothermal transport, weathering, erosion, and sedimentation. In any case, assuming that Rudnick's crustal estimate is correct, the primitive mantle Ba/Rb should lie somewhere between 7 and 11. The lesson from this is that we must be careful when using "canonical" element ratios to make mass-balance estimates for the sizes of different mantle reservoirs.

2.03.3.2 Mass Fractions of Depleted and Primitive Mantle Reservoirs

The simple crust–mantle differentiation model shown in Figure 2 contains three "reservoirs:" continental crust, depleted residue, and oceanic crust. However, the depleted reservoir may well be smaller than the entire mantle, in which case another, possibly primitive reservoir would be needed. Thus, if one assumes that the mantle consists of two reservoirs only, one depleted and one remaining primitive, and if one neglects the oceanic crust because it is thin and relatively depleted in highly incompatible elements, one can calculate the mass fractions of these two reservoirs from their respective isotopic and/or trace-element compositions (Jacobsen and Wasserburg, 1979; O'Nions *et al.*, 1979; DePaolo, 1980; Davies, 1981; Allègre *et al.*, 1983, 1996; Hofmann *et al.*, 1986; Hofmann, 1989a). The results of these estimates have yielded mass fractions of the depleted reservoir ranging from ~30–80%. Originally, the 30% estimate was particularly popular because it matches the mass fraction of the upper mantle above the 660 km seismic discontinuity. It was also attractive because at least some of the mineral physics data indicated that the lower mantle has a different, intrinsically denser, major-element composition. However, more recent data and their evaluations indicate that they do not require such compositional layering (Jackson and Rigden, 1998). Nevertheless, many authors argue that the 660 km boundary can isolate upper-mantle- from lower-mantle convection, either because of the endothermic nature of the phase changes at this boundary, or possibly because of extreme viscosity differences between upper and lower mantle. Although this entire subject has been debated in the literature for many years, there appeared to be good reasons to think that the 660 km seismic discontinuity is the fundamental boundary between an upper, highly depleted mantle and a lower, less depleted or nearly primitive mantle.

The most straightforward mass balance, assuming that we know the composition of the continental crust sufficiently well, can be calculated from the abundances of the most highly incompatible elements, because their abundances in the depleted mantle are so low that even relatively large relative errors do not affect the mass balance very seriously. The most highly enriched elements in the continental crust have estimated crustal abundances (normalized to the primitive mantle abundances given by McDonough and Sun (1995)) of $\mathrm{Cs} = 123$, $\mathrm{Rb} = 97$, and $\mathrm{Th} = 70$ (Rudnick and Fountain, 1995). The estimate for Cs is rather uncertain because its distribution within the crust is particularly heterogeneous, and its primitive-mantle abundance is afflicted by special uncertainties (Hofmann and White, 1983; McDonough *et al.*, 1992). Therefore, a more conservative enrichment factor of 100 (close to the value of 97 for Rb) is chosen for elements most highly enriched in the continental crust. The simple three-reservoir mass balance then becomes

$$X_{\mathrm{lm}} = \frac{1 - C_{\mathrm{cc}} \times X_{\mathrm{cc}} - C_{\mathrm{um}} \times X_{\mathrm{um}}}{C_{\mathrm{lm}}} \quad (9)$$

where C refers to primitive-mantle normalized concentrations (also called "enrichment factors"), X refers to the mass fraction of a given reservoir, and the subscripts cc, lm, and um refer to continental crust, lower mantle, and upper mantle, respectively.

If the lower mantle is still primitive, so that $C_{\mathrm{lm}} = 1$, the upper mantle is extremely depleted, so that $C_{\mathrm{um}} = 0$, and $X_{\mathrm{cc}} = 0.005$, the mass balance yields

$$X_{\mathrm{lm}} = \frac{1 - 100 \times 0.005 - 0 \times X_{\mathrm{um}}}{1} = 0.5 \quad (10)$$

Remarkably, this estimate is identical to that obtained using the amounts of radiogenic argon in the atmosphere, the continental crust and the depleted, upper mantle (Allègre *et al.*, 1996). There are reasons to think that the abundances of potassium and rubidium in BSE used in these calculations have been overestimated, perhaps by as much as 30% (Lassiter, 2002), and this would of course decrease the remaining mass fraction of primitive-mantle material. Thus, we can conclude that at least half, and perhaps 80%, of the most highly incompatible element budget now resides either in the continental crust or in the atmosphere (in the case of argon).

Can we account for the entire silicate earth budget by just these three reservoirs (crust plus atmosphere, depleted mantle, primitive mantle), as has been assumed in all of the above estimates? Saunders *et al.* (1988) and Sun and McDonough (1989) (among others) have shown that this cannot be the case, using global systematics of a single trace-element ratio, Nb/La. Using updated, primitive-mantle normalized estimates for this ratio, namely, $(Nb/La)_n = 0.66$ for the continental crust (Rudnick and Fountain, 1995), and $(Nb/La)_n = 0.81$ for so-called N-type ("normal") MORB (Su, 2002), we see that both reservoirs have lower than primitive Nb/La ratios. Using the additional constraint that niobium is slightly more incompatible than lanthanum during partial melting, we find that the sources of all these mantle-derived basalts must have sources with Nb/La ratios equal to or lower than those of the basalts themselves. This means that all the major mantle sources as well as the continental crust have $(Nb/La)_n \leq 1$. By definition, the entire silicate earth has $(Nb/La)_n = 1$, so there should be an additional, hidden reservoir containing the "missing" niobium. A similar case has more recently been made using Nb/Ta, rather than Nb/La. Current hypotheses to explain these observations invoke either a refractory eclogitic reservoir containing high-niobium rutiles (Rudnick *et al.*, 2000), or partitioning of niobium into the metallic core (Wade and Wood, 2001). Beyond these complications involving special elements with unexpected geochemical "behavior," there remains the question whether ~50% portion of the mantle not needed to produce the continental crust has remained primitive, or whether it is also differentiated into depleted, MORB-source-like, and enriched, OIB-source like subreservoirs. In the past, the occurrence of noble gases with primordial isotope ratios have been used to argue that the lower part of the mantle must still be nearly primitive. However, it will be seen below that this inference is no longer as compelling as it once seemed to be.

2.03.4 MID-OCEAN RIDGE BASALTS: SAMPLES OF THE DEPLETED MANTLE

2.03.4.1 Isotope Ratios of Strontium, Neodymium, Hafnium, and Lead

The long-lived radioactive decay systems commonly used to characterize mantle compositions, their half-lives, and the isotope ratios of the respective radiogenic daughter elements are given in Table 1. The half-lives of ^{147}Sm, ^{87}Sr, ^{186}Hf, ^{187}Re, and ^{232}Th are several times greater than the age of the Earth, so that the accumulation of the radiogenic daughter nuclide is nearly linear with time. This is not the case for the shorter-lived ^{238}U and ^{235}U, and this is in part responsible for the more complex isotopic relationships displayed by lead isotopes in comparison with the systematics of strontium, neodymium, hafnium, and osmium isotopes. The mantle geochemistry of noble gases, although of course an integral part of mantle geochemistry, is treated in Chapter 2.06.

Figures 4–6 show the isotopic compositions of MORBs from spreading ridges in the three major ocean basins. Figures 4(b) and 5(a) also show isotope data for marine sediments, because these are derived from the upper continental crust and should roughly represent the isotopic composition of this crust. In general, the isotopic relationships between the continental and oceanic crust are just what is expected from the elemental parent–daughter relationships seen in Figure 3. The high Rb/Sr and low Sm/Nd and Lu/Hf ratios of continental materials relative to the residual mantle are reflected by high $^{87}Sr/^{86}Sr$ and low $^{143}Nd/^{144}Nd$ and $^{176}Hf/^{177}Hf$ ratios (not shown).

In lead isotope diagrams, the differences are not nearly as clear, and they are expressed primarily by slightly elevated $^{207}Pb/^{204}Pb$ ratios for given values of $^{206}Pb/^{204}Pb$ (Figure 5(a)). This topology in lead-isotope space requires a comparatively complex evolution of the terrestrial U–Pb system. It involves an ancient period of high U/Pb ratios in continental history (with complementary, low ratios in the residual mantle). The higher $^{235}U/^{238}U$ ratios prevailing during that time led to elevated $^{207}Pb/^{206}Pb$ ratios in the crust. This subject is treated more fully in the section on the lead paradox (Section 2.03.6) further below.

Another important observation is that, while strontium, neodymium, and hafnium isotopes all correlate with each other, they form poorer, but still significant, correlations with $^{206}Pb/^{204}Pb$ (or $^{208}Pb/^{204}Pb$, not shown) ratios in the Pacific and Atlantic, but no discernible correlation in the Indian Ocean MORB (Figure 6(a)). Nevertheless, if instead of $^{208}Pb/^{204}Pb$ or $^{206}Pb/^{204}Pb$ ratios one plots the so-called "radiogenic" $^{208}Pb^*/^{206}Pb^*$ ratio, the lead data do correlate with neodymium

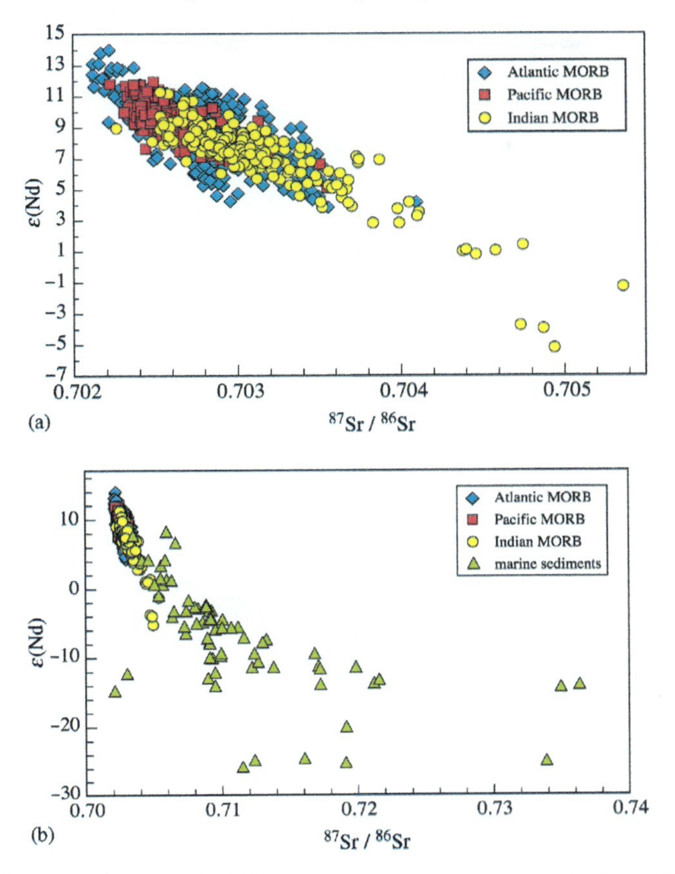

Figure 4 (a) $^{87}Sr/^{86}Sr$ versus ε(Nd) for MORBs from the three major ocean basins. ε(Nd) is a measure of the deviation of the $^{143}Nd/^{144}Nd$ ratio from the chondritic value, assumed to be identical to the present-day value in the bulk silicate earth. It is defined as $\varepsilon(Nd) = 10^4 \times (^{143}Nd/^{144}Nd_{measured} - {}^{143}Nd/^{144}Nd_{Chondrite})/^{143}Nd/^{144}Nd_{Chondrite}$. The chondritic value used is $^{143}Nd/^{144}Nd_{Chondrite} = 0.512638$. The data are compiled from the PETDB database (http://petdb.ldeo.columbia.edu). (b) $^{87}Sr/^{86}Sr$ versus ε(Nd) for MORBs compared with data for turbidites and other marine sediments (Ben Othman *et al.*, 1989; Hemming and McLennan, 2001). This illustrates the complementary nature of continent-derived sediments and MORB expected from the relationships shown in Figure 3.

isotopes in all three ocean basins (Figure 6(b)). This parameter is a measure of the radiogenic additions to $^{208}Pb/^{204}Pb$ and $^{206}Pb/^{204}Pb$ ratios during Earth's history; it is calculated by subtracting the primordial (initial) isotope ratios from the measured values. The primordial ratios are those found in the Th–U-free sulfide phase (troilite) of iron meteorites. Thus, the radiogenic $^{208}Pb^*/^{206}Pb^*$ ratio is defined as

$$\frac{^{208}Pb^*}{^{206}Pb^*} = \frac{^{208}Pb/^{204}Pb\text{-}(^{208}Pb/^{204}Pb)_{init}}{^{206}Pb/^{204}Pb\text{-}(^{206}Pb/^{204}Pb)_{init}} \quad (11)$$

Unlike $^{208}Pb/^{204}Pb$ or $^{206}Pb/^{204}Pb$, which depend on Th/Pb and U/Pb, respectively, $^{208}Pb^*/^{206}Pb^*$ reflects the Th/U ratio integrated over the history of the Earth. The existence of global correlations between neodymium, strontium, and hafnium isotope ratios and $^{208}Pb^*/^{206}Pb^*$, and the absence of such global correlations with $^{208}Pb/^{204}Pb$ or $^{206}Pb/^{204}Pb$, shows that the elements neodymium, strontium, hafnium, thorium, and uranium behave in a globally coherent fashion during crust–mantle differentiation, whereas lead deviates from this cohesion.

Figures 4–6 show systematic isotopic differences between MORB from different ocean basins, reflecting some very large scale isotopic heterogeneities in the source mantle of these basalts. Also, the ranges of ε(Nd) values present in a single ocean basin are quite large. For example, the range of neodymium isotope ratios in Atlantic MORB (~10ε(Nd) units) is somewhat smaller than the respective range of Atlantic OIB values of about $^{14}\varepsilon$(Nd) units (see Section 2.03.5 on OIB), but this difference does not justify calling Atlantic MORB "isotopically homogeneous." This heterogeneity contradicts the widespread notion that the MORB-source mantle reservoir is isotopically nearly uniform, a myth that has persisted through many repetitions in the literature. One can just as easily argue that there is no such thing as a typical "normal"

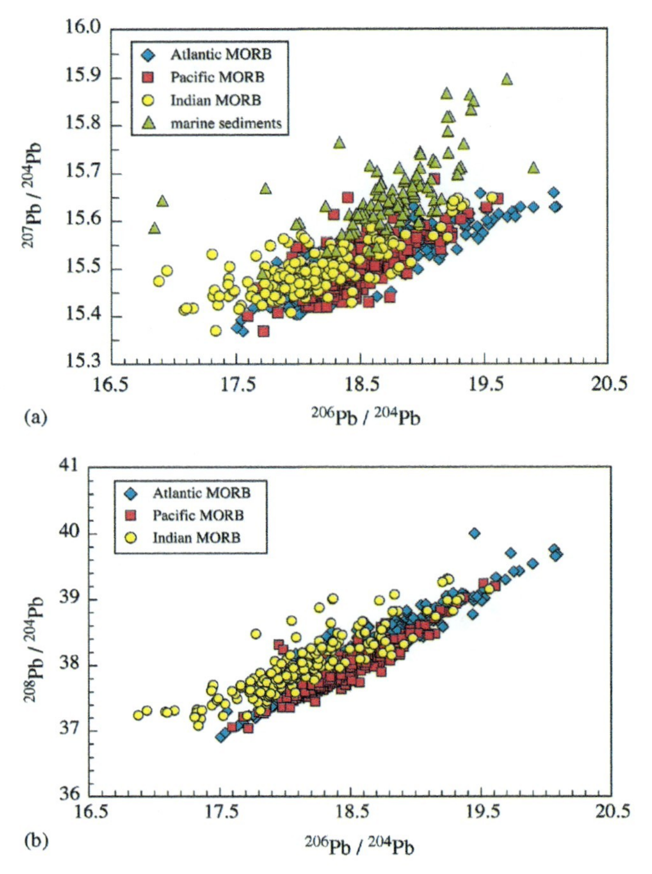

Figure 5 (a) $^{207}Pb/^{204}Pb$ versus $^{206}Pb/^{204}Pb$ for MORB from three major ocean basins and marine sediments. (b) $^{208}Pb/^{204}Pb$ versus $^{206}Pb/^{204}Pb$ for MORB from three major ocean basins. Sediments are not plotted because of strong overlap with the basalt data. For data sources see Figure 4.

(usually called N-type) MORB composition. In particular, the $^{208}Pb/^{204}Pb$ and $^{208}Pb^*/^{206}Pb^*$ ratios of Indian Ocean MORB show very little overlap with Pacific MORB (but both populations overlap strongly with Atlantic MORB). These very large scale regional "domains" were first recognized by Dupré and Allègre (1983) and named DUPAL anomaly by Hart (1984).

The boundary between the Indian Ocean and Pacific Ocean geochemical domains coincides with the Australian–Antarctic Discordance (AAD) located between Australia and Antarctica, an unusually deep ridge segment with several unusual physical and physiographic characteristics. The geochemical transition in Sr–Nd–Hf–Pb isotope space across the AAD is remarkably sharp (Klein *et al.*, 1988; Pyle *et al.*, 1992; Kempton *et al.*, 2002), and it is evident that very little mixing has occurred between these domains. The isotopic differences observed cannot be generated overnight. Rehkämper and Hofmann (1997), using lead isotopes, have estimated that the specific isotopic characteristics of the Indian Ocean MORB must be at least 1.5 Ga old. An important conclusion from this is that convective stirring of the mantle can be remarkably *ineffective* in mixing very large scale domains in the upper mantle (see Chapter 2.12).

When we further consider the fact that the present-day ocean ridge system, though globe encircling, samples only a geographically limited portion of the total, present-day mantle, it is clear that we must abandon the notion that we can characterize the isotopic composition of the depleted mantle reservoir by a single value of any isotopic parameter. What remains is a much broader, nevertheless limited, range of compositions, which, on average, differ from other types of oceanic basalts to be discussed further below. The lessons drawn from Section 2.03.2 (local and regional equilibrium) merely add an additional cautionary note: although it is possible to map the world's ocean ridge system using isotopic compositions of MORB, we cannot be sure how accurately these MORB compositions represent the underlying mantle. The differences between ocean basins are particularly obvious in the $^{208}Pb/^{204}Pb$ versus $^{206}Pb/^{204}Pb$ diagrams, where Indian Ocean MORBs have consistently higher $^{208}Pb/^{204}Pb$ ratios than Pacific Ocean

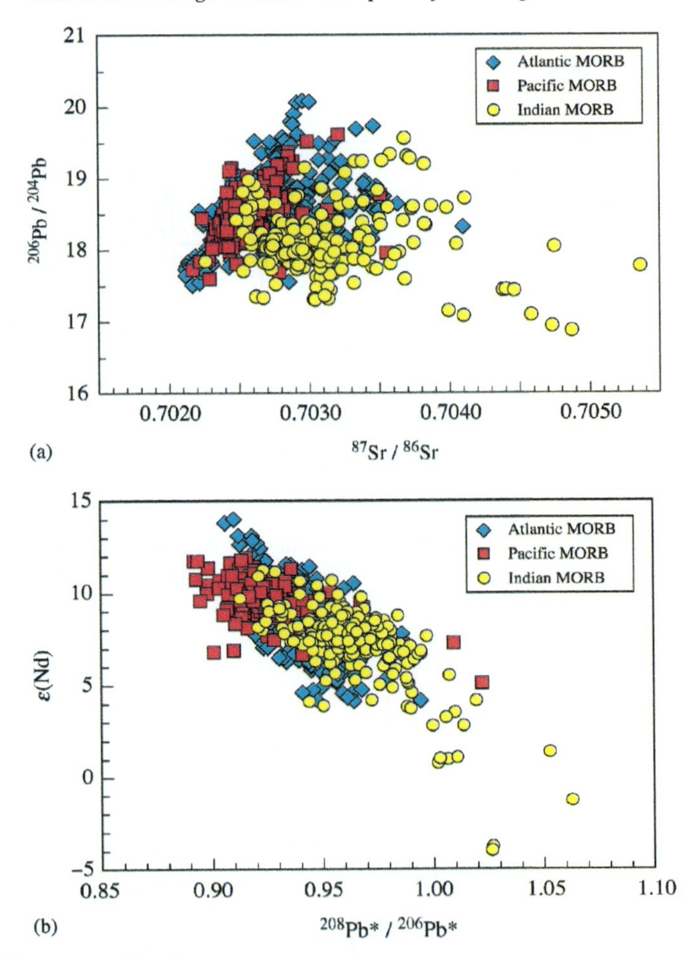

Figure 6 (a) $^{206}Pb/^{204}Pb$ versus $^{87}Sr/^{86}Sr$ for MORB from three major ocean basins. In contrast with the Sr–Nd and the Pb–Pb diagrams (Figures 4 and 5), the $^{206}Pb/^{204}Pb$–$^{87}Sr/^{86}Sr$ data correlate well only for Pacific MORB and not at all for Indian MORB. This indicates some anomalous behavior of the U–Pb decay system during global differentiation. For data sources see Figure 4. (b) $^{208}Pb^*/^{206}Pb^*$ versus ε(Nd) for MORB from three major ocean basins. The overall correlation is similar to the Sr–Nd correlation shown in Figure 4(a). This indicates that the Th/U ratios, which control $^{208}Pb^*/^{206}Pb^*$, do correlate with Sm/Nd (and Rb/Sr) ratios during mantle differentiation. Taken together, (a) and (b) identify Pb as the element displaying anomalous behavior.

MORBs. Diagrams involving neodymium isotopes show more overlap, but Indian Ocean MORBs have many ε(Nd) values lower than any sample from the Pacific Ocean.

An intermediate scale of isotopic variations is shown in Figures 7 and 8, using basalts from the Mid-Atlantic Ridge (MAR). The isotope ratios of strontium and neodymium (averaged over 1° intervals for clarity) vary along the ridge with obvious maxima and minima near the oceanic islands of Iceland, the Azores, and the Bouvet triple junction, and with large scale, relatively smooth gradients in the isotope ratios, e.g., between 20° S and 38° S. In general, the equatorial region between 30° S and 30° N is characterized by much lower strontium isotope ratios than the ridge segments to the north and the south. Some of this variation may be related to the vicinity of the mantle hotspots of Iceland, the Azores, or Bouvet, and the literature contains a continuing debate over the subject of "plume–asthenosphere interaction." Some authors argue the case where excess, enriched plume material spreads into the asthenosphere and mixes with depleted asthenospheric material to produce the geochemical gradients observed (e.g., Hart *et al.*, 1973; Schilling, 1973). Others argue that the hotspot or plume is internally heterogeneous and contains depleted, MORB-like material, which remains behind during normal plume-generated volcanism, spreads out in the asthenosphere, and becomes part of the asthenospheric mantle (Phipps Morgan and Morgan, 1998; Phipps Morgan, 1999). Irrespective of the specific process of plume–ridge interaction, the existence of compositional gradients up to ~2,000 km long implies some rather large-scale mixing processes,

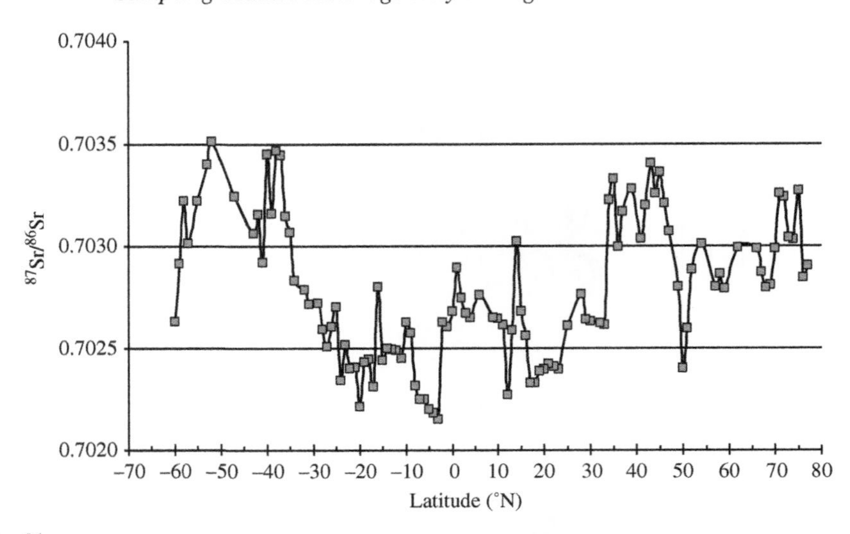

Figure 7 $^{87}Sr/^{86}Sr$ versus latitude variations in MORB along the Mid-Atlantic Ridge. The isotope data have been averaged by 1° intervals. For data sources see Figure 4.

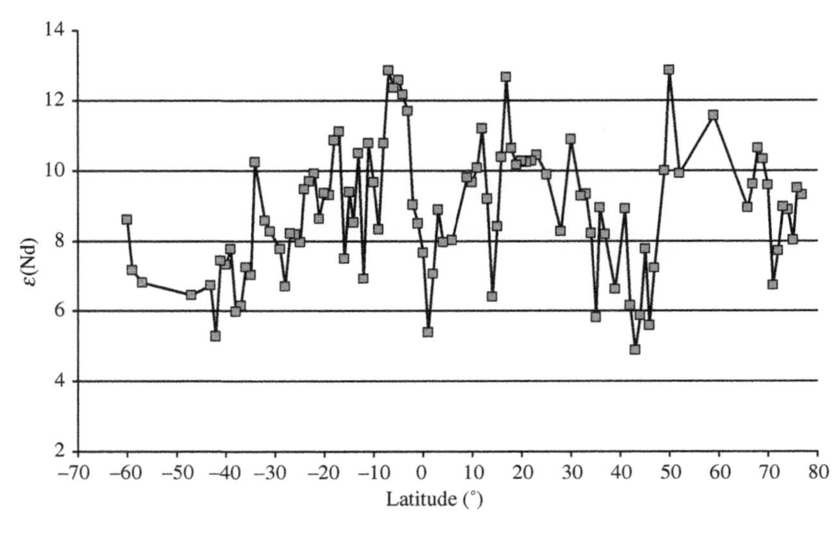

Figure 8 ε(Nd) versus latitude variations in MORB along the MAR. The isotope data have been averaged by 1° intervals. For data sources see Figure 4.

quite distinct from the sharpness of the compositional boundary seen at the AAD.

Isotopic heterogeneities are also observed on much smaller scales than so far discussed. For example, the region around 14 °N on the MAR shows a sharp "spike" in neodymium and strontium isotope ratios (see Figures 7 and 8) with an amplitude in ε(Nd) nearly as large as that of the entire Atlantic MORB variation, even though there is no obvious depth anomaly or other physiographic evidence for the possible presence of a mantle plume.

Finally, work on melt inclusions from phenocrysts has recently shown that sometimes extreme isotopic and trace-element heterogeneities exist within single hand specimens from mid-ocean ridges. Initially, the extreme chemical heterogeneities found in such samples were ascribed to the effects of progressive fractional melting of initially uniform source rocks (Sobolev and Shimizu, 1993), but recent Pb-isotope analyses of melt inclusions from a single MORB sample have shown a large range of isotopic compositions that require a locally heterogeneous source (Shimizu *et al.*, 2003). This phenomenon had previously also been observed in some OIBs (Saal *et al.*, 1998; Sobolev *et al.*, 2000), and future work must determine whether this is the exception rather than the rule. In any case, the normally observed local homogeneity of bulk basalt samples may turn out to be the result of homogenization in magma chambers rather than melting of homogeneous sources.

In summary, it should be clear that the mantle that produces MORB is isotopically heterogeneous on all spatial scales ranging from the size of ocean basins down to kilometers or possibly meters. The often-invoked homogeneity of MORB and MORB sources is largely a myth, and the definition of "normal" or N-type MORB actually applies to the depleted end of the spectrum rather than average MORB. It would be best to abandon these obsolete concepts, but they are likely to persist for years to come. Any unbiased evaluation of the actual MORB isotope data shows unambiguously, e.g., that Indian Ocean MORBs differ substantially from Pacific Ocean MORBs, and that only about half of Atlantic MORB conform to what is commonly referred to as "N-type" with $^{87}Sr/^{86}Sr$ ratios lower than 0.7028 and ε(Nd) values higher than +9. Many geochemists think that MORB samples with higher $^{87}Sr/^{86}Sr$ ratios and lower ε(Nd) values do not represent normal upper mantle but are generated by contamination of the normally very depleted, and isotopically extreme, upper mantle with plume material derived from the deeper mantle. Perhaps this interpretation is correct, especially in the Atlantic Ocean, where there are several hotspots/plumes occurring near the MAR, but its automatic application to all enriched samples (such as those near 14° N on the MAR) invites circular reasoning and it can get in the way of an unbiased consideration of the actual data.

2.03.4.2 Osmium Isotopes

The Re–Os decay system is discussed separately, in part because there are far fewer osmium isotope data than Sr–Nd–Pb data. This is true because, until ~10 years ago, osmium isotopes in silicate rocks were extraordinarily difficult to measure. The advent of negative-ion thermal ionization mass spectrometry has decisively changed this (Creaser *et al.*, 1991; Völkening *et al.*, 1991), and subsequently the number of publications providing osmium isotope data has increased dramatically.

Osmium is of great interest to mantle geochemists because, in contrast with the geochemical properties of strontium, neodymium, hafnium, and lead, all of which are incompatible elements, osmium is a compatible element in most mantle melting processes, so that it generally remains in the mantle, whereas the much more incompatible rhenium is extracted and enriched in the melt and ultimately in the crust. This system therefore provides information that is different from, and complementary to, what we can learn from strontium, neodymium, hafnium, and lead isotopes. However, at present there are still significant obstacles to the full use and understanding of osmium geochemistry. There are primarily three reasons for this:

(i) Osmium is present in oceanic basalts usually at sub-ppb concentration levels. Especially, in low-magnesium basalts, the concentrations can approach low ppt levels. The problem posed by this is that crustal rocks and seawater can have $^{187}Os/^{188}Os$ ratios 10 times higher than the (initial) ratios in mantle-derived melts. Thus, incorporation of small amounts of seawater-altered material in a submarine magma chamber may significantly increase the $^{187}Os/^{188}Os$ ratio of the magma. Indeed, many low-magnesium, moderately to highly differentiated oceanic basalts have highly radiogenic osmium, and it is not easy to know which basalts are unaffected by this contamination.

(ii) The geochemistry of osmium is less well understood than other decay systems, because much of the osmium resides in non-silicate phases such as sulfides, chromite and (possibly) metallic phases, and these phases can be very heterogeneously distributed in mantle rocks. This frequently leads to a "nugget" effect, meaning that a given sample powder is not necessarily representative of the system. Quite often, the reproducibility of concentration measurements of osmium and rhenium is quite poor by normal geochemical standards, with differences of several percent between duplicate analyses, and this may be caused either by intrinsic sample heterogeneity ("nugget effect") or by incomplete equilibration of sample and spike during dissolution and osmium separation.

(iii) There are legitimate doubts whether the osmium-isotopic composition of oceanic basalts are ever identical to those of their mantle source rocks (Section 2.03.2.1).

The point (iii) above is illustrated in Figure 9, which shows osmium isotope ratios and osmium concentrations in abyssal peridotites and in MORB. This diagram shows two remarkable features: (i) The osmium isotope ratios of MORB and abyssal peridotites have very little overlap, the peridotites being systematically lower than those of seafloor basalts. (ii) The MORB data show a strong negative correlation between isotope ratios and osmium concentrations. These results suggest that the basalts may not be in isotopic equilibrium with their source rocks, but we have no proof of this, because we have no samples of specific source rocks for specific basalt samples. Also, the total number of samples represented in Figure 9 is rather small. Nevertheless, the apparently systematically higher $^{187}Os/^{188}Os$ ratios of the basalts compared with the peridotites seem to indicate that unradiogenic

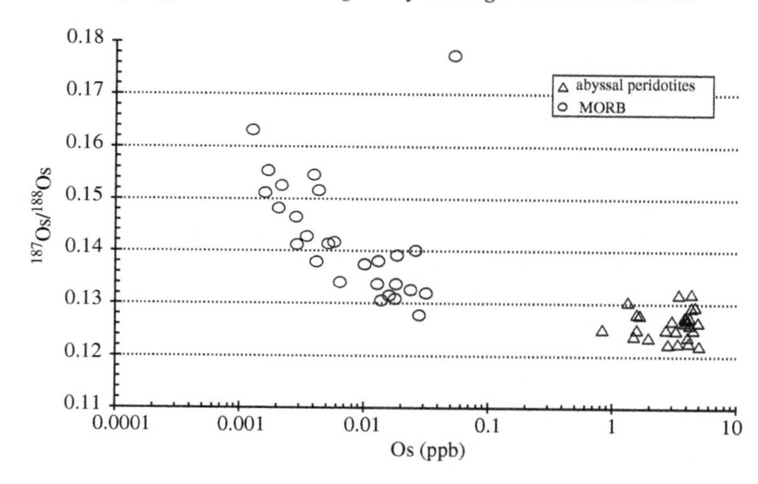

Figure 9 Osmium isotope ratios in MORB and abyssal peridotites. This diagram shows that osmium is generally compatible in peridotites during MORB melting. The systematic differences in $^{187}Os/^{188}Os$ ratios between MORB and peridotites suggest that the melts may not be in isotopic equilibrium with their residual peridotite (sources Martin, 1991; Roy-Barman and Allègre, 1994; Snow and Reisberg, 1995; Schiano *et al.*, 1997; Brandon *et al.*, 2000).

portions of the source peridotites did not contribute to, or react with, the melt. The negative correlation displayed by the MORB data may mean that essentially *all* the melts are contaminated by seawater-derived osmium and that the relative contribution of the contaminating osmium to the measured isotopic compositions is inversely correlated with the osmium concentration of the sample. However, the MORB samples also show a strong positive correlation between $^{187}Os/^{188}Os$ and Re/Os (not shown). Therefore, it is also possible that MORB osmium is derived from heterogeneous sources in such a way that low-osmium, high-Re/Os samples are derived from high-Re/Os portions of the sources (such as pyroxenitic veins), whereas high-osmium, low-Re/Os samples are derived from the peridotitic or even harzburgitic matrix.

To avoid the risk of contamination by seawater, either through direct contamination of the samples or contamination of the magma by assimilation of contaminated material, many authors disregard samples with very low osmium concentrations. Unfortunately, this approach does not remove the inherent ambiguity of interpretation, and it may simply bias the sampling. What is clearly needed are independent measures of very low levels of magma chamber and sample contamination.

2.03.4.3 Trace Elements

The general model of crust–mantle differentiation predicts that, after crust formation, the residual mantle should be depleted in incompatible elements. Melts from this depleted mantle may be absolutely enriched but should still show a relative depletion of highly incompatible elements relative to moderately incompatible elements, as illustrated in Figure 2. Here I examine actual trace-element data of real MORB and their variability. An inherent difficulty is that trace-element abundances in a basalt depend on several factors, namely, the source composition, the degree, and mechanism of melting and melt extraction, the subsequent degree of magmatic fractionation by crystallization, and finally, on possible contamination of the magma during this fractionation process by a process called AFC (assimilation with fractional crystallization). This inherent ambiguity resulted in a long-standing debate about the relative importance of these two aspects. O'Hara, in particular, championed the case of fractional crystallization and AFC processes in producing enrichment and variability of oceanic basalts (e.g., O'Hara, 1977; O'Hara and Mathews, 1981). In contrast, Schilling and coworkers argued that variations in trace-element abundances, and in particular ratios of such abundances, are strongly controlled by source compositions. They documented several cases where REE patterns vary systematically along mid-ocean ridge segments, and they mapped such variations specifically in the vicinity of hotspots, which they interpreted as the products of mantle plumes relatively enriched in incompatible elements. As was the case for the isotopic variations, they interpreted the trace-element variations in terms of mixing of relatively enriched plume material with relatively depleted upper mantle, the asthenospheric mantle (Schilling, 1973; White and Schilling, 1978). Figure 10 shows a compilation of La/Sm ratios (normalized to primitive mantle values) of basalts

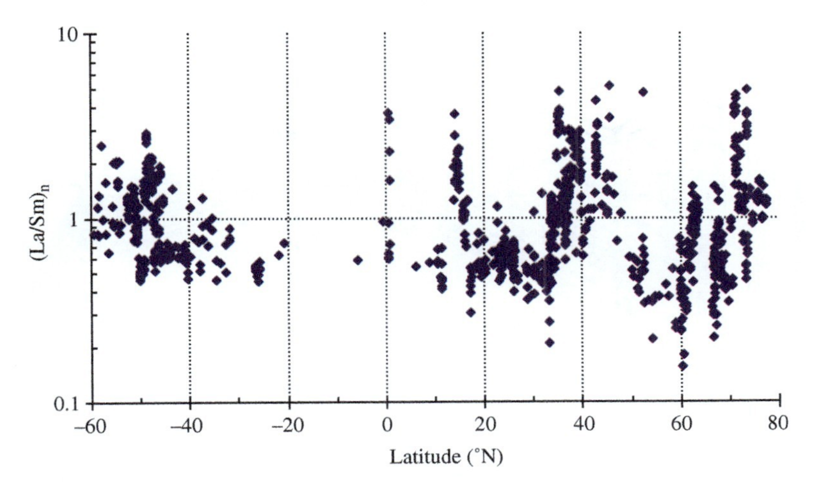

Figure 10 La/Sm ratios in MORB (not smoothed), normalized to primitive-mantle values, as a function of latitude along the MAR.

dredged from the MAR. This parameter has been used extensively by Schilling and co-workers as a measure of source depletion or enrichment, where they considered samples with $(La/Sm)_n < 1.0$ as normal or "N-type" MORB derived from depleted sources with similar or even lower La/Sm ratios. As was found for the isotope ratios, only two-thirds of the MAR shows "typical" or "normal" La/Sm ratios lower than unity. In general, the pattern resembles that of the isotope variations, especially in the North Atlantic, where the coverage for both parameters is extensive. Thus, high La/Sm and $^{87}Sr/^{86}Sr$ values are found near the hotspots of Iceland and the Azores, between 45° S and 50° S, 14° N, and 43° N. Because of these correlations, the interpretation that the trace-element variations are primarily caused by source variations has been widely accepted. Important confirmation for this has come from the study of Johnson *et al.* (1990). They showed that peridotites dredged from near-hotspot locations along the ridge are more depleted in incompatible elements than peridotites from normal ridge segments. This implies that they have been subjected to higher degrees of melting (and loss of that melt). In spite of this higher degree of melting, the near-hotspot lavas are more enriched in incompatible elements, and therefore their initial sources must also have been more enriched.

Trace-element abundance patterns, often called "spidergrams," of MORB are shown in Figure 11 ("spidergram" is a somewhat inappropriate but a convenient term coined by R. N. Thompson (Thompson *et al.*, 1984), presumably because of a perceived resemblance of these patterns to spider webs, although the resemblance is tenuous at best). The data chosen for this plot are taken from le Roux *et al.* (2002) for MORB glasses from the MAR (40–55° S), which encompasses both depleted regions and enriched regions resulting from ridge–hotspot interactions. The patterns are highly divergent for the most incompatible elements, but they converge and become more parallel for the more compatible elements. This phenomenon is caused by the fact that variations in melt fractions produce the largest concentration variations in the most highly incompatible elements in both melts and their residues. This is a simple consequence of Equation (3), which states that for elements with very small values of D the concentration in the melt is inversely proportional to the melt fraction. At the other end of the spectrum, compatible elements, those with D values close to unity or greater, become effectively buffered by the melting assemblage. For an element having $D \gg F$, Equation (2) reduces to

$$C_l \approx \frac{C_0}{D} \tag{12}$$

and for $D = 1$, it reduces to

$$C_l = C_0 \tag{13}$$

In both cases, the concentration in the melt becomes effectively buffered by the residual mineral assemblage until the degree of melting is large enough, so that the specific mineral responsible for the high value of D is exhausted. This buffering effect is displayed by the relatively low and uniform concentrations of scandium (Figure 11). It is caused by the persistence of residual clinopyroxene during MORB melting.

These relationships lead to the simple consequence that the variability of element concentrations in large data sets of basalt analyses are related to the bulk partition coefficients of these elements (Hofmann, 1988; Dupré *et al.*, 1994). This can be verified by considering a set of trace

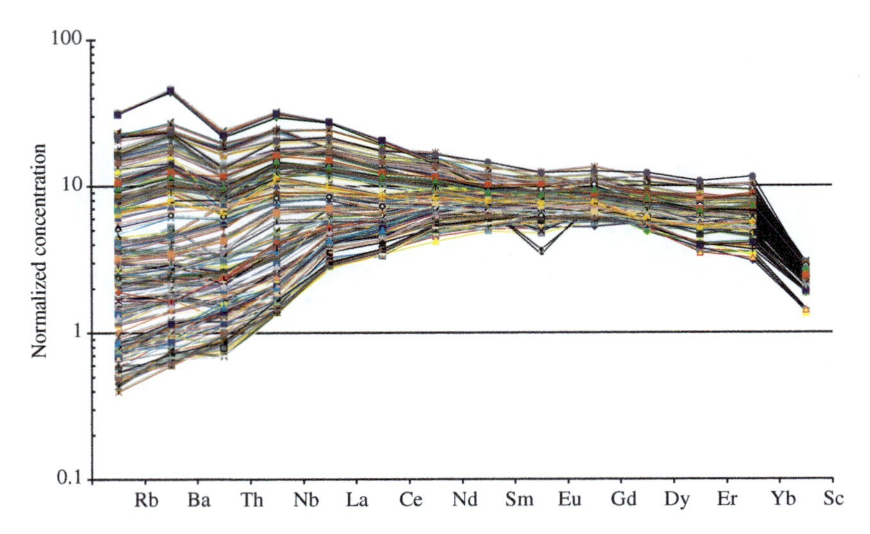

Figure 11 Trace element abundances of 250 MORB between 40° S and 55° S along the Mid-Atlantic Ridge. Each sample is represented by one line. The data are normalized to primitive-mantle abundances of (McDonough and Sun, 1995) and shown in the order of mantle compatibility. This type of diagram is popularly known as "spidergram." The data have been filtered to remove the most highly fractionated samples containing less than 5% MgO (source le Roux *et al.*, 2002).

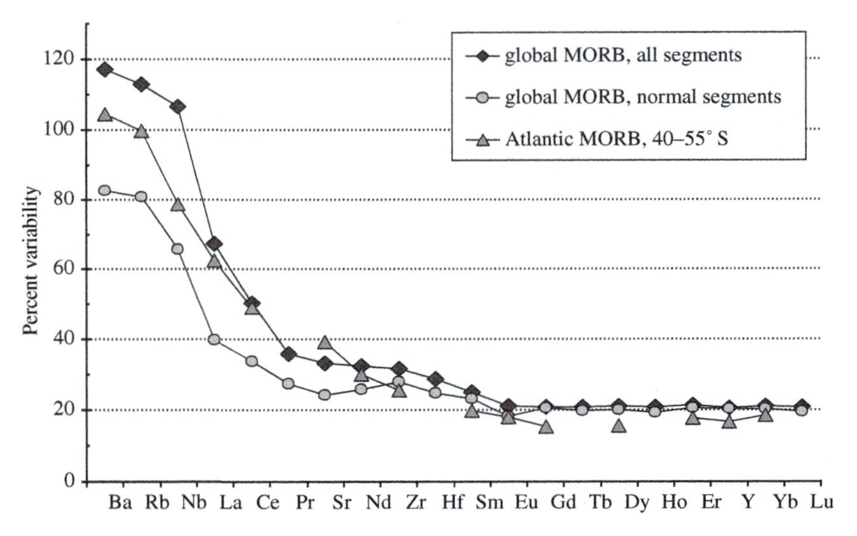

Figure 12 Variability of trace element concentrations in MORB, expressed as 100 * standard deviation/mean concentration. The data for "Global MORB" are from the PETDB compilation of (Su, 2002). "All segments" refers to ~250 ridge segments from all oceans. "Normal segments" refers to ~62 ridge segments that are considered not to represent any sort of "anomalous" ridges, because those might be affected by such factors as vicinity to mantle plumes or subduction of sediments (e.g., back-arc basins and the Southern Chile Ridge). The Atlantic MORB, 40–55° S, from which samples with less than 5% MgO have been removed (source le Roux *et al.*, 2002).

elements for which enough experimental data are available to be confident of the relative solid–melt partition coefficients, namely, the REE. These coefficients decrease monotonically from the heavy to light REEs, essentially because the ionic radii increase monotonically from heavy to light REEs (with the possible exception of europium, which has special properties because of its variable valence). Figure 12 shows three plots of variability of REEs and other trace elements in MORB as a function of mantle compatibility of the elements listed. Variability is defined as the standard deviation of the measured concentrations divided by the respective mean value. Compatibilities of elements other than the REEs are estimated from global correlations of trace-element ratios with absolute abundances as derived from simple partial melting theory (Hofmann *et al.*, 1986; Hofmann, 1988) (see also Figures 17 and 18). Two sets of data are from a new, ridge segment-by-segment compilation made by Su (2002) using the MORB database http://petdb.ldeo.columbia.edu. The third represents data for 270 MORB glasses from the South Atlantic ridge (40–55° S) (le Roux *et al.*, 2002). The qualitative similarity of the three plots is striking. It indicates that the order of variabilities is a robust feature. For example, the variabilities of the heavy REEs (europium to lutetium) are all essentially identical at 20% in all three plots. For the light REEs, the variability increases monotonically from europium to lanthanum, consistent with their decreasing partition coefficients in all mantle minerals. As expected, variabilities are greatest for the very highly incompatible elements (VICEs) niobium, rubidium, and barium. The same is true for thorium and uranium in South Atlantic MORBs (le Roux *et al.*, 2002), which are not shown here because their averages and standard deviations were not compiled by Su (2002). All of this is consistent with the enrichment pattern in the continental crust (Figure 2), which shows the greatest enrichments for barium, rubidium, thorium, and uranium in the continental crust, as well as monotonically decreasing crustal abundances for the REEs from lanthanum to lutetium, with a characteristic flattening from europium to lutetium. Obvious exceptions to this general consistency are the elements niobium and lead. These will be discussed separately below.

Some additional lessons can be learned from Figure 12:

(i) The flattening of the heavy-REE variabilities in MORB are consistent with the flat heavy-REE patterns almost universally observed in MORB, and these are consistent with the flat pattern of HREE partition coefficients in clinopyroxene (see Chapter 2.09). This does not rule out some role of garnet during MORB melting, but it does probably rule out a major role of garnet.

(ii) Strontium, zirconium, and hafnium have very similar variabilities as the REEs neodymium and samarium. Again, this is consistent with the abundance patterns of MORB and with experimental data (Chapter 2.09). Overall, the somewhat tentative suggestion made by Hofmann (1988) regarding the relationship between concentration variability and degree of incompatibility, based on a very small set of MORB data, is strongly confirmed by the very large data sets now available. A note of caution is in order for strontium, which has a high partition coefficient in plagioclase. Thus, when oceanic basalts crystallize plagioclase, the REEs tend to increase in the residual melt, but strontium is removed from the melt by the plagioclase. The net effect of this is that the overall variability of strontium is reduced in data sets incorporating plagioclase-fractionated samples. Such samples have been partly filtered out from the South Atlantic data set. This is the likely reason why strontium shows the greatest inconsistencies between the three plots shown in Figure 12.

2.03.4.4 N-MORB, E-MORB, T-MORB, and MORB Normalizations

It has become a widely used practice to define standard or average compositions of N-MORB, E-MORB, and T-MORB (for normal, enriched, and transitional MORB), and to use these as standards of comparison for ancient rocks found on land. In addition, many authors use "N-MORB" compositions as a normalization standard in trace-element abundance plots ("spidergrams") instead of chondritic or primitive mantle. This practice should be discouraged, because trace-element abundances in MORB form a complete continuum of compositions ranging from very depleted to quite enriched and OIB-like. A plot of global La/Sm ratios (Figure 13) demonstrates this: there is no obvious typical value but a range of lanthanum concentrations covering two orders of magnitude and a range of La/Sm ratios covering about one-and-a-half orders of magnitude. Although the term N-MORB was intended to describe "normal" MORB, it actually refers to depleted MORB, often defined by $(La/Sm)_n < 1$. Thus, while these terms do serve some purpose for characterizing MORB compositions, there is no sound basis for using any of them as normalizing values to compare other rocks with "typical" MORB.

The strong positive correlation seen in Figure 13 is primarily the result of the fact that lanthanum is much more variable than samarium. Still, the overall coherence of this relationship is remarkable. It demonstrates that the variations of the REE abundances are not strongly controlled by variations in the degree of crystal fractionation of MORB magmas, because these would cause similar variability of lanthanum and samarium. Although this reasoning is partly circular because highly fractionated samples containing less than 6% MgO have been eliminated, the total number of such samples in this population of ~2,000 is less than 100. Thus, it is clear that the relationship is primarily controlled either by source or by partial melting effects. Figure 14 shows that the La/Sm ratios are also negatively correlated with $^{143}Nd/^{144}Nd$. Because this isotope ratio is a function of source Sm/Nd (and time), and neodymium is intermediate in bulk partition coefficient between lanthanum and samarium, such a negative correlation is expected if the

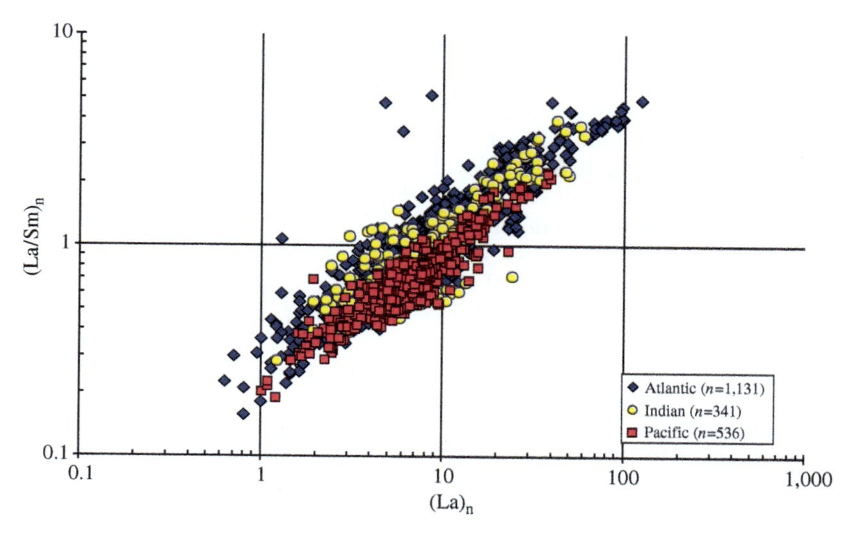

Figure 13 Primitive-mantle normalized La/Sm versus La for MORB from three Ocean basins. Numbers in parentheses refer to the number of samples from each ocean basin. Lanthanum concentrations vary by about two orders of magnitude; La/Sm varies by more than one order of magnitude. Data were extracted from PETDB.

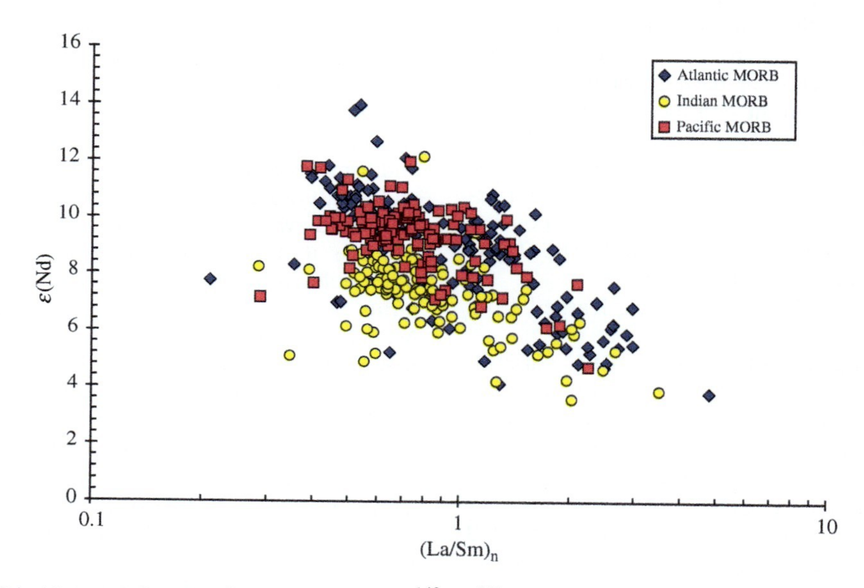

Figure 14 Primitive-mantle normalized La/Sm versus $^{143}Nd/^{144}Nd$ for MORB from three ocean basins. The (weak but significant) negative correlation is consistent with the inference that the variations in La/Sm in the basalts are to a significant part inherited from their mantle sources.

variability of the REE abundances is (at least in part) inherited from the source. Thus, while it would be perfectly possible to generate the relationship seen in Figure 13 purely by variations in partial melting, we can be confident that La/Sm ratios (and other highly incompatible element ratios) do track mantle source variations, as was shown by Schilling many years ago. It is important to realize, however, that such differences in source compositions were originally also produced by melting. These sources are simply the residues of earlier melting events during previous episodes of (continental or oceanic) crust formation.

2.03.4.5 Summary of MORB and MORB-source Compositions

Klein and Langmuir (1987), in a classic paper, have shown that the element sodium is almost uniquely suited for estimating the degree of melting required to produce MORBs from their respective sources. This element is only slightly incompatible at low melt fractions produced at relatively high pressures. As a result, the extraction of the continental crust has reduced the sodium concentration of the residual mantle by no more than ~10% relative to the primitive-mantle value. We therefore know the approximate sodium concentration of all MORB sources. In contrast, this element behaves much more incompatibly during production of oceanic crust, where relatively high melt fractions at relatively low pressures produce ocean floor basalts. This allowed Klein and Langmuir to estimate effective melt fractions ranging from 8% to 20%, with an average of ~10%, from sodium concentrations in MORB. Once the melt fraction is known, the more highly incompatible-element concentrations of the MORB-source mantle can be estimated from the measured concentrations in MORB. For the highly incompatible elements, the source concentrations are therefore estimated at ~10% of their respective values in the basalts. This constitutes a significant revision of earlier thinking, which was derived from melting experiments and the assumption that essentially clinopyroxene-free harzburgites represent the typical MORB residue, and which led to melt fraction estimates of 20% and higher.

The compilation of extensive MORB data from all major ocean basins has shown that they comprise wide variations of trace-element and isotopic compositions and the widespread notion of great compositional uniformity of MORB is largely a myth. An exception to this may exist in helium-isotopic compositions; see Chapter 2.06. From the state of heterogeneity of the more refractory elements it is clear, however, that the apparently greater uniformity of helium compositions is not the result of mechanical mixing and stirring, because this process should homogenize all elements to a similar extent. Moreover, the isotope data of MORB from different ocean basins show that different regions of the upper mantle have not been effectively mixed in the recent geological past, where "recent" probably means approximately the last 10^9 yr.

The general, incompatible-element depleted nature of the majority of MORBs and their sources is well explained by the extraction of the continental crust. Nevertheless, the bulk continental crust and the bulk of the MORB sources are not *exact* chemical complements. Rather, the residual mantle has undergone additional differentiation, most likely involving the generation of OIBs and their subducted equivalents. In addition, there may be more subtle differentiation processes involving smaller-scale melt migration occurring in the upper mantle (Donnelly *et al.*, 2003). It is these additional differentiation processes that have generated much of the heterogeneity observed in MORBs and their sources.

2.03.5 OCEAN ISLAND, PLATEAU, AND SEAMOUNT BASALTS

These basalts represent the oceanic subclass of so-called intraplate basalts, which also include continental varieties of flood and rift basalts. They will be collectively referred to as "OIB," even though many of them are not found on actual oceanic islands either because they never rose above sea level or because they were formed on islands that have sunk below sea level. Continental and island arc basalts will not be discussed here, because at least some of them have clearly been contaminated by continental crust. Others may or may not originate in, or have been "contaminated" by, the subcontinental lithosphere. For this reason, they are not considered in the present chapter, which is concerned primarily with the chemistry of the sublithospheric mantle.

Geochemists have been particularly interested in OIB because their isotopic compositions tend to be systematically different from MORB, and this suggests that they come from systematically different places in the mantle (e.g., Hofmann *et al.*, 1978; Hofmann and Hart, 1978). Morgan's mantle plume theory (Morgan, 1971) thus provided an attractive framework for interpreting these differences, though not quite in the manner originally envisioned by Morgan. He viewed the entire mantle as a single reservoir, in which plumes rise from a lower boundary layer that is not fundamentally different in composition from the upper mantle. Geochemists, on the other hand, saw plumes being formed in a fundamentally different, more primitive, less depleted, or enriched, deeper part of the mantle than MORB sources (e.g., Wasserburg and Depaolo, 1979). The debate about these issues continues to the present day, and some of the mantle models based on isotopic and trace-element characteristics will be discussed below.

2.03.5.1 Isotope Ratios of Strontium, Neodymium, Hafnium, and Lead and the Species of the Mantle Zoo

Radiogenic isotope ratios of OIB are shown on Figure 15. These diagrams display remarkably similar topologies as the respective MORB data shown in Figures 4–6. Strontium isotope ratios are negatively correlated with neodymium and hafnium isotopes, and correlations between strontium, neodymium, and hafnium isotopes and lead

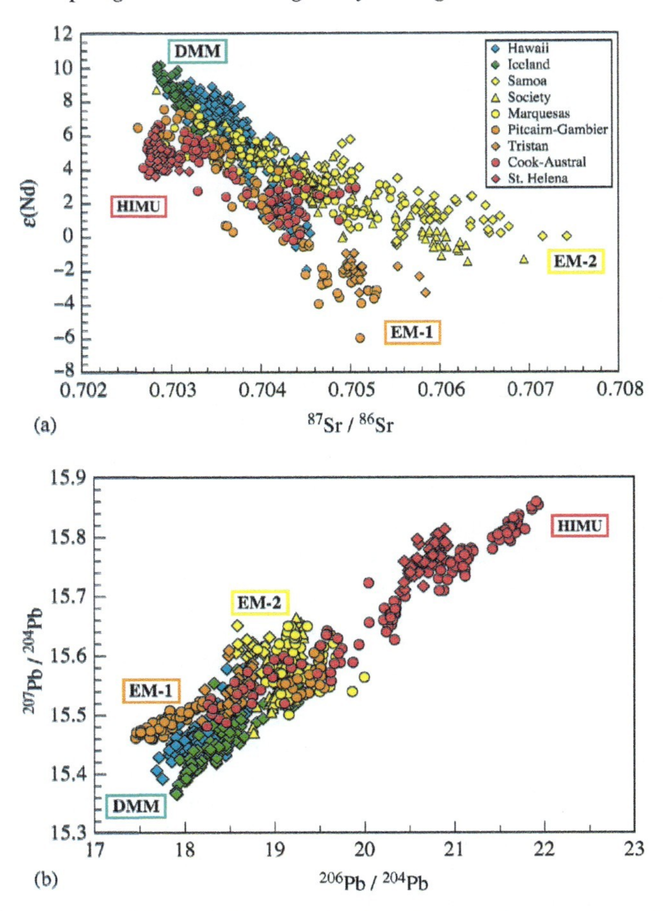

Figure 15 (a) $^{87}Sr/^{86}Sr$ versus ε(Nd) for OIB (excluding island arcs). The islands or island groups selected are chosen to represent extreme isotopic compositions in isotope diagrams. They are the "type localities" for HIMU (Cook-Austral Islands and St. Helena), EM-1 (Pitcairn-Gambier and Tristan), EM-2 (Society Islands, Samoa, Marquesas), and PREMA (Hawaiian Islands and Iceland). See text for explanations of the acronyms. (b) $^{207}Pb/^{204}Pb$ versus $^{206}Pb/^{204}Pb$ for the same OIB as plotted in (a). Note that the $^{207}Pb/^{204}Pb$ ratios of St. Helena and Cook-Australs are similar but not identical, whereas they overlap completely in the other isotope diagrams. (c) $^{208}Pb/^{204}Pb$ versus $^{206}Pb/^{204}Pb$ for the same OIB as plotted in (a). (d) $^{206}Pb/^{204}Pb$ versus $^{87}Sr/^{86}Sr$ for the same OIB as plotted in (a). Note that correlations are either absent (e.g., for the EM-2 basalts from Samoa, the Society Islands and Marquesas) or point in rather different directions, a situation that is similar to the MORB data (Figure 6(a)). (e) $^{208}Pb^*/^{206}Pb^*$ versus ε(Nd) for the same OIB as plotted in (a). Essentially all island groups display significant negative correlations, again roughly analogous to the MORB data. Data were assembled from the GEOROC database (htpp://georoc.mpch-mainz.gwdg.de).

isotopes are confined to $^{208}Pb^*/^{206}Pb^*$, and the ranges of isotope ratios are even greater (although not dramatically so) for OIBs than MORBs. There is one important difference, however, namely, a significant shift in all of these ratios between MORBs and OIBs. To be sure, there is extensive overlap between the two populations, but OIBs are systematically more radiogenic in strontium and less radiogenic in neodymium and hafnium isotopes. In lead isotopes, OIBs overlap the MORB field completely but extend to more extreme values in $^{206}Pb/^{204}Pb$, $^{207}Pb/^{204}Pb$, and $^{208}Pb/^{204}Pb$. As was true for MORBs, OIB isotopic composition can be "mapped," and certain oceanic islands or island groups can be characterized by specific isotopic characteristics. Recognition of this feature has led to the well-known concept of end-member compositions or "mantle components" initially identified by White (1985) and subsequently labeled HIMU, PREMA, EM-1, and EM-2 by Zindler and Hart (1986). These acronyms refer to mantle sources characterized by high μ values (HIMU; $\mu = (^{238}U/^{204}Pb)_{t=0}$), "prevalent mantle" (PREMA), "enriched mantle-1" (EM-1), and "enriched mantle-2" (EM-2). "PREMA" has, in recent years, fallen into disuse. It has been replaced by three new terms, namely, "FOZO" (for "focal zone," Hart *et al.* (1992)), "C" (for "common" component, Hanan and Graham (1996)), or "PHEM"

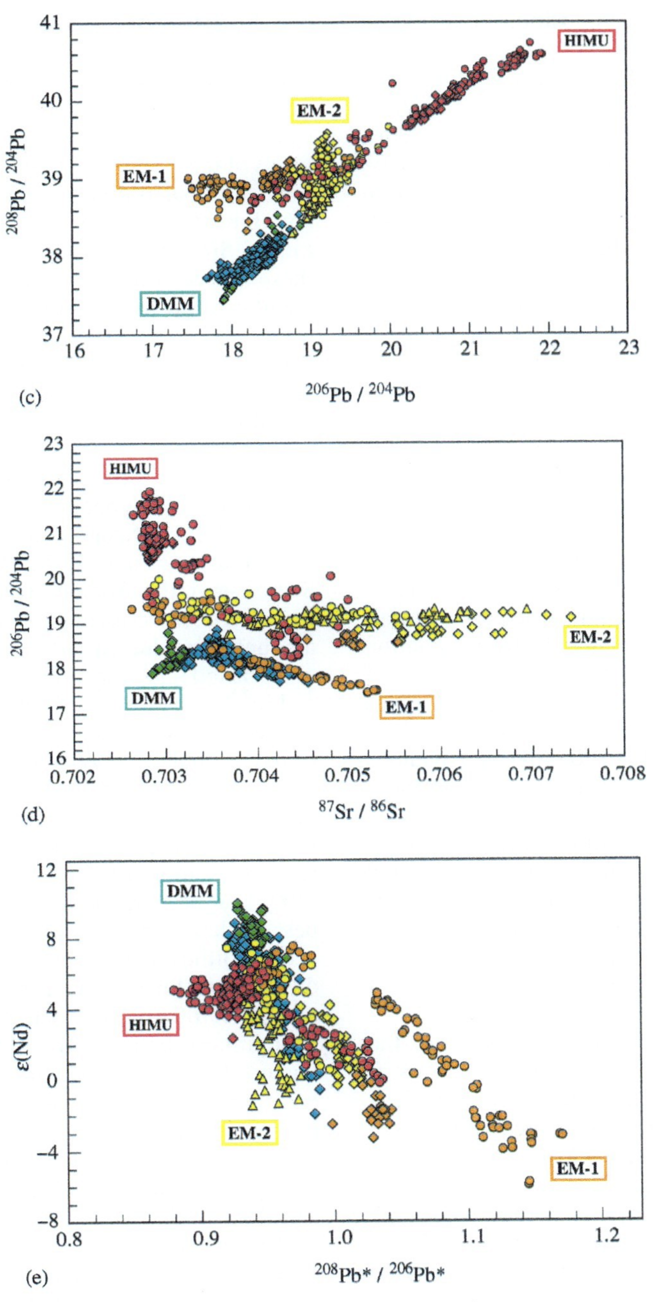

Figure 15 (continued).

(for "primitive helium mantle," Farley *et al.* (1992)), which differ from each other only in detail, if at all. In contrast with the illustration chosen by Hofmann (1997), which used color coding to illustrate how the isotopic characteristics of, e.g., extreme HIMU samples appear in different isotope diagrams, irrespective of their geographic location, here I use the more conventional representation of identifying "type localities" of the various "species" of this mantle isotope zoo.

Two extreme notions about the meaning of these components or end members (sometimes also called "flavors") can be found in the literature. One holds that the extreme isotopic end members of these exist as identifiable "species," which may occupy separate volumes or "reservoirs" in the mantle. In this view, the intermediate compositions found in most oceanic basalts are generated by instantaneous mixing of these species during melting and emplacement of OIBs. The other notion considers them to be merely extremes of a continuum of isotopic compositions existing in mantle rocks.

Apparent support for the "species" hypothesis is provided by the observation that the isotopically extreme compositions can be found in more than a single ocean island or island group, namely, Austral Islands and St. Helena for HIMU, Pitcairn Island and Walvis-Ridge-Tristan Island for EM-1, and Society Islands and Samoa for EM-2 (Hofmann, 1997). Nevertheless, it seems to be geologically implausible that mantle differentiation, by whatever mechanism, would consistently produce just four (or five, when the "depleted MORB mantle" DMM (Zindler and Hart, 1986) is included) species of essentially identical ages, which would then be remixed in variable proportions. It is more consistent with current understanding of mantle dynamics to assume that the mantle is differentiated and remixed continuously through time. Moreover, we can be reasonably certain that a great many rock types with differing chemistries are continuously introduced into the mantle by subduction and are thereafter subjected to variable degrees of mechanical stirring and mixing. These rock types include ordinary peridotites, harzburgites, gabbros, tholeiitic and alkali basalts, terrigenous sediments, and pelagic sediments. Most of these rock types have been affected by seafloor hydrothermal and low-temperature alteration, submarine "weathering," and subduction-related alteration and metasomatism. Finally, it is obvious that, overall, the OIB isotopic data constitute a continuously heterogeneous spectrum of compositions, just as is the case for MORB compositions.

In spite of the above uncertainties about the meaning of mantle components and reservoirs, it is clear that the extreme isotopic compositions represent melting products of sources subjected to some sort of ancient and comparatively extreme chemical differentiation. Because of this, they probably offer the best opportunity to identify the specific character of the types of mantle differentiation also found in other OIBs of less extreme isotopic composition. For example, the highly radiogenic lead isotope ratios of HIMU samples require mantle sources with exceptionally high U/Pb and Th/Pb ratios. At the same time, HIMU samples are among those OIBs with the least radiogenic strontium, requiring source-Rb/Sr ratios nearly as low as those of the more depleted MORBs. Following the currently popular hypothesis of Hofmann and White (1980, 1982); Chase, 1981, it is widely thought that such rocks are examples of recycled oceanic crust, which has lost alkalis and lead during alteration and subduction (Chauvel *et al.*, 1992). However, there are other possibilities. For example, the characteristics of HIMU sources might also be explained by enriching oceanic lithosphere "metasomatically" by the infiltration of low-degree partial melts, which have high U/Pb and Th/Pb ratios because of magmatic enrichment of uranium and thorium over lead (Sun and McDonough, 1989). The Rb/Sr ratios of these sources should then also be elevated over those of ordinary MORB sources, but this enrichment would be insufficient to significantly raise $^{87}Sr/^{86}Sr$ ratios because the initial Rb/Sr of these sources was well below the level where any significant growth of radiogenic ^{87}Sr could occur. Thus, instead of recycling more or less ordinary oceanic crust the enrichment mechanism would involve recycling of magmatically enriched oceanic lithosphere.

The origin of EM-type OIBs is also controversial. Hawkesworth *et al.* (1979) had postulated a sedimentary component in the source of the island of Sao Miguel (Azores), and White and Hofmann (1982) argued that EM-2 basalt sources from Samoa and the Society Islands are formed by recycled ocean crust with an addition of the small amount of subducted sediment. This interpretation was based on the high $^{87}Sr/^{86}Sr$ and high $^{207}Pb/^{204}Pb$ (for given $^{206}Pb/^{204}Pb$) ratios of EM-2 basalts, which resemble the isotopic signatures of terrigenous sediments. However, this interpretation continues to be questioned on the grounds that there are isotopic or trace-element parameters that appear inconsistent with this interpretation (e.g., Widom and Shirey, 1996). Workman *et al.* (2003) argue that the geochemistry of Samoa is best explained by recycling of melt-impregnated oceanic lithosphere, because their Samoa samples do not show the trace-element fingerprints characteristic of other EM-2 suites (see discussion of neodymium below). In addition, it has been argued that the sedimentary signature is present, but it is not part of a deep-seated mantle plume but is introduced as a sedimentary contaminant into plume-derived magmas during their passage through the shallow mantle or crust (Bohrson and Reid, 1995).

The origin of the EM-1 flavor is similarly controversial. Its distinctive features include very low $^{143}Nd/^{144}Nd$ coupled with relatively low $^{87}Sr/^{86}Sr$, and very low $^{206}Pb/^{204}Pb$ coupled with relatively high $^{208}Pb/^{204}Pb$ (leading to high $^{208}Pb^{*}/^{206}Pb^{*}$ values). The two leading contenders for the origin of this are (i) recycling of delaminated subcontinental lithosphere and (ii) recycling of subducted ancient pelagic sediment. The first hypothesis follows a model originally proposed by McKenzie and O'Nions (1983) to explain the origin of OIBs in general. The more specific model for deriving EM-1 type basalts from such a source was developed by Hawkesworth *et al.* (1986), Mahoney *et al.* (1991), and Milner and le Roex (1996). It is based on the observation that mantle xenoliths from Precambrian shields display similar isotopic characteristics. The second hypothesis is based on the observation that many pelagic sediments are characterized by high Th/U and low (U,Th)/Pb ratios (Ben Othman *et al.*, 1989; Plank and Langmuir, 1998), and this will lead to relatively unradiogenic lead with high $^{208}Pb^{*}/^{206}Pb^{*}$ ratios after passage of 1–2 Ga (Weaver, 1991; Chauvel *et al.*, 1992; Rehkämper and Hofmann, 1997; Eisele *et al.*, 2002). Additional support for this hypothesis has come from hafnium isotopes. Many (though not all) pelagic sediments have high Lu/Hf ratios (along with low Sm/Nd ratios), because they are depleted in detrital zircons, the major carrier of hafnium in sediments (Patchett *et al.*, 1984; Plank and Langmuir, 1998). This is expected to lead to relatively high $^{176}Hf/^{177}Hf$ ratios combined with low $^{143}Nd/^{144}Nd$ values, and these relationships have indeed been observed in lavas from Koolau volcano, Oahu (Hawaiian Islands) (Blichert-Toft *et al.*, 1999) and from Pitcairn (Eisele *et al.*, 2002). Gasperini *et al.* (2000) have proposed still another origin for EM-1 basalts from Sardinia, namely, recycling of gabbros derived from a subducted, ancient plume head.

Recycling of subducted ocean islands and oceanic plateaus was suggested by Hofmann (1989b) to explain not the extreme end-member compositions of the OIB source zoo, but the enrichments seen in the basalts forming the main "mantle array" of negatively correlated $^{143}Nd/^{144}Nd$ and $^{87}Sr/^{86}Sr$ ratios. The $^{143}Nd/^{144}Nd$ values of many of these basalts (e.g., many Hawaiian basalts) are too low, and their $^{87}Sr/^{86}Sr$ values too high, for these OIBs to be explained by recycling of depleted oceanic crust. However, if the recycled material consists of either enriched MORB, tholeiitic or alkaline OIB, or basaltic oceanic plateau material, such a source will have the pre-enriched Rb/Sr and Nd/Sm ratios capable of producing the observed range of strontium- and neodymium-isotopic compositions of the main OIB isotope array.

Melt inclusions in olivine phenocrysts have been shown to preserve primary melt compositions, and these have revealed a startling degree of chemical and isotopic heterogeneity occurring in single-hand specimens and even in single olivine crystals (Sobolev and Shimizu, 1993; Sobolev, 1996; Saal *et al.*, 1998; Sobolev *et al.*, 2000; Hauri, 2002). These studies have demonstrated that rather extreme isotopic and chemical heterogeneities exist in the mantle on scales considerably smaller than the melting region of a single volcano, as discussed in Section 2.03.2.2 on "mesoscale" heterogeneities. One of these studies, in particular, demonstrated the geochemical fingerprint of recycled oceanic gabbros in melt inclusions from Mauna Loa Volcano, Hawaii (Sobolev *et al.*, 2000). These rare melt inclusions have trace-element patterns that are very similar to those of oceanic and ophiolitic gabbros. They are characterized by very high Sr/Nd and low Th/Ba ratios that can be ascribed to cumulus plagioclase, which dominates the modes of many of these gabbros. Chemical and isotopic studies of melt inclusions therefore have great potential for unraveling the specific source materials found in oceanic basalts. These inclusions can preserve primary heterogeneities of the melts much better than the bulk melts do, because the latter go through magma chamber mixing processes that attenuate most of the primary melt features.

The origin of FOZO-C-PHEM-PREMA, simply referred to as "FOZO" hereafter, may be of farther-reaching consequence than any of the other isotope flavors, if the inference of (Hart *et al.*, 1992) is correct, namely, that it represents material from the lower mantle that is present as a mixing component in all deep-mantle plumes. The evidence for this is that samples from many individual OIB associations appear to form binary mixing arrays that radiate from this "focal zone" composition in various directions toward HIMU, EM-1, or EM-2. These relationships are shown in Figure 16. The FOZO composition is similar, but not identical, to DMM represented by MORB. It is

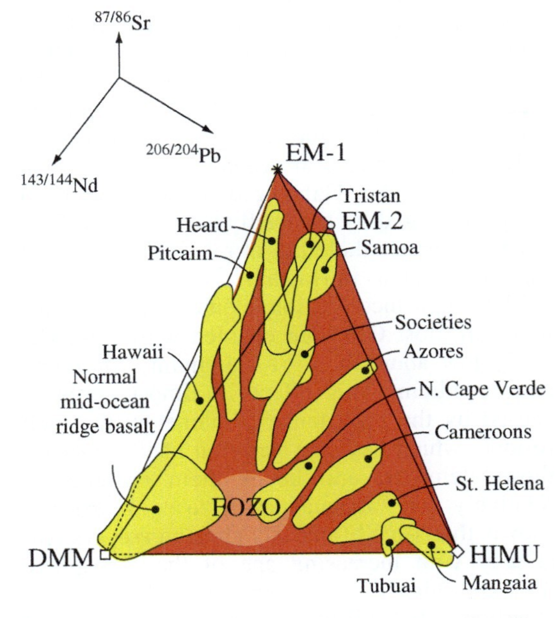

Figure 16 Three-dimensional projection of $^{87}Sr/^{86}Sr$, $^{143}Nd/^{144}Nd$, $^{206}Pb/^{204}Pb$ isotope arrays of a large number of OIB groups (after Hart *et al.*, 1992). Most of the individual arrays appear to radiate from a common region labeled "FOZO" (for focal zone) thought to represent the composition of the deep mantle. The diagram was kindly made available by S. R. Hart.

only moderately more radiogenic in strontium, less radiogenic in neodymium and hafnium, but significantly more radiogenic in lead isotopes than DMM. If plumes originate in the very deep mantle and rise from the core–mantle boundary, rather than from the 660 km seismic discontinuity, they are likely to entrain far more deep-mantle material than upper-mantle material (Griffiths and Campbell, 1990; Hart *et al.*, 1992). It should be noted, however, that the amount of entrained material in plumes is controversial, with some authors insisting that plumes contain very little entrained material (e.g., Farnetani *et al.*, 2002).

2.03.5.2 Trace Elements in OIB

Most OIBs are much more enriched in incompatible trace elements than most MORB, and there are two possible reasons for this: (i) their sources may be more enriched than MORB sources, and (ii) OIBs may be produced by generally lower degrees of partial melting than MORBs. Most likely, both factors contribute to this enrichment. Source enrichment (relative to MORB sources) is required because the isotopic compositions require relative enrichment of the more incompatible of the parent–daughter element ratios. Low degrees of melting are caused by the circumstance that most OIB are also within-plate basalts, so a rising mantle diapir undergoing partial melting encounters a relatively cold lithospheric lid, and melting is confined to low degrees and relatively deep levels. This is also the reason why most OIBs are alkali basalts rather than tholeiites, the predominant rock type in MORBs. Important exceptions to this rule are found primarily in OIBs erupted on or near-ocean ridges (such as on Iceland and the Galapagos Islands) and on hotspots created by especially strong plume flux, which generates tholeiites at relatively high melt fractions, such as in Hawaii.

The high incompatible-element enrichments found in most OIBs are coupled with comparatively low abundances of aluminum, ytterbium, and scandium. This effect is almost certainly caused by the persistence of garnet in the melt residue, which has high partition coefficients for these elements, and keeps them buffered at relatively low abundances. Haase (1996) has shown that Ce/Yb and Tb/Yb increase systematically with increasing age of the lithosphere through which OIBs are erupting. This effect is clearly related to the increasing influence of residual garnet, which is stable in peridotites at depths greater than ~80 km. Allègre *et al.* (1995) analyzed the trace-element abundances of oceanic basalts statistically and concluded that OIBs are more variable in isotopic compositions, but less variable in incompatible element abundances than MORBs. However, their sampling was almost certainly too limited to properly evaluate the effect of lithospheric thickness on the abundances and their variability. In particular, their near-ridge sampling was confined to 11 samples from Iceland and 4 samples from Bouvet. The actual range of abundances of highly incompatible elements, such as thorium and uranium, from Iceland spans nearly three orders of magnitude. In contrast, ytterbium varies by only a factor of 10. This means that either the source of Iceland basalts is internally extremely heterogeneous, or the melt fractions are highly variable, or both. Because of this ambiguity, the REE abundance patterns and most of the moderately incompatible elements, sometimes called "MICE," are actually not very useful to unravel the relative effects of partial melting and source heterogeneity.

The very highly incompatible elements, sometimes called "VICE," are much less fractionated from each other in melts (through normal petrogenetic processes), but they are more severely fractionated in melt residues. This is the reason why their relative abundances vary in the mantle and why these variations can be traced by VICE ratios in basalts, which are in this respect similar to, though not as precise as, isotope ratios. VICE ratios thus enlarge the geochemical arsenal for determining mantle chemical heterogeneities and their origins. These differences are conventionally illustrated either by the "spidergrams" (primitive-mantle normalized element abundance diagrams), or by plotting trace-element abundance ratios.

Spidergrams have the advantage of representing a large number of trace element abundances of a given sample by a single line. However, they can be confusing because there are no standard rules about the specific sequence in which the elements are shown or about the normalizing abundances used. The (mis)use of N-MORB or E-MORBs as reference values for normalizations has already been discussed and discouraged in Section 2.03.4.4. However, there are other pitfalls to be aware of: One of the most widely used normalizations is that given by Sun and McDonough (1989), which uses primitive-mantle estimates for all elements *except* lead, the abundance of which is adjusted by a factor of 2.5, presumably in order to generate smoother abundance patterns in oceanic basalts. The great majority of authors using this normalization simply call this "primitive mantle" without any awareness of the fudge factor applied. Such *ad hoc* adjustments for esthatic reasons should be strongly discouraged. Spidergrams communicate their message most effectively if they are standardized as much as possible, i.e., if they use only one standard for normalization, namely primitive-mantle abundances, and if the sequence of elements used is in the order of increasing compatibility (see also Hofmann, 1988). The methods for determining this order of incompatibility are addressed in the subsequent Section 2.03.5.2.1.

Spidergrams tend to carry a significant amount of redundant information, most of which is useful for determining the general level of incompatible-element enrichment, rather than specific information about the sources. Therefore, diagrams of critical trace-element (abundance) ratios can be very effective in focusing on specific source differences. Care should be taken to use ratios of elements with similar bulk partition coefficients during partial melting (or, more loosely speaking, similar incompatibilities). Otherwise, it may be difficult or impossible to separate source effects from melting effects. Some rather popular element pairs of mixed incompatibility, such as Zr/Nb, which are almost certainly fractionated at the relatively low melt fractions prevailing during intraplate melting, are often used in a particularly confusing manner. For example, in the popular plot of Zr/Nb versus La/Sm, the more incompatible element is placed in the numerator of one ratio (La/Sm) and in the denominator of the other (Zr/Nb). The result is a hyperbolic relationship that looks impressive, but carries little if any useful information other than showing that the more enriched rocks have high La/Sm and low Zr/Nb, and the more depleted rocks have low La/Sm and high Zr/Nb.

Trace-element ratios of similarly incompatible pairs, such as Th/U, Nb/U, Nb/La, Ba/Th, Sr/Nd, or Pb/Nd, tend to be more useful in identifying source differences, because they are fractionated relatively little during partial melting. Elements that appear to be diagnostic of distinctive source types in the mantle are niobium, tantalum, lead, and to a lesser extent strontium, barium, potassium, and rubidium. These will be discussed in connection with the presentation of specific "spidergrams in Section 2.03.5.2.2.

2.03.5.2.1 "Uniform" trace-element ratios

In order to use geochemical anomalies for tracing particular source compositions, it is necessary to establish "normal" behavior first. Throughout the 1980s, Hofmann, Jochum, and co-workers noticed a series of trace-element ratios that are globally more or less uniform in both MORBs and OIBs. For example, the elements barium, rubidium, and caesium, which vary by about three orders of magnitude in absolute abundances, have remarkably uniform relative abundances in many MORBs and OIBs (Hofmann and White, 1983). This became clear only when sufficiently high analytical precision (isotope dilution at the time) was applied to fresh glassy samples. Hofmann and White (1983) argued that this uniformity must mean that the Ba/Rb and Rb/Cs ratios found in the basalts reflect the respective ratios in the source rocks. And because these ratios were so similar in highly depleted MORBs and in enriched OIBs, these authors concluded that these element ratios have not been affected by processes of global differentiation, and they therefore also reflect the composition of the primitive mantle. Similarly, Jochum *et al.* (1983) estimated the K/U ratio of the primitive mantle to be 1.27×10^4, a value that became virtually canonical for 20 years, even though it was based on remarkably few measurements. Other such apparently uniform ratios were Sn/Sm and Sb/Pr (Jochum *et al.*, 1993; Jochum and Hofmann, 1994) and Sr/Nd (Sun and McDonough, 1989). Zr/Hf and Nb/Ta were also thought to be uniform (Jochum *et al.*, 1986), but recent analyses carried out at higher precision and on a greater variety of rock types have shown systematic variations of these ratios.

The above approach of determining primitive-mantle abundances from apparently globally unfractionated trace-element ratios was up-ended by the discovery that Nb/U and Ce/Pb are also rather uniform in MORBs and OIBs the world over, but these ratios are higher by factors of ~5–10 than the respective ratios in the continental crust (Hofmann *et al.*, 1986). This invalidated the assumption that primitive-mantle abundances could be obtained simply from MORB and OIB relations, because the continental crust contains such a large portion of the total terrestrial budget of highly incompatible elements. However, these new observations meant that niobium and lead could potentially be used as tracers for recycled continental material in oceanic basalts. In other words, while ratios such as Nb/U show only limited variation when comparing oceanic basalts as a function of enrichment or depletion on a global or local scale, this uniformity can be interpreted to mean that such a specific ratio is not significantly fractionated during partial melting. If this is true, then the variations that do exist may be used to identify differences in source composition.

Figure 17 shows updated versions of the Nb/U variation diagram introduced by Hofmann *et al.* (1986). It represents an attempt to determine which other highly incompatible trace element is globally most similar to niobium in terms bulk partition coefficient during partial melting. The form of the diagram was chosen because an element ratio will systematically increase as the melt fractions decrease and the absolute concentration of the elements increase. It is obvious that Nb/Th and Nb/La ratios vary systematically with niobium concentration, but Nb/U does not. Extending the comparison to other elements, such as the heavier REEs (not shown), simply increases the slopes of

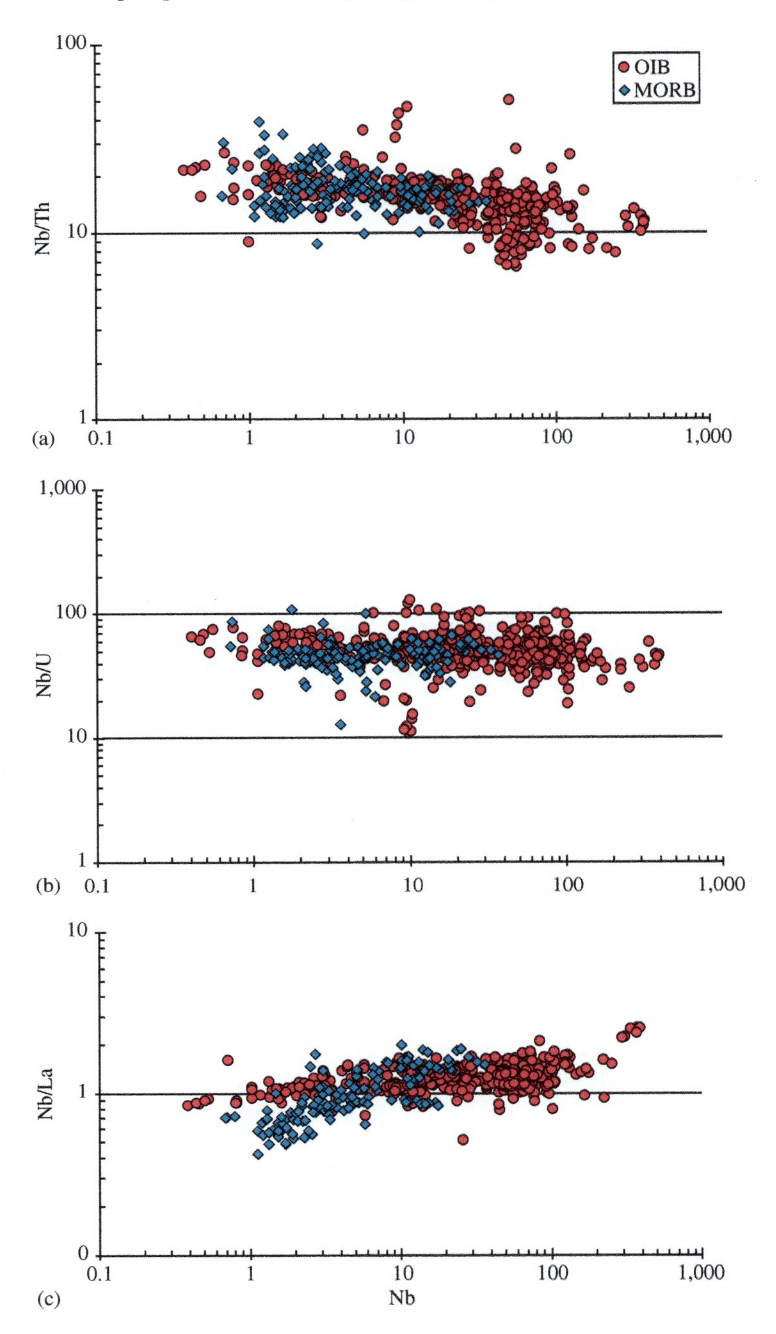

Figure 17 Nb/Th, Nb/U, and Nb/La ratios versus Nb concentrations of global MORB and (non-EM-2-type) OIBs (Hawaiian Isl., Iceland, Australs, Pitcairn, St. Helena, Cnary, Bouvet, Gough Tristan, Ascension, Madeira, Fernando de Noronha, Cameroon Line Isl., Comores, Cape Verdes, Azores, Galapagos, Easter, Juan Fernandez, San Felix). The diagram shows a systematic increase of Nb/Th, approximately constant Nb/U, and systematic decrease of La/Nb as Nb concentrations increase over three orders of magnitude.

such plots. Thus, while Nb/U is certainly *not constant* in oceanic basalts, its variations are the lowest and the least systematic. To be sure, there is possible circularity in this argument, because it is possible, in principle, that enriched sources have systematically lower Nb/U ratios, which are systematically (and relatively precisely) compensated by partition coefficients that are lower for niobium than uranium, thus systematically compensating the lower source ratio during partial melting. Such a compensating mechanism has been advocated by Sims and DePaolo (1997), who criticized the entire approach of Hofmann *et al.* (1986) on this basis. Such fortuitously compensating circumstances, as postulated in the model of Sims and DePaolo (1997), may be *ad hoc* assumptions, but they are not *a priori* impossible.

The model of Hofmann *et al.* (1986) can be tested by examining more local associations of oceanic basalts characterized by large variations in melt fraction. Figure 18 shows Nb–Th–U–La–Nd relationships on Iceland, for volcanic rocks ranging from picrites to alkali basalts, as compiled from the recent literature. The representation differs from Figure 17, which has the advantage of showing the element ratios directly, but the disadvantage pointed out by Sims and DePaolo (1997) that the two variables used are not independent. The simple log–log plot of Figure 18 is less intuitively obvious but in this sense more rigorous. In this plot, a constant concentration ratio yields a slope of unity. The Th–Nb, U–Nb, La–Nb, and Nd–Nb plots show progressively increasing slopes, with the log U–log Nb and the log La–log Nb plots being closest to unity. The data from Iceland shown in Figure 18 are therefore consistent with the global data set shown in Figure 17. This confirms that uranium and niobium have nearly identical bulk partition coefficients during mantle melting in most oceanic environments.

The point of these arguments is not that an element ratio such as Nb/U in a melt will *always* reflect the source ratio very precisely. Rather, because of varying melting conditions, the specific partition coefficients of two such chemically different elements must vary *in detail*, as expected from the partitioning theory of Blundy and Wood (1994). The nephelinites and nepheline melilitites of the Honolulu Volcanic Series, which represent the post-erosional, highly alkalic phase of Koolau Volcano, Oahu, Hawaii, may be an example where the partition coefficients of niobium and uranium are significantly different. These melts are highly enriched in trace elements and must have been formed by very small melt fractions from relatively depleted sources, as indicated by their nearly MORB-like strontium and neodymium isotopic compositions. Their Nb/U ratios average 27, whereas the alkali basalts average Nb/U = 44 (Yang *et al.*, 2003). This may indicate that under melting conditions of very low melt fractions, Nb is significantly more compatible than uranium, and the relationships that are valid for basalts cannot necessarily be extended to more exotic rock types such as nephelinite.

In general, the contrast between Nb/U in most OIBs and MORBs and those in sediments, island arcs and continental rocks is so large that it appears to provide an excellent tracer of recycled continental material in oceanic basalts. A significant obstacle in applying this tracer is the lack of high-quality Nb–U data, partly because of analytical limitations and partly because of sample alteration. The latter can, however, often be overcome by "interpolating" the uranium concentration between thorium and lanthanium (the nearest neighbors in terms of compatibility) and replacing Nb/U by the primitive-mantle normalized Nb/(Th + La) ratio (e.g., Weaver, 1991; Eisele *et al.*, 2002).

Having established that Nb/U or Nb/(Th + La) ratios can be used to trace-mantle source compositions of basalts, this parameter can be turned into a tool to trace recycled continental material in the mantle. The mean Nb/U of 166 MORBs is Nb/U = 47 ± 11, and mean of nearly 500 "non-EM-type" OIBs is Nb/U = 52 ± 15. This contrasts with a mean value of the continental

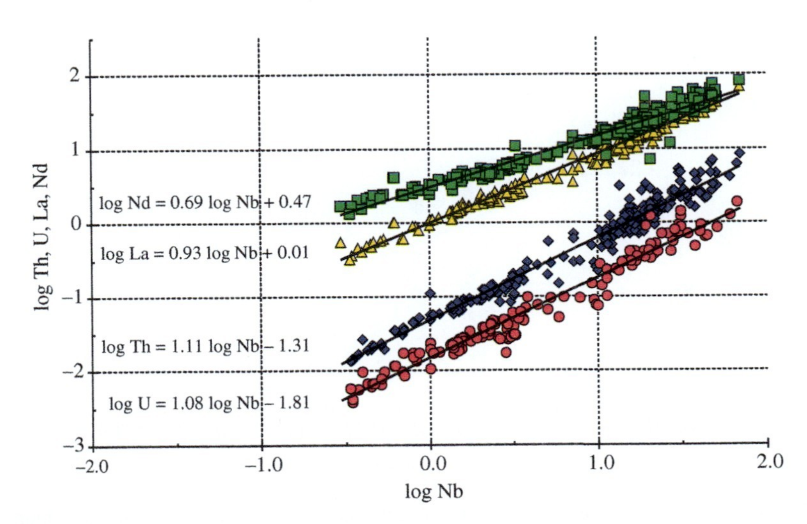

Figure 18 Concentrations of Th, U, La, and Nd versus Nb for basalts and picrites from Iceland. On this logarithmic plot, a regression line of slope 1 represents a constant element concentration ratio (corresponding to a horizontal line in Figure 17). Slopes greater than 1 correspond to positive slopes, and slopes lower than 1 correspond to negative slopes in Figure 17. The correlations of U and La versus Nb yields the slope closest to unity (=1.08 and 0.93).

crust of Nb/U = 8 (Rudnick and Fountain, 1995). As is evident from Figure 4(b), continent-derived sediments also have consistently higher $^{87}Sr/^{86}Sr$ ratios than ordinary mantle rocks; therefore, any OIB containing significant amounts of recycled sediments should be distinguished by high $^{87}Sr/^{86}Sr$ and low Nb/U ratios. Figure 19 shows that this is indeed observed for EM-2 type OIBs and, to a lesser extent, for EM-1 OIBs as well. Of course, this does not "prove" that EM-type OIBs contain recycled sediments. However, there is little doubt that sediments have been subducted in geological history. Much of their trace-element budget is likely to have been short-circuited back into island arcs during subduction. But if any of this material has entered the general mantle circulation and is recycled at all, then EM-type OIBs are the best candidates to show it. Perhaps the greater surprise is that there are so few EM-type ocean islands.

Finding the "constant-ratio partner" for lead has proved to be more difficult. Originally, Hofmann *et al.* (1986) chose cerium because, on average, the Ce/Pb ratio of their MORB data was most similar to their OIB average. However, Sims and DePaolo (1997) pointed out one rather problematic aspect, namely, that even in the original, very limited data set, each separate population showed a distinctly positive slope. In addition, they showed that Ce/Pb ratios appear to correlate with europium anomalies in the MORB population, and this is a strong indication that both parameters are affected by plagioclase fractionation. Recognizing these problems, Rehkämper and Hofmann (1997) argued on the basis of more extensive and more recent data that Nd/Pb is a better indicator of source chemistry than Ce/Pb. Unfortunately, lead concentrations are not often analyzed in oceanic basalts, partly because lead is subject to alteration and partly because it is difficult to analyze, so a literature search tends to yield highly scattered data. Nevertheless, the average MORB value of Pb/Nd = 0.04 is lower than the average continental value of 0.63 by a factor of 15. Because of this great contrast, this ratio is potentially an even more sensitive tracer of continental contamination or continental recycling in oceanic basalts.

But why are Pb/Nd ratios so different in continental and oceanic crust in the first place? An answer to that question will be attempted in the following section.

2.03.5.2.2. Normalized abundance diagrams ("Spidergrams")

The techniques illustrated in Figures 17 and 18 can be used to establish an approximate compatibility sequence of trace elements for mantle-derived melts. In general, this sequence corresponds to the sequence of decreasing (normalized) abundances in the continental crust shown in Figure 2, but this does not apply to niobium, tantalum, and lead for which the results discussed in the previous section demand rather different positions (see also Hofmann, 1988). Here I adopt a sequence similar to that used by Hofmann (1997), but with slightly modified positions for lead and strontium.

Figure 20 shows examples of "spidergrams" for representative samples of HIMU, EM-1, EM-2, and Hawaiian basalts, in addition to average MORB and average "normal" MORB, average

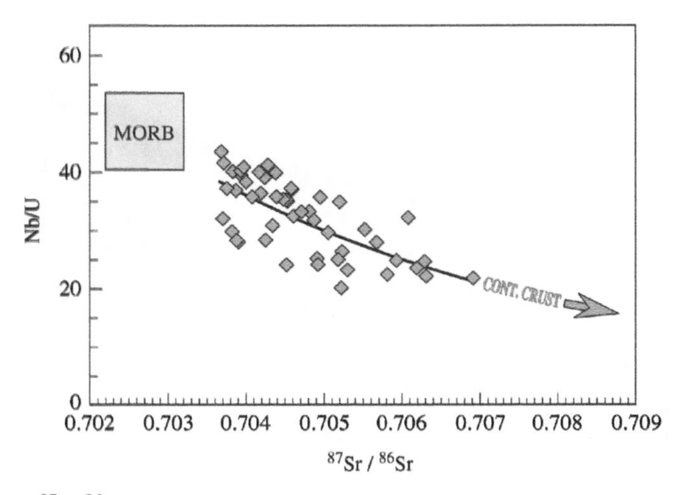

Figure 19 Nb/U versus $^{87}Sr/^{86}Sr$ for basalts from the Society Islands using data of White and Duncan (1996). Two samples with Th/U > 6.0 have been removed because they form outliers on an Nb/Th versus Nb/U correlation and are therefore suspected of alteration or analytical effects on the U concentration. One strongly fractionated trachyte sample has also been removed. This correlation and a similar one of Nd/Pb versus $^{87}Sr/^{86}Sr$ (not shown) is consistent with the addition of a sedimentary or other continental component to the source of the Society Island (EM-2) basalts.

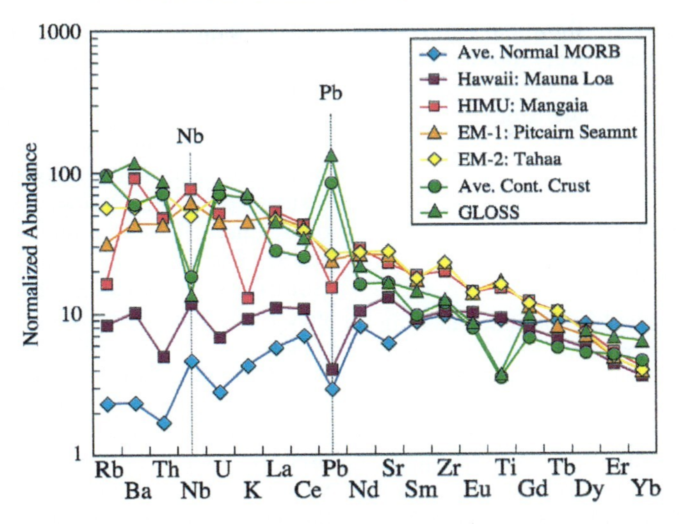

Figure 20 Examples of primitive-mantle normalized trace element abundance diagrams ("spidergrams") for representative samples of HIMU (Mangaia, Austral Islands, sample M-11; Woodhead (1996)), EM-1 (Pitcairn Seamount sample 49DS1; Eisele *et al.* (2002)), EM-2 (Tahaa, Society Islands, sample 73-190; White and Duncan (1996)); Average Mauna Loa (Hawaii) tholeiite (Hofmann, unpublished data), average continental crust (Rudnick and Fountain, 1995), average subducting sediment, GLOSS (Plank and Langmuir, 1998), and Average Normal MORB (Su, 2002). Th, U, and Pb values for MORB were calculated from average Nb/U = 47, and Nd/Pb = 26. All abundances are normalized to primitive/mantle values of McDonough and Sun (1995).

subducting sediment, and average continental crust. Prominent features of these plots are negative spikes for niobium in average continental crust and in sediment, and corresponding positive anomalies in most oceanic basalts except EM-type basalts. Similarly, the positive spikes for lead in average continental crust and in sediments is roughly balanced by negative anomalies in most oceanic basalts. More subtle features distinguishing the isotopically different OIB types are relative deficits for potassium and rubidium in HIMU basalts and high Ba/Th ratios coupled with elevated Sr/Nd ratios in Mauna Loa basalts.

The prominent niobium and lead spikes of continental materials are not matched by any of the OIBs and MORBs reviewed here. They are, however, common features of subduction-related volcanic rocks found on island arcs and continental margins. It is therefore likely that the distinctive geochemical features of the continental crust are produced during subduction, where volatiles can play a major role in the element transfer from mantle to crust. The net effect of these processes is to transfer large amounts of lead (in addition to mobile elements like potassium and rubidium) into the crust. At the same time, niobium and tantalum are retained in the mantle, either because of their low solubility in hydrothermal solutions, or because they are partitioned into residual mineral phases such as Ti-minerals or certain amphiboles. These processes are the subject of much ongoing research, but are beyond the scope of this chapter.

For the study of mantle circulation, these chemical anomalies can help trace the origin of different types of OIBs and some MORBs. Niobium and lead anomalies, coupled with high $^{87}Sr/^{86}Sr$ ratios, seem to be the best tracers for material of continental origin circulating in the mantle. They have been found not only in EM-2-type OIBs but also in some MORBs found on the Chile Ridge (Klein and Karsten, 1995). Other trace-element studies, such as the study of the chemistry of melt inclusions, have already identified a specific recycled rock type, namely, a gabbro, which could be recognized by its highly specific trace-element "fingerprint" (Sobolev *et al.*, 2000).

Until quite recently, the scarceness of high-quality data for the diagnostic elements has been a serious impediment to progress in gaining a full interpretation of the origins of oceanic basalts using complete trace-element data together with complete isotope data. All data compilations aimed at detecting global geochemical patterns are currently seriously hampered by spotty literature data of uncertain quality on samples of unknown freshness. This is now changing, because of the advent of new instrumentation capable of producing large quantities of high-quality data of trace elements at low abundances. The greater ease of obtaining large quantities of data also poses significant risks from lack of quality control. Nevertheless, we are currently experiencing a dramatic improvement in the general quantity and quality of geochemical data, and we can expect significant further improvements in the very near future. These developments offer a bright outlook for the future of deciphering the chemistry and history of mantle differentiation processes.

2.03.6 THE LEAD PARADOX

2.03.6.1 The First Lead Paradox

One of the earliest difficulties in understanding terrestrial lead isotopes arose from the observation that almost all oceanic basalts (i.e., both MORBs and OIBs) have more highly radiogenic lead than does the primitive mantle (Allègre, 1969). In effect, most of these basalts lie to the right-hand side of the so-called "geochron" on a diagram of $^{207}Pb/^{204}Pb$ versus $^{206}Pb/^{204}Pb$ ratios (Figure 21). The Earth was assumed to have the same age as meteorites, so that the "geochron" is identical to the meteorite isochron of 4.56 Ga. If the total silicate portion of the Earth remained a closed system involved only in internal (crust–mantle) differentiation, the sum of the parts of this system must lie on the geochron. The reader is referred to textbooks (e.g., Faure, 1986) on isotope geology for fuller explanations of the construction and meaning of the geochron and the construction of common-lead isochrons.

The radiogenic nature of MORB lead was surprising because uranium is expected to be considerably more incompatible than lead during mantle melting. The MORB source, being depleted in highly incompatible elements, is therefore expected to have had a long-term history of U/Pb ratios lower than primitive ones, just as was found to be the case for Rb/Sr and Nd/Sm. Thus, the lead paradox (sometimes also called the "first paradox") is given by the observation that, although one would expect most MORBs to plot well to the left of the geochron, they actually do plot mostly to the right of the geochron. This expectation is reinforced by a plot of U/Pb versus U, as first used by White (1993), an updated version of which is shown in Figure 22. This shows that the U/Pb ratio is strongly correlated with the uranium concentration, thus confirming the much greater incompatibility of uranium during mantle melting.

Numerous explanations have been advanced for this paradox. The most recent treatment of the subject has been given by Murphy *et al.* (2003), who have also reviewed the most important solutions to the paradox. These include delayed uptake of lead by the core ("core pumping," Allègre *et al.* (1982)) and storage of unradiogenic lead (to balance the excess

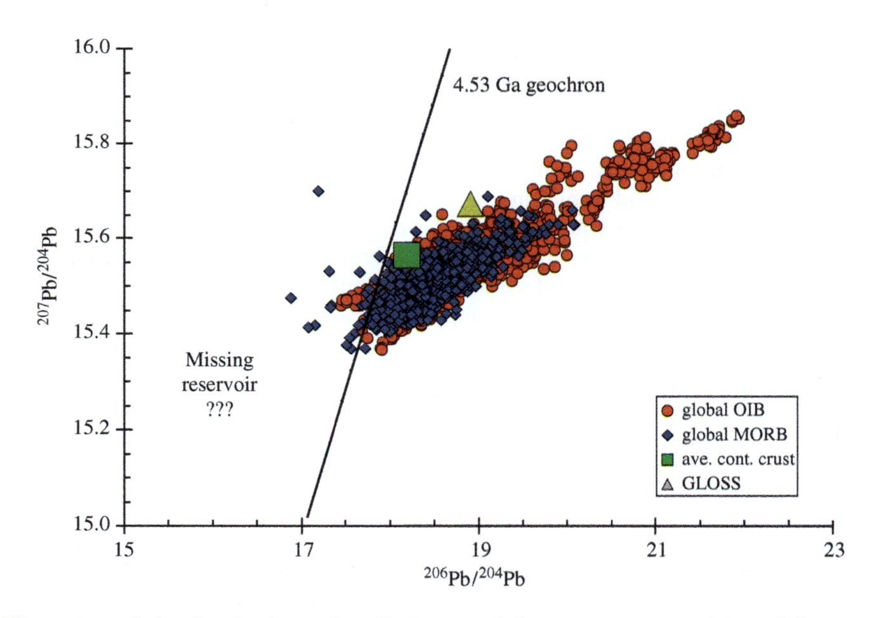

Figure 21 Illustration of the first lead paradox. Estimates of the average composition of the continental crust (Rudnick and Goldstein, 1990), of average "global subducted sediments" (GLOSS, Plank and Langmuir, 1998) mostly derived from the upper continental crust, and a global compilation of MORB and OIB data (from GEOROC and PETDB databases) lie overwhelmingly on the right-hand side of the 4.53 Ga geochron. One part of the paradox is that these data require a hidden reservoir with lead isotopes to the left of the geochron in order to balance the reservoirs represented by the data from the continental and oceanic crust. The other part of the paradox is that both continental and oceanic crustal rocks lie rather close to the geochron, implying that there is surprisingly little net fractionation of the U/Pb ratio during crust–mantle differentiation, even though U is significantly more incompatible than Pb (see Figure 22).

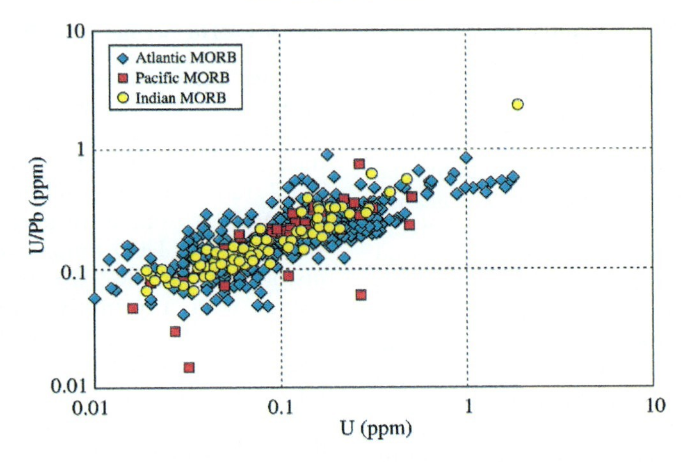

Figure 22 U/Pb versus U concentrations for MORB from three ocean basins. The positive slope of the correlation indicates that U is much more incompatible than Pb during mantle melting. This means that the similarity of Pb isotopes in continental and oceanic crust (see Figure 21) is probably caused by a nonmagmatic transport of lead from mantle to the continental crust.

radiogenic lead seen in MORBs, OIBs, and upper crustal rocks) in the lower continental crust or the subcontinental lithosphere (e.g., Zartman and Haines, 1988; Kramers and Tolstikhin, 1997).

An important aspect not addressed by Murphy *et al.* is the actual position of the geochron. This is the locus of any isotopic mass balance of a closed-system silicate earth. This is not the meteorite isochron of 4.56 Ga, because later core formation and giant impact(s) are likely to have prevented closure of the bulk silicate earth with regard to uranium and lead until lead loss by volatilization and/or loss to the core effectively ended. Therefore, the reference line (geochron) that is relevant to balancing the lead isotopes from the various silicate reservoirs is younger than, and it lies to the right of, the meteorite isochron. The analysis of the effect of slow accretion on the systematics of terrestrial lead isotopes (Galer and Goldstein, 1996) left reasonably wide latitude as to where the relevant geochron should actually be located. This was further reinforced by publication of early tungsten isotope data (daughter product of the short-lived ^{182}Hf), which appeared to require terrestrial core formation to have been delayed by at least 50 Myr (Lee and Halliday, 1995). This would have moved the possible locus of bulk silicate lead compositions closer to the actual positions of oceanic basalts, thus diminishing the magnitude of the paradox, or possibly eliminating it altogether. However, the most recent redeterminations of tungsten isotopes in chondrites by Kleine *et al.* (2002) and Yin *et al.* (2002) have shown the early tungsten data to be in error, so that core and moon formation now appear to be constrained at ~4.53 Ga. This value is not sufficiently lower than the meteorite isochron of 4.56 Ga to resolve the problem.

This means that the lead paradox is alive and well, and the search for the unradiogenic, hidden reservoir continues. The lower continental crust remains (in the author's opinion) a viable candidate, even though crustal xenolith data appear to be, on the whole, not sufficiently unradiogenic (see review of these data by Murphy *et al.* (2003)). It is not clear how representative the xenoliths are, particularly of the least radiogenic, Precambrian lower crust. Another hypothetical candidate is a garnetite reservoir proposed by Murphy *et al.* (2003).

The above discussion, and most of the relevant literature, does not address the perhaps geochemically larger and more interesting question, namely, why are the continental crust and the oceanic basalts so similar in lead isotopes in the first place? It is quite remarkable that most MORBs and most continent-derived sediments cover the same range of ^{206}Pb/^{204}Pb ratios, namely, ~17.5–19.5. This means that MORB sources and upper continental crust, from which these sediments are derived, have very similar U/Pb ratios, when integrated over the entire Earth history. The main offset between lead isotope data for oceanic sediments and MORBs is in terms of ^{207}Pb/^{204}Pb, and even this offset is only marginally outside the statistical scatter of the data. In the previous section we have seen that lead behaves as a moderately incompatible element such as neodymium or cerium, but both uranium and thorium are highly incompatible elements. So the important question remains why lead and uranium are nearly equally enriched in the continental crust, whereas they are very significantly fractionated during the formation of the oceanic crust and ocean islands. Therefore, the unradiogenic reservoir (lower crust or hidden mantle reservoir) needed to balance the existing, slightly radiogenic reservoirs in order to obtain full, bulk-silicate-earth lead isotope values represents only a relatively minor aspect of additional adjustment to this major discrepancy. The answer is, in my opinion, that lead behaves relatively compatibly during MORB–OIB production because it is partially retained in the mantle by a residual phase(s), most likely sulfide(s). This would account for the relationships seen in Figure 22. However, when lead is transferred from mantle to the continental crust, two predominantly nonigneous processes become important: (i) hydrothermal transfer from oceanic crust to metalliferous sediment (Peucker-Ehrenbrink *et al.*, 1994), and (ii) transfer from subducted oceanic crust-plus-sediment into arc magma sources (Miller *et al.*, 1994). This additional, nonigneous transfer enriches the crust to a similar extent as uranium and thorium, and this explains why the $^{206}Pb/^{204}Pb$ ratios of crustal and mantle rocks are so similar to each other and so close to the geochron. Thus, the anomalous geochemical behavior of lead is the main cause of the "lead paradox," the high Pd/Nd ratios of island arc and continental rocks, and the lead "spikes" seen in Figure 20.

The above explanation does not account for the elevated $^{207}Pb/^{204}Pb$ ratios of continental rocks and their sedimentary derivatives relative to mantle-derived basalts (Figures 5(a) and 21). This special feature can be explained by a more complex evolution of continents subsequent to their formation. New continental crust formed during Archean time by subduction and accretion processes must have initially possessed a U/Pb ratio slightly higher than that of the mantle. At that time, the terrestrial $^{235}U/^{238}U$ ratio was significantly higher than today, and this produced elevated $^{207}Pb/^{204}Pb$ relative to $^{206}Pb/^{204}Pb$. Some of this crust was later subjected to high-grade metamorphism, causing loss of uranium relative to lead in the lower crust, and transporting the excess uranium into the upper crust. From there, uranium was lost by the combined action of oxidation, weathering, dissolution, and transport into the oceans. This uranium loss retarded the growth of $^{206}Pb/^{204}Pb$ while preserving the relatively elevated $^{207}Pb/^{204}Pb$ of the upper crust. The net result of this two-stage process is the present position of sediments directly above the mantle-derived basalts in $^{207}Pb/^{204}Pb-^{207}Pb/^{204}Pb$ space. Another consequence of this complex behavior of uranium will be further discussed in the following section. Here it is important to reiterate that lead, not uranium or thorium, is the *major* player in generating the main part of the lead paradox.

2.03.6.2 The Second Lead Paradox

Galer and O'Nions (1985) made the important observation that measured $^{208}Pb^*/^{206}Pb^*$ ratios (see Equation (11)) in most MORBs are higher than can be accounted for by the relatively low Th/U ratios actually observed in MORBs, if the MORB source reservoir had maintained similarly low Th/U ratios over much of Earth's history. A simple two-stage Th/U depletion model with a primitive, first-stage value of $\kappa = (^{232}Th/^{238}U)_{today} = 3.9$ changing abruptly to a second stage value of $\kappa = 2.5$ yields a model age for the MORB source of only ~600 Ma. Because much of the continental mass is much older than this, and because the development of the depleted MORB source is believed to be linked to this, this result presented a dilemma. Galer and O'Nions (1985) and Galer *et al.* (1989) resolved this with a two-layer model of the mantle, in which the upper, depleted layer is in a steady state of incompatible-element depletion by production of continental crust and replenishment by leakage of less depleted material from the lower layer. This keeps the $^{208}Pb^*/^{206}Pb^*$ ratio of the upper mantle at a relatively high value in spite of the low chemical Th/U ratio. However, such two-layer convection models have fallen from grace in recent years, mostly because of the results of seismic mantle tomography (see below). Therefore, other solutions to this second lead paradox have been sought.

A fundamentally different mechanism for lowering the Th/U ratio of the mantle has been suggested by Hofmann and White (1982), namely, preferential recycling of uranium through dissolution of oxidized (hexavalent) uranium at the continental surface, riverine transport into the oceans, and fixation by ridge-crest hydrothermal circulation and reduction to the tetravalent state. The same mechanism was invoked in the quantitative "plumbotectonic" model of Zartman and Haines (1988). Staudigel *et al.* (1995) introduced the idea that this preferential recycling of uranium into the mantle may be connected to a change toward oxidizing conditions at the Earth's surface relatively late in Earth's history. Geological evidence for a rapid atmospheric change toward oxidizing conditions during Early Proterozoic time (i.e., relatively late in Earth's history) has been presented by Holland (1994), and this has been confirmed by new geochemical evidence showing that sulfides and sulfates older than ~2.4 Ga contain non-mass-dependent sulfur isotope fractionations, which can be explained by high-intensity UV radiation in an oxygen-absent atmosphere (Farquhar *et al.*, 2000). Kramers and Tolstikhin (1997) and Elliott *et al.* (1999) have developed quantitative models to resolve the second lead paradox by starting to recycle uranium into the mantle ~2.5 Ga ago. This uranium cycle is an excellent example how mantle geochemistry, surface and atmospheric chemistry, and the evolution of life are intimately interconnected.

2.03.7 GEOCHEMICAL MANTLE MODELS

The major aim of mantle geochemistry has been, from the beginning, to elucidate the structure and evolution of the Earth's interior, and it was clear that this can only be done in concert with observations and ideas derived from conventional field geology and from geophysics. The discussion here will concentrate on the chemical structure of the recent mantle, because the early mantle evolution and dynamics and history of convective mixing are treated in Chapters 2.12 and 2.13.

The isotopic and chemical heterogeneities found in mantle-derived basalts, reviewed on the previous pages, mandate the existence of similar or even greater heterogeneities in the mantle. The questions are: how are they spatially arranged in the mantle? When and how did they originate? Because these heterogeneities have their primary expression in trace elements, they cannot be translated into physical parameters such as density differences. Rather, they must be viewed as passive tracers of mantle processes. These tracers are separated by melting and melt migration, as well as fluid transport, and they are stirred and remixed by convection. Complete homogenization appears to be increasingly unlikely. Diffusion distances in the solid state have ranges of centimeters at best (Hofmann and Hart, 1978), and possibly very much less (Van Orman *et al.*, 2001), and homogenization of a melt source region, even via the movement and diffusion through a partial melt, becomes increasingly unlikely. This is attested by the remarkable chemical and isotopic heterogeneity observed in melt inclusions preserved in magmatic crystals from single basalt samples (Saal *et al.*, 1998; Sobolev *et al.*, 2000). The models developed for interpreting mantle heterogeneities have, with some exceptions, largely ignored the possibly extremely small scale of these heterogeneities. Instead, they have usually relied on the assumption that mantle-derived basalts are, on the whole, representative of some chemical and isotopic average for a given volcanic province or source volume.

In the early days of mantle geochemistry, the composition of the bulk silicate earth, also called "primitive mantle" (i.e., mantle prior to the formation of any crust; see Chapter 2.01) was not known for strontium isotopes because of the obvious depletion of rubidium of the Earth relative to chondritic meteorites (Gast, 1960). The locus of primitive-mantle lead was assumed to be on the meteorite isochron (which was thought to be identical to the geochron until the more recent realization of delayed accretion and core formation; see above), but the interpretation of the lead data was confounded by the lead paradox discussed above. This situation changed in the 1970s with the first measurements of neodymium isotopes in oceanic basalts (DePaolo and Wasserburg, 1976; Richard *et al.*, 1976; O'Nions *et al.*, 1977) and the discovery that $^{143}Nd/^{144}Nd$ was negatively correlated with $^{87}Sr/^{86}Sr$ (see Figures 6 and 15). Because both samarium and neodymium are refractory lithophile elements, the Sm/Nd and $^{143}Nd/^{144}Nd$ ratios of the primitive mantle can be safely assumed to be chondritic. With this information, a primitive-mantle value for the Rb/Sr and $^{87}Sr/^{86}Sr$ was inferred, and it became possible to estimate the size of the MORB source reservoir primarily from isotopic abundances. The evolution of a silicate Earth consisting of three boxes—primitive mantle, depleted mantle, and continental crust—were subsequently modeled by Jacobsen and Wasserburg (1979), O'Nions *et al.* (1979), DePaolo (1980), Allègre *et al.* (1983), and Davies (1981) cast this in terms of a simple mass balance similar to that given above in Equation (11) but using isotope ratios:

$$X_{\text{dm}} = \frac{X_{\text{cc}}C_{\text{cc}}(R_{\text{dm}} - R_{\text{cc}})}{C_{\text{pm}}(R_{\text{dm}} - R_{\text{pm}})} \tag{13}$$

where X are mass fractions, R are isotope or trace-element ratios, C are the concentrations of the denominator element of R, and the subscripts cc, dm, and pm refer to continental crust, depleted mantle, and primitive mantle reservoirs, respectively.

With the exception of Davies, who favored whole-mantle convection all along, the above authors concluded that it was only the upper mantle above the 660 km seismic discontinuity that was needed to balance the continental crust. The corollary conclusion was that the deeper mantle must be in an essentially primitive, nearly undepleted state, and consequently convection in the mantle had to occur in two layers with only little exchange between these layers. These conclusions were strongly reinforced by noble gas data, especially $^{3}He/^{4}He$ ratios and, more recently, neon isotope data. These indicated that hotspots such as Hawaii are derived from a deep-mantle source with a more primordial, high $^{3}He/^{4}He$ ratio, whereas MORBs are derived from a more degassed, upper-mantle reservoir with lower $^{3}He/^{4}He$ ratios. The noble-gas aspects are treated in Chapter 2.06. In the present context, two points must be mentioned. Essentially all quantitative evolution models dealing with the noble gas evidence concluded that, although plumes carry the primordial gas signature from the deep mantle to the surface, the plumes themselves do not originate in the deep mantle. Instead they rise from the base of the upper mantle, where they entrain very small quantities of lower-mantle, noble-gas-rich material. However, all of these models have been constrained by the present-day very low flux of helium from the mantle into the oceans. This flux does not allow the lower mantle to be significantly degassed over the Earth's history. Other authors, who do not consider this constraint on the evolution models to be binding, have interpreted the noble gas data quite differently: they argue that the entire plume comes from a nearly primitive deep-mantle source and rises through the upper, depleted mantle. The former view is consistent with the geochemistry of the refractory elements, which strongly favors some type of recycled, not primitive mantle material to supply the bulk of the plume source. The latter interpretation can be reconciled with the refractory-element geochemistry, if the deep-mantle reservoir is not actually primitive (or close to primitive), but consists of significantly processed mantle with the geochemical characteristics of the FOZO (C, PHEM, etc.) composition, which is characterized by low $^{87}Sr/^{86}Sr$ but relatively high $^{206}Pb/^{204}Pb$ ratios, together with high $^{3}He/^{4}He$ and solar-like neon isotope ratios. The processed nature of this hypothetical deep-mantle reservoir is also evident from its trace-element chemistry, which shows the same nonprimitive (high) Nb/U and (low) Pb/Nd ratios as MORBs and other OIBs. It is not clear how and why the near-primordial noble gas compositions would have survived this processing. So far, except for the two-layer models, no internally consistent mantle evolution model has been published that accounts for all of these observations (see also Chapter 2.12).

The two-layer models have been dealt a rather decisive blow by recent results of seismic mantle tomography. The images of the mantle produced by this discipline appear to show clear evidence for subduction reaching far into the lower mantle (Grand, 1994; van der Hilst *et al.*, 1997). If this is correct, then there must be a counterflow from the lower mantle across the 660 km boundary, which in the long run would surely destroy the chemical isolation between upper- and lower-mantle reservoirs. Most recently, mantle tomography appears to be able to track some of the major mantle plumes (Hawaii, Easter Island, Cape Verdes, Reunion) into the lowermost mantle (Montelli *et al.*, 2003). If these results are confirmed, there is, at least in recent mantle history, no convective isolation, and mantle evolution models reconciling all of the geochemical aspects with the geophysical evidence clearly require new ideas.

As of early 2000s, the existing literature and scientific conferences all show clear signs of a period preceding a significant or even major paradigm shift as described by Thomas Kuhn in his classic work *The Structure of Scientific Revolutions* (Kuhn, 1996): The established paradigm is severely challenged by new observations. While some scientists attempt to reconcile the observations with the paradigm by increasingly complex adjustments of the paradigm, others throw the established conventions overboard and engage in increasingly free speculation. This process continues until a new paradigm evolves or is discovered, which is consistent with all the observations. Examples of the effort of reconciliation are the papers by Stein and Hofmann (1994) and Allègre (1997). These authors point out that the current state of whole-mantle circulation may be episodic or recent, so that whole-mantle chemical mixing has not been achieved.

Examples of the more speculative thinking are the papers by Albarède and van der Hilst (1999) and Kellogg *et al.* (1999), who essentially invent new primitive reservoirs within the deep mantle, which are stabilized by higher chemical density, and may be very irregularly shaped. In the same vein, Porcelli and Halliday (2001) have proposed that the storage reservoir of primordial noble gases may be the core, whereas Tolstikhin and Hofmann (2002) speculate that the lowermost layer of the mantle, called D'' by seismologists, contains the "missing" budget of heat production and primordial noble gases. Finally, an increasing number of convection and mantle evolution modelers are throwing the entire concept of geochemical reservoirs overboard. For example, Phipps Morgan and Morgan (1998) and Phipps Morgan (1999) suggest that the specific geochemical characteristics of plume-type mantle are randomly distributed in the deeper mantle. Plumes rising from the core–mantle boundary constitute the main upward flux balancing the subduction flux. They preferentially lose their enriched components during partial melting, leaving a depleted residue that replenishes the depleted upper mantle.

Starting with the contribution of Christensen and Hofmann (1994), a steadily increasing number of models have recently been published, in which geochemical heterogeneities are specifically incorporated in mantle convection models (e.g., van Keken and Ballentine, 1998; Tackley, 2000; van Keken *et al.*, 2001; Davies, 2002; Farnetani *et al.*, 2002; Tackley, 2002). Thus, while the current state of understanding of the geochemical heterogeneity of the mantle is unsatisfactory, to say the least, the formerly quite separate disciplines of geophysics and geochemistry have begun to interact intensely. This process surely offers the best approach to reach a new paradigm and an understanding of how the mantle really works.

ACKNOWLEDGMENTS

I am most grateful for the hospitality and support offered by the Lamont-Doherty Earth Observatory where much of this article was written. Steve Goldstein made it all possible, while being under duress with the similar project of his own. Cony Class and the other members of the Petrology/Geochemistry group at Lamont provided intellectual stimulation and good spirit, and the BodyQuest Gym and the running trails of Tallman Park kept my body from falling apart.

REFERENCES

Abouchami W., Boher M., Michard A., and Albarède F. (1990) A major 2.1 Ga event of mafic magmatism in West-Africa: an early stage of crustal accretion. *J. Geophys. Res.* **95**, 17605–17629.

Alard O., Griffin W. L., Pearson N. J., Lorand J. P., and O'Reilly S. Y. (2002) New insights into the Re–Os systematics of sub-continental lithospheric mantle from *in situ* analysis of sulphides. *Earth Planet. Sci. Lett.* **203**(2), 651–663.

Albarède F. (1998) Time-dependent models of U–Th–He and K–Ar evolution and the layering of mantle convection. *Chem. Geol.* **145**, 413–429.

Albarède F. and Van der Hilst R. D. (1999) New mantle convection model may reconcile conflicting evidence. *EOS. Trans. Am. Geophys. Union* **80**(45), 535–539.

Allègre C. J. (1969) Comportement des systèmes U–Th–Pb dans le manteau supérieur et modèle d'évolution de ce dernier au cours de temps géologiques. *Earth Planet. Sci. Lett.* **5**, 261–269.

Allègre C. J. (1997) Limitation on the mass exchange between the upper and lower mantle: the evolving convection regime of the Earth. *Earth Planet. Sci. Lett.* **150**, 1–6.

Allègre C. J. and Turcotte D. L. (1986) Implications of a two-component marble-cake mantle. *Nature* **323**, 123–127.

Allègre C. J., Dupré B., and Brévart O. (1982) Chemical aspects of the formation of the core. *Phil Trans. Roy. Soc. London A* **306**, 49–59.

Allègre C. J., Hart S. R., and Minster J.-F. (1983) Chemical structure and evolution of the mantle and continents determined by inversion of Nd and Sr isotopic data: II. Numerical experiments and discussion. *Earth Planet. Sci. Lett.* **66**, 191–213.

Allègre C. J., Schiano P., and Lewin E. (1995) Differences between oceanic basalts by multitrace element ratio topology. *Earth Planet. Sci. Lett.* **129**, 1–12.

Allègre C. J., Hofmann A. W., and O'Nions K. (1996) The argon constraints on mantle structure. *Geophys. Res. Lett.* **23**, 3555–3557.

Ben Othman D., White W. M., and Patchett J. (1989) The geochemistry of marine sediments, island arc magma genesis, and crust–mantle recycling. *Earth Planet. Sci. Lett.* **94**, 1–21.

Ben-Avraham Z., Nur A., Jones D., and Cox A. (1981) Continental accretion: from oceanic plateaus to allochthonous terranes. *Science* **213**, 47–54.

Bird J. M., Meibom A., Frei R., and Nagler T. F. (1999) Osmium and lead isotopes of rare OsIrRu minerals: derivation from the core–mantle boundary region? *Earth Planet. Sci. Lett.* **170**(1–2), 83–92.

Blichert-Toft J., Frey F. A., and Albarède F. (1999) Hf isotope evidence for pelagic sediments in the source of Hawaiian basalts. *Science* **285**, 879–882.

Blundy J. D. and Wood B. J. (1994) Prediction of crystal–melt partition coefficients from elastic moduli. *Nature* **372**, 452–454.

Bohrson W. A. and Reid M. R. (1995) Petrogenesis of alkaline basalts from Socorro Island, Mexico: trace element evidence for contamination of ocean island basalt in the shallow ocean crust. *J. Geophys. Res.* **100**(B12), 24555–24576.

Brandon A. D., Snow J. E., Walker R. J., Morgan J. W., and Mock T. D. (2000) Pt-190–Os-186 and Re-187–Os-187 systematics of abyssal peridotites. *Earth Planet. Sci. Lett.* **177**(3–4), 319–335.

Burton K. W., Schiano P., Birck J.-L., Allègre C. J., and Rehkämper M. (2000) The distribution and behavior of Re and Os amongst mantle minerals and the consequences of metasomatism and melting on mantle lithologies. *Earth Planet. Sci. Lett.* **183**, 93–106.

Calvert A. J. and Ludden J. N. (1999) Archean continental assembly in the southeastern Superior Province of Canada. *Tectonics* **18**(3), 412–429.

Campbell I. H. and Griffiths R. W. (1990) Implications of mantle plume structure for the evolution of flood basalts. *Earth Planet. Sci. Lett.* **99**, 79–93.

Carlson R. W. and Hauri E. (2003) Mantle–crust mass balance and the extent of Earth differentiation. *Earth Planet. Sci. Lett.* (submitted 2002).

Chamorro E. M., Brooker R. A., Wartho J. A., Wood B. J., Kelley S. P., and Blundy J. D. (2002) Ar and K partitioning between clinopyroxene and silicate melts to 8 GPa. *Geochim. Cosmochim. Acta* **66**, 507–519.

Chase C. G. (1981) Oceanic island Pb: two-state histories and mantle evolution. *Earth Planet. Sci. Lett.* **52**, 277–284.

Chauvel C., Hofmann A. W., and Vidal P. (1992) HIMU-EM: the French Polynesian connection. *Earth Planet. Sci. Lett.* **110**, 99–119.

Chauvel C., Goldstein S. L., and Hofmann A. W. (1995) Hydration and dehydration of oceanic crust controls Pb evolution in the mantle. *Chem. Geol.* **126**, 65–75.

Christensen U. R. and Hofmann A. W. (1994) Segregation of subducted oceanic crust in the convecting mantle. *J. Geophys. Res.* **99**, 19867–19884.

Class C., Miller D. M., Goldstein S. L., and Langmuir C. H. (2000) Distinguishing melt and fluid subduction components in Umnak volcanics, Aleutian Arc. *Geochem. Geophys. Geosyst.* **1** #1999GC000010.

Coltice N. and Ricard Y. (1999) Geochemical observations and one layer mantle convection. *Earth Planet. Sci. Lett.* **174**(1–2), 125–137.

Creaser R. A., Papanastasiou D. A., and Wasserburg G. J. (1991) Negative thermal ion mass spectrometry of osmium, rhenium, and iridium. *Geochim. Cosmochim. Acta* **55**, 397–401.

Davidson J. P. and Bohrson W. A. (1998) Shallow-level processes in ocean-island magmatism. *J. Petrol.* **39**, 799–801.

Davies G. F. (1981) Earth's neodymium budget and structure and evolution of the mantle. *Nature* **290**, 208–213.

Davies G. F. (1998) Plates, plumes, mantle convection and mantle evolution. In *The Earth's Mantle-composition, Structure, and Evolution* (ed. I. Jackson). Cambridge University Press, Cambridge, pp. 228–258.

Davies G. F. (2002) Stirring geochemistry in mantle convection models with stiff plates and slabs. *Geochim. Cosmochim. Acta* **66**, 3125–3142.

DePaolo D. J. (1980) Crustal growth and mantle evolution: inferences from models of element transport and Nd and Sr isotopes. *Geochim. Cosmochim. Acta* **44**, 1185–1196.

DePaolo D. J. and Wasserburg G. J. (1976) Nd isotopic variations and petrogenetic models. *Geophys. Res. Lett.* **3**, 249–252.

Donnelly K. E., Goldstein S. L., Langmuir C. H., and Spiegelmann M. (2003) Origin of enriched ocean ridge basalts and implications for mantle dynamics. *Nature* (ms submitted April 2003).

Dupré B. and Allègre C. J. (1983) Pb–Sr isotope variation in Indian Ocean basalts and mixing phenomena. *Nature* **303**, 142–146.

Dupré B., Schiano P., Polvé M., and Joron J.-L. (1994) Variability: a new parameter which emphasizes the limits of extended rare earth diagrams. *Bull. Soc. Geol. Francais* **165**(1), 3–13.

Eisele J., Sharma M., Galer S. J. G., Blichert-toft J., Devey C. W., and Hofmann A. W. (2002) The role of sediment recycling in EM-1 inferred from Os, Pb, Hf, Nd, Sr isotope and trace element systematics of the Pitcairn hotspot. *Earth Planet. Sci. Lett.* **196**, 197–212.

Elliott T., Plank T., Zindler A., White W., and Bourdon B. (1997) Element transport from slab to volcanic front at the Mariana Arc. *J. Geophys. Res.* **102**, 14991–15019.

Elliott T., Zindler A., and Bourdon B. (1999) Exploring the kappa conundrum: the role of recycling on the lead isotope evolution of the mantle. *Earth Planet. Sci. Lett.* **169**, 129–145.

Farley K. A., Natland J. H., and Craig H. (1992) Binary mixing of enriched and undegassed (primitive?) mantle components (He, Sr, Nd, Pb) in Samoan lavas. *Earth Planet. Sci. Lett.* **111**, 183–199.

Farnetani C. G., Legrasb B., and Tackley P. J. (2002) Mixing and deformations in mantle plumes. *Earth Planet. Sci. Lett.* **196**, 1–15.

Farquhar J., Bao H., and Thiemens M. H. (2000) Atmospheric influence of Earth's earliest sulfur cycle. *Science* **290**, 756–758.

Faure G. (1986) *Principles of Isotope Geology*. Wiley, New York.

Foulger G. R. and Natland J. H. (2003) Is "Hotspot" volcanism a consequence of plate tectonics? *Science* **300**, 921–922.

Galer S. J. G. (1999) Optimal triple spiking for high precision lead isotopic measurement. *Chem. Geol.* **157**, 255–274.

Galer S. J. G. and Goldstein S. L. (1996) Influence of accretion on lead in the Earth. In *Earth Processes: Reading the Isotopic Code*, Geophys. Monograph 95 (eds. A. Basu and S. R. Hart). Am. Geophys. Union, Washington, pp. 75–98.

Galer S. J. G. and O'Nions R. K. (1985) Residence time of thorium, uranium and lead in the mantle with implications for mantle convection. *Nature* **316**, 778–782.

Galer S. J. G., Goldstein S. L., and O'Nions R. K. (1989) Limits on chemical and convective isolation in the Earth's interior. *Chem. Geol.* **75**, 252–290.

Gasperini G., Blichert-Toft J., Bosch D., Del Moro A., Macera P., Télouk P., and Albarède F. (2000) Evidence from Sardinian basalt geochemistry for recycling of plume heads into the Earth's mantle. *Nature* **408**, 701–704.

Gast P. W. (1960) Limitations on the composition of the upper mantle. *J. Geophys. Res.* **65**, 1287.

Gast P. W. (1968) Trace element fractionation and the origin of tholeiitic and alkaline magma types. *Geochim. Cosmochim. Acta* **32**, 1057–1086.

Gast P. W., Tilton G. R., and Hedge C. (1964) Isotopic composition of lead and strontium from Ascension and Gough Islands. *Science* **145**, 1181–1185.

Goldschmidt V. M. (1937) Geochemische Verteilungsgesetze der Elemente: IX. Die Mengenverhältnisse der Elemente and Atomarten. *Skr. Nor. vidensk.-Akad. Oslo* **1**, 148.

Goldschmidt V. M. (1954) *Geochemistry*. Clarendon Press, zOxford.

Grand S. P. (1994) Mantle shear structure beneath the Americas and surrounding oceans. *J. Geophys. Res.* **99**, 11591–11621.

Griffiths R. W. and Campbell I. H. (1990) Stirring and structure in mantle starting plumes. *Earth Planet. Sci. Lett.* **99**, 66–78.

Gurenko A. A. and Chaussidon M. (1995) Enriched and depleted primitive melts included in olivine from Icelandic tholeiites: origin by continuous melting of a single mantle column. *Geochim. Cosmochim. Acta* **59**, 2905–2917.

Haase K. M. (1996) The relationship between the age of the lithosphere and the composition of oceanic magmas: constraints on partial melting, mantle sources and the thermal structure of the plates. *Earth Planet. Sci. Lett.* **144**(1–2), 75–92.

Hanan B. and Graham D. (1996) Lead and helium isotope evidence from oceanic basalts for a common deep source of mantle plumes. *Science* **272**, 991–995.

Hanson G. N. (1977) Geochemical evolution of the suboceanic mantle. *J. Geol. Soc. London* **134**, 235–253.

Hart S. R. (1971) K, Rb, Cs, Sr, Ba contents and Sr isotope ratios of ocean floor basalts. *Phil. Trans. Roy. Soc. London Ser. A* **268**, 573–587.

Hart S. R. (1984) A large-scale isotope anomaly in the southern hemisphere mantle. *Nature* **309**, 753–757.

Hart S. R., Schilling J.-G., and Powell J. L. (1973) Basalts from Iceland and along the Reykjanes Ridge: Sr isotope geochemistry. *Nature* **246**, 104–107.

Hart S. R., Hauri E. H., Oschmann L. A., and Whitehead J. A. (1992) Mantle plumes and entrainment: isotopic evidence. *Science* **256**, 517–520.

Hauri E. (2002) SIMS analysis of volatiles in silicate glasses: 2. Isotopes and abundances in Hawaiian melt inclusions. *Chem. Geol.* **183**, 115–141.

Hawkesworth C. J., Norry M. J., Roddick J. C., and Vollmer R. (1979) ^{143}Nd/^{144}Nd and ^{87}Sr/^{86}Sr ratios from the Azores and their significance in LIL-element enriched mantle. *Nature* **280**, 28–31.

Hawkesworth C. J., Mantovani M. S. M., Taylor P. N., and Palacz Z. (1986) Evidence from the Parana of South Brazil for a continental contribution to Dupal basalts. *Nature* **322**, 356–359.

Hedge C. E. and Walthall F. G. (1963) Radiogenic strontium 87 as an index of geological processes. *Science* **140**, 1214–1217.

Hellebrand E., Snow J. E., Dick H. J. B., and Hofmann A. W. (2001) Coupled major and trace elements as indicators of the extent of melting in mid-ocean-ridge peridotites. *Nature* **410**, 677–681.

Hemming S. R. and McLennan S. M. (2001) Pb isotope compositions of modern deep sea turbidites. *Earth Planet. Sci. Lett.* **184**, 489–503.

Hémond C. H., Arndt N. T., Lichtenstein U., Hofmann A. W., Oskarsson N., and Steinthorsson S. (1993) The heterogeneous Iceland plume: Nd–Sr–O isotopes and trace element constraints. *J. Geophys. Res.* **98**, 15833–15850.

Hirschmann M. M. and Stopler E. (1996) A possible role for garnet pyroxenite in the origin of the "garnet signature" in MORB. *Contrib. Mineral. Petrol.* **124**, 185–208.

Hofmann A. W. (1988) Chemical differentiation of the Earth: the relationship between mantle, continental crust, and oceanic crust. *Earth Planet. Sci. Lett.* **90**, 297–314.

Hofmann A. W. (1989a) Geochemistry and models of mantle circulation. *Phil. Trans. Roy. Soc. London A* **328**, 425–439.

Hofmann A. W. (1989b) A unified model for mantle plume sources. *EOS* **70**, 503.

Hofmann A. W. (1997) Mantle geochemistry: the message from oceanic volcanism. *Nature* **385**, 219–229.

Hofmann A. W. and Hart S. R. (1978) An assessment of local and regional isotopic equilibrium in the mantle. *Earth Planet. Sci. Lett.* **39**, 44–62.

Hofmann A. W. and White W. M. (1980) The role of subducted oceanic crust in mantle evolution. *Carnegie Inst. Wash. Year Book* **79**, 477–483.

Hofmann A. W. and White W. M. (1982) Mantle plumes from ancient oceanic crust. *Earth Planet. Sci. Lett.* **57**, 421–436.

Hofmann A. W. and White W. M. (1983) Ba, Rb, and Cs in the Earth's mantle. *Z. Naturforsch.* **38**, 256–266.

Hofmann A. W., White W. M., and Whitford D. J. (1978) Geochemical constraints on mantle models: the case for a layered mantle. *Carnegie Inst. Wash. Year Book* **77**, 548–562.

Hofmann A. W., Jochum K.-P., Seufert M., and White W. M. (1986) Nb and Pb in oceanic basalts: new constraints on mantle evolution. *Earth Planet. Sci. Lett.* **79**, 33–45.

Holland H. D. (1994) Early Proterozoic atmospheric change. In *Early Life on Earth* (ed. S. Bengtson). Columbia University Press, New York, pp. 237–244.

Hurley P. M., Hughes H., Faure G., Fairbairn H., and Pinson W. (1962) Radiogenic strontium-87 model for continent formation. *J. Geophys. Res.* **67**, 5315–5334.

Jackson I. N. S. and Rigden S. M. (1998) Composition and temperature of the mantle: seismological models interpreted through experimental studies of mantle minerals. In *The Earth's Mantle: Composition, Structure and Evolution* (ed. I. N. S. Jackson). Cambridge University Press, Cambridge, pp. 405–460.

Jacobsen S. B. and Wasserburg G. J. (1979) The mean age of mantle and crustal reservoirs. *J. Geophys. Res.* **84**, 7411–7427.

Jochum K. P. and Hofmann A. W. (1994) Antimony in mantle-derived rocks: constraints on Earth evolution from moderately siderophile elements. *Min. Mag.* **58A**, 452–453.

Jochum K.-P., Hofmann A. W., Ito E., Seufert H. M., and White W. M. (1983) K, U, and Th in mid-ocean ridge basalt glasses and heat production, K/U, and K/Rb in the mantle. *Nature* **306**, 431–436.

Jochum K. P., Seufert H. M., Spettel B., and Palme H. (1986) The solar system abundances of Nb, Ta, and Y. and the relative abundances of refractory lithophile elements in differentiated planetary bodies. *Geochim. Cosmochim. Acta* **50**, 1173–1183.

Jochum K. P., Hofmann A. W., and Seufert H. M. (1993) Tin in mantle-derived rocks: constraints on Earth evolution. *Geochim. Cosmochim. Acta* **57**, 3585–3595.

Johnson K. T. M., Dick H. J. B., and Shimizu N. (1990) Melting in the oceanic upper mantle: an ion microprobe study of diopsides in abyssal peridotites. *J. Geophys. Res.* **95**, 2661–2678.

Johnson M. C. and Plank T. (2000) Dehydration and melting experiments constrain the fate of subducted sediments. *Geochem. Geophys. Geosys.* **1**, 1999GC000014.

Kellogg L. H., Hager B. H., and van der Hilst R. D. (1999) Compositional stratification in the deep mantle. *Science* **283**, 1881–1884.

Kempton P. D., Pearce J. A., Barry T. L., Fitton J. G., Langmuir C. H., and Christie D. M. (2002) Sr–Nd–Pb–Hf isotope results from ODP Leg 187: evidence for mantle dynamics of the Australian-Antarctic Discordance and origin of the Indian MORB source. *Geochem. Geophys. Geosys.* **3**, 10.29/2002GC000320.

Kimura G., Ludden J. N., Desrochers J. P., and Hori R. (1993) A model of ocean-crust accretion for the superior province, Canada. *Lithos* **30**(3–4), 337–355.

Klein E. M. and Karsten J. L. (1995) Ocean-ridge basalts with convergent-margin geochemical affinities from the Chile Ridge. *Nature* **374**, 52–57.

Klein E. M. and Langmuir C. H. (1987) Global correlations of ocean ridge basalt chemistry with axial depth and crustal thickness. *J Geophys. Res.* **92**, 8089–8115.

Klein E. M., Langmuir C. H., Zindler A., Staudigel H., and Hamelin B. (1988) Isotope evidence of a mantle convection boundary at the Australian–Antarctic discordance. *Nature* **333**, 623–629.

Kleine T., Munker C., Mezger K., and Palme H. (2002) Rapid accretion and early core formation on asteroids and the terrestrial planets from Hf–W chronometry. *Nature* **418**, 952–955.

Kramers J. D. and Tolstikhin I. N. (1997) Two terrestrial lead isotope paradoxes, forward transport modelling, core formation and the history of the continental crust. *Chem. Geol.* **139**(1–4), 75–110.

Krot A. N., Meibom A., Weisberg M. K., and Keil K. (2002) The CR chondrite clan: implications for early solar system processes. *Meteorit. Planet. Sci.* **37**(11), 1451–1490.

Kuhn T. S. (1996) *The Structure of Scientific Revolutions*. University of Chicago Press, Chicago.

Lassiter J. C. (2002) The influence of recycled oceanic crust on the potassium and argon budget of the Earth. *Geochim. Cosmochim. Acta* **66**(15A), A433–A433(suppl.).

Lassiter J. C. and Hauri E. H. (1998) Osmium-isotope variations in Hawaiian lavas: evidence for recycled oceanic lithosphere in the Hawaiian plume. *Earth Planet. Sci. Lett.* **164**, 483–496.

le Roux P. J., le Roex A. P., and Schilling J.-G. (2002) MORB melting processes beneath the southern Mid-Atlantic Ridge (40–55 degrees S): a role for mantle plume-derived pyroxenite. *Contrib. Min. Pet.* **144**, 206–229.

Lee D. C. and Halliday A. N. (1995) Hafnium–tungsten chronometry and the timing of terrestrial core formation. *Nature* **378**, 771–774.

Mahoney J., Nicollet C., and Dupuy C. (1991) Madagascar basalts: tracking oceanic and continental sources. *Earth Planet. Sci. Lett.* **104**, 350–363.

Martin C. E. (1991) Osmium isotopic characteristics of mantle-derived rocks. *Geochim. Cosmochim. Acta* **55**, 1421–1434.

McDonough W. F. (1991) Partial melting of subducted oceanic crust and isolation of its residual eclogitic lithology. *Phil. Trans. Roy. Soc. London A* **335**, 407–418.

McDonough W. F. and Sun S.-S. (1995) The composition of the Earth. *Chem. Geol.* **120**, 223–253.

McDonough W. F., Sun S.-S., Ringwood A. E., Jagoutz E., and Hofmann A. W. (1992) Potassium, rubidium, and cesium in the Earth and Moon and the evolution of the mantle of the Earth. *Geochim. Cosmochim. Acta* **56**, 1001–1012.

McKenzie D. and O'Nions R. K. (1983) Mantle reservoirs and ocean island basalts. *Nature* **301**, 229–231.

Meibom A. and Frei R. (2002a) Evidence for an ancient osmium isotopic reservoir in Earth (vol. 298, p. 516, 2002). *Science* **297**(5584), 1120–1120.

Meibom A. and Frei R. (2002b) Evidence for an ancient osmium isotopic reservoir in Earth. *Science* **296**(5567), 516–518.

Meibom A., Frei R., Chamberlain C. P., Coleman R. G., Hren M., Sleep N. H., and Wooden J. L. (2002a) OS isotopes, deep-rooted mantle plumes and the timing of inner core formation. *Meteorit. Planet. Sci.* **37**(7), A98.

Meibom A., Frei R., Chamberlain C. P., Coleman R. G., Hren M. T., Sleep N. H., and Wooden J. L. (2002b) Os isotopes, deep-rooted mantle plumes and the timing of inner core formation. *Geochim. Cosmochim. Acta* **66**(15A), A504–A504.

Meibom A., Sleep N. H., Chamberlain C. P., Coleman R. G., Frei R., Hren M. T., and Wooden J. L. (2002c) Re–Os isotopic evidence for long-lived heterogeneity and equilibration processes in the Earth's upper mantle. *Nature* **419**(6908), 705–708.

Meibom A., Anderson D. L., Sleep N. H., Frei R., Chamberlain C. P., Hren M. T., and Wooden J. L. (2003) Are high He-3/He-4 ratios in oceanic basalts an indicator of deep-mantle plume components? *Earth Planet. Sci. Lett.* **208**(3–4), 197–204.

Miller D. M., Goldstein S. L., and Langmuir C. H. (1994) Cerium/lead and lead isotope ratios in arc magmas and the enrichment of lead in the continents. *Nature* **368**, 514–520.

Milner S. C. and le Roex A. P. (1996) Isotope characteristics of the Okenyena igneous complex, northwestern Namibia: constraints on the composition of the early Tristan plume and the origin of the EM 1 mantle component. *Earth Planet. Sci. Lett.* **141**, 277–291.

Montelli R., Nolet G., Masters G., Dahlen F. A., and Hung S.-H. (2003) Global P and PP traveltime tomography: rays versus waves. *Geophys. J. Int.* **142** (in press).

Morgan W. J. (1971) Convection plumes in the lower mantle. *Nature* **230**, 42–43.

Morse S. A. (1983) Strontium isotope fractionation in the Kiglapait intrusion. *Science* **220**, 193–195.

Murphy D. T., Kamber B. S., and Collerson K. D. (2003) A refined solution to the first terrestrial Pb-isotope paradox. *J. Petrol.* **44**, 39–53.

O'Hara M. J. (1973) Non-primary magmas and dubious mantle plume beneath Iceland. *Nature* **243**(5409), 507–508.

O'Hara M. J. (1975) Is there an icelandic mantle plume. *Nature* **253**(5494), 708–710.

O'Hara M. J. (1977) Open system crystal fractionation and incompatible element variation in basalts. *Nature* **268**, 36–38.

O'Hara M. J. and Mathews R. E. (1981) Geochemical evolution in an advancing, periodically replenished, periodically tapped, continuously fractionated magma chamber. *J. Geol. Soc. London* **138**, 237–277.

O'Nions R. K. and Pankhurst R. J. (1974) Petrogenetic significance of isotope and trace element variation in volcanic rocks from the Mid-Atlantic. *J. Petrol.* **15**, 603–634.

O'Nions R. K., Evensen N. M., and Hamilton P. J. (1977) Variations in $^{143}Nd/^{144}Nd$ and $^{87}Sr/^{86}Sr$ ratios in oceanic basalts. *Earth Planet. Sci. Lett.* **34**, 13–22.

O'Nions R. K., Evensen N. M., and Hamilton P. J. (1979) Geochemical modeling of mantle differentiation and crustal growth. *J. Geophys. Res.* **84**, 6091–6101.

Patchett P. J., White W. M., Feldmann H., Kielinczuk S., and Hofmann A. W. (1984) Hafnium/rare earth fractionation in the sedimentary system and crust–mantle recycling. *Earth Planet. Sci. Lett.* **69**, 365–378.

Patterson C. C. (1956) Age of meteorites and the Earth. *Geochim. Cosmochim. Acta* **10**, 230–237.

Peucker-Ehrenbrink B., Hofmann A. W., and Hart S. R. (1994) Hydrothermal lead transfer from mantle to continental crust: the role of metalliferous sediments. *Earth Planet. Sci. Lett.* **125**, 129–142.

Phipps Morgan J. (1999) Isotope topology of individual hotspot basalt arrays: mixing curves of melt extraction trajectories? *Geochem. Geophys. Geosystems* **1**, 1999GC000004.

Phipps Morgan J., Morgan W. J., and Zhang Y.-S. (1995) Observational hints for a plume-fed, suboceanic asthenosphere and its role in mantle convection. *J. Geophys. Res.* **100**, 12753–12767.

Phipps Morgan J. and Morgan W. J. (1998) Two-stage melting and the geochemical evolution of the mantle: a recipe for mantle plum-pudding. *Earth Planet. Sci. Lett.* **170**, 215–239.

Plank T. and Langmuir C. H. (1993) Tracing trace elements from sediment input to volcanic output at subduction zones. *Nature* **362**, 739–742.

Plank T. and Langmuir C. H. (1998) The chemical composition of subducting sediment and its consequences for the crust and mantle. *Chem. Geol.* **145**, 325–394.

Porcelli D. and Halliday A. N. (2001) The core as a possible source of mantle helium. *Earth Planet. Sci. Lett.* **192**(1), 45–56.

Puchtel I. S., Hofmann A. W., Mezger K., Jochum K. P., Shchipansky A. A., and Samsonov A. V. (1998) Oceanic plateau model for continental crustal growth in the Archaean: a case study from the Kostomuksha greenstone belt, NW Baltic Shield. *Earth Planet. Sci. Lett.* **155**, 57–74.

Pyle D. G., Christie D. M., and Mahoney J. J. (1992) Resolving an isotopic boundary within the Australian–Antarctic Discordance. *Earth Planet. Sci. Lett.* **112**, 161–178.

Regelous M., Hofmann A. W., Abouchami W., and Galer J. S. G. (2002/3) Geochemistry of lavas from the Emperor Seamounts, and the geochemical evolution of Hawaiian magmatism 85–42 Ma. *J. Petrol.* **44**, 113–140.

Rehkämper M. and Hofmann A. W. (1997) Recycled ocean crust and sediment in Indian Ocean MORB. *Earth Planet. Sci. Lett.* **147**, 93–106.

Rehkämper M. and Halliday A. N. (1998) Accuracy and long-term reproducibility of Pb isotopic measurements by MC-ICPMS using an external method for the correction of mass discrimination. *Int. J. Mass Spectrom. Ion Processes* **181**, 123–133.

Richard P., Shimizu N., and Allègre C. J. (1976) $^{143}Nd/^{146}Nd$—a natural tracer: an application to oceanic basalt. *Earth Planet. Sci. Lett.* **31**, 269–278.

Roy-Barman M. and Allègre C. J. (1994) $^{187}Os/^{186}Os$ ratios of mid-ocean ridge basalts and abyssal peridotites. *Geochim. Cosmochim. Acta* **58**, 5053–5054.

Rudnick R. L. and Fountain D. M. (1995) Nature and composition of the continental crust: a lower crustal perspective. *Rev. Geophys.* **33**, 267–309.

Rudnick R. L. and Goldstein S. L. (1990) The Pb isotopic compositions of lower crustal xenoliths and the evolution of lower crustal Pb. *Earth Planet. Sci. Lett.* **98**, 192–207.

Rudnick R. L., McDonough W. F., and O'Connell R. J. (1998) Thermal structure, thickness and composition of continental lithosphere. *Chem. Geol.* **145**, 395–411.

Rudnick R. L., Barth M., Horn I., and McDonough W. F. (2000) Rutile-bearing refractory eclogites: missing link between continents and depleted mantle. *Science* **287**, 278–281.

Saal A. E., Hart S. R., Shimizu N., Hauri E. H., and Layne G. D. (1998) Pb isotopic variability in melt inclusions from oceanic island basalts, Polynesia. *Science* **282**, 1481–1484.

Saunders A. D., Norry M. J., and Tarney J. (1988) Origin of MORB and chemically-depleted mantle reservoirs: trace element constraints. *J. Petrol.* (Special Lithosphere Issue), 415–445.

Schiano P., Birck J.-L., and Allègre C. J. (1997) Osmium–strontium–neodymium–lead isotopic covariations in mid-ocean ridge basalt glasses and the heterogeneity of the upper mantle. *Earth Planet. Sci. Lett.* **150**, 363–379.

Schilling J.-G. (1973) Iceland mantle plume: geochemical evidence along Reykjanes Ridge. *Nature* **242**, 565–571.

Schilling J. G. and Winchester J. W. (1967) Rare-earth fractionation and magmatic processes. In *Mantles of Earth and Terrestrial Planets* (ed. S. K. Runcorn). Interscience Publ., London, pp. 267–283.

Schilling J. G. and Winschester J. W. (1969) Rare earth contribution to the origin of Hawaiian lavas. *Contrib. Mineral. Petrol.* **40**, 231.

Shaw D. M. (1970) Trace element fractionation during anatexis. *Geochim. Cosmochim. Acta* **34**, 237–242.

Shimizu N., Sobolev A. V., Layne G. D., and Tsameryan O. P. (2003) Large Pb isotope variations in olivine-hosted melt inclusions in a basalt fromt the Mid-Atlantic Ridge. *Science* (ms. submitted May 2003).

Sims K. W. W. and DePaolo D. J. (1997) Inferences about mantle magma sources from incompatible element concentration ratios in oceanic basalts. *Geochim. Cosmochim. Acta* **61**, 765–784.

Sleep N. H. (1984) Tapping of magmas from ubiquitous mantle heterogeneities: an alternative to mantle plumes? *J. Geophys. Res.* **89**, 10029–10041.

Snow J. E. and Reisberg L. (1995) Os isotopic systematics of the MORB mantle: results from altered abyssal peridotites. *Earth Planet. Sci. Lett.* **133**, 411–421.

Sobolev A. V. (1996) Melt inclusions in minerals as a source of principal petrological information. *Petrology* **4**, 209–220.

Sobolev A. V. and Shimizu N. (1993) Ultra-depleted primary melt included in an olivine from the Mid-Atlantic Ridge. *Nature* **363**, 151–154.

Sobolev A. V., Hofmann A. W., and Nikogosian I. K. (2000) Recycled oceanic crust observed in "ghost plagioclase" within the source of Mauna Loa lavas. *Nature* **404**, 986–990.

Staudigel H., Davies G. R., Hart S. R., Marchant K. M., and Smith B. M. (1995) Large scale isotopic Sr, Nd and O isotopic anatomy of altered oceanic crust: DSDP/ODP sites 417/418. *Earth Planet. Sci. Lett.* **130**, 169–185.

Stein M. and Hofmann A. W. (1994) Mantle plumes and episodic crustal growth. *Nature* **372**, 63–68.

Su Y. J. (2002) Mid-ocean ridge basalt trace element systematics: constraints from database management, ICP-MS analyses, global data compilation and petrologic modeling. PhD Thesis, Columbia University, 2002, 472pp.

Sun S.-S. and McDonough W. F. (1989) Chemical and isotopic systematics of oceanic basalts: implications for mantle composition and processes. In *Magmatism in the Ocean Basins*, Geological Society Spec. Publ. 42 (eds. A. D. Saunders and M. J. Norry). Oxford, pp. 313–345.

Tackley P. J. (2000) Mantle convection and plate tectonics: toward an integrated physical and chemical theory. *Science* **288**, 2002–2007.

Tackley P. J. (2002) Strong heterogeneity caused by deep mantle layering. *Geochem. Geophys. Geosystems* **3**(4), 101029/2001GC000167.

Tatsumoto M., Hedge C. E., and Engel A. E. J. (1965) Potassium, rubidium, strontium, thorium, uranium, and the ratio of strontium-87 to strontium-86 in oceanic tholeiitic basalt. *Science* **150**, 886–888.

Taylor S. R. and McLennan S. M. (1985) *The Continental Crust: its Composition and Evolution*. Blackwell, Oxford.

Taylor S. R. and McLennan S. M. (1995) The geochemical evolution of the continental crust. *Rev. Geophys.* **33**, 241–265.

Taylor S. R. and McLennan S. M. (2001) Chemical composition and element distribution in the Earth's crust. In *Encyclopedia of Physical Sciences and Technology*, Academic Press, vol. 2, pp. 697–719.

Thompson R. N., Morrison M. A., Hendry G. L., and Parry S. J. (1984) An assessment of the relative roles of crust and mantle in magma genesis: an elemental approach. *Phil. Trans. Roy. Soc. London* **A310**, 549–590.

Todt W., Cliff R. A., Hanser A., and Hofmann A. W. (1996) Evaluation of a ^{202}Pb–^{205}Pb double spike for high-precision lead isotope analysis. In *Earth Processes: Reading the Isotopic Code* (eds. A. Basu and S. R. Hart). Am. Geophys. Union, Geophys Monograph 95, Washington, pp. 429–437.

Tolstikhin I. N. and Hofmann A. W. (2002) Generation of a long-lived primitive mantle reservoir during late stages of Earth accretion. *Geochim. Cosmochim. Acta* **66**(15A), A779–A779.

van der Hilst R. D., Widiyantoro S., and Engdahl E. R. (1997) Evidence for deep mantle circulation from global tomography. *Nature* **386**, 578–584.

van Keken P. E. and Ballentine C. J. (1998) Whole-mantle versus layered mantle convection and the role of a high-viscosity lower mantle in terrestrial volatile evolution. *Earth Planet. Sci. Lett.* **156**, 19–32.

van Keken P. E., Ballentine C. J., and Porcelli D. (2001) A dynamical investigation of the heat and helium imbalance. *Earth Planet. Sci. Lett.* **188**(3–4), 421–434.

Van Orman J. A., Grove T. L., and Shimizu N. (2001) Rare earth element diffusion in diopsde: influence of temperature, pressure and ionic radius, and an elastic model for diffusion in silicates. *Contrib. Mineral. Petrol.* **141**, 687–703.

Völkening J., Walczyk T., and Heumann K. G. (1991) Osmium isotope ratio determinations by negative thermal ionization mass spectrometry. *Int. J. Mass Spectrom. Ion Process.* **105**, 147–159.

Vollmer R. (1976) Rb–Sr and U–Th–Pb systematics of alkaline rocks: the alkaline rocks from Italy. *Geochim. Cosmochim. Acta* **40**, 283–295.

Wade J. and Wood B. J. (2001) The Earth's "missing" niobium may be in the core. *Nature* **409**, 75–78.

Wasserburg G. J. and Depaolo D. J. (1979) Models of Earth structure inferred from neodymium and strontium isotopic abundances. *Proc. Natl. Acad. Sci. USA* **76**(8), 3594–3598.

Weaver B. L. (1991) The origin of ocean island basalt end-member compositions: trace element and isotopic constraints. *Earth Planet. Sci. Lett.* **104**, 381–397.

Wedepohl K. H. (1995) The composition of the continental crust. *Geochim. Cosmochim. Acta* **59**(7), 1217–1232.

White W. M. (1985) Sources of oceanic basalts: radiogenic isotope evidence. *Geology* **13**, 115–118.

White W. M. (1993) ^{238}U/^{204}Pb in MORB and open system evolution of the depleted mantle. *Earth Planet. Sci. Lett.* **115**, 211–226.

White W. M. and Duncan R. A. (1996) Geochemistry and geochronology of the Society Islands: new evidence for deep mantle recycling. In *Earth Processes: Reading the Isotopic Code*. (eds. A. Basu and S. R. Hart). Am. Geophys. Union, Geophys. Monograph 95, Washington, pp. 183–206.

White W. M. and Hofmann A. W. (1982) Sr and Nd isotope geochemistry of oceanic basalts and mantle evolution. *Nature* **296**, 821–825.

White W. M. and Schilling J.-G. (1978) The nature and origin of geochemical variation in Mid-Atlantic Ridge basalts from the central North Atlantic. *Geochim. Cosmochim. Acta* **42**, 1501–1516.

Widom E. and Shirey S. B. (1996) Os isotope systematics in the Azores: implications for mantle plume sources. *Earth Planet. Sci. Lett.* **142**, 451–465.

Wood B. J. and Blundy J. D. (1997) A predictive model for rare earth element partitioning between clinopyroxene and anhydrous silicate melt. *Contrib. Mineral. Petrol.* **129**, 166–181.

Wood D. A. (1979) A variably veined suboceanic mantle—genetic significance for mid-ocean ridge basalts from geochemical evidence. *Geology* **7**, 499–503.

Woodhead J. (1996) Extreme HIMU in an oceanic setting: the geochemistry of Mangaia Island (Polynesia), and temporal evolution of the Cook-Austral hotspot. *J. Volcanol. Geotherm. Res.* **72**, 1–19.

Woodhead J. D. and Devey C. W. (1993) Geochemistry of the Pitcairn seamounts: I. Source character and temporal trends. *Earth Planet. Sci. Lett.* **116**, 81–99.

Workman R. K., Hart S. R., Blusztajn J., Jackson M., Kurz M., and Staudigel H. (2003) Enriched mantle: II. A new view from the Samoan hotspot. *Geochem. Geophys. Geosys.* (submitted).

Yang H. J., Frey F. A., and Clague D. A. (2003) Constraints on the source components of lavas forming the Hawaiian north arch and honolulu volcanics. *J. Petrol.* **44**, 603–627.

Yaxley G. M. and Green D. H. (1998) Reactions between eclogite and peridotite: mantle refertilisation by subduction of oceanic crust. *Schweiz. Mineral. Petrogr. Mitt.* **78**, 243–255.

Yin Q., Jacobsen S. B., Yamashita K., Blichert-Toft J., Télouk P., and Albarède F. (2002) A short timescale for terrestrial planet formation from Hf–W chronometry of meteorites. *Nature* **418**, 949–952.

Zartman R. E. and Haines S. M. (1988) The plumbotectonic model for Pb isotopic systematics among major terrestrial reservoirs—a case for bi-directional transport. *Geochim. Cosmochim. Acta* **52**, 1327–1339.

Zindler A. and Hart S. (1986) Chemical geodynamics. *Ann. Rev. Earth Planet. Sci.* **14**, 493–571.

2.04
Orogenic, Ophiolitic, and Abyssal Peridotites

J.-L. Bodinier and M. Godard
CNRS and Université de Montpellier 2, France

2.04.1 INTRODUCTION

"Tectonically emplaced" mantle rocks include subcontinental, suboceanic, and subarc mantle rocks that were tectonically exhumed from the upper mantle and occur:

(i) as dispersed ultramafic bodies, a few meters to kilometers in size, in suture zones and mountain belts (i.e., the "alpine," or "orogenic" peridotite massifs—De Roever (1957), Thayer (1960), Den Tex (1969));

(ii) as the lower ultramafic section of large (tens of kilometers) ophiolite or island arc complexes, obducted on continental margins (e.g., the Oman Ophiolite and the Kohistan Arc Complex—Coleman (1971), Boudier and Coleman (1981), Burg *et al.* (1998));

(iii) exhumed above the sea level in ocean basins (e.g., Zabargad Island in the Red Sea, St. Paul's islets in the Atlantic and Macquarie Island in the southwestern Pacific—Tilley (1947), Melson *et al.* (1967), Varne and Rubenach (1972), Bonatti *et al.* (1981)).

The "abyssal peridotites" are samples from the oceanic mantle that were dredged on the ocean floor, or recovered from drill cores (e.g., Bonatti *et al.*, 1974; Prinz *et al.*, 1976; Hamlyn and Bonatti, 1980).

Altogether, tectonically emplaced and abyssal mantle rocks provide insights into upper mantle compositions and processes that are complementary to the information conveyed by mantle xenoliths (See Chapter 2.05). They provide coverage to vast regions of the Earth's upper mantle that are sparsely sampled by mantle xenoliths, particularly in the ocean basins and beneath passive continental margins, back-arc basins, and oceanic island arcs.

Compared with mantle xenoliths, a disadvantage of some tectonically emplaced mantle rocks for representing mantle compositions is that their original geodynamic setting is not exactly known and their significance is sometimes a subject of speculation. For instance, the provenance of orogenic lherzolite massifs (subcontinental lithosphere versus upwelling asthenosphere) is still debated (Menzies and Dupuy, 1991, and references herein), as is the original setting of ophiolites (mid-ocean ridges versus supra-subduction settings—e.g., Nicolas, 1989). In addition, the mantle structures and mineralogical compositions of tectonically emplaced mantle rocks may be obscured by deformation and metamorphic recrystallization during shallow upwelling, exhumation, and tectonic emplacement. Metamorphic processes range from high-temperature recrystallization in the stability field of plagioclase peridotites (Rampone *et al.*, 1993) to complete serpentinization (e.g., Burkhard and O'Neill, 1988). Some garnet peridotites record even more complex evolutions. They were first buried to, at least, the stability field of garnet peridotites, and, in some cases to greater than 150 km depths (Dobrzhinetskaya *et al.*, 1996; Green *et al.*, 1997; Liou, 1999). Then, they were exhumed to the surface, dragged by buoyant crustal rocks (Brueckner and Medaris, 2000).

Alternatively, several peridotite massifs are sufficiently well preserved to allow the observation of structural relationships between mantle lithologies that are larger than the sampling scale of mantle xenoliths. It is possible in these massifs to evaluate the scale of mantle heterogeneities and the relative timing of mantle processes such as vein injection, melt–rock reaction, deformation, etc... Detailed studies of orogenic and ophiolitic peridotites on centimeter- to kilometer-scale provide invaluable insights into melt transfer mechanisms, such as melt flow in lithospheric vein conduits and wall–rock reactions (Bodinier *et al.*, 1990), melt extraction from mantle sources via channeled porous flow (Kelemen *et al.*, 1995) or propagation of kilometer-scale melting fronts associated with thermal erosion of lithospheric mantle (Lenoir *et al.*, 2001). In contrast, mantle xenoliths may be used to infer either much smaller- or much larger-scale mantle heterogeneities, such as micro-inclusions in minerals (Schiano and Clocchiatti, 1994) or lateral variations between lithospheric provinces (O'Reilly *et al.*, 2001).

The abyssal peridotites are generally strongly affected by oceanic hydrothermal alteration. Most often, their whole-rock compositions are strongly modified and cannot be used straightforwardly to assess mantle compositions (e.g., Baker and Beckett, 1999). However, even in the worst cases the samples generally contain fresh, relic minerals (mainly clinopyroxene) that represent the only available direct information on the oceanic upper mantle in large ocean basins, away from hot-spot volcanic centers. *In situ* trace-element data on clinopyroxenes from abyssal peridotites provide constraints on melting processes at mid-ocean ridges (Johnson *et al.*, 1990).

In this chapter, we review the main inferences on upper mantle composition and heterogeneity that may be drawn from geochemical analyses of the major elements, lithophile trace elements, and Nd–Sr isotopes in tectonically emplaced and abyssal mantle rocks. In addition we emphasize important insights into the mechanisms of melt/fluid transfer that can be deduced from detailed studies of these mantle materials.

2.04.2 TYPES, DISTRIBUTION, AND PROVENANCE

The idea that the "alpine," or "orogenic," peridotites distributed along mountain belts (Benson, 1926; Thayer, 1960, 1967) represent tectonically emplaced mantle rocks rather than crystallized ultramafic magmas or olivine-rich cumulates emerged at the end of the fifties (De Roever, 1957; Hess, 1962; Dietz, 1963). About the same time, a mantle origin was also proposed for the oceanic serpentinized peridotites, mainly known from fault scarps and St. Paul's islets on the Mid-Atlantic Ridge (Hess, 1964). Together with the recognition of the ultramafic composition of the Earth's mantle on the basis of meteorite studies and geophysical constraints (Ringwood, 1966; Harris *et al.*, 1967), these early suggestions have promoted further studies of mantle rock occurrences. Pioneering studies were mainly devoted to unraveling the mechanisms of exhumation and solid emplacement of peridotites and serpentinites (Avé Lallemant, 1967; Jahns, 1967; Lappin, 1967; Ragan, 1967; Den Tex, 1969), and to propose classifications of ultramafic rocks based on their field association, tectonic environment, and metamorphic evolution (O'Hara, 1967a; Wyllie, 1967, 1969).

The principal types of tectonically emplaced mantle rocks were recognized in this period, ranging from the high-pressure garnet peridotites associated with eclogites and granulitic terranes (O'Hara, 1967b; Lappin, 1967)—the orogenic "root-zone" peridotites of Den Tex (1969)—to the peridotites from ophiolite mafic–ultramafic associations (Miyashiro, 1966; Thayer, 1967; Sørensen, 1967). However, the classification of mantle occurrences has slightly evolved since this time, partly because of our better understanding of the significance and origin of mantle rocks. Before Den Tex (1969), the terms "alpine" and "orogenic" peridotites had a broad sense and included almost all types of tectonically emplaced peridotites in mountain chains. Nowadays, these terms are more readily restricted to the lherzolite massifs unrelated with ophiolitic ocean crust (Menzies and Dupuy, 1991). In this respect, the current terminology diverges from the distinction made by Den Tex. The well-known Lherz peridotite body, in the French Pyrenees, is now viewed as a typical "orogenic lherzolite" massif, considered to represent sub-continental lithospheric mantle (Fabriès *et al.*, 1991), whereas it was classified by Den Tex as an "ophiolitic peridotite" (= "truly alpine-type peridotite").

We shall hereafter follow the current terminology. It is convenient to distinguish three main types of mantle occurrences and to examine them in the following sequence: (i) orogenic peridotite massifs, (ii) ophiolitic mantle rocks, and (iii) oceanic peridotites.

2.04.2.1 Orogenic Peridotite Massifs

With few exceptions, orogenic peridotite massifs are characterized by the predominance of lherzolites (Figure 1) equilibrated in any of the garnet-, spinel-, or plagioclase-peridotite facies defined by O'Hara (1967a). Exceptions include the Finero peridotite, in the Italian Alps (Figure 2), which is predominantly composed of amphibole and phlogopite harzburgites (Vogt, 1962; Cawthorn, 1975). However, Finero is otherwise comparable to the orogenic peridotites of the Ivrea Zone in terms of rock association and tectonic setting (Lensch, 1968; Ernst, 1981). The orogenic peridotites vary in size from small blocks (1–10 m) embedded in metamorphic sediments (some of the Pyrenean lherzolites, for example—see table 1 in Fabriès *et al.*, 1991) to large ultramafic exposures (up to 300 km^2) forming the main body of overthrusted tectonic units (e.g., the Ronda massif, southern Spain—Tubia and Cuevas, 1986). However, the vast majority of the lherzolite massifs are intermediate in size between these extremes. In several instances, they occur as strings of several peridotite bodies, a few hundred meters to a few kilometers in size, aligned along thrust sutures or major faults.

The orogenic peridotite massifs normally differ from ophiolitic peridotites in being unrelated to oceanic rock associations. However, some lherzolite massifs contain shallow intrusive rocks interpreted as embryonic ocean crust and are, therefore, considered to be intermediate between true orogenic peridotite massifs and the "lherzolite sub-type" of the ophiolites. The Lanzo plagioclase lherzolite, in the Italian Alps (Figure 2), is an example of such transitional lherzolite massifs (Bodinier *et al.*, 1986, 1991). The orogenic massifs are generally associated with platform sediments or continental rocks recording extreme metamorphic conditions, from high temperature at intermediate or low pressure to high or ultrahigh pressure at variable temperature conditions. Examples of high-temperature metamorphism are observed in the country rocks of the Beni Bousera (northern Morocco) and Ronda (southern Spain) massifs (Loomis, 1972), as well as in the limestones hosting the Pyrenean lherzolites (Goldberg and Leyreloup, 1990). However, several orogenic peridotites from the Alps and the Scandinavian Caledonides (Figure 2) are associated with high- to ultrahigh-pressure terranes (Ernst, 1981; Terry *et al.*, 2000).

The lherzolites of the orogenic massifs are mostly fertile ($\geq$10% clinopyroxene, Figure 1). Subordinate rock types include refractory peridotites (cpx-poor lherzolites, harzburgites, and/or dunites) and mafic layers—a few centimeters to several meters thick—composed of mafic granulites, garnet, spinel, and/or plagioclase clinopyroxenites and websterites. Refractory peridotites may be volumetrically significant (up to 30–40%) while the mafic layers generally do not exceed 5%. Several orogenic peridotites contain accessory lithologies resulting from the segregation and migration of partial melts, or from interaction between mantle rocks and melt/fluids (e.g., Boudier and Nicolas, 1977; Obata and Nagahara, 1987; Bodinier *et al.*, 1988, 1990). Rock types related to melt/fluid processes include

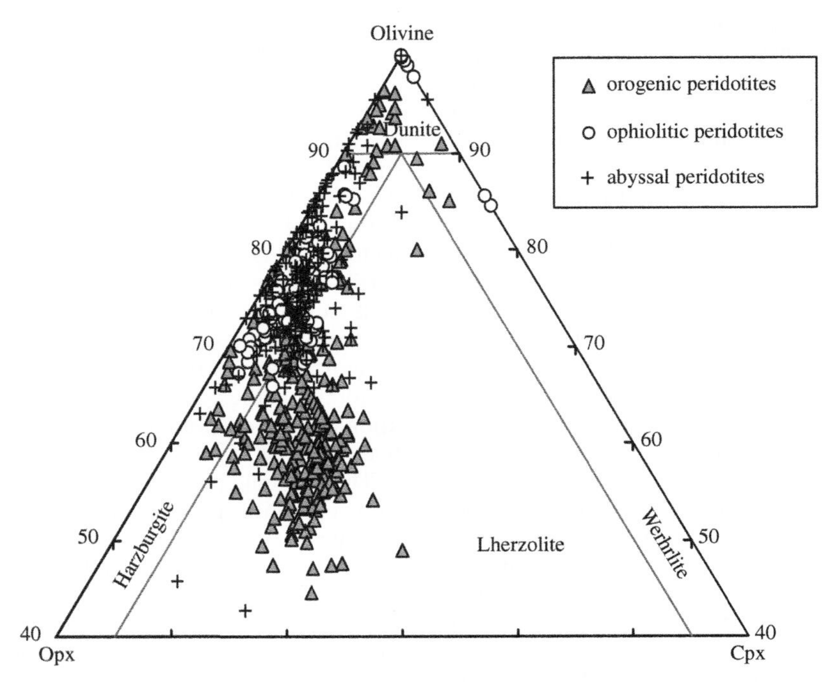

Figure 1 Modal compositions of orogenic, ophiolitic, and abyssal peridotites plotted onto the "peridotite" field of the olivine–orthopyroxene–clinopyroxene diagram (Streckeisen, 1976). The orogenic peridotites (336 samples) are from the Ronda, Pyrenees, and Lanzo massifs (Figure 2); the ophiolitic peridotites (103 samples) are from the Semail ophiolite complex (Figure 3); the abyssal peridotites (185 samples) are from the East Pacific Rise, the Mid-Atlantic Ridge, the American Antarctic Ridge, the Southwest Indian Ridge, the Mariana Trough, the Izu–Bonin–Mariana Forearc, and the South Sandwich Arc-Basin system (Figure 4). Modal compositions for orogenic peridotites and the Semail ophiolitic peridotites were calculated by total inversion of major-element data for whole rocks and minerals. Whole-rock data are mainly from Frey *et al.* (1985), Bodinier (1988 (reproduced by permission of Elsevier from *Tectonophysics* **1988**, *149*, 67–88), 1989), Bodinier *et al.* (1988), Remaïdi (1993), Van der Wal and Bodinier (1996), Fabriès *et al.* (1998), Godard *et al.* (2000) (reproduced by permission of Elsevier from *Earth Planet. Sci. Lett.* **2000**, *180*, 133–148), Lenoir *et al.* (2001), and Gerbert-Gaillard (2002). Mineral compositions were taken from Conquéré and Fabriès (1984), Bodinier *et al.* (1986), Lenoir *et al.* (2001), and from unpublished PhD theses (Boudier, 1976; Conquéré, 1978; Remaïdi, 1993; Garrido, 1995; Lenoir, 2000; Gerbert-Gaillard, 2002). Lherzolites showing variable degrees of re-equilibration in the plagioclase stability field (Lanzo) were recasted as "primary" spinel lherzolites by using appropriate mineral compositions for the mass balance. Modal compositions for abyssal peridotites were obtained by point counting (Dick *et al.*, 1984; Michael and Bonatti, 1985; Johnson and Dick, 1992; Niu and Hékinian, 1997a; Parkinson and Pearce, 1998; Pearce *et al.*, 2000; Ohara *et al.*, 2002).

a variety of dikes, veins and veinlets representing magma conduits, now filled with garnet and/or amphibole pyroxenites, hornblendites, glimmerites (mica-clinopyroxene), orthopyroxenites or gabbros, as well as diffuse facies of pyroxene-rich or amphibole- (±phlogopite-) bearing peridotites resulting from interactions between peridotites and percolating melt/fluids. Some peridotite bodies were almost entirely metasomatized by melts/fluids of mantle or crustal origin. These include, for instance, the Caussou clinopyroxene-rich (±amphibole) peridotite in the eastern Pyrenees (Fabriès *et al.*, 1989), the Finero amphibole (±phlogopite) peridotite in the western Alps (Vogt, 1962; Cawthorn, 1975) and the Ulten amphibole and garnet peridotite in the eastern Alps (Rampone and Morten, 2001).

All lherzolite bodies were deformed at high temperature (≥1,000 °C), as shown by isoclinal folding of mafic layers and systematic olivine lattice orientations (Avé Lallemant, 1967; Nicolas, 1986a). The metasomatic features may partly overprint the deformation. In Lherz, for instance, the amphibole-pyroxenite veins crosscut the mafic layering (Wilshire *et al.*, 1980) and are incompletely transposed in the peridotite foliation (Vétil *et al.*, 1988; Fabriès *et al.*, 1991).

The orogenic peridotites were generally altered by serpentinization during or after their exhumation. The serpentinization degree is extremely variable. Among the 36 Pyrenean peridotite bodies described by Fabriès *et al.* (1991, table 1), 18 show only incipient or weak serpentinization and 12 are strongly to completely serpentinized. The six remaining bodies are intermediate. Similar variations are observed within large lherzolite bodies such as Lanzo and Ronda, where the serpentinization is almost complete

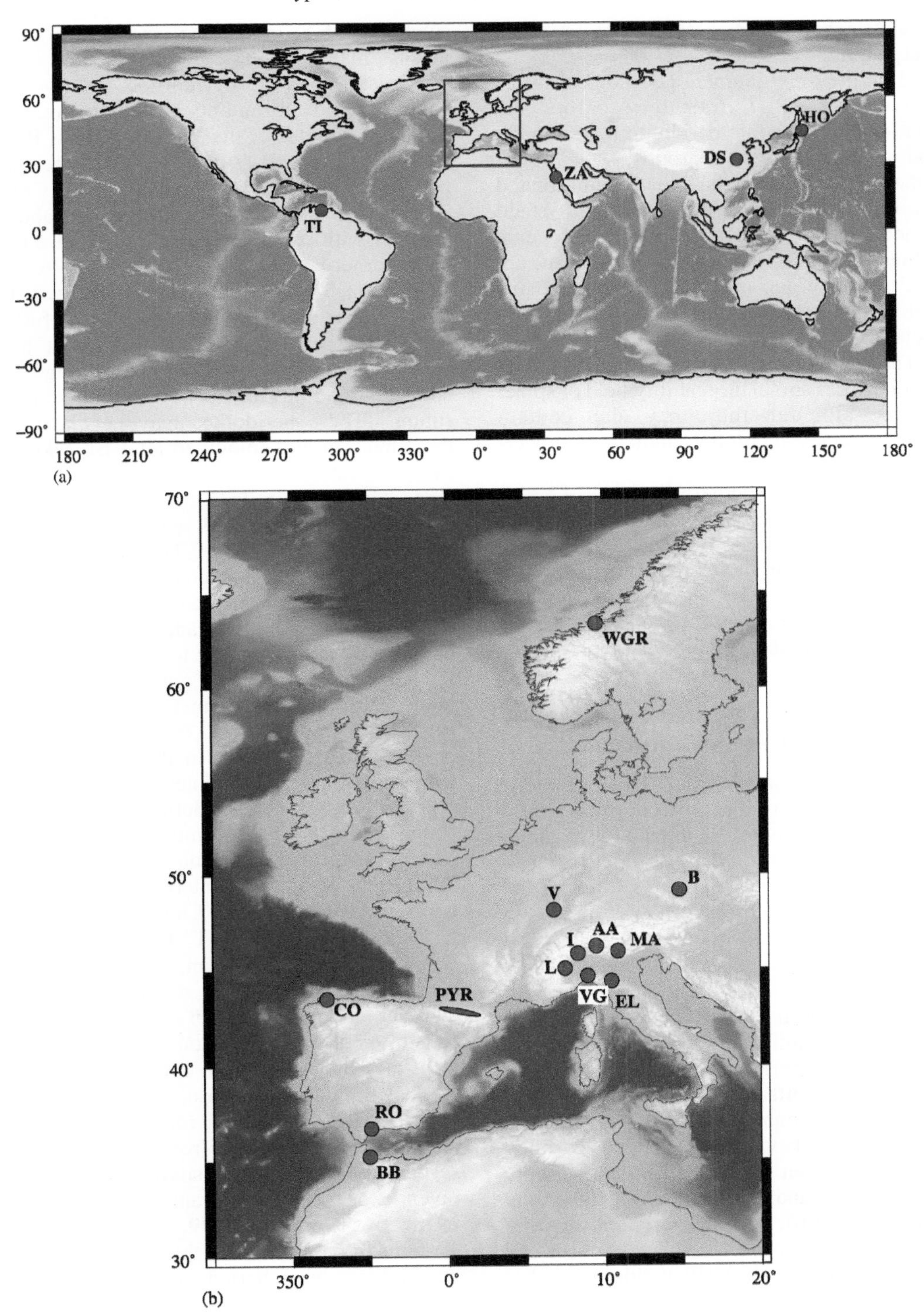

Figure 2 Location of the main orogenic peridotite massifs mentioned in the text: (a) TI—Tinaquillo, northern Venezuela; HO—Horoman, Hokkaido, northern Japan; DS—Dabie Shan garnet peridotites, southwestern China; ZA—Zabargad, Red Sea, Egypt. Inset shows the area enlarged in (b). (b) WGR—garnet peridotites of the Western Gneiss Region, southern Norway; B—Bohemian garnet peridotites, lower Austria; V—Vosges garnet peridotites, eastern France; AA—Alpe Arami, Central Alps, southern Switzerland; M—Val Malenco, Eastern Central Alps, northern Italy; I—Ivrea Zone (Finero, Balmuccia, and Baldissero peridotites), Western Alps, northwestern Italy; L—Lanzo, Western Alps, northwestern Italy; VG—Voltri Group (Erro-Tobbio peridotite), Western Liguria, northwestern Italy; EL—External Ligurides peridotites, Eastern Liguria, northwestern Italy; PYR—Pyrenees, southern France; RO—Ronda, Betic Cordilleras, southern Spain; BB—Beni Bousera, Rif Mountains, northern Morocco; CO—Cabo Ortegal, northwestern Spain.

along faults while some areas contain less than 10% serpentine.

Based on their predominant petrological facies and/or on their $P-T$ trajectory before the exhumation, the orogenic peridotite massifs may be subdivided in three main groups:

(i) the "HP/UHP massifs" chiefly equilibrated at high or ultrahigh pressure, in the stability field of garnet peridotites. The Hokkaido orogenic peridotites, Japan (Horoman and Nikanbetsu) are included in this group in spite of their extensive re-equilibration in the stability fields of spinel and plagioclase.

(ii) The "IP massifs" mostly equilibrated at intermediate pressure in the stability field of spinel peridotites, although they may also contain subordinate facies recording higher- or lower-pressure stages (e.g., Beni Bousera, northern Morocco, and Ronda, southern Spain).

(iii) The "LP massifs," characterized by exhumation at very shallow level in extensional setting, culminating in the tectonic denudation of serpentinized mantle peridotites on the seafloor. The LP massifs often show incipient to complete recrystallization in the stability field of plagioclase peridotites.

2.04.2.1.1 HP/UHP orogenic peridotites

"True" garnet peridotites. These include high-pressure (HP) and ultrahigh-pressure (UHP) ultramafic bodies dominated by garnet peridotites. These rocks have attracted much attention recently, as a potential source of information on relatively deep mantle processes and subduction mechanisms (e.g., Green *et al.*, 1997; Liou, 1999; Brueckner and Medaris, 2000; van Roermund *et al.*, 2001). However, their significance and origin is still a matter of debate. The very high equilibrium pressure estimated for some UHP peridotites (10–24 GPa for Alpe Arami, central Alps, for instance, implying a provenance from 300–650 km deep—Dobrzhinetskaya *et al.*, 1996; Bozhilov *et al.*, 1999; Green *et al.*, 1997) is strongly disputed (Nimis and Trommsdorff, 2001, and references herein). Moreover, contrasting interpretations have been formulated concerning the derivation of the HP/UHP peridotites either from subducted material, cratonic lithosphere, or deep-seated asthenospheric diapirs.

Brueckner and Medaris (2000) reviewed the main occurrences of orogenic garnet peridotites and pointed to their diversity, even within individual HP/UHP terranes. They attributed these differences to different initial environments and different evolutions during the tectonic cycles that ultimately exposed them at the surface, and proposed a classification based on these criteria. The garnet peridotites are first subdivided into two broad categories: "crustal" versus "mantle" types.

"Crustal" garnet peridotites mostly derive from layered ultramafic cumulates (e.g., the Fe–Ti peridotites of the Western Gneiss Region, Norway—Carswell *et al.* (1983)—or the type B peridotites from the Dabie Shan terrane, China—Zhang and Liou (1998)) that were subducted together with continental host rocks and acquired their garnet-bearing assemblages as a result of subduction. These peridotites do not represent true mantle rocks and will not be considered further.

"Mantle" garnet peridotites represent mantle rocks exhumed to the surface along active continental margins or continent–continent sutures. These peridotites may be classified in three main groups (modified after Brueckner and Medaris, 2000):

(i) The "prograde" and "HP–HT" types include garnet peridotites derived from a supra-subduction mantle wedge. They are entrained to the surface by buoyant continental rocks forcefully underthrust in the mantle during continent–continent collision. Prograde peridotites intrude the crustal slab at shallow to intermediate depths (20–50 km) and acquire garnet-bearing assemblages through prograde metamorphism. HP–HT peridotites intrude from the mantle wedge at greater depth (>50 km), in the stability field of garnet peridotites. Examples of the "prograde" type include garnet peridotites of the Swedish Caledonides (van Roermund, 1989), of the Massif Central Variscides (Gardien *et al.*, 1990), and of the Su-Lu terrane, China (type A peridotites of Zhang and Liou (1998); Figure 2). Examples of the "HP–HT" type include peridotites from the Bohemian, Schwarzwald, and Vosges Variscides (Kalt *et al.*, 1995; Altherr and Kalt, 1996; Medaris *et al.*, 1995, 1998) and, possibly, the Alpe Arami peridotite, in the Central Alps (Möckel, 1969; Green *et al.*, 1997; Nimis and Trommsdorff, 2001; Figure 2). Other garnet peridotites from the Central Alps are considered to derive from a mantle protolith serpentinized at very shallow depth, in a juvenile ocean basin (Evans and Trommsdorff, 1978; Evans *et al.*, 1979, 1981; Pfiffner and Trommsdorff, 1998). In this case, these rocks would have been buried down to the stability field of garnet, following a "prograde" trajectory.

(ii) The "relict" orogenic garnet peridotites represent old (cratonic), garnet-bearing subcontinental lithosphere devoid of subduction imprint. The mechanism of their upwelling is not well understood. One possibility is to involve an intracratonic fault developed away from the continent–continent suture zone (i.e., behind the subduction—Brueckner (1998), Brueckner and Medaris (2000)). Lithospheric peridotites would be tectonically intruded in continental rocks driven deep in the mantle along the fault. The Mg–Cr type garnet peridotites of the Western Gneiss region, Norwegian Caledonides, are considered as a typical example of "relict" orogenic peridotites (Carswell *et al.*, 1983; Brueckner and Medaris, 1998, 2000; Figure 2). However, some of these peridotites also record high-temperature conditions ascribed to mantle diapirism before the accretion of the peridotites to cratonic lithosphere (van Roermund *et al.*, 2001; Drury *et al.*, 2001). In this respect, they are somehow akin to the "UHT" group below.

(iii) The "UHT" orogenic garnet peridotites were once equilibrated at ≥1,200 °C in the stability field of spinel before recrystallizing in the garnet stability field. It has been proposed that these peridotites are driven into the garnet field by subduction mechanisms. Then, their upwelling would be triggered by slab break-off or lithospheric delamination (Van der Wal and Vissers, 1993; Davies *et al.*, 1993; Maruyama *et al.*, 1996; Medaris *et al.*, 1998). Examples of "UHT" peridotites include the Mohelno type peridotite, in the Bohemian Variscides (Medaris *et al.*, 1995, 1998; Figure 2). The Beni Bousera and Ronda ultramafic bodies, in the Betico-Rifean Cordilleras, were also included in this group by Brueckner and Medaris (2000). These massifs bear witness of a UHP stage in the form of graphitized diamond pseudomorphs in mafic layers (Pearson *et al.*, 1989, 1993, 1995; Davies *et al.*, 1993). However, Beni Bousera and Ronda are largely dominated by spinel lherzolites (± plagioclase in Ronda), the garnet peridotites being mostly restricted to mylonites near the contact of the massifs with crustal granulites. In addition, the garnet peridotites are spatially related to mafic layers, so that their significance as true garnet-bearing peridotite assemblages was disputed by some authors (Schubert, 1982; Kornprobst *et al.*, 1990). Ronda further differs from the garnet peridotite bodies in being affected by a high-temperature event at shallow depth, shortly before the exhumation (Van der Wal and Vissers, 1993; Lenoir *et al.*, 2001). For these reasons, we classified these massifs in the "IP" group below, notwithstanding that the Ronda massif includes a variety of rock types that encompass almost the whole range of orogenic peridotites worldwide.

When Beni Bousera and Ronda are excluded, the "true" orogenic garnet peridotites have not been much studied from the geochemical point of view. As they witness deep mantle processes that are hardly accessible to the observation otherwise, these rocks represent an exciting field of investigation for the future.

"Symplectite-bearing" peridotites. Although equilibrated in the spinel and plagioclase peridotite facies, the Hokkaido orogenic peridotites, Japan (Horoman and Nikanbetsu, Figure 2) are characterized by the widespread occurrence of symplectitic aggregates that bear evidence for early equilibration of the whole peridotite bodies in the garnet stability field (Tazahi *et al.*, 1972; Takahashi and Arai, 1989; Ozawa and Takahashi, 1995; Takazawa *et al.*, 1996; Takahashi, 2001). The massifs were uplifted from conditions comparable to those inferred for the "HP–HT" garnet peridotites (>1,000 °C and >70 km deep). Similar to the Betico-Rifean peridotites, they followed a high-temperature $P-T$ path during their ascent to the surface, coupled with transient heating, and partial melting in the spinel stability field (Ozawa, 1997; Takahashi, 2001).

The Horoman peridotite has been the subject of detailed geochemical studies devoted to understand the significance of diffusional zoning in minerals and the origin of lithological heterogeneities (Obata and Nagahara, 1987; Frey *et al.*, 1991; Takazawa *et al.*, 1992, 1996, 1999, 2000; Rehkämper *et al.*, 1999; Yoshikawa and Nakamura, 2000; Saal *et al.*, 2001). The massif is well known for its layered structure (Niida, 1974; Obata and Nagahara, 1987; Takahashi, 1992; Takazawa *et al.*, 1992; Toramaru *et al.*, 2001). In the upper part of the massif (the "Upper Zone" after Niida, 1974), the layering is underlined by relatively thin (mostly <1 m) bands of pyroxenites alternating with peridotites. In the "Lower Zone," the layering consists of varied peridotite lithologies showing a symmetrical pattern: elongated lenses of dunites occur at the center of harzburgite bands bounded by layers of refractory spinel lherzolites. Layers of fertile plagioclase lherzolites are intercalated between refractory peridotites. Several hypotheses have been formulated to account for this structure, notably (i) "frozen" compaction waves during partial melting and melt segregation perpendicular to the layering (Obata and Nagahara, 1987), (ii) focusing of partial melt into fractures or refractory porous-flow channels parallel to the layering (Takahashi, 1992; Takazawa *et al.*, 1992), and (iii) stretching and multiple folding of former lithological heterogeneities in the convecting mantle (Toramaru *et al.*, 2001).

2.04.2.1.2 IP orogenic peridotites

The intermediate pressure (IP) orogenic peridotites include several massifs dominantly equilibrated in the spinel (± amphibole, ± phlogopite) peridotite facies, either in the Ariegite (garnet in mafic layers) or in the Seiland (mafic layers devoid of garnet) subfacies (O'Hara, 1967a; Obata, 1980). These massifs were generally preferred to the garnet and plagioclase peridotites for geochemical studies for two main reasons:

(i) their provenance (from subcontinental mantle, at least during the last hundreds of Myr before their exhumation) is less debatable than that of the other types of orogenic peridotites;

(ii) they often show well preserved melt-related mantle structures, textures, and mineralogies whereas in the other orogenic peridotites these features are generally obscured by metamorphic recrystallization in the garnet or plagioclase peridotite facies. In the Ronda massif, which contains the three petrological facies defined by O'Hara (1967a), the melt-related structures (particularly a kilometer-scale melting front—Van der Wal and Bodinier (1996); Lenoir *et al.* (2001)) are much better preserved in the spinel peridotites than in the garnet mylonites or in the plagioclase tectonites.

Studies of orogenic spinel lherzolites have been mainly concentrated on three groups of massifs, all located in western Europe or northern Africa (Figure 2): (i) the ultramafic massifs of the Betico-Rifean belt, in southern Spain and northern Morocco (Ronda and Beni Bousera), (ii) the Pyrenean lherzolites in southern France (mostly Lherz), and (iii) the Ivrea peridotite bodies in the Italian Alps (Baldissero, Balmuccia, and Finero). All these massifs occur above or very close to areas of shallow mantle upwelling indicated by geophysical information (e.g., Torné *et al.*, 2000, for Ronda and Beni Bousera; Daignières *et al.*, 1998, for the Pyrenees; Berckhemer, 1968, and Nicolas *et al.*, 1990, for the Ivrea Zone).

Betico-Rifean belt. The ultramafic massifs of the Betic Cordilleras and of the Rif and Tell ranges of Morocco and Algeria were thrust amidst crustal continental metamorphic rocks in the Neogene (Priem *et al.*, 1979; Zeck *et al.*, 1989). These massifs include the Ronda, Ojén, and Carratraca peridotite bodies in Spain, the Beni Bousera massif in Morocco and the Collo peridotite in Algeria. With an exposure of $\sim 300\ km^2$, Ronda is the largest outcrop of subcontinental mantle exposed at the Earth's surface and probably the orogenic peridotite that has been the most studied from the structural, petrological, and geochemical points of view.

A well-known characteristic of the massif is the existence of a kilometer-scale petrological zoning. From the top of the 1.5 km thick peridotite slab to its base, Obata (1980) distinguished four petrological domains recording decreasing equilibrium pressure: (i) a garnet-lherzolite domain, (ii) a spinel-lherzolite domain of the Ariegite sub-facies, (iii) a spinel-lherzolite domain of the Seiland sub-facies, (iv) a plagioclase lherzolite domain. More recently, Van der Wal and Vissers (1993, 1996) recognized three tectono-metamorphic domains of decreasing age from the top to the base of the peridotite slab:

(i) a "spinel ($\pm$garnet) tectonite" domain interpreted as relict subcontinental lithosphere;

(ii) a "granular peridotite" domain derived from the spinel tectonites by partial melting and a few per cent melt extraction, coupled with annealing recrystallization and grain growth. This occurred across a narrow recrystallization front ($\leq$200 m), which is considered to record the maximum extent of partial melting in the Ronda subcontinental lithospheric mantle during its thermal erosion by upwelling asthenosphere, shortly before the exhumation (Van der Wal and Bodinier, 1996; Garrido and Bodinier, 1999; Lenoir *et al.*, 2001).

(iii) a "plagioclase tectonite" domain related to the crustal emplacement of the massif, which overprints all magmatic and deformation structures observed in the other domains.

The numerous geochemical studies on Ronda include notably Polvé and Allègre (1980), Zindler *et al.* (1983), Menzies (1984), Frey *et al.* (1985), Reisberg and Zindler (1987), Suen and Frey (1987), Hamelin and Allègre (1988), Reisberg *et al.* (1989, 1991), Gervilla and Remaïdi (1993), Reisberg and Lorand (1995), Van der Wal and Bodinier (1996), Garrido and Bodinier (1999), Garrido *et al.* (2000), and Lenoir *et al.* (2001). Several studies were also performed on the Beni Bousera peridotite (Morocco), including pioneering and advanced geochemical studies of mafic layers (Loubet and Allègre, 1979, 1982; Polvé and Allègre, 1980; Kornprobst *et al.*, 1990; Pearson *et al.*, 1989, 1991a,b; 1993; Gueddari *et al.*, 1996; Kumar *et al.*, 1996; Blichert-Toft *et al.*, 1999).

Pyrenees. The Pyrenean lherzolites comprise $\sim$40 separate bodies. Most of them are embedded together with crustal granulites within the "North Pyrenean Metamorphic Zone" (NPMZ). The NPMZ is a narrow (0–5 km wide) band of Mesozoic sediments affected by a low-pressure–high-temperature metamorphism (Goldberg and Leyreloup, 1990). It is ascribed to the transcurrent movement of the Iberian Peninsula relative to the European Plate in the Mid-Cretaceous, which resulted in the opening of elongated asymmetrical pull-apart basins (Choukroune and Mattauer, 1978; Debroas, 1987). Alternative tensional and compressional stress were responsible for crustal thinning and the subsequent ascent of slices of the uppermost lithospheric mantle in the pull-apart basins (Vielzeuf and Kornprobst, 1984).

An overall description of the peridotite bodies may be found in Fabriès *et al.* (1991), completed by Fabriès *et al.* (1998) for the central and western Pyrenees. These papers also synthetize our knowledge on the origin and evolution of Pyrenean peridotites and related rocks. The peridotite bodies vary in size from a few meters to 3 km and are mostly equilibrated in the Ariège subfacies. However, some kelyphized garnet bearing peridotites are observed in the Eastern Pyrenees (Bestiac), where they are closely related to garnet pyroxenite layers (Fabriès and Conquéré, 1983). Conversely, at the other end of the Pyrenean chain to the West ("Pyrénées Atlantiques"), the peridotites are equilibrated in the Seiland subfacies. Some massifs even show incipient recrystallization in the plagioclase stability field (e.g., Turon de Técouère—Fabriès *et al.* (1998)). The westernmost massifs (Saraillé and Col d'Urdach) are strongly serpentinized and the spinel-plagioclase pyroxenites in the Saraillé massif underwent partial rodingitic alteration suggesting that the mantle rocks reached a very shallow level in small oceanic basins. In this respect, the westernmost Pyrenean peridotites are considered to be transitional between the "IP" and "LP" groups of the orogenic lherzolites.

All the Pyrenean massifs are composed of layered spinel lherzolites, where the layering is defined by ubiquitous cm- to dm-thick parallel beds of websterite. Some massifs contain thicker layers (up to 1 m) of garnet-rich pyroxenites. Continuous bands of harzburgites or clinopyroxene-poor lherzolites, up to 20 m thick, lie parallel to the spinel websterite layering (Avé Lallemant, 1967; Conquéré, 1978). In a few massifs of eastern Pyrenees, the layering is cross-cut by a later generation of amphibole-pyroxenite dikes, up to 30 cm thick, and thin hornblendite veins.

The Pyrenean orogenic lherzolite bodies include the well-known Lherz (or "Lers") massif, where the lherzolite rock type was first described (Lacroix, 1917). The Lherz massif is easily accessible and displays the widest variety of rock types of mantle origin in the whole Pyrenees (Lacroix, 1917; Avé Lallemant, 1967; Conquéré, 1971, 1978; Bodinier *et al.*, 1988; Fabriès *et al.*, 1991). Together with a few neighboring lherzolite bodies, Lherz has been the subject of several geochemical studies, including both pioneering studies on mantle rocks and detailed investigations on melt/fluid migration and metasomatic processes (Polvé and Allègre, 1980; Loubet and Allègre, 1982; Bodinier *et al.*, 1987a,b, 1988, 1990, 2004; Hamelin and Allègre, 1988; Lorand *et al.*, 1989, 1999; Chaussidon and Lorand, 1990; Deloule *et al.*, 1991; Downes *et al.*, 1991; Fabriès *et al.*, 1989, 1991, 1998; Lorand, 1991; Mukasa *et al.*, 1991; Reisberg and Lorand, 1995; McPherson *et al.*, 1996; Pattou *et al.*, 1996; Woodland *et al.*, 1996; Zanetti *et al.*, 1996; Burnham, *et al.*, 1998).

Alps. The Baldissero, Balmuccia, and Finero ultramafic bodies are three well-known examples of orogenic spinel peridotites in the Western Alps. They are exposed in the lowermost part of the Ivrea-Verbano Complex, a tilted block of lower continental crust of the Southern Alpine plate uplifted during the Alpine orogenesis (Berckhemer, 1969; Mehnert, 1975). They are generally considered to represent asthenospheric mantle accreted to the lithosphere during the early Paleozoic (Nicolas, 1986a). They were tectonically emplaced into the underplated layered intrusives of the Ivea-Verbano Complex in the late Paleozoic or Early Mesozoic (Voshage *et al.*, 1988, and references herein; Quick *et al.*, 1995). They occur as elongated bodies, a few kilometers long, closely adjacent to the Insubric Line, a major tectonic suture of the Western Alps. In contrast with the Beni Bousera, Ronda, and Pyrenean peridotites, the Ivrea peridotites are equilibrated only in the Seiland subfacies, at pressures of ~1.5 GPa or less.

The Baldissero and Balmuccia peridotite bodies, exposed at the southern tip, and in the middle part of the Ivrea Zone respectively, are dominantly composed of fertile spinel lherzolites, with subordinate harzburgites and dunites (Lensch, 1968, 1971; Sinigoi *et al.*, 1980; Hartmann and Wedepohl, 1993). The massifs also contain mafic layers, among which two suites were distinguished by Shervais (1979) in Balmuccia: (i) a Cr-diopside suite in chemical equilibrium with the host peridotite and (ii) an Al-augite suite in disequilibrium with the peridotite. Except for the western part of the Baldissero massif, along the Insubric Line, which is strongly serpentinized, the two massifs are composed of exceptionally fresh mantle rocks. The Finero massif, exposed at the northern tip of the Ivrea Zone, forms the central part of a antiform elliptic ultramafic–mafic body, ~12 × 2.5 km in size (Vogt, 1962; Lensch, 1968; Cawthorn, 1975; Coltorti and Siena, 1984; Grieco *et al.*, 2001). It is made of amphibole- and phlogopite-bearing harzburgites containing small and diffuse dunite bodies, chomite segregates (in dunites) and minor clinopyroxenite veins. The outer part of the complex comprises, from the core outward, a layered internal zone of ultramafic and mafic rocks (50–130 m thick), an aureole of amphibole peridotite (200–300 m thick) and an external garnet-amphibole gabbro (100–150 m thick).

The Ivrea peridotites and related mafic layers have been the subject of several geochemical studies for more than two decades (Ernst, 1978, 1981; Shervais, 1979; Sinigoi *et al.*, 1983; Ernst and Ottonello, 1984; Garuti *et al.*, 1984; Ottonello *et al.*, 1984a; Voshage *et al.*, 1987, 1988; Shervais and Mukasa, 1991; Hartmann and Wedepohl, 1993; Rivalenti *et al.*, 1995; Zanetti *et al.*, 1999; Grieco *et al.*, 2001).

In addition to the Ivrea peridotites, the Alpine chain contains mantle spinel lherzolites occurring as relicts in major masses of metamorphosed ultramafic rocks, notably at Val Malenco, in the Eastern Central Alps (Müntener and Hermann, 1996). Like the Ivrea peridotites, the Malenco peridotites have a subcontinental lithospheric origin (with mantle–crust relationships exceptionally preserved—Hermann *et al.* (1997)), but they were exhumed during continental rifting and exposed as denuded mantle in the Tethyan ocean, near the continental margin (Trommsdorff *et al.*, 1993; Müntener *et al.*, 2000). Because of this shallow evolution, these spinel peridotites are classified in the "LP" group of the orogenic peridotites, together with the plagioclase and spinel-plagioclase peridotites of Lanzo and Liguria.

Other localities. Away from western Europe and northern Africa, the spinel lherzolite massifs are rare and/or poorly known. A notable exception is the Tinaquillo peridotite, that crops out in the Cordillera de la Costa, in northern Venezuela (MacKenzie, 1960; Seyler and Mattson, 1989, 1993; Figure 2). In part, the Tinaquillo lherzolite is known for having been used as a source material for experimental studies on partial melting of mantle peridotites (Jaques and Green, 1980; Falloon *et al.*, 1988; Robinson and Wood, 1998; Robinson *et al.*, 1998). Like other spinel orogenic lherzolites, the Tinaquillo massif is interpreted as a slice of subcontinental mantle. It was overthrusted amidst the Carribean belt in the Cretaceous, together with crustal granulites (Seyler *et al.*, 1998). It is composed of spinel lherzolites and harzburgites containing an unusual amount of mafic layers (10–15%), including pre-kinematic pyroxenites and syn-kinematic plagioclase and/or hornblende-bearing granulites (Seyler and Mattson, 1993). Only a few major and trace element data have been reported to date for the Tinaquillo massif (Seyler and Mattson, 1989, 1993).

2.04.2.1.3 LP orogenic peridotites

This group includes peridotite massifs dominated by fertile lherzolites that were exhumed in the early stages of continental rifting and exposed as denuded mantle on the seafloor, along passive continental margins.

Western Alps. The LP orogenic peridotites are particularly well exposed in the Alps (Figure 2), where they form several large occurrences of ultramafic rocks (~100 km^2 or more) along the Alpine and North Apennine mountain arc (Figure 2). In general, the massifs underline the suture zone between the southern Alpine and European plates. The most significant bodies include:

(i) Malenco, at the boundary between the Penninic and Austroalpine zones, in eastern central Alps (Trommsdorff and Evans, 1974; Trommsdorff *et al.*, 1993; Müntener and Hermann, 1996);

(ii) Mt Avic, Aosta Valley, and Breithorn, Zermatt-Saas area, in northwestern Alps (Dal Piaz, 1969);

(iii) Lanzo, to the south of the Sesia-Lanzo high-pressure terrane in southwestern Alps (Nicolas, 1969; Boudier and Nicolas, 1972; Ernst, 1978; Pognante *et al.*, 1985; Bodinier *et al.*, 1986, 1991; Bodinier, 1988);

(iv) Erro-Tobbio, in the Voltri group, western Liguria (Bezzi and Piccardo, 1971; Messiga and Piccardo, 1974; Chiesa *et al.*, 1975; Piccardo *et al.*, 1980, 1988a; Hoogerduijn-Strating *et al.*, 1990; Vissers *et al.*, 1991);

(v) Peridotites from the External Ligurides, in eastern Liguria (Galli and Cortesogno, 1970; Bezzi and Piccardo, 1971; Piccardo, 1976).

Some of these massifs were sometimes described as ophiolites in the literature—particularly Lanzo and the External Ligurides (see Section 2.04.2.2). However, they differ from true ophiolite sequences by the absence of a continuous basaltic crust overlying the peridotites. However, they generally differ from the HP/UHP and IP orogenic peridotites in being intruded by small bodies of gabbros, ferro-gabbros, and subordinate acidic differentiates. They are also cross-cut by MORB dikes. These mafic rocks are interpreted as embryonic crust related to the onset of oceanic accretion in the Western Tethys (Bodinier *et al.*, 1986). Ophicarbonates and oceanic sediments are found directly superimposed onto serpentinized peridotites and gabbros. Most of the massifs were strongly affected by hydrothermal alteration, and in several cases the peridotites occur only as relict cores in serpentinite bodies.

To explain the denudation of peridotites on the seafloor before the onset of oceanic accretion, Lemoine *et al.* (1987) and Trommsdorff *et al.* (1993) proposed a model involving extensional exhumation of subcontinental mantle along a major, normal detachment fault rooted in the crust-mantle boundary (Wernicke, 1981, 1985). In most cases, however, the peridotites record a higher-temperature evolution than would be expected from stable lithospheric mantle. This suggests that:

(i) either the mantle exhumation process also involved slices of asthenospheric material;

(ii) or the lithospheric mantle underwent thermomechanical erosion by upwelling asthenosphere before the exhumation.

In this respect, the western Alps LP orogenic peridotites vary between two end-members:

(i) LT (low-temperature) LP peridotites equilibrated in the spinel stability field, which have preserved several lithologies comparable to those observed in "lithospheric" IP orogenic peridotites such as the Ivrea and Pyrenean spinel peridotites.

The Malenco peridotite would represent this end-member. The massif is dominated by spinel lherzolites with no trace of re-equilibration in the plagiocase peridotite facies. In addition, it contains a variety of mafic layers and hydrous veins, including garnet clinopyroxenites and phlogopite hornblendites (Müntener and Hermann, 1996).

(ii) HT (high-temperature) LP peridotites extensively reequilibrated in the plagioclase stability field. These massifs display several features comparable to mildly refractory ophiolitic mantle and contain relatively abundant intrusions of MORB-type gabbros and basalts. Refractory (olivine-rich lherzolites) and dunites are common. A typical example is provided by the southern part of the Lanzo massif (Bodinier, 1988; Bodinier *et al.*, 1991).

The Erro-Tobbio massif (western Liguria), the northern part of the Lanzo massif and the External Ligurian peridotites are transitional between these extremes. These peridotites show variable degrees of re-equilibration in the stability field of plagioclase, from incipient recrystallization (locally enhanced by shearing) in Erro-Tobbio, to ubiquitous but incomplete recrystallization in the northern part of Lanzo and in the External Ligurian peridotites.

Several studies have reported geochemical analyses of LP orogenic peridotites from western Alps. A notable exception is the Malenco massif, which has been poorly studied yet, probably because it was considered until recently as wholly serpentinized (Müntener and Hermann, 1996). However, several pioneering studies on orogenic peridotites included at least one or a few samples from Lanzo (e.g., Loubet *et al.*, 1975; Menzies, 1976; Menzies *et al.*, 1977; Menzies and Murthy, 1978; Richard and Allègre, 1980; Loubet and Allègre, 1982; Ottonello *et al.*, 1984a; Ernst and Ottonello, 1984; Hamelin and Allègre, 1988; Roy-Barman *et al.*, 1996), and several studies were entirely devoted to Lanzo or the Ligurian peridotites (e.g., Ernst and Piccardo, 1979; Ottonello *et al.*, 1979, 1984b; Beccaluva *et al.*, 1984; Bodinier *et al.*, 1986, 1991; Bodinier, 1988; Lorand *et al.*, 1993; Rampone *et al.*, 1993, 1995; Snow *et al.*, 2000; Scambelluri *et al.*, 2001).

Zabargad. The Zabargad peridotite (Red Sea, Egypt, Figure 2) is not involved in a collisional mountain belt and, therefore, should not be considered as an "orogenic" peridotite massif in a strict sense, but rather as an "oceanic" peridotite. However, the massif is virtually indistinguishable from the orogenic lherzolites in several respects, including the predominance of spinel (± amphibole) spinel lherzolites, the occurrence of multiple metasomatic events involving mantle and crustal fluids (Dupuy *et al.*, 1991; Agrinier *et al.*, 1993; Piccardo *et al.*, 1993) and the presence of crustal granulites as host rocks. In addition, similar to the orogenic massifs, the Zabaragad peridotite is considered to derive from subcontinental lithosphere (Brueckner *et al.*, 1988; Bonatti *et al.*, 1990; Dupuy *et al.*, 1991; Piccardo *et al.*, 1988b, 1993), or from upwelling asthenosphere before the onset of oceanic accretion (Bonatti *et al.*, 1986; Nicolas *et al.*, 1987; Bosch and Bruguier, 1998b). In fact, the Zabargad peridotite is especially akin to the LP orogenic peridotites for several reasons:

(i) its location near a passive continental margin (Martinez and Cochran, 1988);

(ii) its exhumation in an extensional tectonic setting during continental rifting and the opening of the Red Sea (Nicolas *et al.*, 1985, 1987; Bonatti *et al.*, 1986);

(iii) the partial re-equilibration of the peridotites in the plagioclase stability field and the existence of gabbro intrusions and MORB dikes (Brueckner *et al.*, 1988; Petrini *et al.*, 1988);

(iv) the possible exhumation of the massif along a low-angle detachment fault associated with the Red Sea rifting (Voggenreiter *et al.*, 1988);

(v) the denudation of mantle rocks on the seafloor, as attested by the hydrothermal alteration of the plagioclase peridotites (Bosch, 1991; Boudier and Nicolas, 1991; Agrinier *et al.*, 1993; Piccardo *et al.*, 1993).

Several geochemical studies were devoted to the Zabargad peridotite, including elemental and isotopic characterization as well as geochronological studies (e.g., Bonatti *et al.*, 1986; Brueckner *et al.*, 1988; Piccardo *et al.*, 1988b, 1993; Villa, 1988, 1990; Bosch, 1991; Dupuy *et al.*, 1991; Jedwab, 1992; Agrinier *et al.*, 1993; Kurat *et al.*, 1993; Vannucci *et al.*, 1991, 1993; Sciuto and Ottonello, 1995; Trieloff *et al.*, 1997; Snow and Schmidt, 1999; Schmidt *et al.*, 2000).

2.04.2.2 Ophiolitic Peridotites

Ophiolites represent slivers of ancient oceanic lithosphere obducted onto continental or oceanic crust. Ophiolite suites are observed in all major orogenic belts, including the Pan-African Belt, the Appalachian–Caledonian–Uralian Belt, the Alpine–Himalayan Belt, and the circum-Pacific orogenic Belt (Figure 3). Understanding the widespread occurrence of oceanic lithosphere in association with continental suture zones is a major issue for the study of plate tectonics and Earth's geodynamics (Abbate *et al.*, 1985; Ishiwatari, 1994; Moores *et al.*, 2000; Yakubchuk *et al.*, 1994). In addition, ophiolites provide a unique opportunity to observe the structure of the oceanic crust and upper mantle. Therefore, the ophiolites have been the subject of multiple studies and reviews for the last 30 years, that is since a world-wide agreement was reached among geologists during the Penrose Conference in 1972

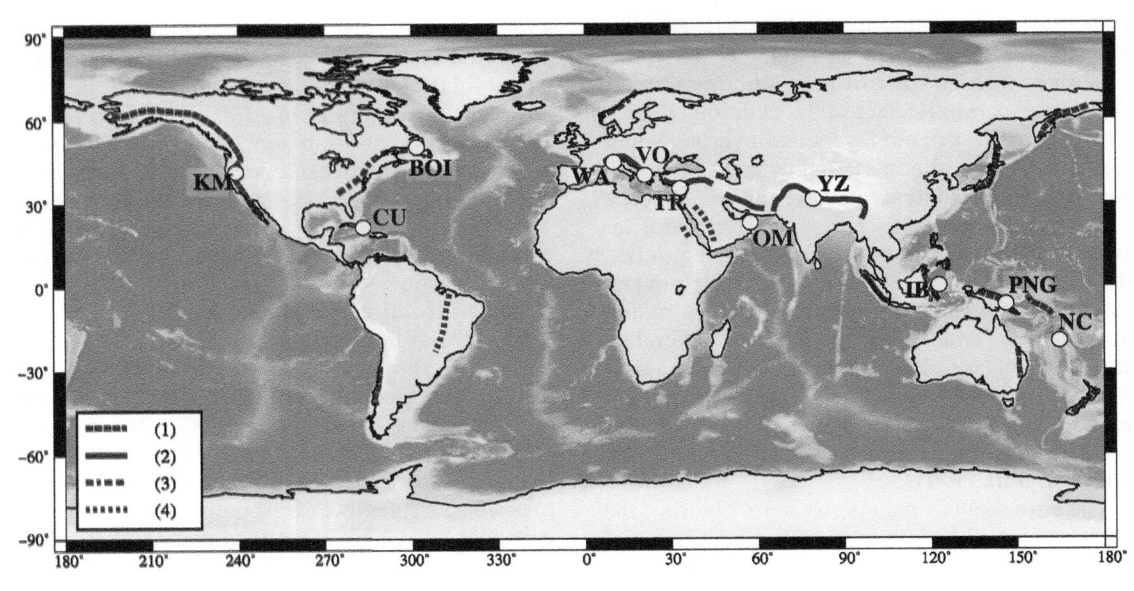

Figure 3 Location of the main ophiolitic belts and ophiolite complexes mentioned in the text (after Coleman, 1977; Ishiwatari, 1994; Moores *et al.*, 2000). (1) *Circum-Pacific Phanerozoic Multiple Ophiolite Belt* (450–0 Ma): KM—Klamath Mountains (Trinity and Josephine, western USA); NC—New Caledonia; PNG—Papua–New Guinea; IB—Indonesian Back-Arc ophiolites. (2) *Himalaya–Alps–Caribbean Mesozoic Belt* (150–80 Ma): YZ—ophiolites of the Yarlung Zangbo Suture Zone, China; OM—Oman (=Semail ophiolite), Sultanate of Oman and Unites Arab Emirates; TR—Troodos, Cyprus; VO—Vourinos and Pindos, Greece; WA—Western Alps (Internal Ligurides and Lanzo): CU—Cuba ophiolites. (3) *Appalachian–Caledonian Belt* (450 Ma): BOI—Bay of Islands, Newfoundland. (4) *Pan-African Belt* (750 Ma).

on the definition of what is an ophiolite (e.g., Coleman, 1977; Dilek *et al.*, 2000; Gass *et al.*, 1984; Ishiwatari *et al.*, 1994; Nicolas, 1989; Parson *et al.*, 1992; Peters *et al.*, 1991a; Vissers and Nicolas, 1995). After years of controversy (see review in Nicolas, 1989), the participants of the Penrose Conference stated that "in a completely developed ophiolite, the rock types occur in the following sequence, starting from the bottom and working up: ultramafic complex (...), gabbroic complex (...), mafic sheeted dike complex (...), mafic volcanic complex, commonly pillowed." Ophiolites are overlain by a "sedimentary section typically including ribbon cherts, thin shale interbeds, and minor limestones." They may contain "podiform bodies of chromite (...) associated with dunite, and sodic felsic intrusive and extrusive rocks." An ophiolite may be incomplete, dismembered, or metamorphosed, and "although ophiolite generally is interpreted to be oceanic crust and upper mantle, the use of the term should be independent of its supposed origin" (Anonymous, 1972). Nevertheless, in the following years, geochemical and isotopic studies established the oceanic origin of several ophiolite massifs (e.g., Allègre *et al.*, 1973; Jacobsen and Wasserburg, 1984; Beccaluva *et al.*, 1984), and nowadays, the term ophiolite refers only to supposed fragments of ancient oceanic lithosphere (e.g., Ishiwatari *et al.*, 1994; Nicolas and Boudier, 2003).

More than 70 ophiolites have been described throughout the world. A number of "completely developed" ophiolitic complexes have been well studied, notably in the Internal Ligurides (northern Italy), Troodos (Cyprus), Semail (Sultanate of Oman and U.A.E.), Xigaze (Tibet), Klamath Mountains (Trinity, California), and Bay of Islands (Newfoundland, Canada). The Internal Ligurides and the Semail ophiolites may be considered as two end-members for the main lithological variations observed in ophiolitic mantle sections:

(i) a mildly refractory mantle section dominated by lherzolites in the Internal Ligurides,

(ii) a strongly refractory mantle section dominated by harzburgites in the Semail ophiolite.

These two end-members provide a simple frame for the classification of the ophiolites in two main groups (e.g., Boudier and Nicolas, 1985; Ishiwatari, 1985; Nicolas and Boudier, in press). They are briefly examined in the following, together with the rare subarc mantle sections exposed at the base of obducted island arcs (e.g., Burg *et al.*, 1998).

2.04.2.2.1 *Ophiolitic lherzolites (Internal Ligurides)*

Ophiolites cropping out in the Western Alps of northern Italy are the remnants of the Jurassic Ligurian Tethys (or "Western Tethys"). The plagioclase peridotite of Lanzo is considered by several authors as part of this ophiolite suite. Situated along the inner arc of the Alps, the Lanzo massif (150 km^2) chiefly consists of fertile plagioclase lherzolites (Boudier, 1978). Small gabbro sills and basaltic dikes are common in the southern part of the massif. They are were interpreted by as embryonic oceanic crust (Boudier and Nicolas, 1985; Bodinier *et al.*, 1986). We classified the Lanzo massif (as well as the External Liguride peridotites) as a LP orogenic lherzolites (see Section 2.04.2.1.3) rather than as a true ophiolite for two reasons: (i) the absence of a continuous basaltic crust overlying the peridotites and (ii) the dominantly fertile composition of the lherzolites. We would nevertheless acknowledge that this distinction is somewhat arbitrary and that there must be a continuum of structures and lithological associations between LP orogenic peridotites and the "lherzolite type" of the ophiolite mantle sections (Boudier and Nicolas, 1985).

However, there is a consensus to consider the ultramafic–mafic association of the Internal Ligurides as a "true" ophiolite. The ophiolite suite represents the basement for the Upper Jurassic to Paleocene sedimentary sequence, characterized by manganiferous radiolarian cherts, silicic limestones, shales with limestones, and ophicalcites. The crustal section is constituted of massive and pillowed basaltic lava flows, but does not show a well-organized sheeted-dike complex. In addition, gabbros occur mainly as intrusive bodies and dikes in the peridotites (Beccaluva *et al.*, 1976; Ferrara *et al.*, 1976; Piccardo, 1977; Bortolotti and Chiari, 1994; Rampone and Piccardo, 2000). The peridotites are strongly serpentinized but nevertheless preserve textural evidence for their derivation from mantle tectonites, as well as rare mineralogical relics of the primary lithologies. Only a few outcrops show small cores of very fresh peridotite. Mantle rocks are dominated by clinopyroxene-poor (5–10%) spinel lherzolites partially reequilibrated in the plagioclase peridotite facies. Geochemical studies on the Internal Ligurides ophiolites were performed by Beccaluva *et al.* (1984), Ottonello *et al.* (1984b), Rampone *et al.* (1996, 1997, 1998), Rampone and Piccardo (2000), and Snow *et al.* (2000).

Only a few ophiolites are comparable to the Internal Ligurides in showing a dominantly lherzolitic mantle sequence. These include, for instance, the Western Dinarides (Albania and Othris; Nicolas and Jackson, 1972; Menzies and Allen, 1974; Nicolas *et al.*, 1999) and—to some degree—the Trinity ophiolite (California—Quick (1981)). Lherzolitic mantle sections in ophiolites are ascribed to lower melting degrees than harzburgitic mantle sections (e.g., Ishiwatari, 1985), a feature which may be explained by slow-spreading rate at mid-ocean ridges (Boudier and Nicolas, 1985; Moores *et al.*, 2000; Nicolas and Boudier, in press). Low melting degrees would also account for the thin, often discontinuous ocean crust observed in these ophiolites. Alternatively, these ophiolites could record the early stages of oceanization. Based on their strongly depleted neodymium isotopic signature, compared with the ophiolitic crustal rocks (Rampone *et al.*, 1996, and Section 2.04.5.2.3), the lherzolites from Internal Ligurides were ascribed to early mantle upwelling and melting during a continental rifting event, in the Permian. This event culminated in the denudation of mantle rocks in the incipent Western Tethys. Thereafter, the ophiolite crust was formed during a later stage of (slow-spreading) ocean accretion, in the Jurassic (Rampone *et al.*, 1998). Notwithstanding that other hypotheses can be possibly envisioned for the mantle–crust isotope decoupling (see Section 2.04.5.2.3), this scenario would reinforce the similarity between the lherzolite-type ophiolites and the LP orogenic peridotites, as the mantle section of this type of ophiolite would be originally of subcontinental origin (Rampone and Piccardo, 2000).

2.04.2.2.2 *Ophiolitic harzburgites (Semail)*

The Semail (=Oman) ophiolite is probably the largest piece of oceanic lithosphere (2×10^4 km^2) exposed at the surface of the Earth. It is considered to represent part of the Neo-Tethys ocean floor obducted onto the Arabian plate at the end of the Cretaceous. It has been the focus of detailed studies by several research groups and is, therefore, one of the best documented ophiolites (e.g., Glennie *et al.*, 1974; Coleman and Hopson, 1981; Lippard *et al.*, 1986; Boudier and Nicolas, 1988; Peters *et al.*, 1991b; Boudier and Juteau, 2000).

The ophiolite consists of several large massifs. Each of them exposes a more or less complete ophiolitic sequence, comprising a thick mantle section (up to 12 km), a gabbro sequence (0.5–6 km), a sheeted dike complex (1–1.5 km), a thick extrusive sequence (0.5–2 km), and interbedded oceanic sediments (e.g., Lippard *et al.*, 1986).

The mantle section represents more than 50% of the exposed Semail ophiolite (Nicolas *et al.*, 2000). It is dominated by harzburgite tectonites with coarse-grain porphyroclastic textures recording high temperature (>1,200 °C) deformation during asthenospheric mantle flow (Boudier and Coleman, 1981). Foliation is generally flat lying and roughly parallel to the Moho, except for 3D vertical structures interpreted as mantle diapirs (Ceuleneer *et al.*, 1988; Ceuleneer and Rabinowicz, 1992). The mantle section contains elongated bodies of dunite that generally crosscut the foliation, except near the base and the top of the sequence where they tend to parallel harzburgite foliation (Kelemen *et al.*, 1995, and references herein). The harzburgites are separated from the overlying gabbros by a "Mantle–Crust Transition Zone" (MTZ), a few meter- to several hundred meter-thick, made of dunites. The MTZ is rich in gabbro sills as well as in diffuse segregates of cpx and plagioclase ("impregnated" dunites). It was often considered as the lower part of the cumulate sequence, but structural evidence indicates that the dunites were deformed at high-temperature mantle conditions and thus belong to the mantle sequence (Boudier and Nicolas, 1995). Mildly fertile facies of cpx-harzburgites or (rarely) lherzolites occur near the base of the mantle section (>10 km below the MTZ—Lippard *et al.*, 1986; Godard *et al.*, 2000; Takazawa *et al.*, 2003). The basal peridotites were affected by relatively low-temperature deformation ascribed to the early obduction stages.

Most of the ophiolites worldwide display similar zoning of their mantle lithologies, i.e., from the top to the base

(i) a dominantly dunitic MTZ, where the dunites may be interlayered with websterites or gabbros, and are often impregnated by melt-derived crystal segregates (cpx and plagioclase),

(ii) a main harzburgitic section including a variety of dunites bodies, layers, and veins,

(iii) a basal section made of mildly refractory peridotites (cpx-harzburgites or lherzolites, sometimes alternating with dunite layers).

These lithological units may vary widely in thickness and composition from one ophiolite to the others, but they are generally observed in all complete ophiolitic mantle sections (Nicolas and Boudier, in press, and references herein). Variations are nevertheless observed in the overall refractory character of the ophiolite mantle sections. The Semail ophiolite (above), the Papua-New Guinea ophiolites (Davies, 1980; Jaques and Chappell, 1980) and the Massif du Sud ophiolite, in New Calenonia (Cassard, 1980; Paris, 1981) provide examples of strongly refractory harzburgitic mantle sections. However, several ophiolites have a mantle sequence composed of mildly refractory (cpx-bearing) harzburgites. This is the case, for example, in the Bay of Islands ophiolite, Newfoundland, where the mantle sequence is mostly composed of cpx-harzburgites and includes spinel and plagioclase lherzolites (Suhr, 1992; Suhr *et al.*, 1998). The main harzburgite sequence of the Xigaze ophiolite, in Tibet, comprises two sub-units with a diffuse transition between them: (i) upper refractory harzburgites, and (ii) lower cpx-harzburgites overlying the basal lherzolites (Girardeau and Mercier, 1988).

Since Miyashiro (1973), who suggested a supra-subduction origin for the Troodos ophiolite (Cyprus), the original tectonic setting of the harzburgite-type ophiolites has been widely debated (e.g., Moores *et al.*, 2000; Nicolas and Boudier, 2003; Shervais, 2001). While there is consensus to ascribe the refractory composition of the harzburgites to a high degree of melt extraction, the opinions diverge on whether this process is achieved by hydrous melting in supra-subduction mantle wedges, or by decompression melting at fast-spreading ridges.

The arguments to ascribe some ophiolites to supra-subduction settings include their situation at active plate margins (e.g., the circum-Pacific ophiolites—Ishiwatari, 1994), their association with tectonic melanges including arc-type volcanoclastic deposits (Moores, 1982) and the "subduction-type" signature of extrusive rocks (e.g., Miyashiro (1973) for Troodos and Pearce *et al.* (1981) for Oman). Suggested supra-subduction settings include back-arc ocean basins (Pearce *et al.*, 1981), juvenile island arcs (Maheo *et al.*, 2000), and forearc settings (Shervais, 2001).

However, several ophiolites show structural and stratigraphic evidence for seafloor spreading in open oceanic environment (e.g., Coleman, 1977; Nicolas, 1989). This led several authors to propose the formation of the harzburgite-type ophiolites at intermediate- to fast-spreading ridges. In this scheme, the variations in the refractory character of the harzburgite mantle sequences are ascribed to differences in spreading-rate (Boudier and Nicolas, 1985; Nicolas and Boudier, in press). To account for the conflicting indications for a suprasudction setting in some of these ophiolites—notably in Oman—Moores *et al.* (2000) suggested that the geochemical signatures of the lavas might reflect a contamination of their mantle sources by ancient subducted slabs.

2.04.2.2.3 Subarc mantle (Kohistan)

A few examples of oceanic island arcs obducted on continental margins have been documented. Some of them (e.g., the Kohistan Complex, northern Pakistan—Bard *et al.*, 1980; Bard, 1983; Coward *et al.*, 1986) include a lower section made of ultramafic rocks. In Kohistan, the "Jijal" ultramafic section is subdivided in four main units, from base to top: (i) a peridotite unit (~2.3 km thick) made of dunites and wehrlites interlayered with pyroxenites, (ii) a pyroxenite unit (~1.5 km) made of chromian websterites, clinopyroxenites, and subordinate dunites, (iii) a garnet-hornblende pyroxenite unit (~0.5 km) composed of garnet pyroxenites and hornblendites, and (iv) a garnet granulite unit (~3.5 km) composed of garnet granulites and subordinate garnet hornblendites topped by rare hornblende gabbro-norites (see, also, Jan *et al.* (1993), for the "Sapat" section). The whole section was generally considered as the lower part of the arc cumulates (e.g., Khan *et al.*, 1993). However, Miller *et al.* (1991) suggested the origin of the ultramafic sequence in the uppermost arc mantle, which was further confirmed by Burg *et al.* (1998) on the basis of field and textural evidence.

Similar to the Jijal sequence, the Cabo Ortegal ultramafic complex (northwestern Spain) contains abundant and thick pyroxenites layers, and might represent an example of subarc mantle in the Varican belt (Gil Ibarguchi *et al.*, 1990; Girardeau and Gil Ibarguchi, 1991).

2.04.2.3 Oceanic Peridotites

In addition to the abyssal peridotites collected on the seafloor, the oceanic peridotites in a broad sense include mantle rocks that were exhumed above sea level by normal faults associated with rifting or by transcurrent movements along transform faults. Examples of "emerged" oceanic peridotites include:

(i) the Zabargad peridotite, in the Red Sea (Figure 2, Bonatti *et al.*, 1981), which is interpreted as passive-margin mantle and is virtually indistinguishable from LP orogenic peridotites (see Section 2.04.2.1.3).

(ii) the St. Peter's and St. Paul's islands, in the Atlantic (Tilley, 1947; Melson *et al.*, 1967; Bonatti, 1990). These peridotites are markedly different from other oceanic peridotites in being modally and chemically metasomatized by alkaline basaltic melts and/or fluid-rich derivatives (Roden *et al.*, 1984; Bonatti *et al.*, 1990). In that sense, they are more akin to metasomatized orogenic lherzolites such as Caussou (Section 2.04.2.1.2), or to metasomatized mantle xenoliths (see Chapter 2.05).

(iii) harzburgites in the Macquarie island, southwest Pacific, that occur at base of a complete section of exhumed oceanic crust. Hence, the Macquarie section is commonly considered as an ophiolite, although it was not obducted (Varne *et al.*, 1969; Griffin and Varne, 1978, 1980; Goscombe and Everard, 2001).

Oceanic peridotites in a strict sense (= abyssal peridotites) were first sampled during the early development of marine geology in the period following the end of World War II—mostly in the sixties (Shand, 1949; Hersey, 1962; Hess, 1964; Fray and Hékinian, 1965; Bowin *et al.*, 1966; Chernysheva and Bezrukov, 1966; Bogdanov and Ploshko, 1967; Bonatti, 1968). Most abyssal peridotites were collected by dredging or using submersibles (Tisseau and Tonnerre, 1995; Karson, 1998; Lagabrielle *et al.*, 1998). Several peridotite samples were also collected in the frame of the international drilling programs Deep Sea Drilling Project (1968–1983) and Ocean Drilling Program (1985–2003) (e.g., Mevel *et al.*, 1996; Karson *et al.*, 1997).

Abyssal peridotite studies provided the first direct information on the uppermost mantle in present-day oceanic settings (e.g., Fisher and Engel, 1969; Miyashiro *et al.*, 1969; Chernysheva and Rudnik, 1970; Cann, 1971; Vinogradov *et al.*, 1971; Ploshko *et al.*, 1973; Jibiki and Masuda, 1974). However, mantle exposures are rare on the seafloor, and limited to specific areas such as the end of ridge segments or fault zones (Dick *et al.*, 1984; Niu and Hékinian, 1997a), deeps (Hékinian *et al.*, 1993), serpentinite domes (Parkinson and Pearce, 1998), or passive continental margins (Cornen *et al.*, 1996). The abyssal peridotite database is, therefore, biased towards these tectonically disturbed areas, and the degree to which they represent the composition of "normal" oceanic mantle has sometimes been questioned (e.g., Nicolas and Boudier, 2003).

Major tectonic windows allowing upper mantle seafloor exposures are most common at slow-spreading mid-ocean ridges (spreading rate $<3\ \mathrm{cm\ yr^{-1}}$). As a result, the majority of the abyssal peridotites come from walls of fracture zones or rift mountains in these ridge systems (Figure 4). They were sampled mainly in the slow-spreading Mid-Atlantic and Indian Ocean ridge systems (Dick and Bullen, 1984; Dick *et al.*, 1984; Johnson *et al.*, 1990; Bonatti *et al.*, 1992, 1993; Elthon, 1992; Casey, 1997; Ross and Elthon, 1997; Seyler and Bonatti, 1997; Stephens, 1997; Hellebrand *et al.*, 2001) and, more recently, at ultra-slow-spreading ridges (spreading rate $<1\ \mathrm{cm\ yr^{-1}}$), such as the South West Indian Ridge (Johnson and Dick, 1992; Seyler *et al.*, 2001, 2003) and the Gakkel Ridge, North Atlantic (Hellebrand *et al.*, 2002). Abyssal peridotites from fast-spreading ridges (spreading rate $>10\ \mathrm{cm\ yr^{-1}}$) are rare, the only tectonic windows on upper mantle in these ridge systems being Hess Deep (Figure 4) and a few transform zones to the south of Hess Deep (Hékinian *et al.*, 1993; Constantin *et al.*, 1995; Allan and Dick, 1996; Dick and Natland, 1996; Niu, 1997; Niu and Hékinian, 1997a,b). In supra-subduction settings, abyssal peridotites were sampled only in three sites (Figure 4): the Tyrrhenian Sea, in the western Mediterranean (Bonatti *et al.*, 1990), the Izu–Bonin–Mariana Forearc (Parkinson and Pearce, 1998), and the South Sandwich Arc Basin (Pearce *et al.*, 2000). Finally, peridotites related to the early stages of the Atlantic opening were also collected on the Galicia passive margin and the Gorringe Bank (Boillot *et al.*, 1988; Abe, 2001; Hébert *et al.*, 2001; Whitmarsh and Wallace, 2001).

All abyssal peridotites are serpentinized, generally to a large extent (e.g., Snow and

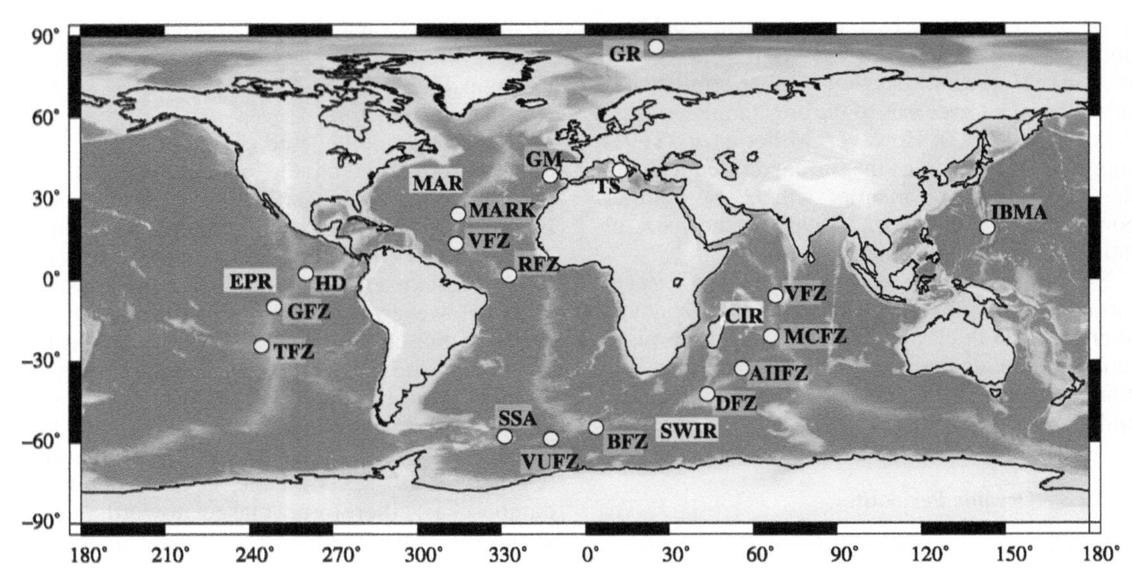

Figure 4 Location of the main occurrences of abyssal peridotites mentioned in the text. *Fast-spreading ridge systems*: EPR—East Pacific Rise (HD—Hess Deep; GFZ—Garret Fracture Zone; TFZ—Terevaka Fracture Zone). *Slow-spreading ridge systems:* MAR—Mid-Atlantic Ridge (MARK—Kane and Atlantis Fracture Zones; VFZ—Vema and 15°20 Fracture Zones; RFZ—Romanche Fracture Zone); CIR—Central Indian Ridge (VFZ—Vema Fracture Zone; MCFZ—Marie Celeste Fracture Zone and Green Rock Hill). *Ultraslow-spreading ridge systems:* SWIR—Southwest Indian Ridge (AIIFZ—Atlantis II Fracture Zone and EDUL; DFZ—Discovery II Fracture Zone; BFZ—Bouvet, Islas Orcadas, and Shaka Fracture Zones); American-Antarctic Ridge (VUFZ—Vulcan and Bullard Fracture Zones); Arctic Ocean (GR—Gakkel Ridge). *Supra-subduction settings:* SSA—South Sandwich Arc-Basin system; IBMA—Izu–Bonin–Mariana Forearc; TS—Tyrrhenian Sea. *Continental Margins:* GM—Galicia Margin and Gorringe Bank.

Dick, 1995). The serpentinite protoliths are mostly harzburgites and, less commonly, lherzolites. Dunites, wehrlites, and pyroxenites are subordinate rock types (e.g., review in Karson, 1998). Most abyssal peridotites are plagioclase-free tectonites with coarse-granular porphyroclastic textures (e.g., Michael and Bonatti, 1985; Johnson *et al.*, 1990). When plagioclase is present (<2%—Dick and Bullen, 1984; Dick *et al.*, 1984), this mineral is generally considered as crystallized from infiltrated melt, or from partial melts incompletely drained from residues after decompression melting (Dick *et al.*, 1984). Plagioclase- and/or cpx-rich peridotites also occur as diffuse lenses and elongated patches cross-cutting the foliation of the host peridotites (e.g., Constantin *et al.*, 1995). These lithologies are very similar to the "melt impregnation" facies observed in the mantle–crust transition zone of the ophiolites where they are ascribed to solidification of interstitial melt (e.g., Boudier and Nicolas, 1995, and references herein).

Partly based on the ophiolite record (see Section 2.04.2.2), the refractory degree of abyssal peridotites was tentatively associated with spreading rate (e.g., Boudier and Nicolas, 1985; Nicolas and Boudier, in press). Mildly refractory peridotites after moderate melt extraction degrees (lherzolites and cpx-harzburgites) would be associated with slow-spreading ridges whereas more refractory residues recording higher melt extraction degrees (harzburgites) would occur at fast-spreading ridges. However, available data on abyssal peridotites do not fit well in this scheme. For instance, harzburgites have been collected at the South West Indian Ridge, an ultra-slow spreading ridge (Johnson *et al.*, 1990; Seyler *et al.*, 2003). Moreover, extremely refractory harzburgites have been sampled between 14° N and 16° N on the slow-spreading Mid-Atlantic Ridge (Sobolev *et al.*, 1992). These peridotites plot at the depleted end-member of the abyssal peridotites in terms of forsterite content in olivine (cationic ratio Mg/(Mg + Fe) = 0.92) and chromium component in spinel (cationic ratio Cr/(Cr + Al) = 0.70).

2.04.3 MAJOR- AND TRACE-ELEMENT GEOCHEMISTRY OF PERIDOTITES

During the last two decades, many studies have reported major- and trace-element data on orogenic, ophiolitic, and oceanic peridotites. Whole-rock compositions representative of the main peridotite occurrences are illustrated on element versus Al_2O_3 diagrams for major elements (Figure 5), minor transition elements (Figure 7),

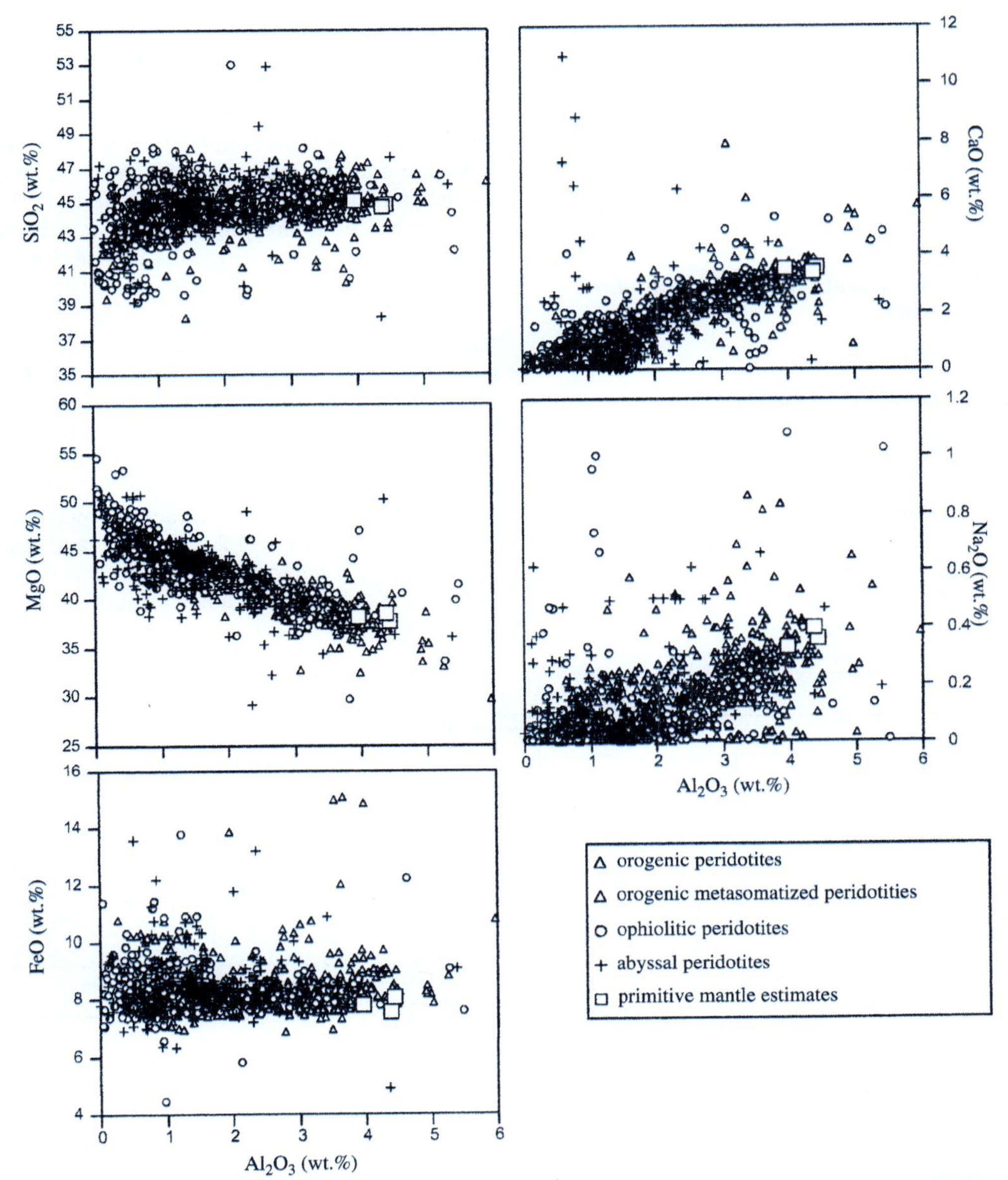

Figure 5 Plots of major-element oxides versus Al_2O_3 (anhydrous wt.%) in whole-rock orogenic, ophiolitic, and abyssal peridotites. *Orogenic peridotites:* Beni Bousera (Pearson *et al.*, 1993, unpublished; Gueddari *et al.*, 1996); Ronda (Frey *et al.*, 1985; Remaïdi, 1993; Gueddari *et al.*, 1996; Van der Wal and Bodinier, 1996; Lenoir, 2000; Lenoir *et al.*, 2001); Pyrenees (Bodinier *et al.*, 1988, 1990, 2004; Bodinier, 1989; Fabriès *et al.*, 1989, 1998; McPherson, 1994; McPherson *et al.*, 1996); Lower Austria Garnet Peridotites (Becker, 1996); Ulten (Obata and Morten, 1987; Morten and Obata, 1990); Finero, Balmuccia, and Baldissero (Ernst, 1978; Sinigoi *et al.*, 1983; Ottonello *et al.*, 1984a; Voshage *et al.*, 1988; Shervais and Mukasa, 1991; Hartmann and Wedepohl, 1993; Rivalenti *et al.*, 1995); Lanzo (Ottonello *et al.*, 1984a; Bodinier, 1988 (reproduced by permission of Elsevier from *Tectonophysics* **1988**, *149*, 67–88), 1989); Voltri Group (Western Liguria) and External Ligurides (Ernst and Piccardo, 1979; Beccaluva *et al.*, 1984; Ottonello *et al.*, 1984b; Rampone *et al.*, 1995); Horoman (Frey *et al.*, 1991; Takazawa *et al.*, 2000). *Ophiolitic peridotites:* Internal Ligurides (Beccaluva *et al.*, 1984; Ottonello *et al.*, 1984b; Rampone *et al.*, 1996, 1998); Antalya (Juteau, 1974); Oman (Lippard *et al.*, 1986; Godard *et al.*, 2000 (reproduced by permission of Elsevier from *Earth Planet. Sci. Lett.* **2000**, *180*, 133–148); Gerbert-Gaillard, 2002; Godard, unpublished); Yarlung Zangbo (Wu and Deng, 1980); Indonesia (Monnier, 1996); Papua–New Guinea (Jaques and Chappell, 1980; Jaques *et al.*, 1983); New Caledonia (Moutte, unpublished); Dun Mountain (Davies *et al.*, 1980); Trinity (Gruau *et al.*, 1991, 1995, 1998); Bay of Islands (Suhr and Robinson, 1994; Edwards and Malpas, 1995; Varfalvy *et al.*, 1997); Polar Urals (Sharma *et al.*, 1995). *Abyssal peridotites:* Iberia Abyssal Plain (Seifert and Brunotte, 1996; Hébert *et al.*, 2001); Mid-Atlantic Ridge (Bonatti *et al.*, 1970, 1971, 1993; Prinz *et al.*, 1976; Shibata and Thompson, 1986; Casey, 1997); South West Indian Ridge and American-Antarctic Ridge (Snow and Dick, 1995); East Pacific Rise (Constantin, 1995; Niu and Hékinian, 1997a); South Sandwich Forearc (Pearce *et al.*, 2000); Izu–Bonin–Mariana Arc-Basin system (Parkinson and Pearce, 1998). Primitive mantle estimates after Green *et al.* (1979), Jagoutz *et al.* (1979), and McDonough and Sun (1995).

REE (Figure 10), thorium and zirconium (Figure 11). Conventional, chondrite-normalized REE patterns are shown in Figures 12–15 for whole rocks and in Figures 18 and 22 for cpx. Primitive mantle-normalized trace-element patterns are shown in Figures 16 and 17 for whole rocks and Figure 21 for minerals.

2.04.3.1 Major Elements and Minor Transition Elements

2.04.3.1.1 Data

In spite of their variable provenance (subcontinental lithosphere, supra-subduction mantle wedge, or oceanic mantle), most of the tectonically emplaced and abyssal peridotites show coherent covariation trends for major elements (Figure 5). These variations reflect their variable modal compositions between a fertile end-member—comparable to proposed estimates for pristine (primitive) mantle compositions (Ringwood, 1975; Jagoutz *et al.*, 1979; McDonough and Sun, 1995)—and a refractory (olivine-rich) end-member. As illustrated in Figure 6, fertile compositions are predominant in the orogenic lherzolite massifs (maxima at 55–60% for modal olivine and 3.0–3.5% for Al_2O_3), whereas refractory compositions are predominant in ophiolitic and abyssal peridotites (maxima around 70–80% for modal olivine and 0.5–1.5% for Al_2O_3).

Similar to the variations observed in individual peridotite bodies (e.g., Frey *et al.*, 1985) as well as in mantle xenoliths (see Chapter 2.05), the whole data set shows a negative correlation between magnesium and aluminum, and a positive correlation between calcium and aluminum (Figure 5). Silicon decreases only slightly from fertile to mildly refractory peridotites (harzburgites with ~1% Al_2O_3), and more markedly from harzburgites to dunites (<0.5% Al_2O_3). FeO decreases rather steadily, but only

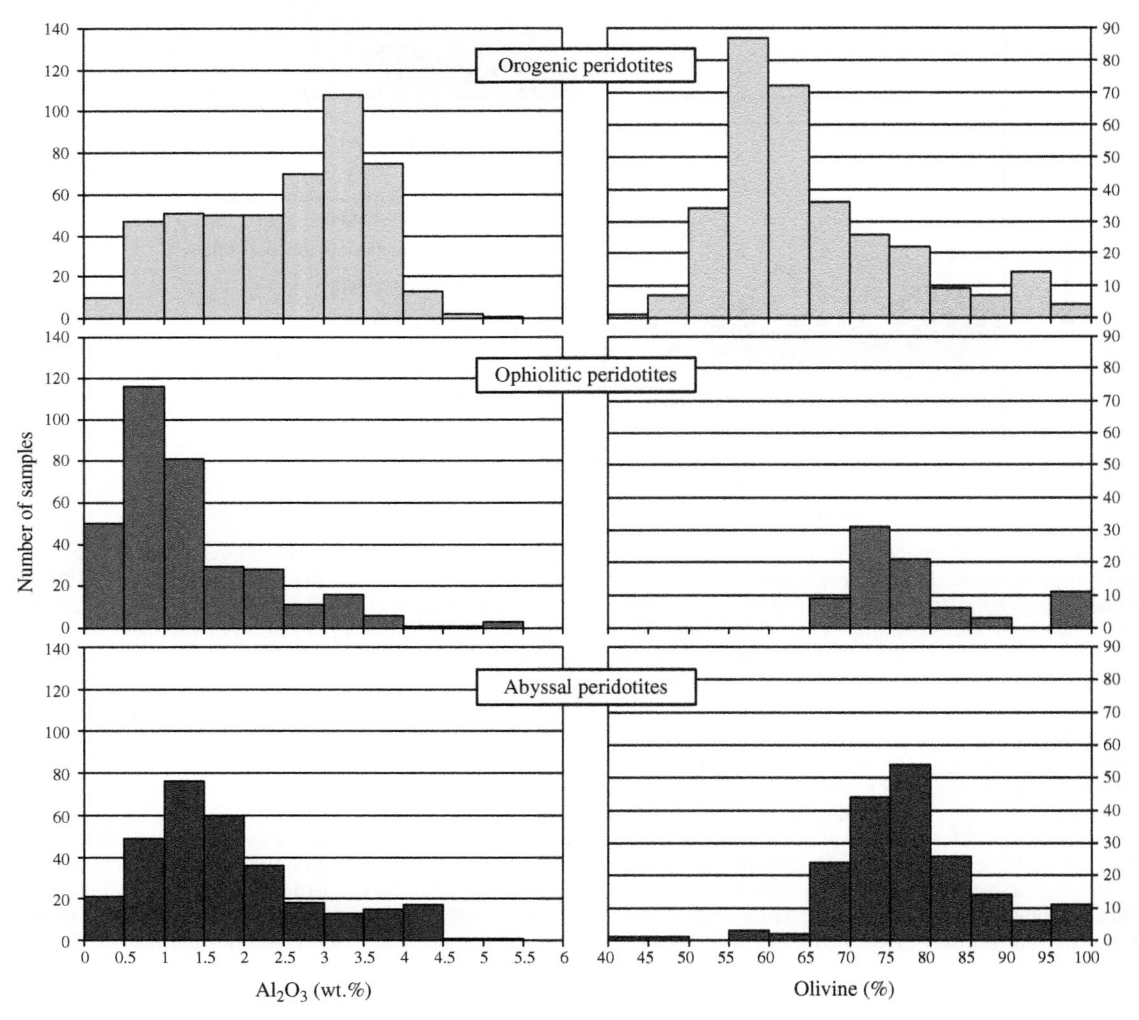

Figure 6 Distribution of Al_2O_3 abundances (anhydrous wt.%) and modal olivine in orogenic, ophiolitic, and abyssal peridotites. Data are the same as in Figure 1 (olivine) and Figure 5 (Al_2O_3).

very slightly, from fertile to refractory peridotites. Sodium is positively correlated with aluminum in the lherzolites (>2% Al_2O_3) but strongly scattered without significant covariation in the more refractory peridotites. Titanium increases with peridotite fertility while nickel and chromium decrease (Figure 7). However, these elements also show scattered variations which are especially marked in the metasomatized orogenic peridotites for titanium and chromium, and in the ophiolitic peridotites for the three elements. The dispersion is particularly important for chromium in orogenic and ophiolitic refractory peridotites (Al_2O_3 <2%).

Some of the deviations from covariation trends in Figures 5 and 7 may be accounted for by small-scale mineralogical heterogeneities. The heterogeneous distribution of spinel in coarse-grain peridotites, for instance, may sometimes affect their chemical composition. This is especially true for chromium in refractory peridotites. The wide range of chromium values reported for metasomatized orogenic peridotites (500–7,000 ppm), for instance, mainly reflects the "nugget" effect of spinel in small-size samples collected along a 65 cm wide harzburgite wall rock from Lherz (Bodinier *et al.*, 1990).

Secondary alteration processes are also responsible for element deviations, particularly the strong depletion of calcium and sodium relative to aluminum in some abyssal peridotites, as well as part of the scattered variations of silicon, magnesium, calcium, and sodium. Serpentinized, orogenic and ophiolitic peridotites may also be strongly depleted in calcium and sodium. However, these samples are rare in the selected database for peridotite massifs, which includes only moderately serpentinized samples.

Other deviations are ascribed to "primary" (mantle) processes in the upper mantle, prior to the tectonic emplacement of the peridotites. Particularly worthy of note is the iron enrichment in several samples (FeO content above 9%, sometimes up to 15%), which is generally coupled with titanium and sodium enrichment as well as with magnesium depletion. Iron enrichment is observed in abyssal peridotites, in refractory peridotites from the MTZ in ophiolites and in some dunites from LP orogenic lherzolites (Bodinier, 1988; Dick and Natland, 1996; Niu and Hékinian, 1997a; Godard *et al.*, 2000). The IP peridotites also contain strongly iron enriched peridotites occurring in wall rocks adjacent to pyroxenite dikes and layers (Bodinier *et al.*, 1988, 1990). Selective iron enrichment is often associated with textural and mineralogical evidence for refertilization or modal metasomatism and is, therefore, ascribed to melt–rock interactions (see discussion below). Likewise, the depletion of nickel (±Cr) in MTZ ophiolitic peridotites and that of chromium in metasomatized IP orogenic peridotites are generally considered as primary signatures related to melt–rock interactions (e.g., Suhr, 1999, for nickel).

2.04.3.1.2 Interpretations

Until the late eighties, the classic interpretation for the elemental covariation trends observed in mantle rocks was to invoke variable degrees of melt extraction during adiabatic partial melting of the asthenospheric mantle (Frey *et al.*, 1985; Bodinier, 1988; McDonough and Frey, 1989). In this scheme, melt extraction degrees calculated by peridotite–melt mass balance varied from nearly zero in the most fertile lherzolites (compositionally similar to PM, such as sample R717 from Ronda—Frey *et al.* (1985)) to >25% in the most refractory peridotites (Figure 8). However, some authors pointed out that the observed covariations were merely the reflection of variable amounts of basaltic component in the peridotites, a feature that could also be accounted for by melt transfer and redistribution in homogeneously molten peridotites (Dobretsov and Ashchepkov, 1991). Obata and Nagahara (1987), for instance, interpreted the compositional layering of the Horoman orogenic peridotite (Japan) with a model involving compaction/porosity waves during segregation of partial melt.

The occurrence of the most refractory peridotites as layers intercalated between more fertile peridotites is indeed hardly consistent with models involving variable degrees of partial melting as these models would imply unrealistic, small-scale, thermal gradients. Layered structures are common in most peridotite massifs (e.g., the harzburgite layers in the Horoman, Ronda, and Lherz orogenic lherzolite bodies, Figure 30, and the "tabular" dunites in the harzburgite mantle sequence of the Oman ophiolite—Avé Lallemant (1967), Niida (1974), Conquéré (1978), Gervilla and Remaïdi (1993), Kelemen *et al.* (1995)). Most orogenic lherzolites show bi- or plurimodal distributions of mineralogical modes and major-element compositions, a feature that cannot be accounted for by a simple model involving variable melting degrees.

A possible interpretation is to ascribe the layering to stretching and multiple folding of older, larger-scale structures by mantle convection. This model is analogous to the "marble-cake" model proposed by Allègre and Turcotte (1986) for the pyroxenite layering (see Section 2.04.4.2.3) and was suggested by Toramaru *et al.* (2001) for the peridotite layering in the Horoman orogenic massif (see also Section 2.04.5.2.1). However, in some instances (e.g., in Ronda and in several ophiolites), the layering overprints the deformation and, therefore, cannot be explained by this model. One may envisage that the peridotites were homogenously molten but

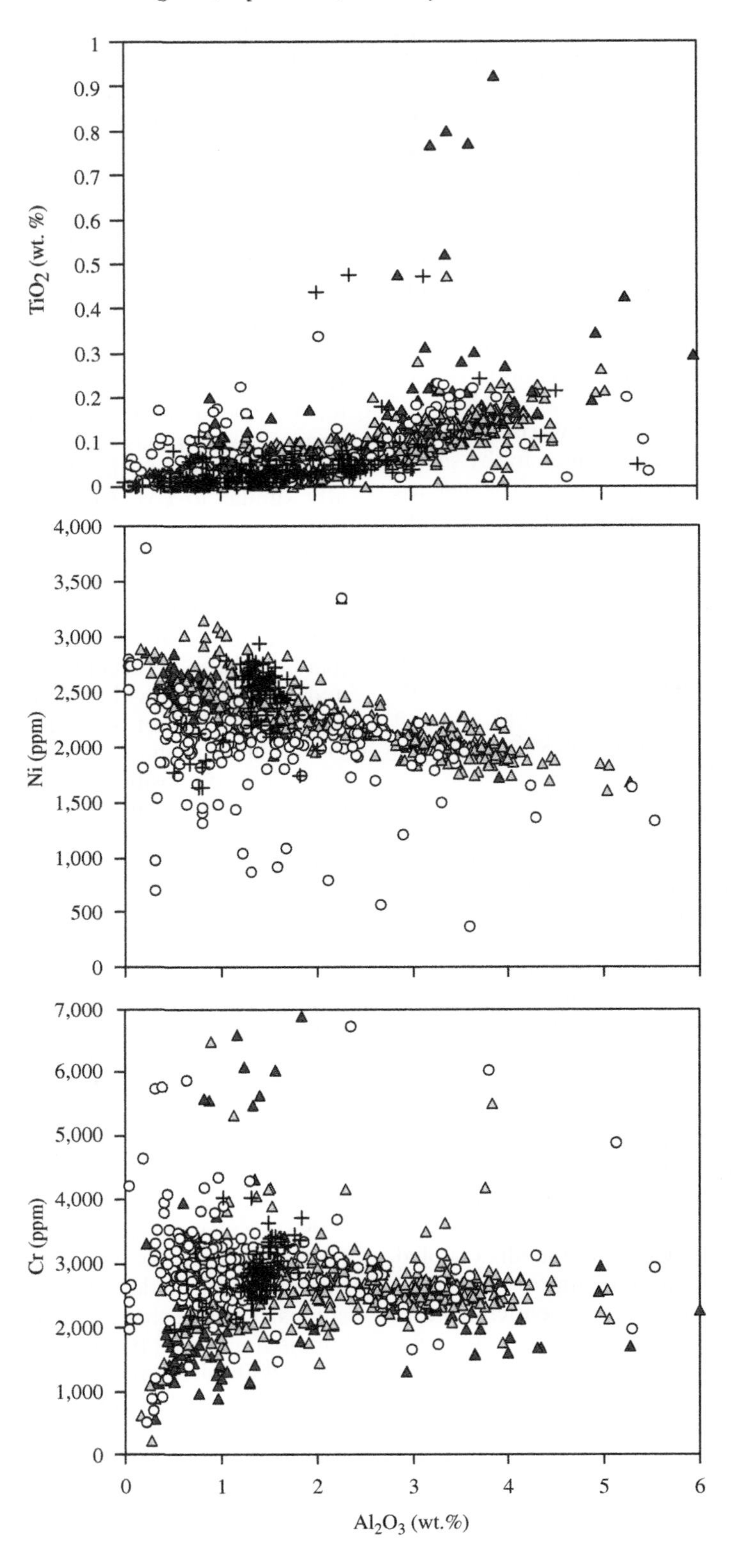

Figure 7 Plots of TiO_2 (anhydrous wt.%), nickel, and chromium (anhydrous ppm) versus Al_2O_3 (anhydrous wt.%) in whole-rock orogenic, ophiolitic, and abyssal peridotites. The symbols and the data sources are the same as in Figure 5.

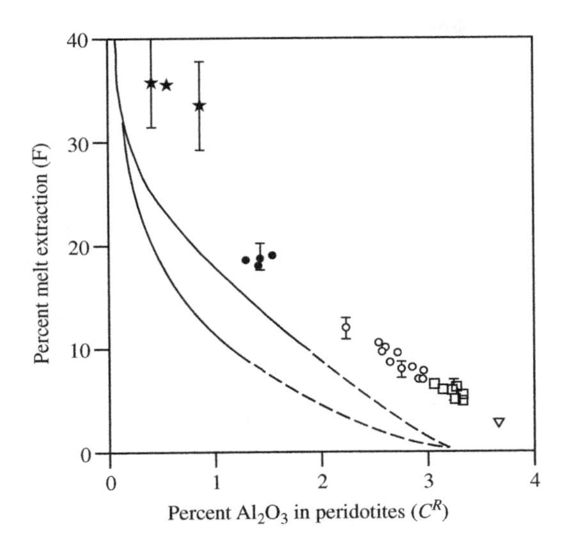

Figure 8 Melt extraction degrees calculated for individual peridotite samples from the Lanzo orogenic peridotite and plotted against Al_2O_3 abundances (anhydrous wt.%) (after Bodinier, 1988) (reproduced by permission of Elsevier from *Tectonophysics* **1988**, *149*, 67–88). Melt extraction degrees were obtained by total inversion of peridotite–melt mass balances, with melt compositions constrained by experimental studies. The symbols for the peridotite facies are as follows: empty triangle—most fertile plagioclase lherzolites ($Al_2O_3 > 3.5\%$); empty squares—mildly fertile plagioclase lherzolites ($3.0\% < Al_2O_3 < 3.5\%$); empty circles—less fertile plagioclase lherzolites ($Al_2O_3 < 3.0\%$); full circles—olivine-rich (refractory) spinel lherzolites; stars—dunites. The continuous lines represent the field of experimental data for the melting of a lherzolite with 3.2% Al_2O_3 in the pressure range 5–15 kbar (Jaques and Green, 1980).

partial melts were preferentially drained from the refractory layers, either due to compaction processes (Obata and Nagahara, 1987), or because they were tapped from wall rocks of magma conduits (Boudier and Nicolas, 1977; Nicolas, 1986b; Takahashi, 1992). However, even these models do not account for all of the mineralogical and geochemical features of the refractory peridotites. In particular, an extremely high melting degree (implying anomalously high temperatures at shallow depth) would be required to produce the dunites. However, these rocks show replacive relationships negating important volume reduction by compaction (Nicolas and Prinzhofer, 1983; Kelemen *et al.*, 1995). Moreover, many refractory peridotites have much lower Mg# values (=Mg/(Mg + Fe) cationic ratio) and nickel content than predicted by melting models (Suhr, 1999).

To account for their distinctive characteristics, several authors (notably Peter Kelemen) suggested the formation of the layered refractory peridotites by olivine-forming, melt–rock reactions, leading to the evolution of "porous-flow channels" (Kelemen, 1990; Bodinier *et al.*, 1991; Kelemen *et al.*, 1990, 1992, 1995, 1997; Gervilla and Remaïdi, 1993; Kelemen and Dick, 1995; Suhr, 1999). Reactive melt transport can result in steep modal/chemical gradients at very small scale (Godard *et al.*, 1995; Suhr, 1999) and is now widely recognized as the predominant mechanism to generate mantle dunites. A similar origin can be envisioned for harzburgites, especially when they are closely associated with dunites (Gervilla and Remaïdi, 1993). Harzburgites may either record transient stages of the dunite-forming melt–rock reaction, or they may represent final reaction products at higher pressure. Alternatively, Kelemen *et al.* (1992) suggested the formation of some harzburgites by opx-forming reactions.

While many refractory peridotites were likely produced by melt–rock reactions at nearly constant or increasing melt mass (Kelemen, 1990; Kelemen *et al.*, 1990), some fertile peridotites were conversely refertilized as the result of the solidification of infiltrated melt (a process also referred to as "percolative fractional crystallization" by Harte *et al.*, 1993). (Re)fertilization processes have been invoked for several types of orogenic, ophiolitic, and oceanic peridotites, such as:

(i) "metasomatized," pyroxene-rich (±amphibole-bearing) spinel lherzolites in the Pyrenean lherzolite bodies (Bodinier *et al.*, 1988; Fabriès *et al.*, 1989);

(ii) "secondary" spinel lherzolites in the granular domain of the Ronda orogenic lherzolite massif (Van der Wal and Bodinier, 1996; Lenoir *et al.*, 2001);

(iii) "fertile" lherzolites in the Horoman orogenic massif (Saal *et al.*, 2001);

(iv) plagioclase lherzolites in the New Caledonia, Western Alps, Ligurian, and Corsican ophiolites (Nicolas and Dupuy, 1984; Rampone *et al.*, 1994, 1997);

(v) plagioclase- and cpx-bearing, and/or iron-rich refractory peridotites in LP orogenic lherzolite massifs, in the mantle–crust transition zone of the ophiolites and in several suites of abyssal peridotites (Dick *et al.*, 1984; Bodinier, 1988; Boudier and Nicolas, 1995; Dick and Natland, 1996; Niu and Hékinian, 1997a); and

(vi) "basal" lherzolites and cpx-harzburgites in the Oman ophiolite (Godard *et al.*, 2000; Takazawa *et al.*, 2003).

Fertilization of mantle rocks may occur in conductively cooled or heated mantle, at the margin of asthenospheric melting domains where infiltrated melts will solidify (e.g., Lenoir *et al.*, 2001). Published examples include transform faults bounding melting regions in an oceanic setting (Nicolas and Dupuy, 1984) and the base of oceanic or continental lithosphere thermally eroded by partially molten, upwelling asthenosphere (Godard *et al.*, 2000; Lenoir *et al.*, 2001).

Fertilization may alternatively result from the fractional solidification of partial melts incompletely drained from residual mantle, upon cooling of melting domains due to mantle convection or tectonic upwelling. In this situation, the refertilized peridotites (sometimes associated with "replacive" pyroxenites—see Section 2.04.4.2.2) may define a layering interpreted in terms of high-porosity compaction waves or porous-flow channels (Obata and Nagahara, 1987; Van der Wal and Bodinier, 1996; Garrido and Bodinier, 1999).

Peridotite fertilization may also result from the fractional solidification of "exotic" (deep-seated) melts infiltrated in wall rocks of translithospheric magma conduits. This process was first described in composite mantle xenoliths (Wilshire and Shervais, 1975; Gurney and Harte, 1980; Irving, 1980; Wilshire *et al.*, 1980; Boivin, 1982; Harte, 1983; Harte *et al.*, 1993; Menzies *et al.*, 1987), where it is referred to as "modal metasomatism" when new (generally hydrous) minerals are precipitated (Dawson, 1984; Kempton, 1987), or "Fe–Ti metasomatism" (Menzies *et al.*, 1987) when the attention is focused on chemical enrichment. In contrast with ultramafic xenoliths, the tectonically emplaced and oceanic peridotites contain only sparse rock types attributable to mantle metasomatism by deep-seated melts. Examples of wall–rock, modal, and Fe–Ti metasomatism were nevertheless described in IP orogenic lherzolites, notably in the Pyrenees (Fabriès *et al.*, 1989; Bodinier *et al.*, 1988, 1990, 2003; McPherson *et al.*, 1996; Woodland *et al.*, 1996).

In several cases, the fertilization is supported by textural and/or mineralogical evidence for crystallization of secondary minerals from infiltrated melt (e.g., interstitial or replacive (coronitic or poikiloblastic) pyroxene ± aluminous phases such as plagioclase, spinel, garnet, or amphibole). However, these textural/mineralogical features may also be erased by recrystallization and deformation so that the fertilization may occur without leaving any trace in terms of texture and mineralogy.

With regard to major and minor transition elements, the chemical signature of fertilization is highly variable depending on the mechanism involved. Two end-member processes may be postulated for simplicity:

(i) Infiltration of equilibrium partial melt in residual peridotite, associated with substantial volume increase and complete solidification of the melt at near-solidus conditions. This process will produce fertile peridotites with compositions that will roughly plot on the element covariation trends (Figures 5 and 7). In the lack of textural evidence, this refertilization process will remain unnoticed.

(ii) Infiltration of melt down a thermal gradient from near- to subsolidus conditions. In this case, the interstitial melt will solidify though successive melt-consuming reactions involving substantial dissolution of pre-existing peridotite minerals. This process is associated with a significant evolution of melt composition, from originally basaltic to progressively volatile-enriched, possibly culminating in low-T small-volume carbonate melts (Bodinier *et al.*, 2004, and references herein). This "fractional" fertilization mechanism results in peridotite compositions that significantly depart from the element covariation trends and may vary widely, depending on several factors:

(a) *Melt composition*. In the Pyrenees, for instance, the selective enrichment of titanium in metasomatized and fertilized peridotites is partly the reflection of the enriched (alkaline) character of the infiltrated melt.

(b) *Mineralogical reactions*. At shallow depth, the fertilized peridotites are often characterized by selective precipitation of cpx at the expense of olivine or opx, resulting in lower Al/Ca ratio than PM (e.g., Fabriès *et al.*, 1989; Godard *et al.*, 2000). Conversely, some peridotites are selectively enriched in aluminous phases (garnet, spinel, or plagioclase) and are, therefore, distinguished by anomalously high Al/Ca ratios.

(c) *Differentiation of interstitial melt*. Melt-consuming reactions in peridotite wall–rock (and mineral segregation in veins) may result in strong iron (±Na, ±Ti) enrichment in fertilized peridotites, that will, therefore, plot well above the element covariation trends (Figures 5 and 7). In their theoretical approach of trace-element redistribution in reactive porous flow systems, Godard *et al.* (1995) predicted that the combination of the source/sink effects of melt–rock reactions and the chromatographic effects of melt transport (Navon and Stolper, 1987) would result in considerable variations of chemical enrichment (or depletions) patterns as a function of peridotite location in a percolation column. Up to a point, this approach can be applied to Mg–Fe variations, suggesting that the iron enrichment is restricted to the domains where the melt/rock ratio was high enough to counterbalance the buffering effect of peridotite minerals. This might explain why in the Pyrenees the wall–rock peridotites closely adjacent to pyroxenite dikes (e.g., in the Lherz and Freychinède bodies) are so strongly enriched in iron whereas the cpx-rich peridotites from Caussou, which are strongly metasomatized but spatially unrelated to dikes, have preserved "normal" mantle Mg# values (Bodinier *et al.*, 1988; Fabriès *et al.*, 1989). Similarly, data on composite mantle xenoliths have led several authors to ascribe the iron-rich peridotites to wall rocks of magma conduits (Menzies *et al.*, 1987). Conversely, the dependence of olivine/melt Mg–Fe partitioning on the magnesium and iron content in melt (e.g., Ulmer, 1989) may result in increasing Mg# values in fertilized peridotites, especially when they are infiltrated by magnesium- and iron-rich melts equilibrated with refractory rocks (Figure 9).

In addition to textural/mineralogical and major-/minor-element chemical arguments, anomalous trace-element signatures were also put forward as evidence for mantle fertilization. Some of these arguments are based on the REE distribution and will be discussed in the next section (e.g., LREE depletion in otherwise fertile lherzolites or europium enrichment indicating plagioclase crystallization in open system). Likewise, suprachondritic $^{187}Re/^{188}Os$ ratios were considered as evidence for the origin of the Horoman fertile lherzolites by refertilization processes (Saal *et al.*, 2001).

There is finally a growing body of evidence that the wide range of modal, major-, and minor-element compositions observed in the orogenic peridotite bodies (notably in the layered ones) is largely contributed by multiple melt transfer and melt–rock reaction processes. The layered granular subdomain of the Ronda orogenic lherzolite (Van der Wal and Bodinier, 1996), for instance, shows a wide dispersion of Mg# values plotted versus olivine proportions, of which only a very limited array can be accounted for by partial melting (Figure 9). Alternatively, melt–rock interactions—varying

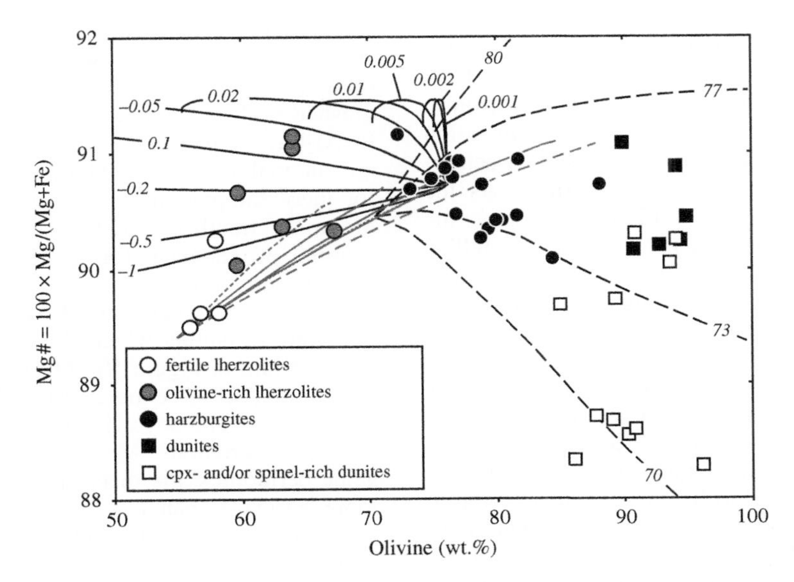

Figure 9 Variation of the Mg# number (= 100 × Mg/(Mg + Fe) cationic ratio) versus modal olivine in the "Layered Granular Subdomain" of the Ronda orogenic peridotite (Gervilla and Remaïdi, 1993; Van der Wal and Bodinier, 1996), compared with modeling of Mg# variations during partial melting and melt–rock interactions. Numerical experiments were performed with the "Plate Model" of Vernières *et al.* (1997), initially designed for trace-element applications. For modeling the Mg# number, we used a total inverse method to constrain the Mg–Fe redistribution between solid and liquid phases at each reaction increment. This allows us to avoid mineral stoichiometry violation, even in the case of complex melt–rock interactions (Bedini *et al.*, 2003). The approach was applied to partial melting in the stability field of spinel peridotites and various melt–rock reactions. Mineral/melt partition coefficients (Ulmer, 1989) are dependent on pressure and temperature for partial melting, and on the MgO content in melt for melt–rock reactions. The results confirm that partial melting accounts only for a small array of the Mg# versus olivine variations observed in the Ronda peridotites. Refertilization produces peridotites with Mg# ratios at a given olivine proportion higher than predicted by melting models. Conversely, olivine- (± cpx-) forming reactions may account for low Mg# values in harzburgites and dunites. The partial melting models (light lines) were calculated with the melting reactions published by Kostopoulos (1991), Baker and Stolper (1994), Walter *et al.* (1995), at 11 kbar and 16 kbar, and Niu *et al.* (1997), at 10 kbar and 15 kbar. The source composition (Mg# = 0.893) was fixed by extrapolation of the Mg# number in fertile Ronda lherzolites down to 55% olivine on a spinel-free basis. The melting models were run up to complete cpx consumption. The refertilization models (bold continuous lines) involve precipitation of cpx + opx at the expense of melt + olivine. As initial peridotite, we took the melting residue obtained after 25% melt extraction with the melting reaction of Niu *et al.* (1997) at 15 kbar and 1,250 °C. The infiltrated melt is the "equilibrium" melt fraction produced with the same model. The refertilization results are shown for *R* values (mass ratio of crystallized minerals to infiltrated melt) ranging from 0.001 to 1—i.e., from conditions essentially controlled by melt percolation to conditions mainly governed by the reaction. The olivine-forming reaction models (bold dashed lines) involve precipitation of olivine at the expense of opx, at nearly constant or decreasing melt mass. As initial peridotite composition for the olivine-forming reaction, we took the melting residue obtained after 20% melt extraction with the melting reaction of Niu *et al.* (1997) at 15 kbar and 1,250 °C. The results are shown for variable infiltrated melt compositions, from Mg# = 70 to Mg# = 80.

from olivine- (± cpx-, ± spinel-) forming reactions at constant or increasing melt mass to cpx- (± opx-, ± spinel-) reactions at decreasing melt mass—may account for the whole range of Mg# versus olivine variation.

Similar to the "layered" refractory peridotites of the orogenic lherzolite massifs, the ophiolitic and abyssal harzburgites recently have been the focus of debates on their significance, with implications for melt generation at ocean ridges (e.g., for abyssal peridotites: Niu, 1997, 1999; Niu *et al.*, 1997; Niu and Hékinian, 1997a; Walter, 1999; Baker and Beckett, 1999). Discussions were especially centered on the extent of the modifications registered by these rocks after melt extraction. Niu (1997) and Niu *et al.* (1997) suggested that the abyssal peridotites from fast-spreading ridges contain far more olivine than would be predicted from melting models. They considered that their composition results from a combination of melting and olivine crystallization that would be both a natural response to ascent of solid and melt beneath an ocean ridge. As the buoyant melts migrate upwards, they encounter the surface thermal boundary layer and crystallize liquidus olivine. The greater the ambient extent of melting of the mantle, the higher the normative olivine content of the melt and the more melt is produced. Hence greater extents of melting would lead to more olivine crystallization at shallow levels. This process is comparable to the olivine-forming melt–rock reaction in increasing the olivine proportion in residual peridotite, but may also be viewed as fertilization mechanisms as it increases the iron and incompatible element content of the peridotites (Niu and Hékinian, 1997a).

However, for the reason exposed above, a significant and widespread decrease of the olivine forsterite content is hardly consistent with the relatively low melt/rock ratios implied by diffuse porous flow. In fact, Baker and Beckett (1999) negated the presence in abyssal peridotites of modal olivine in excess of the values predicted by melting models, as well as the existence of a negative correlation between forsterite and modal olivine. However, they acknowledged that the abyssal peridotites are selectively enriched in incompatible elements, implying a refertilization stage after melt extraction (Niu and Hékininan, 1997a; Asimow, 1999; Walter, 1999). We shall see in Section 2.04.3.2.2 that this enrichment is patent from the trace-element signature of the ophiolitic harzburgites and abyssal peridotites from fast-spreading ridges. This signature can be possibly explained by the combined effects of (reactive) melt transport and melt entrapment. Asimow (1999) found that major- and trace-element data in abyssal peridotites from slow-spreading ridges are consistent with a variety of melting and melt migration histories that include elements or episodes both of near-fractional melting and of reactive porous flow. A component of reactive porous flow explains peridotite compositions better than olivine deposition or refertilization.

2.04.3.2 Lithophile Trace Elements

2.04.3.2.1 Al_2O_3 covariation trends

Al_2O_3 covariation trends. As previously noted by several authors (e.g., McDonough and Frey, 1989, and references herein), the heavy REE (HREE) in mantle peridotites are negatively correlated with magnesium and positively correlated with aluminum. Most tectonically emplaced and abyssal peridotites show a well-defined ytterbium versus Al_2O_3 covariation trend (Figure 10). Only a few, plagioclase-bearing, ophiolitic and orogenic peridotites are distinguished by low ytterbium concentrations relative to Al_2O_3. Conversely, some abyssal peridotites are anomalously enriched in ytterbium. Except for these samples, no significant deviation from the main trend is observed as a function of sample provenance (orogenic, ophiolitic, or abyssal peridotites).

Alternatively, the light REE (LREE) show scattered variations when plotted against Al_2O_3 (e.g., cerium in Figure 10). Overall, the orogenic lherzolites tend to be more enriched in LREE than the ophiolitic and abyssal peridotites, irrespective of Al_2O_3 variations. Moreover, within the orogenic peridotites, the metasomatized samples are more enriched than the unmetasomatized ones. When considered separately, the unmetasomatized orogenic peridotites and the ophiolitic peridotites show rough positive correlations between cerium and Al_2O_3.

The middle REE (MREE) show Al_2O_3 covariation patterns intermediate between those of heavy and LREE. Samarium shows both a rough positive correlation reminiscent of the one observed for ytterbium, and variations irrespective of Al_2O_3 comparable to those observed for cerium (i.e., higher samarium in orogenic peridotites, especially the metasomatized samples, than in ophiolitic and abyssal peridotites).

Among the lithophile trace elements, the highly incompatible elements (HIE—rubidium, caesium, barium, niobium, tantalum, thorium, and uranium) are strongly dispersed on the aluminum covariation diagrams and show variations dominantly governed by peridotite provenance. Thorium, for instance (Figure 11), is not correlated with Al_2O_3 and shows decreasing abundances in the order: metasomatized orogenic peridotites → unmetasomatized orogenic peridotites → ophiolitic and abyssal peridotites. Overall, the less incompatible elements (strontium, yttrium, zirconium, and hafnium) show variations comparable to

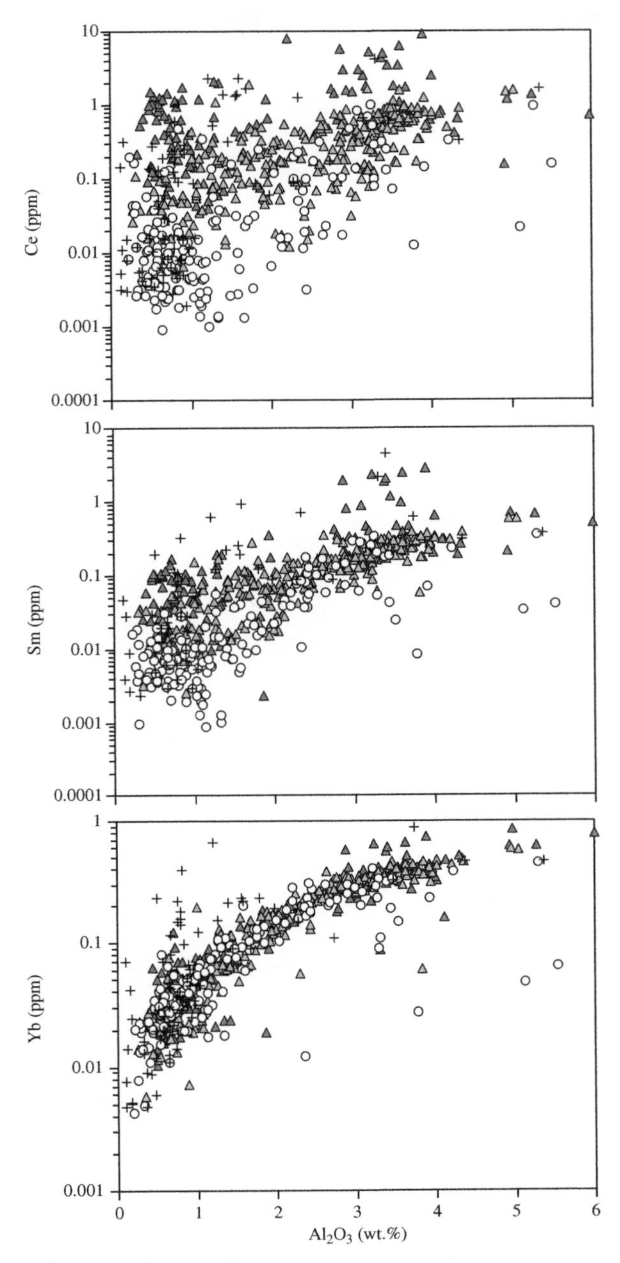

Figure 10 Plots of cerium, samarium, and ytterbium (ppm) versus Al_2O_3 (anhydrous wt.%) in whole-rock orogenic, ophiolitic, and abyssal peridotites. The symbols are the same as in Figure 5. *Orogenic peridotites:* Ronda (Frey *et al.*, 1985; Remaïdi, 1993; Van der Wal and Bodinier, 1996; Garrido *et al.*, 2004; Lenoir, 2000; Lenoir *et al.*, 2001); Pyrenees (Bodinier *et al.*, 1988, 1990, 2004; Bodinier, 1989, unpublished; Fabriès *et al.*, 1989, 1998; McPherson, 1994; McPherson *et al.*, 1996); Lower Austria Garnet Peridotites (Becker, 1996); Ulten (Obata and Morten, 1987; Morten and Obata, 1990); Alpe Arami (Ernst, 1978; Ottonello *et al.*, 1984a); Finero, Balmuccia, and Baldissero (Ernst, 1978; Ottonello *et al.*, 1984a; Hartmann and Wedepohl, 1993; Rivalenti *et al.*, 1995; Grieco *et al.*, 2001); Lanzo (Bodinier, 1988 (reproduced by permission of Elsevier from *Tectonophysics* **1988**, *149*, 67–88), 1989); Voltri Group (Western Liguria) and External Ligurides (Ottonello *et al.*, 1984b; Rampone *et al.*, 1995; Scambelluri *et al.*, 2001); Horoman (Frey *et al.*, 1991; Takazawa *et al.*, 2000); Zabargad (Dupuy *et al.*, 1991). *Ophiolitic peridotites:* Cuba (Proenza *et al.*, 1999); Internal Ligurides (Ottonello *et al.*, 1984b; Rampone *et al.*, 1996, 1998); Oman (Pallister and Knight, 1981; Lippard *et al.*, 1986; Godard *et al.*, 2000 (reproduced by permission of Elsevier from *Earth Planet. Sci. Lett.* **2000**, *180*, 133–148); Gerbert-Gaillard, 2002; Godard, unpublished); Indonesia (Monnier, 1996); New Caledonia (Prinzhofer and Allègre, 1985); Trinity (Gruau *et al.*, 1991, 1995, 1998); Bay of Islands (Edwards and Malpas, 1995; Varfalvy *et al.*, 1997); Polar Urals (Sharma *et al.*, 1995). *Abyssal peridotites:* Iberia Abyssal Plain (Seifert and Brunotte, 1996); Mid-Atlantic Ridge (Casey, 1997); East Pacific Rise (Niu and Hékinian, 1997a; Godard, unpublished); South Sandwich Forearc (Pearce *et al.*, 2000); Izu–Bonin–Mariana Arc-Basin system (Parkinson and Pearce, 1998).

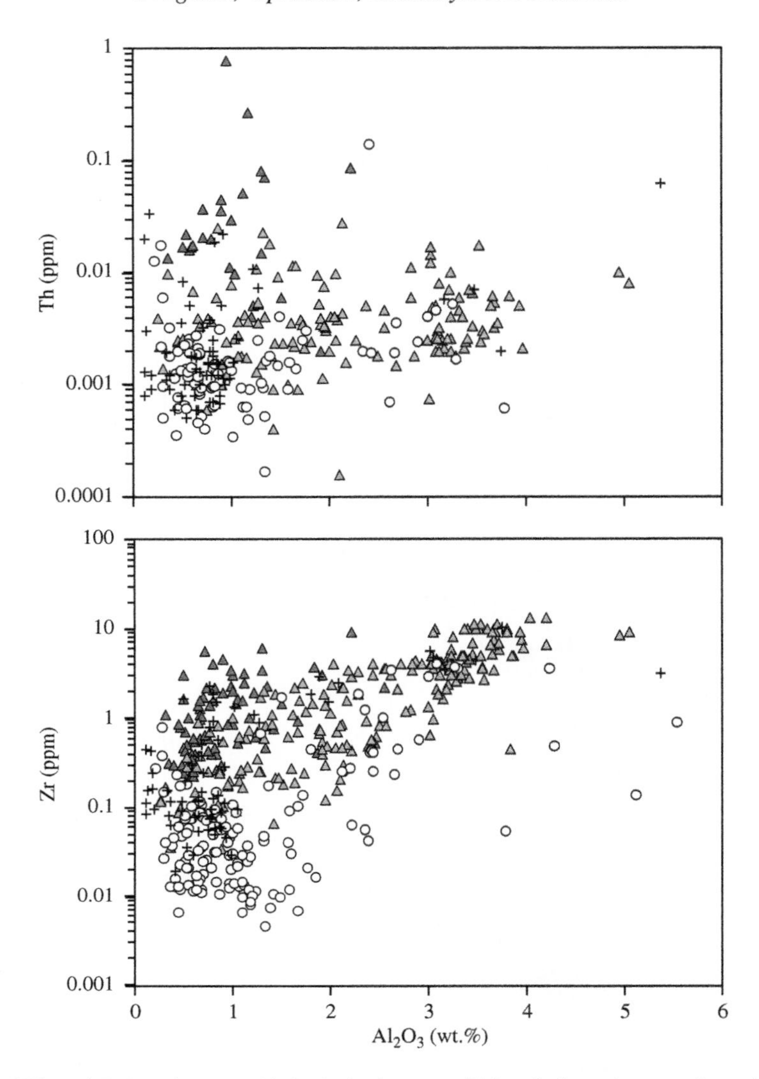

Figure 11 Plots of Th and Zr (ppm) versus Al_2O_3 (anhydrous wt.%) in whole-rock orogenic, ophiolitic, and abyssal peridotites. The symbols are the same as in Figure 5. *Orogenic peridotites:* Ronda (Remaïdi, 1993; Van der Wal and Bodinier, 1996; Garrido *et al.*, 2000; Lenoir, 2000; Lenoir *et al.*, 2001); Pyrenees (McPherson, 1994; McPherson *et al.*, 1996; Bodinier *et al.*, 2004); Finero, Balmuccia, and Baldissero (Hartmann and Wedepohl, 1993; Rivalenti *et al.*, 1995); Voltri Group (Western Liguria) and External Ligurides (Rampone *et al.*, 1995; Scambelluri *et al.*, 2001); Horoman (Frey *et al.*, 1991; Takazawa *et al.*, 2000). *Ophiolitic peridotites:* Cuba (Proenza *et al.*, 1999); Internal Ligurides (Rampone *et al.*, 1996, 1998); Oman (Godard *et al.*, 2000 (reproduced by permission of Elsevier from *Earth Planet. Sci. Lett.* **2000**, *180*, 133–148); Gerbert-Gaillard, 2002; Godard, unpublished); Indonesia (Monnier, 1996); Papua–New Guinea (Jaques and Chappell, 1980; Jaques *et al.*, 1983); Bay of Islands (Varfalvy *et al.*, 1997). *Abyssal peridotites:* Iberia Abyssal Plain (Seifert and Brunotte, 1996); East Pacific Rise (Niu and Hékinian, 1997a; Godard, unpublished); South Sandwich Forearc (Pearce *et al.*, 2000); Izu–Bonin–Mariana Arc-Basin system (Parkinson and Pearce, 1998).

those observed for REE with similar incompatibility degrees. The zirconium versus Al_2O_3 covariation pattern (Figure 11), for instance, is comparable to the one observed for samarium (Figure 10).

Chondrite-normalized REE patterns. With regard to chondrite-normalized REE distributions in whole rocks, the tectonically emplaced and abyssal peridotites may be subdivided in six main groups:

(i) *Unmetasomatized fertile orogenic lherzolites* (Figure 12, continuous lines), that show almost exclusively the classic "N-MORB" REE pattern characterized by a flat HREE segment (Gd–Lu) at ~2 × the chondrite abundances and a selective depletion of LREE, (LREE/MREE and LREE/HREE < chondrites). Lanthanum concentrations are typically in the range 0.1–1× chondrites for these samples. Only a few fertile lherzolites (in the Pyrenees and Lanzo) are

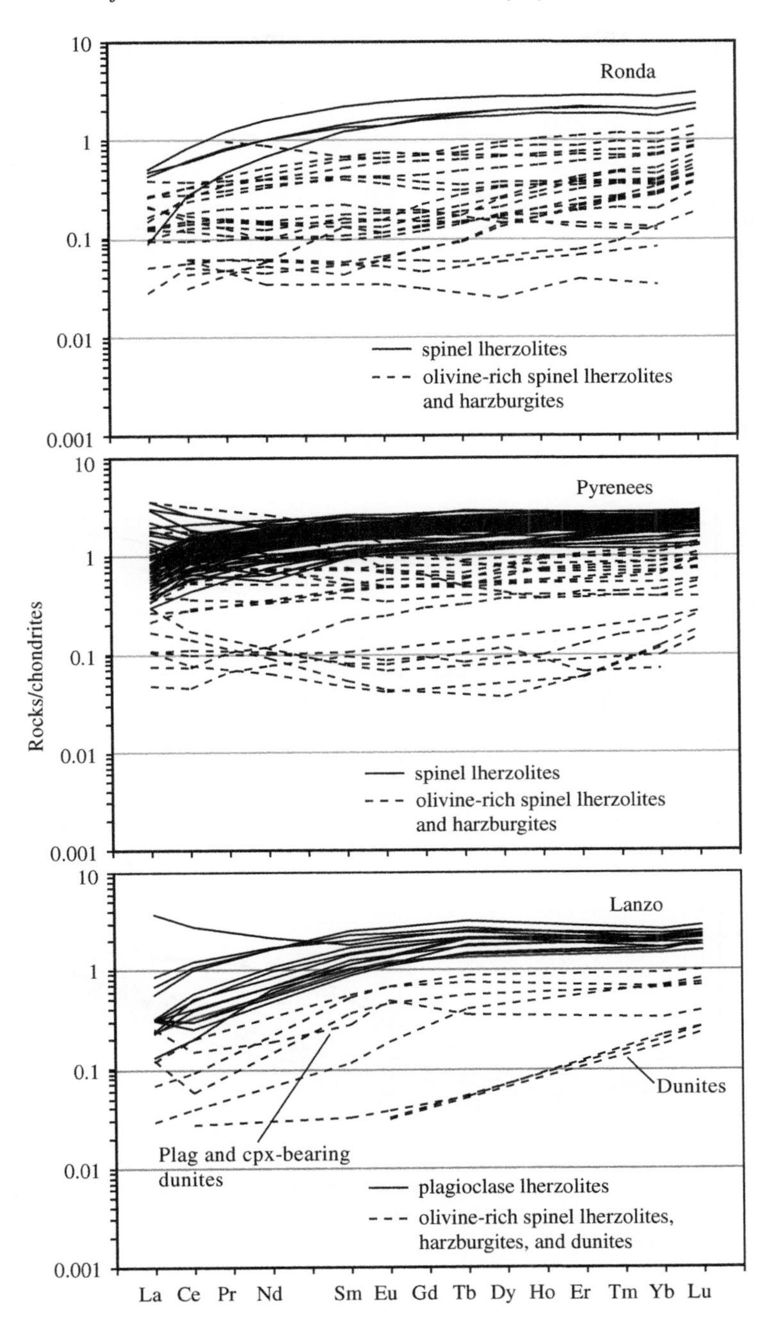

Figure 12 Chondrite-normalized abundances of REEs in representative "unmetasomatized" orogenic peridotites from Ronda, Pyrenees, and Lanzo (whole-rock analyses). Data from Bodinier (1988, 1989, unpublished), Bodinier *et al.* (1988), Remaïdi (1993), Gervilla and Remaïdi (1993), McPherson (1994), McPherson *et al.* (1996), and Fabriès *et al.* (1998). Normalizing values after Sun and McDonough (1989).

distinguished by a slight enrichment of LREE relative to MREE.

(ii) *Ophiolitic lherzolites and cpx-harzburgites* (Figure 13, heavy continuous lines), which are also characterized by LREE-depleted, "N-MORB" patterns. However, these samples differ from the fertile orogenic lherzolites by lower REE concentrations (HREE $= 0.5-1.5 \times$ chondrites) and lower LREE/MREE, MREE/HREE, and LREE/HREE ratios, a difference which is also apparent from cpx compositions (Rivalenti *et al.*, 1996). The lanthanum concentrations are typically in the range $0.005-0.05 \times$ chondrites. Based on their REE distribution in cpx (Figure 18), the abyssal peridotites from slow-spreading ridge systems (<1 cm yr^{-1} half rate) may also be included in this group. The cpx from the American-Antarctic and Southwest Indian Ridges

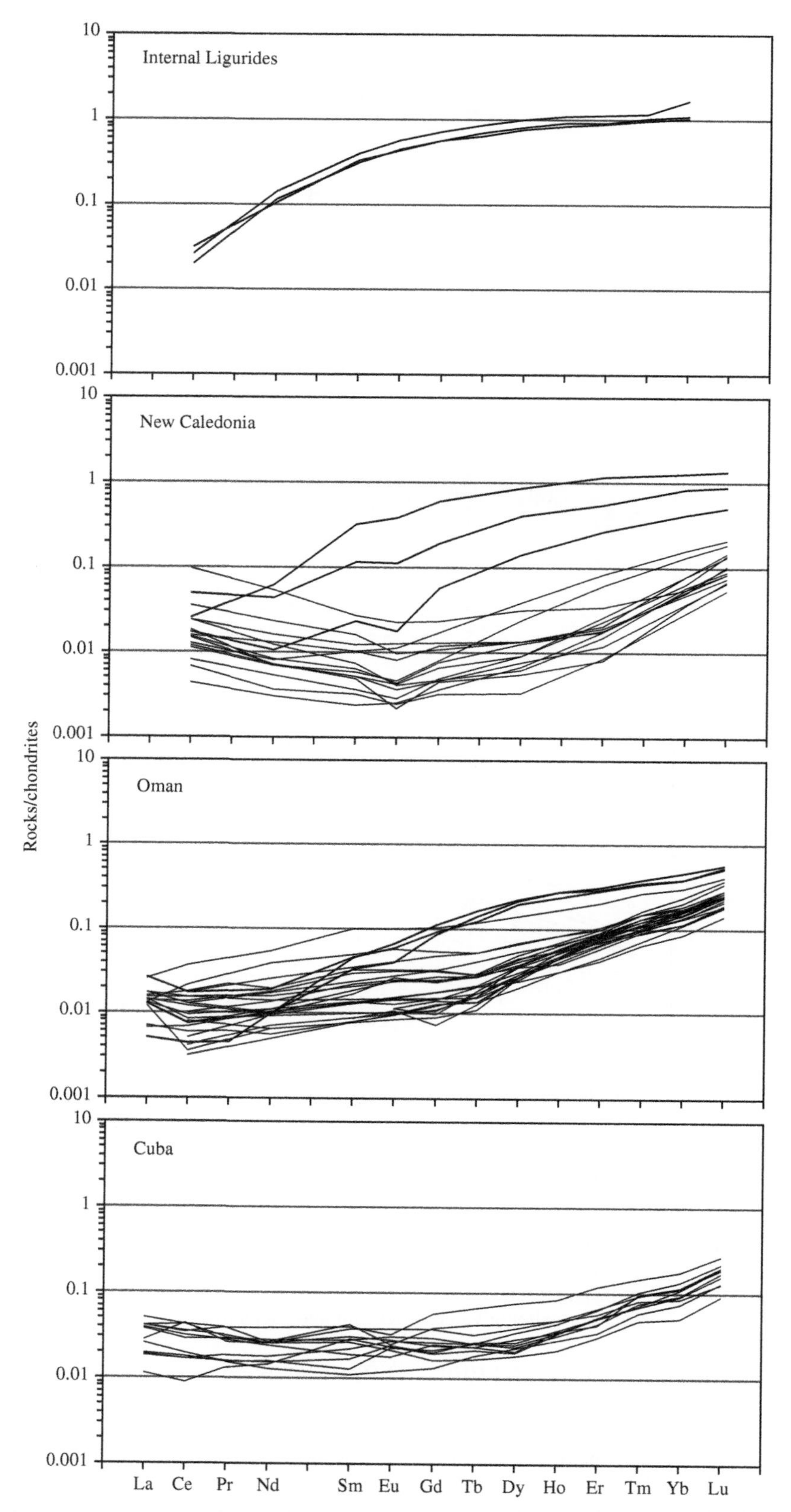

Figure 13 Chondrite-normalized abundances of REEs in representative ophiolitic peridotites from the Internal Ligurides, New Caledonia, Oman, and Cuba (whole-rock analyses). Data from Rampone *et al.* (1996), Prinzhofer and Allègre (1985), Godard *et al.* (2000) (reproduced by permission of Elsevier from *Earth Planet. Sci. Lett.* **2000**, *180*, 133–148), and Proenza *et al.* (1999), respectively. Normalizing values after Sun and McDonough (1989).

systems, for instance (Johnson *et al.*, 1990), are characterized by a strong depletion of LREE relative to HREE, which is a typical feature of this group. In addition, their REE patterns are very similar to those reported for cpx from ophiolitic lherzolites (e.g., the Internal Ligurides—Rampone *et al.*, 1996).

(iii) *Unmetasomatized refractory orogenic peridotites* (Figure 12, dashed lines), characterized by overall REE depletion relative to chondrites. These rocks show a wide compositional range and a variety of REE patterns including:

(a) "N-MORB" patterns observed, for instance, in the olivine-rich spinel lherzolites from Ronda and Lanzo. These samples differ from the fertile orogenic lherzolites (group 1) by lower MREE and HREE concentrations, in the range $0.5-1.5\times$ chondrites. However, they are distinguished from the ophiolitic lherzolites (group 2) by a much lesser degree of LREE depletion.

(b) "HREE-fractionated" patterns, typical of the Ronda harzburgites. These samples are characterized by a progressive depletion from lutetium to europium (MREE/HREE $<$ chondrites) combined with a relatively flat LREE segment, from lanthanum to samarium (LREE/MREE $\sim$ chondrites).

(c) "U-shaped" REE patterns, typical of the Pyrenean harzburgites. These rocks are characterized by a minimum for MREE (Eu–Dy) and relative enrichments of both light and heavy REE (LREE/MREE $>$ chondrites; MREE/HREE $<$ chondrites; LREE/HREE $\sim$ chondrites).

(d) "flat" REE patterns (LREE/MREE $\sim$ MREE/HREE $\sim$ LREE/HREE $\sim$ chondrites), observed mostly in dunites from Ronda and highly refractory harzburgites from Pyrenees.

Within a given orogenic peridotite body, the refractory peridotites are systematically distinguished from the fertile lherzolites by higher LREE/MREE and LREE/HREE ratios (only one exception is noticeable, in Lanzo, Figure 12). McDonough and Frey (1989) have already emphasized the widespread character of this feature in mantle rocks. In addition to the orogenic peridotite massifs, it is observed in ophiolites (e.g., Prinzhofer and Allègre, 1985 and Figure 13) and in several suites of mantle xenoliths (e.g., Song and Frey, 1989). In tectonically emplaced and abyssal peridotites, it is marked by a negative correlation of LREE/MREE and HREE ratios versus HREE abundances (e.g., Ce/Sm versus ytterbium in Figure 20).

(iv) *Ophiolitic and abyssal refractory peridotites* (Figure 13, thin continuous lines) strongly depleted in LREE and MREE. The HREE systematically fractionated (MREE/HREE $<$ chondrites) and the REE patterns vary in shape between two extremes, depending on the LREE/MREE ratio:

(a) "Linear" REE patterns characterized by LREE/MREE ratios $<$ chondrites resulting in almost a continuous decrease of normalized REE abundances from HREE to LREE (e.g., several Oman harzbugites).

(b) "U-shaped" REE patterns characterized by LREE/MREE ratios $>$ chondrites and a minimum for MREE (e.g., the New Caledonia harzburgites and dunites). In addition to the data reported in Figure 13, U-shaped REE patterns have been observed in ophiolitic harzburgites from Troodos, Cyprus (Taylor and Nesbitt, 1988), the Urals, Russia (Sharma and Wasserburg, 1996), and Trinity, California (Gruau *et al.*, 1998).

Between these extremes, several samples (e.g., the harzburgites and dunites from the Cuba ophiolite) show flat LREE segments (LREE/MREE $\sim$ chondrites) and, therefore, resemble the "HREE-fractionated" type recognized in the orogenic refractory peridotites. This REE pattern is also observed in the abyssal peridotites from the Izu-Bonin-Mariana Forearc (Parkinson and Pearce, 1998) and in the less refertilized samples from the East Pacific Rise (Niu and Hékinian, 1997a) (Figure 15). Several refractory orogenic peridotites (Figure 12) have REE patterns comparable in shape to those of the refractory ophiolitic peridotites. However, the ophiolitic samples are distinguished by a more restricted range of composition and a lesser diversity of REE patterns. Above all, they are much more depleted in LREE and MREE than their orogenic counterparts ($0.005-0.05\times$ chondrites, compared with 0.03–0.3).

(v) *Metasomatized orogenic lherzolites*, which are generally enriched in LREE relative to HREE (LREE/HREE $>$ chondrites). In Pyrenees (Figure 14), the modally metasomatized samples (amphibole-bearing or/and cpx-enriched) show a range of REE patterns varying between two extremes:

(a) almost flat (slightly U-shaped to slightly convex upward, or sigmoid) in the amphibole harzburgites, which also have relatively low REE content ($0.2-2\times$ chondrites);

(b) strongly convex upward with a maximum at Nd–Sm in the cpx-rich lherzolites from Caussou (Fabriès *et al.*, 1989), which are strongly enriched in REE ($2-20\times$ chondrites).

The wall–rock harzburgites affected by "cryptic" mantle metasomatism (i.e., incompatible trace-element enrichment with no trace of secondary minerals except for minute amount of apatite—Bodinier *et al.*, 1990, 2004; Woodland *et al.*, 1996) are distinguished by sigmoid REE patterns combining a U-shaped MREE–HREE segment (Eu–Lu) with an enriched, convex-upward LREE segment (maximum at cerium). These samples show extremely high LREE/HREE values (up to $16\times$ chondrites). While it is frequent in mantle xenoliths (see Chapter 2.05), such LREE enrichment is rarely encountered in tectonically

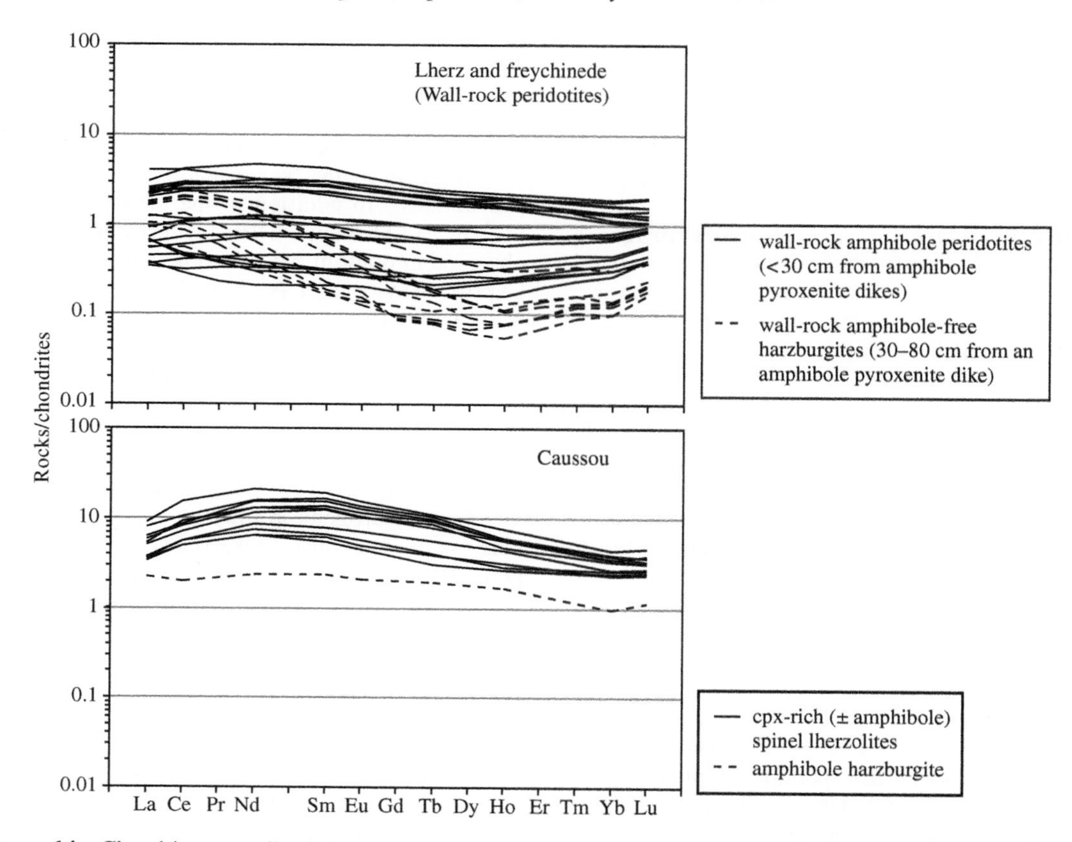

Figure 14 Chondrite-normalized abundances of REEs in "metasomatized" orogenic peridotites from the Eastern Pyrenees (whole-rock analyses). Data from Bodinier *et al.* (1988, 1990) and Fabriès *et al.* (1989). Normalizing values after Sun and McDonough (1989).

emplaced and oceanic peridotites, or it is ascribed to metasomatism by crustal-derived fluids. This origin was advocated, for instance, for the modally metasomatized (amphibole- ± phlogopite) peridotites occurring in the Finero (Western Alps), Ulten (Eastern Alps) and Zabargad (Red Sea) massifs (Dupuy *et al.*, 1991; Piccardo *et al.*, 1993; Zanetti *et al.*, 1999; Rampone and Morten, 2001).

(vi) *Refertilized refractory peridotites*, characterized by relatively flat REE patterns, often associated with a positive europium anomaly (Figure 15). As discussed above (Section 2.04.3.1.2) some refractory peridotites (dunites and harzburgites) from LP orogenic lherzolites (e.g., Lanzo), from the MTZ of the ophiolites, and from abyssal peridotite suites show textural, mineralogical, and/or major-element evidence for refertilization by infiltrated melts. These samples are generally characterized by higher REE abundances and LREE/HREE ratios than their protolith (compare the plagioclase- and cpx-bearing dunite from Lanzo with the other Lanzo dunites in Figure 12). They also show a positive europium anomaly that—intriguingly—was even observed in fertilized abyssal peridotites devoid of plagioclase (e.g., the samples from the Garret Transform reported by Niu and Hékinian, 1997a; Figure 15). Conversely, the most REE-enriched sample from the Garret Transform shows a negative europium anomaly that might result from the entrapment of melt evolved after plagioclase fractionation. However, further assessment of this hypothesis is hampered by secondary alteration (Niu and Hékinian, 1997a).

2.04.3.2.2 *Primitive mantle-normalized trace-element patterns*

The distribution of lithophile trace elements (REE + rubidium, caesium, strontium, barium, yttrium, zirconium, hafnium, niobium, tantalum, thorium, and uranium) normalized to primitive mantle (PM) values are illustrated in Figure 16 for a range of peridotite lithologies from the Ronda orogenic lherzolite massif, and in Figure 17 for ophiolitic and abyssal refractory peridotites.

Most Ronda peridotites show thorium variations consistent with LREE variations. In other words, thorium generally plots on extrapolations of the LREE segments on the PM-normalized diagrams and Th/La is roughly correlated with La/Sm. Both ratios are generally lower than PM values in the lherzolites, close to PM values in the harzburgites and variable in the dunites.

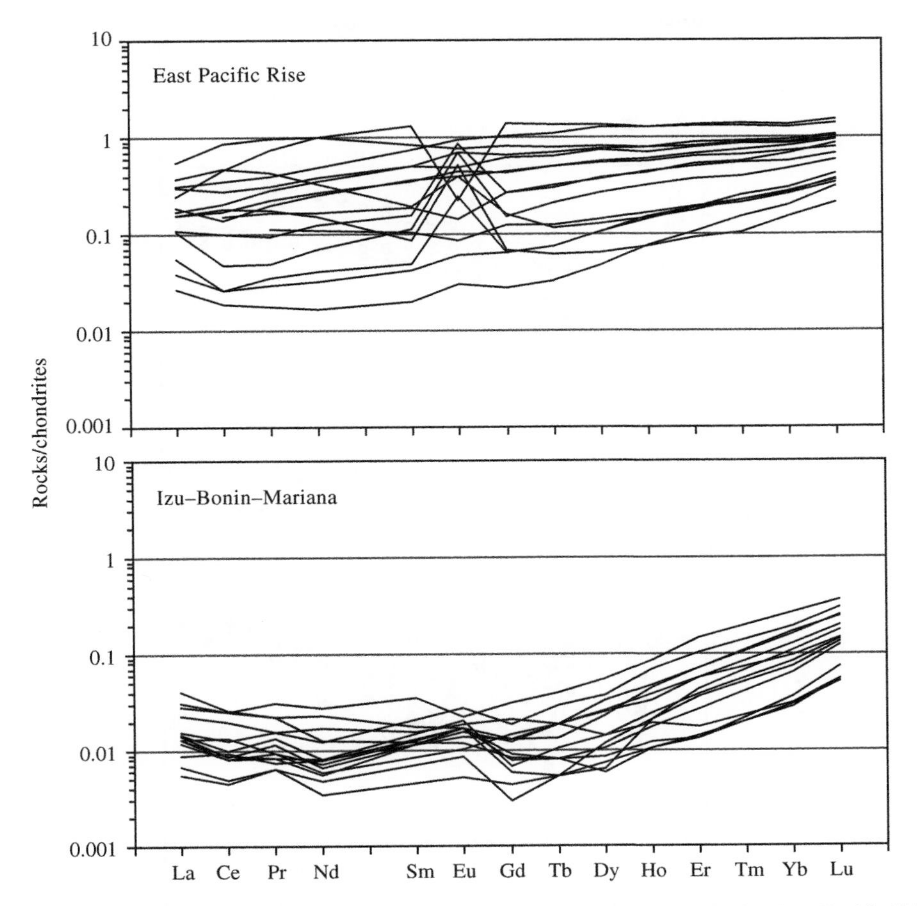

Figure 15 Chondrite-normalized abundances of REEs in abyssal peridotites from the East Pacific Rise and the Izu–Bonin–Mariana Forearc (whole-rock analyses). Data from Niu and Hékinian (1997a) and Godard (unpublished) for the East Pacific Rise, and from Parkinson and Pearce (1998) for the Izu–Bonin–Mariana Forearc. Normalizing values after Sun and McDonough (1989).

A striking feature of thorium distribution in Ronda is the narrow variation range of this element, all thorium concentrations lying within a factor of 5. This range is even more restricted than that of LREE, which vary within a factor of 10 (compared with a factor of 100 for HREE, Figure 12). Garrido *et al.* (2000) have ascribed this feature to melt micro-inclusions trapped in minerals that would govern and homogeneize the concentrations of highly incompatible element in whole rocks.

The other trace elements generally display positive or negative anomalies relative to thorium and REE on PM-normalized diagrams. These anomalies tend to be more marked in the refractory peridotites than in the lherzolites. The U/Th and Ba/Th ratios are generally higher than PM values (mostly in the range (1–5) × PM), without significant variations between rock types. In contrast, rubidium and strontium show variations according to peridotite facies. Most lherzolites are strongly enriched in rubidium relative to other highly incompatible elements, with Rb/Th ratios around 10, but are generally devoid of a strontium anomaly relative to MREE, or show only a subtle positive anomaly. Conversely, the harzburgites and dunites are less enriched in rubidium but generally show a prominent strontium spike.

Niobium and tantalum are roughly interpolated between thorium and lanthanum in the lherzolites, but the Nb/Ta values are variable. Lenoir *et al.* (2001) have revealed an abrupt change in the Nb/Ta ratio across the Ronda recrystallization front (Van der Wal and Vissers, 1993, 1996), from subchondritic values (~10) in the spinel tectonite domain ("ahead" of the front) to near-chondritic (~16) in the granular domain ("behind" the front). They tentatively ascribed this observation to a change from lithospheric conditions, under which Nb–Ta would be dominantly controlled by very small amounts of titanium oxides precipitated from volatile-rich small-volume melts (Bodinier *et al.*, 1996), to more asthenospheric conditions, under which the titanium oxides would be dissolved into basaltic partial melt and Nb–Ta redistributed between silicate phases. The harzburgites are systematically enriched in niobium and tantalum relative to

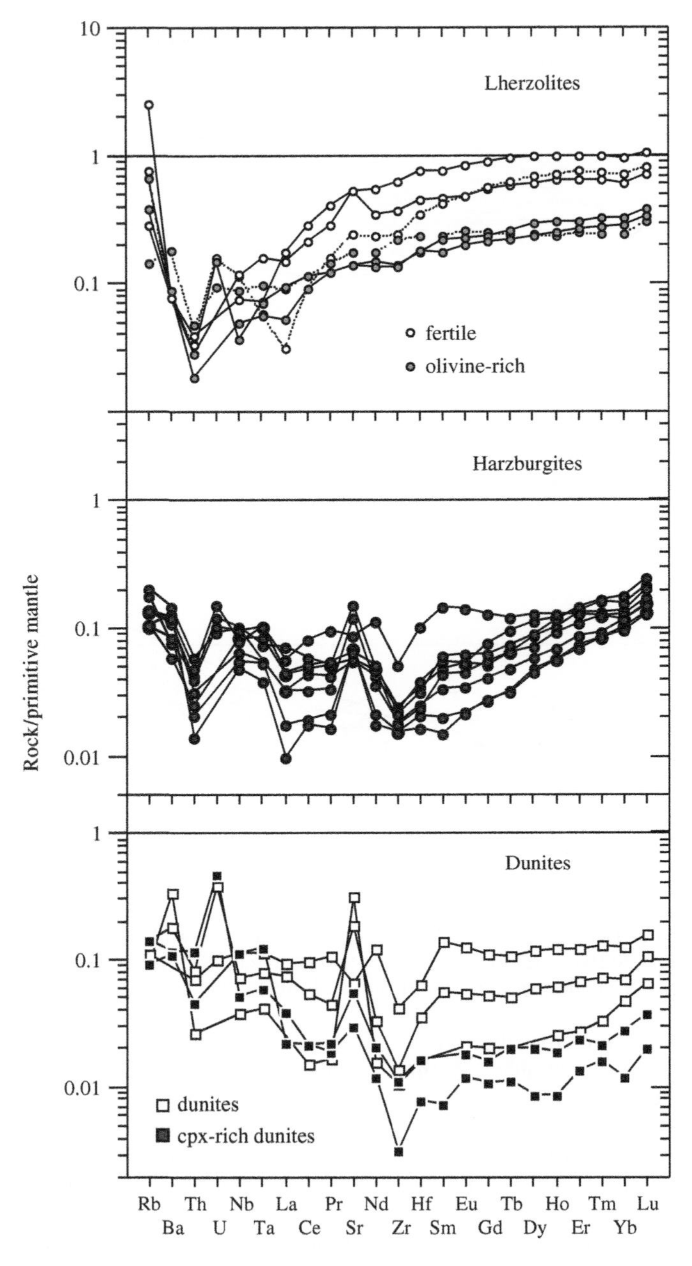

Figure 16 Abundances of lithophile trace elements normalized to PM values in orogenic peridotites from the Ronda massif (whole-rock analyses). Data from Remaïdi (1993). Normalizing values after Sun and McDonough (1989).

thorium and lanthanum, with Ta/La ~ 2–4 × PM. Strong enrichments are also observed in dunites, especially in the samples that are strongly depleted in LREE (Ta/La up to 40 × PM). However, most harzburgites and dunites have nearly chondritic Nb/Ta ratio.

Finally, the Ronda peridotites are generally characterized by negative anomalies of zirconium on PM-normalized diagrams, the amplitude of which increases from lherzolites (which have only subtle anomalies) to dunites. Zr/Sm is mostly in the range (0.5–1) × PM in the lherzolites, compared with 0.2–0.5 in the more refractory rocks. Hafnium also shows negative anomalies relative to MREE, but the hafnium anomalies are smaller than the zirconium anomalies, and the Zr/Hf ratio tends to decrease with increasing refractory character of the peridotites, from 0.6–0.95 × PM in the lherzolites to 0.45–0.8 in the harzburgites and 0.35–0.6 in the dunites.

The harzburgites and dunites from the Oman and Cuba ophiolites show relatively homogenous PM-normalized trace-element patterns characterized by: (i) a subtle (Cuba) to significant (Oman)

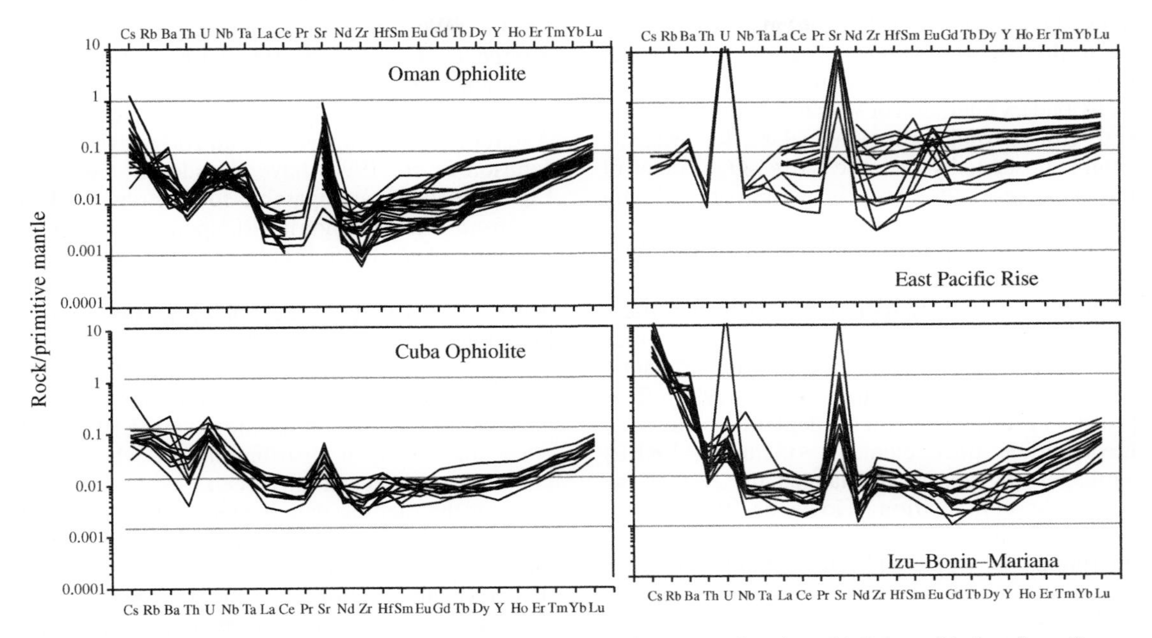

Figure 17 Abundances of lithophile trace elements normalized to PM values in ophiolitic peridotites from Oman and Cuba, and in abyssal peridotites from the East Pacific Rise and the Izu–Bonin–Mariana Forearc (whole-rock analyses). *Ophiolitic peridotites:* data from Godard *et al.* (2000) (reproduced by permission of Elsevier from *Earth Planet. Sci. Lett.* **2000**, *180*, 133–148) for Oman, and from Proenza *et al.* (1999) for Cuba; *abyssal peridotites:* data from Niu and Hékinian (1997a) and Godard (unpublished) for the East Pacific Rise, and from Parkinson and Pearce (1998) for the Izu–Bonin–Mariana Forearc. Normalizing values after Sun and McDonough (1989).

enrichment of thorium relative to lanthanum (Th/La > PM), (ii) a noticeable enrichment of caesium, rubidium, barium, and uranium relative to thorium (in spite of some scatter), and of strontium relative to MREE, (iii) an enrichment of niobium and tantalum relative to lanthanum, which is reminiscent of the one observed in the Ronda harzburgites. The Nb/Ta ratio is clearly supra-chondritic in Cuba and varies from near-chondritic to supra-chondritic in Oman. Godard *et al.* (2000) found that the "diapiric" domains of the Oman mantle sequence (distinguished in the field by lineation dips greater than 30°) differ from the neighboring harzburgites by higher Nb/Ta ratios (≥40, compared with 20 ± 10). A possible interpretation is to consider that, in a late evolutionary stage of the Oman ophiolite, the diapirs focused (volatile-rich?) small melt fractions with the ability to fractionate tantalum from niobium. Finally, the ophiolitic peridotites show only subtle negative anomalies of zirconium and positive anomalies of hafnium.

The abyssal peridotites from the Izu–Bonin–Mariana Forearc resemble the ophiolitic peridotites. They are, nevertheless, distinguished by more prominent enrichments in alkaline and alkaline-earth elements (rubidium, caesium, strontium, and barium), possibly reflecting a supra-subduction imprint (Parkinson and Pearce, 1998). Except for one sample, they lack the niobium enrichment relative to LREE that is observed in the Ronda and ophiolitic refractory peridotites. However, several samples show a positive anomaly of Zr–Hf. Among the East Pacific Rise (EPR) abyssal peridotites, the less refertilized samples (characterized by low REE contents) also resemble the ophiolitic peridotites. They are merely distinguished from their ophiolitic counterparts by a more prominent uranium spike, the lack of niobium enrichment relative to LREE and more important variations of zirconium and hafnium relative to MREE. Some samples show subtle positive anomalies of these elements on the PM-normalized diagram while others are characterized by large negative anomalies. This feature was tentatively ascribed to the uneven precipitation of zirconium-rich microphases (ZrO_2—baddeleyite being the more likely candidate) along with refertilization (Niu and Hékinian, 1997a).

2.04.3.2.3 Chondrite-normalized REE patterns

Ytterbium versus Al_2O_3 covariation trend and fertile orogenic lherzolites. Similar to the correlations observed between major (and minor transition) elements, the good correlation between ytterbium and Al_2O_3 in tectonically emplaced and abyssal peridotites (Figure 10) is classically ascribed to variable degrees of melt extraction. However, any of the alternatives envisioned for the major elements (melt–rock reactions, melt redistribution, and refertilization) may also contribute to the HREE variations. The few samples that plot off the main covariation trend can be explained by secondary alteration or selective addition of minerals from partial melts. Plagioclase, for instance, would substantially increase the Al_2O_3 content of refractory peridotites while increasing only slightly the ytterbium concentration (Bodinier, 1988).

In the orogenic massifs, the fertile lherzolites with depleted LREE compositions ("N-MORB" chondrite-normalized REE patterns) were initially interpreted as resulting from very small degrees (<3%) of melt extraction from a pristine mantle source (~PM) (e.g., Menzies *et al.*, 1977). This process would account for the depletion of the most incompatible trace elements (including LREE) while leaving almost unchanged the rest of the peridotite composition. However, the LREE-depleted fertile lherzolites are now more widely interpreted as "Depleted MORB Mantle (DMM)," i.e., upper mantle rocks representative of the MORB source (e.g., Hartmann and Wedepohl, 1993). In this perspective, they acquired their LREE-depleted character during multiple cycles of convection involving melt extraction at mid-ocean ridges. Their fertile character may result from melt redistribution in the asthenosphere or/and from mixing with recycled crust or deep-seated plume components. Then, these rocks were accreted to the lithosphere before being eventually exhumed to the surface.

An alternative to this scenario is to envisage that some fertile orogenic lherzolites have acquired their geochemical signature as a result of melting and melt–rock interaction processes associated with the thermomechanical erosion of lithospheric mantle by upwelling asthenosphere (e.g., Lenoir *et al.*, 2001). In this scheme, refertilization of lithospheric peridotites by (and re-equilibration with) MORB melts is an alternative to the small degrees of melt extraction to account for LREE depletion in otherwise fertile lherzolites (e.g., Piccardo and Rampone, 2001).

Cpx-poor ophiolitic and abyssal lherzolites. Loubet *et al.* (1975) and Menzies (1976) noted a clear difference between *ophiolitic* and *orogenic* lherzolites in term of LREE abundance, the former being much more depleted in LREE than the latter. The duality of lherzolite compositions was thereafter observed at the scale of a single ocean basin, in Liguria, where the fertile, External Ligurian orogenic lherzolites are moderately depleted in LREE while the less fertile (cpx-poor), Internal Ligurian ophiolitic lherzolites are strongly depleted (Beccaluva *et al.*, 1984; Ottonello *et al.*, 1984b). Based on geological and geochemical arguments, the External Ligurian lherzolites were interpreted as subcontinental mantle exhumed along a passive margin while the Internal Ligurian peridotites would represent suboceanic lithosphere from the Western Tethys (Beccaluva *et al.*, 1984; Rampone *et al.*, 1995, 1996, 1998). Moreover, REE analyses of cpx in abyssal peridotites from slow-spreading ridge systems (e.g., the American-Antarctic and Southwest Indian Ridge systems—Johnson *et al.*, 1990) have revealed a degree of LREE depletion comparable to that observed in cpx from Internal Liguria (Rampone *et al.*, 1996; Rivalenti *et al.*, 1996, Figure 18).

There is a large consensus to ascribe the strong depletion of LREE of the ophiolitic and abyssal lherzolites to a single event of melt extraction during decompression melting of upwelling asthenosphere. Based on some assumptions (such as the mantle source composition), the LREE and other trace-element distributions in whole-rock peridotites or/and cpx have been used to constrain some of the melting parameters, such as depth of melting (garnet or/and spinel stability field), melting degree and melt extraction process (batch versus fractional). Consistent with theoretical and experimental information stating that only very small fractions of melt are needed before segregation can commence, the strong depletion of LREE relative to HREE is considered as evidence for incremental melt extraction approaching the Rayleigh, fractional melting model (Gast, 1968; Shaw, 1970). By modeling REE distribution in cpx, Johnson *et al.* (1990) further suggested that some abyssal peridotites have undergone melting in the spinel stability field alone while others have suffered polybaric melting beginning in the garnet stability field and continuing into the spinel field (Figure 19). They also found that melting associated with slow-spreading ridges commences at greater depth and reaches a higher degree in the vicinity of hot spots.

The predominantly harzburgitic ophiolitic mantle sequences are often underlain by "basal," cpx-poor lherzolites (or cpx-harzburgites) that resemble the lherzolites from the predominantly lherzolitic ophiolitic sequences and the abyssal peridotites from slow-spreading ridge systems in their strong LREE depletion. Typical examples include the Oman ophiolite (Godard *et al.*, 2000; Takazawa *et al.*, 2003—whole-rock cpx-harzburgites shown in Figure 13) and the Bay of Island ophiolite, Newfoundland (Batanova *et al.*, 1998—cpx data shown in Figure 18). In most cases, however, the basal lherzolites are distinguished from the other ophiolitic and abyssal lherzolites by a slight enrichment of lanthanum (± Ce) relative to MREE. These rocks are considered either as mantle residues formed by lower melting degrees than the overlying harzburgites, or as harzburgites refertilized by MORB melts. In the first alternative, the basal lherzolites may be formed in a different tectonic setting and accreted during early ophiolite detachment (Suhr and Batanova, 1998), or they may indicate a steady or transient profile of

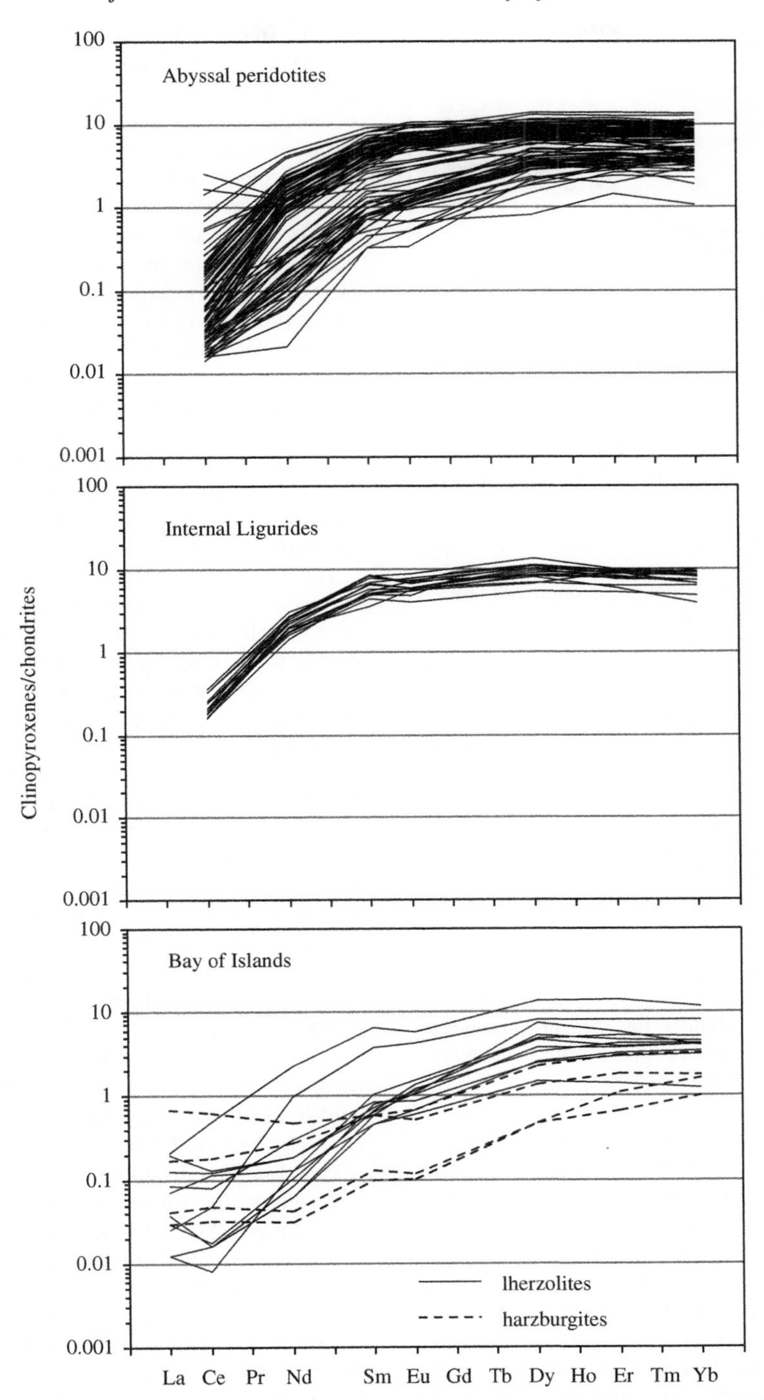

Figure 18 Chondrite-normalized abundances of REEs in clinopyroxene from abyssal peridotites (Mid-Atlantic Ridge and Southwest Indian Ridge) and from the Internal Ligurides and Bay of Islands ophiolitic peridotites. *Abyssal peridotites:* data from Johnson and Dick (1992), Johnson *et al.* (1990) (reproduced by permission of American Geophysical Union from *J. Geophys. Res.* **1990**, *95*, 2661–2678), and Ross and Elthon (1997). *Ophiolitic peridotites:* data from Rampone *et al.* (1996, 1998) for the Internal Ligurides, and from Batanova *et al.* (1998) for the Bay of Islands. Normalizing values after Sun and McDonough (1989).

decreasing melting degree under the ocean ridge (Batanova *et al.*, 1998; Takazawa *et al.*, 2003). In the refertilization scenario, the basal lherzolites are considered to underlie the lithosphere–asthenosphere boundary, along with upwelling partial melts that would freeze through near-solidus, cpx-forming reactions (Godard *et al.*, 2000; Takazawa *et al.*, 2003).

Refractory peridotites: harzburgites and dunites. In contrast with the ophiolitic and abyssal

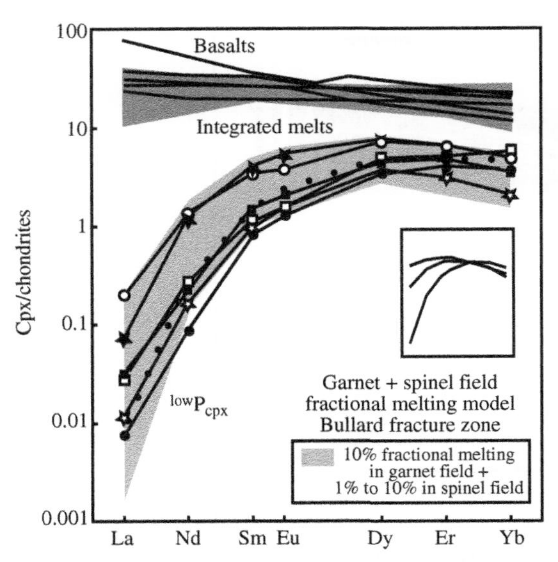

Figure 19 Chondrite-normalized abundances of REEs in clinopyroxene of abyssal peridotites from the Bullard Fracture Zone, American-Antarctic Ridge (symbols), compared with the results of polybaric incremental melting (after Johnson *et al.*, 1990) (reproduced by permission of American Geophysical Union from *J. Geophys. Res.* **1990**, *95*, 2661–2678). Incremental melting (small increments) begins in the garnet stability field (up to 10% total melting in the garnet field) and continues up to an additional 10% in the spinel stability field. Data for the spatially associated basalts (Le Roex *et al.*, 1985) and the integrated composition of melt increments obtained with the melting model are also plotted. The dashed line is a residue model calculated using lower proportions of clinopyroxene in partial melt. The inset is a schematic representation of the evolution of the REE patterns in clinopyroxene with melting from 1% (upper line) to 10% (bolder line) in the presence of garnet. See Johnson *et al.* (1990) for further information on model parameters.

lherzolites, the refractory peridotites (harzburgites and dunites) have LREE distributions that cannot be explained by simple models of melt extraction (i.e., batch, fractional, incremental, and continuous melting—Gast, 1968; Shaw, 1970; Langmuir *et al.*, 1977; Johnson *et al.*, 1990; Johnson and Dick, 1992). As illustrated in Figure 20, all types of refractory peridotites (i.e., orogenic, ophiolitic, and abyssal) diverge from the evolution predicted by the melting models in showing constant or increasing LREE/MREE ratios with decreasing HREE content—whereas the models indicate a strong decrease of the LREE/MREE ratios. The theoretical melting trends shown in Figure 20 were calculated with the fractional melting model of Shaw (1970) and from a "PM" source (Sun and McDonough, 1989). However, changing the melting models or the source composition (e.g., DMM) does not provide a better fit to the data set.

Based on neodymium isotopic signatures, Sharma and Wasserburg (1996), Gruau *et al.* (1998) have attributed the U-shaped patterns observed in refractory ophiolitic peridotites to contamination by crustal fluids during ophiolite obduction, or during their later tectonic involvement in continental suture zones. These authors also suggested that secondary alteration processes involving crustal fluids would account for the LREE abundances—higher in most ophiolitic peridotites than predicted by melting models. However, this interpretation is hardly applicable to all types of refractory peridotites as it fails to account for several observations:

(i) Refractory abyssal peridotites show almost identical REE patterns as their ophiolitic counterparts (compare Figures 13 and 15), although they were not obducted onto continental crust. The abyssal peridotites do not better fit the melting models than the ophiolitic peridotites in the LREE/MREE versus HREE diagram (Figure 20).

(ii) In spite of a similar degree of serpentinization, the ophiolitic lherzolites are generally devoid of the selective LREE enrichment observed in ophiolitic refractory harzburgites. As they are strongly depleted in LREE (e.g., the Ligurian peridotites in Figure 13), they should be highly sensitive to secondary contamination.

(iii) In the orogenic peridotites (Figure 12), LREE enrichment—coupled with low $^{143}Nd/^{144}Nd$ values (Section 2.04.5)—is almost systematically observed in harzburgites and dunites, but rarely in the adjacent lherzolites in spite of their similar degree of serpentinization. Several studies on orogenic peridotites have noted the lack of correlation between LREE enrichment and secondary alteration (Frey *et al.*, 1985; Bodinier, 1988; Bodinier *et al.*, 1988). In Ronda, LREE enrichment is observed in fresh refractory peridotites, virtually devoid of serpentine, as well as in acid-leached mineral separates and in minerals analyzed by LA-ICPMS (Garrido *et al.*, 2000).

(iv) LREE enrichment by crustal fluids has been advocated for several massifs of peridotites (e.g., Finero, Western Alps; Ulten, Eastern Alps; Zabargad, Red Sea) but these massifs show clear evidence for modal metasomatism, as well as peculiar trace-element signature.

Prinzhofer and Allègre (1985) suggested a "disequilibrium melting" mechanism to account for the evolution of the REE patterns in peridotites from the New Caledonia ophiolite, from "N-MORB" in the lherzolites to "U-shaped" in the harzburgites and dunites (Figure 13). However, mineral/melt disequilibrium is unlikely at the high temperatures inferred for decompression mantle melting in the mantle. The good correlations observed between whole-rock and mineral compositions in most suites of mantle rocks bear evidence against significant solid/liquid disequilibrium. Only during relatively low-temperature, melt–rock interactions (in lithospheric

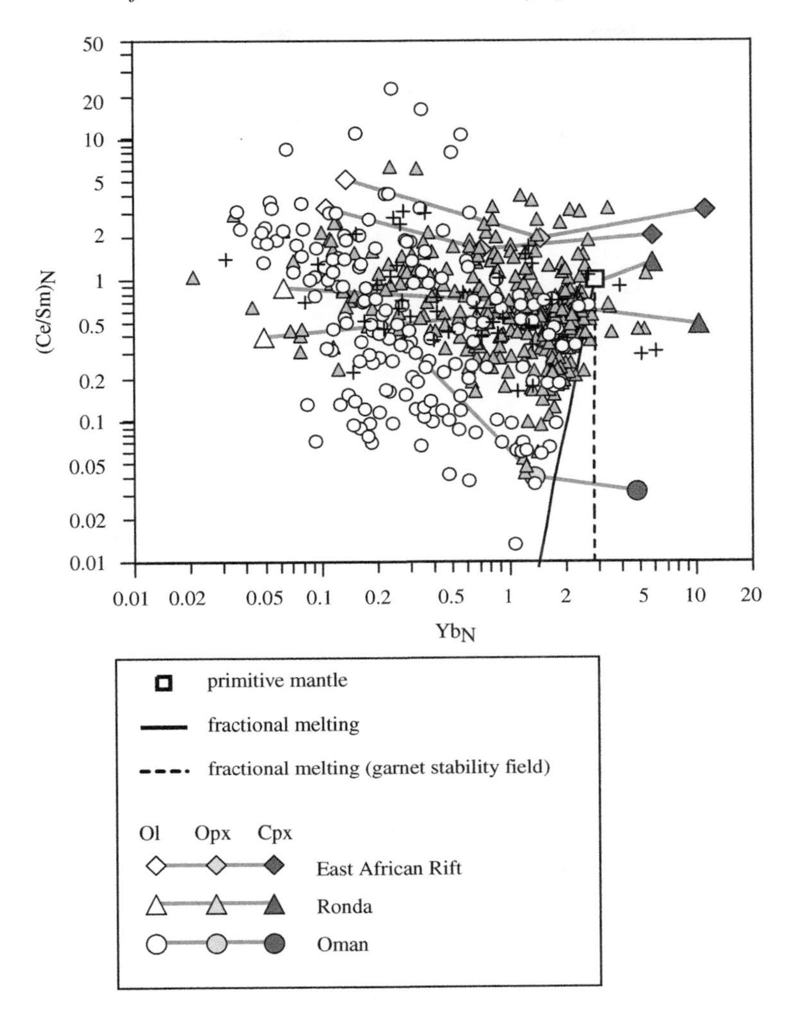

Figure 20 Plot of Ce/Sm versus ytterbium (chondrite-normalized values) for orogenic, ophiolitic, and abyssal peridotites. Whole-rock data are the same as in Figure 10, and the symbols for rock types are the same as in Figure 5. Olivine, orthopyroxene, and clinopyroxene separates from five peridotite samples are also plotted. These include a spinel lherzolite and a harzburgite from the Sidamo suite of mantle xenoliths, East African Rift, Ethiopia (Bedini and Bodinier, 1999), a spinel lherzolite and a harzburgite from the Ronda massif (Garrido *et al.*, 2000) and a harzburgite from the Oman ophiolite (Gerbert-Gaillard, 2002). The composition of fractional melting residues in the spinel and garnet stability field are also shown. The primitive mantle estimate of Sun and McDonough (1989) was taken as source composition for melting. The melting reactions are from Niu (1997) for melting in the stability field of spinel (Equation (6) at 10 kbar) and from Walter *et al.* (1995) for melting in the garnet stability field. Normalizing values after Sun and McDonough (1989).

mantle) can chemical exchange between minerals and melt be slowed down by diffusion in minerals (e.g., Vasseur *et al.*, 1991, and references herein). Moreover, to account for the U-shaped REE patterns of the refractory peridotites, the model proposed by Prinzhofer and Allègre (1985) assumes a "U-shaped" REE distribution in olivine. Mineral analyses have actually confirmed such REE pattern for olivine (Figure 21), but this feature is ascribed to the entrapment of melt micro-inclusions (Garrido *et al.*, 2000).

The divergence of the LREE/MREE ratios from the evolution predicted by the partial melting models may indeed be partly explained by small amounts of equilibrium melt trapped as micro-inclusions in minerals. Because olivine and opx accommodate only small quantities of incompatible elements in their crystal structure, they are more sensitive to the presence of melt inclusions than cpx. Therefore, the presence of micro-inclusions in minerals will result in nearly constant or increasing variations of the LREE/MREE ratios of peridotite minerals in the order cpx → opx → olivine. This feature is illustrated in Figure 20 for minerals separated from an ophiolitic harzburgite (Oman), two orogenic peridotites (Ronda), and two spinel-peridotite mantle xenoliths (East African Rift). Variable modal proportions involving the observed "bulk" mineral compositions (i.e., crystals and their

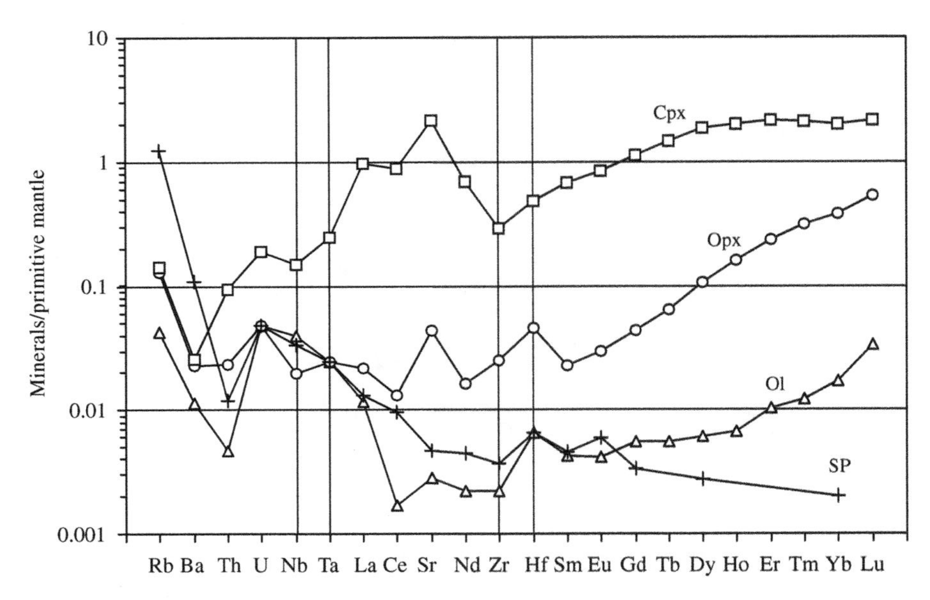

Figure 21 Abundances of lithophile trace elements normalized to PM values in clinopyroxene, orthopyroxene, olivine, and spinel separated from a peridotite from the Ronda massif (Garrido *et al.*, 2000). Normalizing values after Sun and McDonough (1989).

inclusions) would reasonably fit the peridotite data set. Song and Frey (1989) modeled increasing LREE/MREE with decreasing HREE in mantle xenoliths by involving the addition of equilibrium melt to solid residues after partial melting. The presence of melt inclusions in residual minerals is consistent with the suggestion by Niu and Hékinian (1997a) that the composition of refractory abyssal peridotites from fast-spreading ridges is partly governed by trapped melt. Based on a comparison of bulk analyses of mineral separates with *in situ* analyses by LA-ICPMS, Garrido *et al.* (2000) suggested that melt micro-inclusions would account the convergence of mineral compositions towards the most incompatible elements in peridotites from Ronda (Figure 21). They also confirmed with mass-balance calculations that micro-inclusions trapped in minerals are a major host for the most incompatible lithophile trace elements in refractory peridotites (see also Bedini and Bodinier, 1999).

However, the LREE/MREE variations in refractory peridotites cannot be simply modeled by the addition of melt of homogeneous composition to peridotites. This is shown, for instance, by the REE compositions of cpx in coexisting lherzolites and more refractory peridotites, that show a marked increase of the LREE/MREE ratios coupled with a decrease of HREE abundances from lherzolites to refractory peridotites (e.g., the Fontête Rouge orogenic massif, Pyrenees, Figure 22). The strong HREE depletion observed in harzburgite cpx suggests a high extent of melt extraction but, paradoxically, the elevated LREE/MREE ratios indicate equilibration with (± entrapment of) LREE-enriched melt. Conversely, the lherzolites were equilibrated with LREE-depleted melt according to their cpx composition. In fact, it is worthwhile to note that coexisting lherzolites and harzburgites in orogenic massifs and ophiolites are generally characterized by nearly identical LREE contents in cpx (as well as in whole rocks, in several instances), while they strongly differ in their HREE content (Figure 22). As pointed out by Godard *et al.* (2000), this feature may be explained by chromatographic effects associated with porous melt flow. In the frame of the chromatographic theory (Navon and Stolper, 1987), the refractory peridotites would be situated below the LREE chromatographic front—which moves almost as fast as the melt itself due to the strongly incompatible behavior of LREE—but above the much slower HREE chromatographic front. As a result, the LREE are predominantly controlled by the percolating melt and remain roughly constant, both in melt and in peridotite minerals. In contrast, the HREE are governed by the peridotite matrix and are, therefore, expected to vary in response to melting, melt–rock reactions, and mineral segregation (Godard *et al.*, 1995; Ozawa and Shimizu, 1995; Vernières *et al.*, 1997). Melt entrapment as micro-inclusions in minerals would further enhance the LREE enrichment of peridotite minerals and contribute to the divergence of LREE/MREE variations in refractory peridotites from melting models (Figure 20). Numerical experiments (Figure 23) confirm that the REE composition of the Oman harzburgites, for example, can be explained by the combined effects of partial melt

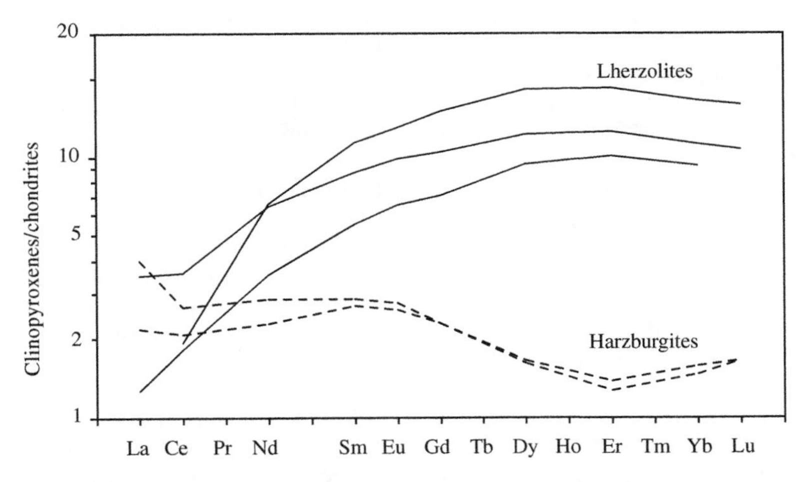

Figure 22 Chondrite-normalized abundances of REEs in clinopyroxene separated from representative peridotite samples from the Fontête Rouge massif, in the Eastern Pyrenees (Downes *et al.*, 1991). Normalizing values after Sun and McDonough (1989).

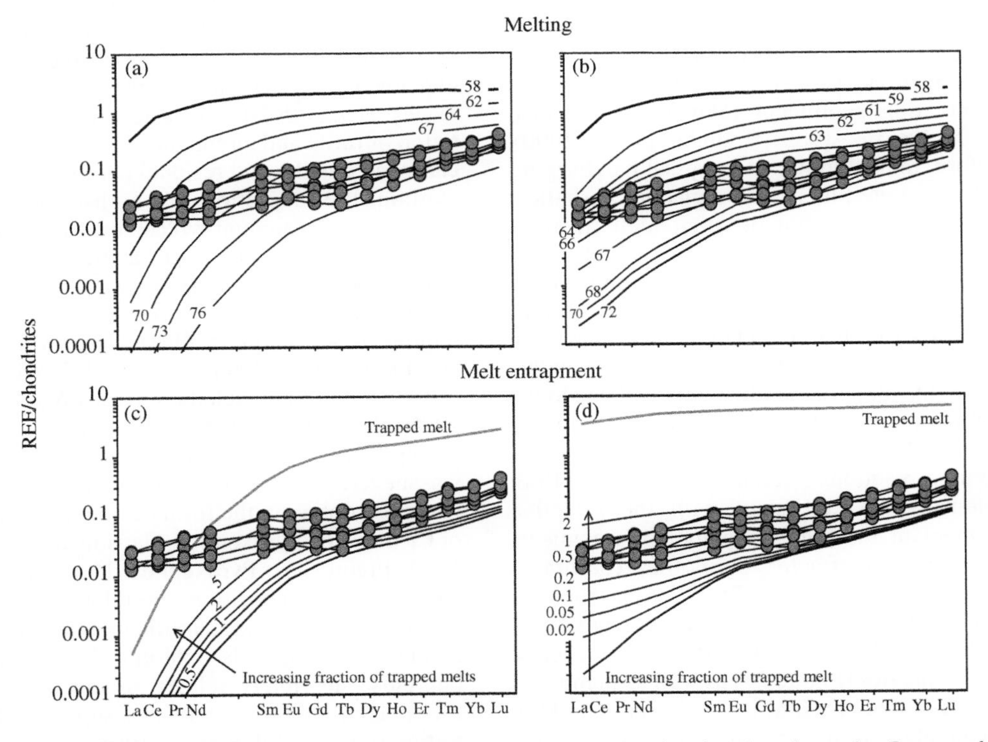

Figure 23 Chondrite-normalized abundances of REEs in representative harzburgites from the Oman ophiolite (symbols—whole-rock analyses), compared with numerical experiments of partial melting performed with the "Plate Model" of Vernières *et al.* (1997), after Godard *et al.* (2000) (reproduced by permission of Elsevier from *Earth Planet. Sci. Lett.* **2000**, *180*, 133–148). *Top:* melting without (a) and with (b) melt infiltration. Model (a) simulates "continuous" melting (Langmuir *et al.*, 1977; Johnson and Dick, 1992), whereas in model (b) the molten peridotites are percolated by a melt of fixed, N-MORB composition. Model (b) is, therefore, comparable to the open-system melting model of Ozawa and Shimizu (1995). The numbers indicate olivine proportions (in percent) in residual peridotites. Bolder lines indicate the REE patterns of the less refractory peridotites. In model (a), the most refractory peridotite (76% olivine) is produced after 21.1% melt extraction. In model (b), the ratio of infiltrated melt to peridotite increases with melting degree, from 0.02 to 0.19. *Bottom:* modification of the calculated REE patterns residual peridotites due to the presence of equilibrium, trapped melt. Models (c) and (d) show the effect of trapped melt on the most refractory peridotites of models (a) and (b), respectively. Bolder lines indicate the composition of residual peridotites without trapped melt. Numbers indicate the proportion of trapped melt (in percent). Model parameters may be found in Godard *et al.* (2000).

extraction, melt transport, and melt entrapment (Godard *et al.*, 2000).

At first sight, the ubiquitous LREE enrichment observed in ophiolitic harzburgites and abyssal peridotites from fast-spreading ridges (compared with values predicted by the melting models) may be viewed as an argument for melt segregation by diffuse porous flow (Niu *et al.*, 1997) rather than by a focused mechanism (e.g., cracks or porous-flow channels—Nicolas, 1986b; Kelemen *et al.*, 1995). However, trace-element data indicate that the percolating melts which ultimately imprinted their REE signatures to the residual peridotites are markedly different from MORB in being enriched in LREE and HIE, but depleted in HREE. Such LREE-enriched, "refractory" melts are probably small volume melts evolved upon incomplete solidification of MORB infiltrated through residual peridotites (e.g., Batanova *et al.*, 1998; Garrido *et al.*, 2000).

It is important to keep in mind that the chromatographic fractionation is a transient process restricted to low melt/rock ratios so that significant REE fractionation can occur only when the melt fraction is small ($<1\%$) or/and the distance of percolation significant (Navon and Stolper, 1987; McKenzie, 1989; Bodinier *et al.*, 1990). Therefore, the ophiolitic and abyssal harzburgites probably acquired their REE signature at shallow level, during the ascent of small melt fractions through conductively cooled, previously depleted residues, rather than in the molten region itself. In the latter, the REE signature of the solid residues may be closer to that predicted by melt extraction models (e.g., Richter and McKenzie, 1984). In this situation, the melt flux is likely reduced due to the precipitation of liquidus phases such as olivine (Niu, 1997; Niu *et al.*, 1997), or as a result of melt-consuming reactions (Godard *et al.*, 2000). Both mechanisms are consistent with the selective enrichment of the highly incompatible elements observed in most refractory peridotites (caesium to tantalum in Figures 16 and 17). These enrichments cannot be related to secondary alteration processes since they are also observed in fresh, acid-leached mineral separates (Figure 21) as well as in mineral analyses run by LA-ICPMS (Garrido *et al.*, 2000). Alternatively, strong enrichments of the most incompatible trace elements are predicted by theoretical modeling of fractional melt solidification combined with melt transport (Godard *et al.*, 1995; Vernières *et al.*, 1997). This mechanism (referred to as "percolative fractional crystallization" by Harte *et al.*, 1993) was advocated to explain ultra-enriched HIE and LREE compositions in metasomatized mantle xenoliths and wall–rock orogenic peridotites (see below and Bedini *et al.*, 1997; Ionov *et al.*, 2002; Bodinier *et al.*, 2003). The same mechanism may account for the selective HIE enrichment in residual peridotites and would further contribute to the "abnormal" LREE/MREE ratios of these rocks.

In addition to HIE, the melts are probably enriched in volatiles, notably in water. Uranium, strontium, and lead may indeed be selectively enriched in water-rich melts/fluids (Gill, 1990; Keppler and Wyllie, 1990; Brenan *et al.*, 1996; Keppler, 1995; You *et al.*, 1996). This might account for the frequent occurrence of positive anomalies of strontium (and lead—not shown) on PM-normalized diagrams, and for the U/Th ratios systematically higher than PM values, both in whole rocks (Figures 16 and 17) and in acid-leached minerals (Figure 21).

Minute amounts of oxide minerals (in the order of a few ppm or less, hence hardly detectable by conventional optical methods) may possibly precipitate from such evolved melts. Minerals such as Nb-rutile (Bodinier *et al.*, 1996) and baddeleyite (Niu and Hékinian, 1997a) would account for the fractionation of the high-field strength elements (niobium, tantalum, zirconium, and hafnium) frequently observed in refractory peridotites (Figures 16 and 17), which is poorly explained by the predominant silicate minerals. Devey *et al.* (2000) suggested that the precipitation of rutile and zircon during the interaction of carbonated melts and residual harzburgites would explain the depletion of high-field strength elements in seamount volcanics from the Juan Fernandez chain (Southeast Pacific).

Because of their low viscosity and solidification temperature, volatile-rich small volume melts may migrate upwards and infiltrate large volumes of peridotites, but will generally solidify before they can segregate and reach the surface (Watson *et al.*, 1990; McKenzie, 1989). Evidence for local segregation of LREE-enriched refractory melts was nevertheless reported from the Ronda orogenic peridotite, in the form of rare and thin dikes of Cr-diopside pyroxenites (Garrido and Bodinier, 1999, Figure 26). Pyroxenites dikes with somewhat similar compositions are relatively common in some ophiolitic harzburgitic mantle suites (e.g., in the Bay of Islands ophiolite, Newfoundland) but those are likely related to the genesis and transport of boninitic melts in a supra-subduction setting (e.g., Varfalvy *et al.*, 1996).

HIE and LREE enrichment due to melt rock interactions were advocated by Parkinson and Pearce (1998), and Pearce *et al.* (2000) for abyssal peridotites from forearc (Izu–Bonin–Mariana) and arc-basin (South Sandwich). In these rocks, however, interaction between arc magmas and pre-existing lithospheric peridotites may significantly contribute to the enrichment, hence resulting in a complex geochemical signature (Pearce *et al.*, 2000).

Metasomatized orogenic peridotites. Compared with mantle xenoliths, the tectonically emplaced, and oceanic peridotites are only rarely affected by metasomatic processes involving deep-seated melts such as alkaline basalts, high-K magmas (kimberlites and lamproites) or their derivatives. This difference probably reflects the distinct provenances of these two types of mantle rocks. In contrast with mantle xenoliths, the massive, and abyssal peridotites do not generally represent lithospheric mantle that has been stabilized for a sufficient period of time to be traversed by intraplate magmatism. Some orogenic peridotites probably include relatively old lithospheric mantle components (see Section 2.04.5.2.1), but they were generally strongly "asthenospherized"—both thermally and chemically—before their exhumation (e.g., Lenoir *et al.*, 2001).

Alternatively, several peridotite massifs were metasomatized by crustal fluids, either pervasively or locally. These peridotites contain secondary, hydrous minerals (amphibole ± phlogopite) and are strongly enriched in LREE and other incompatible elements. Published examples notably include Zabargad, in the Red Sea (Dupuy *et al.*, 1991; Piccardo *et al.*, 1993), Finero, in the Western Alps (Zanetti *et al.*, 1999), and the Ulten garnet peridotites, in the Eastern Alps (Rampone and Morten, 2001). Varied origins were invoked for the fluids, such as crustal dehydration associated with extensional (high-temperature) granulitic metamorphism for Zabargad, fluids released from solidifying migmatites for Ulten and dehydration of a subducted continental slab for Finero.

Most of the orogenic spinel lherzolites also contain small amounts of amphibole (≤1%) texturally equilibrated with the peridotite minerals. These amphiboles have been ascribed to the infiltration of melts/fluids in mantle conditions (Fabriès *et al.*, 1991), but their presence is not associated with noticeable chemical enrichments (Vannucci *et al.*, 1995). A possible explanation is that the liquid/rock ratio was so low that the fluid composition was "buffered" by the LREE-depleted composition of the peridotites. Alternatively, these amphiboles might represent the products of a late crystallization stage during fertilization of the peridotites by LREE-depleted basaltic melts.

However, a few examples of metasomatism and chemical enrichment by alkaline, deep-seated melts have been described, notably in the St. Paul's Island oceanic peridotites, in the Atlantic Ocean (Roden *et al.*, 1984) and in the Lherz orogenic massif and neighboring massifs such as Caussou, in the Eastern Pyrenees (Fabriès *et al.*, 1989; Bodinier *et al.*, 1988, 1990).

As it provides easy access to a variety of metasomatized mantle rocks, the Lherz massif has been recently the focus of detailed geochemical studies—as well as the source of debates—concerning melt infiltration and melt–rock interaction processes in wall rocks of translithospheric melt conduits (Bodinier *et al.*, 1988, 1990, 2004; Downes *et al.*, 1991; Nielson and Wilshire, 1993; McPherson *et al.*, 1996; Woodland *et al.*, 1996; Zanetti *et al.*, 1996). A major advantage of the orogenic peridotites over mantle xenoliths is that there is virtually no limitation to sampling. The massifs provide the opportunity to study spatial and temporal relationships between different metasomatic facies (e.g., modal versus cryptic metasomatism), as well as between metasomatized rocks and their protoliths.

In the Lherz orogenic peridotite, Bodinier *et al.* (1990) studied a 65 cm section of harzburgite wall rock perpendicular to an amphibole-garnet pyroxenite dike. Their results revealed significant mineralogical and chemical zonation of the wall rock. At less than ~30 cm from the dike/wall–rock boundary, hydrous amphibole-bearing peridotite displays convex-upward (0–15 cm) or U-shaped (15–30 cm) normalized REE patterns with relatively unfractionated LREE/HREE ratios (~ chondrites, Figure 14). At >30 cm from the dike, anhydrous harzburgite are strongly LREE-enriched (La/Yb ~3–16 × chondrites) and Woodland *et al.* (1996) noted the presence of apatite in this zone. The REE variation observed in the 65 cm wall rock encompasses almost the whole range of LREE/HREE ratios in metasomatized mantle xenoliths from alkali basalts. Bodinier *et al.* (1990) explained wall–rock zonation with a single-stage process involving progressive chemical evolution of melts infiltrated from the dike into the harzburgite wall rock. In this model, the compatible and moderately incompatible elements are buffered by the wall–rock peridotite proximal to the dike whereas distal to the dike the most incompatible elements remain controlled by the infiltrated melt. This mechanism results in a chromatographic fractionation of REE and accounts for the high LREE/HREE ratios observed at >30 cm from the dike boundary (Figure 24; Zanetti *et al.*, 1996).

However, Nielson and Wilshire (1993) criticized this interpretation. In particular, they questioned the ability of mantle melts to percolate through lithospheric peridotites over distances greater than a few centimeters and negated the very existence of chromatographic fronts in the upper mantle (Navon and Stolper, 1987). They suggested an interpretation for the Lherz wall rock that involved multi-stage metasomatism with a requirement for vein conduits not seen in the peridotite block. At first sight, this interpretation appears to be supported by the isotopic data of Downes *et al.* (1991), indicating that the isotopic signature of the dike does not extend over more than 20 cm into the wall rock. Yet, the strong LREE enrichment typical of the host peridotite at >30 cm from the dike has never been observed in peridotites unrelated to amphibole-pyroxenite

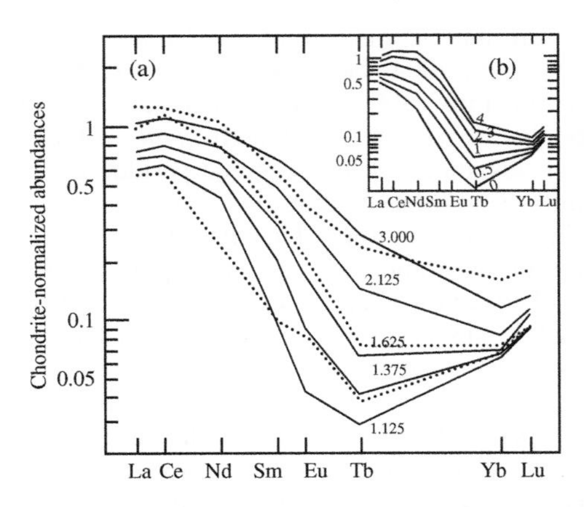

Figure 24 Chondrite-normalized abundances of REEs in a wall-rock harzburgite from Lherz (dotted lines—whole-rock analyses), compared with numerical experiments of 1D porous melt flow, after Bodinier *et al.* (1990). The harzburgite samples were collected at 25–65 cm from an amphibole-pyroxenite dike. In contrast with the 0–25 cm wall-rock adjacent to the dike, they are devoid of amphibole but contain minute amounts of apatite (Woodland *et al.*, 1996). The strong REE fractionation observed in these samples is explained by chromatographic fractionation due to diffusional exchange of the elements between peridotite minerals and advective interstitial melt (Navon and Stolper, 1987; Vasseur *et al.*, 1991). The results are shown in (a) for variable t/t_c ratio, where t is the duration of the infiltration process and t_c the time it takes for the melt to percolate throughout the percolation column (Navon and Stolper, 1987). This parameter is proportional to the average melt/rock ratio in the percolation column. In (b), the results are shown for constant t/t_c but variable proportion of clinopyroxene at the scale of the studied peridotite slices (≤5 cm). All model parameters may be found in Bodinier *et al.* (1990). As discussed in the text, this model was criticized by Nielson and Wilshire (1993). An improved version taking into account the gradual solidification of melt down the wall-rock thermal gradient and the isotopic variations was recently proposed by Bodinier *et al.* (2003).

dikes, neither in Lherz nor in other Pyrenean peridotites (Bodinier *et al.*, 1988; McPherson *et al.*, 1996; Fabriès *et al.*, 1998; Burnham *et al.*, 1998).

The paradox of the spatial decoupling between isotopic contamination and LILE and LREE enrichment was resolved with a numerical simulation of isotopic variations during reactive porous flow (Bodinier *et al.*, 2004). This confirmed that the isotopic contamination (e.g., $^{143}Nd/^{144}Nd$) by the infiltrated melt is restricted to the domain between the melt source (=dike) and the chromatographic front of the element (neodymium)—i.e., ~20 cm from the dike in the studied wall rock. Numerical experiments also showed that the fractional solidification of infiltrated melt (due to amphibole ± cpx precipitation) accounts for the systematic increase in thorium, uranium, LREE, and P_2O_5, that reach a maximum in the distal apatite-bearing wall rock (>50 cm from the dike). It is, therefore, suggested that both the amphibole and the apatite forming metasomatic processes originated from the same dike which produced water-rich and carbonate-rich melts as derivatives from crystal segregation in a magma conduit and wall–rock reactions. In this particular example there is no necessity to invoke different and genetically distinct metasomatic melts/fluids (i.e., a hydrous silicate melt and a carbonate melt). In detail, the envisioned process is as follows:

(i) Emplacement of a "translithospheric" complex of feeder dikes for the alkaline basalts associated with lithospheric thinning in the northern Pyrenees zone during the Cretaceous (ca. 100 Ma), only a few million years before the exhumation of the peridotite bodies in the North Pyrenean basins (Fabriès *et al.*, 1991, 1998).

(ii) Crystal segregation of pyroxene and garnet in the dike conduit, associated with concomitant release of hydrous melts, which resulted in fluid-assisted fracturing of the wall rock and crystallization of hornblendites in branching and anastomosing veins. This was accompanied by more diffuse, intergranular percolation of melt in the dike and the proximal wall rock, that led to the formation of poikiloblastic amphibole in the dike and the precipitation of secondary amphibole (±cpx) in the amphibole harzburgite zone (<25 cm from the dike). Similar processes have been advocated for mantle xenoliths, notably by Harte *et al.* (1993).

(iii) Amphibole ± cpx crystallization resulted in further evolution of the residual melt producing a carbonate-rich melt that was strongly enriched in phosphorus, LREE, and Th–U. The carbonate melt migrated beyond the amphibole-forming reaction front and precipitated apatite in the distal wall rock (>25 cm). The fact that several incompatible elements have their chromatographic fronts nearly coincident with the transition between the amphibole-forming reaction front (i.e., in the range 20–30 cm) suggests that the front corresponds to a significant drop in melt/rock ratio. The amphibole harzburgite reaction zone was very likely bounded by a thermally controlled permeability barrier along which hydrous silicate melts were refracted. The boundary was nevertheless traversed by low-T, carbonated small melt fractions. In that sense, the proximal amphibole peridotite is considered as a small-scale analogue of the thermal boundary layer, at base of lithospheric mantle infiltrated by plume-derived melts (Bedini *et al.*, 1997; Xu *et al.*, 1998).

2.04.4 MAJOR- AND TRACE-ELEMENT GEOCHEMISTRY OF PYROXENITES

2.04.4.1 Data

The tectonically emplaced peridotites and the oceanic suites of mantle rocks contain a variety of mafic rocks. These rocks are often referred to as "pyroxenites" in a broad sense, although they also include eclogites in the HP/UHP orogenic massifs, mafic garnet granulites, hornblendites, and glimmerites (cpx + mica) in the IP orogenic massifs, and olivine gabbros in the LP orogenic massifs, ophiolites, and oceanic suites of mantle rocks.

In addition to variable pressure and temperature equilibrium conditions, the mineralogical diversity of mantle pyroxenites is the reflection of their wide range of major-element compositions. On the SiO_2/MgO versus Al_2O_3 diagram (Figure 25), the pyroxenite array extends from ultramafic compositions close to the peridotite field to silicon- and aluminum-rich compositions approaching the MORB field. In detail, two compositional groups may be distinguished:

(i) *Low-alumina pyroxenites* ($Al_2O_3 < 10\%$) that roughly correspond to the Cr-diopside pyroxenite group of Shervais (1979)—or the type Ib defined by Frey and Prinz (1978) in mantle xenoliths. This group includes clinopyroxenites, websterites (cpx + opx), and orthopyroxenites containing variable amounts of olivine, chromium-spinel, and/or plagioclase.

(ii) *High-alumina pyroxenites* ($Al_2O_3 > 10\%$) that include the Al-augite pyroxenites of Shervais (1979) and the type II pyroxenites of Frey and Prinz (1978), but also contain a variety of amphibole- and/or garnet-bearing mafic rocks.

The chemical diversity of mantle pyroxenites is also patent from REE variations. Figure 26 shows the chondrite-normalized REE patterns of representative pyroxenites from orogenic peridotite massifs in Eastern Pyrenees (Lherz and Freychinède) and the Betico-Rifean Belt (Ronda and Beni Bousera). These rocks show variable REE distributions, but some relationships are observed between the structural and/or mineralogical types of pyroxenites and the REE patterns:

(i) All pyroxenites occurring as cross-cutting dikes and veins (oblique to peridotite foliation and compositional layering) are enriched in LREE and MREE relative to HREE, with convex-upward REE patterns. The amphibole-bearing pyroxenites and hornblendites from the Pyrenees are further distinguished by overall REE enrichment (mostly >10 chondrites). All Pyrenean dikes and veins belong to the high-alumina group defined from major-element compositions. However, the websterite dikes from Ronda are strongly depleted in HREE ($1-2\times$ chondrites) and belong to the low-alumina group of pyroxenites.

(ii) Among the "layered" pyroxenites (parallel to the peridotite foliation), the spinel pyroxenites (±olivine, ±plagioclase) show a wide range of REE concentrations and variable LREE/HREE ratios. The samples that belong to the high-alumina group (aluminum-spinel clinopyroxenites and websterites) are relatively enriched in HREE ($5-10\times$ chondrites) and show either a "N-MORB" REE pattern characterized by a flat HREE segment and substantial LREE depletion, or a convex-upward REE pattern. However, the low-alumina pyroxenites (chromium-spinel clinopyroxenites, websterites, and orthopyroxenites) are depleted in REE (HREE $= 0.3-1\times$

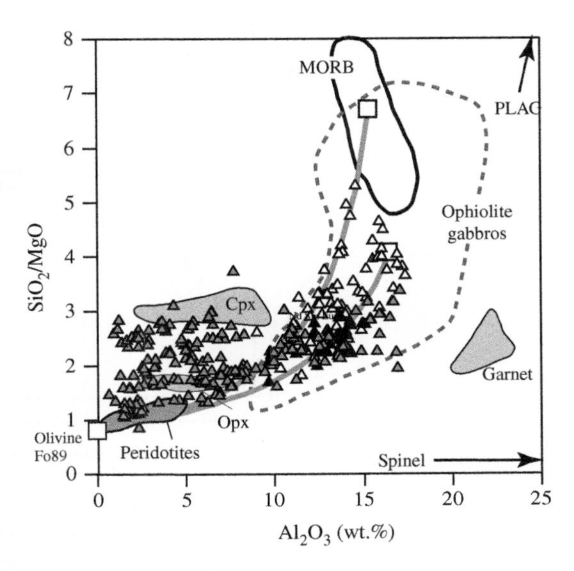

Figure 25 SiO_2/MgO versus Al_2O_3 (anhydrous wt.%) in representative mafic rocks from orogenic and ophiolitic massifs, and from abyssal suites of mantle rocks. The amphibole-bearing dikes and veins are distinguished by darker triangles and the garnet-bearing rocks by empty triangles. The anhydrous, garnet-free pyroxenites are shown by gray triangles. The field of orogenic, ophiolitic, and abyssal peridotites is also shown, as well as the MORB array and the compositional field of gabbros from the Oman ophiolite. Compositions of orthopyroxene, clinopyroxene, and garnet in mafic rocks from orogenic peridotites were also plotted. The figure shows mixing curves between olivine (89% forsterite) and average compositions of MORB and Oman gabbros. *Mafic rocks in orogenic peridotites:* Beni Bousera (Pearson *et al.*, 1993); Ronda (Remaïdi, 1993; Garrido and Bodinier, 1999); Pyrenees (Bodinier *et al.*, 1987a, b; Bodinier, 1989; McPherson *et al.*, 1996); Lanzo (Bodinier, 1988 (reproduced by permission of Elsevier from *Tectonophysics* **1988**, *149*, 67–88), 1989). *Pyroxenites in ophiolites:* Antalya (Juteau, 1974); Thailand (Orberger *et al.*, 1995); New Caledonia (Moutte, unpublished; Bodinier, unpublished); Dun (Davies *et al.*, 1980). *Pyroxenites in abyssal suites:* Iberia Abyssal Plain (Hébert *et al.*, 2001); Mid-Atlantic Ridge (Casey, 1997). Mineral compositions after Bodinier *et al.* (1987b), Takazawa *et al.* (1996), and Woodland *et al.* (1996). Data for orogenic, ophiolitic, and abyssal peridotites are the same as in Figure 5. MORB array from the PetDB database. Oman gabbros after Pallister and Knight (1981), Lippard *et al.* (1986), Ernewein *et al.* (1988), Benoit (1997), Korenaga and Kelemen (1997), Boudier *et al.* (2000), and Garrido and Kelemen (unpublished).

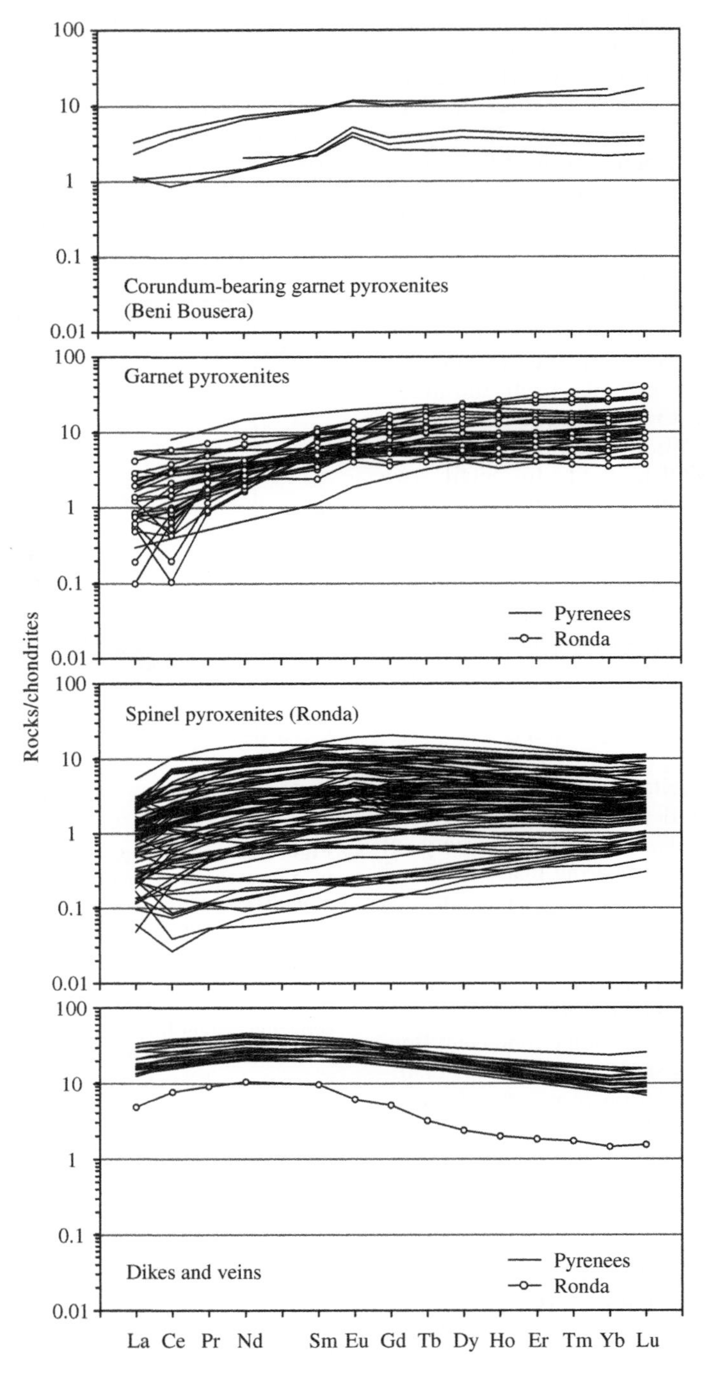

Figure 26 Chondrite-normalized abundances of REEs in representative mafic rocks from the Beni Bousera, Ronda, and Pyrenees orogenic peridotite massifs. Data from Bodinier *et al.* (1987a, 1987b, 1990), Bodinier (1989), Kornprobst *et al.* (1990), Pearson *et al.* (1993), Remaïdi (1993), McPherson (1994), Garrido (1995), McPherson *et al.* (1996), and Garrido and Bodinier (1999). Normalizing values after Sun and McDonough (1989).

chondrites). Similar to the spatially associated refractory peridotites (Figure 12; Gervilla and Remaïdi, 1993) they show either relatively unfractionated, slightly U-shaped REE patterns, or a linear decrease of normalized REE abundances from HREE to LREE.

(iii) Some garnet pyroxenites from Ronda are comparable to the corundum-bearing pyroxenites from Beni Bousera (below) in showing moderately fractionated REE patterns and slight positive europium anomalies. The other garnet pyroxenites have strongly fractionated REE patterns, devoid of europium anomaly. They are depleted in LREE and enriched in HREE relative to MREE. All the garnet pyroxenites belong to the high-alumina group.

(iv) The corundum-bearing garnet pyroxenites (Beni Bousera) are characterized by relatively unfractionated REE patterns, only slightly depleted in LREE relative to HREE and gently sloping from HREE to LREE. They show a positive anomaly of europium, especially marked in the REE-poor samples.

2.04.4.2 Interpretations

Two different hypothesis predominate among the models that have been postulated for the origin of mafic rocks in peridotite massifs:

(i) An origin as high-pressure crystal segregates from mantle melts in magma conduits, comparable to the models envisaged for pyroxenites in mantle xenoliths (e.g., Frey and Prinz, 1978; Irving, 1980). Evidence for this process is based on field, textural and trace-element arguments such as cross-cutting relationships, igneous textures and HREE fractionation indicative of garnet segregation (Shervais, 1979; Wilshire *et al.*, 1980; Bodinier *et al.*, 1987a,b; Sautter and Fabriès, 1990; Garrido and Bodinier, 1999; Takazawa *et al.*, 1999).

(ii) Stretching and recrystallization of subducted oceanic crust in convective mantle (the "marble cake" mantle of Allègre and Turcotte, 1986). This model is based on the extreme diversity of isotopic compositions in mafic rocks and on their isotopic differences with peridotites (Polvé and Allègre, 1980; Hamelin and Allègre, 1988; Roy-Barman *et al.*, 1996; Blichert-Toft *et al.*, 1999). According to Kornprobst *et al.* (1990), the recycling hypothesis is further supported by the existence of positive europium anomalies in garnet pyroxenites from Ronda and Beni Bousera (Figure 26; Pearson *et al.*, 1989; Kornprobst *et al.*, 1990). These anomalies are considered as evidence for a prograde evolution of the pyroxenites.

In both hypotheses, detailed studies of mafic rocks are of prime importance for our understanding of mantle processes. In the first case, these rocks may be used to investigate the geochemical composition of primary mantle melts and the mechanisms of melt segregation from partially molten peridotites (Loubet and Allègre, 1982; Obata and Nagahara, 1987; Suen and Frey, 1987; Pearson *et al.*, 1993; Reisberg and Lorand, 1995), or the subsequent evolution of melt compositions due to high-pressure fractional crystallization and/or reaction with lithospheric peridotites (Bodinier *et al.*, 1987a, b; Downes *et al.*, 1991; Rivalenti *et al.*, 1995; Kumar *et al.*, 1996; Garrido and Bodinier, 1999; Takazawa *et al.*, 1999). In the second hypothesis, the peridotite massifs are considered as analogues for the source region of oceanic basalts, and the mafic layers the cause of the recycling isotopic and trace-element signature in oceanic basalts (e.g., Prinzhofer *et al.*, 1989; Hauri, 1996; Hirschman and Stolper, 1996). Therefore, studies of mafic layers may provide information on the scale, distribution, and extent of geochemical heterogeneities in the convective mantle.

However, between the two end-member processes mentioned above, it is worth noting the alternative suggested by Pearson *et al.* (1993) for the diamond-facies garnet pyroxenites of the Beni Bousera massif. To reconcile the trace-element evidence for high-pressure fractionation in these pyroxenites, and their isotopic signature indicating their derivation from hydrothermally altered crust (±hemi-pelagic sediments—Pearson *et al.*, 1991a,b), the authors have interpreted the pyroxenites as crystal segregates from partial melts derived from subducted oceanic crust. Davies *et al.* (1993) postulated a similar origin for the diamond-facies garnet pyroxenites of the Ronda massif and suggested a geodynamic scenario involving the origin of the pyroxenites in a supra-subduction mantle wedge during an early evolutionary stage of the Ronda and Beni Bousera orogenic peridotites. Conversely, Garrido and Bodinier (1999) found that the majority of the pyroxenites occurring in the granular and plagioclase-tectonite domains of Ronda were formed late in the history of the body, by metasomatic replacement of peridotites. This process occurred during the receding stages of the high-temperature melting event that affected the Ronda peridotite, shortly before its exhumation (Lenoir *et al.*, 2001).

These examples show that the mafic rocks occurring in peridotite massifs are probably extremely variable in origin. It is, therefore, important to constrain their structural and textural relationships with the host peridotite before addressing their petrogenesis and their significance in term of mantle processes. On this ground, the mafic rocks may be classified in three main groups: (i) dikes and veins, (ii) replacive pyroxenites, and (iii) deformed pyroxenites. The main features of these three groups are briefly reviewed below. A number of subgroups can be defined on mineralogical and geochemical grounds, but such distinctions would be beyond the scope of this chapter.

2.04.4.2.1 Dikes and veins

These rocks show clear cross-cutting relationships with the peridotite foliation and compositional layering (Shervais, 1979; Wilshire *et al.*, 1980). They were generally deformed at high temperature, but only partially transposed into the peridotite foliation (Vétil *et al.*, 1988). Hence their igneous origin is hardly disputable. This group of mafic rocks shows a wide variety of mineralogical compositions, including orthopyroxenites, (spinel) websterites, spinel, garnet and/or amphibole clinopyroxenites, hornblendites, glimmerites, and (olivine) gabbros. They are often enriched in LREE and MREE relative to HREE (Figure 26; Bodinier *et al.*, 1987a), although LREE-depleted pyroxenite dikes are also observed (Rivalenti *et al.*, 1995). The olivine gabbro veins occurring in LP orogenic peridotites and ophiolites are distinguished by a prominent positive europium anomaly (Bodinier *et al.*, 1986). Compared with other mafic rock types, the dikes and veins are relatively homogeneous in isotopic compositions within a given peridotite massif, and their isotopic signatures are consistent with crystallization from mantle melts (Voshage *et al.*, 1988; Downes *et al.*, 1991; Mukasa *et al.*, 1991).

The mafic dikes are widely interpreted as crystal segregates precipitated into magma conduits. The parental melts of the high-alumina pyroxenites are generally considered to be of deep-seated origin, and thus unrelated to the host peridotites. This hypothesis was advocated for the Al-augite pyroxenites dikes in Balmuccia, on the basis of isotope data (Voshage *et al.*, 1988). Similarly, in the Eastern Pyrenees, the amphibole pyroxenites dikes are interpreted as "translithospheric" feeder dikes for the alkaline basalts associated with lithospheric thinning in the North Pyrenean zone during the Cretaceous, shortly before the peridotite emplacement (Bodinier *et al.*, 1987a; Fabriès *et al.*, 1991, 1998). The hornblendite veins are ascribed to fluid-assisted fracturing of the dike wall rocks and infiltration of volatile-rich melts in branching and anastomosing cracks. Conversely, the low-alumina pyroxenites from Ronda (Garrido and Bodinier, 1999) and Balmuccia ("Cr-diopside" pyroxenites; Rivalenti *et al.* (1995)) have crystallized from HREE-depleted, refractory melts that were possibly formed within the massifs, together with refractory peridotites (Gervilla and Remaïdi, 1993). Low-alumina pyroxenite dikes are frequent in the ophiolitic harzburgite sequences. Those from the Bay of Island ophiolite (Newfoundland) have been ascribed to the segregation and transport of boninitic melts in a supra-subduction setting (Edwards and Malpas, 1995; Varfalvy *et al.*, 1996, 1997). Similarly, the Nan Uttaradit ophiolite (Northern Thailand) contains low-alumina pyroxenites probably related to boninitic melts (Orberger *et al.*, 1995).

2.04.4.2.2 Replacive pyroxenites

In the Ronda massif, the replacive pyroxenites are distinguished from the other mafic layers by lack of significant deformation, except for some open folds coeval with the development of the plagioclase tectonite domain during the emplacement of the Ronda peridotite into the crust (Van der Wal and Vissers, 1996). They are never isoclinally folded nor boudinaged. They typically occur as swarms of straight layers, or elongated lenses, hosted by harzburgites and dunites (Garrido and Bodinier, 1999). In that sense, they resemble the sill-complexes described in the mantle–crust transition zone of some ophiolites (e.g., Boudier and Nicolas, 1995). However, the outcrops show gradual transitions—both across layering and along strike—from peridotites containing thin and rather diffuse pyroxene layers (<1 cm) to thick websterites lenses (>1 m) containing thin olivine seams. Field observations indicate that the olivine seams derive from relict septa of corroded, host peridotite. Together with petrographic observations showing development of secondary pyroxene after olivine, this feature demonstrates that these pyroxenites were formed by replacement of peridotites via pyroxene-forming melt–rock reaction (e.g., Kelemen *et al.*, 1995). Similar interpretation was recently proposed by Burg *et al.* (1998) for a thick (~1.5 km) websterite sequence containing "relict" olivine flames in the basal mantle section of the obducted Kohistan arc (Northern Pakistan).

In contrast with dikes and veins, the replacive pyroxenites show relatively monotonous mineralogical compositions. Both in Ronda and Jijal, the lithologies vary from low-alumina (olivine) websterites to high-alumina spinel websterites. The samples are characterized by convex-upward ("bell-shaped") REE patterns reminiscent of those observed in the cpx-rich peridotites from the Caussou massif (Eastern Pyrenees, Figure 14), which have been ascribed to cpx-forming reactions between alkaline melts and spinel peridotites (Bodinier *et al.*, 1988; Fabriès *et al.*, 1989). However, the Ronda pyroxenites differ from the Caussou peridotites in being depleted in LREE relative to HREE. LREE depletion is even more accentuated in the Jijal pyroxenites, which are further characterized by an overall depletion in REE (Garrido *et al.*, unpublished data).

The formation of replacive pyroxenites can be explained by melt-consuming reactions at pressure and temperature conditions close to the peridotite solidus. Ronda represents a situation where the reaction was associated with relaxation of thinned and thermally eroded subcontinental lithosphere (Garrido and Bodinier, 1999). Upon cooling of the melting domain developed during the thermal erosion (Lenoir *et al.*, 2001), the interstitial melt became saturated in pyroxene (± aluminous phases) and reacted with olivine to produce secondary cpx, opx, and spinel

(± amphibole upon further cooling). Depending on melt distribution (i.e., pervasive or channeled porous flow), the reaction resulted either in diffuse fertilization of the peridotites or in the formation of websterite layers. The strong melt localization and high melt/rock ratios implied by the formation of the pyroxenites may be accounted for by refraction and channeling of percolating melts below the freezing horizon represented by the base of the eroded lithosphere (Sparks and Parmentier, 1991; Spiegelman, 1993). In this scheme, the websterite layers trace the position of the solidus isotherm—hence of the permeability boundary of the porous flow domain—while the latter was receding down to gradually deeper mantle levels (Garrido and Bodinier, 1999). In contrast, the Jijal complex illustrates a scenario whereby the websterite-forming, melt-consuming reaction occurred because of lithospheric thickening beneath an active island arc (Burg *et al.*, 1998). The thick pyroxenite layers from the Cabo Ortegal ultramafic complex (Northwestern Spain) display some features comparable to Jijal (although strongly obscured by deformation and prograde metamorphism—Gil Ibarguchi *et al.*, 1990; Girardeau and Gil Ibarguchi, 1991) and might also record the formation of replacive pyroxenites in subarc mantle.

2.04.4.2.3 Deformed pyroxenites

The "deformed" pyroxenites are concordant to the foliation at the scale of the outcrop *and* strongly deformed in mantle conditions. Deformation structures commonly observable in the field include multiple isoclinal folding and boudinage (Figure 27). Locally, these pyroxenites form layered sequences where the mafic layers may predominate over the peridotites. These sequences sometimes result from multiple folding of a limited number of layers. In highly deformed peridotites (mylonites), thin pyroxenite layers may be dispersed in the peridotites up to the scale of individual crystal grains, a process that may ultimately result in peridotite fertilization (Kornprobst *et al.*, 1990; Becker, 1996). The deformed pyroxenites display a wide spectrum of mineralogical compositions, ranging from low-alumina olivine westerites and orthopyroxenites, to high-alumina spinel or garnet websterites, garnet clinopyroxenites, and mafic garnet granulites.

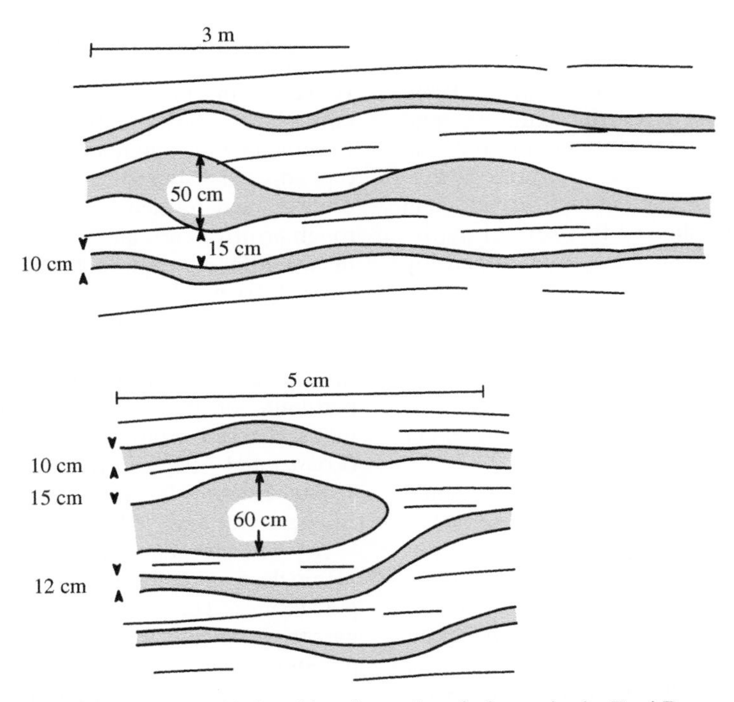

Figure 27 Occurrences of isoclinally folded and boudinaged mafic layers in the Beni Bousera orogenic peridotite, northern Morocco, after Allègre and Turcotte, 1986; reproduced by permission of Nature Publishing Group from *Nature*, **1986**, *323*, 123–127. In their "marble cake" model, Allègre and Turcotte interpret these mafic layers as subducted oceanic lithosphere—modified by partial melting and metamorphism—that was stirred, stretched, and thinned by mantle convection (see, also, Kornprobst *et al.*, 1990; Blichert-Toft *et al.*, 1999). For Pearson *et al.* (1993), these pyroxenites would rather derive from crystallization products of partial melts from subducted crust. They would have been emplaced as dikes in the hanging wall of a subduction zone and deformed during the exhumation of the peridotite (Davies *et al.*, 1993).

The garnet-free websterites and the orthopyroxenites (±olivine, ±spinel, and/or plagioclase, ±amphibole) are widespread in most peridotite massifs where they define a thin and pervasive compositional layering. These rocks are devoid of high-pressure mineral assemblages. Similar to the type Ib defined by Frey and Prinz (1978) in mantle xenoliths, they are generally comparable to the host peridotites in term of mineral compositions. Likewise, the REE patterns of the pyroxenites tend to parallel those of the peridotites at higher concentration levels (Gervilla and Remaïdi, 1993) and their Sr–Nd isotopic compositions are identical (Voshage *et al.*, 1988). In the Western Alps peridotites, these pyroxenites were generally ascribed to the segregation of partial melts from the host peridotites (e.g., Shervais, 1979; Sinigoi *et al.*, 1983; Voshage *et al.*, 1988, for the Cr-diopside suite in Balmuccia, and Bodinier *et al.*, 1986 for similar rock types in Lanzo). To some degree, this observation is sustained by the observation that the wall–rock peridotites adjacent to the pyroxenites are more refractory than the average peridotite, distal to the layers (e.g., Rivalenti *et al.*, 1995). In other peridotite massifs, these pyroxenites are more readily interpreted as residues after partial melting of prexisting, more aluminous pyroxenite layers (Loubet and Allègre, 1982). In Ronda, evidence for melting is provided by the concomitant evolution of pyroxenite and peridotite compositions across a kilometer-scale "melting" front (Garrido and Bodinier, 1999; Lenoir *et al.*, 2001). Partial melting and melt redistribution would account for the homogenization of mineral compositions in pyroxenites and adjacent peridotites. Garrido and Bodinier (1999) also showed that several pyroxenites from Ronda were deeply modified by metasomatic processes *after* the deformation. The metasomatism was responsible for textural annealing and complete mineralogical and chemical transformations. It is ascribed to pervasive percolation of volatile-rich small melt fractions during the receding stages of the melting event that affected the Ronda massif.

Among the deformed garnet pyroxenites, some show textural and/or geochemical features consistent with an origin as high-pressure crystal segregates. They were interpreted as magma conduits crystallized as dikes in the lithospheric mantle or in a mantle wedge above a subduction zone (Bodinier *et al.*, 1987b; Sautter and Fabriès, 1990; Mukasa *et al.*, 1991; Pearson *et al.*, 1993; Takazawa *et al.*, 1999). Thereafter, the pyroxenite dikes would have been deformed and transposed into the peridotite foliation during the tectonic events that led to the exhumation of the peridotite massifs. Textural and petrological features include, for instance, the existence of high-temperature pyroxene megacrysts in garnet pyroxenites from the Eastern Pyrenees (Sautter and Fabriès, 1990). Geochemical evidence for garnet segregation relies on HREE fractionation and includes whole-rock HREE/MREE ratios greater than chondrites ratios, and HREE contents greater than 20 times chondrite abundances (Figure 26). Furthermore, Bodinier *et al.* (1987b) distinguished "primary" (igneous) garnet from "secondary" (metamorphic) garnet on the basis of HREE fractionation in coexisting cpx. The cpx equilibrated with secondary garnet is distinguished by a strong depletion of HREE, which is not observed in cpx equilibrated with primary garnet (see, also, Takazawa *et al.*, 1999). Partial melting and melt extraction leaving residual garnet might explain the high HREE/MREE ratios observed in whole rocks, but this process would hardly account for the elevated HREE abundances.

High- and ultrahigh-pressure equilibrium conditions have been inferred for some garnet pyroxenites. In the Ronda, Beni Bousera, and Horoman orogenic lherzolite bodies, these conditions are constrained by the presence of corundum (±sapphirine) and/or graphitized diamond pseudomorphs (Pearson *et al.*, 1989, 1993, 1995; Kornprobst *et al.*, 1990; Davies *et al.*, 1993; Morishita and Arai, 2001). When these rocks are included, the deformed peridotites show a broad range of major-element compositions on the SiO_2/MgO versus Al_2O_3 diagram (Figure 25). However, in contrast with the replacive pyroxenites that show scattered variations mainly governed by the nature of the segregated minerals (cpx, opx, and spinel), and the amount of residual olivine, the deformed pyroxenites tend to plot on a rough positive correlation that may be interpreted in two different ways (not mutually exclusive):

(i) "Mixing" between peridotites and recycled oceanic crust (MORB or/and gabbros). The mixing process may result from the mechanical/diffusional dispersion of the crustal components in peridotite or from the redistribution of partial melts derived from the recycled mafic rocks (Loubet and Allègre, 1982; Allègre and Turcotte, 1986; Pearson *et al.*, 1993; Becker, 1996; Blichert-Toft *et al.*, 1999).

(ii) The presence of variable amounts of plagioclase in the primary mode of the garnet pyroxenites. This would imply that a number of garnet pyroxenites crystallized primarily as low-pressure, plagioclase-bearing cumulates, and then followed a prograde metamorphic evolution to higher-pressure mantle conditions. This second alternative would be consistent with the positive europium anomalies observed in several samples (Pearson *et al.*, 1989; Kornprobst *et al.*, 1990).

Both hypotheses are consistent with the suggestion that at least some of the deformed pyroxenites represent recycled material. However, it should be noted that these rocks have been found only in strongly deformed orogenic peridotites which have undergone a high- to ultrahigh-pressure evolution in a mantle wedge above a subduction zone, before their exhumation to the surface (Van der Wal and Vissers, 1993; Davies *et al.*, 1993; Becker, 1996; Obata and Niida, 2002). Therefore, the "HP/UHP" deformed pyroxenites may represent crustal material which was first entrained into the subduction, and then exhumed to the surface with their host peridotites (see, e.g., Brueckner and Medaris, 2000, and references herein). In this scenario, no straightforward implications can be drawn on the petrological and geochemical structure of the convective mantle from studies of layered pyroxenites in orogenic peridotite massifs.

2.04.5 Nd–Sr ISOTOPE GEOCHEMISTRY

2.04.5.1 Data

The orogenic peridotites have been the focus of several Nd–Sr isotopic studies that

(i) confirmed their mantle origin (Menzies and Murthy, 1978);

(ii) showed the depleted character of mantle sources for the predominant lherzolite facies (Richard and Allègre, 1980; Zindler *et al.*, 1983),

(iii) revealed significant isotopic discrepancies between pyroxenites and peridotites (Polvé and Allègre, 1980; Voshage *et al.*, 1988; Downes *et al.*, 1991; Mukasa *et al.*, 1991; Becker, 1996), an observation which was further confirmed by lead, osmium, and hafnium isotopes (Hamelin and Allègre, 1988; Roy-Barman *et al.*, 1996; Blicher-Toft *et al.*, 1999),

(iv) reported a wide spectrum of Nd–Sr isotopic variations in several orogenic peridotites, encompassing the whole domain of the oceanic basalts (Figure 28). The variations include fertile lherzolites that are substantially more depleted than average MORB as well as refractory peridotites that are more enriched than Bulk Earth in term of Nd–Sr isotopic compositions (Reisberg and Zindler, 1987; Voshage *et al.*, 1987; Reisberg *et al.*, 1989; Bodinier *et al.*, 1991; Downes *et al.*, 1991; Mukasa *et al.*, 1991; Rampone *et al.*, 1995; Yoshikawa and Nakamura, 2000).

Several Nd–Sr isotopic studies were also devoted to ophiolitic peridotites, and a more restricted number of studies to oceanic peridotites. The early studies have reported anomalously high $^{87}Sr/^{86}Sr$ values (e.g., Bonatti *et al.*, 1970) in these rocks and concluded that they were genetically unrelated to oceanic crust (see references in Menzies and Murthy, 1978). More recent studies

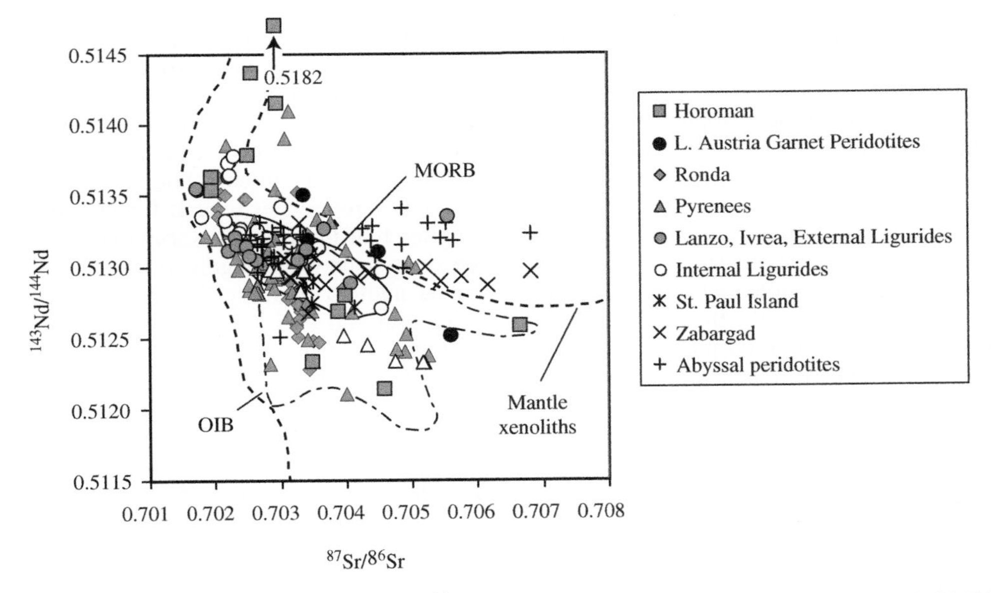

Figure 28 Present-day $^{143}Nd/^{144}Nd$ versus $^{87}Sr/^{86}Sr$ for clinopyroxene separates from orogenic, ophiolitic, and abyssal peridotites and pyroxenites. Data from Yoshikawa and Nakamura (2000) for Horoman, Becker (1996) for the Lower Austria Garnet Peridotites, Reisberg and Zindler (1987) and Reisberg *et al.* (1989) for Ronda, Downes *et al.* (1991), Mukasa *et al.* (1991), McPherson (1994), McPherson *et al.* (1996), and Bodinier *et al.* (2004) for the Pyrenees, Voshage *et al.* (1988), Bodinier *et al.* (1991), and Rampone *et al.* (1995) for the Western Alps (Balmuccia, Lanzo, and the External Ligurides), Brueckner *et al.* (1988) and Bosch (1991, unpublished) for Zabargad, Rampone *et al.* (1996) for the Internal Ligurides, Snow *et al.* (1994), Kempton and Stephens (1997) and Salters and Dick (2002) for the oceanic peridotites. MORB array from the PetDB database. OIB array from the GeoRoc database (data from French Polynesia, Iceland, Hawaii, the Galapagos Islands, and Bouvet).

in abyssal peridotites and in the plagioclase peridotite of the Zabargad Island (Red Sea) demonstrated that the high $^{87}Sr/^{86}Sr$ values measured in oceanic peridotites result from seawater alteration (Bosch, 1991; Agrinier *et al.*, 1993; Snow *et al.*, 1994; Kempton and Stephens, 1997). These samples are distinguished in Figure 28 by $^{143}Nd/^{144}Nd$ and $^{87}Sr/^{86}Sr$ ratios both higher than Bulk Earth. However, several studies on ophiolitic and abyssal peridotites converge on the observation that the "residual" peridotites have too elevated $^{143}Nd/^{144}Nd$ ratios (age-corrected) to be genetically related to the spatially associated oceanic crust by any simple model of partial melting. This is the case, for instance, for the Proterozoic ophiolites from Saudi Arabia (Claesson *et al.*, 1984) as well as for the more recent ophiolites from Trinity, California (Jacobsen *et al.*, 1984; Brouxel and Lapierre, 1988) and Internal Ligurides, Northern Italy (Rampone *et al.*, 1996). Similarly, Göpel *et al.* (1984) concluded from lead isotope data that the mantle rocks of the Xigaze ophiolite (Tibet) are not cogenetic with the crustal rocks. The abyssal peridotites from the Southwest Indian, Atlantic, and American-Antarctic ridge systems appear to be systematically characterized by higher $^{143}Nd/^{144}Nd$ ratios than average MORB values in the same areas (Snow *et al.*, 1994; Kempton and Stephens, 1997; Salters and Dick, 2002). Some exceptions were noted, however, such as the Toad Lake area in the Trinity ophiolite, California (Gruau *et al.*, 1991) and the Urals ophiolite, Russia (Sharma *et al.*, 1995; Sharma and Wasserburg, 1996).

The wide diversity of the isotopic compositions in orogenic peridotites, and the anomalously depleted composition of several ophiolitic and abyssal peridotites have strong implications on the small-scale structure of the convective mantle, as well as on mantle processes such as decompression partial melting of mantle rocks, the formation of oceanic lithosphere and the thermomechanical and chemical erosion of lithospheric mantle by upwelling asthenosphere. We briefly review some of these important issues below.

2.04.5.2 Interpretations

Nd–Sr isotopic studies in orogenic peridotites have emphasized their wide isotopic heterogeneity but also showed that the predominant fertile lherzolite facies are actually fairly homogenous and characterized by depleted, MORB-type isotopic compositions. The low-alumina, deformed (olivine) websterites, that are chemically equilibrated with the peridotites (see Section 2.04.4.2.3), also show a limited range of isotopic variations and are generally indistinguishable from the host peridotites from their $^{143}Nd/^{144}Nd$ and $^{87}Sr/^{86}Sr$ ratios (Voshage *et al.*, 1988).

The contribution of recent mantle metasomatic events to the isotopic heterogeneity of orogenic peridotites is very limited. This is shown, for example, by the Nd–Sr isotopic composition of the amphibole-bearing dikes and hornblendite veins in Pyreneean peridotites (Downes *et al.*, 1991; Mukasa *et al.*, 1991; McPherson *et al.*, 1996). These rocks are relatively homogeneous, slightly more enriched that the lherzolites, but still close to the MORB field. Their isotopic signature is consistent with an origin as crystal segregates in magma conduits and cracks during the Cretaceous alkaline magmatism in Pyrenees (see Section 2.04.4.2.1). In addition, a detailed study of a metasomatized wall rock adjacent to an amphibole-pyroxenite dike has revealed that the contamination in $^{143}Nd/^{144}Nd$ and $^{87}Sr/^{86}Sr$ is spatially limited to a distance of <25 cm from the dike (Bodinier *et al.*, 2003). This distance is interpreted as the chromatographic front of neodymium and strontium since the infiltration of small volume melts probably occurred on a greater distance in the host peridotite, as attested by the enrichment of the highly incompatible elements observed at 25–80 cm from the dike (e.g., lanthanum and cerium, see Section 2.04.3.2.2). This indicates that small volume melts have a limited capacity to modify the Nd–Sr isotopic composition of mantle rocks.

Finally, a careful examination of published data indicates that the isotopic variability of the orogenic peridotites is mostly contributed by two lithologies which are generally subordinate in volume: (i) layers of refractory peridotites (e.g., Downes *et al.*, 1991) and (ii) high-alumina, deformed pyroxenites (Allègre and Turcotte, 1986, and references herein; Becker, 1996). This implies that the isotopic variations occur mostly on short-scale wavelengths (0.1–10 m, typically). Some authors (notably Salters and Dick, 2002) have ascribed to such small-scale variations the isotopic discrepancy observed between abyssal peridotites and ocean crust. In the following, we examine (i) the origin of the isotopic variations from fertile to refractory peridotites, (ii) the significance of deformed pyroxenites in term of mantle veining and crustal recycling, and (iii) the models proposed to explain the isotopic discrepancy between oceanic mantle and crust.

2.04.5.2.1 Harzburgite layering: marble cake of lithospheric strips?

A striking feature of several orogenic peridotites is the existence of a negative correlation between Nd–Sr isotopic enrichment and peridotite fertility. This relationship is illustrated

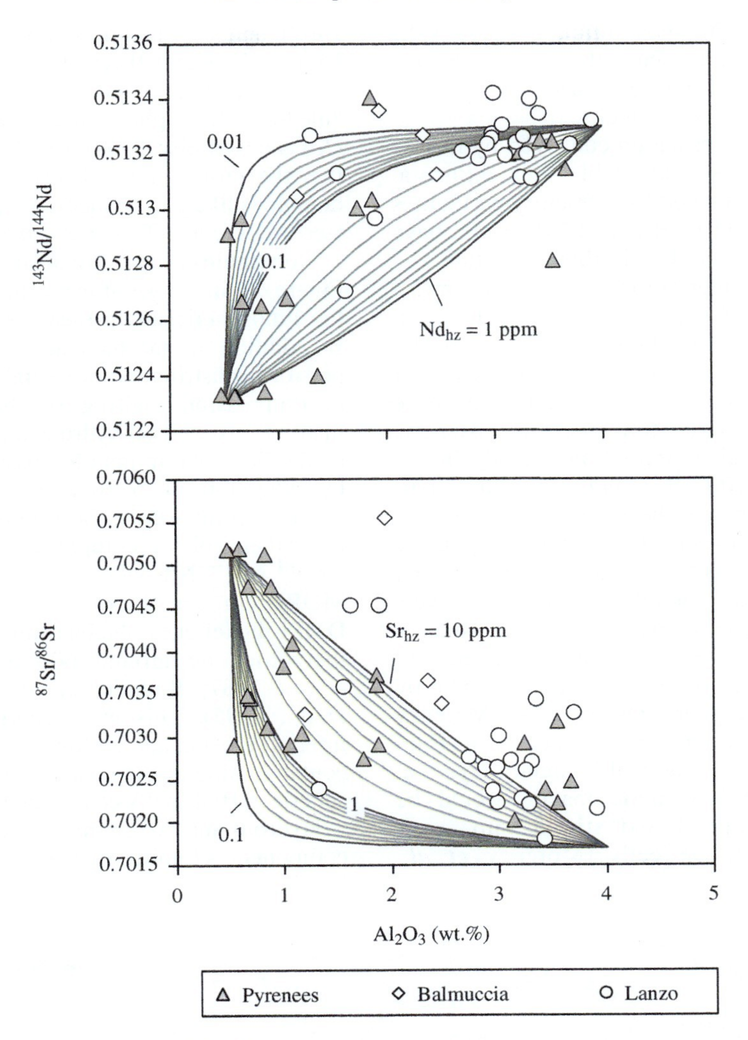

Figure 29 Plots of $^{143}Nd/^{144}Nd$ and $^{87}Sr/^{86}Sr$ versus Al_2O_3 (anhydrous wt.%) for orogenic peridotites from the Pyrenees, Balmuccia, and Lanzo massifs (isotopic ratios = present-day clinopyroxene data). Data from Bodinier (1988, 1989), Bodinier *et al.* (1988, 1991), Voshage *et al.* (1988), Downes *et al.* (1991), McPherson (1994), McPherson *et al.* (1996). Lherzolite–harburgite mixing models are also shown. The elemental and isotopic compositions of the end-members are $Al_2O_3 = 4\%$, Nd = 0.7 ppm, Sr = 12 ppm, $^{143}Nd/^{144}Nd = 0.5133$, $^{87}Sr/^{86}Sr = 0.7017$ for the lherzolite, and $Al_2O_3 = 0.5\%$, Nd varying from 0.01 ppm to 1 ppm, Sr = varying from 0.1 ppm to 10 ppm, $^{43}Nd/^{144}Nd = 0.5123$, $^{87}Sr/^{86}Sr = 0.7052$ for the harzburgite.

in Figure 29 by a positive correlation between the $^{143}Nd/^{144}Nd$ ratio and the Al_2O_3 content of peridotites from the Western Alps (Balmuccia and Lanzo) and the Eastern Pyrenees. Although strongly scattered, a rough, negative correlation is also observed between $^{87}Sr/^{86}Sr$ and Al_2O_3. The correlations tend to be better defined when the massifs are considered individually.

To explain the isotopic enrichment observed in refractory peridotites, Bodinier *et al.* (1991) and Downes *et al.* (1991) suggested an interpretation involving focused percolation of enriched (plume-derived) melts in refractory porous-flow channels. In Lanzo, this process was tentatively ascribed to a "pre-rift" stage of the opening of the Liguro-Piemontese ocean basin (Bodinier *et al.*, 1991). However, this explanation would hardly account for the existence of refractory peridotites with extremely enriched isotopic compositions (particularly very low $^{143}Nd/^{144}Nd$ values, down to 0.5120), as observed in all orogenic peridotites for which a sufficient data set is available. In addition to the Western Alps and the Pyrenees, this includes Ronda (Reisberg and Zindler, 1987; Reisberg *et al.*, 1989) and Horoman (Yoshikawa and Nakamura, 2000). These compositions fall in the lower part of the Ocean Island Basalts (OIB) field, or even below. In conductively cooled, lithosphere mantle, integration of LREE enrichment may contribute to the enriched signature of the harzburgites. However, the Nd/Sm ratios measured in these rocks are too to account for their low $^{143}Nd/^{144}Nd$

values in a reasonable span of time. Moreover, the refractory peridotites generally do not display mineralogical or chemical evidence for interaction with enriched, deep-seated melts.

However, the information conveyed by mantle xenoliths indicates that stable subcontinental lithosphere is dominated by refractory peridotites which are enriched in HIE and LREE and have often acquired an enriched isotopic signature as a result of time integration of their chemical enrichment (see Chapter 2.05). Therefore, an alternative to the porous-flow model is to consider that the harzburgite layers represent strips of lithospheric peridotites embedded into more fertile material derived from the asthenospheric mantle (e.g., the Lherz massif, Figure 30). In this scheme, the $^{143}Nd/^{144}Nd$ and $^{87}Sr/^{86}Sr$ versus Al_2O_3 correlations can be interpreted as incomplete mixing between these two components (Figure 29). A possible mechanism to generate such "marble cake" of relict lithospheric layers within asthenospheric material is suggested by numerical and analogical modeling of thermomechanical erosion of the lithospheric mantle by upwelling asthenosphere (Davies, 1994; Morency *et al.*, 2002). During this process, a thin, conductively heated "thermal boundary layer" at the base of the lithopheric mantle would be periodically entrained and stirred into small (< 10 km) convection cells (Morency *et al.*, 2002), and stretched into asthenospheric material. Upon subsequent thermal relaxation, the mantle domain newly accreted to the lithosphere will be a "blend" of lithospheric and asthenospheric mantle (Xu and Bodinier, 2004). Due to short timescale of the thermomechanical erosion process, this domain is likely to be incompletely homogeneized (see Allègre and Turcotte, 1986).

Nonetheless, the covariations observed in Figure 29 imply significant interactions between the lithospheric and asthenospheric components. In addition to mechanical and diffusional dispersion, redistribution of partial melt produced by decompression melting of the asthenospheric material, as well as hydrous and/or vein melting of the lithospheric mantle, may contribute to the hybridization of peridotites between the end-member components. Decompression melting of the asthenospheric component would account for the $^{143}Nd/^{144}Nd$ values generally higher than MORB in the lherzolites. All neodymium DMM-model ages obtained for the Western Alps lherzolites lie around 600–800 Ma (Voshage *et al.*, 1987; Bodinier *et al.*, 1991; Rampone *et al.*, 1995). This may record the age of the inferred erosion and melting event, but a possible contamination of the neodymium values by lithospheric partial melts cannot be ruled out.

Hydrous and/or vein melting of the lithospheric component is implied by the decoupling of Sm/Nd

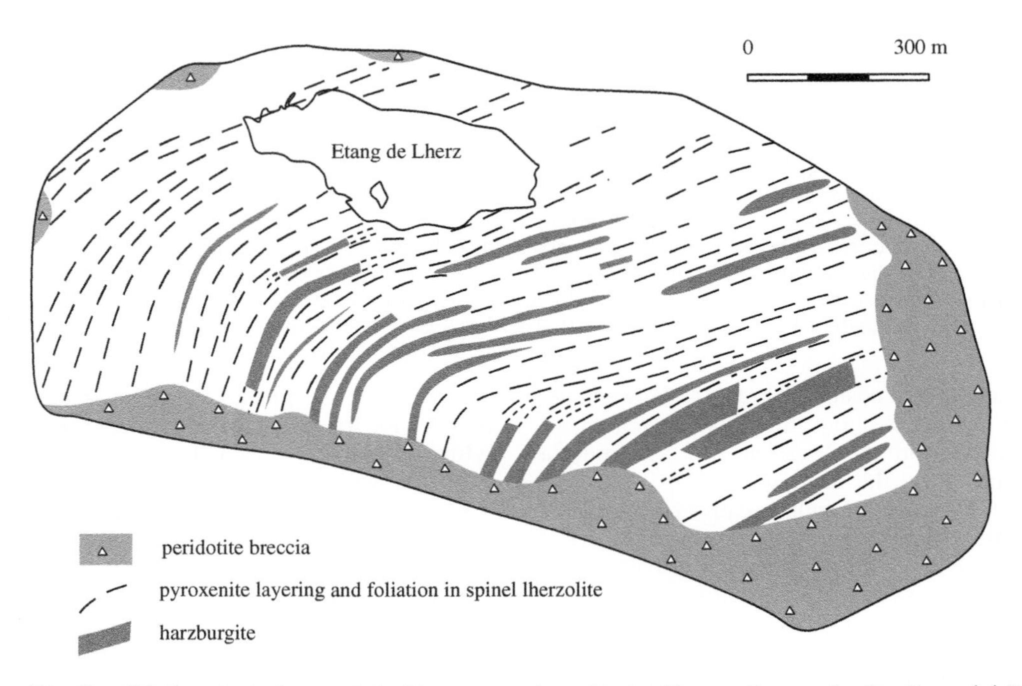

Figure 30 Simplified geological map of the Lherz orogenic peridotite (Eastern Pyrenees), after Conquéré (1978) and Monchoux (unpublished data), showing the hazburgite layering concordant with peridotite foliation. Based on their enriched Nd–Sr isotopic composition (e.g., Downes *et al.*, 1991), the harzburgite layers are interpreted as residual strips of aged, isotopically enriched, but refractory, mantle lithosphere, embedded into younger, depleted but fertile, asthenospheric mantle (spinel lherzolites). This "marble cake" structure is tentatively ascribed to thermomechanical erosion of lithospheric mantle by upwelling asthenosphere.

and $^{143}Nd/^{144}Nd$ ratios in the harzburgites. The $^{143}Nd/^{144}Nd$ versus Al_2O_3 correlation in Figure 29 is best fit with a model assuming neodymium concentrations in harzburgites identical to the values actually observed in the samples (~0.05 ppm in the Pyrenean harzburgites, for instance). However, as noted above, such neodymium values are not high enough to account for the low $^{143}Nd/^{144}Nd$ values in these rocks within a reasonable span of time. Therefore, the original (lithospheric) LREE signature of the harzburgites must have been erased before their entrainment by convection. This may occur via (i) hydrous and/or vein partial melting of the conductively heated lower lithosphere, and (ii) percolation of (lithospheric + asthenospheric) partial melts in the refractory peridotites, before and/or during the mechanical erosion. All these processes were suggested for the Ronda peridotite and ascribed to the erosion of the massif by upwelling asthenosphere shortly before its emplacement (Van der Wal and Bodinier, 1996; Garrido and Bodinier, 1999; Lenoir *et al.*, 2001).

In this scheme, several orogenic lherzolites (particularly in the Western Alps and the Pyrenees) may be interpreted as hybrid mantle containing strips of aged, isotopically enriched, but refractory, lithospheric peridotites, embedded into younger, depleted but fertile, asthenospheric mantle. High-temperature, plagioclase peridotites such as Lanzo would represent strongly "asthenospherized" mantle containing only sparse relicts of lithospheric origin (a few depleted spinel lherzolites with enriched Nd–Sr isotopic compositions—Bodinier *et al.*, 1991). However, lower-temperature spinel peridotites such as Lherz would have preserved a significant lithospheric component in the form of abundant harzburgite layers (Figure 30). This model is consistent with the suggestion by Reisberg and Zindler (1987) that the Nd–Sr isotopic composition of the Ronda massif "was influenced by a component which existed in a LREE-enriched environment for a significant period of time." Similarly, Brueckner *et al.* (1988) suggested that the Nd–Sr covariation in Zabargad requires mixing between a depleted and an enriched reservoir.

A correlation similar to the one observed for the Nd–Sr isotopes versus Al_2O_3 is also noted for osmium in several orogenic peridotites, but is commonly interpreted in term of depletion age (e.g., Reisberg and Lorand, 1995). The mixing hypothesis proposed for Nd–Sr casts some doubt on this interpretation. Saal *et al.* (2001) interpreted the osmium systematics of the Horoman peridotite in term of melt redistribution and fertilization, which is more consistent with the mixing hypothesis.

It is nevertheless worth noting that the marble cake model involving lithospheric strips cannot be straightforwardly applied to all orogenic peridotites, and particularly to the "subduction-related" orogenic peridotites that show:

(i) either Nd–Sr isotopic variations following a pyroxenite control line, rather than the harzburgite control line (e.g., the lower Austria garnet peridotites—Becker, 1996); or

(ii) complex variations that cannot be accounted for by a two-component mixing model (e.g., Horoman-Yoshikawa and Nakamura, 2000; and Ronda, to some degree—Reisberg and Zindler, 1987; Reisberg *et al.*, 1989).

In the first situation, the isotopic compositions are probably dominantly governed by recycled crustal components, as discussed below (Section 2.04.5.2.2). In the second case, they may reflect the heterogeneous composition of the lithospheric component, or the involvement of other components, such as recycled crust, aged (lithospheric) igneous crystal segregates, or refractory mantle from a supra-subduction mantle wedge (Pearson *et al.*, 1991a,b 1993; Davies *et al.*, 1993; Yoshikawa and Nakamura, 2000). Even in the Western Alps and Pyrenean peridotites, the involvement of a third (pyroxenitic) component cannot be ruled out, as it would possibly account for the scattered variation of strontium in Figure 29. Further alternatives to account for complex Nd–Sr isotopic variations in orogenic peridotites involve either diffuse infiltration of subduction-related fluids (Yoshikawa and Nakamura, 2000), or the percolation of melts unrelated to the peridotites, that would re-use the refractory layers as porous-flow channels (Takazawa *et al.*, 1992; Gervilla and Remaïdi, 1993).

2.04.5.2.2 Pyroxenites: crustal recycling versus veined lithosphere

Several studies have confirmed the early observation of Polvé and Allègre (1980) that the deformed high-alumina pyroxenites have a wide spectrum of isotopic variations and are genetically unrelated to the host peridotites. The discrepancy of isotopic compositions between pyroxenites and peridotites led Allègre and Turcotte (1986) to suggest "that the upper mantle contains elongated strips of subducted oceanic lithosphere. These strips are stretched and thinned by the normal and shear strains in the convective mantle, and are destroyed by being reprocessed at ocean ridges or, on the centimeter scale, by dissolution processes; the result is a marble-cake mantle."

At the first sight, this model is supported by the isotopic characterization (Nd–Sr–Pb–O–S) of the corundum and diamond pseudomorph-bearing garnet pyroxenites from Beni Bousera by Pearson *et al.* (1991a,b, 1993), which indicates a derivation from hydrothermally altered oceanic crust

(±hemi-pelagic sediments). However, these authors do not consider these rocks as stretched oceanic crust but as vein conduits of partial melts from subducted crust. The veins would have been emplaced in the hanging wall of a subduction zone (Davies *et al.*, 1993). A similar origin was suggested by Becker (1996) for deformed pyroxenites in Lower Austria garnet peridotites, which are also ascribed to a subduction setting. In Lower Austria, the pyroxenites are isotopically enriched in Nd–Sr and the peridotites show a trend of decreasing $^{143}Nd/^{144}Nd$ and increasing $^{87}Sr/^{86}Sr$ towards the pyroxenite composition. This evolution is explained by mechanical dispersion of the pyroxenite and melt redistribution around pyroxenite layers (Becker, 1996). Partial melting of the pyroxenite layers is also attested in the Beni Bousera pyroxenites by decoupling between isotopic variations and parent/daughter ratios (Pearson *et al.*, 1993). Therefore, even though they cannot be considered as straightforward evidence for a marble cake structure in the upper mantle (see discussion in Section 2.04.4.2.3), the pyroxenites of the subduction-related orogenic peridotites may provide a valuable source of information on partial melting of subducted slabs, and on melt transport and melt–rock interactions in the mantle wedge above subductions.

However, high-pressure crystal segregates (e.g., garnet pyroxenite veins) isolated in the lithospheric mantle during hundreds of million years may develop strong isotopic heterogeneities. This mechanism is supported, for instance, by the huge spectrum of isotopic variations observed in mantle xenoliths and discrete xenocrysts (Figure 28; see Chapter 2.05). A similar origin may be advocated for several high-alumina deformed pyroxenites in orogenic peridotites. In Western Alps and Pyrenees, for instance, these rocks display a wide range of $^{143}Nd/^{144}Nd$ variations that may be accounted for by radiogenic decay following multiples episode of vein injection (Mukasa *et al.*, 1991) and partial melting of pyroxenite veins (Loubet and Allègre, 1982). Such isotopically heterogenous pyroxenites may be entrained together with their host peridotites during the thermomechanical erosion of lithospheric mantle by upwelling asthenosphere. During this process, they may be strongly deformed, boudinaged (Figure 27), and transposed into the peridotite foliation by small-scale convection (Morency *et al.*, 2002). The partial melting event frequently evoked for the deformed pyroxenites since Loubet and Allègre (1982) may be related to this process (e.g., Garrido and Bodinier, 1999).

Finally, the scenario that arises from the isotopic studies of deformed, layered pyroxenites involves two possible origins:

(i) Metamorphosed strips of subducted crust or partial melts from subducted crust crystallized in the hanging wall of subduction zones. Not unexpectedly, these pyroxenites are mostly found in subduction-related orogenic peridotites such as Beni Bousera, Ronda, and the Lower Austria garnet peridotites.

(ii) "Aged" igneous lithospheric veins—the most likely interpretation for the high-alumina, deformed pyroxenites in the orogenic peridotites from the Western Alps and the Pyrenees.

In both cases, however, the pyroxenites may be entrained in convective upper mantle during thermomechanical erosion of supra-subduction mantle wedges (associated with back-arc opening) and of subcontinental lithosphere (during continental rifting).

2.04.5.2.3 Isotope decoupling between oceanic mantle and crust: evidence for marble-cake or veined mantle

As noted above, isotope decoupling between oceanic mantle and crust was observed both in ophiolites (Claesson *et al.*, 1984; Göpel *et al.*, 1984; Jacobsen *et al.*, 1984; Brouxel and Lapierre, 1988; Rampone *et al.*, 1996, 1998) and in abyssal rocks (Snow *et al.*, 1994; Kempton and Stephens, 1997; Salters and Dick, 2002). In most examples, the mantle rocks are distinguished from the oceanic crust by more depleted isotopic compositions, generally reflected by higher $^{143}Nd/^{144}Nd$ values. However, the interpretations proposed for the ophiolites and for the abyssal peridotites are markedly different.

In the ophiolites, the isotope decoupling is ascribed either to a diachronic evolution of the mantle–crust system (Rampone *et al.*, 1996, 1998; Rampone and Piccardo, 2000) or to the juxtaposition of oceanic units of different ages and/or different isotopic signatures (Gruau *et al.*, 1991). In the Internal Ligurides (Northern Italy), the mantle rocks would record early mantle upwelling and melting during a rifting event that culminated in the denudation of peridotites in the incipent Western Tethys, during the Permian. However, the ophiolite crust would be related to a later stage of ocean accretion, in the Jurassic. In the Trinity (California) and Xigaze (Tibet) ophiolites, the authors favor interpretations involving the juxtaposition of oceanic segments of different ages and/or isotopic signatures, in back-arc or inter-arc environments (Göpel *et al.*, 1984; Brouxel and Lapierre, 1988; Gruau *et al.*, 1991).

In contrast, the isotopic studies on abyssal peridotites suggest the presence of a component with a low melting point in the mantle source, which would be more isotopically enriched than the host peridotites. Preferential melting of this component would explain the isotopic discrepancy between "enriched" MORB and "depleted" residual peridotite (Salters and Dick, 2002).

It is probably worthy of note that mantle–crust isotopic decoupling was mostly reported from moderately refractory (cpx-harzburgitic) to mildly fertile (lherzolitic) ophiolitic mantle sequences, as well as from slow-spreading ridge systems—for which low melting degrees are classically inferred. Conversely, the few data available for strongly refractory ophiolitic mantle sequences indicate isotopic equilibrium between residual peridotites and ocean crust (e.g., Sharma *et al.*, 1995; Sharma and Wasserburg, 1996). Although further data will be needed to confirm this difference, the observation that mantle–crust isotopic decoupling would be enhanced by low melting degrees is a solid argument for the existence of enriched small-scale heterogeneities (pyroxenite veins or marble cake) in the MORB upper mantle (Prinzhofer *et al.*, 1989; Hirschman and Stolper, 1996).

These observations also lend support to the models of small-scale isotopic heterogeneity which have been proposed to account for the geochemical heterogeneities observed in OIBs (e.g., Hofmann and White, 1982; Allègre and Turcotte, 1986; Hauri, 1996; see Chapter 2.03). However, the limited extend of the isotopic variations and the lack of trace-element variations indicative of crustal recycling tend to suggest that the heterogeneities involved in MORB genesis derive from igneous lithospheric mantle veining rather than from recycling of oceanic crust. However, further studies will be necessary to constrain the real nature of small-scale heterogeneities in the convective upper mantle.

ACKNOWLEDGMENTS

We are very grateful to Philippe Gaillot, Emmanuel Ball, and Florence Einaudi for their help in the preparation of several figures. We thank Delphine Bosch, Carlos Garrido, and Graham Pearson who contributed unpublished data to our compilation, Rosa Maria Bedini and Jacques Vernières for providing results of numerical experiments on Mg# modeling, and Monique Seyler for her guidance through the literature on abyssal peridotites. Many thanks go to Rick Carlson for his extreme patience. We acknowledge financial support from the Laboratoire de Tectonophysique and the Institut des Sciences de la Terre, de l'Eau et de l'Espace de Montpellier (CNRS and Montpellier University) during the preparation of this chapter.

REFERENCES

Abbate E., Bortolotti V., Passerini P., and Principi G. (1985) The rhythm of Phanerozoic ophiolites. *Ophioliti* **10**, 109–138.

Abe N. (2001) Petrochemistry of serpentinized peridotite from the Iberia Abyssal Plain (ODP Leg 173): its character intermediate between sub-oceanic to sub-continental upper mantle. In *Non-volcanic Rifting of Continental Margins: Evidence From Land and Sea*. Spec. Publ. (eds. R. C. L. Wilson, R. B. Whitmarsh, H. P. J. Taylor, and N. Froitzheim). Geological Society of London, vol. 187, pp. 143–159.

Agrinier P., Mevel C., Bosch D., and Javoy M. (1993) Metasomatic hydrous fluids in amphibole peridotites from Zabargad Island (Red Sea). *Earth Planet. Sci. Lett.* **120**, 187–205.

Allan J. F. and Dick H. J. B. (1996) Cr-rich spinel as a tracer for melt migration and melt-wall rock interaction in the mantle: Hess Deep, Leg 147. In *Proceedings of the Ocean Drilling Program: Scientific Results, Hess Deep Rift Valley: Covering Leg 147 of the Cruises of the Drilling Vessel JOIDES Resolution, San Diego, California, to Balboa Harbor, Panama, Sites 894–895, November 22, 1992–January 21, 1993*. (eds. C. Mevel, K. M. Gillis, J. F. Allan, S. Arai, F. Bouier, B. Celerier, H. J. B. Dick, T. J. Falloon, G. G. L. Frueh, G. J. Iturrino, D. S. Kelley, P. R. Kelso, L. A. Kennedy, E. Kikawa, C. M. Lecuyer, C. J. MacLeod, J. Malpas, C. E. Manning, M. A. MacDonald, D. J. Miller, J. Natland, J. E. Pariso, R. B. Pedersen, H. M. Prichard, H. Puchelt, and C. Richter). Texas A and M University, Ocean Drilling Program, College Station, TX, vol. 1147, pp. 157–172.

Allègre C. J. and Turcotte D. L. (1986) Implications of a two-component marble-cake mantle. *Nature* **323**, 123–127.

Allègre C. J., Montigny R., and Bottinga Y. (1973) Cortège ophiolitique et cortège océanique, géochimie comparée et mode de genèse. *Bull. Soc. Géol. France* **7**, 5–6.

Altherr R. and Kalt A. (1996) Metamorphic evolution of ultrahigh-pressure garnet peridotites from the Variscan Vosges Mts. (France). *Chem. Geol.* **134**, 49–65.

Anonymous (1972) Penrose field conference on ophiolites. *Geotimes* **17**, 24–25.

Asimow P. D. (1999) A model that reconciles major- and trace-element data from abyssal peridotites. *Earth Planet. Sci. Lett.* **169**, 303–319.

Avé Lallemant H. G. (1967) Structural and petrographic analysis of an "Alpine-type" peridotite: the lherzolite of the French Pyrenees. *Leid. Geol. Med.* **42**, 1–57.

Baker M. B. and Beckett J. R. (1999) The origin of abyssal peridotites: a reinterpretation of constraints based on primary bulk compositions. *Earth Planet. Sci. Lett.* **171**, 49–61.

Baker M. B. and Stolper E. M. (1994) Determining the composition of high-pressure mantle melts using diamond aggregates. *Geochim. Cosmochim. Acta* **58**, 2811–2827.

Bard J.-P. (1983) Metamorphism of an obducted island arc: example of the Kohistan sequence (Pakistan) in the Himalayan collided range. *Earth Planet. Sci. Lett.* **65**, 133–144.

Bard J.-P., Maluski H., Matte P., and Proust F. (1980) The Kohistan sequence: crust and mantle of an obducted island arc. *Geol. Bull. Univ. Peshawar* **11**, 87–94.

Batanova V. G., Suhr G., and Sobolev A. V. (1998) Origin of geochemical heterogeneity in the mantle peridotites from the Bay of Islands Ophiolite (Newfoundland, Canada): ion probe study of clinopyroxenes. *Geochim. Cosmochim. Acta* **62**, 853–866.

Beccaluva L., Macciotta G., and Venturelli G. (1976) Differentiation and geochemical characteristics of oceanic basaltic volcanic rocks from the Apennines of Liguria and Emilia-Romagna. *Ofioliti* **1**, 33–65.

Beccaluva L., Maciotta G., Piccardo G. B., and Zeda O. (1984) Petrology of lherzolitic rocks from the Northern Apennine ophiolites. *Lithos* **17**, 299–316.

Becker H. (1996) Geochemistry of garnet peridotite massifs from lower Austria and the composition of deep lithosphere beneath a Paleozoic convergent plate margin. *Chem. Geol.* **134**, 49–65.

Bedini R.-M. and Bodinier J.-L. (1999) Distribution of incompatible trace elements between the constituents of spinel peridotite xenoliths: ICP–MS data from the East African Rift. *Geochim. Cosmochim. Acta* **63**, 3883–3900.

Bedini R.-M., Bodinier J.-L., Dautria J. M., and Morten L. (1997) Evolution of LILE-enriched small melts fractions in the lithospheric mantle: a case study from the East African Rift. *Earth Planet. Sci. Lett.* **153**, 67–83.

Bedini R. M., Bodinier J.-L., and Vernières J. (2003) Numerical simulation of Fe–Mg partitioning during melting and melt-rock interactions in the upper mantle. In *EGS–AGU–EUG Joint Assembly*, Nice, France.

Benoit M. (1997) Caracterisation geochimique (traces, isotopes) d'un systeme de drainage magmatique fossile dans l'ophiolite d'Oman. Paul Sabatier, Toulouse, 265pp.

Benson W. N. (1926) The tectonic conditions accompanying the intrusion of basic and ultrabasic igneous rocks. *Natl. Acad. Sci. Mem.* **19**, 1–90.

Berckhemer H. (1968) Topography of the Ivrea body derived from seismic and gravimetric data. *Schweizer. Mineral. Petrogr. Mitteil.* **48**, 235–246.

Berckhemer H. (1969) Direct evidence for the composition of the lower crust and the Moho. *Tectonophysics* **8**, 97–105.

Bezzi A. and Piccardo G. B. (1971) Structural features of the Ligurian ophiolites: petrologic evidence for the "oceanic" floor of the Northern Apennines geosyncline: a contribution to the problem of the alpinetype gabbro-peridotite associations. *Mem. Soc. Geol. Italiana* **10**, 53–63.

Blichert-Toft J., Albarède F., and Kornprobst J. (1999) Lu–Hf isotope systematics of garnet pyroxenites from Beni Bousera, Morocco: implications for basalt origin. *Science* **283**, 1303–1306.

Bodinier J.-L. (1988) Geochemistry and petrogenesis of the Lanzo peridotite body, Western Alps. *Tectonophysics* **149**, 67–88.

Bodinier J.-L. (1989) *Distribution des terres rares dans les massifs lherzolitiques de Lanzo et de l'Ariège*. Centre Géologique et Géophysique de Montpellier, Montpellier.

Bodinier J.-L., Guiraud M., Dupuy C., and Dostal J. (1986) Geochemistry of basic dykes in the Lanzo Massif (Western Alps): petrogenetic and geodynamic implications. *Tectonophysics* **128**, 77–95.

Bodinier J.-L., Fabriès J., Lorand J.-P., Dostal J., and Dupuy C. (1987a) Geochemistry of amphibole pyroxenite veins from the Lherz and Freychinède ultramafic bodies (Ariège, French Pyrénées). *Bull. Mineral.* **110**, 345–358.

Bodinier J.-L., Guiraud M., Fabriès J., Dostal J., and Dupuy C. (1987b) Petrogenesis of layered pyroxenites from the Lherz, Freychinède and Prades ultramafic bodies (Ariège, French Pyrénées). *Geochim. Cosmochim. Acta* **51**, 279–290.

Bodinier J.-L., Dupuy C., and Dostal J. (1988) Geochemistry and petrogenesis of Eastern Pyrenean peridotites. *Geochim. Cosmochim. Acta* **52**, 2893–2907.

Bodinier J.-L., Vasseur G., Vernières J., Dupuy C., and Fabriès J. (1990) Mechanism of mantle metasomatism: geochemical evidence from the Lherz orogenic peridotite. *J. Petrol.* **31**, 597–628.

Bodinier J.-L., Menzies M. A., and Thirlwall M. (1991) Continental to oceanic mantle transition—REE and Sr–Nd isotopic geochemistry of the Lanzo lherzolite massif. *J. Petrol. (Orogenic lherzolites and mantle processes)* **(sp. vol.)** 191–210.

Bodinier J.-L., Merlet C., Bedini R. M., Simien F., Remaïdi M., and Garrido C. J. (1996) Distribution of Nb, Ta, and other highly incompatible trace elements in the lithospheric mantle: the spinel paradox. *Geochim. Cosmochim. Acta* **60**, 545–550.

Bodinier J.-L., Menzies M. A., Shimizu N., Frey F. A., and McPherson E. (2004) Silicate, hydrous and carbonate metasomatism at Lherz, France: contemporaneous derivatives of silicate melt-harzburgite reaction. *J. Petrol.* (in press).

Bogdanov Y. A. and Ploshko V. V. (1967) Magmaticheskiye i metamorficheskiye porody glubokovodnoy vpadiny Romansh: magmatic and metamorphic rocks of the abyssal Romanche deep. *Doklady Akademii Nauk SSSR* **177**, 909–912.

Boillot G., Winterer E. L., Meyer A. W., Applegate J., Baltuck M., Bergen J. A., Comas M. C., Davies T. A., Dunham K. W., Evans C. A., Girardeau J., Goldberg D., Haggerty J. A., Jansa L. F., Johnson J. A., Kasahara J., Loreau J.-P., Luna E., Moullade M., Ogg J. G., Sarti M., Thurow J., and Williamson M. A. (1988) *Galicia Margin: Covering Leg 103 of the Cruises of the Drilling Vessel JOIDES Resolution. Ponta Delgada, Azores, to Bremerhaven, Germany, April 25, 1985–June 19, 1985. Proceeding of the Ocean Drilling Program, Scientific Results.* Ocean Drilling Program, College Station, TX, USA, Vol. 103, 858pp.

Boivin P. A. (1982) Interactions entre magmas basaltiques et manteau supérieur. Arguments apportés par les enclaves basiques des basaltes alcalins. Exemples du Devès (Massif Central français) et du volcanisme quaternaire de la région de Carthagène (Espagne). Clermont Ferrand.

Bonatti E. (1968) Ultramafic rocks from the mid-Atlantic ridge. *Nature* **219**, 363–364.

Bonatti E. (1990) Subcontinental mantle exposed in the Atlantic Ocean on St. Peter-Paul islets. *Nature (London)* **345**, 800–802.

Bonatti E., Honnorez J., and Ferrara G. (1970) Equatorial mid-Atlantic ridge: petrologic and Sr isotopic evidence for an alpine-type rock assemblage. *Earth Planet. Sci. Lett.* **9**, 247–256.

Bonatti E., Honnorez J., and Ferrara G. (1971) Peridotite-gabbro-basalt complex from the equatorial mid-Atlantic ridge. *Phil. Trans. Roy. Soc. London* **A268**, 385–402.

Bonatti E., Emiliani C., Ferrara G., Honnorez J., and Rydell H. (1974) Ultramafic-carbonate breccias from the equatorial mid Atlantic ridge. *Mar. Geol.* **16**, 83–102.

Bonatti E., Hamlyn P. R., and Ottonello G. (1981) Upper mantle beneath a young oceanic rift; peridotites from the island of Zabargad (Red Sea). *Geology* **9**, 474–479.

Bonatti E., Ottonello G., and Hamlyn P. R. (1986) Peridotites from the island of Zabargad (St. John), Red Sea: petrology and geochemistry. *J. Geophys. Res.* **91**, 599–631.

Bonatti E., Seyler M., Channell J., Giraudeau J., and Mascle G. (1990) Peridotites drilled from the Tyrrhenian Sea, ODP Leg 107. In *Proceedings of Ocean Drilling Program, Scientific Results 107*, College Station, TX, vol. 107, pp. 37–47.

Bonatti E., Peyve A., Kepezhinskas P., Kurentsova N., Seyler M., Skolotnev S., and Udintsev G. B. (1992) Upper mantle heterogeneity below the mid-Atlantic ridge, 0–15°N. *J. Geophys. Res.* **97**, 4461–4476.

Bonatti E., Seyler M., and Sushevskaya N. (1993) A cold suboceanic mantle belt at the Earth's equator. *Science* **261**, 315–320.

Bortolotti V., and Chiari M. (eds.) (1994) *The Northern Apennines: Part 1. The Tuscan Nappe, the Ophiolitic Sequences, the Turbiditic Successions.* Memorie della Societa Geologica Italiana.

Bosch D. (1991) Sr and Nd isotopic evidence for sea water infiltration into the Zabargad mantle diapir, Red Sea. *Comptes Rendus de l'Academie des Sciences, Serie 2* **313**, 49–56.

Bosch D. and Bruguier O. (1998) An early miocene age for a high-temperature event in gneisses from Zabargad Island (Red Sea, Egypt): mantle diapirism? *Terra Nova* **10**, 274–279.

Boudier F. (1976) *Le massif lherzolitique de Lanzo (Alpes piémontaises): etude structurale et pétrologique*. PhD Thesis, Nantes, France, 175pp.

Boudier F. (1978) Structure and petrology of the Lanzo peridotite massif (Piedmont Alps). *Geol. Soc. Am. Bull.* **89**, 1574–1591.

Boudier F. and Coleman R. G. (1981) Cross section through the peridotite in the Samail ophiolite, southeastern Oman mountains. *J. Geophys. Res.* **86**, 2573–2592.

Boudier F. and Juteau T. (eds.) (2000) *The ophiolite of Oman and United Arab Emirates. Mar. Geophys. Res.* **21**, 407pp.

Boudier F. and Nicolas A. (1972) Fusion partielle gabbroïque dans la lherzolite de Lanzo (Alpes piémontaises). *Schweiz. Mineral. Petrog. Mitt.* **52**, 39–56.

Boudier F. and Nicolas A. (1977) Structural controls on partial melting in the Lanzo peridotite. In *Magma Genesis; Proceedings of the American Geophysical Union Chapman Conference on Partial Melting in the Earth's Upper Mantle* (ed. H. J. B. Dick). State Oregon Department of Geology and Mineral Industries, Portland, USA, vol. 96, pp. 63–79.

Boudier F. and Nicolas A. (1985) Harzburgite and lherzolite subtypes in ophiolitic and oceanic environments. *Earth Planet. Sci. Lett.* **76**, 84–92.

Boudier F. and Nicolas A. (eds.) (1988) *The Ophiolites of Oman. Tectono physics* **151**, 390pp.

Boudier F. and Nicolas A. (1991) High-temperature hydrothermal alteration of peridotite, Zabargad Island (Red Sea). In *Orogenic lherzolites and Mantle Processes*, Spec. Vol., J. Petrol. (eds. M. A. Menzies, C. Dupuy, and A. Nicolas). Oxford University Press, Oxford, pp. 243–254.

Boudier F. and Nicolas A. (1995) Nature of the Moho transition zone in the Oman ophiolite. *J. Petrol.* **36**, 777–796.

Boudier F., Godard M., and Armbruster C. (2000) Significance of gabbronorite occurrence in the crustal section of the Semail ophiolite. *Mar. Geophys. Res.* **21**, 307–326.

Bowin C. O., Nalwalk A. J., and Hersey J. B. (1966) Serpentinized peridotite from the north wall of the puerto rico trench. *Geol. Soc. Am. Bull.* **77**, 257–269.

Bozhilov K. N., Green H. W., and Dobrzhinetskaya L. (1999) Clinoenstatite in Alpe Arami peridotite: additional evidence of very high pressure. *Science* **284**, 128–132.

Brenan J.-M., Shaw H. F., and Ryerson F. J. (1995) Experimental evidence for the origin of lead enrichment in convergent-margin magma. *Nature* **378**, 54–56.

Brouxel M. and Lapierre H. (1988) Geochemical study of an early Paleozoic island-arc-back-arc basin system: Part 1. The Trinity ophiolite (northern California). *Geol. Soc. Am. Bull.* **100**, 1111–1119.

Brueckner H. K. (1998) Sinking intrusion model for the emplacement of garnet-bearing peridotites into continent collision orogens. *Geology* **26**, 631–634.

Brueckner H. K. and Medaris L. G. (1998) A tale of two orogens—the contrasting P-T-t history and geological evolution of mantle in ultrahigh pressure (UHP) metamorphic terranes of the Norwegian Caledonides and the Czech Variscides. *Schweizer. Mineral. Petrogr. Mitteil.* **78**, 293–307.

Brueckner H. K. and Medaris L. G. (2000) A general model for the intrusion and evolution of "mantle" garnet peridotites in high-pressure and ultra-high-pressure metamorphic terranes. *J. Metamorph. Geol.* **18**, 123–133.

Brueckner H. K., Zindler A., Seyler M., and Bonatti E. (1988) Zabargad and the isotopic evolution of the sub-Red Sea mantle crust. *Tectonophysics* **150**, 163–176.

Burg J.-P., Bodinier J.-L., Chaudhry S., Hussain S. S., and Dawood H. (1998) Infra-arc mantle-crust transition and intra-arc mantle diapirs in the Kohistan Complex (Pakistani Himalaya): petro-structural evidence. *Terra Nova* **10**, 74–80.

Burkhard D. J. M. and O'Neil J. R. (1988) Contrasting serpentinization processes in the eastern Central Alps. *Contrib. Mineral. Petrol.* **99**, 498–506.

Burnham O. M., Rogers N. W., Pearson D. G., van Calsteren P. W., and Hawkesworth C. J. (1998) The petrogenesis of eastern Pyrenean peridotites: an integrated study of their whole-rock geochemistry and Re–Os isotope composition. *Geochim. Cosmochim. Acta* **62**, 2293–2310.

Cann J. R. (1971) Petrology of basement rocks from Palmer ridge, NE Atlantic: a discussion on the petrology of igneous and metamorphic rocks from the ocean floor. *Phil. Trans. Roy. Soc. London A* **268**, 605–617.

Carswell D. A., Harvey M. A., and Al-Samman A. (1983) The petrogenesis of contrasting Fe–Ti and Mg–Cr garnet peridotite types in the high grade gneiss complex of western Norway. *Bull. Mineral.* **106**, 727–750.

Casey J. F. (1997) Comparison of major- and trace-element geochemistry of abyssal peridotites and mafic plutonic rocks with basalts from the MARK region of the mid-Atlantic ridge. In *Mid-Atlantic Ridge: Leg 153, Sites 920–924* (eds. J. A. Karson, M. Cannat, D. J. Miller, and D. Elthon). College Station, TX vol. 153, pp. 181–241.

Cassard D. (1980) *Structure et origine des gisements de chromite du Massif du Sud (ophiolite de Nouvelle Calédonie)*. University of Nantes.

Cawthorn R. G. (1975) The amphibole peridotite-metagabbro complex, Finero, northern Italy. *J. Geol.* **83**, 437–454.

Ceuleneer G. and Rabinowicz M. (1992) Mantle flow and melt migration beneath oceanic ridges: models derived from observations in ophiolites. In *Mantle Flow and Melt Generation at Mid-Ocean Ridges* (eds. J. P. Morgan, D. K. Blackman, and J. M. Sinton). American Geophysical Union, Washington DC, vol. 71, pp. 123–154.

Ceuleneer G., Nicolas A., and Boudier F. (1988) Mantle flow patterns at an oceanic spreading centre; the Oman peridotites record. *Tectonophysics* **151**, 1–26.

Chaussidon M. and Lorand J.-P. (1990) Sulphur isotope composition of orogenic spinel lherzolite massifs from Ariege (north-eastern Pyrenees, France): an ion microprobe study. *Geochim. Cosmochim. Acta* **54**, 2835–2846.

Chernysheva V. I. and Bezrukov P. L. (1966) Serpentinite from the crest of the Indo-Arabian ridge. *Acad. Sci. USSR, Dokl.: Earth Sci. Sec.: Akad. Nauk SSSR, Dokl.* **166**, 207–210.

Chernysheva V. I. and Rudnik G. B. (1970) Serpentinized varieties of plagioclase lherzolite from the rift zone of the submarine West Indian Ridge. *Trans. Dokl. USSR Acad. Sci.: Earth Sci. Sec.* **194**, 144–145.

Chiesa S., Cortesogno L., Forcella F., Galli M., Messiga B., Pasquare G., Pedemonte G. M., Piccardo G. B., and Rossi P. M. (1975) Structural arrangement and geodynamic interpretation of the Voltri group. *Bollettino della Societa Geologica Italiana* **94**, 555–581.

Choukroune P. and Mattauer M. (1978) Tectonique des plaques et Pyrénées: sur le fonctionnement de la faille transformante nord pyrénéenne: comparaison avec des modèles actuels. *Bull. Soc. Géol. France* **20**, 689–700.

Claesson S., Pallister J. S., and Tatsumoto M. (1984) Samarium–neodymium data on two Late Proterozoic ophiolites of Saudi Arabia and implications for crustal and mantle evolution. *Contrib. Mineral. Petrol.* **85**, 244–252.

Coleman R. G. (1971) Plate tectonic emplacement of upper mantle peridotites along continental edges. *J. Geophys. Res.* **76**, 1212–1222.

Coleman R. G. (1977) *Ophiolites: Ancient Oceanic Lithosphere?*. Springer, Heidelberg, Berlin, New York.

Coleman R. G. and Hopson C. A. (1981) Oman ophiolite. *J. Geophys. Res.* **B86**, 2495–2496.

Coltorti M. and Siena F. (1984) Mantle tectonite and fractionate peridotite at Finero (Italian Western Alps). *Neues Jahrbuch fuer Mineralogie. Abhandlungen* **419**, 244–255.

Conquéré F. (1971) Les pyroxénolites à amphibole et les amphibolites associées aux lherzolites du gisement de Lherz (Ariège, France): un exemple du rôle de l'eau au cours de la cristallisation fractionnée des liquides issus de la fusion partielle de lherzolites. *Contrib. Mineral. Petrol.* **33**, 32–61.

Conquéré F. (1978) *Pétrologie des complexes ultramafiques de l'Ariège*. PhD thesis, Paris, France, 333pp.

Conquéré F. and Fabriès J. (1984) Chemical disequilibrium and its thermal significance in spinel peridotites from the Lherz and Freychinède ultramafic bodies (Ariège, French Pyrénées). In *Kimberlites: II. The Mantle and Crust–Mantle Relationships* (ed. J. Kornprobst). Elsevier, Amsterdam, The Netherlands, pp. 319–331.

Constantin M. (1995) *Petrologie des gabbros et peridotites de la dorsale Est-Pacifique: la transition croute-manteau aux dorsales rapides.* PhD thesis, Brest, France, 286pp.

Constantin M., Hékinian R., Ackermand D., and Stoffers P. (1995) Mafic and ultramafic intrusions into upper mantle peridotites from fast spreading centers of the Easter microsplate (South East Pacific). In *Mantle and Lower Crust Exposed in Oceanic Ridges and in Ophiolites* (eds. R. L. M. Vissers and A. Nicolas). Kluwer Academic, Dordrecht, Boston, London, pp. 71–120.

Cornen G., Beslier M.-O., and Girardeau J. (1996) Petrologic characteristics of the ultramafic rocks from the ocean/continent transition in the Iberia Abyssal Plain. In *Iberia Abyssal Plain; Leg 149, Sites 891–901, March 10, 1993–May 25, 1993, Proceedings of the Ocean Drilling Program; Scientific reports.* Ocean Drilling Program, College Station, TX, USA. vol. 149, pp. 377–396.

Coward M. P., Windley B. F., Broughton R. D., Luff I. W., Petterson M. G., Pudsey C. J., Rex D. C., and Khan M. A. (1986) Collision tectonics in the NW Himalayas. *Geol. Soc. Spec. Publ.* **19**, 203–219.

Daignières M., Gallart J., Hirn A., and Surinach E. (1998) Complementary geophysical surveys along the ECORS Pyrenees line. *Mem. Soc. Geol. France* **173**, 55–80.

Dal Piaz G. V. (1969) Low-temperature rodingite dikes and reaction zones at tectonic contacts between serpentinites and associated rocks in the western Italian Alps. *Rendiconti Soc. Italiana Mineral. Petrol.* **25**, 263–315.

Davies G. (1994) Thermomechanical erosion of the lithosphere by mantle plume. *J. Geophys. Res.* **99**, 5709–15722.

Davies G. R., Nixon P. H., Pearson D. G., and Obata M. (1993) Tectonic implications of graphitized diamonds from the Ronda peridotite massif, southern Spain. *Geology* **21**, 471–474.

Davies H. L. (1980) Crustal structure and emplacement of ophiolite in southern Papua New Guinea. In *Association Mafiques Ultra-mafiques Dans Les.* Orogenes–Orogenic Mafic Ultra-mafic Association (eds. Allegre, J. Claude, Aubouin, and Jean). Centre National de la Recherche Scientifique, Paris, France, vol. 272, pp. 17–33.

Davies T. E., Johnston M. R., Rankin P. C., and Stull R. J. (1979) The Dun Mountain ophiolite belt in East Nelson, New Zealand. In *Ophiolites.* Geological Survey Dept., Cyprus, pp. 480–496.

Dawson J. B. (1984) Contrasting types of upper-mantle metasomatism. In *Kimberlites: II. The Mantle and Crust-mantle Relationships* (ed. J. Kornprobst). Elsevier, Amsterdam, pp. 289–294.

Debroas E. J. (1987) Modèle de bassin triangulaire à l'intersection de décrochements divergents pour le fossé albo-cénomanien de la Ballongue (zone nord-pyrénéenne, France). *Bulletin de la Société Géologique de France* **8**, 887–898.

Deloule E., Albarède F., and Sheppard S. M. F. (1991) Hydrogen isotope heterogeneities in the mantle from ion probe analysis of amphiboles from ultramafic rocks. *Earth Planet. Sci. Lett.* **105**, 543–553.

Den Tex E. (1969) Origin of ultramafic rocks, their tectonic setting and history: a contribution to the discussion of the paper "The origin of ultramafic and ultrabasic rocks" by P. J. Wyllie. *Tectonophysics* **7**, 457–488.

De Roever W. P. (1957) Sind die alpinotypen peridotitmassen vielleicht teknosch verfrachtete bruchstücke der Peridotitschale? *Geol. Rundsch.* **46**, 137–146.

Devey C. W., Hemond C., and Stoffers P. (2000) Metasomatic reactions between carbonated plume melts and mantle harzburgite: the evidence from Friday and Domingo seamounts (Juan Fernandez Chain, SE Pacific). *Contrib. Mineral. Petrol.* **139**, 68–84.

Dick H. J. B. and Bullen T. (1984) Chromian spinel as a petrogenetic indicator in abyssal and alpine-type peridotites and spatially associated lavas. *Contrib. Mineral. Petrol.* **86**, 54–76.

Dick H. J. B. and Natland J. H. (1996) Late-stage melt evolution and transport in the shallow mantle beneath the East Pacific Rise. In *Hess Deep Rift Valley: Leg 147, Sites 894–895.* (eds. C. Mevel, K. M. Gillis, J. F. Allan, and P. S. Meyer) College Station, TX vol. 147, pp. 103–134.

Dick H. J. B., Fisher R. L., and Bryan W. L. (1984) Mineralogic variability of the uppermost mantle along mid-ocean ridges. *Earth Planet. Sci. Lett.* **69**, 88–106.

Dietz R. S. (1963) Alpine serpentinites as oceanic rind segments. *Geol. Soc. Am. Bull.* **74**, 947–952.

Dilek Y., Moores E., Elthon D., and Nicolas A. (eds.) (2000) *Ophiolites and Oceanic Crust: New Insights from Field Studies and the Ocean Drilling Program.* Geological Society of America, Boulder, CO.

Dobretsov N. L. and Ashchepkov I. V. (1991) Melt migration and depletion-regeneration processes in upper mantle of continental and ocean rift zones. *Petrol. Struct. Geol.* **5**, 125–146.

Dobrzhinetskaya L., Green H. W., II, and Wang S. (1996) Alpe Arami: a peridotite massif from depths of more than 300 kilometers. *Science* **271**, 1841–1845.

Downes H., Bodinier J.-L., Thirlwall M. F., Lorand J.-P., and Fabries J. (1991) REE and Sr–Nd isotopic geochemistry of the Eastern Pyrenean peridotite massifs: sub-continental lithospheric mantle modified by continental magmatism. *J. Petrol. (Orogenic lherzolites and mantle processes)* **(sp. vol.)**, 97–115.

Drury M. R., Van Roermund H. L. M., Carswell D. A., De Smet J. H., Van Den Berg A. P., and Vlaar N. J. (2001) Emplacement of deep upper-mantle rocks into cratonic lithosphere by convection and diapiric upwelling. *J. Petrol.* **42**, 131–140.

Dupuy C., Mevel C., Bodinier J.-L., and Savoyant L. (1991) The Zabargad peridotite: evidence for multi-stage metasomatism during Red Sea rifting. *Geology* **19**, 722–725.

Edwards S. J. and Malpas J. (1995) Multiple origins for mantle harzburgites: examples from the Lewis Hills, Bay of Islands ophiolite, Newfoundland. *Can. J. Earth Sci.* **32**, 1046–1057.

Elthon D. (1992) Chemical trends in abyssal peridotites: refertilization of depleted suboceanic mantle. *J. Geophys. Res.* **97**, 9015–9025.

Ernewein M., Pflumio C., and Whitechurch H. (1988) The death of an accretion zone as evidenced by the magmatic history of the Sumail ophiolite (Oman). *Tectonophysics* **151**, 247–274.

Ernst W. G. (1978) Petrochemical study of lherzolitic rocks from the Western Alps. *J. Petrol.* **19**, 341–392.

Ernst W. G. (1981) Petrogenesis of eclogites and peridotites from the Western and Ligurian Alps. *Am. Mineral.* **66**, 443–472.

Ernst W. G. and Ottonello G. (1984) Synthesis of trace element geochemistry for five western Alpine lherzolites. *Ofioliti* **9**, 425–442.

Ernst W. G. and Piccardo G. B. (1979) Petrogenesis of some Ligurian peridotites: 1. Mineral and bulk-rock chemistry. *Geochim. Cosmochim. Acta* **43**, 219–237.

Evans B. W. and Trommsdorff V. (1978) *Petrogenesis of garnet lherzolite, Cima di Gagnone, Lepontinbe Alps.*

Evans B. W., Trommsdorff V., and Richter W. (1979) Petrology of an eclogite-metarodingite suite at Cima di Cagnone, Ticino, Switzerland. *Am. Mineral.* **64**, 15–31.

Evans B. W., Trommsdorff V., and Goles G. G. (1981) Geochemistry of high-grade eclogites and metarodingites from the Central Alps. *Contrib. Mineral. Petrol.* **76**, 301–311.

Fabriès J. and Conquéré F. (1983) Les lherzolites à spinelle et les pyroxénites à grenat associées de Bestiac (Ariège, France). *Bull. Minéral.* **106**, 781–803.

Fabriès J., Bodinier J.-L., Dupuy C., Lorand J.-P., and Benkerrou C. (1989) Evidence for modal metasomatism in the orogenic spinel lherzolite body from Caussou (North-eastern Pyrénées, France). *J. Petrol.* **30**, 199–228.

Fabriès J., Lorand J.-P., Bodinier J.-L., and Dupuy C. (1991) Evolution of the upper mantle beneath the Pyrenees: evidence from orogenic spinel lherzolite massifs. *J. Petrol., (Orogenic lherzolites and mantle processes)* **(sp. vol.)**, 55–76.

Fabriès J., Lorand J.-P., and Bodinier J.-L. (1998) Petrogenetic evolution of orogenic lherzolite massifs in the central and western Pyrenees. *Tectonophysics* **292**, 145–167.

Falloon T. J., Green D. H., Hatton C. J., and Harris K. L. (1988) Anhydrous partial melting of a fertile and depleted peridotite from 2 to 30 Kb and application to basalt petrogenesis. *J. Petrol.* **29**, 1257–1282.

Ferrara G., Innocenti F., Ricci C. A., and Serri G. (1976) Ocean-floor affinity of basalts from north Apennine ophiolites: geochemical evidence. *Chem. Geol.* **17**, 101–111.

Fisher R. L. and Engel C. G. (1969) Ultramafic and basaltic rocks dredged from the nearshore flank of the Tonga trench. *Geol. Soc. Am. Bull.* **80**, 1373–1378.

Fray C. and Hékinian R. (1965) Mafic and ultramafic plutonic rocks from the southeastern Indian Ocean. *Spec. Pap. Geol. Soc. Am.* **82**, 66.

Frey F. A. and Prinz M. (1978) Ultramafic inclusions from San Carlos, Arizona petrologic and geochemical data bearing on their petrogenesis. *Earth Planet. Sci. Lett.* **38**, 139–176.

Frey F. A., Suen C. J., and Stockman H. W. (1985) The Ronda high temperature peridotite: geochemistry and petrogenesis. *Geochim. Cosmochim. Acta* **49**, 2469–2491.

Frey F. A., Shimizu N., Leinbach A., Obata M., and Takazawa E. (1991) Compositional variations within the lower layered zone of the Horoman peridotite, Hokkaido, Japan: constraints on models for melt-solid segregation. In *Orogenic Lherzolites and Mantle Processes*, Spec. Vol. J. Petrol. (eds. M. A. Menzies, C. Dupuy, and A. Nicolas). Oxford University Press, Oxford, UK, pp. 211–227.

Galli M. and Cortesogno L. (1970) Petrography of the ophiolitic formation of the Ligurian Apennines: XIII. Phenomena of low-grade metamorphism in ophiolitic rocks, Ligurian Apennines. *Rendiconti Soc. Italiana Mineral. Petrol.* **26**, 599–647.

Gardien V., Tegyey M., Lardeaux J.-M., Misseri M., and Dufour E. (1990) Crust-mantle relationships in the French Variscan chain: the example of the Southern Monts du Lyonnais unit (eastern French Massif Central). *J. Metamorph. Geol.* **8**, 477–492.

Garrido C. and Bodinier J.-L. (1999) Diversity of mafic rocks in the Ronda peridotite: evidence for pervasive melt/rock reaction during heating of subcontinental lithosphere by upwelling asthenosphere. *J. Petrol.* **40**, 729–754.

Garrido C. J. (1995) Estudio geoquimico de las capas maficas del macizo ultramafico de Ronda (Cordillera Betica, Espana). Granada (Espagne), 220pp.

Garrido C. J., Bodinier J.-L., and Alard O. (2000) Distribution of LILE, REE and HFSE in anhydrous spinel peridotite and websterite minerals from the Ronda massif: insights into the nature of trace element reservoirs in the subcontinental lithospheric mantle. *Earth Planet. Sci. Lett.* **181**, 341–358.

Garuti G., Gorgoni C., and Sighinolfi G. P. (1984) Sulfide mineralogy and chalcophile elements abundances in the Ivrea-Verbano mantle peridotites (western Italian Alps). *Earth Planet. Sci. Lett.* **70**, 69–87.

Gass I. G., Lippard S. J., and Shelton A. W. (eds.) (1984) *Ophiolites and Oceanic Lithosphere*. Geological Society Spec. Publ., London, UK, 413pp.

Gast P. W. (1968) Trace element fractionation and the origin of tholeiitic and alkaline magma types. *Geochim. Cosmochim. Acta*, **32**, 1057–1086.

Gerbert-Gaillard L. (2002) Caractérisation géochimique des peridotites de l'ophiolite d'Oman: processus magmatiques aux limites lithosphere/asthenosphere. In *Geology*. Montpellier 2, Montpellier, 266pp.

Gervilla F. and Remaïdi M. (1993) Field trip to the Ronda ultramafic massif: an example of asthenosphere-lithosphere interaction? *Ofioliti* **18**, 21–35.

Gil Ibarguchi J. I., Mendia M., Girardeau J., and Peucat J.-J. (1990) Petrology of eclogites and clinopyroxene-garnet metabasites from the Cabo Ortegal Complex (northwestern Spain). *Lithos* **25**, 133–162.

Gill R. W. (1990) Th isotope and U series studies of subduction related volcanics rocks. *Geochim. Cosmochim. Acta* **54**, 1427–1442.

Girardeau J. and Gil Ibarguchi J. I. (1991) Pyroxene-rich peridotites of the Cabo Ortegal Complex (Northwestern Spain): evidence for large-scale upper-mantle heterogeneity. *J. Petrol. (Orogenic lherzolites and mantle processes)* **(sp. vol.)**, 135–154.

Girardeau J. and Mercier J. C. C. (1988) Petrology and texture of the ultramafic rocks of the Xigaze Ophiolite (Tibet); constraints for mantle structure beneath slow-spreading ridges. *Tectonophysics* **147**, 33–58.

Glennie K. W., Boeuf M. G. A., Hugues Clark M. W., Moody-Stuart M., Pilaar W. F. H., and Reinhardt B. M. (1974) *Geology of the Oman Mountains*. Verhandelingen Vau het Koninklijk Nederlands Geologisch Mijnbouwkundig Genotschap, Dept. The Netherlands, vol. 31, 432pp.

Godard M., Bodinier J.-L., and Vasseur G. (1995) Effects of mineralogical reactions on trace element redistributions in mantle rocks during percolation processes: a chromatographic approach. *Earth Planet. Sci. Lett.* **133**, 449–461.

Godard M., Jousselin D., and Bodinier J.-L. (2000) Relationships between geochemistry and structure beneath a palaeo-spreading centre: a study of the mantle section in the Oman Ophiolite. *Earth Planet. Sci. Lett.* **180**, 133–148.

Goldberg J.-M. and Leyreloup A. F. (1990) High temperature-low pressure Cretaceous metamorphism related to crustal thinning (Eastern North Pyrenean Zone, France). *Contrib. Mineral. Petrol.* **104**, 194–207.

Göpel C., Allègre C. J., and Xu R.-H. (1984) Lead isotopic study of the Xigaze ophiolite (Tibet): the problem of the relationship between magmatites (gabbros, dolerites, lavas) and tectonites (harzburgites). *Earth Planet. Sci. Lett.* **69**, 301–310.

Goscombe B. D. and Everard J.-L. (2001) Tectonic evolution of Macquarie Island: extensional structures and block rotations in oceanic crust. *J. Struct. Geol.* **23**, 639–673.

Green D. H., Hibberson W. O., and Jacques A. L. (1979) Petrogenesis of mid-ocean basalts. In *The Earth: Its Origin, Structure and Evolution* (ed. H. W. Elhinny). Academic Press, London, pp. 265–299.

Green H. W., II, Dobrzhinetskaya L., Riggs E. M., and Jin Z.-M. (1997) Alpe Arami: a peridotite massif from the mantle transition zone? *Tectonophysics* **279**, 1–21.

Grieco G., Ferrario A., von Quadt A., Koeppel V., and Mathez E. A. (2001) The zircon-bearing chromitites of the phlogopite peridotite of Finero (Ivrea Zone, Southern Alps): evidence and geochronology of a metasomatized mantle slab. *J. Petrol.* **42**, 89–101.

Griffin B. G. and Varne R. (1978) The petrology of the Macquarie Island ophiolite association: mid-tertiary oceanic crust of the Southern Ocean. *Ofioliti* **3**, 230–231.

Griffin B. J. and Varne R. (1980) The Macquarie Island ophiolite complex: mid-tertiary oceanic lithosphere from a major ocean basin. *Chem. Geol.* **30**, 285–308.

Gruau G., Lecuyer C., Bernard-Griffiths J., and Morin N. (1991) Origin and petrogenesis of the Trinity ophiolite complex (California): new constraints from REE and Nd isotope data. In *Orogenic Lherzolites and Mantle Processes*, Spec. Vol. J. Petrol. (eds. M A. Menzies, C. Dupuy, and A. Nicolas). Oxford University Press, Oxford, UK, pp. 229–242.

Gruau G., Bernard-Griffiths J., Lecuyer C., Henin O., Mace J., and Cannat M. (1995) Extreme Nd isotopic variation in the Trinity Ophiolite Complex and the role of melt/rock reaction in the oceanic lithosphere. *Contrib. Mineral. Petrol* **121**, 337–350.

Gruau G., Bernard-Griffiths J., and Lecuyer C. (1998) The origin of U-shaped rare earth patterns in ophiolite peridotites: assessing the role of secondary alteration and melt/rock reaction. *Geochim. Cosmochim. Acta* **62**, 3545–3560.

Gueddari K., Piboule M., and Amosse J. (1996) Differentiation of platinum-group elements (PGE) and of gold during partial melting of peridotites in the lherzolitic massifs of the Betico-Rifean range (Ronda and Beni Bousera). In *Melt Processes and Exhumation of Garnet, Spinel and Plagioclase Facies Mantle* (eds. M. A. Menzies, J. L. Bodinier, F. Frey, F. Gervilla, and P. Kelemen). Elsevier, Amsterdam, Netherlands, vol. 134, pp. 181–197.

Gurney J. and Harte B. (1980) Chemical variations in upper mantle nodules from Southern African Kimberlites. *Phil. Trans. Roy. Soc. London* **A297**, 273–293.

Hamelin B. and Allègre C. J. (1988) Lead isotope study of orogenic lherzolite massifs. *Earth Planet. Sci. Lett.* **91**, 117–131.

Hamlyn P. R. and Bonatti E. (1980) Petrology of mantle-derived ultramafics from the Owen fracture zone, Northwest Indian Ocean: implications for the nature of the oceanic upper mantle. *Earth Planet. Sci. Lett.* **48**, 65–79.

Harris P. G., Reay A., and White I. G. (1967) Chemical composition of the upper mantle. *J. Geophys. Res.* **72**, 6359–6369.

Harte B. (1983) Mantle peridotites and processes—the Kimberlite sample. In *Continental Basalts and Mantle Xenoliths* (eds. C. J. Hawkesworth and M. J. Norry). Shiva Publishing, Nantwich, UK, pp. 46–91.

Harte B., Hunter R. H., and Kinny P. D. (1993) Melt geometry, movement and crystallization, in relation to mantle dykes, veins and metasomatism. *Phil. Trans. Roy. Soc. London A* **342**, 1–21.

Hartmann G. and Wedepohl K. H. (1993) The composition of peridotite tectonites from the Ivrea complex northern Italy: residues from melt extraction. *Geochim. Cosmochim. Acta* **57**, 1761–1782.

Hauri E. H. (1996) Major-element variability in the Hawaiian mantle plume. *Nature (London)* **382**, 415–419.

Hébert R., Gueddari K., Lafleche M. R., Beslier M.-O., and Gardien V. (2001) Petrology and geochemistry of exhumed peridotites and gabbros at non-volcanic margins: ODP Leg 173 West Iberia ocean-continent transition zone. In *Non-volcanic Rifting of Continental Margins: Evidence from Land and Sea*, Spec. Publ. (eds. R. C. L. Wilson, R. B. Whitmarsh, H. P. J. Taylor, and N. Froitzheim). Geological Society of London, vol. 187, pp. 161–189.

Hékinian R., Bideau D., Francheteau J., Cheminée J. L., Armijo R., Lonsdale P., and Blum N. (1993) Petrology of the East Pacific Rise crust and upper mantle exposed in Hess Deep (Eastern Equatorial Pacific). *J. Geophys. Res.* **98**, 8069–8094.

Hellebrand E., Snow J. E., Dick H. J. B., and Hofmann A. W. (2001) Coupled major and trace elements as indicators of the extent of melting in mid-ocean-ridge peridotites. *Nature* **410**, 677–681.

Hellebrand E., Snow J. E., and Muhe R. (2002) Mantle melting beneath Gakkel Ridge (Arctic Ocean): abyssal peridotite spinel compositions. *Chem. Geol.* **182**, 227–235.

Hermann J., Müntener O., Trommsdorff V., Hansmann W., and Piccardo G. (1997) Fossil crust-to-mantle transition, Val Malenco (Italian Alps). *J. Geophys. Res.* **102**, 20123–20132.

Hersey J. B. (1962) Findings made during the June 1961 cruise of Chain to the Puerto Rico trench and Caryn sea mount. *J. Geophys. Res.* **67**, 1109–1116.

Hess H. H. (1962) History of ocean basins. In *Petrologic Studies: a Volume in Honor of A. F. Buddington* (eds. A. E. J. Engle, H. L. James, and B. L. Leonard). Geological Society of America, New York, pp. 599–620.

Hess H. H. (1964) The oceanic crust, the upper mantle, and the Mayaguez serpentinized peridotite. In *A Study of Serpentinite*, Natl. Research Council Publ., Natl. Acad. Sci., vol. 1188, pp. 169–175.

Hirschmann M. M. and Stolper E. M. (1996) A possible role for garnet pyroxenite in the origin of the "garnet signature" in MORB. *Contrib. Mineral. Petrol.* **124**, 185–208.

Hoffmann A. W. and White W. M. (1982) Mantle plumes from ancient oceanic crust. *Earth Planet. Sci.* **57**, 421–436.

Hoogerduijn-Strating E. H., Piccardo G. B., Rampone E., Scambelluri M., and Vissers R. L. (1990) The structure and petrology of the Erro-Tobbio peridotite, Voltri massif, Ligurian Alps: guidebook for a two-day-excursion with emphasis on processes in the upper mantle. *Ofioliti* **15**(1), 119–184.

Ionov D. A., Bodinier J.-L., Mukasa S. B., and Zanetti A. (2002) Mechanisms and sources of mantle metasomatism; major and trace element compositions of peridotite xenoliths from Spitsbergen in the context of numerical modelling. *J. Petrol.* **43**, 2219–2259.

Irving A. J. (1980) Petrology and geochemistry of composite ultramafic xenoliths in alkalic basalts and implications for magmatic processes within the mantle. *Am. J. Sci.* **280**, 389–426.

Ishiwatari A. (1985) Igneous petrogenesis of the Yakuno ophiolite (Japan) in the context of the diversity of ophiolites. *Contrib. Mineral. Petrol* **89**, 155–167.

Ishiwatari A. (1994) Circum-Pacific phanerozoic multiple ophiolite belts. In *Circum Pacific Ophiolites*, Vol. Proc. 29th Int. Geol. Cong. Part D (eds. A. Ishiwatari, J. Malpas, and H. Ishizuka). VSP, Utrecht, The Netherlands, pp. 7–28.

Ishiwatari A., Malpas J., and Ishizuka H. (eds.) (1994) *Circum Pacific Ophiolites.* Vol. Proc. 29th Int. Geol. Cong. Part D, VSP, Utrecht, The Netherlands, 286pp.

Jacobsen S. B. and Wasserburg B. J. (1984) Nd and Sr isotopic study of the Bay of Islands ophiolite complex and evolution of the source of mid-ocean ridge basalts. *J. Geophys. Res.* **84**, 7429–7445.

Jacobsen S. B., Quick J. E., and Wasserburg G. J. (1984) A Nd and Sr isotopic study of the Trinity Peridotite: implications for mantle evolution. *Earth Planet. Sci. Lett.* **68**, 361–378.

Jagoutz E., Palme H., Baddenhausen H., Blum K., Cendales M., Dreibus G., Spettel B., Lorenz V., and Vanke H. (1979) The abundance of major, minor and trace elements in the earth's mantle as derived from primitive ultramafic nodules. *Geochim. Cosmochim. Acta* **11**, 2031–2050.

Jahns R. H. (1967) Serpentinites of the Roxbury District, Vermont. In *Ultramafic and Related Rocks* (ed. P. J. Wyllie). Wiley, NY and London, International (III), pp. 137–160.

Jan M. Q., Khan M. A., and Qazi M. S. (1993) The Sapat mafic-ultramafic complex, Kohistan Arc, North Pakistan. *Geol. Soc. Spec. Publ.* **74**, 113–121.

Jaques A. L. and Chappell B. W. (1980) Petrology and trace element geochemistry of the papuan ultramafic belt. *Contrib. Mineral. Petrol* **75**, 55–70.

Jaques A. L. and Green D. H. (1980) Anhydrous melting of peridotite at 0–15 Kb pressure and the genesis of tholeiitic basalts. *Contrib. Mineral. Petrol.* **73**, 287–310.

Jaques A. L., Chappell B. W., and Taylor S. R. (1983) Geochemistry of cumulus peridotites and gabbros from the Marum ophiolite complex, northern Papua New Guinea. *Contrib. Mineral. Petrol.* **82**, 154–164.

Jedwab J. (1992) Platinum group minerals in ultrabasic rocks and nickeliferous veins from Zabargad Island (Egypt). *Comptes Rendus de l'Academie des Sciences, Serie 2* **314**, 157–163.

Jibiki H. and Masuda A. (1974) Basalts and serpentinite from the Puerto Rico Trench: 2. Rare-earth geochemistry. *Mar. Geol.* **16**, 205–211.

Johnson K. T. M. and Dick H. J. B. (1992) Open system melting and temporal and spatial variation of peridotite and basalt at the Atlantis II fracture zone. *J. Geophys. Res. B: Solid Earth Planets* **97**, 9219–9241.

Johnson K. T. M., Dick H. J. B., and Shimizu N. (1990) Melting in the oceanic upper mantle: an ion microprobe study of diopsides in abyssal peridotites. *J. Geophys. Res.* **95**, 2661–2678.

Juteau T. (1974) Les ophiolites des nappes d'Antalya (Taurides Occidentales): pétrologie d'un fragmentb de l'Ancienne croute océanique téthysienne. Nancy.

Kalt A., Altherr R., and Hanel M. (1995) Contrasting P-T conditions recorded in ultramafic high-pressure rocks from the Variscan Schawarzwald (FRG). *Contrib. Mineral. Petrol.* **121**, 45–60.

Karson J. A. (1998) Internal structure of oceanic lithosphere: a perspective from tectonic windows. In *Faulting and Magmatism at Mid-Ocean Ridges* (eds. W. R. Buck, P. T. Delaney, J. A. Karson, and Y. Lagabrielle). American Geophysical Union, Washington DC, Vol. 106, pp. 177–218.

Karson J. A., Cannat M., Miller D. J., and Elthon D. (eds.) (1997) *Mid-Atlantic Ridge: Leg 153, Sites 920–924. Proceedings of the Ocean Drilling Program, Scientific Result*, Ocean Drilling Program, College Station, TX, USA, Vol. 153, 577pp.

Kelemen P. B. (1990) Reaction between ultramafic rock and fractionating basaltic magma: I. Phase relations, the origin of calc-alkaline magma series, and the formation of discordant dunite. *J. Petrol.* **31**, 51–98.

Kelemen P. B. and Dick H. J. B. (1995) Focused melt flow and localized deformation in the upper mantle; juxtaposition of replacive dunite and ductile shear zones in the Josephine Peridotite, SW Oregon. *J. Geophys. Res. B: Solid Earth Planets* **100**, 423–438.

Kelemen P. B., Joyce D. B., Webster J. D., and Holloway J. R. (1990) Reaction between ultramafic rock and fractionating basaltic magma: II. Experimental investigation of reaction between olivine tholeiite and harzburgite at 1150–1050 degrees C and 5 kb. *J. Petrol.* **31**, 99–134.

Kelemen P. B., Dick H. J. B., and Quick J. E. (1992) Formation of harzburgite by pervasive melt/ rock reaction in the upper mantle. *Nature (London)* **358**, 635–641.

Kelemen P. B., Shimizu, Nobumichi, Salters, and Vincent J. M. (1995) Extraction of mid-ocean-ridge basalt from the upwelling mantle by focused flow of melt in dunite channels. *Nature (London)* **375**, 747–753.

Kelemen P. B., Hirth G., Shimizu N., Spiegelman M., and Dick H. J. B. (1997) A review of melt migration processes in the adiabatically upwelling mantle beneath oceanic spreading ridges. *Phil. Trans. R. Soc. London A.* **355**, 283–318.

Kempton P. D. (1987) Mineralogic and geochemical evidence for differing styles of metasomatism in spinel lherzolite xenoliths: enriched mantle source regions of basalts? In *Mantle Metasomatism* (eds. M. A. Menzies and C. J. Hawkesworth). Academic Press, London, UK, pp. 45–89.

Kempton P. D. and Stephens C. J. (1997) Petrology and geochemistry of nodular websterite inclusions in harzburgite, Hole 920. In *Mid-Atlantic Ridge; Leg 153, sites 920–924. Proceedings of the Ocean Drilling Program, Scientific Results* (eds. J. A. Karson, M. Cannat, D. J. Miller, and D. Elthon). Ocean Drilling Program, College Station, TX, USA, vol. 153, pp. 321–333.

Keppler H. (1996) Constraints from partitioning experiments on the composition of subduction zone fluids. *Nature* **380**, 237–240.

Keppler H. and Wyllie P. J. (1990) Role of fluids in transport and fractionation of uranium and thorium in magmatic processes. *Nature* **348**, 531–533.

Khan M. A., Jan M. Q., and Weaver B. L. (1993) Evolution of the lower arc crust in Kohistan N. Pakistan: temporal arc magmatism through early, mature and intra-arc rift stages. *Geol. Soc. Spec. Publ.* **74**, 123–138.

Korenaga J. and Kelemen P. B. (1997) Origin of gabbro sills in the Moho transition zone of the Oman Ophiolite: implications for magma transport in the oceanic lower crust. *J. Geophys. Res.* **102**, 27729–27749.

Kornprobst J., Piboule M., Roden M., and Tabit A. (1990) Corundum-bearing garnet clinopyroxenites at Beni Bousera (Morocco): original plagioclase-rich gabbros recrystallized at depth within the mantle? *J. Petrol.* **31**, 717–745.

Kostopoulos D. K. (1991) Melting of the shallow upper mantle: a new perspective. *J. Petrol.* **32**, 671–699.

Kumar N., Reisberg L., and Zindler A. (1996) A major and trace element and strontium, neodymium, and osmium isotopic study of a thick pyroxenite layer from the Beni Bousera ultramafic complex of northern Morocco. *Geochim. Cosmochim. Acta* **60**, 1429–1444.

Kurat G., Palme H., Embey-Isztin A., Touret J., Ntaflos T., Spettel B., Brandstaetter F., Palme C., Dreibus G., and Prinz M. (1993) Petrology and geochemistry of peridotites and associated vein rocks of Zabargad Island, Red Sea, Egypt. *Mineral. Petrol.* **48**, 309–341.

Lacroix A. (1917) Les péridotites des Pyrénées et les autres roches intrusives non feldspathiques qui les accompagnent. *C. R. Acad. Sci. Paris* **165**, 381–387.

Lagabrielle Y., Bideau D., Cannat M., Karson J. A., and Mevel C. (1998) Ultramafic-mafic plutonic rock suites exposed along the mid-Atlantic ridge (10°N–30°N): symmetrical-assymetrical distribution and implications for seafloor spreading processes. In *Faulting and Magmatism at Mid-Ocean Ridges* (eds. W. R. Buck, P. T. Delaney, J. Karson, and Y. Lagabrielle). American Geophysical Union, Washington DC, USA, vol. 106, pp. 153–176.

Langmuir C. H., Bender J. F., Bence A. E., Hanson G. N., and Taylor S. R. (1977) Petrogenesis of basalts from the FAMOUS area: mid-Atlantic ridge. *Earth Planet. Sci. Lett.* **36**, 133–156.

Lappin M. A. (1967) Structural and petrofabric studies of the dunites of Almklovdalen, Nordfjord, Norway. In *Ultramafic and Related Rocks* (ed. P. J. Wyllie). Wiley, New York, pp. 183–190.

Le Roex A. P., Dick H. J. B., Reid A. M., Frey F. A., Erlank A. J., and Hart S. R. (1985) Petrology and geochemistry of basalts from the American-Antarctic Ridge, Southern Ocean: implications for the westward influence of the Bouvet mantle plume. *Contrib. Mineral. Petrol.* **90**, 367–380.

Lemoine M., Tricart P., and Boillot G. (1987) Ultramafic and gabbroic ocean floor of the Ligurian Tethys (Alps, Corsica, Apennines): in search of a genetic model. *Geology (Boulder)* **15**, 622–625.

Lenoir X. (2000) *Structure et évolution du mantteau lithosphérique sous-continental en Europe occidentale: étude texturale, pétrologique et géochimique dans le massif de Ronda et les xénolites du Massif Central.* ISTEEM, Montpellier.

Lenoir X., Garrido C. J., Bodinier J.-L., Dautria J.-L., and Gervilla F. (2001) The recrystallization front of the Ronda peridotite: evidence for melting and thermal erosion of subcontinental lithospheric mantle beneath the Alboran basin. *J. Petrol.* **42**, 141–158.

Lensch G. (1968) Die ultramafitite der Zone von Ivrea und ihre geologische interpretation. *Schweizer. Mineral. Petrogr. Mitteil.* **48**, 91–102.

Lensch G. (1971) The occurrence of sapphirine in the peridotite complex of Finero, zone of Ivrea, western Italian Alps. *Contrib. Mineral. Petrol.* **31**, 145–153.

Liou J. G. (1999) Petrotectonic summary of less intensively studied UHP regions. *Int. Geol. Rev.* **41**, 571–586.

Lippard S. J., Shelton A. W., and Gass I. G. (eds.) (1986) *The Ophiolite of Northern Oman. The Geological Society Memiors*, Backwell, London, UK, vol. 11, 178pp.

Loomis T. P. (1972) Contact metamorphism of pelitic rock by the Ronda ultramafic intrusion, southern Spain. *Geol. Soc. Am. Bull.* **83**, 2449–2474.

Lorand J.-P. (1991) Sulphide petrology and sulphur geochemistry of orogenic lherzolites: a comparative study of the Pyrenean bodies (France) and the Lanzo massif (Italy).

In *Orogenic lherzolites and Mantle Processes*, Spec. Vol. J. Petrol. (eds. M. A. Menzies, C. Dupuy, and A. Nicolas). Oxford University Press, Oxford, UK, pp. 77–95.

Lorand J.-P., Bodinier J.-L., Dupuy C., and Dostal J. (1989) Abundances and distribution of gold in the orogenic-type spinel peridotites from Ariège (Northeastern Pyrenees, France). *Geochim. Cosmochim. Acta* **53**, 3085–3090.

Lorand J.-P., Keays R. R., and Bodinier J.-L. (1993) Copper and noble metal enrichments across the lithosphere-asthenosphere boundary of mantle diapirs: evidence from the Lanzo lherzolite massif. *J. Petrol.* **34**, 1111–1140.

Lorand J.-P., Pattou L., and Gros M. (1999) Fractionation of platinum-group elements and gold in the upper mantle: a detailed study in Pyrenean orogenic lherzolites. *J. Petrol.* **40**, 957–981.

Loubet M. and Allègre C. J. (1979) Trace element studies in the Alpine peridotite of Beni-Bouchera (Morocco). *Geochem. J.* **13**, 69–75.

Loubet M. and Allègre C. J. (1982) Trace elements in orogenic lherzolites reveal the complex history of the upper mantle. *Nature* **298**, 809–814.

Loubet M., Shimizu N., and Allègre C. J. (1975) Rare earth elements in alpine peridotites. *Contrib. Mineral. Petrol.* **53**, 1–12.

MacKenzie D. B. (1960) High-temperature Alpine-type peridotite from Venezuela. *Geol. Soc. Am. Bull.* **71**, 303–317.

Maheo G., Bertrand H., Guillot S., Mascle G., Pecher A., Picard C., and de Sigoyer J. (2000) Evidence of an immature Tethyan arc in ophiolites from southern Ladakh, India, northwestern Himalayas. *Comptes Rendus de l'Academie des Sciences, Serie 2* **330**, 289–295.

Martinez F. and Cochran J. R. (1988) Structure and tectonics of the northern Red Sea: catching a continental margin between rifting and drifting. *Tectonophysics* **150**, 1–32.

Maruyama S., Liou J. G., and Terabayashi M. (1996) Blueschists and eclogites of the world and their exhumation. *Int. Geol. Rev.* **38**, 485–594.

McDonough W. F. and Frey F. A. (1989) Rare earth elements in upper mantle rocks. In *Geochemistry and Mineralogy of Rare Earth Elements* (eds. B. R. Lipin and G. A. McKay). The Mineralogical Society of America, Washington, vol. 21, pp. 99–145.

McDonough W. F. and Sun S. S. (1995) The composition of the Earth. *Chem. Geol.* **120**, 223–253.

McKenzie D. P. (1989) Some remarks on the movement of small melt fractions in the mantle. *Earth Planet. Sci. Lett.* **95**, 53–72.

McPherson E. (1994) Geochemistry of silicate melt metasomatism in Alpine peridotite massifs. University of London, London.

McPherson E., Thirlwall M. F., Parkinson I. J., Menzies M. A., Bodinier J. L., Woodland A., and Bussod G. (1996) Geochemistry of metasomatism adjacent to amphibole-bearing veins in the Lherz peridotite massif. In *Melt Processes and Exhumation of Garnet, Spinel, and Plagioclase Facies Mantle* (eds. M. A. Menzies, J. L. Bodinier, F. Frey, F. Gervilla, and P. Kelemen). Elsevier, Amsterdam, The Netherlands, vol. 134, pp. 135–157.

Medaris L. G., Beard B. L., Johnson C. M., Valley J. W., Spicuzza M. J., Jelinek E., and Misar Z. (1995) Garnet pyroxenite and eclogite in the Bohemian Massif: geochemical evidence for Variscan recycling of subducted lithosphere. *Geol. Rundsch.* **84**, 489–505.

Medaris L. G., Fournelle J. H., Ghent E. D., and Jelinek E. (1998) Prograde eclogite in the Gföhl Nappe, Czech Republic: new evidence on Variscan high-pressure metamorphism. *J. Metamorph. Geol.* **16**, 563–576.

Mehnert K. R. (1975) The Ivrea Zone: a model of the deep crust. *Neues Jahrbuch fuer Mineralogie. Abhandlungen* **125**, 156–199.

Melson W. G., Jarosewich E., Bowen V. T., and Thompson G. (1967) Saint Peter and Saint Paul Rocks: a high-temperature, mantle-derived intrusion. *Science* **156**, 1532–1535.

Menzies M. A. (1976) Rare earth geochemistry of fused ophiolitic and alpine lherzolites: I. Othris, Lanzo and Troodos. *Geochim. Cosmochim. Acta* **40**, 645–656.

Menzies M. A. (1984) Chemical and isotopic heterogeneities in orogenic and ophiolitic peridotites. In *Ophiolites and Oceanic Lithosphere* (eds. I. G. Gass, S. J. Lippard, and A. W. Shelton). Blackwell, London, UK, pp. 231–240.

Menzies M. and Allen C. (1974) Plagioclase lherzolite-residual mantle relationships within two eastern Mediterranean ophiolites. *Contrib. Mineral. Petrol.* **45**, 197–213.

Menzies M. A. and Dupuy C. (1991) Orogenic massifs: Protolith, process and provenance. *J. Petrol. (Orogenic lherzolites and mantle processes)* **(sp. vol.)**, 1–16.

Menzies M. A. and Murthy V. R. (1978) Strontium isotope geochemistry of alpine tectonite lherzolites: data compatible with a mantle origin. *Earth Planet. Sci. Lett.* **38**, 346–354.

Menzies M. A., Blanchard D., Brannon J., and Korotev R. (1977) Rare earth geochemistry of fused ophiolitic and alpine lherzolites: II. Beni-Bouchera, Ronda and Lanzo. *Contrib. Mineral. Petrol.* **64**, 53–74.

Menzies M. A., Roggers N. W., Tindle A., and Hawkesworth C. J. (1987) Metasomatic and enrichment processes in lithospheric peridotites: an effect of asthenosphere-lithosphere interaction. In *Mantle Metasomatism* (eds. M. A. Menzies and C. J. Hawkesworth). Academic Press, London, UK, pp. 313–361.

Messiga B. and Piccardo G. B. (1974) Petrographic and structural survey of the Voltri Group: northeastern section: the zone between Mount Tacco and Mount Orditano. *Memorie della Societa Geologica Italiana* **13**, 301–315.

Mevel C., Gillis K. M., Allan J. F., Arai S., Boudier F., Celerier B., Dick H. J. B., Falloon T. J., Frueh G. G. L., Iturrino G. J., Kelley D. S., Kelso P. R., Kennedy L. A., Kikawa E., Lecuyer C. M., MacLeod C. J., Malpas J., Manning C. E., MacDonald M. A., Miller D. J., Natland J., Pariso J. E., Pedersen R. B., Prichard H. M., Puchelt H., and Richter C. (1996) *Proceedings of the Ocean Drilling Program; Scientific Results, Hess Deep Rift Valley; Covering Leg 147 of the Cruises of the Drilling Vessel JOIDES Resolution, San Diego, California, to Balboa Harbor, Panama, Sites 894–895, November 22, 1992–January 21, 1993.* Texas A and M University, Ocean Drilling Program, College Station, TX.

Michael P. J. and Bonatti E. (1985) Peridotite composition from the North Atlantic: regional and tectonic variations and implications for partial melting. *Earth Planet. Sci. Lett.* **73**, 91–104.

Miller D. J., Loucks R. R., and Ashraf M. (1991) Platinum-group element mineralization in the Jijal layered ultramafic-mafic complex, Pakistani Himalayas. *Econ. Geol.* **86**, 1093–1102.

Miyashiro A. (1966) Some aspects of peridotite and serpentinite in orogenic bels. *Japan J. Geol. Geogr. Trans.* **37**, 45–61.

Miyashiro A. (1973) The Troodos complex was probably formed as an island arc. *Earth Planet. Sci. Lett.* **25**, 217–222.

Miyashiro A., Shido F., and Ewing M. (1969) Composition and origin of serpentinites from the mid-Atlantic ridge near 24 degrees and 30 degrees north latitude. *Contrib. Mineral. Petrol.* **23**, 117–127.

Möckel J. R. (1969) Structural petrology of the garnet-peridotite of Alpe Arami (Ticino, Switzerland). *Leidse Geologische Mededelingen* **42**, 61–130.

Monnier C. (1996) Mécanismes d'accrétion des domaines océaniques arrière-arc et géodynamique de l'Asie du Sud-Est; pétrologie et géochimie des ophiolites d'Indonésie (Sulawesi, Haute-Chaine Centrale, Cyclops, Seram et Meratus). Bretagne Occidentale, Nantes, 626pp.

Moores E. (1982) Origin and emplacement of ophiolites. *Rev. Geophys.*, **20**, 735–760.

Moores E., Kellogg L. H., and Dilek Y. (2000) Tethyan ophiolites, mantle convection, and tectonic "historical contingency": a resolution of the "ophiolite conundrum". In *Ophiolites and Oceanic Crust: New Insights from Field Studies and the Ocean Drilling Program*, Special Paper (eds. Y. Dilek, E. M. Moores, D. Elthon, and A. Nicolas). Geological Society of America, Boulder, CO, USA, vol. 349, pp. 3–12.

Morency C., Doin M.-P., and Dumoulin C. (2002) Convective destabilization of a thickened continental lithosphere. *Earth Planet. Sci. Lett.* **202**, 303–320.

Morishita T. and Arai S. (2001) Petrogenesis of corundum-bearing mafic rocks in the Horoman peridotite complex, Japan. *J. Petrol.* **42**, 1279–1299.

Morten L. and Obata M. (1990) Rare earth abundances in the eastern Alpine peridotites, Nonsberg area, northern Italy. *Euro. J. Mineral.* **2**, 643–653.

Mukasa S. B., Shervais J. W., Wilshire H. G., and Nielson J. E. (1991) Intrinsic Nd, Pb and Sr isotopic heterogeneities exhibited by the Lherz alpine peridotite massif, French Pyrenees. In *Orogenic Lherzolites and Mantle Processes.* Spec. Vol. J. Petrol. (eds. M. A. Menzies, C. Dupuy, and A. Nicolas). Oxford University Press, Oxford, UK, pp. 117–134.

Müntener O. and Hermann J. (1996) The Malenco lower crust-to-mantle complex and its field relations. *Schweizer. Mineral. Petrogr. Mitteil.* **76**, 475–500.

Müntener O., Hermann J., and Trommsdorff V. (2000) Cooling history and exhumation of lower-crustal granulite and upper mantle (Malenco, eastern Central Alps). *J. Petrol.* **41**, 175–200.

Navon O. and Stolper E. (1987) Geochemical consequences of melt percolation: the upper mantle as a chromatographic column. *J. Geol.* **95**, 285–307.

Nicolas A. (1969) The structure and metamorphism in the Stura di Lanzo region. *Piedmont Alps. Schweizer. Mineral. Petrogr. Mitteil.* **49**, 359–375.

Nicolas A. (1986a) Structure and petrology of peridotites: clues to their geodynamic environment. *Rev. Geophys.* **24**, 875–895.

Nicolas A. (1986b) A melt extraction model based on structural studies in mantle peridotites. *J. Petrol.* **27**, 999–1022.

Nicolas A. (1989) *Structure of Ophiolites and Dynamics of Oceanic Lithosphere*. Kluwer Academic Press, Dordrecht, Boston, London.

Nicolas A. and Boudier F. (2003) Where ophiolites come from and what do they tell us. Geological Society of America, Special Paper, Boulder, USA, vol. 373 (in press).

Nicolas A. and Dupuy C. (1984Q) Origin of ophiolite and oceanic lherzolites. *Tectonophysics* **110**, 177–187.

Nicolas A. and Jackson E. D. (1972) Répartition en deux provinces des péridotites des chaînes alpines longeant la méditerranée: implications géotectoniques. *Bull. Suisse Miner. Petrol.* **53**, 385–401.

Nicolas A. and Prinzhofer A. (1983) Cumulative or residual origin for the transition zone in ophiolites: structural evidence. *J. Petrol.* **24**, 188–206.

Nicolas A., Boudier F., Lyberis N., Montigny R., and Guennoc P. (1985) Zabargad (Saint-John): a key-witness of early rifting in the Red Sea. *Comptes Rendus de l'Academie des Sciences, Serie 2* **301**, 1063–1068.

Nicolas A., Boudier F., and Montigny R. (1987) Structure of Zabargad Island and early rifting of the Red Sea. *J. Geophys. Res.* **92**, 461–474.

Nicolas A., Hirn A., Nicolich R., Polino R., Bayer R., Lacassin R., Lanza R., Bois C., Damotte B., Roure F., Cazes M., Ravat J., Villien A., Dal Piaz G. V., Guellec S., Tardy M., Gosso G., Marthelot J.-M., Mugnier J.-L., Thouvenot F., Scarascia S., Tabacco I., and Sutter M. (1990) Lithospheric wedging in the Western Alps inferred from the ECORS-CROP traverse. *Geology (Boulder)* **18**, 587–590.

Nicolas A., Boudier F., and Meshi A. (1999) Slow spreading accretion and mantle denudation in the Mirdita ophiolite (Albania). *J. Geophys. Res.* **104**, 15155–15167.

Nicolas A., Boudier F., Ildefonse B., and Ball E. (2000) Accretion of Oman ophiolite and United Emirates ophiolite: discussion of a new structural map. *Mar. Geophys. Res.* **21**, 147–179.

Nielson J. E. and Wilshire H. G. (1993) Magma transport and metasomatism in the mantle: a critical review of current geochemical models. *Am. Mineral.* **78**, 1117–1134.

Niida K. (1974) Structure of the Horoman ultramafic mass of the Hidaka metamorphic belt in Hokkaido, Japan. *J. Geol. Soc. Japan* **80**, 31–44.

Nimis P. and Trommsdorff V. (2001) Revised thermobarometry of Alpe Arami and other peridotites from the central Alps. *J. Petrol.* **42**, 103–115.

Niu Y. (1997) Mantle melting and melt extraction processes beneath ocean ridges: evidence from abyssal peridotites. *J. Petrol.* **38**, 1047–1074.

Niu Y. (1999) Comments on some misconceptions in igneous and experimental petrology and methodology: a reply. *J. Petrol.* **40**, 1195–1203.

Niu Y. and Hékinian R. (1997a) Basaltic liquids and harzburgitic residues in the Garrett Transform: a case study at fast-spreading ridges. *Earth Planet. Sci. Lett.* **146**, 243–258.

Niu Y. and Hékinian R. (1997b) Spreading-rate dependence of the extent of mantle melting beneath ocean ridges. *Nature (London)* **385**, 326–329.

Niu Y., Langmuir C. H., and Kinzler R. J. (1997) The origin of abyssal peridotites: a new perspective. *Earth Planet. Sci. Lett.* **152**, 251–265.

O'Hara M. J. (1967a) Mineral parageneses in ultramafic rocks. In *Ultramafic and Related Rocks* (ed. P. J. Wyllie). Wiley, New York, pp. 393–403.

O'Hara M. J. (1967b) Garniteferous ultrabasic rocks of orogenic regions. In *Ultramafic and Related Rocks* (ed. P. J. Wyllie). Wiley, New York, pp. 167–172.

O'Reilly S. Y., Griffin, W. L., Poudjom-Djomani, Y. H., and Morgan P., (2001) Are lithospheres forever? *GSA Today* **11**, 4–10.

Obata M. (1980) The Ronda peridotite: Garnet-, Spinel-, and Plagioclase-lherzolite facies and the P–T trajectories of a high-temperature mantle intrusion. *J. Petrol.* **21**, 533–572.

Obata M. and Morten L. (1987) Transformation of spinel lherzolite to garnet lherzolite in ultramafic lenses of the Austridic crystalline complex, Northern Italy. *J. Petrol.* **28**, 599–623.

Obata M. and Nagahara N. (1987) Layering of alpine-type peridotite and the segregation of partial melt in the upper mantle. *J. Geophys. Res.* **92**, 3467–3474.

Obata M. and Niida K. (2002) Summary and Synthesis. In *Fourth International Workshop on Orogenic Lherzolites and Mantle Processes Field Guide*. Samani, Hokkaido, pp. 63–64.

Ohara Y., Stern R. J., Ishii T., Yurimoto H., and Yamazaki T. (2002) Peridotites from the Mariana Trough: First look at the mantle beneath an active back-arc basin. *Contrib. Mineral. Petrol.* **143**, 1–18.

Orberger B., Lorand J.-P., Girardeau J., Mercier J.-C. C., and Pitragool S. (1995) Petrogenesis of ultramafic rocks and associated chromitites in the Nan Uttaradit ophilita, Northern Thailand. *Lithos* (in stampa).

Ottonello G., Piccardo G. B., and Ernst W. G. (1979) Petrogenesis of some Ligurian peridotites: 2. Rare earth element chemistry. *Geochim. Cosmochim. Acta* **43**, 1273–1284.

Ottonello G., Ernst W. G., and Joron J. L. (1984a) Rare earth and 3d transition elements geochemistry of peridotitic rocks: 1. Peridotites from the Western Alps. *J. Petrol.* **25**, 343–372.

Ottonello G., Joron J. L., and Piccardo G. B. (1984b) Rare earth and 3d transition elements geochemistry of peridotitic rocks: 2. Ligurian peridotites and associated basalts. *J. Petrol.* **25**, 373–393.

Ozawa K. (1997) P-T history of an ascending mantle peridotite constrained by Al zoning in orthopyroxene: a case study in the Horoman peridotite complex, Hokkaidi, northern Japan. *Mem. Geol. Soc. Japan* **47**, 107–122.

Ozawa K. and Shimizu N. (1995) Open-system melting in the upper mantle: constraints from the Hayachine-Miyamori ophiolite, northeastern Japan. *J. Geophys. Res.* **100**, 22315–22335.

Ozawa K. and Takahashi N. (1995) P-T history of a mantle diapir: the Horoman peridotite complex, Hokkaido, northern Japan. *Contrib. Mineral. Petrol.* **120**, 223–248.

Pallister J. and Knight R. J. (1981) Rare earth element geochemistry of the Samail ophiolite near Ibra, Oman. *J. Geophys. Res.* **86**, 2673–2697.

Paris J. P. (1981) Géologie de la Nouvelle-Calédonie: un essai de synthèse (with geologic map of scale 1:200000), Mémoire du B. R. G. M., Orléans, France.

Parkinson I. J. and Pearce J. A. (1998) Peridotites from the Izu-Bonin-Mariana Forearc (ODP Leg 125): evidence for mantle melting and melt-mantle interaction in a supra-subduction zone setting. *J. Petrol.* **39**, 1577–1618.

Parson L. M., Murton B. J., and Browning P. (eds.) (1992) *Ophiolites and their Modern Analogues*. Geological Society of London, Spec. Publ., London, vol. 60, 325pp.

Pattou L., Lorand J.-P., and Gros M. (1996) Non-chondritic platinum-group element ratios in the Earth's mantle. *Nature* **379**, 712–715.

Pearce J. A., Alabaster T., Shelton A. W., and Searle M. P. (1981) The Oman ophiolite as a cretaceous arc-basin complex: evidence and implications. *Phil. Trans. Roy. Soc. London.* **A300**, 299–317.

Pearce J. A., Barker P. F., Edwards S. J., Parkinson I. J., and Leat P. T. (2000) Geochemistry and tectonic significance of peridotites from the South Sandwich arc-basin system, South Atlantic. *Contrib. Mineral. Petrol* **139**, 36–53.

Pearson D. G., Davies G. R., Nixon P. H., and Milledge H. J. (1989) Graphitized diamonds from a peridotite massif in Morocco and implications for anomalous diamond occurrences. *Nature (London)* **338**, 60–62.

Pearson D. G., Davies G. R., Nixon P. H., Greenwood P. B., and Mattey D. P. (1991a) Oxygen isotope evidence for the origin of pyroxenites in the Beni Bousera peridotite massif, north Morocco: derivation from subducted oceanic lihthosphere. *Earth Planet. Sci. Lett.* **102**, 289–301.

Pearson D. G., Davies G. R., Nixon P. H., and Mattey D. P. (1991b) A carbon isotope study of diamond facies pyroxenites and associated rocks from the Beni Bousera peridotite, north Morocco. In *Orogenic Lherzolites and Mantle Processes*, Spec. Vol. J. Petrol. (eds. M. A. Menzies, C. Dupuy, and A. Nicolas). Oxford University Press, Oxford, UK, pp. 175–189.

Pearson D. G., Davies G. R., and Nixon P. H. (1993) Geochemical constraints on the petrogenesis of diamond facies pyroxenites from Beni Bousera peridotite massif, north Morocco. *J. Petrol.* **34**, 125–172.

Pearson D. G., Davies G. R., and Nixon P. H. (1995) Orogenic ultramafic rocks of UHP (diamond facies) origin. In *Ultrahigh Pressure Metamorphism* (eds. R. G. Coleman and X. Wang). Cambridge University Press, Cambridge, pp. 456–511.

Peters T. J., Nicolas A., and Coleman R. G. (eds.) (1991) *Ophiolite Genesis and Evolution of the Oceanic Lithosphere* Kluwer Academic Press, Dordrecht, Boston, London, 903pp.

Petrini R., Joron J.-L., Ottonello G., Bonatti E., and Seyler M. (1988) Basaltic dykes from Zabargad Island, Red Sea: petrology and geochemistry. *Tectonophysics* **150**, 229–248.

Pfiffner M. and Trommsdorff V. (1998) The high-pressure ultramafic-mafic-carbonate suite of Cima Lunga-Adula, Central Alps: excursions to Cima di Cagnone and Alpe Arami. *Schweizer. Mineral. Petrogr. Mitteil.* **78**, 337–354.

Piccardo G. B. (1976) Petrology of the lherzolite massif of Suvero, La Spezia. *Ofioliti* **1**, 279–317.

Piccardo G. B. (1977) The ophiolite of the Liguria area: petrology and geodynamic environment of its formation. *Rendiconti Soc. Italiana Mineral. Petrol.* **33**, 221–252.

Piccardo G. B. and Rampone E. (2001) Melt impregnation by strongly depleted MORB melts in the ophiolitic peridotites from the Ligurian Tethys (Alps, Apennines and Corsica). In *Geological Society of America, 2001 Annual Meeting.* Abstracts with Programs Geological Society of America, vol. 33, 172pp.

Piccardo G. B., Messiga B., and Cimmino F. (1980) Antigoritic serpentinites and rodingites of the Voltri Massif: some petrological evidences for their evolutive history. *Ofioliti* **5**, 111–114.

Piccardo G. B., Rampone E., and Scambelluri M. (1988a) The Alpine evolution of the Erro-Tobbio Peridotites (Voltri Massif-Ligurian Alps): some field and petrographic constraints. *Ofioliti* **13**, 169–174.

Piccardo G. B., Messiga B., and Vannucci R. (1988b) The Zabargad peridotite-pyroxenite association: petrologic constraints to the evolutive history. *Tectonophysics* **150**, 135–162.

Piccardo G. B., Rampone E., Vannucci R., Shimizu N., Ottolini L., and Bottazzi P. (1993) Mantle processes in the subcontinental lithosphere: the case study of the rifted spIherzolites from Zabargad (Red Sea). *Euro. J. Mineral.* **5**, 1039–1056.

Ploshko V. V., Sidorenko G. A., Knyazeva D. N., and Bogdanov Y. A. (1973) Olivine, pyroxenes, and plagioclase ultrabasites of the Romanche Trench in the Atlantic Ocean. *Oceanology* **13**, 513–518.

Pognante U., Rosli U., and Toscani L. (1985) Petrology of ultramafic and mafic rocks from the Lanzo peridotite body (Western Alps). *Lithos* **18**, 201–214.

Polvé M. and Allègre C. J. (1980) Orogenic lherzolite complexes studied by ^{87}Rb–^{87}Sr: a clue to understand the mantle convection processes? *Earth Planet. Sci. Lett.* **51**, 71–93.

Priem H. N. A., Hebeda E. H., Boelrijk N. A. I. M., Verdurmen T.E.A., and Oen I. S. (1979) Isotopic dating of the emplacement of the ultramafic masses in the Serrania de Ronda, southern Spain. *Contrib. Mineral. Petrol.* **70**, 103–109.

Prinz M., Keil K., Green J. A., Reid A. M., Bonatti E., and Honnorez J. (1976) Ultramafic and mafic dredge samples from the equatorial mid-Atlantic ridge and fracture zones. *J. Geophys. Res.* **81**, 4087–4103.

Prinzhofer A. and Allègre C. J. (1985) Residual peridotites and the mechanisms of partial melting. *Earth Planet. Sci. Lett.* **74**, 251–265.

Prinzhofer A., Lewin E., and Allègre C. J. (1989) Stochastic melting of the marble cake mantle: evidence from local study of the East Pacific Ridge at 12° 50′. *Earth Planet. Sci. Lett.* **92**, 189–206.

Proenza J., Gervilla F., Melgarejo J. C., and Bodinier J.-L. (1999) Al-rich and Cr-rich chromitites from the Mayari-Baracoa ophiolitic belt (Eastern Cuba) as the consequence of interaction between volatile-rich melts and peridotites in suprasubduction mantle. *Econ. Geol.* **94**, 547–566.

Quick J. E. (1981) Petrology and petrogenesis of the Trinity peridotite, an upper mantle diapir in the eastern Klamath mountains, northern California. *J. Geophys. Res.* **86**, 11837–11863.

Quick J. E., Sinigoi S., and Mayer A. (1995) Emplacement of mantle peridotite in the lower continental crust, Ivrea-Verbano Zone, Northwest Italy. *Geology (Boulder)* **23**, 739–742.

Ragan D. M. (1967) The twin sisters dunite, Washington. In *Ultramafic and Related Rocks* (ed. P. J. Wyllie). Wiley, New York, pp. 160–167.

Rampone E. and Morten L. (2001) Records of crustal metasomatism in the garnet peridotites of the Ulten Zone (Upper Austroalpine, Eastern Alps). *J. Petrol.* **42**, 207–219.

Rampone E. and Piccardo G. B. (2000) The ophiolite-oceanic lithosphere analogue: new insights from the Northern Apennines (Italy). In *Ophiolites and Oceanic Crust: New Insights from Field Studies and the Ocean Drilling Program*, Special Paper (eds. Y. Dilek, E. M. Moores, D. Elthon, and A. Nicolas). Geological Society of America, Boulder, CO, USA, vol. 349, pp. 21–34.

Rampone E., Piccardo G. B., Vannucci R., Bottazzi P., and Ottolini L. (1993) Subsolidus reactions monitored by trace element partitioning: the spinel- to plagioclase-facies transition in mantle peridotites. *Contrib. Mineral. Petrol.* **115**, 1–17.

Rampone E., Piccardo G. B., Vannucci R., Bottazzi P., and Zanetti A. (1994) Melt impregnation in ophiolitic peridotites: an ion microprobe study of clinopyroxene and plagioclase. *Min. Mag.* **58A**, 756–757.

Rampone E., Hofmann A. W., Piccardo G. B., Vannucci R., Bottazzi P., and Ottolini L. (1995) Petrology, mineral and isotope geochemistry of the external Liguride peridotites (Northern Apennines, Italy). *J. Petrol.* **36**, 81–105.

Rampone E., Hofmann A. W., Piccardo G. B., Vannucci R., Bottazzi P., and Ottolini L. (1996) Trace element and isotope geochemistry of depleted peridotites from an N-MORB type ophiolite (Internal Liguride N. Italy). *Contrib. Mineral. Petrol.* **123**, 61–76.

Rampone E., Piccardo G. B., Vannucci R., and Bottazzi P. (1997) Chemistry and origin of trapped melts in ophiolitic peridotites. *Geochim. Cosmochim. Acta* **61**, 4557–4569.

Rampone E., Hofmann A. W., and Raczek I. (1998) Isotopic contrasts within the Internal Liguride ophiolite (N. Italy): the lack of a genetic mantle-crust link. *Earth Planet. Sci. Lett.* **163**, 175–189.

Rehkämper M., Halliday A. N., Alt J., Zipfel J., and Takazawa E. (1999) Non-chondritic platinum-group element ratios in oceanic mantle lithosphere: petrogenetic signature of melt percolation? *Earth Planet. Sci. Lett.* **172**, 65–81.

Reisberg L. and Lorand J.-P. (1995) Longevity of sub-continental mantle lithosphere from osmium isotope systematics in orogenic peridotite massifs. *Nature* **376**, 159–162.

Reisberg L. and Zindler A. (1987) Extreme isotopic variations in the upper mantle: evidence from Ronda. *Earth Planet. Sci. Lett.* **47**, 65–74.

Reisberg L., Zindler A., and Jagoutz E. (1989) Further Sr and Nd isotopic results from peridotites of the Ronda ultramafic complex. *Earth Planet. Sci. Lett.* **96**, 161–180.

Reisberg L., Allègre C., and Luck J.-M. (1991) The Re–Os systematics of the Ronda ultramafic complex of southern Spain. *Earth Planet. Sci. Lett.* **96**, 161–180.

Remaïdi M. (1993) Etude géochimique de l'association harzburgite, dunite et pyroxénite de l'Arroyo de la Cala (Massif de Ronda, Espagne). Montpellier 2 (France), Montpellier, 437pp.

Richard P. and Allègre C. J. (1980) Neodymium and strontium isotope study of ophiolite and orogenic lherzolite petrogenesis. *Earth Planet. Sci. Lett.* **59**, 327–342.

Richter F. M. and McKenzie D. (1984) Dynamical models for melt segregation from a deformable matrix. *J. Geol.* **92**, 729–740.

Ringwood A. E. (1966) The chemical composition and origin of the Earth. In *Advances in Earth Sciences* (ed. P. M. Hurley). MIT Press, Cambridge, MA, pp. 287–356.

Ringwood A. E. (1975) *Composition and Petrology of the Earth's Mantle*. McGraw-Hill, New York.

Rivalenti G., Mazzucchelli M., Vannucci R., Hofmann A. W., Ottolini L., Bottazzi P., and Obermiller W. (1995) The relationship between websterite and peridotite in the Balmuccia peridotite massif (NW Italy) as revealed by trace element variations in clinopyroxene. *Contrib. Mineral. Petrol.* **121**, 275–288.

Rivalenti G., Vannucci R., Rampone E., Mazzucchelli M., Piccardo G. B., Piccirillo E. M., Bottazzi P., and Ottolini L. (1996) Peridotite clinopyroxene chemistry reflects mantle processes rather than continental versus oceanic settings. *Earth Planet. Sci. Lett.* **139**, 423–437.

Robinson J. A. C. and Wood B. J. (1998) The depth of the spinel to garnet transition at the peridotite solidus. *Earth Planet. Sci. Lett.* **164**, 277–284.

Robinson J. A. C., Wood B. J., and Blundy J. D. (1998) The beginning of melting of fertile and depleted peridotite at 1.5 GPa. *Earth Planet. Sci. Lett.* **155**, 97–111.

Roden M. K., Hart S. T., Frey F. A., and Melson W. G., Sr (1984) Nd and Pb isotopic and REE geochemistry of St. Paul's Rocks: the metamorphic and metasomatic development of an alkali basalt mantle source. *Contrib. Mineral. Petrol.* **85**, 376–390.

Ross K. and Elthon D. (1997) Extreme incompatible trace-elements depletion of diopside in residual mantle from south of the Kane fracture zone. In *Mid-Atlantic Ridge; Leg 153, Sites 920–924. Proceedings of the Ocean Drilling Program, Scientific Results* (eds. J. A. Karson, M. Cannat., D. J., Miller, and D. Elthon) Ocean Drilling Program, College Station, TX, USA, vol. 153, pp. 277–284.

Roy-Barman M., Luck J.-M., and Allègre C. J. (1996) Os isotopes in orogenic lherzolite massifs and mantle heterogeneities. *Chem. Geol.* **130**, 55–64.

Saal A., Takazawa E., Frey F. A., Shimizu N., and Hart S. R. (2001) Re–Os isotopes in the Horoman peridotite: evidence for refertilization? *J. Petrol.* **42**, 25–37.

Salters V. J. M. and Dick H. J. B. (2002) Mineralogy of the mid-ocean-ridge basalt source from neodymium isotopic composition of abyssal peridotites. *Nature* **418**, 68–72.

Sautter V. and Fabriès J. (1990) Cooling kinetics of garnet websterites from the Freychinède orogenic lherzolite massif, French Pyrenees. *Contrib. Mineral. Petrol.* **105**, 533–549.

Scambelluri M., Rampone E., and Piccardo G. B. (2001) Fluid and element cycling in subducted serpentinite: a trace-element study of the Erro-Tobbio high-pressure ultramafites (Western Alps, NW Italy). *J. Petrol.* **42**, 55–67.

Schiano P. and Clocchiatti R. (1994) Worldwide occurrence of silica-rich melts in sub-continental and sub-oceanic mantle minerals. *Nature (London)* **368**, 621–624.

Schmidt G., Palme H., Kratz K.-L., and Kurat G. (2000) Are highly siderophile elements (PGE, Re and Au) fractionated in the upper mantle of the earth? New results on peridotites from Zabargad. *Chem. Geol.* **163**, 167–188.

Schubert W. (1982) Comments on M. Obata (1980). *J. Petrol.* **23/2**, 293–295.

Sciuto P. F. and Ottonello G. (1995) Water-rock interaction on Zabargad Island, Red Sea: a case study: I. Application of the concept of local equilibrium. *Geochim. Cosmochim. Acta* **59**, 2187–2206.

Seifert K. and Brunotte D. (1996) Geochemistry of serpentinized mantle peridotite from Site 897 in the Iberia Abyssal Plain. *In Iberia Abyssal Plain; Leg 149, Sites 891–901, March 10,1993–May 25, 1993; Proceedings of the Ocean Drilling Program; Scientific Reports* (eds. R. B. Whitmarsh, D. S. Sawyer, A. Klaus, and D. G. Masson). Ocean Drilling Program, College Station, TX, USA, vol. 149, pp. 413–424.

Seyler M. and Bonatti E. (1997) Regional-scale melt-rock interaction in lherzolitic mantle in the Romanche fracture zone (Atlantic Ocean). *Earth Planet. Sci. Lett.* **146**, 273–287.

Seyler M. and Mattson P. H. (1989) Petrology and thermal evolution of the Tinaquillo Peridotite (Venezuela). *J. Geophys. Res.* **94**, 7629–7660.

Seyler M. and Mattson P. H. (1993) Gabbroic and pyroxenite layers in the Tinaquillo, Venezuela, peridotite: succession of melt intrusions in a rising mantle diapir. *J. Geol.* **101**, 501–511.

Seyler M., Paquette J. L., Ceuleneer G., Kienast J. R., and Loubet M. (1998) Magmatic underplating, metamorphic evolution and ductile shearing in a Mesozoic lower crust-upper mantle unit (Tinaquillo, Venezuela) of the Caribbean Belt. *J. Geol.* **106**, 35–58.

Seyler M., Toplis M. J., Lorand J.-P., Luguet A., and Cannat M. (2001a) Clinopyroxene microtextures reveal incompletely extracted melts in abyssal peridotites. *Geology* **29**, 155–158.

Seyler M., Cannat M., and Mével C. (2001b) Major element heterogeneity in the mantle source of abyssal peridotites from the southwest Indian Ridge (52 to 68° East). *Geochem. Geophys. Geosys. G3* **4**, 2002GC000305.

Shand S. J. (1949) Rocks of the mid-Atlantic ridge. *J. Geol.* **57**, 89–92.

Sharma M. and Wasserburg G. J. (1996) The neodymium isotopic compositions and rare earth patterns in highly depleted ultramafic rocks. *Geochim. Cosmochim. Acta* **60**, 4537–4550.

Sharma M., Wasserburg G. J., Papanastassiou D. A., Quick J. E., Sharkov E. V., and Laz'ko E. E. (1995) High 143Nd/144Nd in extremely depleted mantle rocks. *Earth Planet. Sci. Lett.* **135**, 101–114.

Shaw D. M. (1970) Trace element fractionation during anatexis. *Geochim. Cosmochim. Acta* **34**, 237–243.

Shervais J. W. (1979) Ultramafic and mafic layers in the alpine-type lherzolite massif at Balmuccia NW Italy. *Memorie di Scienze Geologiche* **33**, 135–145.

Shervais J. W. (2001) Birth, death, and resurrection: the life cycle of suprasubduction zone ophiolites. *Geochem. Geophys. Geosys. (G3)* **2** 2000GC000080.

Shervais J. W. and Mukasa S. B. (1991) The Balmuccia orogenic lherzolite massif, Italy. In *Orogenic Lherzolites and Mantle Processes*, Spec. vol. J. Petrol. (eds. M. A. Menzies, C. Dupuy, and A. Nicolas). Oxford University Press, Oxford, UK, pp. 155–174.

Shibata T. and Thompson G. (1986) Peridotites from the mid-Atlantic ridge at 43°N and their petrogenetic relation to abyssal tholeiites. *Contrib. Mineral. Petrol.* **93**, 144–159.

Sinigoi S., Comin-Chiaramonti P., and Alberti A. A. (1980) Phase relations in the partial melting of the Baldissero spinel-lherzolite (Ivrea-Verbano Zone, Western Alps, Italy). *Contrib. Mineral. Petrol.* **75**, 111–121.

Sinigoi S., Comin-Chiaramonti P., Demarchi G., and Siena F. (1983) Differentiation of partial melts in the mantle: evidence from the Balmuccia peridotite, Italy. *Contrib. Mineral. Petrol.* **82**, 351–359.

Snow J. E. and Dick H. J. B. (1995) Pervasive magnesium loss by marine weathering of peridotite. *Geochim. Cosmochim. Acta* **59**, 4219–4235.

Snow J. E. and Schmidt G. (1999) Proterozoic melting in the northern peridotite massif, Zabargad Island: Os isotopic evidence. *Terra Nova* **11**, 45–50.

Snow J. E., Hart S. R., and Dick H. J. B. (1994) Nd and Sr isotope evidence linking mid-ocean-ridge basalts and abyssal peridotites. *Nature (London)* **371**, 57–60.

Snow J. E., Schmidt G., and Rampone E. (2000) Os isotopes and highly siderophile elements (HSE) in the Ligurian ophiolites, Italy. *Earth Planet. Sci. Lett.* **175**, 119–132.

Sobolev A. V., Tsamerian G. P., Dmitriev L. V., and Basilev B. (1992) The correlation between the mineralogy of basalt and the associated peridotites: the data from the MAR between 8°–18°N. *EOS* **73**, 584.

Song Y. and Frey F. A. (1989) Geochemistry of peridotite xenoliths in basalt from Hannuoba, Eastern China: implications for subcontinental mantle heterogeneity. *Geochim. Cosmochim. Acta* **53**, 97–113.

Sørensen H. (1967) Metamorphic and metasomatic processes in the formation of ultramafic rocks. In *Ultramafic and Related Rocks* (ed. P. J. Wyllie). Wiley, New York, pp. 204–212.

Sparks D. W. and Parmentier E. M. (1991) Melt extraction from the mantle beneath spreading centers. *Earth. Planet. Sci. Lett.* 105.

Spiegelman M. (1993) Physics of melt extraction: theory, implications and applications. *Phil. Trans. Roy. Soc. London* **A342**, 23–41.

Stephens C. J. (1997) Heterogeneity of oceanic peridotite from the western canyon wall at MARK: results from site 920. In *Mid-Atlantic Ridge; Leg 153, Sites 920–924. Proceedings of the Ocean Drilling Program, Scientific Results* (eds. J. A. Karson, M. Cannat, D. J. Miller, and D. Elthon) Ocean Drilling Program, College Station, TX, USA, vol. 153, pp. 285–301.

Streckeisen A. L. (1976) To each plutonic rock its proper name. *Earth Sci. Rev.* **12**, 1–33.

Suen C. J. and Frey F. A. (1987) Origins of the mafic and ultramafic rocks in the Ronda peridotite. *Earth Planet. Sci. Lett.* **85**, 183–202.

Suhr G. (1992) Upper mantle peridotites in the Bay of Islands Ophiolite, Newfoundland: formation during the final stages of a spreading centre. *Tectonophysics* **206**, 31–53.

Suhr G. (1999) Melt migration under oceanic ridges: inference from reactive transport modeling of upper mantle hosted dunites. *J. Petrol.* **40**, 575–600.

Suhr G. and Batanova V. (1998) Basal lherzolites in the Bay of Islands Ophiolite: origin by detachment-related telescoping of a ridge-parallel melting gradient. *Terra Nova* **10**, 1–5.

Suhr G. and Robinson P. (1994) Origin of mineral chemical stratification in the mantle section of the Table Mountain massif (Bay of Islands Ophiolite, Newfoundland, Canada). *Lithos* **31**, 81–102.

Suhr G., Seck H. A., Shimizu N., Gunther D., and Jenner G. (1998) Infiltration of refractory melts into the lowermost oceanic crust: evidence from dunite- and gabbro-hosted clinopyroxenes in the Bay of Islands Ophiolite. *Contrib. Mineral. Petrol.* **131**, 136–154.

Sun S. S. and McDonough W. F. (1989) Chemical and isotopic systematics of oceanic basalts: implications for mantle composition and processes. In *Magmatism in the Ocean Basins* (eds. A. D. Saunders and M. J. Norry). Geological Society of London, London, vol. 42, pp. 313–345.

Takahashi N. (1992) Evidence for melt segregation towards the fractures in the Horoman mantle peridotite complex. *Nature* **359**, 52–55.

Takahashi N. (2001) Origin of plagioclase lherzolite from the Nikanbetsu peridotite complex, Hokkaido, northern Japan: implications for incipient melt migration and segregation in the partially-molten upper mantle. *J. Petrol.* **42**, 39–54.

Takahashi N. and Arai S. (1989) *Textural and Chemical Features of Chromian Spinel-pyroxene Symplectite in the Horoman Peridotites, Hokkaido, Japan*. Scientific Reports of the Institute of Geosciences of the University of Tsukuba, Section B, **10**, pp. 45–55.

Takazawa E., Frey F. A., Shimizu N., Obata M., and Bodinier J.-L. (1992) Geochemical evidence for melt migration and reaction in the upper mantle. *Nature* **359**, 55–58.

Takazawa E., Frey F., Shimizu N., and Obata M. (1996) Evolution of the Horoman Peridotite (Hokkaido, Japan): implications from pyroxene compositions. In *Melt Processes and Exhumation of Garnet, Spinel and Plagioclase Facies Mantle* (eds. M. A. Menzies, J. L. Bodinier, F. Frey, F. Gervilla, and P. Kelemen). Elsevier, Amsterdam, Netherlands, vol. 134, pp. 3–26.

Takazawa E., Frey F. A., Shimizu N., Saal A., and Obata M. (1999) Polybaric petrogenesis of mafic layers in the Horoman peridotite complex, Japan. *J. Petrol.* **40**, 1827–1851.

Takazawa E., Frey F. A., Shimizu N., and Obata M. (2000) Whole rock compositional variations in an upper mantle peridotite (Horoman, Hokkaido, Japan): are they consistent with a partial melting process? *Geochim. Cosmochim. Acta* **64**, 695–716.

Takazawa E., Okayasu T., Satoh, and Keiichi (2003) Geochemistry and origin of the basal lherzolites from the northern Oman ophiolite (northern Fizh block). *Geochem. Geophys. Geosys. (G3)* **4** 2000GC000080.

Taylor R. N. and Nesbitt R. W. (1988) Light rare-earth enrichment of supra subduction-zone mantle; evidence from the Troodos Ophiolite, Cyprus. *Geology (Boulder)* **16**, 448–451.

Tazahi K., Ito E., and Komatsu M. (1972) Experimental study on a pyroxene-spinel symplectite at high pressures and temperatures. *J. Geol. Soc. Japan* **78**, 347–354.

Terry M. P., Robinson P., and Krogh Ravna E. J. (2000) Kyanite eclogite thermobarometry and evidence for thrusting of UHP over HP metamorphic rocks, Nordøyane, Western Greiss Region, Norway. *Am. Mineral.* **85**, 1637–1650.

Thayer T. P. (1960) *Some Critical differences between Alpine-type and Stratiform Peridotite-gabbro Complexes*. 21st Int. Geol. Congr., Copenhaguen, pp. 247–259.

Thayer T. P. (1967) Chemical and structural relations of ultramafic and felspathic rocks in alpine intrusive complexes. In *Ultramafic and Related Rocks* (ed. P. J. Wyllie). Wiley, New York, pp. 222–239.

Tilley C. E. (1947) The dunite-mylonites of Saint Paul's Rocks (Atlantic). *Am. J. Sci.* **245**, 483–491.

Tisseau C. and Tonnerre T. (1995) Non steady-state thermal model of spreading ridges: implications for melt generation and mantle outcrops. In *Mantle and Lower Crust Exposed in Oceanic Ridges and in Ophiolites* (eds. R. L. M. Vissers and A. Nicolas). Kluwer Academic Press, Dordrecht, Boston, London, pp. 181–214.

Toramaru A., Takazawa E., Morishita T., and Matsukage K. (2001) Model of layering formation in a mantle peridotite (Horoman, Hokkaido, Japan). *Earth Planet. Sci. Lett.* **185**, 299–313.

Torné M., Fernàndez M., Comas M. C., and Soto J. I. (2000) Lithospgeric structure beneath the Alboran Sea Basin: results from 3D modelling and tectonic relevance. *J. Geophys. Res.* **105**, 3209–3228.

Trieloff M., Weber H. W., Kurat G., Jessberger E. K., and Janicke J. (1997) Noble gases, their carrier phases, and argon chronology of upper mantle rocks from Zabargad Island, Red Sea. *Geochim. Cosmochim. Acta* **61**, 5065–5088.

Trommsdorff V. and Evans B. W. (1974) Alpine metamorphhism of peridotitic rocks. *Schweizer. Mineral. Petrogr. Mitteil.* **54**, 333–352.

Trommsdorff V., Piccardo G. B., and Montrasio A. (1993) From magmatism through metamorphism to seafloor emplacement of subcontinental Adria lithosphere during pre-Alpine rifting (Malenco, Italy). *Schweizer. Mineral. Petrogr. Mitteil.* **73**, 191–203.

Tubia J. M. and Cuevas J. (1986) High temperature emplacement of the Los Reales peridotite nappe (Betic Cordillera, Spain). *J. Struct. Geol.* **8**, 473–482.

Ulmer P. (1989) The dependence of the Fe2 + –Mg cation-partitioning between olivine and basaltic liquid on pressure, temperature and composition: an experimental study to 30 kbars. *Contrib. Mineral. Petrol.* **101**, 261–273.

Van der Wal D. and Bodinier J.-L. (1996) Origin of the recrystallization front in the Ronda peridotite by km-scale pervasive porous melt flow. *Contrib. Mineral. Petrol.* **122**, 387–405.

Van der Wal D. and Vissers R. L. M. (1993) Uplift and emplacement of upper mantle rocks in the western Mediterranean. *Geology* **21**, 1119–1122.

Van der Wal D. and Vissers R. L. M. (1996) Structural petrology of the Ronda peridotite, SW Spain: deformation history. *J. Petrol.* **37**, 23–43.

van Roermund H. L. M. (1989) High-pressure ultramafic rocks from the allochtonous nappes of the Swedish Caledonides. In *The Caledonide Geology of Scandinavia* (ed. R. A. Grayer). Grahan and Trotman, London, pp. 205–219.

van Roermund H. L. M., Drury M. R., Barnhoorn A., and de Ronde A. A. (2001) Relict majoritic garnet microstructures from ultra-deep orogenic peridotites in western Norway. *J. Petrol.* **42**, 117–130.

Vanucci R., Shimizu N., Bottazzi P., Ottolini L., Piccardo G. B., and Rampone E. (1991) Rare earth and trace element geochemistry of clinopyroxenes from the Zabargad peridotite-pyroxenite association. *J. Petrol. (Orogenic lherzolites and mantle processes)*, 244–255.

Vanucci R., Shimizu N., Piccardo G. B., Ottolini L., and Botazzi P. (1993) Distribution of trace elements during breakdown of mantle garnete: an example of Zabargad. *Contrib. Mineral. Petrol.* **113**, 437–449.

Vannucci R., Piccardo G. B., Rivalenti G., Zanetti A., Rampone E., Ottolini L., Oberti R., Mazzucchelli M., and Bottazzi P. (1995) Origin of LREE-depleted amphiboles in the subcontinental mantle. *Geochim. Cosmochim. Acta* **59**, 1763–1771.

Varfalvy V., Hébert R., and Bédard J. H. (1996) Interactions between melt and upper-mantle peridotites in the North Arm Mountain massif, Bay of Islands ophiolite, Newfoundland, Canada: implications for the genesis of boninitic and related magmas. *Chem. Geol.* **129**, 71–90.

Varfalvy V., Hébert R., Bédard J. H., and Lafleche M. R. (1997) Petrology and geochemistry of pyroxenite dykes in upper mantle peridotites of the North Arm Mountain Massif, Bay of Islands Ophiolite, Newfoundland: implications for the genesis of boninitic and related magmas. *Can. Mineral.* **35**, 543–570.

Varne R. and Rubenach M. J. (1972) Geology of Macquarie Island and its relationship to oceanic crust. *Antarct. Res. Ser.* **19**, 251–266.

Varne R., Gee R. D., and Quilty P. G. J. (1969) Macquarie island and the cause of oceanic linear magnetic anomalies. *Science* **166**, 230–233.

Vasseur G., Vernières J., and Bodinier J.-L. (1991) Modelling of trace element transfer between mantle melt and heterogranular peridotite matrix. *J. Petrol. (Orogenic lherzolites and mantle processes)*, **(sp. vol.)** 41–54.

Vernières J., Godard M., and Bodinier J.-L. (1997) A plate model for the simulation of trace element fractionation during partial melting and magma transport in the Earth's upper mantle. *J. Geophys. Res.* **102**, 24771–24784.

Vétil J.-Y., Lorand J.-P., and Fabriès J. (1988) Conditions de mise en place des filons des pyroxénites à amphibole du massif ultramafique de Lherz (Ariège, France). *C. R. Acad. Sci. Paris* **307**, 587–593.

Vielzeuf D. and Kornprobst J. (1984) Crustal splitting and the emplacement of Pyrenean lherzolites and granulites. *Earth Planet. Sci. Lett.* **67**, 87–96.

Villa I. M. (1988) (super 40) Ar/ (super 39) Ar analysis of amphiboles from Zabargad Island (Red Sea). *Tectonophysics* **150**, 249.

Villa I. M. (1990) (super 40) Ar/ (super 39) Ar dating of amphiboles from Zabargad Island (Red Sea) is precluded by interaction with fluids. *Tectonophysics* **180**, 369–373.

Vinogradov A. P., Dmitriyev L. V., and Udintsev G. B. (1971) Distribution of trace elements in crystalline rocks of rift zones: a discussion on the petrology of igneous and metamorphic rocks from the ocean floor. *Phil. Trans. Roy. Soc. A* **268**, 487–491.

Vissers R. L. M. and Nicolas A. (ed.) (1995) *Mantle and Lower Crust Exposed in Oceanic Ridges and in Ophiolites*. Kluwer Academic Press, Dordrecht, Boston, London.

Vissers R. L. M., Drury M. R., Hoogerduijn-Strating E. H., and Van der Wal D. (1991) Shear zones in the upper mantle: a case study in an alpine lherzolite massif. *Geology* **19**, 990–993.

Voggenreiter W., Hoetzl H., and Mechie J. (1988) Low-angle detachment origin for the Red Sea Rift system? *Tectonophysics* **150**, 51–75.

Vogt P. (1962) Geologisch-petrographische untersuchungen im peridotistock von Finero. *Schweizer. Mineral. Petrogr. Mitteil.* **49**, 157–198.

Voshage H., Hunziker J. C., Hofmann A. W., and Zingg A. (1987) A Nd and Sr isotopic study of the Ivrea Zone, Southern Alps, N-Italy. *Contrib. Mineral. Petrol.* **97**, 31–42.

Voshage H., Sinigoi S., Mazzucchelli M., Demarchi G., Rivalenti G., and Hofmann A. W. (1988) Isotopic constraints on the origin of ultramafic and mafic dikes in the Balmuccia peridotite (Ivrea Zone). *Contrib. Mineral. Petrol.* **100**, 261–267.

Walter M. J. (1999) Comments on 'Mantle melting and melt extraction processes beneath ocean ridges: Evidence from abyssal peridotites' by Yaoling Niu. *J. Petrol.* **40**, 1187–1193.

Walter M. J., Sisson T. W., and Presnall D. C. (1995) A mass proportion method for calculating melting reactions and application to melting of model upper mantle lherzolite. *Earth Planet. Sci. Lett.* **135**, 77–90.

Watson B. E., Brenan J. M., and Baker D. R. (1990) Distribution of fluids in the continental mantle. In *Continental Mantle* (ed. M. A. Menzies). Clarendon Press, Oxford, pp. 111–125.

Wernicke B. P. (1981) Low-angle normal faults in the Basin and Range Province—nappe tectonics in an extending orogen. *Nature* **291**, 645–648.

Wernicke B. P. (1985) Uniform-sense normal simple shear of the continental lithosphere. *Can. J. Earth Sci.* **22**, 108–125.

Whitmarsh R. B. and Wallace P. J. (2001) The rift-to-drift development of the west Iberia nonvolcanic continental margin: a summary and review of the contribution of Ocean Drilling Program Leg 173. In *Return to Iberia; Covering Leg 173 of the Genesis of the Drilling vessel JOIDES Resolution; Lisbon, Portugal, to Halifass, Nova Scotia; Sites 1065–1070, 15 April–15 June 1997. Proc. ODP, Sci. Results* (eds. M.-O. Beslier, R. B. Whitmarsh, P. J. Wallace, and J. Girardeau) Ocean Drilling Program, College Station, TX, USA, vol. 173, 36pp.

Wilshire H. G. and Shervais J. W. (1975) Al-augite and Cr-diopside ultramafic xenoliths in basaltic rocks from Western United States: structural and textural relationships. *Phys. Chem. Earth* **9**, 257–272.

Wilshire H. G., Nielson J. E., Pike J. E. N., Meyer C. E., and Schwarzman E. C. (1980) Amphibole-rich veins in lherzolite xenoliths, Dish Hill and Deadman Lake, California. *Am. J. Sci.* **280A**, 576–593.

Woodland A. B., Kornprobst J., McPherson E., Bodinier J. L., and Menzies M. A. (1996) Metasomatic interactions in the lithospheric mantle: petrologic evidence from the Lherz Massif, French Pyrenees. In *Melt Processes and Exhumation of Garnet, Spinel and Plagioclase Facies Mantle* (eds. M. A. Menzies, J. L. Bodinier, F. Frey, F. Gervilla, and P. Kelemen). Elsevier, Amsterdam, Netherlands, vol. 134, pp. 83–112.

Wu H. and Deng W. (1979) Basic geological features en the Yarlung Zangbo ophiolite belt, Xizang, China. In *Ophiolites* (ed. P. I. O. Symposium). Geological Survey Department, Republic of Cyprus, pp. 462–472.

Wyllie P. J. (1967) Review. In *Ultramafic and Related Rocks* (ed. P. J. Wyllie). Wiley, New York, pp. 403–416.

Wyllie P. J. (1969) The origin of ultramafic and ultrabasic rocks. *Tectonophysics* **7**, 437–455.

Xu Y.-G. and Bodinier J.-L. (2004) Contrasting enrichments in high and low temperature Mantle Xenoliths from Nushan, Eastern China: results of a single metasomatic event during Lithospheric Accretion? *J. Petrol.* (in press).

Xu Y.- G., Menzies M. A., Bodinier J.-L., Bedini R.-M., Vroon P., and Mercier J.-C. (1998) Melt percolation-reaction at the lithosphere-plume boundary: evidence from the poikiloblastic peridotite xenolith from Borée (Massif Central, Frace). *Contrib. Mineral. Petrol.* 132, 65–84.

Yakubchuk A. S., Nikishin A. M., and Ishiwatari A. (1994) Late Proterozoic ophiolire pulse. In *Circum Pacific Ophiolites*, Vol. Proc. 29th Int. Geol. Cong., Part D (eds. A. Ishiwatari, J. Malpas, and H. Ishizuka). VSP, Utrecht, The Nederlands, pp. 273–286.

Yoshikawa M. and Nakamura E. (2000) Geochemical evolution of the Horoman Peridotite Complex: implications for melt extraction, metasomatism and compositional layering in the mantle. *J. Geophys. Res.* **105**, 2879–2901.

You C. F., Castillo P. R., Gieskes J. M., Chan L. H., and Spivack A. J. (1996) Trace element behavior in hydrothermal experiments: implications for fluid processes at shallow depths in subduction zones. *Earth Planet. Sci. Lett.* **140**, 41–52.

Zanetti A., Vannucci R., Bottazzi P., Oberti R., and Ottolini L. (1996) Infiltration metasomatism at Lherz as monitored by systematic ion-microprobe investigations close to a hornblendite vein. In *Melt Processes and Exhumation of Garnet, Spinel and Plagioclase Facies Mantle* (eds. M. A. Menzies, J. L. Bodinier, F. Frey, F. Gervilla, and P. Kelemen). Elsevier, Amsterdam, Netherlands, vol. 134, pp. 113–133.

Zanetti A., Mazzucchelli M., Rivalenti G., and Vannucci R. (1999) The Finero phlogopite-peridotite massif; an example of subduction-related metasomatism. *Contrib. Mineral. Petrol.* **134**, 107–122.

Zeck H. P., Albat F., Hansen B. T., Torres-Roldan R. L., Garcia-Gasco A., and Martin-Algarra A. (1989) A 21 ± 2 Ma age for the termination of the ductile Alpine deformation in the internal zone of the Betic Cordilleras, South Spain. *Tectonophysics* **169**, 215–220.

Zhang R. Y. and Liou J. G. (1998) Dual origin of garnet peridotites of Dabie-Sulu UHP terrane, eastern-central China. *Episodes* **21**, 229–234.

Zindler A., Staudigel H., Hart S. R., Endres R., and Goldstein S. (1983) Nd and Sr isotopic study of a mafic layer from Ronda ultramafic complex. *Nature* **304**, 226–230.

2.05
Mantle Samples Included in Volcanic Rocks: Xenoliths and Diamonds

D. G. Pearson

Durham University, UK

D. Canil

University of Victoria, BC, Canada

and

S. B. Shirey

Carnegie Institution of Washington, DC, USA

2.05.1 MANTLE XENOLITHS: THE NATURE OF THE SAMPLE

2.05.1.1 Occurrence and Classification

Fragments of the Earth's mantle are frequently transported to the surface via volcanic rocks that are dominantly alkaline in nature. These fragments range up to sizes in excess of 1 m across. The term "mantle xenoliths" or "mantle nodules" is applied to all rock and mineral inclusions of presumed mantle derivation that are found within host rocks of volcanic origin. The purpose of this contribution is to review the geochemistry of mantle xenoliths. For detailed petrological descriptions of individual locations and suites, together with their geological setting, the reader is referred to the major reference work by Nixon (1987).

Despite peridotite xenoliths in basalts being recognized for several centuries and comparisons being made to lherzolite massifs (Lacroix, 1893), it was not until work on garnet peridotites and diamonds in kimberlites by Fermor (1913) and Wagner (1914) that such xenoliths were conceptually associated with a peridotite zone in the Earth beneath the crust, i.e., the zone that we now identify as the mantle. Mantle xenoliths provide snapshots of the lithospheric mantle beneath particular regions at the time of their eruption and hence are crucial direct evidence of the nature of the mantle beneath regions where no samples have been exposed by tectonic activity. As such, xenoliths are an essential compliment to tectonically exposed bodies of mantle (orogenic peridotites and ophiolites) that occur at plate boundaries (see Chapter 2.04). One obvious contrast between the mantle samples provided by xenoliths and those provided by peridotite massifs is the lack of field relationships available for xenoliths. Other drawbacks include the small size of many xenoliths. This makes accurate estimation of bulk compositions difficult and accentuates modal heterogeneities. The frequent infiltration of the host magma also complicates their chemical signature. Despite these drawbacks, xenoliths are of immense value, being the only samples of mantle available beneath many areas. Because they are erupted rapidly, they freeze in the mineralogical and chemical signatures of their depth of origin, in contrast to massifs which tend to re-equilibrate extensively during emplacement into the crust. In addition, many xenolith suites, particularly those erupted by kimberlites, provide samples from a considerably greater depth range than massifs. Over 3,500 mantle xenolith localities are currently known. The location and nature of many of these occurrences are summarized by Nixon (1987). A historical perspective on their study is given by Nixon (1987) and Menzies (1990a). Mantle xenoliths from any tectonic setting are most commonly described from three main igneous/pyroclastic magma types (where no genetic relationships are implied):

(i) Alkalic basalts sensu-lato (commonly comprising alkali basalt-basanites and more evolved derivatives), nephelinites and melilitites.

(ii) Lamprophyres and related magmas (e.g., minettes, monchiquites, and alnoites) and lamproites.

(iii) The kimberlite series (Group I and Group II or orangeites; Mitchell, 1995).

Although mantle xenoliths most commonly occur in primitive members of the above alkaline rocks, rare occurrences have been noted in more evolved magmas such as phonolites and trachytes (e.g., Irving and Price, 1981).

To simplify matters and to circumvent the petrographic complexities of alkaline volcanic rocks in general, we will use the term "alkalic and potassic mafic magmas" to include alkalic basalts, nephelinites, melilitites, and lamprophyres. Occurrence of xenoliths in such magmas can be compared to those occurring in kimberlites and related rocks. As a general rule, the spectrum of mantle xenoliths at a given location varies with host rock type. In particular, alkalic and potassic mafic magmas tend to erupt peridotites belonging predominantly to the spinel-facies, whereas kimberlites erupt both spinel and garnet-facies peridotites (Nixon, 1987; Harte and Hawkesworth, 1989).

Even within either "group" of volcanic rocks the variety of possible xenolith types is great. Table 1 presents a summary of the most common mantle xenolith groups that are found in kimberlitic hosts and within the alkalic and potassic mafic magmas. The significance and abundance of these groups will be discussed below.

Although widespread in the literature, classification of xenoliths on the basis of their host magma is not fully informative. It is more geologically useful to subdivide xenoliths in terms of their tectonic setting. A basic subdivision of xenolith occurrences is between those erupted in oceanic settings and those erupted in continental settings. The continental occurrences far outnumber the oceanic occurrences. The continental occurrences can be further subdivided depending on age of the crust and the tectonic history of the area being sampled. Xenoliths from stable cratonic and circum-cratonic regions are distinctly different in petrology from those occurring in areas that have experienced significant lithospheric rifting, generally in noncratonic crust, in the recent geological past. As such, we will utilize the terms cratonic/circum-cratonic to refer to xenoliths occurring on and around craton margins and the term noncratonic in referring to mantle sampled away from cratons, often in areas that have experienced recent lithospheric thinning. There is a link back to the host rock in that, as a general rule, cratonic and circum-cratonic xenoliths are erupted by kimberlites and noncratonic xenoliths are erupted by alkalic and potassic mafic magmas.

2.05.1.1.1 Mantle xenoliths in continental volcanic rocks

Xenoliths found in Archean cratonic regions are characterized by the lithological types reported in Table 1(a). Garnet-facies peridotites dominate the peridotite xenolith inventory in these locations. In contrast, away from cratons, there is a scarcity of garnet-facies peridotites (Table 1(b)). In addition, cratonic xenolith suites contain samples derived from depths ranging from crustal levels to >200 km, whereas noncratonic xensoliths come from less than 140 km deep. There can be distinct differences between xenoliths erupted on craton and those erupted in stable areas of Proterozoic crust marginal to cratons. For instance, subcalcic-garnet harzburgites occur on most cratons but do not occur in circum-cratonic suites (Boyd *et al.*, 1993). In addition, the maximum depths of equilibration of circum-cratonic peridotite suites are less than for cratonic peridotite suites (e.g., Finnerty and Boyd, 1987). These differences warrant the distinction between "cratonic" and "circum-cratonic" xenoliths. In addition, young rift-related magmatism, marginal to cratons, samples very thin lithosphere compared to cratonic and circum-cratonic lithosphere. The xenoliths sampled in this environment fall into the loose category of "noncratonic" xenoliths. A more detailed and complex tectonic classification is provided by Griffin *et al.* (1999a).

2.05.1.1.2 Mantle xenoliths in oceanic volcanic rocks

The nature of the suboceanic mantle is largely constrained from geochemical studies of its partial melts (see Chapter 2.08) because occurrences of mantle xenoliths in the ocean basins are much rarer than on the continents. The host rocks for these xenoliths are exclusively alkalic and

Table 1 Major groups of mantle xenoliths in kimberlite-related and alkali basalt series volcanic rocks (after Harte and Hawkesworth, 1989). Textural classification follows that of Harte (1977). Terminology for phlogopite-rich mafic mantle xenoliths from Gregoire *et al.* (2002). For supplementary data and classifications see Nixon (1987), table 62.

Type	*Characteristics*	*Examples*	*Mg# olivine*
(A) *Cratonic/circum-cratonic xenoliths erupted by Kimberlite-related volcanics*			
AI: Coarse Mg-rich, low-*T* peridotites	Often abundant. Mostly harzburgites and lherzolites with varying but low modal diopside and garnet. Wide range of orthopyroxene abundance, Kaapvaal examples notably enriched. Crystals typically 0.2 mm with equant or tabular shapes, irregular grain boundaries, rarely granoblastic (Harte, 1977). Bulk compositions typically highly depleted in Fe, Ca, and Al, enriched in Mg. Mineralogy: Cr-rich pyrope, Cr-diopside. Orthopyroxene in garnet facies characterized by >1.0 wt.% Al_2O_3. Cr-spinel sometimes evident. Minor phlogopite common grading into type VIII phlogopite peridotites. Phlogopite often surrounds garnet and is strongly correlated with the presence of diopside. Estimated equilibration temperatures less than 1,100 °C. Equilibration pressures can vary widely within a pipe and range from c. 2 GPa to >6 GPa. Rarely diamondiferous (e.g., Dawson and Smith, 1975), more commonly contain graphite (Pearson *et al.*, 1994).	N. Lesotho (Nixon and Boyd, 1973a), Kaapvaal craton (Gurney and Harte, 1980; Boyd and Nixon, 1978; Boyd and Mertzman, 1987; Nixon, 1987), Siberia (Sobolev, 1974; Boyd *et al.*, 1993); Jericho Slave craton (Kopylova *et al.*, 1999)	Av 92.8 (91–95)
	Subcalcic garnet (high Cr-pyrope; knorringitic) bearing harzburgite varieties scarce but can contain diamond and graphite. Can be megacrystalline. Textures similar to type I. Equilibration temperatures and pressures intermediate between low-*T* and high-*T* lherzolites, i.e., 1,150 °C, 5–6 GPa, but vary widely.	Udachnaya, Siberia (Sobolev *et al.*, 1973; Pokhilenko *et al.*, 1993), Kaapvaal (Boyd *et al.*, 1993)	92–95.5
	Spinel facies widespread but less abundant. Textures as for garnet variety, spinel texture symplectitic or irregular. Equilibration temperatures <800 °C. Can also be orthopyroxene enriched, like garnet facies. Spinel composition can vary widely in Cr# but mostly aluminous. Cr-rich spinels coexist with garnet. Orthopyroxenes in spinel facies have >1.0 wt.% Al_2O_3. Similar range in bulk composition to garnet facies.	Kaapvaal craton (Carswell *et al.*, 1984; Boyd *et al.*, 1999)	91.5–94

AII: coarse, Fe-rich low-*T* peridotites and pyroxenites	Widespread, normally rare but locally abundant. Mainly garnet lherzolites and garnet websterites but also clinopyroxenites and orthopyroxenites ("bronzitites"). Ilmenite can be present in pyroxenites. Coarse grained to "megacrystalline" (at Jericho). Textures and equilibration temperatures as for type I. Sometimes modally layered. Wide ranging bulk and mineral compositions, with high Fe, Ca, Al, and Na relative to type I. Rare fine-grained "quench textured" ilmenite/garnet pyroxenites.	Matsoku, Kaapval craton (Gurney *et al.*, 1975); Jericho, Slave craton (Kopylova *et al.*, 1999); Mzongwana, SE margin Kaapvaal craton (Boyd *et al.*, 1984a)	83–89
AIII: dunites	Widespread, locally common. Two varieties:		
	(i) Highly depleted, coarse to ultracoarse >50 mm olivine (megacrystalline) dunites, often containing chromite or sub calcic high-Cr pyrope and frequently diamondiferous.	Siberia, notably Udachnaya (Pokhilenko *et al.*, 1993)	93–95
	(ii) Often fine to medium grained more Fe-rich dunites, mineral zoning indicates "metasomatism." Mostly deformed textures. Orthopyroxene, garnet, phlogopite, diopside, chromite present.	Kimberley (Boyd *et al.*, 1983; Dawson *et al.*, 1981)	85–93
AIV: deformed low-*T* peridotites and pyroxenites	Widespread, locally common. Porphyroclastic or mosaic-porphyroclastic textures. Modal abundances, chemical characteristics and P–T equilibration conditions similar to those of type I.	Jericho, Slave craton (Kopylova *et al.*, 1999)	91–95
AV: deformed high-*T* peridotites	Widespread but variable abundance in group I kimberlites, absent/scarce in group II kimberlites. Commonly deformed; porphyroclastic and mosaic-porphyroclastic textures with fine neoblasts of olivine. Although generally more depleted than pyrolite, bulk rocks and minerals generally enriched in Fe and Ti compared to type I (low-*T*) and significant compositional overlap of minerals with megacrysts (type X). Equilibration temperatures 1,100 °C to >1,500 °C, equilibration pressures generally 4.5 GPa to >6.5 GPa. Garnets and pyroxenes frequently zoned.	N. Lesotho (Nixon and Boyd, 1973a); Jagersfontein, Kaapvaal craton (Burgess and Harte, 1999); Siberia (Sobolev, 1974; Boyd *et al.*, 1993); Slave (Kopylova *et al.*, 1999); Somerset Island, Churchill Province (Schmidberger and Francis, 1999)	87–92
AVI: phlogopite-rich mafic mantle xenoliths	Widespread and locally common. Olivine poor/absent rocks. Two main subdivisions of this group (Gregoire *et al.*, 2002) are:		
	(i) MARID suite (mica–amphibole–rutile–ilmenite–diopside) with accessory zircon common. Probable genetic link to group II kimberlites. Medium to coarse grained, undeformed to deformed, sometimes modal banding. Amphibole always K-richterite.	Kimberley (Dawson and Smith, 1977; Gregoire *et al.*, 2002)	NA
	(ii) PIC suite (phlogopite–ilmenite–clinopyroxene) with minor rutile. Diopside or Al- and Ti-poor augites. Probable genetic link to group I kimberlites. K-richterite is absent; grade to glimmerites as phlogopite mica reaches >90%. Coarse grained, variably deformed.	Kimberley (Gregoire *et al.*, 2002)	NA

(continued)

Table 1 (continued).

Type	*Characteristics*	*Examples*	*Mg# olivine*
AVII: pyroxenite sheets rich in Fe and Ti	Restricted to Matsoku. Orthopyroxene and clinopyroxene rich rocks with widely variable olivine and garnet compositions, often with ilmenite and phlogopite (the IRPS suite: see type VIII). Bulk compositions Fe and Ti rich. Form magmatic intrusions (<16 cm thick) into type I rocks which become metasomatized.	Matsoku (Gurney *et al.*, 1975; Harte *et al.*, 1975, 1987)	
AVIII: modally metasomatized peridotites	Wide spread, variable abundance. Mostly metasomatized variants of type I. Diverse mineralogies, Two most commonly recognized groups are phlogopite peridotites (PP) and phlogopite–K-richterite–Peridotites (PKP) of Erlank *et al.* (1987). Can be harzburgite or lherzolite, typically coarse grained, undeformed but some display porphyroclastic textures. Assemblages vary with location. Cr-titanate "LIMA" minerals (Lindsleyite–Mathiasite) relatively common at Bultfontein; edenite-phlogopite association at Jagersfontein; ilmenite–rutile–phlogopite–sulfide (IRPS) suite at Matsoku associated with pyroxenitic sheets (type VII). Metasomatic clinopyroxene link to type AI.	Matsoku (Gurney *et al.*, 1975); Kimberley pipes (Erlank *et al.*, 1987; Gregoire *et al.*, 2002); Jagersfontein (Winterburn *et al.*, 1990)	Same as type I, to more Fe-rich.
AIX: eclogites, grospydites, alkremites, and variants	Very widespread, rare to locally abundant. Eclogites (omphacite and pyrope-almandine garnet). Garnet composition widely variable, in grospydites garnet has a large grossular component. At some locations (e.g., Jagersfontein), unusual assemblages of garnet + spinel (Alkremites), garnet + corundum (Corganites) and corundum + garnet + spinel (Corgaspinites) occur (Mazzone and Haggerty 1989). Accessory phases in eclogites include kyanite, corundum, ilmenite, rutile, sanidine, coesite, sulfides, graphite and diamond. Eclogites classified on texture: group I large subhedral to rounded garnets in matrix of omphacite. High Cr, Ca, Fe, and Mn in omphacite. Garnets more Na (avg. 0.1 wt.% Na_2O) and Mg rich. Group II have interlocking texture of anhedral garnet and omphacite and are less altered. Garnets are lower in Na_2O (0.05 wt.%). Common hosts for diamond, especially group I. Not all eclogites of obvious mantle origin and some grade into garnet granulites and pyroxenites of crust origin.	Roberts Victor (MacGregor and Carter, 1970; McCandless and Gurney, 1989); Jagersfontein (Nixon *et al.*, 1978; Mazonne and Haggerty, 1989); Orapa (Robinson *et al.*, 1984), all Kaapvaal craton. Udachnaya, Siberian craton (Sobolev, 1974; Ponomarenko, 1975); Koidu, W. African craton (Tompkins and Haggerty, 1984; Hills and Haggerty, 1989)	NA

AX: megacrysts (discrete nodules)	Single crystals or monominerallic polycrystalline aggregates (sometimes exsolved) weighing up to 15 kg. Rare mutual lamellar or granular intergrowths. Large range in Mg#, Cr, and Ti in a given suite.		
	Cr-poor variety: widespread, locally abundant (e.g., Monastery). Garnets, clino- and orthopyroxenes, phlogopite and ilmenite most common, zircon and olivine rarer. Debatable whether phlogopite and olivine are members of Cr-poor suite. Wide range in chemistry but Cr-poor, Fe–Ti-rich relative to type I (low-*T*) peridotite minerals. Mineral chemistry and estimated equilibration P/Ts overlap those of type V (high-*T*) lherzolites. Some Slave craton "Cr-poor megacrysts" show mineral chemistry links to type II megacrystalline pyroxenite xenoliths. See review of Schulze (1987).	N. Lesotho (Nixon and Boyd 1973b); Monastery (Gurney *et al.*, 1979), Jagersfontein (Hops *et al.*, 1992), Kaapvaal craton; The Malaita megacryst suite (Nixon and Boyd, 1979), occurs in an ocean plateau alnoite, but has many similarities with the kimberlitic low-Cr suite	
	Cr-rich variety: (i) A suite comprising garnet plus ortho- and clinopyroxene, mostly restricted to kimberlites from Colorado-Wyoming. Mineralogically similar to type I lherzolites.	Colorado-Wyoming craton (Eggler *et al.*, 1979)	
	(ii) "Granny Smith" diopsides; bright green Cr-diopside, may contain blebs/intergrowths of ilmenite and phlogopite. Can be polycrystalline.	Kimberley and Jagersfontein (Boyd *et al.*, 1984b)	
	Miscellaneous: mostly garnets and pyroxenes with no clear paragenetic association or links to other megacryst suites. May represent disrupted peridotites/eclogites/pyroxenites.		
AXI: polymict aggregates	Polymict aggregates of peridotite, eclogite and megacrysts, of variable grain size, some containing quenched melt. Mineral assemblages not in elemental or isotopic equilibrium.	Bultfontein, De Beers and Premier mines, Kaapvaal (Lawless *et al.*, 1979). Malaita	Highly variable
AXII: diamond and inclusions in diamonds	Widespread and closely related to cratons. Abundance varies from <1 ppm to 100 ppm by weight. Size $\ll$0.1 g to c. 750 g. Type I diamonds contain abundant N, type II low N (Harris, 1987).	All cratons (Harris, 1987; Meyer, 1987)	93–96
	Inclusion suites divided into peridotitic (P-type) and eclogitic (E-type) parageneses.		
	P-type inclusions: high-Cr, low Ca garnet, Cr-diopside, Fo-rich olivine, orthopyroxene, chromite, wustite, Ni-rich sulfide, have restricted, high Mg, high Ni chemistry. Equilibration temperatures 900–1,100 °C.		
	E-Type inclusions: pyrope-almandine, high Na garnet (>0.1 wt.%), omphacite, coesite, low-Ni sulfide.		

(continued)

Table 1 (continued).

Type	*Characteristics*	*Examples*	*Mg# olivine*
AXIII: ultra-deep peridotites	Rare and restricted to Jagersfontein (Kaapvaal Craton) and Koidu (W. African craton). Four-phase garnet lherzolite. Close association of pyrope-garnet (~70% py; 2 wt.% Cr_2O_3) and jadeite-rich clinopyroxene (20% Jd, & 4% Di). Clinopyroxene forms either orientated rods in garnet host or as discrete grains attached to garnet in cuspate contact. Both pyroxenes exsolved from garnet at 100–150 km depth. Recombination of garnet gives original depths of derivation of 300–400 km. Discrete garnets and "lherzolites" with eclogitic affinities also found (Sautter *et al.*, 1991).	All samples so far from the Jagersfontein kimberlite, S. Africa (Haggerty and Sautter, 1990; Sautter *et al.*, 1991) and Koidu, Sierra Leone (Deines and Haggerty, 2000)	91.6
(**B**): *Non-cratonic xenoliths erupted by alkalic and potassic mafic magmas sensu lato*[a]			
BI: Cr-diopside lherzolite group	Very widespread and common in a variety of tectonic settings, off-craton. Dominantly spinel-facies (Al or Cr-spinel) lherzolites but can be garnet-facies and garnet-spinel facies (e.g., Vitim). Coarse grained, commonly little deformed, sometimes show preferred orientation. Include harzburgites, orthopyroxenites, clinopyroxenites, websterites, and wehrlites. Pargasite and phlogopite may also be common. Both low TiO_2 and high TiO_2 amphiboles can occur at the same locality. Accessory apatite, can be common locally (e.g., Bullenmerri, Victoria). Interstitial silicate glass can be present. Garnet and spinel facies significantly more olivine-rich and orthopyroxene poor than peridotites from cratons such as Kaapvaal and Siberia. Bulk rocks less depleted in Ca, Al, Fe, and lower in Mg than cratonic peridotites. Minerals generally higher Mg# and Cr# and lower Na and Ti than those of the Al-Augite group. Can be subdivided into type IA (LREE depleted clinopyroxene) and type IB (LREE enriched clinopyroxene).	Victoria, SE Australia (Frey and Green, 1974); Vitim (Ionov *et al.*, 1993a); San Carlos and other W. USA localities (Frey and Prinz, 1978; Wilshire and Shervais, 1975); Eifel (Stosch and Seck, 1980); Hawaii (Jackson and Wright, 1970); Scotland (Menzies and Halliday, 1988) Garnet facies: Thumb, Navajo field (Ehrenberg, 1982a,b); Pali-Aike, Patagonia (Stern *et al.*, 1989); Vitim, S. Siberia (Ionov, 1993a,b)	>0.85, Avg. ~90
BII: Al-augite wehrlite-pyroxenite group	Widespread and common. Frequently clinopyroxene-rich rocks but widely variable: wehrlites, clinopyroxenites, dunites, websterites, lherzolites, lherzites, gabbros. Al-spinel is the typical aluminous phase but may contain plagioclase. Kaersutite common along with apatite, Fe–Ti oxides, and phlogopite. Some igneous and metamorphic textures. Composite xenoliths relatively common (in contrast to kimberlite-related xenoliths). Cross-cutting pyroxene-rich veins and layers may occur in olivine-rich hosts. Olivine-rich aggregates also found in pyroxene-rich xenoliths. Minerals generally lower Mg# and Cr#, higher Ti than those of the type I (Cr-diopside group).	Victoria, SE Australia (Frey and Green, 1974); San Carlos and other W. USA localities (Frey and Prinz, 1978; Wilshire and Shervais, 1975; Irving, 1980); Hawaii (Jackson and Wright, 1970; Irving, 1980)	<0.85

BIII: garnet pyroxenite group	Widespread but not abundant. Garnet clinopyroxenites and websterites plus clinopyroxenites and websterites where pyroxenes commonly show exsolution of garnet and/or spinel as well as Ca-rich or Ca-poor pyroxene. Accessory ilmenite and sometimes apatite. Coarse grained, undeformed textures, sometimes layered. "Basaltic" bulk compositions.	Delegate, SE Australia (Lovering and White, 1969; Irving, 1974), Salt Lake Crater, Hawaii (Beeson and Jackson, 1970)	NA
BIV: modal metasomatic group	Widespread varieties of the above groups showing evidence for modal (or "patent") metasomatism. Wehrlite-clinopyroxenites with mica, glimmerites. Typical metasomatic phases include pargasite/kaersutite, phlogopite, apatite, and grain-boundary oxides e.g., rutile. Apatite only in some cases. Silicate glass as melting product of amphibole, clinopyroxene, or phlogopite common. Composite xenoliths occur.	Nunivak, Alaska (Francis, 1976), SE Australia (O'Reilly *et al.*, 1991), Menzies and Murthy, 1980a, Vitim (Ionov *et al.*, 1993a), Loch Roag and Fife Scotland (Menzies *et al.*, 1989)	NA
BV: megacrysts	Widespread with variable abundance. Usually large (>1 cm) single crystals. Large range in Mg#, Cr, and Ti in a given suite. Group A: Al-augite, Al-bronzite, olivine, kaersutite, pyrope, pleonaste, plagioclase; some of which may have crystallized from the host magma Group B: Anorthoclase, Ti-mica, Fe–Na salite, apatite, magnetite, ilmenite, zircon, rutile, sphene, and corundum, all of which are likely xenocrysts. Some coarse crystals are undoubtedly derived from disaggregated type I and type II xenoliths.	SE Australia (Binns *et al.*, 1970; Irving and Frey, 1984; Schulze, 1987), Loch Roag, Scotland (Menzies *et al.*, 1989)	NA

[a] Based on Harte and Hawkesworth (1989) with nomenclature from Frey and Green (1974), Wilshire and Shervais (1975), Frey and Prinz (1978), Irving (1980), and Menzies (1983).

potassic mafic magmas. The xenolith suite of the Hawaiian volcanic chain is perhaps the best characterized (Jackson and Wright, 1970) of the ocean islands, while extensive suites have also been found in the Canary Islands (Neumann *et al.*, 1995), Samoa (Hauri *et al.*, 1993), Grande Comore (Coltorti *et al.*, 1999), and Tahiti. Most of these occurrences sample the oceanic lithosphere directly below the islands and those of type I (Table 1) are proposed to be residues of partial melting that have been variably metasomatized, with carbonatite-like fluids frequently being invoked (Hauri *et al.*, 1993; Coltorti *et al.*, 1999). The Hawaiian suite is more complex. Pyroxenites of type II and type III are common and iron-rich peridotites, some with garnet, are thought to be physical mixtures of spinel lherzolites with the pyroxenite suite (Sen and Leeman, 1991).

Some xenolith localities sample the mantle lithosphere beneath oceanic plateaux. The most extensive and varied xenolith suite in this regard is that from Malaita (Solomon Islands) on the margin of the Ontong Java Plateau (Nixon and Boyd, 1979). This locality is hosted by an alnoite and contains both garnet and spinel-facies lherzolites together with a spectacular megacryst suite. Although in an oceanic setting, the variety of the xenolith suite provided by the Malaita alnoite, in particular the megacrysts, show strong similarities to suites from kimberlites (Nixon and Boyd, 1979).

2.05.1.1.3 Mantle xenoliths in subduction zone environments

Although xenoliths from subduction-related tectonic settings have been known for sometime, their detailed relationship to the subduction zone system has been a matter of debate. Most samples are type-I spinel lherzolites and modal metasomatic variants of type IV, most commonly kaersutite and phlogopite. Among the best-known examples are from Itinome-Gata, Japan (Aoki, 1968) and Simcoe, NW, USA (Brandon *et al.*, 1999), although spinel lherzolites from Grenada, Lesser Antilles, also occur. It is not well established whether these xenoliths actually represent parts of the metasomatized mantle wedge above the subduction zone, or simply mantle lithosphere not intimately related to the subduction zone process. McInnes and Cameron (1994) have reported xenoliths from the Tabar–Lihir–Tanga–Feni arc, Papua New Guinea, that are purported to be mantle wedge compositions.

2.05.1.2 Lithologies

2.05.1.2.1 Nomenclature and abundance

An important point to note in the discussion of mantle xenoliths is the varying petrographic nomenclature used. Workers studying noncratonic peridotite xenoliths and orogenic peridotites use the IUGS nomenclature proposed by Streckeisen (1976), i.e., a harzburgite can contain up to 5 modal % clinopyroxene. In contrast, those who study xenoliths from kimberlites use the more thermodynamically consistent "clinopyroxene-free" definition of a harzburgite. This is based on the reasoning that the presence of even the smallest amount of an extra phase, in this case clinopyroxene, reduces the number of degrees of freedom of the system by one (O'Hara *et al.*, 1975).

The terms "parent" or "modal" metasomatism are usually used to infer the presence of minerals additional to those commonly observed in peridotites (e.g., Harte, 1983). There is a growing body of evidence indicating the addition of certainly clinopyroxene and possibly garnet to peridotites during metasomatic events (Erlank *et al.*, 1987; Pearson *et al.*, 2002; Shimizu, 1999; Simon *et al.*, in press). As suggested by Erlank *et al.* (1987), this possibility and its implications for nomenclature should be kept in mind when discussing metasomatic processes and events.

Both mono- and polymineralic fragments occur as mantle xenoliths. Monomineralic aggregates whose relationship to the host rock is uncertain are generally referred to as "megacrysts" (Table 1), whereas the term "xenolith" is most commonly applied to polymineralic aggregates that are unlikely to have crystallized directly from the host magma. Many megacrysts, sometimes referred to as discrete nodules/xenoliths, are thought to have some genetic/temporal relationship with the host volcanic rock or magmatic activity associated with producing the host rock (Table 1). As such, they do not generally represent fragments of the Earth's mantle that have been accidentally sampled and their geochemistry will not be described in detail here. The reader is referred to Schulze (1987) for a review of megacrysts petrology and to Harte (1983), Irving and Frey (1984), Hops *et al.* (1992), and Nowell *et al.* (in press) for a discussion of megacryst geochemistry.

Studies of mantle xenoliths have confirmed the view from seismology that peridotite is volumetrically the dominant component of the Earth's shallow mantle (<400 km) (see Chapter 2.02). This is because xenolith suites in almost all tectonic environments are dominated by peridotites. Even at localities where other lithologies such as eclogite dominate the intact xenolith suite, mineral concentrate studies show that peridotite dominates the inventory of entrained mantle material (Schulze, 1989). Major- and trace-element studies of mineral concentrates from mined kimberlites have also been used to illustrate how the distribution of lithologies varies with depth within a given mantle section (Griffin *et al.*, 1999a).

2.05.1.2.2 Noncratonic xenoliths

Lithologies most common in this tectonic environment are summarized in Table 1(b) and detailed descriptions and summaries of their petrology can be found in Nixon (1987) and Harte and Hawkesworth (1989). The dominant xenolith lithology is spinel lherzolite belonging to type I but garnet-facies xenoliths also occur, such as at Vitim, Siberia (Ionov *et al.*, 1993a) and Pali-Aike, S. America (Stern *et al.*, 1989). Garnet-facies peridotites carried by the Thumb minette, Arizona (Ehrenberg, 1982a) and the Lashaine ankaramitic scoria and carbonantite tuff in Tanzania (Rudnick *et al.*, 1994) bear close similarities to kimberlite-derived xenoliths. This might be expected for Lashaine given the cratonic setting. Noncratonic xenoliths are generally less than 30 cm in size, although exceptions occur. They are on average smaller than at many kimberlite-bone cratonic xenolith locations. Type I peridotites are commonly lherzolites. Type II or Al-augite wehrlite–pyroxenite groups have lower Mg/(Mg + Fe) and Cr/(Cr + Al) but higher titanium than those of type I.

In contrast to cratonic xenoliths, the main two groups of noncratonic xenoliths commonly have overlapping equilibration temperatures and may occur in close association as composite xenoliths, with the Al-augite group showing intrusive relationships into the Cr-diopside group (e.g., Wilshire and Shervais, 1975; Irving, 1980). The type II pyroxenites are interpreted as intrusive magmatic activity within the mantle. However, their bulk compositions are not melts but are a function of magma–wallrock interaction and crystal plating onto the walls of magma conduits (Wilshire and Shervais, 1975; Irving, 1980). Metasomatism, involving introduction of amphibole and biotite as well as variations in mineral chemistry, are common in the vicinity of Al-augite series intrusions.

The most common and widely studied metasomatic phases in noncratonic type I and II xenoliths are amphibole and mica plus, less commonly, apatite, and carbonate. These phases occur as intergranular material or in veins in association with other metasomatic minerals. Amphibole is frequently contained in melt pockets (glass), e.g., Francis (1976). Primary carbonates have been found in peridotite, pyroxenite, and eclogite xenoliths (Ionov *et al.*, 1993b; Pyle and Haggerty, 1994; Ionov, 1998; Lee *et al.*, 2000b). Experimental evidence suggests that carbonate should be a stable mantle phase in peridotitic assemblages, but it is still relatively scarce considering the geochemical evidence suggesting carbonatite metasomatism in mantle peridotites (see review by Menzies and Chazot, 1995). Canil (1990) interprets the general absence of carbonate in mantle xenoliths to dissociation of carbonate during syn-entrainment decompression.

2.05.1.2.3 Cratonic and Circum-cratonic Xenoliths

Much of our knowledge of cratonic and circum-cratonic xenoliths is based on the extensive studies of samples from southern African kimberlites, particularly the diamondiferous kimberlites of the Kaapvaal craton and surrounding areas (Nixon, 1987; Gurney *et al.*, 1991). Similar xenolith suites have now been described from other cratons such as Siberia (Sobolev, 1974; Boyd *et al.*, 1997), the Slave (Kopylova *et al.*, 1999) and Colorado–Wyoming (Eggler *et al.*, 1979) cratons. In addition, kimberlites located on craton margins (Somerset Island; Mitchell, 1977; Schmidberger and Francis, 1999) or in Proterozoic regions surrounding cratons, e.g., Namibia (Mitchell, 1984) can have excellent peridotite suites that mirror those on craton.

Cratonic xenolith suites provided by kimberlites have two main populations of mantle xenoliths that can be distinguished on the basis of their equilibration pressure, temperature, and textures. Type I, or "low-temperature" peridotites, are generally coarse grained, with low levels of deformation and yield equilibration temperatures generally below 1,100 °C (Table 1; Boyd, 1973; Nixon and Boyd, 1973; Harte, 1983; Finnerty and Boyd, 1987). In contrast, "high-temperature" (type V) peridotites are generally finer grained, highly deformed and yield equilibration temperatures between 1,100 °C and 1,500 °C (Boyd, 1973; Nixon and Boyd, 1973; Harte, 1983; Finnerty and Boyd, 1987). Iron-rich (type II) peridotites and deformed, low-temperature peridotite/pyroxenites (type IV) show intermediate features between type I and type V end-members. In contrast to certain noncratonic xenolith suites, composite xenoliths are much scarcer in cratonic xenolith populations. This may be a function of the differing properties of the contrasting host magmas in the two environments. In particular, no clear examples of xenoliths exhibiting contact relations between common type I peridotites and eclogites (and variants) have been found, despite overlaps in their equilibration conditions (Harte and Hawkesworth, 1989). The Matsoku kimberlite is exceptional in its sampling of composite xenoliths, where iron- and titanium-rich pyroxenite sheets (type VII) show intrusive relationships to low-temperature (type I) peridotites, analogous to the relationships seen in noncratonic type I and type II xenolith groups (Harte *et al.*, 1987).

Harzburgites (clinopyroxene-free assemblages) are relatively common within the low-temperature garnet facies group of kimberlite xenoliths. However, despite a lack of clinopyroxene in hand specimen, mineral chemistry indicates that many of these single-pyroxene garnet peridotite xenoliths equilibrated with clinopyroxene. Kimberlite-derived xenoliths also host a variety of phlogopite-rich mafic xenoliths such as the MARID suite (Table 1; Dawson and Smith, 1977; Gregoire *et al.*, 2002) that show geochemical similarities to kimberlites. Eclogite xenoliths are abundant in some kimberlites (Table 1) and are described in detail in Section 2.05.3.

The megacryst suite is far more common and distinct in kimberlites than in Alkali Basalt Series xenoliths, the one exception being the Malaita megacryst suite (Nixon and Boyd, 1979). Most kimberlitic megacrysts are type X(a) and are believed to be the products of deep crystallization from magma (Nixon and Boyd, 1973b; Menzies, 1983; Harte, 1983). The relationship of this parental megacryst magma to the host kimberlite is debated but must be related in time and space. The mineral chemistry and equilibration conditions of the chromium-poor megacrysts are similar to those of high-temperature type V peridotites (Harte, 1983; Burgess and Harte, 1999). Megacrysts are typically 1, to a few centimeters in diameter. Some localities, such as Monastery, S. Africa, are renown for larger (10 cm and above) samples. Low-chromium garnet megacrysts have been reported from Malaita weighing up to 15 kg (Nixon, 1987).

An unusual variety of ultrahigh pressure xenoliths has been described from the Jagersfontein kimberlite (Haggerty and Sautter, 1990; Sautter *et al.*, 1991). These xenoliths (type XIII, Table 1) comprise four-phase lherzolite, discrete garnets, or iron-rich "lherzolites" with eclogitic affinities (Sautter *et al.*, 1991). Their key characteristic is that the garnets of all varieties contain rods of clinopyroxene that were apparently exsolved from a higher pressure, silicon-rich, majoritic garnet. Equilibration pressures of these samples are 4–5 GPa, but recombination of original garnet compositions before clinopyroxene exsolution gives pressure estimates of 13–15 GPa, equivalent to depths of derivation up to 500 km. Haggerty and Sautter (1990) suggest that similar xenoliths might also be present at other kimberlites such as Koidu, Sierra Leone.

2.05.2 PERIDOTITE XENOLITHS

2.05.2.1 Textures

Mantle xenoliths are metamorphic tectonites whose textures reflect the temperature (T), pressure (P) and differential stress conditions under which they were sampled accidentally by their host magma. The grain size, orientation and interrelationships amongst different minerals with varying strength in peridotites and eclogites will vary depending on the deformation history experienced by the sample, and are used to classify the texture of the rock according to set criteria. Nixon and Boyd (1973a) were some of the first workers to note the relationship between chemistry and texture in peridotite xenoliths. Various classification systems have been devised (e.g., Boullier and Nicolas, 1975; Harte *et al.*, 1975) but the terminology of Harte (1977) is used by most workers (Tables 1 and 2) and will be adopted here.

Xenoliths most commonly show coarse or porphyroclastic textures, or some continuum between the two. The former is defined by a grain size of greater than 2 mm (Table 2) and is commonly equigranular (Figure 1(a)). This texture represents a stable grain size developed under differential stresses that are small and constant over millions of years in the lithosphere. The grain size in olivine, the most abundant but weakest mineral in peridotite xenoliths, can be a measure

Table 2 Brief summary of textural classifications for xenoliths.

Texture	*Grain size*	*Grain boundaries*	*Related terminology*
(1) Coarse	Avg. >2.0 mm	Straight to smoothly curving	Equant, tabular, protogranular,[a] granular[a]
(2) Porphyroclastic (>10% olivine as porhyroclasts)	Two populations: larger p.clasts in finer grained matrix of neoblasts	Straight in neoblasts, irregular in p.clasts	Disrupted, fluidal, sheared[a]
(3) Mosaic porphyroclastic	As in (2)	As in (2)	Fluidal
(4) Granuloblastic	Avg. <2.0 mm	Straight or smooth, polygonal	Equant, tabular

[a] Other terms used in the literature to describe the same texture.

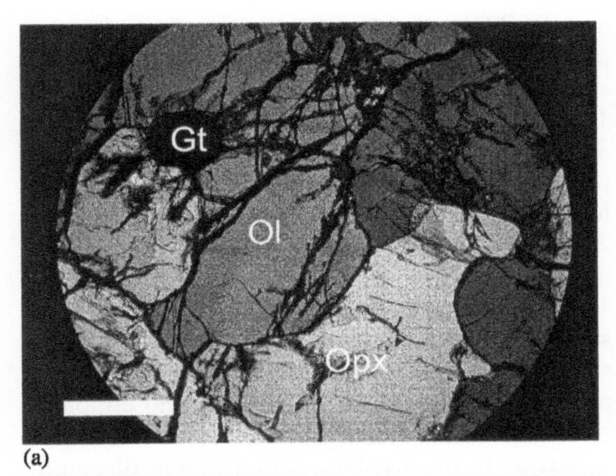

(a)

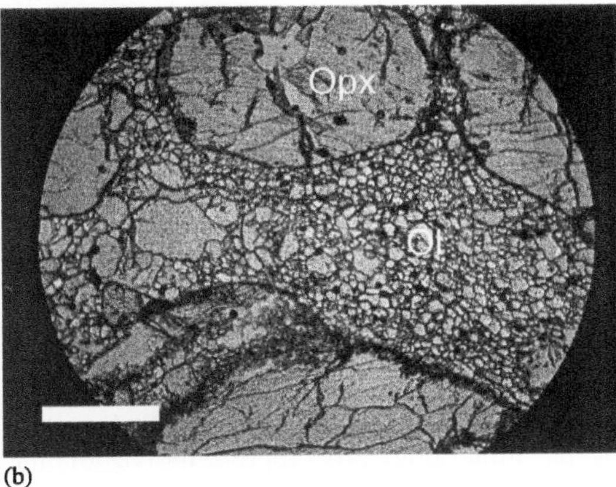

(b)

Figure 1 Common petrographic textures in peridotite xenoliths. Other textural details given in Table 1. (a) Coarse texture in cross-polarized light of garnet peridotite containing olivine (ol), orthopyroxene (opx), and garnet (gt). (b) Mosaic porphyroclastic texture in plane polarized light of garnet peridotite. Both samples are from the Torrie kimberlite, Slave Province (scale bar is 1 mm).

of the state of differential stress. This property of olivine has been calibrated by experiment (Ross *et al.*, 1980) and applied as a grain size "piezometer" to infer states of stress, viscosity, and strain rate in the mantle lithosphere as functions of depth or tectonic setting (Ave Lallement *et al.*, 1980; Mercier, 1980; Ross, 1983). Stress estimates based on this technique are open to interpretation (Twiss, 1977) and may not be entirely representative because the piezometer has been calibrated mainly for aggregates consisting only of olivine, and not multiphase samples.

At higher differential stresses, coarse textures evolve into porphyroclastic, which are defined by the development of smaller neoblasts of olivine amongst stronger pyroxenes, garnet, or spinel (Figure 1(b)). At much higher differential stress or in the presence of melt, fluidal porphyroclastic textures develop, exhibited most classically in garnet peridotite PHN1611. The significance of the latter texture in kimberlite-hosted garnet peridotite xenoliths has been of much interest since its original recognition by Nixon and Boyd (1973a). Proposals for the generation of such textures include shear heating or deformation during related diapiric upwelling. Whatever the cause, porphyroclastic textures cannot be retained for long periods at the high temperatures attendant in the mantle lithosphere, due to the rate at which olivine grains recrystallize and coarsen (Goetze, 1975). Porphyroclastic textures are thus a transient phenomena experienced by samples on a short timescale before entrainment in their host magma, as is also indicated by zoning patterns in minerals from these rocks (Smith and Boyd, 1992).

Gradations between coarse and porphyroclastic textures are observed in some xenolith populations and represent how this recrystallization process can be arrested during sampling by the host magma. In many suites, the change from coarse to porphyroclastic texture occurs as a function of depth, with porphyroclastic textures generally occurring at high temperatures and depths of sampling, although the opposite trend is observed in some suites (Boyd and Nixon, 1978; Franz *et al.*, 1996a).

2.05.2.2 Modal Mineralogy

The modal mineralogy of xenoliths has been used to estimate primitive mantle compositions (Carter, 1970), to constrain the source region of basaltic magmas (Yoder, 1976) and the density of mantle lithosphere (Boyd and McAllister, 1976). Modes in xenoliths are derived by point counting of thin sections or by mass balancing whole-rock compositions with mineral chemical data for all phases present in the xenolith. The former technique is prone to larger uncertainty because standard thin sections (9 cm^2) are not representative of the coarse-grained textures in xenoliths. Larger thin sections can obviate this problem. The mass balance method is more quantitative but uncertainty can arise in coarse rocks if the sample is too small. In addition, minute intergranular secondary phases can contribute significantly to the bulk rock analysis of an element but are not accounted for by the mineral chemistry of the major phases (Boyd and Mertzman, 1987). The mean and median modal mineralogy of cratonic and noncratonic peridotites is summarized in Table 3.

By definition, peridotites contain greater than 40% olivine with lesser amounts of orthopyroxene and clinopyroxene. An aluminous phase, plagioclase, spinel, or garnet may be present depending on the pressure of equilibration and defines the "facies" from which the peridotite xenolith was sampled (Figure 2). Plagioclase-peridotites are generally rare in continental xenolith suites because this facies is stable only at depths shallower than the Moho (~35 km) in most sections of mantle lithosphere sampled by volcanic rocks. Spinel peridotite is very common in off-craton localities. Although garnet peridotites are most frequently studied in on-craton xenolith suites, spinel-facies peridotites can be common (Boyd *et al.*, 1999). In contrast, garnet peridotites are scarce in off-craton mantle xenolith localities. Transitions between the facies are encountered in some xenolith suites. Accessory amounts of ilmenite, amphibole, phlogopite, apatite, monazite, and even zircon are recognized in some xenolith suites as veins, texturally equilibrated grains or grain boundary phases. These grains are usually interpreted as due to chemical modification (metasomatism) of the original peridotite residue.

Xenoliths may also show a spatial association of minerals. Spinel–pyroxene clusters on centimeter-scales are recognized in some samples and are interpreted to reflect conversion of garnet peridotite through the garnet–spinel-facies transition, either by heating or decompression (Smith, 1977). In some cratonic xenoliths, a spatial association of garnet and pyroxenes has been recognized on scales of up to 8 cm^2 (Boyd and Mertzman, 1987; Cox *et al.*, 1987; Saltzer *et al.*, 2001). The spatial association was originally thought to be of exsolution origin (Cox *et al.*, 1987). At high temperatures along the mantle solidus, orthopyroxene is calcium- and aluminum-rich, and on cooling these components exsolve to produce spatially associated garnet and clinopyroxene (Canil, 1991, 1992). The exsolution origin, however, is inconsistent with isotopic disequilibria recorded between pyroxenes and garnet (Gunther and Jagoutz, 1997). Some xenoliths also contain too much clinopyroxene and garnet to account for by this process alone (Canil, 1992), or from a simple melt depletion perspective (Pearson *et al.*, 2002). This evidence argues for a late introduced component of these minerals, as is also evidenced by trace-element studies reviewed below (Shimizu, 1999; Simon *et al.*, 2003). The spatial associations in xenoliths emphasize the potential heterogeneity problem in the chemical analysis of coarser-grained xenoliths. The implication is that xenoliths with small sizes may not be mineralogically representative of the volume of mantle they sample.

Boyd (1989) compiled modes of major minerals in peridotite xenoliths from on and off-craton localities and compared the results with mantle residues from oceanic mantle represented by

Table 3 Mean and median modal mineralogy of xenoliths.

	ol	*opx*	*cpx*	*sp*	*gt*
Off-craton ($n = 98$)					
Average	63.7	21.5	11.7	1.8	9.6
Median	62.9	21.8	12.9	1.7	11.4
1σ	8.4	6.6	5.2	0.8	6.1
On-craton ($n = 210$)					
Average	71.9	20.8	3.3	1.1	6.2
Median	72.0	20.6	2.0	0.8	5.3
1σ	11.6	10.3	4.5	1.0	3.8

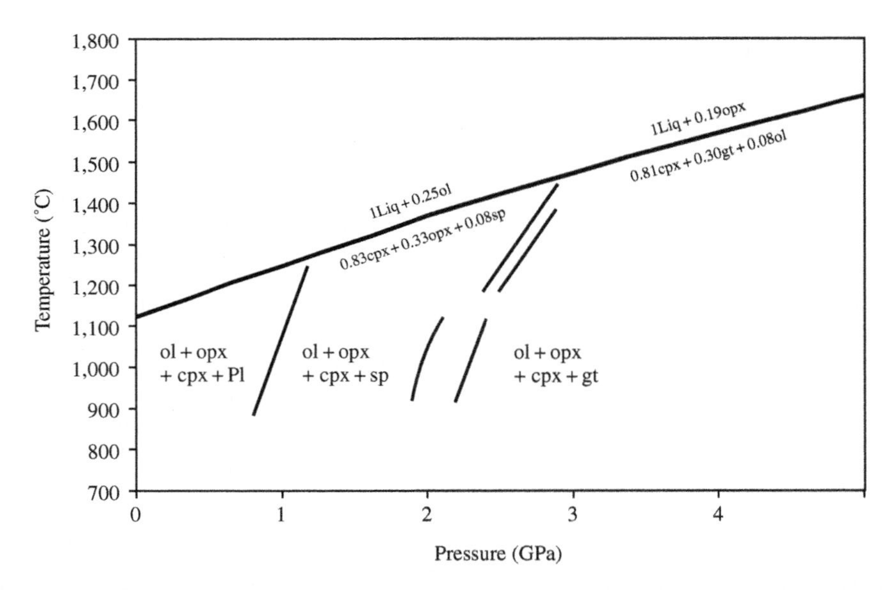

Figure 2 Pressure–temperature diagram showing solidus (after Hirschmann, 2000) and phase boundaries (after Koga *et al.*, 1999) defining the plagioclase-, spinel-, and garnet- facies of peridotite from various experimental studies (Green and Hibberson, 1970; Green and Ringwood, 1970; Nickel, 1986; O'Neill, 1980). Two different versions of the spinel–garnet phase boundary are shown. The stoichiometry of melting reactions on the solidus at different pressures is after Robinson *et al.* (1998) and Walter (1998).

ophiolites and abyssal peridotites. Boyd (1989) showed that in off-craton suites, ophiolites and abyssal peridotites, the amount of olivine shows a good correlation with the degree of depletion in the xenolith, defined by the Mg/(Mg + Fe) (i.e., Mg#) in olivine. In contrast, xenoliths from the Kaapvaal craton have far less olivine and far more orthopyroxene than expected from simple melt depletion (see Chapter 2.08). The latter attribute is reflected by the lower Mg/Si ratio of these rocks (Section 2.05.2.5).

The modal abundances of olivine, orthopyroxene, clinopyroxene and spinel observed for spinel peridotites from six well-characterized off-craton xenolith suites are plotted against a depletion index in Figures 3(a)–(c). The amount of olivine correlates negatively with degree of depletion, as expected because olivine is a product of the reaction that produces melt at the solidus (Figure 2) (see Chapter 2.08). The number of samples compiled ($n = 143$) may not be completely representative, but there is nonetheless a suspicious population gap at ~2 wt.% Al_2O_3. At higher levels of depletion, the rocks are harzburgites (<5% cpx), and the slope of the negative correlation between olivine and Al_2O_3 is steeper (Figure 3(a)), suggesting a change in melting reaction, as is observed in peridotite melting experiments (Baker and Stolper, 1994; Robinson *et al.*, 1998). All five xenolith suites have samples within this group (Figure 3(a)) and thus exhibit bimodality, as noted previously (Shi *et al.*, 1998) but reasons for its origin are unclear.

The trend for orthopyroxene with depletion is more constant, whereas clinopyroxene and spinel show good positive correlations (Figures 3(a)–(c)). In melting reactions on the solidus, clinopyroxene contributes the most to the melt phase (Figure 2) and thus is consumed most rapidly with depletion until it is exhausted after ~20% melting.

Differences between the modes observed in spinel peridotite xenoliths and those predicted by experiment may result because the former are not simple residues, and have had components introduced after their original formation.

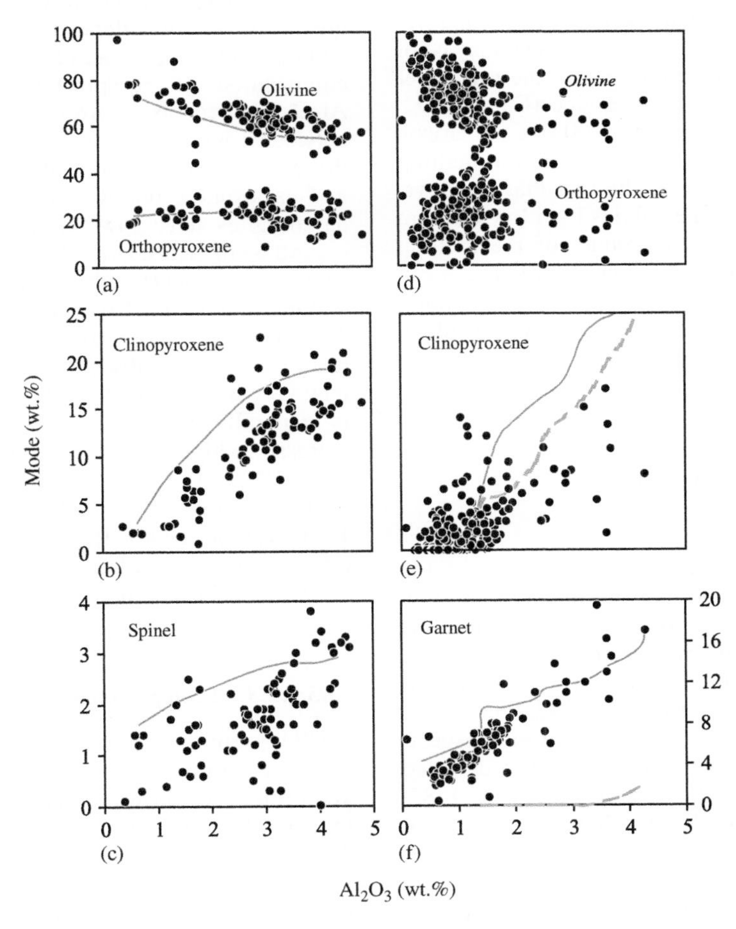

Figure 3 Modal abundances in (a–c) off-craton and (d–f) on-craton xenoliths as a function of depletion (indexed by wt.% Al_2O_3). Shown for illustrative purposes are calculated modal variations in residues as a function of melt fraction for melting at 1.5 GPa (a–c) and 3 GPa and 7 GPa (d–f) using empirical methods in Canil (2002). Data sources in Table 6.

More likely, however, is that modes measured experimentally at high T along the solidus cannot be directly compared with those observed in xenoliths that have cooled hundreds of degrees, changing their modal proportions slightly. Furthermore, the model melting trends and reactions vary in detail depending on the model source mineralogy or melting process (equilibrium versus fractional). Nonetheless, the trends observed in xenoliths generally correlate with those determined in experiment, reiterating the basic tenet that the modal proportions in peridotite xenoliths are due to a source–residue relationship with basaltic/komatiitic melt extraction (Carter, 1970; Walter, 1998; see also Chapter 2.08).

Figures 3(d)–(f) compares modes observed in four well-characterized on-craton xenolith suites ($n = 189$) with degree of depletion. When compared to the off-craton samples, trends are far more scattered for olivine and orthopyroxene, and a significant population of samples are orthopyroxene-rich, as originally remarked by Boyd (1989). The trend for garnet is remarkably regular and uniform, whereas many samples contain far more clinopyroxene than expected for their level of depletion. This excess clinopyroxene may be of exsolution origin, or introduced to the rock after its original formation as a residue (Canil, 1992; Shimizu, 1999; Simon *et al.*, 2003). Comparison with trends expected from peridotite melting models is complicated by the fact that orthopyroxene is replaced at the solidus by a low calcium clinopyroxene, and is a product of the melting reaction at $P > 3$ GPa (Walter, 1998). The mean and median modes of off-craton and on-craton xenoliths from Figure 3 are summarized in Table 2.

2.05.2.3 Mineral Chemistry

2.05.2.3.1 Olivine

Olivine is the main constituent of peridotite xenoliths and is a major host for magnesium, iron, and nickel. The Mg# of olivine reflects that of the whole rock which in turn is related to the degree of melt depletion or enrichment in iron. The Mg# of peridotitic olivine is between 88–92, and 91–94 for off- and on-craton mantle xenoliths, respectively, reflecting the generally more iron-depleted nature of the latter (Boyd and Mertzman, 1987; Boyd, 1997). Xenoliths that have been metasomatized by silicate melt, or that are the crystallization products of silicate melts are generally iron-enriched relative to depleted peridotite and hence have lower olivine Mg# (Table 1). Iron-rich olivines are also found in iron-rich lherzolite xenoliths from Hawaii, where they are interpreted as resulting from physically mixing iron-rich pyroxenites with normal peridotite followed by subsequent re-equilibration (Sen and Leeman, 1991). Carbonatite metasomatism may increase Mg# in secondary olivines due to introduction of magnesium (Hauri and Hart, 1993; Ionov *et al.*, 1993b). Typical analyses are listed in Tables 4 and 9.

2.05.2.3.2 Orthopyroxene

The Mg# of orthopyroxene is similar or slightly greater than that of olivine, due to a relative Fe–Mg partition coefficient (K_D) of ~1 that is independent of P and T (von Seckendorff and O'Neill, 1993). The calcium content of orthopyroxene also varies depending on the temperature of equilibration of the sample and its bulk composition. In both the spinel- and garnet-facies, the CaO content increases with T and varies between 0.2 wt.% and 2.0 wt.%. Orthopyroxenes with very low CaO occur in harzburgites and/or in low-T samples in many kimberlite-borne xenolith suites.

The Al_2O_3 content of orthopyroxene varies greatly in xenoliths depending on the T and P of equilibration of the sample, as well as its bulk composition. In the spinel peridotite facies, the Al_2O_3 content is controlled by T and varies between 1 wt.% and 6 wt.%. In the garnet-facies, the Al_2O_3 content is controlled by both P and T, and is usually below 2 wt.%. Very depleted harzburgites have low Al_2O_3 in orthopyroxene regardless of facies, as evident in samples from the Colorado Plateau (Smith *et al.*, 1999) and in many cratonic xenoliths (Boyd and Mertzman, 1987). The Cr_2O_3 of orthopyroxene is low (<0.6 wt.%); this element can substitute for aluminum and so may vary with T and P (Nickel, 1986). The Fe_2O_3 is generally very low (<0.15 wt.%) but can increase in high-T porphyroclastic xenoliths such as PHN1611 (>0.3 wt.%) and hence affect P–T estimates (Brey and Köhler, 1990; Carswell, 1991; Taylor, 1998). Representative analyses of orthopyroxene in different peridotite facies are shown in Tables 4 and 9.

2.05.2.3.3 Clinopyroxene

Clinopyroxene is a major host for sodium, calcium, chromium, and titanium in mantle xenoliths and shows extensive solid solution toward orthopyroxene and/or garnet at high P and T in the mantle (Boyd, 1969, 1970; Brey and Köhler, 1990). The Mg# of clinopyroxene is usually slightly greater than that of coexisting olivine, due to a K_D greater than 1. The calcium content of clinopyroxene is strongly T-dependent and is between 40 mol.% and 50 mol.% wollastonite component. Subcalcic clinopyroxenes

Table 4 Representative major element mineral chemical data for peridotite xenoliths from different facies (wt.%).

Mineral	*Sample #*	*Facies*	*SiO_2*	*TiO_2*	*Al_2O_3*	*Cr_2O_3*	*FeO*	*Fe_2O_3*	*NiO*	*MnO*	*MgO*	*CaO*	*Na_2O*	*Total*	*Location*	*Data source*
Plagioclase	RH07-217c	Plag-sp	48.64		33.71		0.03					16.28	2.19		Antarctica	Zipfel and Worner (1992)
Clinopyroxene	RH07-217c	Plag-sp	52.32	0.48	5.71	0.84	3.08			0.15	16.02	20.75	1.01	100.12	Antarctica	Zipfel and Worner (1992)
	Pali No. 2 35-H	Plag-sp	50.19	0.65	6.77	1.82	3.3			0.05	16.05	19.81	0.92	99.56	Hawaii	Sen (1988)
	Pali No. 2 36-H	Plag-sp	51.8	0.85	6.37	0.28	4.84				16.61	17.91	1.22	99.88	Hawaii	Sen (1988)
	2905	Sp	52.4	0.39	6.18	0.89	2.35	0.5		0.05	15.3	20.3	1.57		Noorat, Australia	Canil and O'Neill (1996)
	313-37	Sp-gt	52.21	0.45	6.04	1	2.9				15.43	19.12	1.87	99.02	Vitim	Ionov (1996)
	314-74	Sp-gt	52.6	0.28	6.24	1.06	2.72				16.08	19.48	1.47	99.93	Vitim	Ionov (1996)
	313-1	gt	52.82	0.5	5.67	1.29	2.78				15.41	19.04	1.92	99.43	Vitim	Ionov (1996)
	313-3	gt	52.84	0.57	5.81	1.32	3				15.52	18.77	1.99	99.82	Vitim	Ionov (1996)
	FRB 1350	Sp-gt	54.9	0.03	2.04	0.9	1.5	0.23			17.3	22.9	1.23	101	Kaapvaal	Canil and O'Neill (1996)
	BD 1150	gt	54.3	0.18	2.62	1.74	1.97	1.03			16.2	19.7	1.74	99.5	Kaapvaal	Canil and O'Neill (1996)
	PHN 5267	gt	56.8		1.35	0.57	2.89	0.48			21.5	16.8	0.75	101.1	Kaapvaal	Canil and O'Neill (1996)
	PHN 1611	gt	55	0.34	2.46	0.53	5.43				20.9	14.2	1.52	99.7	Kaapvaal	Smith *et al.* (1991)
Orthopyroxene	RH07-217c	Plag-sp	57.32	0.02	3.06	0.31	6.32			0.15	34.64	0.51	0	102.33	Antarctica	Zipfel and Worner (1992)
	Pali No.2 35-H	Plag-sp	55.27	0.14	3.87	0.26	6.72			0.08	33.6	0.61	0.22	100.77	Hawaii	Sen (1988)
	Pali No.2 36-H	Plag-sp	55.5	0.12	2.93	0.15	7.9				33.4	0.2	0.1	100.3	Hawaii	Sen (1988)
	2905	Sp	56	0.05	4.18	0.18	5.9	0.42		0.1	34	0.55	0.1		Noorat, Australia	Canil and O'Neill (1996)
	313-37	Sp-gt	55.53	0.14	3.96	0.49	5.92		0.11		32.51	0.8	0.17	99.63	Vitim	Ionov (1996)
	314-74	Sp-gt	55.61	0.08	4.44	0.52	5.54		0.06		33.21	0.75	0.12	100.33	Vitim	Ionov (1996)

(continued)

Table 4 (continued).

Mineral	*Sample #*	*Facies*	*SiO_2*	*TiO_2*	*Al_2O_3*	*Cr_2O_3*	*FeO*	*Fe_2O_3*	*NiO*	*MnO*	*MgO*	*CaO*	*Na_2O*	*Total*	*Location*	*Data source*
	313-1	gt	55.41	0.19	3.85	0.55	5.88		0.09		32.94	0.75	0.14	99.8	Vitim	Ionov (1996)
	313-3	gt	55.16	0.17	3.73	0.53	5.98		0.03		32.94	0.76	0.11	99.41	Vitim	Ionov (1996)
	FRB 1350	Sp-gt	58	0.03	1.13	0.22	4.89	0.35			36.5	0.27		101.3	Kaapvaal	Canil and O'Neill (1996)
	BD 1150	gt	56.9	0.07	0.72	0.2	5.08	0.63			35.2	0.75	0.1	99.7	Kaapvaal	Canil and O'Neill (1996)
	PHN 5267	gt	56.4		0.91	0.24	4.53	0.44			35	1.38	0.16	99.1	Kaapvaal	Canil and O'Neill (1996)
	PHN 1611	gt	56.1	0.23	1.36	0.22	6.71				32.5	1.59	0.33	99	Kaapvaal	Smith *et al.* (1991)
Spinel	RH07-217c	Plag-sp		0.01	56.86	13.34	10.44			0.1	20.13			100.91	Antarctica	Zipfel and Worner (1992)
	Pali No.2 35-H	Plag-sp		0.2	39.76	25.97	16.96			0.11	17.04	0.78		100.82	Hawaii	Sen (1988)
	Pali No.2 36-H	Plag-sp	0.08	0.62	53.1	8.6	15.6			0.06	19.03	0.06		97.16	Hawaii	Sen (1988)
	2905	Sp		0.5	57.4	11.5	9.1	1.79			20.3					Canil and O'Neill (1996)
	313-37	Sp-gt	0.13	0.43	43.47	22.37	13.42		0.22		18.55			98.59	Vitim	Ionov (1996)
	314-74	Sp-gt	0.12	0.27	47.61	18.25	12.66		0.27		20.12			99.3	Vitim	Ionov (1996)
	313-3	gt	0.11	0.45	44.79	23.05	13.6		0.18		18.92			101.1	Vitim	Ionov (1996)
	FRB 1350	Sp-gt		0.1	19.3	49.4	14.8	2.78			13.4			99.7	Kaapvaal	Canil and O'Neill (1996)
Garnet	313-37	Sp-gt	42.28	0.13	23.42	1.18	7.01			0.33	20.97	5.18		100.5	Vitim	Ionov (1996)
	314-74	Sp-gt	43.41	0.13	23.86	1.18	6.34			0.24	21.47	5.14		101.77	Vitim	Ionov (1996)
	313-1	gt	42.53	0.12	23.62	1.18	7.21			0.24	20.77	4.82		100.49	Vitim	Ionov (1996)
	313-3	gt	42.67	0.19	23.25	1.14	7.51			0.19	20.91	4.9		100.76	Vitim	Ionov (1996)
	FRB 1350	Sp-gt	42	0.03	22.1	2.07	7.95	0.31			20.1	5.43		100	Kaapvaal	Canil and O'Neill (1996)
	BD 1150	gt	41.6	0.02	21.3	2.64	8.09	0.81			19.9	4.54		98.9	Kaapvaal	Canil and O'Neill (1996)
	PHN 5267	gt	42.3	0.03	20.5	3.54	5.63	0.85			22.3	4.93	0.02	100.1	Kaapvaal	Canil and O'Neill (1996)
	PHN 1611	gt	42.6	0.67	21.1	2.22	8.65			0.22	21.1	4.63	0.06	101.3	Kaapvaal	Smith *et al.* (1991)

(Wo <35%) occur in cratonic suites as megacrysts or discrete nodules, and indicate very high *T* of equilibration, perhaps in equilibrium with melt (Boyd, 1969, 1970; Boyd and Nixon, 1978).

The Al_2O_3 content of clinopyroxene varies between 1 wt.% and 7 wt.% and shows a positive correlation with both *T* of equilibration and the bulk composition. Up to 3 wt.% Cr_2O_3 is observed in xenolith clinopyroxenes. The molar Cr/Al in clinopyroxene is nearly identical to that of the bulk rock and correlates positively with the level of depletion. Up to 35% of all iron in clinopyroxene is Fe^{3+} and the partition of this cation between clinopyroxene and garnet may vary with *T* or f_{O_2} (Luth and Canil, 1993; Canil and O'Neill, 1996).

In a given facies, the sodium and titanium content of clinopyroxene shows a negative correlation with depletion (or chromium content) and can increase substantially in metasomatized xenoliths. The substitution of sodium, chromium, and aluminum has a complex relationship with *T*, *P* and degree of depletion but in general, clinopyroxene from higher *P* samples contains more sodium. Representative analyses for different peridotite facies are shown in Table 4.

Many xenoliths can show two generations of clinopyroxene. Second generation clinopyroxene can occur as a minor phase on primary rims or along grain boundaries, and is usually related to infiltration of the host magma. When compared to primary clinopyroxene, secondary clinopyroxene is usually very high in calcium and low in sodium, aluminum, and/or Mg#. The contribution of secondary clinopyroxene to bulk rock sodium, calcium, or iron contents can be considerable (Boyd *et al.*, 1997).

Pyroxenites (as opposed to eclogites) commonly have sodium-rich aluminous augite as their clinopyroxene. Clinopyroxenes from MARID xenoliths are aluminum- and titanium-poor Mg-augites or diopsides (Dawson and Smith, 1977).

2.05.2.3.4 Spinel

The chemical systematics of spinel in peridotite xenoliths and xenocrysts have been reviewed at length (Arai, 1994; Roeder, 1994; Barnes and Roeder, 2001). The Cr/(Cr + Al) and Fe/Mg ratio of spinel in xenoliths shows a strong positive correlation (Figure 4). For a nearly constant mantle olivine composition, the Fe/Fe + Mg) of spinel increases with decreasing *T* of equilibration. The Cr/(Cr + Al) reflects both the degree of depletion of the bulk rock and also the *T* and *P* (and/or facies) of equilibration. Spinels crystallized from depleted xenoliths or at high *P* in the garnet-facies have high Cr/(Cr + Al), whereas spinel from less depleted xenoliths or at lower *T* or *P* in the spinel-facies

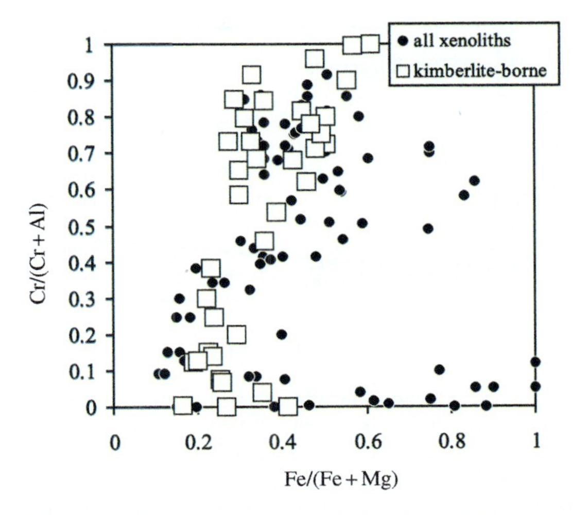

Figure 4 Covariation of Cr/(Cr + Al) and Fe/(Fe + Mg) in spinels from mantle xenoliths in kimberlites and other volcanic rocks. The data points are representative of the extremities of fields for a database of 2.5×10^4 analyses (source Barnes and Roeder, 2001).

have low Cr/(Cr + Al) (Roeder, 1994). The TiO_2 contents of mantle spinels are low (<0.5 wt.%) and correlate with level of depletion in the xenolith.

The Fe^{3+} content in xenolith spinels is large and unlike garnet or pyroxenes can be measured reliably from electron microprobe methods (Wood and Virgo, 1989) and is useful for oxygen barometry of spinel peridotites (Wood *et al.*, 1990; Ballhaus *et al.*, 1991). Representative analyses from different peridotite facies are shown in Table 3.

2.05.2.3.5 Garnet

The major-element systematics of mantle garnets were recently reviewed by Griffin *et al.* (1999a) using a database of over 12,000 analyses. A database of over 900 garnets is used in plots here for illustrative purposes. The Mg# of garnets from peridotite xenoliths ranges from 0.75 to 0.90 and is sensitive to *T*-dependent exchange with coexisting olivine, but these correlations can be obscured by bulk composition effects (Griffin *et al.*, 1999a). Typical analyses are given in Table 4.

The different trends of garnets on a CaO versus Cr_2O_3 plot (Figure 5) are useful for distinguishing the protolith of garnet xenocrysts (Gurney and Switzer, 1973; Sobolev *et al.*, 1973; Schulze, 1995) and garnets from spinel-bearing peridotite (Kopylova *et al.*, 2000). The positive correlation of CaO with Cr_2O_3 in peridotite garnets arises from a complex interplay of *P*, *T*, bulk composition and reciprocal solid solution effects between

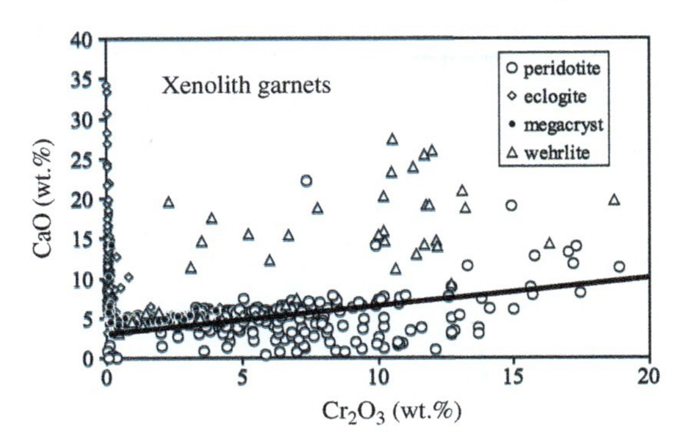

Figure 5 Covariation of CaO with Cr_2O_3 in garnets from a large database ($n = 900$) from a wide variety of xenoliths in kimberlites and other alkaline rocks. Note the positive correlation of calcium and chromium in peridotite garnets. Garnets below the line are "harzburgitic," whereas those above it are lherzolitic or wehrlitic.

Ca–Mg–Fe and Al–Cr end-member components in garnets (Wood and Nicholls, 1978; Nickel, 1986; Girnis and Brey, 1999). To a first order the Cr_2O_3 of peridotitic garnet is controlled by the level of depletion, as shown by a good negative correlation with bulk rock Al_2O_3 (Griffin *et al.*, 1999b), whereas the Ca/Cr ratio depends on $P-T$ of equilibration as well as the Cr/Al ratio of the bulk rock. The high Cr/Al observed in many peridotitic garnets requires a protolith with high Cr/Al, which cannot be achieved by melting in the garnet facies. Protoliths to these chromium-rich garnets must have formed in the spinel peridotite facies (Bulatov *et al.*, 1991; Canil and Wei, 1992; Stachel *et al.*, 1998). Subcalcic (harzburgitic) garnets are confined mainly to lithosphere sampled beneath terrains older than 2.5 Ga and are commonly associated with the presence of diamond (Gurney and Switzer, 1973; Sobolev *et al.*, 1973; Pokhilenko *et al.*, 1993). These subcalcic and chromium-rich garnets are far less abundant in suites from 2.5 Ga to 1 Ga terrains and are absent from mantle beneath terrains less than <1 Ga (Griffin *et al.*, 1999a).

Mantle garnets have $Fe^{3+}/\sum Fe$ ratios of 0.02–0.15 (Luth *et al.*, 1990). The Fe^{3+} content in garnet correlates with T of equilibration and Fe_2O_3 of the bulk rock (Canil and O'Neill, 1996) and is also a useful oxygen barometer for garnet peridotites (Luth *et al.*, 1990; Gudmundsson and Wood, 1995).

2.05.2.3.6 Plagioclase

Plagioclase occurs only in peridotite xenoliths sampled from areas of high geothermal gradient and thinned crust, where the plagioclase-facies is stable and can be sampled from beneath the Moho (Figure 2). It may also occur in shallow lithosphere that has been impregnated by melts (Sen and Leeman, 1991). Plagioclase in peridotite xenoliths is typically calcic (An60–90) due to the low Na_2O levels of most peridotites.

2.05.2.3.7 Amphibole

Amphiboles in most noncratonic xenoliths are essentially (Ti,Cr)-pargasites or, less commonly, kaersutites (Ionov *et al.*, 1997). Those in cratonic xenoliths are typically K-richterites (in MARID/PKP rocks; Gregoire *et al.* (2002); see Table 1 for definitions) or pargasite (Winterburn *et al.*, 1990). Mg# in most cases is close to coexisting pyroxenes (89–91) but vein amphiboles can be lower (Ionov *et al.*, 1997). Mg# for K-richterites in MARID xenoliths is in the range 87.2–92.8. Titanium contents of disseminated amphibole are also generally lower than in vein amphibole. K_2O contents of vein pargasite (1–2%) are higher than disseminated crystals in non-cratonicxenoliths. K-richterites from MARIDS have significantly higher but homogenous K_2O (4.25–4.95 wt.%).

2.05.2.3.8 Mica

Mica in mantle xenoliths is usually phlogopite in composition (Carswell, 1975; Ionov *et al.*, 1997). Mg# usually mimics coexisting pyroxenes, generally being high. MARID phlogopites have Mg# from 87.3 to 88.4 and Al_2O_3 from 10.95 wt.% to 11.2 wt.% (Gregoire *et al.*, 2002). As with amphibole, vein phlogopite can have lower Mg# but higher titanium compared to that disseminated in a peridotite matrix. K_2O contents are generally between 8 wt.% to 10.5 wt.% for phlogopites from veins, metasomatized peridotites or MARID/PIC assemblages (Table 1).

2.05.2.4 Thermobarometry

Defining the distribution of temperature at depth within the Earth as a function of both space and time is the basis for understanding many geological processes (Verhoogen, 1956). Xenoliths have a very central place in this effort because they are pieces of lithosphere and chemical exchange amongst the minerals in these rocks record the physical conditions at the time of their entrainment. In this way xenoliths have become ground-truth information for remote sensing studies of the state of the lithosphere by geophysical techniques.

Thermobarometry is the estimation of the equilibrium temperatures and pressures (depths) recorded by mineral chemical equilibria.

Thermobarometry of mantle xenoliths has been a fruitful endeavour since the pioneering work of Boyd (1973) and serves many purposes:

(i) It provides an estimation of the depth interval and minimum thickness of the lithosphere sampled by a volcanic rock. Because xenoliths are accidentally sampled, not all depth intervals are necessarily represented in a xenolith population.

(ii) The pressures and temperatures provide some spatial context amongst samples within a xenolith population, which may perhaps be divisible according to their textures and/or compositions. This is the basis for the commonly used low (<1,100 °C) versus high (>1,100 °C) T classification of xenoliths from cratonic lithosphere (Boyd and Mertzman, 1987; Boyd, 1989). The depth estimates also provide some insight into the scale of heterogeneity in the textures or bulk chemistry in a xenolith population.

(iii) In many circumstances, xenolith suites may record a regular $P-T$ gradient which can be used to define paleogeothermal gradients in the lithosphere, which vary with tectonic setting and/or age of the crustal province, and provide some insight into the thermal state of the mantle in these environments.

A number of experimentally calibrated thermobarometers are available to estimate the P and T of polyminerallic mantle xenoliths, depending on the mineral assemblage present (Table 5). Brey (1990) estimates that for rocks equilibrated within the P and T range of experiments used to construct the Brey *et al.* (1990) thermobarometer combination, accuracy is ±60 °C and 0.45 GPa. Not all methods necessarily agree for all exchange equilibria amongst all minerals in a given sample. Geobarometry of peridotites has mainly been confined to garnet-facies samples, utilizing the P dependence of chromium or aluminum solubility in pyroxenes coexisting with garnet (Brey *et al.*, 1990; MacGregor, 1974; Nickel and Green, 1985; Nimis and Taylor, 2000). In general, there are few geobarometers available for spinel-facies rocks. The exchange of calcium between olivine and clinopyroxene is P-dependent but is also highly T-sensitive and so late stage heating in xenoliths can be a problem for this method. Currently this geobarometer is the only quantitative method for spinel-facies rocks, but its application can also be analytically challenging. Accurate results, however, are obtainable with careful work (Köhler and Brey, 1990).

The accuracy of xenolith thermobarometry has been critically evaluated several times (Finnerty, 1989; Carswell, 1991), most recently by Smith (1999). Smith (1999) finds that most thermobarometers agree in the range between 900 °C and 1,100 °C and 2–5 GPa. Deviations between thermobarometers below and above that T range partly relate to shortcomings in the experimental calibrations for some exchange equilibria, to the lack of constraints on Fe^{3+} contents of minerals, or to the variable kinetic responses for different cations during chemical exchange amongst minerals (e.g., aluminum exchange versus Fe–Mg exchange). The latter problem can be exploited to unravel the thermal history of individual samples or sections of mantle sampled (Smith *et al.*, 1991; Smith and Boyd, 1992; Franz *et al.*, 1996a,b) but relies on an experimental database for diffusion that is not particularly well-constrained for some cations in mantle minerals (aluminum, chromium). Some limits can be placed on these quantities using

Table 5 Some thermobarometers applicable to xenoliths.[a]

Method[b]	*Assemblage*	*Equilibria*	*Comments/caveats*	*Accuracy*[c]
Barometers				
P_{BKN}	gt–opx	Al exchange	gt-facies only	0.5 GPa
P_{NT}	cpx–gt	Cr/Al exchange	gt-facies only	0.3 GPa
P_{KB}	ol–cpx	Ca exchange	Very T sensitive, analytically challenging	0.7
Thermometers				
T_{WS}	opx–sp	Al exchange	sp-facies only	
T_{Ball}	ol–sp	Fe–Mg exchange	Very T sensitive, fast response	30 °C
T_{BKN}	opx–cpx	Ca–Mg exchange	Insensitive to Fe^{3+}	60 °C
T_{TA}	opx–cpx	Ca–Mg exchange	Insensitive to Fe^{3+}	62 °C
T_{Caopx}	opx	Ca–Mg exchange	Insensitive to Fe^{3+}	60 °C
T_{Kr}	cpx–gt	Fe–Mg exchange	Fe^{3+} sensitive[d]	100 °C
T_{OW}	ol–gt	Fe–Mg exchange	Fe^{3+} sensitive[d]	180 °C
T_{HA}	opx–gt	Fe–Mg exchange	Fe^{3+} sensitive[d]	92 °C
T_{LG}	opx–gt	Fe–Mg exchange	Fe^{3+} sensitive[d]	96 °C

[a] Only the most commonly used are listed. For a more extensive list see Taylor (1998) or Brey *et al.* (1990). [b] P_{BKN}, T_{BKN}, T_{Caopx}: Brey and Köhler (1990); P_{KB}: Kohler and Brey (1990); P_{NT}: Nimis and Taylor (2000); T_{WS}: Witt-Eickschen and Seck (1991); T_{Ball}: Ballhaus *et al.* (1991); T_{TA}: Taylor (1998); T_{Kr}: Krogh (1988); T_{OW}: O'Neill and Wood (1979); T_{HA}: Harley (1984); T_{LG}: Lee and Ganguly (1988). [c] The 2σ accuracies are based on ability to reproduce experimental data (Brey and Köhler, 1990; Taylor, 1998). [d] The sensitivity to Fe^{3+} content of one or both phases is reviewed by Canil and O'Neill (1996).

zoning patterns in natural xenolith minerals (Smith and Barron, 1991).

The number of extant $P-T$ estimates for mantle samples is enormous. The pressures and temperatures recorded by garnet peridotite xenoliths compiled and reviewed recently by Rudnick and Nyblade (1999) are shown with additional data in Figure 6. Many of these suites have regular $P-T$ arrays that are reasonably assumed to record the steady-state paleogeothermal gradient at the time of eruption. The latter information has been central to studies of heat flow in the lithosphere and the distribution of heat producing elements in the crust and mantle (Rudnick and Nyblade, 1999; Russell *et al.*, 2001; Russell and Kopylova, 1999), which in turn constrain models for the bulk composition of the continental crust (Rudnick *et al.*, 1998).

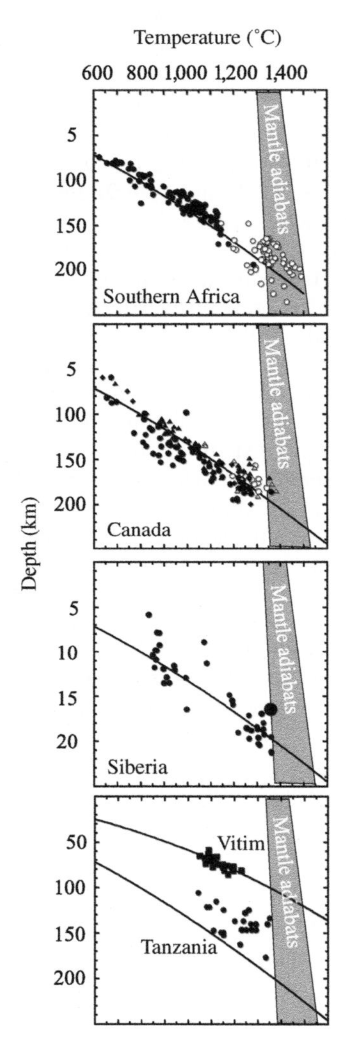

Figure 6 $P-T$ arrays compiled for garnet peridotite xenoliths from several suites using two-pyroxene thermometry and Al-in-orthopyroxene barometry (T_{BKN} and P_{BKN} methods, Table 5). Data sources given in Rudnick and Nyblade with additional data here for Vitim (Ionov *et al.*, 1993a) and Canada (MacKenzie and Canil, 1999; Schmidberger and Francis, 1999). The best-fit line for the Kaapvaal data is plotted in each figure for reference. Intersection of $P-T$ array with mantle adiabats (shaded field) represents an estimate of the thickness of lithosphere at the time of sampling.

The samples shown in Figure 6 reveal a number of features about the lithosphere sampled by xenoliths. The $P-T$ arrays, when projected to a mantle adiabat, can be used to define the maximum depth of mantle lithosphere beneath a locality at the time of eruption. Beneath cratons the lithosphere is at least 220 km deep, whereas below more active tectonic areas (e.g., Vitim) it is much thinner. Paleogeothermal gradients also vary between some cratons, or even within the same craton, as evident in the spread in some data sets (e.g., the Slave samples). Irregularities are evident in some $P-T$ arrays. Some of this variation may be due to lack of mineral equilibrium, perhaps because of recent diopside growth (Brey, 1990; Simon *et al.*, 2003). In other instances the $P-T$ variations signal a transient state of heat advection in the mantle shortly before the sampling process. For example, in many localities, the porphyroclastic xenoliths (open symbols in Figure 4) are displaced to higher temperatures away from the $P-T$ gradient for coarse samples. This is interpreted to be related to heating attended by melt infiltration and recrystallization as recorded by zoning and textural studies of these samples (Smith *et al.*, 1991; Smith and Boyd, 1992; Franz *et al.*, 1996a,b).

2.05.2.5 Bulk Rock Chemistry

The mantle comprises 68% by mass of Earth, and an accurate estimate for its composition is the very basis for unraveling the origin and differentiation of our planet. The bulk chemical analysis of xenoliths has been central to understanding the composition of the Earth's mantle, the genesis of basalt and the physical properties in the lithosphere that bear on its stability in the rigid part of the mantle system.

In view of the inhomogeneities observed for peridotite outcrops in ophiolites and orogenic massifs (Dick and Sinton, 1979; Chapter 2.04), there is some uncertainty as to how representative the bulk analysis of a centimeter-scale sample is of the volume of mantle sampled by a host magma. Furthermore, most peridotite xenoliths are coarse-textured and some contain spatial associations on the centimeter-scale (see above). Compilations discussed or presented here make no distinction for sample size but larger samples (>500 g) are likely to be more representative (Boyd and Mertzman, 1987; Cox *et al.*, 1987).

2.05.2.5.1 Major elements

The database for major-element analyses of mantle xenoliths numbers in the hundreds and has been used in several estimates of average mantle compositions or of the primitive upper mantle (Kuno and Aoki, 1970; Maaloe and Aoki, 1977; Jagoutz *et al.*, 1979; Sun, 1982; Palme and Nickel, 1985; Hart and Zindler, 1986; McDonough and Sun, 1995). Differences in the bulk chemistry of xenoliths as functions of the tectonic environment or tectonothermal age of the crust through which xenolith-bearing magmas have sampled are highlighted in more recent compilations (Boyd, 1989, 1997; Griffin *et al.*, 1999b).

The covariation of Mg/Si and Al/Si for a large database of xenoliths (Table 6) is plotted in Figure 7 and compared with compositions of chondritic meteorites. The xenoliths form an array with negative slope. Melting reactions at low P (<2.5 GPa) along the peridotite solidus produce olivine and liquid at the expense of pyroxenes and spinel (Figure 2). Residues become enriched in magnesium and depleted in silicon and aluminum, explaining the array of compositions in Figure 7 as a partial melting trend to form lithosphere. Scatter in the trend may be due to spatial inhomogeneities, perhaps due to the sample size problem, or to chemical modification (metasomatism) that postdates the original partial melting and depletion process. The chondrite trend shows an opposite slope in Figure 7, which may represent variable compositions in the solar nebula (Jagoutz *et al.*, 1979; Palme and Nickel, 1985). The intersection of the peridotite xenolith and chondrite arrays is interpreted to be "primitive upper mantle" (PUM), or the bulk composition of the silicate earth. This approach, together with studies of other refractory elements known to be in chondritic proportions, has been used to derive some of the various estimates for PUM shown in Table 7 (see Chapter 2.01).

The covariation of Mg/Si with Mg# in a suite of xenoliths from off-craton localities is shown in Figure 8. The negative correlation on this plot parallels that observed for residues from ocean basins and ophiolites and defines the "oceanic" trend of Boyd (1989). With increasing depletion away from "model" primitive mantle compositions (Table 7) more iron is extracted in the residue relative to magnesium and the olivine content of the rock increases, reflected by higher Mg/Si. In the large data set of xenoliths, there is significant scatter in this trend, compared to a much more coherent one which would be observed for orogenic massif or abyssal peridotites (Chapter 2.04). Some of the scatter is due to a subparallel alignment of individual suites on this plot, or to metasomatic effects within each suite, where depleted rocks have had iron added to the rock, shifting it below the oceanic trend (Boyd, 1997). The term "oceanic" for the compositional array of noncratonic xenoliths in this plot may be somewhat misleading, because it is not clear if most of the subcontinental lithosphere originated in an oceanic environment (Griffin *et al.*, 1999b). Noncratonic lithospheric mantle is not as depleted in general and shows different depletion trends in detail than do abyssal peridotites, as will be shown below.

The covariation of Mg/Si and Mg# for xenoliths from six Archean cratons plotted in Figure 8 is quite distinguishable from most of the noncratonic samples by having much higher (Mg#). Despite their higher Mg#, many cratonic xenoliths have low FeO and Mg/Si irrespective of whether they crystallized in the garnet- or spinel-facies. This is especially so for samples from the Kaapvaal and Siberian cratons, but other cratons also have samples within this field. Boyd (1989) and Boyd *et al.* (1997) recognized that cratonic peridotites from the Kaapvaal and Siberian cratons defined a "cratonic" trend. The low Mg/Si in the bulk rock analysis is reflected petrographically by their high modal orthopyroxene contents (compare Figure 3). The origin of the high degrees of depletion but low Mg/Si in cratonic peridotites has been much debated (Boyd, 1989; Herzberg, 1999; Kelemen *et al.*, 1998; Walter, 1999; see Chapter 2.08).

Both calcium and aluminum are refractory lithophile elements during condensation processes in the solar nebula and show relatively constant element-ratios in all classes of chondritic meteorites (Palme and Nickel, 1985). Chondritic meteorites are the presumed starting material from which the mantle is derived (Ringwood, 1966) and so it is surprising that most peridotite xenoliths do not have chondritic Ca/Al ratios, irrespective of their degree of depletion (Figure 9). Some samples have near chondritic ratios but the mean and median for mantle xenoliths are clearly above that (Table 7). In contrast, samples of cratonic mantle have a generally lower than chondritic Ca/Al (Figure 9).

The superchondritic Ca/Al ratio in the upper mantle might be balanced by a slightly differing lower mantle composition, if the latter region accumulated majorite or perovskite during a magma ocean early in Earth evolution (Palme and Nickel, 1985). There is now little evidence to support this suggestion (McDonough and Sun, 1995). Hart and Zindler (1986) interpreted the high Ca/Al of xenoliths as a sampling bias, with preferential sampling and analysis of clinopyroxene-rich (and thus more calcium-rich) xenoliths by most investigators due to their attractive

Table 6 Data sources for major and minor element compositions of xenoliths.

Location	*Authors*
Ocean island	
Samoa	Hauri *et al.* (1993)
Kerguelen	Gregoire *et al.* (2000)
Hawaii	Sen and Leeman (1991)
Oceanic arc	
Lihir, Papua New Guinea	McInnes *et al.* (2001)
Batan, Phillipines	Maury *et al.* (1992)
Bismarck, Papua New Guinea	Franz *et al.* (2002)
Continental rift	
Tariat, Mongolia	Press *et al.* (1986)
East African Rift, Ethiopia	Bedini *et al.* (1997)
Vitim, Russia	Ionov *et al.* (1993a)
Continental intraplate	
Grand Canyon, USA	Smith *et al.* (1999)
Green Knobs, USA	Smith and Levy (1976)
Green Knobs, USA	Aoki (1981)
Bandera, USA	Smith *et al.* (1999)
Puerco Necks, USA	Smith *et al.* (1999)
Thumb, USA	Ehrenberg (1982)
Various locations	Jaguotz *et al.* (1979)
E. Australia	Stolz and Davies (1991)
W. Hungary	Embey-Isztin *et al.* (1989)
Romania	Vaselli *et al.* (1995)
Hannuoba, China	Song and Frey (1999)
Italy	Morten (1987)
N. Africa	Dautria *et al.* (1992)
N. Africa	Dupuy *et al.* (1985)
Sardinia	Dupuy *et al.* (1987)
S.E. Australia	Frey and Green (1974)
S.E. Australia	Yaxley *et al.* (1992)
Yukon, Canada	Francis (1987)
N. British Columbia, Canada	Shi *et al.* (1998)
West Kettle River, Canada	Xue *et al.* (1990)
Massif Central, France	Zangana *et al.* (1999)
Massif Central, France	Zangana *et al.* (1999)
Sahara basin, N. Africa	Dautria *et al.* (1992)
Continental arc	
W. Mexico	Luhr and Aranda-Gomez (1997)
Patagonia	Laurora *et al.* (2001)
Cratonic	
Labait, Tanzania	Lee and Rudnick (1999)
Olmani, Tanzania	Rudnick *et al.* (1994)
Lashaine, Tanzania	Rudnick *et al.* (1993)
Wiedemann, E. Greenland	Bernstein *et al.* (1998)
W. Australia	Jaques *et al.* (1990)
Jericho, Canada	Kopylova and Russell (2000)
Nikos, Canada	Schmidberger and Francis (2000)
Jagersfontein, S, Africa	Winterburn *et al.* (1990)
Kaapvaal, S.Africa	Boyd and Mertzman (1987)
Udachnaya, Siberia	Boyd *et al.* (1997)
Kimberley, S. Africa	Boyd *et al.* (1993)
Premier. S. Africa	Boyd (1999)

appearance in hand sample. This may be part of the problem, but calcium and aluminum are not depleted equably during earlier stages of partial melting (McDonough, 1990). More aluminum is extracted than calcium, and the ratio becomes greater than chondritic with increasing depletion. Furthermore, petrographic or geochemical evidence of metasomatism by carbonatitic melts in some mantle xenoliths show that infiltration of ephemeral, low-melt-fraction carbonatitic melts can greatly elevate Ca/Al in mantle lithosphere (Rudnick *et al.*, 1993; Yaxley *et al.*, 1998).

Neither of these proposals, however, can explain the subchondritic Ca/Al of most cratonic peridotites. If the latter are residues at higher P, more aluminum remains in the residue relative to calcium, the opposite trend of lower P residues. The anomalously low Ca/Al of cratonic peridotites may be unique to the formation of subcratonic mantle, or is a secondary feature with unexplained origin.

Al_2O_3 is a convenient depletion index in peridotites because it is immobile during serpentinization, marine weathering and metamorphic processes that ensue during the exhumation of these rocks (Coleman and Keith, 1971; Snow and Dick, 1995). It thus permits comparison of samples from many different settings in variable states of preservation. Figure 10 shows the mean and median Al_2O_3 content of xenoliths subdivided by sample type or tectonic setting (Table 8). These data are compared with mantle peridotites from orogenic massifs and from known tectonic settings in the ocean basins (abyssal peridotites, pre-oceanic margins and forearcs).

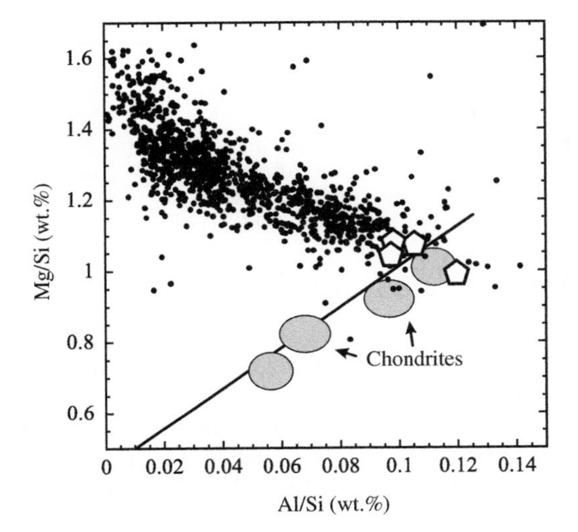

Figure 7 Mg/Si versus Al/Si plot (wt.%) for 593 xenoliths compared with the compositions of chondrites. Data sources in Table 6. Also plotted are various estimates for primitive upper mantle (polygons from Table 7) (source Jagoutz *et al.*, 1979).

On average, the mantle sampled in noncratonic xenolith suites in continental intraplate regions and in young rifts shows a similar limited level of depletion, relative to PUM. Similar, small levels of depletion are also observed for pre-oceanic rifts and orogenic massif peridotites. In contrast, the mean compositions of cratonic, oceanic arc and oceanic island suites are quite depleted. Similar levels of depletion are observed in oceanic forearcs and abyssal peridotites. The 1σ error bars shown in Figure 10 signify considerable overlap between these groups. Unfortunately, there are few published data for ocean island, or continental and oceanic arc suites in the literature (Table 5) to augment the global trends observed in Figure 10.

The Mg# is also another measure of depletion, but is more susceptible to modification by secondary metasomatic processes in the mantle, or to alteration by serpentinization and/or marine weathering. Olivine Mg# is sometimes used to avoid problems of grain-boundary

Table 7 Major and minor elements in xenolith averages and primitive upper mantle (PUM) models.

Sample	*HZ86 PUM*	*J79 PUM*	*PN85 PUM*	*M90 Xeno. avg.*	*M95 PUM*
SiO_2	45.96	45.14	46.20	44.00	44.92
Al_2O_3	4.06	3.97	4.75	2.27	4.44
FeO	7.54	7.82	7.70	8.43	8.05
MgO	37.78	38.30	35.50	41.40	37.80
CaO	3.21	3.50	4.36	2.15	3.54
Na_2O	0.33	0.33	0.40	0.24	0.36
K_2O	0.03	0.03		0.05	0.29
Cr_2O_3	0.47	0.46	0.43	0.39	0.38
MnO	0.13	0.14	0.13	0.14	0.14
TiO_2	0.181	0.217	0.230	0.090	0.201
NiO	0.28	0.27	0.23	0.28	0.25
CoO	0.013	0.013	0.012	0.014	
P_2O_5	0.02			0.06	0.02
Mg/Si	1.06	1.09	0.99	1.21	1.09
Al/Si	0.10	0.10	0.12	0.06	0.11
Mg#	0.90	0.90	0.89	0.90	0.89
Sc (ppm)		17	19	12	16
V (ppm)		77		56	82
Ga (ppm)		3		2	4
References	1	2	3	4	5

References: 1. Hart and Zindler (1986); 2. Jagoutz *et al.* (1979); 3. Palme and Nickel (1985); 4. McDonough (1990); 5. McDonough and Sun (1995).

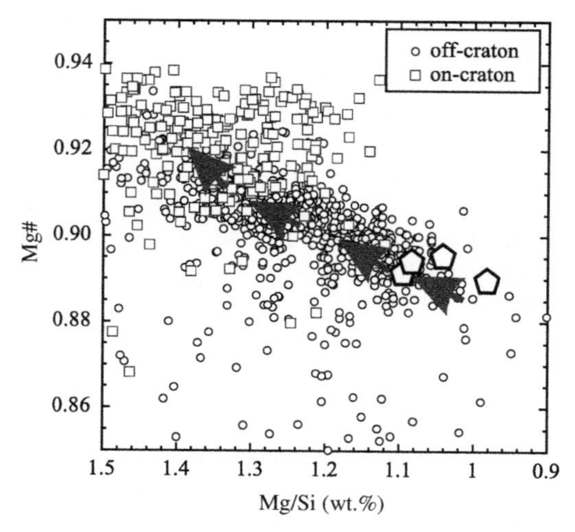

Figure 8 Mg/(Mg + Fe) versus Mg/Si for: (a) off-craton spinel peridotite xenoliths and (b) cratonic garnet and spinel peridotite xenoliths. Arrows mark the "oceanic" trend (Boyd, 1989, 1997) defined by abyssal peridotites. Also shown are various estimates for primitive upper mantle (polygons from Table 7).

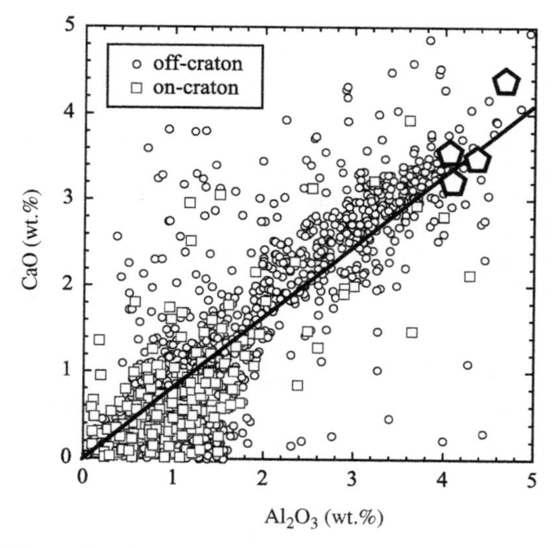

Figure 9 Covariation of CaO and Al_2O_3 in off-craton and on-craton xenoliths. Data sources in Table 6. Shown for reference are various estimates for primitive upper mantle (polygons) and the ratio in chondritic meteorites.

iron-introduction (Boyd, 1989). A summary of the Mg# for xenoliths in different settings shows the same trends as when Al_2O_3 is used as a depletion index (Figure 11). The larger range in Mg# observed for ocean island xenolith suites when compared to other settings is due to secondary melt impregnation in this setting (Boyd, 1997). This plot also emphasizes the depletion of cratonic peridotite xenoliths in FeO.

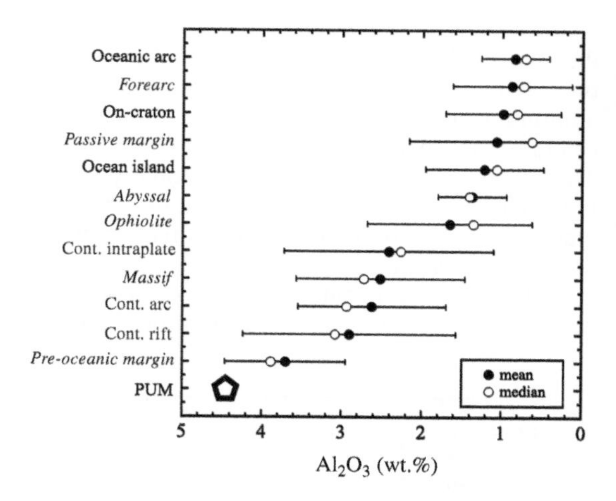

Figure 10 Mean and median Al_2O_3 (wt.%) for peridotite xenoliths from different settings, compared with samples from known modern tectonic settings (in italics; Table 8).

2.05.2.5.2 Minor and trace elements

Compatible trace elements. Both nickel and cobalt in xenoliths show good correlations with depletion parameters such as Al_2O_3 or MgO (Figures 12(a) and (b)). Both elements partition into olivine, which is enriched in the residue during melting and so these elements have bulk distribution coefficients greater than one during partial melting. The Ni/Co ratio in peridotites is nearly constant and approximately chondritic for xenoliths over a range of depletion. This is one of the strongest arguments for heterogeneous accretion of the Earth (Jagoutz *et al.*, 1979). Because nickel is highly siderophile but cobalt only slightly so, a near chondritic Ni/Co ratio requires a late "veneer" be added to Earth's mantle shortly after accretion (O'Neill, 1991). Alternatively, nickel may become less siderophile at extreme T and P during homogeneous accretion (see Chapter 2.10 and references therein).

Unlike nickel and cobalt, chromium shows a nearly constant level of depletion in xenoliths (Figure 12(c)). There is considerable scatter in mantle data sets for chromium that is beyond that attributable to analytical error and much greater than the scatter in nickel or cobalt (Liang and Elthon, 1990). This likely reflects the heterogeneous distribution of chromium in xenoliths, which is contained mainly in a very minor phase (spinel). Small xenolith specimens may over- or under-represent this mineral in peridotite bulk analyses (Liang and Elthon, 1990). The nearly constant chromium levels with varying depletion suggest a bulk distribution coefficient that is ~1 during melting, very different from the bulk distribution coefficient (D) for chromium during crystallization of mafic magmas that are the partial melt complement to the residues represented by

Table 8 Mean and median whole rock compositions of xenoliths.

Sample type	*SiO_2*	*TiO_2*	*Al_2O_3*	*FeO^**	*MnO*	*MgO*	*CaO*	*LOI*	*Al/Si* (wt.%)	*Mg/Si* (wt.%)	*Ca/Al* (wt.%)	*Fe/Al* (wt.%)	*Mg#*	*Ni*	*Co*	*Cr*	*Sc*	*V*	*Cr/Al*	*V/Al*	*n*
Ocean islands																					
Mean	43.72	0.08	1.22	9.27	0.13	44.25	1.10		0.03	1.31	1.36	14.13	0.895	2,410	153	2,794	8	40	0.494	0.0063	16
Median	44.61	0.04	1.07	8.04	0.12	44.73	0.82		0.03	1.30	1.10	10.94	0.908	2,399	141	2,601	9	32	0.433	0.0064	
σ	2.01	0.08	0.74	2.05	0.04	2.34	0.75		0.02	0.11	0.92	9.50	0.022	187	26	806	2	16	0.379	0.0019	
Oceanic arc																					
Mean	44.49	0.02	0.84	8.33	0.14	44.61	0.89	2.67	0.02	1.30	1.75	20.66	0.905	2,262	149	3,326	13	38	0.627	0.0056	21
Median	44.74	0.02	0.71	8.10	0.14	44.66	0.89	2.11	0.02	1.29	1.55	14.97	0.906	2,404	139	2,998	12	40	0.660	0.0000	
σ	1.77	0.02	0.43	0.77	0.01	1.87	0.48	0.90	0.01	0.09	1.32	24.83	0.007	298	33	1,021	5	11	0.677	0.0071	
Continental rifts																					
Mean	44.58	0.10	2.91	7.91	0.13	41.17	2.79	0.13	0.07	1.19	1.28	5.03	0.902	2,152		2,749	11	54	0.246	0.0028	23
Median	44.43	0.11	3.08	7.88	0.13	41.21	2.62	0.13	0.08	1.19	1.17	3.40	0.903	2,160		2,703	10	51	0.159	0.0018	
σ	0.87	0.07	1.33	0.42	0.01	3.43	1.65	0.03	0.03	0.11	0.49	3.71	0.008	220		640	5	20	0.183	0.0030	
Continental intraplate																					
Mean	44.33	0.10	2.41	8.07	0.13	41.84	4.85	0.23	0.06	1.17	1.46	6.10	0.899	2,147	102	2,819	12	59	0.216	0.0037	273
Median	44.42	0.08	2.26	8.12	0.13	41.70	2.57	0.02	0.06	1.20	1.19	4.76	0.900	2,135	104	2,740	12	61	0.182	0.0041	
σ	1.39	0.09	1.31	1.79	0.03	2.96	10.14	0.27	0.03	0.26	0.93	4.51	0.014	263	8	625	4	19	0.190	0.0024	
Continental arc																					
Mean	44.41	0.10	2.62	8.60	0.13	41.14	2.41		0.07	1.20	1.25	5.88	0.895	2,110	108	2,521		53	0.214	0.0041	28
Median	44.71	0.09	2.94	8.20	0.13	40.72	2.52		0.07	1.17	1.24	3.70	0.898	2,066	105	2,552		55	0.179	0.0039	
σ	1.17	0.05	0.92	1.47	0.03	2.80	1.00		0.02	0.10	0.25	6.08	0.018	282	11	499		17	0.110	0.0010	
Cratonic																					
Mean	43.27	0.05	0.99	7.20	0.14	46.19	0.70	3.29	0.03	1.38	0.88	16.62	0.919	2,579	120	3,778	4	29	0.636	0.0043	232
Median	43.39	0.04	0.82	7.01	0.13	46.52	0.49	2.27	0.02	1.39	0.82	11.60	0.921	2,570	116	2,861	4	27	0.456	0.0042	
σ	1.63	0.05	0.72	1.11	0.04	2.94	0.72	2.55	0.02	0.11	0.66	18.18	0.014	324	19	2,720	3	14	0.858	0.0044	
																				Total	593

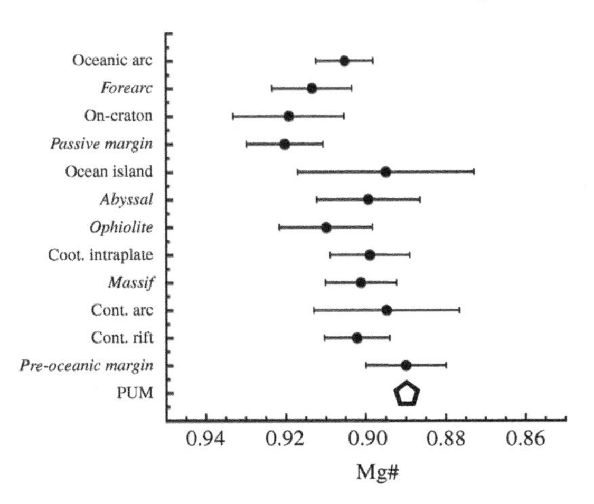

Figure 11 Mean and median Mg/(Mg + Fe) for the same samples in Figure 10 (in italics; Table 8). Error bar represents 1σ of the mean.

xenoliths. The latter feature requires that most magma extracted from the mantle be picritic (Liang and Elthon, 1990). Cratonic peridotite suites show a lower chromium level for a given level of depletion compared to noncratonic xenoliths, but reasons for this are not clear. More than 50% of the cratonic samples plotted are spinel-free and so the reduced scatter in these suites may arise because chromium is more equably partitioned amongst the garnet and pyroxenes in these samples.

The mean and median Cr/Al ratio of various xenoliths varies with tectonic setting, but there are large standard deviations due to the sampling problem for chromium (Figure 13). Despite the scatter, it becomes clear that the high Cr/Al of protoliths required to form low calcium, high chromium harzburgitic garnets in xenoliths occur in the arc or abyssal tectonic setting, at pressures within the spinel-stability field.

Due to the equable partitioning of manganese amongst peridotite minerals, this element shows a much tighter cluster with depletion than does chromium. Like chromium, manganese has a nearly constant abundance in a range of xenolith types and in orogenic massifs (Figure 12(d)). The near parallel trend indicates a bulk D during melting that is ~1. Cratonic peridotites show generally lower manganese contents for a given degree of depletion.

Mildly incompatible elements. The abundance of V in several different mantle residues was recently reviewed by Canil (2002). The trends for abyssal and massif peridotites are compared with on- and off-craton xenolith suites in Figure 14. Abyssal peridotites contain the most V at a given degree of depletion. Massif peridotites are shifted from the abyssal samples and parallel the off-craton samples, but the latter show considerable scatter. Arc xenoliths, although few in number, are the most impoverished in V for a given degree of depletion. Some cratonic peridotites overlap the abyssal samples but many project toward or along the trend for arc xenoliths (Figure 14(b)).

The distinct trends for V abundances in peridotite xenoliths may exist because this element is f_{O_2}-sensitive during melting. Canil (2002) interprets these trends to represent the primary f_{O_2} during melting, with f_{O_2} during partial melting increasing in the following order: abyssal peridotites < massif peridotites ~ off-craton xenoliths < arc and cratonic xenoliths. Scatter in the off-craton xenolith suites is not likely due to metasomatism (Canil and Fedortchouk, 2000) but rather can be attributed to analytical errors or real f_{O_2} variations in residues from a range of tectonic environments. Further effort to apply V abundances as paleo-redox indicators during lithosphere formation will require more analyses for V in xenoliths from a broader range of tectonic settings (arc xenoliths are very under-represented), and an evaluation of interlaboratory accuracy for analysis of this element in peridotites. There are surprisingly many trace-element data sets for xenoliths where V is not measured.

There are few measurements of the Fe^{3+} (Fe_2O_3) abundance in mantle xenoliths (O'Neill *et al.*, 1993; Canil *et al.*, 1994; Canil and O'Neill, 1996). Wet chemical determinations for Fe_2O_3 in peridotites are fraught with errors, but better and more precise results are obtainable with Mossbauer spectroscopy of individual minerals (O'Neill *et al.*, 1993). Fe^{3+} behaves as a mildly incompatible element during melting with a bulk D of ~0.1, similar to that of scandium or V (Figure 12(e)).

Scandium and sodium are very compatible in clinopyroxene. The covariation of Sc or sodium with depletion follows that of V or Fe^{3+} (Figures 12(f) and (h)) and correlates with the decreasing clinopyroxene abundance with increased melting (Figure 3). The kink in the trend for scandium in xenoliths at ~2 wt.% Al_2O_3 may signal exhaustion of clinopyroxene from residues after ~20% melting (Figure 3). Both Sc and sodium show far more scatter in xenoliths than in massif peridotites (Figures 12(f) and (h)), likely because the former samples have experienced more metasomatism, but also possibly because they are contaminated with small amounts of intergranular material (from host volcanic rock) which concentrate these elements and contribute them to the bulk analysis. Scatter due to these effects is particularly pronounced for more depleted xenolith samples with less than ~2 wt.% Al_2O_3. These effects pose even greater problems for the abundances of highly incompatible elements in xenoliths (Zindler and Jagoutz, 1988).

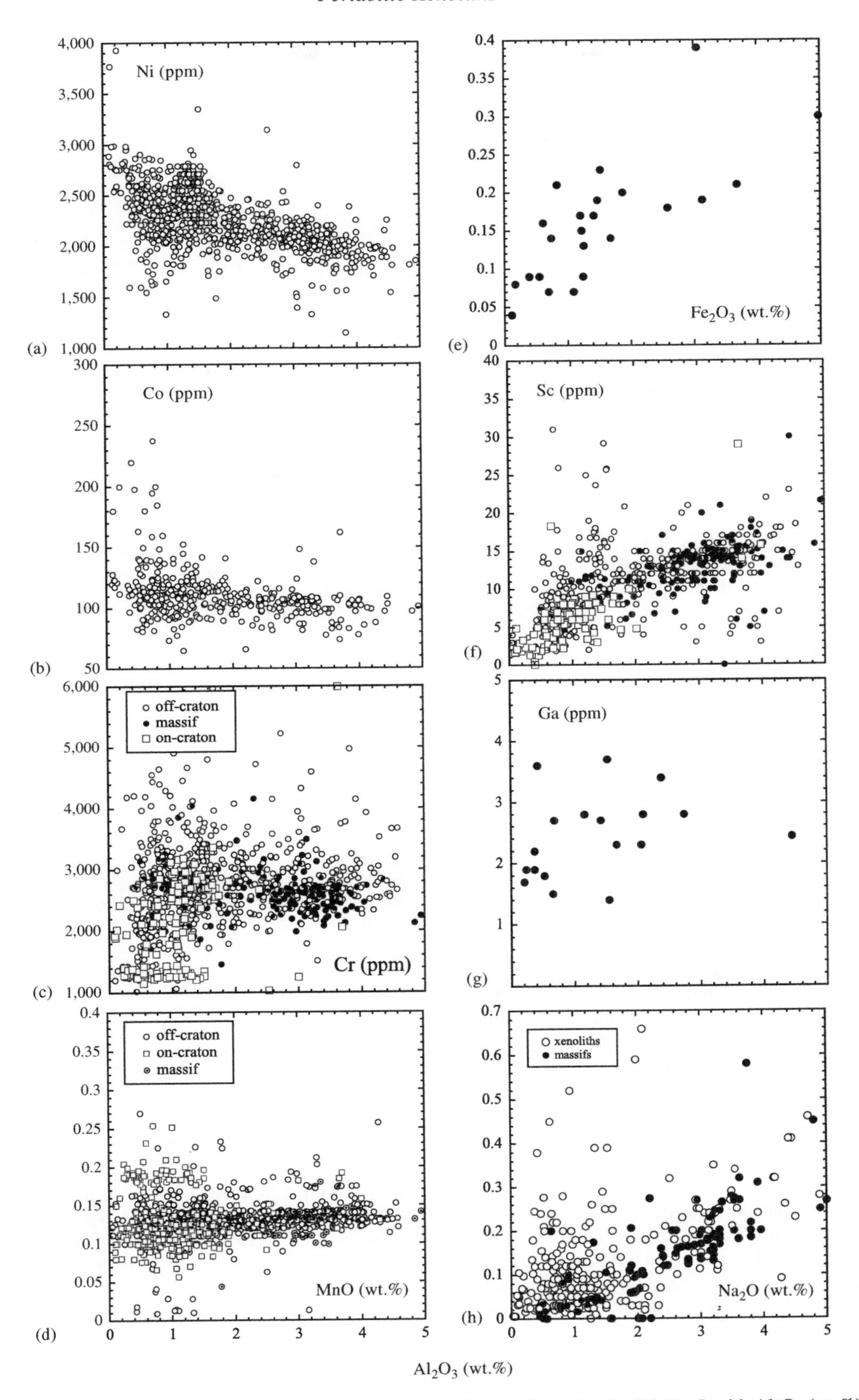

Figure 12 Covariation of: (a) Ni, (b) Co, (c) Cr, (d) MnO, (e) Fe_2O_3, (f) Sc, (g) Ga, (h) Na_2O with Al_2O_3 (wt.%) in on- and off-craton xenolith suites. Samples from all settings are grouped together in the panels where no legend is shown. Data sources in Table 6.

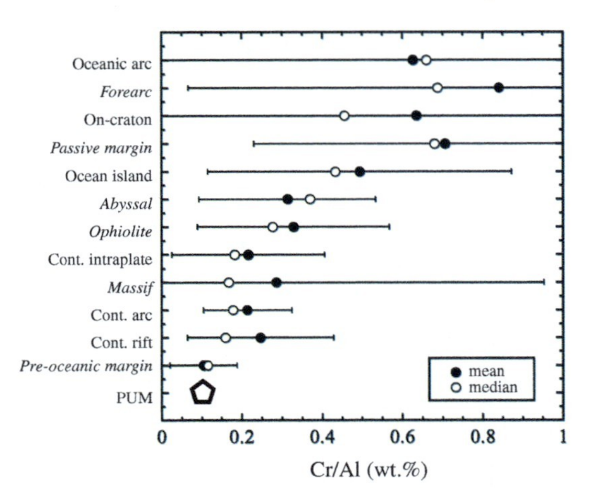

Figure 13 Mean and median Cr/Al for the same samples in Figure 10 (in italics; Table 8). Error bar represents 1σ of the mean.

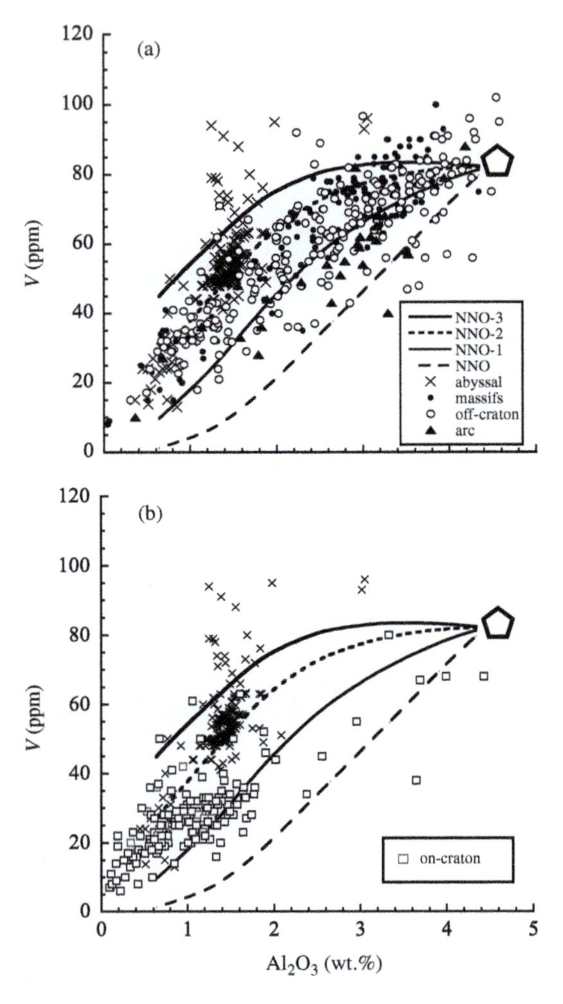

Figure 14 Covariation of V with Al_2O_3 in: (a) abyssal peridotites, orogenic massifs, off-craton xenoliths, arc xenoliths, and (b) cratonic xenoliths (data sources in Table 6). Shown for illustrative purposes are residue trends for partial melting at 1.5 GPa as a functions of f_{O_2} (expressed relative to the nickel-nickel oxide (NNO) buffer) calculated using methods described in Canil (2002).

The few analyses of gallium in xenoliths (Rhodes and Dawson, 1975; Kurat *et al.*, 1980; McKay and Mitchell, 1988) shows that the element has a distribution similar to that of Fe_2O_3, but there are too few data to define any trends (Figure 12(g)). Gallium also partitions itself amongst mantle minerals in the same order as Fe_2O_3.

Incompatible elements (rare-earth elements, large-ion lithophile elements, high-field-strength-elements)

(i) *Peridotites.* The high levels of incompatible element enrichment in almost all mantle xenolith host volcanic rocks results in a high probability that levels of these elements in the bulk xenoliths are compromised by infiltration of the host magma during entrainment and eruption. For this reason and because of the effects of weathering on the whole-rock geochemistry of some highly incompatible elements such as rubidium and strontium, the most reliable data for mantle xenoliths are generally thought to be obtained from mineral studies. However, the possibility that some minerals in certain xenolith suites may have crystallized recently, from the host volcanic rock, complicates the issue further. The effects of weathering and host rock infiltration are discussed further from an isotopic perspective in Section 2.05.2.7. Exceptional to these misgivings are mica- and amphibole-rich rocks such as the MARID suite (Table 1), which have inherently high levels of incompatible elements and are thus more immune from host rock infiltration (Erlank *et al.*, 1987; Gregoire *et al.*, 2002; Pearson and Nowell, 2002).

Detailed studies of the incompatible trace-element budget of mantle xenoliths illustrate the problems. Mass balance studies that use modal analyses and mineral compositions to calculate peridotite bulk compositions show large deficiencies between calculated whole-rocks and those measured (McDonough *et al.*, 1992; Pearson and Nowell, 2002). Schmidberger and Francis (2001) show that calculated whole-rock LREE abundances in peridotites from Somerset Island, Canada, are 70–99% lower than measured whole-rock REE concentrations. The discrepancies can be explained by infiltration of between 0.4 wt.% and 2 wt.% kimberlite (Figure 15). Not all the "interstitial" component in xenoliths is host-rock melt. Bodinier *et al.* (1996) and Bedini and Bodinier (1999) have shown the presence of a very thin (<10 μm thick) reaction layer coating the surfaces of spinel crystals in several suites of

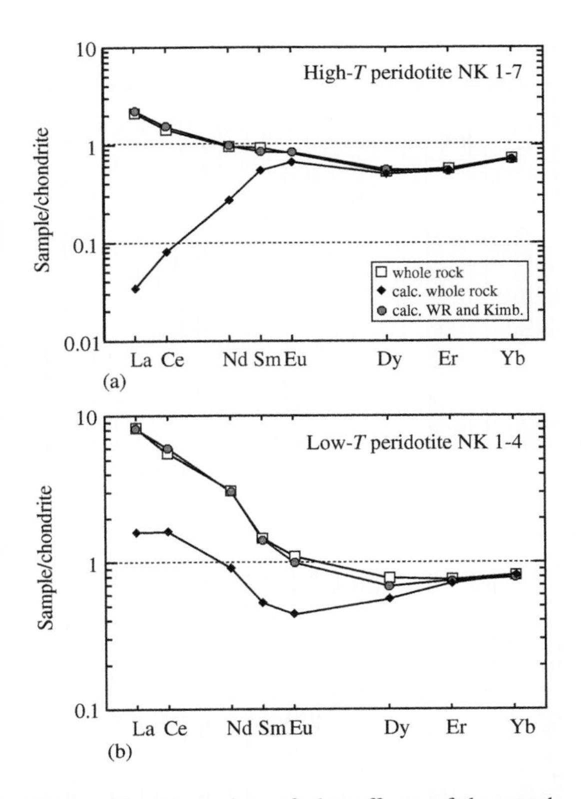

Figure 15 Illustration of the effects of host-rock contamination on whole-rock REE geochemistry. Chondrite-normalized REE patterns of measured whole-rock peridotites compared with REE abundances calculated from modal data plus mineral compositions. Also compared are mixtures of calculated whole-rock and kimberlite for each diagram. Two specimens, (a) and (b) are high-*T* and low-*T* garnet lherzolites from Somerset Island, Nunuvut (Canada) (after Schmidberger and Francis, 2001).

spinel lherzolite xenoliths of noncratonic origin. This layer comprises titanium oxides and phlogopite and can account for 45–90% of the whole-rock budget of rubidium, barium, niobium, and tantalum (Figure 16). Bodinier *et al.* (1996) and Bedini and Bodinier (1999) suggest that spinel rims are likely to be present in most noncratonic spinel lherzolite xenolith suites and attribute the feature to a metasomatic process involving percolation of silica-rich and potassium-rich small melt fractions in the lithospheric mantle. However, a study of peridotite mineral and bulk rock trace-element budgets in two Southeast Australian xenoliths by Eggins *et al.* (1998) did not find evidence for a significant grain boundary component. Detailed studies within individual minerals show that they can contain fluid inclusions that are enriched in uranium, thorium, lead, rubidium, strontium, neodymium, barium, and alkalis. This will significantly affect whole-rock geochemical budgets (Rosenbaum *et al.*, 1996).

Schiano and Clocchiatti (1994) have also demonstrated the ubiquitous presence of volatile, silica- and potassium-rich small melt fractions trapped as inclusions in silicate minerals in noncratonic peridotite xenoliths that affect whole-rock and mineral trace-element budgets. Such inclusions have $SiO_2 > 60$ wt.%, $TiO_2 < 1$ wt.%, $K_2O > 6$ wt.%, $K_2O/Na_2O > 1$ and have high contents of H_2O, CO_2 plus Cl. The inclusions have minimum trapping temperatures of 1,220 °C at 0.7 GPa, corresponding to entrapment at oceanic upper mantle depths. The widespread occurrence of these inclusions in off-craton peridotites is consistent with the continuous infiltration of small melt fractions into the lithospheric mantle as proposed by McKenzie (1989). Recently, it has been shown that infiltration of the host basaltic magma and subsequent reaction may produce glassy veins and pockets whose composition encompasses that of a variety of proposed metasomatic agents (Ciuff *et al.*, 2002). These results suggest caution when drawing inferences about mantle processes from glass pockets and veins in xenoliths.

Mass-balance inversion of a suite of noncratonic type I xenoliths from SE Ethiopia (Bedini and Bodinier, 1999) show that the trace-element composition of some whole-rock peridotites may be controlled by five distinct components:

- The silicate minerals account for all the HREE budget and 50–90% of the LREE, strontium, and Zr–Hf in apatite-free peridotites.
- Mineral hosted fluid inclusions contribute significantly to rubidium (20–25%) and to lesser extents the LILE budgets.
- A pervasive grain-boundary component that is selectively enriched in highly incompatible elements, contributing 25–90% of the whole-rock budget for barium, thorium, uranium, and 10–50% for niobium and LREE in apatite-free samples.
- Thin reaction layers (<10 μm thick) coating spinel surfaces that are composed of titanium oxides and phlogopite. These phases dominate the budget for niobium, tantalum (45–60%), and rubidium plus barium (30–80%) in all the xenoliths studied.
- Apatite dominates the budget of thorium, uranium, strontium, and LREE (25–75%), when present.

Similar findings were made by Kalfoun *et al.* (2002) who suggest that only 1–5% of niobium in spinel lherzolites from S.E. Siberia reside in silicate phases with the remainder hosted in finely disseminated rutile.

Despite these uncertainties and complications, some systematics are evident for whole-rock incompatible element data in xenolith suites.

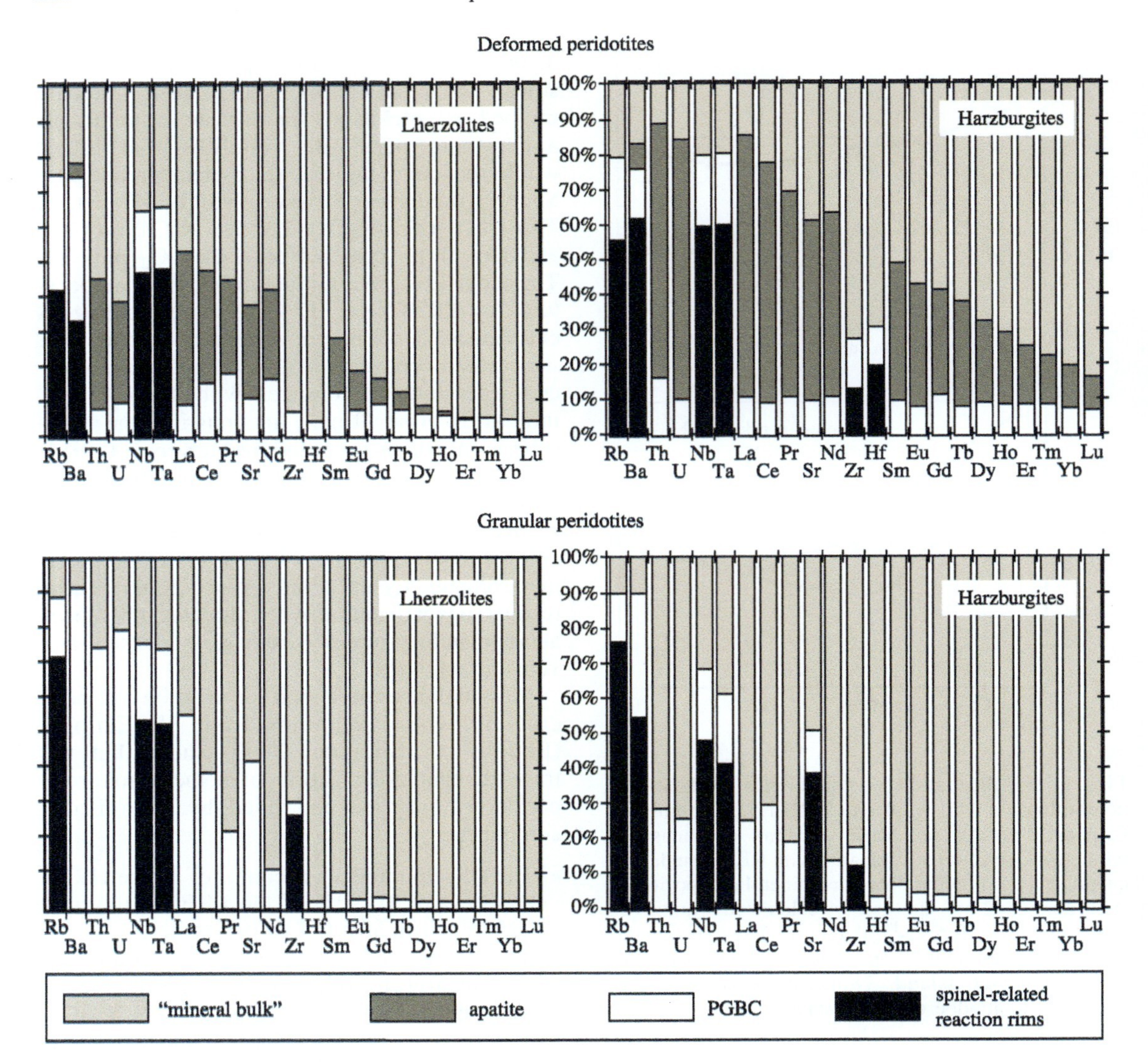

Figure 16 Trace-element mass-balance for deformed and granular lherzolites and harzburgites from E. Africa illustrating the contribution of a pervasive grain boundary component (PGBC), apatite and spinel reaction rims in wt.% of the total elemental whole-rock budget (after Bedini and Bodinier, 1999).

REE and other incompatible element studies of both cratonic (Nixon *et al.*, 1981) and noncratonic xenoliths (Frey and Green, 1974; Stosch and Seck, 1980; McDonough, 1990) have led to the notion that despite their overall major-element depletion, peridotite xenoliths are enriched (relative to chondrites) in incompatible elements, notably LREE. The fact that these samples have higher Mg# and nickel contents than primitive mantle, yet many have LREE enriched signatures, led Frey and Green (1974) to suggest a multistage history whereby the peridotites were initially melt depleted due to basalt extraction and this residue was later enriched by small amounts of incompatible element enriched melt. This basic model is still accepted as commonly applicable to many peridotite xenoliths from a wide variety of tectonic settings.

The concept of incompatible element enrichment has been extended as a general characteristic of the whole lithospheric mantle. While the whole-rock analyses of many peridotites are LREE enriched, the above discussion suggests that this feature could be imposed very recently and that the original character of many of the peridotites and by inference, the lithospheric mantle, is not as incompatible element rich as generally thought. For instance, for a selection of cratonic and circum-cratonic peridotites, Pearson and Nowell (2002) find that the mantle-normalized incompatible element patterns for calculated whole-rock peridotites range from very LREE depleted to slightly LREE enriched, but are generally characterized by very low abundances of the highly incompatible elements barium and lanthanum.

Noncratonic type I (Cr-diopside group) spinel lherzolites (Table 1) have whole-rock REE patterns that are commonly LREE-enriched or less commonly, LREE depleted (Figure 17). This bulk rock chemistry is reflected in their clinopyroxene chemistry such that those with bulk rock LREE-depletion have LREE-depleted clinopyroxene patterns (type IA; Menzies, 1983) and those with bulk rock LREE- enrichment have LREE-enriched clinopyroxene patterns (type IB; Menzies, 1983; Figure 17). The difference between the two groups may well be related to late stage melt/fluid infiltration before/during eruption. Variations in degree of LREE enrichment also correlate with peridotite texture in some spinel lherzolite suites (Downes and Dupuy, 1987). For instance, Xu *et al.* (1998) observed that deformed harzburgites and lherzolites from Wangqing, NE China, were consistently LREE enriched compared to equigranular lherzolites and harzburgites. The type II Al-augite group xenoliths show a wide variety of incompatible element characteristics that are generally enriched, relative to PUM and seem coupled to their relatively fertile major-element compositions.

For cratonic, kimberlite-derived xenoliths, relationships between texture/equilibration temperature and geochemistry also exist. The low-temperature (granular) type I peridotites have generally more LREE enriched compositions than the high-temperature (sheared) type V peridotites (Nixon *et al.*, 1981; Figure 17). These systematics are partly a function of the fact that garnets and diopsides in low-temperature peridotites are more variably LREE enriched than in high temperature peridotites (Figure 17).

Heavy rare earth elements (HREE), in particular ytterbium, are less affected by host rock interaction and show significant positive correlations with CaO, Al_2O_3, scandium, and vanadium contents of noncratonic spinel lherzolites (McDonough, 1990). This is taken as a reflection of the similarity of bulk distribution coefficients of these elements during partial melting. Trends for "hydrous" (amphibole and phlogopite-bearing) spinel peridotites are more scattered than for anhydrous peridotites, probably due to introduction of both aluminum and ytterbium during metasomatism (McDonough, 1990). Cratonic and circum-cratonic peridotites show much less coherency in terms of Al–Ca–Yb correlations (Irvine, 2002), probably due to their more complex metasomatic and magmatic histories. However, Kelemen *et al.* (1998) have argued that low ytterbium levels in cratonic peridotites correlate with low CaO and suggest that this is a function of initial melting in the spinel stability field, releasing ytterbium into the melt, followed by subduction and incorporation into the CLM.

High-quality data for high-field-strength element (HFSE) ratios such as Zr/Hf and Nb/Ta have shown that bulk rock noncratonic spinel lherzolites have ratios similar to primitive mantle (Jochum *et al.*, 1989). This data, together with that from basalts, is used to argue against significant fractionation of these elements during normal mantle melting.

Studies of the HFSE abundances in peridotite xenoliths and their metasomatized variants have also been relevant to arguments concerning flood basalt genesis. Many basaltic rocks erupted through continental areas have distinctive incompatible trace-element signatures compared with basalts erupted in oceanic areas. Continental basalts have characteristic enrichment in large-ion-lithophile (LILE) elements such as barium and depletion of high-field-strength elements (HFSE) such as titanium and niobium (Thompson *et al.*, 1983). This feature has been used to suggest that the CLM is the source region for certain continental basalts (e.g., Hawkesworth *et al.*, 1983; Hergt *et al.*, 1991) although numerous counter arguments have been raised (e.g., Menzies, 1992). McDonough (1990) followed by Arndt and Christensen (1992) examined available trace-element data for off-craton spinel lherzolites and concluded that the CLM, as represented by xenoliths, does not have trace-element characteristics consistent with it being the source of continental basalts because they are not characterized by HFSE depletion. These conclusions have been extended to cratonic peridotites and phlogopite-rich metasomatic mafic xenoliths by Pearson and Nowell (2002), who note that mantle xenoliths containing hydrous minerals such as amphibole and mica have positive HFSE anomalies. Any melt where these phases contribute to the melt fraction will therefore have positive HFSE anomalies. Models involving residual rutile and/or hydrous phases are possible but are of doubtful validity because such phases should be the first to contribute to the melt (e.g., Foley, 1992).

Modeling of the whole-rock and mineral trace-element compositions in xenoliths that have experienced metasomatic infiltration of melts led Ionov *et al.* (2002a) to conclude that the REE and HFSE element compositions of peridotites adjacent to veins bear the chemical fingerprints of the metasomatic agent closest to its source (e.g., a melt vein). Further away, signatures are increasingly dominated by fractionation processes related to melt percolation.

Noncratonic Al-augite group Pyroxenites (type BII). In contrast to the pyroxenites within orogenic massifs (Chapter 2.04) relatively few bulk rock trace-element studies have been performed on noncratonic (or cratonic) pyroxenite xenoliths. The most extensive geochemical studies are of

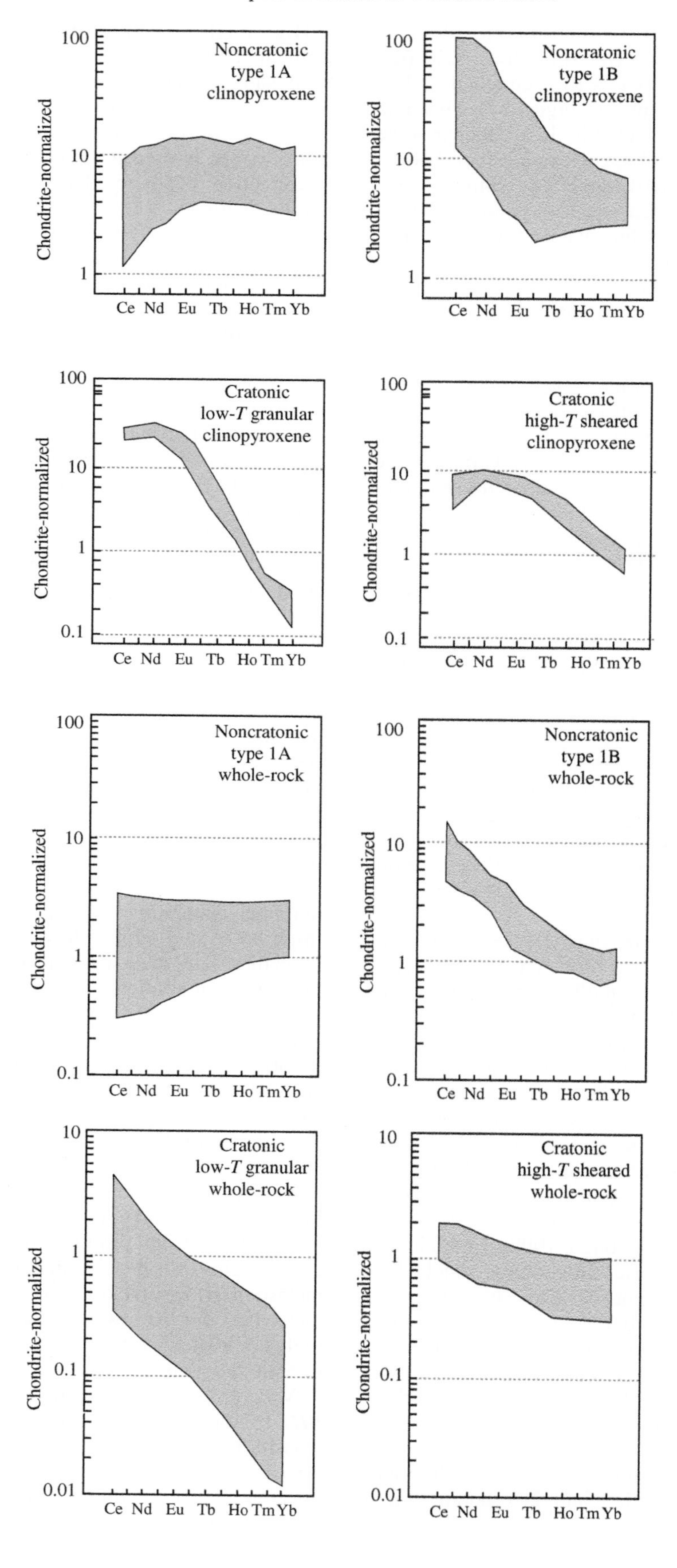

Noncratonic type 1A clinopyroxene
Noncratonic type 1B clinopyroxene
Cratonic low-T granular clinopyroxene
Cratonic high-T sheared clinopyroxene
Noncratonic type 1A whole-rock
Noncratonic type 1B whole-rock
Cratonic low-T granular whole-rock
Cratonic high-T sheared whole-rock
Chondrite-normalized
100
10
1
0.1
0.01
Ce Nd Eu Tb Ho Tm Yb

occurrences in southeastern Australia (Irving, 1974; Griffin *et al.*, 1988) the western USA (Irving, 1980); Hawaii (Frey, 1980). Pyroxene-rich sheets from Kilbourne Hole and San Carlos have greater than chondritic abundances of all REE, with chondrite-normalized patterns that are convex upwards, with maxima between neodymium and samarium (Irving, 1980). Other samples show more variability, ranging from LREE depleted to enriched patterns. Pyroxenites from Salt Lake Crater, Hawaii, have a wide range in nickel contents and Mg together with the convex upward rare earth element distributions relative to chondrites. These features were interpreted by Frey (1980) to be produced by formation of the pyroxenites as "cumulates" from the Honolulu Volcanic Series. As expected, orthopyroxenites and websterites have lower REE abundances and LREE-depleted patterns are common compared with xenoliths dominated by clinopyroxene (McDonough and Frey, 1989). There is general consensus that the REE patterns and abundances reflect pyroxene accumulation on the walls of magma conduits during "flow crystallization" of magma passing through the mantle (Irving, 1980). Clear evidence of host rock contamination is observed in some samples.

Cratonic mica and amphibole-rich "metasomatic" rocks. Incompatible element abundances in these rocks are probably at sufficient levels to minimize the effects of host rock contamination. The phlogopite-rich mafic xenoliths (type VI) are LREE-enriched (10 and 50 times PUM; Figure 18, Erlank *et al.*, 1987; Gregoire *et al.*, 2002; Pearson and Nowell, 2002). Glimmerites (phlogopite-rich rocks—Table 1) have the lowest absolute REE abundances, due to the low levels of REE in phlogopite, but the presence of apatite in glimmerites greatly enhances their REE and strontium content. Virtually all samples display strong positive rubidium, niobium, and tantalum anomalies (Figure 18), making these lithologies unlikely sources for continental basaltic magmatism (Pearson and Nowell, 2002). Variable contents of minor phases such as apatite, allanite and barite cause substantial variations in lanthanum, cerium, barium, and thorium abundances (Gregoire *et al.*, 2002). MARID xenoliths contain ubiquitous K-richterite, which imposes a negative strontium anomaly on most rocks of this type. HFSE ratios such as Zr/Hf, that are normally relatively constant in peridotites, vary widely in these rock types due to the presence of ilmenite and other phases.

2.05.2.5.3 Platinum group elements and rhenium

The platinum group elements (PGEs) comprise osmium, iridium, ruthenium, rhodium, platinum, and palladium. Rhenium is often discussed along with the PGEs because of its similar geochemical behavior and because it forms part of the important Re–Os isotope decay system. The pioneering work on PGE abundances in mantle xenoliths was performed by Morgan *et al.* (1981, 1986). Recent advances in analytical geochemistry allow the routine determination of high-precision PGE data in mantle-derived rocks. In particular, techniques are now available that permit analysis of rhenium and osmium along with all the other PGEs (except rhodium), by isotope dilution (Pearson and Woodland, 2000). Here we discuss recent PGE determinations of mantle xenoliths that contrast the PGE compositions of cratonic and noncratonic xenoliths.

PGEs in fertile, unmelted mantle predominantly reside in base metal sulfides (Jagoutz *et al.*, 1979) and this is supported by positive correlations between bulk rock PGE content and selenium (Lorand and Alard, 2001). Experiments and observation indicate that monosulfide solid solution (mss) preferentially accommodates refractory PGEs (osmium, iridium, ruthenium, rhodium) relative to Cu-sulfides, which concentrate palladium and rhenium (Lorand and Alard, 2001; Pearson *et al.*, 1998b). Once sulfide has been removed from a melt residue, the difference in solid–liquid silicate partition coefficients ($D^{Solid/melt}$) between the I-PGEs (osmium, iridium, ruthenium) and the platinum-group PGEs (P-PGEs; rhodium, platinum, palladium) leads to fractionated PGE patterns, with progressive depletion of P-PGEs, while I-PGEs remain in the residue. This results in a decrease of P-PGE/I-PGE ratios such as Pd/Ir with progressive melting while ratios of I-PGEs, such as Os/Ir, remain close to 1 (Lorand *et al.*, 1999; Handler and Bennett, 1999; Rehkämper *et al.*, 1999; Pearson *et al.*, 2002, 2004). The progressive interelement PGE

Figure 17 Summary fields of chondrite-normalized REE patterns for whole-rock peridotites and clinopyroxenes for peridotite xenoliths. Noncratonic whole-rock peridotites are either LREE-depleted (type IA; least common) or LREE-enriched (type IB; most common). Data sources from Stosch and Seck (1980), Stosch and Lugmair (1986), Menzies *et al.* (1985). Clinopyroxenes from these rocks also show LREE enrichment or depletion. Cratonic peridotite whole rocks are ubiquitously LREE-enriched. Low-*T* (granular) suite show greater LREE/HREE compared to high-*T* (sheared) suite and this is reflected in the more LREE-enriched clinopyroxene compositions in the low-*T* suite. Data sources from Shimizu (1975), Nixon *et al.* (1981), and Irvine (2002). Low-*T* whole-rock suite includes 19 samples from Letseng, N. Lesotho (Irvine, 2002).

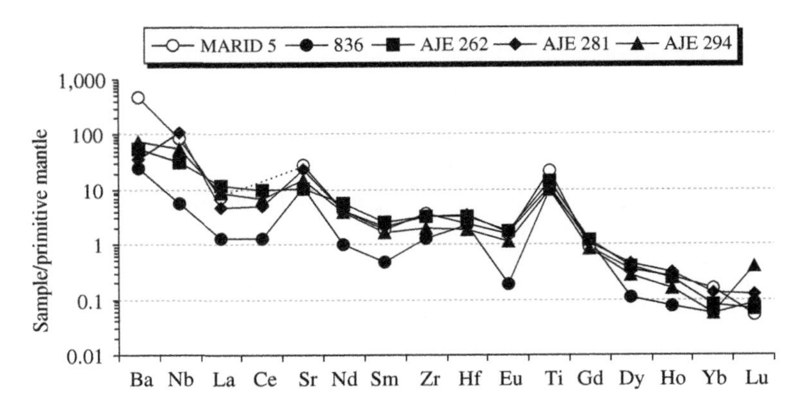

Figure 18 Primitive mantle normalized trace-element abundance patterns for whole-rock MARID xenoliths from kimberlites. Primitive mantle values used for normalisation in this plot and subsequent plots are those of McDonough and Sun (1995) (sources Pearson and Nowell, 2002; Gregoire *et al.*, 2002).

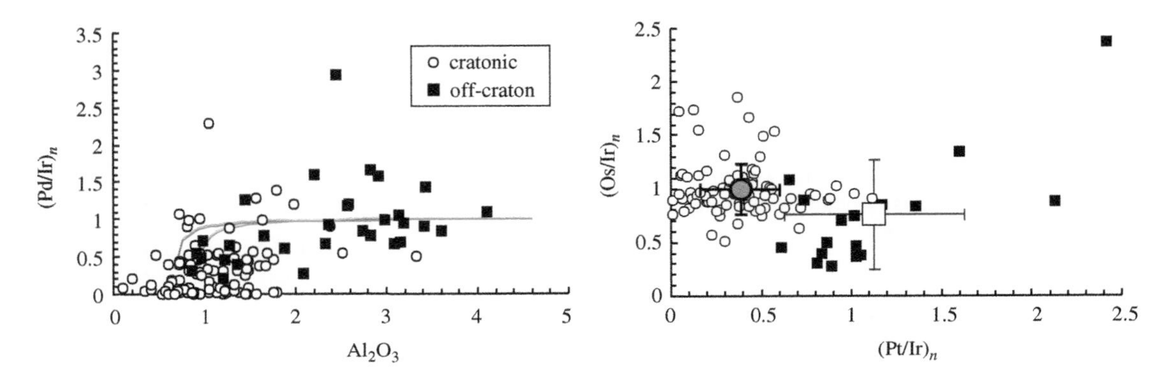

Figure 19 Left: $(Pd/Ir)_n$, where n refers to values normalized to primitive mantle values (McDonough and Sun, 1995) versus bulk rock Al_2O_3. Right: $(Os/Ir)_n$ versus $(Pt/Ir)_n$ plot of cratonic and off-craton whole-rock peridotites. Curves on left-hand plot are trends expected for progressive melting of mantle peridotite. On the right-hand plot, large circle with error bars = mean and 1 SD for cratonic peridotites. Large square with error bars = mean and 1 SD for off-craton peridotites. Cratonic peridotite data are isotope dilution data from Pearson *et al.* (2002), Irvine (2002), Pearson *et al.* (2004), and Irvine *et al.* (2003). Off-craton xenolith data from Handler and Bennett (1999) and Pearson *et al.* (2004).

fractionation during partial melting can be correlated with major-element depletion indices such as Al_2O_3 (Figure 19). If cratonic peridotites represent highly depleted melt residues then they should have very low Pd/Ir ratios to complement the depleted nature of their major-element compositions Pearson *et al.* (in press). Should melt/fluid re-enrichment or some other form of disturbance affect a peridotite subsequent to melting then this should be reflected in PGE patterns and PGE-major-element systematics. Modeling of melt re-enrichment involving the addition of new sulfides to variably depleted melt residues during magma–solid interaction produces elevation of the $(Pd/Ir)_n$ ratio to supra-chondritic levels and elevated P-PGE contents (Rehkämper *et al.*, 1999).

Cratonic peridotite xenoliths from Lesotho, S. Africa, the Jericho kimberlite, Slave craton, and Somerset Island, Churchill Province, northern Canada, along with selected samples from Kimberley, S. Africa, have been analyzed for PGEs plus rhenium (Irvine, 2002; Irvine *et al.*, 2003; Pearson *et al.*, 2002, 2004). Data from the Lesotho suite illustrates the typical within-suite PGE variation (Figure 20; Irvine, 2002; Pearson *et al.*, 2004). The simplest chondrite-normalized PGE-Re patterns to interpret are those of group A, defined by Irvine (2002) as being highly depleted in platinum, palladium, and rhenium but unfractionated in terms of Os–Ir and ruthenium (Figure 20). This signature correlates with the highly depleted major-element characteristics of cratonic xenoliths and reflects the more incompatible nature of the low-melting point or P-PGE members of the PGE group. The P-PGEs and rhenium quantitatively partition into the melt phase during high degrees of melting (>30%) thought to be responsible for generating the major-element characteristics of cratonic peridotites, probably due to the breakdown of

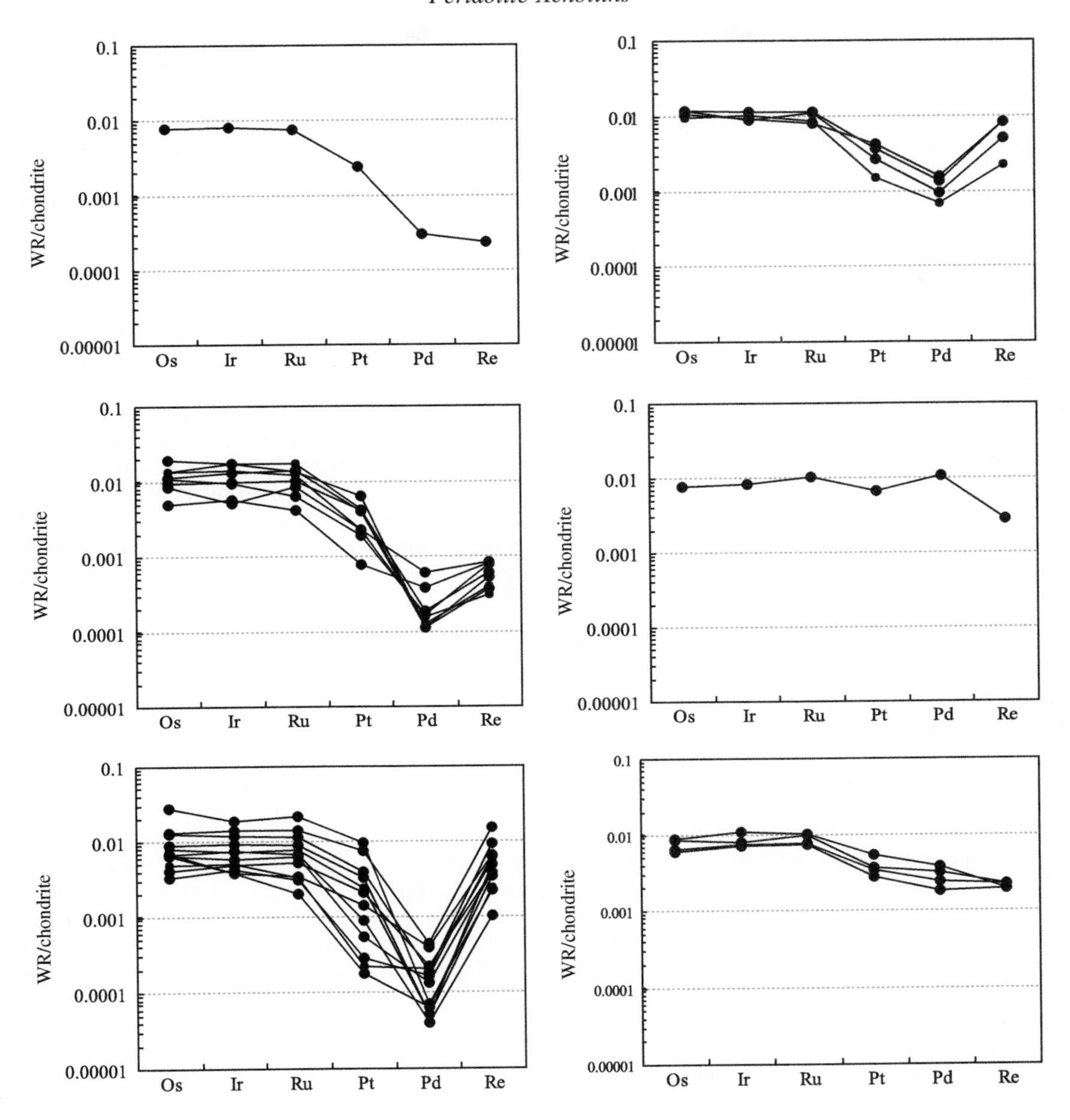

Figure 20 Primitive mantle normalized extended PGE patterns (including rhenium) for cratonic whole-rock garnet peridotite xenoliths from the Letseng kimberlite (Lesotho) (sources Irvine, 2002 and Pearson *et al.*, 2004).

the monosulfide solid solution (mss). In contrast, the higher melting point I-group PGEs remain in the residue and are stabilized in either alloy phases or high melting point residual Ru–Os–Ir sulfides. This produces a highly P-PGE depleted residual peridotite that is considerably more depleted in P-PGEs (Pearson *et al.*, 2004). These peridotites have lower $(Pt/Ir)_n$, and rhenium than fertile, noncratonic spinel lherzolite xenoliths (Figure 19: Handler and Bennett, 1999) or massif peridotites (Lorand *et al.*, 1999). Variations on these scenarios are PGE patterns that still show marked P-PGE depletion, with $(Pd/Ir)_n = 0.007$ to 0.117, but slight rhenium enrichments may occur, probably due to infiltrating melt, that produce small upward kinks at rhenium in the extended PGE pattern (group Bi of Irvine, 2002; Figure 20). Other samples show much greater increases in rhenium abundance with minimal effect on the depleted P-PGEs (e.g., $(Pd/Ir)_n = 0.004$ to 0.102; group Bii of Irvine, 2002; Figure 20). Subgroup Biii has higher Pd abundances and enhanced rhenium. These peridotites have either undergone lower levels of melt depletion or have experienced some palladium re-enrichment following melt depletion. Groups C and D show much less fractionation of P-PGEs from I-PGEs and higher abundances of P-PGEs plus rhenium such that

$(Pd/Ir)_n$ values are greater than 1. Both these groups are likely to have experienced P-PGE enrichment, perhaps by a sulfide melt.

Examination of all data for cratonic peridotites (Figure 19) shows similar features to the Lesotho suite (Irvine, 2002). Highly depleted cratonic peridotites generally have low Al_2O_3 and very low $(Pd/Ir)_n$ values. Variation in Al_2O_3 at very low, constant $(Pd/Ir)_n$ appears characteristic of cratonic peridotites and may be related to the introduction of late-stage diopside by the host kimberlite (Pearson *et al.*, 2002a; Simon *et al.*, 2003). Positive slopes at moderate angles may be related to silicate melt-metasomatism, whereas subvertical trends of increasing Al_2O_3 and $(Pd/Ir)_n$ are indicative of sulfide addition (Rehkämper *et al.*, 1999). Hence, the bulk rock PGE systematics of cratonic peridotites are indicative of their major-element depleted characters, but with metasomatic effects superimposed on this signature to varying degrees.

The best characterized noncratonic, spinel peridotite xenoliths for PGEs are those from southeastern Australia (Handler and Bennett, 1999), the Massif Central, France (Lorand and Alard, 2001), Western USA (Morgan *et al.*, 1981; Lee, 2002) and Vitim, Siberia (Pearson *et al.*, 2004). In general, $(Pd/Ir)_n$ values for noncratonic peridotites are considerably higher than those of cratonic xenoliths (Figure 19), in keeping with their much less depleted nature (Figures 10 and 19). The fertile lherzolites have broadly chondritic PGE interelement ratios with little fractionation of P-PGEs from I-PGEs, in contrast to cratonic peridotites. Some of the more depleted, lower Al_2O_3 peridotites have lower $(Pd/Ir)_n$ but systematics are complicated by metasomatism and alteration processes (e.g., Lee, 2002). Both fertile and depleted peridotites have elevated $(Ir/Os)_n$ that scatter much more than cratonic peridotites (Figure 19). Positive correlation of $(Ir/Os)_n$ with Cu/S suggests late-stage osmium mobility in the southeastern Australian suite (Handler *et al.*, 1999). Similar, but less well developed trends are also seen in western USA xenoliths. In the southeastern Australian and Massif Central peridotites suites, peridotites containing metasomatic phases (amphibole, mica, apatite) have similar PGE systematics to those that do not contain metasomatic phases (Handler *et al.*, 1999; Lorand and Alard, 2001). This has been interpreted as indicating that PGEs were not greatly affected by the melt infiltration events that cause crystallization of mica and amphibole.

Where peridotites show both textural and geochemical evidence for pervasive interaction with percolating melts, PGE and selenium abundances are greatly reduced. PGE abundances decrease by ~80%, probably because of dissolution of intergranular sulfides (Lorand and Alard, 2001). In contrast, peridotites metasomatized by small melt fractions show enrichment in platinum and palladium and elevated $(Pd/Ir)_n$. Bulk mineral separate PGE-Re analyses of two fertile xenoliths from southeastern Australia indicate less than 6% of the whole-rock PGE budget resides in either silicate or oxide phases and further implicates sulfides and alloys as the main controls of PGE-Re abundance. Comparison of sulfide versus whole-rock budgets by Lorand and Alard (2001) demonstrates the dominance of sulfide as the main PGE host in relatively fertile peridotites. This confirms the results of earlier studies of xenolith PGE mass balance (Hart and Ravizza, 1996; Mitchell and Keays, 1981) plus xenolith-derived and diamond inclusion sulfide studies (Jagoutz *et al.*, 1979; Pearson *et al.*, 1998b). As with cratonic xenoliths, sulfur-PGE and major-element-PGE correlations in more depleted noncratonic peridotites indicate that I-PGEs are probably not hosted entirely by sulfide (Lee, 2002).

One striking feature to emerge from the preponderance of high quality rhenium and osmium abundance data resulting from Re–Os isotope work on xenoliths is that the mean osmium concentration for cratonic peridotite xenoliths (3.8 ± 2.6 ppb) is significantly higher than that for noncratonic peridotites (both spinel and garnet facies; 2.0 ± 1.1 ppb; Figure 21 Pearson *et al.* (2004)). Because osmium is compatible, residues from large degrees of melting should have higher osmium than PUM and this is the case for cratonic peridotites. However, the off-craton peridotites are also residual yet they have significantly lower osmium than PUM, with many samples having concentrations below 1 ppb. The reason for this difference is not well understood. Melt percolation through peridotites can dissolve residual sulfides and hence lower osmium and this process may be more prevalent in the off-craton suites studied so far. Lee (2002) suggested multiple processes, including mobilization by aqueous fluids/secondary loss associated with sulfide breakdown were required to account for the osmium (and rhenium, palladium, and platinum) depletion in western USA peridotites. It is clear that noncratonic peridotite xenolith suites are much more susceptible to these processes than cratonic peridotites. The large degrees of melt extraction that have affected cratonic peridotites means that their osmium is likely to be hosted in alloy phases and refractory sulfides that may be more resistant to syn-eruption sulfide breakdown and alteration/leaching. However, we note that nonchondritic Ir/Os values are not a feature of massif peridotites that have experienced similar levels of depletion to noncratonic peridotites (e.g., Burnham *et al.*, 1998).

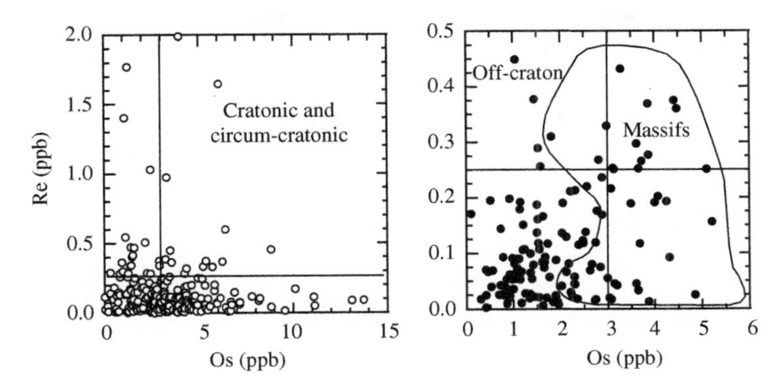

Figure 21 Covariation of rhenium versus osmium in whole-rock cratonic and circum-cratonic peridotite xenoliths (left) and off-craton peridotite xenoliths (right). Off-craton suite is compared to range in Massif peridotites (shown by field). Cratonic values extend to >15 ppb Os and >2 ppb Re. Lines denote Primitive Mantle values of Morgan *et al.* (1986) (source Walker *et al.*, 1989a; Carlson and Erving, 1994; Carlson *et al.*, 1999a; Pearson *et al.*, 1994, 1995a,b, 2002; Reisberg and Lorand, 1995; Handler *et al.*, 1997; Burnham *et al.*, 1998; Chesley *et al.*, 1999; Burton *et al.*, 2000; Peslier *et al.*, 2000; Meisel *et al.*, 2001; Hanghoj *et al.*, 2001; Irvine *et al.*, 2001, 2003; Irvine, 2002).

In summary, PGE studies of peridotite xenoliths from a variety of tectonic settings to date (Morgan *et al.*, 1981; Rehkämper *et al.*, 1997; Lorand and Alard, 2001; Handler and Bennett, 1999; Irvine, 2002; Irvine *et al.*, 2003) reveal a similar history to that indicated from incompatible elements such as REE. The depletion in P-PGE relative to I-PGE correlates broadly with melt depletion indices such as Al_2O_3 and shows that once sulfide-mss has disappeared at high degrees of melting, such as in cratonic peridotites, P-PGE partition quantitatively into the melt. Suprachondritic $(Pd/Ir)_n$ and $(Ru/Ir)_n$, or $(Os/Ir)_n$ values are indicative of metasomatic disturbance of PGEs by later melt infiltration or alteration. However, in contrast to REE studies, where whole-rock peridotites are ubiquitously LREE enriched, whole-rock PGE patterns frequently retain the signature of melt depletion, even when significant LREE enrichments are present in the bulk rock (Lorand and Alard, 2001; Irvine, 2002). This confirms earlier suggestions that PGEs and the Re–Os isotope system are much more robust to the effects of metasomatism than lithophile elements.

2.05.2.6 Mineral Trace-element Geochemistry

2.05.2.6.1 Olivine

Olivine is a major reservoir for divalent transition metal cations in xenoliths. Nickel, manganese, cobalt, and zinc all show correlations with Fo content in olivine from xenoliths. Nickel is positively correlated with Fo content, whereas cobalt, manganese, and zinc are negatively correlated suggesting the latter behave moderately incompatibly during the melting process. The calcium content of olivine is very T sensitive and is often zoned by heating and cooling effects. Typical abundances are listed in Table 9. High magnesium and calcium contents in secondary olivines within some peridotite xenoliths have been suggested to result from carbonatite metasomatism (Hauri *et al.*, 1993; Ionov *et al.*, 1993a).

There are far fewer data for trace-element cations because they are usually at levels below the detection limit of most analytical instruments. Trivalent cations (aluminum, scandium, vanadium, chromium, ytterbium) are all highly correlated with one another in olivine, suggesting crystal chemical effects for their substitution, with general trends of increasing concentration with equilibration temperature (Norman, 1998, 2001). The very few REE analyses in olivine show flat chondrite-normalized profiles that are several orders of magnitude lower in abundance than coexisting clinopyroxene (Figure 22). Li levels in olivine are generally higher than other minerals in mantle peridotite and so olivine will probably dominate the lithium budget in anhydrous peridotites (Eggins *et al.*, 1998; Glaser *et al.*, 1999).

2.05.2.6.2 Orthopyroxene

Orthopyroxene partitions nickel, cobalt, and manganese less than olivine and there are no clear correlations amongst these elements. Although low in abundance, orthopyroxene can be a significant reservoir for the trivalent cations vanadium, scandium plus tetravalent titanium, due to its high modal abundance, especially in depleted xenoliths with little or

Table 9 Representative trace element analyses of minerals from peridotite xenoliths.

Mineral	Setting	Type	Lithology	Notes	Sample no.	Reference	Element													
							Li	*Be*	*B*	*Sc*	*Ti*	*V*	*Rb*	*Sr*	*Y*	*Zr*	*Nb*	*Ba*	*La*	*Ce*
gt	C	AI	gt Lherz	Low-T	8604 gt3	1	0.048		0.25		199			0.272	6.37	14.9	0.005	0.139	0.023	0.220
gt	C	AI	gt Lherz	Low-T	FRB135	1					191			0.511	2.01	37.7	0.665	0.021	0.039	0.984
gt	C	AV	gt Lherz	Low-T	UV128/91	1					958			0.406	3.04	19.4	0.706	0.100	0.060	0.481
gt	C	AI	gt Harz	Low-Ca	RV174	2					73.2			6.13	1.69	7.99		0.010	1.92	7.93
gt	OC	BI	gt-lherz		Vi313-105	3				111	1,260		0.025	28.9		28.9	0.071	0.040	0.009	0.075
cpx	C	AI	gt Lherz	Low-T	8604 Di3	1	0.554	0.289	0.364		401			221	0.884	33.0	0.581	0.168	17.2	41.8
cpx	C	AI	gt Lherz	Low-T	FRB 135	1					104			342	0.205	11.7	0.901	0.686	6.97	27.8
cpx	C	AV	gt Lherz	High-T	UV128/91	1	0.451	0.18	0.324		405			99.1	1.09	4.55	0.232	0.358	2.46	7.79
cpx	OC	AI	gt-lherz		Vi313-105	3				29	3,420		0.011	78.5		21.2	0.50	0.13	1.03	3.58
cpx	OC	BI	sp. Lherz		ET80	4					3,120		0.003	26.9	17.1	13.0	0.027	0.090	0.240	0.950
cpx	OC	BI	sp. Lherz	fertile	2905	9	1.31			31.2	2,704	244	0.008	57.9	21.1	33.4	0.229	0.182	0.774	2.920
cpx	OC	BI	sp. Harz	depleted	84-402	9	1.95			94.4	1,512	219	0.034	155	12.3	50.4	1.46	0.210	5.780	1.340
cpx	OC	BIV	Amph-Sp.	Lherz	DW56-19	5					2,664		0.900	180	20.3	68.1	0.640	0.130	12.0	37.6
cpx	C	AVI	MARID		AJE282	6				43	676	333	0.650	509	7.70	40.0	0.190	11.8	16.0	41.0
opx	C	AI	gt-lherz		8604 En	1					142			0.186	0.017	0.230	0.109	0.039	0.032	0.046
opx	OC	BI	gt. Lherz		Vi313-105	3				8.2	1,050		0.010	0.49		1.58	0.061	0.04	0.010	0.026
opx	OC	BI	sp. Lherz		ET80	4					730		0.006	0.270	0.950	0.650	0.015		0.014	0.020
opx	OC	BI	sp. Lherz		2905	9	0.105			60.2	568	75	0.005	0.129	1.070	1.600	0.006	0.059	0.001	0.010
olv	OC	BI	sp. Lherz		ET80	4					36.1		0.050	0.076	0.040	0.124	0.116	0.1	0.016	
olv	OC	BI	sp. Lherz		2905	9	1.7			3.9	23.6	3	0.023	0.025	0.045	0.118	0.002	0.112	0.001	0.005
olv	OC	BI	gt-lherz		Vi313-105	3				1.5			0.016	0.034		0.044	0.019	0.013	0.0019	0.0037
Spinel	OC	BI	sp. Lherz		ET80	4					416		0.1080	0.0180	0.0005	0.0050	0.0060	0.1700		
Spinel	OC	BI	sp. Lherz		2905	9	0.72			2.1	720	346	0.328		0.200	0.716	0.085	0.312	0.012	0.033
amph	OC	BII	Pxite		SF93803	10	1.16			15.9	26,280	219		249	9.04	98.8	15.9		2.22	10.3
amph	C	AVI	MARID	K-Rich.	AJE288	6				24	978	128	21.0	570	0.650	25.0	3.90	78.0	2.90	7.10
amph	C	AVI	PKP	K-Rich.	BD2346	6					795	51	20.0	294	1.64	67.0	1.50	98.0	3.10	7.10
amph	OC	BIV	Vein		313-103	7				21.2	26,640		3.90	269		123	74.8	88.0	2.34	8.70
amph	OC	BIV	Amph-sp.	Lherz	DW56-19	6					15,300		16.0	470	22.1	82.6	83.9	327	16.1	50.1
phlog.	C	AVI	MARID		FRB836	1	6.90	3.38	3.59		5,036		379	1.14	0.020	1.14	5.36	424	0.127	0.040
phlog.	C	AVI	MARID		AJE288	6					2,385	60	545	4.20	0.280	1.80	14.0	439	0.460	0.200
phlog.	OC	BIV	Phl-sp-	Lherz	Mo4230-16	7				5.1	28,260		97.8	662		3.59	7.08	14,610		0.150
phlog.	OC	AIV	Phl-sp-	Lherz	SG96B11	10	7.7			4.42	27,600	199	154	77.5	0.78	28.1	34.4	824		0.520
Apatite	OC	BI	sp. Lherz		ET49	4							0.400	20,850	147	11.6	4.98	1,240	3,925	4,295
Apatite	OC	BI	Mica vein		4334-lu	4							0.105	4,087	92	0.870	2.66	118	765	623
Zircon	C	AVI	MARID		FRB836	1								14.3	303	58,20,000	10.6	0.253	0.353	18.8
Ilmenite	C	AVI	MARID		AJE288	6					3,20,700	837				53.0	1,318			
Rutile	C	AVI	MARID		AJE262	6					5,64,900	1,236				1,159	1,024			
Lindsleyite	C	AVIII	Phlog.	Lherz	AJE285	6					3,54,720	2,175		20,379		14,952	1,278	20,599		
Carbonate	OC	BIV	Amph-sp-	lherz	4-90-9	8					10.00			1,900	0.71	0.580	0.200	389	0.550	0.770

Element																				
Pr	*Nd*	*Sm*	*Eu*	*Gd*	*Tb*	*Dy*	*Ho*	*Er*	*Yb*	*Lu*	*Hf*	*Ta*	*Pb*	*Th*	*U*	*Rb/Sr*	*Sm/Nd*	*Lu/Hf*	*U/Pb*	*Th/Pb*
0.081	0.921	0.460	0.156	0.718	0.125	1.25	0.211	0.924	1.50	0.342	0.114	0.071					0.499	3		
0.469	0.391	1.5	0.471	1.4	0.113	0.605	0.0865	0.143	0.121		0.159						3.84			
0.156	1.34	0.728	0.19	0.673	0.0849	0.604	0.121		0.906	0.18	0.326						0.543	0.552		
1.21	4.93	0.750	0.200	0.670	0.090	0.350	0.070	0.200	0.210	0.040							0.152			
0.032	0.387	0.572	0.394	2.09	0.54	4.62	1.16	3.62	3.66	0.60	0.48	0.002	0.119	0.001	0.0032	0.0009	1.478	1.260	0.027	0.012
5.01	17.5	2.17	0.495	2.23	0.164	0.678	0.025	0.235	0.359	0.002	1.13	0.039					0.124	0.002		
5.28	28.6	4.27	0.557	0.324	0.087	0.088	0.051	0.163	0.260		0.619						0.149			
1.31	5.83	1.09	0.342	0.543	0.082	0.449	0.044		0.146	0.002	0.233		3.43		0.022		0.187	0.010	0.007	
0.742	4.30	1.50	0.546	1.68	0.232	1.15	0.176	0.358	0.202	0.025	0.99	0.040	0.039	0.022	0.0063	0.0001	0.350	0.025	0.160	0.572
0.280	2.34	1.27	0.560	2.36	0.450	3.33	0.740	2.19	1.96	0.310	0.730	0.002		0.016	0.003	0.0001	0.543	0.425		
	3.69	1.61	0.645	2.52		3.28		2.14	1.9	0.280	1.020	0.041		0.031	0.010	0.0001	0.436	0.275		
	10.1	2.83	0.960	2.85		2.280		1.11	0.893	0.131	1.320	0.215		0.620	0.136	0.0002	0.280	0.099		
6.29	27.2	7.14	1.83	6.22	0.820	4.31	0.970	2.15	1.46	0.254	1.41	0.083				0.0050	0.263	0.180		
6.00	28.0	5.30	1.60	3.60	0.390	1.90	0.350	0.680	0.440	0.040	2.7	0.01	1.5	0.120	0.030	0.0013	0.189	0.015	0.020	0.080
0.012	0.027	0.050	0.018	0.079		0.064	0.006	0.032	0.035	0.003	0.088						1.898	0.035		
0.008	0.048	0.029	0.013	0.052	0.010	0.071	0.015	0.040	0.038	0.006	0.063	0.007	0.035	0.0016	0.0002	0.0199	0.606	0.097	0.007	0.046
0.004	0.033	0.021	0.011	0.052	0.013	0.128	0.037	0.148	0.247	0.051	0.035	0.001		0.002	0.001	0.0222	0.636	1.457		
	0.020	0.019	0.009	0.042		0.116		0.137	0.215	0.041	0.045	0.001			0.001	0.0388	0.950	0.911		
0.001	0.008		0.001	0.004		0.005	0.002	0.009	0.018	0.005	0.004	0.002		0.002		0.6579	0.000	1.190		
	0.002		0.0002	0.001		0.003		0.007	0.018	0.005	0.003	0.007		0.001	0.0004	0.9200		1.660		
	0.003		0.0013	0.004	0.0008	0.0068	0.002	0.0053	0.001	0.0047	0.0012	0.0033	0.0012	0.015	0.0003	0.4838	0.000	3.757	0.251	12.169
0.0001			0.0002			0.0007			0.0003		0.0002	0.0001		0.0001		6.000				
	0.017	0.011	0.004	0.016		0.032		0.024	0.015	0.002	0.020	0.058		0.005	0.001		0.647	0.100		
1.88	10.8	3.360	1.280	3.350	0.540	2.550	0.400	0.790	0.320	0.032	4.85	1.33		0.034	0.001		0.311	0.007		
0.850	3.40	0.500	0.120	0.250	0.025	0.130	0.020	0.050	0.050	0.010	0.990	0.060	3.30	0.300	0.210	0.037	0.147	0.010	0.064	0.091
0.780	3.00	0.740	0.120	0.540	0.060	0.320	0.050	0.120	0.080	0.010	1.800	0.060	4.30	0.670	0.280	0.068	0.247	0.006	0.065	0.156
1.62	9.22	2.88	1.02	2.99	0.39	1.83	0.26	0.46	0.21	0.019	7.54	6.65	0.590	0.072	0.017	0.014	0.312	0.003	0.029	0.122
8.14	32	6.70	2.53	6.53	0.715	4.28	0.762	1.97	1.23	0.265	2.07	2.79				0.034	0.209	0.128		
0.006	0.031	0.026								0.006	0.180		4.38	1.15	0.088	332.5	0.818	0.032	0.020	0.263
																129.8				
0.027	0.29				0.006	0.056	0.011	0.059	0.120	0.026	0.130	0.370	15.4	0.008	0.009	0.148	0.000	0.200	0.001	0.001
											0.780									
315	1035	127	32.3	78.8	10.4	40.2	5.80	11.4	6.1	0.740	0.187	0.120		661	126		0.123	3.957		
	87.0	13.9							5.7		0.022			97.0			0.160			
0.859	9.77	9.28	3.26	19.2	5.88	49.8	12.8		54.5	10.9	14900		18.5	71.2	86.2		0.950	0.001	4.659	3.849
											1.00	40.0								
											28.0	46.0								
											407	90								
0.180	0.900			0.470											0.160					

gt = garnet, cpx = clinopyroxene, opx = orthopyroxene, olv = olivive, amph = amphibole, phlog = phlogopite.
1 = this study; 2 = Stachel *et al.* (1998); 3 = Ionov (1996); 4 = Bedini and Bodinier (1999); 5 = Johnson *et al.* (1996); 6 = Gregoire *et al.* (2002); 7 = Ionov *et al.* (1997); 8 = Ionov (1998); 9 = Eggins *et al.* (1998); 10 = Glaser *et al.* (1999). c = cratonic, oc = off-craton. Type refers to classification of Table 1.

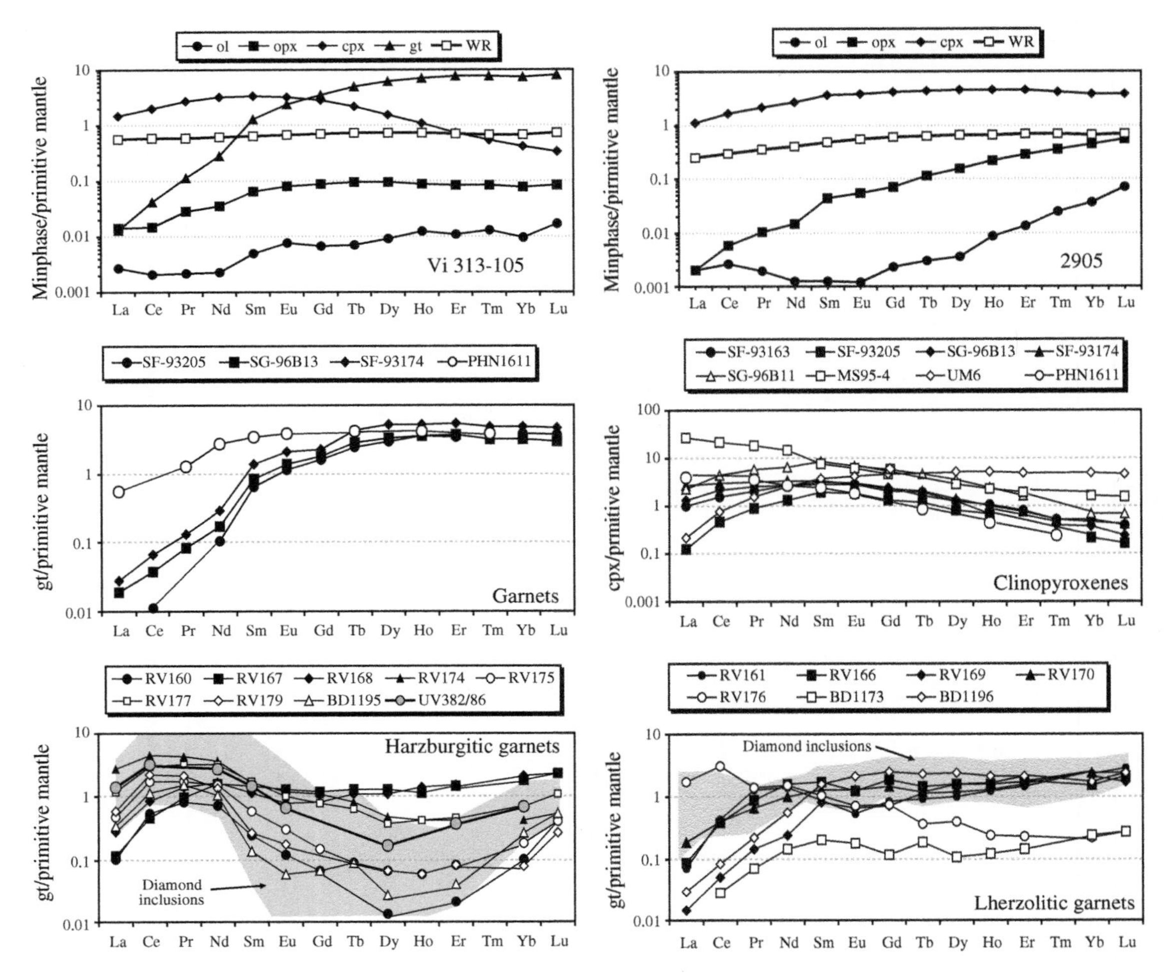

Figure 22 Primitive mantle normalized REE patterns in minerals from peridotite xenoliths and diamond inclusions. Upper two diagrams show coexisiting minerals in a noncratonic garnet lherzolite (Vi 313-105; Vitim suite, Ionov, 1996, 2004; Table 9) and a noncratonic spinel lherzolite (2905, SE Australia; Eggins *et al.*, 1998; see also Table 9). Middle left diagram shows REE data for garnets from noncratonic garnet lherzolites (SF-93205, SG-96B13, SF-93174; Glaser *et al.*, 1999) and high-*T* cratonic garnet lherzolite (PHN1611; Shimizu, 1975). Middle right diagram shows REE data for clinopyroxenes from off-craton (Vitim) garnet lherzolites (SG93163, SF-93205, SG-96B13, SF-93174, SG-96B11; Glaser *et al.*, 1999), off-craton (SE Australia) spinel lherzolites (MS95-4, UM-6; Norman, 1998) and cratonic high-*T* garnet lherzolite (PHN1611; Shimizu, 1975). Lower diagrams show REE patterns for garnets from cratonic lherzolites and harzburgites. Lower left diagram shows garnets in harzburgite xenoliths from the Roberts Victor mine, S. Africa (Stachel *et al.*, 1998). Also shown is a garnet from a diamondiferous low-calcium garnet dunite from Udachnaya, Siberia (UV382/86; Pearson *et al.*, 1995a). Shaded field is for the range of harzburgitic garnet inclusions in diamonds from Akwatia, Guinea (Stachel and Harris, 1997). Lower right diagram shows garnets in lherzolite xenoliths from the Roberts Victor mine, S. Africa (Stachel *et al.*, 1998). Shaded field is for the range of lherzolitic garnet inclusions in diamonds from Akwatia, Guinea (Stachel and Harris, 1997).

no clinopyroxene (Bedini and Bodinier, 1999; Eggins *et al.*, 1998; McDonough *et al.*, 1992). The abundances of strontium, niobium, zirconium, and yttrium in orthopyroxenes are near or below the ppm-level and show no clear correlations. A general feature is enrichment of titanium, zirconium, and niobium relative to coexisting clinopyroxene (Eggins *et al.*, 1998; Norman, 1998, 2001; Simon *et al.*, 2003). The few measurements of the REE patterns in orthopyroxene are typically LREE-depleted, with all REE one to two orders of magnitude below clinopyroxene (Figure 22). There is far more scatter and far less coherent variation amongst many incompatible trace elements in orthopyroxene, likely reflecting more heterogeneity on the micro-scale (Shimizu, 1999) as well as temperature effects that are not well understood with the paucity of data available (Norman, 2001).

2.05.2.6.3 Clinopyroxene

Clinopyroxene is a major host for many incompatible trace elements in peridotite xenoliths (Table 9) and as such its trace-element composition is a useful indicator of chemical modification in the mantle. Clinopyroxene partitions titanium and vanadium equably with spinel but is a major host for strontium and scandium. The partition of the latter element is T sensitive. The trivalent elements all show strong correlations in clinopyroxene from all types of xenoliths. There are significant depletions of zirconium, titanium, and yttrium relative to elements with similar compatibilities (REE), but coexisting orthopyroxene shows a complementary enrichment in these elements (McDonough, 1990). Clinopyroxene generally has lower niobium than garnet or mica/amphiboles (Figure 23). Nb/Ta is close to the PUM value in most instances, despite very variable abundances of both elements. Clinopyroxenes from metasomatic assemblages such as MARID xenoliths show very large negative HFSE anomalies due to equilibration with HFSE-rich phases such as ilmenite, amphibole, mica, and rutile.

Clinopyroxene shows a range of REE patterns from extremely enriched to very depleted LREE signatures (Figure 22). Noncratonic peridotites are subdivided on the basis of clinopyroxene REE patterns into LREE-depleted (type IA) and LREE-enriched (type IB; Menzies, 1983; Figure 17). LREE-enriched type IB pyroxenes are the norm in most suites. LREE-depleted varieties are relatively scarce. Very few clinopyroxenes show simple LREE-depleted REE patterns that can be interpreted solely in terms of melt depletion, i.e., LREE depletion, flat, unfractionated MREE-HREE patterns (e.g., UM-6 or 2905; Figure 22). For peridotites that do have LREE-depleted clinopyroxenes, a correlation of HREE with other incompatible trace elements (e.g., yttrium, strontium, zirconium) in xenoliths suites worldwide requires fractional melting to be the principal means of depletion in the mantle (Norman, 2001).

Clinopyroxenes that have equilibrated with garnet in off-craton garnet peridotites show

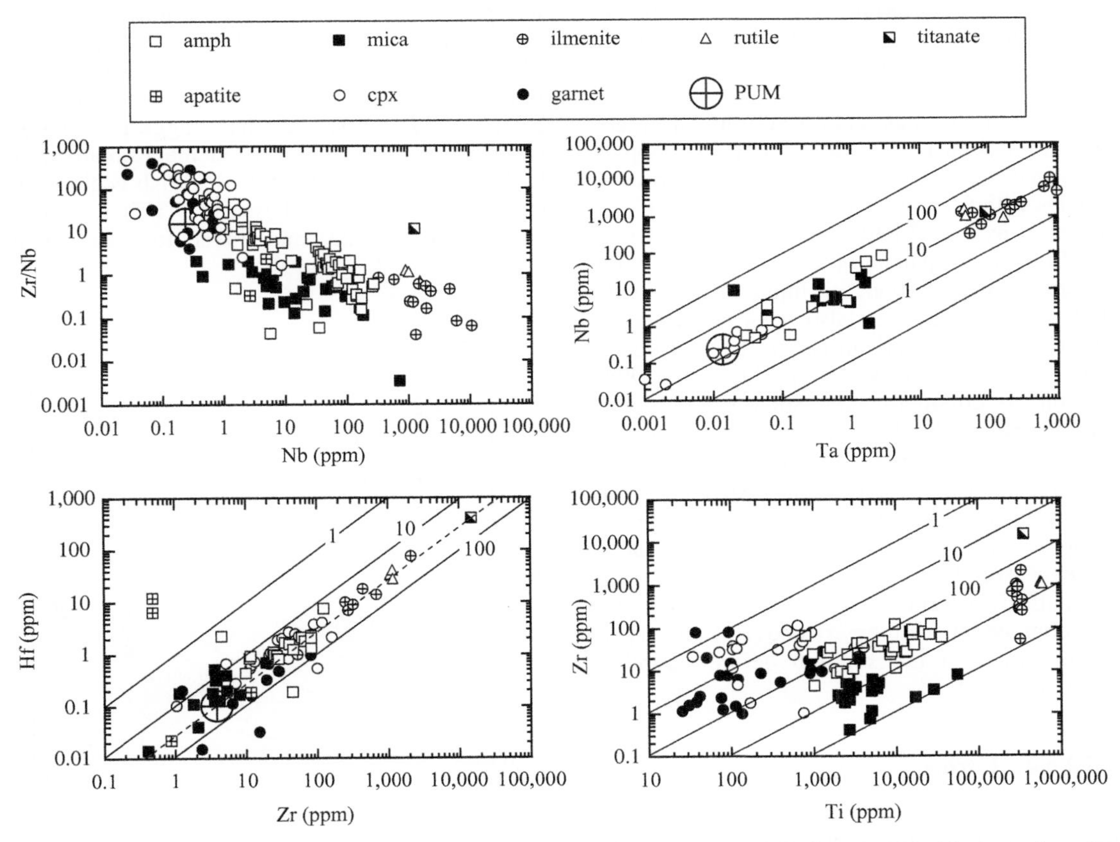

Figure 23 High field strength element (HFSE) variations in mantle xenolith minerals. Primitive mantle values plotted are from McDonough and Sun (1995). Dashed lines extend from PM values at constant ratios, e.g., at Nb/Ta = 17.6, Zr/Hf = 37. Solid lines are constant elemental ratios (1–100 labeled) (sources Moore *et al.*, 1992; Witt-Eickschen and Harte, 1994; Vannucci *et al.*, 1995; Ionov and Hoffman, 1995; Johnson *et al.*, 1996; Chazot *et al.*, 1996; Vaselli *et al.*, 1996; Ionov *et al.*, 1997; Shimizu *et al.*, 1997; Norman, 1998; Glaser *et al.*, 1999; Pearson and Nowell, 2002; Gregoire *et al.*, 2002; Kalfoun *et al.*, 2002; Table 9).

substantially lower levels of HREE than those from spinel-facies peridotites (compare Vi313-105 and 2905; Figure 22). Clinopyroxene REE systematics are linked to texture/equilibration temperature (Figure 17). Clinopyroxene in low-T (granular) peridotites is more LREE enriched and HREE depleted than that from high-T (sheared) peridotites. The relationship between garnet and clinopyroxene in high-T peridotites is close to equilibrium (e.g., PHN1611 shown in Figure 22) and mirrors the relationship shown by off-craton garnet peridotites (e.g., Vi313-105). In contrast, the HREE-depleted clinopyroxenes from the low-T peridotites are often not in equilibrium with their coexisting garnets (Shimizu *et al.*, 1997; Shimizu, 1999; Simon *et al.*, 2003).

LREE enrichment in clinopyroxene is widely linked to metasomatism by either silicate melts or carbonatitic fluids, usually via cryptic (in the sense of no new phase being introduced) metasomatism, although there is a strong possibility that the diopside itself could have precipitated from the melt. This is strongly favored for much of the ubiquitously LREE-enriched diopside found in cratonic peridotites (Figure 17) which are not in trace-element equilibrium with their coexisting garnets and for which equilibrium melts for the diopside closely resemble kimberlite (Simon *et al.*, 2003).

Metasomatism has also fractionated trace-element ratios such as Ti/Eu or Zr/Hf in clinopyroxene that are invariant to the melting process. Samples with fractionated Ti/Eu and or Zr/Hf and possessing extreme LREE-enrichment (near 100 × chondrite) show patent or cryptic evidence for metasomatism by carbonatitic liquid (Yaxley *et al.*, 1998). This process has also been suggested to impart low aluminum (Hauri *et al.*, 1993; Rudnick *et al.*, 1993) and high sodium (Yaxley *et al.*, 1998) to secondary clinopyroxenes. Some xenolith clinopyroxenes, particularly from low-T cratonic xenoliths, show extreme heterogeneity in trace elements. Incompatible element abundances can vary by factors of 2–10 on millimeter scales in clinopyroxenes. Diffusion should erase these gradients over longer time periods at high T in the mantle (Van Orman *et al.*, 2001), requiring that some component of clinopyroxene crystallized late in the history of the xenolith before sampling (Shimizu, 1999).

2.05.2.6.4 Garnet

Garnet can also be a major host for trace elements in peridotite xenoliths, especially in harzburgites. A large data set was recently summarized by Griffin *et al.* (1999a). The abundances of manganese and nickel in peridotitic garnets change due to T-dependent partitioning with coexisting olivine (Sachtleben and Seck, 1981; Griffin *et al.*, 1989; Smith *et al.*, 1991; Canil, 1999). The levels of scandium, titanium, and vanadium in xenolith garnets are commonly correlated with their level of depletion (as measured by the chromium content of the garnet) and to T-sensitive partitioning with coexisting pyroxenes, but the former effect is not well constrained. Trends of zirconium and yttrium in garnets are also correlated with different styles of depletion and/or chemical modification in the mantle (Griffin *et al.*, 1999c). Levels of yttrium in mantle garnet show a strong correlation with depletion (chromium content) and garnets very depleted in this element (<10 ppm) mainly occur in mantle sampled beneath Archean terrains (Griffin *et al.*, 1999a). Because yttrium is compatible in garnet, a low yttrium content implies that this phase was absent during the melting process, but has exsolved during cooling. The positive correlation of yttrium with titanium, gallium, and zirconium, suggests that all of these elements are sensitive to melt depletion. Correlations of the latter three elements with chromium, however, are weak, likely due to the effects of metasomatism. At $T < 1000°C$, metasomatism is interpreted to lead to preferential introduction of zirconium over titanium and yttrium (Figure 24). There are strong correlations of zirconium, titanium, yttrium, and gallium in garnets equilibrated at $T > 1{,}000\ °C$ that suggest a major contribution from melt-related metasomatism at higher T (Griffin *et al.*, 1999a). Zr/Nb is close to, or higher than PUM while Ti/Zr is very variable (Figure 23) and linked to garnet type

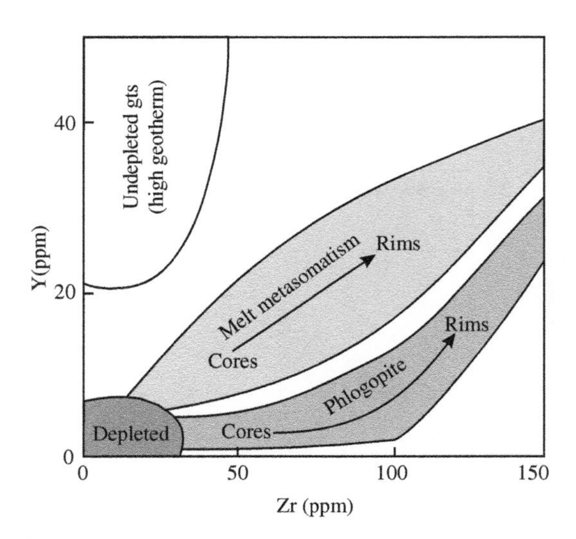

Figure 24 Covariation of yttrium and zirconium in mantle garnets showing fields ascribed to different mantle protoliths (undepleted, depleted) and processes (high-T melt metasomatism, low-T phlogopite metasomatism) and zonation patterns from cores to rims of garnets (after Griffin *et al.*, 1999c).

(Shimizu *et al.*, 1997). Levels of niobium in some garnets can equal those of clinopyroxene.

Two distinct REE patterns are recognized in mantle garnets (Figure 22). Garnets with normal heavy rare earth element (HREE) enriched patterns and garnets with sinuous REE patterns. The normal garnets show depletion in the LREE, steady enrichment from Sm_N to Yb_N and are most frequent in noncratonic xenolith suites or high-T cratonic xenoliths (Shimizu, 1975; Glaser *et al.*, 1999; Figure 22). These garnet REE systematics are compatible with those expected from melt-residues, or garnets that have equilibrated with silicate melts. In contrast, the sinuous REE patterns peak at either neodymium or samarium, Sm_N is greater than Dy_N, and can have fairly flat trends between the HREE or Yb_N is greater than Er_N (Shimizu, 1975; Shimizu and Richardson, 1987; Shimizu *et al.*, 1997; Stachel *et al.*, 1998; Figure 22). This garnet REE pattern is commonly recognized in low-T cratonic xenoliths (harzburgites) or diamond inclusion garnets that are depleted in calcium and have not equilibrated with clinopyroxene (Shimizu and Richardson, 1987; Hoal *et al.*, 1994; Pearson *et al.*, 1995b; Stachel *et al.*, 1998).

Various hypotheses have been proposed to explain the origin of the sinuous REE patterns. Two-stage models of melt depletion followed by metasomatism have been proposed. For instance, Stachel *et al.* (1998) suggest that the high chromium content (>5%) of these garnets is a reflection of the protolith origin as a residue from polybaric melting to create Archean ocean lithosphere. As the oceanic lithosphere is subducted, garnets growing from this depleted protolith were over-printed by a metasomatic fluid with high LREE/HREE. Although attractive, these models have problems. The wide range of HREE shown by subcalcic garnets is difficult to account for. In addition, where overprinting of LREE enrichment occurs in other LREE-depleted minerals such as clinopyroxene or olivine, "spoon-shaped" patterns, with maxima at lanthanum occur rather than sinuous patterns with peaks at neodymium or samarium. Hoal *et al.* (1994) found sinuous REE patterns in calcium-saturated pyrope garnets as well as in subcalcic garnets. These authors propose disequilibrium effects to develop sinuous garnet REE patterns that cause the LREE to diffuse and readjust faster than the HREE during chemical modification of garnet in the mantle. While differences in the diffusion coefficients of LREE and HREE have been measured for diopside (Van Orman *et al.*, 2001), that could account for some of the more complex REE patterns measured in this mineral, experimental determinations have not found any difference in LREE-HREE diffusion coefficients for garnet (Van Orman *et al.*, 2002). Normal and sinuous REE patterns can be present in different parts of the same garnet (Shimizu, 1999; Griffin *et al.*, 1999c), arguing for a disequilibrium feature, arrested shortly before sampling. This is supported by the widely ranging and often unsupported neodymium isotope systematics of these garnets (Pearson *et al.*, 1995a). The origin of the sinuous REE pattern in mantle garnets remains to be unequivocally demonstrated.

The considerable REE heterogeneity in many cratonic garnets is also reflected in other trace elements. Abundances of zirconium, yttrium, hafnium, cerium, and strontium can vary by over a factor of two, and correlate positively, both for multiple analyses of the same garnet grain, or for analyses of different garnet grains from the same sample. The peculiar and varied REE partitioning of mantle garnets leads to highly variable Sm/Nd and Lu/Hf (Figures 25 and 26), many of which are

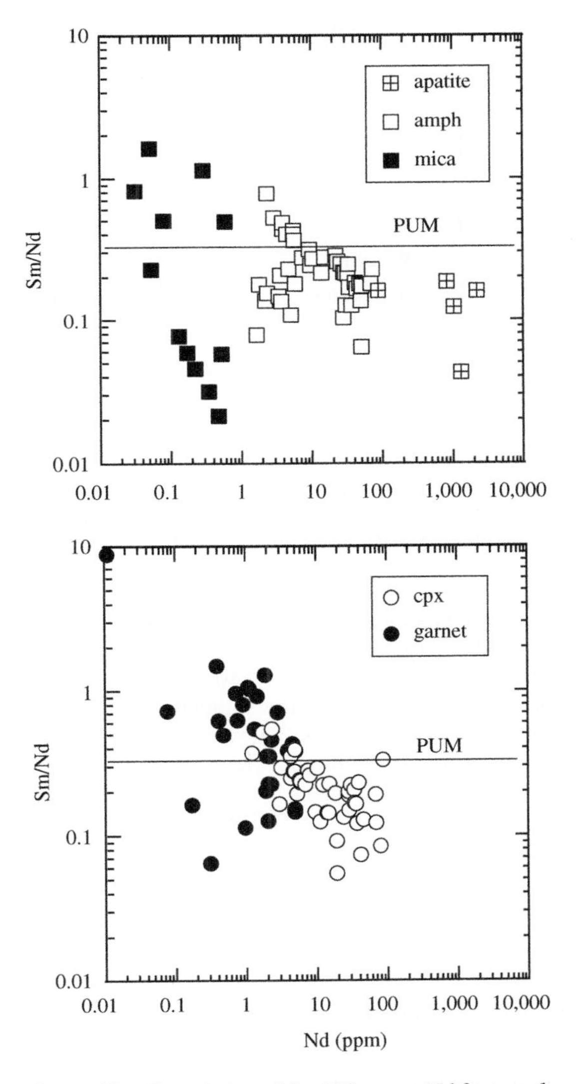

Figure 25 Covariation of Sm/Nd versus Nd for mantle xenolith minerals. Primitive mantle (PUM) line for Sm/Nd marked. Data sources as in Figure 23.

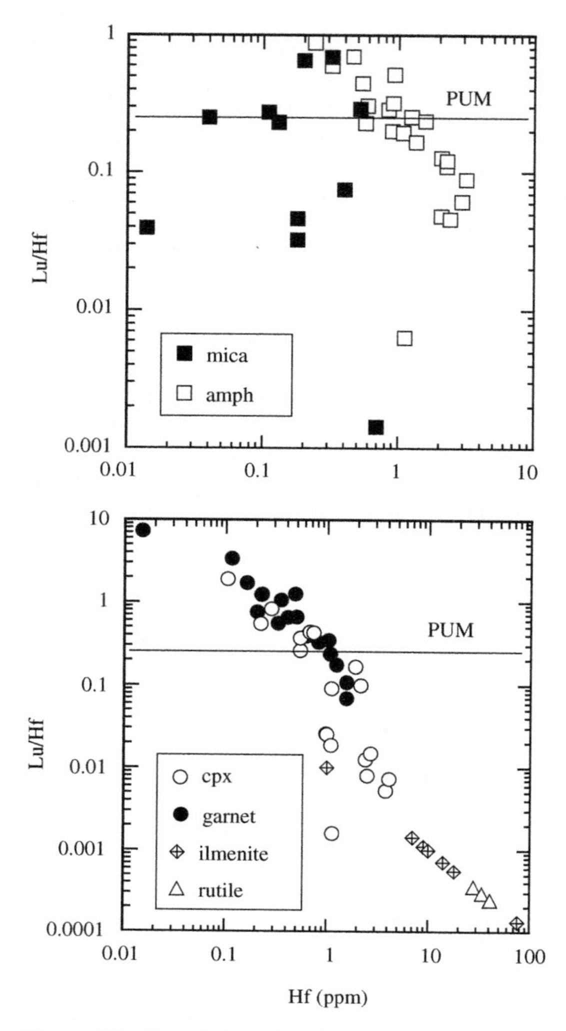

Figure 26 Covariation of Lu/Hf versus Hf for mantle xenolith minerals. Primitive mantle (PUM) line for Lu/Hf marked. Data sources as in Figure 23.

not those expected for equilibrium with a silicate melt. The strontium contents of mantle garnets range widely (0.1 ppm to >40 ppm) and often correlate positively with the sinuosity of the REE pattern (e.g., $(Nd/Yb)_N$). The very high levels of strontium are all from subcalcic, high-chromium garnets of the low-calcium harzburgite/dunite suite or diamond inclusions and have been linked to equilibration with a carbonate-melt fraction possibly during diamond genesis (Pearson *et al.*, 1995c).

2.05.2.6.5 Spinel

Because spinel is a very minor phase in peridotites, it makes up only a very small proportion of the trace-element budget (Eggins *et al.*, 1998; Norman, 2001). Spinel has a limited database for trace cations. The compatible trace-element data for nickel, cobalt, manganese, titanium, vanadium, scandium, zinc, and gallium are shown in Table 9. The nickel and zinc contents are sensitive to *T*-dependent partitioning with olivine, and zinc can be a useful single mineral trace-element geothermometer (Griffin *et al.*, 1993). Experimental work indicates the partition of vanadium into spinel is f_{O_2} sensitive (Horn *et al.*, 1994; Canil, 2002). Moderate concentrations of zirconium and niobium can also be present. Zr–Nb enriched reaction rims have also been found on spinels from numerous off-craton spinel lherzolites (Bedini and Bodinier, 1999; Figure 16).

2.05.2.6.6 Amphibole, phlogopite, apatite, zircon, rutile, ilmenite, titanate, and carbonate

Volatile-bearing minerals (mica, amphibole, apatite) together with titanium-rich phases such as ilmenite, rutile and titanates, are the major manifestation of modal or "patent" metasomatism in mantle xenoliths. The effects of carbonatite metasomatism are widely inferred in many xenolith suites from trace-element systematics (e.g., Hauri *et al.*, 1993; Rudnick *et al.*, 1993; Coltorti *et al.*, 1999) but carbonate of mantle origin is rare in xenoliths (Ionov, 1998). Mantle xenolith studies have allowed direct measurement of both inter-mineral and mineral–melt (through the presence of interstitial glass) trace-element partitioning (Chazot *et al.*, 1996; Ionov *et al.*, 1997). For brevity we will not specifically deal with partitioning data here.

In an extensive review of the geochemistry of volatile-bearing minerals in mantle xenoliths, Ionov *et al.* (1997) have pointed out that although minerals such as mica, amphibole, and apatite are often referred to as "hydrous," in many cases they have very low H_2O contents (Boettcher and O'Neill, 1980). In such cases, these minerals may have significant amounts of fluorine, chlorine and CO_2. Mica, amphibole, and apatite, together with the oxide phases, are important hosts for titanium, potassium, rubidium, strontium, barium, and niobium (Table 9).

Amphibole. Mantle amphibole equilibrated with clinopyroxene generally has lower Rb/Sr (<0.03) than PUM, whereas amphibole within veins and some amphibole-rich peridotites have considerably higher Rb/Sr (Table 9). The weighted mean Rb/Sr (0.036) is very close to the PUM value of 0.03 but values vary widely (Figure 27). Typical strontium contents vary from 200 ppm to 2,000 ppm, averaging close to 500 ppm and are generally higher than those of mica (Figure 27). Ba in mantle xenolith amphiboles typically varies from 100 ppm–1,500 ppm (Table 9;

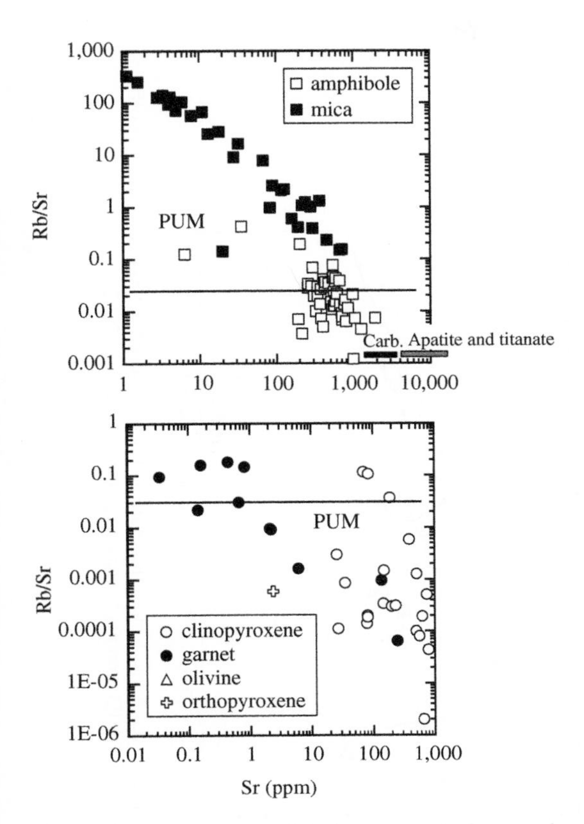

Figure 27 Covariation of Rb/Sr versus Sr for mantle xenolith minerals. Primitive mantle (PUM) line for Rb/Sr marked. Data sources as in Figure 23 (Carb. = carbonate).

Ionov *et al.*, 1997). Both strontium and barium are higher in amphibole from xenoliths than those from massif peridotites (e.g., Vannucci *et al.*, 1995). Numerous ion-probe and laser/solution ICP-MS studies have found wide variation in amphibole REE patterns depending on textural context and tectonic setting (Witt- Eickschen and Harte, 1994; Vannucci *et al.*, 1995; Ionov and Hofmann, 1995; Johnson *et al.*, 1996; Chazot *et al.*, 1996; Vaselli *et al.*, 1995; Ionov *et al.*, 1997; Pearson and Nowell, 2002; Gregoire *et al.*, 2002).

Amphiboles show a wide variety of REE patterns that cannot be uniformly related to modal abundance. Ionov *et al.* (1997) suggest that where the mineral has equilibrated extensively with silicate melt, LREE-enrichment and convex REE patterns occur. Hence, vein amphiboles from noncratonic xenoliths are enriched in LREE and MREE over HREE and thus have convex REE patterns (e.g., Figure 28). Disseminated amphiboles have lower MREE/HREE in some instances but those in xenoliths from the W. Eifel region are much more LREE-enriched than amphibole from other locations. Disseminated amphibole can be either LREE enriched or depleted relative to PUM (Figure 28). The transition from convex-shaped REE patterns in vein amphibole to LREE-depleted disseminated amphibole in vein wall rock has been clearly documented (Vaselli *et al.*, 1995; Zanetti *et al.*, 1996). K-richterites from MARID xenoliths are characteristically LREE-enriched with convex patterns (Gregoire *et al.*, 2002; Figure 28). In contrast to amphibole from xenoliths, disseminated amphibole from massif peridotites is almost exclusively LREE depleted (Vannucci *et al.*, 1995; Zanetti *et al.*, 1996), possibly reflecting more extensive equilibration with other phases. Most xenolith-derived amphibole has Sm/Nd < PUM (Figure 25) and hence, with time, will develop unradiogenic neodymium isotopic signatures with respect to PUM. However, a large proportion of disseminated amphibole has Sm/Nd close to, or above PUM. Hence, the presence of amphibole in a peridotite is no guarantee of the generation of "enriched" neodymium isotopic signatures with time.

Vein amphiboles are generally richer in TiO_2 than disseminated varieties, with Ti/Zr close to the PUM value (112) or above (Figure 23). LREE-enriched disseminated amphiboles from the W. Eifel peridotite xenoliths are low in TiO_2 (<1%) and have lower Ti/Zr than PUM (Witt-Eickschen and Harte, 1994). Multielement patterns are characterized by positive titanium and niobium anomalies (Figure 28). Amphibole has high but variable levels of both zirconium and niobium compared to pyroxenes and garnets. Niobium values vary from <1 ppm to 270 ppm (mean ~25 ppm) and zirconium varies from <1 ppm to 300 ppm (mean ~45 ppm). Zr/Nb is lower than clinopyroxene and generally significantly less than PUM for most amphiboles. This is largely a function of the high niobium contents of amphiboles and means that bulk rock Zr/Nb will be lower than PUM for amphibole-rich samples. Consequently, melts of amphibole-peridotite should have low Zr/Nb.

Amphibole HFSE characteristics are sensitive to the presence of ilmenite and rutile. In MARID samples where rutile dominates over ilmenite, Gregoire *et al.* (2002) note that K-richterites display large negative niobium and tantalum anomalies, whereas when ilmenite dominates, these anomalies are positive. Nb/Ta ratios of amphibole vary from close to the PUM value (17.6) up to ~25 (Figure 23). Zr/Hf values are generally lower than the PUM value of 37 (Figure 23). Differences in HFSE contents between vein and disseminated amphibole (and mica) in spinel peridotites may be explained by a model in which Zr–Nb rich amphibole and mica crystallize close to, or within a melt vein in the mantle. The fractionated, chlorine-rich aqueous residual fluids from the evolved melt then crystallize low Zr–Nb, LREE-depleted amphibole or

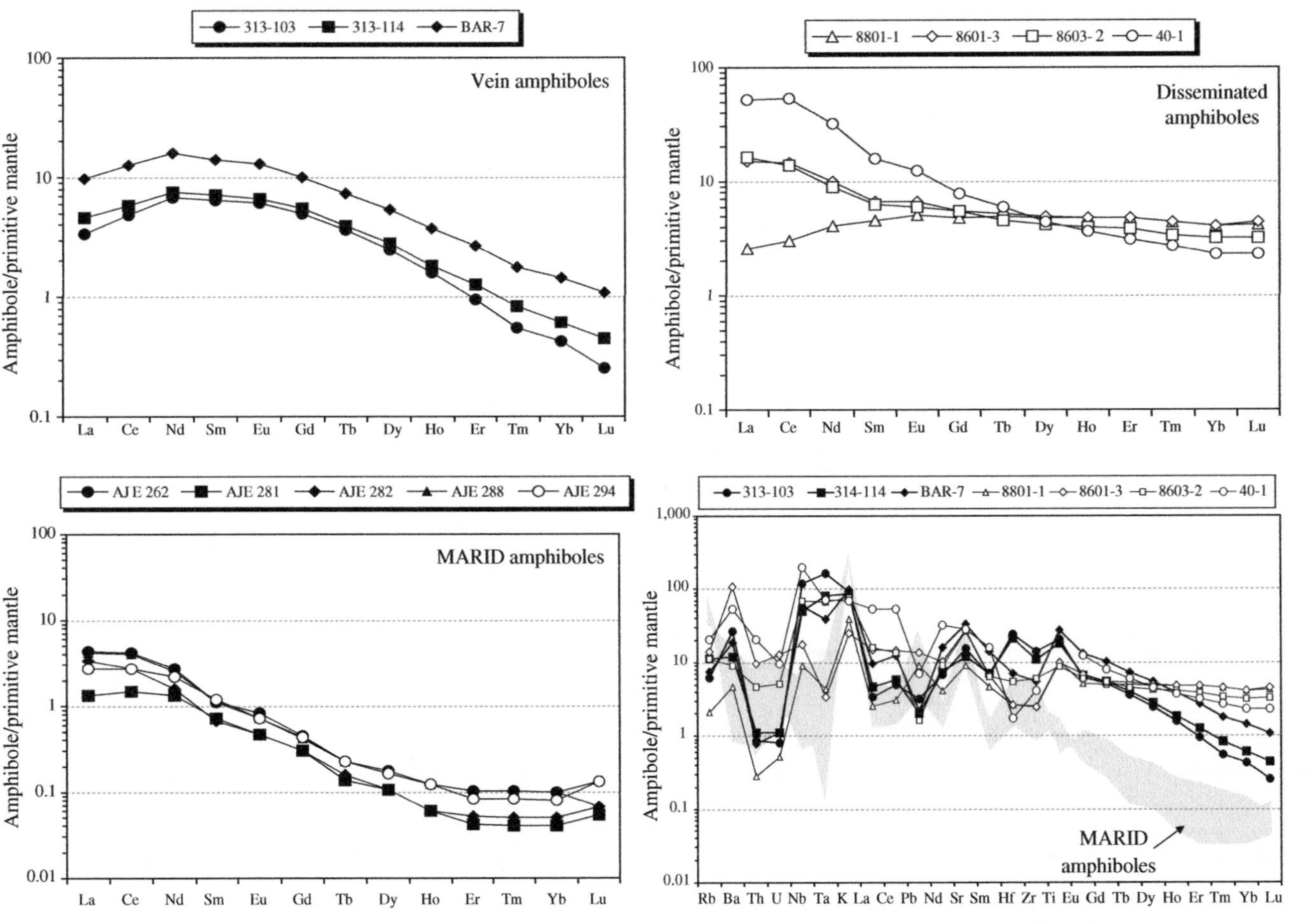

Figure 28 Primitive mantle normalized REE and multi-element patterns for amphiboles from mantle xenoliths. Vein amphibole and disseminated amphibole data from off-craton spinel lherzolite xenoliths (Ionov and Hofmann, 1995; Ionov *et al.*, 1997). MARID amphibole data from Gregoire *et al.* (2002).

mica at a distance from the feeder, as disseminated metasomatic phases within peridotite (Ionov *et al.*, 1997). The residual chlorine-rich fluids are effective at transporting LILE and so disseminated mica crystallising away from the vein in this model is barium and lead rich.

Amphibole in spinel peridotites can have similar zirconium and hafnium contents to clinopyroxene. Lu/Hf ratios span the range of clinopyroxene and can be greater or less than PUM (Figure 26). In the absence of more exotic phases (see below), amphibole and mica dominate the niobium, tantalum, and titanium budget of peridotites (Ionov *et al.*, 1997). To this extent, Pearson and Nowell (2002), have pointed out that any hydrous peridotite proposed as a source for flood basalt volcanism is unlikely to yield melts with the negative HFSE anomalies measured in flood basalts (e.g., Thompson *et al.*, 1983). Amphibole has lead contents in the upper range, to well in excess of those commonly measured in clinopyroxenes (Figure 29), with only mica, of the silicate phases, containing more lead. U/Pb ratios are predominantly greater than PUM but clinopyroxene commonly has a higher U/Pb ratio. Lithium and boron in amphibole are commonly around 1 ppm and thus significantly lower than the levels observed in phlogopite (Table 9; Glaser *et al.*, 1999).

In summary, amphibole, along with mica, is the dominant silicate host for niobium in peridotitic xenoliths. In mica-absent assemblages amphihole also dominates the barium and tantalum budgets (Ionov *et al.*, 1997; Eggins *et al.*, 1998) and its presence strongly affects the bulk rock Zr/Nb ratio.

Mica. Mica, usually phlogopite, is the dominant host for rubidium and barium in the mantle but has highly variable Rb/Sr (0.13–60) and Ba/Sr ratios (Ionov *et al.*, 1997). As with amphibole, the geochemistry of mantle mica varies with texture and hence probably proximity to melt veins (Ionov *et al.*, 1997). Rb contents of micas in peridotites vary from 10 ppm to ~800 ppm. Strontium contents vary from ~1 ppm to almost 1,000 ppm and this variation creates highly variable Rb/Sr that is significantly higher than PUM in most cases (Figure 27). Phlogopites from cratonic type VI and VIII xenoliths, e.g., MARIDs, are exceptionally rubidium-rich compared those found in noncratonic xenoliths (Table 9). Ba contents are very variable and are probably related to the nature of the parental fluid and its evolution away from silicate melt veins. Micas from noncratonic peridotite xenoliths have been reported with 0.5–10 wt.% BaO (Ionov *et al.*, 1997). MARID micas are relatively barium poor, with 300–700 ppm. Along with amphibole, mica dominates the niobium and tantalum budget of silicate-hosted trace elements. Nb/Ta can be close to PUM but shows some scatter (Figure 23) that could be partly related to analytical problems at low tantalum levels. As expected, Ti/Zr is higher than clinopyroxene and largely overlaps the amphibole range, but at lower zirconium contents. As with amphibole, Zr/Nb is low, from a combination of relatively low zirconium and high niobium. Niobium levels are comparable with amphibole. Zr/Hf scatters around PUM values. Mica hafnium levels are lower than amphibole and clinopyroxene (<1 ppm). Lu/Hf ranges widely because of wide variations in both elements (Figure 26). Some of the scatter could be analytical.

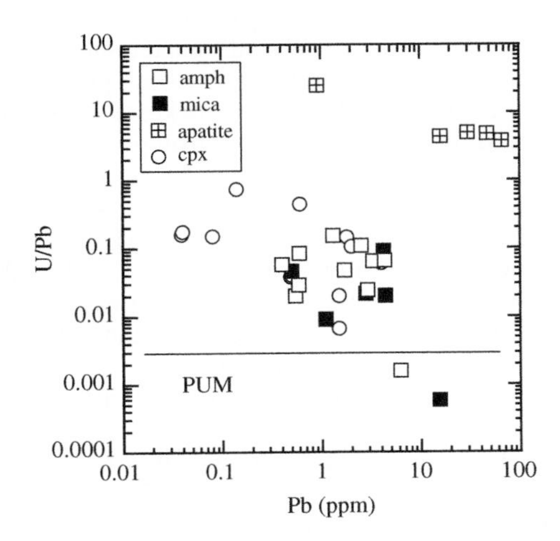

Figure 29 Covariation of U/Pb versus Pb for mantle xenolith minerals. Primitive mantle (PUM) line for U/Pb marked. Data sources as in Figure 23.

Mica from mantle xenoliths is very poor in REE, especially LREE (Table 9). Where reported, Sm/Nd values vary widely both above and below the PUM values (Figure 25); however, the low levels present could be easily influenced by small inclusions or analytical artifacts. Neodymium levels are so low that even the presence of tens of percent mica in a peridotite mineralogy do not significantly influence the neodymium budget of the rock. As such, the presence of mica in a peridotite is not sufficient to generate LREE enrichment and hence its presence alone, is unlikely to cause a bulk peridotite to evolve to "enriched" neodymium isotopic compositions (Pearson and Nowell, 2002).

LREE enrichment relative to PUM in whole-rock peridotites is more likely to be a product of grain-boundary enrichment or clinopyroxenes becoming LREE enriched. Micas are poor in thorium and uranium but high in lead compared to clinopyroxene so that U/Pb is low. Some micas can have up to 20 ppm lead and Rosenbaum (1993) proposes that they are the main mantle lead repository. More recent data have shown that amphiboles can have comparably high lead (Table 9).

Of the volatile-rich minerals, mica has the highest affinity for the light elements lithium and boron, being 3–5 times more enriched than amphibole or pyroxenes (Table 9; Glaser *et al.*, 1999) and hence will affect the light element budget of any peridotite. However, the large modal abundance of olivine together with its moderately high lithium content means that olivine will dominate the budget overall.

Apatite. Apatite is relatively common in noncratonic continental and oceanic peridotite xenoliths (Menzies and Wass, 1983; Ionov *et al.*, 1997). Most apatites are chlorine-rich (2–4.3%) with low F/Cl (0.1–0.3). High fluorine (~5%) and low chlorine (~0.25%) have been reported for apatite in spinel lherzolites from Pacific OIB (Hauri *et al.*, 1993). Extremely high strontium contents, commonly $>2 \times 10^4$ ppm and up to 7 wt.% (Ionov *et al.*, 1997; Table 9) are common in mantle apatites meaning that this phase is a major repository for strontium when present in peridotites at abundances of 0.1% or above. Rb/Sr is very low. Apatites have high levels of REE and are LREE-enriched (Table 9). Lanthanum and cerium concentrations can reach >1 wt.% and neodymium concentrations can be above 1,000 ppm. Sm/Nd is below PUM. HFSE are low and so the presence of this phase does not affect bulk rock HFSE chemistry.

Thorium and uranium contents of apatite vary widely but are normally very high compared to other mantle phases (generally >10 ppm; Table 9; Ionov *et al.*, 1997). Apatites in MARID xenoliths tend to have lower uranium (Kramers *et al.*, 1983), possibly due to uranium partitioning into rutile or zircon. Lead contents are the highest reported for the "common" mantle minerals but U/Pb is generally >>PUM.

Zircon, monazite, rutile, ilmenite, and titanate. Zircons in upper mantle xenoliths have been reported in MARID and PKP xenoliths and glimmerites (Kinny and Dawson, 1992; Konzett *et al.*, 1998; Hamilton *et al.*, 1998; Rudnick *et al.*, 1999). In contrast to rutile, its high blocking temperature and low common Pb have made zircon attractive for U–Pb dating of MARID assemblages. Zircon seems to be found associated with metasomatism characterized by the precipitation of phlogopite and phases such as ilmenite, rutile, monazite, and apatite (Rudnick *et al.*, 1999). Trace-element data are limited but it is apparent that PKP and MARID zircons have higher thorium and uranium contents than zircon megacrysts found in kimberlite (Konzett *et al.*, 1998). Zircons from glimmerites have higher U/Th than zircons from MARID/PKP xenoliths. Recent studies link both the geochemistry and geochronology of MARID xenoliths to the crystallization products of group II kimberlites in the lithospheric mantle (Hamilton *et al.*, 1998; Konzett *et al.*, 2000; Gregoire *et al.*, 2002). Monazite has been reported from metasomatized mantle xenoliths and is also useful for U–Pb geochronology (Rudnick *et al.*, 1993; Carlson and Irving, 1994).

Macrocrystalline rutile is a rare phase in peridotite xenoliths (Haggerty, 1987) but is an essential constituent of MARID xenoliths (Dawson and Smith, 1977). As such, the only published trace-element data for rutiles is from MARIDs (Gregoire *et al.*, 2002; Table 9). In addition to containing substantial vanadium (>1,000 ppm), the other two main elements hosted by rutile are zirconium and niobium, both of which occur at the 1,000 ppm level; Figure 23. In addition, elevated levels of hafnium and tantalum are present. Thus, rutile exerts a huge influence on the HFSE budget of peridotites. Rutile Nb/Ta ratios (~22) are slightly more elevated than PUM (17.6). Lu/Hf values are extremely low (<0.001) such that the mineral should be of great utility in tracing mantle initial hafnium isotope compositions. The potential for U–Pb geochronology using rutile has been hampered in mantle settings by the low closure temperature of the U–Pb system in rutile and its high common (nonradiogenic) lead content. Microcrystalline rutile, plus loveringite and armalcolite, are increasingly found as very finely disseminated phases, sometime coating spinels, in noncratonic lherzolites. These phases are major hosts for niobium and tantalum when compared to silicates (Bodinier *et al.*, 1996; Bedini and Bodinier, 1999; Kalfoun *et al.*, 2002) and have variable Nb/Ta (11–37). Despite high ZrO_2 contents (up to 7 wt.%), in most instances, rutiles do not dominate bulk rock zirconium contents because of the zirconium-rich nature of modally abundant clinopyroxene (Kalfoun *et al.*, 2002).

Ilmenite is a widespread titanium-rich oxide phase in mantle xenoliths, occurring in both metasomatized peridotites, pyroxenites, MARIDs and eclogites in cratonic and noncratonic suites. As with rutile, ilmenite is a HFSE-rich phase and its low Lu/Hf ratio (Figure 26) makes it attractive for determining initial hafnium isotope ratios and for controlling/defining the low Lu/Hf region of Lu–Hf isochrons (Nowell *et al.*, in press). Nb/Ta in ilmenites varies. Those from MARID rocks can be considerably greater than PUM (32; Gregoire *et al.*, 2002; Table 9), whereas megacryst ilmenite Nb/Ta may scatter either side of PUM values (Moore *et al.*, 1992).

Titanates are another group of oxide phases occurring in metasomatized peridotites and MARID assemblages (Jones *et al.*, 1982; Haggerty *et al.*, 1983; Haggerty, 1987). So far, these phases have not been recognized in noncratonic continental or oceanic xenoliths and hence are characteristic of kimberlite-related xenolith sampling. LIMA minerals are members of the lindsleyite (barium-specific) and mathiasite (potassium-specific) series in the crichtonite (strontium-specific) group. As such these phases are hugely enriched in LILE and also HFSE (Table 9). For instance, ZrO_2 contents of up to 4 wt.% and LREE contents of 1–2 wt.% are common in mathiasite (Haggerty, 1989). Strontium concentrations are high (over 2 wt.%), with very low Rb/Sr. Niobium, tantalum, and hafnium are all highly abundant compared to silicate mantle minerals but Nb/Ta values (~15) are close to PUM. The large enrichments of strontium, neodymium, and hafnium in these minerals make them obvious targets for isotopic measurements. K–Ba titanates of the LIMA group are typical of metasomatic associations in cratonic xenoliths of type VI and VII (Table 1) and probably originate from kimberlite-like melts.

Carbonate. Increasing recognition of carbonate within noncratonic peridotite xenoliths in particular has lead to improved geochemical documentation of this phase in the mantle (Ionov, 1998; Lee *et al.*, 2000; Ionov and Harmer, 2002). Of the incompatible elements, carbonate is most enriched in strontium (generally 1,000–5,000 ppm) and Rb/Sr is very low (Figure 30 and Table 9). HFSE are characteristically very low but Ionov and Harmer (2002) report unusual niobium values of ~100 × PUM in carbonates from Mongolian peridotites (Figure 30). U and Th contents are significantly higher than in silicate mantle phases (5–20 ppm) and lead is also generally elevated (10–30 ppm), reflecting the ease of substitution of Pb^{2+} for Ca^{2+}. REE contents are mostly low and not as LREE-enriched as carbonatites. The characteristics of carbonates from mantle xenoliths are similar to early-crystallized calcite phenocrysts in carbonatites (Ionov and Harmer, 2002) and led Ionov (1998), Lee *et al.* (2000) and Ionov and Harmer (2002) to conclude that carbonate grains and pockets in peridotites are crystal "cumulates" from carbonate-rich melts rather than quenched carbonatitic liquids.

2.05.2.7 Radiogenic Isotope Geochemistry

Radiogenic isotope studies of mantle xenoliths have been used to determine their age and to better constrain the origin of the various metasomatic events that have affected them. Extensive reviews on the subject have been published: Menzies (1990a); Menzies and Hawkesworth (1987); Pearson (1999a,b); and Pearson and Nowell (2002). The reader is referred to these works for a more detailed explanation of the isotope systematics.

2.05.2.7.1 Primary and secondary signatures

When ultramafic rocks were initially analyzed for strontium isotopes (Hurley *et al.*, 1964; Roe, 1964; Stueber and Murthy, 1966; Lanphere, 1968) only whole rocks were analyzed. Such data led to numerous erroneous conclusions about the relationship between mantle samples and basalts due to the pervasive alteration suffered by many ultramafic rocks, enhancing the amount of radiogenic strontium present. Subsequent studies have shown that acid-washed mineral separates, particularly diopside and garnet, are the most reliable means of obtaining the primary, unaltered

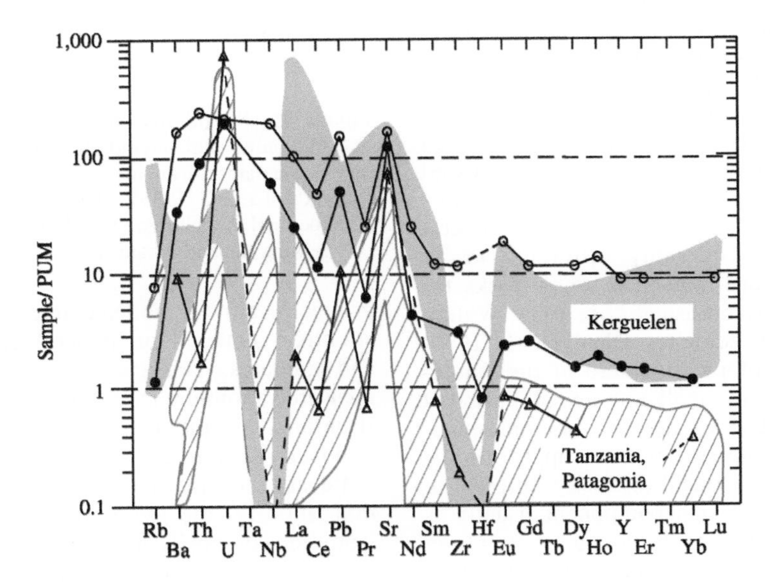

Figure 30 Primitive mantle normalized multielement patterns for carbonates from mantle xenoliths, modified from Ionov and Harmer (2002). Data shown as points are from Mongolia peridotite xenoliths, compared to carbonates in xenoliths from Kerguelen, Tanzania, and Patagonia. See Ionov (1998) and Ionov and Harmer (2002) for data sources.

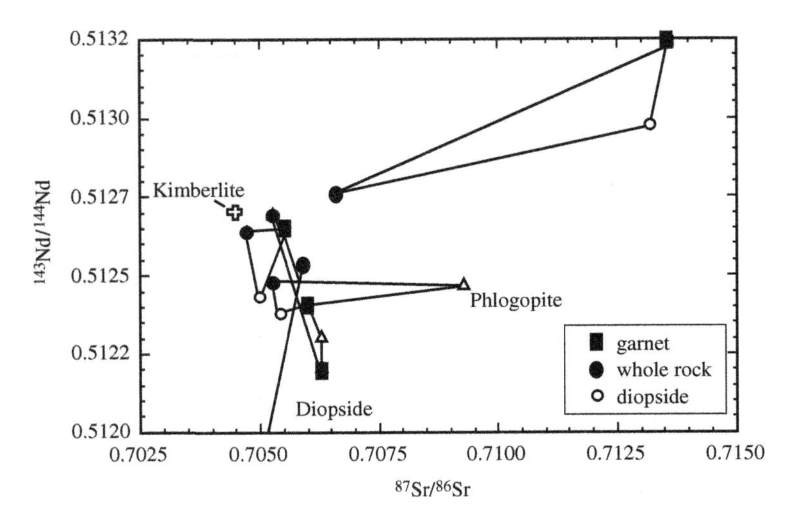

Figure 31 Whole-rock versus mineral separate Nd–Sr isotopic compositions for kimberlite derived xenoliths from Bultfontein (Richardson *et al.*, 1985) and one sample from Jagersfontein (Walker *et al.*, 1989a). Lines connect coexisting phases/whole-rocks. Initial isotopic composition of the Bultfontein kimberlite is also plotted. Jagersfontein cpx plots well off scale on Nd axis.

isotopic composition of mantle material (Bruekner, 1974; Basu and Murthy, 1977; Basu and Tatsumoto, 1980; Jagoutz *et al.*, 1980; Menzies and Murthy, 1980a,b; Richardson *et al.*, 1985; Zindler and Jagoutz, 1988). Even with apparently "clean" mineral separates from mantle rocks, detailed studies have shown that sequential acid leaching procedures are necessary for many minerals in order to obtain the primary (mantle) isotopic composition, unaffected by crustal contamination (Richardson *et al.*, 1985; Zindler and Jagoutz, 1988; Pearson *et al.*, 1993). In general, for isotopic analysis, mineral fragments are selected that are clear, as free of mineral and fluid inclusions as possible and that are bounded by fresh fractures induced by sample preparation, i.e., avoiding grains that retain their primary margins.

In cratonic xenoliths, both low-T secondary alteration and kimberlite contamination have been identified as major concerns to be addressed during the analysis of mineral separates (Richardson *et al.*, 1985). Acid washed mineral separates give considerably different Nd–Sr isotope compositions than their whole-rocks for both peridotites (Richardson *et al.*, 1985; Walker *et al.*, 1989a) and eclogites (Neal *et al.*, 1990; Figure 31), the isotopic signature of the whole-rocks is dominated by small amounts of infiltrated host kimberlite. The whole-rock isotopic signatures of some samples however, appear to be dominated by a component distinctly different to the host kimberlite. This component usually is characterized by much more radiogenic strontium isotopes and appears to be due to the effects of addition of phlogopite mica, which may not be related to the host kimberlite (Figure 31). Also, the nature of kimberlite/xenolith alteration often

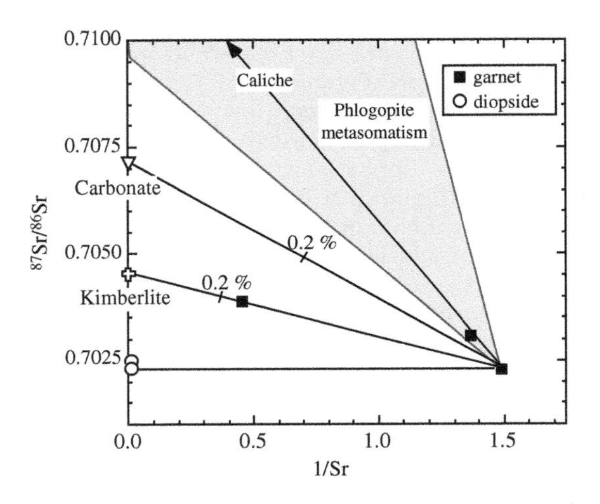

Figure 32 Strontium isotope and abundance systematics of "pure" and altered/metasomatized garnet and diopside separates from the Bultfontein kimberlite (after Richardson *et al.*, 1985). Shaded region indicates the area defined by phlogopite addition. Tick marks on mixing lines illustrate % of component added.

produces almost undetectable films of high-strontium material on mineral grain-boundaries and fractures. This can dramatically affect the measured strontium isotope composition of a low-strontium mineral such as garnet if not properly removed (Figure 32).

Because of the numerous secondary effects that modify the whole-rock elemental and isotopic compositions, our discussion below will focus solely on mineral-separate analyses for the Rb–Sr, Sm–Nd, Lu–Hf, and U–Pb isotope systems. In comparison, the Re–Os system is much more robust to the effects of alteration than the highly incompatible-element-based isotope systems and less affected by invasion of the host rock. For instance, ingress of between 0.5% and 1.5% kimberlite to a host xenolith, as suggested by Schmidberger *et al.* (2001) on the basis of incompatible elements, changes the $^{187}Os/^{188}Os$ ratio of a late Archean peridotite by 0.15%, an insignificant amount. In contrast, the neodymium isotope systematics would be drastically altered because the kimberlite might contribute 70% of the measured whole-rock neodymium (Figure 15). Despite its relatively robust nature, mantle metasomatic events do affect the Re–Os isotope system in mantle xenoliths, just as they do the other radiogenic isotope systems. This will be outlined below.

2.05.2.7.2 Mineral isotope equilibria and parent–daughter fractionation

Peridotites that are residues from melt extraction, i.e., the majority of mantle xenolith peridotites, should contain pyroxenes and garnets that have LREE-depleted compositions such that their low Rb/Sr and high Sm/Nd relative to PUM should plot in the "depleted" field of Figure 33. Clearly, only a small proportion of peridotite xenolith minerals still reflect this depleted history because of either metasomatism, or because some minerals, e.g., certain diopsides in kimberlite-derived xenoliths, have crystallized from melts passing through the peridotite following initial depletion (Shimizu *et al.*, 1997; Shimizu, 1999; Simon *et al.*, 2003). This relationship is also true for the Lu/Hf isotope system where garnets can have either high Lu/Hf, characteristic of melt depletion or low Lu/Hf, indicative of a more complex history (Figure 26). Furthermore, although the minerals within some noncratonic peridotite xenoliths show inter-mineral equilibrium trace-element partitioning relationships (Figure 22), this is not the case for most cratonic peridotites (Shimizu *et al.*, 1997; Shimizu, 1999; Simon *et al.*, 2003).

A consequence of this varied elemental partitioning and hence parent–daughter isotope fractionation, is that the extent to which isotopic equilibria is achieved by xenoliths is also very variable, despite the high equilibration temperatures of mantle rocks. This indicates that many of the processes that affect the incompatible element isotope systematics in mantle xenoliths are of a relatively recent nature compared to the age of the rock. Hence, it is difficult to simply use the measured parent/daughter element ratios of xenoliths and their minerals to predict what their measured isotopic compositions should be.

One of the first studies to show this was performed on Kilbourne Hole spinel lherzolites (Jagoutz *et al.*, 1980). Equilibrated neodymium isotopes in orthopyroxene and diopside defined essentially zero age isochrons, consistent with the very recent eruption age of the host volcanic rocks, while strontium isotopes were un-equilibrated. Stolz and Davies (1988) found varying degrees of equilibration between amphibole, clinopyroxene and apatite in peridotite xenoliths from S.E. Australia. Several samples contained coexisting amphibole and clinopyroxene and had almost reached isotopic equilibrium for strontium but displayed disequilibrium relations for lead and neodymium isotopes. This was taken to indicate more rapid diffusion of strontium than lead and neodymium. Some peridotite and eclogite

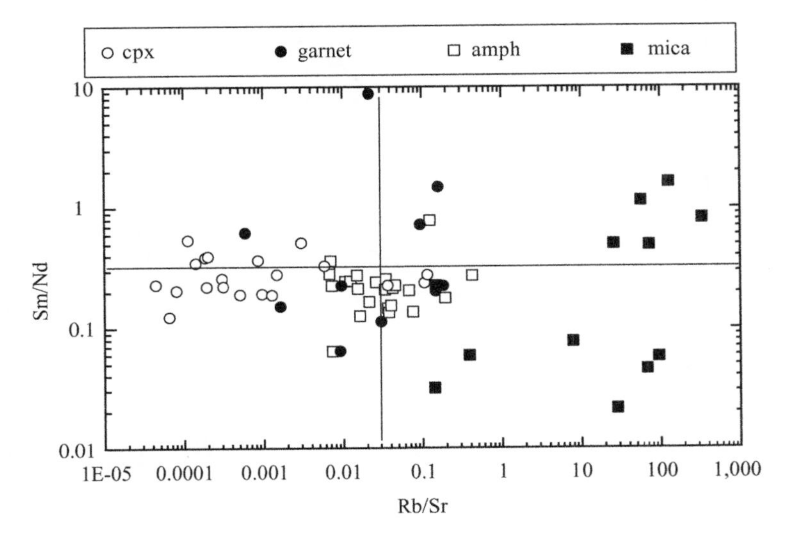

Figure 33 Sm/Nd versus Rb/Sr plot of minerals from mantle xenoliths. Data sources as for Figure 23 and Table 9. Lines denote primitive mantle values.

samples have resided in the mantle at sufficient temperature to equilibrate neodymium isotopes so that the difference in isotopic compositions measured today purely reflects radiogenic decay since eruption (Richardson *et al.*, 1985; Walker *et al.*, 1989; Snyder *et al.*, 1993; Pearson *et al.*, 1995a,b). For instance, Pearson *et al.* (1995b) obtained a relatively precise $1{,}150 \pm 41$ Myr (2σ) clinopyroxene–orthopyroxene–garnet Sm–Nd isochron for a garnet peridotite xenolith from the Proterozoic Premier kimberlite (Figure 34). The equilibration temperature of this xenolith is below 800 °C and the close agreement of the isochron with the eruption age of the Premier kimberlite suggests that the peridotite was above the blocking temperature of the Sm–Nd system in these minerals. Despite the apparent inter-mineral isotopic equilibria, mineral REE studies of the same sample by Shimizu (1999) suggests that they were not in elemental equilibrium and had been disturbed just prior to eruption.

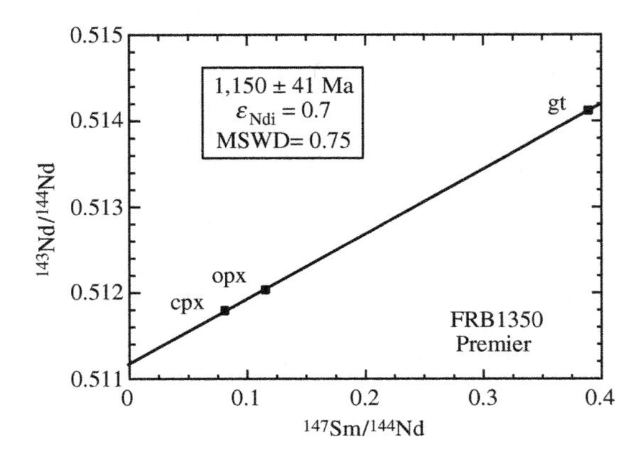

Figure 34 Sm–Nd garnet–orthopyroxene–clinopyroxene mineral isochron for a cratonic garnet lherzolite from Premier, S. Africa. Errors are 2σ (source Pearson *et al.*, 1995b).

More ancient mineral isochrons, significantly in excess of eruption ages, have been determined for diopside-garnet pairs in cratonic peridotites (Gunther and Jagoutz, 1997; McCulloch, 1989; Pearson *et al.*, 1995a,c; Zhuravlev *et al.*, 1991) and eclogites (Jacob *et al.*, 1994; Jagoutz, 1988; Jagoutz *et al.*, 1984). Two-point peridotite mineral isochrons from a single kimberlite pipe can vary widely in slope, giving apparent ages hundreds of Myr different (Figure 35). The geochronological information provided by these isochrons is unclear. The wide variations in xenolith isochron ages from a single kimberlite are unlikely to represent closure ages (Pearson *et al.*, 1995a). Contamination by the host kimberlite can alter peridotite mineral isochrons and this can usually be identified. Gunther and Jagoutz (1997) proposed that the oldest mineral isochrons in Siberian peridotites, of ~2 Ga, represent closure ages. Younger ages represent partial closure/re-equilibration during either lithospheric residence or during eruption.

In situ trace-element measurements of minerals from some of these peridotites provide a different perspective. The frequent presence of both fine-scale (100 μm) zonation and nonequilibrium partitioning behavior for REE between many garnets and clinopyroxenes from Siberian peridotites (Shimizu *et al.*, 1997; Shimizu, 1999) has two possible implications. One is the probable recent

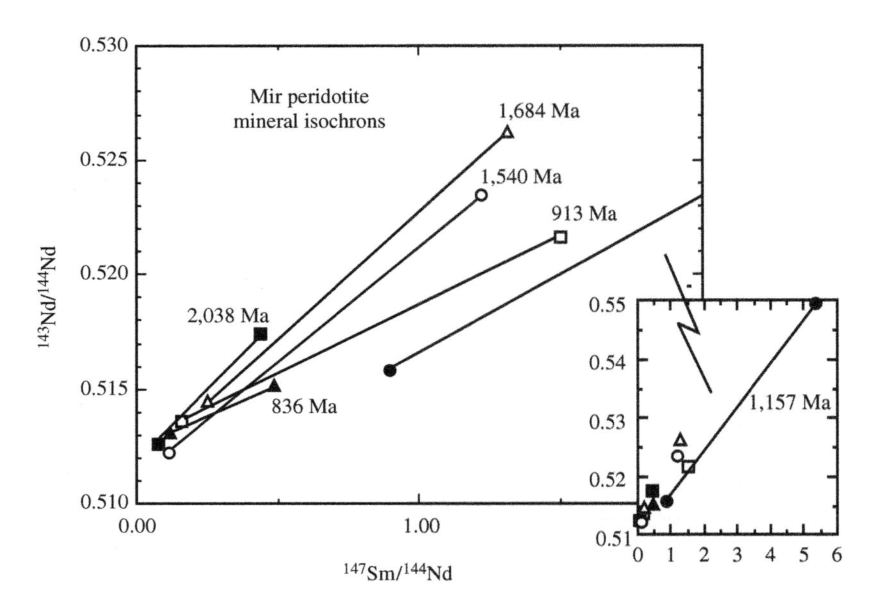

Figure 35 Garnet-diopside Sm–Nd mineral isochrons for peridotites from the Mir kimberlite pipe, Siberia. Ages relating to the slope of each line are given (after Pearson, 1999b).

disruption of isotope systematics in many peridotite minerals due to recent mineral growth. Secondly, in numerous cases it is likely that diopside and garnet have not coexisted long enough to have even partly equilibrated in terms of trace elements and isotopes. A good example of the type of isotopic complexity that this can lead to is shown by reversed mineral "isochrons" in some cratonic peridotites (Figure 36). The Sm–Nd isotope systematics of garnet–orthopyroxene–clinopyroxene clusters in Kimberley peridotites show distinctly more radiogenic neodymium for both clino- and orthopyroxenes compared to the garnet, despite much higher Sm/Nd of the garnet (Figure 36). This feature has been observed in other Kaapvaal low-T peridotites and suggests recent garnet growth from a LREE enriched precursor (Gunther and Jagoutz, 1994) and/or that the clinopyroxene has been newly introduced. Hence, without detailed petrography and mineral elemental chemistry it is unwise to interpret cooling age information from two-point mineral isochrons in mantle rocks.

Carlson and Irving (1994) found correlated Sm–Nd systematics in minerals from Wyoming peridotites that scatter around a 1.8 Ga reference isochron. The neodymium in these cratonic samples is consistently very unradiogenic compared to primitive upper mantle, with an initial ε_{Nd} of -9, indicative of an ancient metasomatic prehistory. The Sm–Nd isochron age agrees well with U–Th–Pb ages for zircon and monazite metasomatic phases in the Wyoming peridotite suite and must clearly reflect the time of a major, regional metasomatic enrichment event in the lithosphere (Carlson and Irving, 1994). This study and others make a clear case for multiple enrichment events following the initial depletion of peridotitic lithosphere sampled by xenoliths (e.g., Pearson *et al.*, 1995a; Pearson, 1999b).

2.05.2.7.3 Rb–Sr, Sm–Nd, Lu–Hf, U–Pb, and Re–Os isotopic signatures

Major-element compositions of most lithospheric peridotites reflect an origin as melt-residues. However, as with parent–daughter isotope ratios, compilation of their strontium and neodymium mineral isotopic compositions reveals that very few samples show the characteristics of ancient melt residues (Figure 37). Osmium isotopes are the exception and dominantly reflect ancient extraction of high Re/Os melts, leaving rhenium-depleted residues to develop time integrated low $^{187}Os/^{188}Os$. The huge isotopic diversity shown by peridotite Nd–Sr isotope systematics (Figure 38) is a reflection of the diverse parent–daughter elemental fractionation shown by mantle silicate minerals coupled with ancient, multiple-stage histories involving melt depletion and subsequent interaction with melts passing through the lithospheric mantle.

Peridotite xenoliths from continental regions show substantial overlap with oceanic mantle for isotopic compositions of lithophile elements such as strontium and neodymium (Menzies, 1990a; Pearson, 1999b; Figure 37). The frequency distribution for neodymium isotopes in peridotite xenoliths that represent the continental lithospheric mantle (CLM) ranges from highly radiogenic compositions (relatively high values of $^{143}Nd/^{144}Nd$; high ε_{Nd}), indicative of long-term parent–daughter depletion, to highly enriched compositions (relatively low values of

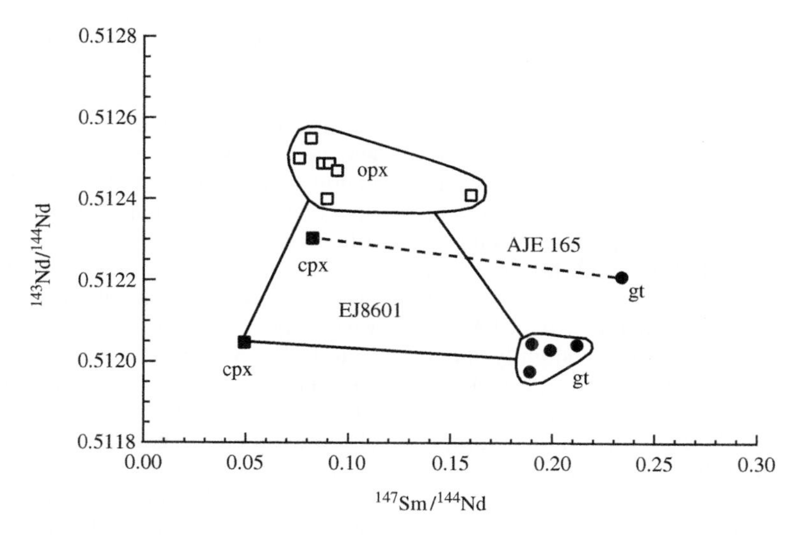

Figure 36 Sm–Nd isotope systematics of minerals from Kimberley peridotites. Data for AJE165 from Richardson *et al.* (1985), and for EJ8601 from Gunther and Jagoutz (1994).

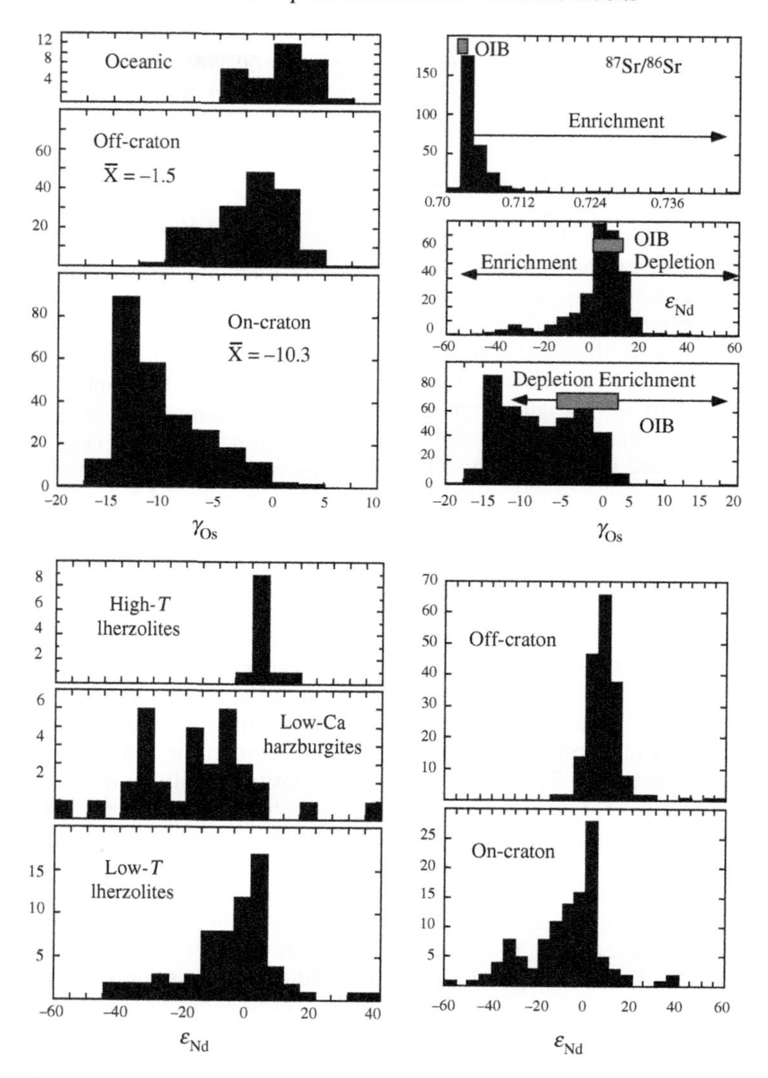

Figure 37 Frequency distribution plots for Os, Nd, and Sr isotope compositions of cratonic and noncratonic peridotite xenoliths. Upper right plots give the range for ocean island basalts (OIB) and arrows show the direction of isotopic evolution for melt depletion and enrichment events. Data compiled from sources cited in Menzies (1990b), Pearson (1999a,b), and those given in Figure 21 (after Pearson and Nowell, 2002).

^{143}Nd/^{144}Nd; low ε_{Nd}) that require ancient parent–daughter enrichment. Despite this wide spread in neodymium isotopic compositions, the remarkable aspect of the frequency distribution for CLM as a whole is that the pronounced mode is within 5ε units of bulk Earth and the mean ε_{Nd} value is 1.8. On this basis, the dominant neodymium isotopic characteristic of CLM is not "enriched" but is close to, or slightly more depleted than bulk Earth. Of course, if Depleted Mantle is used as a reference point then the CLM mode is "enriched." The strontium isotope frequency distribution has a long tail out to very radiogenic ("enriched") compositions but the mean ^{87}Sr/^{86}Sr is 0.7047; very close to estimates of bulk Earth. It is important to bear in mind that this statistical view of CLM geochemistry could be heavily influenced by samples that have geochemical signatures dominated by their host magmas; i.e., the xenolith sample set could be very biased and the real CLM composition grossly different.

If continental peridotite xenoliths are divided into cratonic and noncratonic compositions, the vast majority of highly enriched neodymium isotope compositions originate in cratonic mantle and very few are evident in noncratonic mantle. Enriched neodymium isotope compositions can also be found in mantle sampled by orogenic peridotites (Reisberg and Zindler, 1986; Pearson *et al.*, 1993; Chapter 2.04) but not the extreme values evident in cratonic CLM. This is expected, given the great antiquity of cratonic CLM. Further subdivision of cratonic samples (Figure 37) shows

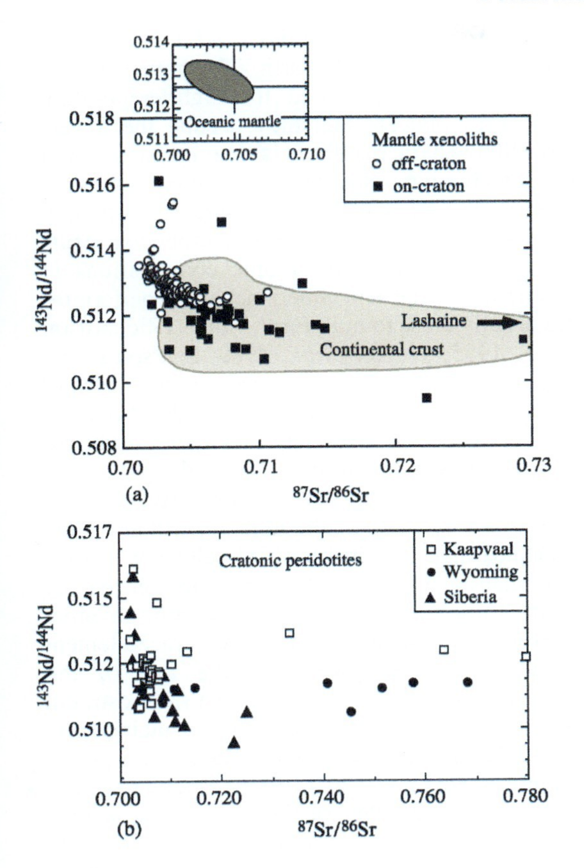

Figure 38 Nd–Sr isotope variation of clinopyroxenes and garnet in peridotite xenoliths. (a) Compares cratonic and noncratonic peridotite xenoliths with continental crust. Inset shows restricted field for oceanic mantle. Arrow points to a peridotite from Lashaine, Tanzania, that lies at an $^{87}Sr/^{86}Sr$ value of 0.83. (b) Compares cratonic peridotites from the Kaapvaal, Wyoming, and Siberian cratons.

that garnets from the low-calcium harzburgite subgroup of type I xenoliths have the most extreme neodymium isotopic variations (Pearson *et al.*, 1995c; Jacob *et al.*, 1998a), whereas the mode for typical low-temperature cratonic lherzolites (type I) is close to bulk Earth and the distribution more restricted. The more tightly clustered lherzolite distribution is probably influenced by the recent formation of diopside, from the host magma, in many of these rocks. Jacob *et al.* (1998a) have shown that low-calcium garnets from an individual kimberlite (Udachnaya) are extremely heterogeneous in their Nd–Sr isotope compositions on a single grain basis, with a range in initial ε_{Nd} of 42 units between grains. The total range for low-calcium garnets, as composites (Pearson *et al.*, 1995a) or single grains from the Udachnaya kimberlite alone is over 60ε units for neodymium, i.e., ~3 times the range found in the convecting mantle, emphasizing the extreme isotopic heterogeneity shown by CLM xenoliths.

Some intracratonic distinctions can be made on the basis of Sr–Nd isotope systematics that reflect differing extents and styles of lithospheric enrichment processes. Peridotites from the Kaapvaal and Wyoming lithospheric roots have strontium isotope compositions that range up to very radiogenic compositions. Probably the most extreme strontium isotope composition of a mantle rock to date comes from a clinopyroxene from a Tanzanian peridotite ($^{87}Sr/^{86}Sr = 0.836$; Cohen *et al.*, 1984), but an insufficient number of Tanzanian samples have been analyzed to make comparisons meaningful. Most of these extreme strontium isotope compositions measured for clinopyroxene and garnet are un-supported by the very low Rb/Sr of these phases (Figure 27). This may indicate that the garnet and/or clinopyroxenes formed from a high Rb/Sr precursor that had sufficient time to evolve radiogenic strontium. The garnet and pyroxene inherited the strontium isotope composition of the precursor but excluded Rb from their lattice. Rubidium-rich phlogopite is a possible candidate to form part of the precursor assemblage. Alternatively, radiogenic strontium could be inherited during H_2O-rich fluid metasomatism via some unspecified mechanism.

In contrast to the Kaapvaal and Wyoming cratons, minerals within Siberian peridotites rarely have $^{87}Sr/^{86}Sr$ over 0.720, with most below 0.710. This possibly correlates with a general paucity of phlogopite in Siberian peridotites (Boyd *et al.*, 1997), indicating that hydrous-fluid-dominated metasomatism has not been as extensive in the Siberian lithosphere. The range in ε_{Nd} for Siberian peridotites is comparable to that of Kaapvaal peridotites (Figure 38(b)), reflecting both ancient LREE depletion and enrichment. Development of very unradiogenic neodymium isotope compositions, relative to bulk Earth, while retaining relatively moderate/low enrichment of $^{87}Sr/^{86}Sr$ is also a characteristic of peridotite and pyroxenite–glimmerite xenoliths from Loch Roag, Scotland, erupted at the margin of the N. Atlantic craton (Menzies and Halliday, 1988). LREE enrichment without marked increase of Rb/Sr is thought to be a characteristic of carbonatite metasomatism, and appears to be a clearly identifiable signature in xenoliths from various cratons, though usually via elemental geochemistry (see Sections 2.05.2.5 and 2.05.2.6).

Xenoliths from the Wyoming craton have consistently very low ε_{Nd} coupled with radiogenic strontium. This radiogenic strontium signature is largely a reflection of the preponderance of mica-peridotites, mica pyroxenites and glimmerites in the xenolith suite analyzed compared to the lherzolite–harzburgite lithologies that dominate xenolith suites from Kaapvaal and Siberia. The radiogenic strontium isotopic signature has been interpreted as the result of shallow slab-derived fluid fluxing during the Archean, soon after formation of the protoliths (Carlson and Irving, 1994).

From the diversity of observed isotopic compositions a number of different metasomatic agents have been inferred, that are thought to produce shifts of Nd–Sr isotope systematics in distinct directions. The three main enrichment processes thought to affect the lithospheric mantle, identifiable from petrographic evidence are silicate melt addition, H_2O-fluid-rich metasomatism, which may result in phlogopite/amphibole metasomatism and may ultimately be linked to silicate melt intrusion (Ionov *et al.*, 1997), and carbonatite metasomatism. These processes operate to different extents in differing tectonic environments and leave a multitude of effects. The reader is referred to Menzies and Hawkesworth (1987) and Menzies and Chazot (1995) for reviews. Recently, Ionov *et al.* (2002a,b) have noted that chromatographic effects related to porous flow during melt percolation can decouple strontium from neodymium isotopes in peridotite xenoliths, creating enrichment in $^{87}Sr/^{86}Sr$ at relatively constant neodymium isotope composition. These authors note that the frequently observed decoupling of strontium from neodymium isotopes, which is often attributed to metasomatism by subduction-related fluids, can be a signature only of silicate melt percolation without the need for subduction zone fluids. Hence, as with trace-element signatures, it may be dangerous to identify the nature of a metasomatic agent solely on isotopic criteria. Coupled petrographic, mineral chemical, elemental and isotopic studies are required to fully understand these processes.

The high-temperature peridotites that appear to occupy the basal portions of CLM, either as a layer (Nixon and Boyd, 1973a), or as local zones surrounding magma intrusions (Gurney and Harte, 1980), have very restricted neodymium isotopic compositions, with a pronounced mode within 1 epsilon unit of bulk Earth. The higher equilibration *P*/*T*'s determined for high-temperature peridotites indicate that they occupy the lowermost lithospheric mantle. It is clear that this region has neodymium isotope compositions indistinguishable from bulk Earth (Figure 37) although the data set is small and data from other continents are needed. Strontium isotopic compositions in southern African and Siberian high-*T* peridotites are usually close to bulk Earth values (Richardson *et al.*, 1985; Walker *et al.*, 1989; Pearson *et al.*, 1995a,b) but clinopyroxenes from high-*T* peridotites at Somerset Island (Churchill Province, northern Canada) have more radiogenic strontium than the low-*T* peridotites (Schmidberger *et al.*, 2001). The generally more primitive Nd–Sr isotope systematics of the high-*T* peridotite suite indicates that their incompatible element isotope systematics have been recently influenced by melts infiltrating from the convecting mantle. This is in agreement with major and trace-element zonation studies. However, the age of the infiltrated protolith has been shown from osmium isotope studies to be much more ancient and probably part of the original Archean lithospheric mantle (Walker *et al.*, 1989; Pearson *et al.*, 1995a,b). A summary of the Sr–Nd–Os isotope characteristics of cratonic and off-craton continental xenoliths and their relation to major-element parameters and density is shown in Figure 39.

Four cratonic, ultradeep xenoliths from S. Africa and Sierra Leone have been analyzed for their strontium and neodymium isotopic compositions (Macdougall and Haggerty, 1999). The neodymium isotopic compositions of minerals from these xenoliths suggest that they were emplaced into the African lithosphere at times ranging from approximately the time of kimberlite eruption to hundreds of millions of years earlier. The samples show a complex history of melt

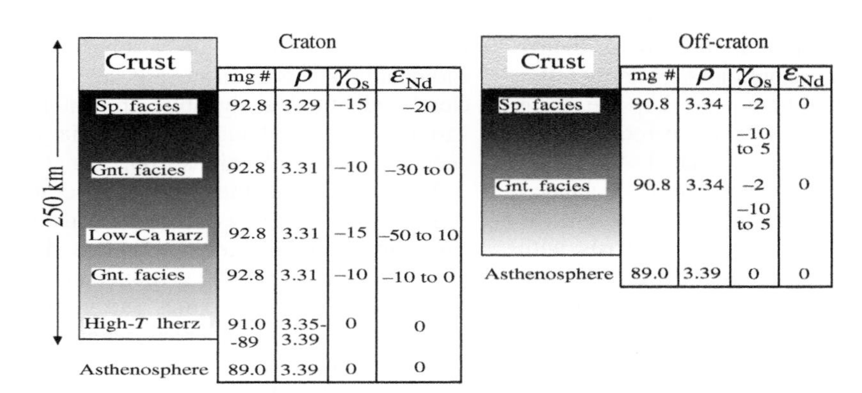

Figure 39 Cross-section of "typical" cratonic and noncratonic lithospheres as sampled by xenoliths showing the variation in olivine Mg#, density (ρ), Os and Nd isotope compositions with depth.

depletion and enrichment. Macdougall and Haggerty (1999) propose that the complex isotope systematics of these samples indicate a heterogeneous deep mantle. Detailed oxygen isotope work on these samples by Deines and Haggerty (2000) concluded that they experienced a series of metasomatic alterations, including an event a few million years before eruption of the host kimberlite. Because of the complicated history experienced by these xenoliths, further studies on additional samples are needed before we can be confident that the data are representative of the Transition zone and deeper.

Few systematic Sr–Nd isotope studies have been performed on ocean island xenolith suites. Ducea *et al.* (2002) analyzed clinopyroxenes from plagioclase-spinel and spinel peridotites from Pali, (Oahu, Hawaii) and found relatively depleted strontium and neodymium isotope systematics that they interpret as representing their evolution as residues from the extraction of Pacific Ocean crust. Consistent with this is a 61 ± 20 Ma errorchron defined by the pyroxene separates that is within error of the 80–85 Ma age of Pacific lithosphere beneath Hawaii.

Compared with neodymium and strontium, there are relatively few studies of the lead isotopic compositions of mantle xenoliths and the systematics are probably biased towards samples that show some degree of patent metasomatism in the form of introduction of amphibole and/or mica. Much of the data come from noncratonic metasomatized peridotites (e.g., Stolz and Davies, 1988) and cratonic MARID xenoliths. Some type I xenoliths that do not have patent metasomatism, from cratonic and noncratonic settings (Kramers, 1977; Galer and O'Nions, 1989; Walker *et al.*, 1989; Lee *et al.*, 1996) together with various pyroxenites and megacrysts have also been analyzed (Ben Othman *et al.*, 1990; Tatsumoto *et al.*, 1992). Only mineral data are considered. As with strontium and neodymium, lead isotopic compositions are extremely variable in mantle xenoliths (Figure 40). Peridotites and their metasomatized variants show most of the heterogeneity while pyroxenites and megacrysts have more constant compositions that relate closely to the host magmas and suggest a genetic link (Ben Othman *et al.*, 1990; Tatsumoto *et al.*, 1992). As expected from their greater antiquity, cratonic xenoliths show considerably more lead isotope variation than their noncratonic counterparts. Both suites define broad positive correlations on a $^{207}Pb/^{204}Pb$ versus $^{206}Pb/^{204}Pb$ diagram that are subparallel to the northern hemisphere reference line (NHRL) defined by Atlantic MORB, but displaced to higher $^{207}Pb/^{204}Pb$ except for two samples from the Cameroon line (Lee *et al.*, 1996). Almost all noncratonic peridotites plot to the right of the Geochron. A number of cratonic peridotites plot well to the left of this reference point, suggesting ancient U/Pb depletion. Cratonic samples also range up to very radiogenic $^{206}Pb/^{204}Pb$. Most noncratonic peridotites have $^{206}Pb/^{204}Pb < 19$ except several samples from the Cameroon Line (Lee *et al.*, 1996) that have HIMU characteristics (Figure 40). Apart from peridotites from the Cameroon Line, Pb–Sr isotope systematics are characterized by heterogeneity. The variability is such that the end-member mantle "components" HIMU, EMI and EMII (Zindler and Hart, 1986) are all strongly developed in some xenolith suites, with those from the Wyoming craton (Carlson and Irving, 1994) and Tanzania (Cohen *et al.*, 1984) showing particularly strong EMII-like characteristics of high $^{87}Sr/^{86}Sr$ at

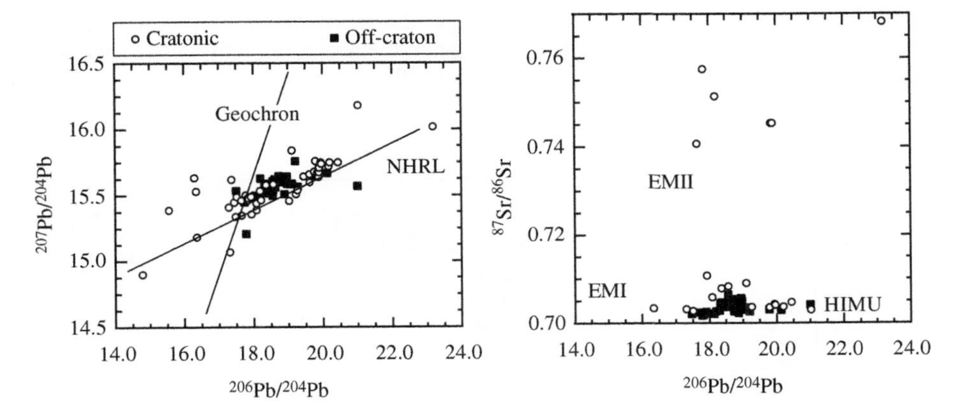

Figure 40 $^{207}Pb/^{204}Pb$ versus $^{206}Pb/^{204}Pb$ isotope plot and $^{87}Sr/^{86}Sr$ versus $^{206}Pb/^{204}Pb$ plot of cratonic and noncratonic peridotite xenoliths minerals. For cratonic peridotites, only clinopyroxenes are plotted. For noncratonic peridotites, amphibole or clinopyroxene are plotted. Metasomatic rocks such as MARIDs from cratons are not plotted. NHRL = Northern Hemisphere Reference Line, EMI = Enriched Mantle I hypothetical mantle component, EMII = Enriched Mantle II hypothetical mantle component and HIMU = the hypothetical mantle component with high time-integrated U/Pb proposed by Zindler and Hart (1986).

moderate $^{206}Pb/^{204}Pb$ (Figure 40). HIMU-type lead isotope characteristics seem to be restricted to cratonic peridotites despite the high measured U/Pb and Th/Pb of some noncratonic peridotites (Meijer *et al.*, 1990). This is further indication of the decoupling of parent–daughter ratios from isotopic compositions in xenoliths, probably due to recent processes. Many noncratonic peridotite xenoliths show well developed EMI and/or EMII characteristics (Tatsumoto *et al.*, 1992). Menzies (1990) has proposed detailed models of lithospheric chemical "domains" related to different times and styles of metasomatism, and suggests that a significant fraction of the lower lithosphere is of HIMU isotopic character.

Improved analytical capabilities have led to the analysis of several hundred xenoliths for osmium isotopic composition. The compatible nature of osmium during mantle melting means that, unlike incompatible-element-based isotope systems, peridotite residues have much higher osmium contents than mantle melts and thus the system is less readily disturbed by later metasomatism (see Section 2.05.2.5.3). This is clearly shown by rhenium and osmium abundances (Figure 21). The vast majority of rhenium contents of both cratonic and noncratonic peridotite xenoliths are below the PUM value proposed by Morgan *et al.* (1981) and many are P-PGE depleted. This contrasts with almost universal LREE enrichment of whole-rock peridotites. That the Re–Os system is not immune from the effects of metasomatism is illustrated by the consideration of extended PGE patterns (Figure 20; Section 2.05.2.5.3; Pearson *et al.*, 2002, 2004). Disruption of both rhenium and osmium in some mantle environments may have occurred (Chesley *et al.*, 1999), especially where sulfide metasomatism is involved (Alard *et al.*, 2000). However, Pearson *et al.* (2002, 2004) and Irvine *et al.* (2003) have shown that coupled PGE and Re–Os isotope analyses can effectively assess the level of osmium isotope disturbance in peridotite suites.

The compatible-element nature of osmium is illustrated by the correlation of osmium isotope composition with indices of melt depletion such as bulk rock Al_2O_3 (Figure 41). Numerous other melt depletion indices show equally good, or sometimes better correlations for a given suite. Meisel *et al.* (2001) use the upper intercept of $^{187}Os/^{188}Os$ at $Al_2O_3 = 4.23\%$ to estimate the primitive mantle osmium isotope composition and find good agreement between seven suites of peridotites (0.1296). More work is required to establish the extent to which these trends are disturbed by melt re-enrichment before accepting this method as the best way to define the PUM osmium isotope composition. Reisberg and Lorand (1995) suggested that the lower intercept of the trends might be used to define the initial $^{187}Os/^{188}Os$ of a given peridotite suite and hence derive a model age. The scatter for the off-craton compilation is then a function of different, subparallel trends for suites of different ages. Pearson (1999a) has pointed out that extrapolation of $^{187}Os/^{188}Os$ to zero Al_2O_3 will lead to an overestimation of model ages because Al_2O_3 is unlikely to be reduced to zero in any melt-residue suite. Sulfur and selenium provide better analogues from this point of view because their abundance should approach zero on the disappearance of sulfide at high levels of melting. However, both these elements suffer from metasomatic disturbance.

Cratonic and circum-cratonic kimberlite-derived xenoliths show poor correlations of $^{187}Os/^{188}Os$ and Al_2O_3 compared to noncratonic peridotites (Figure 41). Although some scatter is present within the total noncratonic data set, individual suites show much better correlations, the suite from Vitim being an excellent example (Pearson *et al.*, 1998a, 2004; Figure 41). Carlson *et al.* (1999b) suggest that the lack of correlation between $^{187}Os/^{188}Os$ and Al_2O_3 in cratonic peridotites is due to a rhenium-rich, aluminum-poor metasomatic component such as a carbonatitic fluid.

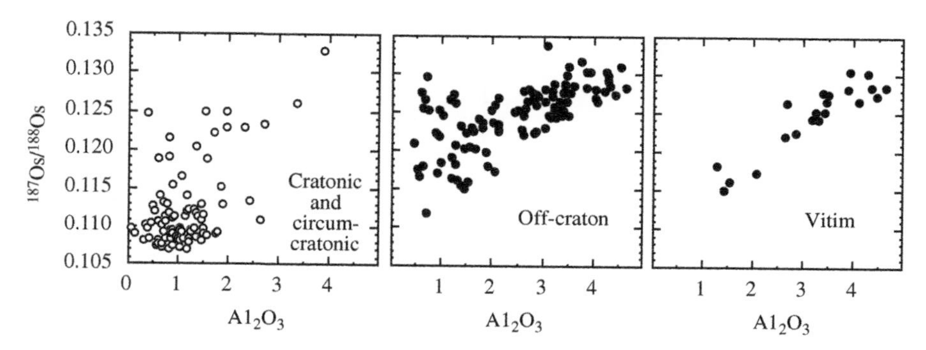

Figure 41 $^{187}Os/^{186}Os$ versus bulk rock Al_2O_3 for cratonic plus circum-cratonic peridotites, off-craton peridotites and the Vitim subset of off-craton peridotites. Data sources as for Figure 21. Vitim data set is from Pearson *et al.* (2002, 2004).

Radiogenic isotope studies of oceanic peridotite xenoliths are relatively few compared to continental peridotite xenoliths. In general, compositions are much more restricted than continental xenoliths, as might be expected for younger lithospheric mantle. However, recent studies are finding examples of very unradiogenic osmium isotopic compositions even in the oceanic environment. One example is the peridotite suite from the Kerguelen plateau, southern Indian Ocean. Unradiogenic osmium isotopes in these peridotites ($^{187}Os/^{188}Os$ ratios as low as 0.119) have been interpreted as indicating that the mantle beneath the Kerguelen plateau is ancient (>1 Ga old) and of continental rather than oceanic origin (Hassler and Shimizu, 1998), although similarly low values have also been reported from some abyssal peridotites (Snow and Reisberg, 1995; Brandon *et al.*, 2001). Even less radiogenic $^{187}Os/^{188}Os$ ratios (as low as 0.113 up to 0.129) have been reported for lherzolites from Salt Lake Crater, Hawaii (Griselin and Lassiter, 2002). These values extend to less radiogenic compositions than reported from abyssal peridotites. Possibilities to explain this data are that the xenoliths sample exotic fragments of subcontinental mantle in the Pacific, or that the xenoliths are derived from a heterogeneous Hawaiian plume that contains ancient, recycled oceanic lithosphere. A further possibility is that the unradiogenic Os isotopes are a result of an melting event that did not lead to lithospheric accretion. The fragment of mantle remained in the convecting mantle to be accreted to lithosphere during some later, more recent differentiation event.

The osmium isotope compositions of oceanic peridotite xenoliths significantly overlap the range for noncratonic continental peridotite xenoliths (Figure 37) although some of the latter suites have less radiogenic compositions that relate to their older ages (see below). The osmium isotopic mode and mean for noncraton peridotite xenoliths ($\gamma_{Os} = -1.5$) are close to chondritic (Figure 37). In contrast, cratonic peridotite xenoliths have extremely unradiogenic osmium isotope compositions with a frequency distribution that is strongly negatively skewed; mean $\gamma_{Os} = -10.3$ and values extending to -17 (Figure 37). Extremely unradiogenic osmium isotope compositions in cratonic xenoliths, even those sampled by Alkali Basalt Series magmas in cratonic settings (Chesley *et al.*, 1999; Hanghoj *et al.*, 2001) set them apart from peridotites from other tectonic settings and are a highly distinctive feature of Archean mantle sampled as peridotite xenoliths (Figure 39). Highly unradiogenic osmium isotope compositions of some cratonic high-*T* peridotites have shown that although they have experienced recent metasomatic disturbance, the protoliths were attached to the mantle root in the Archean (Walker *et al.*, 1989; Pearson *et al.*, 1995a; Carlson *et al.*, 1999). The osmium isotope compositions of almost all lithospheric peridotite suites are completely distinct from those of their host magmas.

Lithologies other than peridotites have not received significant attention from isotope geochemists. Wehrlites and pyroxenites from the off-craton locality of Bullenmerri (southeastern Australia) have strontium, neodymium, and lead isotopic compositions that are generally similar to those found in the regional alkali basalts (Porcelli *et al.*, 1992) indicating a likely origin as the crystallization products of magmas related to the host volcanism. Similarities in the Sr–Nd–Pb isotope data of erupted magma, pyroxenite/amphibole veins and megacrysts in noncratonic xenolith suites support a genetic link between these phenomena (Menzies *et al.*, 1985; Ben Othman *et al.*, 1990).

There are few published Lu–Hf isotope studies of mantle xenoliths because of difficulties in efficient ionization of hafnium by thermal ionization mass spectrometers. Multicollector plasma mass spectrometers are a solution to this problem and data are emerging that promise to be a more revealing tool in mantle environments than neodymium isotopes. The variety of Lu/Hf fractionation displayed by mantle minerals (Figure 42) indicates that, as with other isotope systems, isotopic variation should be considerable and initial results are confirming this. Salters and Zindler (1995) found very radiogenic $^{176}Hf/^{177}Hf$ at relatively unradiogenic neodymium isotope compositions in spinel peridotites from Salt Lake Crater, Hawaii. Radiogenic $^{176}Hf/^{177}Hf$ also characterizes low-*T* circum-cratonic

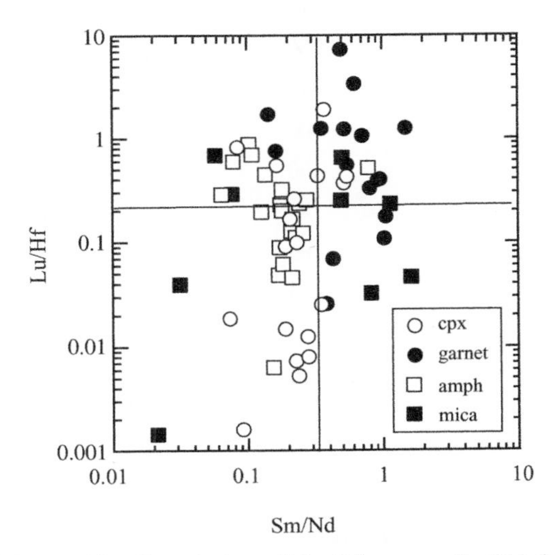

Figure 42 Covariation of Lu/Hf versus Sm/Nd for minerals from mantle peridotite xenoliths. Lines marked are the primitive mantle ratios.

peridotites from the Somerset Island kimberlite (Churchill Province, N. Canada; Schmidberger *et al.*, 2002). ε_{Hf} values range up to +160 at relatively unradiogenic neodymium isotopic compositions, forming a subvertical trend on an Nd–Hf isotope diagram (Figure 43). Radiogenic $^{176}Hf/^{177}Hf$ has also been measured in clinopyroxenes from Vitim peridotites (Blichert-Toft *et al.*, 2000) with compositions that lie consistently on, or well above the mantle Nd–Hf array (Figure 43). Clinopyroxenes from low-*T* Kaapvaal peridotite xenoliths have Hf–Nd isotope systematics that lie within, to just below the terrestrial array (Simon *et al.*, 2002). This is probably consistent with a late, metasomatic origin for many of these clinopyroxenes (Simon *et al.*, 2003). Some clinopyroxenes from Kaapvaal low-*T* lherzolites have ε_{Hf} values up to +70 (Bedini *et al.*, 2002) and are more consistent with a melt depletion origin. In contrast, garnets from cratonic peridotites commonly have highly radiogenic hafnium isotopic compositions that lie well above the mantle Nd–Hf isotope array, with ε_{Hf} values between +100 and +2,500. Thus, it seems clear from the relatively few (<100) data points obtained so far, that continental lithospheric mantle as sampled by xenoliths is characterized by highly radiogenic $^{176}Hf/^{177}Hf$ and by compositions that plot mostly well above the mantle Nd–Hf array, often in subvertical trends. This trend is similar to that found recently for cpx in spinel peridotites from the Beni Bousera orogenic peridotite (Pearson and Nowell, 2003; Figure 43) and hence is not just restricted to xenoliths, but appears a universal characteristic of lithospheric mantle. The highly radiogenic Hf isotopic compositions must be a product of ancient depletion involving residual garnet while the relatively unradiogenic neodymium isotopes are probably a function of more recent metasomatic disturbance that has not greatly affected the Lu–Hf system.

The robustness of the Lu–Hf isotope system in some mantle environments is demonstrated by the precise Lu–Hf isochron of 1,413 ± 67 Myr defined by clinopyroxene separates from the Beni Bousera peridotite massif (Pearson and Nowell, 2003). This age probably dates the time of melt extraction from these rocks and is considerably more precise than the Sm–Nd isochron or the scattered Re–Os isotope systematics of these rocks. This indicates the potential power of this system in dating mantle rocks. The initial results from the Lu–Hf isotope system indicate that of the incompatible element isotope systems, it is the more robust to metasomatic effects, with signatures frequently recording the time-integrated response to melt depletion.

2.05.2.7.4 The age of mantle peridotites

The first modern isotopic studies of continental lithospheric mantle (CLM) revealed that it must have been isolated from the convecting mantle for billion-year timescales (Kramers, 1977; Menzies and Murthy, 1980a; Richardson *et al.*, 1984).

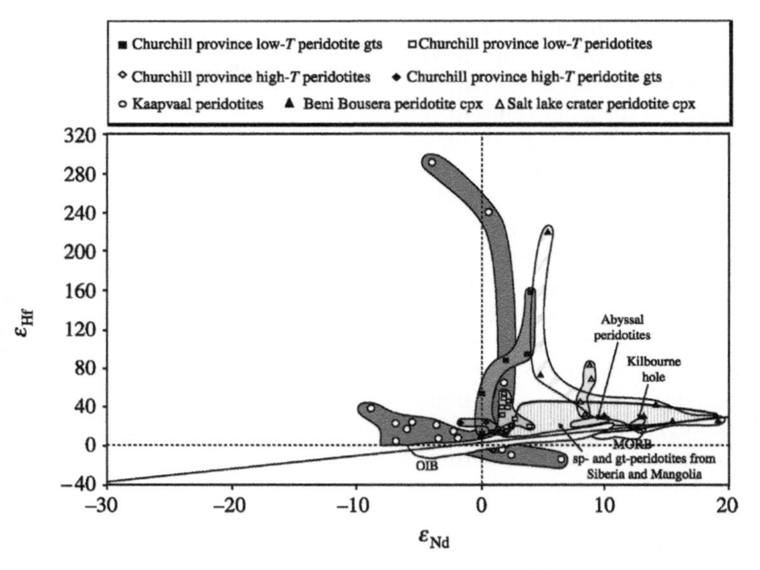

Figure 43 ε_{Hf} versus ε_{Nd} isotope diagrams for lithospheric mantle peridotite minerals. Kaapvaal peridotite data are all garnets and clinopyroxenes (Simon *et al.*, 2002). Slave peridotite data are garnets and whole rocks from Schmidberger *et al.* (2002). Salt Lake Crater peridotites (Hawaii), Kilbourne Hole and Abyssal peridotites, are from Salters and Zindler (1995). Siberian and Mongolian peridotite field are clinopyroxene data from cratonic and off-craton peridotites (field taken from Ionov and Weis, 2002). Fields for MORB (N-MORB) and OIB are from Nowell *et al.* (1998). Field for Beni Bousera peridotites from Pearson and Nowell (2003).

The incompatible element isotope systems used in these studies also showed that the parent-daughter isotope ratios for the respective systems must have been enriched relative to bulk Earth over such timescales. The ages obtained were "enrichment" ages and give rise to the "old enriched mantle" label for CLM. Further detailed work using incompatible-element-based isotopic systems revealed much detailed information concerning the complex, multiphase enrichment history of some peridotites (Richardson *et al.*, 1985; Stosch and Lugmair, 1986; McDonough and McCulloch, 1987; McCulloch, 1989; Gunther and Jagoutz, 1994; Pearson *et al.*, 1995a,b; Xu *et al.*, 1998) but because of these disturbances, the data do not constrain their formation ages.

Although the Re–Os isotope system is not immune from the effects of post-melt depletion processes such as metasomatism (Alard *et al.*, 2000; Burton *et al.*, 2000; Chesley *et al.*, 1999; Pearson *et al.*, 1995c) this system is by far the most reliable we have for estimating the formation ages of peridotites (Pearson, 1999a), although in certain instances, the Lu–Hf isotope system is proving very powerful (Pearson and Nowell, 2003). The Re–Os isotope system has been widely applied to both cratonic and noncratonic peridotite xenoliths. In particular, new developments using either analysis of sulfide inclusions in primary minerals, (Alard *et al.*, 2000; Pearson *et al.*, 2002), chromites (Chesley *et al.*, 1999) or combined osmium-isotope–PGE systematics (Pearson *et al.*, 2002) give us more confidence in the interpretation of Re–Os isotope ages for CLM peridotites.

Cratonic mantle peridotites. Over 230 whole-rock cratonic xenoliths have now been analyzed for Re–Os isotope compositions. Given that many peridotite xenoliths have experienced relatively recent rhenium introduction, it is generally best to use rhenium-depletion model ages (T_{RD}) that do not rely on the measured rhenium content of the rock for model age calculation. For cratonic peridotite xenoliths, the frequency distribution of rhenium-depletion ages shows a wide range, with a pronounced mode at 2.5–2.75 Gyr and some samples that have T_{RD} ages of >3.5 Gyr (Figure 44) With no data filtering, this is a first order observation that the lithospheric mantle underlying Archean cratons is also Archean in age. The Re–Os isotope system has been applied to the dating of seven different cratonic keels; the Kaapvaal craton (Walker *et al.*, 1989; Carlson and Irving, 1994; Pearson *et al.*, 1995b; Carlson *et al.*, 1999b; Irvine *et al.*, 2001), the Siberian craton (Pearson *et al.*, 1995a), the North Atlantic craton (Hanghoj *et al.*, 2001), the Wyoming craton (Carlson and Irving, 1994), the Tanzanian craton (Chesley *et al.*, 1999), the North China craton (Gao *et al.*, 2002), and the

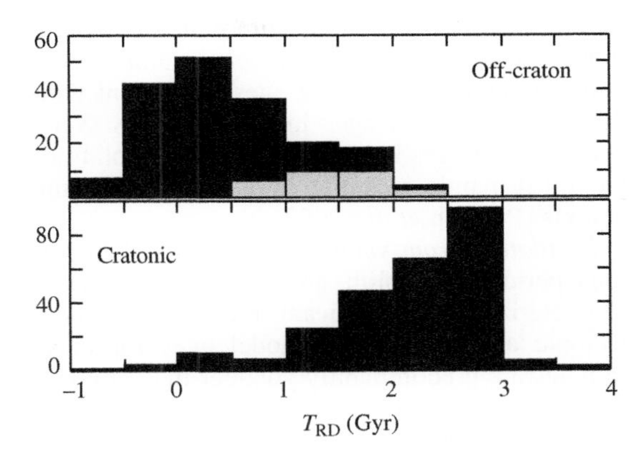

Figure 44 Frequency distribution plots of Re–Os isotope T_{RD} model ages (calculated assuming Re/Os = 0) for cratonic and off-craton peridotite xenoliths and massifs. The light-shaded field in the off-craton plot shows the range for kimberlite-derived peridotites from Namibia and East Griqualand. Data sources as in Figure 21 plus Pearson (unpublished).

Slave craton (Irvine *et al.*, 1999, 2003) which have all produced Archean Re–Os model ages. Some model ages range to in excess of 3.5 Ga but most are in the region between 2.7 Ga and 3.2 Ga (Figure 44). Archean ages are obtained from peridotites derived from well into the diamond stability field.

Re-Os data for peridotite xenoliths erupted by Paleozoic kimberlites of the North China craton indicate the presence of refractory, chemically buoyant Archean lithospheric mantle beneath the craton at that time. Younger, post-Archean Re–Os model ages obtained from peridotite xenoliths erupted by Tertiary alkali basalts in the same craton indicate that the deeper section of the cratonic keel was replaced by more fertile lithospheric mantle sometime after the Palaeozoic (Gao *et al.*, 2002).

Circum-cratonic mantle. The mantle beneath areas adjacent to cratons is also sampled by kimberlite in numerous locations, allowing comparison of the age of CLM beneath circum-cratonic regions with that beneath the cratons. In southern Africa, kimberlites intrude to the West and Southeast of the exposed craton in Namibia and East Griqualand respectively and also to the south, in the Karoo. Rhenium depletion model ages for peridotite xenoliths sampled by these kimberlites are all post-Archean, with maximum ages in the range 2–2.2 Gyr (Pearson *et al.*, 2002a), equivalent to the age of the oldest exposed crust in these areas. Peridotite xenoliths erupted by alkali basalts in the Vitim region of Russia provide samples of CLM adjacent to the Siberian craton (Ionov *et al.*, 1993a). Rhenium depletion ages of up to 2.1 Gyr have been reported from this xenolith suite (Pearson *et al.*, 1998a, 2004).

In all these cases, the circum-cratonic mantle immediately surrounding Archean cratons gives younger rhenium depletion ages, consistent with the younger crustal ages in these regions. Such coherency of ages indicates a general coupling of the crust-mantle systems in and around some cratons (Pearson *et al.*, 2002).

Peridotites from younger lithosphere. Noncratonic peridotite xenoliths are often erupted in areas characterized by significant recent extensional tectonic activity. Re–Os model ages for these samples are predominantly younger than cratonic and circum-cratonic samples (e.g., Peslier *et al.*, 2000), however, peridotites from some localities such as southeastern Australia (Handler *et al.*, 1997) and the Sierra Nevada of the western USA (Lee *et al.*, 2001) have rhenium depletion ages as old as meso-Proterozoic, significantly older than the latest crust generation episodes in these areas. This indicates that old CLM in these cases survives the rifting and crustal differentiation events imposed on them. The old remnant lithospheric mantle found beneath the Sierra Nevada is used to support a model of delamination of deep, unstable lithosphere produced due to collisional thickening (Lee *et al.*, 2000a). Most of the xenoliths analyzed from these areas are spinel-facies xenoliths that are likely to be derived from depths less than 60–70 km deep. Thus, it is only the shallower Proterozoic mantle in general that survives these reworking episodes.

U–Pb ages on high-U/Pb, high-Th/Pb phases. Re–Os model ages record, to varying degrees of precision and accuracy, the timing of melt depletion in peridotites. The presence of high U/Pb and Th/Pb phases such as zircon and monazite in rare peridotite, MARID/PKP and glimmerite xenoliths allow precise U–Th–Pb ages to obtained (Carlson and Irving, 1994; Hamilton *et al.*, 1998; Konzett *et al.*, 1998, 2000; Rudnick *et al.*, 1999). Because the zircon and monazite are associated with metasomatic enrichment events the ages obtained date the timing of melt enrichment in the host rock. Some of these ages are highly precise, e.g., Carlson and Irving (1994) obtain an age of 1,779 ± 2 Myr for a monazite crystal in a glimmerite from the Montana xenolith suite. This age is in agreement with a zircon U–Pb age of 1,784 ± 20 Myr from a peridotite from the same xenolith suite (Rudnick *et al.*, 1999). These ages are suggested to record an enrichment event coincident with the formation of the Great Falls tectonic zone. In contrast, zircons from a phlogopite vein in a harzburgite from the Labait tuff cone, Tanzania record a very young, 400 ± 200 kyr metasomatic event affecting Archean peridotite. Thus, the potential for these phases to yield precise geochronological information in mantle assemblages is clear, if they can be found. These minerals appear to retain their radiogenic Pb because they are the principal sinks for Pb in the host rock. They do not significantly exchange with their surrounding minerals and hence they record ancient ages even though residing at mantle temperatures. In this sense, their "blocking" temperatures in the mantle environment are high, even for monazite.

2.05.2.8 Stable Isotope Chemistry: Oxygen, Carbon, and Sulfur Isotopes

In comparison to radiogenic isotope data, stable isotope data for minerals in peridotites are sparse. As with radiogenic isotopes, alteration processes and invasion by the host rock means that the most reliable data are derived from mineral separate analyses. Initial studies of the variation and inter-mineral fractionation of oxygen isotopes between mantle xenolith minerals were interpreted as due to varying equilibration temperatures (Kyser *et al.*, 1981). Gregory and Taylor (1986) re-interpreted this complex data set in terms of open-system exchange between the minerals and metasomatic fluids. In this model, olivine and spinel exchange oxygen more rapidly than pyroxenes and hence show larger variations in isotope composition. A problem with the Gregory and Taylor (1986) model is how a fluid with a distinct $\delta^{18}O$ can percolate through mantle consisting of 80% olivine without being rapidly buffered to the composition of the ambient mantle. A problem for both models, and for most pre-1990 oxygen isotope data on mantle minerals, is the relatively imprecise data obtained by convention fluorination techniques for mantle minerals, especially olivine, which made interpretation of small inter-mineral isotopic fractionations difficult to do with confidence.

The advent of laser-fluorination techniques dramatically improved capabilities for precise analyses of these phases and led to the observation that olivine from a wide range of peridotites, including hydrous and anhydrous variants, has a very constant $\delta^{18}O$ (Mattey *et al.*, 1994). In addition, olivine and pyroxene data obtained using laser-fluorination show much smaller, more systematic fractionations than conventional fluorination data on the same samples (Figure 45). This observation has been confirmed by subsequent studies (Chazot *et al.*, 1997; Xu *et al.*, 1998).

Because of the buffering effect of mantle olivine on the oxygen isotope composition of metasomatic fluids, oxygen isotopes are not, in most cases, effective tracers of mantle–fluid interaction (Mattey *et al.*, 1994; Menzies and Chazot, 1995). One exception to this is the fluid-process that creates diamond. Comparison of the oxygen isotope compositions of garnets in typical

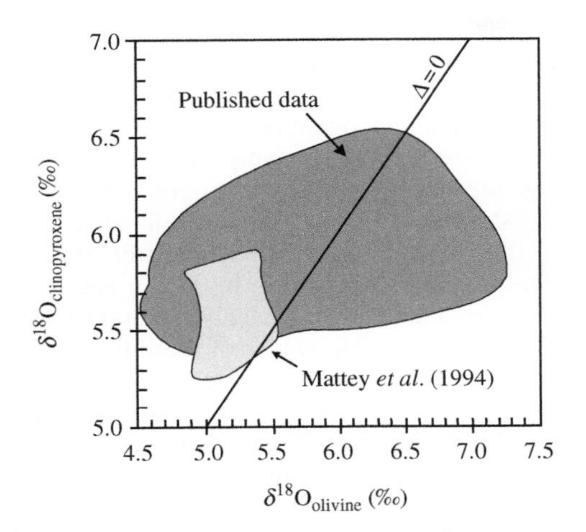

Figure 45 $\delta^{18}O$ clinopyroxene versus $\delta^{18}O$ olivine for peridotite mantle xenoliths. The large-shaded field is for cratonic and noncratonic peridotites analyzed by conventional fluorination techniques cited in Mattey *et al.* (1994). The light-shaded smaller field is for laser-fluorination analyses of a wide-range of cratonic and noncratonic peridotite (source Mattey *et al.*, 1994).

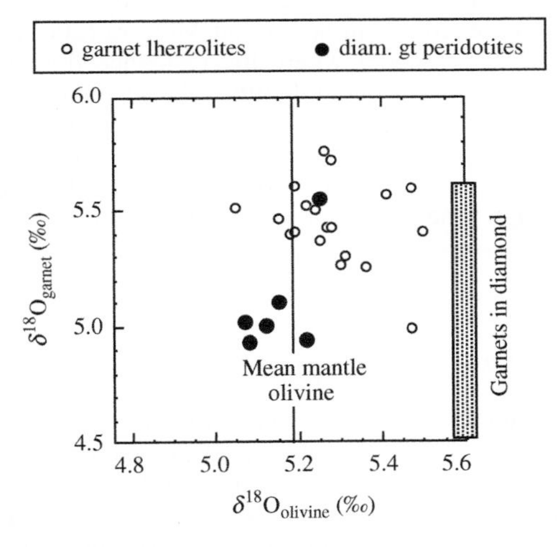

Figure 46 $\delta^{18}O$ garnet versus $\delta^{18}O$ olivine plot for garnet lherzolites (cratonic and noncratonic), subcalcic garnet harzburgite/dunites, and subcalcic garnet inclusions in diamonds. Data from Mattey *et al.* (1994), Lowry *et al.* (1999), and Mattey and Pearson (unpublished). The range for garnets included in diamonds and some xenolith garnets with no corresponding olivine analyses are plotted on the right-hand axis of the diagram. The mean value for mantle olivine is from Mattey *et al.* (1994).

lherzolite xenoliths with peridotitic garnet inclusions in diamonds and garnets from diamondiferous peridotite xenoliths shows that garnets associated with diamonds have significantly lower $\delta^{18}O$ than garnets in nondiamondiferous peridotites (Figure 46; Lowry *et al.*, 1999). Olivines from the diamondiferous peridotites also appear shifted to lower $\delta^{18}O$ than typical for the nondiamondiferous garnet lherzolites but are within the range of mantle olivines from spinel lherzolites (Mattey *et al.*, 1994). In addition, reversed oxygen isotope fractionation between garnet and olivine in both diamond inclusions and diamondiferous peridotites suggests that garnet preserves subtle isotopic disequilibrium related to the genesis of chromium-rich garnet and/or diamond (Lowry *et al.*, 1999). It is also possible to explain the anomalously low $\delta^{18}O$ values of low-calcium harzburgite/dunite assemblages in terms of a subduction model. High bulk rock Cr/Ca and high Cr/Al in garnet in these assemblages may require a low-pressure melting origin, consistent with formation as oceanic lithosphere, followed by subduction (Section 2.05.2.5.2; Schulze, 1986; Canil and Wei, 1992). The $\delta^{18}O$ values are shifted lower than typical mantle, which is consistent with high-T hydrothermal alteration operative in the deep oceanic lithosphere. Subduction of such material could, therefore, account for the low $\delta^{18}O$ values of olivine in subcalcic garnet harzburgites. Further studies are required to better constrain these possibilities.

A detailed laser-fluorination study of individual minerals from ultradeep xenoliths (Deines and Haggerty, 2000) revealed significant oxygen isotopic variation. While evidence for oxygen isotopic equilibrium is observed between some garnet–olivine, garnet–orthopyroxene and clinopyroxene–orthopyroxene pairs, fractionation between garnet and clinopyroxene tends to be out of equilibrium. Deines and Haggerty (2000) concluded from diffusion modeling that the xenoliths had experienced a series of metasomatic alterations, including an event that lowered the $\delta^{18}O$ value of the pyroxenes, a few million years before eruption by the host kimberlites. The mean $\delta^{18}O$ values of garnets in these transition-zone rocks are argued to be unaltered by the metasomatic processes and are indistinguishable from upper-mantle values.

In the oceanic setting, spinel lherzolite xenoliths from Pali (Hawaii) have olivine $\delta^{18}O$ values of 5.09–5.12 per mil, typical of olivines from other oceanic and continental mantle rocks (Ducea *et al.*, 2002). In contrast, olivines from plagioclase peridotites are enriched by ~0.5 per mil. This is interpreted to be due to the formation of plagioclase by reaction with or crystallization from melts intruding the Pacific lithospheric mantle.

Hydrogen isotope data for mantle xenoliths is usually acquired on hydrous minerals such as amphibole and mica. One problem is that early studies did not texturally characterize mica occurrences and so the information is of limited value. Perhaps the best-documented study is that of Boettcher and O'Neil (1980), who analyzed megacrystic phlogopite from kimberlites and phlogopite from peridotites. These authors found a restricted range in δD (-70 per mil to -63 per mil), typical of mantle derived magmas. In megacrysts and veins, amphibole often has lower δD and δ^{18}O than coexisting phlogopite but the reason for this is not well understood (Kyser, 1990).

The systematics of sulfur isotopes in mantle xenoliths have been reviewed by Kyser (1990). Most sulfide data have so far been obtained by *in situ* analyses using SIMS or laser probe and this is less prone to alteration effects than whole-rock analyses. Chaussidon *et al.* (1989) found that considerable sulfur isotope variation exits in mantle minerals (δ^{34}S -5 to 8 per mil), which they attributed to fractionation between residual sulfide and the melt during melting. However, it is megacrysts that show most variation in their data set, possibly due to magmatic processes while sulfide from the garnet peridotites has a much more restricted range, of between -1 per mil and $+4$ per mil, typical of mantle values. Wilson *et al.* (1996) found elevated δ^{34}S in peridotites from Dish Hill, which they proposed was due to metasomatic introduction of subducted crustal sulfur.

2.05.2.9 Noble Gases

2.05.2.9.1 He isotopes

Initial studies of xenoliths from both continental (Porcelli *et al.*, 1986) and oceanic (Vance *et al.*, 1989) settings showed a remarkable consistency of ^{3}He/^{4}He, in contrast to their widely variable strontium isotope compositions. For example, peridotite xenoliths from the Canary Islands, Gough, Kerguelen, and Madeira studied by Vance *et al.* (1989) had a restricted ^{3}He/^{4}He range of 5.2–7R/R_A (where R_A is the atmospheric ^{3}He/^{4}He value), comparable to MORB. In contrast, their clinopyroxene strontium isotope compositions extended over almost the entire range of oceanic basalts. ^{3}He/^{4}He in Hawaiian xenoliths is slightly higher, 8.2–9.5R/R_A, possibly reflecting a sampling bias towards pyroxenites compared to peridotites. The Hawaiian pyroxenites are probably linked to the host magmatism, which has elevated ^{3}He/^{4}He. In contrast, Poreda and Farley (1992) found ^{3}He/^{4}He ratios of up to 21.6 $\pm$ 1 in Samoan peridotite xenoliths. The highest xenolith ^{3}He/^{4}He ratios are comparable to those measured in Samoan lavas and the data are most easily interpreted in a model where the xenoliths trap volatiles from the high ^{3}He/^{4}He lava sources.

The continental samples analyzed by Porcelli *et al.* (1986) had ^{3}He/^{4}He ranging from ~6 to just above 10, with a mean value well within the MORB range. These samples had extremely variable strontium isotope compositions. These initial studies clearly showed that helium and lithophile element isotope compositions in mantle xenoliths from both oceanic and continental settings are decoupled. This was confirmed by a more extensive study of continental peridotites from the Massif Central, The Eifel region, Spitzbergen and Kapfenstein by Dunai and Baur (1995), who analyzed exclusively mineral separates. The mean ^{3}He/^{4}He values of these four suites were 6.5, 6.0, 6.7, and 6.1 times R_A and hence within the MORB range, but lower than the average MORB value of 8. Dunai and Baur (1995) suggested that the helium isotope signature preserved in the xenoliths is inherited from their host magma and reflects a MORB-like source. A recent, extensive review of noble gases from continental peridotite xenoliths and basalts found a mean R/R_A value of 6.1 $\pm$ 0.9 compared to the MORB source ratio of 8 $\pm$ 1 and proposed that this difference is significant (Gautheron and Moreira, 2002). The difference between the continental mantle, as portrayed by the xenoliths, and MORB cannot be accounted for by uranium-decay, leading Gautheron and Moreira (2002) to postulate a model where the residence time for helium in continental lithospheric mantle is around 100 Ma. If such a residence time is correct then the continental mantle cannot be used to track helium isotopes, or other noble gases through time and is completely overprinted by the MORB-source.

2.05.2.9.2 Neon, argon, and xenon isotopes

Neon isotope data from Samoan lavas show elevated ^{20}Ne/^{22}Ne relative to atmosphere. The data are not consistent with a model of mixing between a degassed MORB mantle with high ^{20}Ne/^{22}Ne and a deeper, undegassed plume source with atmospheric ^{20}Ne/^{22}Ne. The similarity of neon isotopes between the Samoan plume-like source and MORB supports the idea that neon isotopes in the mantle as a whole more closely resemble the solar composition than that of the atmosphere (see Chapter 2.06). "Plume-like" neon isotopic signatures have been identified in an apatite from a southeastern Australian spinel lherzolite xenolith (Matsumoto *et al.*, 1997).

^{40}Ar/^{36}Ar in the least contaminated Samoan xenoliths varies from 4,000 to 12,000, with the highest values found in samples with the highest ^{3}He/^{4}He values. There is little evidence for atmosphere-like argon in the source of Samoan lavas, with values likely to be in excess of 5,000 (Poreda and Farley, 1992). These data, combined with ^{129}Xe and ^{136}Xe anomalies 6% greater than the atmospheric value, contradict the idea that high $^3He/^4He$ mantle domains have pristine, near-atmospheric heavy rare gas isotope compositions and suggests that much of the mantle could have non-atmospheric noble gas isotopic compositions.

High uranium and thorium contents of metasomatic minerals in pyroxenites and peridotites from southeastern Australia produce radiogenic ^4He and nucleogenic ^{21}Ne and ^{22}Ne (Matsumoto *et al.*, 2000). Accounting for this and the effects of atmospheric contamination, the heavy noble gas isotope systematics of peridotites from this region and the European continental peridotites analyzed by Dunai and Baur (1995) are MORB-like in character, i.e., with elevated $^{40}Ar/^{36}Ar$, $^{20}Ne/^{22}Ne$, $^{129}Xe/^{130}Xe$, and $^{136}Xe/^{130}Xe$. Thus, the heavy noble gas isotope data from continental lithospheric mantle support models where the volatile budget of this reservoir is dominated by fluxing from the underlying MORB source (Gautheron and Moreira, 2002) and/or that many of the samples formed from MORB-source mantle, accreting to the continental lithosphere during melt depletion. In contrast, Dunai and Baur (1995) have suggested that the He–Ar systematics of some European peridotite xenoliths reflect mixing of recycled crustal material with a MORB-like source during the Variscan orogeny. If data from massif peridotites and peridotite xenoliths are considered together, the continental lithospheric mantle can be argued to have lower $^3He/^4He$ values than the convecting MORB-source mantle but with dominantly MORB-like argon, neon, and xenon isotope systematics.

2.05.3 ECLOGITE XENOLITHS

2.05.3.1 Classification, Mineralogy, and Petrography

All kimberlites within cratonic areas of southern Africa that have been well sampled contain eclogite (Gurney *et al.*, 1991) but at most localities it is a rare xenolith type. Consequently, studies of eclogites have been confined to a few kimberlites, e.g., Newlands, Bellsbank/Bobbejaan, Roberts Victor and Orapa on the Kaapvaal craton (Table 1(a); MacGregor and Carter, 1970; Robinson *et al.*, 1984; Hatton and Gurney, 1987; Taylor and Neal, 1989) and Udachnaya plus Mir from the Siberian craton (Sobolev, 1974; Snyder *et al.*, 1993; Jacob *et al.*, 1994; Beard *et al.*, 1996). Assessing the relative abundance of eclogite within the whole lithospheric mantle is not easy due to considerations of preferential disaggregation of certain types of xenoliths. On the basis on mineral concentrate studies, Schulze (1989) estimates eclogite to be less than 2% on average of a given continental lithospheric mantle section, as sampled by kimberlites. In exceptional circumstances, such as at the Roberts Victor kimberlite (S. Africa), eclogite comprises 95–98% of the mantle xenolith inventory, however, even here, mineral concentrate studies show that peridotite is abundant.

In noncratonic xenolith suites, eclogite is much scarcer and it is notable if this lithology is present. Pyroxenites and garnet pyroxenites are more common in noncratonic xenolith suites.

Eclogites consist of omphacitic clinopyroxene and pyrope-almandine garnet. Their grospydite variants (Table 1) are defined as containing garnets with a large grossular component together with kyanite. Accessory phases vary widely but rutile is perhaps the most common (Table 1). At some localities, such as Roberts Victor, coesite is more common than originally thought (Schulze *et al.*, 2000) and orthopyroxene is sometimes present. Other exotic related rocks such as alkremites contain spinel and/or corundum as major phases associated with garnet (Table 1; type IX). Olivine is unknown. Diamond can be present as a significant component of eclogites at some localities (Gurney *et al.*, 1991).

There are several different classification schemes for eclogites. Roberts Victor eclogites have been divided into group I and group II rocks (MacGregor and Carter, 1970), based on their textures (Table 1). Taylor and Neal (1989) subdivided the Bellsbank/Bobbejaan eclogites into three groups (A–C) on the basis of their mineral chemistry and used this classification in subsequent studies to help assign crust or mantle origins to eclogites. A modified version of the MacGregor and Carter terminology is probably the most widely applied, at least to southern African eclogites. That cratonic eclogites are erupted from mantle depths is confirmed by the presence of diamond in group I variants (Table 1(a)). In addition, the incorporation of relatively high levels of sodium into group I eclogitic garnets and potassium into group I clinopyroxenes (McCandless and Gurney, 1989) requires a high pressure, mantle origin. Eclogites occurring in circum-cratonic and craton-margin kimberlites, often containing feldspar and associated with granulitic assemblages are widely accepted to be of crustal origin (e.g., Gurney *et al.*, 1991). Alkremites (spinel–garnet assemblages; Table 1(a)) probably represent an end-member of the eclogite suite (Nixon *et al.*, 1978).

2.05.3.2 Mineral Chemistry and Equilibration Conditions

Eclogites have widely varying garnet compositions that are less chromium- and magnesium-rich than those of peridotite xenoliths (Figure 47). As a general guideline, eclogite garnets have

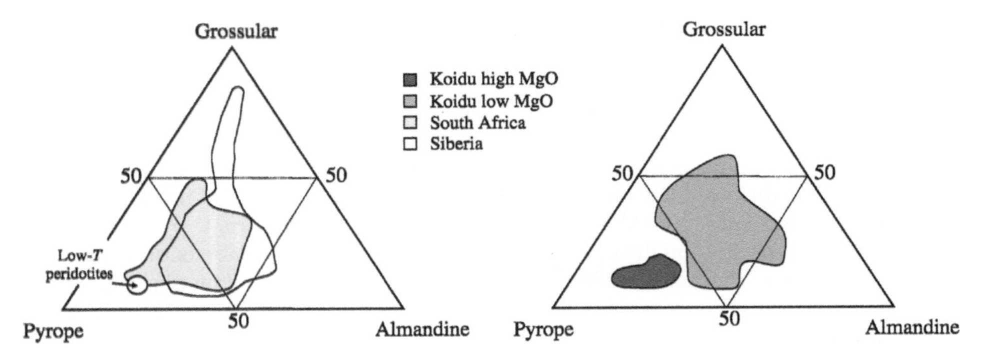

Figure 47 Garnet compositions for eclogites from Siberia, South Africa, and Koidu (Sierra Leone) compared to low-*T* Kaapvaal peridotites, in terms of their pyrope, grossular, and almandine calculated end-members (sources Sobolev, 1974; Hatton and Gurney, 1987; Mazonne and Haggerty, 1989; Taylor and Neal, 1989; Jacob *et al.*, 1994; Hills and Haggerty, 1989; Viljoen *et al.*, 1996; Pyle and Haggerty, 1998; and Barth *et al.*, 2001).

<2 wt.% Cr_2O_3, whereas peridotitic garnets exceed this value (Figure 5; Gurney *et al.*, 1991). The most common garnet compositions are in the pyrope–almandine range but their grossular contents can vary widely. When grossular is dominant over pyrope in the garnet composition in eclogites, the lithology is termed grospydite (Sobolev, 1974). Na_2O contents of garnets from group I eclogites exceed 0.07 wt.% (Schulze *et al.*, 2000) and this feature is interpreted to reflect high-pressure equilibration conditions (>5 GPa; McCandless and Gurney, 1989). Garnets from group II eclogites have <0.07 wt.% Na_2O. Garnets from alkremites and related rocks are generally grossular-rich pyrope–almandine–grossular solid solutions with negligible spressartine (Mazonne and Haggerty, 1989).

Clinopyroxenes are omphacitic, with large jadeite and minor pyroxene quadrilateral components that clearly distinguish them from the clinopyroxene found in pyroxenite xenoliths or pyroxenites from massif peridotites (Figure 48). Omphacite Na_2O contents are typically between 3 wt.% and 8 wt.%. Two-phase assemblages of garnet and clinopyroxene in which the pyroxene is not omphacitic are properly referred to as garnet clinopyroxenites. The high jadeite and low quadrilateral component contents of eclogitic clinopyroxenes are distinct from high-pressure liquidus pyroxenes produced in experiments (Figure 48). This makes an origin for eclogites as high pressure cumulates from magma unlikely. Eclogitic group I clinopyroxenes have >0.08 wt.% K_2O, which is most likely a response to high pressures of equilibration (McCandless and Gurney, 1989).

The presence of coesite and diamond together with high sodium and potassium in garnet and clinopyroxene respectively, make a clear case for the high-pressure origin of cratonic eclogites (Gurney *et al.*, 1991). In contrast to the various quantitative barometers available for peridotites,

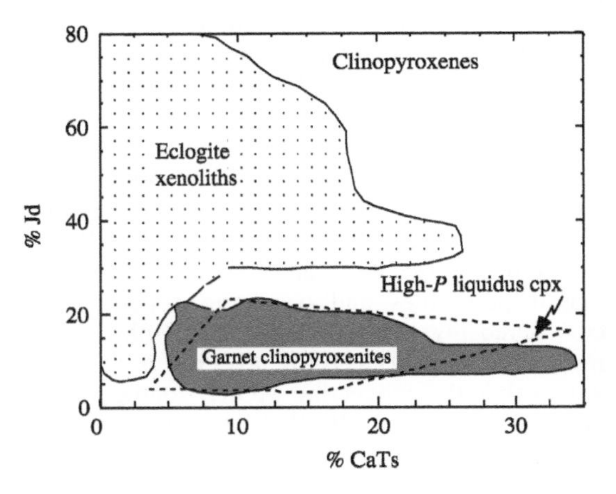

Figure 48 Molecular percent jadeite (Jd) versus Ca-Tschermak's (CaTs) of clinopyroxenes from eclogite xenoliths compared with clinopyroxene from garnet clinopyroxenite xenoliths (Irving, 1974) and from garnet clinopyroxenite layers in massif peridotites (Pearson and Nixon, 1995). Field of high-pressure liquidus clinopyroxenes from Eggins (1992). Data sources as in Figure 47.

the high thermodynamic variance of the garnet–clinopyroxene assemblage means that no barometers for eclogite xenoliths have been developed. Mukhopadhyay (1991) has proposed a barometer for garnet–clinopyroxene assemblages but this is only applicable when clinopyroxenes are enriched in quadrilateral and Tschermak components and hence it is inapplicable to true eclogites.

Equilibration temperatures are commonly calculated using the Fe–Mg exchange reaction between garnet and clinopyroxene, which is slightly pressure dependant and so an equilibration pressure must be assumed. Many thermobarometry studies of eclogites use the assumption that these rocks lie on a suitable geotherm for the region and project equilibration temperatures to the geotherm in order to derive their depth of origin. Variations in equilibration temperatures between different eclogite suites are apparent at a given equilibration pressure. For instance, at 5 GPa the equilibration temperatures of Koidu low-MgO eclogites (Fung and Haggerty, 1995) are much lower (880–930 °C) than for Roberts Victor eclogites (~1,100 °C; Harte and Kirkley, 1997). MacGregor and Manton (1986) suggest that most cratonic eclogites originate from >150 km depth, at the base of the lithosphere. While the high sodium in garnet and high potassium in clinopyroxene, together with the presence of diamonds in group I eclogites support this notion, the absence of these features and the wide range in equilibration temperatures in group II eclogites suggest that some originate from much shallower depths.

2.05.3.3 Bulk Compositions

The bulk compositions of eclogites can be seriously affected by a combination of syn-eruption partial melting, mantle metasomatism (including invasion by their hosts) and low-*T* alteration. Omphacite is generally more affected than garnet. Kyanite-eclogites are usually the most severely altered (Harte and Kirkley, 1997). The geochemical effects are most dramatic for incompatible trace elements but major elements can be affected in some suites also (Ireland *et al.*, 1994), resulting in an increase in Mg# due to invasion of high-magnesium host melts. Accordingly, some authors have used reconstructed bulk compositions calculated from modal analyses and mineral compositions in the study of eclogite bulk rock chemistry. These compositions are subject to errors in the accuracy of modes for coarse-grained rocks. For most suites, effects on major-element compositions are relatively minor and hence their measured bulk compositions can be used to say something about their origin.

The "basaltic" compositions of eclogite xenoliths have lead to two main models of eclogite origin; that they represent the crystallized products of high-pressure mantle melting (O'Hara and Yoder, 1967; Hatton and Gurney, 1987), or that they represent the metamorphosed products of subducted oceanic crust (Helmstaedt and Doig, 1975; Jagoutz *et al.*, 1984; MacGregor and Manton, 1986; Neal *et al.*, 1990; Jacob *et al.*, 1994). A modification of the crystallized mantle melts hypothesis suggests initial crystallization at relatively low pressures (~50 km) and then burial to depths within the diamond stability field (McDade and Harte, 2000). Variations on the subduction end-member model propose that eclogites are residual slab compositions from the extraction of tonalitic magmas during Archean subduction (Ireland *et al.*, 1994; Barth *et al.*, 2001).

In discussing the relevance of eclogite bulk compositions to this argument, it is pertinent to make comparisons of their compositions with Archean basaltic/komatiitic magmatism because of the Archean age of many eclogite xenoliths (Section 2.05.3.5.3). In addition, comparisons of xenoliths with "massif" eclogites exposed in high-pressure orogenic terrains, for which an origin via metamorphism of basalt is widely accepted, are also relevant, despite the Phanerozoic age of most of these occurrences.

The MgO contents of eclogite xenoliths range from ~6% to ~20%, with a mean of 12%. This makes them basaltic to picritic in composition, rather than komatiitic. No samples extend to the very high MgO contents (>25 wt.%) of Archean komatiites (Figure 49). Some eclogite suites can be divided into low-MgO and high-MgO variants (Barth *et al.*, 2001; Barth *et al.*, 2003). Significant overlap between Archean magmatic rocks and eclogite xenoliths exists for CaO, Na_2O, and TiO_2. However, SiO_2 contents for eclogite xenoliths at a given MgO are much lower than Archean magmas while few erupted volcanic rocks reach Al_2O_3 contents of >20% observed in some eclogites. Compared to massif eclogites, there is overlap in CaO and Al_2O_3 contents, but SiO_2, Na_2O, and TiO_2 are, on average, significantly lower for the xenoliths (Figure 49). Eclogites that have been tectonically exposed in the crust in high-pressure metamorphic regions do not appear to contain the alkremite/grospdyite assemblages found in cratonic xenolith suites. This may be a function of the different depths of sampling or due to the differing ages. Orogenic eclogite suites are dominantly post-Archean, whereas cratonic eclogites are Archean in age (Section 2.05.3.5.3). Chromium contents are consistently higher in eclogite xenoliths than massif eclogites. For xenoliths, total alkalis are lower than most magmas at a given MgO content. These differences between eclogite xenolith compositions and those of Archean magmas and massif eclogites pose clear problems for a model where the xenoliths simply represent isochemically metamorphosed basalts.

McDade and Harte (2000) suggest that the major-element compositions of the Roberts Victor eclogite xenoliths are not compatible with recrystallized oceanfloor basalts although their comparisons use exclusively Phanerozoic basalt compositions. McDade and Harte (2000) and Barth *et al.* (2001, 2003) point out that most eclogite compositions are not comparable with the products of peridotite fusion at high pressures. This led McDade and Harte (2000) to appeal to low pressure (spinel-facies) mantle melting and crystallization, followed by burial to greater depths. It is not clear how crustal isotopic

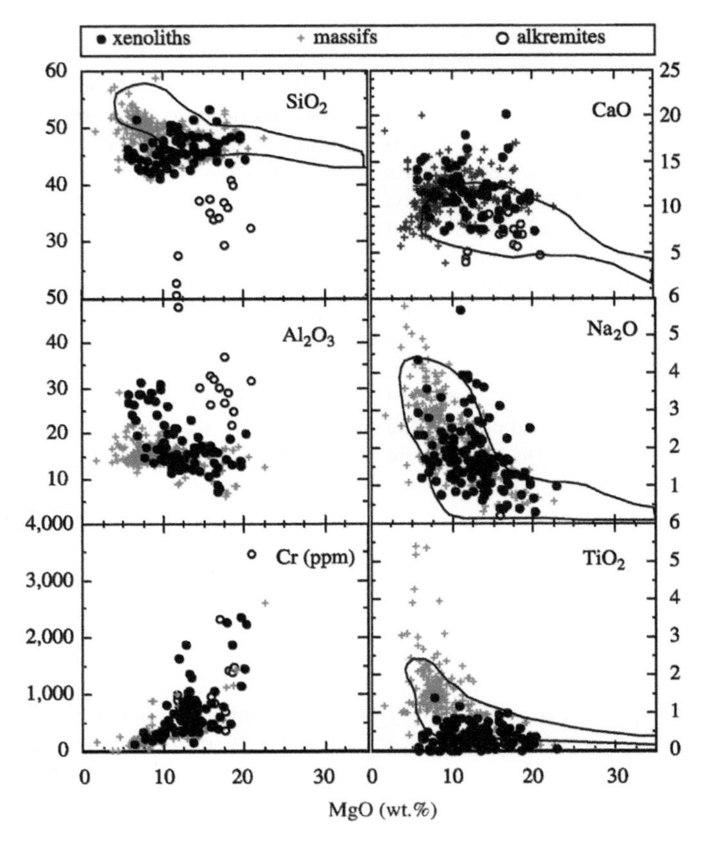

Figure 49 Bulk analyses of eclogite and alkremite (alkremite–corgaspinite–corganite; Table 1) xenoliths from kimberlites compared with massif eclogites and Archean basalts and komatiites (open field). Data sources: Kushiro and Aoki (1968), Sobolev (1974), Mazonne and Haggerty (1989), Hills and Haggerty (1989), Taylor and Neal (1989), Ireland *et al.* (1994), Pyle and Haggerty (1994), Snyder *et al.* (1993), Jacob *et al.* (1994), and Jacob and Foley (1999). Outlined field is for Archean basalts and Komatiites taken from Ireland *et al.* (1994).

signatures are acquired in this model, unless the protoliths are erupted basalts. The moderate-to-low-pressure pyroxenes on the liquidus of mantle melts are not omphacite (Figure 48) and so this must form during burial. A similar model was suggested by Barth *et al.* (2003) to explain the genesis of the Koidu high-MgO eclogite suite. High MgO and Al_2O_3 contents of the eclogites are suggestive of a cumulate origin, either as high-pressure (2–3 GPa) garnet-pyroxene cumulates or low-pressure (<1 GPa) plagioclase–pyroxene–olivine cumulates.

While differences between the major-element composition of eclogite xenoliths versus Phanerozoic *and* Archean magmas certainly exist, and the xenolith compositions also differ with massif eclogites, many of these features can be reconciled with the "residual slab" type model of eclogite xenolith origin (Ireland *et al.*, 1994; Jacob and Foley, 1999; Barth *et al.*, 2001). The relatively low SiO_2 and Na_2O contents of eclogite xenoliths compared to those of massif eclogites and Archean basalts (Figure 49) can be related to the proposed loss of a high-silicon tonalitic melt fraction during subduction. Lowering of SiO_2 during subduction zone melting can also explain the scarcity of quartz (as coesite) in some eclogite xenolith suites (Udachnaya, Koidu, Bellsbank) compared to massif eclogites (Jacob and Foley, 1999). A melting residue model would also explain the lower alkalis and titanium and the higher chromium contents of xenoliths compared with massif eclogites and Archean magmas because of the relative compatibilities of these elements during melting. Schulze *et al.* (2000) suggest that the abundance of free silica (as coesite) in Roberts Victor eclogites indicates that they have not experienced significant partial melting during subduction and represent original ocean-floor type compositions. However, Yaxley and Green (1998) showed in an experimental study that coesite is stable in eclogite until 13% melt loss and so some compositions could be residual. In slab-recycling models, the very aluminous compositions of some eclogite xenoliths, e.g., kyanite-eclogites, has led to the suggestion that they are former plagioclase-rich oceanic cumulates (Jagoutz *et al.*, 1984; Jacob and Foley, 1999).

In support of a crustal origin, Jacob *et al.* (1998b) and Barth *et al.* (2001) point out that the kyanite–eclogite assemblage is not in equilibrium with mantle peridotite at high pressure and would react with olivine to form aluminous pyroxene and garnet, further strengthening an origin via high-grade metamorphism rather than igneous processes.

The bulk compositions of alkremites and related rocks are clearly very anomalous for aluminum and silicon (Figure 49) and are largely a function of the abundance of garnet and spinel in the mode. Nixon *et al.* (1978) explain these unusual rocks in terms of high-pressure phase equilibria and their formation from mantle melts. In contrast, Mazonne and Haggerty (1989) draw attention to similarities with pelitic sediment bulk compositions and suggest an origin as high-pressure metamorphic products of melting residues from the subduction of pelitic sediments.

2.05.3.4 Trace-element Chemistry

The effects of metasomatism and invasion of the host magma alluded to above are much more significant for minor element concentrations in eclogites, particularly highly incompatible elements. For example, Barth *et al.* (2001) demonstrate that addition of ~5% host kimberlite to un-metasomatized protoliths can account for the measured whole-rock trace-element systematics of Koidu eclogites and will significantly affect incompatible element abundances and ratios. As such, our discussion will focus on those studies that have used the trace-element contents of minerals to reconstruct whole-rock trace-element abundances. Ireland *et al.* (1994) compared the trace-element systematics of inclusions in diamond with those of minerals in the host diamondiferous eclogites. They propose that the mineral compositions of the host eclogite are affected by metasomatic processes that veil the true precursor compositions. Other studies (e.g., Taylor *et al.*, 1996) have not yielded such clear results. There are issues concerned with whether the diamond-forming event in diamondiferous eclogites locally affects silicate minerals. These problems aside, there are few samples where studies of inclusions and host phases are possible. Furthermore, the extent of metasomatism by the host rock is much lower in some eclogites suites than others. Hence, we will discuss information that can be obtained from trace-element mineral studies of eclogites in general (Barth *et al.*, 2001; Harte and Kirkley, 1997; Jacob and Foley, 1999; Jerde *et al.*, 1993; Snyder *et al.*, 1997). We will describe mineral trace-element chemistry first followed by reconstructed whole-rock abundances.

Garnets from worldwide eclogite xenolith locations are typically LREE-depleted with values of $(La)_n$ as low as 0.01 (Figure 50) Most garnets have relatively unfractionated HREE abundances with $(Dy/Yb)_n$ of 0.5 to 2.5, $(Yb)_n$ values up to 50 and have small positive europium anomalies. They show variable depletion of zirconium, hafnium, and titanium (e.g., Figure 50). Barth *et al.* (2001) noted that garnets within samples possessing jadeite-poor clinopyroxenes from Koidu (Sierra Leone) show greater HFSE anomalies and less variable REE patterns than those with jadeite-rich clinopyroxenes. Nb–Ta depletions in garnet are variable. Those most deficient in niobium and tantalum are from rutile-bearing samples where these elements preferentially partition into rutile. The LREE-depleted nature of garnets from most worldwide eclogite suites contrasts markedly with the complex REE patterns shown by cratonic peridotites. This suggests that either the eclogites have been subject to much less "cryptic" metasomatism than peridotites and/or the peridotite garnet REE patterns may reflect an origin other than loosely defined metasomatic additions to an originally depleted garnet.

Clinopyroxenes in eclogite xenoliths have convex-upwards REE patterns that are LREE-enriched relative to HREE. $(La)_n$ varies from slightly below 1 to ~15, while $(Yb)_n$ varies between 0.1 and 1.5 (Figure 50). As with garnet, Nb–Ta depletions and, to some extent, other HFSE abundances depend on the coexistence of rutile (Jacob and Foley, 1999; Rudnick *et al.*, 2000; Barth *et al.*, 2001). The relatively homogenous REE patterns of eclogitic clinopyroxene contrast markedly with the wide variety of REE patterns observed in peridotitic clinopyroxene.

Rutile in eclogite xenoliths has highly variable niobium and tantalum concentrations that result in variable Nb/Ta (Rudnick *et al.*, 2000). In common with rutile in peridotites, that found within eclogites shows pronounced enrichment in tungsten, molybdenum, zirconium, and hafnium over silicate phases and this phase is clearly important in controlling the whole-rock budget of these elements plus titanium. Barth *et al.* (2001) note that secondary (metasomatic) rutiles in Koidu eclogites may be distinguished by their very high and heterogeneous niobium and tantalum contents (16,000 ppm Nb, 900–1,700 ppm Ta) plus occasional skeletal textures.

The partitioning of trace elements between clinopyroxene and garnet ($D^{cpx/gt}$) within different eclogite suites can vary by two orders of magnitude, depending on element compatibility. The correlation of $D^{cpx/gt}$ for trace elements with the molar calcium partition coefficient D_{Ca^*} indicates a high level of trace-element equilibrium in most eclogites up to the time of eruption

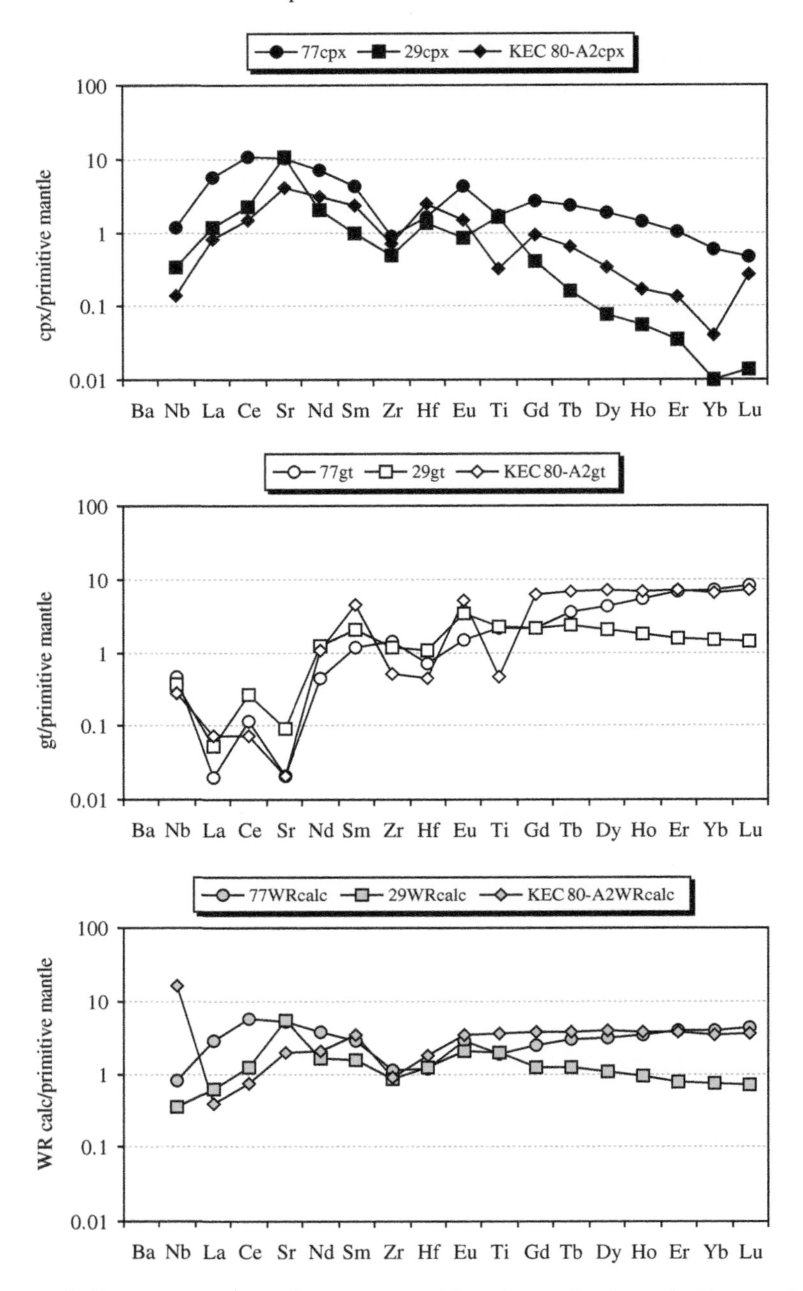

Figure 50 Garnet and clinopyroxene trace-element compositions (normalized to primitive mantle), together with their calculated whole-rock compositions (from modal data). Data sources: 77 and 29 are rutile-free eclogites from Udachnaya, Siberia (Jacob and Foley, 1999); KEC80-A-2 is a high-Nb-rutile eclogite from Koidu, Sierra Leone (Barth *et al.*, 2001).

(Harte and Kirkley, 1997). Differences in garnet–clinopyroxene partitioning behavior between the Roberts Victor and Koidu eclogite suites was attributed to the contrasting equilibration temperatures of the two suites by Barth *et al.* (2001).

Reconstructing whole-rock trace-element concentrations is the best way to avoid the effects of metasomatism, but careful account must be taken of any rutile present that can drastically affect HFSE budgets (Jacob and Foley, 1999; Rudnick *et al.*, 2000; Barth *et al.*, 2001). LREE depletion and very low HFSE abundances seem to be a characteristic of mostly rutile-free eclogites from Udachnaya (Siberia) and suggestive of protoliths originating in an island arc setting (Jacob and Foley, 1999). Positive strontium and europium anomalies and flat HREE profiles in these xenoliths are interpreted to reflect on origin as plagioclase-bearing plutonic rocks originally formed in ocean crust magma chambers

(Jacob and Foley, 1999; Barth *et al.*, 2003). Low-MgO rutile-bearing Koidu eclogites have LREE depleted but niobium-rich (up to 43 ppm) trace-element profiles (Figure 50) that Barth *et al.* (2001) cite as supporting a model where the eclogites are also residues of partial melting of altered, subducted oceanic crust, where the rutile is a residual phase during melting. Modeling of LREE/HREE systematics in the Koidu low-MgO suite suggests that they are the products of 15–40% batch melting of a basaltic protolith similar to Archean basalt. This range of melt extraction agrees with that required to produce Archean tonalitic crustal compositions as determined from eclogite melting experiments (Rapp and Watson, 1995). The relatively large degree of protolith melting also explains the lack/scarcity of coesite in Koidu low-MgO eclogites. Furthermore, the residual model firmly establishes a link between subduction and the generation of Archean continental crust (Ireland *et al.*, 1994; Rudnick *et al.*, 2000; Barth *et al.*, 2001).

For the high-MgO Koidu eclogites, trace-element modeling suggests a low-pressure origin for eclogites with flat HREE patterns and a high-pressure origin for eclogites with fractionated HREE. Flat HREE patterns, the presence of strontium anomalies, and low to moderate transition element contents in the low-P group are consistent with a low-pressure origin as metamorphosed olivine gabbros and troctolites (Barth *et al.*, 2003).

Calculated bulk rock trace-element systematics of eclogites have wider implications for mantle recycling models and bulk silicate earth mass balance. The subchondritic Nb/Ta, Nb/La, and Ti/Zr of both continental crust and depleted mantle require the existence of an additional reservoir with superchondritic ratios to complete the terrestrial mass balance. Rudnick *et al.* (2000) have shown that rutile-bearing eclogites from cratonic mantle have suitably superchondritic Nb/Ta, Nb/La, and Ti/Zr such that if this component formed 1–6% by weight of the bulk silicate earth, this would resolve the mass imbalance. This mass fraction far exceeds the likely mass of eclogite in the continental lithosphere and so the material is proposed to reside in the lower mantle, possibly at the core–mantle boundary.

2.05.3.5 Isotopic Characteristics

2.05.3.5.1 Stable isotopes

The small magnitude of oxygen isotope fractionation at mantle temperatures and pressures, together with the constancy of peridotite oxygen isotope compositions make oxygen isotopes a powerful tool for identifying recycled crustal material in the mantle. Although early studies found that some eclogite xenoliths had highly unusual oxygen isotope compositions (Garlick *et al.*, 1971), it was not until isotopic studies of hydrothermally altered basalts from ophiolite suites that the eclogite signatures were interpreted as an indication of a crustal prehistory for these rocks (Jagoutz *et al.*, 1984). Such data, along with their omphacitic pyroxenes, are some of the strongest evidence against models of eclogite xenoliths as directly crystallized moderate/high-pressure mantle melts. All seven suites of kimberlite-derived eclogites analyzed so far are found to have widely varying mineral $\delta^{18}O$ values that significantly exceed the range shown by olivine (5.18 ± 0.27 per mil) from mantle peridotite (Figure 51). Older analyses performed by conventional fluorination have been confirmed with laser fluorination methods, the laser-data showing more systematic garnet–clinopyroxene fractionations (Figure 51). No other mantle materials have such diverse oxygen isotopic compositions. Thus, while Snyder *et al.* (1997) choose to emphasize the great petrological and geochemical diversity shown by eclogites in general, that make their precise evolution difficult to constrain, a crustal origin for many of these rocks seems clear (Jagoutz *et al.*, 1984; MacGregor and Manton, 1986; Neal *et al.*, 1990; Jacob *et al.*, 1994, 1998b; Jacob and Foley, 1999).

The wide range in $\delta^{18}O$ is shown by both garnets and omphacite. The two minerals show the expected equilibrium fractionation relationship for both diamondiferous and nondiamondiferous samples (Figure 51). This range in $\delta^{18}O$ is considerably larger than that of garnet–diopside pairs from garnet lherzolites (Figure 51). Although it is possible to appeal to metasomatic fluids as the origin of these anomalous signatures, it is notable that metasomatic phases in mantle peridotites have oxygen isotope compositions nearly identical to nonmetasomatic minerals and in the typical mantle range. In addition, it is likely that such fluids would be rapidly buffered to the oxygen isotope composition of mantle peridotite during their movement. Hence, the generally accepted interpretation of the eclogite xenolith oxygen isotope signatures is that they represent a protolith that experienced alteration processes in the crust before subduction into the mantle.

The range in $\delta^{18}O$ values of eclogites is taken to reflect derivation from different portions of the oceanic lithosphere that underwent hydrothermal alteration at varying temperatures. Eclogites with $\delta^{18}O$ below typical mantle values are inferred to be derived from oceanic lithosphere that experienced high-temperature alteration (greenschist to amphibolite conditions; $<450\,°C$), whereas those with $\delta^{18}O$ values above typical mantle values are interpreted as having experienced

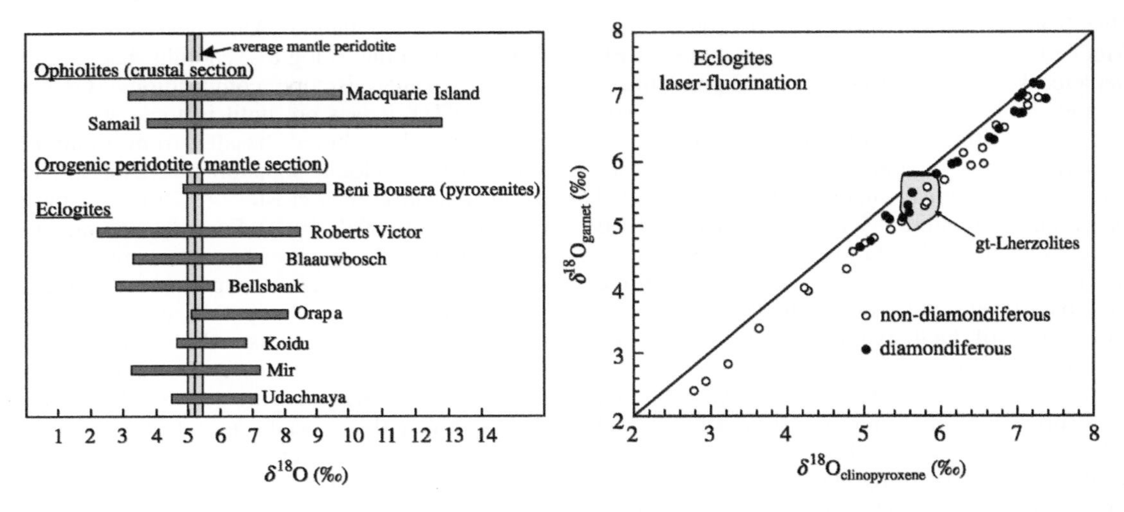

Figure 51 Figure on the left-hand side shows range in $\delta^{18}O$ values for eclogitic minerals from various kimberlites from Africa and Russia (Yakutia) compared with data for pyroxenites from the Beni Bousera peridotite massif (Pearson *et al.*, 1991) and hydrothermally altered rocks from ophiolites (after Jacob and Foley, 1999). The reference line is for average olivine from mantle peridotite (Mattey *et al.*, 1994), shaded zone is 2 SD. Data sources: Udachnaya—Jacob *et al.* (1994), Snyder *et al.* (1997), Mir—Beard *et al.* (1996), Orapa—Viljoen *et al.* (1996), Bellsbank—Neal *et al.* (1990), Blaauwbosch—Schulze *et al.* (2000), Roberts Victor—Garlick *et al.* (1971), Jagoutz *et al.* (1984), Ongley *et al.* (1987), MacGregor and Manton (1986), Schulze *et al.* (2000), Shirey *et al.* (2001), Koidu—Barth *et al.* (2001). Figure on the right-hand side shows $\delta^{18}O$ garnet versus $\delta^{18}O$ clinopyroxene for eclogites analyzed by laser-fluorination compared with the restricted field for garnet lherzolites (after Jacob *et al.*, 1998b). Garnet lherzolite field from Mattey *et al.* (1994).

low-temperature seafloor alteration prior to subduction (MacGregor and Manton, 1986; Jacob and Foley, 1999). Pearson *et al.* (1991) found similarly variable oxygen isotope compositions in garnet pyroxenites from the Beni Bousera orogenic peridotite that they also proposed reflected an origin via high-pressure crystallization from melts of subducted oceanic crust.

Correlations between $\delta^{18}O$ and major or trace elements have been observed at Udachnaya and other eclogite suites (Jacob *et al.*, 1998b) that simulate the variations expected on either side of the mantle $\delta^{18}O$ tie point for seawater alteration processes. Differences in oxygen isotope systematics between eclogite suites exist. For example, eclogites from the Udachnaya kimberlite do not have $\delta^{18}O$ values that extend to the low values found at Roberts Victor or Bellsbank, or even close-by Mir (Figure 51). Those from Orapa are predominantly higher in $\delta^{18}O$ compared to typical mantle. Coesite eclogites from Roberts Victor tend to have $\delta^{18}O$ values closer to "ordinary mantle" and are interpreted to represent the metamorphosed equivalents of deep-oceanic crustal cumulates that escaped significant hydrothermal alteration (Schulze *et al.*, 2000). In this context, Jacob and Foley (1999) point out that "ordinary mantle" oxygen isotope compositions can still be preserved in samples undergoing hydrothermal alteration when the hydrothermal fluid is in equilibrium with the rock, i.e., the system is rock-buffered.

2.05.3.5.2 Radiogenic isotopes

The high modal abundance of both garnet and clinopyroxene in eclogites combined with the appreciable amounts of neodymium that these phases contain means that the composition of both phases must be considered when discussing the Sr–Nd isotope systematics of eclogites. All data discussed here are calculated bulk rock values that account for this mass balance, but we use only the strontium isotope composition of the clinopyroxene because the very low strontium content of garnet make it susceptible to modification by the host magma. Compilation of these calculated bulk rock values show that eclogite xenoliths possess some of the most extreme radiogenic isotope compositions of terrestrial rocks. This feature is testament to their antiquity and to the variety of processes that they have experienced during their evolution. Some samples have the most radiogenic neodymium isotopic compositions of any rocks yet measured (Figure 52), with ε_{Nd} values in excess of 500 for garnets and differences of over 200ε units between coexisting garnet–pyroxene pairs (Jagoutz *et al.*, 1984). Garnet and pyroxene vary from being close to isotopic equilibrium at the time of kimberlite eruption (Snyder *et al.*, 1993) to being in gross disequilibrium. The relatively undisturbed REE systematics observed in garnet and pyroxene for most samples indicates that any isotopic disequilibrium is probably due to radiogenic in-growth following cooling through

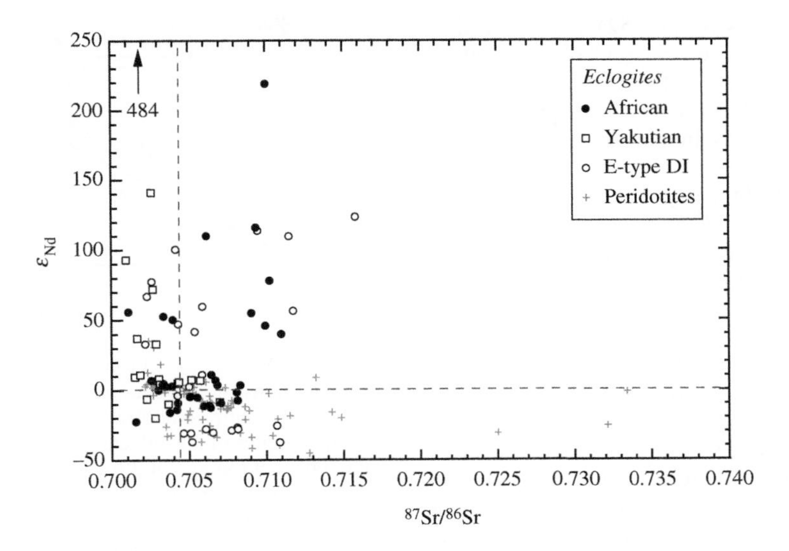

Figure 52 Initial $\varepsilon_{Nd}-\varepsilon_{Sr}$ isotope plot of calculated whole-rock eclogites from Africa (Koidu—Hills and Haggerty, 1989; Orapa—Smith *et al.*, 1989; Viljoen *et al.*, 1996; Roberts Victor—Jagoutz *et al.*, 1984; Smith *et al.*, 1989; Jacob and Jagoutz, 1994; Bellsbank—Neal *et al.*, 1990), Yakutia (Snyder *et al.*, 1993; Jacob *et al.*, 1994; Pearson *et al.*, 1995a; Snyder *et al.*, 1997), and eclogitic inclusions in diamonds (Richardson, 1986; Richardson *et al.*, 1990; 1999; Smith *et al.*, 1991). Eclogite data are initial calculated whole-rock neodymium isotope compositions. Strontium isotope compositions are from clinopyroxene only. Cratonic peridotite minerals and diamond inclusions are plotted for comparison. Arrow points to a sample from Udachnaya that has a calculated initial ε_{Nd} of 484. dashed lines are bulk Earth values.

the Sm–Nd blocking temperature for this system. This must have taken place over 2 Ga ago for some samples.

The Nd–Sr isotope variations of calculated whole-rock African and Yakutian eclogites greatly exceed those of magmas produced from the modern-day convecting mantle and occupy all four quadrants of the diagram (Figure 52). Some samples show very unradiogenic strontium and highly radiogenic neodymium that result from ancient time-integrated low Rb/Sr and high Sm/Nd that are unlike any magmatic rocks. Such characteristics might be compatible with pyroxene–garnet cumulates crystallized from high-pressure melts. However, their mineral chemistry and oxygen isotope composition make a crustal prehistory more likely. The depleted Nd–Sr isotope compositions and their calculated high bulk Sm/Nd ratios could arise if the rocks were metamorphosed oceanic-crustal cumulates, or residues from the melting of basaltic/picritic protoliths during subduction, or both.

A number of eclogites have both radiogenic neodymium isotopes and radiogenic strontium relative to bulk Earth. For these samples, some of the $^{87}Sr/^{86}Sr$ values exceed modern and Archean seawater compositions (see Chapter 6.17). Hence, in a subduction-type model, the radiogenic strontium isotopic composition of the eclogite protolith is not solely inherited from oceanic crust but is likely to be a time-integrated response to Rb/Sr enrichment during hydrothermal alteration in the oceanic crust that did not alter Sm/Nd significantly. This creates radiogenic strontium isotopic compositions without significantly altering neodymium isotopic compositions. Viljoen *et al.* (1996) propose some sediment incorporation during subduction to explain the enriched neodymium and strontium isotopic compositions of eclogites from Orapa, Botswana. Eclogite-suite inclusions in diamonds show similar variation and systematics to eclogite xenoliths. The diamond inclusions show evidence for seawater-like strontium isotope signatures and enriched neodymium isotope systematics that may be evidence of sediment incorporation (Figure 52). In contrast, inclusions in Finsch and Orapa diamonds show considerably less radiogenic neodymium isotopic compositions than xenoliths from the same locality. In these cases, the xenoliths may have been subject to open-system metasomatic behavior compared with the inclusions. Viljoen *et al.* (1996) interpret the radiogenic strontium and highly unradiogenic neodymium isotopic compositions of eclogitic diamond inclusions as a possible result of diamond genesis in a protolith consisting of subducted basalt and sediment. Although not well constrained at the moment, this interpretation is consistent with the osmium isotope systematics of eclogitic diamond inclusion sulfides (Pearson *et al.*, 1998b; Richardson *et al.*, 2001). The strontium isotope compositions of eclogite xenoliths do not extend to the extreme values observed in some peridotites, or peridotitic mineral concentrates (Figure 52), perhaps reflecting the differing nature or extent of metasomatic processes experienced by these lithologies. These differences may also reflect the likely different origins of the phases in peridotite and eclogite xenoliths.

Mazzone and Haggerty (1989) have reported highly radiogenic $^{87}Sr/^{86}Sr$ values in garnets from alkremites of 0.780 to 0.805 that are unsupported by their Rb/Sr ratios. They propose that these compositions are inherited from pelitic crustal protoliths subducted beneath the margin of the Kaapvaal craton and the Rb/Sr lowered by partial melting during subduction. In summary, the Nd–Sr isotope compositions of eclogite xenoliths and related rocks are compatible with a varied history that likely involves crustal protoliths for many and may also reflect the superimposed effects of ancient and/or recent mantle metasomatic effects.

There are few measurements of hafnium isotopes in eclogites. The first data are for samples from the Roberts Victor kimberlite (Jacob *et al.*, 2002) and indicate extreme hafnium isotope variability. Initial ε_{Hf} values (calculated to the time of kimberlite eruption) vary from -9.2 to 166, with correspondingly variable initial ε_{Nd} values (-22 to 484). The samples mostly plot well below the mantle Nd–Hf isotope array. Jacob *et al.* (2002) point out that the good correlation between neodymium and hafnium isotopes in modern terrestrial basalts indicates that the contribution of eclogite melts from the lithosphere to these magmas has been minor. Nowell *et al.* (2003) have reported that some alkremite garnets have extremely radiogenic Hf isotopic compositions. Alkremites from South African kimberlites have ε_{Hfi} ranging from 0 to $+60$ at restricted ε_{Ndi} values of 0 to $+2$. Two alkremites from Udachnaya, Siberia have ε_{Hfi} ranging from $+145$ to $+24{,}960$, the latter value being the most radiogenic Hf isotope composition so far, reported for any mantle mineral. The relatively ordinary ε_{Ndi} values of -7.5 to -11 indicate that significant decoupling of Sm–Nd and Lu–Hf isotope systems has occured during alkremite formation and evolution.

Lead isotope determinations on eclogite minerals are few. Surprisingly, there are no lead isotope analyses reported for rutile within eclogites, despite the high U/Pb of this phase. Over 90% of the lead in most rutile-free eclogites is contained within their clinopyroxenes (Jacob and Foley, 1999) and hence this is the most frequently analyzed phase. $^{206}Pb/^{204}Pb$ ranges from ~15.5 to 18.5 in clinopyroxenes from Udachnaya eclogites, spanning either side of the Geochron on a $^{206}Pb/^{204}Pb$ versus $^{207}Pb/^{204}Pb$ plot (Figure 53). The correlation defines a linear array that can be interpreted as having age significance (see below).

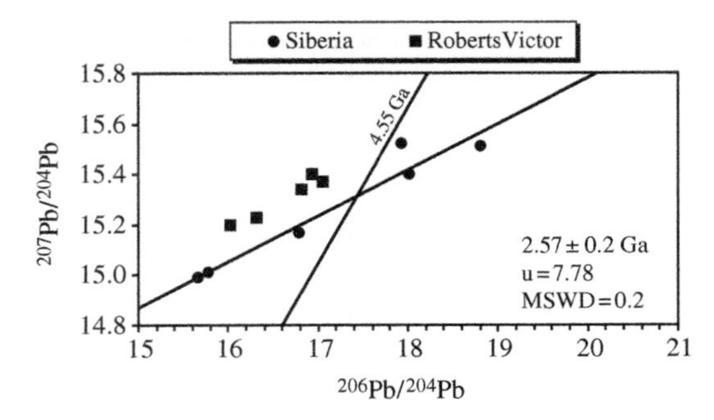

Figure 53 $^{207}Pb/^{204}Pb$ versus $^{206}Pb/^{204}Pb$ isotope compositions of eclogite xenolith clinopyroxenes. The regression line passes through clinopyroxene separates from eclogites of the Udachnaya kimberlite that define a Pb–Pb isochron age of 2.6 ± 0.2 Ga (Jacob and Foley, 1999). Other data points are clinopyroxene separates from Roberts Victor eclogites (Jacob and Jagoutz, 1994). The 4.55 Ga line is the geochron.

Clinopyroxenes from Roberts Victor eclogites generally plot to the left of the Geochron, with very unradiogenic $^{206}Pb/^{204}Pb$, indicative of time-integrated low U/Pb and consistent with the very low time-integrated Rb/Sr suggested by $^{87}Sr/^{86}Sr$ values of <0.702 for some of these samples (Jagoutz *et al.*, 1984). There is no simple relationship between lead and strontium isotopic compositions in the Udachnaya suite analyzed by Jacob *et al.* (1994) and Jacob and Foley (1999), probably reflecting the multistage histories of most of these rocks that may decouple isotope systems with differing geochemical affinities.

In addition to having the most radiogenic neodymium isotopic compositions of any mantle rocks, eclogites commonly have highly radiogenic osmium isotope signatures (Pearson *et al.*, 1992, 1995c; Menzies *et al.*, 1999; Shirey *et al.*, 2001; Barth *et al.*, 2002). Initial γ_{Os} values range from close to chondritic to $>6{,}500$ (Figure 54) are similar to values measured for Archean basalts and komatiites:

$$\gamma_{Os}(t) = \left[({}^{187}Os/{}^{188}Os_{sample(t)}/{}^{187}Os/{}^{188}Os_{chondrite(t)}) - 1\right] \times 100$$

Eclogite osmium isotope compositions are indicative of long-term evolution with basaltic/picritic Re/Os.

2.05.3.5.3 *The age of eclogites*

Pioneering work by Holmes and Paneth (1936) using the U–He system first recognized that eclogites were likely to be much older than their host kimberlites. More recent work has been able to provide firmer constraints. The huge isotopic diversity of eclogites immediately suggests an

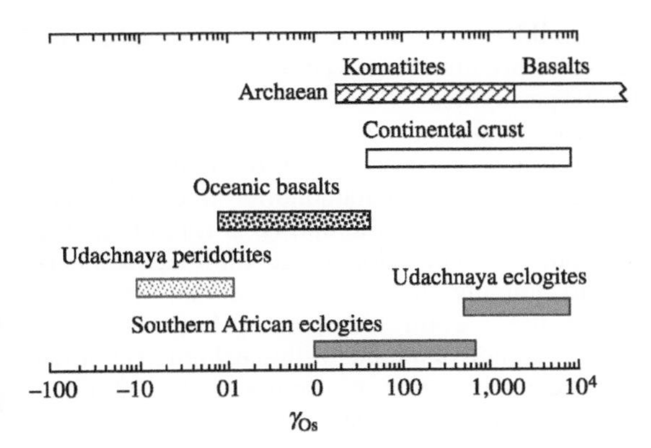

Figure 54 Comparison of present-day osmium isotopic compositions of eclogite xenoliths from Udachnaya, Yakutia (Pearson *et al.*, 1995c) and S. Africa (Pearson *et al.*, 1992; Menzies *et al.*, 1999; Shirey *et al.*, 2001) with continental crust, oceanic basalts (Shirey and Walker, 1998), and Archean komatiites and basalts (Walker *et al.*, 1989b). Udachnaya peridotite data from Pearson *et al.* (1995a).

ancient age for many samples. Jagoutz *et al.* (1984) obtained a 2.7 ± 0.1 Ga isochron age from the calculated bulk compositions of Roberts Victor eclogites. At other localities the Sm–Nd system does not yield easily interpretable age systematics, possibly due to minor metasomatic and partial melting effects during eruption. The large spread in $^{187}Os/^{188}Os$ and Re/Os within eclogite suites makes them amenable to dating using this system. Pearson *et al.* (1995c) found that whole-rock analyses of eclogites from Udachnaya (Yakutia) defined a Re–Os isochron relationship with an age equivalent to 2.9 ± 0.4 Ga. The elevated initial osmium isotope composition of the isochron is further evidence of a crustal prehistory prior to incorporation into the deep lithospheric mantle. This age is within error of the Pb–Pb isochron age defined by clinopyroxene separates from Udachnaya eclogites (Jacob and Foley, 1999; Figure 53) and confirms an Archean age for eclogites from this kimberlite. A less well-defined but clearly Archean age was also determined for eclogites from Newlands (S. Africa) by Menzies *et al.* (1999). Barth *et al.* (2002) also report a mid-Archean Re–Os isochron age for low-MgO Koidu eclogite samples. Recently, Shirey *et al.* (2001) have noted Re–Os isotope systematics of eclogite xenoliths from Roberts Victor and eclogitic-suite sulfide inclusions in diamonds from Koffiefontein, Kimberley and Orapa that are all indicative of Archean ages of ~3 Ga, implying Archean subduction and emplacement of eclogitic components into the lithosphere at this time.

The Archean age of the eclogite xenolith suites that have clear crustal signatures in their elemental, stable and radiogenic isotope compositions makes a very powerful case for plate tectonics operating in the late Archean Earth (Jagoutz *et al.*, 1984; Jacob *et al.*, 1994; Pearson *et al.*, 1995c). Late Archean ages for eclogites, coupled with their major and trace-element systematics indicating loss of a SiO_2-rich melt fraction, have provided a clearer picture of the nature of Archean continental crust generation (Ireland *et al.*, 1994; Barth *et al.*, 2001, 2002).

2.05.4 DIAMONDS

2.05.4.1 Introduction

This section is restricted to the natural occurrences of macrodiamonds from the mantle and their implications for the geochemistry and evolution of the mantle. Diamonds are discrete mantle samples, typically forming in peridotite and eclogite lithologies within the Archean subcontinental lithospheric mantle (Table 1(a)). Diamonds, occurring in kimberlite alongside xenoliths, often contain pristine inclusions of mantle minerals and thus comprise an important source of information about the mineralogy of the lithospheric mantle from which xenoliths are derived (Meyer, 1987; Section 2.05.4.5). This sample is free from the crustal alteration effects that occur in xenolith studies. The relatively recent discovery that some diamonds originate from the Earth's transition zone and the top of the lower mantle (e.g., Scott-Smith *et al.*, 1984; Kerr, 1993; Harte *et al.*, 1999b) holds the promise that diamonds will be the deepest geochemical probes of the mantle (Section 2.05.4.6).

Due to diamond's economic importance and its unique properties, a formidable literature exists with more than 3 times the number of articles published than for mantle xenoliths. Review articles on diamond include articles on diamond geology and genesis (Haggerty, 1986; Gurney, 1989; Kirkley *et al.*, 1991; Harris, 1992; Haggerty, 1999); age dating of diamond (Pearson and Shirey, 1999); diamond inclusions (Harris and Gurney, 1979; Meyer, 1987), and diamond formation in the mantle (Navon, 1999). Reviews of the very extensive literature on the physics, chemistry, economic applications and geology of diamonds can be found in Field (1979, 1992), J. Wilks and E. Wilks (1991), and Harlow (1998).

2.05.4.1.1 Occurrence

Diamond, especially micro- and nanodiamond, is a widely occurring mineral and is probably one of the most abundant forms of carbon in the Galaxy (Alexander, 2001; Allamandola *et al.*, 1993). As such, its geochemistry has implications for both terrestrial and extraterrestrial processes. Diamond is found in meteorites as presolar grains (Fukunaga *et al.*, 1987; Lewis *et al.*, 1987) or shock-related metamorphic minerals (Lipschutz, 1964; Lipschutz and Anders, 1961) and it occurs in high-pressure metamorphic terranes (Rozen *et al.*, 1973; Sobolev and Shatsky, 1990; Xu *et al.*, 1992; Pearson *et al.*, 1995d). Macrodiamonds can occur in polycrystalline aggregates such as framesite or carbonado and as single crystals. Carbonado, a porous, black, irregularly shaped aggregate, occurs alluvially and bears an unclear relationship to the mantle (De *et al.*, 1998; Trueb and Wys, 1969). Framesite, a nonporous, polycrystalline diamond aggregate, occurs in kimberlite and has recently become recognized as a potential recorder of secondary processes in the lithosphere (Burgess *et al.*, 1998; Jacob *et al.*, 2000). Macrodiamond pseudomorphs have been found in orogenic lherzolite (Pearson *et al.*, 1989) and metamorphic eclogite (Shumilova, 2002).

Macrodiamonds occur chiefly as discrete crystals in kimberlite. The association with Archean cratons and kimberlites is strikingly clear on both a local and global scale (Janse, 1985, 1992; Levinson *et al.*, 1992), although there are isolated occurrences of diamondiferous lamproite and diamonds associated with Proterozoic cratons such as the Argyle pipe (e.g., Jaques *et al.*, 1989). Whether isolated diamonds grew in equilibrium with the kimberlite magma (as phenocrysts) or were incorporated from disaggregated lithospheric mantle country rock (as xenocrysts) remained an open question since almost the 1920s, even though the Archean cratonic association for diamond and the young eruptive age of kimberlite suggested a xenocrystal origin. The first radiometric ages on syngenetic inclusions in diamond (Kramers, 1979; Richardson *et al.*, 1984) showed that Archean diamonds occur in Cretaceous kimberlites and firmly established diamonds as xenocrysts. Subsequent work on dating diamonds over the last two decades (Section 2.05.4.5.2) has established that most, but not all, macrodiamonds are xenocrysts and are Archean or Proterozoic in age (e.g., Pearson and Shirey, 1999). An association between Archean cratons and lithospheric carbon in general, including graphite at lower pressures, has been noted by Haggerty (1986, 1999) and Pearson *et al.* (1994).

Peridotitic lithologies comprise more than 95% of the mantle lithosphere (Schulze, 1989). Thus, macrodiamonds in kimberlite are believed to come from mostly peridotitic hosts that break up during the sampling or transport that accompanies kimberlitic volcanism. Sobolev (1974) and Hatton and Gurney (1979) noted the dichotomy that peridotite xenoliths greatly outnumber eclogite xenoliths and peridotitic silicate inclusions in diamond predominate over eclogitic, yet diamondiferous eclogites are much more common than diamondiferous peridotites.

Most of the *in situ* studies of diamonds in xenolith hosts have thus been on diamondiferous eclogites. Occasional diamondiferous eclogites are so diamond-rich it is tempting to ascribe all the diamonds in some eclogite–xenolith-rich kimberlite to eclogite disaggregation alone. However, where the carbon isotopes of both diamond xenocrysts and diamonds within eclogite xenoliths have been analyzed (e.g., Deines *et al.*, 1987) it is clear there are additional noneclogite sources for the diamond xenocrysts, presumably from disaggregated peridotite. Initial studies of diamond in host eclogite concentrated on the inhomogeneities of diamond distribution and the relationship of diamond to metamorphic layering. Recent advances in computer assisted tomographic scanning have allowed diamonds to be located within an eclogite before it is sawed or crushed (e.g., Schulze *et al.*, 1996). In some eclogites, diamond lies preferentially along veins (Taylor *et al.*, 2000), which has sparked a lively debate over the degree to which diamond is a secondary phase introduced into the eclogite or a primary metamorphic phase crystallized from the originally included carbon in the eclogite.

2.05.4.1.2 Diamond morphology

The chemistry of diamond is a strong function of its growth form. The systematics of diamond growth and the factors influencing it have been reviewed by Gurney (1989), J. Wilks and E. Wilks (1991), Harris (1992), Bulanova (1995), Mendelssohn and Milledge (1995), and Harlow (1998).

Single-crystal macrodiamonds can take many forms such as the cube, the cuboctahedron, the dodecahedron and the trisoctahedron but the octahedron is the most common (Harlow, 1998). Experimental diamond synthesis has established that diamond morphology is a function of the pressure and temperature of growth. Octahedra are the high-temperature growth form, whereas cubo-octahedra grow at lower temperatures (J. Wilks and E. Wilks, 1991). An important determinant of external morphology is dissolution, presumably in the lithosphere or in kimberlitic magma during transport to the surface, which produces rounded variants of the above forms and numerous etch features on the diamond surface such as trigons (Mendelssohn and Milledge, 1995; J. Wilks and E. Wilks, 1991).

Study of the internal morphology of macrodiamond with X-ray topography (e.g., J. Wilks and E. Wilks, 1991, figures 5.32–5.36), optical microscopy and especially cathodoluminescence (e.g., Bulanova, 1995, figures 2–9) reveals a complicated growth/resorption history in many gem-quality stones. Knowledge of such complexity has become crucial over the years as analytical sensitivity has improved and spot analyses have become possible for many spectroscopic (FTIR), crystallographic (analytical SEM), elemental (EPMA, PIXE), and isotopic (SIMS) analytical techniques. Such detail will lead to more accurate geochemical models of natural macrodiamond growth that will account for local changes in diamond-forming fluid composition (Sections 2.05.4.3 and 2.05.4.4).

2.05.4.1.3 Diamond types and classification

Diamonds can be classified by morphologic shape, size, and color (e.g., Harris, 1992). For geochemistry, the most useful classifications are those based on the content and lattice-substitutional state of the most abundant impurity in diamond, nitrogen, or the affinity of the diamond to either of the two dominant diamond host lithologies peridotite (lherzolite–harzburgite) or eclogite (Table 1(a)). The classification of diamond types by their nitrogen content is covered succinctly by Harlow (1998) and, in more detail, by Evans (1992) and J. Wilks and E. Wilks (1991). Low-nitrogen diamonds (<10 ppm) are termed type II and higher nitrogen diamonds (10–3,000 ppm) are termed type I. Type I diamonds are termed Ia if nitrogen is distributed in the crystal lattice as polyatomic aggregates and Ib if it is distributed as single atoms. Natural macrodiamonds are about 98% type Ia, 0.1% type Ib, and 2% type II. In microdiamond populations, type II is more common (Harlow, 1998). The nitrogen classification system can be applied to virtually any diamond.

To relate diamonds to their host lithologies in the lithospheric mantle, the diamond classification based on inclusion paragenesis has been the most useful. Inclusions in diamond can grow at the same time as the diamond (syngenetic) or predate the diamond (protogenetic, see Meyer, 1987; Bulanova, 1995; Pearson and Shirey, 1999). Most work has been done on inclusions thought to be syngenetic and discussions concerning inclusions in this chapter will be confined to this type. Silicates and sulfides comprise the two most abundant inclusion mineral groups and each inclusion group has affinities with the minerals in eclogite or peridotite xenoliths in kimberlite (Table 1). Diamonds are classified as peridotitic or "P-type" (or U-type by some workers, e.g., Sobolev *et al.*, 1998) when they contain Cr-pyrope, diopside, enstatite, chromite, or olivine (Harris and Gurney, 1979; Meyer, 1987; Meyer and Boyd, 1972). P-type diamonds can be further subdivided into harzburgitic or lherzolitic by the degree of depletion indicated by the Cr_2O_3 and CaO content of their included garnet (e.g., Harris and Gurney, 1979). Diamonds are classified as eclogitic or "E-type" when they contain pyrope-almandine garnet, omphacite, coesite, or kyanite. A similar E versus P paragenetic distinction has been applied to sulfide inclusions but with additional uncertainties stemming from sulfide exsolution, identification within the diamond (sulfides are opaque and also create internal fractures in the host diamond that obscure the inclusion) and rarity of coexisting silicates and sulfides. Nonetheless peridotitic sulfides from the Siberian craton typically contain nickel contents greater than 22.8 wt.%, whereas eclogitic sulfides typically have less than 8 wt.% (Yefimova *et al.*, 1983). Such a clear break between E- and P-type sulfides has not yet been established for inclusions in Kaapvaal craton diamonds (Deines and Harris, 1995; Harris and Gurney, 1979). Sulfides from pyroxenite assemblages also complicate the situation (Pearson *et al.*, 1999b).

2.05.4.2 Nitrogen in Diamond

The chief geological significance of nitrogen in diamond is because its abundance and distribution contains a record of the partitioning of nitrogen during diamond growth from the original diamond-forming fluid. Furthermore, the nitrogen aggregation state reflects a time-temperature history that relates to diamond storage in the lithospheric mantle. Diamond's resistance to metamorphism, coupled with the ancient age of many diamonds, may provide an estimate of the nitrogen content of the early Earth's mantle.

Specialized sample preparation techniques such as etching have long revealed complex zoning and resorption in natural macrodiamonds. Cathodoluminescence imaging (CL) and recent improvements in FTIR and SIMS sensitivity and resolution suggest that most macrodiamonds are zoned in nitrogen content and nitrogen aggregation (Bulanova, 1995; Bulanova *et al.*, 2002) and potentially in carbon and nitrogen isotopic composition (Hauri *et al.*, 2002; Bulanova *et al.*, 2002).

2.05.4.2.1 Nitrogen abundance

Nitrogen abundance in diamonds is an integrated reflection of the nitrogen content of the diamond-forming fluids in the mantle, changes in fluid evolution during diamond growth and possible changes in the diamond/fluid partition coefficient. Detailed studies of nitrogen abundances have been made for diamond populations from individual kimberlite mines in southern Africa, particularly such as the De Beers Pool, Finsch, Jagersfontein, Jwaneng, Koffiefontein, Orapa, Premier, and Roberts Victor mines. These studies have shown, in general, that diamonds with E-type silicate inclusions have higher nitrogen content than diamonds with P-type silicate inclusions (Cartigny *et al.*, 1998a, 2001; Deines *et al.*, 1987, 1991a, 1993, 1989). The combined data from these mines are shown in Figure 55. Kaapvaal P-type diamonds rarely have more than 400 ppm nitrogen (Figure 55), whereas Kaapvaal E-type diamonds have almost a bimodal distribution with at least a third of the population having nitrogen concentrations above 400 ppm. Similarly, mines that have diamonds where E-types predominate have systematically lower proportions of type II diamonds (Shirey *et al.*, 2002). This difference could be due to a higher nitrogen content in the diamond source fluid of E-type diamonds. However, the high nitrogen content of fibrous diamonds (average = 900 ppm N) compared to P-type diamonds has been used to argue for strong dependence on nitrogen partitioning on diamond growth rate (Cartigny *et al.*, 2001). Using this rationale, the higher nitrogen

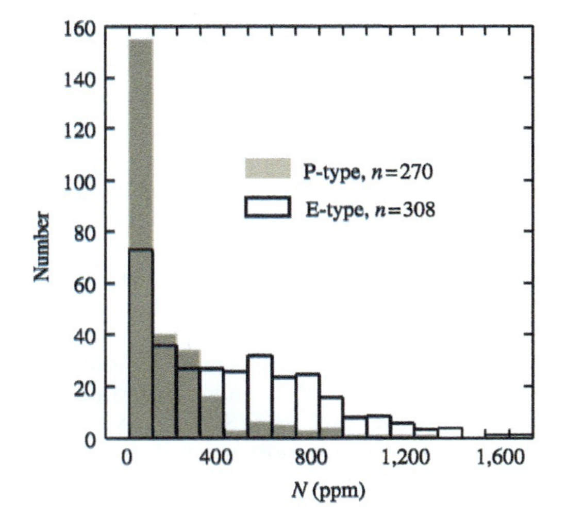

Figure 55 The nitrogen content (ppm) of diamonds of P-type and E-type parageneses. The number of total specimens is given for each group ($n = 578$). P-type diamonds (number analyzed in parens) are from these kimberlites: DO-27 (15), Finsch (66), Jagersfontein (13), Jwaneng (17), KanKan (13), Koffiefontein (27), Orapa (21), Premier (33), and Roberts Victor (65). E-type diamonds are from the following kimberlites: Argyle (20), DO-27 (10), Finsch (9), Jagersfontein (13), Jwaneng (74), KanKan (15), Koffiefontein (19), Orapa (56), Premier (76), Roberts Victor (14), and Sao Luiz (2) (sources Cartigny *et al.*, 1998a, 1999; Davies *et al.*, 1999; Deines *et al.*, 1987, 1989, 1991a,b, 1993, 1997; van Heerden *et al.*, 1995; Stachel *et al.*, 2002).

contents of E-type diamonds would be a result of their faster growth rate compared with P-type diamonds.

Current debate centers on whether nitrogen abundance variations in diamonds reflect the primary nitrogen content of the diamond-forming fluid or are controlled by effects more local to the site of diamond growth in the mantle, such as diamond growth rate or *in situ* changes in nitrogen content of the fluids by differentiation (see also Section 2.05.4.3.3). If nitrogen abundance is dominantly a source effect then the nitrogen content of diamonds might relate to geological processes and could lead to an understanding of the nitrogen content of the ancient mantle. If kinetic and crystallographic effects dominate then detailed nitrogen studies will lead to a better understanding of how diamonds form and grow.

In studies where nitrogen abundance has been determined by both FTIR and SIMS techniques, the agreement between the resulting concentrations is variable (Harte *et al.*, 1999a; Bulanova *et al.*, 2002; Hauri *et al.*, 1999, 2002). The agreement depends on how complex growth zonation and nitrogen distribution are and how well the sample preparation method (polished plate or fragment/chip) samples these variations. The FTIR technique provides an integrated signal through the entire thickness of the sample plate. The SIMS technique measures the signal from material sputtered from a few microns, at most, of the diamond surface. The SIMS technique thus has the potential to make more detailed studies of nitrogen distribution, but cannot measure nitrogen aggregation.

2.05.4.2.2 Nitrogen aggregation

The crystallographic distribution of nitrogen in the diamond lattice and its optical effects are described in detail by J. Wilks and E. Wilks (1991). Briefly, diamonds quenched rapidly in experiments contain single nitrogen atoms substituting randomly for single carbon atoms (diamond type Ib; Evans and Zengdu, 1982). This is extremely rare in natural diamonds. In the mantle, aggregation starts with paired nitrogen atoms (A-centers, diamond type IaA) and progresses ultimately to fourfold clusters of nitrogen atoms with an accompanying vacancy (B-centers, diamond type IaB; Evans, 1992; Evans and Zengdu, 1982; Taylor *et al.*, 1990). Many diamonds are intermediate between these two end-members and show a mix of A and B centers (diamond type IaA/B) which may be accompanied by N3 centers (threefold nitrogen and a vacancy) and platelets (Evans, 1992; Evans and Zengdu, 1982). Nitrogen aggregation is driven by time, temperature, and nitrogen content. For diamond populations of similar nitrogen content and age, those with a lesser percentage of highly aggregated nitrogen (e.g., IaB diamonds) are more likely to have resided in cooler lithosphere, whereas diamonds with a higher type IaB nitrogen aggregation state are more likely to have resided in warmer lithosphere. Navon (1999) has made the case that, because temperature is the most important variable in nitrogen aggregation, the process of nitrogen aggregation in diamonds makes an excellent geothermometer for residence times in the upper mantle of greater than 200 Ma.

In general, temperatures for diamond storage in the lithospheric mantle obtained with nitrogen aggregation studies agree surprisingly well with traditional mineral geothermometers in the temperature range of 1,050–1,200 °C (Evans and Harris, 1989; Harris, 1992; Navon, 1999). Similarly, in studies of Archean diamonds where ages were determined on inclusions in diamond and nitrogen aggregation measured on the diamond hosting the inclusions (Pearson *et al.*, 1998, 1999a; Richardson and Harris, 1997) the ages obtained are typically supported by the advanced aggregation state of the nitrogen in the host diamond.

This applies to diamonds where inclusions give both ancient and young ages. For instance, a young P-type diamond dated from the Koffiefontein kimberlite had poorly aggregated nitrogen (Pearson *et al.*, 1998b), consistent with a short mantle residence time. Furthermore, fibrous diamond coats, thought to be much younger than their octahedral cores also have less aggregated nitrogen (Boyd *et al.*, 1987).

The chief problems with using nitrogen aggregation in natural diamonds come in accurately determining the activation energy of transition from A to B centre coupled with the time/temperature/N content tradeoffs (e.g., Evans and Harris, 1989). Also, FTIR measurements of complexly zoned diamonds, even in carefully prepared polished plates, may still integrate signals across different growth zones (e.g., Bulanova, 1995); Taylor *et al.* (1996) have quantified the kinetics of the Ib to IaA nitrogen aggregation step but further progress is required in this area.

2.05.4.3 Isotope Systematics of Diamond and its Impurities

2.05.4.3.1 Carbon isotopes

The carbon isotopic composition of diamonds has been studied for many years (e.g., Galimov *et al.*, 1978; Deines, 1980; Galimov, 1984). These early studies attempted to correlate diamond properties such as color and form with carbon isotopic composition in some form of classification scheme. These efforts have given way to using carbon isotopes in diamond to understand diamond formation and especially to examine the carbon geochemical cycle in the mantle focussing on the relative roles of recycled crustal carbon versus primordial carbon (Deines, 1980; Galimov, 1991; Gurney, 1989; Haggerty, 1999; Harris, 1992; Kirkley *et al.*, 1991; Navon, 1999).

There is a general relationship between carbon isotopic composition and diamond paragenesis as designated by inclusion chemistry (Sobolev *et al.*, 1979). Most P-type silicate inclusion-bearing diamonds show a narrow range in carbon isotopic composition of $\delta^{13}C = -5‰ \pm 4‰$ (Figure 56), a value that overlaps the carbon isotopic composition of the upper mantle as given by oceanic basalts, carbonatites, and kimberlitic carbonate (e.g., Cartigny *et al.*, 1998b). E-type diamonds show a wider range in $\delta^{13}C$ of $+3‰$ to $-34‰$ with a mode in their frequency distribution that is identical to the P-type diamonds at every locality. The carbon isotope frequency distribution of E-type diamonds at most localities is markedly skewed relative to that of P-type diamonds, usually towards isotopically light carbon values

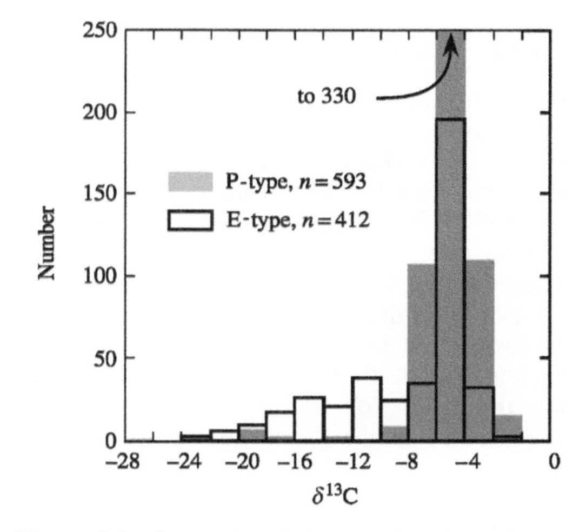

Figure 56 Comparison of the carbon isotopic composition ($\delta^{13}C$ in ‰) of diamonds of P-type and E-type parageneses. The total number of specimens is given for each group ($n = 1005$). Note that the -4 to -6 bar for P-type diamonds is off-scale at a number of 330. P-type diamonds are from these kimberlites: DeBeers Pool (125), DO-27 (14), Dokolwayo (41), Finsch (70), Jagersfontein (14), Jwaneng (20), KanKan (13), Koffiefontein (35), Orapa (22), Premier (34), Roberts Victor (65), and Venetia (140). E-type diamonds are from the following kimberlites: Argyle (24), DeBeers Pool (66), DO-27 (14), Dokolwayo (30), Finsch (10), Jagersfontein (18), Jwaneng (47), KanKan (15), Koffiefontein (20), Orapa (56), Premier (77), Roberts Victor (14), Sao Luiz (2), and Venetia (19). Data are taken from the references cited in Figure 55 caption and the following: Cartigny *et al.* (1998a), Daniels and Gurney (1999), Deines *et al.* (2001), and Hutchison *et al.* (1999).

but more rarely to isotopically heavy values. The carbon isotopic range for E-type diamonds spans a large part of the isotopic range seen for terrestrial carbon, from very light $\delta^{13}C$ values, characteristic of organic matter to heavy values similar to marine carbonate (Cartigny *et al.*, 1998b).

The carbon isotopic composition of diamonds has been explained by a variety of processes:

- intramantle isotopic fractionation during fluid transport and diamond growth (Deines, 1980; Galimov, 1991; Cartigny *et al.*, 2001);
- primoridal mantle heterogeneities (e.g., Deines *et al.*, 1987); and
- introduction of heterogeneous subducted carbon (e.g., Sobolev *et al.*, 1979; Kirkley *et al.*, 1991) as recently discussed by Navon (1999).

Carbon isotopic studies alone are not able to resolve the source of such carbon isotopic variability and the relative importance of these competing models. Thus, current approaches are to combine carbon isotopic studies with nitrogen abundance, aggregation, isotopic composition and even age determinations on the same diamonds (e.g., Hauri *et al.*, 1999, 2002).

2.05.4.3.2 *Nitrogen isotopes*

The application of nitrogen isotopic analyses to diamonds is well established by the early work of Javoy and co-workers (Javoy *et al.*, 1984; Javoy *et al.*, 1986), Boyd and co-workers (Boyd *et al.*, 1987, 1992; Boyd and Pillinger, 1994) and Cartigny and co-workers (Cartigny, 1998; Cartigny *et al.*, 1998a,b). The data set for nitrogen is less extensive than that for carbon, chiefly due to the lower abundance of nitrogen (nitrogen is a trace element at <10–1,000 ppm in diamond) and hence the need for lower analytical blanks (nitrogen is the major gaseous component of the atmosphere).

The nitrogen isotopic composition of diamonds shows more isotopic heterogeneity and does not display the clear paragenetic distinction evident in the frequency distribution of carbon isotopes. Except for the unusually large nitrogen isotopic range seen in P-type diamonds from Pipe 50, China (Cartigny *et al.*, 1997) and the "low $\delta^{13}C$ group" of Boyd and Pillinger (1994), P-type and E-type diamonds from many localities have an overlapping nitrogen isotopic range from $\delta^{15}N$ of −12‰ to +4‰ (Cartigny *et al.*, 1998b; Figure 57). This range is isotopically identical to that of the modern terrestrial depleted upper mantle as measured in ocean ridge basalts (Javoy *et al.*, 1986; Marty and Zimmerman, 1999). Close examination of the diamond nitrogen data allows a crude paragenetic distinction to be made. Nearly all diamonds that are light in carbon

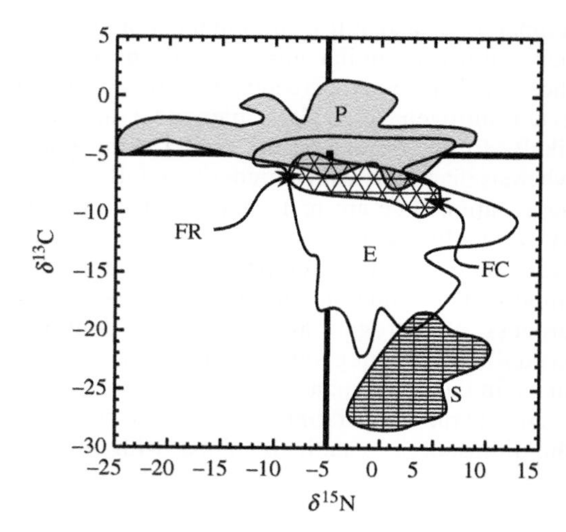

Figure 57 Carbon ($\delta^{13}C$ in ‰) and nitrogen ($\delta^{15}N$ in ‰) isotopic composition of P-type (P) and E-type (E) diamonds compared to the mantle (thick solid lines), modern sediments (S), and fibrous diamonds (FR = fibrous diamond rim; FC = fibrous diamond, non-fibrous core). Sources of data as follows: P-type diamonds (DeBeers Pool, Orapa, Pipe 50) from Cartigny *et al.* (1997, 1998a, #65; 1999), E-type diamonds (Argyle, Jwaneng, Orapa, DeBeers Pool) from Cartigny *et al.* (1998a,b, 1999) and van Heerden *et al.* (1995). Note that the "high $\delta^{13}C$" group (Finsch, Premier, Williamson, Jagersfontein, Arkansas) of Boyd and Pillinger (1994) is overlapped by the field for P-type diamonds and their "low $\delta^{13}C$" group (Argyle, Jagersfontein, Arkansas) is overlapped by the field for E-Type diamonds. Fibrous diamonds (Zaire) are from Boyd *et al.* (1987). Field for modern sediments after Navon (1999) and references therein.

isotopic composition are E-type and it is these diamonds that comprise the bulk of diamonds with isotopically heavy nitrogen that extends towards values characteristic of sediments and granites derived from sedimentary protoliths (+5‰ to +10‰).

2.05.4.3.3 *C–N–S isotopic systematics*

The combined systematics of $\delta^{13}C$–$\delta^{15}N$ and nitrogen abundance in P-type and E-type diamonds provide the strongest constraints on the abundance and ultimate sources of carbon and nitrogen in the mantle. Cartigny *et al.* (2001) estimate a diamond growth—medium C/N of 200–500 (mantle C = 400 ppm; mantle N = 1–2 ppm). C/N ratios in this medium could be as low as 10 depending on assumptions about the partitioning behavior of nitrogen. The former values would put continental lithospheric mantle C/N ratios in line with estimates for the modern depleted upper mantle from MORB analyses (Cartigny *et al.*, 2001). Based on their mantle-like carbon isotopic composition, P-type diamonds must have predominantly mantle-derived nitrogen. Collectively, when including the diamonds from Pipe 50, China, the entire P-type diamond population shows a large range in $\delta^{15}N$ from −25‰ to +6‰ (Figure 57) that has been ascribed to primordial mantle heterogeneity in N, inherited from the large isotopic range seen in meteorites (Cartigny *et al.*, 1997). To explain the diamonds with anomalously low-$\delta^{13}C$ and high-$\delta^{15}N$ that are outside the normal mantle range (Figure 57), Cartigny and co-workers (1998a, 1999, 2001) argue for a fractionation process in the diamond-forming fluid because mixing between mantle and sedimentary end-members cannot reproduce key features of the data. Specifically, recycling of crustal material is not consistent with the occurrence of E-type diamonds with negative $\delta^{15}N$. In addition, concave upward data arrays in the $\delta^{13}C$-$\delta^{15}N$ plot would be expected for mantle–crustal fluid mixing because of by the high C/N crustal fluids. Other features that are inconsistent with a recycling model include the drop in nitrogen content of a diamond when $\delta^{15}N$ increases and the positive correlation of nitrogen content with $\delta^{13}C$, seen in some eclogitic diamond suites.

Navon (1999) objectively considered arguments against recycling models and advanced a model whereby the C–N isotope systematics of E-type diamonds can be explained as a mixture of mantle and subducted components. This model overcomes the objections of Cartigny *et al.* (1997, 1998a,b) to subduction scenarios by using mantle and sedimentary mixing end-members that have variable compositions, a variable C/N ratio of the diamond-forming fluid, and possible nitrogen isotopic fractionations during diamond growth. Such growth-related nitrogen isotopic fractionations are indicated by *in situ* nitrogen isotope determinations at high spatial resolution (Hauri *et al.*, 1999, 2002; Bulanova *et al.*, 2002). Indeed, the huge range for nitrogen isotopic composition of P-type diamonds (Cartigny *et al.*, 1997) and the lack of data on nitrogen isotopes in Archean crustal rocks require variable end-members to be considered. Future resolution of the subduction versus mantle fractionation debate for C–N isotope systematics in diamonds perhaps lies in detailed *in situ*, high spatial resolution C–N isotopic studies on typically complex zoned diamonds using cathodoluminescence, FTIR, and SIMS (Boyd and Pillinger, 1994; Bulanova, 1995; Bulanova *et al.*, 2002; Fitzsimons *et al.*, 1999; Harte *et al.*, 1999a; Hauri *et al.*, 2002). These nascent studies show large covariations in coupled C–N isotopic compositions that correlate with diamond growth zones. (e.g., Bulanova *et al.*, 2002; Fitzsimons *et al.*, 1999). Furthermore, they suggest decoupling of the carbon and nitrogen isotopic compositions and point toward temporal changes in fluid composition and/or isotopic fractionation during diamond growth (Bulanova *et al.*, 2002; Cartigny *et al.*, 2001; Fitzsimons *et al.*, 1999).

Convincing evidence of recycling of volatiles to the mantle and their incorporation into diamonds has recently been reported from sulfur isotope studies of sulfide inclusions within diamond (Farquhar *et al.*, 2002). SIMS analysis of sulfide inclusions has revealed anomalous mass-independently fractionated sulfur isotopic compositions from the Orapa kimberlite, which are explained as a consequence of the recycling of surface sulfur produced through photolytic chemistry in the Archean atmosphere. Such studies need to be extended to other diamond populations and combined with C–N isotope studies on the host diamonds.

2.05.4.4 Volatiles and Fluids in Diamonds

2.05.4.4.1 Fluid inclusions in diamonds

The first indication of the presence of fluid inclusions within diamonds was reported by Chrenko *et al.* (1967), who used infrared spectroscopy to identify the presence of water and carbonate in the fibrous coating of an octahedral diamond. Subsequent studies have recognized that fibrous diamonds contain very high densities of fluid trapped during diamond growth and hence this type of diamond offers an insight into the chemistry of deep mantle fluids from which fibrous diamonds grow (Navon *et al.*, 1988). Fibrous diamonds commonly occur either as cube-shaped crystals, where the entire diamond is fibrous throughout, or as coatings on octahedral diamonds. Fibrous diamonds are widespread and have been reported from Sierra Leone, Botswana, Zaire (Congo), India, and Yakutia (Schrauder and Navon, 1994). Although generally forming less than 1 wt.% of mine production, fibrous diamonds can represent up to 8 wt.% of production at Jwaneng, Botswana (Harris, 1992). Recently, fluid inclusions have also been found in "cloudy" monocrystalline diamonds (Izraeli *et al.*, 2001).

Analyses of the fluid inclusions within fibrous diamonds from Botswana show that the composition of fluids within individual diamonds is uniform but that significant variation exists between different diamonds. The fluids range in composition from carbonatitic to hydrous end-members, with intermediate compositions (Schrauder and Navon, 1994). The carbonatitic fluid is rich in carbonate, CaO, FeO, MgO, and P_2O_5 and the hydrous fluid is rich in SiO_2 and Al_2O_3 (Schrauder and Navon, 1994; Figure 58). K_2O contents are high and Mg# low in both end-members. In contrast, fluid inclusions from

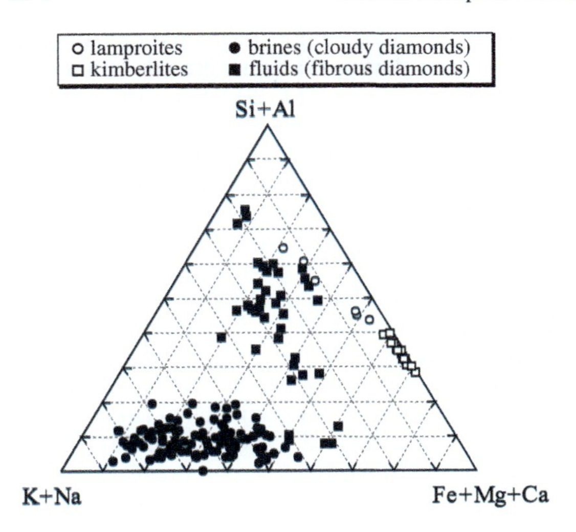

Figure 58 Major-element compositions of fluid inclusions within fibrous diamonds (squares) compared to cloudy diamonds and kimberlites/lamproites (after Izraeli *et al.*, 2001).

cloudy diamonds contain much higher chlorine contents and are classified as brines, being distinct from the other fluid types found in fibrous diamonds (Izraeli *et al.*, 2001; Figure 58). The brines carry very little SiO_2 (3–4 wt.%), possibly because of the low water content restricting the solvating capacity of the fluid.

The concentration of incompatible elements of varying geochemical affinity (potassium, sodium, bromine, rubidium, strontium, zirconium, caesium, barium, hafnium, tantalum, thorium, uranium, and LREEs) in the fluid inclusions from fibrous diamonds is higher than in typical mantle-derived magmas and melt inclusions (Schrauder *et al.*, 1996). The concentrations of most trace elements decrease by a factor of two from the carbonate-rich fluids to the hydrous fluids. REE contents of the fluid inclusions are higher than those of kimberlites and lamproites but the fluids show very similar levels of LREE/HREE fractionation to these rock types. Hence, they may be related to kimberlite-like magmas via fractionation of carbonate plus other phases such as rutile, apatite and zircon (Schrauder *et al.*, 1996). The amount of fractionation of these latter minerals must be small otherwise MREE and HREE systematics of the fluids would show distinctive fractionation effects. The link between the fluids in fibrous diamonds and kimberlite-like magmas is further supported by strontium isotope (Akagi and Masuda, 1988) and carbon isotope measurements (Boyd *et al.*, 1987, 1992) of fibrous diamonds. The incompatible-element-rich nature of the fluids in fibrous diamonds illustrate that carbonate-rich and hydrous deep mantle fluids are efficient carriers of incompatible elements.

2.05.4.4.2 *Noble gases in diamonds*

Diamonds provide a valuable sample of deep mantle volatiles and thus are obvious targets for noble gas analysis. Although monocrystalline (nonfibrous) diamonds have been analyzed (e.g., Honda *et al.*, 1987), noble gas abundances are very low and uncertainties very large. Far more reliable data comes from fibrous diamonds due to their high fluid inclusion contents that lead to high noble gas contents and precise analyses. Indeed, their gas-rich nature has allowed some of the most precise noble gas isotope ratio determinations of mantle materials to date. An important note here is that several studies have cited the old age of diamonds as a means of comparing mantle noble gas systematics over time. However, the very unaggregated nature of nitrogen within fibrous diamonds, which are typically analyzed in preference to monocrystalline diamonds, indicates a very young age is likely for fibrous stones, that is not much older than the host kimberlite eruption event (see Section 2.05.4.2.2). As such, fibrous diamonds provide an additional view of modern mantle to that available from volcanic rocks and not a view of the ancient mantle, as claimed by some authors. A complication to account for in the analysis of fibrous (cubic) diamonds is that the high uranium contents of their inclusions, and the high uranium-content of the host kimberlite mean that nucleogenic contributions to noble gas inventories are most likely. Cosmogenic ^3He may also be produced from carbon within the diamond itself.

The relatively gas-rich nature of fibrous diamonds has allowed precise analysis of helium, neon, and xenon isotopes (e.g., Ozima and Zashu, 1988, 1991; Wada and Matsuda, 1998). $^3He/^4He$ values are generally higher than the typical MORB value of 1×10^{-6} (Wada and Matsuda, 1998) but overlap the range of helium isotope compositions found in continental mantle xenoliths (e.g., Dunai and Baur, 1995; Section 2.05.2.9; see Chapter 2.06). Lowering of $^3He/^4He$ from a MORB-like value by direct decay of uranium trapped within the fluid inclusions requires time periods in excess of 1 Ga if uranium abundances are low (~0.5 ppb, Wada and Matsuda, 1998). This contrasts with the young age of fibrous diamonds indicated by nitrogen-aggregation studies. To reconcile these differences, Wada and Matsuda (1998) propose differentiation of the diamond fluid source from the mantle over 1 Ga ago followed by recent crystallization of the diamond from fluid, just before kimberlite sampling. The timescale required for modifying $^3He/^4He$ ratios via uranium decay is dramatically reduced if higher uranium abundances of up to 50 ppm, suggested by INAA studies (Schrauder *et al.*, 1996) are more applicable.

Fibrous diamonds have provided some of the more precise analyses of mantle neon and xenon isotopic composition and, together with MORB glass data, they have clearly revealed that the mantle is elevated in $^{20}Ne/^{22}Ne$ and $^{21}Ne/^{22}Ne$ (11–14) relative to atmosphere (9.8; Ozima and Zashu, 1988). High $^{20}Ne/^{22}Ne$ values in fibrous diamonds relative to atmosphere (9.8) are taken to indicate a primordial, Solar-like neon component in the mantle (Ozima and Zashu, 1988, 1991). Scatter in $^{21}Ne/^{22}Ne$ could be either source heterogeneity or due to nucleogenic ^{21}Ne. Wada and Matsuda (1998) have noted that many of the data points on $^{20}Ne/^{22}Ne$ versus $^{21}Ne/^{22}Ne$ correlation plots for fibrous diamonds are below the MORB regression line. This indicates greater amounts of excess ^{21}Ne (referred to as $^{21}Ne^*$) in cubic diamonds than the MORB source. Enrichment in ^{21}Ne due to nuclear reactions is consistent with $^{3}He/^{4}He$ values lower than MORB, since ^{21}Ne and ^{4}He are both derived from decay of uranium and thorium, via spallation. However, $^{21}Ne^*/^{4}He$ in cubic diamonds is significantly higher than the production rate estimated for average mantle material.

Fibrous diamonds show positively correlated excesses of ^{129}Xe and $^{131-136}Xe$ of up to 10% relative to atmospheric values, similar to the systematics shown by MORB (Figure 59). These data provide some of the best-determined mantle xenon isotope measurements. The xenon isotope excesses are attributed to decay of extinct ^{129}I and ^{244}Pu. $^{136}Xe/^{130}Xe$ versus $^{129}Xe/^{130}Xe$ correlations are identical to MORB glasses (Figure 59) and suggest that the fluid source of fibrous diamonds is very similar to the MORB/convecting mantle source (Ozima and Zashu, 1988, 1991; Wada and Matsuda, 1998). This is further

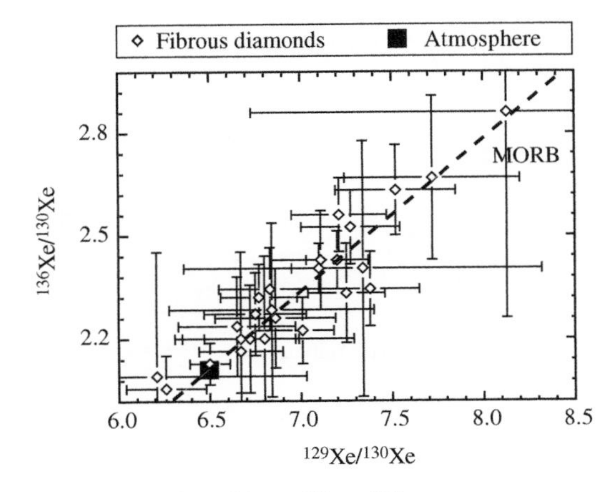

Figure 59 $^{136}Xe/^{130}Xe-^{129}Xe/^{130}Xe$ correlation plot of fibrous diamonds compared with the MORB correlation line and the atmospheric composition. Data and correlation line from Ozima and Zashu (1991) and Wada and Matsuda (1998). Error bars are 1σ and include blank correction.

supported by the uniformity and normal mantle-like carbon and nitrogen isotopic compositions of the fibrous coats of coated diamonds (Boyd *et al.*, 1987, 1992).

$^{40}Ar/^{36}Ar$ values of up to 35,000 are found in fibrous diamonds and are similar to estimates of the MORB-source mantle (e.g., Wada and Matsuda, 1998), further supporting the carbon and nitrogen isotope evidence for the involvement of this reservoir in the genesis of fibrous diamonds. Slightly elevated $^{38}Ar/^{36}Ar$ may be a product of the $^{35}Cl(\alpha,p)^{38}Ar$ reaction in the fluid phase. This is possible because of the high chlorine contents of some fibrous diamonds (>100 ppm) although levels are normally <50 ppm (Burgess *et al.*, 2002). Wada and Matsuda (1998) suggest that it is difficult for all the excess ^{38}Ar to be of nucleogenic origin and do not rule out the possibility that mantle $^{38}Ar/^{36}Ar$ is greater than atmospheric. Resolution of this possibility is needed because it has important implications for planetary–solar system noble gas reservoir requirements.

The likely young ages of fibrous diamonds mean that their isotopic compositions are directly comparable to the MORB source. When nucleogenic effects are accounted for, the isotopic signature of fibrous diamonds (as cubes or coats on octahedra) shares many similarities with the source of MORB, i.e., $\delta^{13}C$ -4 per mil to -8 per mil, $\delta^{15}N$ -2 per mill to -9 per mil, $^{40}Ar/^{36}Ar \leq 4 \times 10^4$, $^{3}He/^{4}He$ 4–8R_A, $^{20}Ne/^{22}Ne$ from 10 to 14 and $^{136}Xe/^{130}Xe$ plus $^{129}Xe/^{130}Xe$ up to 10% > atmosphere. This suggests that the origin of the parental fluid forming fibrous diamonds is in the convecting mantle.

2.05.4.4.3 *Halogen contents of diamonds*

Determination of the halogen content of the Earth's mantle is a difficult problem because the largest proportion of these elements is now at the Earth's surface and because seawater/surficial contamination affects recently erupted lavas such as MORB (see Chapter 2.07). Most of these problems are overcome when studying diamonds because of their deep origin and the ability to chemically remove surficial contamination before analysis. As with the noble gases, it is fluid-inclusion-rich fibrous diamonds, in the form of coated diamonds or cubes, that are the most attractive for analysis because of their high volatile contents (Burgess *et al.*, 2002). The very high chlorine contents of cloudy diamonds also make them amenable to study (Izraeli *et al.*, 2001). The Ar–Ar analytical technique can be extended for halogen analysis and enables correlation of argon isotope data with halogen contents. Burgess *et al.* (2002) have recently used this technique to study fibrous diamonds from Siberia,

Africa, and Canada. In the Siberian samples, three components have been identified with different Ar isotope compositions. Atmospheric blank is a small component present in most analyses. The major component is characterized by high $^{40}Ar/^{36}Ar$ (>11,000) and constant $^{40}Ar^*/Cl$ ($527 \pm 22 \times 10^{-6}$), $Br/Cl = 1.7 \times 10^{-3}$ and $I/Cl = 22 \times 10^{-6}$, indicative of a mantle fluid phase. This component is also characteristic of African and Canadian fibrous diamonds and leads to estimated mantle halogen abundances in the fluid source region of 3 ppm Cl, 11 ppb Br, and 0.4 ppb I (Burgess *et al.*, 2002). As with noble gas systematics, these data are very comparable to estimates for MORB-source mantle and implies that the mantle is >90% degassed of its halogens. A third Ar component in coated diamonds appears to be the result of ^{40}K decay since the time of kimberlite eruption.

2.05.4.5 Solid Inclusions in Diamonds

2.05.4.5.1 Geochemistry of inclusions

The P-type versus E-type paragenetic classification of diamonds based on their inclusions is introduced in Section 2.05.4.1.3. The geochemical basis for this fundamental difference between inclusion types is discussed in review articles by Meyer (1987), Harris and Gurney (1979), Gurney (1989), and Kirkley *et al.* (1991), summarized in brief here and discussed in the context of newer SIMS trace-element data on inclusions. Meyer (1987) points out the importance of inclusions in diamonds for the study of the mantle. First, inclusions are the chief way to understand the relationship of diamonds to their mantle host lithologies. Second, inclusions often represent pristine, geochemically unaltered samples that are not subject to the chemical re-equilibration and alteration that affects the minerals in xenoliths and macrocrysts.

Meyer (1987) documents nearly 20 minerals that have been proposed as protogenetic (pre-existing diamond growth) or syngenetic (forming with diamond) on the basis of modern analytical methods (X-ray or electron probe). An additional 10 minerals have been proposed to be of epigenetic origin (forming after diamond growth). In the P-type paragenesis, six minerals are typically found (sulfide, olivine, orthopyroxene, garnet, chromite, and clinopyroxene), whereas in the E-type paragenesis five minerals are typically found (sulfide, garnet, clinopyroxene, and rarely orthopyroxene and coesite; Gurney, 1989).

Olivine and orthopyroxene are generally associated with the P-type paragenesis. However, it is difficult to relate either of these two minerals specifically to a lherzolitic or harzurgitic facies because of their wide range in ($Mg/(Mg + Fe^{2+})$; 0.90–0.96; e.g., Meyer, 1987). As a general rule, highly magnesian olivines are likely to be related to the harzburgitic facies and those richer in iron to the lherzolitic facies. The difference between harzburgitic and lherzolitic garnet compositions is most easily distinguished using the relationship between calcium and chromium (e.g., Sobolev *et al.*, 1973; Gurney and Switzer, 1973) where harzburgitic garnets have much higher chromium and lower calcium contents (Figure 60). The P-type versus E-type distinction for garnet and clinopyroxene is fundamental and evident on key major-element plots (Meyer, 1987). P-type garnet inclusions have much lower FeO, CaO, and TiO_2 and much higher MgO compared to E-type garnet. P-type clinopyroxene has much lower Na_2O, Al_2O_3, and FeO and much higher MgO and Cr_2O_3 compared to E-type clinopyroxene. Sulfides that are dominantly Fe–S with minor copper and nickel follow similar paragenetic distinctions. P-type sulfides have higher nickel content, typically >13 wt.%, and E-type sulfides have lower nickel content, typically <10 wt.% (Yefimova *et al.*, 1983; Deines and Harris, 1995) but some complications exist that can be clarified using Re–Os systematics (Pearson *et al.*, 1999b). E-type sulfide inclusions have low osmium (e.g., <200 ppb to 300 ppb) and high rhenium that give combined high Re/Os compared to P-type sulfides which have lower rhenium and much higher osmium (2–30 ppm; Pearson *et al.*, 1999b).

While the major-element compositions of inclusions have served to link them to harzburgite, lherzolite or eclogite, it is their trace-element content, determined via laser ICP-MS or SIMS analysis, that has served to reveal how these

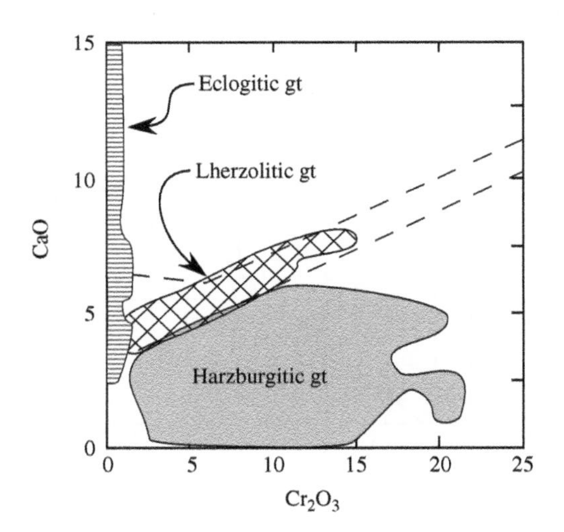

Figure 60 Cr_2O_3 versus CaO (wt.%) content of garnet inclusions in diamond. Original plot and lherzolite field (dashed) from Sobolev *et al.* (1973). This figure redrafted from plots and database of Stachel *et al.* (1997, 1998, 2000).

inclusions were formed in the mantle and the complex metasomatic processes to which they have been subjected. The first Sm–Nd isotopic analyses (Richardson *et al.*, 1984) and the first SIMS trace-element analyses on harzburgitic garnet from Finsch and Kimberley (Shimizu and Richardson, 1987) confirmed extreme light rare earth element enrichment paired with extreme depletion in major-element composition. This was recognized as a signature of ancient mantle metasomatism that preceded incorporation of the garnets in diamond (Shimizu and Richardson, 1987) and is a feature of nearly all garnets in harzburgitic xenolith suites analyzed for Sm–Nd isotopic ages (e.g., Pearson *et al.*, 1995a; Pearson and Shirey, 1999; Figure 61). Subsequent work on Udachnaya (Pearson and Milledge, 1998), Akwatia (Stachel and Harris, 1997; Stachel *et al.*, 1998), and Mwadui (Stachel *et al.*, 1999) has shown a whole range of complex (e.g., sinuous) REE patterns. Almost none are simply depleted in the LREE, as their major-element composition would indicate they should be (Figure 62). These trace-element complexities have supported the following processes in forming and metasomatising the continental lithospheric mantle:

(i) depletion by polybaric fractional melting;

(ii) enrichment during mantle metasomatism; and

(iii) slight light rare earth element depletion (e.g., La, Ce $\pm$ Pr) due to re-equilibration of the garnet with clinopyroxene (Stachel and Harris, 1997; Stachel *et al.*, 1998), all followed by diamond formation.

Similar trace-element patterns between harzburgitic garnets from xenoliths (diamondiferous and nondiamondiferous) and harzburgitic garnet inclusions (Stachel *et al.*, 1998) show that host lithologies reflect the same processes as seen in the inclusions. Lherzolitic garnets show much less complexity in their REE patterns. They are LREE depleted (Figure 62) and could have formed in equilibrium with kimberlitic/lamproitic or carbonatitic fluids (Stachel and Harris, 1997; Stachel *et al.*, 1998). The lherzolitic garnets are proposed to be distinctly different than harzburgitic garnets and are thought to be associated with additional metasomatic enrichment of the lithosphere that introduced additional calcium, iron, and silicon (Stachel *et al.*, 1998).

Eclogitic garnet and clinopyroxene inclusions studied from Udachnaya (Ireland *et al.*, 1994; Taylor *et al.*, 1996), Mwadui (Stachel *et al.*, 1999), and Kankan (Stachel *et al.*, 2000) are characterized by more "regular" REE patterns that reflect typical equilibrium distribution of the REEs between the two minerals. As is the case with the P-type paragenesis, similar trace-element patterns between garnets and clinopyroxene from eclogite xenoliths and those occurring as

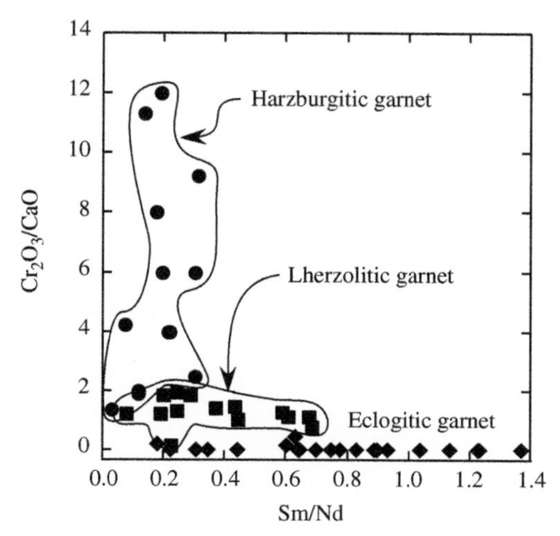

Figure 61 Cr_2O_3/CaO versus Sm/Nd for garnet inclusions in diamonds from Akwatia, Argyle, Finsch, KanKan, Kimberley Pool, Mir, Orapa, Premier, and Udachnaya (sources Richardson *et al.*, 1984, 1986, 1990; Richardson and Harris, 1993, 1997; Stachel *et al.*, 2000a, 1997; Smith *et al.*, 1991; Taylor *et al.*, 1996).

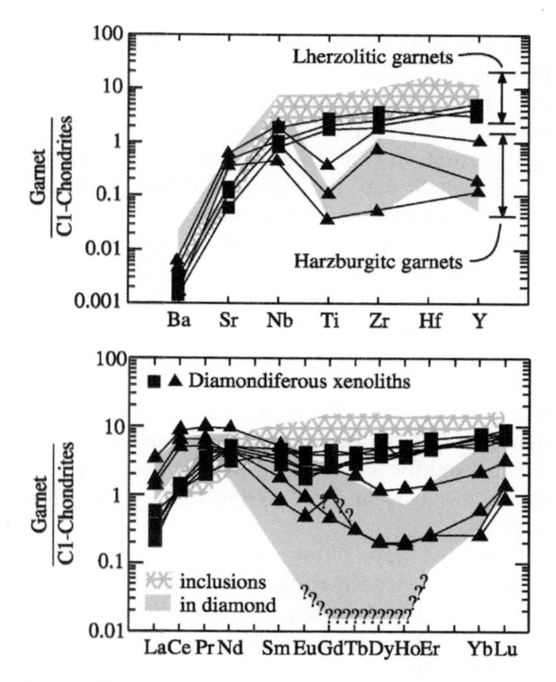

Figure 62 Trace-element plots for harzburgitic and lherzolitic diamond inclusions from Akwatia Ghana compared to harzburgitic and lherzolitic garnets from rare (<0.6% of the Roberts Victor xenolith population) diamondiferous peridotites from Roberts Victor. Ion microprobe analyses and original figures from Stachel *et al.* (1997, 1998). Labeling applies to both figure panels. Solid triangles and squares are for harzburgitic and lherzolitic (respectively) garnets in diamondiferous peridotites. Shaded and cross-hatched patterns are for harzburgitic and lherzolitic (respectively) garnet inclusions in diamond.

inclusions in diamond (Taylor *et al.*, 1996) show that the xenolith host lithologies reflect the same processes as seen in the inclusions. Since the mineral trace-element patterns are regular and eclogites are roughly 50 : 50 garnet : clinopyroxene, meaningful bulk-rock trace-element patterns can be readily constructed from trace-element analyses of the individual minerals (Ireland *et al.*, 1994; Stachel *et al.*, 1999; Taylor *et al.*, 1996). These reconstructed whole-rocks show strong light REE depletions that are interpreted by Ireland *et al.* (1994) as being consistent with those of subducted basalt residues from tonalite-trondhjemite melt extraction in the garnet stability field.

2.05.4.5.2 Ages of inclusions and their diamond hosts

The ages of diamond inclusions have been summarized by Harris (1992), Pearson and Shirey (1999), and Navon (1999). Analyses typically are performed on inclusions, thought to be syngenetic, of both E-type and P-type paragenesis in otherwise gem-quality diamonds using the following systems: Sm–Nd in garnet and clinopyroxene, Re–Os in sulfide, Ar–Ar in clinopyroxene and Pb–Pb in sulfide. Supportive age constraints can be obtained with the Rb–Sr system and nitrogen aggregation. In exploring the literature, the reader is cautioned that the ages determined do not always sample inclusion suites in the same manner and this may lead to apparent age discrepancies. For example Sm–Nd isotope studies on silicate inclusion suites typically require composites of many inclusions and provide a statistical average for many diamonds, whereas Re–Os on sulfides can be performed on single diamonds (Pearson *et al.*, 1998b). The diamond size fraction is also important. The small stones used for Sm–Nd composites need not be of the same age as the large stones used for Re–Os in sulfides.

Until the application of the Re–Os isotopic system to whole-rock peridotites (Section 2.05.2.7.4) the Archean and Proterozoic Pb–Pb and Sm–Nd model ages and isochrons obtained on diamonds were the chief constraints on the antiquity of the continental lithospheric mantle (Kramers, 1979; Richardson *et al.*, 1984). Early work, most of which was on diamonds from the Kaapvaal–Zimbabwe craton because of sample availability, suggested a simple difference from Meso-Archean P-type (harzburgitic) diamonds and Proterozoic E-type diamonds (Richardson, 1986; Richardson *et al.*, 1984, 1993). More recent Re-Os work on single sulfide inclusions from the Kaapvaal–Zimbabwe craton has underscored the importance of a Neo-Archean E-type diamond suite (Shirey *et al.*, 2001) and has resolved Proterozoic diamond populations (Pearson *et al.*, 1998b). The Neo-Archean E-type diamonds appear to be related to assembly of the Kaapvaal-Zimbabwe craton, whereas Proterozoic diamonds appear to be related to magmatic/metasomatic process by which the Archean lithosphere was modified in the Proterozoic (Shirey *et al.*, 2002).

The accuracy of diamond ages based on inclusions has sparked a lively debate in the literature (Navon, 1999; Pidgeon, 1989; Richardson, 1989; Taylor *et al.*, 2000). The main issue is to what extent the diamond ages on silicate composites are averages of many ages, and to what extent younger diamonds can encapsulate older, pre-existing inclusions. Even though some diamonds occur in fracture systems within their host lithology which must post-date the rock itself (Taylor *et al.*, 2000), there can be little doubt that diamonds can be Archean and Proterozoic in age although some are as young as the Phanerozoic age of most kimberlites (Pearson *et al.*, 1998b). Evidence for Proterozoic and Archean diamonds includes the occurrence of Proterozoic kimberlites such as Premier that carry abundant xenocrystic diamonds (Navon, 1999) and Re–Os ages of lithospheric peridotitic hosts that agree with diamond ages (Shirey *et al.*, 2001). In addition, type Ia diamonds show advanced aggregation that would take billions of years to achieve at lithospheric storage temperatures (Richardson and Harris, 1997; Pearson *et al.*, 1999a). Neodymium and strontium isotopic differences between concentrate garnets and garnets encapsulated in diamond suggest encapsulation for billion-year timescales (Richardson *et al.*, 1984; Richardson, 1989; Pearson and Shirey, 1999). Finally, diamonds occur as detrital grains in the later Archean/neo-Proterozoic Wits basin (Pearson and Shirey, 1999). The answer to the question of age accuracy must be sought in independent corroboration of the measured age (Navon, 1999). Such corroboration would come from obtaining the same age on different isotopic systems and the systematics of multiple age determinations with the same isotopic system: (i) relation of age and inclusion composition; (ii) repetition of ages in different diamonds and pipes; and (iii) agreement of ages with geological events recorded in the lithosphere or crust.

2.05.4.6 Ultradeep Diamonds

An exciting finding of the last two decades is that rare macrodiamonds occurring in kimberlites originate well below the lithosphere (Scott-Smith *et al.*, 1984; Harris, 1992) and can come from the lower mantle (McCammon, 2001). From the first reported lower mantle assemblage in a diamond

(Scott-Smith *et al.*, 1984), recognition of these ultradeep diamonds has extended to more than 12 localities on eight cratons (McCammon, 2001). The most thoroughly studied include the Slave (Davies *et al.*, 1999), Brazilian (Harte *et al.*, 1999b; Hutchison *et al.*, 1999, 2001; Kaminsky *et al.*, 2001; McCammon *et al.*, 1997; Wilding *et al.*, 1991), West African (Joswig *et al.*, 1999; Stachel *et al.*, 2000, 2002), and Kaapvaal (Deines *et al.*, 1991; McDade and Harris, 1999; Moore and Gurney, 1985; Scott-Smith *et al.*, 1984) cratons. Ultradeep mantle xenoliths are known to exist (Haggerty and Sautter, 1990; Sautter *et al.*, 1991) but ultradeep diamonds are perhaps the more revealing probes of the deep mantle. This is because, apparently, they are more widely distributed, they suffer less from retrogression and they derive from both the transition zone and the lower mantle (Harte *et al.*, 1999). Ultradeep diamonds encapsulate bonafide lower mantle mineral assemblages and allow a direct estimation of lower-mantle oxygen fugacity, carbon isotopic composition, and nitrogen abundance. However, it is important to keep in mind the potential and likely overprint of the diamond forming environment on the chemistry of these inclusions, i.e., they may be atypical samples of the ultradeep mantle environment.

Ferropericlase is the dominant ultradeep inclusion found in diamond (McCammon, 2001). Because it can be stable under upper-mantle conditions its presence alone does not signify a lower-mantle origin. The coexistence of ferropericlase with enstatite (Mg-silicate perovskite) and Ca-silicate perovskite is taken as a strong indication of lower mantle formation conditions (Stachel *et al.*, 2000). These additional ultradeep phases are the low-pressure polymorphs or exsolved multiphase chemical equivalents of their high-pressure precursors such as enstatite for $MgSiO_3$-perovskite, calcium silicates for $CaSiO_3$-perovskite and quartz/coesite for stishovite (Figure 63). A garnet composition phase, tetragonal almandine pyrope phase (TAPP; Harris *et al.*, 1997), is thought to exist at high pressure but may be a retrograde garnet (Harte *et al.*, 1999b). Nickel contents of $MgSiO_3$ inclusions in equilibrium with ferropericlase are an order of magnitude lower than typical upper-mantle enstatite and provide a ready way to establish a lower-mantle heritage for $MgSiO_3$ when it occurs in isolation (Stachel *et al.*, 2000). Recently, corundum inclusions were discovered in association with Al_2O_3-rich $MgSiO_3$-perovskite and ferropericlase, which has established that a free aluminum phase can exist in the lower mantle (Hutchison *et al.*, 2001).

The trace-element content of the ultradeep inclusion suite has been characterized by SIMS. $CaSiO_3$-perovksite is observed to have a strongly

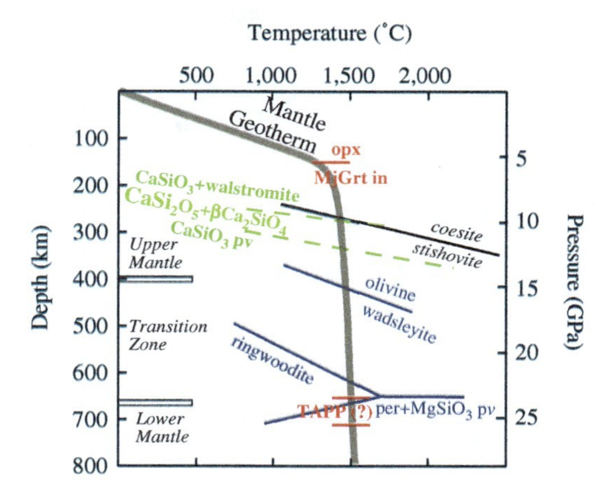

Figure 63 Mantle geotherm and *P–T* diagram for phases included in transition zone and lower mantle diamonds. Figure modified by Joswig *et al.* (1999) to include stability fields in the Mg–Al–Si–O system for the tetragonal almandine pyrope phase (TAPP) and majorite garnet (MjGrt) using occurrences (Harte *et al.*, 1999; Stachel *et al.*, 2000), experimental data (Gasparik and Hutchison, 2000; Irifune *et al.*, 1996) and summaries of experimental work (Fei and Bertka, 1999). Si–O system in black italics; Mg–Si–O system in blue lower case; Mg–Al–Si–O system in red capitals; and the Ca–Si–O system in green boldface. Note that a higher temperature geotherm than that shown recently has been proposed (Gasparik and Hutchison, 2000).

light rare earth element enriched pattern and high strontium content, whereas $MgSiO_3$-perovskite is observed to have a flat to slightly depleted rare earth element pattern (Harte *et al.*, 1999b; Hutchison *et al.*, 2001; Kaminsky *et al.*, 2001; Stachel *et al.*, 2000). These patterns have been interpreted as evidence for the presence of substantial lower-mantle/transition-zone source heterogeneity (Hutchison *et al.*, 2001) that includes nonprimitive LREE enriched sources (Stachel *et al.*, 2000). Caution must be exercised, however, in extrapolating from minute single inclusions to the entire lower mantle or transition zone. The effects of the complete mantle mineral assemblage on fractionating the trace-element patterns prior to incorporation in the diamond at these pressures and temperatures are not well-known and potential effects from the diamond-forming event also complicate interpretation.

Nearly all lower-mantle diamonds have such low nitrogen that they are classified type II (McCammon, 2001). These features are very distinctive from upper-mantle diamonds, which are chiefly type Ia and have widely variable percentages of A, A/B, and B centers. The carbon isotopic composition of ultradeep, lower-mantle diamonds is surprisingly homogeneous, most samples having typical upper-mantle values of

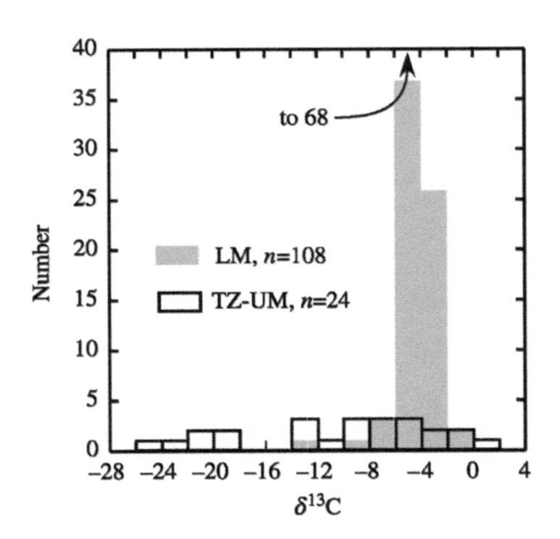

Figure 64 Comparison of the carbon isotopic composition of diamonds containing inclusions ascribed to transition zone and sublithospheric depths (TZ-UM; clear) to diamonds with inclusions derived from lower mantle depths (LM; gray). The number of specimens is given for each group ($n = 132$). Note that the -4 to -6 bar for lower mantle diamonds is off-scale at a number of 68. Transition zone and upper mantle diamonds are from these kimberlites: Jagersfontein (6), Juina (1), KanKan (5), Orapa (1), Premier (1), and Sao Luiz (10). Lower mantle diamonds (and the number analyzed from each) are from these kimberlites: Dokolwayo (1), DO-27 (5), Juina (30), KanKan (36), Koffiefontein (3), Letseng-la-Terai (1), and Sao Luiz (33) (sources Daniels and Gurney, 1999; Davies *et al.*, 1999; Deines *et al.*, 1989, 1991a, 1993; Hutchison *et al.*, 1999; Kaminsky *et al.*, 2001; McDade and Harris, 1999; Stachel *et al.*, 2000, 2002).

$\delta^{13}C = (-5 \pm 4)$‰ (Figure 64). In contrast, diamonds originating in the transition zone show considerable carbon isotopic variability, with the lightest isotopic compositions originating in the shallowest specimens, from the top of the transition zone (Davies *et al.*, 1999; Deines *et al.*, 1991; Kaminsky *et al.*, 2001; Stachel *et al.*, 2002; Wilding *et al.*, 1991).

The typical low nitrogen content of lower-mantle diamonds does not support an enhanced reservoir of nitrogen in the lower mantle (McCammon, 2001), although little is known about nitrogen partitioning during growth of diamond under these conditions. While the LREE-enriched patterns of $CaSiO_3$ perovskite could indicate a protolith subducted to transition zone depths (Stachel *et al.*, 2000), these signatures may also arise from interaction with very enriched alkaline magmas/liquids. Furthermore, the primitive and mantle-like carbon isotopic composition of many transition-zone and lower-mantle diamonds is not diagnostic of a subducted origin. Of the few ultradeep diamonds with isotopically light $\delta^{13}C$ (e.g., Deines *et al.*, 1991), most are from the transition zone as opposed to the lower mantle. Lower-mantle diamonds are characterized by restricted and heavier, mantle-like $\delta^{13}C$. This fact and a propensity for a majorite garnet and jadeitic diopside association in transition zone inclusions versus the ferropericlase-$MgSiO_3$ perovskite association with lower-mantle inclusions has led Hutchison *et al.* (2001) to suggest that the transition-zone inclusion suite is dominated by eclogitic assemblages, whereas the lower mantle is dominated by peridotitic assemblages. The ultradeep inclusion suites studied so far support a layered mantle convection model, which allows plume penetration upward through the transition zone and the accumulation of subducted slab material in the transition zone (Harte *et al.*, 1999b; Hutchison *et al.*, 2001; Stachel *et al.*, 2000) as first envisioned by Ringwood (1991). If any future lower-mantle inclusions are discovered that have isotopically light $\delta^{13}C$, then this would provide direct evidence for slab penetration into the lower mantle.

ACKNOWLEDGMENTS

The authors are very grateful to Geoff Nowell for enthusiastically helping to create many of the diagrams in this manuscript and to Dave Dowall for logistical help. Pete Nixon gave guidance with Table 1. Dmitri Ionov, Pierre Cartigny, Gordon Irvine, Roberta Rudnick, and Thomas Stachel provided assistance by donating unpublished data, database compilations, figures, and preprints. Dorrit Jacob reviewed the section on eclogites. Martin Menzies gave advice on scope and content. Comments by Nina Simon, Rick Carlson, and Tony Irving helped to create a much more readable manuscript. We acknowledge the inspiration provided by Pete Nixon and Joe Boyd in the pursuit of xenolith studies across the world.

REFERENCES

Akagi T. and Masuda A. (1988) Isotopic and chemical evidence for a relationship between kimberlite and Zaire cubic diamonds. *Nature* **336**, 665–667.

Alard O., Griffin W. L., Lorand J. P., Jackson S. E., and O'Reilly S. Y. (2000) Non-chondritic distribution of the highly siderophile elements in mantle sulphides. *Nature* **407**, 891–894.

Alexander C. M. O. D. (2001) Inherited material from the protosolar cloud: composition and origin. *Phil. Trans. Roy. Soc. London A* **359**, 1973–1989.

Allamandola L. J., Sandford S. A., Tielens A. G. G. M., and Herbst T. M. (1993) Diamonds in dense molecular clouds: a challenge to the standard interstellar medium paradigm. *Science* **260**, 64–66.

Aoki K. (1968) Petrogenesis of ultrabasic and basic inclusions in Iki Island, Japan. *Am. Mineral.* **53**, 241–256.

Arai S. (1994) Characterization of spinel peridotites by olivine-spinel compositional relationships: review and interpretation. *Chem. Geol.* **113**, 191–204.

Arndt N. T. and Christensen U. (1992) The role of lithospheric mantle in flood volcanism: thermal and geochemical constraints. *J. Geophys. Res.* **97**, 10967–10981.

Ave Lallement H. L., Carter N. L., Mercier J. C., and Ross J. V. (1980) Rheology of the uppermost mantle: inferences from peridotite xenoliths. *Tectonophysics* **70**, 221–234.

Baker M. B. and Stolper E. M. (1994) Determining the composition of high-pressure mantle melts using diamond aggregates. *Geochim. Cosmochim. Acta* **58**, 2811–2827.

Ballhaus C., Berry R. F., and Green D. H. (1991) High pressure experimental calibration of the olivine-orthopyroxene-spinel oxygen barometer: implications for the oxidation state of the upper mantle. *Contrib. Mineral. Petrol.* **107**, 27–40.

Barnes S. J. and Roeder P. L. (2001) The range of spinel compositions in terrestrial mafic and ultramafic rocks. *J. Petrol.* **42**, 2279–2302.

Barth M. G., Rudnick R. L., Horn I., McDonough W. F., Spicuzza M. J., Valley J. W., and Haggerty S. E. (2001) Geochemistry of xenolithic eclogites from West Africa: Part I. A link between low MgO eclogites and Archean crust formation. *Geochim. Cosmochim. Acta* **65**, 1499–1527.

Barth M. G., Rudnick R. L., Carlson R. W., Horn I., and McDonough W. F. (2003) Re–Os and U–Pb geochronological constraints on the eclogite-tonalite connection in the Archean Man Shield, West Africa. *Precamb. Res.* **118**, 267–283.

Barth M. G., Rudnick R. L., Horn I., McDonough W. F., Spicuzza M. J., Valley J. W., and Haggerty S. E. (2002) Geochemistry of xenolithic eclogites from West Africa: Part II. Origins of the high MgO eclogites. *Geochim. Cosmochim. Acta* **66**, 4325–4345.

Basu A. R. and Murthy V. R. (1977) Ancient lithospheric lherzolite xenoliths in alkali basalt from Baja California. *Earth Planet. Sci. Lett.* **35**, 239–246.

Basu A. R. and Tatsumoto M. (1980) Nd isotopes in selected mantle-derived rocks and minerals and their implications for mantle evolution. *Contrib. Mineral. Petrol.* **75**, 43–54.

Beard B. L., Fraracci K. N., Taylor L. A., Snyder G. A., Clayton R. N., Mayeda T., and Sobolev N. V. (1996) Petrography and geochemistry of eclogites from the Mir kimberlite, Yakutia, Russia. *Contrib. Mineral. Petrol.* **125**, 293–310.

Bedini R. M. and Bodinier J. L. (1999) Distribution of incompatible trace elements between the constituents of spinel peridotite xenoliths: ICP–MS data from the east African rift. *Geochim. Cosmochim. Acta* **63**, 3883–3900.

Bedini R. M., Blichert-Toft J., Boyet M., and Albarede F. (2002) Lu–Hf isotope geochemistry of garnet-peridotite xenoliths from the Kaapvaal craton and the thermal regime of the lithosphere. *Geochim. Cosmochim. Acta (Spec. Suppl.)* **66S1**, A61.

Beeson M. H. and Jackson E. D. (1970) Origin of the garnet pyroxenite xenoliths from Salt Lake Crater Oahu. *Spec. Pap. Min. Soc. Am.* **3**, 95–112.

Ben Othman D., Tilton G. R., and Menzies M. A. (1990) Pb–Nd–Sr isotopic investigations of kaersutite and clinopyroxene from ultramafic nodules and their host basalts: the nature of the subcontinental mantle. *Geochim. Cosmochim. Acta* **54**, 3449–3460.

Binns R. A., Duggan M. B., and Wilkinson J. F. G. (1970) High pressure megacrysts in alkaline lavas from northeastern New South Wales. *Am. J. Sci.* **269**, 132–168.

Blichert-Toft J., Ionov D. A., and Albarede F. (2000) The nature of the sub-continental lithospheric mantle: Hf isotope evidence from garnet peridotite xenoliths from Siberia. *J. Conf. Abstr.* **5**, 217.

Bodinier J. L., Merlet C., Bedini R. M., Simien F., Remaidi M., and Garrido C. J. (1996) Distribution of niobium, tantalum, and other highly incompatible trace elements in the lithospheric mantle: the spinel paradox. *Geochim. Cosmochim. Acta* **60**, 545–550.

Boettcher A. L. and O'Neill J. R. (1980) Stable isotope, chemical, and petrographic studies of high-pressure amphiboles and micas: evidence for metasomatism in the mantle source regions for alkali basalts and kimberlites. *Am. J. Sci.* **280A**, 594–621.

Boullier A. M. and Nicolas A. (1975) Classification of textures and fabrics of peridotite xenoliths from South African kimberlites. *Phys. Chem. Earth* **9**, 467–475.

Boyd F. R. (1969) Electron-probe study of diopside inclusions from kimberlite. *Am. J. Sci.* **267A**, 50–69.

Boyd F. R. (1970) Garnet peridotites and the system $CaSiO_3$–$MgSiO_3$–Al_2O_3. *Min. Soc. Am. Spec. Publ.* **3**, 63–75.

Boyd F. R. (1973) A pyroxene geotherm. *Geochim. Cosmochim. Acta* **37**, 2533–2546.

Boyd F. R. (1989) Compositional distinction between oceanic and cratonic lithosphere. *Earth Planet. Sci. Lett.* **96**, 15–26.

Boyd F. R. (1997) Origins of peridotite xenoliths: major element considerations. In *Short Course on High P and T Research on the Lithosphere*. U. Siena. pp. 89–106.

Boyd F. R. and McAllister R. H. (1976) Densities of fertile and sterile garnet peridotites. *Geophys. Res. Lett.* **3**, 509–512.

Boyd F. R. and Mertzman S. A. (1987) Composition and structure of the Kaapvaal lithosphere, Southern Africa. In *Magmatic Processes: Physicochemical Principles* (ed. B. O. Mysen). The Geochemical Society, Houston, vol. 1, pp. 3–12.

Boyd F. R. and Nixon P. H. (1978) Ultramafic nodules from the Kimberley pipes, South Africa. *Geochim. Cosmochim. Acta* **42**, 1367–1382.

Boyd F. R., Jones R. A., and Nixon P. H. (1983) Mantle metasomatism: the Kimberley dunites. *Carneg. Inst. Wash. Yearb.* **82**, 330–336.

Boyd F. R., Dawson J. B., and Smith J. V. (1984a) Granny Smith diopside megacrysts from the kimberlites of the Kimberley area and Jagersfontein, South Africa. *Geochim. Cosmochim. Acta* **48**, 381–384.

Boyd F. R., Nixon P. H., and Boctor N. Z. (1984b) Rapidly crystallised garnet pyroxenite xenoliths possibly related to discrete nodules. *Contrib. Mineral. Petrol.* **86**, 119–130.

Boyd F. R., Pearson D. G., Nixon P. H., and Mertzman S. A. (1993) Low Ca garnet harzburgites from southern Africa: their relation to craton structure and diamond crystallisation. *Contrib. Mineral. Petrol.* **113**, 352–366.

Boyd F. R., Pokhilenko N. P., Pearson D. G., Mertzman S. A., Sobolev N. V., and Finger L. W. (1997) Composition of the Siberian cratonic mantle: evidence from Udachnaya peridotite xenoliths. *Contrib. Mineral. Petrol.* **128**, 228–246.

Boyd F. R., Pearson D. G., and Mertzman S. A. (1999) Spinel-facies peridotites from the Kaapvaal root. In *Proc. 7th Int. Kimb. Conf.* (eds. J. J. Gurney, J. L. Gurney, M. D. Pascoe, and S. H. Richardson). National Book Printers, Red Roof Design, Cape Town, vol. 1, pp. 40–48.

Boyd S. R. and Pillinger C. T. (1994) A preliminary study of $^{15}N/^{14}N$ in octahedral growth form diamonds. *Chem. Geol.* **116**(1–2), 43–59.

Boyd S. R., Mattey D. P., Pillinger C. T., Milledge H. J., Mendelssohn M., and Seal M. (1987) Multiple growth events during diamond genesis: an integrated study of carbon and nitrogen isotopes and nitrogen aggregation state in coated stones. *Earth Planet. Sci. Lett.* **86**(2), 341–353.

Boyd S. R., Pillinger C. T., Milledge H. J., and Seal M. J. (1992) C and N isotopic composition and the infrared absorption spectra of coated diamonds: evidence for the regional uniformity of CO (sub 2)-H (sub 2) O rich fluids in lithospheric mantle. *Earth Planet. Sci. Lett.* **108**, 139–150.

Brandon A. D., Becker H., Carlson R. W., and Shirey S. B. (1999) Isotopic constraints on timescales and mechanisms of slab material transport in the mantle wedge: evidence from the Simcoe mantle xenoliths, Washington USA. *Chem. Geol.* **160**, 387–407.

Brandon A. D., Snow J. E., Walker R. J., Morgan J. W., and Mock T. D. (2001) ^{190}Pt–^{186}Os and ^{187}Re–^{187}Os systematics of abyssal peridotites. *Earth Planet. Sci. Lett.* **177**, 319–335.

Brey G. P. (1990) Geothermobarometry for lherzolites: experiments from 10 kb to 60 kb, new thermobarometers and application to natural rocks. Habil. Thesis, University of Aarmstadt, Germany, 227pp.

Brey G. and Köhler T. (1990) Geothermobarometry in four-phase lherzolites: II. New thermobarometers, and practical assessment of existing thermobarometers. *J. Petrol.* **31**, 1353–1378.

Brey G., Köhler T., and Nickel K. G. (1990) Geothermobarometry in four-phase lherzolites: I. Experimental results from 10 to 60 kb. *J. Petrol.* **31**, 1313–1352.

Bruekner H. K. (1974) Mantle Rb/Sr and $^{87}Sr/^{86}Sr$ ratios from clinopyroxenes from Norwegian garnet peridotites and pyroxenites. *Earth Planet. Sci. Lett.* **24**, 26–32.

Bulanova G. P. (1995) The formation of diamond. Diamond Exploration into the 21st Century. *J. Geoch. Explor.* **53**(1–3), 1–23.

Bulanova G. P., Pearson D. G., Hauri E. H., and Griffin William L. (2002) Carbon and nitrogen isotope systematics within a sector-growth diamond from the Mir kimberlite, Yakutia. *Chem. Geol.* **188**, 105–123.

Bulatov V., Brey G. P., and Foley S. F. (1991) Origin of low-Ca, high-Cr garnets by recrystallization of low-pressure harzburgites. *Fifth International Kimberlite Conference, Extended Abstracts,* CPRM Spec. Publ., 2/91, CPRM, Brasilia, pp. 29–31.

Burgess R., Johnson L. H., Mattey D. P., Harris J. W., and Turner G. (1998) He, Ar and C isotopes in coated and polycrystalline diamonds. *Chem. Geol.* **146**(3–4), 205–217.

Burgess R., Layzelle E., Turner G., and Harris J. W. (2002) Constraints on the age and halogen composition of mantle fluids in Siberian coated diamonds. *Earth Planet. Sci. Lett.* **197**, 193–203.

Burgess S. R. and Harte B. (1999) Tracing lithosphere evolution through the analysis of heterogeneous G9/G10 garnets in peridotite xenoliths: I. Major element analysis. In *Proc. 7th Int. Kimberlite Conf.* (eds. J. J. Gurney, J. L. Gurney, M. D. Pascoe, and S. H. Richardson). Red Roof Design, Cape Town, vol. 1, pp. 66–80.

Burnham O. M., Rogers N. W., Pearson D. G., van calsteren P. W., and Hawkesworth C. J. (1998) The petrogenesis of the eastern Pyrenean peridotites: an integrated study of their whole-rock geochemistry and Re–Os isotope composition. *Geochim. Cosmochim. Acta* **62**, 2293–2310.

Burton K. W., Schiano P., Birck J.-L., Allegre C. J., Rehkämper M., Halliday A. N., and Dawson J. B. (2000) The distribution and behaviour of rhenium and osmium amongst mantle minerals and the age of the lithospheric mantle beneath Tanzania. *Earth Planet. Sci. Lett.* **183**, 93–106.

Canil D. (1990) Experimental study bearing on the absence of carbonate in mantle-derived xenoliths. *Geology* **18**, 1011–1013.

Canil D. (1991) Experimental evidence for the exsolution of cratonic peridotite from high temperature harzburgite. *Earth Planet. Sci. Lett.* **106**, 64–72.

Canil D. (1992) Orthopyroxene stability along the peridotite solidus and the origin of cratonic lithosphere beneath southern Africa. *Earth Planet. Sci. Lett.* **111**, 83–95.

Canil D. (1999) The Ni-in-garnet geothermometer: calibration at natural abundances. *Contrib. Mineral. Petrol.* **136**, 240–246.

Canil D. (2002) Vanadium in peridotites, mantle redox and tectonic environments: Archean to present. *Earth Planet. Sci. Lett.* **195**, 75–90.

Canil D. and Fedortchouk Y. (2000) Clinopyroxene-liquid partitioning for vanadium and the oxygen fugacity during formation of cratonic and oceanic mantle lithosphere. *J. Geophys. Res.* **105**, 26003–26016.

Canil D. and O'Neill H. S. C. (1996) Distribution of ferric iron in some upper mantle assemblages. *J. Petrol.* **37**, 609–635.

Canil D. and Wei K. (1992) Constraints on the origin of mantle-derived low Ca garnets. *Contrib. Mineral. Petrol.* **109**, 421–430.

Canil D., O'Neill H. S. C., Pearson D. G., Rudnick R. L., McDonough W. F., and Carswell D. A. (1994) Ferric iron in peridotites and mantle oxidation states. *Earth Planet. Sci. Lett.* **123**, 205–220.

Carlson R. W. and Irving A. J. (1994) Depletion and enrichment history of sub-continental lithospheric mantle: Os, Sr, Nd and Pb evidence for xenoliths from the Wyoming Craton. *Earth Planet. Sci. Lett.* **126**, 457–472.

Carlson R. W., Irving A. J., and Hearn B. C., Jr. (1999a) Chemical and isotopic systematics of peridotite xenoliths from the Williams kimberlite, Montana: clues to processes of lithospheric formation, modification and destruction. In *Proc. 7th Int. Kimberlite Conf.* (eds. J. J. Gurney, J. L. Gurney, M. D. Pascoe, and S. H. Richardson). Red Roof Design, Cape Town, vol. 1, pp. 90–98.

Carlson R. W., Pearson D. G., Boyd F. R., Shirey S. B., Irvine G., Menzies A. H., and Gurney J. J. (1999b) Regional age variation of the southern African mantle: significance for models of lithospheric mantle formation. In *Proc. 7th Int. Kimberlite Conf.* (eds. J. J. Gurney, J. L. Gurney, M. D. Pascoe, and S. H. Richardson). Red Roof Design, Cape Town, vol. 1, pp. 99–108.

Cartigny P. (1998) Carbon isotopes in diamond. PhD Thesis, Universite de Paris VII.

Cartigny P., Boyd S. R., Harris J. W., and Javoy M. (1997) Nitrogen isotopes in peridotitic diamonds from Fuxian, China: the mantle signature. *Terra Nova* **9**(4), 175–179.

Cartigny P., Harris J. W., and Javoy M. (1998a) Eclogitic diamond formation at Jwaneng: no room for a recycled component. *Science* **280**, 1421–1424.

Cartigny P., Harris J. W., Phillips D., Girard M., and Javoy M. (1998b) Subduction-related diamonds? The evidence for a mantle-derived origin from coupled delta (^{13}C-delta ^{15}N determinations. In *The Degassing of the Earth* (eds. M. R. Carroll, S. C. Kohn, and B. J. Wood). Elsevier, Amsterdam, vol. 147(1–2), pp. 147–159.

Cartigny P., Harris J. W., and Javoy M. (1999) Eclogitic, peridotitic and metamorphic diamonds and the problems of carbon recycling: the case of Orapa (Botswana). In *Proc. 7th Int. Kimberlite Conf.* (eds. J. J. Gurney, J. L. Gurney, M. D. Pascoe, and S. H. Richardson). Red Roof Design, Cape Town, vol. 1, pp. 117–124.

Cartigny P., Harris J. W., and Javoy M. (2001) Diamond genesis, mantle fractionations and mantle nitrogen content: a study of $d^{13}C$–N concentrations in diamonds. *Earth Planet. Sci. Lett.* **185**, 85–98.

Carswell D. A. (1975) Primary and secondary phlogopites and clinopyroxenes in garnet lherzolite xenoliths. *Phys. Chem. Earth* **9**, 417–429.

Carswell D. A. (1991) The garnet-orthopyroxene Al barometer: problematic application to natural garnet lherzolite assemblages. *Mineral. Mag.* **55**, 19–31.

Carswell D. A., Griffin W. L., and Kresten P. (1984) Peridotite nodules from the Ngopetsoeu and Lipelaneng kimberlites. Lesotho: a crustal or mantle origin. In *Kimberlites II: The Mantle and Crust-mantle Relationships* (ed. J. Kornprobst). Elsevier, Amsterdam, pp. 229–243.

Carter J. L. (1970) Mineralogy and chemistry of the earth's upper mantle based on partial fusion-partial crystallisation model. *Geol. Soc. Am. Bull.* **81**, 2021–2034.

Chaussidon M., Albarède F., and Sheppard S. M. F. (1989) Sulfur isotope variations in the mantle from ion-microprobe analyses of micro-sulfide inclusions. *Earth Planet. Sci. Lett.* **92**, 144–156.

Chazot G., Menzies M. A., and Harte B. (1996) Determination of partition coefficients between apatite, clinopyroxene, amphibole and melt in natural spinel lherzolites from Yemen: implications for wet melting of the lithospheric mantle. *Geochim. Cosmochim. Acta* **60**, 423–437.

Chazot G., Lowry D., Menzies M. A., and Mattey D. (1997) Oxygen isotopic composition of hydrous and anhydrous mantle peridotites. *Geochim. Cosmochim. Acta* **61**, 161–169.

Chesley J. T., Rudnick R. L., and Lee C. T. (1999) Re–Os systematics of mantle xenoliths from the East African Rift: age, structure and history of the Tanzanian craton. *Geochim. Cosmochim. Acta* **63**, 1203–1217.

Chrenko R. M., McDonald R. S., and Darrow K. A. (1967) Infra-red spectrum of diamond coat. *Nature* **214**, 474–476.

Ciuff S., Rivalenti G., Vannucci R., Zanetti A., Mazzucchelli M., and Cingolani C. A. (2002) Are the glasses in mantle xenoliths witness of the metasomatic agent composition? *Geochim. Cosmochim. Acta (Spec. Suppl.)* **66S1**, A143.

Cohen R. S., O'Nions R. K., and Dawson J. B. (1984) isotope geochemistry of xenoliths from East Africa: implications for development of mantle reservoirs and their interaction. *Earth Planet. Sci. Lett.* **68**, 209–220.

Coleman R. G. and Keith T. E. (1971) A chemical study of serpentinization—Burro Mountain, California. *J. Petrol.* **12**, 311–329.

Coltorti M., Bonadiman C., Hinton R. W., Siena F., and Upton B. G. J. (1999) Carbonatite metasomatsim of the oceanic upper mantle: evidence from clinopyroxenes and glasses in ultramafic xenoliths of Grande Comore, Indian Ocean. *J. Petrol.* **40**, 133–165.

Cox K. G., Smith M. R., and Beswetherick S. (1987) Textural studies of garnet lherzolites: evidence of exsolution origin from high-temperature harzburgites. In *Mantle Xenoliths* (ed. P. H. Nixon). Wiley, Chichester, pp. 537–550.

Daniels L. R. M. and Gurney J. J. (1999) Dokolwayo diamond carbon isotopes. In *The J. B. Dawson Volume: Proceedings of the VIIth International Kimberlite Conference* (eds. J. J. Gurney, J. L. Gurney, M. D. Pascoe, and S. H. Richardson). Red Roof Designs, Cape Town, vol. 1, pp. 143–147.

Davies R. M., Griffin W. L., Pearson N. J., Andrew A. S., Doyle B. J., and O'Reilly S. Y. (1999) Diamonds from the deep: pipe DO-27, Slave Craton, Canada. In *The J. B. Dawson Volume: Proceedings of the VIIth International Kimberlite Conference* (eds. J. J. Gurney, J. L. Gurney, M. D. Pascoe, and S. H. Richardson). Red Roof Designs, Cape Town, vol. 1, pp. 148–155.

Dawson J. B. and Smith J. V. (1975) Occurrence of diamond in a mica-garnet lherzolite xenolith from kimberlite. *Nature* **254**, 580–58.

Dawson J. B. and Smith J. V. (1977) The MARID (mica-amphibole-rutile-ilmenite-diopside) suite of xenoliths in kimberlite. *Geochim. Cosmochim. Acta* **41**, 309–323.

De S., Heaney P. J., Hargraves R. B., Vicenzi E. P., and Taylor P. T. (1998) Microstructural observations of polycrystalline diamond: a contribution to the carbonado conundrum. *Earth Planet. Sci. Lett.* **164**(3–4), 421–433.

Deines P. (1980) The carbon isotopic composition of diamonds: relationship to diamond shape, color, occurrence and vapor composition. *Geochim. Cosmochim. Acta* **44**(7), 943–962.

Deines P. and Haggerty S. (2000) Small-scale oxygen isotope variations and petrochemistry of ultradeep (>300 km) and transition zone xenoliths. *Geochim. Cosmochim. Acta* **64**, 117–131.

Deines P. and Harris J. W. (1995) Sulfide inclusion chemistry and carbon isotopes of African diamonds. *Geochim. Cosmochim. Acta* **59**(15), 3173–3188.

Deines P., Harris J. W., and Gurney J. J. (1987) Carbon isotopic composition, nitrogen content and inclusion composition of diamonds from the Roberts Victor Kimberlite, South Africa: evidence for ^{13}C depletion in the mantle. *Geochim. Cosmochim. Acta* **51**, 1227–1243.

Deines P., Harris J. W., Spear P. M., and Gurney J. J. (1989) Nitrogen and ^{13}C content of Finsch and Premier diamonds and their implications. *Geochim. Cosmochim. Acta* **53**, 1367–1378.

Deines P., Harris J. W., and Gurney J. J. (1991a) The carbon isotopic composition and nitrogen content of lithospheric and asthenospheric diamonds from the Jagersfontein and Koffiefontein Kimberlite, South Africa. *Geochim. Cosmochim. Acta* **55**, 2615–2625.

Deines P., Harris J. W., Robinson D. N., Gurney J. J., and Shee S. R. (1991b) Carbon and oxygen isotope variations in diamond and graphite eclogites from Orapa, Botswana, and the nitrogen content of their diamonds. *Geochim. Cosmochim. Acta* **55**(2), 515–524.

Deines P., Harris J. W., and Gurney J. J. (1993) Depth-related carbon isotope and nitrogen concentration variability in the mantle below the Orapa Kimberlite, Botswana, Africa. *Geochim. Cosmochim. Acta* **57**, 2781–2796.

Deines P., Harris J. W., and Gurney J. J. (1997) Carbon isotope ratios, nitrogen content and aggregation state, and inclusion chemistry of diamonds from Jwaneng, Botswana. *Geochim. Cosmochim. Acta* **61**(18), 3993–4005.

Deines P., Viljoen F., and Harris J. W. (2001) Implications of the carbon isotope and mineral inclusion record for the formation of diamonds in the mantle underlying a mobile belt: Venetia, South Africa. *Geochim. Cosmochim. Acta* **65**(5), 813–838.

Dick H. J. B. and Sinton J. M. (1979) Compositional layering in Alpine peridotites: evidence for pressure solution creep in the mantle. *J. Geol.* **87**, 403–416.

Downes H. and Dupuy C. (1987) Textural, isotopic and rare earth variations in spinel peridotites xenoliths, Massif Central, France. *Earth Planet. Sci. Lett.* **82**, 121–135.

Ducea M., Sen G., Eiler J., and Fimbres J. (2002) Melt depletion and subsequent metasomatism in the shallow mantle beneath Koolau volcano (Hawaii). *Geochem. Geophys. Geosys.* **3** 10.1029/2001GC000184.

Dunai T. J. and Baur H. (1995) Helium, neon, and argon systematics of the European subcontinental mantle: implications for its geochemical evolution. *Geochim. Cosmochim. Acta* **59**, 2767–2783.

Eggins S. M. (1992) Petrogenesis of Hawaiin tholeiites: 1. Phase equilibria constraints. *Contrib. Mineral. Petrol.* **110**, 387–397.

Eggins S. M., Rudnick R. L., and McDonough W. F. (1998) The composition of peridotites and their minerals: a laser-ablation ICP–MS study. *Earth Planet. Sci. Lett.* **154**, 53–71.

Eggler D. H., McCallum M. E., and Smith C. B. (1979) Megacryst assemblages in kimberlite from northern Colorado and southern Wyoming: petrology, geothermometry–barometry and areal distribution. In *Kimberlites, Diatremes, and Diamonds: Their Geology, Petrology, and Geochemistry* (eds. F. R. Boyd and H. O. A. Meyer). American Geophysical Union, Washington, DC, vol. 1, pp. 213–226.

Ehrenberg S. N. (1982a) Petrogenesis of garnet lherzolite and megacrystalline nodules from the Thumb, Navajo Volcanic Field. *J. Petrol.* **23**, 507–47.

Ehrenberg S. N. (1982b) Rare earth element geochemistry of garnet lherzolite and megacrystalline nodules from minette of the Colorado Plateau province. *Earth Planet. Sci. Lett.* **57**, 191–210.

Erlank A. J., Waters F. G., Hawkesworth C. J., Haggerty S. E., Allsopp H. L., Rickard R. S., and Menzies M. A. (1987) Evidence for mantle metasomatism in peridotite nodules from the Kimberley pipes, South Africa. In *Mantle Metasomatism* (eds. C. J. Hawkesworth and M. A. Menzies). Academic Press, London, pp. 221–311.

Evans T. (1992) Aggregation of nitrogen in diamond. In *The Properties of Natural and Synthetic Diamond* (ed. J. E. Field). Academic Press, London, pp. 259–290.

Evans T. and Harris J. W. (1989) Nitrogen aggregation, inclusion equilibration temperatures and the age of diamonds. In *Kimberlites and Related Rocks* (ed. N. Ross). Spec. Publ. Geol. Soc. Austral. No. 14, IGSA, Perth, vol. 2, pp. 1001–1006.

Evans T. and Zengdu Q. (1982) The kinetics of the aggregation of nitrogen atoms in diamond. *Proc. Roy. Soc. London, Ser. A: Math. Phys. Sci.* **381**(1780), 159–178.

Farquhar J., Wing B. A., McKeegan K. D., Harris J. W., Cartigny P., and Thiemens M. H. (2002) Anomalous sulfur isotope compositions of inclusions from diamonds: evidence for recycling of sulfur to the mantle on early Earth. *Science* **298**, 2369–2372.

Fei Y. and Bertka C. M. (1999) Phase transitions in the Earth's mantle and mantle mineralogy. In *Mantle Petrology: Field Observations and High-pressure Experimentation; a Tribute to Francis R. (Joe) Boyd* (eds. Y. Fei, M. Bertka Constance, and O. Mysen Bjorn). Geochemical Society, Houston, vol. 6, pp. 189–207.

Fermor L. L. (1913) Preliminary note on garnet as a geological barometer and on an infra-plutonic zone in the Earth's crust. *Geol Surv. India, Records* **43**, 41–47.

Field J. E. (1979) *The Properites of Diamond.* Academic Press, London, 674pp.

Field J. E. (1992) *The Properties of Natural and Synthetic Diamond.* Academic Press, London 710pp.

Finnerty A. A. (1989) Xenolith-derived mantle geotherms: whither the inflection? *Contrib. Mineral. Petrol.* **102**, 367–375.

Finnerty A. A. and Boyd F. R. (1987) Thermobarometry for garnet peridotite xenoliths: a basis for upper mantle stratigraphy. In *Mantle Xenoliths* (ed. P. H. Nixon). Wiley, Chichester, pp. 381–402.

Fitzsimons I. C. W., Harte B., Chinn I. L., Gurney J. J., and Taylor W. R. (1999) Extreme chemical variation in complex diamonds from George Creek, Colorado: a SIMS study of carbon isotope composition and nitrogen abundance. *Min. Mag.* **63**(6), 857–878.

Foley S. F. (1992) Vein-plus-wall-rock melting mechanisms in the lithosphere and the origin of potassic alkaline magmas. *Lithos* **28**, 435–453.

Francis D. (1976) The origin of amphibole in lherzolite xenoliths from Nunivak Island, Alaska. *J. Petrol.* **17**, 357–78.

Franz L., Brey G., and Okrusch M. (1996a) Steady state geotherm, thermal disturbances and tectonic development of the lower lithosphere underneath the Gibeon Kimberlite Province, Namibia. *Contrib. Mineral. Petrol.* **126**, 181–198.

Franz L., Brey G. P., and Okrusch M. (1996b) Re-equilibration of ultramafic xenoliths from Namibia by metasomatic processes at the mantle boundary. *J. Geol.* **104**, 599–615.

Frey F. A. (1980) The origin of pyroxenites and garnet pyroxenites from Salt Lake Crater, Oahu, Hawaiii: trace element evidence. *Am. J. Sci* **280**, 427–449.

Frey F. A. and Green D. H. (1974) The mineralogy, geochemistry and origin of lherzolite inclusions in Victorian basanites. *Geochim. Cosmochim. Acta* **38**, 1023–1059.

Frey F. A. and Prinz M. (1978) Ultramafic inclusions from San Carlos, Arizona: petrological and geochemical data bearing on their petrogenesis. *Earth Planet. Sci. Lett.* **38**, 129–176.

Fukunaga K., Matsuda J. i., Nagao K., Miyamoto M., and Ito K. (1987) Noble-gas enrichment in vapour-growth diamonds and the origin of diamonds in ureilites. *Nature* **328**(6126), 141–143.

Fung A. T. and Haggerty S. E. (1995) Petrography and mineral composition of eclogites from the Koidu kimberlite complex, Sierra Leone. *J. Geophys. Res. B.* **100**(10), 20451–20473.

Galer S. J. G. and O'Nions R. K. (1989) Chemical and isotopic studies of ultramafic inclusions from the San Carlos Volcanic Field, Arizona: a bearing on their petrogenesis. *J. Petrol.* **30**, 1033–1064.

Galimov E. M. (1984) The relation between formation conditions and variations in isotope composition of diamonds. *Geochemistry* **8**, 1091–1118.

Galimov E. M. (1991) Isotope fractionation related to kimberlite magmatism and diamond formation. *Geochim. Cosmochim. Acta* **55**(6), 1697–1708.

Galimov E. M., Kaminskiy F. V., and Ivanovskaya I. N. (1978) Carbon isotope compositions of diamonds from the Urals, Timan, Sayan, the Ukraine, and elsewhere. *Geochem. Int.* **15**(2), 11–18.

Garlick G. D., MacGregor I. D., and Vogel D. E. (1971) Oxygen isotope ratios in eclogites from kimberlites. *Science* **172**, 1025–1027.

Gao S., Rudnick R. L., Carlson R. W., McDonough W. F., and Liu Y.-S. (2002) Re–Os evidence for replacement of ancient mantle lithosphere beneath the North China craton. *Earth Planet. Sci. Lett.* **198**, 307–322.

Gasparik T. and Hutchison M. T. (2000) Experimental evidence for the origin of two kinds of inclusions in diamonds from the deep mantle. *Earth Planet. Sci. Lett.* **181**(1–2), 103–114.

Gautheron C. and Moreira M. (2002) Helium isotopic signature of the subcontinental lithospheric mantle. *Earth Planet. Sci. Lett.* **199**, 39–47.

Girnis A. V. and Brey G. P. (1999) Garnet-spinel-olivine-orthopyroxene equilibria in the $FeO-MgO-Al_2O_3-SiO_2-Cr_2O_3$ system: II. Thermodynamic analysis. *Euro. J. Mineral.* **11**, 619–636.

Glaser S. M., Foley S. F., and Gunther D. (1999) Trace element compositions of minerals in garnet and spinel peridotite xenoliths from the Vitim volcanic field, Transbaikalia, eastern Siberia. *Lithos* **48**, 263–285.

Goetze C. (1975) Sheared lherzolites: from the point of view of rock mechanics. *Geology* **3**, 172–173.

Green D. H. and Hibberson W. (1970) The instability of plagioclase in peridotite at high pressure. *Lithos* **3**, 209–221.

Green D. H. and Ringwood A. E. (1970) Mineralogy of peridotitic compositions under upper mantle conditions. *Phys. Earth Planet. Int.* **3**, 359–371.

Gregoire M., Bell D. R., and Roux A. P. L. (2002) Trace element geochemistry of phlogopite-rich mafic mantle xenoliths: their classification and their relationship to phlogopite-bearing peridotites and kimberlites revisited. *Contrib. Mineral. Petrol.* **142**, 603–625.

Gregory R. T. and Taylor H. P. (1986) Non-equilibrium metasomatic $^{18}O/^{16}O$ effects in upper mantle mineral assemblages. *Contrib. Mineral. Petrol.* **93**, 124–135.

Griselin M. and Lassiter J. C. (2002) Extreme unradiogenic Os isotopes in Hawaiin mantle xenoliths: implications for mantle convection. *Geochim. Cosmochim. Acta. (Spec. Suppl.)* **66(S1)**, A292.

Griffin W. L., Cousens D. R., Ryan C. G., and Suter G. F. (1989) Ni in chrome garnet: a new geothermometer. *Contrib. Mineral. Petrol.* **103**, 199–202.

Griffin W. L., Sobolev N. V., Ryan C. G., Pokhilenko N. P., Win T. T., and Yefimova E. S. (1993) Trace elements in garnets and chromites: diamond formation in the Siberian lithosphere. *Lithos* **29**, 235–256.

Griffin W. L., O'Reilly S. Y., and Stabel C. G. (1988) Mantle metasomatism beneath western Victoria, Australia: II. isotopic geochemistry of Cr-diopside lherzolites and Al-augite pyroxenites. *Geochim. Cosmochim. Acta.* **52**, 449–459.

Griffin W. L., Fisher N. I., Friedman J., Ryan C. G., and O'Reilly S. Y. (1999a) Cr-pyrope garnets in the lithospheric mantle: I. Compositional systematics and relations to tectonic setting. *J. Petrol.* **40**, 679–704.

Griffin W. L., O'Reilly S. Y., and Ryan C. G. (1999b) The composition and origin of sub-continental lithospheric mantle. In *Mantle Petrology: Field Observations and High Pressure Experimentation* (eds. Y. Fei, C. M. Bertka, and B. O. Mysen). The Geochemical Society, Houston, vol. 6, pp. 13–45.

Griffin W. L., Shee S. R., Ryan C. G., Win T. T., and Wyatt B. A. (1999c) Harzburgite to lherzolite and back again: metasomatic processes in ultramafic xenoliths from the Wesselton kimberlite. Kimberley, South Africa. *Contrib. Mineral. Petrol.* **134**, 232–250.

Gudmundsson G. and Wood B. J. (1995) Experimental tests of garnet peridotite oxygen barometry. *Contrib. Mineral. Petrol.* **199**, 56–67.

Gunther M. and Jagoutz E. (1994) isotopic disequilibria (Sm/Nd, Rb/Sr) between minerals of coarse grained, low temperature peridotites from Kimberely floors, southern Africa. In *Kimberlites, Related Rocks and Mantle Xenoliths, Proc. 5th Int. Kimberlite Conf., Araxa, Brazil. CPRM.* Spec.

Publ. 1/a (eds. H. O. A. Meyer and O. H. Leonardos). CPRM, Brasilia, pp. 354–356.

Gunther M. and Jagoutz E. (1997) The meaning of Sm/Nd apparent ages from kimberlite-derived coarse grained low temperatures peridotites from Yakutia. *Russian Geol. Geophys.* **38**, 229–239.

Gurney J. J. (1989) Diamonds. *Kimberlites and Related Rocks*. Spec. Publ. Geol. Soc. Austral., No. 14, vol. 2, pp. 935–965.

Gurney J. J. and Harte B. (1980) Chemical variations in upper mantle nodules from southern African kimberlites. *Phil. Trans. Roy. Soc. London* **A297**, 273–293.

Gurney J. J. and Switzer G. S. (1973) The discovery of garnets closely related to diamonds in the Finsch pipe, South Africa. *Contrib. Mineral. Petrol.* **39**, 103–116.

Gurney J. J., Harte B., and Cox K. G. (1975) Mantle xenoliths in the Matsoku kimberlite pipe. *Phys. Chem. Earth* **9**, 507–523.

Gurney J. J., Jacob W. R. O., and Dawson J. B. (1979) Megacrysts from the Monastery kimberlite pipe, South Africa. In *Kimberlites, Diatremes and Diamonds: Their Geology, Petrology and Geochemistry* (eds. F. R. Boyd and H. O. A. Meyer). American Geophysical Union, Washington, DC, vol. 1, pp. 222–243.

Gurney J. J., Moore R. O., Otter M. L., Kirkley M. B., Hops J. J., and McCandless T. E. (1991) Southern African kimberlites and their xenoliths. In *Magmatism in Extensional Structural Settings* (eds. A. B. Kampunzu and R. T. Lubala). Springer, Berlin, pp. 495–536.

Haggerty S. E. (1986) Diamond genesis in a multiply-constrained model. *Nature* **320**, 34–38.

Haggerty S. E. (1987) Metasomatic mineral titanates in upper mantle xenoliths. In *Mantle Xenoliths*. (ed. P. H. Nixon), Wiley, Chichester, pp. 671–690.

Haggerty S. E. (1989) Upper mantle opaque mineral stratigraphy and the genesis of metasomites and alkali-rich melts. In *Kimberlites and Related Rocks*. Geological Society of Australia Special Publication No. 14 (ed. J. Ross). Blackwell, Perth, vol. 2, pp. 687–699.

Haggerty S. E. (1999) A diamond trilogy: superplumes, supercontinents, and supernovae. *Science* **285**, 851–860.

Haggerty S. E. and Sautter V. (1990) Ultradeep (greater than 300 kilometers), ultramafic upper mantle xenoliths. *Science* **248**, 993–996.

Haggerty S. E., Smyth J. R., Erlank A. J., Rickard R. S., and Danchin R. V. (1983) Lindsleyite (Ba) and Mathiasite (K): two new chromium titanates in the crichtonite series from the upper mantle. *Am. Mineral.* **68**, 494–505.

Hamilton M. A., Pearson D. G., Stern R. A., and Boyd F. R. (1998) Constraints on MARID petrogenesis: SHRIMP II U–Pb zircon evidence for pre-eruption metasomatism at Kampfersdam. *7th Int. Kimberlite. Conf. (Ext. Abstr.)* Cape Town, 296–298.

Handler M. R. and Bennett V. C. (1999) Behaviour of platinum-group elements in the subcontinental mantle of eastern Australia during variable metasomatism and melt depletion. *Geochim. Cosmochim. Acta* **63**, 3597–3618.

Handler M. R., Bennett V. C., and Esat T. Z. (1997) The persistence of off-cratonic lithospheric mantle: Os isotopic systematics of variably metasomatised southeast Australian xenoliths. *Earth Planet. Sci. Lett.* **151**, 61–75.

Handler M. R., Bennett V. C., and Dreibus G. (1999) Evidence from correlated Ir/Os and Cu/S for late-stage Os mobility in peridotite xenoliths: implications for Re–Os systematics. *Geology* **27**, 75–78.

Hanghoj K., Kelemen P. B., Bernstein S., Blustztajn J., and Frei R. (2001) Osmium isotopes in the Wiedemann Fjord mantle xenoliths: a unique record of cratonic mantle formation by melt depletion in the Archaean. *Geochem. Geophys. Geosys.* **2** (20010109): 2000GC000085.

Harley S. L. (1984) An experimental study of the partitioning of Fe and Mg between garnet and orthopyroxene. *Contrib. Mineral. Petrol.* **86**, 359–373.

Harlow G. E. (1998) *The Nature of Diamonds*. Cambridge University Press, Cambridge, UK.

Harris J., Hutchison M. T., Hursthouse M., Light M., and Harte B. (1997) A new tetragonal silicate mineral occurring as inclusions in lower-mantle diamonds. *Nature (London)* **387**(6632), 486–488.

Harris J. W. (1987) Recent physical, chemical and isotopic research of diamond. In *Mantle Xenoliths* (ed. P. H. Nixon). Wiley, Chichester, pp. 477–500.

Harris J. W. (1992) Diamond geology. In *The Properties of Natural and Synthetic Diamond* (ed. J. E. Field). Academic Press, London, pp. 345–393.

Harris J. W. and Gurney J. J. (1979) Inclusions in diamond. In *The Properties of Diamond* (ed. J. E. Field). Academic Press, London, pp. 555–591.

Hart S. R. and Ravizza G. E. (1996) Os partitioning between phases in lherzolite and basalt. In *Earth Processes: Reading the isotopic Code*, Geophysical Monograph 95 (eds. A. Basu and S. R. Hart). American Geophysical Union, Washington, DC, pp. 123–134.

Hart S. R. and Zindler A. (1986) In search of a bulk Earth composition. *Chem. Geol.* **57**, 247–267.

Harte B. (1977) Rock nomenclature with particular relation to deformation and recrystallisation textures in olivine-bearing xenoliths. *J. Geology* **85**, 279–288.

Harte B. (1983) Mantle peridotites and processes—the kimberlite sample. In *Continental Basalts and Mantle Xenoliths* (eds. C. J. Hawkesworth and M. J. Norry). Shiva, Nantwich, pp. 46–91.

Harte B. and Hawkesworth C. J. (1989) Mantle domains and mantle xenoliths. In *Kimberlites and Related Rocks,* Geol. Soc. Austral. Spec. Publ., No. 14 (ed. J. Ross). Blackwell, Perth, vol. 2, pp. 649–686.

Harte B. and Kirkley M. B. (1997) Partitioning of trace elements between clinopyroxene and garnet: data from mantle eclogites. *Chem. Geol.* **136**, 1–24.

Harte B., Cox K. G., and Gurney J. J. (1975) Petrography and geological history of upper mantle xenoliths from the Matsoku kimberlite pipe. *Phys. Chem. Earth* **9**, 477–506.

Harte B., Winterburn P. A., and Gurney J. J. (1987) Metasomatic and enrichment phenomena in garnet peridotite facies mantle xenoliths from the Matsoku kimberlite pipe, Lesotho. In *Mantle Metasomatism* (eds. C. J. Hawkesworth and M. A. Menzies). Academic Press, London, pp. 145–220.

Harte B., Fitzsimons I. C. W., Harris J. W., and Otter M. L. (1999a) Carbon isotope ratios and nitrogen abundances in relation to cathodoluminescence characteristics for some diamonds from the Kaapvaal Province S. Africa. *Mineral. Mag.* **63**(6), 829–856.

Harte B., Harris J. W., Hutchison M. T., Watt G. R., and Wilding M. C. (1999b) Lower mantle mineral associations in diamonds from Sao Luiz, Brazil. In *Mantle Petrology; Field Observations and High-pressure Experimentation: A Tribute to Francis R. (Joe) Boyd* (eds. Y. Fei, M. Bertka Constance, and O. Mysen Bjorn). Geochemical Society, Houston, vol. 6, pp. 125–153.

Hassler D. R. and Shimizu N. (1998) Osmium isotopic evidence for ancient sub-continental lithospheric mantle beneath the Kerguelen Islands, southern Indian Ocean. *Science* **280**, 418–420.

Hatton C. J. and Gurney J. J. (1979) Eclogites and peridotites from kimberlites. In *The Mantle Sample: Inclusions in Kimberlites and Other Volcanics* (eds. F. R. Boyd and H. O. A. Meyer). American Geophysical Union, Washington, DC, pp. 29–36.

Hatton C. J. and Gurney J. J. (1987) Roberts Victor eclogites and their relation to the mantle. In *Mantle Xenoliths* (ed. P. H. Nixon). Wiley, Chichester, pp. 453–463.

Hauri E. and Hart S. R. (1993) Re–Os isotope systematics in HIMU and EMII Ocean island basalts. *Earth Planet. Sci. Lett.* **114**, 253–271.

Hauri E. H., Shimizu N., Dieu J. J., and Hart S. R. (1993) Evidence for hotspot-related carbonatite metasomatism in the oceanic upper mantle. *Nature* **365**, 221–227.

Hauri E. H., Pearson D. G., Bulanova G. P., and Milledge H. J. (1999) Microscale variations in C and N isotopes within mantle diamonds revealed by SIMS. *Proc. 7th Int. Kimberlite Conf.* (eds. J. J. Gurney, J. L. Gurney, M. D. Pascoe, and S. H. Richardson). Red Roof Design, Cape Town, vol. 1, pp. 341–347.

Hauri E. H., Wang J., Pearson D. G., and Bulanova G. P. (2002) Microanalysis $\delta^{15}C$, $\delta^{15}N$, and N abundances in diamonds by secondary ion mass spectrometry. *Chem. Geol.* **185**(1–2), 149–163.

Hawkesworth C. J., Erlank A. J., Marsh J. S., Menzies M. A., and van Calsteren P. (1983) Evolution of the continental lithosphere: evidence from volcanics and xenoliths from Southern Africa. In *Continental Basalts and Mantle Xenoliths* (eds. C. J. Hawkesworth and M. J. Norry). Shiva, Nantwich, pp. 111–138.

Helmstaedt H. and Doig R. (1975) Eclogite nodules from kimberlite pipes in the Colorado plateau-samples of subducted Franciscan type oceanic lithosphere. In *Phys. Chem. Earth* (eds. L. H. Ahrens, J. B. Dawson, A. R. Duncan, and A. J. Erlank). Pergamon, Oxford, vol. 9, pp. 95–111.

Hergt J. M., Peate D. W., and Hawkesworth C. J. (1991) The petrogenesis of Mesozoic Gondwana low-Ti flood basalts. *Earth Planet. Sci. Lett.* **105**, 134–148.

Herzberg C. (1999) Phase equilibrium constraints on the formation of cratonic mantle. In *Mantle Petrology: Field Observations and High Pressure Experimentation*, Spec. Publ. Geochem. Soc. (eds. Y. Fei, C. Bertka, and B. O. Mysen) Geochemical Society, Houston, vol. 6, pp. 241–250.

Hills D. V. and Haggerty S. E. (1989) Petrochemistry of eclogites from the Koidu Kimberlite Complex, Sierra Leone. *Contrib. Mineral. Petrol.* **103**, 397–422.

Hirschmann M. M. (2000) Mantle solidus: experimental constraints and the effects of peridotite composition. *Geochem. Geophys. Geosys.* **1** (20001024).

Hoal K. E. O., Hoal B. G., Erlank A. J., and Shimizu N. (1994) Metasomatism of the mantle lithosphere recorded by rare earth elements in garnets. *Earth Planet. Sci. Lett.* **126**, 303–313.

Holmes A. and Paneth F. A. (1936) Helium ratios of rocks and minerals from the diamond pipes of South Africa. *Proc. Roy. Soc. Ser.* **154A**, 385–413.

Honda M., Reynolds J. H., Roedder E., and Epstein S. (1987) Noble gases in diamonds: occurrences of solarlike helium and neon. *J. Geophys. Res.* **92**, 12507–12521.

Hops J., Gurney J. J., and Harte B. (1992) The Jagersfontein Cr-poor megacryst suite-towards a model for megacryst petrogenesis. *J. Volcanol. Geotherm. Res.* **50**, 143–160.

Horn I., Foley S. F., Jackson S. E., and Jenner G. A. (1994) Experimentally determined partitioning of high field strength and selected transition elements between spinel and basaltic melt. *Chem. Geol.* **117**, 193–218.

Hutchison M. T., Cartigny P., and Harris J. W. (1999) Carbon and nitrogen compositions and physical characteristics of transition zone and lower mantle diamonds from Sao Luiz, Brazil. In *The J. B. Dawson Volume: Proceedings of the VIIth International Kimberlite Conference* (eds. J. J. Gurney, J. L. Gurney, M. D. Pascoe, and S. H. Richardson), Red Root Designs, Cape Town, vol. 1, pp. 372–382.

Hutchison M. T., Hursthouse M. B., and Light M. E. (2001) Mineral inclusions in diamonds: associations and chemical distinctions around the 670-km discontinuity. *Contrib. Mineral. Petrol.* **142**(1), 119–126.

Hurley P. M., Fairburn H. W., and Pinson W. H. (1964) Rb–Sr relationships in serpentine from Mayaguez, Puerto Rico and dunite from St. Paul's rocks: a progress report. In *A Study of Serpentinite: The AMSOC Core Hole near Mayaguez, Puerto Rico*. (ed. C. A. Burke) Natl. Tes. Counc. vol. 1118, pp. 149–151.

Ionov D. (1998) Trace element composition of mantle-derived carbonates and coexisting phases in peridotite xenoliths from alkali basalts. *J. Petrol.* **39**, 1931–1941.

Ionov D. A. (1996) Distribution and residence of lithophile trace element in minerals of garnet and spinel peridotites: an ICP–MS study. *J. Conf. Abstr.* **1**, 278.

Ionov D. A. (2004) Chemical variations in peridotite xenoliths from Vitim, Siberia: inferences for REE and Hf behaviour in the garnet facies upper mantle. *J. Petrol.* **45**.

Ionov D. A. and Harmer R. E. (2002) Trace element distribution in calcite-dolomite carbonatites from Spitskop: inferences for differentiation of carbonatite magmas and the origin of carbonates in mantle xenoliths. *Earth Planet. Sci. Lett.* **198**, 495–510.

Ionov D. A. and Hofmann A. W. (1995) Nb–Ta–rich mantle amphiboles and micas: implications for subduction-related metasomatic trace element fractionations. *Earth Planet. Sci. Lett.* **131**, 341–356.

Ionov D. A., and Weis D. (2002) Hf isotope composition of mantle peridotites: first results and inferences for the age and evolution of the lithospheric mantle. *Abstract, 4th Int. Workshop on Orogenic Lherzolites and Mantle Processes*, Samani, Japan, pp. 56–57.

Ionov D. A., Ashchepkov I. V., Stosch H.-G., Witt-Eickschen G., and Seck H. A. (1993a) Garnet peridotite xenoliths from the Vitim volcanic field, Baikal Region: the nature of the garnet-spinel peridotite transition zone in the continental mantle. *J. Petrol.* **34**, 1141–1175.

Ionov D. A., Dupuy C., O'Reilly S., Kopylova M. G., and Genshaft Y. S. (1993b) Carbonated peridotite xenoliths from Spitsbergen: implications for trace element signature of mantle carbonate metasomatism. *Earth. Planet. Sci. Lett.* **119**, 283–297.

Ionov D. A., Griffin W. L., and O'Reilly S. Y. (1997) Volatile-bearing minerals and lithophile trace elements in the upper mantle. *Chem. Geol.* **141**, 153–184.

Ionov D. A., Bodinier J.-L., Mukasa S. B., and Zanetti A. (2002a) Mechanisms and sources of mantle metasomatism: major and trace element compositions of peridotite xenoliths from Spitsbergen in the context of numerical modelling. *J. Petrol.* **43**, 2219–2259.

Ionov D. A., Mukasa S. B., and Bodinier J.-L. (2002b) Sr–Nd–Pb isotopic compositions of peridotite xenoliths from Spitsbergen: numerical modelling indicates Sr–Nd decoupling in the mantle by melt percolation metasomatism. *J. Petrol.* **43**, 2261–2278.

Ireland T. R., Rudnick R. L., and Spetsius Z. (1994) Trace elements in diamond inclusions from eclogites reveal link to Archean granites. *Earth Planet. Sci. Lett.* **128**, 199–213.

Irifune T., Koizumi T., and Ando J.-I. (1996) An experimental study of the garnet-perovskite transformation in the system $MgSiO_3$–$MgAl_2SiO1_2$. *Phys. Earth Planet. Inter.* **96**, 147–157.

Irvine G. J. (2002) Time constraints on the formation of lithospheric mantle beneath cratons: a Re–Os isotope and Platinum Group Element study of peridotite xenoliths from Northern Canada and Lesotho, PhD Thesis, University of Durham.

Irvine G. J., Carlson R. W., Kopylova M. G., Pearson D. G., Shirey S. B., and Kjarsgaard B. A. (1999) Age of the lithospheric mantle beneath and around the Slave craton: a rhenium-osmium isotopic study of peridotite xenoliths from the Jericho and Somerset Island kimberlites. In *Abst. Ninth Annual V. M. Goldschmidt Conference* LPI contribution no. 971. Lunar and Planetary Institute, Houston, pp. 134–135.

Irvine G. J., Pearson D. G., and Carlson R. W. (2001) Lithospheric mantle evolution in the Kaapvaal craton: A Re-Os isotope study of peridotite xenoliths from Lesotho kimberlites. *Geophys. Res. Lett.* **28**, 2505–2508.

Irvine G. J., Pearson D. G., Carlson R. W., Kjarsgaard B. A., Dreibus G. E., and Kopylova M. G. (2003) A PGE and

Re–Os isotope study of mantle xenoliths from Somerset Island. *Lithos.*

Irving A. J. (1974) Geochemical and high pressure experimental studies of garnet pyroxenite and pyroxene granulite xenoliths from the Delegate basaltic pipes. *Australia. J. Petrol.* **15**, 1–40.

Irving A. J. (1980) Petrology and geochemistry of composite ultramafic xenoliths in alkalic basalts and implications for magmatic processes in the mantle. *Am. J. Sci.* **280A**, 389–426.

Irving A. J. and Frey F. A. (1984) Trace element abundances in megacrysts and their host basalts: constraints on partition coefficients and megacryst genesis. *Geochim. Cosmochim. Acta* **48**, 1201–1221.

Irving A. J. and Price R. C. (1981) Geochemistry and evolution of high pressure phoonolitic lavas from Nigeria,Australia, Eastern Germany and New Zealand. *Geochim. Cosmochim. Acta* **45**, 1309–1320.

Izraeli E. S., Harris J. W., and Navon O. (2001) Brine inclusions in diamonds: a new upper mantle fluid. *Earth Planet. Sci. Lett.* **187**, 323–332.

Jackson E. D. and Wright T. L. (1970) Xenoliths in the Honolulu volcanic series, Hawaii. *J. Petrol.* **11**, 405–430.

Jacob D. and Jagoutz E. (1994) A diamond-graphite bearing eclogite xenolith from Roberts Victor (South Africa): implications for petrogenesis from Pb-, Nd-, and Sr isotopes. In *Kimberlites, Related Rocks and Mantle Xenoliths* (eds. H. O. A. Meyer and O. H. Leonardos) CPRM Spec. Publ. CPRM, Brasilia, vol. 1, pp. 304–317.

Jacob D., Jagoutz E., Lowry D., Mattey D., and Kudrjavtseva G. (1994) Diamondiferous eclogites from Siberia: remnants of Archean oceanic crust. *Geochim. Cosmochim. Acta* **58**, 5191–5207.

Jacob D. E., Jagoutz E., and Sobolev N. V. (1998a) Neodymium and strontium isotopic measurements on single subcalcic garnet grains from Yakutian kimberlites. *Neus. Jahrb. Mineral. Abh.* **172**, 357–379.

Jacob D., Jagoutz E., Lowry D., and Zinngrebe E. (1998b) Comments on "The origins of Yakutian Eclogite Xenoliths" by G. A. Snyder L. A. Taylor G. Crozaz A. N. Halliday B. L. Beard V. N. Sobolev and N. V. Sobolev. *J. Pet.* **39**, 1527–1533.

Jacob D. E. and Foley S. F. (1999) Evidence for Archean ocean crust with low high field strength element signature from diamondiferous eclogite xenoliths. *Lithos* **48**, 317–336.

Jacob D. E., Viljoen K. S., Grassineau N., and Jagoutz E. (2000) Remobilization in the cratonic lithosphere recorded in polycrystalline diamond. *Science* **289**, 1182–1185.

Jacob D. E., Bizinis M., and Salters V. J. M. (2002) Lu–Hf isotope systematics of subducted ancient oceanic crust: Roberts Victor eclogites. *Geochim. Cosmochim. Acta (Spec. Suppl.)* **66S1**, A360.

Jagoutz E. (1988) Nd and Sr systematics in an eclogite xenolith from Tanzania: evidence for frozen mineral equilibria in the continental lithsphere. *Geochim. Cosmochim. Acta* **52**, 1285–1293.

Jagoutz E., Palme H., Baddenhausen H., Blum K., Cendales M., Dreibus G., Spettel B., Lorenz V., and Wanke H. (1979) The abundances of major, minor and trace elements in the Earth's mantle as derived from primitive ultramafic nodules. In *Proc. 10th Lunar and Planetary Science Conference*, Pergamon, Houston, pp. 2031–2050.

Jagoutz E., Carlson R. W., and Lugmair G. W. (1980) Equilibrated Nd-, unequilibrated Sr isotopes in mantle xenoliths. *Nature* **296**, 708–710.

Jagoutz E., Dawson J. B., Hoernes S., Spettel B., and Wanke H. (1984) Anorthositic oceanic crust in the Archean Earth. In *Abstr. 15th Lunar and Planetary Science Conf.* pp. 395–396. Lunar and Planetary Science Institute, Houston.

Janse A. J. A. (1985) Kimberlites; where and when. In *Kimberlite Occurrence and Origin; a Basis for Conceptual Models in Exploration* (eds. J. E. Glover and P. G. Harris). University of Western Australia, Geology Department and Extension Service, Perth, West, vol. 8, pp. 19–61.

Janse A. J. A. (1992) New ideas in subdividing cratonic areas. *Russian Geol. Geophys.* **33**(10), 9–25.

Jaques A. L., Haggerty S. E., Lucas H., and Boxer G. L. (1989) Mineralogy and petrology of the Argyle (AK1) lamproite pipe, Western Australia. In *Kimberlites and Related Rocks* (eds. J. Ross, A. L. Jaques, J. Ferguson, D. H. Green, O'Reilly S. Y., R. V. Danchin, and A. J. A. Janse). Geological Society of Australia, Perth, vol. 14(12), pp. 153–188.

Javoy M., Pineau F., and Demaiffe D. (1984) Nitrogen and carbon isotopic composition in the diamonds of Mbuji Mayi (Zaire). *Earth Planet. Sci. Lett.* **68**(3), 399–412.

Javoy M., Pineau F., and Delorme H. (1986) Carbon and nitrogen isotopes in the mantle. In *Isotopes in Geology: Picciotto Volume* (eds. S. Deutsch and A. W. Hofmann). Elsevier, Amsterdam, vol. 57(1–2), pp. 41–62.

Jerde E. A., Taylor L. A., Crozaz G., Sobolev N. V., and Sobolev V. S. (1993) Diamondiferous eclogites from Yakutia, Siberia: evidence for a diversity of protoliths. *Contrib. Mineral. Petrol.* **114**, 189–192.

Jochum K. P., McDonough W. F., Palme H., and Spettel B. (1989) Compositional constraints on the continental lithospheric mantle from trace elements in spinel peridotite xenoliths. *Nature* **340**, 548–550.

Johnson K. E., Davis A. M., and Bryndzia L. T. (1996) Contrasting styles of hydrous metasomatism in the upper mantle: an ion microprobe investigation. *Geochim. Cosmochim. Acta* **60**, 1367–1385.

Jones A. P., Smith J. V., and Dawson J. B. (1982) Mantle metasomatism in 14 veined peridotites from Bultfontein Mine, South Africa. *J. Geology* **90**, 435–453.

Joswig W., Stachel T., Harris J. W., Baur W. H., and Brey G. P. (1999) New Ca-silicate inclusions in diamonds-tracers from the lower mantle. *Earth Planet. Sci. Lett.* **173**(1–2), 1–6.

Kalfoun F., Ionov D., and Merlet C. (2002) HFSE residence and Nb/Ta ratios in metasomatised rutile-bearing mantle peridotites. *Earth Planet. Sci. Lett.* **199**, 49–65.

Kaminsky F. V., Zakharchenko O. D., Davies R., Griffin W. L., Khachatryan B. G. K., and Shiryaev A. A. (2001) Superdeep diamonds form the Juina area, Mato Grosso State, Brazil. *Contrib. Mineral. Petrol.* **140**(6), 734–753.

Kelemen P. B., Hart S. R., and Bernstein S. (1998) Silica enrichment in the continental upper mantle via melt/rock reaction. *Earth Planet. Sci. Lett.* **164**, 387–406.

Kerr R. A. (1993) Bits of the lower mantle found in Brazilian diamonds. *Science* **261**(5127), 1391.

Kinny P. D. and Dawson J. B. (1992) A mantle metasomatic injection event linked to late Cretaceous kimberlite magmatism. *Nature* **360**, 726–728.

Kirkley M. B., Gurney J. J., and Levinson A. A. (1991) Age, origin, and emplacement of diamonds; scientific advances in the last decade. *Gems and Gemology* **27**, 2–25.

Koga K. T., Shimizu N., and Grove T. L. (1999) Disequilibrium trace element redistribution during garnet to spinel facies transformation. In *Proceedings of the VIIth International Kimberlite Conference* (eds. J. Gurney John, L. Gurney James, D. Pascoe Michelle, and H. Richardson Stephen). Red Roof Designs, Cape Town, vol. 1, pp. 444–451.

Köhler T. and Brey G. P. (1990) Calcium exchange between olivine and clinopyroxene calibrated as a geothermobarometer for natural peridotites from 2–60 kb with applications. *Geochim. Cosmochim. Acta* **54**, 2375–2388.

Konzett J., Armstrong R. A., Sweeney R. J., and Compston W. (1998) The timing of MARID metasomatism in the Kaapvaal mantle: an ion probe study of zircons from MARID xenoliths. *Earth Planet. Sci. Lett.* **160**, 133–145.

Konzett J., Armstrong R. A., and Gunther D. (2000) Modal metasomatism in the Kaapvaal craton lithosphere: constraints on timing and genesis from U–Pb zircon dating of metasomatised peridotites and MARID xenoliths. *Contrib. Mineral. Petrol.* **139**, 704–719.

Kopylova M. G. and Russell J. K. (2000) Chemical stratification of cratonic lithosphere: constraints from the Northern Slave craton, Canada. *Earth Planet. Sci. Lett.* **181**, 71–87.

Kopylova M. G., Russell J. K., and Cookenboo H. (1999) Petrology of peridotite and pyroxenite xenoliths from the Jericho kimberlite: implications for the thermal state of the mantle beneath the Slave craton, northern Canada. *J. Petrol.* **40**, 79–104.

Kopylova M. G., Russell J. K., Stanley C., and Cookenboo H. (2000) Garnet from Cr- and Ca-saturated mantle: implications for diamond exploration. *J. Geochem. Explor.* **68**, 183–199.

Kramers J. D. (1977) Lead and strontium isotopes in Cretaceous kimberlites and mantle-derived xenoliths from Southern Africa. *Earth Planet. Sci. Lett.* **34**, 419–431.

Kramers J. D. (1979) Lead, uranium, strontium, potassium and rubidium in inclusion-bearing diamonds and mantle-derived xenoliths from southern Africa. *Earth Planet. Sci. Lett.* **42**(1), 58–70.

Kramers J. D., Roddick J. C. M., and Dawson J. B. (1983) Trace element and isotope studies on veined, metasomatic and "MARID" xenoliths from Bultfontein, South Africa. *Earth Planet. Sci. Lett.* **65**, 90–106.

Krogh E. J. (1988) The garnet-clinopyroxene Fe–Mg geothermometer—a reinterpretation of existing experimental data. *Contrib. Mineral. Petrol.* **99**, 44–48.

Kuno H. and Aoki K. (1970) Chemistry of ultramafic nodules and their bearing on the origin of basaltic magmas. *Phys. Earth Planet. Int.* **3**, 273–301.

Kurat G., Palme H., Spettel B., Baddenhausen H., Hofmeister H., Palme C., and Wanke H. (1980) Geochemistry of ultramafic xenoliths from Kapfenstein, Austria: evidence for a variety of upper mantle processes. *Geochim. Cosmochim. Acta* **44**, 45–60.

Kushiro I. and Aoki K. (1968) Origin of some eclogite inclusions in kimberlite. *Am. Mineral.* **53**, 1347–1367.

Kyser T. K. (1990) Stable isotopes in the continental lithospheric mantle. In *Continental Mantle* (ed. M. A. Menzies). Clarendon Press, Oxford, pp. 127–156.

Kyser T. K., O'Neill J. R., and Carmichael S. E. (1981) Oxygen isotope thermometry of basic lavas and mantle nodules. *Contrib. Mineral. Petrol.* **77**, 11–23.

Lacroix A. (1893) Les enclaves des roches volcaniques. Protat Freres, Macon, 710pp.

Lanphere M. (1968) Sr–Rb–K and Sr isotopic ratios in ultramafic rocks S. E. Alaska. *Earth Planet. Sci. Lett.* **9**, 247–256.

Lawless P. J., Gurney J. J., and Dawson J. B. (1979) Polymict peridotites from the Bultfontein and De Beers mines, Kimberley, South Africa. In *The Mantle Sample: Inclusions in Kimberlites and other Volcanics* (eds. F. R. Boyd and H. O. A. Meyer). American Geophysical Union, Washington, DC, pp. 144–155.

Lee C., Rudnick R. L., and Jacobsen S. (2001) Preservation of ancient and fertile lithospheric mantle beneath the southwestern United States. *Nature* **411**, 69–73.

Lee C. T. (2002) Platinum-group element geochemsitry of peridotite xenoliths from the Sierra Nevada and the Basin and Range, California. *Geochim. Cosmochim. Acta* **66**, 3987–4006.

Lee C. T., Yin Q., Rudnick R. L., Chesley J. T., and Jacobsen S. B. (2000a) Osmium isotopic evidence for Mesozoic removal of lithospheric mantle beneath the Sierra Nevada. *Science* **289**, 1912–1916.

Lee C. T., Rudnick R. L., McDonough W. F., and Horn I. (2000b) Petrologic and geochemical investigation of carbonates in peridotite xenoliths from northeastern Tanzania. *Contrib. Mineral. Petrol.* **139**, 470–484.

Lee D.-C., Halliday A. N., Davies G. R., Essene E. J., Fitton J. G., and Temdjim R. (1996) Melt enrichment of shallow depleted mantle: a detailed petrological, trace element and isotopic study of mantle-derived xenoliths and megacrysts from the Cameroon Line. *J. Petrol.* **37**, 415–441.

Lee H.-Y. and Ganguly J. (1988) Equilibrium composition of coexisting garnet and orthopyroxene: experimental determinations in the system $FeO–MgO–Al_2O_3–SiO_2$ and applications. *J. Petrol.* **29**, 93–113.

Levinson A. A., Gurney J. J., and Kirkley M. B. (1992) Diamond sources and production; past, present, and future. *Gems and Gemology* **28**(4), 234–254.

Lewis R. S., Ming T., Wacker J. F., Anders E., and Steel E. (1987) Interstellar diamonds in meteorites. *Nature (London)* **326**(6109), 160–162.

Liang Y. and Elthon D. (1990) Evidence from chromium abundances in mantle rocks for extraction of picrite and komatiite melts. *Nature* **343**, 551–553.

Lipschutz M. E. (1964) Origin of diamonds in the ureilites. *Science* **143**(3613), 1431–1434.

Lipschutz M. E. and Anders E. (1961) The record in the meteorites: Part 4. Origin of diamonds in iron meteorites. *Geochim. Cosmochim. Acta* **24**, 83–105.

Lorand J.-P. and Alard O. (2001) Platinum-group element abundances in the upper mantle: new constraints from *in situ* and whole-rock analyses of Massif Central xenoliths. *Geochim. Cosmochim. Acta* **65**, 2789–2806.

Lorand J.-P., Pattou L., and Gros M. (1999) Fractionation of platinum-group elements and gold in the upper mantle: a detailed study in Pyrenean orogenic lherzolites. *J. Petrol.* **40**, 957–981.

Lovering J. F. and White A. J. R. (1969) Granulitic and eclogitic inclusions from basic pipes at Delegate, Australia. *Contrib. Mineral. Petrol.* **21**, 9–52.

Lowry D., Mattey D. P., and Harris J. W. (1999) Oxygen isotope composition of syngenetic inclusions in diamond from the Finsch Mine, RSA. *Geochim. Cosmochim. Acta* **63**, 1825–1836.

Luth R. W. and Canil D. (1993) Ferric iron in mantle-derived pyroxenes and a new oxybarometer for the upper mantle. *Contrib. Mineral. Petrol.* **113**, 236–248.

Luth R. W., Virgo D., Boyd F. R., and Wood B. J. (1990) Ferric iron in mantle derived garnets: implications for thermobarometry and for the oxidation state of the mantle. *Contrib. Mineral. Petrol.* **104**, 56–72.

Maaloe S. and Aoki K. (1977) The major element composition of the upper mantle estimated from the composition of lherzolites. *Contrib. Mineral. Petrol.* **63**, 161–173.

Macdougal J. D. and Haggerty S. F. (1999) Ultradeep xenoliths from African kimberlites: Sr and Nd isotopic compositions. *Earth Planet. Sci. Lett.* **170**, 73–82.

MacGregor I. D. (1974) The system $MgO–Al_2O_3–SiO_2$: solubility of Al_2O_3 in enstatite for spinel and garnet peridotite compositions. *Am. Mineral.* **59**, 110–119.

MacGregor I. D. and Carter J. L. (1970) The chemistry of clinopyroxenes and garnets of eclogite and peridotite xenoliths from the Roberts Victor Mine, South Africa. *Phys. Earth Planet. Int.* **3**, 391–397.

MacGregor I. D. and Manton W. I. (1986) Roberts Victor Eclogites: ancient oceanic crust. *J. Geophys. Res.* **91**, 14063–14079.

MacKenzie J. M. and Canil D. (1999) Composition and thermal evolution of cratonic mantle beneath the central Archean Slave Province, NWT, Canada. *Contrib. Mineral. Petrol.* **134**, 313–324.

Marty B. and Zimmerman L. (1999) Volatiles (He, C, N, Ar) in mid-ocean ridge basalts: assessment of shallow-level fractionation and characterization of source composition. *Geochim. Cosmochim. Acta* **63**, 3619–3633.

Mattey D. P., Lowry D., and Macpherson C. (1994) Oxygen isotope composition of mantle peridotite. *Earth Planet. Sci. Lett.* **128**, 231–241.

Matsumoto T., Honda M., McDougall I., Yatsevich I., and O'Reilly S. Y. (1997) Plume-like neon in a metasomatic apatite from the Australian lithospheric mantle. *Nature* **388**, 162–164.

Matsumoto T., Honda M., McDougall I., O'Reilly S. Y., Norman M., and Yaxley G. (2000) Noble gases in pyroxenites and metasomatised peridotites from the Newer Volcanics, southeastern Australia: implications for mantle metasomatism. *Chem. Geol.* **168**, 49–73.

Mazonne P. and Haggerty S. E. (1989) Peraluminous xenoliths in kimberlite: metamorphosed restites produced by partial melting of pelites. *Geochim. Cosmochim. Acta* **53**, 1551–1561.

McCammon C. (2001) Geophysics—Deep diamond mysteries. *Science* **293**(5531), 813–814.

McCammon C. A., Hutchinson M., and Harris J. (1997) Ferric iron content of mineral inclusions in diamonds from Sao Luiz: a view into the lower mantle. *Science* **278**(5337), 434–436.

McCandless T. E. and Gurney J. J. (1989) Sodium in garnet and potassium in clinopyroxene: criteria for classifying mantle eclogites. In *Kimberlites and Related Rocks*, Geol. Soc. Austral. Spec. Publ. No. 14 (ed. J. Ross). Blackwell, Perth, vol. 2, pp. 827–832.

McCulloch M. T. (1989) Sm–Nd systematics in eclogite and garnet peridotite nodules from kimberlites: implications for the early differentiation of the earth. In *Kimberlites and Related Rocks*, Geol. Soc. Austral. Spec. Publ. No. 14 (ed. J. Ross). Blackwell, Perth, vol. 2, pp. 864–876.

McDade P. and Harris J. W. (1999) Syngenetic inclusion bearing diamonds from Letseng-la-Terai, Lesotho. In *Proceedings of the VIIth International Kimberlite Conference* (eds. J. Gurney John, L. Gurney James, D. Pascoe Michelle, and H. Richardson Stephen). Red Roof Designs, Cape Town, vol. 2, pp. 557–565.

McDade P. and Harte B. (2000) Roberts Victor eclogites with a spinel-facies mantle signature. *J. Conf. Abstr.* **5**, 690.

McDonough W. F. (1990) Constraints on the composition of the continental lithospheric mantle. *Earth Planet. Sci. Lett.* **101**, 1–18.

McDonough W. F. and Frey F. A. (1989) Rare earth elements in upper mantle rocks. In *Geochemistry and Mineralogy of Rare Earth Elements*. Mineral. Soc. Am.—Rev. Mineral. Ser. (eds. B. R. Liplin and G. A. McKay) AGU, Washington, DC, vol. 21, pp. 99–145.

McDonough W. F. and McCulloch M. T. (1987) The southeast Australian lithospheric mantle: isotopic and geochemical constraints on its growth and evolution. *Earth. Planet. Sci. Lett.* **86**, 327–340.

McDonough W. F. and Sun S. S. (1995) The composition of the earth. *Chem. Geol.* **120**, 223–253.

McDonough W. F., Stosch H. G., and Ware N. G. (1992) Distribution of titanium and rare earth elements between peridotitic minerals. *Contrib. Mineral. Petrol.* **110**, 321–328.

McInnes B. I. A. and Cameron E. M. (1994) Carbonated alkaline hybridising melts from a sub-arc environment: mantle wedge samples from the Tabar-Lihir-Tanga-Feni arc, Papua New Guinea. *Earth Planet. Sci. Lett.* **122**, 125–141.

McKay D. B. and Mitchell R. H. (1988) Abundance and distribution of gallium in some spinel and garnet lherzolites. *Geochim. Cosmochim. Acta* **52**, 2867–2870.

McKenzie D. (1989) Some remarks on the movement of small melt fractions in the mantle. *Earth. Planet. Sci. Lett.* **95**, 53–72.

Meijer A., Kwon S. T., and Tilton G. R. (1990) U–Th–Pb partitioning behavior during partial melting in the upper mantle; implications for the origins of high mu components and the "Pb-paradox". *J. Geophys. Res.* **95**, 433–448.

Meisel T., Walker R. J., Irving A. J., and Lorand J.-P. (2001) Osmium isotopic compositions of mantle xenoliths: a global perspective. *Geochim. Cosmochim. Acta* **65**, 1311–1323.

Mendelssohn M. J. and Milledge H. J. (1995) Morphological characteristics of diamond populations in relation to temperature-dependent growth and dissolution rates. *Int. Geol. Rev.* **37**(4), 285–312.

Menzies A. H., Shirey S. B., Carlson R. W., and Gurney J. J. (1999) Re–Os systematics of Newlands peridotite xenoliths: implications for diamond and lithosphere formation. In *Proc. 7th Int. Kimberlite Conf.* (eds. J. J. Gurney, J. L. Gurney, M. D. Pascoe, and S. H. Richardson). Red Roof Design, Cap Town, pp. 566–583.

Menzies M. and Chazot G. (1995) Fluid processes in diamond to spinel facies shallow mantle. *J. Geodyn.* **20**, 387–415.

Menzies M. and Halliday A. N. (1988) Lithospheric mantle domains beneath the Archean and Proterozoic crust of Scotland. In *Oceanic and Continental Lithosphere: Similarities and Differences*, J. Petrology (Spec. volume) (eds. M. A. Menzies and K. G. Cox). Oxford University Press, Oxford, pp. 275–302.

Menzies M. and Murthy V. R. (1980a) Enriched mantle: Nd and Sr isotopes in diopsides from kimberlite nodules. *Nature* **283**, 634–636.

Menzies M. and Murthy V. R. (1980b) Nd and Sr isotope geochemistry of hydrous mantle nodules and their host alkali basalts; implications for local heterogeneities in metasomatically veined mantle. *Earth Planet. Sci. Lett.* **46**, 323–334.

Menzies M. and Wass S. Y. (1983) CO_2 rich mantle beneath eastern Australia: REE, Sr and Nd isotopic study of Cenozoic alkaline magmas and apatite-rich xenoliths, southern Highlands Province, New South Wales, Australia. *Earth Planet. Sci. Lett.* **65**, 287–302.

Menzies M. A. (1983) Mantle ultramafic xenoliths in alkaline magmas: evidence for mantle heterogeneity modified by magmatic activity. In *Continental Basalts and Mantle Xenoliths* (eds. C. J. Hawkesworth and M. J. Norry). Shiva, Nantwich, pp. 92–110.

Menzies M. A. (1990a) Archaean, proterozoic and phanerozoic lithospheres. In *Continental Mantle*. (ed. M. A. Menzies). Clarendon Press, Oxford, pp. 67–86.

Menzies M. A. (1990b) Petrology and geochemistry of the continental mantle: an historical perspective. In *Continental Mantle* (ed. M. A. Menzies). Clarendon press, Oxford, pp. 31–54.

Menzies M. A. (1992) The lower lithosphere as a major source for continental flood basalts: a re-appraisal. In *Magmatism and the Causes of Continental Break-up*, Geol. Soc. Spec. Publ. (eds. B. C. Storey, T. Alabaster, and R. J. Pankhurst). The Geological Society, London, vol. 68, pp. 31–40.

Menzies M. A. and Hawkesworth C. J. (1987) *Mantle Metasomatism*. Academic Press, London, 472pp.

Menzies M. A., Kempton P. D., and Dungan M. (1985) Interaction of continental lithosphere and asthenospheric melts below the Geronimo volcanic field, Arizona USA. *J. Petrol.* **26**, 663–693.

Menzies M. A., Halliday A. N., Hunter R. H., MacIntyre R. M., and Upton B. J. G. (1989) The age, composition and significance of a xenolith-bearing monchiquite dike, Lewis, Scotland. In *Kimberlites and Related Rocks*, Geol. Soc. Austral. Spec. Publ. No. 14 (ed. J. Ross). Blackwell, Perth, vol. 2, pp. 843–852.

Mercier J. C. (1980) Magnitude of continental lithospheric stresses inferred from rheomorphic petrology. *J. Geophys. Res.* **85**, 6293–6303.

Meyer H. O. A. (1987) Inclusions in diamond. In *Mantle Xenoliths* (ed. Peter H. Nixon). Wiley, Chichester, pp. 501–523.

Meyer H. O. A. and Boyd F. R. (1972) Composition and origin of crystalline inclusions in natural diamonds. *Geochim. Cosmochim. Acta* **36**(11), 1255–1273.

Mitchell R. H. (1977) Ultramafic xenoliths from the Elwin bay kimberlite: the first Canadian paleogeotherm. *Can. J. Earth Sci.* **14**, 1202–1210.

Mitchell R. H. (1984) Garnet lherzolites from the Hanaus-1 and Louwrensia kimberlites of Namibia. *Contrib. Mineral. Petrol.* **86**, 178–188.

Mitchell R. H. and Keays R. R. (1981) Abundance and distribution of gold, palladium and iridium in some spinel and garnet lherzolites: implications for the nature and origin of precious metal-rich intergranular components in the upper mantle. *Geochim. Cosmochim. Acta* **45**, 2425–2442.

Mitchell R. M. (1995) *Kimberlites, Orangeites and Related Rocks*. Plenum, New York.

Moore R. O. and Gurney J. J. (1985) Pyroxene solid solution in garnets included in diamond. *Nature (London)* **318**(6046), 553–555.

Moore R. O., Griffin W. L., Gurney J. J., Ryan C. G., Cousens D. R., Shee S. H., and Suter G. F. (1992) Trace element geochemistry of ilmenite megacrysts from the Monastery kimberlite, South Africa. *Lithos* **29**, 1–16.

Morgan J. W. (1986) Ultramafic xenoliths: clues to Earth's late accretionary history. *J. Geophys. Res.* **91**(B12), 12375–12387.

Morgan J. W., Wandless G. A., Petrie R. K., and Irving A. J. (1981) Composition of the Earth's upper mantle: I. Siderophile trace elements in ultramafic nodules. *Tectonophysics* **75**, 47–67.

Mukhopadhyay B. (1991) Garnet-clinopyroxene geobarometry: the problems, a prospect and an approximate solution. *Am. Mineral.* **76**, 512–529.

Navon O. (1999) Diamond formation in the Earth's mantle. In *The P. H. Nixon Volume* (eds. J. J. Gurney, J. L. Gurney, M. D. Pascoe, and S. H. Richardson). Red Roof Design, Cape Town, pp. 584–604.

Navon O., Hutcheon I., Rossman G. R., and Wasserburg G. J. (1988) Mantle-derived fluids in diamond micro-inclusions. *Nature* **335**, 784–789.

Neal C. R., Taylor L. A., Davidson J. P., Holden P., Halliday A. N., Nixon P. H., Paces J. B., Clayton R. N., and Mayeda T. (1990) Eclogites with oceanic crustal and mantle signatures from the Bellsbank kimberlite, South Africa: Part 2. Sr, Nd and O isotope geochemistry. *Earth Planet. Sci. Lett.* **99**, 362–379.

Neumann E. R., Wulff P. E., Johnsen K., Andersen T., and Krogh E. (1995) Petrogenesis of spinel harzburgite and dunite suite xenoliths from Lanzarote, eastern Canary Islands: implications for the upper mantle. *Lithos* **35**, 83–107.

Nickel K. G. (1986) Phase equilibria in the system SiO_2–MgO–Al_2O_3–CaO–Cr_2O_3 (SMACCR) and their bearing on the spinel/garnet lherzolite relationships. *Neus. Jahrb. Mineral. Abh.* **155**, 259–287.

Nickel K. G. and Green D. H. (1985) Empirical geothermobarometry for garnet peridotites and implications for the nature of the lithosphere, kimberlites and diamonds. *Earth Planet. Sci. Lett.* **73**, 158–170.

Nimis P. and Taylor W. R. (2000) Single clinopyroxene thermobarometry for garnet peridotites: Part 1. Calibration and testing of a Cr-in-cpx barometer and enstatite-in-cpx thermometer. *Contrib. Mineral. Petrol.* **139**, 541–554.

Nixon P. H. (1987) *Mantle Xenoliths*. Wiley, Chichester.

Nixon P. H. and Boyd F. R. (1973a) Petrogenesis of the granular and sheared ultrabasic nodule suite in kimberlite. In *Lesotho Kimberlites* (ed. P. H. Nixon). Cape and Transvaal, Maseru, pp. 48–56.

Nixon P. H. and Boyd F. R. (1973b) The discrete nodule association in kimberlites from northern Lesotho. In *Lesotho Kimberlites*, (ed. P. H. Nixon) Lesotho National Development Corporation, Maseru, Lesotho, 67–75.

Nixon P. H. and Boyd F. R. (1979) Garnet bearing lherzolites and discrete nodules from the Malaita alnoite, Solomon Islands S. W. Pacific, and their bearing on oceanic mantle composition and geotherm. In *The Mantle Sample: Inclusions in Kimberlites and Other Volcanics* (eds. F. R. Boyd and H. O. A. Meyer). American Geophysical Union, Washington, DC, pp. 400–423.

Nixon P. H., Chapman N. A., and Gurney J. J. (1978) Pyrope-Spinel (Alkremite) xenoliths from kimberlite. *Contrib. Mineral. Petrol.* **65**, 314–346.

Nixon P. H., Rogers N. W., Gibson I. L., and Grey A. (1981) Depleted and fertile mantle xenoliths from southern African kimberlites. *Ann. Rev. Earth Planet. Sci.* **9**, 285–309.

Norman M. (2001) Applications of laser-ablation ICPMS to the trace element geochemistry of basaltic magmas and mantle evolution. In *Min. Assoc. Can. Short Course Series* (ed. P. Sylvester). Ottawa, vol. 29, pp. 163–184.

Norman M. D. (1998) Melting and metasomatism in the continental lithosphere: laser ablation ICPMS analysis of minerals in spinel lherzolites from eastern Australia. *Contrib. Mineral. Petrol.* **130**, 240–255.

Nowell G. M., Kempton P. D., Noble S. R., Fitton J. G., Saunders A. D., Mahoney J. J., and Taylor R. N. (1998) High precision Hf isotope measurements of MORB and OIB by thermal ionisation mass spectrometry: insights into the depleted mantle. *Chem. Geol.* **149**, 211–233.

Nowell G. M., Pearson D. G., Jacob D. J., Spetsius Z. V., Nixon P. H., and Haggerty S. E. (2003). The origin of alkremites and related rocks: Lu–Hf, Rb–Sr and Sm–Nd isotope study. Abstr. 8th Int. Kimberlite Conference, Victoria, FLA 0271.

Nowell G. M., Pearson D. G., Bell D. R., Carlson R. W., Smith C. B., Kempton P. D., and Noble S. R. (2004) Hf isotope systematics of kimberlites and their megacrysts: new constraints on their source regions. *J. Petrol.* (in press).

O'Hara M. J. and Yoder H. S. (1967) Formation and fractionation of basic magmas at high pressures. *Scott. J. Geol.* **3**, 67–117.

O'Hara M. J., Saunders M. J., and Mercy E. L. P. (1975) Garnet-peridotite, primary ultrabasic magma and eclogite: interpretation of upper mantle processes in kimberlite. *Phys. Chem. Earth* **9**, 571–604.

O'Neill H. S. C. (1980) The transition between spinel lherzolite and garnet lherzolite and its use as a geobarometer. *Contrib. Mineral. Petrol.* **77**, 185–194.

O'Neill H. S. C. (1991) The origin of the Moon and early history of the Earth—a chemical model: Part 2. The Earth. *Geochim. Cosmochim. Acta* **55**, 1159–1172.

O'Neill H. S. C. and Wood B. J. (1979) An experimental study of Fe–Mg partitioning between garnet and olivine and its calibration as a geothermometer. *Contrib. Mineral. Petrol.* **70**, 59–70.

O'Neill H. S. C., Rubie D. C., Canil D., Geiger C. A., Ross C. R. I., Seifert F., and Woodland A. B. (1993) Ferric iron in the upper mantle and in transition zone assemblages: implications for relative oxygen fugacities in the mantle. In *Evolution of the Earth and Planets*, Geophys. Monogr. 74 (eds. E. Takahashi, R. Jeanloz, and D. Rubie). American Geophysical Union, Washington, DC, pp. 73–87.

Ongley J. S., Basu A. R., and Kyser T. K. (1987) Oxygen isotopes in coexisting garnets, clinopyroxenes and phlogopites of Roberts Victor eclogites: implications for petrogenesis and mantle metasomatism. *Earth Planet. Sci. Lett.* **83**, 80–84.

O'Reilly S. Y., Griffin W. L., and Ryan C. G. (1991) Residence of trace elements in metasomatized spinel lherzolite xenoliths: a proton-microprobe study. *Contrib. Mineral. Petrol.* **109**, 98–113.

Ozima M. and Zashu S. (1988) Solar-type neon in Zaire cubic diamonds. *Geochim. Cosmochim. Acta* **52**, 19–25.

Ozima M. and Zashu S. (1991) Noble gas state of the ancient mantle as deduced from noble gases in coated diamonds. *Earth Planet. Sci. Lett.* **105**, 13–27.

Palme H. and Nickel K. (1985) Ca/Al ratio and composition of the Earth's mantle. *Geochim. Cosmochim. Acta* **49**, 2123–2132.

Pearson D. G. (1999a) The age of continental roots. *Lithos* **48**, 171–194.

Pearson D. G. (1999b) Evolution of cratonic lithospheric mantle: an isotopic perspective. In *Mantle Petrology: Field Observations and High Pressure Experimentation* (eds. Y. Fei, C. M. Bertka, and B. O. Mysen). The Geochemical Society, Houston, vol. 6, pp. 57–78.

Pearson D. G. and Milledge H. J. (1998) Diamond growth conditions and preservation: inferences from trace elements in a large garnet inclusion in a Siberian diamond. Extended Abstracts of 7th Int. Kimb. Conf., Cape Town, pp. 667–669.

Pearson D. G. and Nixon P. H. (1995) Diamonds in young orogenic belts: graphitised diamonds from Beni Bousera N. Morocco, a comparison with kimberlite-derived diamond occurrences and implications for diamond genesis and exploration. *Africa Geosci. Rev.* **3**, 295–316.

Pearson D. G. and Nowell G. M. (2002) The continental lithospheric mantle: characteristics and significance as a mantle reservoir. *Proc. Roy. Soc. Ser. A.* **360**, 2383–2410.

Pearson D. G. and Nowell G. M. (2003) Dating mantle differentiation: a comparison of the Lu–Hf, Re–Os and Sm–Nd isotope systems in the Beni Bousera peridotite massif and constraints on the Nd–Hf composition of the lithospheric mantle. *Geophys. Res. Abstr.* **5**, 05430.

Pearson D. G. and Shirey S. B. (1999) Isotopic dating of diamonds. In *Application of Radiogenic isotopes to Ore Deposit Research and Exploration* (eds. D. D. Lambert and Ruiz J.). Society of Economic Geologists, Boulder, vol. 12, pp. 143–172.

Pearson D. G. and Woodland S. J. (2000) Carius tube digestion and solvent extraction/ion exchange separation for the analysis of PGEs (Os, Ir, Pt, Pd, Ru) and Re–Os isotopes in geological samples by isotope dilution ICP-mass spectrometry. *Chem. Geol.* **165**, 87–107.

Pearson D. G., Davies G. R., Nixon P. H., and Milledge H. J. (1989) Graphitized diamonds from a peridotite massif in Morocco and implications for anomalous diamond occurrences. *Nature (London)* **338**(6210), 60–62.

Pearson D. G., Davies G. R., Nixon P. H., Greenwood P. B., and Mattey D. P. (1991) Oxygen isotope evidence for the origin of pyroxenites in the Beni Bousera peridotite massif N. Morocco: derivation from subducted oceanic lithosphere. *Earth Planet. Sci. Lett.* **102**, 289–301.

Pearson D. G., Shirey S. B., Carlson R. W., and Taylor L. R. (1992) Os isotope constraints on the origin of eclogite xenoliths. *EOS, Trans. Am. Geophys. Union* **73**(14), 376.

Pearson D. G., Davies G. R., and Nixon P. H. (1993) Geochemical constraints on the petrogenesis of diamond facies pyroxenites from the Beni Bousera peridotite massif, north Morocco. *J. Petrology* **34**, 125–172.

Pearson D. G., Boyd F. R., Haggerty S. E., Pasteris J. D., Field S. W., Nixon P. H., and Pokhilenko N. P. (1994) The characterisation and origin of graphite in cratonic lithospheric mantle: a petrological carbon isotope and Raman spectroscopic study. *Contrib. Mineral. Petrol.* **115**, 449–466.

Pearson D. G., Shirey S. B., Carlson R. W., Boyd F. R., Pokhilenko N. P., and Shimizu N. (1995a) Re–Os, Sm–Nd and Rb–Sr isotope evidence for thick Archaean lithospheric mantle beneath the Siberia craton modified by multi-stage metasomatism. *Geochim. Cosmochim. Acta* **59**, 959–977.

Pearson D. G., Carlson R. W., Shirey S. B., Boyd F. R., and Nixon P. H. (1995b) The stabilisation of Archaean lithospheric mantle: a Re–Os isotope study of peridotite xenoliths from the Kaapvaal craton. *Earth Planet. Sci. Lett.* **134**, 341–357.

Pearson D. G., Snyder G. A., Shirey S. B., Taylor L. A., Carlson R. W., and Sobolev N. V. (1995c) Archaean Re–Os age for Siberian eclogites and constraints on Archaean tectonics. *Nature* **374**, 711–713.

Pearson D. G., Davies G. R., and Nixon P. H. (1995d) Orogenic ultramafic rocks of UHP (Diamond Facies) origin. In *Ultrahigh Pressure Metamorphism* (eds. R. G. Coleman and X. Wang). Cambridge University Press, Cambridge, pp. 456–510.

Pearson D. G., Ionov D., Carlson R. W., and Shirey S. B. (1998a) Lithospheric evolution in circum-cratonic settings: a Re–Os isotope study of peridotite xenoliths from the Vitim region, Siberia. *Mineral. Mag.* **62A**, 1147–1148.

Pearson D. G., Shirey S. B., Harris J. W., and Carlson R. W. (1998b) A Re–Os isotope study of sulfide diamond inclusions from the Koffiefontein kimberlite S. Africa: constraints on diamond crystallisation ages and mantle Re–Os systematics. *Earth Planet. Sci. Lett.* **160**, 311–326.

Pearson D. G., Shirey S. B., Bulanova G. P., Carlson R. W., and Milledge H. J. (1999a) Re–Os isotope measurements of single sulfide inclusions in a Siberian diamond and its nitrogen aggregation systematics. *Geochim. Cosmochim. Acta* **63**(5), 703–711.

Pearson D. G., Shirey S. B., Bulanova G. P., Carlson R. W., and Milledge H. J. (1999b) Dating and paragenetic distinction of diamonds using the Re–Os isotope system. In *Proc. 7th Int. Kimberlite Conf.* (eds. J. J. Gurney, J. L. Gurney, M. D. Pascoe, and S. H. Richardson). Red Roof Design, Cape Town, vol. 2, pp. 637–643.

Pearson D. G., Irvine G. J., Carlson R. W., Kopylova M. G., and Ionov D. A. (2002) The development of lithospheric mantle keels beneath the earliest continents: time constraints using PGE and Re–Os isotope systematics. In *The Early Earth* (eds. M. Fowler, C. J. Ebinger, and C. J. Hawkesworth). Geological Society of London Special Publication, London, vol. 199, pp. 65–90.

Pearson D. G., Irvine G. J., Ionov D. A., Boyd F. R., and Dreibus G. E. (2004) Re–Os isotope systematics and platinum group element fractionation during mantle melt extraction: a study of massif and xenolith peridotite suites. *Chem. Geol.* (special volume): *Highley Siderophile Elements* (in press).

Pearson N. J., Alard O., Griffin W. L., Jackson S. E., and O'Reilly S. (2002) *In situ* measurement of Re–Os isotopes in mantle sulfides by laser ablation multicollector inductively coupled plasma mass spectrometry: anaytical methods and preliminary results. *Geochim. Cosmochim. Acta* **66**, 1037–1050.

Peslier A. H., Reisberg L. R., Ludden J., and Francis D. (2000) Os isotope systematics in mantle xenoliths: age constraints on the Canadian Cordillera lithosphere. *Chem. Geol.* **166**, 85–101.

Pidgeon R. T. (1989) Archaean diamond xenocrysts in kimberlites: how definitive is the evidence?. In *Kimberlites and Related Rocks* (eds. J. Ross, A. L. Jaques, J. Ferguson, D. H. Green, S. Y. O'Reilly, R. V. Danchin, and A. J. A. Janse). Geological Society of Australia, Perth, vol. 14, pp. 1006–1011.

Pokhilenko N. P., Sobolev N. V., Boyd F. R., Pearson D. G., and Shimizu N. (1993) Megacrystalline pyrope peridotites in the lithosphere of the Siberian platform: mineralogy, geochemical peculiarities and the problem of their origin. *Russian J. Geol. Geophys.* **34**(1–12), 56–67.

Porcelli D., Nions R. K. O., and O'Reilly S. Y. (1986) Helium and strontium isotopes in ultramafic xenoliths. *Chem. Geol.* **54**, 237–250.

Porcelli D., O'Nions R. K., Galer S. J. G., Cohen A. S., and Mattey D. P. (1992) isotopic relationships of volatile and lithophile trace elements in continental ultramafic xenoliths. *Contrib. Mineral. Petrol.* **110**, 528–538.

Poreda R. J. and Farley K. A. (1992) Rare gases in Samoan xenoliths. *Earth Planet. Sci. Lett.* **113**, 129–144.

Pyle J. M. and Haggerty S. E. (1994) Silicate-carbonate liquid immiscibility in upper mantle eclogites: evidence of natrosilicic and carbonatitic conjugate melts. *Geochim. Cosmochim. Acta* **58**, 2997–3011.

Pyle J. M. and Haggerty S. E. (1998) Eclogites and the metasomatism of eclogites from the Jagersfontein kimberlite: punctuated transport and implications for alkali magmatism. *Geochim. Cosmochim. Acta* **62**, 1207–1231.

Rapp R. P. and Watson E. B. (1995) Dehydration melting of metabasalt at 8–32 kbar: implications for continental growth and crust-mantle recycling. *J. Petrol.* **36**, 891–932.

Rehkämper M., Halliday A. N., Barford D., Fitton J. G., and Dawson J. B. (1997) Platinum-group element abundance patterns in different mantle environments. *Science* **278**, 1595–1598.

Rehkämper M., Halliday A. N., Alt J., Fitton J. G., Zipfel J., and Takazawa E. (1999) Non-chondritic platinum-group element ratios in oceanic mantle lithosphere: petrogenetic signature or melt percolation? *Earth Planet. Sci. Lett.* **172**, 65–81.

Reisberg L. and Lorand J. P. (1995) Longevity of sub-continental mantle lithosphere from osmium isotope systematics in orogenic peridotite massifs. *Nature* **376**, 159–162.

Reisberg L. and Zindler A. (1986) Extreme isotopic variations in the upper mantle: evidence from Ronda. *Earth Planet. Sci. Lett.* **81**, 29–45.

Rhodes J. M. and Dawson J. B. (1975) Major and trace element chemistry of peridotite inclusions from the Lashaine volcano, Tanzania. *Phys. Chem. Earth* **11**, 545–555.

Richardson S. H. (1986) Latter-day origin of diamonds of eclogitic paragenesis. *Nature* **322**, 623–626.

Richardson S. H. (1989) As definitive as ever: a reply to Archean diamond xenocrysts in kimberlites: how definitive is the evidence? by R. T. Pidgeon. In *Kimberlites and Related Rocks* (eds. J. Ross, A. L. Jaques, J. Ferguson, D. H. Green, S. Y. O'Reilly, R. V. Danchin, and A. J. A. Janse). Geological Society of Australia, Perth, vol. 14, pp. 1070–1072.

Richardson S. H. and Harris J. W. (1997) Antiquity of peridotitic diamonds from the Siberian Craton. *Earth Planet. Sci. Lett.* **151**(3–4), 271–277.

Richardson S. H., Gurney J. J., Erlank A. J., and Harris J. W. (1984) Origin of diamonds in old enriched mantle. *Nature* **310**, 198–202.

Richardson S. H., Erlank A. J., and Hart S. R. (1985) Kimberlite-borne garnet peridotite xenoliths from old enriched subcontinental lithosphere. *Earth Planet Sci. Lett.* **75**, 116–128.

Richardson S. H., Erlank A. J., Harris J. W., and Hart S. R. (1990) Eclogitic diamonds of Proterozoic age from Cretaceous kimberlites. *Nature* **346**, 54–56.

Richardson S. H., Harris J. W., and Gurney J. J. (1993) Three generations of diamonds from old continental mantle. *Nature* **366**, 256–258.

Richardson S. H., Chinn I. L., and Harris J. W. (1999) Age and origin of eclogitic diamonds from the Jwaneng kimberlite, Botswana. *Proc. 7th Int. Kimberlite Conf.* (eds. J. J. Gurney, J. L. Gurney, M. D. Pascoe, and S. H. Richardson). Red Roof Design, Cape Town, vol. 2, pp. 709–713.

Richardson S. H., Shirey S. B., Harris J. W., and Carlson R. W. (2001) Archean subduction recorded by Re–Os isotopes in eclogitic sulfide inclusions in Kimberley diamonds. *Earth Planet. Sci. Lett.* **191**, 257–266.

Ringwood A. E. (1966) The chemical composition and origin of the Earth. In *Advances in Earth Science* (ed. P. M. Hurley). MIT press, Cambridge, Massachusetts, pp. 287–356.

Ringwood A. E. (1991) Phase transformations and their bearing on the constitution and dynamics of the mantle. *Geochim. Cosmochim. Acta* **55**, 2083–2110.

Robinson D. N., Gurney J. J., and Shee S. R. (1984) Diamond eclogite and graphite eclogite xenoliths from Orapa. Botswana. In *Kimberlites: II. The Mantle and Crust-Mantle Relationships* (ed. J. Kornprobst). Elsevier, Amsterdam, pp. 11–24.

Robinson J. A. C., Wood B. J., and Blundy J. D. (1998) The beginning of melting of fertile and depleted peridotite at 1.5 GPa. *Earth Planet. Sci. Lett.* **155**, 97–111.

Roe G. D. (1964) Rubidium-strontium analyses of ultramafic rocks and the origin of peridotites. DPhil. Thesis, Massachusetts Institute of Technology.

Roeder P. L. (1994) Chromite: from the fiery rain of chondrules to the Kilauea Iki lava lake. *Can. Mineral.* **32**, 729–746.

Rosenbaum J. M. (1993) Mantle phlogopite: a significant lead repository? *Chem. Geol.* **106**, 475–483.

Rosenbaum J. M., Zindler A., and Rubenstone J. L. (1996) Mantle fluids: evidence from fluid inclusions. *Geochim. Cosmochim. Acta* **60**, 3229–3252.

Ross J. V. (1983) The nature and rheology of the Cordilleran upper mantle of British Columbia: inferences from peridotite xenoliths. *Tectonophysics* **100**, 321–357.

Ross J. V., Ave Lallement H. G., and Carter N. L. (1980) Stress dependence of recrystallized grain and subgrain size in olivine. *Tectonophysics* **70**, 147–158.

Rozen O. M., Zorin Y. M., and Zayachkovskiy A. A. (1973) A diamond find in Precambrian eclogite of the Kokchetav Block. *Trans. (Doklady) USSR Acad. Sci.: Earth Sci. Sec.* **203**(1–6), 163–165.

Rudnick R. L. and Nyblade A. A. (1999) The thickness and heat production of Archean lithosphere: constraints from xenolith thermobarometry and surface heat flow. In *Mantle Petrology: Field Observations and High Pressure Experimentation* (eds. Y. Fei, C. Bertka, and B. O. Mysen). The Geochemical Society, Houston, vol. 6, pp. 3–12.

Rudnick R. L., McDonough W. F., and Chappell B. W. (1993) Carbonatite metasomatism in the Northen Tanzanian mantle: petrographic and geochemical characteristics. *Earth. Planet. Sci. Lett.* **114**, 463–475.

Rudnick R. L., McDonough W. F., and Orpin A. (1994) Northern Tanzanian peridotite xenoliths: a comparison with Kaapvaal peridotites and inferences on metasomatic interactions. In *Proc. 5th Int. Kimberlite Conf.* (eds. H. O. A. Meyer and O. H. Leonardos). CRPM, vol. 1, pp. 336–353.

Rudnick R. L., McDonough W. F., and O'Connell R. J. (1998) Thermal structure, thickness, and composition of continental lithosphere. *Chem. Geol.* **145**, 395–411.

Rudnick R. L., Ireland T. R., Gehrels G., Irving A. J., Chesley J. T., and Hanchar J. M. (1999) Dating mantle metasomatism: U–Pb geochronology of zircons in cratonic mantle xenoliths from Montana and Tanzania. In *Proc. 7th Int. Kimberlite Conf.* (eds. J. J. Gurney, J. L. Gurney, M. D. Pascoe, and S. H. Richardson). Red Roof Design, Cape Town, vol. 2, pp. 728–735.

Rudnick R. L., Barth M. G., Horn I., and McDonough W. F. (2000) Rutile-bearing refractory eclogites: missing link between continents and depleted mantle. *Science* **287**, 278–281.

Russell J. K. and Kopylova M. G. (1999) A steady state conductive geotherm for the north central Slave, Canada: inversion of petrological data from the Jericho kimberlite pipe. *J. Geophys. Res.* **104**, 7089–7101.

Russell J. K., Dipple G. M., and Kopylova M. G. (2001) Heat production and heat flow in the mantle lithosphere, Slave craton, Canada. *Phys. Earth Planet. Int.* **123**, 27–44.

Sachtleben T. H. and Seck H. A. (1981) Chemical control of Al-solubility in orthopyroxene and its implications on pyroxene geothermometry. *Contrib. Mineral. Petrol.* **78**, 157–165.

Salters V. J. M. and Zindler A. (1995) Extreme $^{176}Hf/^{177}Hf$ in the sub-oceanic mantle. *Earth Planet. Sci. Lett.* **129**, 13–30.

Saltzer R. L., Chatterjee N., and Grove T. L. (2001) The spatial distribution of garnets and pyroxenes in mantle peridotites: pressure-temperature history of peridotites from the Kaapvaal craton. *J. Petrol.* **42**, 2215–2229.

Sautter V., Haggerty S. E., and Field S. W. (1991) Ultradeep (>300 kilometers) ultramafic xenoliths: petrological evidence from the transition zone. *Science* **252**, 827–830.

Schiano P. and Clocchiatti R. (1994) Worldwide occurrence of silicate-rich melts in sub-continental and sub-oceanic mantle minerals. *Nature* **368**, 621–624.

Schmidberger S. S. and Francis D. (1999) Nature of the mantle roots beneath the North American craton: mantle xenolith evidence from Somerset Island kimberlites. *Lithos* **48**, 195–216.

Schmidberger S. S. and Francis D. (2001) Constraints on the trace element composition of the Archean mantle root beneath Somerset Island, Arctic Canada. *J. Petrol.* **42**, 1095–1117.

Schmidberger S. S., Simonetti A., and Francis D. (2001) Sr–Nd–Pb isotope systematcis of mantle xenoliths from Somerset Island kimberlites: evidence for lithosphere stratification beneath Arctic Canada. *Geochim. Cosmochim. Acta* **65**, 4243–4255.

Schmidberger S. S., Simonetti A., Francis D., and Gariepy C. (2002) Probing Archean lithosphere using the Lu–Hf isotope systematics of peridotite xenoliths from Somerset Island kimberlites, Canada. *Earth Planet. Sci. Lett.* **197**, 245–259.

Schrauder M. and Navon O. (1994) Hydrous and carbonatitic mantle fluids in fibrous diamonds from Jwaneng, Botswana. *Geochim. Cosmochim. Acta* **58**, 761–771.

Schrauder M., Koeberl C., and Navon O. (1996) Trace element analyses of fluid-bearing diamonds from Jwaneng, Botswana. *Geochim. Cosmochim. Acta* **60**, 4711–4724.

Schulze D. J. (1986) Calcium anomalies in the mantle and a subducted metaserpentinite origin for diamonds. *Nature* **319**, 483–485.

Schulze D. J. (1987) Megacrysts from alkalic volcanic rocks. In *Mantle Xenoliths* (ed. P. H. Nixon). Wiley, Chichester, pp. 433–451.

Schulze D. J. (1989) Constraints on the abundance of eclogite in the upper mantle. *J. Geophys. Res.* **94**, 4205–4212.

Schulze D. J. (1995) Low-Ca garnet harzburgites from Kimberley, South Africa: abundance and bearing on the structure and evolution of the lithosphere. *J. Geophys. Res.* **100**(12), 513–526.

Schulze D. J., Wiese D., and Steude J. (1996) Abundance and distribution of diamonds in eclogite revealed by volume visualization of CT X-ray scans. *J. Geol.* **104**(1), 109–113.

Schulze D. J., Valley J. W., and Spicuzza M. (2000) Coesite eclogites from the Roberts Victor kimberlite, South Africa. *Lithos* **54**, 23–34.

Scott-Smith B. H., Danchin R. V., Harris J. W., and Stracke K. J. (1984) Kimberlites near Orroroo, South Australia. In *Kimberlites II: the Mantle and Crust-mantle Relationships (Proc. Third Int.l Kimberlite Conf.)* (ed. J. Kornprobst). Elsevier, Amsterdam, pp. 121–142.

Sen G. (1988) Petrogenesis of spinel lherzolite and pyroxenite suite xennoliths from the Koolau shield, Oahu, Hawaii: implications for petrology of the post-eruptive lithosphere beneath Oahu. *Contrib. Mineral. Petrol.* **100**, 61–91.

Sen G. and Leeman W. P. (1991) Iron-rich lherzolite xenoliths from Oahu: origin and implications for Hawaiian magma sources. *Earth Planet. Sci. Lett.* **102**, 45–57.

Shi L., Francis D., Ludden J., Frederiksen A., and Bostock M. (1998) Xenolith evidence for lithospheric melting above anomalously hot mantle under the northern Canadian Cordillera. *Contrib. Mineral. Petrol.* **131**, 39–53.

Shimizu N. (1975) Rare earth elements in garnets and clinopyroxenes from garnet lherzolite nodules in kimberlites. *Earth Planet. Sci. Lett.* **25**, 26–35.

Shimizu N. (1999) Young geochemical features in cratonic peridotites from southern Africa and Siberia. In *Mantle Petrology: Field Observations and High Pressure Experimentation* (eds. Y. Fei, C. M. Bertka, and B. Mysen). The Geochemical Society, Houston, vol. 6, pp. 47–55.

Shimizu N. and Richardson S. H. (1987) Trace element abudance patterns of garnet inclusions in peridotite suite diamonds. *Geochim. Cosmochim. Acta* **51**, 755–758.

Shimizu N., Sobolev N. V., and Yefimova E. S. (1997) Chemical heterogeneities of inclusion garnets and juvenile character of peridotitic diamonds from Siberia. *Russian J. Geol. Geophys.* **38**(2), 356–372.

Shirey S. B. and Walker R. J. (1998) The Re–Os isotope system in cosmochemistry and high-temperature geochemistry. *Ann. Rev. Earth Planet. Sci.* **26**, 423–500.

Shirey S. B., Carlson R. W., Richardson S. H., Menzies A., Gurney J. J., Pearson D. G., Harris J. W., and Wiechert U. (2001) Archean emplacement of eclogitic components into the lithospheric mantle during formation of the Kaapvaal craton. *Geophys. Res. Lett.* **28**, 2509–2512.

Shirey S. B., Harris J. W., Richardson S. H., Fouch M. J., James D. E., Cartigny P., and Viljoen F. (2002) Seismic structure, diamond geology and evolution of the Kaapvaal-Zimbabwe craton. *Science* **297**, 1683–1686.

Shumilova T. G. (2002) Carbynoid carbon and its pseudomorphs after diamond in the eclogitization zone—(Shumikha complex, central Urals). *Doklady—Earth Sciences* **383**(2), 222–224.

Simon N. S. C., Carlson R. W., Pearson D. G., and Davies G. R. (2002) The Lu–Hf isotope composition of cratonic lithosphere: disequilibrium between garnet and clinopyroxene in kimberlite xenoliths. *Geochim. Cosmochim. Acta* **6651**, A717.

Simon N. S. C., Irvine G. J., Davies G. R., Pearson D. G., and Carlson R. W. (2003) The origin of garnet and clinopyroxene in "depleted" Kaapvaal peridotites. *Lithos* (in press).

Smith C. B., Gurney J. J., Harris J. W., Robinson D. N., Shee S. R., and Jagoutz E. (1989) Sr and Nd isotopic systematics of diamond-bearing eclogite xenoliths and eclogite inclusions in diamond from southern Africa. *Kimberlites and Related Rocks.* Spec. Pub. Geol. Soc. Austral., no. 14, vol. 2, pp. 853–863.

Smith C. B., Gurney J. J., Harris J. W., Otter M. L., Kirkley M. B., and Jagoutz E. (1991) Neodymium and strontium isotope systematics of eclogite and websterite paragenesis inclusions from single diamonds, Finsch and Kimberley Pool, RSA. *Geochim. Cosmochim. Acta* **55**, 2579–2590.

Smith D. (1977) The origin and interpretation of spinel-pyroxene clusters in peridotite. *J. Geol.* **85**, 476–482.

Smith D. (1999) Temperatures and pressures of mineral equilibration in peridotite xenoliths: review, discussion and implications. In *Mantle Petrology: Field Observations and High Pressure Experimentation* (eds. Y. Fei, C. Bertka, and B. O. Mysen). The Geochemical Society, Houston, vol. 6, pp. 171–188.

Smith D. and Barron B. R. (1991) Pyroxene-garnet equilibration during cooling in the mantle. *Am. Mineral.* **76**, 1950–1963.

Smith D. and Boyd F. R. (1992) Compositional zonation in garnets of peridotite xenoliths. *Contrib. Mineral. Petrol.* **112**, 134–147.

Smith D., Griffin W. L., Ryan C. G., and Sie S. H. (1991) Trace element zonation in garnets from the thumb: heating and melt infiltration below the Colorado Plateau. *Contrib. Mineral. Petrol.* **107**, 60–79.

Smith D., Riter J. C. A., and Mertzman S. A. (1999) Water rock interactions, orthopyroxene growth and Si-enrichment in the mantle: evidence in xenoliths from the Colorado Plateau, southwestern United States. *Earth Planet. Sci. Lett.* **167**, 347–356.

Snow J. E. and Dick H. J. B. (1995) Pervasive magnesium loss by marine weathering of peridotite. *Geochim. Cosmochim. Acta* **59**, 4219–4235.

Snow J. E. and Reisberg L. C. (1995) Erratum of "Os isotopic systematics of the MORB mantle: results from altered abyssal peridotites. *Earth Planet. Sci. Lett.* **136**, 723–733.

Snyder G. A., Jerde E. A., Taylor L. A., Halliday A. N., Sobolev V. N., and Sobolev N. V. (1993) Nd and Sr isotopes from diamondiferous eclogites, Udachnaya kimberlite pipe, Yakutia, Siberia: evidence of differentiation in the early Earth? *Earth. Planet. Sci. Lett.* **118**, 91–100.

Snyder G. A., Taylor L. A., Crozaz G., Halliday A. N., Beard B., Sobolev V., and Sobolev N. (1997) The origins of Yakutian eclogite xenoliths. *J. Petrol.* **38**, 85–113.

Sobolev N. V. (1974) Deep-seated Inclusions in Kimberlites and the Problem of the Composition of the Upper Mantle. Nauka, Novosibirsk, 264pp (in Russian).

Sobolev N. V. and Shatsky V. S. (1990) Diamond inclusions in garnets from metamorphic rocks. *Nature* **343**, 742–746.

Sobolev N. V., Lavrent'yev Y. G., Pokhilenko N. P., and Usova L. V. (1973) Chrome-rich garnets from the kimberlites of Yakutia and their paragenesis. *Contrib. Mineral. Petrol.* **40**, 39–52.

Sobolev N. V., Galimov E. M., Ivanovskaya I. N., and Yefimova E. S. (1979) isotopic composition of carbon of diamonds containing crystalline inclusions. *Akad. Nauk SSSR Doklady* **249**, 1217–1220.

Sobolev N. V., Yefimova E. S., Channer D. M. D., Anderson P. F. N., and Barron K. M. (1998) Unusual upper mantle beneath Guaniamo, Guyana Shield, Venezuela: evidence from diamond inclusions. *Geology* **26**, 971–974.

Stachel T. and Harris J. W. (1997) Diamond precipitation and mantle metasomatism: evidence from the trace element chemistry of silicate inclusions in diamonds from Akwatia, Ghana. *Contrib. Min. Petrol.* **129**(2–3), 143–154.

Stachel T., Viljoen K. S., Brey G., and Harris J. W. (1998) Metasomatic processes in lherzolitic and harzburgitic domains of diamondiferous lithospheric mantle. *Earth Planet. Sci. Lett.* **159**, 1–12.

Stachel T., Harris J. W., and Brey G. P. (1999) REE patterns of peridotitic and eclogitic inclusions in diamonds from Mwadui (Tanzania). In *Proceedings of the VIIth International Kimberlite Conference* (eds. J. Gurney, J. L. Gurney, M. D. Pascoe, and S. H. Richardson), Red Roof Designs, Cape Town, vol. 2, pp. 829–835.

Stachel T., Harris J. W., Brey G. P., and Joswig W. (2000) Kankan diamonds (Guinea): II. Lower mantle inclusion parageneses. *Contrib. Min. Petrol.* **140**(1), 16–27.

Stachel T., Harris J. W., Aulbach S., and Deines P. (2002) Kankan diamonds (Guinea): III. $\delta^{13}C$ and nitrogen characteristics of deep diamonds. *Contrib. Min. Petrol.* **142**(4), 465–475.

Stern C. R., Saul S., Skewes M. A., and Futa K. (1989) Garnet peridotite xenoliths from the Pali-Aike alkali basalts of southernmost South America. In *Kimberlites and Related Rocks*. Geol. Soc. Austral. Spec. Publ. No. 14 (ed. J. Ross). Blackwell, Perth, vol. 2, pp. 735–744.

Stolz A. J. and Davies G. R. (1988) Chemical and isotopic evidence from spinel lherzolite xenoliths for episodic metasomatism of the upper mantle beneath southeast Australia. In *Oceanic and Continental Lithosphere: Similarities and Differences* J. Petrology (Spec. volume) (eds. M. A. Menzies and K. G. Cox). Oxford University Press, Oxford, pp. 303–330.

Stosch H. G. and Lugmair G. W. (1986) Trace element and Sr and Nd isotope geochemistry of peridotite xenoliths from the Eifel (West Germany) and their bearing on the evolution of the sub-continental lithosphere. *Earth Planet. Sci. Lett.* **80**, 281–298.

Stosch H. G. and Seck H. A. (1980) Geochemistry and mineralogy of two spinel peridotite suites from Dreiser-Weiher, West Germany. *Geochim. Cosmochim. Acta* **44**, 457–470.

Streckeisen A. (1976) To every plutonic rock its proper name. *Earth Sci. Rev.* **12**, 1–33.

Stueber A. M. and Murthy V. R. (1966) Sr isotope and alkali element abundances in ultramafic rocks. *Geochim. Cosmocim. Acta* **33**, 543–553.

Sun S. S. (1982) Chemical composition and origin of the Earth's primitve mantle. *Geochim. Cosmochim. Acta* **46**, 179–192.

Tatsumoto M., Basu A. R., Wankang H., Junwen W., and Guanhong X., Sr (1992) Nd and Pb isotopes of ultramafic xenoliths in volcanic rocks of eastern China: enriched components, EMI and EMII in sub-continental lithosphere. *Earth Planet. Sci. Lett.* **113**, 107–128.

Taylor L. A. and Neal C. R. (1989) Eclogites with oceanic crustal and mantle signatures from the Bellsbank kimberlite, South Africa: Part 1. Mineralogy, Petrography, and whole rock chemistry. *J. Geol.* **97**, 551–567.

Taylor L. A., Snyder G. A., Crozaz G., Sobolev V. N., Yefimova E. S., and Sobolev N. V. (1996) Eclogitic inclusions in diamonds: evidence of complex mantle processes over time. *Earth Planet. Sci. Lett.* **142**(3–4), 535–551.

Taylor L. A., Keller R. A., Snyder G. A., Wang W. Y., Carlson W. D., Hauri E. H., McCandless T., Kim K. R., Sobolev N. V., and Bezborodov S. M. (2000) Diamonds and their mineral inclusions, and what they tell us: a detailed "pull-apart" of a diamondiferous eclogite. *Int. Geol. Rev.* **42**(11), 959–983.

Taylor W. R. (1998) An experimental test of some geothermometer and geobarometer formulations for upper mantle peridotites with application to the thermobarometry of fertile lherzolite and garnet websterite. *Neus. Jahrb. Mineral. Abh.* **172**, 381–408.

Taylor W. R., Jaques A. L., and Ridd M. (1990) Nitrogen-defect aggregation characteristics of some Australasian diamonds: time-temperature constraints on the source regions of pipe and alluvial diamonds. *Am. Min.* **75**, 1290–1310.

Thompson R. N., Morrison M. A., Hendry G. L., and Parry S. L. (1983) Continental flood basalts...arachnids rule OK? In *Continental Flood Basalts and Mantle Xenoliths* (eds. C. J. Hawkesworth and M. J. Norry). Shiva, Nantwich, pp. 158–185.

Tompkins L. A. and Haggerty S. E. (1984) The Koidu Kimberlite Complex, Sierra Leone: geological setting, petrology and mineral chemistry. In *Kimberlites II: the Mantle and Crust-Mantle Relationships* (ed. J. Kornprobst). Elsevier, Amsterdam, pp. 83–105.

Trueb L. F. and Wys E. C. d. (1969) Carbonado: natural polycrystalline diamond. *Science* **165**(2895), 799–802.

Twiss R. J. (1977) Theory and applicability of a recrystallized grain size paleopiezometer. *Pure Appl. Geophys.* **115**, 227–244.

van Heerden L. A., Boyd S. R., Milledge H. J., and Pillinger C. T. (1995) The carbon- and nitrogen isotope characteristics of the Argyle and Ellendale diamonds, Western Australia. *Int. Geol. Rev.* **37**(1), 39–50.

Van Orman J. A., Grove T. L., and Shimizu N. (2001) Rare earth element diffusion in diopside: influence of temperature, pressure, and ionic radius, and an elastic model for diffusion in silicates. *Contrib. Min. Petrol.* **141**, 687–703.

Van Orman J. A., Grove T. L., Shimizu N., and Layne G. D. (2002) Rare Earth element diffusion in a natural pyrope single crystal at 2.8 GPa. *Contrib. Min. Petrol.* **142**, 416–424.

Vance D., Stone J. O. H., and O'Nions R. K. (1989) He, Sr and Nd isotopes in xenoliths from Hawaii and other oceanic islands. *Earth Planet. Sci. Lett.* **96**(1–2), 147–160.

Vannucci R., Piccardo G. B., Rivalenti G., Zanetti A., Rampone E., Ottolini L., Oberti R., Mazzucchelli M., and Bottazzi P. (1995) Origin of LREE-depleted amphiboles in the sub-continental mantle. *Geochim. Cosmochim. Acta* **59**, 1763–1771.

Vaselli O., Downes H., Thirwall M. F., Dobosi G., Coradossi N., Seghedi I., Szakacs A., and Vannucci R. (1995) Ultramafic xenoliths in Plio-pleistocene alkali basalts from the eastern Transylvanian Basin: depleted mantle enriched by vein metasomatism. *J. Petrol.* **36**, 23–53.

Verhoogen J. (1956) Temperatures within the earth. *Phys. Chem. Earth* **1**, 17–43.

von Seckendorff V. and O'Neill H. S. C. (1993) An experimental study of Fe–Mg partitioning between olivine and orthopyroxene at 1173, 1273 and 1423 K and 1.6 GPa. *Contrib. Min. Petrol.* **113**, 196–207.

Viljoen K. S., Smith C. B., and Sharp Z. D. (1996) Stable and radiogenic isotope study of eclogite xenoliths from the Orapa kimberlite, Botswana. *Chem. Geol.* **131**, 235–255.

Wada N. and Matsuda J. (1998) A noble gas study of cubic diamonds from Zaire: constraints on their mantle source. *Geochim. Cosmochim. Acta* **63**, 2335–2345.

Wagner P. A. (1914) *The Diamond Fields of Southern Africa*. Transvaal Leader, Johannesburg 347pp.

Walker R. J., Carlson R. W., Shirey S. B., and Boyd F. R. (1989a) Os, Sr, Nd, and Pb isotope systematics of southern African peridotite xenoliths: implications for the chemical evolution of sub-continental mantle. *Geochim. Cosmochim. Acta* **53**, 1583–1595.

Walker R. J., Shirey S. B., Horan M. F., and Hanson G. N. (1989b) Re—Os, Rb–Sr, and O isotopic systematics of the Archean Kolar schist belt, Karnataka, India. *Geochim. Cosmochim. Acta.* **53**, 3005–3013.

Walter M. J. (1998) Melting of garnet peridotite and the origin of komatiite and depleted lithosphere. *J. Petrol.* **39**, 29–60.

Walter M. J. (1999) Melting residues of fertile peridotite and the origin of cratonic lithosphere. In *Mantle Petrology: Field Observations and High Pressure Experimentation* (eds. Y. Fei, C. Bertka, and B. O. Mysen). The Geochemical Society, Houston, vol. 6, pp. 225–240.

Wilding M. C., Harte B., and Harris J. W. (1991) Evidence for a deep origin for Sao Luiz diamonds. In *Fifth International Kimberlite Conference; Extended Abstracts* (ed. Anonymous), vol. 5, pp. 456–458 [publisher varies].

Wilks J. and Wilks E. (1991) *Properties and Applications of Diamond.* Heinemann, Oxford.

Wilshire H. G. and Shervais J. W. (1975) Al-augite and Cr-diopside ultramafic xenoliths in basaltic rocks from the Western United States. *Phys. Chem. Earth* **9**, 257–272.

Wilson M. R., Kyser T. K., and Fagan R. (1996) Sulfur isotope systematics and platinum group element behavior in REE-enriched metasomatic fluids: a study of mantle xenoliths from Dish Hill, California, USA. *Geochim. Cosmochim. Acta* **60**, 1933–1942.

Winterburn P. A., Harte B., and Gurney J. J. (1990) Peridotite xenoliths from the Jagersfontein kimberlite pipe: I. Primary and primary-metasomatic mineralogy. *Geochim. Cosmochim. Acta* **54**, 329–341.

Witt-Eickschen G. and Harte B. (1994) Distribution of trace elements between amphibole and clinopyroxene from mantle peridotites of the Eifel (western Germany): an ion microprobe study. *Chem. Geol.* **117**, 235–250.

Witt-Eickschen G. and Seck H. A. (1991) Solubility of Ca and Al in orthopyroxene from spinel peridotite: an improved version of an empirical thermometer. *Contrib. Mineral. Petrol.* **106**, 431–439.

Wood B. J. and Nicholls J. (1978) The thermodynamic properties of reciprocal solid solutions. *Contrib. Mineral. Petrol.* **66**, 389–400.

Wood B. J. and Virgo D. (1989) Upper mantle oxidation state: ferric iron contents of lherzolite spinels by ^{57}Fe Mössbauer spectroscopy and resultant oxygen fugacities. *Geochim. Cosmochim. Acta* **53**, 1277–1291.

Wood B. J., Bryndzia L. T., and Johnson K. E. (1990) Mantle oxidation state and its relationship to tectonic environment and fluid speciation. *Science* **248**, 337–345.

Xu S., Okay A. I., Ji S., Sengor A. M. C., Su W., Liu Y., and Jiang L. (1992) Diamond from the Dabie Shan metamorphic rocks and its implication for tectonic setting. *Science* **256**(5053), 80–82.

Xu Y., Menzies M. A., Vroon P., Mercier J. C., and Lin C. (1998) Texture-temperature-geochemistry relationships in the upper mantle as revealed from spinel peridotite xenoliths from Wanqing, NE China. *J. Petrol.* **39**, 469–493.

Yaxley G. M. and Green D. H. (1998) Reactions between eclogite and peridotite: mantle refertilisation by subduction of oceanic crust. *Schweitz. Mieral. Petrogr. Mitt.* **78**, 243–255.

Yaxley G. M., Green D. H., and Kamenetsky V. (1998) Carbonatite metasomatism in the southeastern Australian lithosphere. *J. Petrol.* **39**, 1917–1930.

Yefimova E. S., Sobolev N. V., and Pospelova L. N. (1983) Sulfide inclusions in diamond and specific features of their paragenesis. *Zap. Vsesoy. Mineral. Obsh.* **112**, 300–310.

Yoder H. S. J. (1976) *Generation of Basaltic Magma.* National Academy of Sciences, Washington, DC.

Zanetti E., Vannucci R., Botazzi P., Oberi R., and Ottolini L. (1996) Infiltration metasomatism at Lherz as monitored by systematic ion-microprobe investigations close to a hornblendite vein. *Chem. Geol.* **134**, 113–133.

Zhuravlev A. Z., Laz'ko Y. Y., and Ponomarenko A. I. (1991) Radiogenic isotopes and REE in garnet peridotite xenoliths from the Mir kimberlite pipe, Yakutia. *Geokhimiya* **7**, 982–994.

Zipfel J. and Wörner G. (1992) Thermobarometry on four- and five-phase periodotites from a continental rift system: evidence for upper mantle uplift and cooling at the Ross Sea margin (Antartica). *Contrib. Mineral. Petrol.* **111**, 24–36.

Zindler A. and Hart S. R. (1986) Chemical geodynamics. *Ann. Rev. Earth Plan. Sci.* **14**, 493–571.

Zindler A. and Jagoutz E. (1988) Mantle cryptology. *Geochim. Cosmochim. Acta.* **52**, 319–333.

2.06
Noble Gases as Mantle Tracers

D. R. Hilton

University of California San Diego, La Jolla, CA, USA

and

D. Porcelli

University of Oxford, UK

2.06.1 INTRODUCTION

The study of the noble gases has been associated with some of the most illustrious names in experimental science, and some of the most profound discoveries. Fundamental advances in nuclear chemistry and physics—including the discovery of isotopes—have resulted from their study, earning Nobel Prizes for a number of early practitioners (Rutherford in 1908; Soddy in 1921; Aston in 1922) as well as for their discoverers (Ramsay and Rayleigh in 1904). Within the Earth Sciences, the noble gases found application soon after discovery—helium was used as a chronometer to estimate formation ages of various minerals (Strutt, 1908). In more recent times, the emphasis of noble gas research has shifted to include their exploitation as inert tracers of geochemical processes. In large part, this shift stems from the realization that primordial volatiles have been stored within the Earth since the time of planetary accretion and are still leaking to the surface today. In this introduction, we give a brief overview of the discovery of the noble gases and their continuing utility in the Earth Sciences, prior to setting into perspective the present contribution, which focuses on noble gases in the Earth's mantle.

2.06.1.1 Historical Overview

The first noble gas to be discovered was helium (Lockyer, 1869). Spectroscopic investigations of the Sun's chromosphere during a solar eclipse in India in 1868 revealed a previously unobserved line, close in wavelength to the D_1 and D_2 Fraunhofer lines of sodium. The new line was designated D_3 and the element it represented helium (=sun).

The chemical and physical characterization of the noble gases commenced in the 1890s with the discovery of argon. Rayleigh (1892) found a consistent density difference between nitrogen prepared from ammonia and "nitrogen" prepared from air. By passing the "air-nitrogen" over heated magnesium, he and Ramsey succeeded in isolating a new gas of greater density than nitrogen and having a different spectrum from any of the known elements. They named the new gas argon (=not working, lazy) for its unreactive nature (Rayleigh and Ramsay, 1892). Pursuing further research on argon, Ramsay undertook acid treatment experiments on uraninite in the hope of liberating argon. Spectroscopic examination of the gas, however, revealed the D_3 line observed in India by Lockyer. Ramsay reported the discovery of terrestrial helium (Ramsay, 1895).

The discovery of the first two noble gases led Ramsay to conclude that they were members of a new column in the Periodic Table and to search for the remaining members by a new distillation technique. The method involved (i) taking samples of liquid air, (ii) allowing them to boil away almost completely, and then (iii) examining the residue, or further distillate of the residue, spectroscopically. Krypton (=hidden) was the first of the remaining gases to be discovered by this method (Ramsay and Travers, 1898a). Similar experiments on argon yielded two portions, the lighter of which was found to be neon (=new) (Ramsay and Travers, 1898b). The same technique on liquid krypton led to the discovery of xenon (=stranger) (Ramsay and Travers, 1898c).

At the beginning of the twentieth century, research turned to the newly discovered radioactive substances. Ramsay and Soddy (1903) showed that helium was derived by radioactive disintegration of radium, the first demonstration that one element was derived from another. Once the decay constant of radium had been determined, the door was open for the first application of the noble gases: geochronology (Strutt, 1908), a methodology that has been pursued ever since.

In an attempt to explain why the atomic weights of the elements were almost, but not quite, whole integers, Soddy (1910) suggested the existence of chemically identical substances of different atomic weight. J. J. Thomson (1912), using his positive ray apparatus, showed that neon had a line in its spectrum corresponding to mass 22 (as well as mass 20) that could not be matched to any known line at the time. Perceiving the need for a specific name for such substances, Soddy suggested the term "isotopes" (iso = same; topes = place) because they occupied the same place in the Periodic Table.

Following the development of the first mass spectrographs (Aston, 1919; Dempster, 1918) to succeed the parabolic mass analyzer of Thomson, Aston demonstrated the existence, and estimated the relative proportions, of ^{20}Ne and ^{22}Ne (Aston, 1920a), ^{36}Ar and ^{40}Ar (Aston, 1920b, 1920c), all six isotopes of krypton (Aston, 1920b) and five of the isotopes of xenon (Aston, 1920c). He reported the proportions of the remaining xenon isotopes in 1927 (Aston, 1927). Of the remaining noble gas isotopes, ^{21}Ne and ^{38}Ar were found by Hogness and Kvalnes (1928) and Zeeman and De Gier (1934), respectively (see citation by Nier (1936) to the work of Zeeman and De Gier, 1934). Alvarez and Cornog (1939) discovered the one remaining noble gas isotope, namely ^{3}He, using a cyclotron as a mass spectrometer.

The dawn of the modern era of terrestrial noble gas studies can be traced to a suggestion by Suess and Wanke (1965) that the ocean floor should be characterized by a steady-state loss of helium equal to its production in the mantle. Although they measured a 6% excess helium saturation anomaly in Pacific deep water, doubts were expressed due to the possibility of air entrainment. To circumvent these difficulties, Clarke *et al.* (1969) measured the $^3He/^4He$ ratio expecting to distinguish between air helium and mantle helium, which conventional wisdom at the time dictated should have low (radiogenic) helium isotopic ratios (Morrison and Pine, 1955). Measured helium, however, showed excess ^3He (~22% above air) that was immediately attributed to the presence of primordial ^3He leaking from the mantle. At the same time, Mamyrin *et al.* (1969) reported extremely high $^3He/^4He$ ratios ($\sim 10^{-5}$) for hydrothermal fluids from the southern Kuril Islands. Subsequent studies have shown that high $^3He/^4He$ ratios are characteristic of all samples of recent mantle origin, reflecting the transport of mantle volatiles through the crust in a variety of tectonic environments. This realization has led to a remarkable broadening in scope of terrestrial noble gas studies into areas such as oceanography, hydrology, and volcanology, to name but a few. However, as will be demonstrated in this work, understanding mantle-related processes through the study of the noble gases remains a key research area in isotope geochemistry and a continuing challenge to the noble gas community.

2.06.1.2 Scope of Present Contribution

In this work, we present an overview of the current status of noble gas geochemistry as it relates to understanding the terrestrial mantle. We first review the intrinsic properties of the noble gases that make them so useful as geochemical tracers. This leads to a description of the main isotopic and relative abundance features of the noble gases in mantle-derived materials. We then concentrate on three aspects of the utility of noble gases in mantle-related studies: (i) as a tracer of recycling between the Earth's exosphere (hydrosphere, atmosphere, and crust) and mantle, and how noble gases are crucial in assessing the state of volatile mass balance between these reservoirs; (ii) as a fundamental constraint on a host of different classes of models aimed at describing the structure of the mantle (e.g., whether it is layered) and the related topic of the origin of ocean island basalts (OIBs); and (iii) as a means to understand depletion/enrichment processes and volatile transport in the subcontinental mantle. It should be noted that noble gases have also found utility in other related areas of mantle geochemistry not covered in this chapter: the reader is referred to Chapters 4.11 and 4.12 for respective background on (i) noble gases in planetary accretion process and the acquisition of the terrestrial volatile inventory, and (ii) mantle–atmosphere coupling and the formation of the Earth's atmosphere.

This work follows on from many earlier reviews of the noble gas geochemistry of the Earth's mantle. For more details on this topic, the reader is referred to the following references: Craig and Lupton (1981); Farley and Neroda (1998); Lupton (1983); Mamyrin and Tolstikhin (1984); McDougall and Honda (1998); Ozima (1994); Ozima and Podosek (1983); and Porcelli *et al.* (2002).

2.06.2 NOBLE GASES AS GEOCHEMICAL TRACERS

Noble gases have been widely exploited as geochemical tracers. They have three attributes that make them particularly sensitive as tracers of mantle processes. Specifically, noble gases are (i) composed of different components each having diagnostic isotope characteristics, (ii) incompatible, and hence depleted in the solid Earth, and (iii) chemically inert. Each of these features of the noble gas family is discussed in turn.

2.06.2.1 Component Structures of the Noble Gases

The principal sources of terrestrial noble gases are primordial gases, inherited from the solar nebula at the time of planetary accretion, and nucleogenic gases generated by specific nuclear reactions subsequent to accretion. Interactions with cosmic rays produce a third but relatively minor source of noble gases (cosmogenic noble gases) mainly by spallation reactions at or close to the Earth's surface. Identification of the isotopic characteristics of these "end-member" compositions is an essential prerequisite to attempts to resolve terrestrial noble gas isotope variations into component structures, and hence in their exploitation as tracers of the origin, development, and structure of the Earth's mantle.

Models of planetary evolution assume that at the time of planetary formation the solar system had a single universal and well-mixed composition from which all parts of the solar system were derived (see Podosek, 1978). Information as to the elemental and isotopic characteristics of this primordial composition is presently available from the Sun, meteorites, and the atmospheres of the giant planets (Wieler, 2002). In the case of the Sun, distinction is usually made between the present-day composition, which is available via spectral analysis of the solar atmosphere and capture of the solar wind, either directly in space or by using metallic foil targets, and the proto-Sun (the composition at the time of planetary accretion) whereby the lunar regolith and/or meteorites are utilized as archives of ancient solar wind. As discussed below, the distinction is only really important for helium due to production of ^3He by deuterium burning.

Although the Sun represents the obvious choice for defining the primordial composition of the solar nebula, the initial discovery of primordial noble gases was made using meteorites. Gerling and Levskii (1956) claimed the honor by arguing that observed noble gas abundances in a gas-rich achondrite could not be supported by *in situ* nuclear processes (radiogenic or spallation reactions). Subsequent investigations of other gas-rich meteorites (e.g., Suess *et al.*, 1964) showed that the relative abundances of the trapped noble gases are similar to those in the Sun, earning the label "solar" gases for the trapped component. Direct measurements of the relative abundances and isotopic structure of noble gases in the solar wind (see Wieler, 2002) have confirmed the origin of the solar component in meteorites as implanted solar wind. Therefore, noble gas measurements of both gas-rich meteorites and the solar wind serve to define the primordial noble gas signature of the solar system. In Table 1 we highlight some of the principal characteristics of "solar" noble gases.

A problem with adopting the solar wind $^3He/^4He$ ratio as representative of the solar nebula is the production of ^3He from deuterium very early in solar system history: consequently, the solar wind value ($\sim 4.4 \times 10^{-4}$) is too high by a factor between ~2.5 and ~3 relative to the proto-Sun (Geiss and Reeves, 1972). To circumvent this difficulty, recourse has been made to analyzing the giant planets whose atmospheres are expected to reflect proto-solar values (Wieler, 2002). Jupiter is the only giant planet whose atmospheric $^3He/^4He$ ratio has been determined (Mahaffy *et al.*, 1998). Its value of 1.66×10^{-4} ($\sim 120\,R_A$, where R_A = air $^3He/^4He$)—remarkably similar to measurements on primitive meteorites (see below), is now adopted as most representative of primordial (proto-solar) helium.

It should be noted at this point that studies of the most primitive class of meteorites—CI chondrites—have found that many contain trapped gases in a distinctly different pattern to the "solar" signature. Indeed, the relative abundances of the noble gases resemble those in the Earth's atmosphere (Figure 1) so they have been dubbed "planetary" gases (Signer and Suess, 1963). Whereas all meteorites show a marked depletion in noble gases compared to solar abundances (see next section), the "planetary" pattern shows a greater relative depletion in the lighter noble gases. In addition, the isotopic composition of the planetary component differs from that of solar gas (Table 1). For example, the $^3He/^4He$ ratio of carbonaceous material in primitive chondrites (the so-called Q-gases; see discussion by Ott, 2002) is 1.23×10^{-4}, significantly lower than that of solar helium. At this stage, firm genetic relationships between "planetary" and solar noble gases remain to be established but it appears probable that, with the exception of helium, the solar noble gas component better approximates primordial characteristics at the time of Earth accretion. This follows from considerations of the origins of noble gases on the planets, as discussed in Chapter 4.12. Indeed, the noble gases seen in the Earth's and other planetary atmospheres may record the effects of a number of additional processes, e.g., Rayleigh distillation, hydrodynamic loss, adsorption on planetesimal surface grains, or any combination thereof, associated with the accretionary process itself and/or subsequent

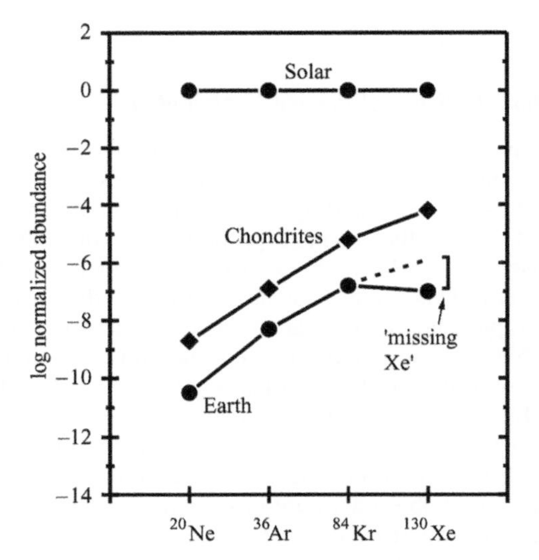

Figure 1 The noble gas abundance patterns of the Earth and CI chondrites normalized to silicon and the solar composition (reproduced by permission of Mineralogical Society of America from *Rev. Mineral. Geochem.*, **2002**, *47*, 418).

Table 1 Solar system isotopic compositions of He, Ne, and Ar.

Reservoir	$^3He/^4He$ ($\times 10^{-6}$)	R/R_A [a]	$^{20}Ne/^{22}Ne$	$^{21}Ne/^{22}Ne$	$^{38}Ar/^{36}Ar$	$^{40}Ar/^{36}Ar$
Solar	457	326	13.8	0.0328	0.1825	$\sim 3 \times 10^{-4}$
Planetary	143	102	8.2	0.024	0.188	$\sim 3 \times 10^{-4}$
Earth atmosphere	1.4	1	9.8	0.0290	0.1880	295.5

After McDougall and Honda (1998).
[a] $R/R_A = (^3He/^4He)_{observed}/(^3He/^4He)_{air}$, where air $^3He/^4He = 1.4 \times 10^{-6}$.

Table 2 Half-lives of parent nuclides for noble gases.

Nuclide	*Half-life*	*Daughter*	*Yield* (atoms/decay)	*Comments*
^{3}H	12.26 yr	^{3}He	1	Continuously produced in atm
^{238}U	4.468 Gyr	^{4}He	8	
		^{136}Xe	3.6×10^{-8} $(4.4\pm0.1)\times10^{-8}$	Spontaneous fission
^{235}U	0.7038 Gyr	^{4}He	7	$^{238}U/^{235}U = 137.88$
^{232}Th	14.01 Gyr	^{4}He	6	Th/U = 3.8 in bulk Earth
		^{136}Xe	$<4.2\times10^{-11}$	No significant production in Earth
^{40}K	1.251 Gyr	^{40}Ar	0.1048	$^{40}K = 0.01167\%$ total K
^{244}Pu	80.0 Myr	^{136}Xe	7.00×10^{-5}	$^{244}Pu/^{238}U = 6.8\times10^{-3}$ at 4.56 Ga
^{129}I	15.7 Myr	^{129}Xe	1	$^{129}I/^{127}I = 1.1\times10^{-4}$ at 4.56 Ga

Source: Porcelli *et al.* (2002).

Table 3 Some nuclear reactions producing noble gases.

Reaction	*Upper crust production ratios*[a]
$^{6}Li(n,\alpha)^{3}H(\beta-)^{3}He$	$^{3}He/^{4}He = 1\times10^{-8}$
$^{17}O(\alpha,n)^{20}Ne$	$^{20}Ne/^{4}He = 4.4\times10^{-9}$
$^{18}O(\alpha,n)^{21}Ne$	$^{21}Ne/^{4}He = 4.5\times10^{-8}$
$^{24}Mg(n,\alpha)^{21}Ne$	$^{21}Ne/^{4}He = 1\times10^{-10}$
$^{25}Mg(n,\alpha)^{22}Ne$	Combined crustal production:
$^{19}F(\alpha,n)^{22}Na(\beta+)^{22}Ne$	$^{22}Ne/^{4}He = 9.1\times10^{-8}$
$^{35}Cl(n,\gamma)^{36}Cl(\beta-)^{36}Ar$	$^{40}Ar/^{36}Ar = 1.5\times10^{7}$

Source: Porcelli *et al.* (2002).
[a] See Ballentine and Burnard (2002) for details and production rates in different compositions.

planetary growth (Pepin, 1992), and these processes have resulted in distinctive isotopic compositions and relative abundances that are distinctive from those found on the planets (see Chapter 4.12). There now appears to be a general consensus that the "planetary" component of noble gases found in meteorites is unrelated to the noble gases seen in planetary atmospheres.

Radiogenic and nucleogenic noble gases have been produced within the Earth subsequent to planetary accretion. They are produced as a result of the spontaneous decay of a parent radionuclide or as a consequence of specific nuclear reactions. The principal noble gas isotope daughters produced by radioactive decay are given in Table 2, whereas the most common nuclear reactions producing noble gas isotopes are summarized in Table 3 (see review by Ballentine and Burnard, 2002). Modification of the (primordial) isotopic composition of a noble gas within the Earth occurs by addition of a radiogenic/nucleogenic daughter product.

The most conspicuous daughter isotope produced by nuclear methods is ^{4}He. It originates through α-particle decay of members of the ^{238}U, ^{235}U, and ^{232}Th decay series. Uranium-238 also undergoes spontaneous fission and produces small but sufficient quantities of xenon isotopes to perturb the primordial pattern within the Earth. The short-lived nuclides ^{129}I and ^{244}Pu undergo spontaneous decay and are now both extinct on Earth. They have produced measurable variations in the xenon spectra of the mantle as sampled by oceanic basalts (Staudacher and Allègre, 1982). The only other major noble gas daughter produced by spontaneous decay is ^{40}Ar, which results from electron capture by a ^{40}K nucleus.

Specific nuclear reactions capable of producing noticeable quantities of noble gas daughters in the Earth (^{3}He and ^{21}Ne in particular) are initiated by alpha and fission activities of the natural radioelements. Helium-3 is produced through a neutron capture reaction involving ^{6}Li (Hill, 1941), whereas ^{21}Ne production occurs through a number of α-induced reactions (Wetherill, 1954). In the case of helium, the $^{3}He/^{4}He$ ratio produced is of the order 10^{-8} and primarily reflects the lithium abundance at the site of production (Mamyrin and Tolstikhin, 1984). For neon, the only conspicuous isotope produced is ^{21}Ne due to its low natural abundance. The present-day $^{21}Ne/^{4}He$ production ratio in the mantle has been calculated at 4.5×10^{-8} (Yatsevich and Honda, 1997) (see Ballentine and Burnard, 2002 for discussion regarding calculation of this parameter).

Spallation reactions are the only other means of producing significant abundances of noble gases on Earth. Interactions of cosmic rays with the Earth's atmosphere produce high-energy particles that can interact with nuclei in the atmosphere and at the Earth's surface. Sizeable quantities of ^{3}He, ^{21}Ne, and ^{38}Ar can be produced in this manner (see Ballentine and Burnard, 2002). In the context of the present work on the Earth's mantle, however, the primary concern regarding noble gases of cosmic origin relates to the suggestion of Anderson (1993) that interplanetary dust particles (IDPs) rich in ^{3}He (and ^{21}Ne) could account for the origin of the high $^{3}He/^{4}He$ ratios observed in mantle emanations. This idea has been criticized on two fronts: first, the rate of IDP deposition is many orders of magnitude too low to sustain the flux of ^3He (and ^{21}Ne) released via the ridge system (Stuart, 1994; Trull, 1994). Second, the diffusivity of helium in IDP material is too high for it to be effectively transported into the mantle (Hiyagon, 1994). We are left with the conclusion, therefore, that the range of isotopic compositions observed for the noble gases in mantle-derived materials primarily reflects admixture—to varying degrees—of primordial and radiogenic and/or nucleogenic components, with the superimposition of other effects (e.g., degassing, crustal assimilation) onto one or both components.

2.06.2.2 Partitioning of the Noble Gases

A major reason for the success of the noble gases as geochemical tracers is their depletion in the terrestrial realm (Figure 1). For example, ^{36}Ar is nearly nine orders of magnitude less abundant in the Earth relative to solar abundances (normalized to silicon) (Ozima and Podosek, 1983). The available evidence further indicates that a substantial proportion of the terrestrial inventory of various noble gas species (e.g., ^{130}Xe; Pepin and Porcelli, 2002) is now resident in the atmosphere, implying that the Earth has retained noble gases in low abundance following atmosphere formation (see Chapter 4.11). Herein lies the particular advantage of the noble gases as mantle tracers: abundances of residual (primordial) noble gases are sufficiently low that significant (i.e., measurable) shifts in isotopic ratios are produced by addition of radiogenic and nucleogenic noble gases. The key to understanding the depletion of noble gases in the solid Earth lies with knowledge of the partitioning behavior of noble gases between melt and vapor, melt and crystal, and liquid iron and silicate solid.

There are three facets of the partitioning behavior of the noble gases that work together to deplete the gases from the silicate Earth. First, during mantle melting and the production of partial melt, the noble gases show a strong affinity for the melt phase. Early experimental work on synthetically made olivines (e.g., Hiyagon and Ozima, 1982) showed that D-values (where D is the weight concentration in mineral/weight concentration in melt) are generally low ($D_{He} \leq 0.07$; $D_{Ar} \leq 0.05-0.15$; $D_{Xe} \leq 0.3$). Although there was some debate whether D values are consistently low for all noble gases and all minerals phases (e.g., Broadhurst *et al.*, 1990, 1992), work on naturally occurring glass–mineral pairs consistently found low D values for olivine–melt partitioning (e.g., $D_{He} \leq 0.008$; $D_{Ar} \leq 0.003$; Marty and Lussiez, 1993). All these values are considered minima. More recent experimental work on clinopyroxene-silicate melt (Chamorro *et al.*, 2002) has found that D_{Ar} is low ($\sim 4 \times 10^{-4}$) and constant to pressures of at least 80 kbar. Brooker *et al.* (2003) have demonstrated that such low values extend to all the noble gases. Therefore, during even minor and/or relatively deep episodes of mantle melting, the expectation is that noble gases will partition strongly from the solid (mantle) into the melt phase.

Second, the solubility of noble gases in basaltic melt is generally low: Jambon *et al.* (1986) reports values of 56, 25, 5.9, 3.0, and 1.7 (units of cm^3 STP g^{-1} $bar^{-1} \times 10^3$) for helium through xenon for basaltic liquid at 1,400 °C. The ionic porosity of the melt—controlled by its composition—dictates the solubility of a particular species (Carroll and Draper, 1994). Due to low noble gas solubilities, any exsolution of a vapor phase (such as CO_2) during decompression of a melt as it moves toward the surface will result in the rapid removal of the noble gases from the melt phase. In Figure 2, the partitioning of the noble gases between melt and vapor is illustrated for a vesiculated basaltic magma (Carroll and Draper, 1994). Note that for the range of vesicularities observed in submarine basalts, partitioning of the noble gases into the vesicle phase is almost complete. In this way, any melt phase can be effectively stripped of its volatile inventory. Therefore, both mantle melting and vapor exsolution act together to remove noble gases and other volatiles from the silicate Earth and to transfer them to the atmosphere and hydrosphere.

Finally, a related but more speculative means to deplete the silicate Earth of its noble gas inventory is if noble gases were sequestered by Fe–Ni and transferred to the Earth's core. The proposition has been evaluated for helium and found to be plausible notwithstanding significant uncertainties in solid-silicate/liquid-metal partitioning behavior (Porcelli and Halliday, 2001). However, the proportion of helium lost to the core in this way is likely to be relatively small

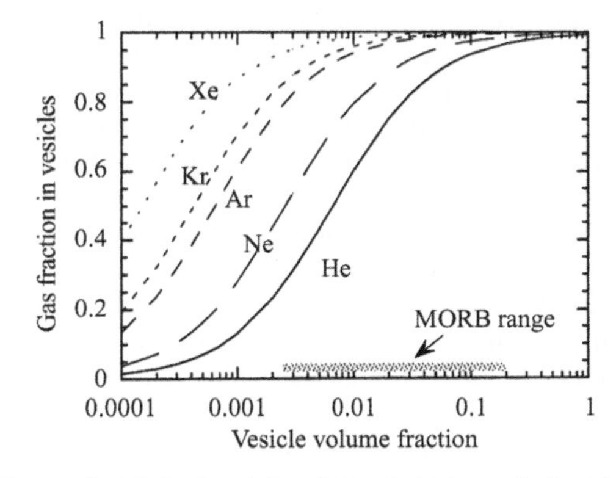

Figure 2 Calculated fraction of total gas in basaltic magma which is present in vesicles as a function of vesicularity (reproduced by permission of Elsevier from *Chem. Geol*, **1994**, *117*, 37–56).

(<2% of the original mantle inventory; Porcelli and Halliday, 2001). Insufficient data preclude this analysis being extended to the other noble gases.

In sum, the partitioning behavior of the noble gases has acted to deplete the silicate Earth of its noble gas inventory. As the mantle comprises the bulk of the silicate Earth, noble gases are relatively scarce in this terrestrial reservoir. A consequence of this is that noble gases are particularly sensitive tracers of events occurring within the mantle. The scarcity of noble gases in the solid Earth has also given rise to their common pseudonym, the rare gases, although this is far from true in a solar sense.

2.06.2.3 Chemical Inertness of the Noble Gases

The scarcity of the noble gases on Earth is a direct result of one of their most obvious characteristics—their chemical inertness, or failure to react with other species under almost all circumstances. As discussed above, a major consequence of this inertness is that noble gases pass easily into the gas phase, and are efficiently lost from the solid Earth. However, there are situations where the physical chemistry of the noble gases may be important for explaining experimental observations. In such cases, the effects will be almost entirely due to van der Waal's forces, which increase in strength with number of electrons, i.e., Xe > Kr > Ar > Ne > He. Following this line of reasoning, a significant effort has been expended since the 1980s evaluating continental shales and glacial ice as possible reservoirs for the so-called "missing-Xe" (see Figure 1), based upon the potential of xenon to interact more strongly with possible sorbent material. Significantly, no terrestrial reservoir has been found as the repository of the extraneous xenon (Ozima and Podosek, 1983). Indeed, the current consensus for the low Xe/Kr ratio in the terrestrial atmosphere is that it is not indicative of xenon loss but is an inherited feature characteristic of planetary formation (Porcelli and Ballentine, 2002).

Another corollary of the failure of the noble gases to react chemically is that they generally behave coherently as a group, with systematic and often predictable variations or responses from helium through xenon. Again, this is advantageous in a geochemical context because such responses can yield information on particular physical processes or perturbations to natural systems. An example of such a process is vapor partitioning, with the distribution of noble gases between a melt and vesicle phase useful for tracing extent and mode of degassing (e.g., Moreira and Sarda, 2000). It is important to note that using the noble gases as a group allows for redundancy in the tracing process: this can be significant if, e.g., any one species (e.g., ^{36}Ar) is particularly susceptible to modification by other means (e.g., air contamination). Likewise, the coherence of the noble gases facilitates estimation of unknown parameters. For example, the diffusivity of the heavy noble gases under mantle $P-T$ conditions is poorly constrained: however, this is not the case for helium (Trull and Kurz, 1993). Therefore, by scaling to helium it becomes possible to place reasonable limits on unknown quantities.

Taken together, the properties of wide-ranging isotopic systematics, incompatibility in the solid Earth and chemical inertness act to make noble gases tracers "par excellence" in geochemical research. In the following discussion, we show that noble gases are particularly adept at tracing processes associated with the origin and development of the Earth's mantle.

2.06.3 MANTLE NOBLE GAS CHARACTERISTICS

In this section we present an overview of the principal means available to sample mantle-derived noble gases, followed by a summary of their main isotope and relative abundance characteristics in the mantle. Mostly, the mantle shows a wide range in noble gas isotope variations serving to impart information on a variety of topics. The only exception is krypton whose isotopic composition is steadfastly air-like in mantle materials. Consequently, we do not consider krypton in this review.

2.06.3.1 Localities and Sampling Media

Given their geochemical incompatibility and affinity for the liquid silicate phase, noble gases of mantle origin are overwhelmingly transferred to the Earth's surface during mantle melting. Consequently, the distribution of locations worldwide exploited to yield information on the noble gases characteristics of the mantle is heavily skewed towards regions undergoing active melting which results in magmatic or volcanic activity. The global-encircling mid-ocean ridge system allows relatively straightforward access to the oceanic mantle at points worldwide. Ocean islands, be they intraplate in origin or associated with arc volcanism, also facilitate loss of noble gases from the oceanic mantle—albeit on a smaller scale—and are consequently prime sampling targets. On the continents, regions of active volcanism (e.g., continental arcs) provide similar access to volatiles from the underlying mantle, as do areas associated with rifting and nascent magmatism. Although crustal fluids such as groundwaters and natural gases may influence their distribution, the primary control on the location of mantle-derived noble gases in continental regions is still likely to be the presence of underlying melts.

Mantle noble gases can be sampled through a wide variety of fluid and rock types. For example, fumarolic gas discharges, bubbling hot springs, natural gases, and groundwaters are prime sampling media for noble gases. Silicate hosts for noble gases include quenched glasses (submarine and subglacial in origin) as well as various minerals phases (e.g., olivine, pyroxene) erupted as phenocrysts or xenocrysts. In mineral phases, noble gases are usually trapped within fluid or melt inclusions. In the following, we discuss how these various sampling media are exploited in different tectonic environments to provide a picture of the noble gas signature of the mantle. For simplicity, we consider sampling in the submarine and subaerial environments, which roughly maps to sampling of the oceanic and continental segments of the mantle.

In the submarine environment, dredged vitreous glass is a widely exploited medium for obtaining noble gas isotopic compositions and abundances. Most of the mid-ocean ridge basalt (MORB) database has been obtained using pillow-rim glasses erupted on or close to the spreading axes of diverging plate boundaries. Rapid quenching of lavas as they are extruded onto the seafloor traps mantle-derived volatile phases in the glassy rinds. The vesicular and granular interior sections of the basalt pillows usually show atmospheric-like volatile characteristics, indicating contamination with ambient seawater and/or air during sample recovery or subsequent processing (Ballentine and Barfod, 2000). Submarine hydrothermal fluids or vents also provide a means to access mantle-derived noble gases. The usual sampling scenario involves collecting high- and low-enthalpy hydrothermal fluids directly at the vent orifice using leak-tight containers manipulated into place using submersible vehicles. Alternatively, the vent-derived noble gases can be sampled at various depths in the water column by means of hydrocasting: in this case, the mantle noble gas signature is usually heavily diluted by ambient seawater. Noble gas data collected at the same locations using hydrothermal fluids and lavas (e.g., Loihi Seamount; Hilton *et al.*, 1998) generally show good agreement.

A variety of sampling opportunities for mantle noble gases also arise in subaerial regions. Recent volcanism at ocean islands, active margins or other continental regions provides ready access to magmatic (and presumably mantle-derived) volatiles through hydrothermal manifestations (fumaroles, hot springs, bubbling mudpots) and various eruptive products such as ash, tephra, lavas, and xenoliths. In addition, crustal fluids, notably groundwaters and natural gases, can also be exploited as carriers of mantle-derived noble gases to the Earth's surface—this is particularly the case in tectonically active areas. It should be noted that the potential for contamination of the mantle noble gas component is probably greater in subaerial regions: the major contaminants are either atmospheric gases or various daughter products (i.e., radiogenic/nucleogenic noble gases) produced *in situ* in the crust. Fortunately, contamination effects can be recognized through isotopic considerations and (in some cases) corrections applied.

2.06.3.2 Helium Isotopes

MORBs provide a global perspective on the helium isotope systematics of the shallow (uppermost) mantle. Numerous compilations of the mean MORB $^3He/^4He$ value have been produced (see summary in Hilton *et al.* (2000a)) with the general consensus being that MORB-mantle is characterized by a $^3He/^4He$ ratio of $8 \pm 1\ R_A$ (R_A = air $^3He/^4He$). Remarkably, in all compilations produced to date (also see Graham, 2002), there appears no statistical evidence of differences in $^3He/^4He$ between MORBs erupted in the Atlantic, Pacific, and Indian oceans. The observation that mantle $^3He/^4He$ ratios are greater than air has been interpreted as reflecting the fact that the mantle still retains primordial helium (enriched in ^3He) since accretion of the planet, and that it is still leaking to the surface today albeit heavily diluted by time-integrated radiogenic helium grown-in throughout Earth's history (Clarke *et al.*, 1969). Alternative ideas as to the origin of the high $^3He/^4He$ ratios in mantle-derived materials (e.g., due to subduction of cosmogenic dust; Anderson, 1993) have not gained widespread acceptance by the noble gas community (see Porcelli and Ballentine, 2002).

The conclusion that the MORB mantle is characterized by such a narrow range in $^3He/^4He$ (Figure 3) is crucially predicated on the definition of what constitutes MORBs, and various authors have invoked different criteria to either accept or reject samples from the MORB database. The primary and most widely adopted criteria appear to be that samples should be "depleted" in large-ion lithophile elements (LILEs) and have unradiogenic strontium and lead (and radiogenic neodymium) isotope systematics. Although difficult to apply for each and every sample, these criteria are aimed at identifying MORB erupted at shallow levels proximal to obvious plume influences and/or showing geochemical evidence of mantle plume involvement in their petrogenesis The most obvious example where difficulties arise

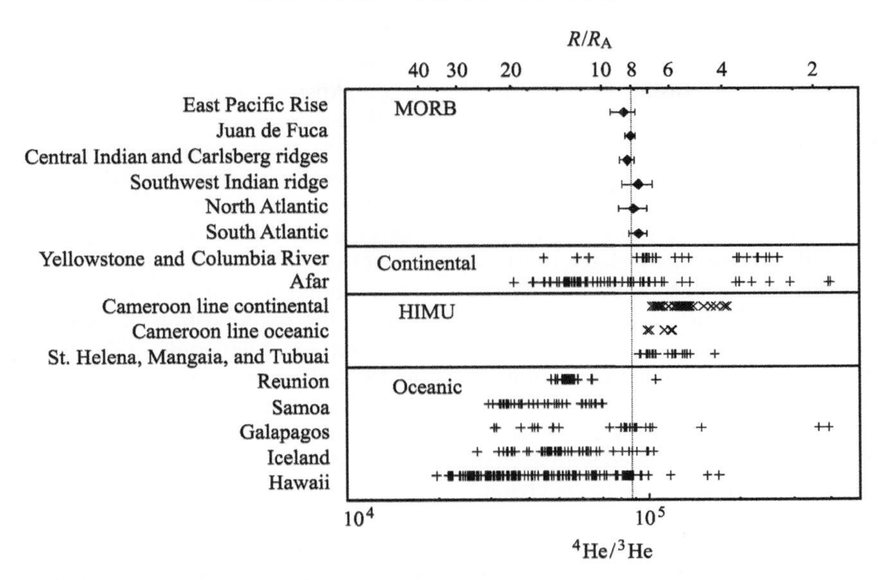

Figure 3 Compilation of helium isotope data from MORBs, continental hotspots, ocean island basalts, and HIMU sources (reproduced by permission of Mineralogical Society of America from *Rev. Mineral. Geochem.*, **2002**, *47*, 419).

is for submarine MORB erupted close to Iceland (the location of the highest global $^3He/^4He$ ratios—see below). Hilton *et al.* (1993a) excluded MORB samples erupted along the Reykjanes and Kolbeinsey ridges and produced an estimate of $8.29 \pm 0.78\,R_A$ for North Atlantic MORB—a value very close to their global mean estimate of $8.18 \pm 0.73\,R_A$ (1σ). In contrast, Graham (2002)) included all oceanic basalts close to Iceland, and derived a mean $^3He/^4He$ ratio (for the whole Atlantic) of $9.58 \pm 2.94\,R_A$—in comparison to a global estimate of $8.75 \pm 2.14\,R_A$. Clearly, in this latter case, the mean $^3He/^4He$ value for Atlantic (and global) MORB helium was influenced by the inclusion of data from the Reykjanes ridge ($11.0-17.6\,R_A$). Although the above approaches differ somewhat in their selection of samples, the two estimates for MORB helium do not diverge markedly indicating that the number of contentious samples is small in comparison to the total number of MORB. The observation that a major part of the mantle is characterized by helium with a $^3He/^4He$ of $8 \pm 1R_A$ is a fundamental constraint on models of mantle structure and evolution (see Section 2.06.4.2). It also provides an important benchmark to contrast with other regions of the mantle that have evolved differently to MORB mantle.

The most obvious suites of samples to search for noble gas differences in the mantle are OIBs given their distinct trace element and isotopic geochemistry compared to MORB (see Hofmann, 1997; see Chapter 2.03). The first indication that mantle $^3He/^4He$ values did not fall within the canonical range of $8 \pm 1R_A$ now ascribed to MORB came with the publication of a $^3He/^4He$ ratio of $15R_A$ for fumarolic gas from Kilauea volcano in Hawaii (Craig and Lupton, 1976). Given notions of a mantle plume origin for the Hawaiian islands (Morgan, 1971), the high $^3He/^4He$ ratio was ascribed to a lower mantle source possessing a higher time-integrated $^3He/(U+Th)$ ratio. Clearly, the source of OIB and MORB mantle must have remained isolated for a significant period of Earth's history to maintain differences in respective $^3He/^4He$ ratios. Other OIB localities with $^3He/^4He \gg 8R_A$ (termed "high-^3He" hotspots) include Iceland, Samoa, Reunion, Heard Island, Easter Island, and Jan Fernandez (see compilation by Farley and Neroda, 1998; Graham, 2002). The highest $^3He/^4He$ value reported to date is $38R_A$ for Tertiary basalts from Iceland (Hilton *et al.*, 1999).

In addition to OIB with $^3He/^4He$ values greater than the MORB range, there is subpopulation of OIB with $^3He/^4He$ values less than MORB. The islands of St. Helena, Mangaia, Tubuaii, and Tristan da Cunha fall into this category, with their $^3He/^4He$ values falling between $5R_A$ and $8R_A$. It is noteworthy that these so-called "low-^3He" hotspot islands tend to have extreme radiogenic isotope characteristics (Zindler and Hart, 1986) prompting the suggestion that OIB $^3He/^4He <$ MORB represents addition of radiogenic helium from ancient crustal material recycled back into the mantle (Graham *et al.*, 1993). In order to explain the relatively high $^3He/^4He$ values of these OIBs, in comparison to significant lower (i.e., pure radiogenic) $^3He/^4He$ ratios anticipated for old recycled crust, models of diffusive exchange of helium between the recycled material and ambient

(MORB) mantle have been invoked (e.g., Hanyu and Kaneoka, 1997). As an alternative explanation, it has been suggested that crustal–magma interaction coupled with prior degassing of the magmatic helium component has resulted in $^3He/^4He$ values <MORB for a number of OIBs (Hilton *et al.*, 1995). A test of this suggestion at La Palma (an active oceanic island in the Canaries archipelago with HIMU-like trace element characteristics) found $^3He/^4He$ values as high as $9.6R_A$—at the transition between the MORB range and that of "high-^{3}He" hotspots (Hilton *et al.*, 2000a). If this observation is generally applicable then notions of a bimodal distribution of OIB $^3He/^4He$ on either side of the MORB range will need revising. However, if OIB genesis is associated with ancient recycled crust then such material must be characterized by production of (some) radiogenic helium which will be superimposed upon helium acquired from ambient mantle: the wide range of $^3He/^4He$ values seen in OIBs, therefore, may reflect variations in the balance of helium from these different sources.

Another prime locality to gain information on the helium isotope systematics of the mantle is convergent margins. Arc magmatism at such localities can sample volatiles derived from the mantle wedge, the subducting slab (both crustal basement and its sedimentary veneer) and the arc lithosphere through which magmas are erupted. In the vast majority of arcs, however, $^3He/^4He$ ratios lie coincident with, or slightly lower than, the range found in MORBs (Poreda and Craig, 1989). The range of measured $^3He/^4He$ values, $(5-8)R_A$, indicates that the mantle wedge is the main contributor to arc-derived helium. Discriminating between the subducting slab and arc lithosphere as the source of the small but discernible contribution of radiogenic helium is difficult and relies on circumstantial evidence. For example, Hilton *et al.* (2002) pointed out that arc segments with both significant volumes of subducting sediment (e.g., Alaska, the Aleutian Islands, Java) or old oceanic basement (e.g., the Marianas, Japan) still record $^3He/^4He$ values within the canonical MORB range. In contrast, arc segments that erupt through thick continental crust (e.g., the Andes, Kamchatka) tend to have $^3He/^4He < 7R_A$; thereby implicating the arc lithosphere as the source of radiogenic helium. As in the case of "low-^{3}He" hotspots, it appears that prior degassing of the magmatic component followed by magma–crust interaction could be the responsible mechanism (Hilton *et al.*, 1993b).

Although most arcs are dominated by MORB-like helium, segments of two arcs (the Banda arc of eastern Indonesia and the Campanian Magmatic Province of southern Italy) are highly unusual in that they emit predominantly radiogenic helium (Hilton and Craig, 1989; Marty *et al.*, 1994). In both cases, the regions are characterized by the subduction of continental crust—the leading edge of the Australian plate in the former case and the northward-moving African plate in the latter. Therefore, the helium budget in these two arcs is dominated by the input of radiogenic helium from the subducting slab with mantle wedge helium taking a subordinate role in the helium inventory. The observation of radiogenic helium in these two arc systems has clear tectonic implications, e.g., for tracing the juxtaposition of two colliding plates; however, more generally, it illustrates the fact that under certain circumstances helium can be subducted into the mantle and may therefore contribute to some of the heterogeneity observed in mantle $^3He/^4He$ ratios.

In the western Pacific Ocean, back-arc basin volcanism is often associated with subduction zone activity: as such, it is also a useful source of helium data on that portion of the mantle that might have been modified by subducted oceanic crust and sediments. To date, $^3He/^4He$ ratios are available for four back-arc regions: the Lau Basin, Mariana Trough, North Fiji Basin and Manus Basin (see summary in Hilton *et al.*, 2002). A wide range of $^3He/^4He$ values characterize basalts erupted in these areas—from highs of $22R_A$ and $15R_A$ in the Lau and Manus basins respectively, through the MORB range (all four basins) to predominantly radiogenic values ($<4R_A$)—again in all four basins. In the case of the high $^3He/^4He$ ratios (>MORB), their origin can safely be ascribed to a mantle plume source. The observation of MORB-like helium is not unexpected given the fact that mature back-arc basin basalts also show MORB-like trace element and radiogenic isotope characteristics (Saunders and Tarney, 1991). Debate continues on the origin of the radiogenic helium: is it slab-derived and a tracer of recycled material into the mantle, or is a shallow-level contaminant introduced into the magma source by assimilation of arc crust immediately prior to eruption? Answering this question has far-reaching ramifications for understanding geochemical recycling in general and the helium isotope evolution of the mantle in particular (see discussion in Hilton *et al.*, 2002).

Mantle noble gases are also observed in continental localities. However, the principal difficulty is that these volatiles must traverse a major geochemical reservoir that is characterized by a $^3He/^4He$ ratio $\leq 0.05R_A$ (Andrews, 1985). In spite of this obstacle, mantle helium is recognized throughout the continents (e.g., Oxburgh *et al.*, 1986; Poreda *et al.*, 1986) based on the fact that there is a large isotopic contrast between radiogenic helium and mantle-derived helium as defined by the $^3He/^4He$ ratio of MORB mantle (Section 2.06.2.1). For example, a 5% addition of MORB-like helium ($8R_A$) to radiogenic helium gives a resultant $^3He/^4He$ value $>0.1R_A$—a value significantly higher than can be produced by crustal lithologies. In this way, relatively small additions of mantle helium to crustal reservoirs can be recognized and quantified. Of course, some continental regions record high (MORB-like) $^3He/^4He$ ratios implying that dilution with radiogenic helium is low or nonexistent. The main regions of the crust where mantle helium is found includes areas undergoing extension (e.g., the Rhine Graben, East African Rift), subduction-type volcanoes (e.g., the Andes), major fault systems (e.g., the San Andreas Fault) as well as kimberlite pipes and other xenolith-bearing localities (Ballentine and Burnard, 2002; Dunai and Porcelli, 2002). It should be noted also that there are two "high-^3He" hotspots ($^3He/^4He \gg$ MORB) found in continental locations—at Yellowstone and along the Ethiopian Rift (see Graham, 2002).

2.06.3.3 Neon Isotopes

Determining the neon isotopic composition of the mantle has proved to be more difficult than for helium since neon is not highly depleted in the atmosphere due to losses to space and is present in low abundances in mantle-derived rocks, resulting in the almost inevitable contamination of all mantle samples with (some) air-derived neon. Furthermore, the analytical challenges are more severe for neon due to the need to correct for isobaric inferences on masses 20 and 22 due to doubly charged interfering species. Therefore, although there are early reports (e.g., Craig and Lupton, 1976) that the neon isotopic composition of the mantle was different from air, it was not until the mid-to-late 1980s that work on diamonds (Honda *et al.*, 1987) and MORB (Sarda *et al.*, 1988) demonstrated conclusively that the mantle had higher $^{20}Ne/^{22}Ne$ and $^{21}Ne/^{22}Ne$ ratios compared to the Earth's atmosphere.

In Figure 4 the neon isotope systematics of MORBs (and OIBs) are illustrated on a traditional three-isotope neon plot. It can be seen that MORB samples define an array between the air value ($^{20}Ne/^{22}Ne = 9.8$; $^{21}Ne/^{22}Ne = 0.029$) and an end-member enriched in both $^{20}Ne/^{22}Ne$ and $^{21}Ne/^{22}Ne$. This array is usually interpreted as a binary mixing trajectory between air and MORB mantle, with individual samples contaminated to varying extents with air neon. The enrichment of MORB in $^{20}Ne/^{22}Ne$ compared to air can be explained by the presence of a solar neon component with $^{20}Ne/^{22}Ne = 13.6$ (Wieler, 2002) or 12.5 (Trieloff *et al.*, 2000) trapped within the mantle (like so-called primordial helium) since accretion (see discussion in Chapter 4.12). Extrapolation of the MORB array to the solar $^{20}Ne/^{22}Ne$ ratio (in effect removing the air contaminant from the mixture) allows an estimate of the MORB mantle $^{21}Ne/^{22}Ne$ ratio of 0.074 (using solar $^{20}Ne/^{22}Ne = 13.6$). This relative enrichment in ^{21}Ne can be attributed to the so-called Wetherill reactions—$^{18}O(\alpha,n)^{21}Ne$ and $^{24}Mg(n,\alpha)^{21}Ne$ (Wetherill, 1954), producing nucleogenic ^{21}Ne in the Earth's mantle over the past 4.55 Ga.

In contrast to MORB samples, OIBs (Hawaii, Iceland, and Reunion Island) plot with steeper trajectories in three-isotope neon space i.e., for a given $^{21}Ne/^{22}Ne$ ratio OIB would have a higher $^{20}Ne/^{22}Ne$ value than MORBs (Figure 4). Following the same extrapolation methodology to

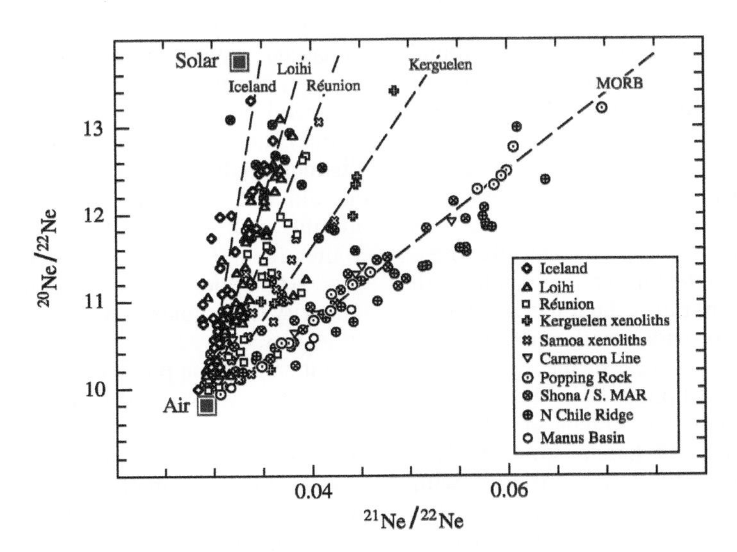

Figure 4 Three-isotope neon plot showing isotopic composition of air and solar neon and the MORB correlation line plus trajectories for various OIBs (reproduced by permission of Mineralogical Society of America from *Rev. Mineral. Geochem.*, **2002**, *47*, 280).

correct for atmospheric contamination as described above for MORBs, then OIB mantle would have a lower $^{21}Ne/^{22}Ne$ ratio when extrapolated to a solar $^{20}Ne/^{22}Ne$ ratio. Therefore, evolution of the OIB $^{21}Ne/^{22}Ne$ ratio has been retarded with respect to that of MORB mantle. This observation is consistent with the notion that the Earth accreted with solar neon, and production of nucleogenic ^{21}Ne has simply moved OIB and MORB mantle to higher $^{21}Ne/^{22}Ne$ ratios with time (Honda *et al.*, 1991). The observation, however, that OIBs have experienced less of a shift in $^{21}Ne/^{22}Ne$ than MORBs implies that the OIB source reservoir has a higher time-integrated $^{21}Ne/(U + Th)$ ratio than the MORB mantle, consistent with the idea that it has suffered a lower degree of primordial (solar) gas loss over time.

In the solid Earth, production of nucleogenic ^{21}Ne is coupled to that of radiogenic ^{4}He. This is because production of ^{21}Ne is directly proportional to the α-particle production ratio from the uranium and thorium series. The $^{21}Ne/^{4}He$ production ratio is constant and has been estimated at a value of 4.5×10^{-8} (Yatsevich and Honda, 1997). In this way, if the Earth accreted with solar helium and neon and initial ratios were modified by production of ^{21}Ne and ^{4}He in a fixed proportion then the present-day $^{3}He/^{4}He$ and $^{21}Ne/^{22}Ne$ ratios in the mantle should be correlated. Honda *et al.* (1993) noted a strong correlation between OIB helium and neon isotopes such that steeper trajectories in three-isotope neon space were characterized by samples with high $^{3}He/^{4}He$ ratios. Indeed, they showed that it was possible to estimate the $^{3}He/^{4}He$ ratio of a suite of OIBs based solely on measurements of the neon isotope composition.

More recent work has found that the correlation between helium and neon breaks down for certain localities. For example, Dixon *et al.* (2000) found solar-like neon isotopic compositions for a suite of Icelandic basalts even through their $^{3}He/^{4}He$ ratios were $<30R_A$. In this case, the gradient of the correlation on the three-isotope neon plot was steeper than anticipated from the $^{3}He/^{4}He$ ratios. Such decoupling of helium and neon was explained by source heterogeneity, and the preservation of domains in the OIB source mantle with high $[Ne_{solar}]/(U + Th)$ ratios (Dixon *et al.*, 2000). Shaw *et al.* (2001) provide another example of He–Ne decoupling: however, in the case of the Manus Basin, the decoupling is in the opposite sense to that of Iceland. The Manus sample suite (with $^{3}He/^{4}He >$ MORB) has a gradient less than that of MORB (Figure 4) consistent with preferential loss of neon relative to helium (the $^{3}He/^{22}Ne$ ratio—an index of degassing—is high for the Manus Basin compared to MORB: 23 versus 10). While it is possible that the fractionation could have occurred not long ($\sim$10 Ma) before generation of the Manus basalts, it is also possible that the fractionation is a remnant of early Earth degassing from a magma ocean (Shaw *et al.*, 2001). The relationship between the isotopes of helium and neon, therefore, bears fundamental information on the nature of the mantle source regions and the degassing history of the planet.

The other main source of information on the neon isotope systematics of the oceanic mantle comes from arc- and back-arc basalts. With the exception of the Manus Basin samples discussed above, the available back-arc basalt database (from the Lau Basin and Mariana Trough) show coherent He–Ne isotope systematics such that the neon isotope data fall on the MORB array of the three-isotope neon plot and the samples have $^{3}He/^{4}He$ ratios within the canonical MORB range. From a neon perspective, therefore, it seems that normal MORB-type mantle underlies these back-arc regions of the western Pacific. The few arc mantle samples analyzed for neon generally convey the same information as back-arc basalts (Figure 5) except for one striking difference: arc phenocrysts from the Aeolian arc and Campanian Magmatic Province have highly nucleogenic neon (i.e., enrichments in ^{21}Ne and, to a lesser extent, ^{22}Ne relative to air). This observation has been ascribed to the influence of subducted crustal material in the region, contributing nucleogenic neon to the source of arc magmas (Tedesco and Nagao, 1996). Curiously, these samples have arc-like $^{3}He/^{4}He$ ratios between $7R_A$ and $8R_A$, giving another and hitherto unexplained example of He–Ne decoupling.

The identification of mantle-like neon isotopic compositions in crustal materials is often difficult due to potentially significant production of nucleogenic neon in crustal lithologies and/or overwhelming contamination by air neon in a subaerial environment. However, deconvolving mantle, crustal, and atmospheric contributions to the neon budget is possible because the isotopic compositions of the three end-members are so distinct (Ballentine and O'Nions, 1992). Using helium and neon data of natural gases from a number of continental regions worldwide, Ballentine (1997) showed that after subtracting the atmospheric neon component, the vast majority of samples plot on a well-defined mixing trajectory between mantle- and crustal-neon end-members. This approach allows an estimate of the $^{3}He/^{22}Ne$ ratio of the subcontinental mantle assuming that mantle volatiles are introduced into tectonically active areas of the continents by emplacement of small degrees of asthenospheric melts. Surprisingly, the range of $^{3}He/^{22}Ne$ ratios observed for the mantle end-member component lies between 0.6 and 1.9, or approximately one-half the value estimated for MORB mantle. The explanation for this apparent discrepancy lies with

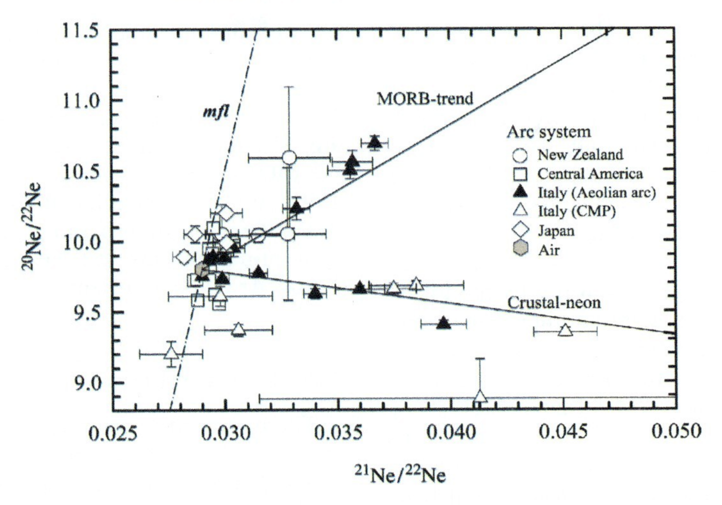

Figure 5 Three-isotope neon plot showing arc-related neon isotope variations (reproduced by permission of Mineralogical Society of America from *Rev. Mineral. Geochem.*, **2002**, *47*, 331).

the realization that solubility-controlled degassing of basaltic melt preferentially releases neon to helium (Section 2.06.2.2). Consequently, although volatiles from the subcontinental mantle are trapped (and can be sampled) in regions of continental extension, these volatiles are characterized by elemental abundances modified from source values.

2.06.3.4 Argon Isotopes

Argon is composed of three isotopes: ^{36}Ar, ^{38}Ar, and ^{40}Ar. The former two isotopes are primordial in origin with no significant contributions by production within the Earth: ^{40}Ar, however, is produced by decay of ^{40}K ($t_{1/2} = 1.40 \times 10^9$ yr). The initial (or primordial) $^{40}Ar/^{36}Ar$ ratio of the solar system lies between 10^{-4} and 10^{-3} (Begemann *et al.*, 1976). In contrast, the terrestrial atmosphere has a $^{40}Ar/^{36}Ar$ ratio of 296, and all other terrestrial materials have higher values. Terrestrial $^{40}Ar/^{36}Ar$ ratios therefore are essentially the result of mixing of radiogenic ^{40}Ar with primordial ^{36}Ar.

A large range in $^{40}Ar/^{36}Ar$ ratios has been measured in MORBs. The variation is generally interpreted as due to mixing of variable proportions of atmospheric argon (with $^{40}Ar/^{36}Ar = 296$) with a single, much more radiogenic, mantle composition. The minimum value for this mantle composition is represented by the highest measured values of 2.8×10^4 (Staudacher *et al.*, 1989) to 4×10^4 (Burnard *et al.*, 1997). From correlations between $^{20}Ne/^{22}Ne$ and $^{40}Ar/^{36}Ar$ in step-heating results of a gas-rich MORB sample, a maximum value of $^{40}Ar/^{36}Ar = 4.4 \times 10^4$ has been obtained (Moreira and Allègre, 1998). This (isotopic) composition is an important constraint on the degassing of the mantle (see Section 2.06.4.2).

It is difficult to determine if any argon isotopic variations exist in the upper mantle due to the pervasive contamination of mantle-derived samples by atmospheric argon. This is an important topic, as a range in mantle $^{40}Ar/^{36}Ar$ values can potentially be used to trace the impact of either variable degassing/contamination or other processes such as subduction of argon. It has been suggested that there are indeed heterogeneities in the $^{40}Ar/^{36}Ar$ ratio of the MORB source, based on a correlation between lead isotopes and $^{40}Ar/^{36}Ar$ ratios for samples with $^3He/^4He \leq 9.5R_A$ (Sarda *et al.*, 1999). This was ascribed to subduction of a component with relatively radiogenic lead and low $^{40}Ar/^{36}Ar$, implying a substantial flux of argon from the surface to the upper mantle. However, Burnard (1999) pointed out that lead isotope compositions are known to be more radiogenic in shallow eruptive environments, and that the lower $^{40}Ar/^{36}Ar$ values correlate equally well with depth of eruption. Therefore, the shallower samples may have lower $^{40}Ar/^{36}Ar$ ratios due to greater gas depletion and so proportionally more air contamination. Also, Ballentine and Barfod (2000) showed that the amount of air contamination is related to basalt vesicularity, which in turn is related to eruption depth and volatile content, further explaining correlations of atmosphere-derived noble gases with radiogenic lead (Sarda *et al.*, 1999). Overall, the debate has highlighted the difficulty of separating true mantle $^{40}Ar/^{36}Ar$ variations from contamination effects which occur either within the magma chamber (Burnard *et al.*, 1994; Farley and Craig, 1994), through assimilation of crustal material during melt transit to the surface (Hilton *et al.*, 1993a), by equilibration with seawater during eruption (Patterson *et al.*, 1990), or via sample vesicularity (Ballentine and Barfod, 2000).

Considerable effort has also been expended determining whether OIBs—with distinctive $^3He/^4He$ (>$10R_A$), have $^{40}Ar/^{36}Ar$ ratios that can be distinguished from those of the MORB source. It is generally expected that high $^3He/^4He$ ratios, reflecting high $^3He/(U + Th)$ ratios in the source region(s), would be accompanied by low $^{40}Ar/^{36}Ar$ ratios, reflecting similarly high $^{36}Ar/K$ ratios. Surprisingly, the first $^{40}Ar/^{36}Ar$ values reported for Loihi Seamount glasses were found to be similar to the values for air, prompting suggestions of a complementary relationship between argon in the OIB source and the atmosphere (Allègre *et al.*, 1983; Kaneoka *et al.*, 1986; Staudacher *et al.*, 1986). At that time, ideas of air contamination (e.g., Fisher, 1985; Patterson *et al.*, 1990) were dismissed. However, a study of basalts from Juan Fernandez (Farley and Craig, 1994) found that the atmospheric argon component resided within phenocrysts—an observation consistent with introduction of air argon into the magma chamber. This study provided a ready explanation for the prevalence of air contamination of OIBs. More recent measurements of Loihi samples have reported significantly higher $^{40}Ar/^{36}Ar$ values: between 2,600 and 2,800 (Hiyagon *et al.*, 1992; Valbracht *et al.*, 1997), and up to 8,000 (Trieloff *et al.*, 2000)—all on samples with $^3He/^4He$ as high as $24R_A$ (and so midway between MORB and the highest OIB). Additionally, Poreda and Farley (1992) found values of $^{40}Ar/^{36}Ar \leq 1.2 \times 10^4$ in Samoan xenoliths that have intermediate $^3He/^4He$ ratios ($9-20R_A$). Other attempts at characterizing OIB $^{40}Ar/^{36}Ar$ have used associated neon isotope variations and the (debatable) assumption that the contaminant Ne/Ar ratio is constant in a strategy designed to remove the effects of air contamination. Using this approach, Sarda *et al.* (2000) reported that $^{40}Ar/^{36}Ar$ ratios in the high $^3He/^4He$ OIB source are >3,000 but still considerably lower than that of the MORB source (see also Matsuda and Marty, 1995). It must be concluded, therefore, that due to the difficulties of separating atmospheric argon contamination, present-day estimates of OIB $^{40}Ar/^{36}Ar$ ratios represent a lower limit only.

2.06.3.5 Xenon Isotopes

There are nine isotopes of xenon. Isotopic variations have been generated in the Earth due to the β decay of ^{129}I ($t_{1/2} = 16$ Ma) producing ^{129}Xe, and the spontaneous fission of both ^{244}Pu ($t_{1/2} = 80$ Ma) and ^{238}U ($t_{1/2} = 4.47$ Ga). Since ^{129}I and ^{244}Pu are short-lived nuclides, they were only present early in Earth history. MORB $^{129}Xe/^{130}Xe$ and $^{136}Xe/^{130}Xe$ ratios lie on a correlation extending from atmospheric ratios to higher values (Kunz *et al.*, 1998; Staudacher and Allègre, 1982), and likely reflect mixing of variable proportions of air contaminant xenon with a single upper mantle component having more radiogenic $^{129}Xe/^{130}Xe$ and $^{136}Xe/^{130}Xe$ ratios (Figure 6). The highest measured values thus provide lower limits for the MORB source. As with argon isotopes, due to the pervasiveness of atmospheric xenon contamination, it has not been possible to resolve isotopic variations in the upper mantle that can serve as useful tracers.

While the origin of the high $^{129}Xe/^{130}Xe$ ratios is clearly due to decay of ^{129}I, an important issue regarding mantle xenon is the origin of

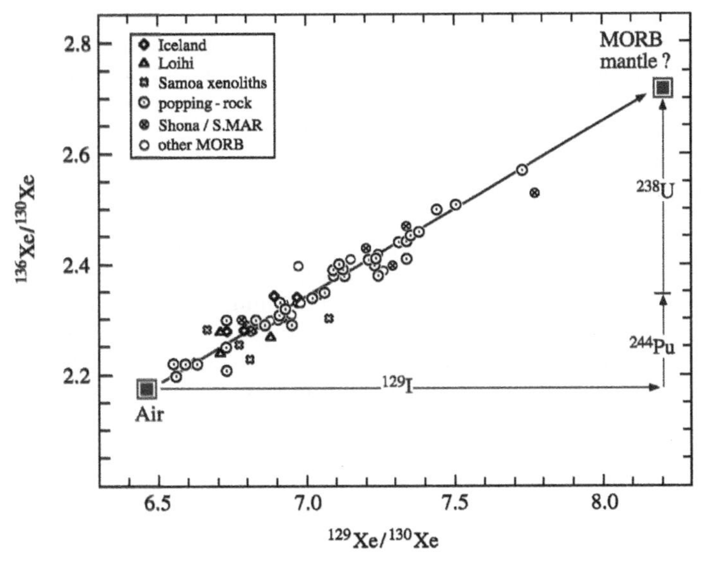

Figure 6 Xenon isotope plot for MORBs and OIBs. A proportion of 32% for fissiogenic ^{136}Xe (from ^{224}Pu decay) is shown (reproduced by permission of Mineralogical Society of America from *Rev. Mineral. Geochem.*, **2002**, *47*, 291).

radiogenic ^{136}Xe. Contributions to ^{136}Xe enrichments in MORBs by decay of ^{238}U or ^{244}Pu in theory can be distinguished based on the spectrum of contributions to other xenon isotopes, although analyses have typically not been sufficiently precise to identify the parent nuclide. Precise measurements of xenon in CO_2 well gases found ^{129}Xe and ^{136}Xe enrichments similar to those found in MORBs, and so it is likely that this xenon is derived from the MORB source (Staudacher, 1987). These indicate that ^{244}Pu has contributed <10–20% of the ^{136}Xe that is in excess of the atmospheric composition (Caffee *et al.*, 1999; Phinney *et al.*, 1978). An error-weighted best fit to recent precise MORB data (Kunz *et al.*, 1998) yielded a value of 32 ± 10% for the fraction of ^{136}Xe excesses relative to atmosphere that are ^{244}Pu-derived (Figure 6), although there is considerable scatter in the data (see Marti and Mathew, 1998). Clearly, further work is warranted on the proportion of plutonium-derived heavy xenon in the mantle, although it appears that the fissiogenic xenon is dominantly derived from uranium.

Measurements of xenon in high $^{3}He/^{4}He$ OIB samples have often found atmospheric isotopic ratios (e.g., Allègre *et al.*, 1983) that appear to be due to overwhelming air xenon contamination (Harrison *et al.*, 1999; Patterson *et al.*, 1990) rather than reflecting mantle xenon with an air composition. Although Samoan samples with intermediate (9–20R_A) helium isotope ratios have been found with xenon isotopic ratios distinct from those of the atmosphere (Poreda and Farley, 1992), the xenon in these samples may have been derived largely from the MORB source. Harrison *et al.* (1999) have found slight ^{129}Xe excesses in Icelandic samples with $^{129}Xe/^{3}He$ ratios that are compatible with the ratio in gas-rich MORB, but due to the uncertainties in the data it cannot be determined whether there are indeed differences between the MORB and OIB sources. Trieloff *et al.* (2000) reported xenon isotope compositions in Loihi dunites and Icelandic glasses that were on the MORB correlation line and had values up to $^{129}Xe/^{130}Xe = 6.9$. These were accompanied by $^{3}He/^{4}He$ ratios up to 24R_A, and so may contain noble gases from both MORB (~8R_A) and the highest $^{3}He/^{4}He$ ratio (37R_A) OIB source. From these data it appears that the OIB source may have xenon that is different from the terrestrial atmosphere but (marked) differences between OIB and MORB still await discovery.

2.06.3.6 Mantle Noble Gas Abundances

In order to use noble gases as tracers for the involvement of different mantle components in a particular environment, the absolute concentrations are required to determine how each noble gas signature will have an impact. The concentrations of helium are most easily inferred, and those of the other noble gases can be determined from the elemental abundance patterns.

2.06.3.6.1 Helium concentrations in the upper mantle

An estimate of the ^{3}He concentration of the upper mantle can be obtained by noting that there is a clearly recognizable flux of ^{3}He into the oceans from degassing of MORB. The mantle ^{3}He flux from mid-ocean ridges (Clarke *et al.*, 1969; Craig *et al.*, 1975) has been obtained by combining seawater ^{3}He concentrations in excess of dissolved air helium with seawater mixing and ocean-atmosphere exchange models (see review by Schlosser and Winckler, 2002). A value of 1,060 ± 250 mol yr^{-1} ^{3}He (Lupton and Craig, 1975) has been confirmed with more recent ocean circulation models (Farley *et al.*, 1995). This value represents an average over the last 1,000 years. The concentration of ^{3}He in the mantle can be determined by dividing the flux of ^{3}He into the oceans by the rate of production of melt that is responsible for carrying this ^{3}He from the mantle, which is equivalent to the rate of ocean crust production of 20 km^3 yr^{-1} (Parsons, 1981). MORBs that degas quantitatively to produce a ^{3}He flux of 1,060 mol yr^{-1} then must have an average ^{3}He content of 1.96×10^{-14} mol g^{-1} or 4.4×10^{-10} cm^3 STP g^{-1}. Note that the validity of combining a flux estimate based upon a millennium-scale record of ocean ventilation with a magma production rate derived using paleomagnetic reversals occurring on considerably longer timescales has been questioned recently (Ballentine *et al.*, 2002). It is significant, however, that the derived ^{3}He concentration lies within a factor of 2 of that obtained for the most gas rich basalt glass yet found (the so-called "popping-rock"), with $\sim 10.0 \times 10^{-10}$ cm^3 STP ^{3}He g^{-1} (Moreira *et al.*, 1998; Sarda *et al.*, 1988). This sample is noteworthy in having sample vesicle size distributions consistent with closed system formation during magma ascent (Sarda and Graham, 1990) and individual vesicles with He/Ar, C/He, and C/N ratio variations consistent with closed system formation during ascent (Javoy and Pineau, 1991). An independent estimate of the carbon content of undegassed MORB of 900–1,800 ppm C (Holloway, 1998), combined with a MORB-source mantle $CO_2/^{3}He$ ratio of 2×10^9 (Marty and Jambon, 1987) gives a ^{3}He concentration in undegassed MORB of $(8.4–17.0) \times 10^{-10}$ cm^3 STP ^{3}He g^{-1}. Assuming 10% partial melting and quantitative extraction of helium from the solid phase into the melt

(Section 2.06.2.2), these undegassed melt values give a MORB-source mantle concentration of $(0.44-1.7) \times 10^{-10}$ cm^3 STP ^{3}He g^{-1} or $1.2-4.6 \times 10^9$ atoms ^{3}He g^{-1} (Porcelli and Ballentine, 2002).

2.06.3.6.2 Helium concentrations in the OIB source

It is generally assumed that there is a single mantle reservoir that supplies the helium of ocean islands with high ^{3}He/^{4}He ratios. There are two approaches to estimating the concentration of ^{3}He in this mantle source: (i) through using model calculations, and (ii) via direct measurements of OIB volcanics. Helium with high ^{3}He/^{4}He ratios is often assumed to be stored in a mantle reservoir that has evolved approximately as a closed system for noble gases and has bulk silicate Earth (BSE) parent nuclide concentrations. Since this reservoir must be isolated from degassing at mid-ocean ridges and subduction zones, it is often placed within the deeper, or lower mantle. Assigning the highest OIB ^{3}He/^{4}He ratios to this reservoir, a comparison between the total production of ^{4}He and the shift in ^{3}He/^{4}He from the initial terrestrial value to the present value provides an estimate of the ^{3}He concentration in this reservoir. Using BSE values of 21 ppb U and Th/U = 3.8 (Doe and Zartman, 1979; Rocholl and Jochum, 1993), 1.02×10^{15} atoms ^{4}He g^{-1} are produced over 4.55 Ga. Using the highest Iceland value of ^{3}He/^{4}He = 37R_A to represent this reservoir and an initial value of ^{3}He/^{4}He = 120R_A (Section 2.06.2.1) then the reservoir has 7.6×10^{10} atoms ^{3}He g^{-1}. This is over an order of magnitude greater than the value derived above for the MORB source mantle. Note that other explanations have been put forth regarding the nature of the reservoir supplying high ^{3}He/^{4}He hotspots (see discussion in Section 2.06.4.2), although a gas-rich source has been the most commonly assumed.

The flux of ^{3}He from intra-plate volcanic systems is dominantly subaerial and so it is not possible to directly obtain the time-integrated value for even a short geological period. It is noteworthy that OIB samples typically have substantially lower helium concentrations than MORB samples. However, extensive degassing during eruption can account for the strong depletion of helium in many samples. Therefore, it is often assumed that OIB source ^{3}He concentrations are high, but have been highly degassed prior to or during eruption (see Marty and Tolstikhin, 1998). There may be a number of factors promoting more extensive degassing for OIB than MORB. OIB source may be more volatile-rich than MORB and therefore may degas more effectively (Dixon and Stolper, 1995). Also, the water content of basalts reduces CO_2 solubility: therefore, water-rich basalts (with high ^{3}He/^{4}He ratios e.g., along the shallow Reykjanes Ridge) can have significantly lower CO_2 (and helium) compared to water-poor MORB erupted at greater confining pressures/depths (Hilton *et al.*, 2000b).

Concentrations of ^{3}He in OIB sources can be estimated in several ways. The ^{3}He flux can be estimated from volcanic CO_2 fluxes and magma generation rates. The summit CO_2 flux of Kilauea volcano, Hawaii, can be combined with observed CO_2/^{3}He values and estimated magma production rates to obtain an undegassed plume magma concentration of 4.7×10^{-10} cm^3 STP ^{3}He g^{-1} (Hilton *et al.*, 1997). Assuming magma production by 7% melting, this corresponds to a source concentration of 3.3×10^{-11} cm^3 STP ^{3}He g^{-1}. Prior plume degassing (at Loihi Seamount?) was invoked to explain such a low value (Hilton *et al.*, 1997). Moreira and Sarda (2000) assessed the extent of prior basalt degassing from the magnitude of air-corrected helium, neon, and argon fractionations in filtered MORB and OIB data sets. Using an open system Rayleigh degassing model, they calculated a minimum OIB source concentration of 1.1×10^{-10} cm^3 STP ^{3}He g^{-1}, which is comparable to that of the MORB source. Furthermore, they suggested that measured OIB helium concentrations can also be subject to variable post-eruption gas loss that occurs with no elemental fractionation (e.g., by vesicle rupture during sample handling), and so the source may have had even higher concentrations.

OIB source helium concentrations can also be derived from isotopic mixing patterns at various locations. For example, where plumes have strong interactions with mid-ocean ridges, such as along the Reykjanes Ridge, south of Iceland, basalt characteristics can be ascribed to mixing between a MORB source and a plume component. Hilton *et al.* (2000b) found that ^{3}He/^{4}He and ^{206}Pb/^{204}Pb are linearly correlated along the Reykjanes Ridge and may reflect mixing between components with similar He/Pb ratios. Eiler *et al.* (1998) found a similar relationship in Hawaiian lavas. In the limiting case where the lead concentration of the plume is the same as that of MORB-source mantle, the ^{3}He concentration in the plume source is four times higher than that of the MORB-source mantle, taking into account differences in end-member ^{3}He/^{4}He ratios. Higher plume lead concentrations would increase this value. For example, adopting Pb = 0.05 ppm for the MORB-source mantle and 0.185 ppm for the plume source (the BSE value), Hilton *et al.* (2000b) calculated that the ^{3}He concentration of the plume source component is 15 times that of MORB-source mantle.

Further, they suggested that the data might support a mixing line involving an even higher plume source helium concentration. Predegassing of the plume source prior to mixing with MORB-source mantle also remains a possibility, thus allowing a higher helium concentration in the plume component.

It should be emphasized that, in many locations, the high $^{3}He/^{4}He$ source component comprises a small fraction of the sample source. A very gas-rich helium source need only contribute a small mass fraction of the source region to dominate the helium budget. The presence of the OIB source in such a case would be much less discernible in other chemical signatures, and so helium isotopes would provide the most sensitive indicator of mixing. Alternatively, if the OIB source has concentrations that are much lower than that of the MORB source, then clearly it must constitute a much greater proportion of ocean island source regions, and this should be evident in other chemical signatures as well.

2.06.3.6.3 *Noble gas abundance patterns*

Noble gas abundance patterns measured in MORBs and OIBs scatter greatly. This is due to post-eruptive alteration processes as well as fractionation during noble gas partitioning between basaltic melts and a vapor phase. Moreover, the vapor phase may be preferentially gained or lost by a particular sample. However, MORB Ne/Ar and Xe/Ar ratios that are greater than the air values are common. This pattern was found in a gas-rich MORB (popping-rock) sample with high $^{40}Ar/^{36}Ar$ and $^{129}Xe/^{130}Xe$ ratios (and so containing relatively little air contamination) and $^{4}He/^{40}Ar$ ratios that are near the expected production ratio of the upper mantle (and so not fractionated) (Moreira *et al.*, 1998; Staudacher *et al.*, 1989).

A noble gas abundance pattern for the mantle that is not dependent upon unraveling fractionation effects due to transport and eruption can be calculated by assuming that the radiogenic nuclides of the different elements are present in the mantle in relative abundances that are equal to their production ratios: in this case, the relative proportions of radiogenic ^{4}He, ^{21}Ne, ^{40}Ar, and ^{136}Xe are known. This process assumes that the changes in the $^{3}He/^{4}He$, $^{21}Ne/^{22}Ne$, $^{40}Ar/^{36}Ar$, and $^{136}Xe/^{130}Xe$ ratios from their initial values can be related to the relative abundances of ^{3}He, ^{22}Ne, ^{36}Ar, and ^{130}Xe (see Porcelli and Ballentine, 2002). Using the MORB isotopic compositions discussed above, and $K/U = 1.27 \times 10^{4}$ (Jochum *et al.*, 1983) and $Th/U = 3.8$ for parent element relative ratios, then $^{3}He/^{22}Ne_{MORB} = 11$, $^{3}He/^{36}Ar_{MORB} = 1.7$ (and so $^{22}Ne/^{36}Ar_{MORB} = 0.15$), and $^{130}Xe/^{36}Ar_{MORB} = 3.3 \times 10^{-4}$. This is consistent with the pattern discussed above, with $^{22}Ne/^{36}Ar$ and $^{130}Xe/^{36}Ar$ ratios greater than those of the atmosphere (0.05×10^{-4} and 1.1×10^{-4}, respectively). Note that these arguments change qualitatively only if some noble gases are removed from the mantle with much shorter time constants, thereby maintaining ratios between radiogenic isotopes that are different from the production ratios.

Unfortunately, it is not possible to similarly calculate the abundance pattern of OIB source regions, since the heavy element isotope ratios are not constrained. A survey of helium and neon isotope compositions of OIB and MORB has allowed calculations of the $^{3}He/^{22}Ne$ ratios for the source region of each type, with $^{3}He/^{22}Ne = 6.0 \pm 1.4$ for Hawaii and 10.2 ± 1.6 for MORB, with a mantle average of 7.7 (Honda and McDougall, 1998). This approach suggests that there may be a systematic difference between the MORB and OIB sources. In contrast, it has been argued that Loihi and gas-rich MORB samples have similar $^{3}He/^{22}Ne$ ratios (Moreira and Allègre, 1998), and so the issue remains open.

2.06.4 NOBLE GASES AS MANTLE TRACERS

In this section, we focus on three key areas of noble gas studies of the mantle. First, we consider how noble gases (^{3}He in particular) are used to assess volatile fluxes involving the mantle as a principal terrestrial reservoir. It is possible to both understand the cycling history of the noble gases between the mantle and the atmosphere and hydrosphere using this approach and to constrain the provenance of major volatiles (such as CO_2 and N_2) thereby leading to a better understanding of their evolution in the mantle. Second, we illustrate how noble gases can be used to place constraints on the different classes of model that have been proposed to account for the structure of the mantle. A related topic also covered in this section is the utility of noble gases to trace the origin of plumes and the formation of ocean islands. Finally, we assess the noble gas systematics of mantle-derived xenoliths as a means to gain insight into the development and evolution of the subcontinental mantle, and to trace the transport of volatiles from the convecting mantle into the lithosphere.

2.06.4.1 Volatile Fluxes and Geochemical Recycling

Noble gases have been at the forefront of studies defining volatile fluxes between the mantle and other terrestrial reservoirs. This stems from the fact that in the case of ^{3}He there is no question or ambiguity regarding its origin: the mantle ^{3}He

budget is overwhelmingly dominated by the primordial component (Craig and Lupton, 1981). Furthermore, it is implicitly assumed that the flux of ^{3}He is unidirectional from the mantle to the exosphere (hydrosphere, atmosphere, and crust) based upon the hypothesis that subduction of ^{3}He into the mantle occurs—if at all—at a level negligible relative to the mantle inventory (Hilton *et al.*, 1992; Hiyagon, 1994; Trull, 1994). Consequently, mantle ^{3}He fluxes can be estimated with a fair degree of certainty, and exploited as a flux monitor for other volatiles of geochemical interest (e.g., CO_2 and N_2). In this way, ^{3}He is integral to assessing the state of volatile mass balance involving the mantle and other terrestrial reservoirs.

2.06.4.1.1 ^{3}He fluxes from the mantle

As discussed in Section 2.06.3.6.1 a value of 1,060 ± 250 mol yr^{-1} is taken as the canonical value for the upper mantle ^{3}He degassing flux (Lupton and Craig, 1975). This estimate is based upon knowledge of the ^{3}He concentration anomaly in the oceans (relative to saturation) coupled with ocean–atmosphere exchange models (see review by Schlosser and Winckler, 2002). Estimating the flux of ^{3}He associated with OIB volcanism is more problematic, however. The parameters necessary to make an estimate of the OIB ^{3}He flux are (i) rates of OIB magmatism, and (ii) the helium concentration of the source region(s).

Estimates of the rate of intraplate magma production vary considerably: 0.6×10^{15} g yr^{-1} (Reymer and Schubert, 1984), 4.4×10^{15} g yr^{-1} (Schilling *et al.*, 1978), $(5-7) \times 10^{15}$ g yr^{-1} (including seamounts) (Batiza, 1982), and $(0.4-5) \times 10^{15}$ g yr^{-1} when considering magma emplacement estimates for the last 180 Ma (Crisp, 1984). These values fall between 1% and 12% of the MORB production rate.

Information on the ^{3}He content of OIB sources is available from both modeling and concentration measurements of helium in intraplate volcanics (Section 2.06.3.6.2). For example, Hilton *et al.* (2000b) estimated that the high ^{3}He/^{4}He component erupted along the Reykjanes Ridge had a ^{3}He content 15 times that of the MORB source. Unfortunately, it is not clear what fraction of such source material supplying OIBs worldwide has this concentration (Porcelli and Ballentine, 2002). However, using the highest estimate of source ^{3}He content over a range of assumed OIB magma production rates, Porcelli and Ballentine (2002) obtained a global flux of $8.4 \times 10^4 - 1.5 \times 10^6$ cm^3 STP ^{3}He yr^{-1} (38–670 mol ^{3}He yr^{-1}) for the OIB ^{3}He flux from the mantle. It should be noted, however, that helium characteristics of various OIB sources are likely to vary considerably. The range of measured ^{3}He/^{4}He ratios indicates that several components may be involved in differing proportions, and these are likely to have different helium concentrations. Therefore, detailed studies of individual hotspots may not necessarily be combined to yield a representative estimate.

Subduction zones also represent a primary escape route for primordial ^{3}He from the mantle. There are two general approaches for estimating the arc ^{3}He flux. First, assuming that the ^{3}He content of magma in the mantle wedge is the same as that beneath spreading ridges, Torgersen (1989) combined the mid-ocean ridge degassing flux (~1,000 mol ^{3}He yr^{-1}; Craig *et al.*, 1975) with a figure of 20% as the assumed magma production rate of arcs relative to MOR (Crisp, 1984) to yield an arc ^{3}He flux of ~200 ± 40 mol yr^{-1}. The second method relies on actual ^{3}He (and some other normalizing species) measurements of arc-related and other subaerial volcanoes. For example, Allard (1992) derived an estimate of 240–310 mol yr^{-1} for the total flux of ^{3}He into the atmosphere by subaerial volcanism (based upon integrating the CO_2 flux from 23 individual volcanoes worldwide and coupling this flux with measurements of the CO_2/^{3}He ratios). Of the total subaerial ^{3}He flux, he suggested that approximately 70 mol yr^{-1} was arc-related. Adopting a similar approach, Marty and LeCloarec (1992) used ^{210}Po as the flux indicator (along with the ^{210}Po/^{3}He ratio) to estimate a total subaerial volcanic ^{3}He flux of 150 mol yr^{-1}—of which over half (>75 mol yr^{-1}) was due to arc volcanism. Hilton *et al.* (2002) derived an estimate of 92 mol yr^{-1} based on integrating SO_2 fluxes (combined with SO_2/^{3}He ratios) from individual arc segments worldwide.

It is important to point out that neither approach attempts a direct measurement of the arc ^{3}He flux. To estimate magma production rates, scaling is used in the first instance, whereas the second methodology relies on knowledge of an absolute flux of some chemical species from volcanoes together with a measurement of the ratio of that species to ^{3}He. The most widely used species to derive absolute chemical fluxes from subaerial volcanoes is SO_2 using the correlation spectrometer technique (COSPEC) (Stoiber *et al.*, 1983). Carbon dioxide (see Brantley and Koepenick, 1995) and ^{210}Po (Marty and LeCloarec, 1992) have also been used in an analogous manner.

Continental settings provide a small but significant flux of ^{3}He from the mantle. In the Pannonian Basin (4,000 km^2 in Hungary), the flux of ^{3}He has been estimated at 8×10^4 atoms m^{-2} s^{-1} (Martel *et al.*, 1989) and $0.8-5 \times 10^4$ atoms m^{-2} s^{-1} (Stute *et al.*, 1992). Making the assumption that the area of the continents is 2×10^{14} m^2 and that 10% is under extension, this yields a total ^{3}He flux of 8.4–84 mol ^{3}He yr^{-1}. This value compares with <3 mol ^{3}He yr^{-1} from the stable continental crust (see O'Nions and Oxburgh, 1988).

In summary, based upon the range of values reported for the different regions (MOR, OIB, arcs, and continents), the total rate of contemporary degassing of mantle-derived ^{3}He lies in the range of 1,039–2,270 mol yr^{-1}. The oceanic ridge system dominates the flux with a contribution between 57% and 78 % of the total.

2.06.4.1.2 *Major volatile fluxes*

The establishment of the ^{3}He flux from the mantle allows scaling to other volatiles of geochemical interest so that their mantle fluxes can also be quantified. The seminal work in this respect is that of Marty and Jambon (1987), who combined measured $CO_2/^3He$ ratios in MORB ($\sim 2 \times 10^9$) with the MOR ^{3}He flux estimate of 1,000 mol yr^{-1} (Craig *et al.*, 1975) to yield a value of 2×10^{12} mol yr^{-1} for the flux of CO_2 from the upper mantle. Although there is evidence of fractionation between helium and CO_2 during magmatic degassing (e.g., Hilton *et al.*, 1998), it has been argued that similar solubilities of helium and CO_2 in, and restricted compositional range typical of, most tholeiites erupted along oceanic ridges results in little or negligible modification of the source $CO_2/^3He$ value: in any case, a correction is possible using measured He/Ar ratios (Marty and Zimmermann, 1999). In addition to allowing assessment of volatile mass balance (see below), quantifying the flux of magmatic carbon at ridges provides an independent check on the carbon content of the upper mantle (Marty and Tolstikhin, 1998).

In the case of arc volcanism, the critical observation is that the $CO_2/^3He$ ratio of arc-related volatiles is significantly greater than that of MORBs (Marty *et al.*, 1989; Sano and Marty, 1995; Sano and Williams, 1996; Varekamp *et al.*, 1992). In a compilation of subduction zone type gases, Marty and Tolstikhin (1998) report a median value of $11.0 \pm 3.3 \times 10^9$, or $\sim$5 times that of MORBs. Consequently, an estimate of the proportion of carbon from nonmantle sources ($\sim$80%), presumably the subducted slab, can be inferred by scaling to the (upper mantle) $CO_2/^3He$ value.

Marty and Tolstikhin (1998) detail how knowledge of the arc $CO_2/^3He$ ratio can be used to calculate the arc-related CO_2 flux. They used the following relationship:

$$\varnothing_{C,arc} = \varnothing_{arc} \times [^3He]_{um} \times (CO_2/^3He)_{arc} \times 1/r_{arc}$$

where $\varnothing_{arc}$ is the flux of arc magma and r_{arc} is the mean partial melting rate at arcs. Adopting values of 2.5–8.0 km^3 yr^{-1} for the rate of magma emplacement at accreting plate margins (Crisp, 1984) and 10–25% for the range in the extent of partial melting (Plank and Langmuir, 1988), then a CO_2 arc flux of $\sim 2.5 \times 10^{12}$ mol yr^{-1} can be calculated (Marty and Tolstikhin, 1998). This estimate compares well with other values produced using different approaches: for example, Sano and Williams (1996) combined the arc $CO_2/^3He$ with the arc ^{3}He flux to estimate an arc CO_2 flux of 3.1×10^{12} mol yr^{-1} whereas Hilton *et al.* (2002) integrated CO_2 fluxes from arc segments worldwide to produce a global arc estimate of 1.6×10^{12} mol yr^{-1}.

The flux of CO_2 from plumes can be computed in an analogous manner to that described above for arcs. Marty and Tolstikhin (1998) estimated that the plume CO_2 flux at 3×10^{12} mol yr^{-1} by assuming plume $CO_2/^3He = 3 \times 10^9$, $r = 0.05$, $\varnothing_{plume} = 5 \times 10^{15}$ g yr^{-1} and $[^3He]_{plume} = 10^{-14}$ mol g^{-1}. The flux is comparable to that at mid-ocean ridges in spite of the significantly lower magma production rates at hotspots compared to ridges. It should be appreciated, however, that plume flux estimates derived using this approach are only as good as the input parameters, and there is significant uncertainty in the ^{3}He content of plume source material (Section 2.06.3.6.2).

The above examples illustrate how knowledge of ^{3}He fluxes from various mantle reservoirs can be used to estimate concomitant fluxes of CO_2 from the same reservoirs. The methodology of using ^{3}He in this manner can also be extended to other volatiles of geochemical interest—nitrogen being a prime example. Marty (1995) estimated fluxes of $1.2–3.2 \times 10^9$ mol yr^{-1} for the N_2 flux from mid-ocean ridges based on measured $N_2/^{40}Ar$ ratios, assumed $^4He/^{40}Ar$ mantle production ratios and knowledge of the mid-ocean ridge ^{3}He (and ^{4}He) flux. Similarly, Sano *et al.* (2001) produced estimates of the total mantle nitrogen flux from mid-ocean ridges, plumes, and arcs (2.2×10^9, 0.004×10^9, and 0.6×10^9, in units of mol yr^{-1}, respectively) by using measured $N_2/^3He$ ratios and normalizing to ^{3}He flux estimates at each locality.

2.06.4.1.3 *The question of volatile provenance*

Volatile flux estimates derived by the approach described in the previous section make no distinction as to the source or provenance of the volatiles. However, for arc-related regions especially, quantitatively constraining the origin of the volatiles is important as they may be derived from different sources, vis-à-vis the mantle, subducting slab and arc lithosphere. In order to assess the state of volatile mass balance of the various terrestrial reservoirs, particularly involving outputs from the mantle and inputs associated with the subducting slab, the total arc output flux must be resolved into its component structures. As we show in this section, helium has proven remarkably sensitive in discerning volatile provenance. We use CO_2 and N_2 to illustrate the case.

As discussed above, the $CO_2/^3He$ ratio is significantly higher in arc-related terrains compared to mid-ocean ridge spreading centers. Such high values have been used to argue for addition of slab carbon to the source region of arc volcanism. However, in addition to the slab (both its sedimentary veneer and underlying oceanic basement), the mantle wedge and/or the arc crust through which magmas traverse en route to the surface may also contribute to the total carbon output. Distinguishing between these various sources is possible by considering carbon and helium together (both isotopic variations and relative abundances).

The carbon output at arcs can be resolved into end-member components involving MORB mantle (M), and slab-derived marine carbonate/limestone (L) and (organic) sedimentary components (S). Sano and Marty (1995) used the following mass balance equations:

$$(^{13}C/^{12}C)_o = f_M(^{13}C/^{12}C)_M + f_L(^{13}C/^{12}C)_L + f_S(^{13}C/^{12}C)_S$$

$$1/(^{12}C/^3He)_o = f_M/(^{12}C/^3He)_M + f_L/(^{12}C/^3He)_L + f_S/(^{12}C/^3He)_S$$

$$f_M + f_S + f_L = 1$$

where o = observed, and f is the fraction contributed by L, S, and M to the total carbon output. It is possible to determine the relative proportions of M-, L-, and S-derived carbon in individual samples of arc-related geothermal fluids by measuring the isotopic composition of carbon and the $CO_2/^3He$ ratio (which involves measuring the $^3He/^4He$ ratio). Appropriate end-member compositions must be selected, and both Sano and Marty (1995) and Sano and Williams (1996) suggest $\delta^{13}C$ values of -6.5‰, 0‰, and -30‰ (relative to PDB) with corresponding $CO_2/^3He$ ratios of 1.5×10^9, 1×10^{13}, and 1×10^{13} for M, L, and S, respectively.

Based on the analysis of arc-related geothermal samples from 30 volcanic centers worldwide and utilizing high, medium, and low temperature fumaroles, Sano and Williams (1996) estimated that between 10% and 15% of the arc-wide global CO_2 flux is derived from the mantle wedge—the remaining 85–90% coming from decarbonation reactions involving subducted marine limestone, slab carbonate, and pelagic sediment. Subducted marine limestone and slab carbonate supply the bulk of the nonmantle carbon— ~70–80% of the total carbon—the remaining ~10–15% is contributed from subducted organic (sedimentary) carbon. It should be noted, however, that most studies ignore the arc crust as a potential source of carbon: although this omission may not be significant in intra-oceanic settings, this is unlikely to be the case at all localities (van Soest *et al.*, 1998). With this combined C–He approach, therefore, it is possible to more accurately constrain (i) mantle fluxes of various volatile species, and (ii) volatile mass balances at arcs i.e., input via the trench versus output via the arc.

Using an approach analogous to that for carbon, Sano *et al.* (1998, 2001) have attempted to understand the nitrogen cycle at subduction zones. Again, the problem is to identify and quantify the various contributory sources to the volcanic output. At subduction zones, there are three major sources of nitrogen: the mantle (M), atmosphere (A), and subducted sediments (S), and each has a diagnostic $\delta^{15}N$ value and $N_2/^{36}Ar$ ratio. Therefore, observed (o) variations in these two parameters for individual samples can be resolved into their component structures using the following equations (Sano *et al.*, 1998):

$$(\delta^{15}N)_o = f_M(\delta^{15}N)_M + f_A(\delta^{15}N)_A + f_S(\delta^{15}N)_S$$

$$1/(N_2/^{36}Ar)_o = f_M/(N_2/^{36}Ar)_M + f_A/(N_2/^{36}Ar)_A + f_S/(N_2/^{36}Ar)_S$$

$$f_M + f_S + f_A = 1$$

where f_M is the fractional contribution of mantle-derived nitrogen, etc. Note that the noble gas isotope ^{36}Ar is used in this case since degassing is not expected to fractionate the $N_2/^{36}Ar$ ratio due to the similar solubilities of nitrogen and argon in basaltic magma—this is not the case for helium and nitrogen, so that degassing corrections may be necessary if the $N_2/^3He$ ratio is used (Sano *et al.*, 2001). In the above scheme, end-member compositions are generally well constrained: the mantle and sedimentary end-members both have $N_2/^{36}Ar$ ratios of 6×10^6 (air is 1.8×10^4) but their $\delta^{15}N$ values are distinct. The upper mantle has a $\delta^{15}N$ value of -5 ± 2‰ (Marty and Humbert, 1997; Sano *et al.*, 1998) whereas sedimentary nitrogen is assumed to be $+7 \pm 4$‰ (Bebout, 1995; Peters *et al.*, 1978) (air has $\delta^{15}N = 0$‰). The wide difference in $\delta^{15}N$ between the potential end-members makes this approach a sensitive tracer of N_2 provenance.

Gas discharges from island arc volcanoes and associated hydrothermal systems have $N_2/^{36}Ar$ ratios which reach a maximum of 9.7×10^4, with $\delta^{15}N$ values up to $+4.6$‰ (Sano *et al.*, 2001).

This would indicate that a significant proportion (up to 70%) of the N_2 could be derived from a subducted sedimentary or crustal source. The situation is reversed in the case of back-arc basin glasses, which have significantly lower $\delta^{15}N$ values (−2.7‰ to +1.9‰): this implies that up to 70% of the nitrogen could be mantle-derived (Sano *et al.*, 2001). A first-order conclusion from these observations is that N_2 is efficiently recycled from the subducting slab to the atmosphere and hydrosphere through arc (and back-arc volcanism) (see also Fischer *et al.*, 2002).

There are two major issues of concern with the approach of using $CO_2/^3He$ and $N_2/^{36}Ar$ (or $N_2/^3He$) ratios in combination with $\delta^{13}C$ and $\delta^{15}N$ values to constrain the sources of volatiles at arcs. The first issue is the selection of representative end-member isotopic and relative elemental abundances—this factor has a profound effect on the deduced provenance of the volatile of interest. The second is the assumption that various elemental (and isotopic) ratios observed in the volcanic products are representative of the magma source. Both have the potential to compromise the accuracy of the output flux estimates.

In the case of CO_2, the methodology of Sano and co-workers assumes that subducted marine carbonate and sedimentary organic matter can be distinguished as potential input parameters based solely on their perceived carbon isotopic compositions prior to subduction (0‰ versus −30‰, respectively). However, this approach ignores the anticipated evolution of organic-derived CO_2 to higher $\delta^{13}C$ as a function of diagenetic and/or catagenetic changes experienced during subduction (Ohmoto, 1986). In the southern Lesser Antilles, for example, van Soest *et al.* (1998) calculated that >50% of the total carbon would be assigned to an organic, sedimentary origin if an end-member *S*-value of −10‰ were chosen—as opposed to <20% for an adopted end-member value of −30‰. In this scenario, it was suggested that the large sedimentary input implied by adopting a heavier $\delta^{13}C$ sedimentary end-member could be accommodated by loss of CO_2 from the arc crust, which is particularly thick in the southern Antilles arc. This example illustrates the point that a realistic mass balance at arcs is impossible without taking into account (i) the effect of subduction on the evolution of the carbon isotopic signature of the sedimentary input, and (ii) the possibility of an additional input from the arc crust. Note, however, that not all arcs require a crustal input of volatiles. For example, Fischer *et al.* (1998) showed that the volatiles discharged from the Kuril Island arc (75 $t\,d^{-1}$ $volcano^{-1}$) could be supplied from subducted oceanic crust and mantle wedge alone.

The second potential complication is the possible fractionation of CO_2, N_2, and helium during subduction and/or subsequent magma degassing. Little has been reported on elemental fractionation during the subduction process but studies at Loihi Seamount have shown that magma degassing can exert a strong control on resultant $CO_2/^3He$ ratios as sampled in hydrothermal fluid discharges. Hilton *et al.* (1998) reported large variations in $CO_2/^3He$ ratios at Loihi Seamount that were correlated with the composition of the magma undergoing degassing. For example, as helium is more soluble in tholeiitic basalt than CO_2 (i.e., $S_{He}/S_{CO_2} > 1$ where S = solubility), the $CO_2/^3He$ ratio in the melt phase will evolve to lower values as a function of fractionation style (Rayleigh or Batch) and extent of degassing. Measured $CO_2/^3He$ ratios in fluids during periods of tholeiitic volcanism were low ($\sim 5 \times 10^8$). In contrast, $CO_2/^3He$ ratios $\sim 10^{10}$ were recorded in other active periods, which is consistent with degassing of alkalic magmas (in this case—$S_{He}/S_{CO_2} < 1$). In order to obtain a meaningful estimate of the carbon budget in arcs, therefore, the initial (pre-degassing) $CO_2/^3He$ ratio is of prime importance, and it is often assumed that the measured $CO_2/^3He$ ratio equates to the initial magmatic value—as shown above for Loihi Seamount, this may not necessarily be the case. The same issues of degassing-induced changes to elemental ratios apply also to the N_2–He–Ar systematics used to resolve the provenance of nitrogen in arcs (Sano *et al.*, 2001).

A related concern is that of isotopic fractionation of carbon (or nitrogen) during subduction and/or magma degassing. Sano and Marty (1995) have concluded that arc-related high-temperature fluids are likely to preserve the $\delta^{13}C$ values of the (magmatic) source based on comparisons of $\delta^{13}C$ values in fluids and phenocrysts. Furthermore, they cite evidence of overlapping $\delta^{13}C$ values between high- and medium-to-low-enthalpy hydrothermal fluids, leading to the general conclusion that any fractionation induced by degassing and/or interactions within the hydrothermal system must be minimal. On the other hand, Snyder *et al.* (2001) have argued that geothermal fluids in Central America have experienced 1–2‰ shifts in $\delta^{13}C$ resulting from removal of bicarbonate during slab dewatering and/or by precipitation of calcite in the hydrothermal system. It should be noted, however, that even if observed values of $\delta^{13}C$ in arc-fluids are fractionated, the magnitude of the isotopic shifts proposed by Snyder *et al.* (2001) would make a minor difference only to calculations involving the source of carbon. To date, there is no evidence of nitrogen isotopic fractionation during magmatic degassing (Marty and Humbert, 1997).

2.06.4.1.4 Volatile inputs and mass balance

With realistic estimates of the volatile output from the mantle, particularly the mantle wedge, it is possible to assess the state of volatile mass balance for the mantle—comparing inputs via subduction zones with outputs via arc, mid-ocean ridge, and (possibly) plume-related magmatism. Understanding the volatile systematics of the mantle is a key component in defining its structure and evolutionary history.

Hilton *et al.* (2002) compared the major volatile input via the trench (based on the lithology of drilled sedimentary sequences adjacent to arcs worldwide plus assumed volatile characteristics of underlying basaltic basement) with the magmatic output—both for arcs individually and for the mantle globally. In the case of CO_2, for example, it was found that the input of limestone-derived CO_2 exceeds its output at all arcs except Japan. This reinforces the notion that subduction zones act as conduits for the transfer of carbon into the mantle beyond the zone of magma generation (Kerrick and Connolly, 2001). Furthermore, on a global basis, carbon has a mean degassing duration (ratio of its total surface inventory to its mantle output flux; Marty and Dauphas, 2002) less than the age of the Earth implying that it is recycled rapidly between the mantle and the exospheric reservoirs. In contrast, nitrogen has a mean degassing duration significantly greater than that of carbon suggesting it is characterized by a large surficial inventory compared to the mantle flux, or that it is inefficiently recycled into the mantle at subduction zones (Fischer *et al.*, 2002).

In addition to considering major volatiles, it is useful to address the question of volatile mass balance for the noble gases themselves. Porcelli and Wasserburg (1995b) reviewed the sedimentary budget and found, for example, that pelagic sediments have a wide range of measured xenon concentrations ($0.05-7 \times 10^{10}$ atoms ^{130}Xe g^{-1} with a geometric mean of 6×10^9 atoms g^{-1}; Matsuda and Nagao, 1986; Podosek *et al.*, 1980; Staudacher and Allègre, 1988). The amount of sediment subducted has been estimated at 3×10^{15} g yr^{-1} (von Huene and Scholl, 1991). Holocrystalline basalts have been found to have $(4-42) \times 10^7$ atoms ^{130}Xe g^{-1} (Dymond and Hogan, 1973; Staudacher and Allègre, 1988), with a mean of 2×10^8 atoms ^{130}Xe g^{-1}. Low temperature enrichment of alkalies seems to extend ~600 m into the ocean crust (Hart and Staudigel, 1982) and this may apply to xenon as well—7×10^{15} g yr^{-1} of this material is subducted. In total, these numbers result in 2×10^{25} atoms ^{130}Xe g^{-1} (33 mol yr^{-1}) reaching subduction zones. In a similar exercise, Staudacher and Allègre (1988) assumed that 40–80% of the subducting flux of oceanic crust (6.3×10^{16} g yr^{-1}) is altered and contains atmosphere-derived noble gases, and that up to 18% of this mass is ocean sediment. They obtained noble gas fluxes of 13.8–39 mol ^{130}Xe yr^{-1} (similar to the above estimate), along with $1.9-2.5 \times 10^6$ mol ^{4}He yr^{-1}, $1.9-5.9 \times 10^3$ mol ^{20}Ne yr^{-1}, $5.8-21.9 \times 10^3$ mol ^{36}Ar yr^{-1}, and $3.0-12.3 \times 10^3$ mol ^{84}Kr yr^{-1}. Subduction of noble gases at these rates over 10^9 yr would have resulted in 1%, 90%, 110%, and 170% of the ^{20}Ne, ^{36}Ar, ^{84}Kr, and ^{130}Xe, respectively estimated to be in the upper mantle (Allègre *et al.*, 1986). The effects of dewatering and melting are unknown but may remove a large fraction of these noble gases from subducting materials, although silica appears to retain xenon to high temperatures (Matsuda and Matsubara, 1989). It should be emphasized that these numbers are highly uncertain and can only be used for order-of-magnitude comparisons.

Subduction zone processing and volcanism may return much of the noble gases to the atmosphere. Hilton *et al.* (2002) calculated that the output via arc-related magmatism for helium and argon greatly exceeded the potential input via the subducting slab, implying that the output fluxes are dominated by mantle wedge and/or arc crust contributions. Staudacher and Allègre (1988) argued that subducting argon and xenon is almost completely lost back to the atmosphere during subduction zone volcanism (the so-called "subduction barrier") to preserve the high ^{129}Xe/^{130}Xe and ^{136}Xe/^{130}Xe in the upper mantle throughout Earth history. However, it is possible that the total noble gas abundances reaching subduction zones is sufficiently high that subduction into the mantle of only a small fraction may have a considerable impact upon the composition of argon and xenon in the upper mantle. This contrary viewpoint produces the upper mantle xenon composition through mixing between subducted noble gases and nonrecycled, mantle-derived xenon (Porcelli and Wasserburg, 1995a,b). For example, the ^{128}Xe/^{130}Xe ratio measured in mantle-derived xenon trapped in CO_2 well gases may be interpreted as a mixture of ~90% subducted xenon with 10% trapped solar xenon. The same proportions could be applied to subducted and trapped radiogenic ^{129}Xe. Since mantle evolution models can have nonradiogenic noble gases in the upper mantle that are either primordial or largely dominated by subducted gases, isotopic evolution arguments alone cannot preclude subduction of atmospheric argon and xenon. Furthermore, direct input of the subducted slab into a gas-rich deeper reservoir that has ^{40}Ar/^{36}Ar values significantly lower than the MORB source mantle is also a possibility (e.g., Trieloff *et al.*, 2000).

2.06.4.2 Mantle Structure and the Source of Ocean Islands

The range of $^3He/^4He$ ratios in materials from the convecting mantle, and in particular the OIB ratios that are substantially higher than those in MORBs, provide a tracer for the source of the plumes supplying hotspots such as Hawaii and Iceland. However, in order to relate the surface expressions of these features to mantle structure, the nature of the reservoir storing the helium with high $^3He/^4He$ ratios must be identified. There have been various models addressing this topic. The reservoir must have a distinctly higher $^3He/(U + Th)$ ratio than the upper mantle in order to generate higher $^3He/^4He$ ratios, and the main issue is how this reservoir can be preserved in the mantle over sufficient time to generate the distinctive helium isotope composition. Also, material from this reservoir must be preferentially incorporated into mantle plume material. This is generally, but not universally, taken to link the source of the plumes with the reservoir, so that models explaining the high $^3He/^4He$ ratios also incorporate an explanation for the generation of plumes. There are three general categories of models for the OIB helium reservoir: (i) a mantle with a deep layer, with plumes arising from the boundary with the upper, MORB-source mantle; (ii) the core, with plumes derived from the core–mantle boundary; and (iii) heterogeneities within some whole mantle circulation pattern. The distribution of helium isotope compositions found at the surface reflects the conditions of plume generation and transport of helium with high $^3He/^4He$ ratios within the context of each model. It should be noted that the different models also have implications for the degassing of the mantle and the formation of the atmosphere. These are discussed in Chapter 4.11 along with the planetary budgets for radiogenic isotopes and the overall constraints on the extent of solid Earth degassing.

2.06.4.2.1 Isolated lower mantle models

There are several possible descriptions of a layered mantle. The possibilities that have been incorporated into noble gas models include a boundary layer at the 670 km seismic discontinuity, a deeper layer of variable thickness, and a boundary layer at the core–mantle boundary.

The first mantle degassing models that incorporated considerations of mantle structure were developed with the discovery of mantle noble gas isotope heterogeneities (Allègre *et al.*, 1986, 1983; Hart *et al.*, 1979; Kurz *et al.*, 1982). These models, following earlier interpretations of lithophile element isotope composition variations (DePaolo, 1979; O'Nions *et al.*, 1979), divide the mantle into two convectively isolated layers with a boundary at 670 km. Such models incorporate the degassing of the upper mantle reservoir to the atmosphere. In order to explain the high OIB $^3He/^4He$ ratios, the underlying gas-rich reservoir is isolated from the degassing upper mantle. Therefore, these layered mantle models can be considered to incorporate two separate systems: the upper mantle–atmosphere and the lower mantle. There is no interaction between these two systems, and the lower mantle is completely isolated except for a minor flux to OIB that marks its existence. It is further assumed that the mantle was initially uniform in noble gas and parent isotope concentrations, so that both systems had the same starting conditions. Note that various modifications to this basic scheme have been proposed, and are discussed below (Allègre *et al.*, 1983, 1986).

This class of model has been used to calculate the degassing history of the atmosphere. The initial atmospheric abundances were assumed to have been zero, and all noble gases were once contained within the upper mantle. The atmosphere has progressively degassed from the upper mantle. The evolution of noble gases in the upper mantle is then a function of radiogenic production of ^{21}Ne, ^{40}Ar, ^{129}Xe, and ^{136}Xe within the mantle and gas loss to the atmosphere. The degassing history of the atmosphere can be calculated from the isotopic compositions of the present atmosphere, the initial mantle, and the present upper mantle. A key conclusion of these models is that high $^{40}Ar/^{36}Ar$ and $^{129}Xe/^{130}Xe$ ratios require both early (even before ^{129}I was extinct) and extensive mantle degassing to set high ratios of parent elements to noble gases in the mantle. More complicated degassing calculations can be used to take into account factors such as the depletion history of the parent by extraction to the continental crust, but the conclusions do not change significantly (e.g., Hamano and Ozima, 1978).

The lower mantle supplies OIB through plumes that originate through instabilities at the boundary, and so the high $^3He/^4He$ ratios trace the rising of lower mantle material. This reservoir has evolved essentially as a closed system; the fluxes to OIB are sufficiently low that the closed system approximation remains valid. The high $^3He/^4He$ ratios seen in OIBs have been interpreted as reflecting closed system evolution, and can be used to calculate the reservoir $^3He/(U + Th)$ ratio. It is generally assumed that the initial concentrations of the parent elements are equal to those of the bulk silicate Earth. The present isotopic compositions of argon and xenon are equal to those of the BSE and so those of the atmosphere. This is because noble gases and parent elements were initially uniformly distributed in the Earth, and since the upper mantle is highly degassed, the atmosphere contains essentially all of the noble gases in the upper portion of the Earth, and so has BSE noble gas isotope ratios. Note that early observations supporting atmospheric argon and xenon ratios in the lower mantle have been shown to reflect only contamination (e.g., Fisher, 1985; Patterson *et al.*, 1990), and so the lower mantle ratios are calculated, rather than observed, values. Using closed system calculations for the lower mantle helium isotope evolution, the model predicts that the lower mantle is enriched in helium by $\sim 10^2$ relative to the present upper mantle concentration.

A variation of the residual upper mantle model by Zhang and Zindler (1989) considers degassing of MORBs by partitioning of noble gases into CO_2 vapor for transport to the surface, with the remaining noble gases in MORBs returned to the mantle. Due to different solubilities, there is fractionation between noble gases in the residual MORBs and those lost to the atmosphere. The net result of this partial gas loss from a MORB is fractionation during mantle degassing. Another variation has been presented by Kamijo *et al.* (1998) and Seta *et al.* (2001). Here two additional factors are considered: depletion of the lower mantle reservoir by plume activity, and the subduction of parent elements into both the upper and lower mantle reservoirs. As with the other models, an initially uniform distribution of noble gases and parent elements in the mantle was assumed. Using mass fluxes between reservoirs and a linear continental growth function, it was concluded that the lower mantle could contain as little as 13% of its original noble gas inventory, and have $^{40}Ar/^{36}Ar$ ratios up to 3×10^4, much higher than that of the atmosphere, due to progressive depletion.

There are several observations that require that this type of mantle model be modified. The relationship between the upper mantle and the atmosphere cannot be simply related. The difference between MORB and air $^{20}Ne/^{22}Ne$ ratios indicates either that neon isotopes were not initially uniformly distributed in the Earth or that neon in the atmosphere has been modified by losses to space after degassing from the mantle. Regardless of the reason, the assumption that the atmosphere and upper mantle together form a closed system does not hold. Also, the higher $^{20}Ne/^{22}Ne$ and Ne/Ar ratios of the upper mantle limit the contribution that presently degassing volatiles can have made to the atmosphere (Marty and Allé, 1994). These observations suggest that neon has been lost by fractionating processes from the early atmosphere, a feature that can be appended to the model of mantle structure.

A major objection that cannot be satisfied through simple modification of the model is that atmospheric and MORB xenon are not complementary; that is, extraction of the xenon now seen in the atmosphere from the upper mantle will not leave a mantle reservoir with the xenon isotope characteristics presently observed there. Due to the greater half-life of ^{244}Pu, the ratio of ^{244}Pu-derived ^{136}Xe to radiogenic ^{129}Xe within a residual upper mantle reservoir must be greater than that of the atmosphere extracted from that reservoir (see Porcelli and Ballentine, 2002). This does not appear to be the case, since the ratio of radiogenic ^{136}Xe (derived from ^{244}Pu) to radiogenic ^{129}Xe in MORB (Kunz *et al.*, 1998) and mantle-derived xenon in CO_2 well gas (Phinney *et al.*, 1978) is clearly below that of the atmosphere. There are also difficulties with maintaining a highly depleted upper mantle from a gas-rich underlying reservoir. Distinctive radiogenic xenon isotope ratios in the upper mantle, established by degassing early in Earth history, would be obliterated by contamination from plumes rising from the gas-rich lower mantle (Porcelli *et al.*, 1986b). The requirement that all rising gas-rich material is completely degassed at hotspots is probably not reasonable, and the mixing of even a small fraction of incompletely degassed material into the surrounding mantle will have an impact.

Overall, using the basic principles of this model, the basic relationships between the isotopic compositions of the atmosphere and upper mantle cannot be explained. However, the preservation of noble gas isotope heterogeneities in convectively isolated mantle layers remains appealing.

2.06.4.2.2 Models with transfer between mantle reservoirs

To overcome some of the problems with the isolated mantle models (above), a new set of models allowed interaction between the gas-rich lower mantle below 670 km and the upper mantle. The first of this type sought to explain ^4He fluxes (O'Nions and Oxburgh, 1983) and assumed that highly incompatible elements in the upper mantle were (i) introduced in material entrained from the lower mantle, and (ii) in steady state concentrations. The model was then applied to the U–Pb system (Galer and O'Nions, 1985), and finally to the other noble gases (Kellogg and Wasserburg, 1990; O'Nions and Tolstikhin, 1994; Porcelli and Wasserburg, 1995a, 1995b). The central focus of the model is not degassing of the upper mantle to form the atmosphere, but rather mixing in the upper mantle, which has noble gas inventories that are presently not continually depleting, but rather are the result of continuing inputs from surrounding reservoirs. In addition to the upper mantle inputs by radiogenic production from decay of uranium, thorium, and potassium, atmospheric argon and xenon are subducted into the upper mantle, and lower mantle noble gases are transported into the upper mantle within fluxes of upwelling material. Noble gases from these sources comprise the outflows at mid-ocean ridges. The isotopic systematics of the different noble gases are linked by the assumption that transfer of noble gases from the upper mantle to the atmosphere by volcanism, as well as the transfer from the lower into the upper mantle by the rising of plumes, occurs without elemental fractionation.

It is assumed that upper mantle concentrations are in steady state, so that the inflows and outflows are equal; therefore, there are no time-dependent functions (and so additional parameters) determining upper mantle concentrations. Degassing of the upper mantle occurs by essentially complete removal of noble gases in the melting zone of MORB; therefore, the noble gas residence time in the upper mantle is determined by the rate at which upper mantle material is processed at ridges. The residence time is then the rate of mantle processing at ridges divided by the total mass of the reservoir. For the upper mantle above 670 km and the current rate of MORB production (from 10% melting), this is ~1.4 Ga (Kellogg and Wasserburg, 1990; Porcelli and Wasserburg, 1995b). It is important to note that the assumption that upper mantle concentrations are in steady state is only a simplification in the model for interaction between mantle reservoirs. Nevertheless, as discussed below, in this way all of the presently observed broad noble gas characteristics of MORB and OIB can be explained (Porcelli and Wasserburg, 1995a,b). Additional parameters required to calculate time-dependent characteristics would only create under-constraint in the calculated model solutions, and therefore the introduction of non-steady-state conditions are justified only when new constraints are found. However, modification of this model by increasing the size of the upper mantle, as discussed below and so extending the reservoir to a large fraction of the age of the Earth, then would require time-dependent solutions. Tolstikhin and Marty (1998) have explored the evolution of the mantle prior to the establishment of steady state concentrations, and some effects of the time dependence of noble gas transfers into the upper mantle.

The atmosphere generally has no *a priori* connection with other reservoirs. It simply serves as a source for subducted gases, and no assumptions are made about its origin. While the radiogenic noble gases presently found in the atmosphere were clearly degassed from the solid Earth, and the nonradiogenic noble gases may also have been originally incorporated within the solid Earth, this degassing necessarily occurred prior to establishment of the present upper mantle characteristics. The history of degassing cannot be obtained from the noble gases presently found in the upper mantle, which are largely derived from the deeper mantle. Also, if upper mantle noble gases are assumed to be in steady state, by definition their characteristics do not contain any historical information regarding the establishment of present conditions.

The lower mantle reservoir evolves as an approximately closed system, and so the concentrations of ^{4}He, ^{21}Ne, ^{40}Ar, ^{129}Xe, and ^{136}Xe that have been produced over Earth history can be calculated. The lower mantle ^{20}Ne, ^{36}Ar, and ^{130}Xe concentrations, and so isotope compositions, can be calculated from the balance of fluxes into the upper mantle. The MORB ^{3}He/^{4}He ratio is a result of mixing between lower mantle helium and production of ^{4}He in the upper mantle (O'Nions and Oxburgh, 1983). Since the flux of the latter is fixed from the concentrations of parent nuclides, the rate of helium transfer from the lower mantle can be calculated. The lower mantle mass flux into the upper mantle that is calculated from helium isotopes (Kellogg and Wasserburg, 1990) can then be used to obtain the fluxes of other daughter nuclides from the lower mantle from their lower mantle concentrations (Porcelli and Wasserburg, 1995a,b). Production of other daughter noble gases in the upper mantle, ^{21}Ne, ^{40}Ar, and ^{136}Xe, can also be calculated from parent nuclide concentrations. Using the upper mantle neon, argon, and xenon isotope compositions, the lower mantle isotopic compositions of neon, argon, and xenon can be calculated. If subduction of any argon and xenon occurs, then a fraction of the nonradiogenic nuclides is introduced into the upper mantle, and so more radiogenic argon and xenon isotope compositions are calculated for the lower mantle.

The model is consistent with the isotopic evidence that upper mantle xenon does not have a simple direct relationship to atmospheric xenon. The radiogenic xenon presently seen in the atmosphere was degassed from the upper portion of the solid Earth prior to the establishment of the present upper mantle steady state xenon isotope compositions and concentrations. The lower mantle ratios are established early in Earth history by decay of ^{129}I and ^{244}Pu; ^{238}U decay produces a relatively small fraction of fissiogenic nuclides (Porcelli and Wasserburg, 1995b). The xenon daughters (now in the upper mantle) of the short-lived parents are supplied from the lower mantle. The MORB ^{129}Xe/^{130}Xe ratio (when corrected for air contamination) has no radiogenic contributions from present production in the upper mantle, and so is simply due to mixing between lower mantle and subducted xenon. Since the MORB $^{129}Xe/^{130}Xe$ value is greater than that of the atmosphere, the lower mantle ratio must be equal to that of MORB (if there is no xenon subduction) or higher. Once lower mantle xenon is transported into the upper mantle, it is significantly augmented by uranium-derived ^{136}Xe, since the U/Xe ratio is much greater in this gas-depleted reservoir. The ^{130}Xe flux from the lower mantle cannot be determined without a constraint on the amount subducted, but is inversely proportional to the lower mantle $^{129}Xe/^{130}Xe$ and $^{136}Xe/^{130}Xe$ ratios (Porcelli and Wasserburg, 1995b). There is therefore a trade-off between two presently unknown quantities; the lower mantle ^{130}Xe concentration (i.e., how high are the lower mantle $^{136}Xe/^{130}Xe$ and $^{129}Xe/^{130}Xe$ ratios) and the flux of subducted xenon (how much is mixed with lower mantle xenon to lower these ratios). Alternatively, if a further assumption is made regarding a specific Xe/He ratio for the lower mantle (e.g., the solar ratio), then a xenon concentration can be calculated from the helium concentration, along with the amount of subducted xenon. In the model, ~50% of the radiogenic ^{136}Xe in the upper mantle is derived from the lower mantle, where it was produced largely by ^{244}Pu.

Overall, this type of model has been successful at explaining the general noble gas characteristics of the mantle. The difficulties with relating the atmosphere to the upper mantle are solved through decoupling of the histories of these two reservoirs, and the origins of the atmosphere constituents can be explained by a separate, early history of the upper part of the planet. Modifications of the model for interacting mantle reservoirs are of course possible, although in some cases the attainment of upper mantle steady state isotope concentrations may not be reasonable and time-dependent solutions may be necessary.

2.06.4.2.3 Difficulties with mantle layering

The principal objection to the various versions of a layered mantle model has been geophysical arguments advanced against maintaining a distinctive deep mantle reservoir. The mass flux from the lower mantle into the upper mantle in the interacting mantle model is ~50 times less than the rate of ocean crust subduction, greatly limiting the fraction of subducted material that could cross into the lower mantle (O'Nions and Tolstikhin, 1996). While there are a variety of different lines of evidence for the lack of a boundary layer at 670 km (see Porcelli and Ballentine, 2002; Tackley, 2000), the clearest has been seismic tomographic evidence that subducting slabs have penetrated the 670 km discontinuity (Creager and Jordan, 1986; Grand, 1987; van der Hilst *et al.*, 1997), and so greater mass exchange occurs with the lower mantle below 670 km. It appears that a mantle-wide boundary separating distinct mantle reservoirs is not present at this level. Many geophysical studies have therefore assumed that the mantle is stirred by whole mantle convection.

An important constraint on mantle reservoirs comes from the relationship between the radiogenic ^{4}He and heat fluxes at the surface. The Earth's global heat loss amounts to 44 TW (Pollack *et al.*, 1993). Subtracting the heat production from the continental crust (4.8–9.6 TW), and the core (3–7 TW; Buffett *et al.*, 1996) leaves 9.6–14.4 TW to be accounted for by present day radiogenic heating and 17.8–21.8 TW as a result of secular cooling. O'Nions and Oxburgh (1983) pointed out that the present day ^{4}He mantle flux was produced along with only 2.4 TW of heat, a factor of up to six times lower than the total radiogenic mantle heat flux associated with present radiogenic production. If a portion of the heat from secular cooling is from past radiogenic production, this must also be associated with ^{4}He production, and so the imbalance is even greater. Therefore, a mechanism is required for allowing the escape of heat, but not ^{4}He, from the reservoir that contains the bulk of the heat-producing elements uranium and thorium in the mantle. O'Nions and Oxburgh (1983) suggested that this could be achieved by a boundary layer in the mantle through which heat could pass, but behind which helium was trapped. As an alternative, van Keken *et al.* (2001) used a secular cooling model of the Earth to investigate the possibility that heat and helium are separated due to the different mechanisms that extract heat and helium from the mantle. The rates of release of heat and helium over a 4 Ga model run varied substantially, but the ratio of the surface ^{4}He flux to heat flow equaled that of the present Earth only during infrequent periods of very short duration. It is unlikely that the present day Earth happens to correspond to such a period. However, Ballentine *et al.* (2002) pointed out that if the long-term helium flux from the mantle were ~3 times greater than presently inferred, and so the upper mantle helium concentration were correspondingly higher, then the heat-helium balance could be satisfied. In addition, such concentrations throughout most of the mantle would also account for the mantle ^{40}Ar budget (see Chapter 4.11). Further data are required to substantiate this argument. Morgan (1998) suggested that heat and helium are transported to the surface at hotspots, where the bulk of the helium is lost, while the heat is lost subsequently at ridges. However, evidence for such a hotspot helium flux at the present day or in the past, and the formulation of a mantle noble gas model incorporating this suggestion, is unavailable. At present, there appears to be no convincing alternative to maintaining much of the mantle uranium and thorium behind a boundary across which heat, but not ^{4}He, can efficiently pass. However, this does not need to be constrained to a depth of 670 km.

Allègre (1997) has argued that geochemical models for long-term mantle layering can be reconciled with geophysical observations for present-day whole mantle convection if the mode of mantle convection changed less than 1 Ga ago from layered to whole mantle convection. A difficulty with this hypothesis is the expected consequences for the thermal history of the upper mantle. The principle process driving changes in the mode of mantle convection is the development of thermal instabilities in the lower layer that eventually result in either massive or episodic mantle overturn (Davies, 1995; Tackley *et al.*, 1993). Reviewing the consequences for models that reproduce mantle layering, Silver *et al.* (1988) noted that a >1,000 K temperature difference develops between the two portions of the mantle, and so overturns will cause large variations in upper mantle temperatures. However, the available geological record suggests that there has been a relatively uniform mantle cooling rate of ~50–57 K Ga^{-1} (Abbott *et al.*, 1993; Galer and Mezger, 1998). Further, numerical models simulating ^{3}He/^{4}He variations in a mantle that has undergone this transition do not reproduce the observed ^{3}He/^{4}He distributions seen today (van Keken and Ballentine, 1999).

While the lack of geophysical evidence for layered mantle convection poses the greatest problem for maintaining separate reservoirs with distinct helium isotope compositions, the intra-reservoir relationships involved in this model may be adaptable to other configurations of noble gas reservoirs that might be found to be compatible with geophysical data. It might be possible to reformulate the model for a larger upper mantle, or greater mass fluxes between reservoirs, although in some cases non-steady-state upper mantle concentrations may be required.

2.06.4.2.4 Deeper layer reservoirs

Kellogg *et al.* (1999) have developed and numerically tested a model in which mantle below ~1,700 km has a composition, and so density, that is sufficiently different from that of the shallower mantle to largely avoid being entrained and homogenized in the overlying convecting mantle. This model has generated a great deal of interest because of its ability to preserve a region in the mantle behind which the radioelements and primitive noble gases can be preserved, while accommodating many geophysical observations. For example, it is argued that the depth of the compositional change varies, allowing slab penetration to the CMB in some locations while providing a barrier into the lower mantle elsewhere. Supporting tomographic evidence for a significant number of slabs being disrupted at 1,700 km is given by van der Hilst and Karason (1999). More recently, this supporting evidence has been questioned because of the loss of tomographic resolution in this portion of the mantle (Kàrason and van der Hilst, 2001). Although Kellogg *et al.* (1999) suggest that this layer will be hard to detect seismically because of its neutral thermal buoyancy, irregular shape and small density contrast, Vidale *et al.* (2001) argue that seismic scattering nevertheless should be resolvable, and yet is not observed. Also, if the overlying mantle has the composition of the MORB source, then the abyssal layer must contain a large proportion of the heat-producing elements, and must efficiently remove heat from the core. It is not yet clear what this effect would have on the thermal stability of the layer or temperature contrast with the overlying mantle.

The boundary layer at the CMB has been explored as a reservoir for high ^{3}He/^{4}He ratio helium, in the context of whole mantle convection. It has been suggested that subducted oceanic crust could accumulate there and form a distinct chemical boundary layer, accounting for the properties of the D'' layer (Christensen and Hofmann, 1994). Altered ocean crust is strongly depleted in ^{3}He and may be enriched in uranium, and so is likely to have relatively radiogenic helium. However, a complementary harzburgitic lithosphere that is depleted in both uranium and helium also develops during crust formation, and a layer of such material above the CMB has been surmised as a high ^{3}He/^{4}He reservoir (Coltice and Ricard, 1999). This reservoir preserves a high initial helium content, and addition of subsequent uranium- and thorium-depleted material simply dilutes this. The model presently lacks a mechanism for establishing the layer that is then modified by the incorporation of subducted material (Porcelli and Ballentine, 2002). More generally, it has not been extended to explain the other noble gases, and would require substantial modification to explain xenon isotope systematics and the radiogenic nuclide budgets. Also, since such a layer could not contain a dominant fraction of the heat-producing elements, another mechanism must be evoked to explain the separation of heat and helium.

Overall, if a deep mantle layer is found to be geophysically viable, it must be incorporated into a comprehensive noble gas model. This will inevitably include some of the features described in the above layered mantle models.

2.06.4.2.5 Heterogeneities in the convecting mantle

A different approach to the interpretation of isotopic variations of noble gases, as well as of lithophile trace elements, has been to hypothesize the preservation of heterogeneities within the convecting mantle. The central issue is whether this can explain the highest ratios in OIBs by creating domains imbedded in the convecting MORB-source mantle, so that there is no requirement for convective isolation of mantle with high $^3He/^4He$ ratios. Two end-member models have been postulated, one in which "blobs" or "plums" of enriched material are passively entrained in the convecting mantle to provide OIB-source material (Davies, 1984). The other has been called "penetrative convection" (Silver *et al.*, 1988), in which downgoing cold material drops into a compositionally different lower mantle layer. The slabs, on heating at the CMB, regain positive buoyancy and on return to the surface entrain a small portion of the deeper reservoir. In this way, lower mantle material is provided to either OIB-source or is mixed into the MORB-source mantle. Regarding the first case, numerical models have shown that more viscous "blobs" can be preserved in a convective regime if they are at least 10–100 times more viscous than the surrounding mantle (Manga, 1996). Becker *et al.* (1999) have investigated the dynamical, rheological, and thermal consequences of such blobs containing high radioelement and noble gas concentrations to account for the bulk Earth uranium budget and the high $^3He/^4He$ ratios seen in OIB. The higher heat production and resulting thermal buoyancy within these blobs must be offset by a combination of increased density (~1%) and small size (<160 km diameter) to avoid both seismic detection and thermal buoyancy that will result in the blobs rising into the upper mantle. These high viscosity zones must fill 30–65% of the mantle to satisfy geochemical mass balance constraints, and must be surrounded by low viscosity mantle to transport the resulting heat away. The thermal gradient generated through the blobs will result in a high viscosity shell around a lower viscosity core that controls the dynamical mixing behavior. For a viscosity contrast between blobs and host mantle of 100, the average viscosity of the lower mantle is predicted to be greater, by a factor of 5, than that of the upper mantle. The mechanism creating the more dense "blob" material is uncertain, but the higher density may be explained by a different perovskite/magnesiowüstite ratio than the surrounding mantle (Becker *et al.*, 1999). Manga (1996) noted that clumping of "blobs" is likely to occur and form large-scale heterogeneities, and this process is expected to work against the careful size balance required to avoid increased thermal-driven buoyancy and/or seismic detection.

The heterogeneities are often envisioned as containing gas-rich material preserved since early Earth history. It is generally assumed that noble gases are the most highly incompatible elements, so that any mantle reservoir that undergoes melting would be preferentially depleted in noble gases relative to the parent elements uranium, thorium, and potassium. Therefore, melt residues are expected to have low He/U ratios and so develop relatively radiogenic helium isotope compositions. Alternatively, it has been suggested (Anderson, 1998; Graham *et al.*, 1990; Helffrich and Wood, 2001) that uranium may be more incompatible than helium, so that melting can leave a residue with a higher He/U ratio than the original source material. In this case, high $^3He/^4He$ ratios seen in OIB can be generated in mantle domains that have been previously depleted by melting and can survive for some time. To date, this idea has not been developed into a coherent mantle model that examines the evolution of the MORB- and OIB-source mantle domains, so that it cannot be evaluated rigorously. However, several issues that must be incorporated in such a model can be considered. The source of the domains must have high $^3He/^4He$ ratios at the time of depletion. Also, the source must have started out with sufficiently high ^3He to be able to supply OIBs after depletion. Melting of highly depleted MORB-source mantle will leave a component that will not readily impart a distinctive isotopic signature to OIBs. This is a particular problem when it is considered that such high $^3He/^4He$ components must dominate the helium signature in OIB source regions where there is clear evidence for recycled oceanic crust that likely contains high ^4He concentrations, such as in Hawaii (Hauri *et al.*, 1994). Therefore, the initial mantle source must be more gas-rich. The need to preserve a gas-rich precursor is contrary to the original motivation for hypothesizing these gas-poor heterogeneities as an alternative to long-term storage of gas-rich material from early Earth history.

Note that if such a scheme is viable, the evolution of mantle noble gases, reflecting degassing and interaction between reservoirs, may still follow those calculated in layered mantle models. An important issue that must be resolved is that there must be a mechanism for the preferential involvement of this previously melted material at OIB. This would link the involvement of these reservoirs in OIB source regions with where such material is stored. In this case, high $^3He/^4He$ ratios at the surface will trace the transfer of material from this region. Also, helium would provide information on the generation of such depleted material. However, this process has yet to be substantiated.

2.06.4.2.6 *The core*

The core has often been suggested as a source for ^{3}He presently found in plumes. If the core served as the long-term storage reservoir for helium in OIBs with distinctively high ^{3}He/^{4}He ratios, then the difficulties of maintaining an isolated mantle helium reservoir separate from the source of MORB are removed. The possibility of trapping helium into the core and releasing it into the overlying mantle has been systematically evaluated by Porcelli and Halliday (2001). Appealing to the core as a source of noble gases necessarily evokes specific conditions of terrestrial noble gas acquisition and core formation, as well as core composition characteristics. Matsuda *et al.* (1993) obtained partitioning data for liquid silicate and liquid metal for pressures up to 100 kbar (corresponding to depths of up to ~300 km). Liquid–metal/liquid–silicate partition coefficients of ~(0.01–3) $\times 10^{-4}$ were obtained. While such low values have been taken to indicate that substantial quantities of noble gases cannot be in the core, the amounts partitioned into core-forming metal depend upon the concentrations initially available in the mantle. Since noble gases are incompatible during silicate melting, liquid–metal/solid–silicate partition coefficients will be greater, and so the conditions under which core formation occurred must be considered. It was shown that there might have been sufficient gas present in the mantle during core segregation to supply a substantial quantity of helium and neon to the core (Porcelli and Halliday, 2001) although this is dependent upon how noble gases are incorporated into the early Earth. In order to supply the ^{3}He found in OIBs, transfer from the core to the mantle by either bulk entrainment of core material or chemical interaction at the CMB may provide a mechanism for supplying relatively unfractionated noble gases to plumes. Using presently available data, these scenarios may not involve unreasonable amounts of core material. It was emphasized that the concentrations of noble gases within the early Earth, the partition coefficients between silicates and metal, and the concentrations presently required in the core to supply OIBs are all highly uncertain. Nonetheless, calculations using the presently available constraints indicate that the core remains a plausible source of the ^{3}He found in OIBs. Measured high ^{3}He/^{4}He ratios would then be the result of mixtures of helium from the core that has a solar nebula ^{3}He/^{4}He ratio with radiogenic ^{4}He from the mantle. However, the implications for other geochemical parameters also must be considered. Other elements that would be affected by bulk transfer of core material into the mantle include the platinum group elements and volatiles that may be relatively abundant in the core, such as hydrogen and carbon. While some limits on the amount of transferred core material are provided by these (Porcelli and Halliday, 2001), it should be noted that the amount needed to sustain the ^{3}He flux is still uncertain, and it is possible that upward revisions in the amount of noble gases within the early Earth, or in the metal/silicate partition coefficients, could reduce the required flux of core material by an order of magnitude or more.

Note that storage of helium in the core remains only one component of a noble gas model that can describe the range of noble gas observations. The core has only been evaluated as a possible storage of ^{3}He. The incorporation in the core of other noble gases, and their relative fractionations, cannot be clearly evaluated without more data. Also, the distribution of radiogenic nuclides such as ^{40}Ar, ^{129}Xe, and ^{136}Xe that are produced within the mantle must be explained with a model that fully describes the mantle reservoirs. While these issues may be tractable, a comprehensive model that incorporates a core reservoir remains to be formulated. It should be emphasized that the core does not completely explain the distribution of helium isotopes, since the issue of the ^{4}He-heat imbalance is not addressed at all by this model. It appears that even if high ^{3}He/^{4}He ratios are the signature of involvement of core material in the source of mantle plumes, several mantle reservoirs are still required.

2.06.4.3 Mantle Volatiles in the Continental Lithosphere

While the convecting upper mantle is expected to be well mixed, with MORB providing representative samples for noble gas characterization, a significant portion of the upper mantle is stabilized against convection beneath the continents as continental lithospheric mantle. This region may supply magmas to continental volcanics, but these are generally erupted subaerially and have been extensively degassed and so can only provide limited noble gas information of this mantle region. However, these magmas can also contain mantle xenoliths—portions of the lithosphere that have been entrained and brought to the surface without melting. The thickness of the lithosphere varies, and reaches the greatest depths under Archean cratons, with estimates of ~250–400 km based on seismic data (Grand, 1994; Jordan, 1975; Nolet *et al.*, 1994; Polet and Anderson, 1995; Ricard *et al.*, 1996; Simons *et al.*, 1999; Su *et al.*, 1994) and on xenolith thermobarometry data (e.g., O'Reilly and Griffin, 2000; Rudnick and Nyblade, 1999). Xenolith samples are found in kimberlites and related rocks allowing access to the subcontinental mantle: however, with the exception of diamonds (which unfortunately has provided ambiguous information (see Dunai and Porcelli, 2002), little work has been performed on mantle noble gases in major silicate phases from kimberlitic xenoliths. Younger continental areas such as western and central Europe typically have lithospheric thicknesses up to 100 km (Blundell *et al.*, 1992; Sobolev *et al.*, 1997). The lithosphere can be much thinner locally, especially in rifted areas. Rifted areas also contain the majority of volcanic centers that have yielded xenoliths (e.g., O'Reilly and Griffin, 1984) thus making the underlying mantle accessible to study. It is worth noting that if the lithosphere has an average thickness of 150 km and underlies 30% of the Earth's surface area then it constitutes ~7% of the upper mantle, and may therefore be a significant reservoir of mantle noble gases.

2.06.4.3.1 Chemistry of the lithosphere

Major element data of xenoliths suggest that the lithospheric mantle has been largely depleted by melt extraction, while many trace element and isotopic characteristics show evidence of subsequent modification and both ancient and recent enrichment (e.g., Carlson and Irving, 1994; Frey and Green, 1974; Hawkesworth *et al.*, 1990, 1984; Menzies and Chazot, 1985; Richardson *et al.*, 1984; Wass *et al.*, 1980; see Chapter 2.05). This is due to infiltration by H_2O-rich fluids, silicate melts, or carbonatitic melts that act to re-enrich depleted rocks in LREE and other trace elements (e.g., Downes, 2001; Hawkesworth *et al.*, 1990; Pearson, 1999b; Wilson and Downes, 1991). These enriching fluids were likely introduced from a variety of sources with different characteristics, including fluids above subduction zones, fluids akin to the basaltic and kimberlitic volcanics that reach the surface, and highly mobile carbonate melts (see e.g., Menzies and Chazot, 1995). The fluid source regions may involve different mantle domains, such as those seen in oceanic and island arc volcanism, or lithospheric material that has undergone heating or uplift. Also, multiple transits of fluid through the lithosphere can result in the generation of distinctive lithospheric compositions. The resulting enrichments evolve isotopically over various times, so that while some xenolith compositions clearly show the influence of subducted components or fluids related to host magmas, a great diversity of compositions that reflect complex lithospheric histories is usually observed (see Carlson, 1995). It should be noted, however, that it is not clear to what extent such trace-element-enriched samples are representative of the bulk lithosphere (McDonough, 1990; Rudnick *et al.*, 1998). For example, Rudnick and Nyblade (1999) found that the geotherm below the Kalahari craton recorded in xenoliths was most compatible with no heat production within the lithosphere, implying minimal concentrations of uranium and thorium. In contrast, O'Reilly and Griffin (2000) argue that a considerable amount of uranium and thorium is stored in apatite and other secondary phases within the Phanerozoic lithospheric mantle, and this generates a significant amount of heat as well as radiogenic noble gases.

The age of depletion and subsequent enrichment of the subcontinental lithosphere is related to the age of the overlying crust and the timing of tectonic processes that lead to its consolidation. Archean cratons may have lithospheric mantle roots that were depleted up to 3.3 Ga ago, as most clearly seen in diamond inclusion Sm–Nd isochrons and model age data (e.g., Richardson *et al.*, 1984; Richardson and Harris, 1997), Re–Os model depletion ages of mantle xenoliths (e.g., Pearson, 1999a,b; Walker *et al.*, 1989), and lead isotope data for lithosphere-derived lavas and xenoliths (e.g., Rodgers *et al.*, 1992). The lithosphere underlying younger, more tectonically active continental regions such as Wyoming, southeast Australia, and western Europe may have been depleted ≤ 2 Ga ago (e.g., Burnham *et al.*, 1998; Carlson and Irving, 1994; Handler *et al.*, 1997). While these ages may pertain to the initial stabilization of the lithosphere, subsequent enrichment events and so open system chemical behavior can extend up to rifting events and the deformation related to the eruption of the host magma. There is, therefore, a range of timescales for lithospheric trace element enrichment over which production of radiogenic noble gases may occur.

Noble gases, in particular helium, have the potential for identifying the sources of metasomatising fluids, as long as subsequent evolution in the lithosphere does not significantly alter its isotope composition. In this respect, an important issue is the location of noble gases and parent elements in xenoliths. The noble gases are not readily incorporated into crystal lattices, and strongly partition into the CO_2-rich fluids that typically occupy mantle fluid inclusions (Section 2.06.2.2). This has been confirmed by studies that showed trapped helium can be released by crushing and separated from lattice-hosted noble gases (Hilton *et al.*, 1993b; Kurz *et al.*, 1990; Scarsi, 2000). Note that fluids and noble gases may also reside along grain boundaries at depth, and are likely to be completely lost by movement along the boundaries during transport, incipient decompression melting, and sample preparation. Therefore, the quantities stored in grain boundaries at depth cannot be constrained.

2.06.4.3.2 Helium isotopic variations

Early measurements of ultramafic xenoliths from alkali basalts found $^3He/^4He$ ratios up to $10R_A$ (e.g., Kyser and Rison, 1982; Tolstikhin *et al.*, 1974). These data indicated that helium with significant fractions of primordial ^3He, rather than just radiogenic helium, was present in xenoliths. Porcelli *et al.* (1986a) found that recently erupted samples from a number of locations worldwide had isotopic compositions that fell within the restricted range of 6.2–$10.6R_A$, and suggested that some of the samples studied earlier with lower values had been altered by post-eruption radiogenic production of ^4He. The first-order observation (Porcelli *et al.*, 1986a) is that the data fall into a relatively narrow range that is close to that of typical or "normal" MORBs (N-MORBs), and is distinctive from both radiogenic helium that characterizes crustal rocks ($\sim 0.05R_A$) and mantle helium typical of OIBs with high $^3He/^4He$ ratios (Figure 7). Therefore, it appears that helium in the lithosphere sampled by xenoliths is similar to that in the underlying convecting mantle, and does not contain large fractions of radiogenic helium. In detail, however, the samples do not always correspond completely with MORB values. A common feature of some regions is that the $^3He/^4He$ ratios are somewhat lower, i.e., more radiogenic, than the depleted N-MORB reservoir (Dunai and Baur, 1995).

The isotopic composition of helium can be compared with that of other tracers of mantle sources. Available data for measured $^3He/^4He$ ratios are compared with those of $^{87}Sr/^{86}Sr$ in Figure 8. A range of $^{87}Sr/^{86}Sr$ ratios have been obtained for these samples, from values within the range of MORB to the highest value measured in a mantle sample of 0.8360 for a Tanzanian garnet lherzolite (Cohen *et al.*, 1984). These values represent a range of mantle sources for the lithophile elements as well as examples of very long isolation times from asthenospheric mantle in a lithospheric region with high Rb/Sr ratios. Since xenoliths that are enriched in trace elements are not particularly depleted in uranium (e.g., Cohen *et al.*, 1984), radiogenic ^{87}Sr is expected to be accompanied by ^4He, even though it is not known how Rb/Sr relates to U/He ratios. However, in contrast to the wide spread in strontium isotope compositions, helium isotopes fall within a restricted range. The lack of correlation between these ratios indicates that there generally has been complete decoupling of the two isotopic systems (Porcelli *et al.*, 1986a). Similar decoupling between helium and strontium isotopes has also

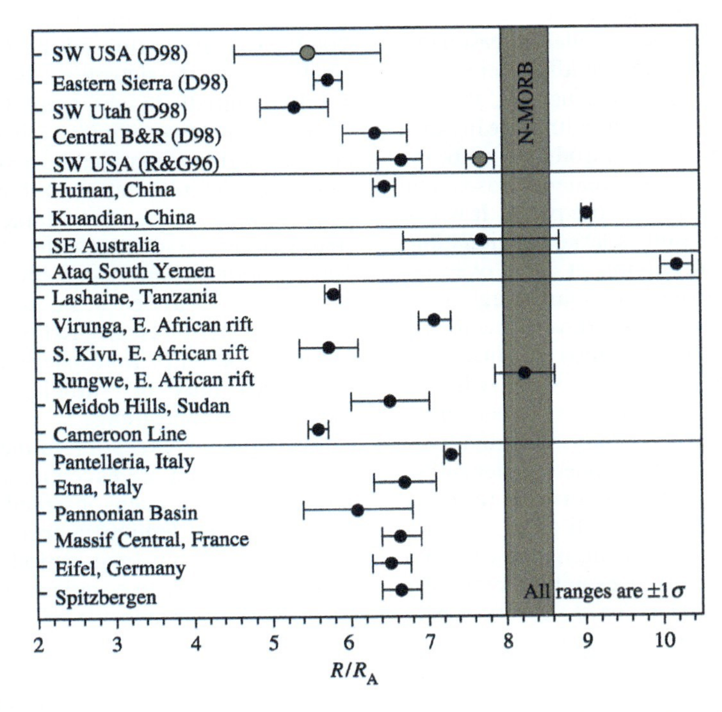

Figure 7 Published $^3He/^4He$ for phenocrysts and ultramafic xenoliths from various continental localities worldwide (reproduced by permission of Mineralogical Society of America from *Rev. Mineral. Geochem.*, **2002**, *47*, 380).

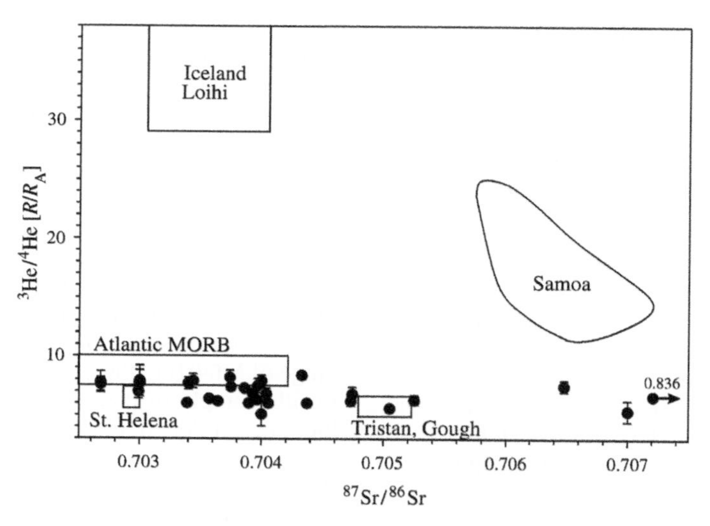

Figure 8 He–Sr isotope relationships in xenoliths (reproduced by permission of Mineralogical Society of America from *Rev. Mineral. Geochem.*, **2002**, *47*, 383).

been observed for xenoliths from ocean islands (Vance *et al.*, 1989), as well as between helium and neodymium isotopes (Stone *et al.*, 1990). It is possible that the decoupling of helium and strontium ratios reflects processes that fractionate rubidium from strontium but not uranium from helium, and so the He/U ratio of the lithosphere is very low. Since substantial amounts of helium are likely to have been lost during eruption due to decrepitation of helium-bearing fluid inclusions, measured He/U ratios are generally gross underestimates of the ratio within the lithosphere. However, while it might be considered reasonable that the formation of migrating fluids preferentially concentrated both uranium and helium, there is likely to be strong fractionations within the invaded mantle as the uranium is introduced into the mineral lattices or newly created phases and the helium remains in the volatile phase. It would be expected that the result is large fractionations between uranium and helium, at least over the limited distances between the source regions of different xenoliths, and so the development of a much wider range of $^{3}He/^{4}He$ ratios than observed. The more likely explanation is that while a dominant fraction of the helium has been introduced into the samples recently, the dominant fraction of strontium has been added earlier and so has had time to isotopically evolve to more radiogenic compositions (Porcelli *et al.*, 1986a). It has been shown that neodymium, and certainly many other trace elements, does not preferentially enter CO_2-rich fluids (Meen *et al.*, 1989). Where these elements are enriched, the transporting agent must have been a silicate or carbonatitic melt. Therefore, the introduction of the volatiles found in the fluid inclusions may have been the result of fluid migration that occurred much more recently than the events responsible for the enrichment of other trace elements.

2.06.4.3.3 Helium transport

Available evidence suggests that CO_2-rich fluids invade the source regions of the xenoliths not long before their entrainment. Trapping of the volatiles and associated noble gases from the host magma degassing during ascent is generally not likely (see discussion in Dunai and Porcelli, 2002). The generation of the CO_2-rich fluids is likely related to upwelling of mantle material, resulting initially in very small degrees of melting. These fluids must efficiently invade the xenolith source region, completely overwhelming any radiogenic helium that might have been expected to accompany the radiogenic strontium and neodymium. It is possible that while He/U fractionation occurs in the lithosphere, this may occur on a restricted scale so that the regional average is not unusually radiogenic. In this case, homogenization of lithospheric helium by invading fluids may be sufficient to erase more radiogenic compositions.

An important exception to the general trend in the xenolith data are exceptionally high $^{3}He/^{4}He$ ratios that have been observed in samples that have very low helium concentrations and have been exposed for long periods of time at the surface. Exposure to cosmic rays can result in the production of cosmogenic ^{3}He (Craig and Poreda, 1986; Kurz, 1986), and this can dominate over the mantle ^{3}He contained in the samples (Porcelli *et al.*, 1987). Subsequent studies have endeavored to avoid any samples that may be susceptible to such alterations.

2.06.4.3.4 Tracing mantle sources

Various studies have examined whether the mantle sources of helium in particular regions can be identified. An extensive study of the Eifel, Germany, and Massif Central, France (Dunai and Baur, 1995) found a restricted range of $^3He/^4He$ ratios. The samples had sufficiently high helium that corrections for post-eruption additions of radiogenic ^4He and cosmogenic ^3He, using intermineral data, were small. Samples from the Massif Central had a mean $^3He/^4He$ ratio of $6.53 \pm 0.25R_A$, while two xenoliths from the Eifel had a value of $6.03 \pm 0.14R_A$. The samples include a range of lithologies, including pyroxenites, dunites, and lherzolites, and often contained hydrous phases. Lithophile isotopes and trace elements also provided evidence for both depletion and subsequent trace element enrichment events (e.g., Stosch *et al.*, 1980). The general uniformity of the values over many vents and regardless of lithology is remarkable (Dunai and Baur, 1995). The somewhat lower values at two other Eifel vents may reflect some radiogenic production in the source (Dunai and Baur, 1995). A review of neodymium, strontium, and lead data indicates that the volcanics at each of a number of localities across Europe, including the Eifel and Massif Central, fall on correlation lines that extend from a common component that has the characteristics of HIMU mantle (Wilson and Downes, 1991). The correlations have been interpreted as mixing between HIMU mantle plume material with small variable contributions from a local lithospheric source. Seismic imaging suggests that small plume upwellings supply volcanic centers throughout continental Europe, and the geochemical similarities suggest a common deep plume source (Granet *et al.*, 1995; Hoernle *et al.*, 1995). Recent seismic data suggests that broader upwelling occurs from depths of up to 2,000 km, and this single source may feed smaller plumes that are responsible for the surface volcanics (Goes *et al.*, 1999). In such a case, the overlap of helium isotopic compositions in the xenoliths with that of some HIMU basalts ($6.8 \pm 0.9\,R_A$; Hanyu and Kaneoka, 1997) suggests that the xenolith helium is dominated by this asthenospheric mantle component, and the local lithospheric components that add other trace elements are an insignificant source of helium. It appears that at many locations across Europe, helium in the xenoliths was likely supplied by melts from the same source as the host magmas that invaded the lithospheric mantle in advance of xenolith entrainment, and provided exsolved CO_2 and other volatiles.

A number of noble gas studies have focused on xenoliths from the Pliocene to Recent alkali basalts of the Newer Volcanics in SE Australia (Matsumoto *et al.*, 1997, 1998, 2000; Porcelli *et al.*, 1986a, 1987, 1992). These basalts appear to be associated with plume-related volcanism (Wellman and McDougall, 1974; Zhang *et al.*, 2001), and xenoliths from a range of depths have been found in a number of young vents. The xenoliths represent a range of petrologic types (O'Reilly and Griffin, 1984) and have undergone differing degrees of metasomatic addition of trace elements, with the addition in some samples of phases such as amphibole and apatite (Griffin *et al.*, 1988; O'Reilly and Griffin, 1988). A range of values for $^3He/^4He$ ratios of $7.1-9.8R_A$ has been found for anhydrous spinel lherzolites, with no systematic difference between vents (Matsumoto *et al.*, 1998). These values clearly overlap the MORB range, and are accompanied by clearly defined neon isotope systematics that also overlap the MORB range. Analyses of peridotites containing amphibole and apatite, as well as garnet pyroxenites, found values of $6-10\,R_A$ (Matsumoto *et al.*, 2000; Porcelli *et al.*, 1986a, 1992), excluding those that have higher ratios clearly due to cosmogenic inputs (Porcelli *et al.*, 1987). Note that values below $7.2R_A$ are generally due to radiogenic inputs that are more evident in low temperature step-heating extraction steps. Therefore, helium in all these xenoliths could have been derived from the MORB-source region. The petrology and noble gas characteristics of the xenoliths clearly require a series of events well after formation of the lithosphere. The garnet pyroxenites appear to be cumulates from intruding melts that crosscut the peridotites (Griffin *et al.*, 1988). Lithophile isotopes indicate that multiple metasomatic episodes occurred in the source (Griffin *et al.*, 1988; McDonough and McCulloch, 1987). The amphiboles, derived from volatile-rich fluids, were formed ~300–500 Ma, based on Sr–Nd systematics (Griffin *et al.*, 1988). The noble gases, derived from a source with MORB characteristics—presumably the convecting upper mantle, were introduced either very recently, overwhelming any helium encountered in the lithosphere, or in such high concentrations that isotopic compositions were not substantially altered by radiogenic growth of ^4He in the lithosphere.

Extensive volcanism can be found in eastern Africa associated with the East African rift system. Helium isotope ratios measured in basalts from the Ethiopian Rift Valley and Afar, near the triple junction of Red Sea, Aden, and Ethiopian rift systems, are $6-17R_A$ (Marty *et al.*, 1993, 1996; Scarsi and Craig, 1996). These values include MORB ratios as well as indications of both contributions from radiogenic ^4He and high $^3He/^4He$ hotspots. To the south, the rift system divides into eastern and western branches. Xenoliths from Tanzanian vents in the eastern branch have $^3He/^4He = 5.8–7.3R_A$ (Porcelli *et al.*, 1986a, 1987). The olivines in these samples have $^3He/^4He$ $\sim 7R_A$, and so coincide with MORB values, so lower values in this study may be due to post-eruption radiogenic production in the uranium-rich phases. In contrast, osmium isotope data of xenoliths from the nearby Labait vent suggests the involvement of plume material (Chesley *et al.*, 1999). Therefore, the complex involvement of different mantle sources may be involved. Kivu and Virunga volcanics of the western branch of the East African rift show highly variable trace element and isotopic compositions that are indicative of a lithospheric mantle source that is heterogeneous on a small scale (Furman and Graham, 1999), and have $^3He/^4He$ ratios which are significantly different from each other (Graham *et al.*, 1995). The range appears to reflect different mixing proportions of asthenospheric sources with lithospheric components with lower $^3He/^4He$ ratios that presumably have developed within the lithosphere itself.

The relative importance of lithospheric and asthenospheric contributions to the helium in basalts in the southwestern USA has been investigated by Reid and Graham (1996) and Dodson *et al.* (1998). Reid and Graham (1996) found that the most radiogenic helium ratios (5.5 R_A) were associated with basalts that exhibit thorium and lead isotope evidence for derivation from an enriched source in the lithospheric mantle that has remained unmodified since $\sim$1.7 Ga. Furthermore, a strong positive correlation between $^3He/^4He$ ratios and ε_{Nd} was found, suggesting an affinity between light rare earth enrichment and time-integrated $(U + Th)/^3He$ ratios. The range in observed $^3He/^4He$ ratios (5.5–7.7 R_A) most likely represents a mixture between asthenospheric and lithospheric melts. Reid and Graham (1996) concluded that the Proterozoic lithospheric mantle in the southwestern US is not a highly degassed reservoir contaminated by helium derived from the underlying asthenosphere, but rather a reservoir with only slightly elevated $(U + Th)/^3He$ ratios (and so slightly lower $^3He/^4He$ ratios) compared to the depleted upper mantle MORB source. Another conclusion was that processes that enrich the lithospheric mantle in LILEs also add helium, so that there is no significant decoupling between LILE and helium. Dodson *et al.* (1998) significantly expanded the database of Reid and Graham (1996) and found a much larger range in $^3He/^4He$ ratios extending to lower, i.e., more radiogenic, $^3He/^4He$ ratios (2.8–7.8 R_A). This was interpreted as reflecting variations in the age of the lithospheric mantle and the degree of degassing. It has been suggested that lower $^3He/^4He$ ratios could be explained by degassing and radiogenic in-growth during magmatic differentiation with the superimposed effects of crustal contamination (Gasparon *et al.*, 1994; Hilton *et al.*, 1993a, 1995; Parello *et al.*, 2000). Overall, the question of whether the low $^3He/^4He$ ratios represent a source feature or are due to processes occurring at shallower depths can only be settled by studying more samples from each location, especially the locations that are closest to end-member compositions. A convergence of $^3He/^4He$ ratios in multiple samples from a single locality and with high helium concentrations is needed to conclusively confirm source region characteristics.

It is worth noting that although not seen in xenoliths, helium from hotspots with high $^3He/^4He$ ratios have sometimes found their surface expression in continental areas (e.g., Basu *et al.*, 1993, 1995; Dodson *et al.*, 1997; Graham *et al.*, 1998; Kirstein and Timmerman, 2000; Marty *et al.*, 1996; Scarsi and Craig, 1996) and consequently must have influenced the affected continental lithosphere. This includes 250 Ma old Siberian flood basalts (with 13 R_A; Basu *et al.*, 1995), 42 Ma old proto-Iceland plume basalts in Ireland (Kirstein and Timmerman, 2000), 380 Ma old Kola Peninsula carbonatites (with 19 R_A; Marty *et al.*, 1998), and Archean komatiites (Richard *et al.*, 1996). However, there are few indications of $^3He/^4He$ ratios greater than MORB in young xenoliths. It might be noted that the products of the earliest stages of the plume-related volcanism in a given continental area seem to have the largest chance of picking up relatively unaltered asthenospheric high $^3He/^4He$ ratios, and the isotopic expression of high $^3He/^4He$ hotspots wanes relatively quickly as the continental plate moves over the hotspot (Dunai and Porcelli, 2002). Basalts erupting subsequently in the same area may then fail to pick up any high $^3He/^4He$ ratios (Basu *et al.*, 1993, 1995; Dodson *et al.*, 1997). Therefore, input of helium with high $^3He/^4He$ ratios related to hotspots may be a transient feature in the lithospheric mantle and so has not left an enduring imprint on lithospheric xenoliths that have been brought to the surface subsequently.

2.06.5 CONCLUDING REMARKS

Although noble gases have contributed much to the development of ideas concerning the Earth's mantle, there remain large gaps in our knowledge that limit further progress. In terms of understanding volatile recycling involving the mantle, there is still poor control on the input parameters of the noble gases (and other volatiles of interest), as there is on fluxes from output localities such as fore-arc and back-arc regions. Furthermore, there are large uncertainties in arc output fluxes—both for various arc segments individually as well as arcs globally. We will become less reliant on the limited datasets currently available only through collection of more (high quality) data on targeted areas and samples. A related concern pertains to our understanding of elemental and isotopic fractionation processes during metamorphic devolatilization (subduction) and during magma storage, crystallization, and degassing. Progress in these areas may come from vapor–melt partitioning studies allied to volatile studies specific to particularly relevant magmatic systems.

In the case of understanding mantle structure, the main obstacle to a comprehensive description of mantle noble gas evolution is in finding a configuration for distinct mantle domains that is compatible with geophysical observations. In this regard, a major difficulty is the heat-helium imbalance. Models involving mantle stratification can account for this, but have been discounted on other geophysical grounds. No other adequate explanation for this imbalance has been proposed. Another problem requiring resolution is the nature of the high $^3He/^4He$ OIB source region. Most models equate this region with undepleted, undegassed mantle, although some models invoke depletion mechanisms. However, none of these have matched the end-member components seen in OIB lithophile isotope correlations. It remains to be demonstrated that a primitive component is present and so can dominate the helium and neon isotope signatures in OIBs. Further data is also required regarding the heavy noble gas isotope compositions of OIBs, the rates of noble gas subduction, and the mechanisms of noble gas incorporation within the nascent Earth. A final comprehensive model may involve the core and several mantle reservoirs, with the dispositions of the mantle reservoirs and the fluxes between them defined by geophysical considerations.

The decoupling of helium from other isotopic signatures in samples from the subcontinental mantle is likely due to relatively recent addition of helium to the xenolith in fluids with high concentrations of helium relative to other trace elements, either within the mantle or during transport. The introduction of helium isotopes into ultramafic xenoliths clearly occurs not long before eruption and does not reflect the longer lithospheric history of the rest of the rocks. One avenue of future research is relating the introduction of volatiles into the lithosphere to local tectonic and volcanic history. If the metasomatic history of the xenoliths can be related to regional rifting and magmatic activity, then a greater understanding of the history and composition of the lithosphere can be obtained from the more completely explored surface geologic history.

There are few constraints on the heavier noble gases in ultramafic xenoliths. While obtaining credible heavy isotope compositions is hampered by low concentrations, more refined analyses and the collection of more gas-rich samples may obtain compositions that provide information on whether separation of the noble gases occurs within the subcontinental lithosphere. Further progress on the origin and nature of noble gases in the lithospheric mantle may be obtained from studies that combine evidence from fluid inclusions, textures, lithophile trace element isotope systematics, and major element variations to understand how noble gas concentrations and isotopic compositions are introduced and modified within the lithosphere.

ACKNOWLEDGMENTS

Discussions over many years with friends and colleagues contributed to ideas presented in this review. The authors thank Rick Carlson for editorial patience and remarks on the manuscript, and Chris Ballentine for a detailed and constructive review.

REFERENCES

Abbott D. A., Burgess L., Longhi J., and Smith W. H. F. (1993) An empirical thermal history of the Earth's upper mantle. *J. Geophys. Res.* **99**, 13835–13850.

Allard P. (1992) Global emissions of He-3 by subaerial volcanism. *Geophys. Res. Lett.* **19**, 1479–1481.

Allègre C. J. (1997) Limitation on the mass exchange between the upper and lower mantle: the evolving convection regime of the Earth. *Earth Planet. Sci. Lett.* **150**, 1–6.

Allègre C. J., Staudacher T., Sarda P., and Kurz M. (1983) Constraints on the evolution of Earth's mantle from the rare gas systematics. *Nature* **303**, 762–766.

Allègre C. J., Staudacher T., and Sarda P. (1986) Rare gas systematics: formation of the atmosphere, evolution and structure of the Earth's mantle. *Earth Planet. Sci. Lett.* **87**, 127–150.

Alvarez L. W. and Cornog R. (1939) He-3 in helium. *Phys. Rev.* **56**, 379.

Anderson D. L. (1993) Helium-3 from the mantle: primordial signal or cosmic dust? *Science* **261**, 170–176.

Anderson D. L. (1998) The helium paradoxes. *Proc. Natl Acad. Sci.* **95**, 4822–4827.

Andrews J. N. (1985) The isotopic composition of radiogenic helium and its use to study groundwater movement in confined aquifers. *Chem. Geol.* **49**, 339–351.

Aston F. W. (1919) A positive ray spectrograph. *Phil. Mag.* **38**, 707–714.

Aston F. W. (1920a) The constitution of atmospheric neon. *Phil. Mag.* **39**, 449–455.

Aston F. W. (1920b) The constitution of the elements. *Nature* **105**, 8.

Aston F. W. (1920c) The mass spectra of chemical elements. *Phil. Mag.* **39**, 611–625.

Aston F. W. (1927) Bakerian lecture: a new mass-spectrograph and the whole number rule. *Proc. Roy. Soc. London* **A115**, 487–514.

Ballentine C. J. (1997) Resolving the mantle He/Ne and crustal Ne-21/Ne-22 in well gases. *Earth Planet. Sci. Lett.* **152**, 233–249.

Ballentine C. J. and Barfod D. N. (2000) The origin of air-like noble gases in MORB and OIB. *Earth Planet. Sci. Lett.* **180**, 39–48.

Ballentine C. J. and Burnard P. G. (2002) Production, release, and transport of noble gases in the continental crust. *Rev. Mineral. Geochem.* **47**, 481–538.

Ballentine C. J. and O'Nions R. K. (1992) The nature of mantle neon contributions to Vienna basin hydrocarbon reservoirs. *Earth Planet. Sci. Lett.* **113**, 553–567.

Ballentine C. J., van Keken P. E., Porcelli D., and Hauri E. H. (2002) Numerical models, geochemistry and the zero-paradox noble-gas mantle. *Phil. Trans. Roy. Soc. London* **A360**, 2611–2631.

Barfod D. N., Ballentine C. J., Halliday A. N., and Fitton J. G. (1999) Noble gases in the Cameroon line and the He, Ne, and Ar isotopic compositions of high mu (HIMU) mantle. *J. Geophys. Res.* **104**, 29509–29527.

Basu A. R., Renne P. R., Dasgupta D. K., Teichmann F., and Poreda R. J. (1993) Early and late alkali igneous pulses and a high ^{3}He plume origin for the Deccan flood basalts. *Science* **261**, 902–906.

Basu A. R., Poreda R. J., Renne P. R., Teichmann F., Vasiliev Y. R., Sobolev N. V., and Turrin B. D. (1995) High-He-3 plume origin and temporal spatial evolution of the Siberian flood basalts. *Science* **269**, 822–825.

Batiza R. (1982) Abundances, distribution and sizes of volcanoes in the Pacific Ocean and implications for the origin of non-hotspot volcanoes. *Earth Planet. Sci. Lett.* **60**, 195–206.

Bebout G. E. (1995) The impact of subduction-zone metamorphism on mantle-ocean chemical cycling. *Chem. Geol.* **126**, 191–218.

Becker T. W., Kellogg J. B., and O'Connell R. J. (1999) Thermal constraints on the survival of primitive blobs in the lower mantle. *Earth Planet. Sci. Lett.* **171**, 351–365.

Begemann R., Weber H. W., and Hintenberger H. (1976) On the primordial abundance of argon-40. *Astrophys. J.* **203**, L155–L157.

Blundell D., Freeman R., and Müller S. (1992) *A Continent Revealed: The European Geotraverse*. Cambridge University Press, Cambridge.

Brantley S. L. and Koepenick K. W. (1995) Measured carbon dioxide emissions from Oldoinyo Lengai and the skewed distribution of passive volcanic fluxes. *Geology* **23**, 933–936.

Broadhurst C. L., Drake M. J., Hagee B. E., and Bernatowicz T. J. (1990) Solubility and partitioning of Ar in anorthite, diopside, forsterite, spinel, and synthetic basaltic liquids. *Geochim. Cosmochim. Acta* **54**, 299–309.

Broadhurst C. L., Drake M. J., Hagee B. E., and Bernatowicz T. J. (1992) Solubility and partitioning of Ne, Ar, and Xe in minerals and synthetic basaltic melts. *Geochim. Cosmochim. Acta* **56**, 709–723.

Brooker R. A., Du Z., Blundy J. D., Kelley S. P., Allan N. L., Wood B. J., Chamorro E. M., Wartho J. A., and Purton J. A. (2003) 'Zero charge' partitioning behaviour of noble gases during mantle melting. *Nature* **473**, 738–741.

Buffett B. A., Huppert H. E., Lister J. R., and Woods A. W. (1996) On the thermal evolution of the Earth's core. *J. Geophys. Res.* **101**, 7989–8006.

Burnard P. (1999) Origin of argon–lead isotopic correlations in basalts. *Science* **286**, 871.

Burnard P., Graham D., and Turner G. (1997) Vesicle-specific noble gas analyses of "popping rock": implications for primordial noble gases in Earth. *Science* **276**, 568–571.

Burnard P. G., Stuart F. M., Turner G., and Oskarsson N. (1994) Air contamination of basaltic magmas—implications for high ^{3}He/^{4}He mantle Ar isotopic composition. *J. Geophys. Res.* **99**, 17709–17715.

Burnham O. M., Rogers N. W., Pearson D. G., van Calsteren P. W., and Hawkesworth C. J. (1998) The petrogenesis of the eastern Pyrenean peridotites: an integrated study of their whole rock geochemistry and Re–Os isotope composition. *Geochim. Cosmochim. Acta* **62**, 2293–2310.

Caffee M. W., Hudson G. B., Velsko C., Huss G. R., Alexander E. C., and Chivas A. R. (1999) Primordial noble gases from Earth's mantle: identification of a primitive volatile component. *Science* **285**, 2115–2118.

Carlson R. W. (1995) Isotopic inferences on the chemical structure of the mantle. *J. Geodynam.* **20**, 365–386.

Carlson R. W. and Irving A. J. (1994) Depletion and enrichment history of subcontinental lithospheric mantle—an Os, Sr, Nd, and Pb isotopic study of ultramafic xenoliths from the northwestern Wyoming craton. *Earth Planet. Sci. Lett.* **126**, 457–472.

Carroll M. R. and Draper D. S. (1994) Noble gases as trace elements in magmatic processes. *Chem. Geol.* **117**, 37–56.

Chamorro E. M., Brooker R. A., Wartho J. A., Wood B. J., Kelley S. P., and Blundy J. D. (2002) Ar and K partitioning between clinopyroxene and silicate melt to 8 GPa. *Geochim. Cosmochim. Acta* **66**, 507–519.

Chesley J. T., Rudnick R. L., and Lee C.-T. (1999) Re–Os systematics of mantle xenoliths from the East African rift: age, structure, and history of the Tanzanian craton. *Geochim. Cosmochim. Acta* **63**, 1203–1217.

Christensen U. R. and Hofmann A. W. (1994) Segregation of subducted oceanic crust in the convecting mantle. *J. Geophys. Res.* **99**, 19867–19884.

Clarke W. B., Beg M. A., and Craig H. (1969) Excess He-3 in the sea: evidence for terrestrial primordial helium. *Earth Planet. Sci. Lett.* **6**, 213–220.

Cohen R. S., O'Nions R. K., and Dawson J. B. (1984) Isotope geochemistry of xenoliths from east Africa: implications for development of mantle reservoirs and their interaction. *Earth Planet. Sci. Lett.* **68**, 209–220.

Coltice N. and Ricard Y. (1999) Geochemical observations and one layer mantle convection. *Earth Planet. Sci. Lett.* **174**, 125–137.

Craig H. and Lupton J. E. (1976) Primordial neon, helium, and hydrogen in oceanic basalts. *Earth Planet. Sci. Lett.* **31**, 369–385.

Craig H. and Lupton J. E. (1981) Helium-3 and mantle volatiles in the ocean and oceanic crust. In *The Sea* (ed. C. Emiliani). Wiley, New York, vol. 7, pp. 391–428.

Craig H. and Poreda R. J. (1986) Cosmogenic ^{3}He in terrestrial rocks: the summit lavas of Maui. *Proc. Natl Acad. Sci. USA* **83**, 1970–1974.

Craig H., Clarke W. B., and Beg M. A. (1975) Excess ^{3}He in deep water on the East Pacific rise. *Earth Planet. Sci. Lett.* **26**, 125–132.

Creager K. C. and Jordan T. H. (1986) Slab penetration into the lower mantle beneath the Marianas and other island arcs of the northwest Pacific. *J. Geophys. Res.* **91**, 3573–3589.

Crisp J. A. (1984) Rates of magma emplacement and volcanic output. *J. Volcanol. Geotherm. Res.* **20**, 177–211.

Davies G. F. (1984) Geophysical and isotopic constraints on mantle convection: an interim synthesis. *J. Geophys. Res.* **89**, 6017–6040.

Davies G. F. (1995) Punctuated tectonic evolution of the Earth. *Earth Planet. Sci. Lett.* **136**, 363–379.

Dempster A. J. (1918) A new method of positive ray analysis. *Phys. Rev.* **11**, 316–325.

DePaolo D. J. (1979) Implications of correlated Nd and Sr isotopic variations for the chemical evolution of the crust and mantle. *Earth Planet. Sci. Lett.* **43**, 201–211.

Dixon E. T., Honda M., McDougall I., Campbell I. H., and Sigurdsson I. (2000) Preservation of near-solar neon isotopic ratios in Icelandic basalts. *Earth Planet. Sci. Lett.* **180**, 309–324.

Dixon J. E. and Stolper E. M. (1995) An experimental study of water and carbon dioxide solubilities in mid-ocean ridge basaltic liquids: 2. Applications to degassing. *J. Petrol.* **36**, 1633–1646.

Dodson A., Kennedy B. M., and DePaolo D. J. (1997) Helium and neon isotopes in the Imnaha basalt, Columbia River Basalt Group: evidence for a Yellowstone plume source. *Earth Planet. Sci. Lett.* **150**, 443–451.

Dodson A., DePaolo D. J., and Kennedy B. M. (1998) Helium isotopes in lithospheric mantle: evidence from Tertiary basalts of the western USA. *Geochim. Cosmochim. Acta* **62**, 3775–3787.

Doe B. R. and Zartman R. E. (1979) Plumbotectonics: I. The Phanerozoic. In *Geochemistry of Hydrothermal Ore Deposits* (ed. H. L. Barnes). Wiley, New York, pp. 22–70.

Downes H. (2001) Formation and modification of the shallow sub-continental lithospheric mantle: a review of geochemical evidence from ultramafic xenolith suites and tectonically emplaced ultramafic massifs of western and central Europe. *J. Petrol.* **42**, 233–250.

Dunai T. J. and Baur H. (1995) Helium, neon, and argon systematics of the European subcontinental mantle—implications for its geochemical evolution. *Geochim. Cosmochim. Acta* **59**, 2767–2783.

Dunai T. J. and Porcelli D. (2002) Storage and transport of noble gases in the subcontinental lithosphere. *Rev. Mineral. Geochem.* **47**, 371–409.

Dymond J. and Hogan L. (1973) Noble gas abundance patterns in deep sea basalts—primordial gases from the mantle. *Earth Planet. Sci. Lett.* **20**, 131–139.

Eiler J. M., Farley K. A., and Stolper E. M. (1998) Correlated helium and lead isotope variations in Hawaiian lavas. *Geochim. Cosmochim. Acta* **62**, 1977–1984.

Farley K. A. and Craig H. (1994) Atmospheric argon contamination of ocean island basalt olivine phenocrysts. *Geochim. Cosmochim. Acta* **58**, 2509–2517.

Farley K. A. and Neroda E. (1998) Noble gases in the Earth's mantle. *Ann. Rev. Earth Planet. Sci.* **26**, 189–218.

Farley K. A., Maier-Reimer E., Schlosser P., and Broecker W. S. (1995) Constraints on mantle ^{3}He fluxes and deep-sea circulation from an oceanic general circulation model. *J. Geophys. Res.* **100**, 3829–3839.

Fischer T. P., Giggenbach W. F., Sano Y., and Williams S. N. (1998) Fluxes and sources of volatiles discharged from Kudryavy, a subduction zone volcano Kurile Islands. *Earth Planet. Sci. Lett.* **160**, 81–96.

Fischer T. P., Hilton D. R., Zimmer M. M., Shaw A. M., Sharp Z. D., and Walker J. A. (2002) Subduction and recycling of nitrogen along the central American margin. *Science* **297**, 1154–1157.

Fisher D. E. (1985) Noble gases from oceanic island basalt do not require an undepleted mantle source. *Nature* **316**, 716–718.

Frey F. A. and Green D. H. (1974) The mineralogy, geochemistry, and origin of lherzolite inclusions in Victorian basanites. *Geochim. Cosmochim. Acta* **38**, 1023–1059.

Furman T. and Graham D. W. (1999) Erosion of lithospheric mantle beneath the East African rift system: geochemical evidence from the Kivu volcanic province. *Lithos* **48**, 237–262.

Galer S. J. G. and Mezger K. (1998) Metamorphism, denudation and sea level in the Archean and cooling of the Earth. *Precamb. Res.* **92**, 389–412.

Galer S. J. G. and O' Nions R. K. (1985) Residence time of thorium, uranium and lead in the mantle with implications for mantle convection. *Nature* **316**, 778–782.

Gasparon M., Hilton D. R., and Varne R. (1994) Crustal contamination processes traced by helium isotopes—examples from the Sunda arc, Indonesia. *Earth Planet. Sci. Lett.* **126**, 15–22.

Geiss J. and Reeves H. (1972) Cosmic and solar system abundances of deuterium and helium-3. *Astron. Astrophys.* **18**, 126–132.

Gerling E. K. and Levskii L. K. (1956) On the origin of the rare gases in stony meteorites. *Geokhimiya* **7**, 59–64.

Goes S., Spakman W., and Bijwaard H. (1999) A lower mantle source for central European volcanism. *Science* **286**, 1928–1931.

Graham D. W. (2002) Noble gas isotope geochemistry of mid-ocean ridge and ocean island basalts: characterization of mantle source reservoirs. *Rev. Mineral. Geochem.* **47**, 247–317.

Graham D. W., Lupton J., Albarède F., and Condomines M. (1990) Extreme temporal homogeneity of helium isotopes at Piton de la Fournaise, Reunion Island. *Nature* **347**, 545–548.

Graham D. W., Christie D. M., Harpp K. S., and Lupton J. E. (1993) Mantle plume helium in submarine basalts from the Galapagos platform. *Science* **262**, 2023–2026.

Graham D. W., Furman T. H., Ebinger C. J., Rogers N. W., and Lupton J. E. (1995) Helium, lead, strontium, and neodymium isotope variations in mafic volcanic rocks from the western branch of the East African rift. *EOS Trans.: Am. Geophys. Union* **76**, F686.

Graham D. W., Larsen L. M., Hanan B. B., Storey M., Pedersen A. K., and Lupton J. E. (1998) Helium isotope composition of the early Iceland mantle plume inferred from the tertiary picrites of West Greenland. *Earth Planet. Sci. Lett.* **160**, 241–255.

Grand S. P. (1987) Tomographic inversion of sear velocity beneath the North American plate. *J. Geophys. Res.* **92**, 14065–14090.

Grand S. P. (1994) Mantle shear structure beneath the Americas and surrounding oceans. *J. Geophys. Res.* **99**, 66–78.

Granet M., Wilson M., and Achauer U. (1995) Imaging a mantle plume beneath the French Massif Central. *Earth Planet. Sci. Lett.* **136**, 281–296.

Griffin W. L., O'Reilly S. Y., and Stabel A. (1988) Mantle metasomatism beneath western Victoria, Australia: II. Isotopic geochemistry of Cr-diopside lherzolites and Al-augite pyroxenites. *Geochim. Cosmochim. Acta* **52**, 449–459.

Hamano Y. and Ozima M. (1978) Earth-atmosphere evolution model based on Ar isotopic data. In *Terrestrial Rare Gases* (eds. M. Ozima and E. C. Alexander, Jr.). Japan Scientific Societies Press, Tokyo, pp. 155–171.

Handler M. R., Bennett V. C., and Esat T. M. (1997) The persistence of off-cratonic lithospheric mantle: Os isotopic systematics of variably metasomatised southeast Australian xenoliths. *Earth Planet. Sci. Lett.* **151**, 61–75.

Hanyu T. and Kaneoka I. (1997) The uniform and low ^{3}He/^{4}He ratios of HIMU basalts as evidence for the origin as recycled materials. *Nature* **390**, 273–276.

Harrison D., Burnard P., and Turner G. (1999) Noble gas behaviour and composition in the mantle: constraints from the Iceland plume. *Earth Planet. Sci. Lett.* **171**, 199–207.

Hart R., Dymond J., and Hogan L. (1979) Preferential formation of the atmosphere-sialic crust system from the upper mantle. *Nature* **278**, 156–159.

Hart S. R. and Staudigel H. (1982) The control of alkalis and uranium in seawater by ocean crust alteration. *Earth Planet. Sci. Lett.* **58**, 202–212.

Hauri E. H., Whitehead J. A., and Hart S. R. (1994) Fluid dynamic and geochemical aspects of entrainment in mantle plumes. *J. Geophys. Res.* **99**, 24275–24300.

Hawkesworth C. J., Rogers N. W., van Calsteren P. W. C., and Menzies M. A. (1984) Mantle enrichment processes. *Nature* **311**, 331–334.

Hawkesworth C. J., Kempton P. D., Rogers N. W., Ellam R. M., and van Calsteren P. W. (1990) Continental mantle lithosphere, and shallow level enrichment processes in the Earth's mantle. *Earth Planet. Sci. Lett.* **96**, 256–268.

Helffrich G. R. and Wood B. J. (2001) The Earth's mantle. *Nature* **412**, 501–507.

Hill R. D. (1941) Production of helium-3. *Phys. Rev.* **59**, 103.

Hilton D. R. and Craig H. (1989) A helium isotope transect along the Indonesian archipelago. *Nature* **342**, 906–908.

Hilton D. R., Hoogewerff J. A., van Bergen M. J., and Hammerschmidt K. (1992) Mapping magma sources in the east Sunda-Banda arcs, Indonesia: constraints from helium isotopesd. *Geochim. Cosmochim. Acta* **56**, 851–859.

Hilton D. R., Hammerschmidt K., Loock G., and Friedrichsen H. (1993a) Helium and argon isotope systematics of the central Lau Basin and Valu Fa ridge: evidence of crust–mantle interactions in a back-arc basin. *Geochim. Cosmochim. Acta* **57**, 2819–2841.

Hilton D. R., Hammerschmidt K., Teufel S., and Friedrichsen H. (1993b) Helium isotope characteristics of Andean geothermal fluids and lavas. *Earth Planet. Sci. Lett.* **120**, 265–282.

Hilton D. R., Barling J., and Wheller G. E. (1995) Effect of shallow-level contamination on the helium isotope systematics of ocean-island lavas. *Nature* **373**, 330–333.

Hilton D. R., McMurtry G. M., and Kreulen R. (1997) Evidence for extensive degassing of the Hawaiian mantle plume from helium–carbon relationships at Kilauea volcano. *Geophys. Res. Lett.* **24**, 3065–3068.

Hilton D. R., McMurtry G. M., and Goff F. (1998) Large variations in vent fluid $CO_2/^3He$ ratios signal rapid changes in magma chemistry at Loihi Seamount, Hawaii. *Nature* **396**, 359–362.

Hilton D. R., Gronvold K., Macpherson C. G., and Castillo P. R. (1999) Extreme $^3He/^4He$ ratios in northwest Iceland: constraining the common component in mantle plumes. *Earth Planet. Sci. Lett.* **173**, 53–60.

Hilton D. R., Macpherson C. G., and Elliott T. R. (2000a) Helium isotope ratios in mafic phenocrysts and geothermal fluids from La Palma, the Canary Islands (Spain): implications for HIMU mantle sources. *Geochim. Cosmochim. Acta* **64**, 2119–2132.

Hilton D. R., Thirlwall M. F., Taylor R. N., Murton B. J., and Nichols A. (2000b) Controls on magmatic degassing along the Reykjanes ridge with implications for the helium paradox. *Earth Planet. Sci. Lett.* **183**, 43–50.

Hilton D. R., Fischer T. P., and Marty B. (2002) Noble gases and volatile recycling at subduction zones. *Rev. Mineral. Geochem.* **47**, 319–370.

Hiyagon H. (1994) Retentivity of solar He and Ne in IDPs in deep sea sediment. *Science* **263**, 1257–1259.

Hiyagon H. and Ozima M. (1982) Noble gas distribution between basalt melt and crystals. *Earth Planet. Sci. Lett.* **58**, 255–264.

Hiyagon H., Ozima M., Marty B., Zashu S., and Sakai H. (1992) Noble gases in submarine glasses from mid-oceanic ridges and Loihi Seamount—constraints on the early history of the Earth. *Geochim. Cosmochim. Acta* **56**, 1301–1316.

Hoernle K., Zhang Y. S., and Graham D. (1995) Seismic and geochemical evidence for large-scale mantle upwelling beneath the eastern Atlantic and western and central Europe. *Nature* **374**, 34–39.

Hofmann A. W. (1997) Mantle geochemistry: the message from oceanic volcanism. *Nature* **385**, 219–229.

Hogness T. R. and Kvalnes H. M. (1928) The ionization processes in methane interpreted by the mass spectrograph. *Phys. Rev.* **32**, 942–945.

Holloway J. R. (1998) Graphite-melt equilibria during mantle melting: constraints on CO_2 in MORB magmas and the carbon content of the mantle. *Chem. Geol.* **147**, 89–97.

Honda M. and McDougall I. (1998) Primordial helium and neon in the Earth—a speculation on early degassing. *Geophys. Res. Lett.* **25**, 1951–1954.

Honda M., Reynolds J., Roedder E., and Epstein S. (1987) Noble gases in diamonds: occurrences of solar like helium and neon. *J. Geophys. Res.* **92**, 12507–12521.

Honda M., McDougall I., Patterson D. B., Doulgeris A., and Clague D. (1991) Possible solar noble gas component in Hawaiian basalts. *Nature* **349**, 149–151.

Honda M., McDougall I., and Patterson D. (1993) Solar noble gases in the Earth: the systematics of helium–neon isotopes in mantle derived samples. *Lithos* **30**, 257–265.

Jambon A., Weber H., and Braun O. (1986) Solubility of He, Ne, Ar, Kr, and Xe in a basalt melt in the range 1,250–1,600 °C. Geochemical implications. *Geochim. Cosmochim. Acta* **50**, 401–408.

Javoy M. and Pineau F. (1991) The volatile record of a "popping" rock from the Mid-Atlantic Ridge at 14° N: chemical composition of the gas trapped in the vesicles. *Earth Planet. Sci. Lett.* **107**, 598–611.

Jochum K. P., Hofmann A. W., Ito E., Seufert H. M., and White W. M. (1983) K, U, and Th in mid-ocean ridge glasses and heat production, K/U and K/Rb in the mantle. *Nature* **306**, 431–436.

Jordan T. H. (1975) The continental tectosphere. *Rev. Geophys. Space Phys.* **13**, 1–2.

Kamijo K., Hashizume K., and Matsuda J. (1998) Noble gas constraints on the evolution of the atmosphere mantle system. *Geochim. Cosmochim. Acta* **62**, 2311–2321.

Kaneoka I., Takaoka N., and Upton B. G. J. (1986) Noble gas systematics in basalts and a dunite nodule from Reunion and Grand Comore Islands, Indian Ocean. *Chem. Geol.* **59**, 35–42.

Kàrason H. and van der Hilst R. D. (2001) Tomographic imaging of the lowermost mantle with differential times of refracted and diffracted core phases (PKP, P-diff). *J. Geophys. Res.* **106**, 6569–6587.

Kellogg L. H., Hager B. H., and van der Hilst R. D. (1999) Compositional stratification in the deep mantle. *Science* **283**, 1881–1884.

Kellogg L. H. and Wasserburg G. J. (1990) The role of plumes in mantle helium fluxes. *Earth Planet. Sci. Lett.* **99**, 276–289.

Kerrick D. M. and Connolly J. A. D. (2001) Metamorphic devolatilization of subducted marine sediments and the transport of volatiles into the Earth's mantle. *Nature* **411**, 293–296.

Kirstein L. A. and Timmerman M. J. (2000) Evidence of the proto-Iceland plume in northwestern Ireland at 42 Ma from helium isotopes. *J. Geol. Soc. London* **157**, 923–927.

Kunz J., Staudacher T., and Allègre C. J. (1998) Plutonium-fission xenon found in Earth's mantle. *Science* **280**, 877–880.

Kurz M. D. (1986) Cosmogenic helium in a terrestrial igneous rock. *Nature* **320**, 435–439.

Kurz M. D., Jenkins W. J., and Hart S. R. (1982) Helium isotope systematics of ocean islands and mantle heterogeneity. *Nature* **297**, 43–47.

Kurz M. D., Colodner D., Trull T. W., Moore R. B., and O'Brien K. (1990) Cosmic ray exposure dating with *in situ* produced cosmogenic He-3—results from young Hawaiian lava flows. *Earth Planet. Sci. Lett.* **97**, 177–189.

Kyser T. K. and Rison W. (1982) Systematics of rare gas isotopes in basic lavas and ultramafic xenoliths. *J. Geophys. Res.* **87**, 5611–5630.

Lockyer J. N. (1869) Spectroscopic observations of the sun. *Phil. Trans.* **159**, 425–444.

Lupton J. E. (1983) Terrestrial inert gases: isotope tracer studies and clues to primordial components in the mantle. *Ann. Rev. Earth Planet. Sci.* **11**, 371–414.

Lupton J. E. and Craig H. (1975) Excess ^3He in oceanic basalts: evidence for terrestrial primordial helium. *Earth Planet. Sci. Lett.* **26**, 133–139.

Mahaffy P. R., Donahue T. M., Atreya S. K., Owen T. C., and Niemann H. B. (1998) Galileo probe measurements of D/H and $^3He/^4He$ in Jupiter's atmosphere. *Space Sci. Rev.* **84**, 251–263.

Mamyrin B. A. and Tolstikhin I. N. (1984) *Helium Isotopes in Nature*. Elsevier, Amsterdam.

Mamyrin B. A., Tolstikhin I. N., Anufriev G. S., and Kamensky I. L. (1969) Anomalous isotopic composition of helium in volcanic gases. *Dokl. Akad. Nauk. SSSR* **184**, 1197–1199.

Manga M. (1996) Mixing of heterogeneities in the mantle: effect of viscosity differences. *Geophys. Res. Lett.* **23**, 403–406.

Martel D. J., Deak J., Dovenyi P., Horvath F., O'Nions R. K., Oxburgh E. R., Stegena L., and Stute M. (1989) Leakage of helium from the Pannonian basin. *Nature* **342**, 908–912.

Marti K. and Mathew K. (1998) Noble-gas components in planetary atmospheres and interiors in relation to solar wind and meteorites. *Proc. Indian Acad. Sci.: Earth Planet. Sci.* **107**, 425–431.

Marty B. (1995) Nitrogen content of the mantle inferred from N_2/Ar correlations in oceanic basalts. *Nature* **377**, 326–329.

Marty B. and Allé P. (1994) Neon and argon isotopic constraints on Earth-atmosphere evolution. In *Noble Gas Geochemistry and Cosmochemistry* (ed. J.-I. Matsuda). Terra Scientific Publishing Company, Tokyo, pp. 191–204.

Marty B. and Dauphas N. (2002) Formation and early evolution of the atmosphere. *J. Geol. Soc. London Spec. Publ.: Early History of the Earth* (eds. C. M. R. Fowler, C. J. Ebinger, and C. J. Hawkesworth). Geological Society of London, London, UK, vol. 199, pp. 213–219.

Marty B. and Humbert F. (1997) Nitrogen and argon isotopes in oceanic basalts. *Earth Planet. Sci. Lett.* **152**, 101–112.

Marty B. and Jambon A. (1987) C/^{3}He in volatile fluxes from the solid Earth: implication for carbon geodynamics. *Earth Planet. Sci. Lett.* **83**, 16–26.

Marty B. and LeCloarec M. F. (1992) Helium-3 and CO_2 fluxes from subaereal volcanoes estimated from Polonium-210 emissions. *J. Volcanol. Geotherm. Res.* **53**, 67–72.

Marty B. and Lussiez P. (1993) Constraints on rare gas partition coefficients from analysis of olivine glass from a picritic mid-ocean ridge basalt. *Chem. Geol.* **106**, 1–7.

Marty B. and Tolstikhin I. N. (1998) CO_2 fluxes from mid-ocean ridges, arcs and plumes. *Chem. Geol.* **145**, 233–248.

Marty B. and Zimmermann L. (1999) Volatiles (He, C, N, Ar) in mid-ocean ridge basalts: assessment of shallow-level fractionation and characterization of source composition. *Geochim. Cosmochim. Acta* **63**, 3619–3633.

Marty B., Jambon A., and Sano Y. (1989) Helium isotopes and CO_2 in volcanic gases of Japan. *Chem. Geol.* **76**, 25–40.

Marty B., Appora I., Barrat J. A. A., Deniel C., Vellutini P., and Vidal P. (1993) He, Ar, Sr, Nd, and Pb isotopes in volcanic rocks from Afar—evidence for a primitive mantle component and constraints on magmatic sources. *Geochem. J.* **27**, 219–228.

Marty B., Trull T., Lussiez P., Basile I., and Tanguy J. C. (1994) He, Ar, O, Sr, and Nd isotope constraints on the origin and evolution of Mount Etna magmatism. *Earth Planet. Sci. Lett.* **126**, 23–39.

Marty B., Pik R., and Gezahegn Y. (1996) Helium isotopic variations in Ethiopian plume lavas—nature of magmatic sources and limit on lower mantle contribution. *Earth Planet. Sci. Lett.* **144**, 223–237.

Marty B., Tolstikhin I., Kamensky I. L., Nivin V., Balagans-kaya E., and Zimmermann J. L. (1998) Plume-derived rare gases in 380 Ma carbonatites from the Kola region (Russia) and the argon isotopic composition in the deep mantle. *Earth Planet. Sci. Lett.* **164**, 179–192.

Matsuda J. and Marty B. (1995) The $^{40}Ar/^{36}Ar$ ratio of the undepleted mantle: a reevaluation. *Geophys. Res. Lett.* **22**, 1937–1940.

Matsuda J. and Matsubara K. (1989) Noble gases in silica and their implication for the terrestrial "missing" Xe. *Geophys. Res. Lett.* **16**, 81–84.

Matsuda J. and Nagao K. (1986) Noble gas abundance in a deep-sea sediment core from eastern equatorial Pacific. *Geochem. J.* **20**, 71–80.

Matsuda J., Sudo M., Ozima M., Ito K., Ohtaka O., and Ito E. (1993) Noble gas partitioning between metal and silicate under high pressures. *Science* **259**, 788–790.

Matsumoto T., Honda M., McDougall I., Yatsevich I., and O'Reilly S. Y. (1997) Plume-like neon in a metasomatic apatite from the Australian lithospheric mantle. *Nature* **388**, 162–164.

Matsumoto T., Honda M., McDougall I., and O'Reilly S. Y. (1998) Noble gases in anhydrous lherzolites from the Newer volcanics, southeastern Australia: a MORB like reservoir in the subcontinental mantle. *Geochim. Cosmochim. Acta* **62**, 2521–2533.

Matsumoto T., Honda M., McDougall I., O'Reilly S. Y., Norman M., and Yaxley G. (2000) Noble gases in pyroxenites and metasomatized peridotites from the Newer volcanics, southeastern Australia: implications for mantle metasomatism. *Chem. Geol.* **168**, 49–73.

McDonough W. F. (1990) Constraints on the composition of the continental lithospheric mantle. *Earth Planet. Sci. Lett.* **101**, 1–18.

McDonough W. F. and McCulloch M. T. (1987) The southeast Australian lithospheric mantle: isotopic and geochemical constraints on its growth and evolution. *Earth Planet. Sci. Lett.* **86**, 327–340.

McDougall I. and Honda M. (1998) Primordial solar gas component in the Earth: consequences for the origin and evolution of the Earth and its atmosphere. In *The Earth's Mantle: Composition, Structure, and Evolution* (ed. I. Jackson). Cambridge University Press, Cambridge, pp. 159–187.

Meen J. K., Eggler D. H., and Ayers J. C. (1989) Experimental evidence for very low solubility of rare earth elements in CO_2-rich fluids at mantle conditions. *Nature* **340**, 301–303.

Menzies M. and Chazot G. (1985) Fluid processes in diamond to spinel facies shallow mantle. *J. Geodynam.* **20**, 387–415.

Menzies M. and Chazot G. (1995) Fluid processes in diamond to spinel facies shallow mantle. *J. Geodynam.* **20**, 387–415.

Moreira M. and Allègre C. J. (1998) Helium–neon systematics and the structure of the mantle. *Chem. Geol.* **147**, 53–59.

Moreira M. and Sarda P. (2000) Noble gas constraints on degassing processes. *Earth Planet. Sci. Lett.* **176**, 375–386.

Moreira M., Kunz J., and Allègre C. (1998) Rare gas systematics in popping rock: isotopic and elemental compositions in the upper mantle. *Science* **279**, 1178–1181.

Morgan J. P. (1998) Thermal and rare gas evolution of the mantle. *Chem. Geol.* **145**, 431–445.

Morgan W. J. (1971) Convection plumes in the lower mantle. *Nature* **230**, 42–43.

Morrison P. and Pine J. (1955) Radiogenic origin of the helium isotopes in rock. *Ann. NY Acad. Sci.* **62**, 69–92.

Nier A. O. (1936) The isotopic constitution of rubidium, zinc, and argon. *Phys. Rev.* **49**, 272.

Nolet G., Grand S. P., and Kennet B. L. N. (1994) Seismic heterogeneity in the upper mantle. *J. Geophys. Res.* **99**, 23753–23776.

Ohmoto H. (1986) Stable isotope geochemistry of ore deposits. *Rev. Mineral.* **16**, 491–560.

O'Nions R. K. and Oxburgh E. R. (1983) Heat and helium in the Earth. *Nature* **306**, 429–431.

O'Nions R. K. and Oxburgh E. R. (1988) Helium, volatile fluxes and the development of continental crust. *Earth Planet. Sci. Lett.* **90**, 331–347.

O'Nions R. K. and Tolstikhin I. N. (1994) Behaviour and residence times of lithophile and rare gas tracers in the upper mantle. *Earth Planet. Sci. Lett.* **124**, 131–138.

O'Nions R. K. and Tolstikhin I. N. (1996) Limits on the mass flux between lower and upper mantle and stability of layering. *Earth Planet. Sci. Lett.* **139**, 213–222.

O'Nions R. K., Evenson N. M., and Hamilton P. J. (1979) Geochemical modeling of mantle differentiation and crustal growth. *J. Geophys. Res.* **84**, 6091–6101.

O'Reilly S. Y. and Griffin W. L. (1984) A xenolith-derived geotherm for southeastern Australia and its geophysical implications. *Tectonophysics* **111**, 41–63.

O'Reilly S. Y. and Griffin W. L. (1988) Mantle metasomatism beneath western Victoria, Australia: I. Metasomatic processes in Cr-diopside lherzolites. *Geochim. Cosmochim. Acta* **52**, 433–447.

O'Reilly S. Y. and Griffin W. L. (2000) Apatite in the mantle: implications for metasomatic processes and high heat production in Phanerozoic mantle. *Lithos* **53**, 217–232.

Ott U. (2002) Noble gases in meteorites-trapped components. *Rev. Mineral. Geochem.* **47**, 71–100.

Oxburgh E. R., O'Nions R. K., and Hill R. L. (1986) Helium isotopes in sedimentary basins. *Nature* **324**, 632–635.

Ozima M. (1994) Noble gas state in the mantle. *Rev. Geophys.* **32**, 405–426.

Ozima M. and Podosek F. A. (1983) *Noble Gas Geochemistry*. Cambridge University Press, Cambridge.

Parello F., Allard P., D'Alessandro W., Federico C., Jean-Baptiste P., and Catani O. (2000) Isotope geochemistry of Pantelleria volcanic fluids, Sicily Channel rift: a mantle volatile end-member for volcanism in southern Europe. *Earth Planet. Sci. Lett.* **180**, 325–339.

Parsons B. (1981) The rates of plate creation and consumption. *Geophys. J. Roy. Astron. Soc.* **67**, 437–448.

Patterson D. B., Honda M., and McDougall I. (1990) Atmospheric contamination—a possible source for heavy noble gases in basalts from Loihi Seamount, Hawaii. *Geophys. Res. Lett.* **17**, 705–708.

Pearson D. G. (1999a) The age of continental roots. *Lithos* **48**, 171–194.

Pearson D. G. (1999b) Evolution of cratonic lithospheric mantle: an isotopic perspective. In *Mantle Petrology: Field Observations and High Pressure Experimentation: A Tribute to Francis R. (Joe) Boyd*, The Geochemical Society Special Publication (eds. Y. Fei, C. M. Bertka, and B. O. Mysen). The Geochemical Society, Houston, Texas, vol. 6, pp. 57–78.

Pepin R. O. (1992) Origin of noble gases in the terrestrial planets. *Ann. Rev. Earth Planet. Sci.* **20**, 389–430.

Pepin R. O. and Porcelli D. (2002) Origin of noble gases in the terrestrial planets. *Rev. Mineral. Geochem.* **47**, 191–246.

Peters K. E., Sweeney R. E., and Kaplan I. R. (1978) Correlation of carbon and nitrogen stable isotope ratios in sedimentary organic matter. *Limnol. Ocean.* **23**, 598–604.

Phinney D., Tennyson J., and Frick U. (1978) Xenon in CO_2 well gas revisited. *J. Geophys. Res.* **83**, 2313–2319.

Plank T. and Langmuir C. H. (1988) An evaluation of the global variations in the major element chemistry of arc basalts. *Earth Planet. Sci. Lett.* **90**, 349–370.

Podosek F. A. (1978) Isotopic structure in solar system materials. *Ann. Rev. Astron. Astrophys.* **16**, 293–334.

Podosek F. A., Honda M., and Ozima M. (1980) Sedimentary noble gases. *Geochim. Cosmochim. Acta* **44**, 1875–1884.

Polet J. and Anderson D. L. (1995) Depth extension of cratons as inferred from tomographic studies. *Geology* **23**, 205–208.

Pollack H. N., Hurter S. J., and Johnson J. R. (1993) Heat flow from the Earth's interior: analysis of the global data set. *Rev. Geophys.* **31**, 267–280.

Porcelli D. and Ballentine C. J. (2002) Models for the distribution of terrestrial noble gases and the evolution of the atmosphere. *Rev. Mineral. Geochem.* **47**, 412–480.

Porcelli D. and Halliday A. N. (2001) The core as a possible source of mantle helium. *Earth Planet. Sci. Lett.* **192**, 45–56.

Porcelli D. and Wasserburg G. J. (1995a) Mass transfer of helium, neon, argon, and xenon through a steady state upper mantle. *Geochim. Cosmochim. Acta* **59**, 4921–4937.

Porcelli D. and Wasserburg G. J. (1995b) Mass transfer of xenon through a steady-state upper mantle. *Geochim. Cosmochim. Acta* **59**, 1991–2007.

Porcelli D., O'Nions R. K., and O'Reilly S. Y. (1986a) Helium and strontium isotopes in ultramafic xenoliths. *Chem. Geol.* **54**, 237–249.

Porcelli D., Stone J. O. H., and O'Nions R. K. (1986b) Rare gas reservoirs and Earth degassing. *Lunar Planet. Sci.* **XVII**, 674–675.

Porcelli D., Stone J. O. H., and O'Nions R. K. (1987) Enhanced $^3He/^4He$ ratios and cosmogenic helium in ultramafic xenoliths. *Chem. Geol.* **64**, 25–33.

Porcelli D., Ballentine C. J., and Wieler R. (eds.) (2002) In *Noble Gases in Geochemistry and Cosmochemistry, Reviews in Mineralogy and Geochemistry*, Mineralogical Society of America, Washington, DC, vol. 47.

Porcelli D. R., O'Nions R. K., Galer S. J. G., Cohen A. S., and Mattey D. P. (1992) Isotopic relationships of volatile and lithophile trace elements in continental ultramafic xenoliths. *Contrib. Mineral. Petrol.* **110**, 528–538.

Poreda R. and Craig H. (1989) Helium isotope ratios in circum-Pacific volcanic arcs. *Nature* **338**, 473–478.

Poreda R. J. and Farley K. A. (1992) Rare gases in Samoan xenoliths. *Earth Planet. Sci. Lett.* **113**, 129–144.

Poreda R. J., Jenden P. D., Kaplan I. R., and Craig H. (1986) Mantle helium in Sacramento basin natural gas wells. *Geochim. Cosmochim. Acta* **50**, 2847–2853.

Ramsay W. (1895) Discovery of helium. *Chem. News* **71**, 151.

Ramsay W. and Soddy F. (1903) Experiments in radioactivity and the production of helium from radium. *Proc. Roy. Soc. London* **A72**, 204.

Ramsay W. and Travers M. W. (1898a) On a new constituent of atmospheric air. *Proc. Roy. Soc. London* **A63**, 405–408.

Ramsay W. and Travers M. W. (1898b) On the companions of argon. *Proc. Roy. Soc. London* **A63**, 437–440.

Ramsay W. and Travers M. W. (1898c) On the extraction from air of the companions of argon and on neon. *Br. Ass. Rept.* 828–830.

Rayleigh L. (1892) Density of nitrogen. *Nature* **46**, 512–513.

Rayleigh L. and Ramsay W. (1892) Argon, a new constituent of the atmosphere. *Proc. Roy. Soc. London* **A57**, 265.

Reid M. R. and Graham D. W. (1996) Resolving lithospheric and sub-lithospheric contributions to helium isotope variations in basalts from the southwestern US. *Earth Planet. Sci. Lett.* **144**, 213–222.

Reymer A. and Schubert G. (1984) Phanerozoic addition rates to the continental crust. *Tectonics* **3**, 63–77.

Ricard Y., Nataf H.-C., and Montagner J.-P. (1996) The three-dimensional seismological model a priori constrained: confrontation with seismic data. *J. Geophys. Res.* **101**, 8457–8472.

Richard D., Marty B., Chaussidon M., and Arndt N. (1996) Helium isotopic evidence for a lower mantle component in depleted Archean komatiite. *Science* **273**, 93–95.

Richardson S. H. and Harris J. W. (1997) Antiquity of peridotitic diamonds from the Siberian craton. *Earth Planet. Sci. Lett.* **151**, 271–277.

Richardson S. H., Gurney J. J., Erlank A. J., and Harris J. W. (1984) Origin of diamonds in old enriched mantle. *Nature* **310**, 198–202.

Rocholl A. and Jochum K. P. (1993) Th, U, and other trace elements in carbonaceous chondrites—implications for the terrestrial and solar system Th/U ratios. *Earth Planet. Sci. Lett.* **117**, 265–278.

Rodgers N. W., De Mulder M., and Hawkesworth C. J. (1992) An enriched mantle source for potassic basanites: evidence from Karisimbi volcano, Virunga volcanic province, Rwanda. *Contrib. Mineral. Petrol.* **111**, 543–556.

Rudnick R. L. and Nyblade A. A. (1999) The thickness and heat production of Archean lithosphere: constraints from xenolith thermobarometry and surface heat flow. In *Mantle Petrology: Field Observations and High Pressure Experimentation: A Tribute to Francis R. (Joe) Boyd*, The Geochemical Society Special Publication (eds. Y. Fei, C. M. Bertka, and B. O. Mysen). The Geochemical Society, Houston, Texas, vol. 6, pp. 3–12.

Rudnick R. L., McDonough W. F., and O'Connell R. J. (1998) Thermal structure, thickness and composition of continental lithosphere. *Chem. Geol.* **145**, 399–415.

Sano Y. and Marty B. (1995) Origin of carbon in fumarolic gas from island arcs. *Chem. Geol.* **119**, 265–274.

Sano Y. and Williams S. N. (1996) Fluxes of mantle and subducted carbon along convergent plate boundaries. *Geophys. Res. Lett.* **23**, 2749–2752.

Sano Y., Takahata N., Nishio Y., and Marty B. (1998) Nitrogen recycling in subduction zones. *Geophys. Res. Lett.* **25**, 2289–2292.

Sano Y., Takahata N., Nishio Y., Fischer T. P., and Williams S. N. (2001) Volcanic flux of nitrogen from the Earth. *Chem. Geol.* **171**, 263–271.

Sarda P. and Graham D. (1990) Mid-ocean ridge popping rocks: implications for degassing at ridge crests. *Earth Planet. Sci. Lett.* **97**, 268–289.

Sarda P., Staudacher T., and Allègre C. J. (1988) Neon isotopes in submarine basalts. *Earth Planet. Sci. Lett.* **91**, 73–88.

Sarda P., Moreira M., and Staudacher T. (1999) Argon–lead isotopic correlation in Mid-Atlantic ridge basalts. *Science* **283**, 666–668.

Sarda P., Moreira M., Staudacher T., Schilling J. G., and Allègre C. J. (2000) Rare gas systematics on the southernmost Mid-Atlantic ridge: constraints on the lower mantle and the Dupal source. *J. Geophys. Res.* **105**, 5973–5996.

Saunders A. and Tarney J. (1991) Back-arc basins. In *Oceanic Basalts* (ed. P. A. Floyd). Blackie, Glasgow, pp. 219–263.

Scarsi P. (2000) Fractional extraction of helium by crushing of olivine and clinopyroxene phenocrysts: effects on the He-3/He-4 measured ratio. *Geochim. Cosmochim. Acta* **64**, 3751–3762.

Scarsi P. and Craig H. (1996) Helium isotope ratios in Ethiopian rift basalts. *Earth Planet. Sci. Lett.* **144**, 505–516.

Schilling J. G., Unni C. K., and Bender M. L. (1978) Origin of chlorine and bromine in the oceans. *Nature* **273**, 631–636.

Schlosser P. and Winckler G. (2002) Noble gases in ocean waters and sediments. *Rev. Mineral. Geochem.* **47**, 701–730.

Seta A., Matsumoto T., and Matsuda J.-I. (2001) Concurrent evolution of $^3He/^4He$ ratio in the Earth's mantle reservoirs for the first 2 Ga. *Earth Planet. Sci. Lett.* **188**, 211–219.

Shaw A. M., Hilton D. R., Macpherson C. G., and Sinton J. M. (2001) Nucleogenic neon in high $^3He/^4He$ lavas from the Manus back-arc basin: a new perspective on He–Ne decoupling. *Earth Planet. Sci. Lett.* **194**, 53–66.

Signer P. and Suess H. E. (1963) Rare gas in the sun, in the atmosphere, and in meteorites. In *Earth Science and Meteorites* (eds. J. Geiss and E. D. Goldberg). Wiley, New York, pp. 241–272.

Silver P. G., Carlson R. W., and Olson P. (1988) Deep slabs, geochemical heterogeneity, and the large-scale structure of mantle convection—investigation of an enduring paradox. *Ann. Rev. Earth Planet. Sci.* **16**, 477–541.

Simons F. J., Zielhuis A., and van der Hilst R. D. (1999) The deep structure of the Australian continent from surface wave tomography. *Lithos* **48**, 17–43.

Snyder G., Poreda R., Hunt A., and Fehn U. (2001) Regional variations in volatile composition: isotopic evidence for carbonate recycling in the central American volcanic arc. *Geochem. Geophys. Geosyst.* **2**, U1–U32.

Sobolev S. V., Zeyen H., Granet M., Achauer U., Bauer C., Werling F., Altherr R., and Fuchs K. (1997) Upper mantle temperatures and lithosphere-asthenosphere system beneath the French Massif Central constrained by seismic, gravity, petrologic, and thermal observations. *Tectonophysics* **275**, 143–164.

Soddy F. (1910) Radioactivity. *Ann. Rep. Chem. Soc.* **7**, 256–286.

Staudacher T. (1987) Upper mantle origin for Harding County well gases. *Nature* **325**, 605–607.

Staudacher T. and Allègre C. J. (1982) Terrestrial xenology. *Earth Planet. Sci. Lett.* **60**, 389–406.

Staudachèr T. and Allègre C. J. (1988) Recycling of oceanic crust and sediments: the noble gas subduction barrier. *Earth Planet. Sci. Lett.* **89**, 173–183.

Staudacher T., Kurz M. D., and Allègre C. J. (1986) New noble gas data on glass samples form Loihi Seamount and Huahalai and on dunite samples from Loihi and Reunion Island. *Chem. Geol.* **56**, 193–205.

Staudacher T., Sarda P., Richardson S. H., Allégre C. J., Sagna I., and Dmitriev L. V. (1989) Noble gases in basalt glasses from a mid-Atlantic ridge topographic high at 14-degrees-N—geodynamic consequences. *Earth Planet. Sci. Lett.* **96**, 119–133.

Stoiber R. E., Malinconico L. L. J., and Williams S. N. (1983) Use of the correlation spectrometer at volcanoes. In *Forecasting Volcanic Events* (eds. H. Tazieff and J. Sabroux). Elsevier, Amsterdam, vol. 1, pp. 425–444.

Stone J., Porcelli D., Vance D., Galer S., and O'Nions R. K. (1990) Volcanic traces. *Nature* **346**, 228.

Stosch H. G., Carlson R. W., and Lugmair G. W. (1980) Episodic mantle differentiation: Nd and Sr isotopic evidence. *Earth Planet. Sci. Lett.* **47**, 263–271.

Strutt R. J. (1908) The accumulation of helium in geologic time. *Proc. Roy. Soc. London* **A81**, 272–277.

Stuart F. M. (1994) Speculations about the cosmic origin of He and Ne in the interior of the Earth—comment. *Earth Planet. Sci. Lett.* **122**, 245–247.

Stute M., Sonntag C., Deak J., and Schlosser P. (1992) Helium in deep circulating groundwater in the Great Hungarian Plain—flow dynamics and crustal and mantle helium fluxes. *Geochim. Cosmochim. Acta* **56**, 2051–2067.

Su W.-J., Woodward R. L., and Dziewonski A. M. (1994) Degree 12 model of shear velocity heterogeneity in the mantle. *J. Geophys. Res.* **99**, 6945–6980.

Suess H. E. and Wanke H. (1965) On the possibility of a helium flux through the ocean floor. In *Progress in Oceanography* (ed. M. Sears). Pergamon, Oxford, vol. 3, pp. 347–353.

Suess H. E., Wanke H., and Wlotzka F. (1964) On the origin of gas rich meteorites. *Geochim. Cosmochim. Acta* **28**, 595–607.

Tackley P. J. (2000) Mantle convection and plate tectonics: toward an integrated physical and chemical theory. *Science* **288**, 2002–2007.

Tackley P. J., Stevenson D. J., and Glatzmaier G. A. (1993) Effects of an endothermic phase transition at 670 km depth on spherical mantle convection. *Nature* **361**, 699–704.

Tedesco D. and Nagao K. (1996) Radiogenic ^4He, ^{21}Ne, and ^{40}Ar in fumarolic gases on vulcano: implication for the presence of continental crust beneath the island. *Earth Planet. Sci. Lett.* **144**, 517–528.

Thomson J. J. (1912) Further experiments on positive rays. *Phil. Mag.* **24**, 209–253.

Tolstikhin I. N. and Marty B. (1998) The evolution of terrestrial volatiles: a view from helium, neon, argon, and nitrogen isotope modelling. *Chem. Geol.* **147**, 27–52.

Tolstikhin I. N., MaMain B. A., Khabarin L. B., and Erlikh E. N. (1974) Isotope composition of helium in ultrabasic xenoliths from volcanic rocks of Kamchatka. *Earth Planet. Sci. Lett.* **22**, 73–84.

Torgersen T. (1989) Terrestrial helium degassing fluxes and the atmospheric helium budget: implications with respect to the degassing processes of continental crust. *Chem. Geol.* **79**, 1–14.

Trieloff M., Kunz J., Clague D. A., Harrison D., and Allègre C. J. (2000) The nature of pristine noble gases in mantle plumes. *Science* **288**, 1036–1038.

Trull T. (1994) Influx and age constraints on the recycled cosmic dust explanation for high $^3He/^4He$ ratios at hotspot volcanoes. In *Noble Gas Geochemistry and Cosmochemistry* (ed. J. Matsuda). Terra Publishers, Tokyo, pp. 77–88.

Trull T. W. and Kurz M. D. (1993) Experimental measurements of ^3He and ^4He mobility in olivine and clinopyroxene at magmatic temperatures. *Geochim. Cosmochim. Acta* **57**, 1313–1324.

Valbracht P. J., Staudacher T., Malahoff A., and Allègre C. J. (1997) Noble gas systematics of deep rift zone glasses from Loihi Seamount, Hawaii. *Earth Planet. Sci. Lett.* **150**, 399–411.

Vance D., Stone J. O. H., and O'Nions R. K. (1989) He, Sr, and Nd isotopes in xenoliths from Hawaii and other oceanic islands. *Earth Planet. Sci. Lett.* **96**, 147–160.

van der Hilst R. D. and Karason H. (1999) Compositional heterogeneity in the bottom 1000 kilometers of Earth's mantle: toward a hybrid convection model. *Science* **283**, 1885–1888.

van der Hilst R. D., Widiyantoro S., and Engdahl E. R. (1997) Evidence for deep mantle circulation from global tomography. *Nature* **386**, 578–584.

van Keken P. E. and Ballentine C. J. (1999) Dynamical models of mantle volatile evolution and the role of phase transitions and temperature-dependent rheology. *J. Geophys. Res.* **104**, 7137–7151.

van Keken P. E., Ballentine C. J., and Porcelli D. (2001) A dynamical investigation of the heat and helium imbalance. *Earth Planet. Sci. Lett.* **188**, 421–434.

van Soest M. C., Hilton D. R., and Kreulen R. (1998) Tracing crustal and slab contributions to arc magmatism in the Lesser Antilles island arc using helium and carbon relationships in geothermal fluids. *Geochim. Cosmochim. Acta* **62**, 3323–3335.

Varekamp J. C., Kreulen R., Poorter R. P. E., and Vanbergen M. J. (1992) Carbon sources in arc volcanism with implications for the carbon cycle. *Terra Nova* **4**, 363–373.

Vidale J. E., Schubert G., and Earle P. S. (2001) Unsuccessful initial search for a mid-mantle chemical boundary with seismic arrays. *Geophys. Res. Lett.* **28**, 859–862.

von Huene R. and Scholl D. W. (1991) Observations at convergent margins concerning sediment subduction, subduction erosion, and the growth of continental crust. *Rev. Geophys.* **29**, 279–316.

Walker R. J., Carlson R. W., Shirey S. B., and Boyd F. R. (1989) Os, Sr, Nd, and Pb isotope systematics of southern African peridotite xenoliths: implications for the chemical evolution of subcontinental mantle. *Geochim. Cosmochim. Acta* **53**, 1583–1595.

Wass S. Y., Henderson P., and Elliot C. J. (1980) Chemical heterogeneity and metasomatism in the upper mantle: evidence from rare earth and other elements in apatite-rich xenoliths in basaltic rocks from eastern Australia. *Phil. Trans. Roy. Soc. London* **A297**, 333–346.

Wellman P. and McDougall I. (1974) Cainozoic igneous activity in eastern Australia. *Tectonophysics* **23**, 49–65.

Wetherill G. W. (1954) Variations in the isotopic abundance of neon and argon extracted from radioactive minerals. *Phys. Rev.* **96**, 679–683.

Wieler R. (2002) Noble gases in the solar system. *Rev. Mineral. Geochem.* **47**, 21–70.

Wilson M. and Downes H. (1991) Tertiary–Quaternary extension-related alkaline magmatism in western and central Europe. *J. Petrol.* **32**, 811–849.

Yatsevich I. and Honda M. (1997) Production of nucleogenic neon in the Earth from natural radioactive decay. *J. Geophys. Res.* **102**, 10291–10298.

Zeeman P. and De Gier J. (1934) Third preliminary note on some experiments concerning the isotope of hydrogen. *Proc. K. Akad. Amsterdam* **37**, 1–3.

Zhang M., Stephenson J., O'Reilly S. Y., McCulloch M. T., and Norman M. (2001) Petrogenesis of Late Cenozoic basalts in North Queensland and its geodynamic implications: trace element and Sr–Nd–Pb isotope evidence. *J. Petrol.* **42**, 685–719.

Zhang Y. and Zindler A. (1989) Noble gas constraints on the evolution of Earth's atmosphere. *J. Geophys. Res.* **94**, 13710–13737.

Zindler A. and Hart S. R. (1986) Chemical geodynamics. *Ann. Rev. Earth Planet. Sci.* **14**, 493–571.

2.07
Mantle Volatiles—Distribution and Consequences

R. W. Luth
University of Alberta, Edmonton, Canada

NOMENCLATURE

f_{O_2}	oxygen fugacity
Fo_{xx}	olivine, with composition given as subscript xx = 100*Mg/(Mg + Fe) on cation basis
P	pressure
T	temperature
$X(H_2O)$	mole fraction H_2O
θ	dihedral angle

2.07.1 INTRODUCTION

Volatiles in the mantle have, for many years, been the subject of intensive study from a number of perspectives. They are of interest because of their potential effects on melting relationships, on transport of major and trace elements, and on the rheological and other physical properties of the mantle. By convention, "volatiles" in this context are constituents that are liquid or gaseous at normal Earth surface conditions. This review will look at the behavior of C–O–H–S–halogen volatiles, beginning with H_2O and C–O volatiles.

There have been tremendous strides made recently towards understanding how volatiles in general and water in particular is transported and stored in the mantle. This progress is based on research on a number of fronts: studies of mantle-derived samples have provided insight into the nature and occurrence of hydrous phases such as amphibole, mica, and chlorite, and have provided constraints on the capacity of nominally anhydrous minerals (NAMs) such as olivine, pyroxenes, and garnet to contain "water" by a variety of substitution mechanisms. Experimental studies on mantle-derived magmas have provided constraints on volatile contents in their source regions. Other studies have constrained the pressure, temperature, and composition conditions over which hydrous phases are stable in the mantle.

Fundamental questions remain about the geochemical cycling of volatiles in the mantle, and between the mantle and the surface. Much attention has focused on the capability of hydrous phases such as amphibole, mica, serpentine, chlorite, and a family of "dense hydrous magnesian silicates" (DHMSs) to act as carriers of water in subducting slabs back into the mantle. It has been clear since the work of Ito *et al.* (1983) that there is a discrepancy between the amount of volatiles subducted into the mantle and those returned to the surface by arc magmatism. A recent overview of volatile cycling in subduction systems by Bebout (1996) suggests that 5–15% of the H_2O and 10–44% of the CO_2 that is subducted is returned to the surface in arc magmatism. He emphasized that the "missing" volatiles may have multiple fates, including incorporation into the mantle wedge, large-scale fluid flow up along the interface between the subducting slab and overlying mantle, and transport into the deeper mantle.

Because of the hydrous nature of arc magmatism, a common hypothesis is that there is a hydrous phase that breaks down at subarc conditions to trigger melting in the overlying mantle wedge to produce arc magmas. A key research goal has been to identify this phase, or phases. For example, serpentine in peridotite will break down during subduction to produce olivine + orthopyroxene + fluid or, in cooler slabs, a progression of DHMSs, the last of which may survive into the transition zone.

At some point, however, because of the limited thermal or pressure stability of the hydrous phases, water will be liberated from the slab into the surrounding mantle. At this point, the water will either exist as a fluid, a melt—or something intermediate if we are above the second critical end point in the relevant system (Wyllie and Ryabchikov, 2000)—or it may dissolve into nominally anhydrous phases.

The understanding of the relevant phase relations for the other volatiles is not as advanced. For carbon, we have a reasonable understanding of its phase stability in the mantle, but there is still no good understanding of the relative importance of carbonates, elemental carbon, and other forms as hosts for carbon in the mantle. In the upper mantle, sulfur resides primarily in sulfides; their behavior during partial melting will play a major role in the geochemical cycling of sulfur as well as of chalcophile elements. The halogens are rare (and rarely studied) in mantle-derived samples; more insight into their behavior is currently coming from the study of mantle-derived magmas.

This review will first consider the evidence from mantle-derived magmas pertaining to volatiles in the mantle, then turn to mantle-derived samples, looking at each major volatile in turn. The experimental studies on questions of volatile behavior will be considered briefly where appropriate. For further information, the reader is referred to the recent reviews on this topic (Ulmer and Trommsdorff, 1999; Frost, 1999; Luth, 1999; Schmidt and Poli, 1998; Wyllie and Ryabchikov, 2000; Williams and Hemley, 2001; Poli and Schmidt, 2002).

2.07.2 EVIDENCE FROM MANTLE-DERIVED MAGMAS

2.07.2.1 Introduction

Mantle-derived magmas provide both qualitative and quantitative indications of the presence and nature of volatiles in the mantle. Qualitative indications of the presence of volatiles in the mantle include the existence of volatile-rich magmas such as kimberlites, lamproites, lamprophyres, and carbonatites. The generally accepted interpretation of these magmas is that they are low-degree partial melts that contain high concentrations of incompatible elements, including volatiles. More quantitative information may be gleaned from the lavas, especially from the volatiles trapped in glassy groundmass or in inclusions, and from considerations of the dependence of the stability of hydrous phases on water content of the melt. It is important to realize that much of the information derived by these techniques addresses the volatile content of the magma in the near surface, pre-eruptive state, and even then has a number of pitfalls, as reviewed by Johnson *et al.* (1994). To go from pre-eruptive concentrations in the magma to concentrations in the source region requires a quantitative understanding of what happens to the magma during ascent to the surface, and of the fine details of the melting process itself. The pre-eruptive concentrations will, however, allow some indication of the variability of volatile concentrations in the source regions.

2.07.2.2 Basalts

Mid-ocean ridge basalts (MORBs), because they erupt subaqueously with a confining pressure imposed by the overlying water column, are excellent candidates to preserve pre-eruptive volatile contents. There is an extensive body of data on H_2O contents of quenched glasses from MORBs (Danyushevsky *et al.*, 2000; Michael, 1995; Johnson *et al.*, 1994; and references therein). The measured H_2O contents are typically less than the amount of H_2O required to saturate the melt at the depth of eruption, and there is no good correlation of H_2O content with depth. Many of these lavas are vesicular, implying saturation with an H_2O-poor vapor phase during emplacement. Given that CO_2 is the second most abundant volatile in MORBs, insight into this problem may be gained by considering the vapor phase produced by vesiculation as a CO_2–H_2O vapor. Dixon and Stolper (1995) analyzed this question in detail, based on their solubility data for H_2O and CO_2 in basaltic liquids (Dixon *et al.*, 1995). They concluded that for relatively H_2O-poor magmas such as MORBs and oceanic island basalts (OIBs), water contents of quenched glasses are good estimates of the water content of the melt for samples erupted at depths greater than 1,000 m. For more H_2O-rich magmas, such as arc or back-arc basalts, exsolution of an H_2O-rich vapor would occur even at water depths >2,000 m.

With these caveats in mind, measured water contents of MORBs range from 0.05 wt.% to 1.40 wt.% (Figure 1) in a skewed distribution with a mean of ~0.33 wt.%. Of the 328 analyses shown in Figure 1, 95% have ≤0.65 wt.% H_2O. These results are corroborated by studies of melt inclusions in phenocrysts from MORBs. Sobolev and Chaussidon (1996) found H_2O contents of 0.07–0.66 wt.%, with enriched mid-ocean ridge basalts (E-MORBs) having consistently higher H_2O contents (0.34–0.66 wt.%) compared to normal mid-ocean ridge basalts (N-MORBs) (0.07–0.19 wt.%).

It has been noted in a number of these studies that H_2O correlates well with other incompatible elements such as lanthanum or caesium (e.g., Michael, 1988; Dixon *et al.*, 1988; and many subsequent studies). This correlation has led to the identification of geographical variation in the H_2O/Ce in the basalts, and by inference in their source regions (e.g., Michael, 1995; Danyushevsky *et al.*, 2000). These studies have

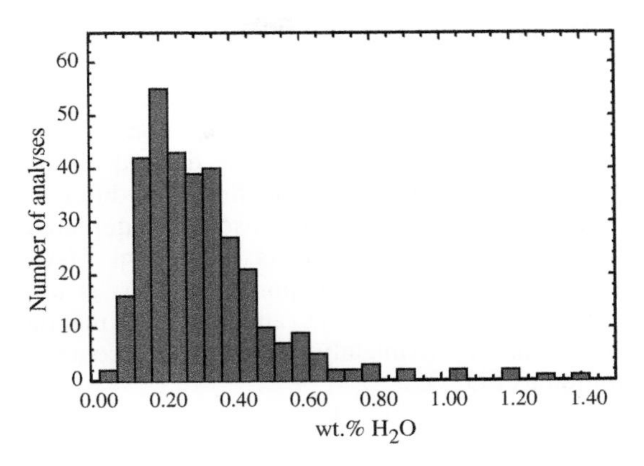

Figure 1 Measured water contents for MORB glasses (sources Johnson *et al.*, 1994; Michael, 1995; Danyushevsky *et al.*, 2000).

used these correlations to estimate the water contents of the source regions (e.g., Byers *et al.*, 1986; Michael, 1988, 1995; Dixon *et al.*, 1988, 2002; Sobolev and Chaussidon, 1996; Danyushevsky *et al.*, 2000). The estimated H_2O contents of the source regions for N-MORBs are on the order of 80–180 ppm, whereas E-MORB source regions are damper, with estimated H_2O concentrations of 200–950 ppm.

There is considerable interest in the water content of plume-related magmas and their source regions. Schilling *et al.* (1980) proposed that hot-spots were also wet-spots; subsequent work confirms this hypothesis (e.g., Dixon and Clague, 2001; Nichols *et al.*, 2002; Dixon *et al.*, 2002). For example, Nichols *et al.* (2002) measured water contents of up to 1.1 wt.% in glasses from submarine and subglacial basaltic eruptions along the Reykjanes Ridge and Iceland, respectively. They evaluated the effects of crystallization, which reduced the maximum water content to ~0.6 wt.%, and used a batch melting model to infer concentrations of H_2O in the source region. Their inferred concentrations of H_2O in the source region increase from 165 ppm at the southern end of the Reykjanes Ridge (furthest from the Icelandic mantle plume) to 620–920 ppm underneath Iceland.

The situation in any single plume appears to be complex; the Hawaii case is illustrative. Dixon *et al.* (1997) estimated H_2O concentrations of 525 ± 75 ppm in the source for North Arch basalts, a seafloor field of alkalic basalts north of Oahu. Wallace (1998) estimated 450 ± 190 ppm H_2O in the source for Kilauea basalts, and Dixon and Clague (2001) inferred 400 ± 30 ppm H_2O in the source for Loihi seamount basalts. Dixon and Clague (2001) argue that these differences are consistent with a mantle plume with multiple components, such as a wet rim and dry core plume model as proposed by Sen *et al.* (1996).

Dixon *et al.* (2002) take this approach one step further, and use new data on water contents, trace element and isotope geochemistry of Atlantic basalts to argue that mantle endmember "reservoirs" or components, previously defined on the basis of their isotope geochemistry, have different water contents. They propose that depleted MORB mantle (DMM) has 100 ppm H_2O, FOZO ("focus zone" component of Hart *et al.*, 1992) has 750 ppm H_2O, and enriched mantle (EM) has approximately half the H_2O concentration as FOZO.

Clearly, there is a heterogenous distribution of H_2O in the mantle, one that appears to correlate with other geochemical and isotopic evidence of heterogeneity (see Chapter 2.03). The implications for the nature of the host phase(s) for water depend very much on the scale of this heterogeneity. If there is 100 ppm H_2O in the MORB source region, and it is distributed relatively homogeneously, the storage capacity of NAMs would suffice to host most, if not all, of the water. However, if water is concentrated in geochemically enriched, low-abundance lithologies such as pyroxenites, a hydrous phase such as amphibole, phlogopite, or a fluid would be required (cf. discussion in Williams and Hemley, 2001).

Because of the questions concerning the fate of subducted water, constraining the water content in the source region of arc magmas is of considerable interest. Analysis of groundmass glass of arc basalts is more problematic than for MORBs, because the eruptions are often subaerial, which are likely to exsolve and degass more efficiently. In addition, as pointed out by Dixon and Stolper (1995), the higher water contents of these magmas produce exsolution and degassing at higher pressures, such that extensive degassing could occur in crustal magma chambers prior to eruption.

Because of these difficulties, the water contents of melt inclusions trapped in phenocrysts in arc volcanic rocks have been used to constrain magmatic water contents. Such analyses provide a snapshot of the melt composition, including volatile content, at one point during the crystallization of the phenocryst. Complications arise because the melt in the inclusion can react with the host mineral, modifying its composition (Danyushevsky *et al.*, 2002)—including volatile content by concentrating the volatiles into a smaller volume of melt—and because volatiles can be lost during experimental rehomogenization of the melt inclusion (Massare *et al.*, 2002). In addition, a number of processes, such as fractional crystallization, assimilation, degassing, or magma mixing, could have affected the volatile content of the melt prior to entrapment of the inclusion. Therefore, extrapolation from the measured volatile concentration back to the concentration in the parental melt may be necessary. Such extrapolations are minimized by judicious choice of samples, such as looking at the most primitive lavas and at early-crystallizing phenocrysts for melt inclusions. Of course, such studies are constrained by the phases that crystallize under conditions that trap melt inclusions in the first place.

Sisson and Layne (1993) analyzed melt inclusions in olivines in samples from four arc volcanoes. Inclusions in olivines in basalts from the 1974 eruption of Fuego volcano, Guatemala, contain 2.1–4.6 wt.% H_2O. Inclusions in olivines in basaltic andesites from the same eruption have a wider range, from 1.0 wt.% to 6.2 wt.% H_2O. Inclusions hosted in olivines in basalts from three centers in the Southern Cascades have lower water contents, ranging from 0.2 wt.% to 1.4 wt.% H_2O.

Sobolev and Chaussidon (1996) summarize an extensive body of data on olivine-hosted melt inclusions from basalts from a variety of tectonic settings. They concentrated on inclusions hosted in magnesium-rich ($f_{o_{87}}$–$f_{o_{94}}$) olivines, with the logic that these olivines would be crystallizing from the most primitive basaltic magmas, and hence provide the best estimate of the water content of the parental, least-modified magma. For arc-related tholeiitic basalts, they report a range of 1.2–2.5 wt.% H_2O. They suggest that the higher values found by Sisson and Layne (1993) result from 30–40% fractional crystallization of more primitive magnesium-rich basalt.

Sobolev and Chaussidon (1996) found a range of 1.0–2.9 wt.% H_2O for boninites. Kamenetsky *et al.* (2002) determined H_2O contents of 0.8–1.9 wt.% in melt inclusions in chromites from low-calcium boninites from Cape Vogel, Papua New Guinea. They suggest that these are minimum values because of low-pressure degassing, and that the H_2O content of the original magma might have been as high as 3 wt.%.

Dobson *et al.* (1995) measured water contents in melt inclusions in pyroxenes in boninites from the Bonin Islands. These melt inclusions contained 2.8–3.2 wt.% H_2O, whereas quenched glass from pillow lava rims had 2.2–2.4 wt.% H_2O, which Dobson *et al.* interpreted as a result of degassing between the time of entrapment of the melt inclusions and the eruption on the seafloor. They also measured the water content of the orthopyroxene, which contained 80–120 ppm H_2O.

Jamtveit *et al.* (2001) measured water contents in olivines from basalts and picrites from the North Atlantic Volcanic Province using Fourier-transform infrared spectroscopy (FTIR). They found H_2O contents from <0.5 ppm to ~18 ppm, and suggested that olivines with

3–7 ppm H_2O crystallized from melts containing a minimum of 0.2–0.5 wt.% H_2O. They infer that the mantle source region for these magmas had >300 ppm H_2O, higher than the H_2O content inferred for the source region of N-MORB (see discussion above). This technique holds great promise for tracing the volatile evolution of evolving magmatic systems, as well as for constraining concentrations of volatiles in the source region. Quantitative insights into the latter will require better understanding both of the partitioning behavior of H_2O between melt and NAMs, and of the nature of the melting process.

Other evidence for the presence of volatile-bearing phases in the source region of mantle-derived magmas comes from modeling the behavior of their minor and trace elements (e.g., Oxburgh, 1964; Sun and Hanson, 1975; Clague and Frey, 1982; Wright, 1984; Hoernle and Schmincke, 1993; Francis and Ludden, 1995; Class and Goldstein, 1997; among many others). For example, Francis and Ludden (1995) used the systematics of the variations in incompatible (large ion lithophile, light rare earth, and high field strength) elements in Late Tertiary to Recent alkaline basalts to argue for residual amphibole in the source region.

Another example of this line of reasoning is given by Class and Goldstein (1997), who argue amphibole and/or phlogopite are residual phases in the source regions for alkaline basaltic magmas on oceanic islands (Grande Comore, Indian Ocean and Oahu, Hawaii). Models of the melting process are used to constrain the amount of these hydrous phases as well; Class and Goldstein (1997) suggest 2% amphibole in the source region for the Grande Comore lavas, for example. Such estimates of modal abundance of hydrous phases are critically dependent on the assumed melting model, however, both in terms of the style of melting (fractional, accumulated fractional, etc.) and the assumed nature of the source region, such as whether the hydrous phase is distributed in veins or homogeneously throughout the mantle peridotite (cf. Foley, 1992 for a discussion of this issue).

Additional insight into the water contents of arc magmas and their source regions comes from experimental studies. In "forward" studies, the effects of H_2O on the composition of partial melts of a postulated source lithology are studied. Initial experiments were conducted in H_2O-saturated systems (e.g., Kushiro *et al.*, 1968; Green, 1973, 1976; Nicholls, 1974; Mysen and Boettcher, 1975). More recently, attention has focused on fluid-undersaturated partial melting (e.g., Kushiro, 1990; Hirose and Kawamoto, 1995; Gaetani and Grove, 1998; Falloon and Danyushevsky, 2000). Ulmer (2001) provides an extensive discussion of the effects of P, T, and water content on the composition of partial melts of peridotitic source lithologies.

More directly relevant to constraining water contents of source regions of specific arc-related rocks are "inverse" experiments. In these experiments, the P, T conditions at which a "primary" liquid is in equilibrium with peridotitic residual phases are determined. The liquid is normally selected on geochemical criteria such as Mg# and nickel concentration, and the residue can be olivine and orthopyroxene, with additional clinopyroxene and an aluminous phase if small degrees of partial melting are considered appropriate. Recent such studies include Tatsumi (1981, 1982), Tatsumi *et al.* (1983, 1994), Foden and Green (1992), Baker *et al.* (1994), and Pichavant *et al.* (2002). Complications in the interpretation of these experiments include (i) the assumption that the liquid did not change composition after formation and before sampling, (ii) the inability of this approach to deal with phases in reaction relation with the liquid, (iii) the lack of any insight into the host phase(s) for water in the source region, and (iv) the exploration of a limited range of P–T–$X(H_2O)$ space (unlike earlier work on more silicic systems by Wyllie and co-workers and others, where T–$X(H_2O)$ sections at fixed pressures were studied for particular bulk compositions). The study of Pichavant *et al.* (2002), in which multiple saturation points for two H_2O contents were determined, is a promising step towards addressing this issue.

2.07.2.3 Kimberlites and Lamproites

The presence of volatile-bearing phases such as phlogopite, apatite, and carbonates in kimberlites testify to the volatile-rich nature of the parental magma (e.g., Mitchell, 1986). The ubiquitous serpentization present in kimberlites cannot be used as evidence of magmatic water, with the exception of groundmass serpentine that is interpreted to be primary in nature. As discussed by Mitchell (1986), there are limited stable isotopic data consistent with a meteoric origin for some of the water in the serpentine. However, it is unclear if these results could be attributed to post-emplacement exchange of deuteric serpentine with meteoric fluids.

Placing constraints on the concentration of volatiles in the original melt or source region is difficult. To date, there have been no studies of melt inclusions in phenocrysts in kimberlites, nor any experimental studies that would allow constraints to be placed on the volatile content of either the kimberlitic parent magma, or its source region.

2.07.2.4 Lamprophyres

Lamprophyres are characteristically volatile-rich, containing hydrous phases such as amphibole or mica. They are usually extensively altered, either by deuteric processes during cooling, or by interaction with other fluids, so there is no groundmass glass present. As of early 2000s, the only study that constrains the water content of lamprophyre melts is that of Righter and Carmichael (1996), who determined that 3–5 wt.% H_2O is required to duplicate experimentally the observed phenocryst assemblages in olivine- and augite-bearing minettes from Mexico.

2.07.2.5 Carbonatites

The presence of carbonatites in both oceanic and continental settings can be interpreted as evidence of widespread carbon-bearing phases in the mantle. Like other indicators, this one is ambiguous. One significant problem is that most natural carbonatites in the geological record are calcium-rich; reconciling this observation with the sodic nature of the active carbonatite magmatism at Oldoinyo Lengai on the one hand (Dawson *et al.*, 1990), and with the dolomitic character of carbonatitic magmas expected from melting of carbonate-bearing peridotite (Dalton and Presnall, 1998; Wyllie and Ryabchikov, 2000) on the other remains problematic. It is possible that the Oldoinyo Lengai natrocarbonatite is simply anomalous and has nothing to do with "normal" carbonatites. This possibility is supported by the radiogenic isotope geochemistry. Many calcium-carbonatites have OIB-like isotopic compositions (Hoernle *et al.*, 2002 and references therein), unlike Oldoinyo Lengai (Bell and Simonetti, 1996 and references therein). Bell and Simonetti (1996) developed a model for the formation of Oldoinyo Lengai and other East African carbonatites in which plume-derived melts or fluids interact with the subcontinental lithosphere. Subsequent melting of this metasomatized lithosphere produces nephelinitic melts, which may either be accompanied by carbonatitic melts, or produce such melts by liquid immiscibility upon ascent (e.g., Kjarsgaard *et al.*, 1995). In the latter case, the sodic nature of the Oldoinyo Lengai carbonatites says nothing about the nature of primary, mantle-derived carbonatites.

The commonly used model for calcium-carbonatite genesis involves partial melting of a carbonate-bearing peridotite, forming a carbonate-rich silicate melt that upon ascent produces immiscible calcium-rich carbonate and magnesium-rich silicate melts (e.g., Lee and Wyllie, 1998 and references therein). Alternatively, calcium-rich carbonatites may be produced by melting of carbonated wehrlites, or by reaction of a precursor dolomitic carbonatite melt with peridotite during ascent (Dalton and Wood, 1993; Lee and Wyllie, 2000). In this model, the wehrlites themselves would be the product of metasomatic reaction of a carbonate-rich melt derived from greater depths. It is safe to say that there is no consensus on the formation of carbonatite, but there is consensus on their mantle origin. Their existence means that there must be a carbon-bearing phase in the mantle, but clarification of the identity of this phase awaits further constraints on the depth of origin of carbonatites, and on the nature of the source-region mineralogy.

For example, an unexplored possibility is that carbonatites form by melting of a source region containing carbon, not carbonate. Melting in the presence of graphite to produce MORB was discussed by Holloway (1998), who found that melts produced in such a fluid-absent process have restricted values of melt Fe^{3+}/Fe^{2+} and CO_2 content. Such a process would require a source of oxygen, which could be the reduction of Fe^{3+}. The primitive upper mantle has been modeled as containing 0.2 wt.% Fe_2O_3 on a whole-rock basis (O'Neill *et al.*, 1993). This relatively low abundance of Fe^{3+} would be a limiting factor in generating carbonatitic melts. However, O'Neill *et al.* (1993) argue that the transition zone would have a minimum Fe_2O_3 content (at metal saturation) of ~0.44 wt.%. They then explore the mechanisms by which this increase in Fe_2O_3 content might occur in going from the upper mantle to the transition zone, primarily by a reaction

$$3Fe^{2+}O_{(silicate)} = Fe^{0}_{(sulfide)} + 2Fe^{3+}O_{1.5(silicate)}$$

essentially liberating oxygen by dissolving Fe^0 in sulfide to produce an Fe–Ni–S alloy phase. This ~100% difference between the Fe_2O_3 contents of the upper mantle and the transition zone poses an intriguing possibility for generation of carbonatites. Geophysical evidence for whole-mantle convective flow (cf. review by van Keken *et al.*, 2002) implies that the transition zone is not an impermeable boundary for mantle flow. Therefore, there must be convective upwellings where mantle moves from the transition zone upwards into the upper mantle, producing a situation in which the amount of Fe_2O_3 cannot be accommodated in the upper mantle olivine + pyroxenes + garnet mineral assemblage. If the Fe–Ni–S alloy of O'Neill *et al.* (1993) is still in contact with this volume of mantle, the excess oxygen could dissolve by reduction of Fe^{3+} by the reverse of the reaction above. Alternatively, if the mantle assemblage contains diamond, it is possible that the following reaction takes place instead:

$$4Fe^{3+}O_{1.5(silicate)} + C_{(diamond)} = 4Fe^{2+}O_{(silicate)} + CO_{2(carbonate\ melt)}$$

producing a carbonate melt by an oxidative-melting process, but in the absence of a fluid, which is required for the redox melting model of Green *et al.* (1987).

Once formed by whatever mechanism, carbonate melts will wet an olivine and pyroxene matrix, so the melt will be interconnected and mobile at extremely low melt fraction (Hunter and McKenzie, 1989; Minarik and Watson, 1995), rendering it a potential metasomatic agent.

2.07.3 EVIDENCE FROM MANTLE-DERIVED SAMPLES

In this section, we review the evidence for volatiles in the mantle obtained from the study of mantle-derived samples. The primary focus is on xenoliths brought to the surface in alkali basalts and kimberlite host magmas, but some relevant results obtained from study of alpine peridotites are also discussed.

These samples may provide a biased perspective of the Earth's mantle because of the association of the host magmas with certain tectonic environments. For example, xenolith-bearing kimberlites are usually found on cratons, which have been shown to be underlain by mantle with anomalous geophysical, and presumably geochemical, nature (see Chapter 2.05). Another caveat is that at many localities, samples containing hydrous minerals may be uncommon, even if they are disproportionately represented in the literature published on a given site—Dish Hill, California, is a good example of this phenomenon.

An additional complication is that the xenoliths are a limited resource, and are not replenished except by mining in the case of kimberlites, or by erosion of the primary deposit, such as mass wasting of an alkali basalt cinder cone. It is possible that the earlier workers, at a given location, if they collect extensively—or exclusively—samples containing hydrous phases, would bias the population observed by later workers, which will be under-represented in these hydrous samples.

It is equally important to acknowledge that the host magmas are unlikely to sample representatively the mantle through which they transit to the surface. Nevertheless, these xenoliths are direct samples of regions of the Earth's interior; their diversity illustrates the variety of lithologies and compositions present in the mantle. The length-scales of these heterogeneities cannot be well constrained from the xenolith record.

2.07.3.1 Hydrogen-bearing Phases

Hydrogen-bearing phases in mantle-derived samples include fluid and melt inclusions trapped in host minerals, hydrous minerals such as amphibole, mica, chlorite, serpentine, and apatite, and NAMs that contain hydrogen at the sub-1 wt.% level. The mode of occurrence of hydrogen in the mantle will profoundly affect the implications of its presence. For example, if fluid is present, melting may occur at lower temperatures than otherwise. If hydrous phases are present, dehydration melting reactions will dictate the conditions at which the mantle will begin to melt. In addition, such phases will impact the trace-element geochemistry and partitioning in the mantle. Even if hydrogen is only hosted in NAMs, it could have significant effects on mantle rheology and melting behavior.

2.07.3.1.1 Fluids

Although shallow-mantle xenoliths, hosted in alkali basalts, commonly contain CO_2-rich fluid inclusions (see below), there have been no reports, to the author's knowledge, of H_2O-rich fluid inclusions in these samples. The CO_2-rich fluid inclusions are commonly attributed to late, possibly magma-derived, metasomatism of the samples. If such metasomatism was produced by silicate- or carbonate-rich melts, ascent of such a melt could produce saturation in a CO_2-rich vapor, but H_2O would partition strongly into either residual melt or hydrous phases such as phlogopite or amphibole. Thus, the absence of H_2O in the fluid inclusions in these samples cannot be taken as evidence that the metasomatic agent was anhydrous.

In contrast to the fluid inclusions in shallow-mantle minerals, diamonds have been found with inclusions of H_2O-bearing fluids. Most studies of fluids have concentrated on fibrous or cubic diamonds, which contain abundant inclusions (Chrenko *et al.*, 1967; Navon *et al.*, 1988; Akagi and Masuda, 1988; Turner *et al.*, 1990; Schrauder and Navon, 1994; Schrauder *et al.*, 1996; Johnson *et al.*, 2000). Schrauder and Navon (1994) showed the major-element compositions of these fluids were quite variable but could be considered in terms of two end-members, a carbonate-rich one enriched in CaO, FeO, MgO, and P_2O_5; the other a hydrous fluid enriched in SiO_2 and Al_2O_3. Most of the fluids were potassium-rich (Schrauder and Navon, 1994) and high in argon and the halogens (Johnson *et al.*, 2000). It is unclear, however, if these fluids are representative of fluids present during growth of "normal," nonfibrous diamond (Izraeli *et al.*, 2001).

For this reason, the study of Izraeli *et al.* (2001) may be more relevant to constraining fluids present during crystallization of "normal" diamonds. They studied "cloudy" diamonds from the Koffiefontein kimberlite in South Africa. Typically, <20% of the diamond was cloudy, sometimes in the central part of the diamond, sometimes forming a mantle around a clear core. These cloudy regions contain large numbers of inclusions of ~1 μm size. The diamonds also contained silicate mineral inclusions, some of the eclogitic paragenesis, other diamonds containing peridotitic mineral inclusions. They found a very silica-poor, chlorine-rich brine contained in the fluid inclusions in diamond of both the eclogite and peridotite suites. Izraeli *et al.* (2001) determined an average brine composition of $(K,Na)_8(Ca,Fe,Mg)_4SiO(CO_3)_4Cl_{10}(H_2O)_{28-44}$ and proposed that diamonds in both suites precipitated in a single event from this brine.

There is considerable interest in the mobility of possible hydrous fluids in the mantle. The controlling factor is the efficiency with which the fluid wets grain boundaries. The criterion often used to judge this efficiency is the dihedral angle (θ), formed at the intersection of a grain boundary with two fluid–crystal phase boundaries. A dihedral angle of 60° marks the transition between an interconnected fluid phase ($\theta < 60°$) and a fluid phase isolated at grain boundary junctions ($\theta > 60°$). Experiments by Watson and co-workers (Watson and Brenan, 1987; Watson *et al.*, 1990) document $\theta > 60°$ for H_2O at $T \leq 1,100°$ at 1 GPa, but $\theta < 60°$ at higher temperatures. Watson *et al.* (1990) document a negative dependence of θ in the olivine–H_2O system at pressure between 0.5 GPa and 2 GPa. At 1,200 °C, θ becomes <60° at pressures above ~0.7 GPa. Similar dunite–H_2O experiments at higher pressures confirm the negative pressure and temperature dependence of θ (Mibe *et al.*, 1998, 1999). Indeed, Mibe *et al.* (1999) propose that the transition between non-interconnected and interconnected fluid networks in the subducting slab control the position of the volcanic front in island arcs, rather than a specific dehydration reaction controlling its position.

Interestingly, there is a strong mineralogical control on the observed values of θ. Watson and Lupulescu (1993) report $\theta > 60°$ in clinopyroxenite–H_2O system at 1.5 GPa, 900–950 °C. Ono *et al.* (2002) studied pyrope-H_2O at 4–13 GPa and 900–1,200 °C. They found that θ increased with pressure up to 9 GPa. At 9 GPa and above, θ was >60°. They suggested that in the metabasaltic portion of a subducting slab, lawsonite or phengite dehydration at ≥9 GPa would produce an H_2O-rich fluid, which would be trapped in the garnet-rich matrix and subducted further into the mantle.

The alternative to such a scenario is that the fluid phase triggers melting. Experimental studies of H_2O-present melting in peridotite (Figure 2) document that the H_2O-saturated solidus is at ~1,000°, lower than the mantle adiabat any pressure. Whether the H_2O-saturated solidus of metabasalt or garnetite is similar is not clear at present, although the results of Sumita and Inoue (1996) on the H_2O-present melting of pyrope suggest a higher solidus: they found the solidus at 1,157 °C at 4.2 GPa and 1,237 °C at 12 GPa. Although these values are higher, they imply that melting would occur once the slab warmed to ambient mantle temperatures.

An open question at present concerns the nature of the fluid phase, and the pressure at which there is a second critical end point on the solidus, above which there is complete miscibility between hydrous silicate melt and silicate-bearing hydrous fluid, such that the concept of a "solidus" requires rethinking. Complete miscibility between melt and fluid was observed in diamond-anvil experiments in albite–H_2O (Shen and Keppler, 1997) and in other simple systems (Bureau and Keppler, 1999), although this crest of the solvus between melt and fluid is not the same as a second critical end point. Quench experiments by Stalder *et al.* (2000) are consistent with a second critical end point in albite–H_2O above 1.5 GPa.

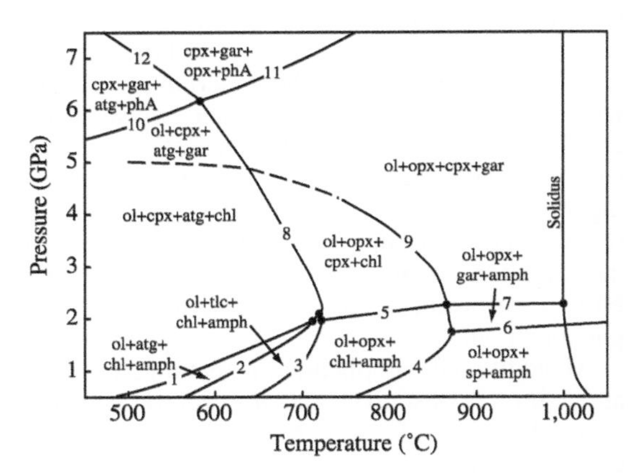

Figure 2 Phase relationships for lherzolite + H_2O after Kawamoto and Holloway (1997) and Schmidt and Poli (1998). Solidus location from Kawamoto and Holloway (1997). Subsolidus phase relationships from Schmidt and Poli (1998), with phase abbreviations modified for consistency with other figures. Numbered reactions: (1) atg + chl + cpx = amph + ol + fl, (2) atg = tlc + ol + fl, (3) tlc + ol = opx + fl, (4) chl = opx + ol + sp + fl, (5) chl + opx + cpx = amph + ol + fl, (6) gar lhz = sp lhz, (7) cpx + gar + opx + fl = amph + ol, (8) atg = opx + ol + fl, (9) chl + opx = ol + gar + fl, (10) atg + phA = ol + fl, (11) phA + opx = ol + fl, (12) atg = phA + opx + fl.

In experiments in the Mg_2SiO_4–$Mg_2Si_2O_6$–H_2O system, Stalder *et al.* (2001) found that melts and fluids remained immiscible to 9 GPa, although their compositions converge with increasing pressure. They extrapolated their data, and predicted complete miscibility at 12–13 GPa. Wyllie and Ryabchikov (2000) provide an extensive discussion of some of the possible consequences of the appearance of a second critical end point on the solidus of mantle peridotite.

2.07.3.1.2 Amphibole and mica

Amphibole and mica are the most common hydrous minerals in mantle-derived xenoliths. In mantle samples, the presence of these minerals is commonly considered a hallmark and a result of metasomatic processes. In the literature, there is a distinction drawn between the so-called "primary" hydrous phases, which formed during residence in the mantle and typically form discrete grains appearing to be in textural equilibrium with the surrounding anhydrous silicates, and the "secondary" hydrous phases, which form overgrowths or reaction products on primary minerals. These latter phases are often interpreted to result from interaction with the host magma just before or during ascent. It is important to note that even the "primary" phases are usually interpreted to be the result of interaction of a metasomatic agent with the host rock.

There is an abundant literature on these minerals, beginning with the first examples of micas in peridotite xenoliths with unambiguously "primary" textures (Dawson and Powell, 1969). There are two excellent review volumes (Menzies and Hawkesworth, 1987; Nixon, 1987) that bear specifically on this topic. Given the voluminous literature on this subject, it is impossible to provide more than an overview, with sufficient references to point the interested reader to the literature. Chapter 2.05 provides additional background on the mineralogy and geochemistry of mantle xenoliths.

For ease of presentation, we consider spinel-facies samples first, then the deeper garnet-facies samples. The key points to keep in mind for the purpose of this chapter are: (i) amphibole and mica are the products of melt or fluid interaction with the host mantle rocks, and (ii) once introduced, both amphibole and mica are stable at the pressure–temperature conditions from which the xenoliths were sampled; hence they are viable hosts for volatiles at (upper) mantle conditions.

Spinel-facies samples. These samples are usually hosted in alkali basalts and are typically represented by two groups, the Cr-diopside series and the Al-augite series of Wilshire and Shervais (1975). Other classification schemes exist (e.g., type I/type II of Frey and Prinz (1978)), but for our purposes the Wilshire/Shervais classification is satisfactory.

The Cr-diopside series is the most abundant type of xenolith found in alkali basalts. Amphibole is uncommon in samples of this series, but rare examples have been found from locations across the world (see review by Kempton, 1987). The amphibole is typically a chromium-rich pargasite and has been observed to constitute up to 6% of the mode. Commonly, these amphiboles have partially broken down, a process interpreted to be a response to the incorporation of the xenolith into the ascending host magma. Phlogopite seems to be less commonly observed in spinel peridotites, but is present along with amphibole in some suites (Kempton, 1987 and references therein). In other suites, phlogopite is the only hydrous phase present (Francis, 1987; Canil and Scarfe, 1989).

The Al-augite series occurs either as discrete xenoliths, or as veins cross-cutting Cr-diopside series peridotite. They are usually interpreted to be high-pressure segregations from hydrous mafic melts. At many locations, the Al-augite series characteristically contains a kaersutitic amphibole and phlogopite (see review by Kempton, 1987).

Subsequent to the studies that established amphibole and phlogopite as stable phases in spinel-facies peridotites, albeit introduced by fluids or melts, attention turned to determining the trace-element composition of these phases (O'Reilly *et al.*, 1991; Witt-Eickschen and Harte, 1994; Ionov and Hofmann, 1995; Vannucci *et al.*, 1995; Johnson *et al.*, 1996; Chazot *et al.*, 1996; Ionov *et al.*, 1997). Ionov *et al.* (1997) combined their data on amphibole and mica, along with apatite and clinopyroxene, with results from previous studies and found that mica hosts most of the rubidium and barium, and that both amphibole and mica were major hosts for niobium. They concluded that the trace-element and radiogenic isotopic characteristics of the lithospheric mantle would be strongly affected by the presence of these minerals. They further argued that the behavior of these phases during partial melting events would have a major impact on the trace-element characteristics of any melts formed in the lithosphere. In support of this argument, they referred to the vein-plus-wall-rock model of Foley (1992), in which veins containing hydrous minerals (phlogopite, amphibole, apatite) melt at temperatures lower than the surrounding peridotitic wall rock, and that melt subsequently reacts with the wall-rock peridotite. This interaction has a greater effect on major elements than on incompatible trace elements, because of the paucity of incompatible elements in "normal" peridotite. Hence, the vein minerals would exert a disproportionate effect on the trace-element characteristics of the derived magma.

Garnet-facies samples. Samples of peridotite brought to the surface by kimberlites may contain phlogopite, edenite-pargasite, or a potassium richterite. Numerous studies on African examples, beginning with reports at the First International Kimberlite Conference in 1973, have extensively documented that "primary" hydrous phases are present and represent phases stable in the mantle prior to entrainment in the kimberlite. Papers that document the evolving interpretation of the micas, for example, include Carswell (1975) and Delaney *et al.* (1980).

In a classic paper, Erlank *et al.* (1987) documented four types of metasomatized peridotites in samples from Kimberley: garnet peridotites (GPs), which lack primary phlogopite; garnet phlogopite peridotites (GPPs), which contain primary phlogopite; phlogopite peridotites (PPs), which contain primary phlogopite but lack garnet; phlogopite K-richterite peridotites (PKPs), which contain primary phlogopite and K-richterite but usually lack garnet. They demonstrated that these peridotite types correspond to increasing interactions with metasomatic fluids, in the sequence GP–GPP–PP–PKP (least to most). They were careful to stress that although the PKP suite was the most metasomatized, they may not have formed from GP precursors, but may have had harzburgitic precursors. They found textural evidence for replacement of garnet by phlogopite $\pm$ clinopyroxene, as well as phlogopite replacing all anhydrous silicates. Similarly, K-richterite appears to replace all pre-existing silicate minerals, including phlogopite. They proposed a generalized overall reaction (Erlank *et al.*, 1987, p. 247):

$$\begin{aligned}&\text{olivine} + \text{enstatite} + \text{garnet} + \text{diopside}_1\\&\quad + \text{Cr-spinel}_1 + \text{fluid}\,(\text{K}_2\text{O etc., H}_2\text{O}, \text{CO}_2)\\&\qquad \rightarrow \text{phlogopite} + \text{K-richterite} \pm \text{diopside}_2\\&\qquad\quad + \text{Cr-spinel}_2 \pm \text{calcite} \pm \text{LIMA, ilmenite,}\\&\qquad\qquad \text{rutile, armalcolite, sulphides}\end{aligned}$$

In this reaction, LIMA refers to the titanite minerals lindsleyite and mathiasite.

There are many other studies of phlogopite-bearing mantle peridotites (see reviews by Harte, 1987; Menzies and Chazot, 1995), all of which support the concept that phlogopite is formed by interaction of a metasomatic agent with a pre-existing anhydrous peridotite.

Two additional suites of mantle-derived samples found in kimberlites provide evidence for amphibole and/or mica stability in the lithospheric mantle. These are the ilmenite–rutile–phlogopite–sulphide (IRPS) and mica–amphibole–rutile–ilmenite–diopside (MARID) suites. The IRPS suites (Harte *et al.*, 1987 and references therein) result from metasomatism of the pre-existing peridotites, and are related to pyroxenite veins and sheets that appear to be magmatic in origin. Harte *et al.* (1987) proposed that the relationship is genetic, with the same agent being responsible for both the pyroxenites and the modal metasomatism in the peridotites. This suite has been described only in samples from the Matsoku kimberlite in Lesotho.

Xenoliths belonging to the MARID suite have been found in kimberlites from southern Africa and Russia (Dawson and Smith, 1977; Dawson, 1987) and in monchiquites and olivine nephelinites in Morocco (Wagner *et al.*, 1996). Dawson and Smith (1977) proposed that the MARID suite crystallized from a kimberlite-like magma at depths less than ~100 km. They also proposed that fluids migrating away from crystallizing MARID intrusions would metasomatize the surrounding lithosphere. Dawson (1987), building on this and subsequent work, suggested that such fluids would be responsible for the type of metasomatism described by Erlank and co-workers (e.g., Erlank *et al.*, 1987).

Waters (1987), while accepting an igneous origin for these rocks, proposed that they were the product of crystallization at high pressure of magnesium-rich ultrapotassic magmas such as lamproites, which had formed by small degrees of partial melting of phlogopite-bearing peridotite. Waters (1987) reconciled the lower incompatible element composition in MARIDs relative to lamproites by invoking a late-stage escape of volatile- and incompatible-element-rich melt from the crystallizing MARID that could act to metasomatize the surrounding lithospheric mantle.

Based on their phase equilibrium study of a natural MARID sample, Sweeney *et al.* (1993) ruled out crystallization of MARID from a hydrous melt with the same bulk composition, even if water-saturated, because of high liquidus temperatures. They proposed two alternative models, one in which MARIDs form by interaction of alkali-rich fluid with peridotite, the other in which MARIDs result from fractionation of olivine and exsolution of carbonatite from a group II kimberlitic parent magma. A connection between MARID formation and group II kimberlites is supported by U–Pb dates of some zircons from Kimberley MARIDs, which gave ages similar to the period of group II kimberlite magmatism in the Kaapvaal (Konzett *et al.*, 1998). Their first model is supported by the experiments of Odling (1995), who formed an assemblage including phlogopite and K-richterite via interaction of a hydrous fluid with peridotite.

Grégoire *et al.* (2002) revisited the MARID issue by re-examining mica-rich mafic xenoliths from the Kimberly kimberlites. They differentiated these xenoliths into two groups, one corresponding to the previously described MARID group, the other characterized by the mineral assemblage phlogopite–ilmenite–clinopyroxene with minor rutile, which they termed the phlogopite + ilmenite + clinopyroxene (PIC) group. They used major- and trace-element chemistry and extant strontium- and neodymium-isotopic data to propose genetic relationships between, and common parental magmas for the PIC samples and group I kimberlites on the one hand, and for MARIDs and group II kimberlites on the other. They further propose that these two distinctly different melts metasomatized the mantle, producing the PKP and PP phlogopite-bearing peridotites of Erlank *et al.* (1987), respectively.

Experimental constraints on the stability of amphibole and mica. Calcic amphiboles are stable in peridotites at pressures <2.5 GPa (Figure 2). Schmidt and Poli (1998) and Poli and Schmidt (2002) reviewed this topic. These references, along with Ulmer and Trommsdorff (1999), should be consulted for a more detailed discussion. Phlogopite can be stable in peridotite to higher pressure. Trønnes (2002) determined the stability of synthetic phlogopite using the bulk composition $K_2Mg_6Al_2Si_6O_{20}(OH)_2$ (Figure 3). Sato *et al.* (1997) determined the stability of natural magnesium-rich phlogopite and phlogopite + enstatite mixtures to 8 GPa. This study focused on the high-temperature stability of the two compositions, which were found to be ~1,350 °C and ~1,300 °C, respectively at 4–6 GPa. At 8 GPa, phlogopite was still stable in both bulk compositions at temperatures <1,300 °C. A possible problem with these experiments is that they were run in platinum capsules, and it is possible that iron was lost to the capsule. The analyses of the micas in the run products reported by Sato *et al.* (1997) are very iron-poor relative to the starting phlogopite; no analyses of the coexisting phases were reported.

Konzett and Ulmer (1999) bracketed the phlogopite-out reaction in a natural peridotite at 6–7 GPa at 1,150 °C and in a subalkaline iron-free bulk composition at 8–9 GPa at 1,150 °C and at <8 GPa at 1,200 °C (Figure 3). Phlogopite becomes unstable with increasing pressure relative to a potassium-rich amphibole; the stability of this amphibole has been studied by a number of workers (Figure 3).

Modreski and Boettcher (1973), Sudo and Tatsumi (1990) and Luth (1997) determined the stability of phlogopite coexisting with clinopyroxene (Figure 4). With increasing pressure, phlogopite breaks down to form a potassic amphibole, which in turn decomposes at higher pressure to form another hydrous potassium magnesium silicate, termed "phase X" by Trønnes (1990).

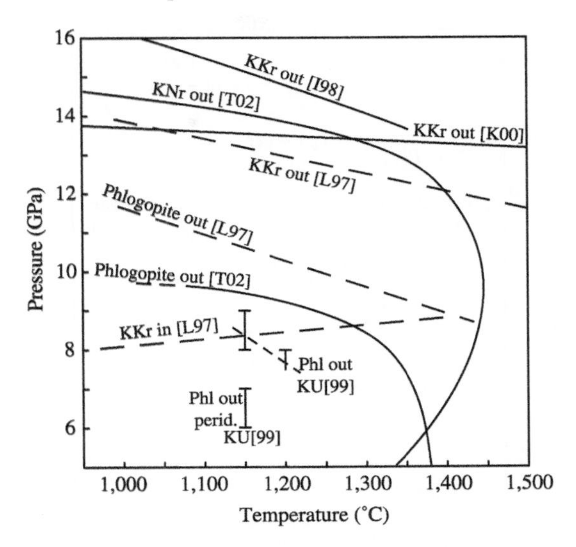

Figure 3 Stability of phlogopite and potassic amphibole in various subalkaline compositions. "Phl out perid. KU[99]" bracket is from Konzett and Ulmer (1999) for phlogopite decomposition in a natural peridotite bulk composition. "Phl out KU[99]" marks the upper pressure limit of phlogopite in a subalkaline bulk composition in the K_2O–Na_2O–CaO–MgO–Al_2O_3–SiO_2–H_2O system (in wt.%, SiO_2 47.28, Al_2O_3 11.50, MgO 27.26, CaO 6.58, Na_2O 1.22, K_2O 4.94, H_2O 1.22) from the same study. "Phlogopite out [T02]" curve is for $K_2Mg_6Al_2Si_6O_{20}(OH)_2$ bulk composition from Trønnes (2002). "KKr in [L97]" is where K-amphibole becomes stable in the phlogopite–diopside system (Luth, 1997). "Phlogopite out [L97]" and "KKr out [L97]" mark the high-pressure limit to phlogopite and K-amphibole stability, respectively, in the phlogopite–diopside system (Luth, 1997). "KKr out [K00]" is the high-pressure stability limit for K-amphibole in the subalkaline KNCMASH bulk composition of Konzett and Ulmer (1999) as delimited by Konzett and Fei (2000). "KKr out [I98]" is the high-pressure stability limit of K-amphibole found by Inoue *et al.* (1998a) for a bulk composition of $K_2CaMg_5Si_8O_{22}(OH)_2$. The "KNr out [T02]" curve is the corresponding stability limit for the sodium-bearing analogue, $KNaCaMg_5Si_8O_{22}(OH)_2$, from Trønnes (2002).

Phase X has been observed in a number of studies on hydrous potassium-bearing systems (Trønnes, 1990, 2002; Inoue *et al.*, 1998a; Luth, 1997). Its stability relations have been studied by Konzett and Fei (2000), who found that it breaks down between 20 GPa and 23 GPa at 1,500–1,700 °C. Its breakdown products were reported by Konzett and Fei (2000) to be K-hollandite, γ-Mg_2SiO_4, majorite, Ca-perovskite, and fluid. Hence, phase X is not succeeded by another hydrous potassic solid phase, and is therefore the hydrous potassic (solid) phase with the highest-pressure stability. The crystal structures of phase X and some related phases were determined by Yang *et al.* (2001).

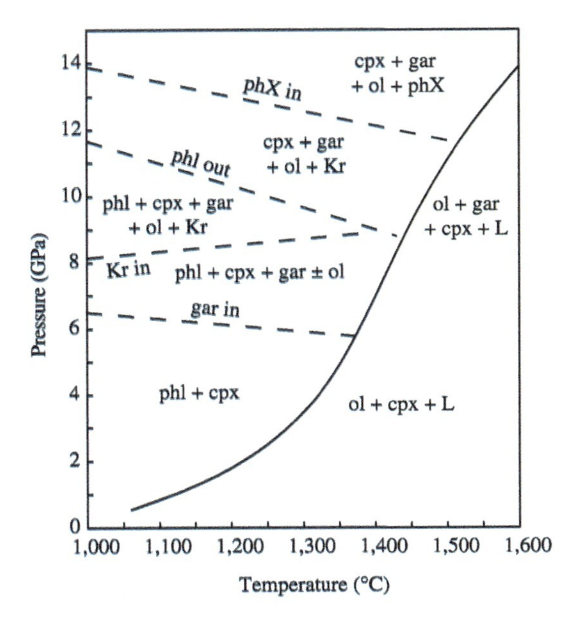

Figure 4 Stability of hydrous phases in the system $K_2Mg_6Al_2Si_6O_{20}(OH)_2$–$CaMgSi_2O_6$ (phlogopite–diopside). At pressures below ~8 GPa, phlogopite is stable. It becomes unstable at higher pressures relative to K-richterite (Kr). Kr in turn becomes unstable at ~13–14 GPa relative to phase X (phX), a hydrous potassic magnesian silicate. The high-temperature stability limit for all three hydrous phases is defined by the vapor-absent solidus. Phase relationships after Modreski and Boettcher (1973), Sudo and Tatsumi (1990), and Luth (1997).

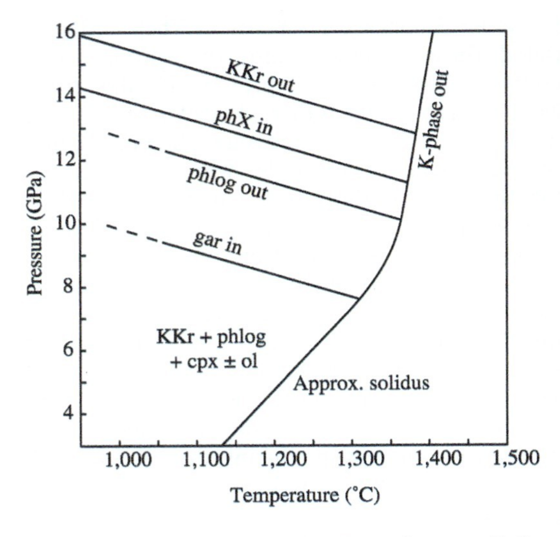

Figure 5 Stability of hydrous phases in a peralkaline bulk composition, after Konzett *et al.* (1997) and Konzett and Fei (2000). Bulk composition is (wt.%) SiO_2 47.61, Al_2O_3 6.67, MgO 23.79, CaO 7.61, Na_2O 1.94, K_2O 7.62, H_2O 4.76. Abbreviation "phX" refers to phase X, a hydrous potassium–magnesium silicate (see text for details).

The general sequence of stability of hydrous phases with increasing pressure (phlogopite → K-amphibole → phase X) has been seen consistently in studies of various bulk compositions relevant to peridotites (Figure 3). The same sequence was also seen in studies of a peralkaline bulk composition (Figure 5), in which the stability fields of potassium-rich phases should be maximized (Konzett *et al.*, 1997; Konzett and Fei, 2000).

The abundances of these potassic hydrous phases, and hence the amount of water they can host, are limited by the potassium content of the mantle, but they are of considerable interest because of their high thermal stability. Depending on the precise mantle geotherm (adiabat) assumed, these hydrous phases may be stable in convecting mantle as well as in the cooler portions of the mantle such as the subcontinental lithosphere. This stability to high temperatures provides a mechanism not only to transport water into the mantle by subduction, but also to retain it. In this respect, the potassic hydrous phases contrast with the hydrous magnesium-rich phases, as outlined below.

2.07.3.1.3 Serpentine

Serpentinized peridotite is a major constituent of the oceanic lithosphere, and is most frequently, but not exclusively, found on slow-spreading ridges. Serpentinized ultramafic rocks crop out in a number of tectonic environments associated with mid-ocean ridge magmatism: fracture zones and transform faults (e.g., Dick *et al.*, 1984; Dick, 1989; Cannat *et al.*, 1990; Snow and Dick, 1995; Escartín and Cannat, 1999), rift valleys (Cannat *et al.*, 1992; Cannat, 1993; Bougault *et al.*, 1993), and discrete massifs (Bougault *et al.*, 1993). Abyssal peridotites typically are highly (70–80%) serpentinized (Dick *et al.*, 1984).

A key question is the extent to which the oceanic lithospheric mantle is serpentinized, because it is a significant factor in the amount of water present in a subducting slab. O'Hanley (1996) argues that "...serpentinization of the lower oceanic crust and upper mantle is minimal..." (p. 228) and that the extent of serpentinization is inversely correlated to spreading rate. Ernst (1999) pointed out that fracture zones and transform faults, the most common source of samples of abyssal peridotites, would be characterized by unusually high fluid flow and hence may be more serpentinized than average oceanic lithospheric mantle. Based on density and seismic-velocity arguments, he argued that only 10–20% of the subducting peridotite would be serpentinized. Ulmer and Trommsdorff (1995) point out that completely serpentinized peridotites contain ~13 wt.% H_2O, and argue that such peridotites would be more important in transporting H_2O into the mantle than would be hydrated mafic oceanic crust. Reducing the volume of serpentinized peridotite in accord with the estimate of Ernst (1999), the subducting ultramafic section would contain ~1.3–3 wt.% H_2O, compared to 5–6 wt.% H_2O in the (thinner) basaltic crust (Schmidt and Poli, 1998). Clearly, the relative contribution of each depends on the extent of hydration of each, which may vary from slab to slab, and laterally along a slab.

Ulmer and Trommsdorff (1999) reviewed thermodynamic and field constraints from Alpine ultramafic rocks, which document that lizardite and crysotile, the other two serpentine minerals, break down to form antigorite via reactions such as chrysotile + talc = antigorite, chrysotile + tremolite = antigorite + diopside, and chrysotile = brucite + antigorite. Therefore, antigorite is the appropriate serpentine mineral to study experimentally to determine the maximum pressure–temperature stabilities for serpentine *sensu lato*.

Antigorite, and its breakdown products, may be considered in the system $MgO-SiO_2-H_2O$ (Table 1, Figure 6). Experiments by Ulmer and Trommsdorff (1995) show that antigorite breaks down along a negatively-sloped reaction boundary, at ~720 °C at 2 GPa, ~690 °C at 3 GPa, and ~620 °C at 5 GPa. The precise products of the decomposition of antigorite depend on pressure and temperature; at higher temperatures, antigorite reacts to form forsterite + enstatite + fluid. At lower temperatures and higher pressures, the products produced depend on the bulk composition (Figures 7 and 8), in particular, on the extent of serpentinization. Both the 10 Å phase and phase A illustrated on these diagrams are DHMS and have not been observed in nature.

The thermal stability of antigorite is clearly insufficient to stabilize it at "normal" mantle temperatures and hence limits it to a role of supplying water in subducting slabs (Ulmer and Trommsdorff, 1995, 1999; Peacock, 2001),

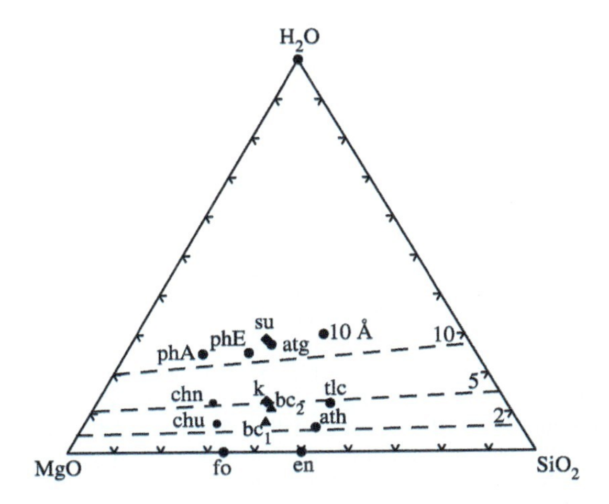

Figure 6 Composition of selected phases (circles) and bulk compositions (triangles, diamonds) plotted in the system $MgO-SiO_2-H_2O$. Abbreviations: fo = forsterite, en = enstatite, ath = anthophyllite, tlc = talc, 10 Å = 10 Å phase, atg = antigorite, phE = phase E, phA = phase A, chn = phondrodite, chu = clinohumite. Compositions of phases are given in Table 1. Dashed lines represent compositions with 2 wt.%, 5 wt.%, and 10 wt.% H_2O, respectively. Triangles bc_1 and bc_2 are bulk compositions of serpentinized harzburgite (60% olivine, 40% orthopyroxene) with 40% and 60% serpentinization of the olivine, respectively. The phase relationships for these two bulk compositions are shown in Figures 7 and 8, respectively. Diamond labeled "su" is the bulk composition used by Stalder and Ulmer (2001) (Figure 10). Diamond labeled "k" is bulk composition used by Khodyrev *et al.* (1990) (Figure 11) projected into $MgO-SiO_2-H_2O$ system.

Table 1 Compositions of phases in the $MgO-Al_2O_3-SiO_2-H_2O$ system.

Name	*Phase*	*Formula*	wt.% H_2O
Forsterite	fo	Mg_2SiO_4	0.00
Enstatite	en	$Mg_2Si_2O_6$	0.00
Phase A	phA	$Mg_7Si_2O_8(OH)_6$	11.84
Phase E	phE	$Mg_{2.3}Si_{1.25}H_{2.4}O_6$	11.41
Chrysotile	ctl	$Mg_3Si_2O_5(OH)_4$	13.00
Antigorite	atg	$Mg_{48}Si_{34}O_{85}(OH)_{62}$	12.31
10 Å phase	10 Å	$Mg_3Si_4O_{10}(OH)_2 \times 2H_2O$	13.01
Talc	tlc	$Mg_3Si_4O_{10}(OH)_2$	4.75
Anthophyllite	ath	$Mg_7Si_8O_{22}(OH)_2$	2.31
Chondrodite	chn	$Mg_5Si_2O_8(OH)_2$	5.30
Clinohumite	chu	$Mg_9Si_4O_{16}(OH)_2$	2.90
Chlorite (clinochlore)	chl	$Mg_5Al_2Si_3O_{10}(OH)_8$	12.97
Pyrope	prp	$Mg_3Al_2Si_3O_{12}$	0.00
Spinel	spl	$MgAl_2O_4$	0.00
Mg-sursassite	Mg-s	$Mg_5Al_5Si_6O_{21}(OH)_7$	7.17

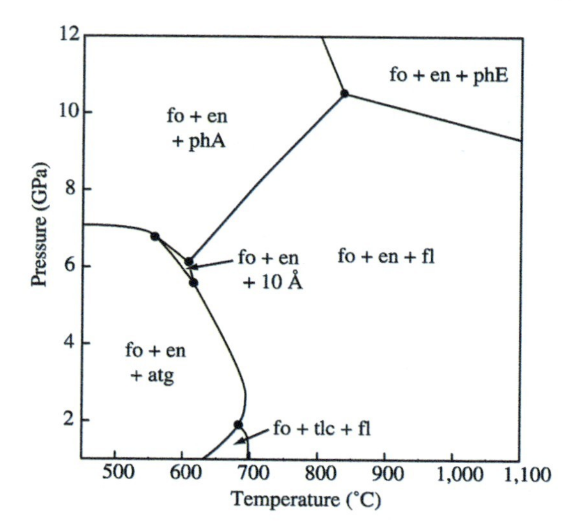

Figure 7 Phase relationships in the system MgO–SiO_2–H_2O for bulk composition bc_1 (Figure 6: 60% olivine, 40% orthopyroxene, with 40% serpentization of the olivine), showing the stable mineral assemblages as a function of pressure and temperature. Stability fields and topology of reaction boundaries after Ulmer and Trommsdorff (1999).

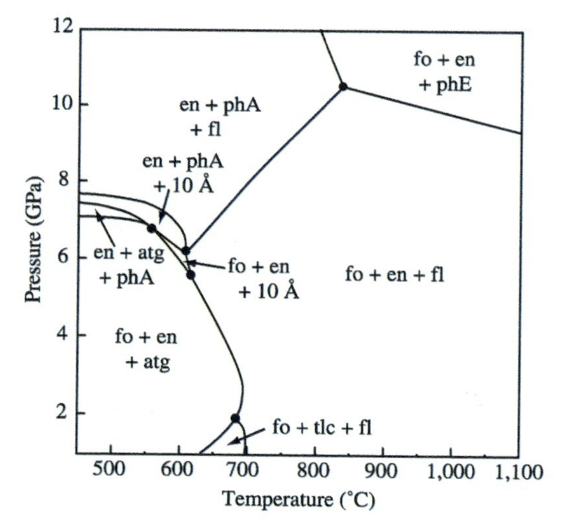

Figure 8 Phase relationships in the system MgO–SiO_2–H_2O for bulk composition bc_2 (Figure 6: 60% olivine, 40% orthopyroxene, with 60% serpentization of the olivine), showing the stable mineral assemblages as a function of pressure and temperature. Stability fields and topology of reaction boundaries after Ulmer and Trommsdorff (1999). The stability fields for en + atg + phA and en + phA + 10 Å are present for this bulk composition but not for bc_1 because bc_1 and bc_2 lie on opposite sides of the en–phA tieline (Figure 6).

and potentially in hydrating the fore-arc mantle wedge (Peacock and Hyndman, 1999; Bostock *et al.*, 2002). The low thermal stability of antigorite also explains why it is not observed as a primary mineral in most mantle-derived xenoliths. An exception is the observation of Smith (1979) of texturally primary antigorite in low-temperature peridotite xenoliths from the Buell Park and Green Knobs diatremes (Colorado Plateau, southwestern USA).

2.07.3.1.4 Clinohumite

Titanium-rich clinohumite is found in ultramafic massifs in Europe and Asia (Trommsdorff and Evans, 1980; Evans and Trommsdorff, 1983; Scambelluri *et al.*, 1991; Okay, 1994; Zhang *et al.*, 1995; Liou *et al.*, 1998; Gil Ibarguchi *et al.*, 1999; Medaris, 1999). In many of these occurrences, it is interpreted to form at "ultrahigh" (metamorphically speaking) pressures and temperatures. Titanian clinohumite has been reported in ultramafic breccia and as inclusions in pyrope-rich garnets from heavy-mineral concentrates from the Mose Rock dike, Colorado Plateau, USA by McGetchin *et al.* (1970). Subsequently, Aoki *et al.* (1976) discovered both titanoclinohumite and titanochondrodite in heavy-mineral concentrates from the Buell Park kimberlite, also on the Colorado Plateau. They were unable to find either mineral in thin sections of the kimberlite. Their interpretation was that both phases crystallized from the kimberlitic magma at ~100 km depth and ~1,000 °C (Aoki *et al.*, 1976; Aoki, 1977). This interpretation was disputed by Smith (1977) based on his observations of hydrous minerals at the Green Knobs diatreme. He subsequently published detailed textural evidence for the formation of amphibole, chlorite, titanoclinohumite, magnesite, and possibly antigorite prior to incorporation of the peridotitic xenoliths into the ultramafic host magma at both Buell Park and Green Knobs diatremes (Smith, 1979). He proposed that the reactions that produced the hydrated mineral assemblages occurred at 600–700 °C at 45–60 km depth.

In natural samples, clinohumite is invariably titanium-rich and often fluorine-bearing (in those cases where fluorine analyzed). There is no apparent systematic composition difference (e.g., Mg/Fe, Ti content) between clinohumites found in massifs and those found in kimberlites and xenoliths. Ulmer and Trommsdorff (1999) argue that titanium-saturated clinohumite has the maximum thermal stability, with titanium-undersaturated clinohumite breaking down at lower temperatures in a divariant reaction to form a titanium-saturated clinohumite + olivine + ilmenite + H_2O, based on experiments by Weiss (1997) (Figure 9). Because of the dependence of the stability of clinohumite on TiO_2, Ulmer and Trommsdorff (1999) argue that the amount of titanoclinohumite in the peridotitic mantle would be controlled by the low TiO_2 content

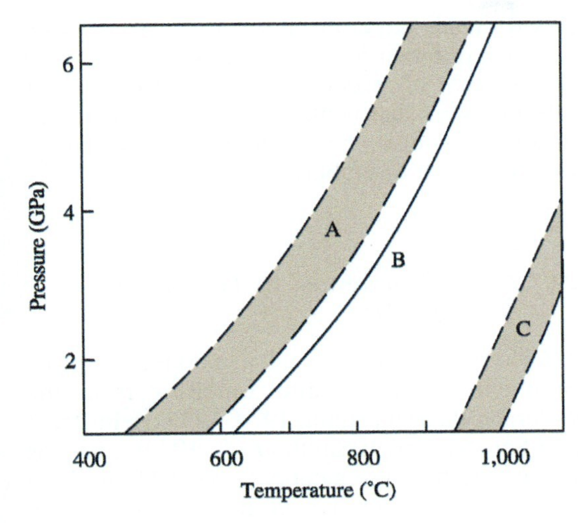

Figure 9 Location of the breakdown reaction clinohumite = olivine + ilmenite + fluid after Weiss (1997), as reported in Ulmer and Trommsdorff (1999). Experiments were conducted on three clinohumites of different composition. Abbrevations: $X_{Fe} = Fe/(Fe + Mg)$, $X_{Ti} = 2Ti/(2Ti + OH)$, and $X_F = F/(F + OH)$. Shaded areas represent divariant fields over which clinohumite coexists with olivine + ilmenite + fluid because the composition of the clinohumite is changing. Curves bounding shaded area A represent breakdown of a clinohumite with $X_{Fe} = 0.04$, $X_{Ti} = 0.28$, and $X_F = 0$. Curve B represents maximum stability limit of natural, F-free clinohumite with $X_{Fe} = 0.19$, $X_{Ti} = 0.46$, and $X_F = 0$. Curves bounding shaded area C represent breakdown of a clinohumite with $X_{Fe} = 0.03$, $X_{Ti} = 0.47$, and $X_F = 0.47$. Increasing X_{Ti}, increasing X_F, and decreasing X_{Fe} all tend to stabilize clinohumite to higher temperatures.

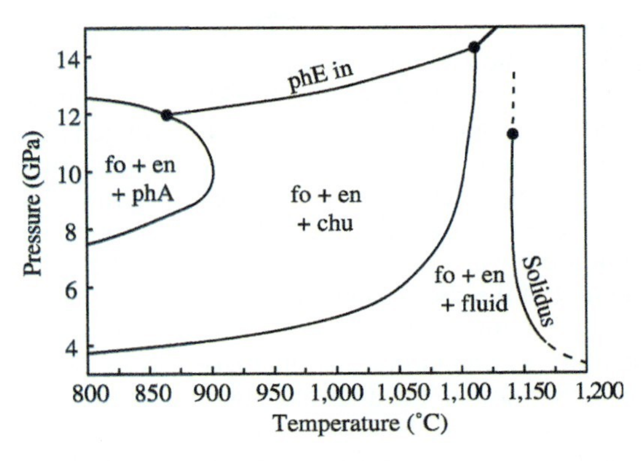

Figure 10 Stability of hydrous phases in hydrated harzburgite based on experiments on a serpentine bulk composition $Mg_3Si_2O_5(OH)_4$ with ~1 wt.% FeO, 0.5 wt.% Al_2O_3, 0.1 wt.% F, and 0.02 wt.% TiO_2. This bulk composition is that given by the diamond labeled "su" on Figure 6. The dot at the top of the solidus represents the approximate location of second critical end point, above which there is no first-order phase transition separating fluid from melt. Note that the phase assemblages shown on the phase diagram are appropriate for bulk compositions within the fo–en–chu three-phase triangle in the $MgO-SiO_2-H_2O$ system (Figure 6), although the experiments used a more H_2O-rich starting composition and hence different phase assemblages were observed in the run products. Abbreviations as in Table 1 (after Stalder and Ulmer, 2001).

of the peridotite. They calculated that the maximum amount of H_2O hosted by clinohumite to be on the order of 1,000–1,500 ppm.

The subsequent work of Stalder and Ulmer (2001), however, is not in agreement with this line of reasoning. These authors conducted experiments on a mixture of synthetic talc and brucite with a nominal composition of chrysotile serpentine [$Mg_3Si_2O_5(OH)_4$]. Their starting material also had ~0.02 wt.% TiO_2, 0.5 wt.% Al_2O_3, 1 wt.% FeO, and 0.1 wt.% F. They found a large subsolidus P, T region over which clinohumite was stable coexisting with enstatite (Figure 10), a region much larger than had previously been observed in the $MgO-SiO_2-H_2O$ system. They analyzed some of their clinohumites, and only the low-pressure ones had >0.1 wt.% TiO_2 and >1 wt.% F. They explain the large increase in size of the clinohumite + enstatite stability field at the expense of forsterite + fluid by the stabilizing effect of fluorine and titanium on the clinohumite. Given the low concentrations of these elements in their analyzed clinohumites, this effect is quite extraordinary. Their results support, however, the earlier work of Khodyrev *et al.* (1990), who found a large stability field for clinohumite under both subsolidus and supersolidus conditions in experiments on a pyrolite + 5% H_2O bulk composition (Figure 11).

Critical questions that must be addressed to make progress in reconciling the experimental results to date include: (i) distinguishing quench phases from stable phases, (ii) interlaboratory P, T calibrations, (iii) reversals of reaction boundaries (keeping in mind that metastable reactions can equally well be reversed), and (iv) careful evaluation of the effects of minor elements such as titanium and fluorine on phase stability. The discussion of Stalder and Ulmer (2001) concerning the interpretation of run products that are zoned both in their composition and their phase assemblage is a welcome beginning to addressing some of these questions.

2.07.3.1.5 Chondrodite

Titanium-bearing chondrodite appears to be much less common than is titanoclinohumite in either ophiolitic or xenolithic occurrences. It has been reported from meta-gabbroic dikelets in serpentinized ophiolite of Liguria (Italy) by

Scambelluri and Rampone (1999). Aoki *et al.* (1976) found it in mineral separates from the Buell Park diatreme, and Smith (1995) reported it in xenoliths in the Moses Rock diatreme. In general, chondrodite is more titanium-rich than coexisting clinohumite (if present). This is best illustrated by the analyses of Aoki *et al.* (1976) of intergrown clinohumite and chondrodite; they found the TiO_2 content of the chondrodite to be ~1.8 times that of the clinohumite. If the effect of titanium on the stability of chondrodite is analogous to its effect on clinohumite, the thermal stability of natural chondrodite may be significantly enhanced relative to that inferred from studies of OH-chondrodite in the $MgO-SiO_2-H_2O$ system. To date, there have been no reported quantitative analyses for fluorine or chlorine in chondrodite, although Fujino and Takéuchi (1978) stated that there was no detectable fluorine or chlorine in the chondrodites they studied, which were taken from those collected by Aoki *et al.* (1976). No detection limit was given by Fujino and Takéuchi (1978). Given the strong effect of fluorine on the stability of clinohumite and of hydrous minerals in general, constraining the fluorine-content of natural samples might prove enlightening.

Based on their experimental study of a serpentine composition at high pressures and temperatures, Stalder and Ulmer (2001) argue that chondrodite would only be stable in ultramafic compositions with >2 wt.% H_2O.

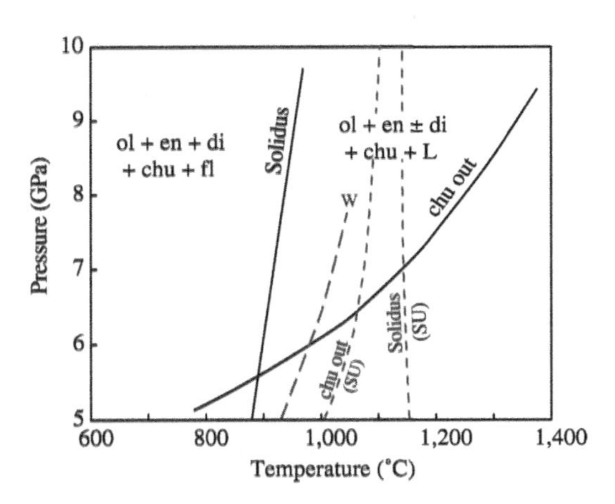

Figure 11 Experimentally determined phase relationships for pyrolite + ~5 wt.% H_2O by Khodyrev *et al.* (1990). Their bulk composition (in wt.%) was SiO_2 45.15, TiO_2 0.64, Al_2O_3 3.7, FeO 7.8, MnO 0.13, MgO 37.88, CaO 3.25, Na_2O 0.66, K_2O 0.14, P_2O_5 0.05, Cr_2O_3 0.4, NiO 0.2. For comparison, this composition is projected into the $MgO-SiO_2-H_2O$ system in Figure 6. For comparison, the breakdown reaction for clinohumite with $X_{Fe} = 0.19$, $X_{Ti} = 0.46$, and $X_F = 0$ (Figure 9) from Weiss (1997) is shown by the dashed line labeled "W." The solidus and clinohumite-out curves for serpentinite from Stalder and Ulmer (2001) (Figure 10) are shown for comparison as well.

Based on his experiments in $MgO-SiO_2-H_2O$, and on a different Schreinemakers analysis of the arrangement of reactions in *P*, *T* space, Wunder (1998) argues that neither clinohumite nor chondrodite would be stable in olivine + orthopyroxene assemblages.

2.07.3.1.6 Chlorite and its breakdown products

Chlorite has been found in peridotite, garnet pyroxenite, and eclogite xenoliths from the Colorado Plateau (Smith, 1979, 1995; Helmstaedt and Schulze, 1979; Smith *et al.*, 1999), and was interpreted by these authors to be stable in these samples prior to entrainment and eruption. Chlorite is an example of a hydrous phase that is stabilized in ultramafic and mafic systems by the presence of aluminum, and the system $MgO-Al_2O_3-SiO_2-H_2O$ is a useful model system in which to examine the relevant phase relationships. Figure 12 shows the compositional relationships between phases in this system, projected from H_2O onto the $MgO-Al_2O_3-SiO_2$ ternary. In spinel lherzolites, the reaction chlorite = forsterite + enstatite + spinel + H_2O would be expected, and indeed has been determined experimentally between an invariant point at ~0.3 GPa, 720 °C involving chlorite, cordierite, forsterite, enstatite, spinel, and fluid and a higher-pressure invariant point at ~2 GPa, ~894 °C involving chlorite, forsterite, enstatite, spinel, pyrope, and fluid (Fawcett and Yoder, 1966; Staudigel and Schreyer, 1977; Jenkins and Chernosky, 1986). Based on

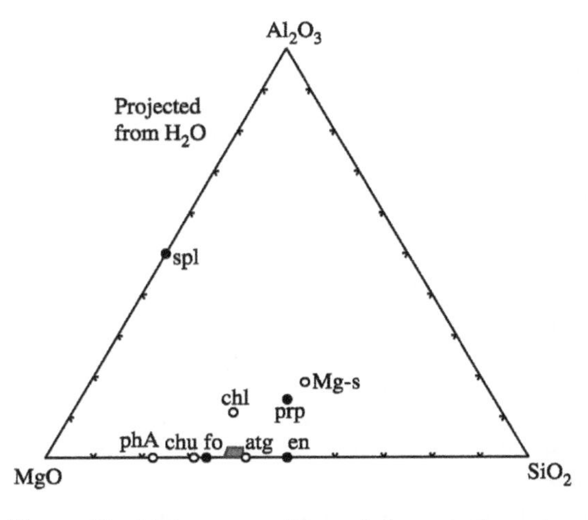

Figure 12 Molar compositions of phases in the system $MgO-Al_2O_3-SiO_2-H_2O$, projected from H_2O into the $MgO-Al_2O_3-SiO_2$ ternary. Solid circles are anhydrous phases, open circles are hydrous phases. Abbreviations as in Table 1. The shaded box represents plausible mantle peridotite compositions, with the upper bound being at ~5 wt.% Al_2O_3.

the chemography illustrated, the analogous decomposition reaction, "chlorite = pyrope + forsterite + spinel + H_2O" (Staudigel and Schreyer, 1977) would only be relevant to bulk compositions that plot to the aluminum-rich side of the forsterite–pyrope join in the $MgO-Al_2O_3-SiO_2$ ternary.

Mantle peridotites, however, plot on the aluminum-poor side of the forsterite–pyrope join (gray-shaded trapezoidal area in Figure 12). The relevant reaction by which chlorite would be expected to decompose for these bulk compositions would be "chlorite + enstatite = forsterite + pyrope + H_2O." Such bulk compositions would have the phase assemblage chlorite + forsterite + enstatite coexisting with either H_2O or pyrope, depending on whether the bulk composition lies to the water-rich or water-poor side of the forsterite–enstatite–chlorite plane in the system $MgO-Al_2O_3-SiO_2-H_2O$. Ulmer and Trommsdorff (1999) argue that this reaction will not be displaced greatly from the position of the "chlorite = pyrope + forsterite + spinel + H_2O" reaction, based on the small difference between the lower-pressure reactions "chlorite = forsterite + cordierite + spinel + H_2O" and "chlorite + enstatite = forsterite + cordierite + H_2O" found by Jenkins and Chernosky (1986). This argument is corroborated by the position of the two reaction boundaries calculated using the most recent version of THERMOCALC (version 3.21) (Holland and Powell, 1998) (Figure 13). The calculated breakdown reactions of chlorite and of chlorite + enstatite differ by 51 °C at 4 GPa and by 73 °C at 5 GPa. Although this difference is larger than Ulmer and Trommsdorff's estimate as illustrated in Figure 13, their point—that the stability of chlorite in peridotite will not be significantly less than the stability of chlorite in a bulk composition of its own stoichiometry—is still valid. This prediction has been confirmed by the new experimental results of Pawley (2003) (solid line, Figure 13).

Another constraint on the ability of chlorite to transport H_2O into the mantle results from its aluminous composition. Chlorites in natural ultramafic samples are close to stoichiometric clinochlore [$Mg_5Al_2Si_3O_{10}(OH)_8$] in composition, which contains 12.97 wt.% H_2O and 18.35 wt.% Al_2O_3 (Table 1). Given that even fertile peridotites contain $\leq$4 wt.% Al_2O_3, a simple mass balance calculation yields a maximum of 2.8 wt.% H_2O being contained in chloritized peridotite, if all Al_2O_3 is present in chlorite.

With increasing temperature in the 3–5 GPa pressure range, chlorite (+enstatite) in peridotite will break down to forsterite + pyrope + H_2O. With increasing pressure at lower temperatures, however, a new hydrous magnesium aluminosilicate is stabilized (Mg-s in Figure 14). This phase is of interest because its stability bridges the pressure range between the low-pressure region where chlorite and antigorite are stable

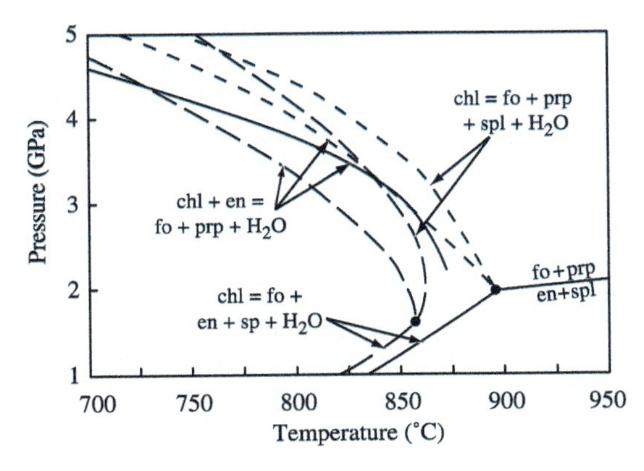

Figure 13 Experimental, estimated, and calculated phase relationships for chlorite stability. Thin solid lines: experimentally determined reactions chl = fo + prp + spl + H_2O and chl = fo + en + sp + H_2O as summarized by Ulmer and Trommsdorff (1999). Short dashed line for chl + en = fo + prp + H_2O as estimated by Ulmer and Trommsdorff (1999). Solid line for chl + en = fo + prp + H_2O from Pawley (2003). Long dashed lines: calculated position of the same three reaction boundaries. Calculated with version 3.21 of THERMOCALC (source Holland and Powell, 1998).

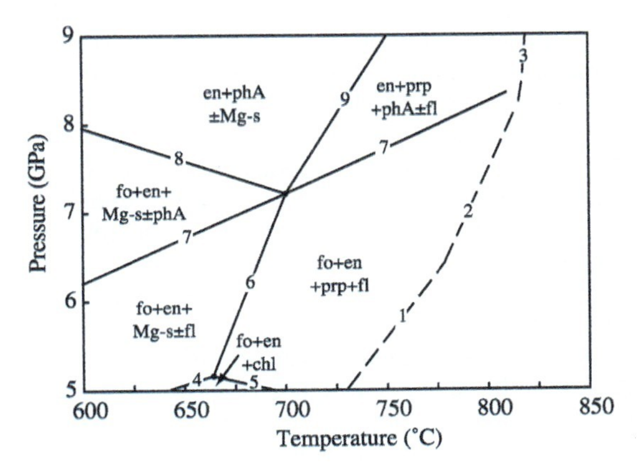

Figure 14 Phase relationships of Mg-sursassite ($Mg_5Al_5Si_6O_{21}(OH)_7$) coexisting with forsterite and enstatite after Bromiley and Pawley (2002). The dashed lines are the breakdown reactions of Mg-sursassite in its own bulk composition after Fockenberg (1998). Numbered reactions are: (1) Mg-s = prp + ky + cs + fl, (2) Mg-s = prp + toz-OH + cs + fl, (3) Mg-s = prp + toz-OH + stv + fl, (4) Mg-s + fo + fl = chl + en, (5) chl + en = prp + fo + fl, (6) Mg-s + fo + en = prp + fl, (7) phA + en = fo + fl, (8) Mg-s + fo = prp + phA + en, (9) Mg-s + phA + en = prp + fl. Abbreviations: as per Kretz (1983) with additions: Mg-s = Mg-sursassite, cs = coesite, toz-OH = Topaz-OH, stv = stishovite, phA = phase A, fl = fluid.

and the higher pressures at which the dense hydrous magnesium silicate phase A is stable. This new phase was originally synthesized by Schreyer *et al.* (1986), who identified it as a calcium-free pumpellyite based on its chemistry and powder X-ray diffractogram. They called this new phase Mg–Mg–Al pumpellyite. Subsequent preliminary work by Schreyer (1988) and Liu (1989), and a detailed study by Fockenberg (1998) delineated the stability of this phase in a bulk composition of its own stoichiometry. Two crystal structure refinements on this phase have come up with differing interpretations of the crystal structure. Artioli *et al.* (1999) described the structure as "pumpellyite-type structure having a certain amount of sursassite domains." Gottschalk *et al.* (2000) determined the structure to be isostructural with sursassite, rather than with pumpellyite, and found none of the stacking disorder that Artioli *et al.* (1999) found, and concluded that "significant differences exist between the samples investigated here and those studied by Artioli *et al.* (1999)." As of early 2000s, it appears that this phase is best described as Mg-sursassite, following the usage of Gottschalk *et al.* (2000), Grevel *et al.* (2001), Bromiley and Pawley (2002), and Wunder and Gottschalk (2002).

In their chemographic analyses of the $MgO-Al_2O_3-SiO_2-H_2O$ system, both Ulmer and Trommsdorff (1999) and Artioli *et al.* (1999) concluded that in a peridotitic assemblage, Mg-sursassite would break down with increasing temperature according to the reaction "forsterite + enstatite + Mg-sursassite = pyrope + H_2O." Ulmer and Trommsdorff (1999) show this reaction with a steep positive slope, whereas Artioli *et al.* (1999) argue for a negative slope for this reaction. Bromiley and Pawley (2002) experimentally determined the stability of Mg-sursassite coexisting with forsterite and enstatite. Their results (Figure 14) require a steep positive slope to the fo + en + Mg-s = prp + H_2O reaction, and define the ability of Mg-sursassite to act as an intermediate water host between chlorite and phase A. As can be seen from Figure 14, if temperatures do not exceed ~660 °C until >7 GPa, Mg-sursassite would be stable, and remain stable until the reaction Mg-sursassite + forsterite = phase A + enstatite + pyrope. The reaction that forms Mg-sursassite from chlorite with increasing pressure is

$$\underset{\text{chlorite}}{10Mg_5Al_2Si_3O_{10}(OH)_8} + \underset{\text{enstatite}}{9Mg_2Si_2O_6} = \underset{\text{Mg-sursassite}}{4Mg_5Al_5Si_6O_{21}(OH)_7} + \underset{\text{forsterite}}{24Mg_2SiO_4} + \underset{\text{fluid}}{26H_2O}$$

This reaction transfers a significant amount of the structural OH in chlorite into a fluid, rather than into the Mg-sursassite phase. Following through with the assumed maximum of 2.8 wt.% H_2O in chlorite peridotite, this would produce a Mg-sursassite bearing peridotite with 0.98 wt.% H_2O bound in the Mg-sursassite, with the remaining H_2O being lost to the fluid. The higher-pressure reaction by which Mg-sursassite breaks down to form phase A is one in which no fluid is involved:

$$6Mg_5Al_5Si_6O_{21}(OH)_7 + 41Mg_2SiO_4 = 7Mg_7Si_2O_8(OH)_6 + 15Mg_3Al_2Si_3O_{12} + 9Mg_2Si_2O_6$$

and hence there need be no change in water content in the hydrated peridotite across this reaction boundary, just a change in hydrous host from Mg-sursassite to phase A.

The most significant difference between the MASH model system and the natural one is the absence of FeO. As of early 2003, there are no experimental data on how these reaction boundaries will shift in the iron-bearing system; the shifts will depend on the relative partitioning of iron between the hydrous phases chlorite, Mg-sursassite, and phase A, and the coexisting anhydrous ferromagnesian phases olivine, orthopyroxene, and garnet. Recently, Wunder and Gottschalk (2002) reported successful syntheses of Fe–Mg sursassites, an essential first step towards constraining the behavior of the iron-bearing system.

It is worth emphasizing that the temperatures over which many of these hydrous phases in the MASH system are stable are sufficiently low that their presence is only possible in cold subducting slabs or subcontinental lithosphere. They will not be present in "normal" mantle, given what we believe to be the case concerning mantle geothermal gradients.

Because this review is intended to focus on the observational evidence for volatiles in the mantle, we shall stop at this stage, and not discuss the higher-pressure reactions involving the various DHMS. The interested reader is directed to the reviews by Frost (1999) and Williams and Hemley (2001) for more information on this topic.

2.07.3.1.7 *Nominally anhydrous minerals*

Since the 1960s, there has been accumulating evidence that mantle-derived NAMs contain measurable amounts of "water"—on the ppm to thousands of ppm level. This leads to the intriguing possibility that the major hosts for water in the Earth's mantle are these nominally anhydrous phases, rather than "conventional" hydrous minerals with stoichiometric OH (Bell and Rossman, 1992a).

Some of the original researches in this area were experimental studies. Sclar and co-workers (Sclar *et al.*, 1967a,b; Sclar, 1970) reported synthesis of "hydroxylated pyroxenes" at conditions of 9.5–18 GPa and 700–1,050 °C. They reported a deduced formula $Mg_2SiH_4O_6$ (with 20.4 wt.% H_2O) and calculated densities of 2.83 gm cm^{-3} and 2.82 gm cm^{-3} (Sclar *et al.*, 1967a). They proposed that the water enters the structure via the hydrogarnet substitution of H_4O_4 for SiO_4. No crystal structure refinement of the "hydroxylated pyroxenes" has been published, nor have more recent experimental studies in the extensively studied $MgO-SiO_2-H_2O$ system reported these phases.

Fyfe (1970) proposed that the hydrogarnet substitution could be of general significance, allowing mantle minerals to host water. Martin and Donnay (1972) disagreed that this substitution would occur at high pressures, and proposed instead that OH substitutes for O in NAMs, with charge balance provided by heterovalent substitution or vacancies. They cited a number of lines of evidence, including published infrared (IR) spectroscopic data on forsteritic olivine, pyroxene, the Al_2SiO_5 polymorphs, garnet, and quartz.

Wilkins and Sabine (1973) used IR spectroscopy to determine water contents of kyanite, andalusite, sillimanite, grossular, andradite, pyrope, diopside, olivine, and feldspars. They found low water contents: 0.008 wt.% in olivine, 0.02 wt.% in diopside, and 0.009 wt.% in pyrope. Zemann, Beran, and co-workers published a series of papers on IR spectroscopy of both hydrous minerals and NAMs (e.g., Tillmanns and Zemann, 1965; Beran and Zemann, 1971, 1986; Beran, 1971, 1986, 1987; Beran and Götzinger, 1987; Beran *et al.*, 1993). For the most part, these contributions were focused on the substitutional mechanisms by which hydrogen entered the crystal structure, rather than on the absolute amount of hydrogen in the crystal structure.

Since the application of IR spectroscopy to water contents in mantle-derived minerals by Rossman and co-workers (summarized in Bell and Rossman, 1992a), it is clear that nominally anhydrous phases are potentially the largest water reservoir in the mantle, and have the potential to be hosts for water that are stable in the convecting mantle. The range of water content for all phases is large: mantle-derived olivines have <1–140 ppm H_2O, orthopyroxenes have 60–650 ppm H_2O, garnets have <1–200 ppm H_2O, and clinopyroxenes have 100–1,300 ppm H_2O (cf. review by Ingrin and Skogby, 2000).

Analytical methods. IR spectroscopy is the most common method of detecting OH in minerals (Rossman, 1988, 1996), and is able to distinguish structurally incorporated OH from water in fluid inclusions, cracks, and hydrous alteration products. Polarized IR spectroscopy provides, in addition, information on the orientation of the OH bond in the crystal structure. Determining the amount of water (as OH) in the mineral of interest requires calibration of the relationship between the intensity of the absorption and the amount of water present; an independent technique to determine the amount of water in standard materials is therefore required. Unfortunately, this has proved to be a stumbling block, in part because there seems to be considerable variation in the molar absorption coefficient depending on the mineral group, and even on composition within a given mineral group. In effect, there is a significant matrix correction to these calibration coefficients. There are generalized relationships based on glasses and quartz (e.g., Paterson, 1982) and others based on nominally hydrous minerals (Libowitzky and Rossman, 1997). As pointed out explicitly by the latter authors, this relationship may not be applicable to NAMs because the absorption characteristics of OH or hydrogen, if present as point defects in low concentration, may differ from its characteristics when present as an essential structural constituent of hydrous minerals. This possibility, in turn, leads to a very difficult situation, where the water content must be measured as a function of composition for each of the minerals of interest—and at the concentrations of interest. Further, there has to be some assurance that the standard materials that are analyzed by the independent technique do not have additional contributions to their total water content from water that is adsorbed, present in cracks, or present in fluid inclusions. Given these requirements, it is clear that there is a large step—and much effort required—in going from answering the qualitative question "is there OH present?" to the quantitative question "how much?"

A number of other techniques hold potential. Hydrogen manometry, where the hydrogen is extracted from the sample and its volume precisely measured, is an absolute technique. It requires, however, large samples (e.g., 4 g of garnet for the pyrope calibration of Bell *et al.*, 1995). It is, therefore, a bulk sample, and could only be useful in studies of structurally incorporated trace amounts of water if there was absolutely no doubt that there was no contribution to the hydrogen measured from water present in fluid inclusions, in cracks, as adsorbed water, or in alteration minerals.

Another "bulk" technique that holds promise is ^{1}H MAS NMR (Yesinowski *et al.*, 1988; Cho and Rossman, 1993; Kohn, 1996), which has the advantages that the method is intrinsically quantitative, so that the area of the spectrum is directly proportional to the amount of hydrogen in the sample independent of matrix, and that the measured chemical shifts are characteristic of the environment surrounding the hydrogen atom. Kohn's results on olivine and pyroxene were higher than those expected based on IR measurements, and he suggested that IR measurements may systematically underestimate the OH concentration. Keppler and Rauch (2000) addressed this divergence, and demonstrated that Kohn's results, measured on powdered samples, were probably affected by adsorbed water and fluid inclusion water, again demonstrating the difficulty of analyzing a "fugitive" component like OH in a bulk sample, particularly when the sample is powdered. For example, Keppler and Rauch (2000) measured 199 ppm water in clear single crystals of enstatite by polarized FTIR. In polycrystalline enstatite aggregates equilibrated at the same conditions, FTIR yielded 2,000–4,000 ppm. They note that these aggregates were turbid, and the spectrum contained broad, nearly isotropic absorptions that they interpreted to result from water in grain boundaries, growth defects, and fluid or melt inclusions. After grinding enstatite powder in air, they measured total water contents of 5,300 ppm, and even higher contents (12,600 ppm) after grinding under water, even though this sample was dried at 150 °C before the measurement.

Another technique is the nuclear reaction analysis of Lanford *et al.* (1976), which was applied to NAMs by Rossman *et al.* (1988), Skogby *et al.* (1990), Maldener *et al.* (2001), and Bell *et al.* (2003). This technique has the advantage of yielding absolute hydrogen concentrations. It is a near-surface technique in which the hydrogen concentration is measured as a function of depth. The spatial resolution of the technique is at the millimeter level; Bell *et al.* (2003) prepared polished surfaces 5×5 mm in area. Maldener *et al.* (2001) state that 1 mm diameter samples can be analyzed. This technique has been used to calibrate absorption coefficients for IR spectroscopy (e.g., Maldener *et al.*, 2001; Bell *et al.*, 2003). Bell *et al.* (2003) applied this technique to hydrogen in olivine, and found some of the previous estimates of hydrogen concentration in olivine need to be revised upward by factors between 2 and 4. Such a correction cannot be applied uniformly to all previous studies, because their new calibrations are specific to polarized spectra.

Clearly, what is needed is a technique that is readily calibrated and can analyze with spatial resolution of a few microns. Two possibilities, aside from micro-FTIR, appear to exist. Secondary ion mass spectrometry (SIMS) (ion microprobe) is one possibility (Kurosawa *et al.*, 1992, 1993, 1997; Hauri, 2002; Koga *et al.*, 2003), although, like FTIR, it has significant issues with the matrix dependence of the correction factors required. Another possibility that merits further exploration is the elastic recoil detection analysis (ERDA) technique (Barbour and Doyle, 1995) that was applied to silicates by Sweeney *et al.* (1997). This latter paper also gives a useful overview of the advantages and disadvantages of the various microanalytical techniques. At present, it appears there are several potential microanalytical techniques that may be useful; what would be helpful now are single-crystal standards of known water content that are free of all spurious sources of hydrogen-contamination that can be used for calibration and comparison purposes.

At this point, it seems useful to review the current state of the science with respect to the major minerals of the mantle. Different authors report "water" contents in different ways: atoms $H/10^6$ atoms Si, ppm OH, ppm H_2O, etc. To ease comparison, the authors' original values have been converted, where necessary, to ppm H_2O.

Olivine. Olivine, as the most abundant mineral in the peridotitic upper mantle, has been the subject of numerous studies of natural samples and experimental studies designed to characterize the amount of hydrogen that could be stored in olivine. The former include Wilkins and Sabine (1973), Beran and Putnis (1983), Kitamura *et al.* (1987), Miller *et al.* (1987), Bell and Rossman (1992a), Libowitzky and Beran (1995), Kohn (1996), Kurosawa *et al.* (1997), and Jamtveit *et al.* (2001). The latter include Mackwell and Kohlstedt (1990), Bai and Kohlstedt (1992, 1993), Kohlstedt *et al.* (1996), and Matveev *et al.* (2001). Additional insight has been provided by computer simulations (Wright and Catlow, 1994; Brodholt and Refson, 2000).

The "wettest" mantle-derived olivine is an iron-rich olivine megacryst from Monastery kimberlite, South Africa, that contained 140 ppm H_2O (Bell and Rossman, 1992a). This sample may constrain the crystallization conditions of minerals of the megacryst suite. Indeed, the high water content supports an origin by crystallization from a volatile-enriched magma (Harte and Gurney, 1981), similar to the proposed "mantle pegmatite" origin of megacrysts in alkali basalts (Righter and Carmichael, 1993).

Peridotitic olivines from Kimberley and Monastery kimberlites studied by Miller *et al.* (1987) contain the equivalent of ~16 ppm H_2O, although they report one from Kimberley "including molecular water" as having the equivalent to ~60 ppm H_2O. A subsequent SIMS study by Kurosawa *et al.* (1997) on a variety of mantle-derived samples found hydrogen contents from 10–60 ppm H_2O. Given these results, and figure 2 of Bell and Rossman (1992a), which showed only the single sample discussed above at >70–80 ppm H_2O, it appears that peridotitic olivine contains $\leq$60 ppm H_2O.

From available experimental data, it appears that these values do not reflect saturation with a fluid phase. For example, Kohlstedt *et al.* (1996) synthesized $f_{o_{90}}$ olivines with water contents that increased with pressure from 135 ppm H_2O at 2.5 GPa, 1,100 °C to 1,510 ppm at 12 GPa, 1,100 °C and 1,090 ppm at 13 GPa, 1,100 °C.

There are two additional issues to consider:

- how is the hydrogen substituting in the olivine structure?
- does the hydrogen content of the olivine change en route to the surface?

These two issues are linked; the substitution mechanism of hydrogen, along with diffusion rates, will control the loss or gain of hydrogen during ascent. Most models of hydrogen substitution have involved vacancies on the tetrahedral sublattice or on the metal (M1, M2) sublattice, although some authors have invoked oxygen interstitials (see review by Ingrin and Skogby (2000), and further discussion in Matveev *et al.* (2001)). As discussed by Ingrin and Skogby (2000), it will be the mobility of vacancies, especially on the metal sublattice, that control the loss of hydrogen during ascent. They calculate timescales for 95% re-equilibration of olivine crystals as a function of size and temperature. For example, according to their figure 6, a 10 mm olivine crystal will reequilibrate in <50 h at 1,000 °C, whereas a 1 mm crystal would take <1 h. They argue, based on these calculations, that it is unlikely that measured olivine OH contents truly reflect the amount of OH present in the olivine at mantle conditions. That olivine may lose much of its hydrogen during ascent is supported by some authors (Mackwell and Kohlstedt, 1990; Bai and Kohlstedt, 1992, 1993) but not some others (Rossman, 1996; Kurosawa *et al.*, 1997).

A further complication is provided by Khisina *et al.* (2001) and Khisina and Wirth (2002), who observed nanometer-scale lamellar- and hexagonal-shaped OH-bearing inclusions in a peridotitic olivine from Udachnaya kimberlite, Siberia. These inclusions were lower in (Mg + Fe)/Si than the host olivine, and they used a variety of TEM-based techniques to provide supporting evidence that these inclusions were hydrous olivine, with the general formula $(Mg_{1-y}Fe_y^{2+})_{2-x}v_xSiO_4H_{2x}$, where v is a vacancy on the metal sublattice. They propose that these inclusions exsolved from a precursor olivine that was saturated with OH-bearing point defects while the olivine was still in the mantle. If this exsolution is a general mechanism, it may act to stabilize the hydrous phase and decrease the extent of hydrogen loss during ascent.

Kudoh (2002) proposed a possible structure for hydrous olivine that is a recombination structure derived from the olivine structure. This generates a crystal structure similar to that of clinohumite, and Kudoh proposes that the humite group minerals and his HyM-α hydrous olivine form a homologous series as recombination structures. The formula of HyM-α ($Mg_9Si_5H_2O_{20}$), if expressed on a four-oxygen basis ($Mg_{1.8}SiO_4H_{0.40}$) is the same as that of the hydrous olivine of Khisina and Wirth (2002) for $x = 0.20$.

Orthopyroxene. Orthopyroxene has been studied less than olivine. There were no mantle-derived orthopyroxenes in the study of Skogby and Rossman (1989); that of Skogby *et al.* (1990) included an enstatite (Mg/(Mg + Fe) = 0.90) from the Premier kimberlite, South Africa, which contained 500 ppm OH, or 265 ppm H_2O. According to Ingrin and Skogby (2000), these values should be reduced by a factor of 0.4 to correspond to the new calibration for orthopyroxene of Bell *et al.* (1995). Bell and Rossman (1992a) report a range from 40–460 ppm H_2O in mantle-derived orthopyroxene. Peslier *et al.* (2002) found water contents of 39–265 ppm H_2O for orthopyroxenes in spinel peridotite xenoliths from Mexico and Washington state, which they argue come from the sub-arc mantle wedge.

Peslier *et al.* (2002) also found systematic correlations between the water contents and major-element compositional data for their pyroxenes, and between the water contents and the oxidation state of the xenolith. They argue that these correlations require that the water contents correspond to mantle values, and the samples have not been subjected to loss of hydrogen during ascent. Ingrin and Skogby (2000), based on the diffusion data for pyroxene, considered that pyroxenes would be significantly slower than olivine to re-equilibrate during ascent, consistent with the findings of Peslier *et al.* (2002).

The amount of water that can dissolve in orthopyroxene depends on pressure, temperature, and composition. In a study at 1,100 °C, Rauch and Keppler (2002) found that water content in pure enstatite increased from 55 ppm H_2O at 0.2 GPa to 867 ppm at 7.5 GPa, then decreased slightly to 714 ppm at 10 GPa. They also looked at the effect of changing pyroxene composition; the most impressive change occurred with increasing aluminum content of the pyroxene. The addition of ~1 wt.% Al_2O_3 increased water solubility from 199 ppm to 1,102 ppm H_2O at 1.5 GPa and 1,100 °C. They suggest this implicates a substitution of $Al^{3+} + H^+$ for Si^{4+}, and argue that examining the aluminum distribution between octahedral and tetrahedral sites in mantle-derived pyroxenes would allow estimation of original water contents even for samples that have lost hydrogen during ascent.

Clinopyroxene. Studies on water contents of natural clinopyroxenes include Wilkins and Sabine (1973), Ingrin *et al.* (1989), Skogby and Rossman (1989), Skogby *et al.* (1990), Smyth *et al.* (1991), Bell and Rossman (1992a), and Bell *et al.* (1995). Experimental studies by Skogby (1994) and Ingrin *et al.* (1995) provide insight into solubilities and diffusivities, respectively. In general, clinopyroxene has the highest measured water contents of any upper-mantle NAM, ranging from 100–600 ppm H_2O for augite and diopside, and up to ~1,300 ppm H_2O for omphacite from mantle eclogite. Ingrin and Skogby (2000) caution that the omphacite values may be overestimated because of the systematic differences in the absorption spectra of augite and omphacite, and the lack of an omphacite-specific calibration. The measured abundances are likely to be mantle values, again based on the timescale calculations of Ingrin and Skogby (2000).

Garnet. There are many studies that report water contents of natural garnets (Wilkins and Sabine, 1973; Ackermann *et al.*, 1983; Aines and Rossman, 1984a,b; Rossman *et al.*, 1988, 1989; Lager *et al.*, 1989; Rossman and Aines, 1991; Bell and Rossman, 1992a,b; Langer *et al.*, 1993; Bell *et al.*, 1995; Amthauer and Rossman, 1998). Experimental studies include those of Geiger *et al.* (1991), Khomenko *et al.* (1994), L. Wang *et al.* (1996), Lu and Keppler (1997), and Withers *et al.* (1998). Mantle-derived pyropic garnets usually contain <60 ppm H_2O (Bell and Rossman, 1992a), and the range in values considered reliable by Ingrin and Skogby (2000) is <1 ppm to 200 ppm H_2O.

Geiger *et al.* (1991) synthesized pyropes with 200–700 ppm H_2O at 2–5 GPa, 800–1,200 °C. They proposed that OH was entering the structure in the hydrogarnet substitution based on their X-ray and IR data. They noted that the more complicated IR spectra of natural water-bearing pyropic garnets implicates different substitution mechanism(s) for these garnets compared to the synthetic ones.

Lu and Keppler (1997) experimentally determined the solubility of water in a natural pyrope from Dora Maira in a series of experiments from 1.5 GPa to 10 GPa, mostly at 1,000 °C. They found the water content to systematically increase from 73.5 ppm H_2O at 1.5 GPa, 1,000 °C to 198.9 ppm H_2O at 10 GPa, 1,000 °C. Lu and Keppler's results contrast with the 1,000 °C results of Withers *et al.* (1998), who found a systematic increase in water contents from 140 ppm H_2O at 2 GPa to 1,010 ppm at 4 GPa. At 6 GPa, the water content had decreased slightly to 960 ppm H_2O. At 7 GPa, 8 GPa, and 13 GPa, the water content of garnets synthesized by Withers *et al.* were below the detection limit of their spectrometer. Withers *et al.* (1998) explain their results by suggesting that the water is entering into the crystal structure as the hydrogarnet substitution, and arguing that the estimated molar partial volume of H_2O in garnet in this substitution becomes greater than the molar volume of H_2O as a discrete fluid at the pressure where the maximum solubility occurs. In effect, they verified the prediction of Martin and Donnay (1972).

The apparently discrepant results of Lu and Keppler (1997) and Withers *et al.* (1998) may be reconciled if different substitution mechanisms were at work. Supporting this idea, Lu and Keppler (1997) showed that their results were not consistent with the hydrogarnet substitution favored by Withers *et al.* (1998). Lu and Keppler (1997) proposed that OH substituted for oxygen with charge balance provided by boron substituting for silicon on the tetrahedral site and lithium substituting for magnesium, on the dodecahedral site. They calculated that the amount of lithium and boron in their natural garnet would charge-balance 466 ppm and 54 ppm H_2O, respectively. Given their maximum water content was ~200 ppm H_2O, this mechanism, and this explanation for the difference in their results and those of Withers *et al.* (1998), seems quite reasonable.

Higher-pressure phases. Among NAMs, stable at higher pressures, much attention focused on wadsleyite, following the original suggestion on crystal chemical grounds by Smyth (1987) that wadsleyite could be a significant host for hydrogen in the mantle. Electrostatic calculations by Downs (1989) corroborated the Smyth proposal, and further suggested another oxygen site to be a candidate for protonation as well. Subsequently, McMillan *et al.* (1991) reported the presence of hydroxyl groups in β-Mg_2SiO_4 and estimated a water content of ~300 ppm H_2O, even though their samples were not synthesized under conditions designed to have a high water fugacity. Young *et al.* (1993) subsequently synthesized β-phase with 618–4,001 ppm H_2O.

Smyth (1994) proposed a hypothetical end-member for hydrous wadsleyite with stoichiometry $Mg_7Si_4O_{14}(OH)_2$ and a maximum water content of 3.3 wt.%.

This proposal was confirmed by Inoue *et al.* (1995), Kudoh *et al.* (1996), and Smyth *et al.* (1997). Kohlstedt *et al.* (1996) synthesized wadsleyite with ~24,000 ppm H_2O (2.4 wt.%) at 14–15 GPa and 1,100 °C, but noted that the absence of an H_2O fluid at the end of their experiments implied that saturation conditions were not established at the run conditions.

The ringwoodite polymorph ((γ-Mg,Fe)$_2SiO_4$) has also been found to contain water. Kohlstedt *et al.* (1996) synthesized ringwoodite with ~27,000 ppm H_2O (2.7 wt.%) at 19.5 GPa and 1,100 °C in experiments that they interpret to be fluid-undersaturated. Subsequent studies confirm high water contents in ringwoodite; e.g., Inoue *et al.* (1998b) synthesized samples with 2.2 wt.% H_2O. Kudoh *et al.* (2000) determined the crystal structure for a sample of hydrous ringwoodite of composition $Mg_{1.89}Si_{0.98}H_{0.30}O_4$ and derived the site occupancies, which show vacancies on both tetrahedral and octahedral sites, as well as a small amount of cation disorder: $(Mg_{1.84}Si_{0.05}v_{0.11})(Si_{0.93}Mg_{0.05}v_{0.02})H_{0.30}O_4$, where v = vacancy.

Bolfan-Casanova *et al.* (2000) studied partitioning of water between phases in the system $MgO-SiO_2-H_2O$ at 15–24 GPa and 1,200–1,600 °C. They found that $MgSiO_3$ perovskite did not contain detectable H_2O, and that the water solubility in wadsleyite and ringwoodite were ~10 times greater than in the high-pressure polymorphs of $MgSiO_3$. Stishovite was found to contain 15–72 ppm H_2O. At 15 GPa and 1,300 °C, coexisting wadsleyite and clinoenstatite contained 2,212 ppm and 583 ppm H_2O, respectively. At 19 GPa and 1,300 °C, coexisting ringwoodite and akimotoite (ilmenite-structure $MgSiO_3$) contained 7,817 ppm and 352 ppm H_2O, respectively.

Murakami *et al.* (2002) studied a natural peridotite composition (with 7.5–13.5 wt.% H_2O) at 25.5 GPa and 1,600–1,650 °C. They measured water contents in their run products by SIMS. They found magnesium-rich perovskite and ferropericlase to have ~2,000 ppm H_2O and calcium-rich perovskite to have ~4,000 ppm H_2O. A lower mantle consisting of 79 wt.% Mg-perovskite, 16 wt.% ferropericlase, and 5 wt.% Ca-perovskite could contain 2,100 ppm H_2O, which when integrated over the mass of the lower mantle yields a reservoir ~5 times greater than the oceans. They compare this to the transition zone, which can store nearly six oceans worth of water, despite its smaller volume, because of the greater solubility of water in wadsleyite and ringwoodite (~3.3×10^4 ppm and 2.2×10^4 ppm, respectively). The great contrast in water solubility between the transition zone phases and those of the lower mantle suggest the possibility of release of water from slabs, and from convecting mantle, upon going from the transition zone down into the lower mantle.

2.07.3.2 Carbon-bearing Species

Carbon is present in a variety of modes in the mantle, both in oxidized form as fluid, carbonate, and carbonatitic melt, and as neutral species, such as graphite and diamond. The carbon cycle is complex because of the multiplicity of forms in which carbon may be present, and the consequences of carbon presence are very different depending on the nature of the host. For example, at pressures where carbonate is stable in peridotite, the temperature of the solidus is lower than in the volatile-free case, or indeed in the lower-pressure situation where a CO_2 fluid, rather than carbonate, coexists with the peridotite. Carbonate-rich melts have been implicated in a number of mantle processes, from metasomatism to growth media for diamonds. Alternatively, at more reducing conditions in which diamond is stable rather than carbonate, such metasomatic activity by carbonate-rich melts would seem to be unlikely.

2.07.3.2.1 Fluid inclusions

Shallow-mantle xenoliths, hosted in alkali basalts, commonly contain CO_2-rich fluid inclusions (e.g., Roedder, 1965, 1984; Frey and Prinz, 1978; Murck *et al.*, 1978; Miller and Richter, 1982; Pasteris, 1987; Frezzotti *et al.*, 1994; Burnard *et al.*, 1998; Ertan and Leeman, 1999; Andersen and Neumann, 2001). In most cases, these fluid inclusions are related to metasomatic processes to which the samples were subjected. As outlined previously, the CO_2-rich nature of these inclusions may be a natural consequence of degassing from an ascending melt that contains both H_2O and CO_2, because the greater solubility of H_2O in silicate melts would allow it to remain in solution in the residual melt. Alternatively, as proposed by Andersen and Neumann (2001), the high CO_2 content in these inclusions could be the result of removal of water via reactions between the original fluid and the host mineral surrounding the inclusion.

2.07.3.2.2 Carbon

The presence of diamond and graphite in mantle-derived samples such as kimberlites and the xenoliths they host is *prima facie* evidence that neutral carbon is stable in the Earth's mantle. Outstanding questions remain, however, concerning the stability of neutral carbon in areas other than those beneath continental cratons, and concerning the mechanism by which diamond forms. To a large extent, the latter question revolves around the unresolved problem of the oxidation state of the mantle, and how—and if—the oxidation state is controlled.

The geographical distribution of diamond is intrinsically biased, because only a few magma types (kimberlites, lamproites, and possibly a few other members of the lamprophyre clan) have depths of origin within the field of diamond stability. If the conditions that produce such magmas—defining these conditions is itself an unanswered question—are restricted to mantle beneath cratons, either for thermal or chemical reasons, then the biased geographical distribution of diamond-bearing rocks cannot be used to infer the spatial distribution of diamond in the mantle. A comprehensive understanding of the conditions of formation of natural diamond is required to make progress on this problem. Toward this end, recent experimental studies in which diamond has nucleated and grown in the presence of C–O–H fluids or carbonate melts, or from carbonate–silicate interaction, rather than in the normal molten metal catalysts, provide pathways for diamond nucleation and growth that may be appropriate to the natural environment (e.g., Akaishi *et al.*, 1990, 2000; Taniguchi *et al.*, 1996; Pal'yanov *et al.*, 1999a,b, 2002a,b; Sato *et al.*, 1999; Akaishi and Yamaoka, 2000; Kumar *et al.*, 2000; Sun *et al.*, 2000; Sokol *et al.*, 2001).

Neutral carbon can occur dissolved in minerals, as interstitial atoms, in defects, or even substituting for silicon. The latter was proposed by Fyfe (1970), who speculated on the possibility of Mg_2CO_4 spinel as a structural analogue to ringwoodite. Green (1972, 1985) and Green and Guegen (1983) suggested one of the first two mechanisms might be operative in olivine. Freund *et al.* (1980) determined high solubilities (and diffusivities) of carbon in olivine, but their results could not be reproduced by other workers (Mathez *et al.*, 1984; Tsong *et al.*, 1985; Tsong and Knipping, 1986). Experimental studies by Tingle *et al.* (1988) determined that carbon solubility must be <30 ppm in olivine at pressures to 3 GPa and temperatures of 1,180–1,529 °C. They discussed in detail the analytical pitfalls, including carbon-rich surface layers, concentration of carbon along cracks, and quench artifacts. At this point in time, it is unclear if carbon can dissolve in high-pressure silicates to any significant extent.

2.07.3.2.3 Carbonates

Carbonate is rarely found in mantle-derived xenoliths. It has been found in mantle-derived garnets (McGetchin and Besançon, 1975; Smith, 1987), clinopyroxenes (Hervig and Smith, 1981) and in rare xenoliths (e.g., Ionov *et al.*, 1993a, 1996; Lee *et al.*, 2000; Laurora *et al.*, 2001).

Ionov *et al.* (1993a, 1996) found carbonate in spinel lherzolite xenoliths as interstitial crystals and as aggregates with calcium-rich olivine and aluminum- and titanium-rich clinopyroxene. They interpreted the former to be primary and the latter as evidence for metasomatism by a carbonate-rich melt. Subsequently, Ionov (1998) measured trace-element abundances in the carbonates and coexisting phases, and proposed the aggregate carbonates were formed by crystal fractionation from a carbonate melt. That these carbonates represent crystallized cumulates, rather than quenched melts, is supported by subsequent studies of Lee *et al.* (2000), Ionov and Harmer (2002), and Laurora *et al.* (2001), as well as by the extensive experimental studies by Wyllie and co-workers (e.g., Lee and Wyllie, 1998, 2000, and references therein).

The rarity of carbonate may be attributed to one of two possibilities: either carbonate is rare in the mantle sampled by such xenoliths, or carbonate does not survive transit to the surface because it reacts readily with silicates to produce CO_2 by reactions such as $2MgCO_3 + Mg_2Si_2O_6 \rightarrow 2Mg_2SiO_4 + 2CO_2$ (Wyllie *et al.*, 1983; Canil, 1990). This process, if it occurs, would also be a very efficient mechanism for disaggregating the xenoliths (Boyd and Gurney, 1986, and references therein). Alternatively, Berg (1986) interpreted brucite–calcite intergrowths in peridotite xenoliths to be the result of reaction of dolomite with a hydrous fluid during ascent.

Despite the paucity of evidence from natural samples for carbonate being stable in the mantle, there has been much experimental work since the 1970s devoted to understanding the behavior of oxidized carbon (as CO_2 and carbonate) at pressure and temperature conditions of the Earth's mantle (see reviews by Wyllie, 1995; Luth, 1999; Wyllie and Ryabchikov, 2000). As outlined by Eggler and Baker (1982) and by Luth (1993), understanding the stability of oxidized carbon provides fundamental constraints on the stability of reduced carbon in mantle mineral assemblages.

Relative oxidation state is often referenced to the chemical potential or fugacity of oxygen (f_{O_2}). The maximum f_{O_2} at which carbon can exist under any circumstances is defined by the reaction $C + O_2 = CO_2$. In the presence of silicates, however, that maximum f_{O_2} is reduced because reactions such as $Mg_2Si_2O_6 + 2MgCO_3 = 2Mg_2SiO_4 + 2C + O_2$ (for peridotitic assemblages; Eggler and Baker, 1982) or reactions involving clinopyroxene or garnet for eclogitic assemblages (Luth, 1993) occur at lower f_{O_2}. Our understanding of where these reactions lie in P–T–f_{O_2} space relative to "normal" mantle assemblages is still at an unsatisfactory level.

Carbonate has been observed infrequently as inclusions in diamond. Calcite was described by Meyer and McCallum (1986) and Brenker *et al.* (2002), magnesite by A. Wang *et al.* (1996), and dolomite by Stachel *et al.* (1998). Of the three, the magnesite is least surprising, in that magnesite is the carbonate that is stable in peridotitic assemblages in the diamond stability field.

Meyer and McCallum (1986) concluded that the calcite in diamonds from the Sloan kimberlites was epigenetic, not primary, in origin. Brenker *et al.* (2002) found $CaCO_3$ as submicron precipitates in olivine in a multiphase inclusion from Kankan, Guinea. Their interpretation was that the carbonate formed by exsolution from the olivine during cooling, with the carbonate provided possibly from the same reservoir that formed the diamond.

The dolomite observed by Stachel *et al.* (1998) was a single-phase inclusion. The presence of dolomite within the stability field of diamond requires a protolith in which the exchange reaction $Mg_2Si_2O_6 + CaMg(CO_3)_2 = CaMgSi_2O_6 + 2\ MgCO_3$ is prevented from occurring by the absence of orthopyroxene.

2.07.3.2.4 *Evidence for carbonate metasomatism*

As discussed above, many of the occurrences of carbonate in spinel-facies mantle xenoliths are interpreted to be the product of metasomatism by a carbonate-rich melt. Less obvious effects of this style of metasomatism have been documented in a number of cases, in which a sodic clinopyroxene replaces primary orthopyroxene, accompanied by the crystallization of apatite $\pm$ amphibole $\pm$ phlogopite, or where specific trace-element geochemical signatures are observed in the peridotitic minerals, with no mineralogical manifestation of alteration. These studies include Green and Wallace (1988), O'Reilly and Griffin (1988), Yaxley *et al.* (1991, 1998), Dautria *et al.* (1992), Dalton and Wood (1993), Hauri *et al.* (1993), Ionov *et al.* (1993a), Rudnick *et al.* (1993b), Coltorti *et al.* (1999), Chalot-Prat and Arnold (1999), Ionov *et al.* (2002), and Neumann *et al.* (2002). Where there is an overt change in mineralogy, the process is usually interpreted to result from ascending carbonatitic melts crossing an analogue of the reaction orthopyroxene + dolomite = forsterite + clinopyroxene + CO_2. This destabilizes the carbonate liquid and generates clinopyroxene and a CO_2-rich fluid at the expense of orthopyroxene. The accompanying hydrous phases would be explained by water in solution in the original carbonate-rich liquid. More subtle effects attributed to carbonatitic fluids are where only changes in trace-element geochemistry are observed (cf. review by Menzies and Chazot, 1995). These effects typically include enrichment in elements such as rubidium, barium, thorium, uranium, strontium, REEs, and depletion in elements such as niobium, tantalum, zirconium, hafnium, and titanium. However, the attribution of such signatures to carbonatitic metasomatism is often done rather uncritically; the study of Bedini *et al.* (1997) provides a cautionary note in this regard (see Chapter 2.04). They demonstrated that chromatographic infiltration of an OIB-type melt can produce these same effects, given reasonable assumptions concerning the nature of the melt–mantle interaction.

In deeper samples, the trace-element geochemistry observed in garnet inclusions in diamonds has been attributed to carbon-bearing fluids (e.g., Stachel and Harris, 1997; Wang *et al.*, 2000; Dobosi and Kurat, 2002), although the oxidation state of the fluid (CO_2- versus CH_4-rich) remains open to debate.

2.07.3.2.5 *Moissanite*

There have been a number of reports of moissanite, α-SiC, in diamond inclusions and in concentrates from kimberlite. The earlier reports were controversial, because of the possibility of contamination during crushing. *In situ* observation of moissanite and β-SiC as inclusions in diamond (Leung, 1990; Leung *et al.*, 1996) and studies that excluded any contact of bulk kimberlite samples with carbide (Leung *et al.*, 1990; Mathez *et al.*, 1995) have confirmed the rare presence of SiC in diamond and kimberlite. Stability of SiC requires very reducing conditions, such that iron would be present as native metal (Mathez *et al.*, 1995; Ulmer *et al.*, 1998). Such conditions contrast with the generally accepted view of relatively oxidizing conditions in most regions of the upper mantle. Mathez *et al.* (1995) suggested that SiC may have formed by metamorphism of reduced, carbon-bearing sediments during subduction, whereas Ulmer *et al.* (1998) argue for a bias in our understanding of mantle oxidation state because of the changes induced by magmatism that brought the samples to the surface. They propose that there are local regions at >300 km depth that are sufficiently reduced to stabilize SiC.

Another possibility, somewhat speculative, exists. A region of the Earth in which reduced iron is generally accepted to exist is the core. There have been recent proposals, based on Re–Os systematics, that argue for incorporation of core material into the mantle (Walker *et al.*, 1995; Brandon *et al.*, 1998, 1999; D. Walker, 2000). SiC would be stable under core conditions. Haggerty (1994) proposed that protokimberlite magma and entrainment of xenolithic material originates at the core–mantle boundary—citing the presence of SiC, among other lines of evidence, as supporting this deep origin.

2.07.3.3 Sulfur

Sulfur is almost always present in mantle-derived magmas and mantle samples as sulfide, which has been documented from mantle xenolith suites, abyssal peridotites, peridotite massifs, and diamonds (Meyer and Brookins, 1971; Desborough and Czamanske, 1973; Frick, 1973; Vakhrushev and Sobolev, 1973; Bishop *et al.*, 1975; De Waal and Calk, 1975; Meyer and

Boctor, 1975; Peterson and Francis, 1977; Harris and Gurney, 1979; Tsai *et al.*, 1979; Mitchell and Keays, 1981; Lorand and Conquéré, 1983; Garuti *et al.*, 1984; Gurney *et al.*, 1984; Morgan, 1986; Andersen *et al.*, 1987; Dromgoole and Pasteris, 1987; Distler *et al.*, 1987; Lorand, 1987, 1988, 1989, 1990, 1991; Fleet and Stone, 1990; Ionov *et al.*, 1992; Rudnick *et al.*, 1993a; Deines and Harris, 1995; Szabó and Bodnar, 1995; Bulanova *et al.*, 1996; Guo *et al.*, 1999; Luguet *et al.*, 2001). As emphasized by a number of authors, sulfides of the compositions expected to be present in the mantle have low melting temperatures, such that they will be liquid at temperatures above ~1,200–1,300 °C at 1–2 GPa. Thus, in many cases, the phase that is trapped is a sulfide melt, or a monosulfide solid solution coexisting with a sulfide melt.

One point of controversy in the literature stems from the high reactivity of sulfides, which leads to the potential for re-equilibration and exsolution during ascent and emplacement, and for post-emplacement alteration (Eggler and Lorand, 1993; Lorand, 1993; Ionov *et al.*, 1993b).

It is noteworthy that even in oxidized samples (ΔFMQ ~2) from a fore-arc environment (Lihir, Papua New Guinea), the sulfur is present as Fe–Ni sulfides rather than as sulfate (McInnes *et al.*, 2001).

The usual interpretation of sulfides in peridotitic samples is that they were trapped in the residual assemblage upon partial melting, implying that the degree of partial melting experienced by these samples was insufficient to exhaust the sulfide in the source region (e.g., Dromgoole and Pasteris, 1987; Szabó and Bodnar, 1995; Guo *et al.*, 1999). This would also imply that the melt produced is sulfide-saturated, consistent with observations on primitive basalts (Czamanske and Moore, 1977; Mathez, 1976; Peach *et al.*, 1990; Stone and Fleet, 1991; Roy-Barman *et al.*, 1998). Sulfides found in pyroxenite samples are interpreted to have precipitated from a basaltic liquid in the mantle, along with the rest of the pyroxenite mineral assemblage. Again, this is consistent with sulfide saturation of basaltic melts (e.g., Dromgoole and Pasteris, 1987; Szabó and Bodnar, 1995; Guo *et al.*, 1999).

The probability that sulfide is molten in some situations within the mantle raises the possibility that this melt may be an effective metasomatic agent, and that sulfides observed in mantle-derived samples have been introduced in metasomatic events (e.g., Fleet and Stone, 1990), rather than being trapped in peridotitic residue of partial melting events. Given the ability of sulfide melt to mobilize chalcophilic elements, a significant amount of experimental work has been conducted to evaluate the mobility of liquid sulfide at high pressures and temperatures, both to constrain element mobility in the present mantle and models of core formation (Ballhaus and Ellis, 1996; Minarik *et al.*, 1996; Gaetani and Grove, 1999; Rose and Brenan, 2001).

Another mode of occurrence of sulfide is in polyphase inclusions, coexisting with silicate glass and/or hydrous fluid (Kovalenko *et al.*, 1987; Hansteen *et al.*, 1991; Schiano *et al.*, 1992, 1994, 1995; Wulff-Pedersen *et al.*, 1996). Kogarko *et al.* (1995) found Fe–Cu-rich sulfides coexisting with syenitic silicate glass and calcic carbonate (interpreted to be quenched liquid—but see Section 2.07.3.2.3) in wehrlitic alteration zones in a metasomatized harzburgite from the Canary Islands. These zones also contained secondary olivine, clinopyroxene, and spinel. They interpreted the metasomatic agent to have been coexisting immiscible dolomitic, silicate, and sulfide liquids.

Finally, sulfide is a common inclusion in diamonds (cf. Chapter 2.05), and sulfide-bearing fluids/melts have been suggested to be the medium from which diamond crystallizes (e.g., Bulanova, 1995).

2.07.3.4 Halogens

There is ample evidence that seafloor hydrothermal alteration involves a variety of fluids, including Na–Ca-rich brines (Von Damm, 1995; Kelley and Früh-Green, 2000). The presence of brine-filled fluid inclusions in altered oceanic crust (e.g., Vanko, 1988; Kelley and Früh-Green, 2000, 2001) testifies to the potential for chlorine to be subducted. Extensive studies in eclogite-facies rocks have documented that some of the fluids that are released during subduction are highly saline, with salinities >50 wt.% NaCl equivalent (Philippot and Selverstone, 1991; Selverstone *et al.*, 1992; Philippot *et al.*, 1995, 1998; Scambelluri *et al.*, 1997, 1998; Scambelluri and Philippot, 2001).

What happens to the chlorine with subduction? Scambelluri *et al.* (1997) traced a sequence of events in an Alpine peridotite that started with hydrothermal alteration of an original peridotite by seawater-derived fluids to form serpentine with 0.35 wt.% Cl, brucite with 0.2 wt.% Cl, and phyllosilicates with ~0.2 wt.% Cl. These phases were preserved as relics in a higher-pressure olivine + titanoclinohumite + antigorite assemblage that formed at 550–600 °C and 2.5 GPa. These minerals lack chlorine but coexist with salt-saturated fluid inclusions. In other cases, chloride-rich fluid inclusions in anhydrous eclogites are interpreted to be the result of breakdown of hydrous phases during prograde metamorphism (Philippot *et al.*, 1998). These authors estimate that the fluid inclusions correspond to ~2,000–3,000 ppm H_2O and 110–167 ppm Cl in these rocks.

Some of the subducted chlorine would be expected to be present in the slab-derived fluid that participates in subduction-zone magmatism, and hence arc magmas should have elevated chlorine contents relative to MORBs or OIBs. Melt inclusion studies indicate that some primitive arc melts have >1,000 ppm Cl (Garcia *et al.*, 1979; Harris and Anderson, 1984; Sisson and Layne, 1993; Sisson and Bronto, 1998; Kent and Elliott, 2002). In contrast, primitive N-MORB glasses have 20–50 ppm Cl (Schilling *et al.*, 1980; Michael and Schilling, 1989; Michael and Cornell, 1998). Higher chlorine contents found in MORBs and submarine OIBs have been attributed to assimilation of hydrothermally altered wall rocks or incorporation of chlorine-rich brines into the magma prior to eruption (Michael and Schilling, 1989; Michael and Cornell, 1998; Kent *et al.*, 1999a,b; Hauri, 2002).

In a study of the Lau Basin back-arc system, Kent *et al.* (2002) was able to resolve contributions from assimilation and from slab-derived fluids. Based on their data and models of the melting process, they proposed that the slab-derived fluids are chlorine-rich, with equivalent salinities of up to 19 wt.% NaCl. For their Mariana Trough samples, they inferred an equivalent salinity of 2.5–4.5 wt.% NaCl, in excellent agreement with the value of 4.3 ± 0.4 derived for the "H_2O-rich subduction component" for the Mariana Trough glasses by Stolper and Newman (1994). Based on the Cl/H_2O, Cl/K_2O, and Cl/P_2O_5 they derive for the slab component, Kent *et al.* (2002) rule out amphibole, phlogopite, and apatite as the primary host for chlorine. They argue for an additional chlorine-rich phase, possibly a chlorine-rich fluid, in the mantle wedge source region for the back-arc magmas. Along the same lines, Kamenetsky *et al.* (1997) had earlier proposed that a LILE-, lead-, and chlorine-rich supercritical fluid was released by dehydration reactions in the subducting slab and added to the mantle wedge. The presence of such a fluid would also be consistent with the presence of chlorine-rich fluid inclusions in the anhydrous eclogite minerals discussed above.

What happens to chlorine at depths below those supplying fluid for subduction-related magmatism? There are two contrasting views at present. Philippot *et al.* (1998) argue that 70% of the chlorine in the altered crust is recycled into the mantle in order to produce an isotopic composition of chlorine equal to that in the source region for MORBs. Alternatively, Lassiter *et al.* (2002) argue that the consistency of the Cl/K_2O ratio of MORBs, Austral Island and other OIBs requires that the Cl/K_2O of the mantle does not vary significantly, despite variability in the nature and quantity of recycled, altered oceanic crust in some parts of the mantle, such as the source regions for plume-related basalts. They use the lack of variability in this ratio in plume-related basalts as evidence that most of the chlorine added during seafloor alteration must be removed during subduction, and not recycled into the mantle. Evidence in support of this point of view comes from the study of Becker *et al.* (2000), who found potassium is depleted by 95–98% during dehydration of oceanic basalts during subduction. Because of the constancy in Cl/K in the mantle, this result would require a similar depletion in chlorine during subduction.

Based on the work of Philippot *et al.* (1998), one might expect to observe a certain proportion of chlorine-rich fluid inclusions in mantle-derived xenoliths, but inclusions in these xenoliths are overwhelmingly CO_2-rich, and chlorine-rich inclusions have not been reported (cf. reviews by Roedder, 1984; Pasteris, 1987; Andersen and Neumann, 2001), with the intriguing exception of the brines reported as inclusions in some diamonds (Johnson *et al.*, 2000; Izraeli *et al.*, 2001). The lack of direct observation of chlorine-rich fluid inclusions in mantle-derived xenoliths may be a result of the lack of examination of appropriate samples that record a previous history as subducted oceanic crust, an absence of these fluids in deeper samples because of participation of these fluids in other petrological processes, such as melt production, or because such fluids do not survive subduction below the slab dehydration limit. Conversely, the presence of chlorine in fluid inclusions in diamonds argues for the existence of chlorine-rich fluids at least in some circumstances in the mantle in the pressure range of diamond stability.

Potential insight into the fate of a chlorine-bearing fluid came from the study of Andersen *et al.* (1984) of xenoliths from Bullenmerri and Gnotuk maars in southwestern Australia that contained abundant CO_2-rich fluid inclusions and vugs up to 1.5 cm in diameter. They found the trapped fluids had reacted with the host minerals to produce secondary carbonates and amphiboles, such that the original composition of the fluid was inferred to be a chlorine- and sulfur-bearing CO_2–H_2O fluid. The evidence for chlorine was the presence of a chlorine peak in the energy-dispersive spectrum of the amphibole; unfortunately, no quantitative analyses were possible on these amphiboles. This does pose the possibility that this sort of reaction is common, and that the normal host for chlorine in the mantle is a mineral phase, such as apatite, amphibole, and mica.

Mantle-derived amphibole is typically chlorine-poor; Garcia *et al.* (1980) measured values from

0.001 wt.% Cl to 0.010 wt.% Cl, Smith *et al.* (1981) report a range of ~30 ppm to 0.03 wt.% Cl, Vannucci *et al.* (1995) and Johnson *et al.* (1996) reported chlorine ≤0.03 wt.%, and Laurora *et al.* (2001) report values of 0.07 wt.% Cl. Only in the Zabargad (Red Sea) peridotites did amphiboles contain greater chlorine contents; up to 0.55 wt.% (Agrinier *et al.*, 1993). These authors interpret many of the high-chlorine amphiboles as having formed during or after diapiric uplift, and hence do not reflect "mantle" values.

Mantle micas are also chlorine-poor; Smith *et al.* (1981) report ranges of 0.04–0.11 wt.% for chlorine and 0.15–2.2 wt.% for fluorine in primary-textured micas. Matson *et al.* (1986) report 0.35–0.53 wt.% F and 0.04–0.07 wt.% Cl in primary micas from peridotites. They found a similar range in chlorine in secondary micas in peridotites but a wider range in fluorine, from 0.02 wt.% to 0.54 wt.%. Subsequent analyses (e.g., Wagner *et al.*, 1996) have not found micas richer in either halogen. In contrast, mantle apatites are often halogen-rich. Smith *et al.* (1981) describe an apatite included in a dolomite inclusion in a Cr-diopside megacryst with 0.87 wt.% Cl and 1.4 wt.% F. Exley and Smith (1982) analyzed apatites from a variety of mantle-derived samples and found 0.008–0.58 wt.% Cl and 1.3–2.37 wt.% F. Ionov *et al.* (1997) found 0.27–2.68 wt.% Cl and 0.00–1.46 wt.% F in mantle-derived apatites. O'Reilly and Griffin (2000) analyzed a variety of apatites hosted in lherzolite xenoliths from Australia and Nunivak Island, Alaska. These apatites are present either as dispersed grains within the lherzolite, or in veins coexisting with Cr-diopside or amphibole + mica. These apatites contain 1.43–2.48 wt.% Cl and 0.00–1.26 wt.% F. They attribute the apatite to metasomatism and calculate that the coexisting fluid would contain 500–3,500 ppm chlorine. Qualitative support for an involvement of a metasomatic fluid comes from the earlier study of Andersen *et al.* (1984), as discussed above.

The presence of apatite in subcontinental mantle samples raises the question of the thermal stability of apatite, and whether it would be stable in mantle other than cool lithospheric roots. A limited amount of experimental work has been done that addresses this question; Vukadinovic and Edgar (1993) determined the solidus for phlogopite–apatite mixtures at 2 GPa in the K_2O–CaO–MgO–Al_2O_3–SiO_2–P_2O_5–H_2O–F system. They looked at two bulk compositions, one with the hydroxy end-members and the other where F/OH = 1. The solidus was 1,225 °C for the fluorine-free system, and 1,260 °C for the fluorine-bearing system. It seems likely that adding iron to the system will decrease solidus temperatures. Given that the average current mantle adiabat is ~1,300 °C at 2 GPa (McKenzie and Bickle, 1988), it appears unlikely that apatite would survive as a possible host for halogens in the convecting mantle, away from lithospheric roots. In another study, Foley *et al.* (1999) found the solidus for a K-richterite + clinopyroxene + phlogopite + ilmenite + apatite to be 1,025–1,075 °C at 1.5 GPa and ~1,300 °C at 5 GPa. Apatite was not stable at temperatures greater than ~50 °C above the solidus. Again, these temperatures are lower than would be expected in the convecting mantle at this depth, arguing that apatite's role as host for halogens is limited to subcontinental mantle.

2.07.4 SUMMARY AND CONCLUSIONS

We have a reasonable picture of the geochemical cycle of water and to a lesser extent that of carbon. To return to the question of the fate of volatiles in subducting slabs, it is clear that there are a number of viable hosts for volatiles in the mantle. The crux is whether they actually contain H_2O or CO_2, or whether essentially all the volatiles are stripped out of the slab during subduction, leaving a desiccated slab to continue down into the mantle. There is no consensus on this issue. One extreme is represented by Dixon *et al.* (2002), who argue that water is extracted from the subducting slab with 92–97% efficiency.

Some insight into this issue may be obtained by combining the experimentally determined phase equilibria with internally consistent thermodynamic databases, which allow interpolation and extrapolation of the extant data. In a series of papers, Kerrick and Connolly (1998, 2001a,b) calculated phase equilibria and modal mineralogies

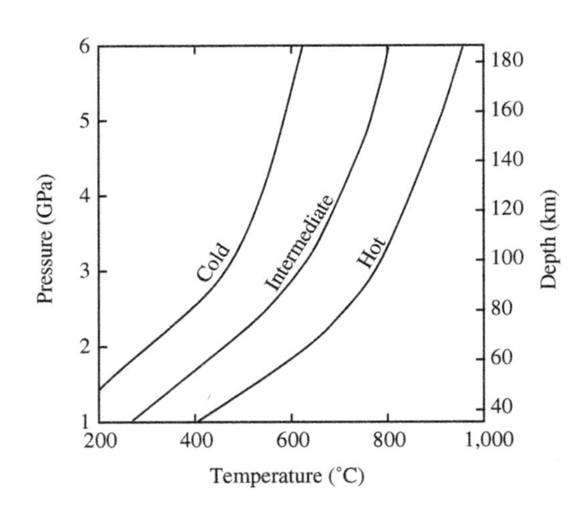

Figure 15 Geotherms used in contructing Figures 16–18 (after Kerrick and Connolly, 2001a; Peacock and Wang, 1999).

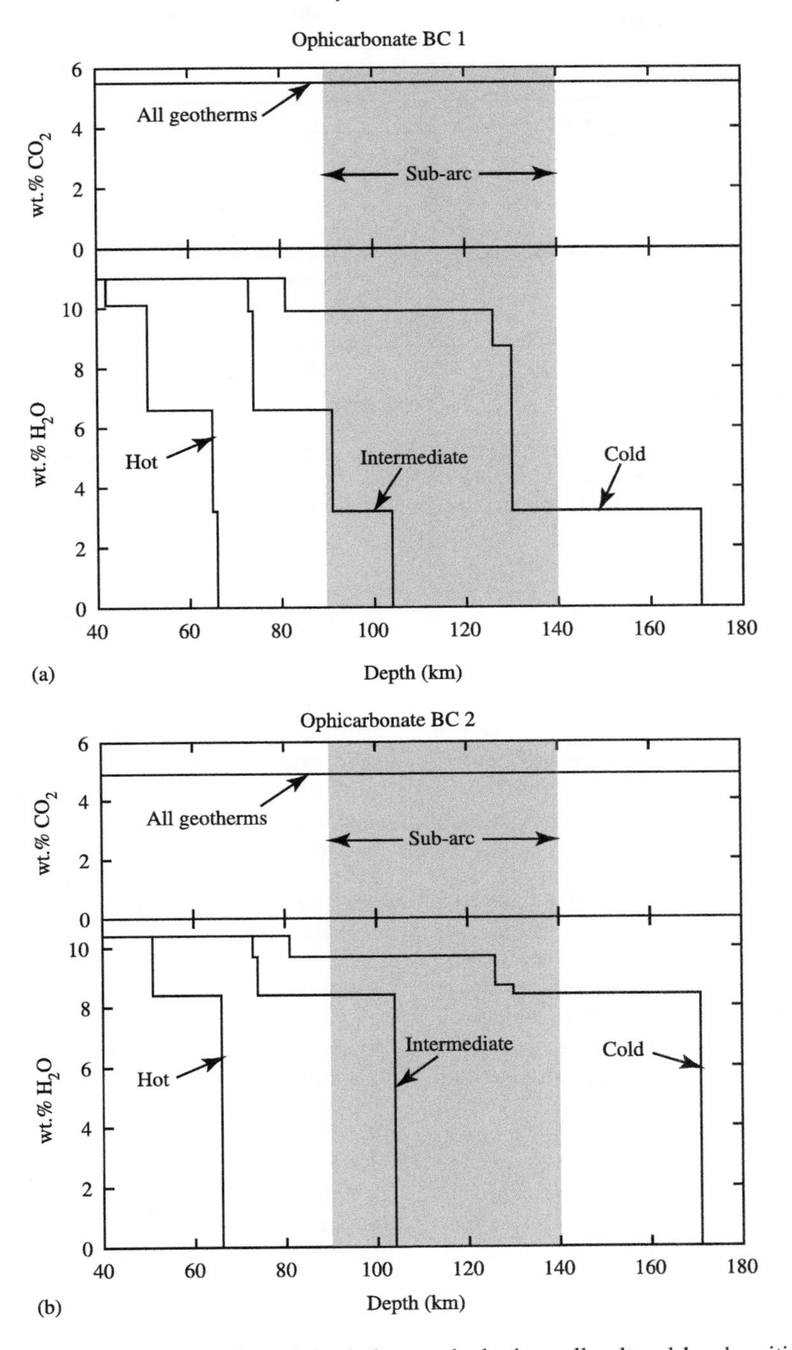

Figure 16 Volatile content as a function of depth for two hydrothermally altered harzburgitic (ophicarbonate) protoliths, shown for three geotherms. Constructed from *P–T* projections in Kerrick and Connolly (1998). Protolith mineralogy is: (a) antigorite + brucite + calcite and (b) antigorite + talc + calcite. "Hot," "intermediate," and "cold" refer to geotherms illustrated in Figure 15. Shaded "sub-arc" region after Kerrick and Connolly (2001a).

as functions of pressure and temperature for bulk compositions representative of different components of the subducting lithosphere. Their calculations assumed closed-system behavior, and thus represent one end-member of a range of possible behaviors. These calculations provide the basis to estimate the amount of H_2O and CO_2 that are present as a function of depth in the slab, which they present in all their papers in *P–T* projections that also show the relevant mineral assemblage. A more useful representation for our purposes, used in Kerrick and Connolly (2001a), is a plot of

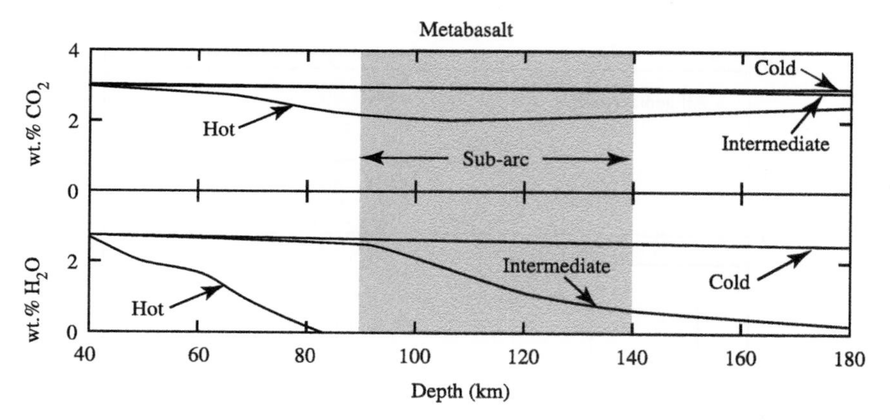

Figure 17 Volatile content as a function of depth for average hydrothermally altered metabasalt, shown for three geotherms (after Kerrick and Connolly, 2001a). "Hot," "intermediate," and "cold" refer to geotherms illustrated in Figure 15. Shaded "sub-arc" region after Kerrick and Connolly (2001a).

volatile content against depth for three slab geotherms. They used the upper surface geotherms for the high- and low-T cases from Peacock and Wang (1999), and chose an intermediate geotherm half way between the other two (Figure 15). For this chapter, I derived plots of volatile content versus depth for the same three geotherms for the other bulk compositions in Kerrick and Connolly (1998, 2001b).

Starting with the ultramafic part of the subducting slab, Figure 16 shows the H_2O and CO_2 content of two altered ultramafic protoliths as a function of depth (Kerrick and Connolly, 1998). In both cases, carbonate would remain stable to depths of at least 180 km on all three geotherms. Water, however, is liberated with increasing depth by dehydration reactions involving brucite and talc at shallow depths, and antigorite at greater depths. The stepwise decrease in H_2O content reflects the univariant nature of these dehydration reactions. For both compositions, significant sub-arc dehydration occurs only for intermediate geotherms, but all water is ultimately driven off by 170 km depth even for the coolest geotherm.

Moving up-section in the slab, Kerrick and Connolly (2001a) considered the phase stability in a metabasalt bulk composition, with an initial mineralogy of glaucophane + chlorite + clinozoisite + aragonite + dolomite + quartz + sphene, containing 2.68 wt.% H_2O and 2.95 wt.% CO_2. In this case, CO_2 decreases only in the hot geotherm case, and only slightly (Figure 17). The behavior of H_2O is strongly dependent on the geotherm; the metabasalt completely dehydrates by ~90 km on a hot geotherm, but requires >180 km on the intermediate geotherm. Because lawsonite remains stable on the cold geotherm, little H_2O is lost over the depth range of their calculation. In contrast to the ultramafic case, dehydration of the metabasalt is continuous over a large depth interval, a result of the high variance of the dehydration reactions. This point is also made by Poli and Schmidt (2002), who suggest that dehydration of the subducting lithosphere is a continuous process, extending over 300 km of the slab–mantle interface.

The behavior of H_2O and CO_2 in subducted sediments was addressed by Kerrick and Connolly (2001b), who calculated phase stabilities and volatile contents for four bulk compositions designed to span the observed range of oceanic sediments. The devolatilization behavior is a sensitive function of bulk composition and geothermal gradient (Figure 18). It is noteworthy that carbonate can survive subduction to >180 km depth in many situations. This observation, combined with their results on metabasalts (Kerrick and Connolly, 2001a), led them to note that there is a significant imbalance between the amount of carbon subducted and that returned to the surface in arc magmatism, reinforcing the point made earlier by Ito *et al.* (1983) and Bebout (1996). Indeed, the stability of carbonate in subducting lithologies led Kerrick and Connolly (2001b) to propose that decarbonation occurred by infiltrating H_2O-rich fluids at sub-arc conditions in order to provide any CO_2 to arc magmatism.

At greater depths, hydrous phases such as chlorite and serpentine are succeeded in cooler subduction zones by DHMSs. These DHMSs can transport water to the transition zone, where they may break down because of their lack of stability at high temperature. At these depths, however, wadsleyite would be able to accommodate the water released by these dehydration reactions. Further subduction may trigger H_2O release at the wadsleyite → ringwoodite transition, because of

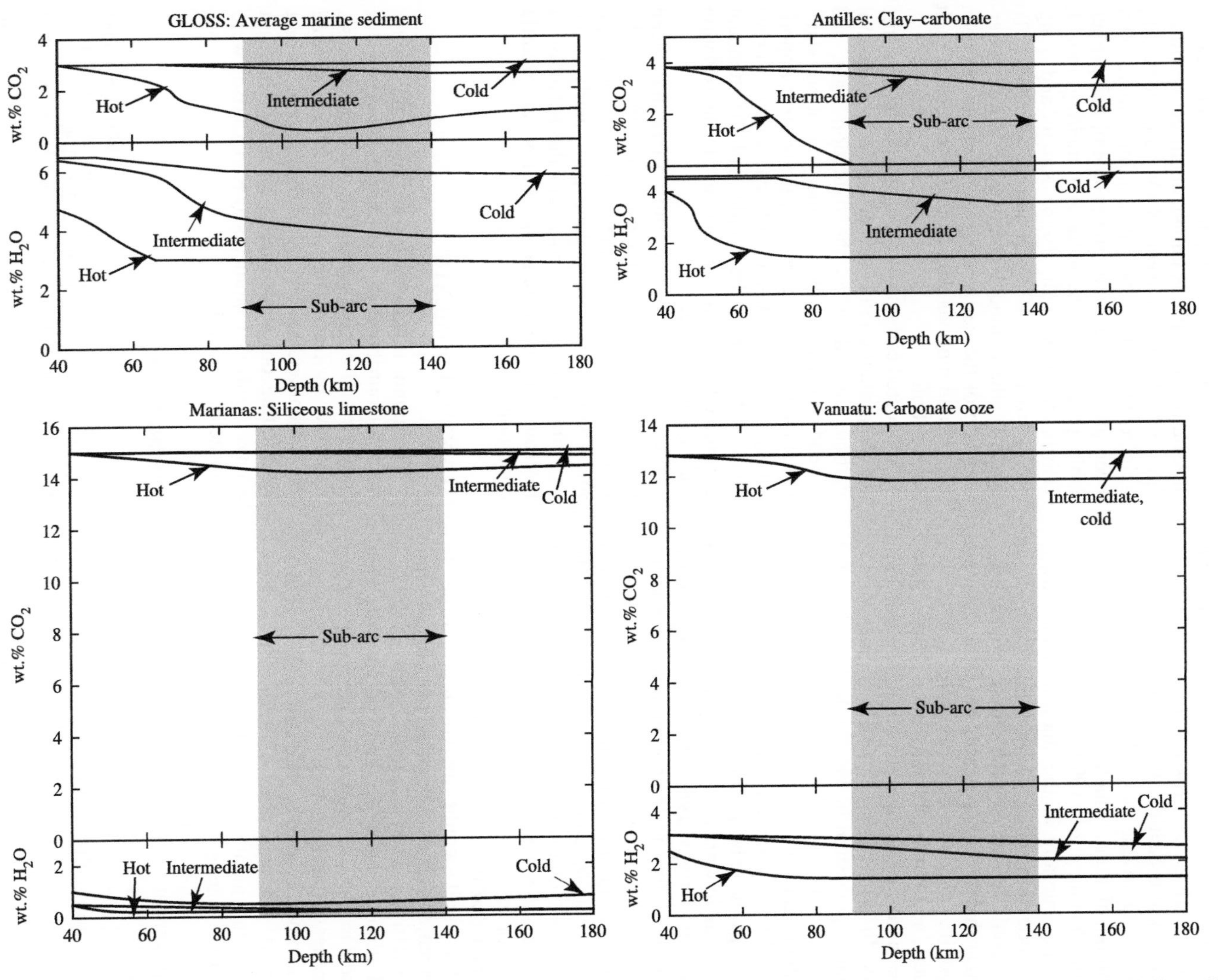

Figure 18 Volatile content as a function of depth for four oceanic sediment bulk compositions. Constructed from P–T projections in Kerrick and Connolly (2001b). "Hot," "intermediate," and "cold" refer to geotherms illustrated in Figure 15. Shaded "sub-arc" region after Kerrick and Connolly (2001a).

the somewhat lower water solubility in ringwoodite compared to wadsleyite, or at the ringwoodite → perovskite + ferropericlase transition at 660 km, where there is a pronounced decrease in the saturation water content (from wt.% to ppm level). At the other side of a convection cell, as upwelling mantle containing hydrous wadsleyite crosses the wadsleyite → olivine phase transition, the difference in solubility of H_2O in the two phases could lead to exsolution of fluid, which might in turn trigger melting (Young *et al.*, 1993).

Moving laterally in the shallow mantle away from subduction zones, where nominally hydrous phases could be stable (if the slab is cool enough), it is clear that the only potential hosts for water are phlogopite and its high-pressure breakdown products, and nominally anhydrous phases. Phlogopite and its successors require potassium; NAMs do not. Given current estimates of the amount of potassium in the mantle, it is clear that NAMs are the most likely hosts for water in nonsubduction zone settings.

The effect of water on physical properties, such as rheology, have been recognized since the 1960s and the classic works on hydrolytic weakening of quartz (e.g., Griggs, 1967). For the geodynamist, water is by far the most important volatile. For that reason, much attention has focused on the role of water in deformation of olivine (Mei and Kohlstedt, 2000, and references therein). The effect of water on rheology has important potential consequences for mantle flow and melt segregation. For example, Hirth and Kohlstedt (1996) suggested that the efficient partitioning of water into the melt would dehydrate, and hence strengthen, the residual mantle during decompression melting. They suggest that it is this dehydration-produced increase in viscosity that is critical to focusing melt into a narrow upwelling zone beneath the ridge axis.

In another example, Richard *et al.* (2002) simulated the transport of water in a two-dimensional mantle convection model. They found that mantle flow, not diffusion, was the primary control on water distribution, which led to a homogeneous distribution of water in the mantle. If this is the case, the transition zone may contain less water than could be dissolved into the nominally anhydrous phases present there. Because of the low solubility of water in lower-mantle nominally anhydrous phases (Bolfan-Casanova *et al.*, 2000), Richard *et al.* proposed that there might be a water-rich fluid phase in the lower mantle. They did not, however, consider the possibility of water-induced partial melting, leading to a melt rather than a fluid.

In terms of magma genesis, water again is the most important volatile, at least in the upper mantle. The implications of magma production in the presence of a stoichiometric hydrous phase such as amphibole or phlogopite have been explored experimentally to different degrees (cf. Schmidt and Poli, 1998; Ulmer and Trommsdorff, 1999). The effect of water present in NAMs on melt production has yet to be explored experimentally, and little is known about the partitioning of hydrogen between NAMs and coexisting silicate melt. That notwithstanding, there have been attempts to frame the problem, using plausible partition coefficients and melting models (Hirth and Kohlstedt, 1996; Karato and Jung, 1998; Ito *et al.*, 1999; Braun *et al.*, 2000). To quantify these models requires a better understanding of the partitioning of hydrogen between NAMs and melt, as well as the processes of melt segregation.

Subducted carbon as carbonate can survive into the deep mantle either as carbonate or as neutral carbon, depending on the oxidation state, and indeed on what controls the oxidation state. Outstanding questions include the relative importance—and abundance—of carbonate versus elemental carbon in the mantle, with the attendant unresolved question of what controls the oxidation state of the mantle.

For the halogens, recently there has been considerable progress in the understanding of the geochemical cycles for these elements, although much work remains to be done, particularly on the experimental front, to constrain the behavior of hydrous, halogen-bearing fluids at mantle conditions.

ACKNOWLEDGMENTS

The author thanks Rick Carlson for the invitation to contribute to this chapter, and for his patience and encouragement during the process. A review by David Bell substantially improved the content and readability of the chapter, in part by pointing out some biases the author had not even recognized. Comments by Rick Carlson were very helpful as well.

REFERENCES

Ackermann L., Cemic L., and Langer K. (1983) Hydrogarnet substitution in pyrope: a possible location for "water" in the mantle. *Earth Planet. Sci. Lett.* **62**, 208–214.

Agrinier P., Mével C., Bosch D., and Javoy M. (1993) Metasomatic hydrous fluids in amphibole peridotites from Zabargad Island (Red Sea). *Earth Planet. Sci. Lett.* **120**, 187–205.

Aines R. D. and Rossman G. R. (1984a) The water content of mantle garnets. *Geology* **12**, 720–723.

Aines R. D. and Rossman G. R. (1984b) The hydrous component in garnets: pyralspites. *Am. Mineral.* **69**, 1116–1126.

Akagi T. and Masuda A. (1988) Isotopic and elemental evidence for a relationship between kimberlite and Zaire cubic diamonds. *Nature* **336**, 665–667.

Akaishi M. and Yamaoka S. (2000) Crystallization of diamond from C–O–H fluids under high-pressure and high-temperature conditions. *J. Cryst. Growth* **213**, 999–1003.

Akaishi M., Kanda H., and Yamaoka S. (1990) Synthesis of diamond from graphite-carbonate systems under very high temperature and pressure. *J. Cryst. Growth* **104**, 578–581.

Akaishi M., Kumar M. D. S., Kanda H., and Yamaoka S. (2000) Formation process of diamond from supercritical H_2O–CO_2 fluid under high pressure and high temperature conditions. *Diam. Relat. Mater.* **9**, 1945–1950.

Amthauer G. and Rossman G. R. (1998) The hydrous component in andradite garnet. *Am. Mineral.* **83**, 835–840.

Andersen T. and Neumann E.-R. (2001) Fluid inclusions in mantle xenoliths. *Lithos* **55**, 301–320.

Andersen T., O'Reilly S. Y., and Griffin W. L. (1984) The trapped fluid phase in upper mantle xenoliths from Victoria, Australia: implications for mantle metasomatism. *Contrib. Mineral. Petrol.* **88**, 72–85.

Andersen T., Griffin W. L., and O'Reilly S. Y. (1987) Primary sulphide melt inclusions in mantle-derived megacrysts and pyroxenites. *Lithos* **20**, 279–295.

Aoki K. (1977) Titanochondrodite and titanoclinohumite derived from the upper mantle in the Buell Park kimberlite, Arizona, USA. A reply. *Contrib. Mineral. Petrol.* **61**, 217–218.

Aoki K., Fujino K., and Akaogi M. (1976) Titanochondrodite and titanoclinohumite derived from the upper mantle in the Buell Park kimberlite, Arizona, USA. *Contrib. Mineral. Petrol.* **56**, 243–253.

Artioli G., Fumagalli P., and Poli S. (1999) The crystal structure of $Mg_8(Mg_2Al_2)Al_8Si_{12}(O,OH)_{56}$ pumpellyite and its relevance in ultramafic systems at high pressure. *Am. Mineral.* **84**, 1906–1914.

Bai Q. and Kohlstedt D. L. (1992) Substantial hydrogen solubility in olivine and implications for water storage in the mantle. *Nature* **357**, 672–674.

Bai Q. and Kohlstedt D. L. (1993) Effects of chemical environment on the solubility and incorporation mechanism for hydrogen in olivine. *Phys. Chem. Mineral.* **19**, 460–471.

Baker M. B., Grove T. L., and Price R. (1994) Primitive basalts and andesites from the Mt. Shasta region, N. California: products of varying melt fraction and water content. *Contrib. Mineral. Petrol.* **118**, 111–129.

Ballhaus C. and Ellis D. J. (1996) Mobility of core melts during Earth's accretion. *Earth Planet. Sci. Lett.* **143**, 137–145.

Barbou J. C. and Doyle B. L. (1995) Elastic recoil detection: ERD (or forward recoil spectrometry: FRES). In *Handbook of Modern Ion Beam Analysis* (ed. J. R. Tesmer). Materials Res. Soc. Warren Hale, PA, pp. 83–138.

Bebout G. E. (1996) Volatile transfer and recycling at convergent margins: mass-balance and insights from high-*P/T* metamorphic rocks. In *Subduction: Top to Bottom*, Geophysical Monograph 96 (eds. G. E. Bebout, D. W. Scholl, S. H. Kirby, and J. P. Platt). American Geophysical Union, Washington, DC, pp. 179–193.

Becker H., Jochum K. P., and Carlson R. W. (2000) Trace element fractionation during dehydration of eclogites from high-pressure terranes and the implications for element fluxes in subduction zones. *Chem. Geol.* **163**, 65–99.

Bedini R. M., Bodinier J.-L., Dautria J.-M., and Morten L. (1997) Evolution of LILE-enriched small melt fractions in the lithospheric mantle: a case study from the East African Rift. *Earth Planet. Sci. Lett.* **153**, 67–83.

Bell D. R. and Rossman G. R. (1992a) Water in Earth's mantle: the role of nominally anhydrous minerals. *Science* **255**, 1391–1397.

Bell D. R. and Rossman G. R. (1992b) The distribution of hydroxyl in garnets form the subcontinental mantle of southern Africa. *Contrib. Mineral. Petrol.* **111**, 161–178.

Bell D. R., Ihinger P. D., and Rossman G. R. (1995) Quantitative analysis of trace OH in garnet and pyroxenes. *Am. Mineral.* **80**, 465–474.

Bell D. R., Rossman G. R., Maldener J., Endisch D., and Rauch F. (2003) Hydroxide in olivine: a quantitative determination of the absolute amount and calibration of the IR spectrum. *J. Geophys. Res.* **108**(B2), 2105, doi: 10.1029/2001JB000679.

Bell K. and Simonetti A. (1996) Carbonatite magmatism and plume activity: implications from the Nd, Pb and Sr isotope systematics of Oldoinyo Lengai. *J. Petrol.* **37**, 1321–1339.

Beran A. (1971) Messung des Ultrarot-Pleochroismus von Mineralen: XII. Der Pleochroismus der OH-Streckfrequenz in Disthen. *Tschermaks Min. Petr. Mitt.* **16**, 129–135.

Beran A. (1986) A model of water allocation in alkali feldspar, derived from infrared-spectroscopic investigations. *Phys. Chem. Mineral.* **13**, 306–310.

Beran A. (1987) OH groups in nominally anhydrous framework structures: an infrared spectroscopic investigation of danburite and labradorite. *Phys. Chem. Mineral.* **14**, 441–445.

Beran A. and Götzinger M. A. (1987) The quantitative IR spectroscopic determination of structural OH groups in kyanites. *Mineral. Petrol.* **36**, 41–49.

Beran A. and Putnis A. (1983) A model of the OH positions in olivine, derived from infrared-spectroscopic investigations. *Phys. Chem. Mineral.* **9**, 57–60.

Beran A. and Zemann J. (1971) Messung des Ultrarot-Pleochroismus von Mineralen: XI. Der Pleochroismus der OH-Streckfrequenz in Rutil, Anatas, Brookit und Cassiterit. *Tschermaks Min. Petr. Mitt.* **15**, 71–80.

Beran A. and Zemann J. (1986) The pleochroism of a gem-quality enstatite in the region of the OH stretching frequency, with a stereochemical interpretation. *Tschermaks Min. Petr. Mitt.* **35**, 19–25.

Beran A., Langer K., and Andrut M. (1993) Single crystal infrared spectra in the range of OH fundamentals of paragenetic garnet, omphacite and kyanite in an eklogitic mantle xenolith. *Mineral. Petrol.* **48**, 257–268.

Berg G. W. (1986) Evidence for carbonate in the mantle. *Nature* **324**, 50–51.

Bishop F. C., Smith J. V., and Dawson J. B. (1975) Pentlandite-magnetite intergrowth in De Beers spinel lherzolite: review of sulfide in nodules. *Phys. Chem. Earth* **9**, 323–337.

Bolfan-Casanova N., Keppler H., and Rubie D. C. (2000) Water partitioning between nominally anhydrous minerals in the MgO–SiO_2–H_2O system up to 24 GPa: implications for the distribution of water in the Earth's mantle. *Earth Planet. Sci. Lett.* **182**, 209–221.

Bostock M. G., Hyndman R. D., Rondenay S., and Peacock S. M. (2002) An inverted continental Moho and serpentinization of the forearc mantle. *Nature* **417**, 536–538.

Bougault H., Charlou J.-L., Fouquet Y., Needham H. D., Vaslet N., Appriou P., Baptiste P. J., Rona P. A., Dimitriev L., and Silantiev S. (1993) Fast and slow spreading ridges: structure and hydrothermal activity, ultramafic topographic highs and CH_4 output. *J. Geophys. Res.* **98**, 9643–9651.

Boyd F. R. and Gurney J. J. (1986) Diamonds and the African lithosphere. *Science* **232**, 472–477.

Brandon A. D., Walker R. J., Morgan J. W., Norman M. D., and Prichard H. M. (1998) Coupled ^{186}Os and ^{187}Os evidence for core–mantle interaction. *Science* **280**, 1570–1573.

Brandon A. D., Norman M. D., Walker R. J., and Morgan J. W. (1999) ^{186}Os–^{187}Os systematics of Hawaiian picrites. *Earth Planet. Sci. Lett.* **174**, 25–42.

Braun M. G., Hirth G., and Parmentier E. M. (2000) The effect of deep damp melting on mantle flow and melt generation beneath mid-ocean ridges. *Earth Planet. Sci. Lett.* **176**, 339–356.

Brenker F. E., Stachel T., and Harris J. W. (2002) Exhumation of lower mantle inclusions in diamond: ATEM investigation of retrograde phase transitions, reactions and exsolution. *Earth Planet. Sci. Lett.* **198**, 1–9.

Brodholt J. P. and Refson K. (2000) An *ab initio* study of hydrogen in forsterite and a possible mechanism for hydrolytic weakening. *J. Geophys. Res.* **105**, 18977–18982.

Bromiley G. D. and Pawley A. R. (2002) The high-pressure stability of Mg-sursassite in a model hydrous peridotite: a possible mechanism for the deep subduction of significant volumes of H_2O. *Contrib. Mineral. Petrol.* **142**, 714–723.

Bulanova G. P. (1995) The formation of diamond. *J. Geochem. Explor.* **53**, 1–23.

Bulanova G. P., Griffin W. L., Ryan C. G., Shestakova O. Y., and Barnes S.-J. (1996) Trace elements in sulfide inclusions from Yakutian diamonds. *Contrib. Mineral. Petrol.* **124**, 111–125.

Bureau H. and Keppler H. (1999) Complete miscibility between silicate melts and hydrous fluids in the upper mantle: experimental evidence and geochemical implications. *Earth Planet. Sci. Lett.* **165**, 187–196.

Burnard P. G., Farley K. A., and Turner G. (1998) Multiple fluid pulses in a Samoan harzburgite. *Chem. Geol.* **147**, 99–114.

Byers C. D., Garcia M. O., and Muenow D. W. (1986) Volatiles in basaltic glasses from the East Pacific Rise at 21°N: implications for MORB sources and submarine lava flow morphology. *Earth Planet. Sci. Lett.* **79**, 9–20.

Canil D. (1990) Experimental study bearing on the absence of carbonate in mantle-derived xenoliths. *Geology* **18**, 1011–1013.

Canil D. and Scarfe C. M. (1989) Origin of phlogopite in mantle xenoliths from Kostal Lake, Wells Gray Park, British Columbia. *J. Petrol.* **30**, 1159–1179.

Cannat M. (1993) Emplacement of mantle rocks in the seafloor at mid-ocean ridges. *J. Geophys. Res.* **98**, 4163–4172.

Cannat M., Bideau D., and Hébert R. (1990) Plastic deformation and magmatic impregnation in serpentinized ultramafic rocks from the Garrett Transform Fault, East Pacific Rise. *Earth Planet. Sci. Lett.* **101**, 216–232.

Cannat M., Bideau D., and Bougault H. (1992) Serpentinized peridotites and gabbros in the Mid-Atlantic Ridge axial valley at 15°37′N and 16°52′N. *Earth Planet. Sci. Lett.* **109**, 87–106.

Carswell D. A. (1975) Primary and secondary phlogopites and clinopyroxenes in garnet lherzolite xenoliths. *Phys. Chem. Earth* **9**, 417–430.

Chalot-Prat G. and Arnold M. (1999) Immiscibility between calciocarbonatitic and silicate melts and related wall rock reactions in the upper mantle: a natural case study from Romanian mantle xenoliths. *Lithos* **46**, 627–659.

Chazot G., Menzies M., and Harte B. (1996) Determination of partition coefficients between apatite, clinopyroxene, amphibole and melt in natural spinel lherzolites from Yemen: implications for wet melting of the lithospheric mantle. *Geochim. Cosmochim. Acta* **60**, 423–437.

Cho H. and Rossman (1993) Single-crystal NMR studies of low-concentration hydrous species in minerals: grossular garnet. *Am. Mineral.* **78**, 1149–1164.

Chrenko R. M., McDonald R. S., and Darrow K. A. (1967) Infra-red spectrum of diamond coat. *Nature* **214**, 474–476.

Clague D. A. and Frey F. A. (1982) Petrology and trace element geochemistry of the Honolulu Volcanic Series. *J. Petrol.* **23**, 447–504.

Class C. and Goldstein S. L. (1997) Plume-lithosphere interactions in the ocean basins: constraints from the source mineralogy. *Earth Planet. Sci. Lett.* **150**, 245–260.

Coltorti M., Bonadiman C., Hinton R. W., Siena F., and Upton B. G. J. (1999) Carbonatite metasomatism of the oceanic upper mantle: evidence from clinopyroxenes and glasses in ultramafic xenoliths of Grande Comore, Indian Ocean. *J. Petrol.* **40**, 133–165.

Czamanske G. K. and Moore J. G. (1977) Composition and phase chemistry of sulfide globules in basalt from the Mid-Atlantic Ridge rift valley near 37°N lat. *Geol. Soc. Am. Bull.* **88**, 587–599.

Dalton J. A. and Presnall D. C. (1998) Carbonatitic melts along the solidus of model lherzolite in the system $CaO-MgO-Al_2O_3-SiO_2-CO_2$ from 3 to 7 GPa. *J. Petrol.* **39**, 1953–1964.

Dalton J. A. and Wood B. J. (1993) The compositions of primary carbonate melts and their evolution through wall-rock reaction in the mantle. *Earth Planet. Sci. Lett.* **119**, 511–525.

Danyushevsky L. V., Eggins S. M., Falloon T. J., and Christie D. M. (2000) H_2O abundance in depleted to moderately enriched mid-ocean ridge magmas: Part I. Incompatible behavior, implications for mantle storage, and origin of regional variations. *J. Petrol.* **41**, 1329–1364.

Danyushevsky L. V., McNeill A. W., and Sobolev A. V. (2002) Experimental and petrological studies of melt inclusions in phenocrysts from mantle-derived magmas: an overview of techniques, advantages and complications. *Chem. Geol.* **183**, 5–24.

Dautria J. M., Dupuy C., Takherist D., and Dostal J. (1992) Carbonate metasomatism in the lithospheric mantle: the peridotitic xenoliths from a melilititic district of the Sahara Basin. *Contrib. Mineral. Petrol.* **111**, 37–52.

Dawson J. B. (1987) The MARID suite of xenoliths in kimberlite: relationship to veined and metasomatised peridotite xenoliths. In *Mantle Xenoliths* (ed. P. H. Nixon). Wiley, Chichester, pp. 465–473.

Dawson J. B. and Powell D. G. (1969) Mica in the upper mantle. *Contrib. Mineral. Petrol.* **22**, 233–237.

Dawson J. B. and Smith J. V. (1977) The MARID (mica-amphibole-rutile-ilmenite-diopside) suite of xenoliths in kimberlite. *Geochim. Cosmochim. Acta* **41**, 309–323.

Dawson J. B., Pinkerton H., Norton G. E., and Pyle D. M. (1990) Physicochemical properties of alkali carbonatite lavas: data from the 1988 eruption of Oldoinyo Lengai, Tanzania. *Geology* **18**, 260–263.

Deines P. and Harris J. W. (1995) Sulfide inclusion chemistry and carbon isotopes of African diamonds. *Geochim. Cosmochim. Acta* **59**, 3173–3188.

Delaney J. S., Smith J. V., Carswell D. A., and Dawson J. B. (1980) Chemistry of micas from kimberlites and xenoliths: II. Primary- and secondary-textured micas from peridotite xenoliths. *Geochim. Cosmochim. Acta* **44**, 857–872.

Desborough G. A. and Czamanske G. K. (1973) Sulfides in eclogite nodules from a kimberlite pipe, South Africa, with comments on violarite stoichiometry. *Am. Mineral.* **58**, 195–202.

De Waal S. A. and Calk L. C. (1975) The sulfides in the garnet pyroxenite xenoliths from Salt Lake Crater, Oahu. *J. Petrol.* **16**, 134–153.

Dick H. J. B. (1989) Abyssal peridotites, very slow spreading ridges and ocean ridge magmatism. In *Magmatism in the Ocean Basins*, Geological Society Special Publication No. 42 (eds. A. D. Saunders and M. J. Norry). Geological Society of London, pp. 71–105.

Dick H. J. B., Fisher R. L., and Bryan W. B. (1984) Mineralogic variability of the uppermost mantle along mid-ocean ridges. *Earth Planet. Sci. Lett.* **69**, 88–106.

Distler V. V., Ilupin I. P., and Laputina I. P. (1987) Sulfides of deep-seated origin in kimberlites and some aspects of copper–nickel mineralization. *Int. Geol. Rev.* **29**, 456–464.

Dixon J. E. and Clague D. A. (2001) Volatiles in basaltic glasses from Loihi Seamount, Hawaii: evidence for a relatively dry plume component. *J. Petrol.* **42**, 627–654.

Dixon J. E. and Stolper E. M. (1995) An experimental study of water and carbon dioxide solubilities in mid-ocean ridge basaltic liquids: Part II. Applications to degassing. *J. Petrol.* **36**, 1633–1646.

Dixon J. E., Stolper E., and Delaney J. R. (1988) Infrared spectroscopic measurements of CO_2 and H_2O in Juan de Fuca Ridge basaltic glasses. *Earth Planet. Sci. Lett.* **90**, 87–104.

Dixon J. E., Stolper E. M., and Holloway J. R. (1995) An experimental study of water and carbon dioxide solubilities in mid-ocean ridge basaltic liquids: Part I. Calibration and solubility models. *J. Petrol.* **36**, 1607–1631.

Dixon J. E., Clague D. A., Wallace P., and Poreda R. (1997) Volatiles in alkalic basalts from the North Arch Volcanic Field, Hawaii: extensive degassing of deep submarine-erupted alkalic series lavas. *J. Petrol.* **38**, 911–939.

Dixon J. E., Leist L., Langmuir C., and Schilling J.-G. (2002) Recycled dehydrated lithosphere observed in plume-influenced mid-ocean-ridge basalt. *Nature* **420**, 385–389.

Dobosi G. and Kurat G. (2002) Trace element abundances in garnets and clinopyroxenes from diamondites—a signature of carbonatitic fluids. *Mineral. Petrol.* **76**, 21–38.

Dobson P. F., Skogby H., and Rossman G. R. (1995) Water in boninite glass and coexisting orthopyroxene: concentration and partitioning. *Contrib. Mineral. Petrol.* **118**, 414–419.

Downs J. W. (1989) Possible sites for protonation in β-Mg_2SiO_4 from an experimentally derived electrostatic potential. *Am. Mineral.* **74**, 1124–1129.

Dromgoole E. L. and Pasteris J. D. (1987) Interpretation of the sulfide assemblages in a suite of xenoliths from Kilbourne Hole, New Mexico. *Geol. Soc. Am. Spec. Pap.* **215**, 25–46.

Eggler D. H. and Baker D. R. (1982) Reduced volatiles in the system C–O–H: implications to mantle melting, fluid formation and diamond genesis. In *High-pressure Research in Geophysics* (eds. S. Akimoto and M. H. Manghnani). Center for Academic Publications Japan, Toyko, pp. 237–250.

Eggler D. H. and Lorand J. P. (1993) Mantle sulfide geobarometry. *Geochim. Cosmochim. Acta* **57**, 2213–2222.

Erlank A. J., Waters F. G., Hawkesworth C. J., Haggerty S. E., Allsopp H. L., Rickard R. S., and Menzies M. (1987) Evidence for mantle metasomatism in peridotite nodules from the Kimberley Pipes, South Africa. In *Mantle Metasomatism* (eds. M. A. Menzies and C. J. Hawkesworth). Academic Press, London, pp. 221–310.

Ernst W. G. (1999) H_2O and ultra-high pressure subsolidus phase relations for mafic and ultramafic systems. *Int. Geol. Rev.* **41**, 886–894.

Ertan I. E. and Leeman W. P. (1999) Fluid inclusions in mantle and lower crustal xenoliths from the Simcoe volcanic field, Washington. *Chem. Geol.* **154**, 83–95.

Escartín J. and Cannat M. (1999) Ultramafic exposures and the gravity signature of the lithosphere near the fifteen–twenty fracture zone (Mid-Atlantic Ridge, 14°–16.5° N). *Earth Planet. Sci. Lett.* **171**, 411–424.

Evans B. and Trommsdorff V. (1983) Fluorine hydroxyl titanian clinohumite in Alpine recrystallized garnet peridotite: compositional controls and petrologic significance. *Am. J. Sci.* **283A**, 355–369.

Exley R. A. and Smith J. V. (1982) The role of apatite in mantle enrichment processes and in the petrogenesis of some alkali basalt suites. *Geochim. Cosmochim. Acta* **46**, 1375–1384.

Falloon T. J. and Danyushevsky L. V. (2000) Melting of refractory mantle at 1.5, 2 and 2.5 GPa under anhydrous and H_2O-undersaturated conditions: implications for the petrogenesis of high-Ca boninites and the influence of subduction components on mantle melting. *J. Petrol.* **41**, 257–283.

Fawcett J. J. and Yoder H. S. (1966) Phase relationships of chlorites in the system $MgO–Al_2O_3–SiO_2–H_2O$. *Am. Mineral.* **51**, 353–380.

Fleet M. E. and Stone W. E. (1990) Nickeliferous sulfides in xenoliths, olivine megacrysts and basaltic glass. *Contrib. Mineral. Petrol.* **105**, 629–636.

Fockenberg T. (1998) An experimental study of the pressure–temperature stability of MgMgAl-pumpellyite in the system $MgO–Al_2O_3–SiO_2–H_2O$. *Am. Mineral.* **83**, 220–227.

Foden J. D. and Green D. H. (1992) Possible role of amphibole in the origin of andesites: some experimental and natural evidence. *Contrib. Mineral. Petrol.* **109**, 479–493.

Foley S. (1992) Vein-plus-wall-rock melting mechanisms in the lithosphere and the origin of potassic alkaline magmas. *Lithos* **28**, 435–453.

Foley S. F., Musselwhite D. S., and van der Laan S. R. (1999) Melt compositions from ultramafic vein assemblages in the lithospheric mantle: a comparison of cratonic and non-cratonic settings. In *Proceedings of the International Kimberlite Conference 7, Cape Town, South Africa* (eds. J. J. Gurney, J. L. Gurney, M. D. Pascoe, and S. H. Richardson), vol. 1, pp. 238–246.

Francis D. (1987) Mantle–melt interaction recorded in spinel lherzolite xenoliths from the Alligator Lake Volcanic Complex, Yukon, Canada. *J. Petrol.* **28**, 569–597.

Francis D. and Ludden J. (1995) The signature of amphibole in mafic alkaline lavas, a study in the Northern Canadian Cordillera. *J. Petrol.* **36**, 1171–1191.

Freund F., Kathrein H., Wengeler H., Knobel R., and Heinen H. J. (1980) Carbon in solid solution in forsterite—a key to the untractable nature of reduced carbon in terrestrial and cosmogenic rocks. *Geochim. Cosmochim. Acta* **44**, 1319–1333.

Frey F. A. and Prinz M. (1978) Ultramafic inclusions from San Carlos, Arizona: petrologic and geochemical data bearing on their petrogenesis. *Earth Planet. Sci. Lett.* **38**, 129–176.

Frezzotti M. L., Touret J. L. R., Lustenhouwer W. J., and Neumann E. R. (1994) Melt and fluid inclusions in dunite xenoliths from La Gomera, Canary Islands: tracking the mantle metasomatic fluids. *Euro. J. Mineral.* **6**, 805–817.

Frick C. (1973) The sulphides in griquaite and garnet-peridotite xenoliths in kimberlite. *Contrib. Mineral. Petrol.* **39**, 1–16.

Frost D. J. (1999) The stability of dense hydrous magnesium silicates in Earth's transition zone and lower mantle. In *Mantle Petrology: Field Observations and High Pressure Experimentation: A Tribute to Francis R. (Joe) Boyd*, Geochemical Society Special Publication No. 6 (eds. Y. Fei, C. M. Bertka, and B. O. Mysen). Geochemical Society, Houston, pp. 283–296.

Fujino K. and Takéuchi Y. (1978) Crystal chemistry of titanian chondrodite and titanian clinohumite of high-pressure origin. *Am. Mineral.* **63**, 535–543.

Fyfe W. S. (1970) Lattice energies, phase transformations, and volatiles in the mantle. *Phys. Earth Planet. Int.* **3**, 196–200.

Gaetani G. A. and Grove T. L. (1998) The influence of water on melting of mantle peridotite. *Contrib. Mineral. Petrol.* **131**, 323–346.

Gaetani G. A. and Grove T. L. (1999) Wetting of mantle olivine by sulfide melt: implications for Re/Os ratios in mantle peridotite and late-stage core formation. *Earth Planet. Sci. Lett.* **169**, 147–163.

Garcia M. O., Liu N. W. K., and Muenow D. W. (1979) Volatiles in submarine volcanic rocks from the Mariana Island arc and trough. *Geochim. Cosmochim. Acta* **43**, 305–312.

Garcia M. O., Muenow D. W., and Liu N. W. K. (1980) Volatiles in Ti-rich amphibole megacrysts, southwest USA. *Am. Mineral.* **65**, 306–312.

Garuti G., Gorgoni C., and Sighinolfi G. P. (1984) Sulfide mineralogy and chalcophile and siderophile element abundances in the Ivrea-Verbano mantle peridotites (western Italian Alps). *Earth Planet. Sci. Lett.* **70**, 69–87.

Geiger C. A., Langer K., Bell D. R., Rossman G. R., and Winkler B. (1991) The hydroxide component in synthetic pyrope. *Am. Mineral.* **76**, 49–59.

Gil Ibarguchi J. I., Abalos B., Azcarraga J., and Puelles P. (1999) Deformation, high-pressure metamorphism and exhumation of ultramafic rocks in a deep subduction/collision setting (Cabo Ortegal, NW Spain). *J. Metamorph. Geol.* **17**, 747–764.

Gottschalk M., Fockenberg T., Grevel K.-D., Wunder B., Wirth R., Schreyer W., and Maresch W. V. (2000) Crystal structure of the high-pressure phase $[Mg_4(MgAl)Al_4Si_6O_{21}/(OH)_7]$: an analogue of sursassite. *Euro. J. Mineral.* **12**, 935–945.

Green D. H. (1973) Experimental melting studies on a model upper mantle composition at high pressure under water-saturated and water-undersaturated conditions. *Earth Planet. Sci. Lett.* **19**, 37–53.

Green D. H. (1976) Experimental testing of equilibrium partial melting of peridotite under water saturated, high pressure conditions. *Can. Mineral.* **14**, 255–268.

Green D. H. and Wallace M. E. (1988) Mantle metasomatism by ephemeral carbonatite melts. *Nature* **336**, 459–462.

Green D. H., Falloon T. J., and Taylor W. R. (1987) Mantle-derived magmas—roles of variable source peridotite and variable C–H–O fluid compositions. In *Magmatic Processes and Physicochemical Principles,* Geochemical Society Special Publication No. 1 (ed. B. O. Mysen). Geochemical Society, Universtiy Park, PA, pp. 139–154.

Green H. W. (1972) A CO_2-charged asthenosphere. *Nat. Phys. Sci.* **238**, 2–5.

Green H. W. (1985) Coupled exsolution of fluid and spinel from olivine: evidence for O^- in the mantle? In *Point Defects in Minerals*, Geophysical Monograph Series 31 (ed. R. N. Schock). American Geophysical Union, Washington, DC, pp. 226–232.

Green H. W. and Guegen Y. (1983) Deformation of peridotite in the mantle and extraction by kimberlite: a case history documented by fluid and solid precipitates in olivine. *Tectonophysics* **92**, 71–92.

Grégoire M., Bell D. R., and Le Roex A. P. (2002) Trace element geochemistry of phlogopite-rich mafic mantle xenoliths: their classification and their relationship to phlogopite-bearing peridotites and kimberlites revisited. *Contrib. Mineral. Petrol.* **142**, 603–625.

Grevel K.-D., Navrotsky A., Kahl W. A., Fasshauer D. W., and Majzlan J. (2001) Thermodynamic data of the high-pressure phase $Mg_5Al_5Si_6O_{21}(OH)_7$ (Mg-sursassite). *Phys. Chem. Mineral.* **28**, 475–487.

Griggs D. (1967) Hydrolytic weakening of quartz and other silicates. *Geophys. J. Roy. Astro. Soc.* **14**, 19–31.

Guo J., Griffin W. L., and O'Reilly S. Y. (1999) Geochemistry and origin of sulphide minerals in mantle xenoliths: Qilin, southeastern China. *J. Petrol.* **40**, 1125–1149.

Gurney J. J., Harris J. W., and Rickard R. S. (1984) Minerals associated with diamonds from the Roberts Victor Mine. In *Kimberlites II: The Mantle and Crust–Mantle Relationships* (ed. J. Kornprobst). Elsevier, Amsterdam, pp. 25–32.

Haggerty S. E. (1994) Superkimberlites: a geodynamic diamond window to the Earth's core. *Earth Planet. Sci. Lett.* **122**, 57–69.

Hansteen T. H., Andersen T., Neumann E.-R., and Jelsma H. (1991) Fluid and silicate glass inclusions in ultramafic and mafic xenoliths from Hierro, Canary Islands: implications for mantle metasomatism. *Contrib. Mineral. Petrol.* **107**, 242–254.

Harris D. M. and Anderson A. T., Jr. (1984) Volatiles H_2O, CO_2, and Cl in a subduction related basalt. *Contrib. Mineral. Petrol.* **87**, 120–128.

Harris J. W. and Gurney J. J. (1979) Inclusions in diamond. In *The Properties of Diamond* (ed. J. E. Field). Academic Press, London, pp. 555–591.

Hart S. R., Hauri E. H., Oschmann L. A., and Whitehead J. A. (1992) Mantle plumes and entrainment: isotopic evidence. *Science* **256**, 517–520.

Harte B. (1987) Metasomatic events recorded in mantle xenoliths: an overview. In *Mantle Xenoliths* (ed. P. H. Nixon). Wiley, Chichester, pp. 625–640.

Harte B. and Gurney J. J. (1981) The mode of formation of chromium-poor megacryst suites from kimberlites. *J. Geol.* **89**, 749–753.

Harte B., Winterburn P. A., and Gurney J. J. (1987) Metasomatic and enrichment phenomena in garnet peridotite facies mantle xenoliths from the Matsoku kimberlite pipe, Lesotho. In *Mantle Metasomatism* (eds. M. A. Menzies and C. J. Hawkesworth). Academic Press, London, pp. 145–220.

Hauri E. (2002) SIMS analysis of volatiles in silicate glasses: 2. Isotopes and abundances in Hawaiian melt inclusions. *Chem. Geol.* **183**, 115–141.

Hauri E., Shimizu N., Dieu J. J., and Hart S. R. (1993) Evidence for hotspot-related carbonatite metasomatism in the oceanic uper mantle. *Nature* **365**, 221–227.

Helmstaedt H. and Schulze D. J. (1979) Garnet clinopyroxenite—chlorite eclogite transition in a xenolith from Moses Rock: further evidence for metamorphosed ophiolites under the Colorado Plateau. In *The Mantle Sample: Inclusions in Kimberlites and other Volcanics* (eds. F. R. Boyd and H. O. A. Meyer). American Geophysical Union, Washington, DC, pp. 357–365.

Hervig R. L. and Smith J. V. (1981) Dolomite-apatite inclusion in chrome-diopside crystal, Bellsbank kimberlite, South Africa. *Am. Mineral.* **66**, 246–249.

Hirose K. and Kawamoto T. (1995) Hydrous partial melting of lherzolite at 1 GPa: the effect of H_2O on the genesis of basaltic magmas. *Earth Planet. Sci. Lett.* **133**, 463–473.

Hirth G. and Kohlstedt D. L. (1996) Water in the oceanic upper mantle: implications for rheology, melt extraction and the evolution of the lithosphere. *Earth Planet. Sci. Lett.* **144**, 93–108.

Hoernle K., Tilton G., Le Bas M. J., Duggen S., and Garbe-Schönberg D. (2002) Geochemistry of oceanic carbonatites compared with continental carbonatites: mantle recycling of oceanic crustal carbonate. *Contrib. Mineral. Petrol.* **142**, 520–542.

Hoernle K. A. J. and Schmincke H.-U. (1993) The petrology of the tholeiites through melilite nephelinites on Gran Canary, Canary Island: crystal fractionation, accumulation, and depth of melting. *J. Petrol.* **34**, 573–597.

Holland T. J. B. and Powell R. (1998) An internally consistent thermodynamic data set for phases of petrological interest. *J. Metamorph. Geol.* **16**, 309–343.

Holloway J. R. (1998) Graphite-melt equilibria during mantle melting: constraints on CO_2 in MORB magmas and the carbon content of the mantle. *Chem. Geol.* **147**, 89–97.

Hunter R. H. and McKenzie D. (1989) The equilibrium geometry of carbonate melts in rocks of mantle composition. *Earth Planet. Sci. Lett.* **92**, 347–356.

Ingrin J. and Skogby H. (2000) Hydrogen in nominally anhydrous upper-mantle minerals: concentration levels and implications. *Euro. J. Mineral.* **12**, 543–570.

Ingrin J., Latrous K., Doukhan J. C., and Doukhan N. (1989) Water in diopside: an electron microscopy and infrared spectroscopy study. *Euro. J. Mineral.* **1**, 327–341.

Ingrin J., Hercule S., and Charton T. (1995) Diffusion of hydrogen in diopside: results of dehydration experiments. *J. Geophys. Res.* **100**, 15489–15499.

Inoue T., Yurimoto H., and Kudoh Y. (1995) Hydrous modified spinel, $Mg_{1.75}SiH_{0.5}O_4$: a new water reservoir in the mantle transition region. *Geophys. Res. Lett.* **22**, 117–120.

Inoue T., Irifune T., Yurimoto H., and Miyagi I. (1998a) Decomposition of K-amphibole at high pressures and implications for subduction zone volcanism. *Phys. Earth Planet. Int.* **107**, 221–231.

Inoue T., Weidner D. J., Northrup P. A., and Parise J. B. (1998b) Elastic properties of hydrous ringwoodite (γ-phase) in Mg_2SiO_4. *Earth Planet. Sci. Lett.* **160**, 107–113.

Ionov D. (1998) Trace element composition of mantle-derived carbonates and coexisting phases in peridotite xenoliths from alkali basalts. *J. Petrol.* **39**, 1931–1941.

Ionov D. A. and Harmer R. E. (2002) Trace element distribution in calcite–dolomite carbonatites from Spitskop: inferences for differentiation of carbonatite magmas and the origin of carbonates in mantle xenoliths. *Earth Planet. Sci. Lett.* **198**, 495–510.

Ionov D. A. and Hofmann A. W. (1995) Nb–Ta-rich mantle amphiboles and micas: implications for subduction-related metasomatic trace element fractionation. *Earth Planet. Sci. Lett.* **131**, 341–356.

Ionov D. A., Hoefs J., Wedepohl K. H., and Wiechert U. (1992) Content and isotopic composition of sulphur in ultramafic xenoliths from central Asia. *Earth Planet. Sci. Lett.* **111**, 269–286.

Ionov D. A., Dupuy C., O'Reilly S. Y., Kopylova M. G., and Genshaftet Y. S. (1993a) Carbonated peridotite xenoliths from Spitzbergen: implications for trace element signature of mantle metasomatism. *Earth Planet. Sci. Lett.* **119**, 283–297.

Ionov D. A., Hoefs J., Wedepohl K. H., and Wiechert U. (1993b) Content of sulphur in different mantle reservoirs: reply to comment on Ionov *et al.* (1992). *Earth Planet. Sci. Lett.* **119**, 635–640.

Ionov D. A., O'Reilly S. Y., Genshaft Y. S., and Kopylova M. G. (1996) Carbonate-bearing mantle peridotite xenoliths from Spitsbergen: phase relationships, mineral compositions and trace-element residence. *Contrib. Mineral. Petrol.* **125**, 375–392.

Ionov D. A., Griffin W. L., and O'Reilly S. Y. (1997) Volatile-bearing minerals and lithophile trace elements in the upper mantle. *Chem. Geol.* **141**, 153–184.

Ionov D. A., Bodinier J.-L., Mukasa S. B., and Zanetti A. (2002) Mechanisms and sources of mantle metasomatism: major and trace element compositions of peridotite xenoliths from Spitsbergen in the context of numerical modelling. *J. Petrol.* **43**, 2219–2259.

Ito E., Harris D. M., and Anderson A. T., Jr. (1983) Alteration of oceanic crust and geologic cycling of chlorine and water. *Geochim. Cosmochim. Acta* **47**, 1613–1624.

Ito G., Shen Y., Hirth G., and Wolfe C. J. (1999) Mantle flow, melting, and dehydration of the Iceland mantle plume. *Earth Planet. Sci. Lett.* **165**, 81–96.

Izraeli E. S., Harris J. W., and Navon O. (2001) Brine inclusions in diamonds: a new upper mantle fluid. *Earth Planet. Sci. Lett.* **187**, 323–332.

Jamtveit B., Brooker R., Brooks K., Larsen L. M., and Pedersen T. (2001) The water content of olivines from the North Atlantic volcanic province. *Earth Planet. Sci. Lett.* **186**, 401–415.

Jenkins D. M. and Chernosky J. V., Jr. (1986) Phase equilibria and crystallochemical properties of Mg-chlorite. *Am. Mineral.* **71**, 925–936.

Johnson K. E., Davis A. M., and Bryndzia L. T. (1996) Contrasting styles of hydrous metasomatism in the upper mantle: an ion probe investigation. *Geochim. Cosmochim. Acta* **60**, 1367–1385.

Johnson L. H., Burgess R., Turner G., Milledge H. J., and Harris J. W. (2000) Noble gas and halogen geochemistry of mantle fluids: comparison of African and Canadian diamonds. *Geochim. Cosmochim. Acta* **64**, 717–732.

Johnson M. C., Anderson A. T., Jr., and Rutherford M. J. (1994) Pre-eruptive volatile contents of magmas. In *Volatiles in Magmas*, Reviews in Mineralogy (eds. M. R. Carroll and J. R. Holloway). Mineralogical Society of America, Washington, DC, vol. 20, pp. 281–330.

Kamenetsky V. S., Crawford A. J., Eggins S., and Mühe R. (1997) Phenocryst and melt inclusion chemistry of near-axis seamounts, Valu Fa Ridge, Lau Basin: insight into mantle wedge melting and the addition of subduction components. *Earth Planet. Sci. Lett.* **151**, 205–223.

Kamenetsky V. S., Sobolev A. V., Eggins S. M., Crawford A. J., and Arculus R. J. (2002) Olivine-enriched melt inclusions in chromites from low-Ca boninites, Cape Vogel, Papua New Guinea: evidence for ultramafic primary magma, refractory mantle source and enriched components. *Chem. Geol.* **183**, 287–303.

Karato S.-I. and Jung H. (1998) Water, partial melting and the origin of the seismic low velocity and high attenuation zone in the upper mantle. *Earth Planet. Sci. Lett.* **157**, 193–207.

Kawamoto T. and Holloway J. R. (1997) Melting temperature and partial melt chemistry of H_2O-saturated mantle peridotite to 11 Gigapascals. *Science* **276**, 240–243.

Kelley D. S. and Früh-Green G. L. (2000) Volatiles in mid-ocean ridge environments. In *Ophiolites and Oceanic Crust: New Insights from Field Studies and the Ocean Drilling Program*, Geological Society of America Special Paper 349 (eds. Y. Dilek, E. M. Moores, D. Elthon, and A. Nicolas). Geological Society of America, Boulder, Colorado, pp. 237–260.

Kelley D. S. and Früh-Green G. L. (2001) Volatile lines of descent in submarine plutonic environments: insights from stable isotope and fluid inclusion analyses. *Geochim. Cosmochim. Acta* **65**, 3325–3346.

Kempton P. D. (1987) Mineralogic and geochemical evidence for differing styles of metasomatism in spinel lherzolite xenoliths: enriched mantle source regions of basalts? In *Mantle Metasomatism* (eds. M. A. Menzies and C. J. Hawkesworth). Academic Press, London, pp. 45–89.

Kent A. J. R. and Elliott T. R. (2002) Melt inclusions from Marianas arc lavas: implications for the composition and formation of island arc magmas. *Chem. Geol.* **183**, 263–286.

Kent A. J. R., Clague D. A., Honda M., Stolper E. M., Hutcheon I. D., and Norman M. D. (1999a) Widespread assimilation of a seawater-derived component at Loihi Seamount, Hawaii. *Geochim. Cosmochim. Acta* **63**, 2749–2761.

Kent A. J. R., Norman M. D., Hutcheon I. D., and Stolper E. M. (1999b) Assimilation of seawater-derived components in an oceanic volcano: evidence from matrix glasses and glass inclusions from Loihi seamount, Hawaii. *Chem. Geol.* **156**, 299–319.

Kent A. J. R., Peate D. W., Newman S., Stolper E. M., and Pearce J. A. (2002) Chlorine in submarine glasses from the Lau Basin: seawater contamination and constraints on the composition of slab-derived fluids. *Earth Planet. Sci. Lett.* **202**, 361–377.

Keppler H. and Rauch M. (2000) Water solubility in nominally anhydrous minerals measured by FTIR and ^1H MAS NMR: the effect of sample preparation. *Phys. Chem. Mineral.* **27**, 371–376.

Kerrick D. M. and Connolly J. A. D. (1998) Subduction of ophicarbonates and recycling of CO_2 and H_2O. *Geology* **26**, 375–378.

Kerrick D. M. and Connolly J. A. D. (2001a) Metamorphic devolatilization of subducted oceanic metabasalts: implications for seismicity, arc magmatism and volatile recycling. *Earth Planet. Sci. Lett.* **189**, 19–29.

Kerrick D. M. and Connolly J. A. D. (2001b) Metamorphic devolatilization of subducted marine sediments and transport of volatiles into the Earth's mantle. *Nature* **411**, 293–296.

Khisina N. R. and Wirth R. (2002) Hydrous olivine $(Mg_{1-y}Fe_y^{2+})_{2-x}v_xSiO_4H_{2x}$—a new DHMS phase of variable composition observed as nanometer-sized precipitations in mantle olivine. *Phys. Chem. Mineral.* **29**, 98–111.

Khisina N. R., Wirth R., Andrut M., and Ukhanov A. V. (2001) Extrinsic and intrinsic mode of hydrogen occurrence in natural olivines: FTIR and TEM investigation. *Phys. Chem. Mineral.* **28**, 291–301.

Khodyrev O. Y., Agoshkov V. M., and Slutskiy A. B. (1990) The system peridotite—aqueous fluid at upper mantle pressures. *Trans. USSR Acad. Sci.: Earth Sci. Sec.* **312**, 255–258.

Khomenko V. M., Langer K., Beran A., Koch-Müller M., and Fehr T. (1994) Titanium substitution and OH-bearing defects in hydrothermally grown pyrope crystals. *Phys. Chem. Mineral.* **20**, 483–488.

Kitamura M., Kondoh S., Morimoto N., Miller G. H., Rossman G. R., and Putnis A. (1987) Planar OH-bearing defects in mantle olivine. *Nature* **328**, 143–145.

Kjarsgaard B. A., Hamilton D. L., and Peterson T. D. (1995) Peralkaline nephelinite/carbonatite liquid immiscibility: comparison of phase compositions in experiments and natural lavas from Oldoinyo Lengai. In *Carbonatite Volcanism: Oldoinyo Lengai and the Petrogenesis of Natrocarbonatites* (eds. K. Bell and J. Keller). Springer, Berlin, pp. 163–190.

Koga K., Hauri E., Hirschmann M., and Bell D. (2003) Hydrogen concentration analyses using SIMS and FTIR: comparison and calibration for nominally anhydrous minerals. *Geochem. Geophys. Geosys.* **4**(2), 1019, doi: 10.1029/2002GC000378.

Kogarko L. N., Henderson C. M. B., and Pacheco H. (1995) Primary Ca-rich carbonatite magma and carbonate-silicate-sulphide liquid immiscibility in the upper mantle. *Contrib. Mineral. Petrol.* **121**, 267–274.

Kohlstedt D. L., Keppler H., and Rubie D. C. (1996) Solubility of water in the α, β, and γ phases of $(Mg,Fe)_2SiO_4$. *Contrib. Mineral. Petrol.* **123**, 345–357.

Kohn S. C. (1996) Solubility of H_2O in nominally anhydrous mantle minerals using ^1H MAS NMR. *Am. Mineral.* **81**, 1523–1526.

Konzett J. and Fei Y. (2000) Transport and storage of potassium in the Earth's upper mantle and transition zone: an experimental study to 23 GPa in simplified and natural bulk compositions. *J. Petrol.* **41**, 583–603.

Konzett J. and Ulmer P. (1999) The stability of hydrous potassic phases in lherzolitic mantle—an experimental study to 9.5 GPa in simplified and natural bulk compositions. *J. Petrol.* **40**, 629–652.

Konzett J., Sweeney R. J., Thompson A. B., and Ulmer P. (1997) Potassium amphibole stability in the upper mantle: an experimental study in a peralkaline KNCMASH system to 8.5 GPa. *J. Petrol.* **38**, 537–568.

Konzett J., Armstrong R. A., Sweeney R. J., and Compston W. (1998) The timing of MARID metasomatism in the Kaapvaal mantle: an ion probe study of zircons from MARID xenoliths. *Earth Planet. Sci. Lett.* **160**, 133–145.

Kovalenko V. I., Solovova I. P., Ryabchikov I. D., Ionov D. A., and Bogatikov O. A. (1987) Fluidized CO_2-sulphide-silicate media as agents of mantle metasomatism and megacrysts formation: evidence from a large druse in a spinel-lherzolite xenolith. *Phys. Earth Planet. Int.* **45**, 280–293.

Kretz R. (1983) Symbols for rock-forming minerals. *Am. Mineral.* **68**, 277–279.

Kudoh Y. (2002) Predicted model for hydrous modified olivine (HyM-α). *Phys. Chem. Mineral.* **29**, 387–395.

Kudoh Y., Inoue T., and Arashi H. (1996) Structure and crystal chemistry of hydrous wadsleyite, $Mg_{1.75}SiH_{0.5}O_4$: possible hydrous magnesium silicate in the mantle transition zone. *Phys. Chem. Mineral.* **23**, 461–469.

Kudoh Y., Kuribayashi T., Mizobata H., and Ohtani E. (2000) Structure and cation disorder of hydrous ringwoodite, γ-$Mg_{1.89}Si_{0.98}H_{0.30}O_4$. *Phys. Chem. Mineral.* **27**, 474–479.

Kumar M. D. S., Akaishi M., and Yamaoka S. (2000) Formation of diamond from supercritical H_2O–CO_2 fluid at high pressure and high temperature. *J. Cryst. Growth* **213**, 203–206.

Kurosawa M., Yurimoto H., Matsumoto K., and Sueno S. (1992) Hydrogen analysis of mantle olivine by secondary ion mass spectrometry. In *High-pressure Research in Mineral Physics: Applications to Earth and Planetary Sciences* (eds. Y. Syono and M. H. Manghnani). Terra Publishers, Tokyo and American Geophysical Union, Washington, DC, pp. 283–287.

Kurosawa M., Yurimoto H., Matsumoto K., and Sueno S. (1993) Water in the Earth's mantle: hydrogen analysis of mantle olivine, pyroxenes and garnet using the SIMS. *Proc. Lunar Planet. Sci. Conf. 24th*, Lunar and Planetary Institute, Houston, pp. 839–840.

Kurosawa M., Yurimoto H., and Sueno S. (1997) Patterns in the hydrogen and trace element compositions of mantle olivines. *Phys. Chem. Mineral.* **24**, 385–395.

Kushiro I. (1990) Partial melting of mantle wedge and evolution of island arc crust. *J. Geophys. Res.* **95**, 15929–15939.

Kushiro I., Syono Y., and Akimoto S. (1968) Melting of a peridotite nodule at high pressures and high water pressures. *J. Geophys. Res.* **73**, 6023–6029.

Lager G. A., Armbruster T., Rotella F. J., and Rossman G. R. (1989) OH substitution in garnets: X-ray and neutron diffraction, infrared, and geometric-modelling studies. *Am. Mineral.* **74**, 840–851.

Lanford W. A., Trautvetter H. P., Ziegler J. F., and Keller J. (1976) New precision technique for measuring the concentration versus depth of hydrogen in solids. *Appl. Phys. Lett.* **28**, 554–570.

Langer K., Robarick E., Sobolev N. V., Shatsky V. S., and Wang W. (1993) Single-crystal spectra of garnets from diamondiferous high-pressure metamorphic rocks from Kazakhstan: indications for OH^-, H_2O, and FeTi charge transfer. *Euro. J. Mineral.* **5**, 1091–1100.

Lassiter J. C., Hauri E. H., Nikogosian I. K., and Barsczus H. G. (2002) Chlorine–potassium variations in melt inclusions from Raivavae and Rapa, Austral Islands: constraints on chlorine recycling in the mantle and evidence for brine-induced melting of oceanic crust. *Earth Planet. Sci. Lett.* **202**, 525–540.

Laurora A., Mazzucchelli M., Rivalenti G., Vannucci R., Zanetti A., Barbieri M. A., and Cingolani C. A. (2001) Metasomatism and melting in carbonated peridotite xenoliths from the mantle wedge: the Gobernador Gregores case (southern Patagonia). *J. Petrol.* **42**, 69–87.

Lee C.-T., Rudnick R. L., McDonough W. F., and Horn I. (2000) Petrologic and geochemical investigation of carbonates in peridotite xenoliths from northeastern Tanzania. *Contrib. Mineral. Petrol.* **139**, 470–484.

Lee W.-J. and Wyllie P. J. (1998) Process of crustal carbonatite formation by liquid immiscibility and differentiation, elucidated by model systems. *J. Petrol.* **39**, 2005–2013.

Lee W.-J. and Wyllie P. J. (2000) The system CaO–MgO–SiO_2–CO_2 at 1 GPa, metasomatic wehrlites, and primary carbonatite magmas. *Contrib. Mineral. Petrol.* **138**, 214–228.

Leung I. S. (1990) Silicon carbide cluster entrapped in a diamond from Fuxian, China. *Am. Mineral.* **75**, 1110–1119.

Leung I. S., Guo W., Friedman I., and Gleason J. (1990) Natural occurrence of silicon carbide in a diamondiferous kimberlite from Fuxian. *Nature* **346**, 352–354.

Leung I. S., Taylor L. A., Tsao C. S., and Han Z. (1996) SiC in diamond and kimberlites: implications for nucleation and growth of diamond. *Int. Geol. Rev.* **38**, 595–606.

Libowitzky E. and Beran A. (1995) OH defects in forsterite. *Phys. Chem. Mineral.* **22**, 387–392.

Libowitzky E. and Rossman G. R. (1997) An IR absorption calibration for water in minerals. *Am. Mineral.* **82**, 1111–1115.

Liou J. G., Zhang R. Y., Ernst W. G., Rumble D., III, and Maruyama S. (1998) High-pressure minerals from deeply-subducted metamorphic rocks. In *Ultrahigh-pressure Mineralogy: Physics and Chemistry of the Earth's Deep Interior*, Reviews in Mineralogy (ed. R. J. Hemley). Mineralogical Society of America, Washington, DC, vol. 37, pp. 33–96.

Liu L.-G. (1989) Stability fields of Mg-pumpellyite composition at high pressures and temperatures. *Geophys. Res. Lett.* **16**, 847–849.

Lorand J. P. (1987) Cu–Fe–Ni–S mineral assemblages in mantle peridotites from the Table Mountain and Blow-Me-Down Mountain ophiolite massifs (Bay-of-Islands area, Newfoundland): relationships with silicate melts and fluids. *Lithos* **20**, 59–77.

Lorand J. P. (1988) The Cu–Fe–Ni sulfide assemblages of tectonic peridotites from the Maqsad district, Sumail ophiolite, southern Oman: implications for the origin of the sulfide component in the oceanic upper-mantle. *Tectonophysics* **151**, 57–74.

Lorand J. P. (1989) Mineralogy and chemistry of Cu–Fe–Ni sulfides in mantle-derived spinel peridotite bodies from Ariège (northeastern Pyrenees, France). *Contrib. Mineral. Petrol.* **103**, 335–345.

Lorand J. P. (1990) Are spinel lherzolite xenoliths representative of the sulfur content of the upper mantle? *Geochim. Cosmochim. Acta* **54**, 1487–1492.

Lorand J. P. (1991) Sulphide petrology and sulphur geochemistry of orogenic lherzolites: a comparative study of the

Pyrenean bodies (France) and the Lanzo Massif (Italy). *J. Petrol.* Special Lherzolites Issue, pp. 77–95.

Lorand J. P. (1993) Comment on Ionov *et al.* (1992). *Earth Planet. Sci. Lett.* **119**, 627–634.

Lorand J. P. and Conquéré F. (1983) Contribution à l'étude des paragenèses sulfurées dans les enclaves de basalte alcalin du Massif Central et du Languedoc (France). *Bull. Minér.* **106**, 585–606.

Lu R. and Keppler H. (1997) Water solubility in pyrope to 100 kbar. *Contrib. Mineral. Petrol.* **129**, 35–42.

Luguet A., Alard O., Lorand J. P., Pearson N. J., Ryan C., and O'Reilly S. Y. (2001) Laser-ablation microprobe (LAM)-ICPMS unravels the highly siderophile element geochemistry of the oceanic mantle. *Earth Planet. Sci. Lett.* **189**, 285–294.

Luth R. W. (1993) Diamonds, eclogites, and the oxidation state of the Earth's mantle. *Science* **261**, 66–68.

Luth R. W. (1997) Experimental study of the system phlogopite-diopside from 3.5 to 17 GPa. *Am. Mineral.* **82**, 1198–1209.

Luth R. W. (1999) Carbon and carbonates in the mantle. In *Mantle Petrology: Field Observations and High Pressure Experimentation: A Tribute to Francis R. (Joe) Boyd*, Geochemical Society Special Publication No. 6 (eds. Y. Fei, C. M. Bertka, and B. O. Mysen). Geochemical Society, Houston, pp. 297–322.

Mackwell S. J. and Kohlstedt D. L. (1990) Diffusion of hydrogen in olivine: implications for water in the mantle. *J. Geophys. Res.* **95**, 11319–11333.

Maldener J., Rauch F., Gavranic M., and Beran A. (2001) OH absorption coefficients of rutile and cassiterite deduced from nuclear reaction analysis and FTIR spectroscopy. *Mineral. Petrol.* **71**, 21–29.

Martin R. F. and Donnay G. (1972) Hydroxyl in the mantle. *Am. Mineral.* **57**, 554–570.

Massare D., Métrich N., and Clocchiatti R. (2002) High-temperature experiments on silicate melt inclusions in olivine at 1 atm: inferences on temperatures of homogenization and H_2O concentrations. *Chem. Geol.* **183**, 87–98.

Mathez E. A. (1976) Sulfur solubility and magmatic sulfides in submarine basalt glass. *J. Geophys. Res.* **81**, 4269–4276.

Mathez E. A., Blacic J. D., Beery J., Maggiore C., and Hollander M. (1984) Carbon abundances in mantle minerals determined by nuclear reaction analysis. *Geophys. Res. Lett.* **11**, 947–950.

Mathez E. A., Fogel R. A., Hutcheon I. D., and Marshintsev V. K. (1995) Carbon isotopic composition and origin of SiC from kimberlites of Yakutia, Russia. *Geochim. Cosmochim. Acta* **59**, 781–791.

Matson D. W., Muenow D. W., and Garcia M. O. (1986) Volatile contents of phlogopite micas from South African kimberlite. *Contrib. Mineral. Petrol.* **93**, 399–408.

Matveev S., O'Neill H. St. C., Ballhaus C., Taylor W. R., and Green D. H. (2001) Effect of silica activity on OH^- IR spectra of olivine: implications for low-$a$$SiO_2$ mantle metasomatism. *J. Petrol.* **42**, 721–729.

McGetchin T. R. and Besançon J. R. (1975) Carbonate inclusions in mantle-derived pyropes. *Earth Planet. Sci. Lett.* **18**, 408–410.

McGetchin T. R., Silver L. T., and Chodos A. A. (1970) Titanoclinohumite: a possible mineralogical site for water in the upper mantle. *J. Geophys. Res.* **75**, 255–259.

McInnes B. I. A., Gregoire M., Binns R. A., Herzig P. M., and Hannington M. D. (2001) Hydrous metasomatism of oceanic sub-arc mantle, Lihir, Papua New Guinea: petrology and geochemistry of fluid-metasomatized mantle wedge xenoliths. *Earth Planet. Sci. Lett.* **188**, 169–183.

McKenzie D. and Bickle M. J. (1988) The volume and composition of melt generated by extension of the lithosphere. *J. Petrol.* **29**, 625–679.

McMillan P. F., Akaogi M., Sato R. K., Poe B., and Foley J. (1991) Hydroxyl groups in β-Mg_2SiO_4. *Am. Mineral.* **76**, 354–360.

Medaris L. G., Jr. (1999) Garnet peridotites in Eurasian high-pressure and ultrahigh-pressure terranes: a diversity of origins and thermal histories. *Int. Geol. Rev.* **41**, 799–815.

Mei S. and Kohlstedt D. L. (2000) Influence of water on plastic deformation of olivine aggregates: 1. Diffusion creep regime. *J. Geophys. Res.* **105**, 21457–21469.

Menzies M. and Chazot G. (1995) Fluid processes in diamond to spinel facies shallow mantle. *J. Geodynam.* **20**, 387–415.

Menzies M. A. and Hawkesworth C. J. (eds.) (1987) Mantle *Metasomatism*. Academic Press, London.

Meyer H. O. A. and Boctor N. Z. (1975) Sulfide-oxide minerals in eclogite from Stockdale kimberlite, Kansas. *Contrib. Mineral. Petrol.* **52**, 57–68.

Meyer H. O. A. and Brookins D. G. (1971) Eclogite xenoliths from Stockdale Kimberlite, Kansas. *Contrib. Mineral. Petrol.* **34**, 60–72.

Meyer H. O. A. and McCallum M. E. (1986) Mineral inclusions in diamonds from the Sloan kimberlites, Colorado. *J. Geol.* **94**, 600–612.

Mibe K., Fujii T., and Yasadu A. (1998) Connectivity of aqueous fluids in the Earth's upper mantle. *Geophys. Res. Lett.* **25**, 1233–1236.

Mibe K., Fujii T., and Yasadu A. (1999) Control of the location of the volcanic front in island arcs by aqueous fluid connectivity in the mantle wedge. *Nature* **401**, 259–262.

Michael P. J. (1988) The concentration, behavior and storage of H_2O in the suboceanic upper mantle: implications for mantle metasomatism. *Geochim. Cosmochim. Acta* **52**, 555–566.

Michael P. J. (1995) Regionally distinctive sources of depleted MORB: evidence from trace elements and H_2O. *Earth Planet. Sci. Lett.* **131**, 301–320.

Michael P. J. and Cornell W. C. (1998) Influence of spreading rate and magma supply on crystallization and assimilation beneath mid-ocean ridges: evidence from chlorine and major element chemistry of mid-ocean ridge basalts. *J. Geophys. Res.* **103**, 18325–18356.

Michael P. J. and Schilling J.-G. (1989) Chlorine in mid-ocean ridge magmas: evidence for assimilation of seawater-influenced components. *Geochim. Cosmochim. Acta* **53**, 3131–3143.

Miller C. and Richter W. (1982) Solid and fluid phases in lherzolite and pyroxenite inclusions from Hoggar, Central Sahara. *Geochem. J.* **16**, 263–277.

Miller G. H., Rossman G. R., and Harlow G. E. (1987) The natural occurrence of hydroxide in olivine. *Phys. Chem. Mineral.* **14**, 461–472.

Minarik W. G. and Watson E. B. (1995) Interconnectivity of carbonate melts at low melt fraction. *Earth Planet. Sci. Lett.* **133**, 423–437.

Minarik W. G., Ryerson F. J., and Watson E. B. (1996) Textural entrapment of core-forming melts. *Science* **272**, 530–533.

Mitchell R. H. (1986) *Kimberlites: Mineralogy, Geochemistry, and Petrology*. Plenum Press, New York, 442p.

Mitchell R. H. and Keays R. R. (1981) Abundance and distribution of gold, palladium and iridium in some spinel and garnet lherzolites: implications for the nature and origin of precious metal rich intergranular components in the upper mantle. *Geochim. Cosmochim. Acta* **45**, 2425–2445.

Modreski P. J. and Boettcher A. L. (1973) Phase relationships of phlogopite in the system K_2O–MgO–CaO–Al_2O_3–SiO_2–H_2O to 35 kilobars: a better model for micas in the interior of the Earth. *Am. J. Sci.* **273**, 385–414.

Morgan J. W. (1986) Ultramafic xenoliths: clues to Earth's late accretional history. *J. Geophys. Res.* **91**, 12375–12387.

Murakami M., Hirose K., Yurimoto H., Nakashima S., and Takafuji N. (2002) Water in the Earth's lower mantle. *Science* **295**, 1885–1887.

Murck B. W., Burruss R. C., and Hollister L. S. (1978) Phase equilibria in fluid inclusions in ultramafic xenoliths. *Am. Mineral.* **63**, 40–46.

Mysen B. O. and Boettcher A. L. (1975) Melting of a hydrous mantle: I. Phase relations of natural peridotite at high pressures and temperatures with controlled activities of water, carbon dioxide, and hydrogen. *J. Petrol.* **16**, 520–548.

Navon O., Hutcheon I. D., Rossman G. R., and Wasserburg G. L. (1988) Mantle-derived fluids in diamond micro-inclusions. *Nature* **335**, 784–789.

Neumann E.-R., Wulff-Pedersen E., Pearson N. J., and Spencer E. A. (2002) Mantle xenoliths from Tenerife (Canary Islands): evidence for reactions between mantle peridotites and silicic carbonatite melts inducing Ca metasomatism. *J. Petrol.* **43**, 825–857.

Nicholls I. A. (1974) Liquids in equilibrium with peridotitic mineral assemblages at high water pressure. *Contrib. Mineral. Petrol.* **45**, 289–316.

Nichols A. R. L., Carroll M. R., and Höskuldsson Á. (2002) Is the Iceland hot spot also wet? Evidence from the water contents of undegassed submarine and subglacial pillow basalts. *Earth Planet. Sci. Lett.* **202**, 77–87.

Nixon P. H. (ed.) (1987) *Mantle Xenoliths.* Wiley, Chichester.

Odling N. W. A. (1995) An experimental replication of upper-mantle metasomatism. *Nature* **373**, 58–60.

O'Hanley D. S. (1996) *Serpentinites: Records of Tectonic and Petrological History.* Oxford University Press, New York, 277p.

Okay A. I. (1994) Sapphirine and Ti-clinohumite in ultra-high-pressure garnet-pyroxenite and eclogite from Dabie Shan, China. *Contrib. Mineral. Petrol.* **116**, 145–155.

O'Neill H. St. C., Rubie D. C., Canil D., Geiger C. A., Ross C. R., II, Seifert F., and Woodland A. B. (1993) Ferric iron in the upper mantle and in transition zone assemblages: implications for relative oxygen fugacities in the mantle. In *Evolution of the Earth and Planets*, Geophysical Monograph 74 (eds. E. Takahashi, R. Jeanloz, and D. Rubie). International Union of Geodesy and Geophysics and American Geophysical Union, Washington, DC, vol. 14, pp. 73–88.

Ono S., Mibe K., and Yoshino T. (2002) Aqueous fluid connectivity in pyrope aggregates: water transport into the deep mantle by a subducted oceanic crust without any hydrous minerals. *Earth Planet. Sci. Lett.* **203**, 895–903.

O'Reilly S. Y. and Griffin W. L. (1988) Mantle metasomatism beneath Victoria, Australia: I. Metasomatic processes in Cr-diopside lherzolites. *Geochim. Cosmochim. Acta* **52**, 433–447.

O'Reilly S. Y. and Griffin W. L. (2000) Apatite in the mantle: implications for metasomatic processes and high heat production in Phanerozoic mantle. *Lithos* **53**, 217–232.

O'Reilly S. Y., Griffin W. L., and Ryan C. G. (1991) Residence of trace elements in metasomatized spinel lherzolite xenoliths: a proton microprobe study. *Contrib. Mineral. Petrol.* **109**, 98–113.

Oxburgh E. R. (1964) Petrological evidence for the presence of amphibole in the upper mantle and its petrogenetic and geophysical implications. *Geol. Mag.* **101**, 1–19.

Pal'yanov Y. N., Sokol A. G., Borzdov Y. M., Khokhryakow A. F., and Sobolev N. V. (1999a) Diamond formation from mantle carbonate fluids. *Nature* **400**, 417–418.

Pal'yanov Y. N., Sokol A. G., Borzdov Y. M., Khokhryakow A. F., Shatsky A. F., and Sobolev N. V. (1999b) The diamond growth from Li_2CO_3, Na_2CO_3, K_2CO_3, and Cs_2CO_3 solvent-catalysts at $P = 7$ GPa and $T = 1700–1750$ °C. *Diam. Relat. Mater.* **8**, 1118–1124.

Pal'yanov Y. N., Sokol A. G., Borzdov Y. M., and Khokhryakow A. F. (2002a) Fluid-bearing alkaline-carbonate melts as the medium for the formation of diamonds in the Earth's mantle: an experimental study. *Lithos* **60**, 145–159.

Pal'yanov Y. N., Sokol A. G., Borzdov Y. M., Khokhryakow A. F., and Sobolev N. V. (2002b) Diamond formation through carbonate–silicate interaction. *Am. Mineral.* **87**, 1009–1013.

Pasteris J. D. (1987) Fluid inclusions in mantle xenoliths. In *Mantle Xenoliths* (ed. P. H. Nixon). Wiley, pp. 691–707.

Paterson M. (1982) The determination of hydroxyl by infrared absorption in quartz, silicate glasses, and similar materials. *Bull. Mineral.* **105**, 20–29.

Pawley A. (2003) Chlorite stability in mantle peridotite: the reaction clinochlore + enstatite = forsterite + pyrope + H_2O. *Contrib. Mineral. Petrol.* **144**, 449–456.

Peach C. L., Mathez E. A., and Keays R. R. (1990) Sulfide melt-silicate melt distribution coefficients for noble metals and other chalcophile elements as deduced from MORB: implications for partial melting. *Geochim. Cosmochim. Acta* **54**, 3379–3389.

Peacock S. M. (2001) Are the lower planes of double seismic zones caused by serpentine dehydration in subducting oceanic mantle? *Geology* **29**, 299–302.

Peacock S. M. and Hyndman R. D. (1999) Hydrous minerals in the mantle wedge and the maximum depth of subduction thrust earthquakes. *Geophys. Res. Lett.* **26**, 2517–2520.

Peacock S. M. and Wang K. (1999) Seismic consequences of warm versus cool subduction metamorphism: examples from southwest and northeast Japan. *Science* **286**, 937–939.

Peslier A. H., Luhr J. F., and Post J. (2002) Low water contents in pyroxenes from spinel-peridotites of the oxidized, sub-arc mantle wedge. *Earth Planet. Sci. Lett.* **201**, 69–86.

Peterson R. and Francis D. (1977) The origin of sulfide inclusions in pyroxene megacrysts. *Am. Mineral.* **62**, 1049–1051.

Philippot P. and Selverstone J. (1991) Trace-element-rich brines in eclogitic veins: implications for fluid composition and transport during subduction. *Contrib. Mineral. Petrol.* **106**, 417–430.

Philippot P., Chevallier P., Chopin C., and Dubessy J. (1995) Fluid composition and evolution in coesite-bearing rocks (Dora-Maira massif, Western Alps): implications for element recycling during subduction. *Contrib. Mineral. Petrol.* **121**, 29–44.

Philippot P., Agrinier P., and Scambelluri M. (1998) Chlorine cycling during subduction of altered oceanic crust. *Earth Planet. Sci. Lett.* **161**, 33–44.

Pichavant M., Mysen B. O., and MacDonald R. (2002) Source and H_2O content of high-MgO magmas in island arc settings: an experimental study of a primitive calc-alkaline basalt from St. Vincent, Lesser Antilles arc. *Geochim. Cosmochim. Acta* **66**, 2193–2209.

Poli S. and Schmidt M. W. (2002) Petrology of subducted slabs. *Ann. Rev. Earth Planet. Sci.* **30**, 207–235.

Rauch M. and Keppler H. (2002) Water solubility in orthopyroxene. *Contrib. Mineral. Petrol.* **143**, 525–536.

Richard G., Monnereau M., and Ingrin J. (2002) Is the transition zone an empty water reservoir? Inferences from numerical model of mantle dynamics. *Earth Planet. Sci. Lett.* **205**, 37–51.

Righter K. and Carmichael I. S. E. (1993) Mega-xenocrysts in alkali olivine basalts: fragments of disrupted mantle assemblages. *Am. Mineral.* **78**, 1230–1245.

Righter K. and Carmichael I. S. E. (1996) Phase equilibria of phlogopite lamprophyres from western Mexico: biotite-liquid equilibria and $P–T$ estimates for biotite-bearing igneous rocks. *Contrib. Mineral. Petrol.* **123**, 1–21.

Roedder E. (1965) Liquid CO_2 inclusions in olivine-bearing nodules and phenocrysts from basalts. *Am. Mineral.* **50**, 1746–1782.

Roedder E. (1984) *Fluid Inclusions*, Reviews in Mineralogy. Mineralogical Society of America, vol. 12, 646pp.

Rose L. A. and Brenan J. M. (2001) Wetting properties of Fe–Ni–Co–Cu–O–S melts against olivine: implications for sulfide melt mobility. *Econ. Geol.* **96**, 145–157.

Rossman G. R. (1988) Vibrational spectroscopy of hydrous components. In *Spectroscopic Methods in Mineralogy*,

Reviews in Mineralogy (ed. F. C. Hawthorne). Mineralogical Society of America, Washington, DC, vol. 18, pp. 193–206.

Rossman G. R. (1996) Studies of OH in nominally anhydrous minerals. *Phys. Chem. Mineral.* **23**, 299–304.

Rossman G. R. and Aines R. D. (1991) The hydrous components in garnets: grossular–hydrogrossular. *Am. Mineral.* **76**, 1153–1164.

Rossman G. R., Rauch F., Livi R., Tombrello T. A., Shi C. R., and Zhou Z. Y. (1988) Nuclear reaction analysis of hydrogen in almandine, pyrope, and spessartite garnets. *Neus Jahrb. Mineral. Mh.* **1988**(H.4), 172–178.

Rossman G. R., Beran A., and Langer K. (1989) The hydrous component of pyrope from the Dora Maira Massif, Western Alps. *Euro. J. Mineral.* **1**, 151–154.

Roy-Barman M., Wasserburg G. J., Papanastassiou D. A., and Chaussidon M. (1998) Osmium isotopic compositions and Re–Os concentrations in sulfide globules from basaltic glasses. *Earth Planet. Sci. Lett.* **154**, 331–347.

Rudnick R. L., Eldridge C. S., and Bulanova G. P. (1993a) Diamond growth history from *in situ* measurement of Pb and S isotopic compositions of sulfide inclusions. *Geology* **21**, 13–16.

Rudnick R. L., McDonough W. F., and Chappell B. C. (1993b) Carbonatite metasomatism in the northern Tanzanian mantle. *Earth Planet. Sci. Lett.* **114**, 463–475.

Sato K., Katsura T., and Ito E. (1997) Phase relations of natural phlogopite with and without enstatite up to 8 GPa: implications for mantle metasomatism. *Earth Planet. Sci. Lett.* **146**, 511–526.

Sato K., Akaishi M., and Yamaoka S. (1999) Spontaneous nucleation of diamond in the system $MgCO_3$–$CaCO_3$–C at 7.7 GPa. *Diam. Relat. Mater.* **8**, 1900–1905.

Scambelluri M. and Philippot P. (2001) Deep fluids in subduction zones. *Lithos* **55**, 213–227.

Scambelluri M. and Rampone E. (1999) Mg-metasomatism of oceanic gabbros and its control on Ti-clinohumite formation during eclogitization. *Contrib. Mineral. Petrol.* **135**, 1–17.

Scambelluri M., Hoogerduijn-Strating E. H., Piccardo G. B., Vissers R. L. M., and Rampone E. (1991) Alpine olivine- and titanian clinohumite-bearing assemblages in the Erro-Tobbio peridotite (Voltri Massif, NW Italy). *J. Metamorph. Geol.* **9**, 79–91.

Scambelluri M., Piccardo G. B., Philippot P., Robbiano A., and Negretti L. (1997) High salinity fluid inclusions formed from recycled seawater in deeply subducted alpine serpentinite. *Earth Planet. Sci. Lett.* **148**, 485–499.

Scambelluri M., Pennacchioni G., and Philippot P. (1998) Salt-rich aqueous fluids formed during eclogitization of metabasites in the Alpine continental crust (Austroalpine Mt. Emilius unit, Italian western Alps). *Lithos* **43**, 151–167.

Schiano P., Clocchiatti R., and Joron J. L. (1992) Melt and fluid inclusions in basalts and xenoliths from Tahaa Island, Society archipelago: evidence for a metasomatized upper mantle. *Earth Planet. Sci. Lett.* **111**, 69–82.

Schiano P., Clocchiatti R., Shimizu N., Weis D., and Mattielli N. (1994) Cogenetic silica-rich and carbonate-rich melts trapped in mantle minerals in Kerguelen ultramafic xenoliths: implications for metasomatism in the oceanic upper mantle. *Earth Planet. Sci. Lett.* **123**, 167–178.

Schiano P., Clocchiatti R., Shimizu N., Maury R. C., Jochum K. P., and Hofmann A. W. (1995) Hydrous, silica-rich melts in the sub-arc mantle and their relationship with erupted arc lavas. *Nature* **377**, 595–600.

Schilling J.-G., Bergeron M. B., and Evans R. (1980) Halogens in the mantle beneath the North Atlantic. *Phil. Trans. Roy. Soc. London* **297**, 147–178.

Schmidt M. W. and Poli S. (1998) Experimentally based water budgets for dehydrating slabs and consequences for arc magma generation. *Earth Planet. Sci. Lett.* **163**, 361–379.

Schrauder M. and Navon O. (1994) Hydrous and carbonatitic mantle fluids in hydrous diamonds from Jwaneng, Botswana. *Geochim. Cosmochim. Acta* **58**, 761–771.

Schrauder M., Koeberl C., and Navon O. (1996) Trace element analyses of fluid-bearing diamonds from Jwaneng, Botswana. *Geochim. Cosmochim. Acta* **60**, 4711–4724.

Schreyer W. (1988) Experimental studies on metamorphism of crustal rocks under mantle pressures. *Min. Mag.* **52**, 1–26.

Schreyer W., Maresch W. V., Medenbach O., and Baller T. (1986) Calcium-free pumpellyite, a new synthetic hydrous Mg–Al-silicate formed at high pressures. *Nature* **321**, 510–511.

Sclar C. B. (1970) High-pressure studies in the system MgO–SiO_2–H_2O. *Phys. Earth Planet. Int.* **3**, 333.

Sclar C. B., Carrison L. C., and Stewart O. M. (1967a) High-pressure synthesis and stability of hydroxylated clinoenstatite in the system MgO–SiO_2–H_2O. *Geol. Soc. Amer. Program with Abstr.* 198p.

Sclar C. B., Carrison L. C., and Stewart O. M. (1967b) High-pressure synthesis of a new hydroxylated pyroxene in the system MgO–SiO_2–H_2O. *Trans. Am. Geophys. Union* **48**, 226.

Selverstone J., Franz G., Thomas S., and Getty S. (1992) Fluid variability in 2 GPa eclogites as an indicator of fluid behavior during subduction. *Contrib. Mineral. Petrol.* **112**, 341–357.

Sen G., Macfarlane A., and Srimal N. (1996) Significance of rare hydrous alkaline melts in Hawaiian xenoliths. *Contrib. Mineral. Petrol.* **122**, 415–427.

Shen A. H. and Keppler H. (1997) Direct observation of complete miscibility in the albite-H_2O system. *Nature* **385**, 710–712.

Sisson T. W. and Bronto S. (1998) Evidence for pressure-release melting beneath magmatic arcs from basalt at Galunggung, Indonesia. *Nature* **391**, 883–886.

Sisson T. W. and Layne G. D. (1993) H_2O in basalt and basaltic andesite glass inclusions from four subduction-related volcanoes. *Earth Planet. Sci. Lett.* **117**, 619–635.

Skogby H. (1994) OH incorporation in synthetic clinopyroxene. *Am. Mineral.* **79**, 240–249.

Skogby H. and Rossman G. R. (1989) OH- in pyroxene: an experimental study of incorporation mechanisms and stability. *Am. Mineral.* **74**, 1059–1069.

Skogby H., Bell D. R., and Rossman G. R. (1990) Hydroxide in pyroxene: variations in the natural environment. *Am. Mineral.* **75**, 664–764.

Smith D. (1977) Titanochondrodite and titanoclinohumite derived from the upper mantle in the Buell Park kimberlite, Arizona, USA. A discussion. *Contrib. Mineral. Petrol.* **61**, 213–215.

Smith D. (1979) Hydrous minerals and carbonates in peridotite inclusions from the Green Knobs and Buell Park kimberlitic diatremes on the Colorado Plateau. In *The Mantle Sample: Inclusions in Kimberlites and other Volcanics* (eds. F. R. Boyd and H. O. A. Meyer). American Geophysical Union, Washington, DC, pp. 345–356.

Smith D. (1987) Genesis of carbonate in pyrope from ultramafic diatremes on the Colorado Plateau, southeastern United States. *Contrib. Mineral. Petrol.* **97**, 389–396.

Smith D. (1995) Chlorite-rich ultramafic reaction zones in Colorado Plateau xenoliths: recorders of sub-Moho hydration. *Contrib. Mineral. Petrol.* **121**, 185–200.

Smith D., Riter J. C. A., and Mertzman S. A. (1999) Water-rock interactions, orthopyroxene growth, and Si-enrichment in the mantle: evidence in xenoliths from the Colorado Plateau, southwestern United States. *Earth Planet. Sci. Lett.* **165**, 45–54. (also cf. erratum: *Earth Planet. Sci. Lett.* **167**, 347–356).

Smith J. V., Delaney J. S., Hervig R. L., and Dawson J. B. (1981) Storage of F and Cl in the upper mantle: geochemical implications. *Lithos* **14**, 133–147.

Smyth J. R. (1987) β-Mg_2SiO_4: a potential host for water in the mantle? *Am. Mineral.* **72**, 1051–1055.

Smyth J. R. (1994) A crystallographic model for hydrous wadsleyite (β-Mg_2SiO_4): an ocean in the Earth's interior? *Am. Mineral.* **79**, 1021–1024.

Smyth J. R., Bell D. R., and Rossman G. R. (1991) Incorporation of hydroxyl in upper-mantle clinopyroxenes. *Nature* **351**, 732–735.

Smyth J. R., Kawamoto T., Jacobsen S. D., Swope R. J., Hervig R. L., and Holloway J. R. (1997) Crystal structure of monoclinic hydrous wadsleyite [β-$(Mg,Fe)_2SiO_4$]. *Am. Mineral.* **82**, 270–275.

Snow J. E. and Dick H. J. B. (1995) Pervasive magnesium loss by marine weathering of peridotite. *Geochim. Cosmochim. Acta* **59**, 4219–4235.

Sobolev A. V. and Chaussidon M. (1996) H_2O concentrations in primary melts from supra-subduction zones and mid-ocean ridges: implications for H_2O storage and recycling in the mantle. *Earth Planet. Sci. Lett.* **137**, 45–55.

Sokol A. G., Pal'yanov Y. N., Pal'yanova G. A., Khokhryakov A. F., and Borzdov Y. M. (2001) Diamond and graphite crystallization from C–O–H fluids. *Diam. Relat. Mater.* **10**, 2131–2136.

Stachel T. and Harris J. W. (1997) Diamond precipitation and mantle metasomatism—evidence from the trace element chemistry of silicate inclusions in diamonds from Akwatia, Ghana. *Contrib. Mineral. Petrol.* **129**, 143–154.

Stachel T., Harris J. W., and Brey G. P. (1998) Rare and unusual mineral inclusions in diamonds from Mwadui, Tanzania. *Contrib. Mineral. Petrol.* **132**, 34–57.

Stalder R. and Ulmer P. (2001) Phase relations of a serpentine composition between 5 and 14 GPa: significance of clinohumite and phase E as water carriers into the transition zone. *Contrib. Mineral. Petrol.* **140**, 670–679. (also cf. errata: Contrib. Mineral. Petrol. **140**, 754).

Stalder R., Ulmer P., Thompson A. B., and Günther D. (2000) Experimental approach to constrain second critical end points in fluid/silicate systems: near-solidus fluids and melts in the system albite-H_2O. *Am. Mineral.* **85**, 68–77.

Stalder R., Ulmer P., Thompson A. B., and Günther D. (2001) High pressure fluids in the system $MgO–SiO_2–H_2O$ under upper mantle conditions. *Contrib. Mineral. Petrol.* **140**, 607–618.

Staudigel H. and Schreyer W. (1977) The upper thermal stability of clinochlore, $Mg_5Al[AlSi_3O_{10}](OH)_8$, at 10–35 kb P_{H_2O}. *Contrib. Mineral. Petrol.* **61**, 187–198.

Stolper E. M. and Newman S. (1994) The role of water in the petrogenesis of the Mariana Trough magmas. *Earth Planet. Sci. Lett.* **121**, 293–325.

Stone W. E. and Fleet M. E. (1991) Nickel–copper sulfides from the 1959 eruption of Kilauea volcano, Hawaii: contrasting compositions and phase relations in eruption pumice and Kilauea Iki lava lake. *Am. Mineral.* **76**, 1363–1372.

Sudo A. and Tatsumi Y. (1990) Phlogopite and K-amphibole in the upper mantle: implication for magma genesis in subduction zones. *Geophys. Res. Lett.* **17**, 29–32.

Sumita T. and Inoue T. (1996) Melting experiments and thermodynamic analyses on silicate-H_2O systems up to 12 GPa. *Phys. Earth Planet. Int.* **96**, 187–200.

Sun L., Akaishi M., and Yamaoka S. (2000) Formation of diamond in the system of Ag_2CO_3 and graphite at high pressure and high temperatures. *J. Cryst. Growth* **213**, 411–414.

Sun S.-S. and Hanson G. N. (1975) Origin of Ross Island basanitoids and limitations upon the heterogeneity of mantle sources for alkali basalts and nephelinites. *Contrib. Mineral. Petrol.* **52**, 77–106.

Sweeney R. J., Thompson A. B., and Ulmer P. (1993) Phase relations of a natural MARID composition and implications for MARID genesis, lithospheric melting and mantle metasomatism. *Contrib. Mineral. Petrol.* **115**, 225–241.

Sweeney R. J., Prozesky V. M., and Springhorn K. A. (1997) Use of the elastic recoil detection analysis (ERDA) microbeam technique for the quantitative determination of hydrogen in materials and hydrogen partitioning between olivine and melt at high pressures. *Geochim. Cosmochim. Acta* **61**, 101–113.

Szabó C. and Bodnar R. J. (1995) Chemistry and origin of mantle sulfides in spinel peridotite xenoliths from alkaline basaltic lavas, Nógrád-Gömör Volcanic Field, northern Hungary and southern Slovakia. *Geochim. Cosmochim. Acta* **59**, 3917–3927.

Taniguchi T., Dobson D., Jones A. P., Rabe R., and Milledge H. J. (1996) Synthesis of cubic diamond in the graphite–magnesium carbonate and graphite–$K_2Mg(CO_3)_2$ systems at high pressure of 9–10 GPa region. *J. Mater. Res.* **11**, 2622–2632.

Tatsumi Y. (1981) Melting experiments on a high-magnesium andesite. *Earth Planet. Sci. Lett.* **54**, 357–365.

Tatsumi Y. (1982) Origin of high-magnesium andesites in the Setouchi volcanic belt, southwest Japan: II. Melting phase relations at high pressures. *Earth Planet. Sci. Lett.* **60**, 305–317.

Tatsumi Y., Sakuyama M., Fukuyama H., and Kushiro I. (1983) Generation of arc basalt magmas and thermal structure of the mantle wedge in subduction zones. *J. Geophys. Res.* **88**, 5815–5825.

Tatsumi Y., Furukawa Y., and Yamashita S. (1994) Thermal and geochemical evolution of the mantle wedge in the northeast Japan arc: I. Contribution from experimental petrology. *J. Geophys. Res.* **99**, 22275–22283.

Tillmanns E. and Zemann J. (1965) Messung des Ultrarot-Pleochroismus von Mineralen: I. Der Pleochroismus der OH-Streckfrequenz in Azurit. *Neus Jahrb. Mineral. Mh.* **1965**, 228–231.

Tingle T. N., Green H. W., and Finnerty A. A. (1988) Experiments and observations bearing on the solubility and diffusivity of carbon in olivine. *J. Geophys. Res.* **93**, 15289–15304.

Trommsdorff V. and Evans B. W. (1980) Titanian hydroxyl-clinohumite: formation and breakdown in antigorite rocks (Malenco, Italy). *Contrib. Mineral. Petrol.* **72**, 229–242.

Trønnes R. G. (1990) Low-Al, high-K amphiboles in subducted lithosphere from 200 to 400 km depth: experimental evidence (abstr.). *EOS* **71**, 1587.

Trønnes R. G. (2002) Stability range and decomposition of potassic richterite and phlogopite end members at 5–15 GPa. *Mineral. Petrol.* **74**, 129–148.

Tsai H. M., Shieh Y. N., and Meyer H. O. A. (1979) Mineralogy and 34S/32S ratios of sulfides associated with kimberlites, xenoliths and diamonds. In *The Mantle Sample: Inclusions in Kimberlites and other Volcanics* (eds. F. R. Boyd and H. O. A. Meyer). American Geophysical Union, Washington, DC, pp. 87–103.

Tsong I. S. T. and Knipping U. (1986) Comment on "Solute carbon and carbon segregation in magnesium oxide single crystals—a secondary ion mass spectrometry study" by F. Freund. *Phys. Chem. Mineral.* **13**, 277–279.

Tsong I. S. T., Knipping U., Loxton C. M., Magee C. W., and Arnold G. W. (1985) Carbon on surfaces of magnesium oxide and olivine single crystals; diffusion from the bulk or surface contamination? *Phys. Chem. Mineral.* **12**, 261–270.

Turner G., Burgess R., and Bannon M. (1990) Volatile-rich mantle fluids inferred from inclusions in diamond and mantle xenolith. *Nature* **344**, 653–655.

Ulmer G. C., Grandstaff D. E., Woermann E., Göbbels M., Schönitz M., and Woodland A. B. (1998) The redox stability of moissanite (SiC) compared with metal–metal oxide buffers at 1773 K and at pressures up to 90 kbar. *Neus Jahrb. Mineral. Abh.* **172**(2/3), 279–307.

Ulmer P. (2001) Partial melting in the mantle wedge—the role of H_2O in the genesis of mantle-derived "arc-related" magmas. *Phys. Earth Planet. Int.* **127**, 215–232.

Ulmer P. and Trommsdorff V. (1995) Serpentine stability to mantle depths and subduction-related magmatism. *Science* **268**, 858–861.

Ulmer P. and Trommsdorff V. (1999) Phase relations of hydrous mantle subducting to 300 km. In *Mantle Petrology: Field Observations and High Pressure Experimentation: A Tribute to Francis R. (Joe) Boyd,* Geochemical Society

Special Publication No. 6 (eds. Y. Fei, C. M. Bertka, and B. O. Mysen). Geochemical Society, Houston, pp. 259–281.

Vakhrushev V. A. and Sobolev N. V. (1973) Sulfidic formations in deep xenoliths from kimberlite pipes in Yakutia. *Int. Geol. Rev.* **15**, 103–110.

van Keken P. E., Hauri E. H., and Ballentine C. J. (2002) Mantle mixing: the generation, preservation, and destruction of chemical heterogeneity. *Ann. Rev. Earth Planet. Sci.* **30**, 493–525.

Vanko D. A. (1988) Temperature, pressure, and composition of hydrothermal fluids, with their bearing on the magnitude of tectonic uplift at mid-ocean ridges, inferred from fluid inclusions in Oceanic Layer 3 rocks. *J. Geophys. Res.* **93**, 4595–4611.

Vannucci R., Piccardo G. B., Rivalenti G., Zanetti A., Rampone E., Ottolini L., Oberti R., Mazzucelli M., and Bottazzi P. (1995) Origin of LREE-depleted amphiboles in the subcontinental mantle. *Geochim. Cosmochim. Acta* **59**, 1763–1771.

Von Damm K. L. (1995) Controls on the chemistry and temporal variability of seafloor hydrothermal fluids. In *Seafloor Hydrothermal Systems: Physical, Chemical, Biological, and Geological Interactions*, Geophysical Monograph 91 (eds. S. E. Humphris, R. A. Zierenberg, L. S. Mullineaux, and R. E. Thomson). American Geophysical Union, Washington, DC, pp. 222–247.

Vukadinovic D. and Edgar A. D. (1993) Phase relations in the phlogopite–apatite system at 20 kbar: implications for the role of fluorine in mantle melting. *Contrib. Mineral. Petrol.* **114**, 247–254.

Wagner C., Deloule E., and Mokhtari A. (1996) Richterite-bearing peridotites and MARID-type inclusions in lavas from North Eastern Morocco: mineralogy and D/H isotopic studies. *Contrib. Mineral. Petrol.* **124**, 406–421.

Walker D. (2000) Core participation in mantle geochemistry: Geochemical Society Ingerson Lecture, GSA Denver, October 1999. *Geochim. Cosmochim. Acta* **64**, 2897–2911.

Walker R. J., Morgan J. W., and Horan M. F. (1995) 187Os enrichment in some plumes: evidence for core–mantle interaction. *Science* **269**, 819–822.

Wallace P. J. (1998) Water and partial melting in mantle plumes: inferences from the dissolved H_2O concentrations of Hawaiian basaltic magmas. *Geophys. Res. Lett.* **25**, 3639–3642.

Wang A., Pasteris J. D., Meyer H. O. A., and Delduboi M. L. (1996) Magnesite-bearing inclusion assemblage in natural diamond. *Earth Planet. Sci. Lett.* **141**, 293–306.

Wang L., Zhang Y., and Essene E. (1996) Diffusion of the hydrous component in pyrope. *Am. Mineral.* **81**, 706–718.

Wang W., Sueno S., Takahashi E., Yurimoto H., and Gasparik T. (2000) Enrichment processes at the base of the Archean lithospheric mantle: observations from trace element characteristics of pyropic garnet incluisons in diamonds. *Contrib. Mineral. Petrol.* **139**, 720–733.

Waters F. G. (1987) A suggested origin of MARID xenoliths in kimberlites by high pressure crystallization of an ultrapotassic rock such as lamproite. *Contrib. Mineral. Petrol.* **95**, 523–533.

Watson E. B. and Brenan J. M. (1987) Fluids in the lithosphere: 1. Experimentally-determined wetting characteristics of CO_2–H_2O fluids and their implications for fluid transport, host-rock physical properties, and fluid inclusion formation. *Earth Planet. Sci. Lett.* **85**, 497–515.

Watson E. B. and Lupulescu A. (1993) Aqueous fluid connectivity and chemical transport in clinopyroxene-rich rocks. *Earth Planet. Sci. Lett.* **117**, 279–294.

Watson E. B., Brenan J. M., and Baker D. R. (1990) Distribution of fluids in the continental mantle. In *Continental Mantle* (ed. M. A. Menzies). Clarendon Press, Oxford, pp. 111–125.

Weiss M. (1997) Clinohumites: a field and experimental study. PhD Dissertation, Swiss Federal Institute of Technology (ETH), Zurich, 168pp (unpublished). (Not seen; referenced in Ulmer and Trommsdorff 1999.).

Wilkins R. W. T. and Sabine W. (1973) Water content of some nominally anhydrous silicates. *Am. Mineral.* **58**, 508–516.

Williams Q. and Hemley R. J. (2001) Hydrogen in the deep Earth. *Ann. Rev. Earth Planet. Sci.* **29**, 365–418.

Wilshire H. G. and Shervais J. W. (1975) Al-augite and Cr-diopside ultramafic xenoliths in basaltic rocks from the western United States. *Phys. Chem. Earth* **9**, 257–272.

Withers A. C., Wood B. J., and Carroll M. R. (1998) The OH content of pyrope at high pressure. *Chem. Geol.* **147**, 161–171.

Witt-Eickschen G. and Harte B. (1994) Distribution of trace elements between amphibole and clinopyroxene from mantle peridotites of the Eifel (Western Germany): an ion-microprobe study. *Chem. Geol.* **117**, 235–250.

Wright K. and Catlow C. R. A. (1994) A computer simulation study of OH defects in olivine. *Phys. Chem. Mineral.* **20**, 515–518.

Wright T. L. (1984) Origin of Hawaiian tholeiite: a metasomatic model. *J. Geophys. Res.* **89**, 3233–3252.

Wulff-Pedersen E., Neumann E.-R., and Jensen B. B. (1996) The upper mantle under La Palma, Canary Islands: formation of a Si–K–Na-rich melt and its importance as a metasomatic agent. *Contrib. Mineral. Petrol.* **125**, 113–139.

Wunder B. (1998) Equilibrium experiments in the system MgO–SiO_2–H_2O (MSH): stability fields of clinohumite-OH [$Mg_9Si_4O_{16}(OH)_2$], chondrodite-OH [$Mg_5Si_2O_8(OH)_2$] and phase A [$Mg_7Si_2O_8(OH)_6$]. *Contrib. Mineral. Petrol.* **132**, 111–120.

Wunder B. and Gottschalk M. (2002) Fe–Mg solid solution of sursassite, $(Fe,Mg)_4(Mg,Fe,Al)_2Al_4[Si_6O_{21}/(OH)_7]$. *Euro. J. Mineral.* **14**, 575–580.

Wyllie P. J. (1995) Experimental petrology of upper mantle materials, processes and products. *J. Geodynam.* **20**, 429–468.

Wyllie P. J. and Ryabchikov I. D. (2000) Volatile components, magmas, and critical fluids in upwelling mantle. *J. Petrol.* **41**, 1195–1206.

Wyllie P. J., Huang W.-L., Otto J., and Byrnes A. P. (1983) Carbonation of peridotites and decarbonation of siliceous dolomites represented in the system CaO–MgO–SiO_2–CO_2 to 30 kbar. *Tectonophysics* **100**, 359–388.

Yang H., Konzett J., and Prewitt C. T. (2001) Crystal structure of phase X, a high pressure alkali-rich hydrous silicate and its anhydrous equivalent. *Am. Mineral.* **86**, 1483–1488.

Yaxley G. M., Crawford A. J., and Green D. H. (1991) Evidence for carbonatite metasomatism in spinel peridotite xenoliths from western Victoria, Australia. *Earth Planet. Sci. Lett.* **107**, 305–317.

Yaxley G. M., Green D. H., and Kamenetsky V. (1998) Carbonatite metasomatism in the southeastern Australian lithosphere. *J. Petrol.* **39**, 1917–1930.

Yesinowski J. P., Eckert H., and Rossman G. R. (1988) Characterization of hydrous species in minerals by high-speed ^1H MAS NMR. *J. Am. Chem. Soc.* **110**, 1367–1375.

Young T. E., Green H. W., II, Hofmeister A. M., and Walker D. (1993) Infrared spectroscopic investigation of hydroxyl in β-$(Mg,Fe)_2SiO_4$ and coexisting olivine: implications for mantle evolution and dynamics. *Phys. Chem. Mineral.* **19**, 409–422.

Zhang R. Y., Liou J. G., and Cong B. L. (1995) Talc-, magnesite- and Ti-clinohumite-bearing ultrahigh-pressure meta-mafic and ultramafic complex in the Dabie Mountains, China. *J. Petrol.* **36**, 1011–1037.

2.08
Melt Extraction and Compositional Variability in Mantle Lithosphere

M. J. Walter

Okayama University, Misasa, Japan

2.08.1 INTRODUCTION

Samples of peridotite from the uppermost mantle exhibit considerable variation in mineralogic and chemical composition (e.g., Dawson *et al.*, 1980; McDonough, 1990; Boyd, 1989; Griffin *et al.*, 1999b; Chapters 2.04 and 2.05). Systematic depletion in clinopyroxene (cpx), aluminous phases (e.g., plagioclase, spinel, or garnet), and magmaphile major and trace elements strongly implicates partial melt extraction as the predominant process for producing compositional variation in these rocks.

In the modern tectonic environment, melt extraction from the uppermost mantle occurs in numerous settings that include oceanic and continental spreading centers, "hot-spots," and subduction zones, with the vast majority of melt extraction occurring at mid-ocean ridges (~20 $km^3\ yr^{-1}$) (e.g., Wilson, 1989). Primary melt compositions produced at these locations are controlled by temperature, pressure, and bulk composition of the source, and although in detail the mineral–melt equilibria that govern partial melt composition show a rich variation, melts extracted from the upper mantle are predominantly basaltic in composition (e.g., Basaltic Volcanism Study Project (BVSP), 1981). Based on the abundance of greenstone belts in Precambrian terrains, mafic melt extraction from the mantle has apparently been dominant since the Archean, although relative to the Phanerozoic, melt extraction in the Precambrian also included volumetrically significant amounts of more magnesium-rich picritic and komatiitic melts (e.g., Condie, 1981).

In this chapter the mineral–melt phase equilibria that control the compositions of partial melts are examined on the basis of experimental and thermodynamic databases, and this information is used to predict the effects of partial melt extraction from fertile upper mantle on residual mineralogy and major-element chemistry.

Direct evidence for the compositional effects of partial melt extraction is preserved in samples of upper-mantle lithosphere with a range of ages, including Archean cratonic mantle, Proterozoic subcontinental mantle, and modern oceanic mantle. Samples of upper mantle are collected as xenoliths, peridotites dredged from oceanic fracture zones, and slices of upper mantle tectonically exposed at the surface, and extensive samples exist from both oceanic and continental settings (see Chapters 2.04 and 2.05). Here, data sets are assembled for oceanic and subcontinental mantle lithosphere, and compositional trends are compared to those predicted for partial melt extraction from fertile peridotite in order to deduce the role that melt extraction has played in producing compositional variability in upper-mantle lithosphere, and to place constraints on the thermal evolution of the mantle.

2.08.2 PHASE EQUILIBRIUM AND MELT EXTRACTION

The effects of melt extraction on the chemistry and mineralogy of the residual mantle are controlled by the phase equilibrium of melting. The intensive variables, pressure, temperature, and bulk composition, exert primary control on the complimentary compositions of extracted melts and residues. The process of melt extraction, which is controlled by such factors as residual porosity and the extent of equilibration between migrating melts and wall rock, also effects melt and residue composition. Here, partial melt extraction is modeled generally as a batch process, wherein a given amount of melt equilibrates with the residue at a prescribed final pressure and temperature, irrespective of the initial pressure of melting or the melt migration path. The batch of melt can be large or small. When the amount is small, say less than 2% or so, the process is often referred to as incremental batch melting and is used as an approach to modeling a fractional melting process. Fractional melting is a process in which an infinitesimal amount of melt is extracted and isolated from the residue as soon as it is formed without any subsequent re-equilibration. Both batch and fractional melting can be modeled as isobaric or polybaric processes. As will be discussed in some detail in Section 2.08.4, in nature a spectrum from batch to fractional melting likely exists, with important effects on residual chemistry and mineralogy (see Kelemen *et al.* (1997) and Asimow (1999)).

The phase relations for melting of fertile mantle peridotite at upper-mantle conditions have been studied extensively using experimental and thermodynamic techniques (e.g., O'Hara, 1968; Presnall *et al.*, 1979; Takahashi and Kushiro, 1983; Ghiorso, 1985; Falloon *et al.*, 1988; Kinzler and Grove, 1992a, 1993; Hirose and Kushiro, 1993; Walter and Presnall, 1994; Zhang and Herzberg, 1994; Baker *et al.*, 1995; Hirschmann *et al.*, 1998; Kushiro, 1998; Walter, 1998; Gudfinnsson and Presnall, 2000; Pickering-Witterer and Johnston, 2000; Ghiorso *et al.*, 2002). The information gained from these and other studies about how melt compositions change with intensive variables has been used with great success in modeling the conditions at which mafic and ultramafic melts are produced in the upper mantle. Although outstanding problems and controversies remain, generally speaking it is well established that (i) basalt generation at oceanic spreading centers occurs by ~5–20% partial melting of fertile peridotite primarily in the depth range between 10 km and 70 km, (ii) basalt and basaltic andesite melt generation at convergent margins occurs by partial melting of hydrated peridotite at similar degrees of melting and over a similar range of depths as mid-ocean ridge basalts (MORBs), (iii) alkalic basalts typical of within-plate settings are low-degree melts generated over a range of depths typically less than 100 km, and (iv) rare picritic and komatiitic rocks are generally considered to have been generated by relatively high degrees of melting (>30%) of fertile peridotite at greater depths than basaltic melts (~100–200 km).

In this section a general review is presented of the melting phase relations for fertile upper-mantle peridotite, concentrating on how variations in the depth and degree of partial melt extraction impart compositional variability to the residual source rock.

2.08.2.1 Melting Phase Relations of Fertile Mantle Peridotite

Model fertile peridotite compositions such as those given in Table 1 crystallize lherzolite mineral assemblages at uppermost mantle conditions, with olivine the most abundant mineral constituting some 50–60% by weight, the pyroxenes, orthopyroxene (opx), and clinopyroxene (cpx), constituting some 30–40%, and a fourth, pressure-dependent aluminous mineral constituting the remainder. Figure 1 is a generalized pressure–temperature phase diagram for fertile mantle peridotite (Table 1) compiled from experimental data for subsolidus and batch melting phase relations (Takahashi and

Table 1 Model fertile upper-mantle compositions.

	1	*2*	*3*	*4*	*5*	*6*	*7*	*8*
SiO_2	45.0	45.2	45.96	46.12	45.1	44.7	45.4	44.90
TiO_2	0.20	0.22	0.18	0.18	0.2	0.15	0.21	0.13
Cr_2O_3	0.38	0.46	0.47	0.38	0.4		0.37	0.4
Al_2O_3	4.45	3.97	4.06	4.09	3.3	3.9	4.49	3.49
FeO	8.05	7.82	7.54	7.49	8.0	8.5	8.1	8.27
MgO	37.8	38.3	37.78	37.77	38.1	38.0	36.78	38.59
CaO	3.55	3.50	3.21	3.23	3.1	3.2	3.7	3.19
MnO	0.14	0.13	0.13	0.15	0.15	0.14	0.14	
NiO	0.25	0.27	0.28	0.25	0.2		0.23	0.26
Na_2O	0.36	0.33	0.33	0.36	0.4	0.34	0.35	0.24
K_2O	0.03	0.03	0.03	0.03	0.03	0.03	0.03	
Total	100.2	100.2	100.0	100.0	100.0	99.0	99.8	99.5
MgO/SiO_2	0.84	0.85	0.82	0.82	0.85	0.85	0.81	0.86
CaO/Al_2O_3	0.80	0.88	0.79	0.79	0.94	0.82	0.82	0.91
Mg#[a]	89.2	89.6	89.8	89.8	89.3	88.7	89.0	89.3
Olivine	54.2	54.8	49.8	49.6	55.7	56.8	49.4	56.8
Opx	22.7	23.0	29.8	29.8	24.4	22.8	27.6	23.6
Cpx	19.9	19.3	17.4	17.6	17.6	17.9	19.8	17.0
Spinel	3.2	2.9	3.0	2.9	2.4	2.6	3.3	2.6
Cpx/Opx[b]	0.88	0.84	0.58	0.59	0.72	0.79	0.72	0.72

(1) Pyrolite model of McDonough and Sun (1995); (2) least depleted ultramafic xenolith model of Jagoutz *et al.* (1979); (3) LOSIMAG C1 model of Hart and Zindler (1986); (4) PRIMA model of Allegre *et al.* (1995); (5) pyrolite model of Ringwood (1979); (6) average Zabargad fertile peridotite from Bonatti *et al.* (1986); (7) primitive mantle of Palme and O'Neill, Chapter 2.01 this volume; and (8) melt extraction model, this study.

[a] Mg# calculated as molar Mg/(Mg + Fe) × 100. [b] Normative cpx/opx calculated from the spinel lherzolite norm of Kelemen *et al.* (1992).

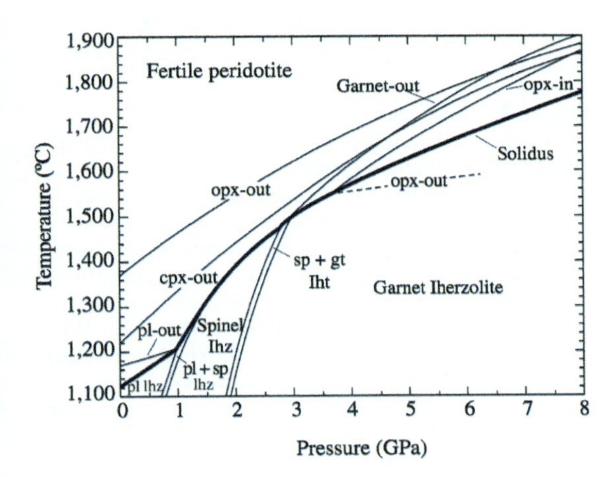

Figure 1 Generalized phase relations for fertile peridotite mantle from 1 atm to 8 GPa. The diagram has been constructed on the basis of experimental data for melting of two model fertile mantle compositions: KLB-1 (Takahashi and Kushiro, 1983; Takahashi, 1986; Canil, 1991; Takahashi *et al.*, 1993; Hirose and Kushiro, 1993; Zhang and Herzberg, 1994) and KR4003 (Walter, 1998). Subsolidus phase relations based on experimental studies of Walter and Presnall (1994), Gudfinnsson and Presnall (2000), Robinson and Wood (1998), and Walter *et al.* (2002). Curves labeled "out" mean that phases are stable at temperatures below the curve. pl = plagioclase, cpx = clinopyroxene (augitic and pigeonitic), opx = orthopyroxene, and lhz = lherzolite.

Kushiro, 1983; Takahashi, 1986; Takahashi *et al.*, 1993; Canil, 1991; Hirose and Kushiro, 1993; Zhang and Herzberg, 1994; Walter and Presnall, 1994; Robinson and Wood, 1998; Walter, 1998; Gudfinnsson and Presnall, 2000). The phase relations exhibited among various fertile peridotite compositions are generally consistent and are displayed adequately on this diagram, although in detail there are differences in the solidi and phase stability curves that relate to subtle differences in major and minor element components (e.g., Hirschmann *et al.*, 1999b; Hirschmann, 2000; Pickering-Witterer and Johnston, 2000). At pressures less than ~1 GPa, plagioclase is the stable aluminous phase at the solidus, giving way to spinel (~1–2.5 GPa) and garnet (>2.5 GPa) at higher pressures. At low pressures, plagioclase is the first phase consumed during batch partial melting, followed progressively by cpx, opx, and olivine with increase in temperature. Melting of spinel lherzolite is similar, except that spinel becomes progressively more refractory with increase in temperature due to an increase in Cr/Al, but is generally exhausted at a degree of melting similar to cpx. In general, for both plagioclase and spinel lherzolite, cpx is exhausted after ~20–25% batch melting leaving a harzburgite residue, with opx remaining stable up to 40–50% melting.

Melting phase relations at higher pressures (>3 GPa) are more complex (Figure 1). Garnet is the stable aluminous phase at 3 GPa and is the first phase consumed after only ~10% melting. However, garnet becomes progressively more stable in the residue at higher pressure, being replaced by cpx as the first phase consumed at 4–5 GPa, and replaced by opx as the second phase consumed around 7 GPa, where it is stable with olivine up to ~50% melting. As will be explained in more detail below, cpx becomes progressively enriched in enstatite component as pressure and temperature increase along the lherzolite solidus, resulting in an increase in the cpx/opx ratio. Eventually, opx is eliminated at the solidus of fertile lherzolite depending sensitively on the bulk composition (Takahashi, 1986; Canil, 1992; Walter, 1998; Weng and Presnall, 2001). However, as shown in Figure 1, opx reappears within the melting interval as a reaction product with melt over some pressure interval. Above ~10 GPa, opx is no longer stable anywhere in the melting interval, and the solidus mineral assemblage is calcium-poor cpx, majorite garnet, and olivine (e.g., Takahashi, 1986, Zhang and Herzberg, 1994).

2.08.2.2 Melting Reactions and Residual Mineral Modes

Extraction of partial melt from fertile peridotite induces systematic changes in the modal abundances and compositions of mineral solid solutions in the residue. During melting, minerals in the source rock either contribute an extract to the melt causing a reduction in modal abundance, or consume melt by crystallization causing an increase in modal abundance. Changes in the stoichiometry of mantle melting reactions with pressure and temperature are due to changes in the compositions of coexisting minerals and melt (e.g., Kinzler and Grove, 1992a; Bertka and Holloway, 1993; Walter and Presnall, 1994; Walter *et al.*, 1995; Longhi and Bertka, 1996; Kinzler, 1997; Gudfinnsson and Presnall, 2000). Figure 2 shows schematic diagrams that illustrate chemographic relationships among coexisting minerals and melt as a function of pressure along the lherzolite solidus. These diagrams are projections from the aluminous mineral onto a plane comprised of olivine and pyroxene components. At lower pressures in the spinel lherzolite field, partial melts are relatively depleted in olivine component and cpx is enriched in diopside component. The melt lies outside the area circumscribed by olivine + opx + cpx (a volume when the aluminous phase is included) so the melting reaction is not eutectic-type. Tie lines reveal a reaction where cpx + opx + spinel combine to form olivine + melt. As pressure increases, the melt becomes progressively enriched in olivine component while cpx becomes progressively enriched in enstatite component. At some pressure

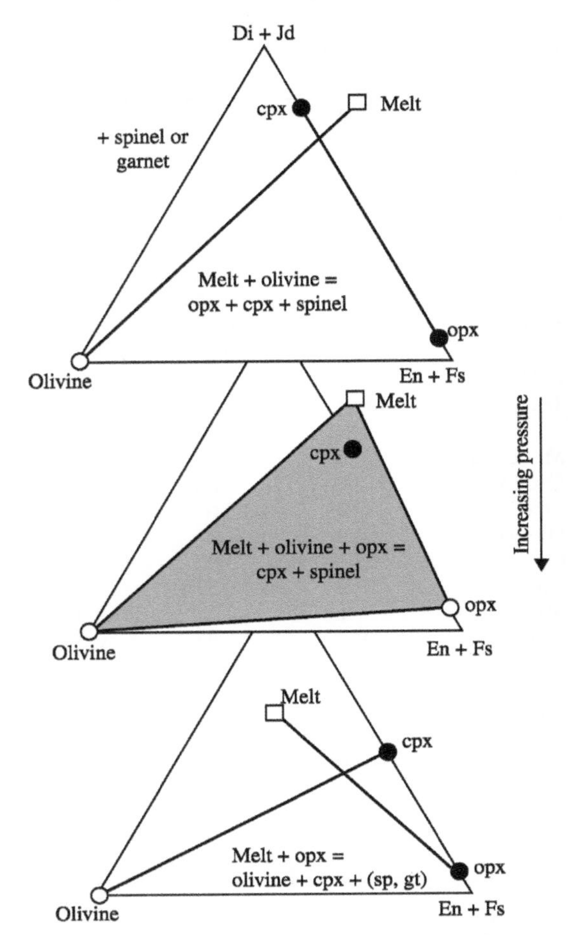

Figure 2 Schematic diagrams illustrating changes in chemographic relationships among coexisting minerals and melt as a function of pressure along the solidus of spinel and garnet lherzolite. Compositions are projected from the aluminous mineral, spinel or garnet, onto a plane comprised of olivine and pyroxene components. Di + Jd = diopside + jadeite, En + Fs = enstatite + ferrosilite. At low pressures the peritectic reaction is one in which olivine crystallizes from melt. With increase in pressure, cpx becomes enriched in enstatite component as melt becomes enriched in olivine component, resulting first in opx crystallization from melt along with olivine, and finally with only opx crystallizing from melt.

close to 1.7 GPa for fertile mantle peridotite, cpx crosses the melt + olivine tie line and the reaction changes to one in which both olivine and opx crystallize during melting (Kinzler, 1997). With further increase in pressure, melt and cpx components continue to become enriched in olivine and enstatite, respectively, eventually resulting in a reaction where only opx crystallizes during melting (see also Weng and Presnall (2001) for ternary phase relations). Figure 3 shows the progressive changes in the enstatite component in cpx and the olivine component in the melt as a function of pressure along the solidus of fertile

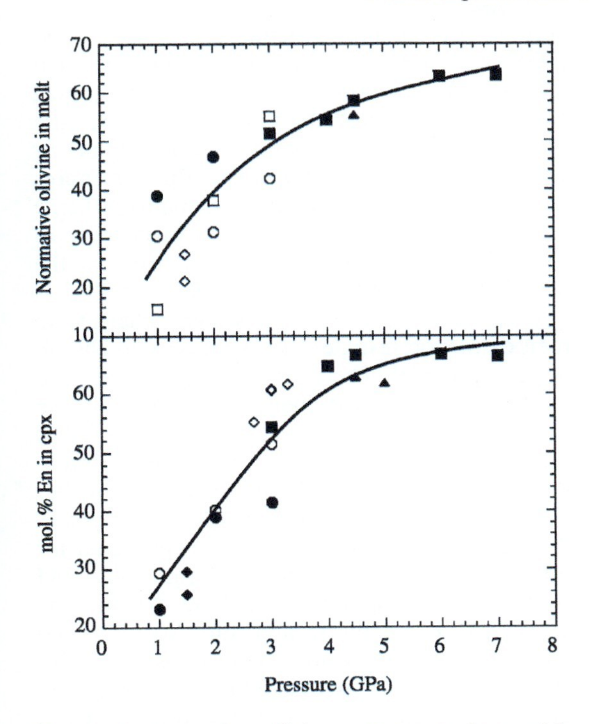

Figure 3 Normative olivine component in partial melt and mol.% enstatite component in cpx as a function of pressure along the fertile peridotite solidus, based on experimental data and pMELTS (sources: pMELTS (solid circles, Ghiorso *et al.*, 2002), CMASN (open circles, Walter and Presnall, 1994), CMASF (open squares, Gudfinnsson and Presnall, 2000), and for natural experimental compositions (filled triangles, Canil, 1992; open diamonds, Robinson and Wood, 1998; filled diamonds, Robinson *et al.*, 1998; Walter, 1998)).

peridotite as constrained by experiments and the pMELTS thermodynamic model (Ghiorso *et al.*, 2002). While the melt becomes progressively enriched in olivine component, cpx changes progressively from an augitic composition to a pigeonitic composition, and apparently reaches a maximum enstatite content of ~70 mol.%.

Melting reactions can be quantified by observing the rate of change in modal abundances of phases (e.g., Walter *et al.*, 1995; Baker *et al.*, 1995), and Table 2 gives a list of melting reactions that have been deduced for melting of spinel and garnet lherzolite using phase equilibrium data from simplified and natural model mantle systems. There is remarkable agreement in the predicted stoichiometry of the peritectic-type spinel lherzolite melting reactions at low pressures (~1 GPa), with olivine crystallizing from melt. As pressure increases, the rate at which opx is consumed decreases as does the rate at which olivine crystallizes and eventually, opx begins to crystallize from melt and olivine begins to be consumed by melt. Based on Table 2, these changes occur generally at pressures in the range of 1.7–2.0 GPa for mantle compositions, and at lower pressures in more iron-rich compositions (Bertka and Holloway, 1993). Melting of garnet lherzolite at 3 GPa is peritectic-type with opx crystallizing from melt. Experimental data at higher pressure show that if opx is absent from the solidus the melting reaction is eutectic-type, with olivine, garnet, and cpx being consumed (Walter, 1998). However, opx crystallizes over some portion of the melting interval, probably up to ~10 GPa, where it is no longer stable above the

Table 2 Comparison of mantle melting reactions.

P (GPa)	*Melting interval*	*System*	*Melting reaction* (wt.%)	*Ref.*
Spinel lherzolite				
1.1	Invariant	CMAS	36 opx + 55 cpx + 9 sp = 77 liq + 23 ol	1
1.1	12–22%	CMASN	34 opx + 56 cpx + 10 sp = 75 liq + 25 ol	2
1.15	12–22%	CMASF	26 opx + 63 cpx + 11 sp = 75 liq + 25 ol	3
1.0	~7–18%	Natural	31 opx + 58 cpx + 11 sp = 82 liq + 18 ol	4
1.0	Mg#75–67	Natural	35 opx + 59 cpx + 5 sp = 78 liq + 22 ol	5
1.0	0–20%	Natural	31 opx + 60 cpx + 9 sp = 85 liq + 15 ol	6
1.7	$\Delta = 29\%$	Natural	7 opx + 84 cpx + 9 sp = 93 liq + 7 ol	5
1.7	$\Delta = 17\%$	Natural	91 cpx + 9 sp = 68 liq + 8 ol + 24 opx	5
2.0	0–20%	CMASN	5 opx + 83 cpx + 12 sp = 84 liq + 16 ol	2
2.0	1–20%	CMASF	90 cpx + 10 sp = 86 liq + 13 ol + 1 opx	3
2.0	0–20%	Natural	25 ol + 69 cpx + 6 sp = 81 liq + 19 opx	6
Garnet lherzolite				
3.0	0–14%	Natural	4 ol + 84 cpx + 12 gt = 87 liq + 13 opx	7
5.1	Invariant	CMAS	7 ol + 75 cpx + 18 gt = 60 liq + 40 opx	8
7.0	0–35%	Natural	27 ol + 48 cpx + 25 gt = 100 liq	7

(1) Gudfinnsson and Presnall (1996); (2) Walter and Presnall (1994); (3) Gudfinnsson and Presnall (2000); (4) Baker and Stolper (1994); (5) Kinzler and Grove (1992a); (6) pMELTS (Ghiorso *et al.*, 2002; Asimow, private communication); (7) Walter (1998); (8) Presnall *et al.* (2002).

fertile peridotite solidus (Takahashi, 1986; Zhang and Herzberg, 1994; Walter, 1998).

The net effect of changes in mineral chemistry and melting reactions are systematic changes in the modal abundances of minerals in peridotite residues as a function of pressure. For example, phase equilibrium data from simplified mantle systems as well as pMELTS show that cpx/opx and olivine/opx ratios increase markedly along the solidus, which is a consequence of enstatite component dissolving into cpx (Walter *et al.*, 1995; Walter, 1999a). As melting proceeds above the solidus, the cpx/opx ratio in the residue decreases as cpx is consumed at a high rate and opx is consumed at a lower rate or crystallizes from melt at higher pressure. Melt extraction at low pressures leads to higher olivine/opx ratios in residues as olivine crystallizes during melting and opx is consumed. The opposite is true at higher pressure as olivine is consumed during melting and opx crystallizes. Thus, lower-pressure melting residues are relatively enriched in olivine component but depleted in pyroxene components, whereas higher-pressure residues become progressively more enriched in opx component. Due to the increased stability of garnet with increase in pressure, higher-pressure residues are also relatively enriched in aluminous components. These melt extraction characteristics are very useful for diagnosing the depth of melt extraction recorded in mantle residues.

Lastly, it is important to note that experimentally predicted changes in mineral modes that occur above the solidus are not directly comparable to melting residues from the mantle. This is because samples from the mantle invariably have re-equilibrated at conditions different from those during melt extraction, usually at considerably lower temperatures, resulting in changes in mineral composition or mineral exsolution. The important point is that re-equilibration leads to changes in modal mineralogy unrelated to melt extraction (Walter, 1999a). This problem can be circumvented by normalizing experimental and natural residues to a common set of mineral components or by making direct comparison of major-element oxide concentrations between mantle and experimental residues, and both methods are employed here.

2.08.2.3 Compositional Trends in Melt Extraction Residues

Recasting the major-element compositions of experimental residues and mantle samples into a common set of mineral components effectively eliminates differences in temperatures and pressures of final equilibration, as the normalization essentially "re-equilibrates" all compositions to a common temperature and pressure. Here, we use the "high $P-T$" spinel and garnet lherzolite normative calculation schemes of Keleman *et al.* (1992) to compare mantle rocks to experimental residues. These norms are designed to reflect equilibration of mantle peridotite at high-pressure and high-temperature conditions where spinel or garnet lherzolite would crystallize. But it is important to emphasize that no fixed normative scheme can recapture the complex history of changes in mineral chemistry and modes that occur at the actual melting conditions (e.g., Walter, 1999a).

In order to predict compositional changes in melt extraction residues, it is necessary to choose a melting model. There are numerous alternative models available for melting of peridotite in the shallow upper mantle (i.e., <3 GPa), including models based on phase equilibrium data in simplified mantle systems (e.g., Walter and Presnall, 1994; Gudfinnsson and Presnall, 2000), parametrizations of experimental data for melting natural compositions (e.g., Niu and Batiza, 1991; Langmuir *et al.*, 1992; Kinzler and Grove, 1992a, 1993), and the MELTS and pMELTS thermodynamic algorithms (Ghiorso, 1985; Hirschmann *et al.*, 1998; Ghiorso *et al.*, 2002). Overall there is a remarkably good correspondence among these models (for some comparisons see Niu *et al.* (1997)), as exemplified by the similarity in melting reactions listed in Table 2. The biggest difference among the models is that MELTS and pMELTS tend to predict partial melts that are more enriched in olivine component and depleted in opx component than other models, especially at higher pressures (i.e., >2 GPa). Here, the model of Kinzler and Grove (1992a, 1993) is chosen as representative of experimentally based models due to its inherent flexibility in dealing with changes in bulk composition. At 3 GPa and above, there are much less experimental data available, and the data of Walter (1998, 1999b) for melting of fertile peridotite will be used to model high-pressure melt extraction.

Figure 4 shows whole-rock Mg# (molar $[Mg/(Mg + Fe)] \times 100$) versus spinel lherzolite normative mineral abundances for up to 25% batch melt extraction from fertile peridotite at 1 GPa and 2 GPa. In general, whole-rock Mg# serves as a good proxy for degree of melt depletion, although the rate of change in Mg# with degree of melt extraction is also pressure dependent. For a given amount of melt extraction, higher-pressure residues have higher Mg#s, a feature that when coupled with changes in mineral abundances is very useful for deducing the depth of melt extraction. Normative olivine increases and normative cpx decreases (as does spinel, not shown) nearly linearly with degree of melt extraction, although there is some curvature at higher Mg#s. Normative opx decreases at a lesser rate than cpx, and at 2 GPa is nearly constant up to

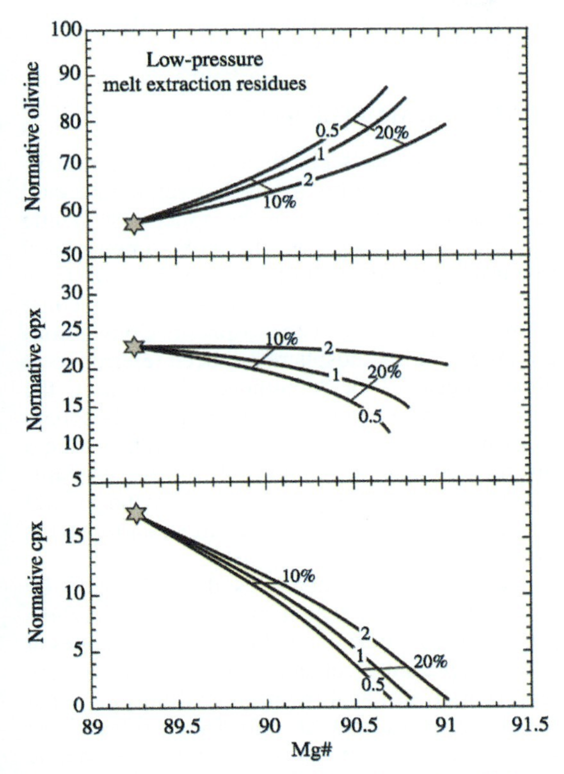

Figure 4 Normative spinel lherzolite mineral abundances (wt.%) in batch partial melt extraction residues (0–25%) from fertile peridotite (composition 8, Table 1) as a function of Mg# (molar Mg/(Fe + Mg)) at 0.5 GPa, 1 GPa, and 2 GPa, based on the melting model of Kinzler and Grove (1992a, 1993). Normative mineral compositions are calculated using the spinel lherzolite normative algorithm of Kelemen *et al.*, (1992).

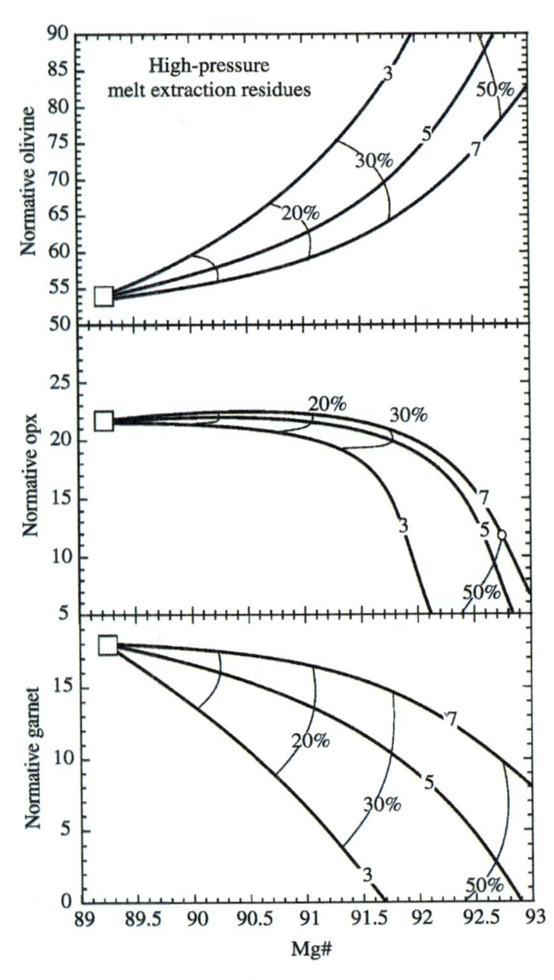

Figure 5 Normative garnet lherzolite mineral abundances (wt.%) in batch partial melt extraction residues (0% to >50%) from fertile peridotite as a function of Mg# from 3 GPa to 7 GPa (based on the experimental data of Walter (1998) as parametrized by Walter (1999b) for melting of fertile peridotite KR4003; normative mineral compositions are calculated using the garnet lherzolite normative algorithm of Kelemen (1992)).

~20% melt extraction due to opx crystallization in the melting reaction.

Figure 5 shows whole-rock Mg# versus garnet lherzolite normative mineral abundances for batch melt extraction from fertile peridotite at 3–7 GPa based on the experimental data of Walter (1998). As pressure increases, the amount of normative olivine decreases for a given degree of melt extraction, but as with lower pressure melt extraction, at constant olivine content the Mg# is greater. The amount of normative opx increases slightly or remains nearly constant in residues up to ~30% melt extraction. Notice that even though opx crystallizes in a peritectic-type melting reaction during melting of garnet lherzolite, the opx-component in the residue does not increase markedly. This is because a considerable amount of the source rock enstatite-component is held in cpx as discussed above, and cpx is rapidly consumed at all pressures during melting (not shown in Figure 5). Normative garnet is a good melting barometer as lower-pressure residues become depleted rapidly during melt extraction, whereas the rate of garnet depletion is suppressed at higher pressures (see Figure 1).

A complimentary perspective of compositional variation in melting residues can be made from major-element oxide variation diagrams. Figure 6 shows the variation in major-element oxides versus FeO content as a function of pressure (1–7 GPa) and degree of batch melt extraction. FeO was chosen as the abscissa because of its large relative variation in partial melts and residues as a function of pressure. At 1 GPa the FeO content in the residue shows a mild increase with degree of melt extraction, but by 2 GPa FeO shows a mild decrease. This reflects the fact that low-pressure partial melts have FeO contents less than that of the bulk rock, but as pressure increases so does the FeO

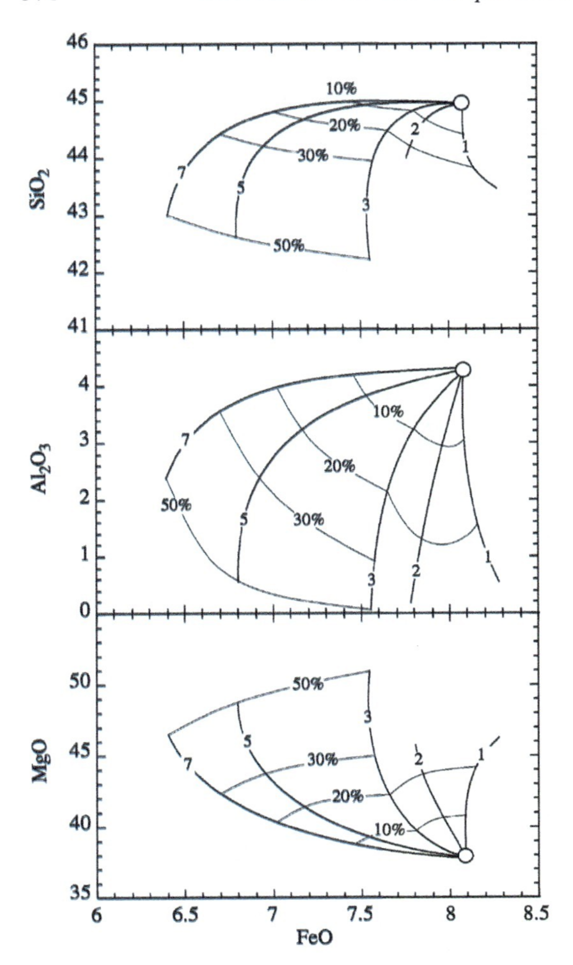

Figure 6 Major-element oxides (wt.%) versus FeO as a function of pressure (GPa) and degree of batch melt extraction (sources: the 1 GPa and 2 GPa trends are based on the Kinzler and Grove (1992a, 1993) model for melting of primitive mantle of McDonough and Sun (1995) (composition 1, Table 1), and the trends at higher pressures are based on the data of Walter (1998) for melting of fertile peridotite KR4003).

content of the melt at a given degree of melting. The exact pressure and melt fraction at which the FeO contents of partial melts begin to exceed the bulk rock depends on the bulk composition. SiO_2 and Al_2O_3 exhibit similar pressure-dependent behavior in that both become depleted in residues at a higher rate at low pressures than at higher pressures. This reflects the increased stability of opx and garnet, respectively, in high-pressure melting residues. MgO increases at a nearly constant rate during melt extraction at all pressures, and because FeO decreases in residues more rapidly at high pressures, for a given degree of melting, high-pressure residues have higher Mg#s than low-pressure residues. The dependence of these major elements on both pressure and degree of melting make oxide and normative mineral variation diagrams very powerful tools for deducing the conditions of melt extraction in samples of depleted mantle.

2.08.3 THE MANTLE SAMPLE

Mineral and bulk chemical analyses are available in the literature for well over a thousand samples of peridotite from the uppermost mantle (<~250 km). Based on thermobarometric and textural evidence, most of these samples are from rigid lithospheric mantle, but some also come from the deforming asthenosphere. Conventionally, the mantle lithosphere has been referred to as either "oceanic" or "continental," reflecting the nature of the crustal portion of the lithosphere from which samples were collected rather than a discrete genetic separation of mantle domains (e.g., Menzies and Dupuy, 1991). Here the general classification of oceanic and continental mantle is maintained.

Modern oceanic mantle is defined solely by abyssal peridotites, which are samples of harzburgite and lherzolite collected from fracture zones at oceanic spreading centers, and these samples are representative of shallow oceanic lithosphere that has been processed at mid-ocean ridges (see Chapter 2.04). Two types of continental mantle lithosphere are considered: (i) cratonic mantle, which refers to xenoliths collected from kimberlites that sample portions of mantle beneath stable, Archean cratons and (ii) off-craton mantle, which refers to xenoliths collected from alkalic basalts that have sampled portions of the subcontinental mantle adjacent to ancient cratonic mantle (see Chapter 2.05). Also included with off-craton lithosphere are orogenic lherzolites and ophiolites, which are slices of mantle tectonically emplaced typically at convergent margins.

In this section the normative modal and major-element compositions of a large number of samples collected from both oceanic and continental mantle lithosphere are presented. Compositional trends exhibited by oceanic and off-craton mantle are used to develop a model fertile upper-mantle composition. The reader is referred to Chapters 2.04 and 2.05 in this volume, and to reviews by McDonough (1990), McDonough and Rudnick (1998), and Griffin *et al.* (1999b) for alternative and detailed perspectives of the mantle sample.

2.08.3.1 Mantle Heterogeneity

Before presenting a detailed account of the compositional variability of mantle lithosphere and the role that melt extraction has played in producing that variation, it is important to acknowledge the likelihood of intrinsic heterogeneity in the mantle that is not a direct consequence of primary melt extraction, and to assess whether such heterogeneity is likely to be preserved as a distinguishable mineralogic or major-element signal in samples of mantle lithosphere.

Gross chemical heterogeneity could conceivably have been imparted to the primordial mantle shortly after accretion by large-scale mineral layering due to crystal/melt fractionation in a deep magma ocean (e.g., Agee and Walker, 1988). There are, however, several converging lines of evidence that are unsupportive of large-scale layering in the modern mantle. First, many workers have noted that refractory lithophile elements are apparently present in near-chondritic relative proportions in primitive upper-mantle rocks (e.g., Hart and Zindler, 1986; McDonough and Sun, 1995; Allegre *et al.*, 1995; O'Neill and Palme, 1998; Chapter 2.01). Geochemical models based on experimental mineral/melt partitioning data predict that large-scale layering involving peridotite liquidus phases would cause gross fractionations in refractory lithophile element ratios that should be readily apparent both geochemically and isotopically (e.g., Kato *et al.*, 1988; McFarlane *et al.*, 1994; Corgne and Wood, 2002). Second, seismic tomographic observations indicate mass exchange between the upper and lower mantle (i.e., across the 670 km seismic discontinuity), implying flow of mantle material between the upper and lower mantle (e.g., Kennett and van der Hilst, 1998). Numerical models indicate that solid-state convective mixing is expected to be a relatively efficient process over the lifetime of the Earth (i.e., ~4.5 Gyr), such that gross layering could not remain isolated (e.g., van Keken and Zhong, 1999; see Chapter 2.12). Third, seismic discontinuities at 410 km and 660 km are best explained by mineral phase transformations (e.g., in $(Mg,Fe)Si_2O_4$) rather than as chemical boundaries (e.g., Helffrich and Wood, 2001). Finally, the overall physical properties of the mantle are consistent with a generally uniform and broadly "pyrolitic" composition throughout (see reviews by Bina (1998), Jackson and Rigden (1998), and Chapter 2.02). For these reasons it is commonly held that at least for major elements (e.g., SiO_2, MgO, FeO) the mantle is broadly homogeneous (e.g., McDonough and Sun, 1995; Allegre *et al.*, 1995; see Chapter 2.01).

Kellogg *et al.* (1999), however, have suggested, on the basis of a transition in seismic heterogeneity observed at ~1,600 km depth, the possibility of a very deep layer extending hundreds of kilometers above the core–mantle boundary. One possibility is that a relic layer of dense, primordial crystalline differentiates (e.g., magnesium- and calcium-silicate perovskite) may have remained buried in the deep lower mantle until the present. Such a layer is a potential storehouse for trace elements, including radioactive heat-producing elements, and potentially could provide an important reservoir for bulk silicate Earth chemical mass balance as well as a source for missing heat required to satisfy surface heat flux (e.g., Kellogg *et al.*, 1999; Albarede and van der Hilst, 1999). The geochemical, geophysical, and geodynamic effects of such a layer are current subjects of investigation.

Even though the mantle may be considered broadly "homogeneous," the continuous process of extraction of crust from the mantle followed by subduction of oceanic crust and depleted lithosphere back into the mantle for as much as 4 Gyr has undoubtedly resulted in preserved mantle heterogeneity (see Helffrich and Wood, 2001). Nonperidotite lithologies such as eclogite and pyroxenite comprise several percent of the population of uppermost mantle samples (e.g., Schulze, 1989), and pyroxenite veins can be pervasive features in slices of mantle peridotite that are exposed at the surface (e.g., Menzies and Dupuy, 1991; see Chapter 2.04). The modern upper mantle is not homogeneous in terms of trace element or isotopic composition, and this diversity can be directly related to crustal recycling (e.g., Zindler and Hart, 1986; Carlson, 1994; Hofmann, 1997; Albarede, 2001; Hauri, 2002; see Chapter 2.03). Domains of eclogite or pyroxenite are inferred to exist in the source regions of magmas generated in plumes and mid-oceanic ridges (e.g., Hauri, 1996, Hirschmann and Stolper, 1996; Takahashi *et al.*, 1998; Lundstrom *et al.*, 1999; Salters and Dick, 2002).

The extent and nature of mantle chemical heterogeneity due to material recycling is a matter of much speculation and depends on the scale of interest (e.g., Hart and Zindler, 1989). There is unequivocal isotopic evidence from erupted magmas from oceanic island or "hot-spot" locations that subducted crust and lithosphere are stored in the mantle for long periods (see reviews by Carlson (1994), Hofmann (1997), and van Keken *et al.* (2002)). Subducted lithologies may be preserved in the mantle over a wide range of scales, from large, kilometer-sized chunks or blobs to highly dispersed fragments of ancient material. Scattered seismic waves indicate that bodies less than 10 km in size may be pervasive features in the mantle (e.g., Hedlin *et al.*, 1997). Meter- to kilometer-sized eclogite domains are postulated to exist in the source regions of plume-related magmas (Hauri, 1996; Takahashi *et al.*, 1998; Takahashi and Nakajima, 2002). Ancient subducted materials may be highly dispersed or become completely homogenized into mantle peridotite by diffusive equilibration, which occurs at a scale of a few centimeter over Gyr timescales. Highly dispersed small-scale fragments (centimeter to meter) of oceanic crust may be efficiently recycled and consumed at oceanic spreading centers (Helffrich and Wood, 2001).

In order for small-scale mantle heterogeneities to be detected geochemically or isotopically in erupted lavas requires that partial melts are sampled at similar scales in the source region and remain isolated. This is probably not the general case for melt extraction from the upper mantle, which typically involves segregation and pooling of melts from a melt zone that is large relative to the scale of heterogeneity. However, direct evidence for small-scale heterogeneity in the upper mantle is provided by extreme isotopic heterogeneities in melt inclusions preserved in crystals in "hot-spot" and MORB lavas (e.g., Saal *et al.*, 1998; Shimizu and Sobolev, 1998).

In recent years the effects of melting nonperidotite lithologies in mantle peridotite have begun to be systematically investigated and a rich detail of phase relations is emerging from these studies (Kogiso *et al.*, 1998; Yaxley and Green, 1998; Kogiso and Hirschmann, 2001; Bulatov *et al.*, 2002; Pertermann and Hirschmann, 2003; Hirschmann *et al.*, 2003). A thermal divide apparently exists that is straddled by natural eclogite and garnet pyroxenite compositions, with more silica-rich eclogite with compositions like average MORB generating silica-normative melts, whereas silica-poor pyroxenites produce nephaline-normative melts (e.g., Yaxley and Green, 1998; Hirschmann *et al.*, 2003).

Discrete eclogite domains in peridotite melt at considerably lower temperatures than pyrolitic peridotite at uppermost mantle conditions, and the siliceous partial melts react and refertilize the surrounding peridotite, most notably by increasing the opx/olivine ratio (e.g., Yaxley and Green, 1998; Pertermann and Hirschmann, 2003). At temperatures above the normal peridotite solidus the eclogitic component would be entirely molten, and the phase relations are dictated by the enriched peridotite bulk composition (Yaxley and Green, 1998; Bulatov *et al.*, 2002). In contrast, some natural garnet pyroxenite compositions have solidi that are close to mantle peridotite and partial melts are silica-poor (Hirschmann *et al.*, 2003). If eclogite or pyroxenite are intimately mixed with normal pyrolitic peridotite, the hybrid or "refertilized" composition generally melts at a moderately lower temperature and has a higher melt productivity, and melt products are generally enriched in FeO, TiO_2, and alkalis relative to melts from "normal" mantle (e.g., Kogiso *et al.*, 1998). Compositional features of partial melts from nonperidotite lithologies are inferred to exist in OIB lavas, although it is unclear how discrete melts of eclogite or pyroxenite in a peridotite matrix can reach the surface without being consumed by reaction with peridotite (e.g., Hauri, 1996; Kogiso *et al.*, 1998; Lundstrom *et al.*, 2000; Hirschmann *et al.*, 2003).

In the following presentation, samples of mantle lithosphere are considered from the perspective of mineralogic and major-element systematics. It will be shown that the general trends in data sets compiled for oceanic and off-craton subcontinental lithosphere are consistent with melt extraction from a generally common, fertile peridotite protolith. The melting regimes that produced the melt extraction trends involved, on average, relatively extensive melting of peridotite. In such regimes, small-scale, nonperidotite lithologies would generally be melted completely, and in terms of major-element chemistry or mineralogy all traces of their existence in residual peridotite would be wiped out by extraction of the higher-temperature peridotite melts, although an isotopic signal could linger (Salters and Dick, 2002). If such heterogeneity did exist in fertile mantle protoliths of depleted oceanic and subcontinental mantle, it likely contributes an irresolvable component of noise to the observed melt extraction trends. It is conceivable, however, that in low-degree melting regimes the relative contribution from pyroxenite may be proportionally larger and refractory pyroxenitic residues may remain in the source.

2.08.3.2 Oceanic Mantle

Peridotite samples collected in dredge hauls and as drill cores from oceanic rift valleys and fracture zones are pieces of the uppermost oceanic lithosphere and typically consist of cpx-poor lherzolite and harzburgite (e.g., Dick *et al.*, 1984; Dick, 1989; Bonatti and Michael, 1989; see Chapter 2.04). Due to their residence in an oceanic and hydrothermal environment, the samples are highly serpentinized (~75% on average) so bulk chemical analyses are not reliable for obtaining primary compositions (e.g., Dick *et al.*, 1984). However, successful reconstructions of major-element compositions have been achieved using carefully obtained mineral modal abundances and analyses of one or more relict minerals (Dick, 1989; Niu *et al.*, 1997; Baker and Beckett, 1998; Asimow, 1999). On the basis of both mineral chemistry and reconstructed bulk compositions, it is well established that abyssal peridotites are the complimentary residues of MORB melt extraction (e.g., Dick and Bullen, 1984; Dick, 1989; Bonatti and Michael, 1989; Johnson *et al.*, 1990). Further, detailed examinations of trace- and minor-element compositions have led to insights regarding the melt extraction process itself. For example, the highly depleted LREE abundance patterns in cpx from abyssal peridotites have been used as primary evidence that the melting process involved incremental extraction of small-degree melts, or "near-fractional" melting (Johnson *et al.*, 1990), and incompatible minor element trends have led to inferences regarding possible olivine accumulation (Niu *et al.*, 1997) or mixed-mode melt extraction processes (Asimow, 1999).

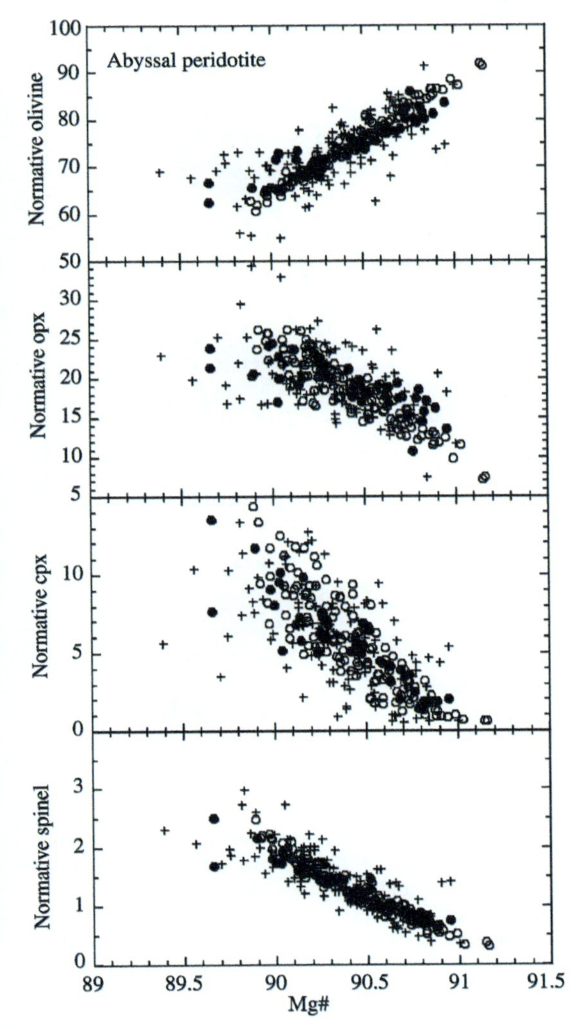

Figure 7 Normative spinel lherzolite mineral abundances (wt.%) versus rock Mg# for a set of reconstructed abyssal peridotite compositions. Reconstructions were made using mineral modes and phase composition data (see text Niu *et al.* (1997), Baker and Beckett (1998), and Asimow (1999) for details). Circles are from Baker and Beckett (1998), and filled circles are reconstructions based on site-averages, whereas open circles are reconstructions based on single thin section modes. Crosses are from Asimow (1999), and are reconstructions based on single thin section modes after the method of Niu *et al.* (1997).

Shown on Figure 7 are normative mineral modes versus whole-rock Mg# for two sets of reconstructed abyssal peridotite compositions, both of which are based on mineral modes and mineral compositions reported in the literature. All major ocean basins are represented in these reconstructions, but data are highly skewed toward the heavily sampled, slow-spreading Southwest Indian Ridge. Crosses are bulk compositions reported in Asimow (1999), and reconstructions were made using single thin section modes together with analyzed mineral composition data following the method of Niu *et al.* (1997). The data set includes 37 fully constrained compositions, meaning compositional data for all four primary lherzolite minerals were available, and 85 compositions where from one to three mineral compositions were available. When a given mineral composition was not available, its composition was estimated on the basis of major-element correlations between cpx and other primary minerals among other samples (see also Niu *et al.* (1997)). Filled circles are bulk compositions reported in Baker and Beckett (1998), and reconstructions were made using site-averaged mineral modes rather than single thin section modes, together with published mineral compositions. The data set includes 11 fully constrained compositions and 32 compositions where from one to three mineral compositions were available. In contrast to the method of Niu *et al.* (1997) and Asimow (1999), Baker and Beckett (1998) rely on correlations between mineral compositions and olivine modal abundance to estimate mineral compositions. Open circles are 131 reconstructions using the method of Baker and Beckett (1998), but modes are based on single thin sections rather than site averages.

All normative mineral abundances are well correlated with Mg# on Figure 7. As Mg# increases, normative olivine increases, and normative opx, cpx, and spinel decrease, trends that are consistent with melt extraction at shallow levels from a fertile mantle protolith (Figure 4). The reconstructed compositions of Baker and Beckett (1998) show a higher correlation between normative mineralogy and Mg# than those of Asimow (1999), but an important feature is that the slopes of regression lines are nearly identical in all reconstructions. Trends for both site-averaged and single thin section modes using the Baker and Beckett (1998) method are highly correlated and, interestingly, correlations actually increase for all normative minerals except opx when single thin section modes are used rather than site-averages. Thus, the correlations produced by the Baker and Beckett (1998) method apparently reflect a high degree of correlation between olivine modal abundance and the Mg# of the primary minerals.

Any method for reconstructing abyssal peridotite bulk compositions involves assumptions and uncertainties. Rather than choosing a single reconstruction method as superior, we consider that in total these reconstructions are highly representative of depleted oceanic mantle beneath the axis of mid-ocean ridges.

2.08.3.3 Off-craton Subcontinental Mantle

Circum-cratonic mantle beneath continental crust is sampled as xenoliths in alkalic basalt magmas and as orogenic lherzolite massifs (see also Chapters 2.04 and 2.05). Xenoliths typically range from a few to tens of centimeters in diameter and exhibit a wide range of mineralogy and chemistry, but are dominated by spinel lherzolite with rare garnet and plagioclase lherzolite. Samples exhibit a range in alteration from serpentinized to extremely fresh, but on average are much more pristine than abyssal peridotites. Although the limited size and presumably random nature of sampling precludes detailed spatial reconstructions of the mantle beneath a given collection site, xenoliths often are sampled over a range of pressures and temperatures, and because they are transported to the surface with little or no chance for re-equilibration, they have proved to be very useful for estimating lithospheric geotherms and thickness (e.g., Boyd *et al.*, 1985; Rudnick *et al.*, 1998; MacKenzie and Canil, 1999), and in some cases provide evidence for a vertical stratigraphy (e.g., Griffin *et al.*, 1999a; Lee and Rudnick, 1999; Carbno and Canil, 2002). Final equilibration temperatures are variable, but generally fall in the range of 600–1,200 °C. In contrast, orogenic lherzolite bodies are typically exposed over km to tens of km surface area and provide a window into mantle processes at this scale. Like xenoliths, these rocks range from highly serpentinized to remarkably fresh (e.g., Menzies and Dupuy, 1991). Off-craton mantle samples typically yield Late-Proterozoic to Early-Phanerozoic isotopic model ages, but examples of Early-Proterozoic off-craton mantle have also been reported. The antiquity and tectonic location of continental mantle has enhanced its chances of subjection to one or more stages of metasomatic enrichment by fluids or melts subsequent to melt depletion, which can obscure an original melt depletion signature (e.g., Fabriès *et al.*, 1989; Downes, 2001).

Global data sets including large numbers of xenolith and orogenic lherzolite samples have been assembled in previous studies in order to place constraints on the compositional variability and average composition of subcontinental mantle (e.g., McDonough, 1990; Griffin *et al.*, 1999b). Here, we are interested specifically in the role that melt extraction has played in producing compositional variability. However, the scatter found in large data sets makes deducing the melt extraction component of variability an ambiguous task as several competing processes have operated in the upper mantle, the affects of which are more prevalent in some sampling localities than others. For example, Figure 8 shows a global data set of 802 individual bulk analyses of off-craton peridotite xenolith (575) and orogenic lherzolite (227) samples in plots of normative olivine, opx, and cpx versus Mg#. The scatter in the data is much more pronounced than for reconstructed abyssal

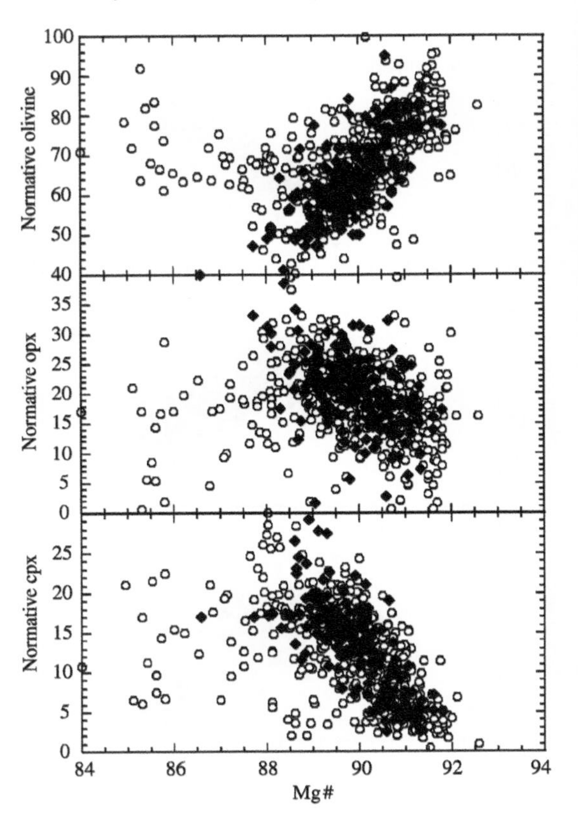

Figure 8 Normative olivine and opx (wt.%) versus rock Mg# for a set of 770 group 1 off-craton xenoliths (open circles) and orogenic lherzolites (filled diamonds). Data sources for xenoliths are: Beard and Glazner (1995), Beccaluva *et al.* (2001a,b), Brandon and Draper (1996), Cao and Zhu (1987), Chen *et al.* (2001), Dautria and Girod (1987), Downes *et al.* (1992), Downes and Dupuy (1987), Ehrenberg (1982), Francis (1987), Frey and Prinz (1978), Girod *et al.* (1987), Heinrich and Besch (1992), Hunter and Upton (1987), Ionov *et al.* (1993, 1996), Jakes and Vokurka (1987), Liang and Elthon (1990), Luhr and Aranda-Gomez (1997), Morten (1987), Nimz *et al.* (1995), O'Reilly and Griffin (1987, 1988), Press *et al.* (1986), Qi *et al.* (1995), Rampone *et al.* (1995), Reid and Woods (1978), Rivalenti *et al.* (2000), Roden *et al.* (1988), Shi *et al.* (1998), Siena *et al.* (1991), Song and Frey (1987), Stern *et al.* (1999), Vaselli *et al.* (1995), Wiechert *et al.* (1997), Xu *et al.* (1988), Xue *et al.* (1990), and Zangana *et al.* (1999). Data sources for orogenic lherzolites are: Bodinier *et al.* (1988), Bonatti *et al.* (1986), Fabries *et al.* (1989), Frey *et al.* (1985, 1991), Godard *et al.* (2000), Gruau *et al.* (1991), Hartmann and Wedepohl (1993), Lenoir *et al.* (2001), Lugovic *et al.* (1991), Rampone and Morten (2001), Rivalenti *et al.* (1995), and Shervais and Mukasa (1991).

peridotite compositions, although similar trends are apparent in the data with olivine increasing and pyroxenes decreasing with increase in Mg#, generally consistent with melt extraction. Although the bulk of the samples have Mg#s greater than 88, numerous samples have lower Mg#s and among these the scatter in the data is pronounced, indicating that such samples may generally be products of random, secondary enrichment (e.g., Fabriès *et al.*, 1989). There is essentially no difference in the fields occupied by the bulk of the xenolith and the orogenic lherzolite compositions confirming the close affinity between these two types of off-craton mantle samples, except that there are fewer orogenic lherzolites with Mg#s less than 89.

In order to focus more closely on the melt extraction component in off-craton mantle, a subset of data has been selected. A single criterion was found to be very efficient for culling the data. That is, data from any individual locality with five or more samples that show a linear correlation between Mg# and normative olivine abundance with a correlation coefficient, R^2, of greater than 0.60 was included. The assumption here is that the strong, generally linear correlation between these two factors produced during melt extraction (see Figure 4) has remained nearly intact in some data sets, whereas random processes, such as enrichment by melts or fluids, have smeared out melt extraction trends in other data sets. This simple criterion produced a data subset with 292 compositions representing nine xenolith localities from four continents (174 samples) and six orogenic lherzolite localities from two continents (118 samples). It is noteworthy that all but one sample with Mg# less than 88 were eliminated in the procedure.

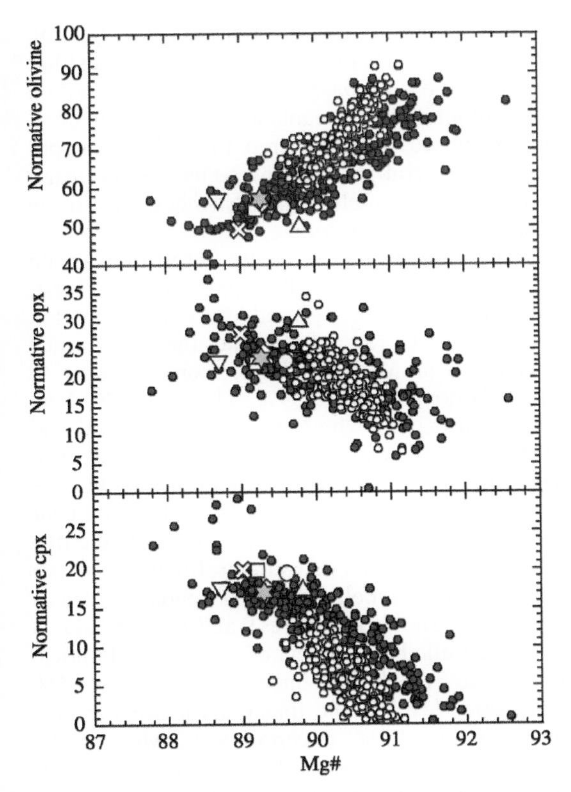

Figure 9 Normative spinel lherzolite mineral abundances (wt.%) versus rock Mg# for a subset of 292 off-craton mantle compositions (shaded circles). Data from a given locality were included if compositions had a correlation between Mg# and normative olivine with a correlation coefficient (R^2) of 0.6 or better. Data sources for xenoliths are: Beccaluva *et al.* (2001a,b), Reid and Woods (1978), Rivalenti *et al.* (2000), Stern *et al.* (1999), Vaselli *et al.* (1995), Wiechert *et al.* (1997), Xu *et al.* (1988), and Zangana *et al.* (1999). Data sources for orogenic lherzolites are: Bodinier *et al.* (1988), Frey *et al.* (1985, 1991), Hartmann and Wedepohl (1993), and Lugovic *et al.* (1991). Open circles are the reconstructed abyssal peridotite compositions from Figure 7. Also shown are estimates of primitive mantle from Table 2: white square = 1, circle = 2, triangle = 3, diamond = 5, inverted triangle = 6, ex = 7, and shaded star = 8.

Figure 9 shows normative minerals versus Mg# for the subset of off-craton mantle together with reconstructed abyssal peridotite compositions and estimates of primitive mantle from Table 1. The scatter in the data is considerably reduced in the off-craton subset relative to the entire off-craton data set (Figure 8), and correlations exist between all mineral components and rock Mg#. There is overlap between the off-craton and abyssal data sets for all mineral components, but some distinguishing features are apparent. First, abyssal peridotites are restricted to a narrower range in Mg# than off-craton mantle and probably do not contain any examples of fertile upper mantle. Second, the slopes of trends for all mineral components are shallower in off-craton mantle, being especially apparent for olivine and cpx, features that are consistent with a somewhat higher average pressure of melt extraction for off-craton mantle (see Section 2.08.4.2.1).

2.08.3.4 Fertile Upper-mantle Composition

Figure 10 shows major-element oxides versus Mg# for off-craton and oceanic mantle, as well as some estimated compositions for primitive mantle (Table 1). As expected from the normative plots, the two sets of mantle compositions have distinct trends for all oxides. Previous models for primitive upper mantle have a range in Mg# from ~89 to 90, and Figures 9 and 10 show that the oceanic and off-craton trends also converge within this range. Assuming that the off-craton and abyssal mantle trends are due primarily to melt extraction from a common protolith, then the intersection of the trends should provide a good estimate for the composition of fertile upper mantle for major elements.

Based on the incompatible trace element depleted isotopic character of MORB (e.g., Hofman, 1988), the MORB source has a long-term

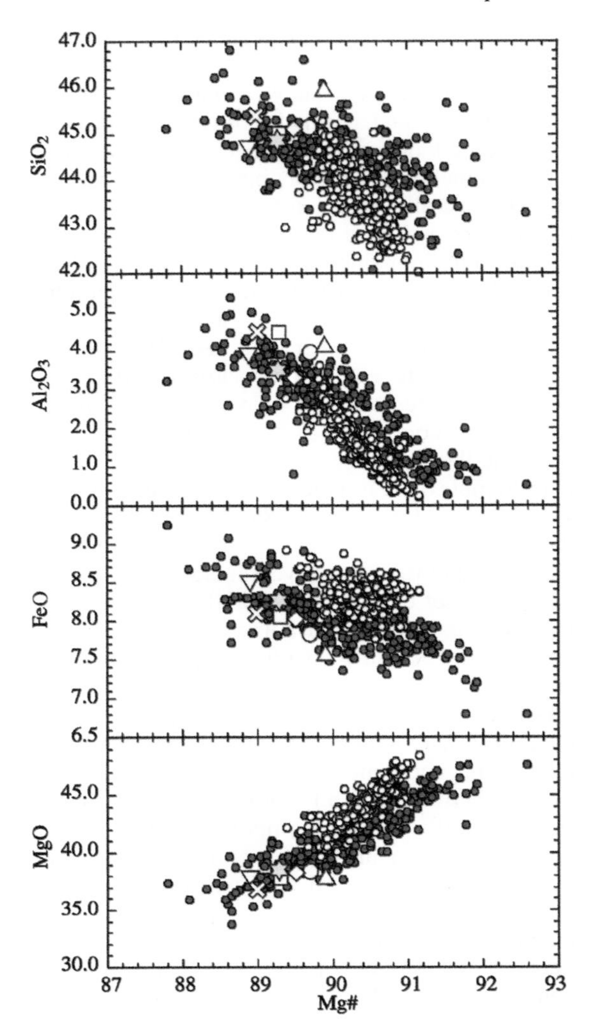

Figure 10 Major-element oxides versus Mg# for the off-craton mantle subset (shaded circles) and reconstructed abyssal peridotite compositions (open circles). Primitive mantle compositions from Table 2 are also shown with symbols as in Figure 9.

depletion in magmaphile components (see Chapter 2.03). Samples of off-craton mantle have isotopic compositions with depletion characteristics similar to modern depleted MORB mantle, indicating a common protolith (e.g., Hartmann and Wedepohl, 1993; Wiechert *et al.*, 1997; Rivalenti *et al.*, 2000). Thus, both oceanic and off-craton mantle show incompatible element depletions related to early crust extraction. The continental crust, which is permanently separated from the mantle, comprises only ~0.6% of bulk silicate Earth, so the effect of its extraction on major elements is negligible. Most of the mantle may have been processed by continuous extraction and subduction of oceanic crust since the Archean (e.g., Helffrich and Wood, 2001). Depleted MORB mantle may represent depleted mantle plus recycled oceanic crust. Mixing of these materials may be efficient over long

Table 3 Model fertile upper mantle with uncertainties at 95% confidence level.

	Best	*High*	*Low*
Mg#	89.28	89.49	88.91
SiO_2	44.90	45.19	44.74
TiO_2	0.13	0.15	0.12
Cr_2O_3	0.40	0.42	0.38
Al_2O_3	3.49	4.00	3.20
FeO	8.27	8.41	8.04
MgO	38.59	39.25	37.46
NiO	0.26	0.28	0.24
CaO	3.19	3.66	2.92
Na_2O	0.24	0.28	0.22

timescales, at least at the spatial scale of sampling in upper-mantle melting regimes (see Section 2.08.3.1 and Chapter 2.12). The composition derived below is based on the melt extraction trends exhibited by oceanic and off-craton mantle, and is subsequently referred to as fertile upper mantle. This composition will be used as a starting composition for melt extraction models for these types of mantle lithosphere.

The following procedure was used to estimate the composition of fertile upper mantle. First, correlation between Mg# and MgO is very good in both data sets ($R^2 \sim 0.8$), and the two trends intersect at an Mg# of ~89.27, corresponding to an MgO content of 38.59%. This fixes the FeO content at 8.27%, which is poorly correlated and nearly constant with Mg# in the abyssal data set. The abundance of each remaining oxide was calculated at an Mg# of 89.27 based on linear regressions to Mg# versus oxide trends in the off-craton data subset. The estimate for fertile upper mantle based on melt extraction trends is given in Table 1 and shown in Figures 9 and 10, and Table 3 gives a detailed assessment of oxide uncertainties. Errors are based on uncertainty at the 95% confidence level in the linear fits to Mg# versus oxide trends. First, uncertainty in Mg# at the intersection of the abyssal and off-craton trends was determined. This uncertainty was then combined with uncertainty in the linear fit to each oxide.

The best estimate for fertile upper mantle deduced from melt extraction trends is generally consistent with previous primitive mantle models, but with some notable differences in detail. The composition is most similar to the average of five fertile samples from Zabargad (Bonatti *et al.*, 1986), which is a nascent rift exposing a section that includes presumably pristine upper mantle. SiO_2 is very similar to pyrolite, but lower than in models of primitive mantle determined by Hart and Zindler (1986) and Allegre *et al.* (1995). The higher value of ~46% in these models is difficult to reconcile with the off-craton mantle subset. The Al_2O_3 content estimated here (3.5%) is close to but on the low side of previous estimates. FeO is higher than most previous estimates, closer to the Zabargad average, whereas other previous estimates tend to fall along the lower bound of the field of mantle samples.

Many of the samples in the off-craton data subset have Mg# less than 89.3. This raises a question regarding the status of samples that are apparently more fertile than the model fertile upper-mantle composition. However, processes other than melt extraction no doubt contribute to the spread of data on oxide versus Mg# plots. Primary among these are small-scale heterogeneity in small xenolith samples due to mineral segregations, and secondary enrichments due to fluid metasomatic processes or melt infiltration/reaction (see also Chapters 2.04 and 2.05). The data filtration method used to cull the off-craton data set was designed to eliminate these effects, but it could not have been entirely effective. It was noted above that the filter eliminated all of the samples with Mg# less than 88, suggesting that the "fertile" nature of these samples is likely due to random secondary processes. Even in the off-craton subset samples with Mg# less than 89.3 tend to show scatter that is at least as great or greater than other samples. There is considerable ambiguity in choosing the most "fertile" samples as representative of the upper mantle. In the approach used here, the presumption is that bulk melt extraction from a common protolith is the dominant feature preserved in the oceanic and off-craton data sets, and that the fitted trends in Figures 9 and 10 adequately recapture this feature.

Refractory lithophile major-element ratios in model fertile upper mantle are nonchondritic. The MgO/SiO_2 ratio is ~20% greater than the chondritic (primitive solar) ratio, as in nearly all model compositions of upper mantle and bulk silicate Earth. It is likely that this is a primary feature of the mantle, due either to cosmochemical fractionation or solution of silicon into Earth's core (e.g., Ringwood, 1977, 1979; Hart and Zindler, 1986; Allegre *et al.*, 1995; Gessmann *et al.*, 2001; see also Chapter 2.15). The CaO/Al_2O_3 ratio in model fertile upper mantle is 0.91, higher than the chondritic ratio of 0.8. Previous estimates of primitive upper mantle based on fertile xenoliths compositions, as well as pyrolite reconstructions, have also yielded superchondritic CaO/Al_2O_3 ratios (Jagoutz *et al.*, 1979; Ringwood, 1979; Palme and Nickel, 1985), whereas the constraint of a chondritic ratio has been effectively imposed in other models for bulk silicate Earth (Hart and Zindler, 1986; McDonough and Sun, 1995; Allegre *et al.*, 1995; O'Neill and Palme, 1998; Chapter 2.01). Figure 11 shows a plot of CaO/Al_2O_3 versus normative cpx/opx showing off-craton mantle and abyssal peridotite compositions together with estimates of

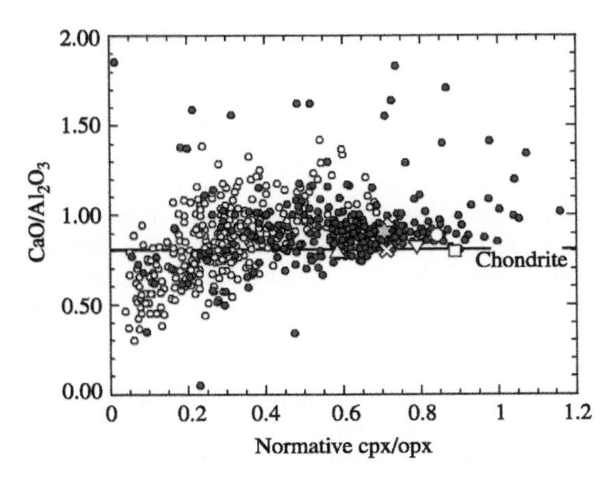

Figure 11 CaO/Al_2O_3 versus spinel lherzolite normative cpx/opx for the off-craton mantle subset (shaded circles) and reconstructed oceanic mantle (open circles). Primitive mantle compositions from Table 2 are also shown and the symbols are the same as in Figure 9.

primitive mantle (Table 1). The chondritic models of Hart and Zindler (1986) and Allegre *et al.* (1995) have low cpx/opx and low CaO/Al_2O_3 that do not make good candidates as protoliths of upper-mantle samples. Although there is much scatter in the mantle CaO/Al_2O_3 ratio, the fertile end of off-craton samples (i.e., high cpx/opx) is more consistent with a superchondritic ratio, tending to funnel to a value of ~0.9. This could mean that bulk silicate Earth has a superchondritic CaO/Al_2O_3 ratio, but this is unlikely as these elements are among the most refractory in terms of their cosmochemical condensation temperatures.

If the fertile mantle protolith to oceanic and subcontinental mantle is nonchondritic in refractory elements, and if the bulk silicate Earth is chondritic in refractory elements, then a complimentary reservoir must exist elsewhere, presumably buried in the deep lower mantle and isolated from mantle convection (e.g., Anderson, 1989; Kellogg *et al.*, 1999; Albarede and van der Hilst, 1999). The primitive upper mantle could have acquired a superchondritic ratio as a consequence of crystal fractionation in a magma ocean, or perhaps by extraction of an early crust with low CaO/Al_2O_3.

2.08.3.5 Cratonic Mantle

The mantle roots of Archean–Proterozoic cratons extend to depths in the range of 200–400 km (e.g., Jordan, 1978; Rudnick *et al.*, 1998), and are sampled as xenoliths erupted in kimberlitic lavas. Based on thermobarometric and textural evidence, a twofold classification of cratonic xenoliths has been made that includes (i) shallower, coarse-grained, low-temperature xenoliths (<~1,100 °C), and (ii) deeper, sheared, high-temperature xenoliths (>~1,100 °C).

The composition of cratonic mantle as deduced from low-T xenoliths indicates lithosphere that is highly depleted in FeO, Al_2O_3, and CaO, presumably from extraction of mafic and ultramafic melts (e.g., Boyd and Mertzman, 1987; Boyd, 1989; Chapter 2.05). This buoyant, depleted mantle is essentially welded onto continental crust, acting as a keel stabilizing the craton. High-T xenoliths are typically less depleted and, based on their compositions and textures, many workers have suggested that these samples may represent cratonic asthenosphere (e.g., Boyd, 1987; Walter, 1998), or subducted and metasomatized oceanic lithosphere accreted to the lithosphere (Kesson and Ringwood, 1989). However, on the basis of Re–Os depletion ages that relate high-T samples to Archean lithosphere (e.g., Pearson, 1999), as well the geochemistry of garnets (Griffin *et al.*, 1999b), it is most likely that these samples are cratonic lithosphere that has been pervasively metasomatized by asthenospheric melts (e.g., Griffin *et al.*, 1989; Smith *et al.*, 1993; Griffin *et al.*, 1999b). Thus, these samples will not be given further consideration here.

Figure 12 shows normative minerals versus rock Mg# for a set of coarse-grained, low-temperature cratonic mantle xenoliths. Relative to abyssal peridotite and off-craton mantle, cratonic mantle xenoliths typically have considerably higher rock Mg#s, mostly in the range of 91–94. Mg# and normative mineralogy are poorly correlated, showing a wide range in olivine and opx contents at a given Mg#. A large number of xenoliths have high normative opx of greater than 30%, most notably xenoliths from the Kaapvaal and Siberian cratons. The high Mg#s and extreme depletions in CaO and Al_2O_3 have been interpreted as evidence for a large amount of melt extraction from fertile mantle (e.g., Boyd and Mertzman, 1987; Takahashi, 1990; Boyd *et al.*, 1997; Walter, 1999b). However, opx contents above ~30% and in some cases negatively correlated FeO and SiO_2 cannot be reconciled directly with partial melt extraction (e.g., Boyd, 1989; Kelemen *et al.*, 1992; Canil, 1992; Herzberg, 1993; Boyd *et al.*, 1997; Walter, 1999b), and several competing models involving secondary processes have been proposed to account for these features. However, as more data have become available, it is becoming clear that more than one breed of cratonic mantle exists. This is evident from Figure 12, where it can be seen that xenoliths from Canada, Greenland, and Tanzania tend to be more enriched in olivine and poorer in opx.

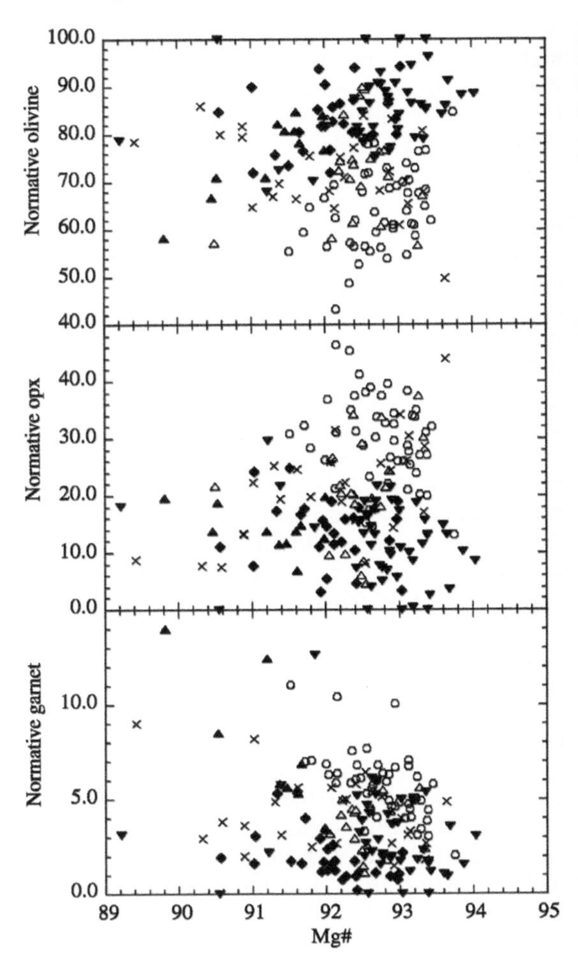

Figure 12 Normative garnet lherzolite mineral abundances (wt.%) versus rock Mg# for low-temperature cratonic mantle. Data are: open circles, Kaapvaal craton (Boyd and Mertzman, 1987 and references therein); filled diamonds, Tanzanian craton (Lee and Rudnick, 1999); open triangle, Siberian craton (Boyd *et al.*, 1997); cross, Slave craton, northwest Canada (Kopylova and Russell, 2000); filled triangle, northern Canadian craton (Schmidberger and Francis, 1999); and inverted triangle, central Greenland craton (Bernstein *et al.*, 1998).

2.08.4 THE ROLE OF MELT EXTRACTION

In this section the role that melt extraction has played in producing the chemical and mineralogic variability in samples from the upper mantle is explored. This is achieved by comparing trends for melt extraction from fertile peridotite that are predicted on the basis of experimental data with lithospheric mantle compositions as presented in the previous section. Such comparisons always have an element of nonuniqueness because assumptions are required regarding the source composition, the choice of melting model, and details of the melt extraction process.

Here, the fertile upper-mantle composition derived in the previous section is assumed for oceanic and off-craton mantle, and the model of Kinzler and Grove (1992a, 1993) is used to model melt extraction at pressures of ≤ 2.5 GPa.

Melt extraction at higher pressures as applied to cratonic mantle is modeled using the experimental data of Walter (1998) for a composition that is very similar to the primitive mantle of McDonough and Sun (1995).

2.08.4.1 Oceanic Mantle

Figure 13 shows abyssal peridotite trends relative to predicted residue trends for isobaric batch melting of fertile peridotite at 0.5 GPa, 1 GPa, and 2 GPa on plots of normative minerals versus Mg#. The abyssal peridotite trends are generally well reproduced by 5–25% batch melt extraction in this pressure range, with the bulk of the abyssal compositions falling between the 1 GPa and 2 GPa trends. This amount of melt extraction is similar in extent to that predicted in melting models to produce MORB magmas (e.g., Klein and Langmuir, 1987; Kinzler and Grove, 1992b; Presnall *et al.*, 2002), and on average is sufficient to produce the average oceanic crustal thickness of ~6 km. Isobaric batch melt extraction is not consistent with the modern paradigm for melting beneath mid-oceanic ridges to generate MORB, which dictates a polybaric, near-fractional melting process. Polybaric melt extraction is expected as a consequence of adiabatic decompression of mantle peridotite during passive upwelling beneath a spreading center, and some approach to fractional melting is expected if melt migration from the source occurs due to buoyancy driven, porous flow (e.g., Ahern and Turcotte, 1979; McKenzie, 1984; McKenzie and Bickle, 1988). Global systematics in the major-element chemistry of MORB are best explained by polybaric melting (Klein and Langmuir, 1987), and LREE depletions in cpx from abyssal peridotites require an element of fractional melting (Johnson *et al.*, 1990; Kelemen *et al.*, 1997).

Figure 14 shows normative mineral modes versus rock Mg# for abyssal peridotite trends relative to calculated trends for polybaric, near-fractional melting. These models are constructed

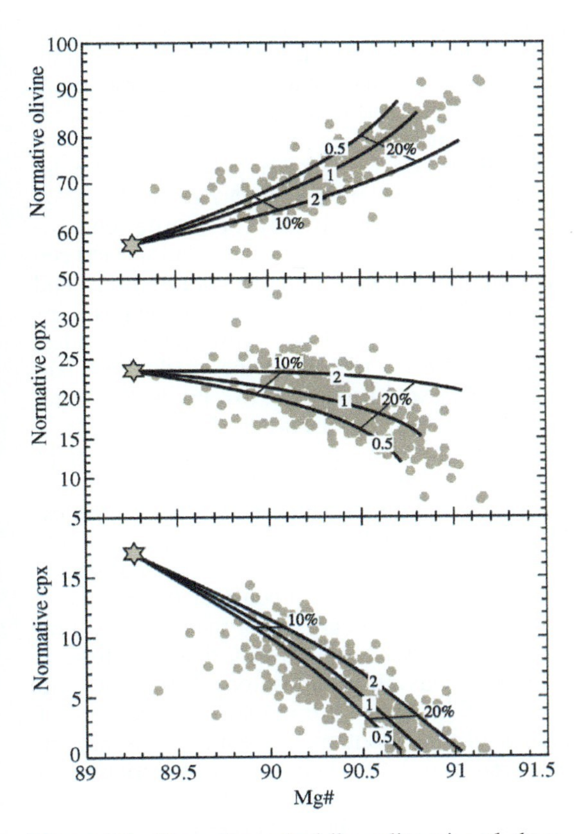

Figure 13 Normative spinel lherzolite mineral abundances (wt.%) versus rock Mg# for oceanic mantle (as in Figure 7) relative to trends for 0–25% batch melt extraction at 0.5–2 GPa. The starting composition is fertile upper mantle as determined in this study (Table 1, #8), and residues are calculated using the melting model of Kinzler and Grove (1992a, 1993).

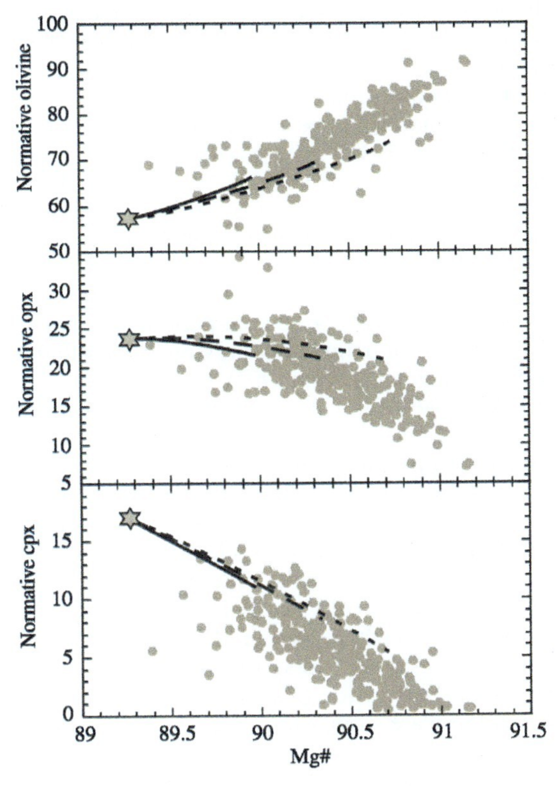

Figure 14 Normative spinel lherzolite mineral abundances (wt.%) versus rock Mg# for abyssal peridotite trends relative to calculated trends for polybaric, near-fractional melting. Three melt extraction models are shown over pressure ranges of 2.5–0.4 (short dashed line), 2.0–0.4 (long dashed line), and 1.5–0.4 GPa (solid line). One percent melting occurs per 0.1 GPa of decompression, and 90% of the melt is extracted at each pressure. Starting mantle is fertile upper mantle (Table 1, #8).

in the same way as those reported in Kinzler and Grove (1992b, 1993) to model MORB generation, in that melt extraction occurs over a range of pressures (2.5–0.4 GPa, 2.0–0.4 GPa, and 1.5–0.4 GPa), 1% melting occurs per 0.1 GPa of decompression, and 90% of the melt is extracted at each pressure. These melt extraction characteristics are not unique, as neither the exact amount of melt extraction per 0.1 GPa (i.e., melt productivity) nor the residual porosity is precisely known. However, the assumed melt productivity is generally consistent with recent theoretical and petrologic estimates, although it is not expected to be constant with pressure (Asimow *et al.*, 1997; Yang *et al.*, 1998), and could increase considerably around the plagioclase to spinel transition (Presnall *et al.*, 2002).

Predicted residues of polybaric near-fractional melting are not well reconciled with the abyssal peridotite compositional trends (Figure 14). This is true for individual residue paths as well as for the combination of final residues of all paths. In general, abyssal compositions are enriched in olivine and depleted in opx and cpx relative to the model melt extraction trends, especially at higher Mg#s. Although not shown, if the primitive mantle composition of McDonough and Sun (1995) is chosen as the starting composition the fractional trends are even poorer in olivine and richer in opx than abyssal peridotites. This observation also holds if the MELTS model is used, rather than the Kinzler and Grove model (e.g., Asimow, 1999).

The failure of polybaric, near-fractional melting residues to account for abyssal peridotite compositions has been pointed out previously by Elthon (1992), Langmuir *et al.* (1992), Niu *et al.* (1997), and Asimow (1999) based on major and minor element variation diagrams. Three models have been developed to explain why MORB can be successfully modeled as fractional melting products, but abyssal peridotites apparently cannot: (i) refertilization due to crystallization of trapped melts at shallow levels (Elthon, 1992), (ii) crystallization of olivine from melts that are channeled through peridotite (Niu *et al.*, 1997), and (iii) polybaric mixed-mode (fractional + batch) melt extraction (Asimow, 1999). The first two of these are similar in suggesting that abyssal peridotite trends do not faithfully reproduce melt extraction residues, and that a secondary process has led to changes in bulk composition and mineralogy unrelated to melt extraction. In contrast, the third model relies on melt extraction to account for abyssal compositions.

Low-pressure trapped melts crystallize plagioclase, and it has been documented that the relatively much less abundant plagioclase-bearing abyssal peridotites are products of refertilization by trapped melt (e.g., Dick, 1989). For this reason, Asimow (1999) avoided plagioclase-bearing samples and Baker and Beckett (1998) established an arbitrary but low cutoff point for plagioclase abundance in samples included in reconstructions. Thus, refertilization is not considered to contribute significantly to the reconstructed major-element compositions in either of these data sets.

Based on two main lines of evidence, Niu *et al.* (1997) concluded that abyssal peridotites are the end products of melt extraction followed by variable amounts of olivine crystallization. First, in their set of reconstructed compositions they found that model fractional and batch melt extraction trends could not reproduce major and minor element variations in their data set. Most importantly, they found that melt extraction models failed to account for the strong positive correlation between FeO and MgO, as well as incompatible minor-element concentrations. Specifically, at a given Na_2O or TiO_2 content, abyssal peridotites are enriched in MgO relative to model melt extraction residues. Niu *et al.* (1997) showed that these compositional anomalies can be reconciled by a model of melt extraction followed by olivine crystallization, with more MgO-enriched samples having more accumulated olivine. If correct, this model has important implications for understanding melt extraction at oceanic ridges, and it has recently been the focus of re-evaluation.

Baker and Beckett (1998), employing a fundamentally different method than Niu *et al.* (1997) for reconstructing abyssal peridotite compositions, found no significant correlation between FeO and MgO in site-averaged compositions, and only a slight positive correlation in compositions reconstructed from single thin-section modes. Asimow (1999), following the method of Niu *et al.* (1997), was also unable to reproduce a strong positive correlation between FeO and MgO. Although there is the appearance of a slight positive correlation between FeO and MgO in the combined data sets (Figure 10, open circles, $R^2 = 0.2$), the highly correlated, strong positive trend found by Niu *et al.* (1997) does not occur. Baker and Beckett (1998) consider the Niu *et al.* (1997) trend to be an artifact of the reconstruction method, the critical point being that the Mg#s of calculated olivine compositions using the Niu *et al.* (1997) method are virtually independent of modal olivine content, a feature contrary to observed correlation between these variables in abyssal peridotites in which olivine contents and mineral modes have been analyzed.

In order to explore the role of melt extraction more fully in generating abyssal peridotite compositions, Asimow (1999) used the MELTS model to investigate a variety of melt extraction scenarios including polybaric batch and mixtures of polybaric near-fractional and batch melting, and found that such models are more successful in reproducing abyssal peridotite compositions than are fractional melting models. Figure 15 shows oxide versus FeO variation diagrams in which melt extraction trends from isobaric batch, polybaric near-fractional, polybaric batch, and mixed-mode melting models are plotted relative to abyssal peridotite compositions. Again, polybaric fractional melt extraction cannot reproduce the abyssal compositions. Polybaric batch and mixed-mode melt extraction models produce trends that largely overlap abyssal compositions, and unlike the fractional model, can reproduce the apparent negative correlation between FeO and SiO_2 and the slight positive correlation between FeO and MgO. On the basis of modeling using MELTs, Asimow (1999) used similar arguments to conclude that abyssal peridotites record a melt extraction process that included a large component of batch melting. Kelemen *et al.* (1997) discussed the possibility for a mixed-mode style of melt extraction beneath ridges in which near-fractional melt extraction in a porous flow regime is mixed with batch melt extraction in an equilibrium, channel flow regime. They showed that the LREE depleted patterns in cpx from abyssal peridotites can be accounted for by mixed-mode melt extraction with a minimum of 3% fractional melting.

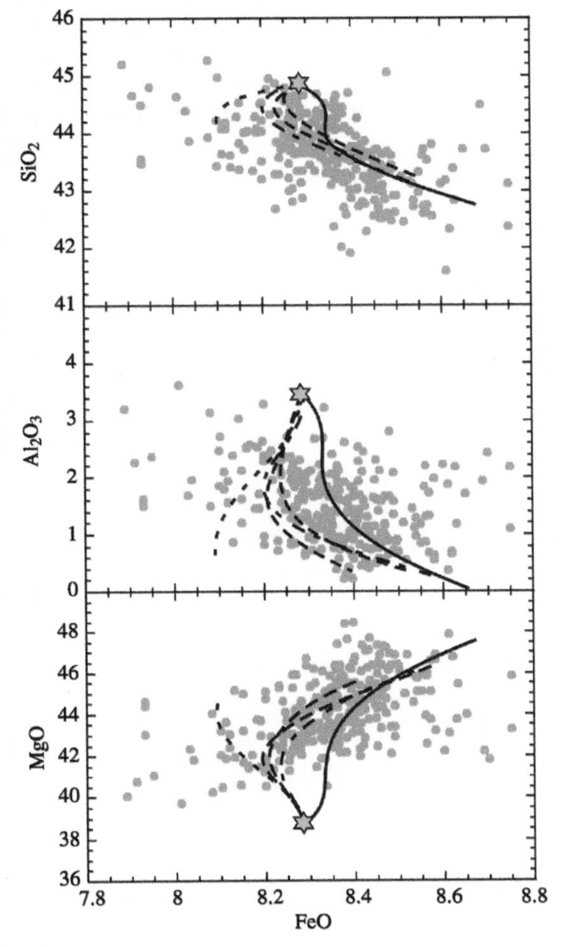

Figure 15 Major-element oxides versus FeO variation diagrams showing reconstructed abyssal peridotite compositions relative to trends produced by a variety of melt extraction models. Solid line is 0–25% batch melt extraction at 1 GPa. Short-dashed line is polybaric, near-fractional melting from 2.5 GPa to 0.4 GPa as described in Figure 14. Long–short dashed line is for polybaric batch melt extraction from 2.5 GPa to 0.4 GPa, in which the degree of melting at a given pressure increases by 1% per 0.1 GPa of decompression. Long-dash lines are for two mixed-mode polybaric melt extraction models from 2.5 GPa to 0.4 GPa. In one model (shorter curve), the first 8% melting is extracted in a polybaric, near-fractional manner as described in Figure 14, whereas in the other only the first 3% is near-fractional. The remainder in both models is extracted in a polybaric batch manner as described above. Starting mantle is fertile upper mantle (Table 1, #8).

Figure 16 shows Na_2O versus normative olivine content for abyssal peridotites and melt extraction models. Consistent with the observations of Niu *et al.* (1997) and Asimow (1999), polybaric fractional melting residues are too enriched in olivine at a given Na_2O content to account for abyssal compositions. Conversely, combinations of polybaric near-fractional and batch melt extraction can provide a relatively good match to the abyssal data. Although not shown, combinations of polybaric batch + fractional melting can also reproduce the TiO_2 contents of abyssal peridotites (see, e.g., Figure 4 in Asimow (1999)). Because the Na_2O and TiO_2 contents of the fertile

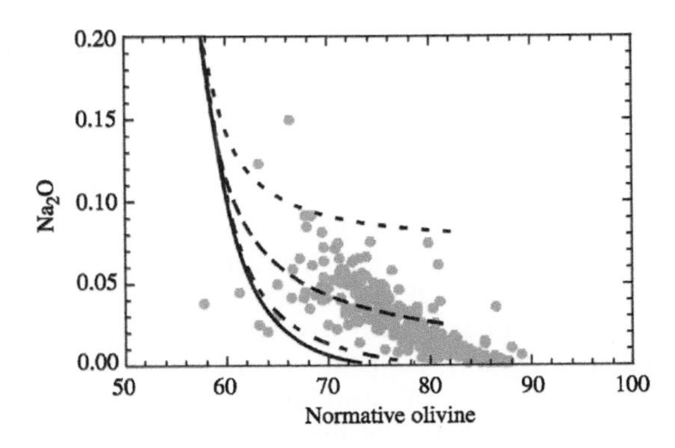

Figure 16 Na_2O versus normative olivine content for reconstructed abyssal peridotites and melt extraction models. Short-dashed line is 0–25% batch melt extraction at 1 GPa. Solid line is polybaric, near-fractional melting from 2.5 GPa to 0.4 GPa as described in Figure 14. Long dash and long-short dashed lines are for mixed-mode melting with 3% polybaric fractional and 8% fractional melt extraction, respectively, followed by polybaric batch melting as described in Figure 15.

MORB source are not well constrained, and because of variations in the partitioning behavior of these elements among melting models, it is not possible to determine accurately the relative proportions of batch and fractional melting from this type of analysis. However, if the abyssal trend is due to melt extraction and not olivine addition, then it is clear that the batch component of melting must be considerable, and may well dominate over fractional melting.

Asimow (1999) provided further tests for discerning between melt extraction and olivine addition by looking at variations among ratios that are sensitive to melt extraction but insensitive to olivine crystallization, such as CaO/Al_2O_3, Al_2O_3/TiO_2, and modal cpx/opx, and found that polybaric fractional melting could not account for the array of abyssal compositions, whereas polybaric batch and mixed-mode melting could. Figure 11 shows CaO/Al_2O_3 versus normative cpx/opx for abyssal and off-craton mantle compositions. There is considerable spread in the abyssal data, as well as a general trend toward decreasing CaO/Al_2O_3 with decreasing cpx/opx, which *a priori* cannot be attributed to olivine addition. Melt extraction produces a decrease in cpx/opx at all pressures as cpx is always consumed at a greater rate than opx (see melting reactions in Table 2). The Kinzler and Grove (1992a, 1993) model predicts that variation in CaO/Al_2O_3 in residues during batch melt extraction of spinel lherzolite is sensitive to pressure, but also very sensitive to the starting CaO/Al_2O_3 ratio in the source. Regardless, a spread in CaO/Al_2O_3 with decreasing cpx/opx is expected for polybaric batch and mixed-mode melting, whereas fractional melting alone produces only a limited range of CaO/Al_2O_3 (see Asimow, 1999). The abyssal data could still be accounted for by melt extraction followed by olivine addition, but this would require that the MORB source was extremely heterogeneous in CaO/Al_2O_3 ratio.

In summary, reconstructed abyssal peridotite bulk compositions are most consistent with a large component of batch melt extraction at shallow levels (e.g., ~1 GPa). Presuming that melting beneath mid-ocean ridges is polybaric, as seems required by any realistic melting scenario for adiabatic ascent of upwelling mantle, abyssal peridotites indicate that the melting process was nearer to batch melting than fractional melting, in contrast to the modern paradigm for MORB generation (Asimow, 1999). A wide range of possible melt extraction scenarios exists that can include elements of both equilibrium and near-fractional melting (Kelemen *et al.*, 1997; Asimow, 1999), and the range of pressures over which melt is extracted is not necessarily large (Presnall *et al.*, 2002).

2.08.4.2 Subcontinental Mantle

2.08.4.2.1 Off-craton mantle

The off-craton mantle subset is shown relative to isobaric batch melt extraction curves on plots of normative olivine and major-element oxides versus Mg# in Figure 17. Generally speaking, off-craton mantle compositions are consistent with effectively ~0–30% melt extraction from fertile upper mantle in the range of 1–5 GPa. The chemical signature recorded in off-craton mantle mimics closely the maximum degree of melt extraction recorded in oceanic mantle, but in contrast there are many samples that show little or

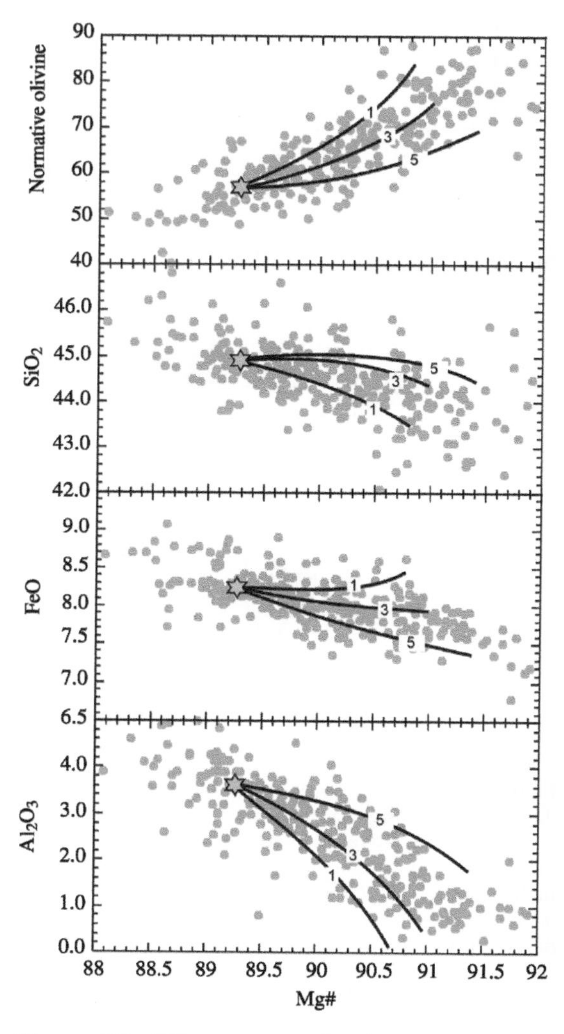

Figure 17 Off-craton mantle shown relative to isobaric batch melt extraction curves on plots of normative olivine and major-element oxides versus Mg#. Batch melt extraction at 1 GPa modeled using the model of Kinzler and Grove (1992a, 1993), and at 3 GPa and 5 GPa using the model of Walter (1998, 1999b). Starting composition is fertile upper mantle (Table 1, #8).

no melt extraction. This may be due to a sampling bias, because abyssal peridotites presumably sample the most depleted portions of the mantle at the axis of the mid-ocean ridge melting zone, whereas off-craton mantle provides a more random sampling of ancient melting zones. Off-craton mantle is best matched by melt extraction at higher average pressure than oceanic mantle. Notice especially the negative trend between FeO and Mg# that is characteristic of higher-pressure melt extraction (Walter, 1999b), in contrast with the slightly positive trend exhibited for oceanic mantle (see Figure 10) that is characteristic of lower-pressure melt extraction. Using oceanic mantle as an analog, it is reasonable to speculate that off-craton mantle formed by similar melt extraction processes. That is, some combination of polybaric fractional and batch melt extraction, with a large component of batch melting. For example, Frey *et al.* (1985) concluded from trace-element modeling that the melt extraction events that formed residual Ronda peridotite are best modeled by batch melt extraction, and that fractional melt extraction is limited to less than ~5%. This possibility is supported by the observation that on a plot of CaO/Al_2O_3 versus normative cpx/opx (Figure 11), off-craton mantle and abyssal peridotites show a similar range of compositions, which as discussed above are better modeled with a large component of batch melt extraction (Asimow, 1999). Further, the off-craton data mimic the abyssal data in that, at a given normative olivine content, the Na_2O contents are high relative to polybaric fractional melting residues, although there is much scatter in the Na_2O contents of off-craton samples that is probably due to some extent to the mobility of Na_2O in fluids.

Samples of off-craton mantle are typically much older than oceanic mantle, and unlike oceanic mantle it is essentially impossible to relate samples from a given locality to any specific set of complimentary basaltic rocks. Off-craton xenoliths typically have highly variable and often enriched incompatible element abundances as well as isotopic characteristics that preclude them as source rocks for the basalts that transport them to the surface (e.g., McDonough, 1990). Thus, it is difficult to relate any specific set of off-craton mantle samples to melt extraction in a particular tectonic setting. A general similarity in both major and trace element depletion characteristics imply a similar mode of origin for off-craton and oceanic mantle (e.g., Frey *et al.*, 1985; Boyd, 1989; Hartmann and Wedepohl, 1993), indicating that off-craton lithosphere may represent former oceanic lithosphere that was plastered against cratons. However, a firm connection cannot be established unequivocally because for a common fertile source, residues of polybaric melt extraction of shallow upper mantle will be chemically indistinguishable irrespective of tectonic setting (e.g., mid-ocean ridge, continental rift, hot-spot, subduction zone). In any case, normative mineral and major-element compositional trends show that off-craton mantle records higher average pressures of melt extraction than does modern oceanic mantle, indicating a larger proportion of picritic melt extraction due either to higher mantle potential temperature or elevated H_2O contents (e.g., Frey *et al.*, 1985; Fabries *et al.*, 1991; Shi *et al.*, 1998).

2.08.4.2.2 *Cratonic mantle*

Figure 18 is a plot of normative olivine and opx (garnet lherzolite norm) versus rock Mg# showing low-*T* cratonic mantle relative to batch melt extraction trends from 3 GPa to 7 GPa. Open circles are compositions from the Kaapvaal, Siberian, and Slave cratons, and the filled circles are compositions from the Tanzanian, Canadian, and Greenland cratons. Taken at face value, most of the data require high pressures and high degrees of melt extraction. The combination of high Mg#s with normative olivine generally in the range of

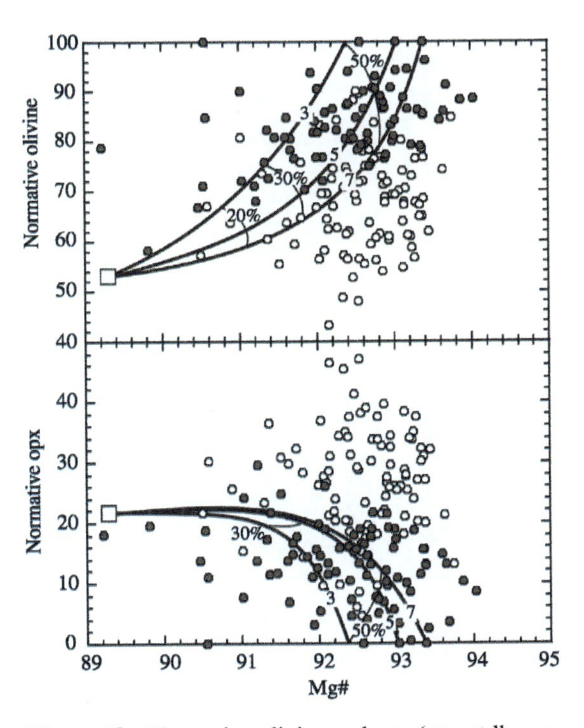

Figure 18 Normative olivine and opx (garnet lherzolite norm) versus rock Mg# showing low-*T* cratonic mantle relative to batch melt extraction trends from 3 GPa to 7 GPa (see Figure 6). Open circles are compositions from the Kaapvaal, Siberian, and Slave cratons, and the filled circles are compositions from the Tanzanian, Canadian, and Greenland cratons.

50–90% and opx from 10% to 40% cannot be generated by low-pressure melt extraction. There is a well-defined segregation between the locations signified by open and filled circles. Xenoliths from the Kaapvaal, Siberian, and Slave cratons (open circles) are generally depleted in olivine and enriched in opx at a given Mg# relative to xenoliths from other cratons, and the opx-rich nature of these xenoliths is not likely to result from melt extraction. The maximum opx component in melting residues from primitive mantle compositions like those in Table 1 will reach a maximum in the range of ~20–25% normative opx (garnet lherzolite norm), and this is not likely to increase much at pressures above 7 GPa as the SiO_2 content of partial melts remain similar to that of the bulk composition. Xenoliths from the Tanzanian, Canadian, and Greenland cratons (filled circles) can be interpreted as melt extraction residues generally in the pressure range of 3–7 GPa and 30% to more than 50% melt extraction.

Figure 19 is a plot of SiO_2 and Al_2O_3 versus FeO showing low-*T* cratonic mantle relative to batch melt extraction trends from 3 GPa to 7 GPa. Samples from the Kaapvaal, Siberian, and Slave cratons (open circles) have considerably higher SiO_2 and somewhat higher Al_2O_3 at a given FeO content relative to samples from the Tanzanian, Canadian, and Greenland cratons (filled circles). There is also a negative correlation between SiO_2 and FeO among the samples from the Kaapvaal, Siberian, and Slave cratons, a correlation that is especially well developed in the Siberian samples (Boyd *et al.*, 1997). This trend toward high SiO_2 and Al_2O_3 at low FeO is in marked contrast to that expected for high-pressure melt extraction, regardless of the melting process (Walter, 1999b). In contrast, samples from the Tanzanian, Canadian, and Greenland cratons can be explained directly by high average degrees of melt extraction at high average pressures.

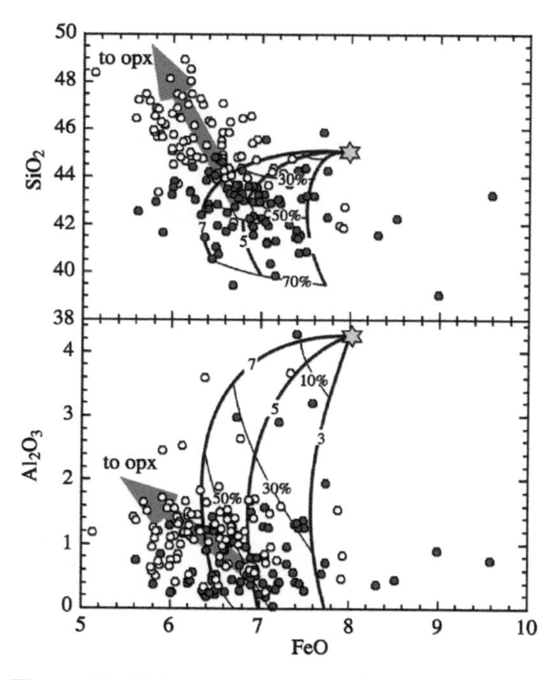

Figure 19 SiO_2 and Al_2O_3 versus FeO showing low-*T* cratonic mantle relative to batch melt extraction trends from 3 GPa to 7 GPa. Small open circles are samples from the Kaapvaal, Siberian, and Slave cratons, and filled circles are samples from the Tanzanian, Canadian, and Greenland cratons. Large open circle shows composition after 50% melt extraction at 5 GPa. Arrow shows the effect of addition of an opx composition that would be in equilibrium with melt at high *P* and *T*.

Extensive compositional data for low-temperature cratonic mantle were first reported from the Kaapvaal and Siberian cratons, so it was initially thought that enrichment in SiO_2 (i.e., opx) was a fundamental feature of cratonic lithosphere (e.g., Boyd and Mertzman, 1987; Boyd, 1989, Boyd *et al.*, 1997). The failure of melt extraction from pyrolite-like mantle to account for the SiO_2-rich nature of these cratons led to numerous ingenious models for their origin, including: (1) melt extraction from a fertile mantle protolith considerably richer in SiO_2 than present day upper mantle, perhaps more akin to chondrite (Herzberg, 1993); (2) high-SiO_2 upper mantle formed directly as residual melt after extensive majorite fractionation, or as cumulates from high-SiO_2 melts produced by extensive melting in a large plume (Herzberg *et al.*, 1988; Herzberg, 1993); (3) melt extraction from hydrous peridotite (Ohtani *et al.*, 1996); (4) melt extraction followed by metamorphic differentiation resulting in opx- and olivine-rich segregations (Boyd, 1989; Takahashi, 1990; Boyd *et al.*, 1997); and (5) melt extraction from pyrolitic mantle followed by secondary enrichment of opx, either as a cumulate from ultramafic magmas or as a product of reaction with SiO_2-rich melts related to post-Archean subduction zones (e.g., Kelemen *et al.*, 1992, 1998; Rudnick *et al.*, 1994; Herzberg, 1999; Walter, 1999b).

As samples have become available from more cratons, it is now apparent that SiO_2 enrichment is not a global feature of cratonic mantle, and samples from Kaapvaal and Siberia now appear to have the most anomalously high SiO_2 contents among all cratons sampled to date. Indeed, the majority of samples from the Tanzanian, Canadian, and Greenland cratons, as well as many samples from the Siberian and Slave cratons, can be explained in terms of high-pressure melt extraction from nominally anhydrous pyrolitic mantle. Models calling for enrichments in upper-mantle SiO_2 due to accretionary mechanisms or crystallization events in an Early-Hadean magma ocean can generally be excluded because model

Re–Os ages of cratons are similar to the ages of overlying Archean crust (e.g., Pearson, 1999), and further, such processes would be expected to be global mechanisms for craton formation. A cumulate origin for cratonic mantle in gigantic Archean plumes is viable, but unattractive due to the vast amounts of melt required to crystallize a cratonic volume of peridotite with a limited range of Mg# (Boyd *et al.*, 1997). It is also not clear why some plumes would crystallize opx-rich cumulates (e.g., Kaapvaal), and others olivine-rich cumulates (e.g., Greenland). Formation of SiO_2-rich residues due to melt extraction from hydrous peridotite is an interesting possibility that has been difficult to test fully due to lack of experimental data. However, the SiO_2 contents of partial melts from pyrolitic peridotite with 1–2% H_2O at 6.5 GPa as reported by Ohtani *et al.* (1996) are still higher than the bulk composition, and so could not cause SiO_2 enrichment in residues. Further, unless the Archean mantle was heterogeneously enriched in H_2O, this model cannot not explain olivine-rich cratonic mantle that can be produced by melt extraction from nominally anhydrous mantle. Metamorphic differentiation and sorting into olivine-rich and opx-rich domains has probably occurred on some scale and led to sampling bias at all localities. However, it is not obvious why such a process should be so prominent in Kaapvaal and Siberian samples, but apparently much less evident in samples from other cratons. Presently, the most plausible explanation for variable SiO_2 enrichment is that a secondary, selective process has acted to various degrees in some cratons, and the currently favored mechanism is secondary addition of opx (Boyd and Mertzman, 1987; Kelemen *et al.*, 1998; Walter, 1999b; Herzberg, 1999).

A general scenario for melt extraction followed by opx addition is illustrated in Figure 19. The large open circle represents residual mantle depleted by 50% melt extraction at 5 GPa, leaving a harzburgite residue. High-degree melt extraction leaving a garnet-free, harzburgite residue is consistent with the observation that the Ca/Yb systematics in cratonic peridotites are not consistent with residual garnet (Kelemen *et al.*, 1998), and that garnet in cratonic harzburgites may have an exsolution origin (Canil, 1991). The large, shaded arrow shows a compositional vector for up to 50% addition of opx with a composition that could be in equilibrium with melt at these conditions. The vector is consistent with a negative correlation between SiO_2 and FeO, and the generally higher Al_2O_3 at a given FeO content that is characteristic of samples from the Kaapvaal, Siberian and Slave cratons (e.g., Walter, 1999b).

Further evidence for secondary enrichment in opx comes from the systematics of nickel in olivine. Kelemen *et al.* (1998) first pointed out the existence of a positive correlation between nickel in olivine versus modal opx, which they observed in samples from the Premier kimberlite in southern Africa, as well as in average craton compositions. Figure 20 shows a highly correlated positive trend ($R^2 = 0.86$) between craton-averaged nickel in olivine versus modal opx, although correlations are not observed among individual samples from a given location. Kelemen *et al.* (1998) explored three possible models to explain this positive trend including: (i) melt extraction; (ii) metamorphic differentiation and re-equilibration upon cooling (see also Herzberg, 1999); and (iii) melt/rock reaction. Kelemen *et al.* (1998) found that melt extraction is clearly unable to produce the positive trend, whereas metamorphic-differentiation/re-equilibration and melt/rock reaction can both produce positive trends. However, of the two, only melt/rock reaction can effectively reproduce the magnitude of the positive slope. These authors concluded that reaction between residual harzburgitic peridotite and FeO-poor siliceous melts, such as might be derived by partial melting of eclogite in a subducting slab, can account for the opx enrichment in cratonic mantle. This model is also appealing because such enrichment would naturally be selective, occurring only in cratons, or portions of cratons, that were subjected to subduction zone processes.

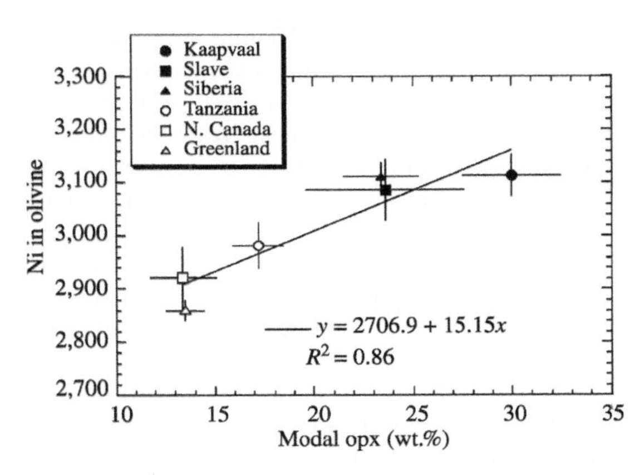

Figure 20 Craton averaged Ni in olivine versus modal opx (wt.%). Error bars are one standard error of the mean. A least-squares linear fit yields an R^2 of 0.86.

2.08.5 PERSPECTIVE ON MANTLE THERMAL EVOLUTION

Mineralogic and major-element variations in samples from the upper mantle provide strong evidence that melt extraction has played a predominant role in the origin of mantle lithosphere of all ages and provenance. There are compositional distinctions among samples of ancient Archean cratonic mantle, younger off-craton subcontinental mantle, and modern MORB mantle that can be related to average depth and temperature of melt extraction if source regions were compositionally similar (i.e., pyrolitic). However, in addition to the potential for major-element and mineralogic variation among primary source regions, variation in volatile content can also lead to ambiguous or erroneous interpretations regarding the pressures and especially the temperatures of melt extraction.

Experimental studies show that volatile components such as H_2O and CO_2 can have profound effects on melting temperature and melt composition (e.g., Kushiro, 1972; Hirose and Kawamoto, 1995; Kawamoto and Holloway, 1997; Dalton and Presnall, 1998; Gaetani and Grove, 1998; Hirschmann *et al.*, 1999a; Lee *et al.*, 2000; Asahara and Ohtani, 2001). It has been implicitly assumed in the assessment above that melt extraction occurred in nominally anhydrous mantle. This assumption is most robust in the case of MORB mantle, which has been shown to have a low volatile content (e.g., Michael, 1988; Saal *et al.*, 2002; Chapter 2.07). Inasmuch as off-craton mantle has isotopic characteristics that indicate similar long-term incompatible element depletion to the MORB source, and considering that volatiles are very incompatible elements, a low volatile content at the time of melt extraction from off-craton mantle is implied. Indeed, off-craton mantle may be genetically related to modern MORB mantle.

The volatile content in the mantle protolith of cratonic lithosphere is not known. The generally high degree of melting required to form cratonic mantle implies komatiite melt extraction. In recent years the water content of Archean komatiites has become a hotly debated subject and it seems apparent that at least some komatiites were produced in the presence of water, perhaps in a subduction setting (e.g., Ohtani *et al.*, 1996; Parman *et al.*, 1997; Stone *et al.*, 1997; Shimizu *et al.*, 2001). However, the compositions of both aluminum-depleted and aluminum-undepleted komatiites are well matched by high-degree, high-pressure melts of nominally anhydrous peridotite (e.g., Herzberg and Zhang, 1996; Walter, 1998; Arndt *et al.*, 1998). The role of water as a primary agent in komatiite and craton formation is unresolved.

As an initial premise, it is assumed in the following that in general water played a relatively minor role in the melt extraction events that led to the primary depletion characteristics of upper-mantle lithosphere. In this case, compositional distinctions among oceanic and continental lithosphere of variable age can be related directly to average depth and temperature of melt extraction, helping to constrain models for the thermal evolution of the mantle.

Figure 21 shows site-averaged and grand-average compositions for oceanic, off-craton, and cratonic mantle on plots of normative garnet lherzolite minerals versus rock Mg#. Also shown

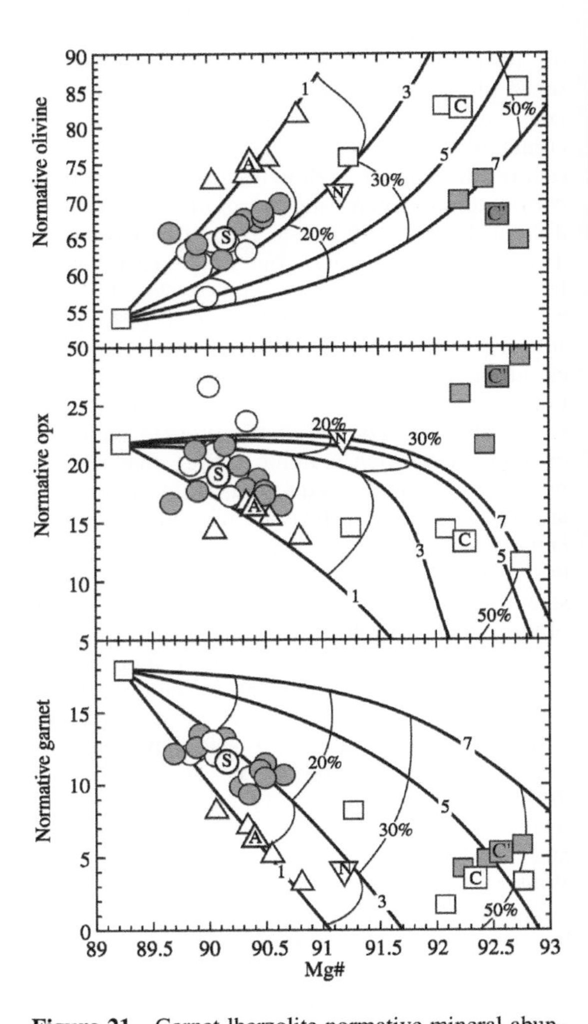

Figure 21 Garnet lherzolite normative mineral abundances versus Mg# for averaged mantle compositions relative to batch melt extraction trends from 1 GPa to 7 GPa. Triangles show site-averaged abyssal peridotite compositions as reported in Dick (1989). Triangle with A gives the average of all reconstructed abyssal peridotite compositions from Figure 7. Shaded circles show site-averaged xenolith compositions and open circles site-averaged orogenic lherzolite compositions for off-craton mantle subset from Figure 9. Circle with S is off-craton mantle subset average. Shaded squares show site-averaged compositions from the Kaapvaal, Siberian, and Slave cratons, and open squares site-averages from the Tanzanian, Canadian, and Greenland cratons. Square with C is average of Tanzanian, Canadian, and Greenland cratons. Square with C′ is average of Kaapvaal, Siberian, and Slave cratons. Inverted triangle with N is the average mode of Early-Proterozoic xenoliths from Namibia (source Pearson *et al.*, 1994).

are batch melt extraction trends from 1 GPa to 7 GPa for mantle of pyrolitic composition. This figure illustrates the distinct compositional features among the three types of upper-mantle lithosphere. Using the phase relations displayed on Figure 1 as a guide, generalized conditions of melt extraction are deduced for each type of lithosphere as shown on Figure 22.

Relatively young (<250 Ma) oceanic upper mantle is formed at the lowest average pressures and temperatures by melt extraction at mid-ocean ridges. Melt extraction is expected to be a polybaric process, and likely involves a combination of near-fractional and batch melt extraction. Assuming an average pressure of melt extraction of ~1 GPa and 15–20% melt extraction at the ridge axis, an average temperature of 1,315 ± 50 °C is estimated.

Off-craton, late Proterozoic–Phanerozoic subcontinental lithosphere records higher average pressures of melt extraction generally in the range of 1–3 GPa (Figure 21). Off-craton mantle apparently is generated on average by lower degrees of melt extraction than modern oceanic mantle, but this likely reflects a sampling bias because abyssal peridotites sample the most-depleted mantle near the axis of the mid-ocean ridge melt zone, whereas off-craton mantle provides a more random sampling of ancient melting zones. Figure 17 shows that the maximum extent of melt extraction preserved in samples of oceanic and off-craton mantle is essentially the same, at ~25–30%. It is also worth noting that on the basis of modeling of incompatible trace elements in MORB, an average of ~8% and a maximum of ~10% melt extraction is indicated (Hofmann, 1988), slightly lower than the average of 12% estimated here for off-craton mantle.

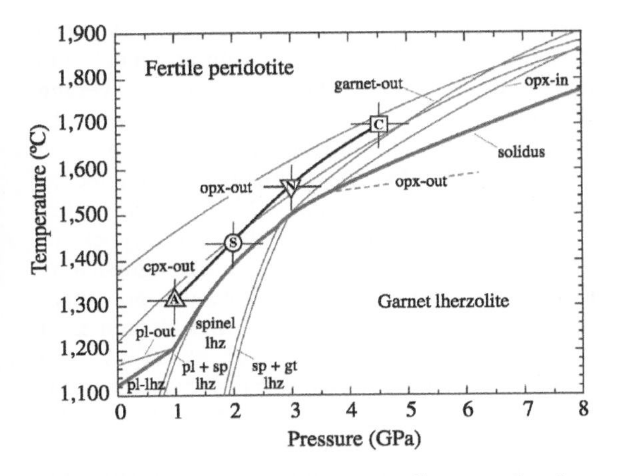

Figure 22 Pressure–temperature diagram showing estimated conditions of melt extraction for average lithosphere compositions. Estimates are based on generalized phase relations for batch melting of fertile mantle as depicted in Figure 1 and shown in the background. Lithosphere labels are as in Figure 21.

Assuming an average pressure of melt extraction of ~2 GPa for off-craton mantle, with a maximum of ~30% and an average of 12% melt extraction, an average temperature of 1,440 ± 50 °C is estimated.

Average cratonic mantle shows two distinct types, opx-rich and opx-poor, as indicated by C and C′ in Figure 21, respectively. Opx-rich cratonic mantle is likely to be derived by melt extraction followed by secondary addition of opx, as discussed above. Assuming that average opx-poor cratonic mantle directly reflects a melt extraction signature, then cratonic mantle is generated on average by ~45% melt extraction at 4.5 GPa, yielding an estimated temperature of 1,700 ± 50 °C.

There is a considerable gap between the average compositions of Archean cratonic mantle, and late Proterozoic–Phanerozoic off-craton and modern oceanic mantle. Early-Proterozoic mantle would be expected to fall within this gap given a model for secular change in average depth of melting. Samples of Early-Proterozoic mantle are rare, but Pearson *et al.* (1994) have reported average modes from a set of xenoliths from Namibia that yield Re–Os model extraction ages of ~2 Ga, and found that they are compositionally intermediate between Archean and oceanic lithosphere in terms of Mg# and mineral mode. The average modal mineralogy of Namibian xenoliths is also plotted on Figure 21, although these are actual modes and not normative modes. However, on the basis of comparisons between actual garnet lherzolite modes from xenolith samples and normative garnet lherzolite modes calculated from bulk compositions using the method of Kelemen *et al.* (1992), differences are expected to be less than 5% for a given mineral abundance and, of course, Mg# remains unchanged. The Early-Proterozoic Namibian peridotites apparently are compositionally intermediate between younger and older lithospheric mantle, indicating an average pressure and degree of melt extraction of ~3 GPa and 25%, respectively, or an average temperature of ~1,560 ± 50 °C.

Figure 22 illustrates that on the basis of average lithospheric mantle compositions, secular variation in mantle potential temperature (i.e., melting temperature decompressed adiabatically to 1 atm) on the order of 350 ± 50 °C is indicated from Archean to present. Griffin *et al.* (1998) have documented secular variation in the composition of garnets from subcontinental mantle and relate these variations to changes in mantle melting and crust generation processes. Using a large global database with over 8,000 samples, these authors showed that garnets from Archean xenoliths have distinctly high chromium and low calcium, consistent with derivation from subcalcic harzburgite residues. Proterozoic and younger garnets have chromium and calcium contents indicating derivation from calcic harzburgites and lherzolites, and also show a decrease in mean chromium and Zr/Y with decreasing crustal age. Griffin *et al.* (1998) relate these changes in garnet chemistry to increases in (cpx + garnet) and cpx/garnet, and a decrease in Mg# in mantle residues from the Proterozoic to the present, and suggest that subcontinental mantle shows a continuous evolution to less depleted compositions.

Secular variation in the average pressure and temperature of melt extraction near the surface is a natural consequence of a cooling Earth. Modern accretion theory dictates that much of Earth's mass was added in series of large impacts, and an inescapable consequence of this violent, earliest period of Earth evolution is that the primordial "Hadean" mantle was hot, likely passing through at least one magma ocean stage (e.g., Melosh, 1990). Models of mantle thermal evolution show that after a transient period of relatively rapid cooling lasting for several hundred million years after the Hadean, the mantle heat budget is dominated by radioactive decay and cooling is achieved by convective heat transport (see review in Davies (1998)). Figure 23 shows two models for thermal evolution in the upper mantle, one based on whole mantle convection and another for mantle undergoing transient layered convection due to a phase barrier between the upper and lower mantle (e.g., spinel-perovskite transition in $(MgFe)_2SiO_4$), resulting in periodic mantle overturns (Davies, 1995). Whole-mantle convection models indicate that mantle potential temperature would have declined since the Archean by at most 200–300 °C. Secular evolution in average melt extraction temperatures predicted for mantle lithosphere indicates a higher-temperature cooling trajectory with a difference from Archean to present of ~350 °C, perhaps more consistent with the trace of peak upper-mantle temperatures predicted for mantle overturn when hot lower mantle replaces cooler upper mantle. The transient layered convection model predicts that overturns would have been prevalent in the Archean and Early Proterozoic, but would have diminished in the last billion years or so as whole mantle convection becomes dominant (e.g., Davies, 1995, 1998). Mantle overturns have also been related to the periodicity of major Precambrian crust formation events (e.g., Condie, 1998). An alternative explanation is that water played a more important role in formation of ancient lithosphere than at present, lowering the temperature required to form depleted lithosphere. In this sense, variation in lithosphere composition could reflect secular dehydration of the upper mantle.

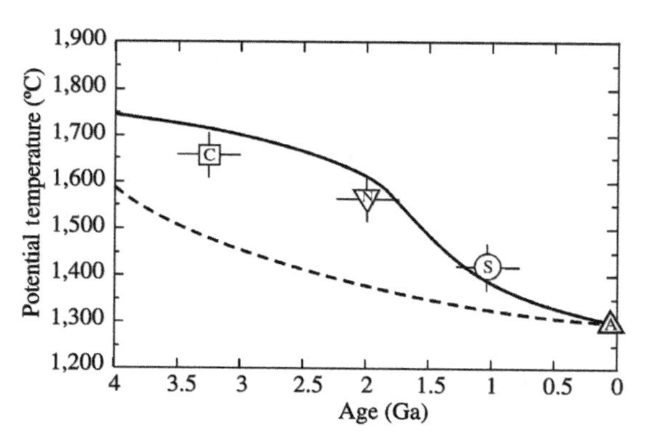

Figure 23 Mantle potential temperature (°C) versus age (Ga) showing thermal evolution models for the upper mantle. The dashed line is a model for whole-mantle convection, and the solid line shows the trace of maximum upper-mantle temperatures in a model of transient layered convection with periodic mantle overturn (Davies, 1995, 1998). The large circles show estimates for average mantle lithosphere. Lithosphere labels are as in Figure 21.

2.08.6 SUMMARY

Samples of mantle lithosphere beneath oceanic and continental crust exhibit mineralogic and chemical variability resulting from extraction of partial melt. Chemical trends from oceanic and off-craton subcontinental lithosphere intersect in normative mineral and major-element composition space, indicating a common lherzolite protolith. Fertile upper-mantle composition calculated from these trends has nonchondritic refractory lithophile major-element ratios. Melt extraction trends that are based on phase relations for melting of fertile peridotite are used to constrain the conditions at which mantle lithosphere formed. Reconstructed abyssal peridotite compositions are successfully modeled by moderate degrees (~5–25%) of combined batch and near-fractional melt extraction at an average pressure of ~1 GPa. Proterozoic off-craton subcontinental mantle lithosphere, which may be genetically akin to oceanic mantle, is modeled by similar extents of melting at a somewhat higher average pressure of ~2 GPa. Two types of Archean cratonic lithosphere are distinguishable. Low-SiO_2 cratonic lithosphere has compositional features that are explicable by high-degree melt extraction (~45% on average) at an average pressure of ~4.5 GPa. High-SiO_2 cratonic lithosphere is modeled by melt extraction followed by addition of opx in a secondary process, most likely reaction of depleted lithosphere with siliceous melts produced in a subduction setting. Estimates for average mantle potential temperature at the time of lithosphere formation indicate secular cooling of the mantle by ~350 ± 50 °C since the Archean. This amount of cooling is generally greater than predicted in models for mantle cooling based on whole-mantle convection, and could be indicative of transient periods of layered convection and mantle overturn during the first several billion years of Earth's history.

ACKNOWLEDGMENTS

Special thanks are extended to P. Asimow and M. Baker for providing data tables of reconstructed abyssal peridotite compositions, P. Asimow for providing tables of MELTS data, G. Gudfinnsson for providing tabulated data for CMASF melting models, and T. Grove for providing software for the Kinzler and Grove melting model. Reviews by D. Presnall and R. Carlson helped to improve the manuscript. Thanks are also extended to M. Hirschmann for comments on short notice, and to Y. Nakano for assistance. And finally, thanks to Rick Carlson for encouragement and an amazing amount of patience.

REFERENCES

Agee C. B. and Walker D. (1988) Mass balance and phase density constraints on early differentiation of chondritic mantle. *Earth Planet. Sci. Lett.* **90**, 144–156.

Ahern J. L. and Turcotte D. L. (1979) Magma migration beneath an oceanic ridge. *Earth Planet. Sci. Lett.* **45**, 115–122.

Albarede F. (2001) Radiogenic ingrowth in systems with multiple reservoirs: applications to the differentiation of the mantle–crust system. *Earth Planet. Sci. Lett.* **189**, 59–73.

Albarede F. and Hilst R. D. V. d. (1999) New mantle convection model may reconcile conflicting evidence. *EOS* **80**, 535, 537–539.

Allegre C. J., Poirier J. P., Humbler E., and Hofmann A. W. (1995) The chemical composition of the Earth. *Earth Planet. Sci. Lett.* **134**, 515–526.

Anderson D. L. (1989) Composition of the Earth. *Science* **243**, 367–370.

Arndt N., Ginibre C., Chauvel C., Albarede F., Cheadle M., Herzberg C., Jenner G., and Lahaye Y. (1998) Were komatiites wet? *Geology* **26**, 739–742.

Asahara Y. and Ohtani E. (2001) Melting relations of the hydrous primitive mantle in the CMAS-H_2O system at high pressures and temperatures, and implications for generation of komatiites. *Phys. Earth Planet. Inter.* **125**, 31–44.

Asimow P. D. (1999) A model that reconciles major- and trace-element data from abyssal peridotites. *Earth Planet. Sci. Lett.* **169**, 303–319.

Asimow P. D., Hirschmann M. M., and Stolper E. M. (1997) An analysis of variations in isentropic melt productivity. *Phil. Trans. Roy. Soc. London* **355**(A), 255–281.

Baker M. B. and Beckett J. R. (1998) The origin of abyssal peridotites: a reinterpretation of constraints based on primary bulk compositions. *Earth Planet. Sci. Lett.* **171**, 49–61.

Baker M. B. and Stolper E. M. (1994) Determining the composition of high-pressure mantle melts using diamond aggregates. *Geochim. Cosmochim. Acta* **58**, 2811–2827.

Baker M. B., Hirschmann M. M., Ghiorso M. S., and Stolper E. M. (1995) Compositions of near-solidus peridotite melts from experiments and thermodynamic calculations. *Nature* **375**, 308–311.

Beard B. L. and Glazner A. F. (1995) Trace element and Sr and Nd isotopic composition of the mantle xenoliths from the big pine volcanic field, California. *J. Geophys. Res.* **100**(B3), 4169–4179.

Beccaluva L., Bianchini G., Coltorti M., Perkins W. T., Siena F., Vaccaro C., and Wilson M. (2001a) Multistage evolution of the European lithospheric mantle: new evidence from Sardinia peridotite xenoliths. *Contrib. Mineral. Petrol.* **142**, 284–297.

Beccaluva L., Bonadiman C., Coltorti M., Salvini L., and Siena F. (2001b) Depletion events, nature of metasomatizing agent and timing of enrichment processes in lithospheric mantle xenoliths from the Veneto volcanic province. *J. Petrol.* **42**, 173–187.

Bernstein S., Kelemen P. B., and Brooks C. K. (1998) Depleted spinel harzburgite xenoliths in Tertiary dykes from East Greenland: restites from high degree melting. *Earth Planet. Sci. Lett.* **154**, 221–235.

Bertka C. M. and Holloway J. R. (1993) Pigeonite at solidus temperatures: implications for partial melting. *J. Geophys. Res.* **98**, 19755–19766.

Bina C. R. (1998) Lower mantle mineralogy and the geophysical perspective. In *Ultrahigh-Pressure Mineralogy: Physics and Chemistry of the Earth's Deep Interior* (ed. R. J. Hemley). Mineralogical Society of America, Washington, DC, vol. 37, pp. 205–240.

Bodinier J. D., Dupuy C., and Dostal J. (1988) Geochemistry and petrogenesis of Eastern Pyrenean peridotites. *Geochim. Cosmochim. Acta* **52**, 2893–2907.

Bonatti E. and Michael J. P. (1989) Mantle peridotites from continental rifts to ocean basins to subduction zones. *Earth Planet. Sci. Lett.* **91**, 297–311.

Bonatti E., Ottonello G., and Hamlyn P. R. (1986) Peridotites form the island of Zabargad (St. John), Red Sea: petrology and geochemistry. *J. Geophys. Res.* **91**, 599–631.

Boyd F. R. (1987) High- and low-temperature garnet peridotite xenoliths and their possible relation to the lithosphere-asthenosphere boundary beneath southern Africa. In *Mantle Xenoliths* (ed. P. H. Nixon). Wiley, London, pp. 403–412.

Boyd F. R. (1989) Compositional distinction between oceanic and cratonic lithosphere. *Earth Planet. Sci. Lett.* **96**, 15–26.

Boyd F. R. and Mertzman S. A. (1987) Composition and structure of the Kaapvaal lithosphere, southern Africa. In *Magmatic Processes: Physicochemical Principles*, The Geochemical Society Special Publication No. 1 (ed. B. O. Mysen), pp. 13–24.

Boyd F. R., Gurney J. J., and Richardson S. H. (1985) Evidence for a 150–200 km thick Archean lithosphere from diamond inclusion thermobarometry. *Nature* **315**, 387–389.

Boyd F. R., Pokhilenko N. P., Pearson D. G., Mertzman S. A., Sobolev N. V., and Finger L. W. (1997) Composition of the Siberian cratonic mantle: evidence from Udachnaya peridotite xenoliths. *Contrib. Mineral. Petrol.* **128**, 228–246.

Brandon A. D. and Draper D. S. (1996) Constraints on the origin of the oxidation state of mantle overlying subduction zones: an example from Simcoe, Washington, USA. *Geochim. Cosmochim. Acta* **60**(10), 1739–1749.

Bulatov V. K., Girnis A. V., and Brey G. P. (2002) Experimental melting of a modally heterogeneous mantle. *Mineral. Petrol.* **75**, 131–152.

Basaltic Volcanism Study Project (1981). Basaltic Volcanism on the Terrestrial Plants, Permagon, New York, 1286pp.

Canil D. (1991) Experimental evidence for the exsolution of cratonic peridotite from high-temperature harzburgite. *Earth Planet. Sci. Lett.* **106**, 64–72.

Canil D. (1992) Orthopyroxene stability along the peridotite solidus and the origin of cratonic lithosphere beneath southern Africa. *Earth Planet. Sci. Lett.* **111**, 83–95.

Cao R. and Zhu S. (1987) Mantle xenoliths and alkali-rich host rocks in eastern China. In *Mantle Xenoliths* (ed. P. H. Nixon). Wiley, London, pp. 167–180.

Carbno G. B. and Canil D. (2002) Mantle structure beneath the SW Slave craton, Canada: constraints from garnet geochemistry in the Drybones bay kimberlite. *J. Petrol.* **43**, 129–142.

Carlson R. W. (1994) Mechanisms of earth differentiation-consequences for the chemical-structure of the mantle. *Rev. Geophys.* **32**, 337–361.

Chen S., O'Relly S. Y., Zhou X., Griffin W. L., Zhang G., Sun M., Feng J., and Zhang M. (2001) Thermal and petrological structure of the lithosphere beneath Hannuoba, Shino-Korean Craton, China: evidence from xenoliths. *Lithos* **56**, 267–301.

Condie K. C. (1981) *Archean Greenstone Belts*. Elsevier, Amsterdam.

Condie K. C. (1998) Episodic continental growth and super-continents: a mantle avalanche connection? *Earth Planet. Sci. Lett.* **163**, 97–108.

Corgne A. and Wood B. (2002) $CaSiO_3$ and $CaTiO_3$ perovskite-melt partitioning of trace elements: implications for gross mantle differentiation. *Geophys. Res. Lett.* **29**, (article no. 1933), doi: 10.1029/2001GL014398.

Dalton J. A. and Presnall D. C. (1998) The continuum of primary carbonatitic-kimberlitic melt compositions in equilibrium with lherzolite: data from the system $CaO-MgO-Al_2O_3-SiO_2-CO_2$ at 6 GPa. *J. Petrol.* **39**, 1953–1964.

Dautria J. M. and Girod M. (1987) Cenozoic volcanism associated with swells and rifts. In *Mantle Xenoliths* (ed. P. H. Nixon). Wiley, London, pp. 195–214.

Davies G. F. (1995) Punctuated tectonic evolution of the earth. *Earth Planet. Sci. Lett.* **36**, 363–380.

Davies G. F. (1998) Plates, plumes, mantle convection, and mantle evolution. In *The Earth's Mantle: Composition, Structure and Evolution* (ed. I. Jackson). Cambridge University Press, Cambridge, pp. 228–258.

Dawson J. B., Smith J. V., and Hervig R. L. (1980) Heterogeneity in upper mantle lherzolites and harzburgites. *Phil. Trans. Roy. Soc. London* **297**(A), 323–331.

Dick H. J. B. (1989) Abyssal peridotites, very slow spreading ridges and ocean ridge magmatism. In *Magmatism in the Ocean Basins*, Geological Society of London Special Publication 42 (eds. A. D. Saunder and M. J. Norry). Geological Society of London, 71pp.

Dick H. J. B. and Bullen T. (1984) Chromian spinel as a petrogenetic indicator in abyssal and alpine-type peridotites and spatially associated lavas. *Contrib. Mineral. Petrol.* **86**, 54–76.

Dick H. J. B., Fisher R. L., and Bryan W. B. (1984) Mineralogic variability of the uppermost mantle along mid-ocean ridges. *Earth Planet. Sci. Lett.* **69**, 88–106.

Downes H. (2001) Formation and modification of the shallow sub-continental lithospheric mantle: a review of geochemical evidence from ultramafic suites and tectonically emplaced ultramafic massifs of western and central Europe. *J. Petrol.* **42**, 233–250.

Downes H. and Dupuy C. (1987) Textual, isotopic and REE variations in spinel peridotite xenoliths, Massif Central, France. *Earth Planet. Sci. Lett.* **82**, 121–135.

Downes H., Embey-Isztin A., and Thirwall M. F. (1992) Petrology and geochemistry of spinel peridotite xenoliths from the western Pannonian Basin (Hungary): evidence for an association between enrichment and texture in the upper mantle. *Contrib. Mineral. Petrol.* **109**, 340–354.

Ehrenberg S. N. (1982) Petrogenesis of garnet lherzolite and megacrystalline nodules from the Thumb, Navajo volcanic field. *J. Petrol.* **23**(4), 507–547.

Elthon D. (1992) Chemical trends in abyssal peridotites. *J. Geophys. Res.* **97**, 9015–9025.

Fabriès J., Bodinier J., Dupuy C., Lorand J., and Benkerrou C. (1989) Evidence for model metasomatism in the orogenic spinel lherzolite body from Caussou (Northeastern Pyrenees, France). *J. Petrol.* **30**(1), 199–228.

Fabriès J., Lorand J.-P., Bodinier J.-L., and Dupuy C. (1991) Evolution of the upper mantle beneath the Pyrenees: evidence from orogenic spinel lhezolite massifs. *J. Petrol.* (Special Lherzolites Issue) 55–76.

Falloon T. J., Green D. H., Hatton C. J., and Harris K. L. (1988) Anhydrous partial melting of a fertile and depleted peridotite from 2 to 30 kb and application to basalt petrogenesis. *J. Petrol.* **29**, 1257–1282.

Francis D. (1987) Mantle-melt interaction recorded in spinel lherzolite xenoliths from the Alligator Lake volcanic complex, Yukon, Canada. *J. Petrol.* **28**, 569–597.

Frey F. A. and Prinz M. (1978) Ultramafic inclusions from San Carlos, Arizona: petrologic and geochemical data bearing on their petrogenesis. *Earth Planet. Sci. Lett.* **38**, 129–176.

Frey F. A., Suen C. J., and Stockman H. W. (1985) The Ronda high temperature peridotite: geochemistry and petrogenesis. *Geochim. Cosmochim. Acta* **49**, 2469–2491.

Frey F. A., Shimizu N., Leinbach A., Obata M., and Takazawa E. (1991) Compositional variations within the low layered zone of the Horoman Peridotite, Hokkaido, Japan: constraints on models for melt-solid segregation. *J. Petrol.* (Special Lherzolites Issue) 211–227.

Gaetani G. A. and Grove T. L. (1998) The influence of water on melting of mantle peridotite. *Contrib. Mineral. Petrol.* **131**, 323–346.

Gessmann C. K., Wood B. J., Rubie D. C., and Kilburn M. R. (2001) Solubility of silicon in liquid metal at high pressure: implications for the composition of the Earth's core. *Earth Planet. Sci. Lett.* **184**, 367–376.

Ghiorso M. S. (1985) Chemical mass transfer in magmatic processes: I. Thermodynamic relations and numerical algorithms. *Contrib. Mineral. Petrol.* **90**, 107–120.

Ghiorso M. S., Hirschmann M. M., Reiners P. W., and III V. C. Kress (2002) The pMELTS: a revision of MELTS for improved calculation of phase relations and major-element partitioning related to partial melting of the mantle to 3 GPa. *Geochem. Geophys. Geosys.* **3**, 1030, doi: 10.1029/2001GC000217.

Girod M., Dautria J. M., and Giovanni R. D. (1987) A first insight into the constitution of the upper mantle under the Hoggar area (Southern Algeria): the lherzolite xenoliths in the alkali-basalts. *Mantle Xenoliths* **77**, 66–73.

Godard M., Jousselin D., and Bodinier J. (2000) Relationships between geochemistry and structure beneath a paleo-spreading centre: a study of the mantle section in the Oman ophiolite. *Earth Planet. Sci. Lett.* **180**, 133–148.

Griffin W. L., Smith D., Boyd F. R., Cousens D. R., Ryan C. G., Sie S. H., and Suter G. F. (1989) Trace element zoning in garnets from sheared mantle xenoliths. *Geochim. Cosmochim. Acta* **53**, 561–567.

Griffin W. L., O'Reilly S. Y., Ryan C. G., Gaul O., and Ionov D. A. (1998) Secular variation in the composition of subcontinental lithospheric mantle. In *Structure and Evolution of the Australian Continent*, Geodynamic Series (eds. J. Braun, J. C. Dooley, B. R. Goleby, R. D. V. d. Hilst, and C. T. Klootwijk). American Geophysical Union, Washington, DC, vol. 26, pp. 1–26.

Griffin W. L., Doyle B. J., Ryan C. G., Pearson N. J., O'Reilly S. Y., Davies R., Kivi K., Achterbergh E. V., and Natapov L. M. (1999a) Layered mantle lithosphere in the Lac de Gras Area, Slave craton: composition, structure and origin. *J. Petrol.* **40**, 705–727.

Griffin W. L., O'Reilly S. Y., and Ryan C. G. (1999b) The composition and origin of sub-continental lithospheric mantle. In *Mantle Petrology: Field Observations and High-Pressure Experimentation*: A Tribute to Francis R. (Joe) Boyd. Vol. Special Publication No. 6 (eds. Y. Fei, C. Bertka, and B. Mysen). The Geochemical Society, pp. 13–46.

Gruau G., Lecuyer C., Bernard-Griffiths J., and Morin N. (1991) Origin and petrogenesis of the Trinity Ophiolite complex (California): new constraints from REE and Nd isotope data. *J. Petrol.* (Special Lherzolite Issue) 229–242.

Gudfinnsson G. H. and Presnall D. C. (1996) Melting relations of model lherzolite in the system $CaO-MgO-Al_2O_3-SiO_2$

at 2.4 to 3.4 GPa and the generation of komatiites. *J. Geophys. Res.* **101**, 27701–27709.

Gudfinnsson G. H. and Presnall D. C. (2000) Melting behavior of model lherzolite in the system $CaO-MgO-Al_2O_3-SiO_2-FeO$ at 0.7 to 2.8 GPa. *J. Petrol.* **41**, 1241–1269.

Hart S. R. and Zindler A. (1986) In search of a bulk-Earth composition. *Chem. Geol.* **57**, 247–267.

Hart S. R. and Zindler A. (1989) Constraints on the nature and development of chemical heterogeneities in the mantle. In *Mantle Convection: Plate Tectonics and Global Dynamics* (ed. W. R. Peltier). Gordon and Breach, vol. 4, pp. 261–387.

Hartmann G. and Wedepohl K. H. (1993) The composition of peridotite tectonites from the Ivrea complex, northern Italy: residues from melt extraction. *Geochim. Cosmochim. Acta* **57**, 1761–1782.

Hauri E. H. (1996) Major-element variability in the Hawaiian mantle plume. *Nature* **382**, 415–419.

Hauri E. H. (2002) Osmium isotopes and mantle convection. *Phil. Trans. Roy. Soc. London* **360**(A), 2371–2382.

Hedlin M. A. H., Shearer P. M., and Earle P. S. (1997) Seismic evidence for small-scale heterogeneity throughout the Earth's mantle. *Nature* **387**, 145–150.

Heinrich W. and Besch T. (1992) Thermal history of the upper mantle beneath a young back-arc extensional zone: ultramafic xenoliths from the San Luis potosì, Central Mexico. *Contrib. Mineral. Petrol.* **111**, 126–142.

Helffrich G. R. and Wood B. J. (2001) The Earth's mantle. *Nature* **412**, 501–507.

Herzberg C. (1999) Phase equilibrium constraints on the formation of cratonic mantle. In *Mantle Petrology: Field Observations and High-Pressure Experimentation*: A Tribute to Francis R. (Joe) Boyd. Vol. Special Publication No. 6 (eds. Y. Fei, C. Bertka, and B. Mysen). The Geochemical Society, pp. 241–258.

Herzberg C. and Zhang J. (1996) Melting experiments on anhydrous peridotite KLB-1: compositions of magmas in the upper mantle and transition zone. *J. Geophys. Res.* **101**, 8271–8295.

Herzberg C., Feigenson M., Skuba C., and Ohtani E. (1988) Majorite fractionation recorded in the geochemistry of peridotites from South Africa. *Nature* **332**, 823–826.

Herzberg C. T. (1993) Lithosphere peridotites of the Kaapvaal craton. *Earth Planet. Sci. Lett.* **120**, 13–29.

Hirose K. and Kushiro I. (1993) Partial melting of dry peridotites at high pressures: determination of compositions of melts segregated from peridotite using aggregates of diamond. *Earth Planet. Sci. Lett.* **114**, 477–489.

Hirose K. and Kawamoto T. (1995) Hydrous partial melting of lherzolite at 1 GPa: the effect of H_2O on the genesis of basaltic magmas. *Earth Planet. Sci. Lett.* **133**, 463–473.

Hirschmann M. M. (2000) Mantle solidus: experimental constraints and the effects of peridotite composition. *Geochem. Geophys. Geosys.* **1**, paper number 2000GC000070.

Hirschmann M. M. and Stolper E. M. (1996) A possible role for garnet pyroxenite in the origin of the "garnet signature" in MORB. *Contrib. Mineral. Petrol.* **124**, 185–208.

Hirschmann M. M., Ghiorso M. S., Wasylenki L. E., Asimow P. D., and Stolper E. M. (1998) Calculation of peridotite partial melting from thermodynamic models of minerals and melts: I. Review of methods and comparison with experiments. *J. Petrol.* **39**, 1091–1115.

Hirschmann M. M., Asimow P. D., Ghiorso M. S., and Stolper E. M. (1999a) Calculation of peridotite partial melting from thermodynamic models of minerals and melts: III. Controls on isobaric melt production and the effect of water on melt production. *J. Petrol.* **40**, 831–851.

Hirschmann M. M., Ghiorso M. S., and Stolper E. M. (1999b) Calculation of peridotite partial melting from thermodynamic models of minerals and melts: II. Isobaric variations in melts near the solidus owing to variable source composition. *J. Petrol.* **40**, 297–313.

Hirschmann M. M., Kogiso T., Baker M. B., and Stolper E. M. (2003) Alkalic magmas generated by partial melting of garnet pyroxenite. *Geology* **31**, 481–484.

Hofmann A. W. (1988) Chemical differentiation of the Earth: the relationship between mantle, continental crust, and oceanic crust. *Earth Planet. Sci. Lett.* **90**, 297–314.

Hofmann A. W. (1997) Mantle geochemistry: the message from oceanic volcanism. *Nature* **385**, 219–229.

Hunter R. H. and Upton B. G. J. (1987) The British Isles—a Paleozoic mantle sample. In *Mantle Xenoliths* (ed. P. H. Nixon). Wiley, London, pp. 107–118.

Ionov D. A., Ashchepkov I. V., Stosch H.-G., Witt-Eickscen G., and Seck H. A. (1993) Garnet peridotite xenoliths from the Vitim volcanic field, Baikal region: the nature of the garnet-spinel peridotite transition zone in the continental mantle. *J. Petrol.* **34**(6), 1141–1175.

Ionov D. A., O'Reilly S. Y., Genshaft Y. S., and Kopylova M. G. (1996) Carbonate-bearing mantle peridotite xenoliths from Spitsbergen: phase relationships, mineral compositions and trace-element residence. *Contrib. Mineral. Petrol.* **125**, 375–392.

Jackson I. and Rigden S. M. (1998) Composition and temperature of the Earth's mantle: seismological models interpreted through experimental studies of Earth materials. In *The Earth's Mantle: Composition, Structure and Evolution* (ed. I. Jackson). Cambridge University Press, Cambridge, pp. 405–460.

Jagoutz E., Palme H., Baddenhausen H., Blum K., Cendales K., Dreibus G., SPettel B., Lorenz V., and Wanke H. (1979) The abundances of major, minor and trace elements in the Earth's mantle as derived from primitive ultramafic nodules. *Proc. 10th Lunar Planet Sci. Conf.*, 2031–2050.

Jakes P. and Vokurka K. (1987) Central Europe. In *Mantle Xenoliths* (ed. P. H. Nixon). Wiley, London, pp. 149–154.

Johnson K. T. M., Dick H. J. B., and Shimizu N. (1990) Melting in the oceanic upper mantle: an ion microprobe study of diopsides in abyssal peridotites. *J. Geophys. Res.* **95**, 2661–2678.

Jordan T. H. (1978) Composition and development of the continental tectosphere. *Nature* **274**, 544–548.

Kato T., Ringwood A. E., and Irifune T. (1988) Experimental determination of element partitioning between silicate perovskites, garnets, and liquids: constraints on early differentiation of the mantle. *Earth Planet. Sci. Lett.* **89**, 123–145.

Kawamoto T. and Holloway J. R. (1997) Melting temperature and partial melt chemistry of H_2O-saturated mantle peridotite to 11 GPa. *Science* **276**, 240–243.

Kelemen P. B., Dick H. J. B., and Quick J. E. (1992) Formation of harzburgite by pervasive melt/rock reaction in the upper mantle. *Nature* **358**, 635–641.

Kelemen P. B., Hirth G., Shimizu N., Spiegelman M., and Dick H. J. B. (1997) A review of melt migration processes in the adiabatically upwelling mantle beneath oceanic spreading ridges. *Phil. Trans. Roy. Soc. London* **355**(A), 283–318.

Kelemen P. B., Hart S. R., and Bernstein S. (1998) Silica enrichment in the continental upper mantle via melt/rock reaction. *Earth Planet. Sci. Lett.* **164**, 387–406.

Kellogg L. H., Hager B. H., and Hilst R. D. V. d. (1999) Compositional stratification in the deep mantle. *Science* **283**, 1881–1884.

Kennett B. L. N. and Hilst R. D. V. d. (1998) Seismic structure of the mantle: from subduction zone to craton. In *The Earth's Mantle: Composition, Structure and Evolution* (ed. I. Jackson). Cambridge University Press, Cambridge, pp. 381–404.

Kesson S. E. and Ringwood A. E. (1989) Slab-mantle interactions: 1. Sheared and refertilized garnet peridotite xenoliths-samples of wadati-benioff zones? *Chem. Geol.* **78**, 89–96.

Kinzler R. J. (1997) Melting of mantle peridotite at pressures approaching the spinel to garnet transition: application to mid-ocean ridge basalt petrogenesis. *J. Geophys. Res.* **102**, 853–874.

Kinzler R. J. and Grove T. L. (1992a) Primary magmas of mid-ocean ridge basalts: 1. Experiments and methods. *J. Geophys. Res.* **97**(B5), 6885–6906.

Kinzler R. J. and Grove T. L. (1992b) Primary magmas of mid-ocean ridge basalts: 2. Applications. *J. Geophys. Res.* **97**(B5), 6907–6926.

Kinzler R. J. and Grove T. L. (1993) Corrections and further discussion of the primary magmas of mid-ocean ridge basalts, 1 and 2. *J. Geophys. Res.* **98**(B12), 22339–22347.

Klein E. M. and Langmuir C. H. (1987) Global correlations of mid-ocean ridge basalt chemistry with axial depth and crustal thickness. *J. Geophys. Res.* **92**, 8089–8115.

Kogiso T. and Hirschmann M. M. (2001) Experimental study of clinopyroxenite partial melting and the origin of ultra-calcic melt inclusions. *Contrib. Mineral. Petrol.* **142**, 347–360.

Kogiso T., Hirose K., and Takahashi E. (1998) Melting experiments on homogeneous mixtures of peridotite and basalt: application to the genesis of ocean island basalts. *Earth Planet. Sci. Lett.* **162**, 45–61.

Kopylova M. G. and Russell J. K. (2000) Chemical stratification of cratonic lithosphere: constraints from the Northern Slave craton, Canada. *Earth Planet. Sci. Lett.* **181**, 71–87.

Kushiro I. (1972) Effect of water on the composition of magmas formed at high pressures. *J. Petrol.* **13**(part 2), 311–334.

Kushiro I. (1998) Compositions of partial melts formed in mantle peridotites at high pressure and their relation to those of primitive MORB. *Phys. Earth Planet. Inter.* **107**, 103–110.

Langmuir C. H., Klein E. M., and Plank T. (1992) Petrological systematics of mid-ocean ridge basalts: constraints on melt generation beneath oceanic ridges. In *Mantle Flow and Melt Generation at Mid-Ocean Ridges*, Vol. Geophysical Monograph 71 (eds. J. P. Morgan, D. K. Blackman, and J. M. Sinton). American Geophysical Union, Washington, DC, pp. 183–280.

Lee C.-T. and Rudnick R. L. (1999) Compositionally stratified cratonic lithosphere: petrology and geochemistry of peridotite xenoliths from the Labait tuff cone, Tanzania. In *The Nixon Volume, Proc. 7th Int. Kimberlite Conference* (eds. J. J. Gurney, J. L. Gurney, M. D. Pascoe, and S. R. Richardson). Red Roof Designs, pp. 503–521.

Lee W. J., Huang W. L., and Wyllie P. (2000) Melts in the mantle modeled in the system $CaO-MgO-SiO_2-CO_2$ at 2.7 GPa. *Contrib. Mineral. Petrol.* **138**, 199–213.

Lenoir X., Garrido C. J., Bodnier J., Dautria J., and Gervilla F. (2001) The recrystallization front of the Ronda peridotite: evidence for melting and thermal erosion of subcontinental lithospheric mantle beneath the Alboran basin. *J. Petrol.* **42**(1), 141–158.

Liang Y. and Elthon D. (1990) Geochemistry and petrology of spinel lherzolite xenoliths from Xalapasco de la Joya, San Luis Potosi, Mexico: partial melting and mantle metasomatism. *J. Geophys. Res.* **95**(10), 15859–15877.

Longhi J. and Bertka C. M. (1996) Graphical analysis of pigeonite-augite liquidus equilibria. *Am. Mineral.* **81**, 685–695.

Lugovic B., Alther R., Raczek I., Hofmann A. W., and Majer V. (1991) Geochemistry of peridotites and mafic igneous rocks from the Central Dinaric Ophiolite Belt, Yugoslavia. *Contrib. Mineral. Petrol.* **106**, 201–216.

Luhr J. F. and Aranda-Gomez J. J. (1997) Mexican peridotite xenoliths and tectonic terranes: correlations among vent location, texture, temperature, pressure, and oxygen fugacity. *J. Petrol.* **38**(38), 1075–1112.

Lundstrom C. C., Sampson D. E., Perfit M. R., Gill J., and Williams Q. (1999) Insights into mid-ocean ridge basalt petrogenesis: u-series disequilibria from the Siqueiros transform, Lamont Seamounts, and East Pacific Rise. *J. Geophys. Res.* **104**, 13035–13048.

Lundstrom C. C., Gill J., and Williams Q. (2000) A geochemically consistent hypothesis for MORB generation. *Chem. Geol.* **162**, 105–126.

MacKenzie J. M. and Canil D. (1999) Composition and thermal evolution of cratonic mantle beneath the central Archean Slave province, NWT, Canada. *Contrib. Mineral. Petrol.* **134**, 313–324.

McDonough W. F. (1990) Constraints on the composition of the continental lithospheric mantle. *Earth Planet. Sci. Lett.* **101**, 1–18.

McDonough W. F. and Rudnick R. L. (1998) Mineralogy and composition of the upper mantle. In *Ultrahigh-Pressure Mineralogy: Physics and Chemistry of the Earth's Deep Interior* (ed. R. J. Hemley). Mineralogical Society of America, Washington, DC, vol. 37, pp. 139–164.

McDonough W. F. and Sun S.-S. (1995) The composition of the Earth. *Chem. Geol.* **120**, 223–253.

McFarlane E. A., Drake M. J., and Rubie D. C. (1994) Element partitioning between Mg-perovskite, magnesiowustite, and silicate melt at conditions of the Earth's mantle. *Geochim. Cosmochim. Acta* **58**, 5161–5172.

McKenzie D. (1984) The generation and compaction of partially molten rock. *J. Petrol.* **25**, 713–765.

McKenzie D. and Bickle M. J. (1988) The volume and composition of melt generated by extension of the lithosphere. *J. Petrol.* **29**, 625–679.

Melosh H. J. (1990) Giant impacts and the thermal state of the early earth. In *Origin of the Earth* (eds. H. E. Newsom and J. H. Jones). Oxford University Press, New York, pp. 69–83.

Menzies M. A. and Dupuy C. (1991) Orogenic massifs: protolith, process and provenance. *J. Petrol.* (Special Lherzolites Issue) 1–16.

Michael P. J. (1988) The concentration, behavior and storage of H_2O in the suboceanic upper mantle-implications for mantle metasomatism. *Geochim. Cosmochim. Acta* **52**, 555–566.

Morten L. (1987) A review of xenolithic occurrences and their comparison with Alpine peridotites. In *Mantle Xenoliths* (ed. P. H. Nixon). Wiley, London.

Nimz G. J., Cameron K. L., and Niemeyer S. (1995) Formation of mantle lithosphere beneath northern Mexico: chemical and Sr–Nd–Pb isotopic systematics of peridotite xenolith from La Olivina. *J. Geophys. Res.* **100**(B3), 4181–4196.

Niu Y. and Batiza R. (1991) An empirical method for calculating melt compositions produced beneath mid-oceanic ridges: application for axis and off-axis (seamount) melting. *J. Geophys. Res.* **96**, 21753–21777.

Niu Y., Langmuir C. H., and Kinzler R. J. (1997) The origin of abyssal peridotites: a new perspective. *Earth Planet. Sci. Lett.* **152**, 251–265.

O'Hara M. J. (1968) The bearing of phase equilibria studies in synthetic and natural systems on the origin and evolution of basic and ultrabasic rocks. *Earth Sci. Rev.* **4**, 69–134.

Ohtani E., Mibe K., and Kato T. (1996) Origin of cratonic peridotite and komatiite: evidence for melting in the wet Archean mantle. *Proc. Japan Acad.* **72**(B), 113–117.

O'Neill H. S. C. and Palme H. (1998) Composition of the silicate earth: implications for accretion and core formation. In *The Earth's Mantle: Composition, Structure and Evolution* (ed. I. Jackson). Cambridge University Press, Cambridge, pp. 3–126.

O'Reilly S. Y. and Griffin W. L. (1987) Eastern Australia-4000 kilometres of mantle samples. In *Mantle Xenoliths* (ed. P. H. Nixon). Wiley, London, pp. 267–280.

O'Reilly S. Y. and Griffin W. L. (1988) Mantle metasomatism beneath western Victoria, Australia: I. Metasomatic processes in Cr-diopside lherzolites. *Geochim. Cosmochim. Acta* **52**, 433–447.

Palme H. and Nickel K. G. (1985) Ca/Al ratio and composition of the earths upper mantle. *Geochim. Cosmochim. Acta* **49**, 2123–2132.

Parman S. W., Dann J. C., Grove T. L., and Wit M. J. d. (1997) Emplacement conditions of komatiite magmas from the 3.49 Ga komati formation, Barberton Greenstone belt, South Africa. *Earth Planet. Sci. Lett.* **150**, 303–323.

Pearson D. G. (1999) The age of continental roots. *Lithos* **48**, 171–194.

Pearson D. G., Boyd F. R., Hoal K. E. O., Hoal B. G., Nixon P. H., and Rogers N. W. (1994) A Re–Os isotopic and petrological study of Namibian peridotites: contrasting petrogenesis and composition of on- and off-craton lithospheric mantle. *Min. Mag.* **58** A, 703–704.

Pickering-Witterer J. and Johnston A. D. (2000) The effects of variable bulk composition on the melting systematics of fertile peridotitic assemblages. *Contrib. Mineral. Petrol.* **140**, 190–211.

Pertermann M. and Hirschmann M. M. (2003) Anhydrous partial melting experiments on MORB-like eclogite: phase relations, phase compositions and mineral/melt partitioning of major elements at 2–3 GPa. *J. Geophys. Res.* **108**, doi: 10.1029/2000JB000118.

Presnall D. C., Dixon T. H., O'Donell T. H., and Dixon S. A. (1979) Generation of mid-ocean ridge tholeiites. *J. Petrol.* **20**, 3–35.

Presnall D. C., Gudfinnsson G. H., and Walter M. J. (2002) Generation of mid-ocean ridge basalts at pressures from 1 to 7 GPa. *Geochim. Cosmochim. Acta* **66**, 2073–2090.

Press S., Witt G., Seck H. A., Eonov D., and Kovalenko V. I. (1986) Spinel peridotite xenoliths from the Tariat depression, Mongolia: I. Major element chemistry and mineralogy of a primitive mantle xenolith suite. *Geochim. Cosmochim. Acta* **50**, 2587–2599.

Qi Q., Taylor L. A., and Zhou X. (1995) Petrology and geochemistry of mantle peridotite xenoliths from SE China. *J. Petrol.* **36**(1), 55–79.

Rampone E. and Morten L. (2001) Records of crustal metasomatism in the garnet peridotites of the Ulten zone (Upper Austroalpine, Eastern Alps). *J. Petrol.* **42**(1), 207–219.

Rampone E., Hofmann A. W., Piccardo G. B., Vannucci R., Bottazzi P., and Ottolini L. (1995) Petrology, mineral and isotope geochemistry of the external Liguride peridotites (Northern Apennines, Italy). *J. Petrol.* **36**(1), 81–105.

Reid J. B. and Woods G. A. (1978) Oceanic mantle beneath the southern Rio Grande Rift. *Earth Planet. Sci. Lett.* **41**, 303–316.

Ringwood A. E. (1977) Composition of the core and implications for the origin of the earth. *Geochem. J.* **11**, 111–135.

Ringwood A. E. (1979) *Origin of the Earth and Moon*. Springer, New York, 295pp.

Rivalenti G., Mazzucchelli M., Vannucci R., Hofmann A. W., Ottolini L., Bpttazzi P., and Obermiller W. (1995) The relationship between websterite and peridotite in the Balmuccia peridotite massif (NW Italy) as revealed by trace element variations in clinopyroxene. *Contrib. Mineral. Petrol.* **121**, 275–288.

Rivalenti G., Mazzucchelli M., Girardi V. A. V., Vannucci R., Barbieri M. A., Zanetti A., and Goldstein S. L. (2000) Composition and processes of the mantle lithosphere in northeastern Brazil and Fernando de Noronha: evidence from mantle xenoliths. *Contrib. Mineral. Petrol.* **138**, 308–325.

Robinson J. A. C. and Wood B. J. (1998) The depth of the spinel to garnet transition at the peridotite solidus. *Earth Planet. Sci. Lett.* **164**, 277–284.

Robinson J. A. C., Wood B. J., and Blundy J. D. (1998) The beginning of melting of fertile and depleted peridotite at 1.5 GPa. *Earth Planet. Sci. Lett.* **155**, 97–111.

Roden M. F., Irving A. J., and Murthy V. R. (1988) Isotopic and trace element composition of the upper mantle beneath a young continental rift: results from Kilbourne hole, New Mexico. *Geochim. Cosmochim. Acta* **52**, 461–473.

Rudnick R. L., McDonough W. F., and Orpin A. (1994) Northern Tanzanian peridotite xenoliths: a comparison with Kaapvaal peridotites and inferences on metasomatic interactions. In *Kimberlites, Related Rocks, and Mantle Xenoliths*, Proc 5th Int. Kimberlite Conf., Vol. 1, CPRM Spec. Publ. 1A (eds. H. O. A. Meyer and O. H. Leonardas). CPRM, Brasilia, vol. 1, pp. 336–353.

Rudnick R. L., McDonough W. F., and O'Connell R. J. (1998) Thermal structure, thickness and composition of continental lithosphere. *Chem. Geol.* **145**, 395–411.

Saal A. E., Hart S. R., Shimizu N., Hauri E. H., and Layne G. D. (1998) Pb isotopic variability in melt inclusions from oceanic island basalts, Polynesia. *Science* **282**, 1481–1484.

Saal A. E., Hauri E. H., Langmuir C. H., and Perfit M. R. (2002) Vapour undersaturation in primitive mid-ocean-ridge basalt and the volatile content of Earth's upper mantle. *Nature* **419**, 451–455.

Salters V. J. M. and Dick H. J. B. (2002) Mineralogy of the mid-ocean-ridge basalt source from neodymium isotopic composition of abyssal peridotites. *Nature* **418**, 68–72.

Schmidberger S. S. and Francis D. (1999) Nature of the mantle roots beneath the North American craton: mantle xenolith evidence from Somerset Island kimberlites. *Lithos* **48**, 195–216.

Schulze D. J. (1989) Constraints on the abundance of eclogite in the upper mantle. *J. Geophys. Res.* **94**, 4205–4212.

Shervais J. W. and Mukasa S. B. (1991) The Balmuccia orogenic lherzolite massif, Italy. *J. Petrol.* (Special Lithosphere Issue) 155–174.

Shi L., Francis D., Ludden J., Frederiksen A., and Bostok M. (1998) Xenolith evidence for lithospheric melting above anomalously hot mantle under the northern Canadian Cordillera. *Contrib. Mineral. Petrol.* **131**, 39–53.

Shimizu N. and Sobolev A. (1998) *In-situ* Pb isotopic analysis of olivine-hosted melt inclusions from mid-ocean ridges. *EOS Trans*, **79**, f790.

Shimizu K., Komiya T., Hirose K., Shimizu N., and Maruyama S. (2001) Cr-spinel, an excellent micro-container for retaining primitive melts-implications for a hydrous plume origin for komatiites. *Earth Planet. Sci. Lett.* **189**, 177–188.

Siena F., Beccaluva L., Coltorti M., Marchesi S., and Morra V. (1991) Ridge to hot-spot evolution of the Atlantic lithospheric mantle: evidence from Lanzarote peridotite xenoliths (Canary Islands). *J. Petrol.* (Special Lherzolites Issue) 271–290.

Smith D., Griffin W. L., and Ryan C. G. (1993) Compositional evolution of high-temperature sheared lherzolite PHN1611. *Geochim. Cosmochim. Acta* **57**, 605–613.

Song Y. and Frey F. A. (1987) Geochemistry of peridotite xenoliths in basalt from Hannuoba, Eastern China: implications for subcontinental mantle heterogeneity. *Geochim. Cosmochim. Acta* **53**, 97–113.

Stern C. R., Kilian R., Olker B., Hauri E. H., and Kyser T. K. (1999) Evidence from mantle xenoliths for relatively thin (< 100 km) continental lithosphere below the Phanerozoic crust of southernmost South America. *Lithos* **48**, 217–235.

Stone W. E., Deloule E., Larson M. S., and Lesher C. M. (1997) Evidence for hydrous high-MgO melts in the Precambrian. *Geology* **25**, 143–146.

Takahashi E. (1986) Melting of a dry peridotite KLB-1 up to 14 GPa: implications on the origin of peridotitic upper mantle. *J. Geophys. Res.* **91**, 9367–9382.

Takahashi E. (1990) Speculations on the Archean mantle: missing link between komatiite and depleted garnet peridotite. *J. Geophys. Res.* **95**, 15941–15954.

Takahashi E. and Kushiro I. (1983) Melting of a dry peridotite at high pressures and basalt magma genesis. *Am. Mineral.* **68**, 859–879.

Takahashi E. and Nakajima K. (2002) Melting process in the Hawaiian plume. In *Hawaiian Volcanoes: Deep Underwater Perspectives*, Vol. Geophysical Monograph 128 (eds. E. Takahashi, P. W. Lipman, M. O. Garcia, J. Naka, and S. Aramaki). American Geophysical Union, Washington, DC, pp. 403–418.

Takahashi E., Shimazaki T., Tsuzaki Y., and Yoshida H. (1993) Melting study of peridotite KLB-1 to 6.5 GPa, and the origin of basaltic magmas. *Phil. Trans. Roy. Soc. London* **342**(A), 105–120.

Takahashi E., Nakajima K., and Wright T. L. (1998) Origin of the Columbia river basalts: melting model of a heterogeneous plume head. *Earth Planet. Sci. Lett.* **162**, 63–80.

van Keken P. E. and Zhong S. J. (1999) Mixing in a 3D spherical model of present-day mantle convection. *Earth Planet. Sci. Lett.* **171**, 533–547.

van Keken P. E., Hauri E. H., and Ballentine C. J. (2002) Mantle mixing: the generation, preservation, and destruction of chemical heterogeneity. *Ann. Rev. Earth Planet. Sci.* **30**, 493–525.

Vaselli O., Downes H., Thirwall M., Dobosi G., Coradossi N., Seghedi I., Szakacs A., and Vannucci R. (1995) Ultramafic xenoliths in plio-pleistocene alkali basalts from the eastern Transylvanian basin: depleted mantle enriched by vein metasomatism. *J. Petrol.* **36**(1), 23–53.

Walter M. J. (1998) Melting of garnet peridotite and the origin of komatiite and depleted lithosphere. *J. Petrol.* **39**, 29–60.

Walter M. J. (1999a) Comments on 'Mantle melting and melt extraction process beneath ocean ridges: evidence from abyssal peridotites' by Yaoling Niu. *J. Petrol.* **40**, 1187–1193.

Walter M. J. (1999b) Melting residues of fertile peridotite and the origin of cratonic lithosphere. In *Mantle Petrology: Field Observations and High-pressure Experimentation*: A Tribute to Francis R. (Joe) Boyd, Special Publication No. 6 (eds. Y. Fei, C. Bertka, and B. Mysen). The Geochemical Society, pp. 225–240.

Walter M. J. and Presnall D. C. (1994) Melting behavior of simplified lherzolite in the system $CaO-MgO-Al_2O_3-SiO_2-Na_2O$ from 7 to 35 kbar. *J. Petrol.* **35**, 329–359.

Walter M. J., Sisson T. W., and Presnall D. C. (1995) A mass proportion method for calculating melting reactions and application to melting of model upper mantle lherzolite. *Earth Planet. Sci. Lett.* **135**, 77–90.

Walter M., Katsura T., Kubo A., Shinmei T., Nishikawa O., Ito E., Lesher C., and Funakoshi K. (2002) Spinel-garnet lherzolite transition in the system $CaO-MgO-Al_2O_3-SiO_2$ revisited: an *in situ* X-ray study. *Geochim. Cosmochim. Acta* **66**, 2109–2121.

Weng Y. H. and Presnall D. C. (2001) The system diopside-forsterite-enstatite at 5.1 GPa: a ternary model for melting of the mantle. *Can. Mineral.* **39**, 299–308.

Wiechert U., Ionov D. A., and Wedepohl K. H. (1997) Spinel peridotite xenoliths from the Atsagin-Dush volcano, Dariganga lava plateau, Mongolia: a record of partial melting and cryptic metasomatism in the upper mantle. *Contrib. Mineral. Petrol.* **126**, 346–364.

Wilson M. (1989) *Igneous Petrogenesis: A Global Tectonic Approach*. Kluwer, Dordrecht.

Xu Y., Menzies M. A., Vroon P., Mercier J., and Lin C. (1988) Texture-temperature-geochemistry relationships in the upper mantle as revealed from spinel peridotite xenoliths from Wangqing, NE China. *J. Petrol.* **39**(3), 469–493.

Xue X., Baadsgaard H., Irving A. J., and Scarfe C. M. (1990) Geochemical and isotopic characteristics of lithospheric mantle beneath West Kettle river, British Columbia: evidence from ultramafic xenoliths. *J. Geophys. Res.* **95**(B10), 15879–15891.

Yang H.-J., Sen G., and Shimizu N. (1998) Mid-ocean ridge melting: constraints from lithospheric xenoliths at Oahu, Hawaii. *J. Petrol.* **39**, 277–295.

Yaxley G. M. and Green D. H. (1998) Reactions between eclogite and peridotite: mantle refertilisation by subduction of oceanic crust. *Schweiz. Mineral. Petrograph. Mitteil.* **78**, 243–255.

Zangana N. A., Downes H., Thirlwall M. F., Marriner G. F., and Bea F. (1999) Geochemical variation in peridotite xenoliths and their constituent clinopyroxenes from ray pic (French Massif Central): implications for the compositions of the shallow lithospheric mantle. *Chem. Geol.* **153**, 11–35.

Zhang J. and Herzberg C. (1994) Melting experiments on anhydrous peridotite KLB-1 from 5.0 to 22.5 GPa. *J. Geophys. Res.* **99**(B9), 17729–17742.

Zindler A. and Hart S. (1986) Chemical geodynamics. *Ann. Rev. Earth Planet. Sci.* **14**, 493–571.

2.09
Trace Element Partitioning under Crustal and Uppermost Mantle Conditions: The Influences of Ionic Radius, Cation Charge, Pressure, and Temperature

B. J. Wood and J. D. Blundy

University of Bristol, UK

2.09.1 INTRODUCTION

The controls on partitioning of trace elements between crystals and silicate melts were initially the subject of crystal-chemical, rather than petrogenetic interest. Goldschmidt (1937) systematized his observations of elemental concentrations in minerals as a means of understanding and predicting element behavior during crystallization from liquids or gases. Thus, he proposed his three "rules" of element partitioning, which may be summarized as follows: (i) Any two ions of the same charge and very similar ionic radius have essentially the same crystal–liquid partition coefficient ($D = [i]_{xtl}/[i]_{liq}$, where $[i]$ refers to the concentration of element i). (ii) If there is a small difference of ionic radius, the smaller ion enters the crystal preferentially, e.g., $D_{Mg^{2+}} > D_{Fe^{2+}}$, $D_{K^+} > D_{Rb^+} > D_{Cs^+}$. (iii) For ions of similar radius but different charges, the ion with the higher charge enters the crystal preferentially, i.e., $D_{Sc^{3+}} > D_{Mg^{2+}} > D_{Li^+}$, $D_{Ca^{2+}} > D_{Na^+}$, and $D_{Ba^{2+}} > D_{K^+}$. These principles were taught to generations of students and, as we will show below, under certain circumstances, retain a degree of validity. They are neither, however, universally correct nor do they have any quantitative applicability. The aim of this chapter is to summarize the ways in which Goldschmidt's work has been amplified through a combination of theory and experimental measurement in order to quantify crystal–liquid partitioning behavior.

Since the development of accurate methods of determining element concentration at the ppm level, the trace-element contents of igneous rocks have frequently been used to model their chemical evolution. These studies use estimated crystal–liquid partition coefficients together with solutions for the differential equations describing, e.g., fractional crystallization or fractional melting (Schilling and Winchester, 1967; Gast, 1968; Shaw, 1970) to model evolution of the melt during precipitation or dissolution of the crystalline phases. Generally, because of lack of data, the crystal–liquid partition coefficients are assumed to be constant during differentiation (e.g., McKenzie and O'Nions, 1991) an assumption which is thermodynamically implausible and which contradicts observed variations in partition coefficients (e.g., Wilson, 1988; Jones, 1995). Table 1 gives part of the summary of experimental crystal–melt partition coefficients presented by Wilson (1988). As can be seen, values for most elements in the important phases—plagioclase, clinopyroxene, and garnet—vary by more than an order of magnitude, in some cases by two orders of magnitude. So, modeling of fractional crystallization or partial melting involving these phases presents problems in both the proportions of phases involved and in the appropriate partition coefficients to be used.

Normally, modelers use the simplest possible approach when applying trace-element partition coefficients. They take average values and assume that these do not vary during the petrogenetic process. Both of these assumptions can lead to large errors, however. For example, the mean partition coefficient for barium entering clinopyroxene is, according to Table 1, ~0.3. This is a dramatic overestimate for all clinopyroxene–liquid partitioning, however, probably arising from analytical errors in some of the data summarized by Wilson (1988). Our results

Table 1 Summary of experimental solid–melt partition coefficients $D_i = [i]_{crystal}/[i]_{melt}$.

	K	*Ti*	*Ni*	*Cr*	*Sr*	*Ba*	*Rb*	*Yb*
Plagioclase	0.02–0.35		0.01–0.4	0.01–0.3	0.3–3	0.2–2	0.01–0.5	0.01–0.06
Clinopyroxene	0.01–0.3	0.35–0.8	0.2–5	2.0–5.0	0.01–0.7	0.01–0.6	0.01–0.06	0.3–0.9
Garnet	0.01		0.4	0.06–2	0.008	0.008–0.04		>4

	Eu	*Sm*	*Ce*	*La*	*Ta*	*Hf*	*Zr*	*U*	*Th*
Plagioclase	0.2–0.7	0.02–0.08	0.04–0.2	0.05–0.4	0.01–0.2	0.01–0.4	0.01–0.45	0.01–0.3	0.01–0.35
Clinopyroxene	0.2–2	0.15–0.4	0.1–0.4	0.05–0.3	0.05–0.5	0.2–0.7	0.2–0.7	0.005–0.15	0.005–0.4
Garnet	0.3–2	0.06–1.5	0.02–0.05	0.005–0.07	0.2	0.2–0.6		<0.005	<0.005

Source: Wilson (1988).

indicate that D_{Ba}^{cpx} is generally less than 10^{-2}. This means, e.g., that measured barium concentrations of differentiated rocks would generally lead to overestimates of the extent of fractional crystallization. Turning to the more common problem arising from the assumption of constant partition coefficients throughout the petrogenetic process, Blundy and Wood (1991) showed how, during cotectic fractional crystallization of plagioclase plus pyroxene, the partition coefficients of barium and strontium between plagioclase and liquid ($D = Sr_{pl}/Sr_{liq}$) increase substantially, causing, in the case of strontium, a rise and then a fall in the strontium concentration of the liquid. This happens because strontium is more compatible in the albitic plagioclases which crystallize in the later stages of fractionation than in earlier anorthitic feldspar (Figure 1). Cotectic crystallization can, therefore, explain peaks in strontium concentration in derivative liquids such as those shown by fractionated melts from the Aden Main Cone (Hill, 1974). In this case bulk $D_{Sr} < 1$ for a 60 : 40 mixture of clinopyroxene plus plagioclase when the latter phase is anorthitic, and $D_{Sr} > 1$ when it is albitic. If constant partition coefficients are assumed, then strontium concentration is calculated to either rise monotonically or fall monotonically and peaks in liquid concentration must be ascribed either to a change in cotectic proportions or to the appearance of another crystalline phase such as apatite with $D_{Sr} > 1.0$.

Another example of an important effect of bulk composition on partitioning behavior is provided by uranium and thorium partitioning into clinopyroxene. The ^{238}U decay series is frequently used as an indicator of the nature and timescales of magmatic processes (e.g., Condomines *et al.*, 1981; McKenzie, 1985). The parent isotope ^{238}U decays to the intermediate short-lived isotopes ^{230}Th and ^{226}Ra which have half-lives of 75,380 yr, and 1,600 yr, respectively. Mid-ocean ridge and ocean island lavas commonly exhibit secular disequilibrium in the decay series, with excesses of ^{230}Th and ^{226}Ra over ^{238}U being most frequently observed. This means that melting and melt transport is rapid with respect to the half-lives of the daughter isotopes and that uranium is partitioned more strongly into the residual solids (i.e., it is retained in the mantle) than are the daughter elements thorium and radium. The retention of uranium in the mantle requires the presence of a solid phase which exhibits D_U/D_{Th} and D_U/D_{Ra}, both greater than 1.0. Furthermore, the bulk solid–liquid partition coefficient for uranium, D_U^{bulk}, must be large with respect to the threshold porosity at which melt is extracted, which probably means greater than 10^{-5} (McKenzie, 1989). Data of Beattie (1993a,b), Hauri *et al.* (1994), La Tourrette *et al.* (1993), and Lundstrom *et al.* (1994) indicate that the only phase with D_U/D_{Th} greater than 1.0 and large absolute D_{Th} values is garnet. Clinopyroxene generally shows $D_U/D_{Th} < 1.0$, while olivine and orthopyroxene have very low absolute D_{Th} values. Measured ^{230}Th excesses in lavas are, therefore, frequently interpreted as arising from the presence of garnet in the source region.

The inference that ^{230}Th excesses require the presence of residual garnet was challenged by Wood *et al.* (1999). These authors showed that variations in the crystal chemistry of mantle phases could result in the generation of excess ^{230}Th at lower pressures, in the stability field of spinel lherzolite. The reason is that D_U/D_{Th} for clinopyroxene is very sensitive to the size of the VIII-coordinate crystal site (M2) which the large U^{4+} (radius 0.98 Å) and Th^{4+} (1.04 Å) ions enter. Thus, for calcium-rich clinopyroxenes such as those which crystallize from basalt or are stable in peridotite at low pressure, the site size is >1.02 Å and D_U/D_{Th} is less than 1, as shown in Figure 2. Under these circumstances, in the stability fields of plagioclase and spinel lherzolite, excess ^{230}Th cannot be produced by partial melting. With increasing pressure on the spinel lherzolite solidus, however, the clinopyroxene becomes increasingly subcalcic and aluminous, and both these compositional effects act to reduce the size of the M2 site such that, at pressures greater than ~1.5 GPa, D_U/D_{Th} becomes greater than 1.0 (Figure 2). Thus, partial melting of the assemblage olivine, orthopyroxene, and clinopyroxene ± spinel can generate ^{230}Th excesses at pressures ~1.3 GPa lower than the first appearance of garnet. Furthermore, the progressive reduction of M2 site size with increasing pressure on the mantle solidus means that the

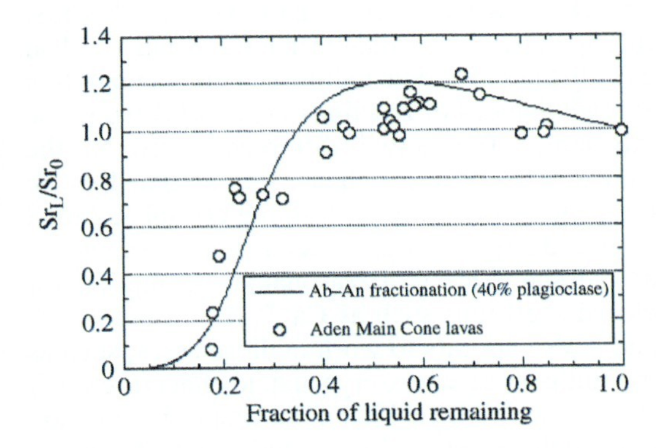

Figure 1 Variation in Sr content of Aden Main Cone volcanic rocks during differentiation (Hill, 1974). The solid curve is calculated for a fractionating assemblage containing 40% plagioclase. Liquid and solid Ab contents are based on the Ab–An binary loop. Note that, because D_{Sr} depends on Ab content of feldspar, Sr content of the liquid increases and then decreases as feldspars become more albitic.

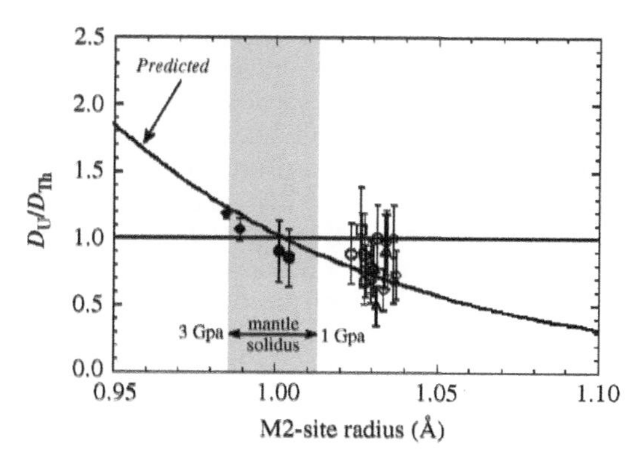

Figure 2 Calculated and observed values of D_U/D_{Th} for clinopyroxene–liquid partitioning under low f_{O_2} conditions. Solid line corresponds to predictions based on r_o expression of Wood and Blundy (1997). Note clinopyroxenes on the mantle solidus at high pressure are predicted to have $D_U/D_{Th} > 1.0$.

extent of ^{230}Th excess depends on the depth at which melting originates, a conclusion completely consistent with the observed relationship between ridge axis depth and ^{230}Th excess (Bourdon *et al.*, 1996). Thus, the conclusion that melting must begin in the presence of garnet has been shown to be vitiated by the assumption of constant partition coefficients. When compositional effects on partitioning are considered, garnet presence is shown not to be required.

These examples serve to illustrate fundamental limitations in the use of trace-element partition coefficients in petrogenetic modeling. Until the effects of pressure, temperature, and composition on element partitioning are known, the results, independent of the type of fractionation process involved, are bound to be subject to large uncertainty. The aim of this work, therefore, is to develop methods of systematizing and predicting the ways in which partition coefficients vary with ionic radius and cation charge, and with pressure, temperature, and the compositions of the host phases. It is, in essence, an attempt to update Goldschmidt's rules using recent theoretical developments and experimental data.

2.09.2 IONIC RADIUS AND LATTICE STRAIN THEORY

We start with the influence of ionic radius on partitioning. Consider a hypothetical silicate $Y^{2+}Si_xO_z$ where Y^{2+} is a cation of radius r_o which fits exactly into the largest cation site without straining the structure. The partitioning of Y^{2+} between crystal and coexisting silicate melt should be determined principally by the free energy of its end-member melting reaction, defined as follows:

$$\underset{\text{crystal}}{YSi_xO_z} = \underset{\text{melt}}{YSi_xO_z} \tag{1}$$

$$\Delta G_Y^o = (G_{YSi_xO_z}^o)_{\text{melt}} - (G_{YSi_xO_z}^o)_{\text{crystal}}$$

The relationship between partitioning and free energy of melting can be seen by considering the equilibrium constant:

$$K_0 = \frac{a_{YSi_xO_z}^{\text{melt}}}{a_{YSi_xO_z}^{\text{cryst}}} = \exp\left(\frac{-\Delta G_Y^o}{RT}\right) \tag{2}$$

where $a_{YSi_xO_z}^{\text{cryst}}$ and $a_{YSi_xO_z}^{\text{melt}}$ refer to the thermodynamic activity of the YSi_xO_z component in the crystal and melt phases, respectively. Given activity-composition models for melt and crystal, it should be possible to relate K_0 to the measured Nernst partition coefficient D_o where D_o is defined as

$$D_o = \frac{[Y]_{\text{cryst}}}{[Y]_{\text{melt}}} \tag{3}$$

here $[Y]_{\text{cryst}}$ and $[Y]_{\text{melt}}$ refer to weight fractions of cation Y in crystal and melt phases, respectively.

Before connecting D_o to K_0 let us consider the effect of replacing Y^{2+} in crystal and melt by very small amounts of cation I^{2+} of the same charge, but different radius r_i. Equilibrium of I^{2+} between melt and crystal can be represented in an analogous way to that outlined for YSi_xO_z above:

$$\underset{\text{crystal}}{ISi_xO_z} = \underset{\text{melt}}{ISi_xO_z} \tag{4}$$

In this case, however, the ISi_xO_z component is at infinite dilution in a host of essentially pure YSi_xO_z. Now we assume that Goldschmidt's first rule applies, i.e., we assume that if I^{2+} and Y^{2+} had exactly the same ionic radius then the standard free energy changes of reactions (1) and (4) would be the same. The actual difference between the standard free energy changes is assumed to be due to the work done in straining crystal and melt by introducing a cation which is not the same size as the site. This is a reasonable assumption for closed-shell ions such as Ca^{2+}, Sr^{2+}, and Mg^{2+} and it also appears to work in those cases, such as the lanthanides, where crystal field effects are small (Blundy and Wood, 1994). For first row transition ions such as Ni^{2+}, Co^{2+}, and Cu^{2+}, however, crystal field effects provide significant and in some cases major contributions to the free energy of exchange (e.g., Burns, 1970). These ions must be excluded from the following treatment.

Given the assumption that strain energy is the sole additional contribution when I^{2+} is introduced, the standard free energy change of reaction

(4) $\Delta G_{\mathrm{I}}^{\mathrm{o}}$, with I infinitely dilute in both phases, is related to that of reaction (1) by

$$\Delta G_{\mathrm{I}}^{\mathrm{o}} = \Delta G_{\mathrm{Y}}^{\mathrm{o}} + \Delta G_{\mathrm{strain}}^{\mathrm{melt}} - \Delta G_{\mathrm{strain}}^{\mathrm{crystal}} \quad (5)$$

In Equation (5), $\Delta G_{\mathrm{strain}}^{\mathrm{melt}}$ and $\Delta G_{\mathrm{strain}}^{\mathrm{crystal}}$ refer to the strain energies generated when replacing 1 mol of Y^{2+} by 1 mol of I^{2+} in an infinite volume of pure YSi_xO_z melt and crystal, respectively.

The strain energy generated by placing a sphere in a hole of different size in a continuous medium depends on the elastic properties of the medium, specifically the Young's modulus (E) and Poisson's ratio (σ). Since, for all liquids, Young's modulus is zero and Poisson's ratio 0.5, the strain energy of substitution into a melt should be zero. This yields the following relationship between the equilibrium constants for reactions (1), K_0, and (4), K_{i} :

$$K_{\mathrm{i}} = \frac{a_{\mathrm{ISi}_x\mathrm{O}_z}^{\mathrm{melt}}}{a_{\mathrm{ISi}_x\mathrm{O}_z}^{\mathrm{crystal}}} = \exp\left(\frac{-\Delta G_{\mathrm{Y}}^{\mathrm{o}} + \Delta G_{\mathrm{strain}}^{\mathrm{crystal}}}{RT}\right) = K_0 \exp\left(\frac{\Delta G_{\mathrm{strain}}^{\mathrm{crystal}}}{RT}\right) \quad (6)$$

Equation (6) shows how the free energy relationships for trace-element melting reactions should be related to those for the host cation, given the assumption that strain energy terms dominate. In order to relate the equilibrium constant K_{i} to partition coefficient D_{i} for trace-element I, we take the activity coefficient for YSi_xO_z in the crystal to be a constant near 1.0 (Raoult's law region) and assume that the activity coefficients for ISi_xO_z and YSi_xO_z in the melt are identical. The latter seems reasonable since I^{2+} and Y^{2+} have the same charge, enter the same sites and the replacement of one by the other involves no strain energy in the melt. These assumptions yield a proportionality between D_{o} and K_0 which depends on $\gamma_{\mathrm{YSi}_x\mathrm{O}_z}$, the activity coefficient in the melt and a correction factor for mean molecular weights of melt and crystal:

$$D_{\mathrm{o}} = \frac{\gamma_{\mathrm{YSi}_x\mathrm{O}_z}\{\mathrm{melt}\}}{K_{\mathrm{o}}\{\mathrm{crystal}\}} \quad (7)$$

where {melt} and {crystal} refer to the mean molecular weights of melt and crystalline phases, respectively, on the basis of the same number (z) of oxygen atoms. Given these assumptions and a few lines of algebra the relationship between K_0 and K_{i} can be converted into a relationship between D_{o} and D_{i} as follows:

$$D_{\mathrm{i}} = D_{\mathrm{o}} \exp\left(\frac{-\Delta G_{\mathrm{strain}}^{\mathrm{crystal}}}{RT}\right) \quad (8)$$

Nagasawa (1966) was the first to propose that the crystal–melt partition coefficients for ions of the same charge could be related to one another by an equation of the form of (8). He calculated the form of the strain energy for the case where a spherical ion is forced into a smaller site. Because $\Delta G_{\mathrm{strain}}^{\mathrm{crystal}}$ is always greater than zero this produces partition coefficients which decrease with increasing radius, as predicted by Goldschmidt's second rule. The treatment, discussed in the geochemical literature by Onuma *et al.* (1968) and Beattie (1994) only applies for the case where the substituting ion is larger than r_{o}. Brice (1975) investigated the case where the substituting ion may be smaller or larger than the site radius r_{o}. By treating the medium as elastically isotropic he obtained a simplified form of the strain energy as follows:

$$\Delta G_{\mathrm{strain}}^{\mathrm{crystal}} = -4\pi E_{\mathrm{s}} N_{\mathrm{A}}\left(\frac{r_{\mathrm{o}}}{2}(r_{\mathrm{o}} - r_{\mathrm{i}})^2 - \frac{1}{3}(r_{\mathrm{o}} - r_{\mathrm{i}})^3\right) \quad (9)$$

where N_{A} is Avogadro's number and E_{s} is the effective Young's modulus of the crystal site. Despite Brice's simplification, this equation yields virtually identical results to that of Nagasawa (1966) for the case where r_{i} is greater than r_{o} and contains one fewer parameter. We, therefore, consider it a reasonable approximation. Substituting Equation (9) into (8) yields an approximately parabolic dependence of D_{i} on ionic radius, with the maximum value of D corresponding to D_{o}, the partition coefficient for strain-free substitution into the crystal site:

$$D_{\mathrm{i}} = D_{\mathrm{o}} \times \exp\left(\frac{-4\pi E_{\mathrm{s}} N_{\mathrm{A}}\left((r_{\mathrm{o}}/2)(r_{\mathrm{o}} - r_{\mathrm{i}})^2 - (1/3)(r_{\mathrm{o}} - r_{\mathrm{i}})^3\right)}{RT}\right) \quad (10)$$

The theoretical dependence of D_{i} on ionic radius, shown in Figure 3, means that Goldschmidt's second rule should only apply if r_{i} is greater than r_{o} and that for small ions D_{i} should decrease with *decreasing* radius. Obviously, r_{o}, the radius of the ion which fits without strain into the lattice, should bear a close relationship to the radius of the site in the crystal. The Young's modulus of the site E_{s} controls the tightness of the partitioning parabola. Very rigid sites exclude ions of the "wrong" size and have high values of E_{s}. The height of the parabola D_{o} depends predominantly, as shown in Equation (7), on the free energy of the fusion reaction involving the cation which enters the site without straining it.

Figures 4–6 show that Equation (10) is in excellent agreement with the partitioning behavior of 1+, 2+, and 3+ ions substituting into the large sites in plagioclase, clinopyroxene, and garnet. The only ions which deviate noticeably from the predicted behavior are those such as Pb^{2+} with

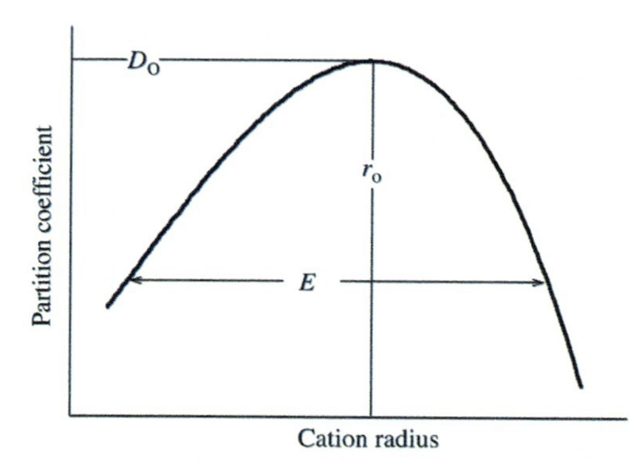

Figure 3 A schematic showing the effects on partition coefficient of the parameters in the Brice Equation (10). The "best-fit" ion has radius r_o and a corresponding partition coefficient D_o at the peak of the parabola. Partition coefficients decrease as r_i deviates in either positive or negative direction away from r_o. The derivative $(\partial D/\partial r)$ (or the "tightness" of the parabola) increases with increasing E, the Young's modulus of the site D_o, and to a much lesser extent E, varies as a function of pressure and temperature. r_o varies primarily with crystal composition.

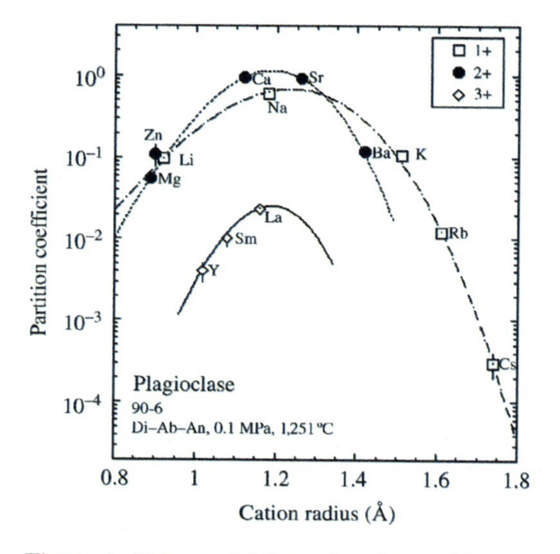

Figure 4 Brice model fits to 1+, 2+, and 3+ cation partitioning between plagioclase (An_{89}) and silicate melt in the system diopside–albite–anorthite (Blundy and Wood, 1994). Note that the best-fit curve becomes tighter as the charge on the cation increases from 1+ through 3+. This is due to increase in effective Young's modulus of the site with increasing charge (see text).

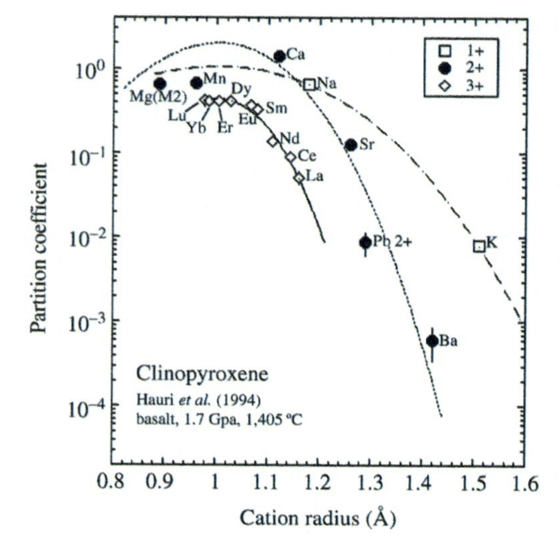

Figure 5 Brice model fits to the experimental data of Hauri *et al.* (1994) for clinopyroxene–melt partitioning. Note that the dependence of partitioning on charge corresponds to that found by Blundy and Wood (1994) (Figure 3). Deviation of Pb^{2+} from the best-fit curve may be related to its electronic structure (see text).

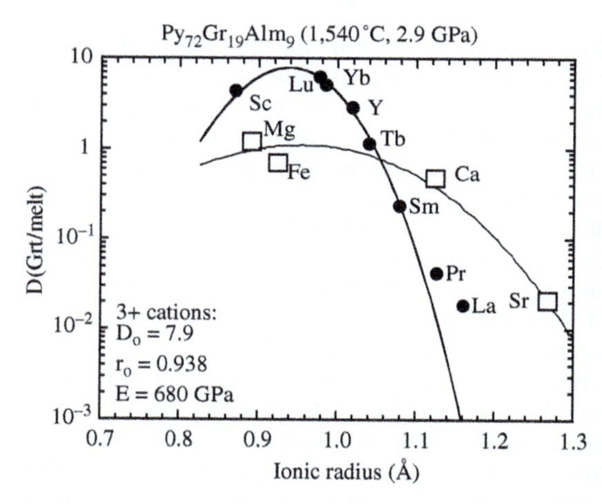

Figure 6 Brice model fits to garnet–melt partitioning of Van Westrenen *et al.* (2000b) for 2+ and 3+ ions.

lone pairs of electrons or those such as Ni^{2+} where crystal field effects are significant. Given that the theory is broadly correct, this means that partition coefficients of ions of the same charge can be related to one another through differences in their ionic radii. Blundy and Wood (1994) showed, from Equation (10), that the partition coefficient of ion i of radius r_i (D_i) can be calculated from that of ion "a" of radius r_a (D_a) as follows:

$$D_i = D_a \times \exp\left(\frac{-4\pi E_s N_A[(r_o/2)(r_a^2 - r_i^2) + (1/3)(r_i^3 - r_a^3)]}{RT}\right) \quad (11)$$

where r_o is, as before, the radius of the site or the radius of an ion which enters the site without strain. Equation (11) provides a direct and simple way of predicting partition coefficient ratios such as the D_U/D_{Th} for clinopyroxene shown in Figure 2. In this case, as discussed above, the dependence of D_U/D_{Th} on r_o is particularly important.

Since we observe that the theory embodied in Equations (9)–(11) is in good agreement with partitioning behavior of ions of different charge into several major silicate phases (Figures 4–6), the question now is how to develop the framework of lattice strain theory into a quantitative model. Specifically, we need to know how D_o, r_o, and E_s depend on charge, major element composition of the crystal, pressure, and temperature. We begin by considering E_s and r_o.

2.09.3 DETERMINATION OF E_s AND r_o

After establishing that the Brice model provides a good description of partitioning behavior for major silicate phases, our initial approach (Blundy and Wood, 1994) was to treat E_s, r_o, and D_o as unknowns and to fit experimentally measured partition coefficients to Equation (10). We adopted the ionic radii of Shannon (1976) and generated Brice model curves such as those shown in Figures 4–6. This approach immediately showed some interesting systematics. For plagioclase, r_o was found to decrease with increasing anorthite content of the crystal. The trend is less clear for 1+ ions than for 2+, because fitted r_o values have higher uncertainties for ions of low charge. For 2+ ions, however, we found the following dependence of r_o on composition:

$$r_o^{2+} = 1.258 - 0.057X_{An}\,\text{Å} \tag{12}$$

This means that albite has an effectively larger M-site than does anorthite, a result which agrees with crystallographic data. One implication is that Sr^{2+} with ionic radius 1.26 Å fits more readily into the albite structure than into anorthite, something which had been deduced from partitioning data by Blundy and Wood (1991). Second, the elastic behavior of the site depends on the charge on the trace ion, not on the nature of the major ion which it is replacing. This can be seen in Figures 4–6 where the parabolas clearly become tighter as the charge on the trace ion increases. For plagioclase, the best-fit values of E_s^{2+} are, as shown in Figure 7, twice the apparent values of E_s^{1+} and ~0.6 times the values of E_s^{3+}. These effective Young's moduli are (as can be seen from Figure 7) virtually independent of the anorthite content of the plagioclase. Furthermore, the average fitted value of E_s^{2+} (116 GPa) is very close to the bulk measured value for anorthite (107 GPa), while average fitted E_s^{1+} of 61 GPa is virtually identical to the bulk albite value (62 GPa). This implies that the bulk elastic behavior of the plagioclases is controlled by that of the M-site and, perhaps, that bulk elastic behavior may, in some cases be used to estimate E_s and hence the form of the partitioning parabola.

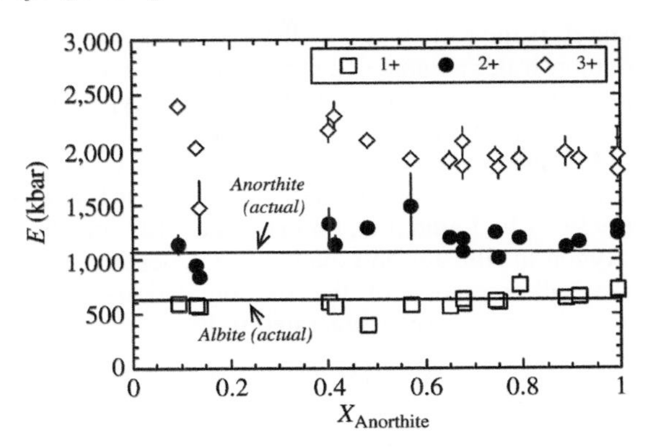

Figure 7 Apparent Young's modulus E for 1+, 2+, and 3+ ions entering plagioclase. Note that E is independent of plagioclase composition, but depends on cation charge with $E^{1+} \approx (1/2)E^{2+} \approx (1/3)E^{3+}$ (see text).

Crystal–melt partition coefficients for cations entering the clinopyroxene M2-site (Figure 5) show dependence on ionic radius and charge which are similar to those observed in the plagioclase series. Some deviation from the Brice curve, noted by Blundy and Wood (1994), can be seen in the cases of nonspherical ions such as Pb^{2+}, with its lone pair of electrons, and in transition elements such as manganese and nickel where electronic effects may be important. In general, however, closed-shell ions and lanthanides partition in accordance with the Brice model. The $D-r_i$ curves become tighter as cation charge increases from 1+ to 3+ demonstrating, as for plagioclase, that $E_{M2}^{1+} < E_{M2}^{2+} < E_{M2}^{3+}$. The exact relationship is more difficult to determine, however, because at r_i values below ~0.95 Å (in VIII co-ordination), cations start to enter the pyroxene M1-site. This means that unequivocal M2-liquid partitioning data cannot be obtained for cations such as Li^+, Mg^{2+}, and Sc^{3+} and hence that the $E-r_i$ curves cannot be as well constrained as for plagioclase, which has only one large cation site. The clear result for both plagioclase M-site and clinopyroxene M2 is, however, that apparent E_s values are in the order $E_s^{1+} < E_s^{2+} < E_s^{3+}$. This observation can be understood from a consideration of the physical principles underlying compressibility in simple oxides and silicates.

D. L. Anderson and O. L. Anderson (1970) showed that, for an ionic crystal with electrostatic attractive forces and a Born power-law repulsive potential, the bulk modulus K is given by

$$K = \frac{AZ_aZ_ce^2(n-1)}{9d_oV_o}V \tag{13}$$

In Equation (13), V_o is the molecular volume, A is the Madelung constant, Z_a and Z_c are the anion and cation valences, respectively, e is the charge on the electron, n is the Born power-law exponent, and d_o is the cation–anion distance. This equation demonstrates a linear dependence of bulk modulus on cation charge for fixed structure type and molecular volume.

Extension of this approach to individual cation–anion polyhedra was made by Hazen and Finger (1979), who found that, for cations coordinated by oxygen in silicates and oxides, the dependence of the bulk modulus of the site on cation–anion distance and charge (illustrated in Figure 8) is

$$K = 750Z_c d^{-3}\ \text{GPa} \qquad (14)$$

where d is the cation–anion distance in Å, equivalent to the cation radius r_o plus the radius of the oxygen anion, 1.38 Å (Shannon, 1976), Z_c is the cation charge and 750 is an empirical constant with an uncertainty of ±20. K is therefore observed to be a linear function of cation charge in oxides and in oxygen polyhedra of the kind we are concerned with in clinopyroxene. To extend the observed correlation to Young's modulus E, we need to consider the identity relating E and K:

$$E = 3K(1 - 2\sigma)$$

where σ is Poisson's ratio. Most minerals approximate Poisson solids and in this case, when σ is ~0.25 we obtain

$$E \approx 1.5K \qquad (15)$$

Therefore, the value of E_s in Equation (10) should, like bulk modulus, be an approximately linear function of cation charge. This is exactly what we observed previously (Blundy and Wood, 1994) for plagioclases, where 1+, 2+, and 3+ values of the

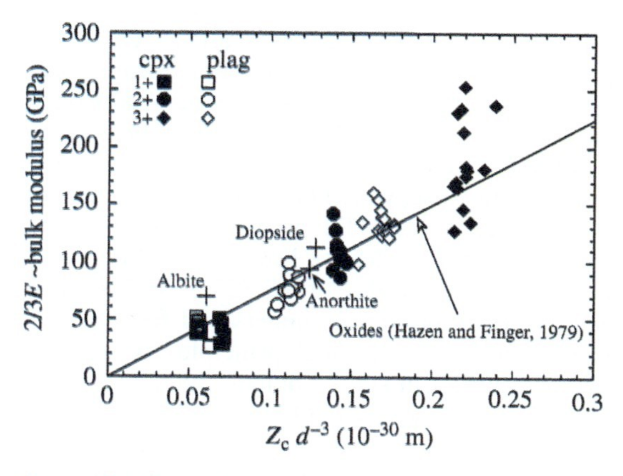

Figure 8 Plot of bulk modulus versus cation charge (Z_c) divided by metal–oxygen distance cubed (Å^3) for cation polyhedra in oxides (Hazen and Finger, 1979) and for 1+, 2+, and 3+ ions in the M2-site of clinopyroxene (Wood and Blundy, 1997) and the cation site in plagioclase (Blundy and Wood, 1994). Note that the linear correlation derived from compressibility data by Hazen and Finger holds reasonably well in our results independently estimated from the r_i-dependence of cation partition coefficient.

Young's modulus of the large cation site were found to be 67 GPa, 128.5 GPa, and 190 GPa, respectively, for anorthite and 52 GPa, 104 GPa, and 198 GPa for albite. When converted to site bulk modulus using Equation (15) we found, as shown in Figure 8, that the elastic properties of the large cation site in plagioclase calculated from trace element partitioning are in very good agreement with the empirical relationship derived by Hazen and Finger (1979) from observed compressibilities of oxide polyhedra in minerals. Furthermore, as noted earlier, the values obtained from trace-element partitioning are, for 1+ and 2+ ions, in excellent agreement with the bulk properties of albite and anorthite, respectively, suggesting that the bulk elastic properties are controlled, in this case, by the large cation site.

For clinopyroxene, partitioning of 1+, 2+, and 3+ ions fitted to Equation (10) also produces acceptable agreement with the Hazen–Finger relationship (Wood and Blundy, 1997; Figure 8). We believe that the scatter at high charge (Figure 8) arises predominantly from the increasing uncertainty in our fit values as E_s increases. Some may, however, indicate the beginning of deviation from the Hazen–Finger relationship which we find with garnet and the smaller clinopyroxene M1-site (see below).

Wood and Blundy (1997) used the agreement of E_s values with the Hazen–Finger relationship and imposed temperature and pressure dependences which are consistent with those of bulk diopside. Fits to the partitioning data then yielded for E^{3+}_{M2}:

$$E^{3+}_{M2} = 318.6 + 6.9P - 0.036T\ \text{GPa} \qquad (16)$$

where P is in GPa and T in K.

Given Equation (16) we derived, by nonlinear least-squares fitting, values of r_o from a total of 82 experiments in which three or more clinopyroxene–liquid REE partition coefficients were measured and in which the compositions of crystal and liquid phases were well constrained. Adopted radii values for individual REE were those of Shannon (1976) for 3+ ions in VIII coordination. Resulting values of r_o range from 0.979 Å to 1.055 Å. We performed stepwise linear regression of the derived values of r_o against all major compositional parameters, pressure and temperature. The result was that the only important parameters appeared to be the octahedral (M1) Al content and the calcium content of the M2-site. Fitting r_o to these parameters yields an equation which reproduces the 82 points with a standard deviation of 0.009 Å (i.e., less than the difference in ionic radius between adjacent REE):

$$r_o = 0.974 + 0.067X^{M2}_{Ca} - 0.051X^{M1}_{Al}\ \text{Å} \qquad (17)$$

where the X terms denote cation atomic fractions. The derived dependence of r_o on

$X_{Al}^{M1}(-0.051$ Å) is consistent with crystal-chemical data on the change in M2-site radius on going from $CaMgSi_2O_6$ to $CaAl_2SiO_6$ (Smyth and Bish, 1988) and with REE partitioning data in the system $CaO-MgO-Al_2O_3-SiO_2$ (Blundy *et al.*, 1996) which give −0.04 Å per Al atom in M1. The form of Equation (17) provides the explanation for the observed pressure dependence of D_U/D_{Th} for clinopyroxene at the mantle solidus shown in Figure 2. With increasing pressure the solidus clinopyroxene becomes less calcic and more aluminous leading to a smaller effective M2-site, as shown in Figure 9. Since U^{4+} has an ionic radius 0.05 Å smaller than Th^{4+}, increasing pressure leads to increasing preference of the crystal for uranium over thorium. The work discussed by Wood and Blundy (1997) and Wood *et al.* (1999) has been expanded by Landwehr *et al.* (2001) with a detailed study of partitioning into model mantle-solidus clinopyroxenes. Using compositions in the system $Na_2O-CaO-MgO-Al_2O_3-SiO_2$, these authors further elucidated the effects of pyroxene composition on partitioning relationships. They confirmed the crossover of D_U/D_{Th} from values <1.0 to >1.0 with increasing pressure and found that the apparent M2-site radius agrees with Equation (17) for all clinopyroxenes except those (rare in nature) containing greater than 6 wt.% Na_2O.

Thus far, we have shown that the large cation sites in clinopyroxene and plagioclase have elastic properties which follow, quite closely, the oxygen polyhedral trend of Equation (14). Their "best-fit" cation radii r_o vary with bulk composition in ways which bear reasonable relationships to crystallographic data (e.g., Equation (17)). Now, we turn to some additional complexities which have arisen with more detailed study. Law *et al.* (2000) determined the partitioning of a wide range of trace cations between wollastonite and carbonated silicate melt at 3 GPa and 1,420 °C. Values of r_o, D_o, and E_s for partitioning into the octahedrally coordinated calcium sites were obtained by fitting the Brice model to partitioning of 1+ (lithium, sodium, potassium, rubidium, and caesium), 2+ (magnesium, calcium, strontium, and barium), 3+ (REE plus chromium), and 4+ (zirconium, uranium, and thorium) cations. The most surprising observation is that r_o depends strongly on the charge of the substituent cation (Figure 10), decreasing from 1.17 Å for 1+ cations to 0.79 Å for 4+ ions. Much less surprising are the dependences of D_o and E_s on cation charge. The latter agrees reasonably well with the Hazen–Finger relationship and shows the anticipated relationship $E^{1+} < E^{2+} < E^{3+}$. D_o is a maximum for 2+ ions replacing the host Ca^{2+} and exhibits an approximately parabolic dependence on charge (Figure 10). We will return to this observation in a later section. The strong dependence of r_o on charge shown in Figure 10 means that both the elastic properties of the site and its effective size depend on the substituent charge. The latter is partly due to the effect of cation charge on the polarization of the nearest-neighbor oxygen anions. Some of the observed dependence of r_o on charge, however, is an artifact caused by the assumption that the melt contributes nothing to the exchange free energy. The most important conclusion, from the partitioning standpoint, is that r_o is generally a function of charge. Therefore we cannot, in general, fix r_o at the value determined from REE partitioning and assume that this must apply for 1+, 2+, and 4+ ions.

The dependence of r_o on cation charge was not found in our earlier studies of plagioclase and

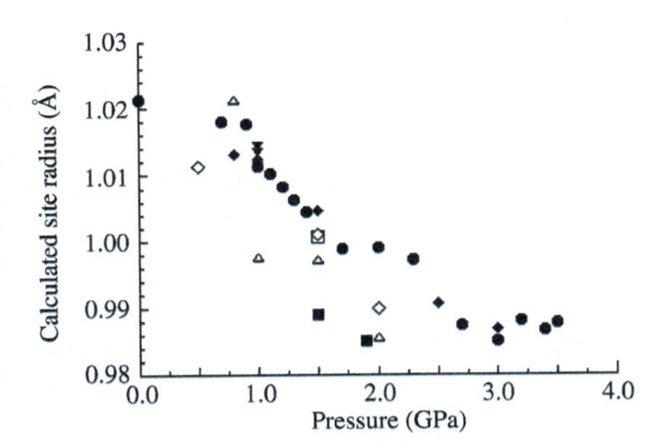

Figure 9 Calculated (from Equation (17)) M2 site radius for experimentally produced clinopyroxenes in peridotite close to the solidus. Note that, despite scatter due to differences in bulk composition, the trend to lower radius with increasing pressure is clear and reproduced in all of the polybaric data sets. This leads to crossover from D_U/D_{Th} less than 1.0 at low pressure to D_U/D_{Th} greater than 1.0 above ~1.5 GPa (●: Walter and Presnall (1994); ◆: Falloon and Green (1987); △: Takahashi and Kushiro (1983); ▼: Baker and Stolper (1994); □: Robinson *et al.* (1998); ■: Wood *et al.* (1999); and ◇: Hirose and Kushiro (1998)).

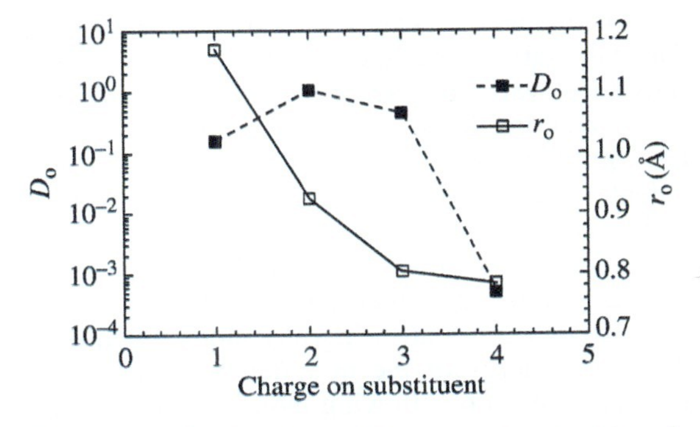

Figure 10 Fitted values of D_o and r_o as a function of charge for wollastonite-melt partitioning from Law *et al.* (2000). Note parabolic dependence of D_o on cation charge and decreasing r_o with increasing charge.

clinopyroxene, mainly because the experiments were not designed to test for it and the data did not cover a wide enough range of ionic radii for all charges. Our results on plagioclase (Figure 4), for example, only yielded accurate r_o values for 1+ and 2+ ions and these are, within uncertainty, indistinguishable from one another. Similarly, for clinopyroxene (Figure 5), partitioning of small ions such as Li^+ and Mg^{2+} into the M1-site means that r_o for M2 is not well constrained except for 3+ ions and then only for fairly calcic clinopyroxene in which M2 is large (Equation (17)).

In order better to understand the apparent dependence of r_o on cation charge and to elucidate the dependence of E_s on charge of the substituent, we have performed computer modeling studies of trace substitution into clinopyroxene, wollastonite, and garnet. We summarize the results for garnet in order to illustrate the kinds of information which can be obtained.

2.09.4 SIMULATIONS OF TRACE-ELEMENT SUBSTITUTION INTO GARNET

Van Westrenen *et al.* (2000a) employed the General Utility Lattice Program (GULP; Gale, 1997) for the simulations. A Born (ionic) model was used for all atoms, meaning that the normally accepted integral ionic charges were assigned, 2+ for magnesium and calcium, 4+ for silicon, 2− for oxygen, and so on. Short-range interactions between cations and oxygen were described using Buckingham potentials (Purton *et al.*, 1996, 1997a)

Static simulations of perfect lattices give the lattice energy and crystal structure of the garnets at 0 K. In the static limit, the lattice structure is determined by the condition $\partial U/\partial X_i = 0$, where U is the internal energy, and the variables $\{X_i\}$ define the structure (i.e., the lattice vectors, the atomic positions in the garnet unit cell, and the oxygen shell displacements).

Simulated structures for pyrope, almandine, spessartine, and grossular were used as the basis for calculations of the energies required to introduce various trace-element "defects." In every computational run, one or more defects are introduced into the crystal, e.g., for homovalent (same-charge) substitution, one divalent cation at the X-site of a perfect garnet lattice is replaced by one trace-element divalent cation. Initial, unrelaxed defect energies $U_{def}(i)$ were calculated without allowing any atoms to move. The total energy of the new defect system was then minimized by allowing the surrounding ions to relax in order to accommodate the misfit cation(s).

In the case of heterovalent (different charge) substituents, such as Yb^{3+} replacing Mg^{2+} at the X-site, it is necessary to ensure overall charge balance in the structure. This was achieved by simultaneously inserting a second trace element into the crystal, as near as possible to the first defect. Previous work (Purton *et al.*, 1997a) had demonstrated the importance of defect association in forsterite and diopside, and so we concentrated on results for associated defects in garnet. For trivalent trace cations entering the X-site, therefore, a sodium or lithium cation was placed on an adjacent X-site, or an aluminum cation replaced silicon on an adjacent Z-site (YAG-type substitution). For univalent trace cations, compensation was achieved by a trivalent cation on the X-site or a silicon replacing aluminum on the Y-site (majorite-type substitution). For more details on defect energy calculations in silicates, the reader is referred to Purton *et al.* (1996, 1997a).

Our results (VanWestrenen *et al.*, 2000a) demonstrate that the energetics of both homovalent and heterovalent substitution can be understood and parametrized using the Brice model. The simulations indicate, however, that r_o should depend on charge (as observed experimentally) and that apparent r_o and E_s values for the crystal also depend on the environments of ions in the melt. Thus, r_o and E_s values obtained from trace-element partitioning data will, in general, not be identical to those derived from crystallographic and elastic property measurements on the crystals. In the case of plagioclase, therefore, the excellent agreement between E_s^{1+} and E_s^{2+} and the bulk Young's moduli of albite and anorthite, respectively (Figure 7), is coincidental.

2.09.5 DEVIATIONS FROM SIMPLE BULK MODULUS SYSTEMATICS

In view of the dependence of fitted E_s on melt properties discussed in the previous section, it is not surprising that the effective values of E_s derived from partitioning data deviate markedly, in some cases, from the Hazen–Finger relationship (Equation (14)). The X-site in garnet is a clear example (VanWestrenen *et al.*, 1999). Figure 11 shows that values of E_{gt}^{3+} obtained by fitting the Brice equation to garnet–melt partition coefficients for REE, yttrium, and scandium are markedly higher (by up to a factor of 2) than predicted from Equation (14), especially at high values of Z_c/d^3 (i.e., at small X-site volumes). The partitioning data appear to show a power-law dependence of K and E on V ($= d^3$, see the dotted line in Figure 11). Such a dependence broadly agrees with the observations of Anderson (1972), who showed that KV^y is approximately constant for isostructural groups of elements, oxides, halides, and sulfides.

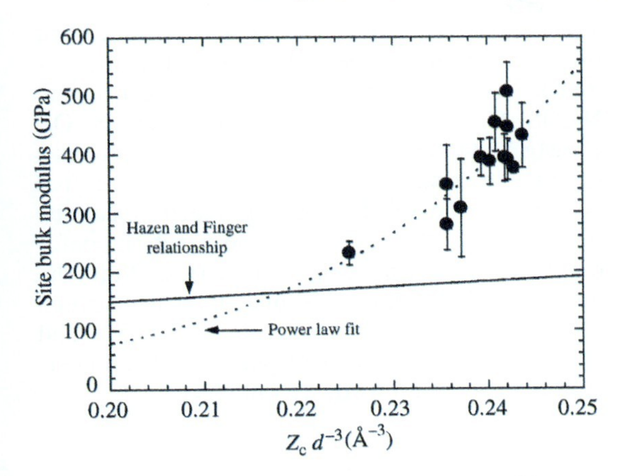

Figure 11 Comparison of fitted values of E^{3+} from garnet–melt partitioning with the Hazen–Finger relationship of Equations (14) and (15). Note that the X-site is much "stiffer" than would be anticipated (see text).

Hill *et al.* (2000) obtained a similar result for the clinopyroxene M1-site to that shown in Figure 11 for garnet. Apparent Young's modulus of the site increases with increasing charge, as anticipated from Equation (14). The apparent site values of K and E are, however, much larger than would be predicted from Equations (14) and (15). As with garnet, the clinopyroxene M1-site appears to be very "stiff" and hence excludes ions of larger and smaller radius than r_o much more strongly than anticipated. This leads us to wonder whether any agreement with the Hazen–Finger relationship is purely fortuitous.

Our computer simulations discussed above clearly show that the apparent site modulus E_s derived from partitioning data must depend on the environment of the trace element in the melt. Specifically, the lower the coordination number of the trace element in the melt, the higher the apparent E derived from partitioning studies involving garnet and that melt. The "melt effect" cannot, however, explain all deviations from the Hazen–Finger relationship. We observe, for example, that while E_{gt}^{3+} values derived from REE, yttrium, and scandium partition coefficients deviate strongly from the Hazen–Finger relationship, apparent E^{3+} for plagioclase and the clinopyroxene M2-site derived using the same elements actually obey it quite closely. Thus, the dominant control on apparent elastic properties must still be the crystal. Melt effects are not negligible but are of secondary importance.

In conclusion, we consider that E, D_o, and r_o can, for ions of constant charge partitioning into a given crystalline site, be parametrized using the Brice model. Our simulations and the preceding discussion serve to make us very cautious about trying to predict E and r_o values, however and we prefer to treat them as "fit parameters" which can be broadly understood in terms of the elastic properties of the crystal and the actual radius of the site. Parametrization in these terms has the important advantage that relatively few data serve to produce a generally applicable equation. We now turn to a consideration of the relationship between D_o and the free energy of fusion of the trace element component of the crystal. This is shown in terms of the equilibrium constant for the fusion reaction, K_o in Equation (7). By relating D_o to K_o we are able, in principle, to parameterize D_o in terms of enthalpy, entropy, and volume of fusion and hence develop an expression for the temperature and pressure dependences of D_o.

2.09.6 TEMPERATURE AND PRESSURE DEPENDENCES OF D_O AND PARTITIONING

For illustration of our approach we use the partitioning of REE into the M2- site in clinopyroxene and into the X-site in garnet. The aim is to relate D_o to the free energy of fusion of appropriate REE pyroxene and garnet end-members. In this case we use "REE" to denote that rare earth element which is of radius r_o and hence which enters the crystalline phase without strain. The main difficulty arises from the fact that no thermodynamic or melting data exists for these end-member REE-bearing phases. In place of such data we need, as a minimum, a good understanding of the mechanisms by which REE^{3+} substitute for the major Ca^{2+} and Mg^{2+} ions.

Consider the replacement of Ca^{2+} on M2-sites by REE^{3+}. In order to maintain charge balance, this substitution may be coupled to the replacement of a tetrahedral silicon atom by aluminum, leading to the hypothetical REE pyroxene end-member $REEMgAlSiO_6$. The charge-balancing mechanism is not, however, unique. Additional ways of replacing Ca by REE and maintaining charge balance are to replace 2Ca by REE plus Na or to replace 3Ca by 2REE plus a vacancy. These lead to the hypothetical end-members $Na_{0.5}REE_{0.5}MgSi_2O_6$ and $REE_{0.667}MgSi_2O_6$, respectively. In practice, all three substitutions can be rigorously described thermodynamically and are hence equally valid. Practically speaking, however, the experimental difficulties involved in obtaining the vacancy concentration on M2 make the activity of the third component $REE_{0.667}MgSi_2O_6$ almost impossible to define, and uncertainties in sodium partitioning mean that use of the sodium end-member leads to appreciable scatter. Wood and Blundy (1997), therefore, confined themselves principally to consideration of the component $REEMgAlSiO_6$.

Consider a melting reaction analogous to reaction (1), but involving the component $REEMgAlSiO_6$:

$$\underset{\text{clinopyroxene}}{REEMgAlSiO_6} = \underset{\text{melt}}{REEMgAlSiO_6} \quad (18)$$

The free energy change (ΔG_f^o) of this reaction is related to the equilibrium constant K_o as follows (Equation (2)):

$$K_o = \frac{a_{REEMgAlSiO_6}^{melt}}{a_{REEMgAlSiO_6}^{cpx}} = \exp\left(\frac{-\Delta G_f^o}{RT}\right)$$

If we take logarithms of both sides and expand ΔG_f^o in terms of enthalpy (ΔH_f^o), entropy (ΔS_f^o), and volume (ΔV_f^o) of fusion, we obtain

$$\Delta H_f^o - T\Delta S_f^o + \int_1^P \Delta V_f^o \mathrm{d}P = -RT \ln\left(\frac{a_{REEMgAlSiO_6}^{melt}}{a_{REEMgAlSiO_6}^{cpx}}\right) \quad (19)$$

In Equation (19), the enthalpy and entropy changes refer to values at 1 bar and the volume of fusion is pressure dependent. The main difficulty with the practical application of (19) to partitioning relations is, as discussed earlier, finding appropriate ways of connecting the weight partition coefficient D_o to its thermodynamic expression K_o. This requires making assumptions about the relationship between composition and activity of the component $REEMgAlSiO_6$.

Although there is no general model for activity-composition relationships in silicate melts, several studies have shown that, under certain circumstances, a semi-empirical approach works reasonably well. Blundy *et al.* (1995) investigated the partitioning behavior of sodium between clinopyroxene and silicate melts over wide ranges of pressure and temperature. They showed that the crystal–liquid partition coefficient D_{Na} bears a very simple relationship to the equilibrium constant K_{Na} for the melting reaction:

$$\underset{\text{crystal}}{NaAlSi_2O_6} = \underset{\text{melt}}{NaAlSi_2O_6}$$

Taking standard states of crystal and melt phase to be pure $NaAlSi_2O_6$ crystal and liquid, respectively, at the pressure and temperature of interest, Blundy *et al.* then converted crystal activities to composition using the observation of Holland (1990) that $a_{NaAlSi_2O_6}^{cpx}$ at low $X_{NaAlSi_2O_6}^{cpx}$ is approximately equal to X_{Na}^{M2} the mole fraction of sodium on the M2-site in clinopyroxene, i.e., $a_{NaAlSi_2O_6}^{cpx} = X_{Na}^{M2}$ = number of Na atoms per six oxygens.

For the melt, Blundy *et al.* (1995) used exactly the same assumption:

$$a_{NaAlSi_2O_6}^{melt} = \text{number of Na atoms per six oxygens}$$

The latter is equivalent to a quasicrystalline melt model (Burnham, 1981) in which the melt is treated as if it consisted of six-oxygen units, $NaAlSi_2O_6$, $CaAl_2SiO_6$, $(Mg,Fe)_3Si_{1.5}O_6$, Si_3O_6, $Ca(Mg)Si_2O_6$, and so on. If this simplification is a reasonable assumption, the standard state free energy change of the melting reaction can be related to D_{Na}, the crystal–liquid partition coefficient through Equation (7). Given that the difference in mean molecular weight between crystal and liquid (generally less than 3% and hence insignificant with respect to other uncertainties) can be ignored, we obtain

$$K_{Na} = \frac{1}{D_{Na}}$$

$$\Delta G_f^o = -RT \ln\left(\frac{1}{D_{Na}}\right) \quad (20)$$

Although the melt model is a gross simplification, Blundy *et al.* (1995) found that the standard state free energy change of the melting reaction derived from calorimetric and phase equilibrium data on pure $NaAlSi_2O_6$ is in excellent agreement with the sodium partitioning data for a wide range of synthetic and natural compositions. Wood and Blundy (1997) followed this up by showing that the quasicrystalline model of the melt could be successfully extended to the partitioning of diopside $CaMgSi_2O_6$ component between liquid and solid. As with $NaAlSi_2O_6$, the simple activity-composition relations produce good agreement between the fusion curve of pure diopside and the partitioning of $CaMgSi_2O_6$ component between clinopyroxene and silicate melt for a wide range of bulk compositions. This means that, for liquids which are precipitating clinopyroxene, the quasicrystalline model based on six-oxygen units provides a reasonable representation of activity-composition relations. This does not imply however, that the model is generally applicable to all melts or that the melt has pyroxene structure.

For REE partitioning, we relied on the results for $NaAlSi_2O_6$ and $CaMgSi_2O_6$ partitioning discussed above and assumed that liquid and solid phases obey simple activity-composition relations similar to those exhibited by $NaAlSi_2O_6$ and $CaMgSi_2O_6$ components. This means treating the pyroxene as if there were complete disorder on M2- and M1-sites but that the tetrahedral Al–Si substitution is locally coupled to the M-sites and contributes nothing to the partial molar entropy of mixing. This gives, with ideal solution on each individual sublattice

$$a_{REEMgAlSiO_6}^{cpx} = X_{REE}^{M2} X_{Mg}^{M1} \quad (21)$$

Similarly, for the melt, an analogous model to that used for $CaMgSi_2O_6$ component is

$$a^{\text{melt}}_{\text{REEMgAlSiO}_6} = [\text{REE}]_{\text{melt}}\left(\frac{\text{Mg}}{\text{Mg}+\text{Fe}}\right)_{\text{melt}} \quad (22)$$

where $[\text{REE}]_{\text{melt}}$ refers to the atomic concentration of the (strain-free) REE in the melt on a six-oxygen basis. These assumptions lead to the following expression for equilibrium constant:

$$\Delta H^{\text{o}}_{\text{f}} - T\Delta S^{\text{o}}_{\text{f}} + \int_1^P \Delta V^{\text{o}}_{\text{f}} \text{d}P = -RT\ln\left(\frac{[\text{REE}]_{\text{melt}}(\text{Mg}/(\text{Mg}+\text{Fe}))_{\text{melt}}}{X^{\text{M2}}_{\text{REE}}\, X^{\text{M1}}_{\text{Mg}}}\right) \quad (23)$$

In Equation (23), $X^{\text{M2}}_{\text{REE}}$ and $[\text{REE}]_{\text{melt}}$ are atomic concentrations on a six-oxygen basis in crystal and melt phases, respectively. If, as they usually are, the molecular weights of crystal and melt (on a six-oxygen basis) are within a few percent of one another, then the ratio of REE atomic concentrations approximates D^{3+}_{o} :

$$D^{3+}_{\text{o}} \cong \frac{X^{\text{M2}}_{\text{REE}}}{[\text{REE}]_{\text{melt}}}$$

On the left-hand side of Equation (23) the 1 bar enthalpy and entropy of fusion are both temperature dependent, but because of the opposite signs in the equation their temperature dependence tend to cancel one another (Wood and Fraser, 1976, p. 29). This means that $\Delta G^{\text{o}}_{\text{f}}$ can generally be calculated reasonably accurately over a wide temperature range with fixed values of enthalpy ($\Delta H^{\text{o}}_{\text{f}}$) and entropy ($\Delta S^{\text{o}}_{\text{f}}$). However, the volume of fusion $\Delta V^{\text{o}}_{\text{f}}$ is, because of the compressibility of the liquid, rather pressure dependent. Assuming that the pressure dependence is linear, Equation (23) yields

$$\Delta H^{\text{o}}_{\text{f}} - T\Delta S^{\text{o}}_{\text{f}} + \Delta V^{\text{o}}_{\text{f}}(P-1) + 0.5\Delta V^{\text{a}}_{\text{f}}(P^2-1) = +RT\ln\left(\frac{D^{3+}_{\text{o}} X^{\text{M1}}_{\text{Mg}}}{(\text{Mg}/(\text{Mg}+\text{Fe}))_{\text{melt}}}\right) \quad (24)$$

where D^{3+}_{o} refers to the partition coefficient of the "strain-free" REE substituent. Note that the term in brackets on the right-hand side has been inverted by changing the sign.

Equation (24) now provides a means of parametrizing measured REE partition coefficients in a thermodynamically consistent manner. The bulk composition of the crystal yields $X^{\text{M1}}_{\text{Mg}}$ from the method of Wood and Banno (1973). Similarly the composition of the melt yields $(\text{Mg}/(\text{Mg}+\text{Fe}))_{\text{melt}}$ directly. For any measured rare earth partition coefficient D_{REE}, a value of D^{3+}_{o} can be derived from the Brice equation using r_{o} from Equation (17) and E^{3+}_{M2} from Equation (16). Thus, for any given experimental measurement, the right-hand side of Equation (24) is defined. Wood and Blundy (1997) used 481 experimental data points on REE (plus Y) clinopyroxene–melt partition coefficients and for each one calculated the right-hand side as described above. They then fit the data set by treating the left-hand side as having four unknowns: $\Delta H^{\text{o}}_{\text{f}}$, $\Delta S^{\text{o}}_{\text{f}}$, $\Delta V^{\text{o}}_{\text{f}}$, and $\Delta V^{\text{a}}_{\text{f}}$. This yielded

$$88{,}750 - 65.644T + 7{,}050P - 770P^2 = +RT\ln\left(\frac{D^{3+}_{\text{o}} X^{\text{M1}}_{\text{Mg}}}{(\text{Mg}/(\text{Mg}+\text{Fe}))_{\text{melt}}}\right) \quad (25)$$

Figure 12 shows the left-hand side of Equation (25) evaluated for all 481 points available to Wood and Blundy, plotted against temperature. As can be seen, the simple activity-composition relations appear to remove most melt-compositional dependence of partition coefficient. Using normal error propagation we found that uncertainties in the strain-radius correction contribute 2,762 J to the standard error of the fit and that the errors introduced by the thermodynamic approximations are 2,219 J, equivalent to about a 20% error in D. This seems rather minor when compared to the order of magnitude, or more, uncertainties in some partition coefficients (Table 1).

2.09.7 GARNET–MELT PARTITIONING OF REE

The approach embodied in Equation (19), in which partition coefficients are related to the free energy of fusion of a fictive, in this case clinopyroxene, component, has been found to be applicable to other solid phases. For example, Van Westrenen *et al.* (2001a) found that the partitioning of REE between garnet and melt could be treated in an analogous way. These authors considered the melting of a fictive $\text{REEMg}_2\text{Al}_3\text{Si}_2\text{O}_{12}$ component, where "REE" refers to a rare earth element which enters the garnet X-site without straining the lattice:

$$\underset{\text{garnet}}{\text{REEMg}_2\text{Al}_3\text{Si}_2\text{O}_{12}} = \underset{\text{melt}}{\text{REEMg}_2\text{Al}_3\text{Si}_2\text{O}_{12}} \quad (26)$$

Instead of using six-oxygen units for the melt, as Wood and Blundy (1997) had done for clinopyroxene–melt partitioning, the liquid phase was treated as a mixture of 12-oxygen "garnet-like" units. These were assumed to mix ideally with similar entropy of mixing to the solid garnet, an assumption which produces agreement between the fusion curve of pure pyrope and the partitioning of magnesium between garnet and melt in complex compositions. Van Westrenen *et al.* (2001a) then fitted D^{3+}_{o} values to an equation of similar form to (24). This approach reproduces experimentally measured REE partitioning data within ~15% (1σ).

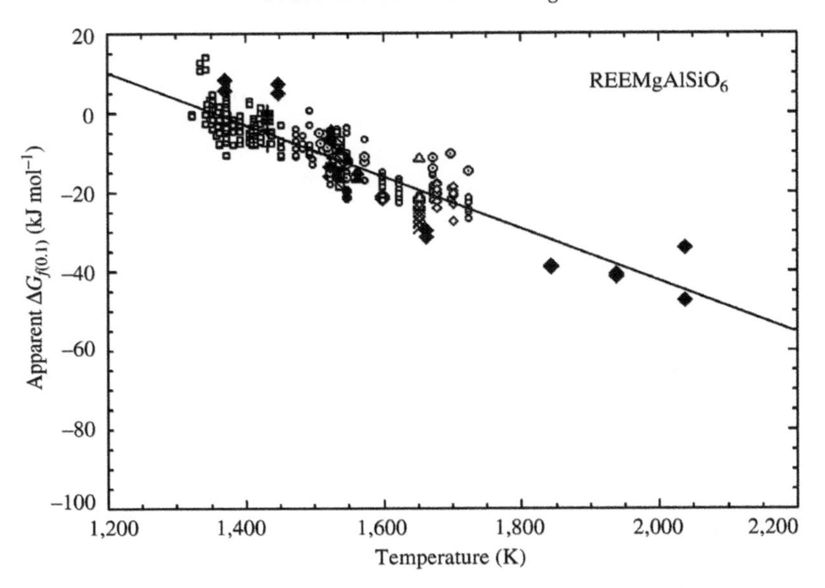

Figure 12 The apparent 1 bar free energy of fusion of hypothetical $REEMgAlSiO_6$ clinopyroxene plotted as a function of temperature. Experimental data for all REE and Y were used and D values were corrected to D_0^{3+}. Results were fitted to Equation (24). Key: □ Gallahan and Nielsen (1992), • Watson *et al.* (1987), ◆ Blundy (unpublished data), × Hart and Dunn (1993), ○ Hack *et al.* (1994), Grutzeck *et al.* (1974), · Gaetani and Grove (1995), ◇ Hauri *et al.* (1994), ▲ Dunn (1986), ■ Jones and Burnett (1987), McKay *et al.* (1986), + McKay *et al.* (1994), △ Nicholls and Harris (1980).

The examples discussed above show that the pressure and temperature dependence of D_o can, in many cases, be expressed in terms of the thermodynamic variables ΔH_f, ΔS_f, and ΔV_f. In cases where data for pure phases are available (jadeite, diopside, and pyrope, for example) values of ΔH_f, ΔS_f, and ΔV_f obtained from fitting partition coefficients agree well with measured thermodynamic properties. Given that partition coefficients for ions of radii other than r_o are predictable from D_o using lattice strain theory, the approach discussed here enables partition coefficients for many ions into clinopyroxene, garnet, and (potentially) plagioclase to be predicted over wide ranges of pressure, temperature, and composition. The next step is to consider how the D_o values for ions of different charge in a single mineral may be related to one another.

2.09.8 DEPENDENCE OF D_O ON IONIC CHARGE

Consider a trace ion entering the clinopyroxene M2-site. In general, the ion has a charge which is different from that of the major ion which it is replacing (e.g Th^{4+} replacing Ca^{2+}), and it therefore forms a charged defect in the crystal lattice. In order to maintain charge balance, the substitution of Th^{4+} for Ca^{2+} must be compensated by other substitutions or by the creation of cation vacancies, e.g.,

$$ThMgAl_2O_6 \Leftrightarrow CaMgSi_2O_6$$

$$Th^{4+} + V'' \Leftrightarrow 2Ca^{2+}$$

where V'' is a vacancy on the M2-site. Activity-composition relations for clinopyroxene solid solutions requiring charge-balanced substitution, e.g., $CaMgSi_2O_6{-}NaAlSi_2O_6$, indicate that there is considerable local order of the charge-balancing cations, i.e., sodium and aluminum and calcium and magnesium tend to be associated in clinopyroxene (Wood *et al.*, 1980). At very high dilution, however, the entropic effects of dissociation of the charge balancing cations must eventually overwhelm the energy of association. Under these conditions the trace ion Th^{4+} enters a site of charge -2 (vacated by Ca^{2+}) leading to a net positive charge of $+2$ and requiring that electrostatic work is done. This electrostatic energy acts to discriminate against the entry of Th^{4+} into the crystal and could reasonably be an important part of the reason why $D_{Th} \ll D_{Ca}$ for clinopyroxene–melt partitioning.

The work done in placing a charged sphere in a medium of dielectric constant ε was addressed by Born (1920) with respect to the hydration energies of ions in solution. In the case of crystals, similar treatments can be applied when explicit provision is made for the electrostatic forces exerted by atoms adjacent to the defect (Mott and Littleton, 1938). For illustration and to gain a qualitative understanding of the "charge effect," Wood and Blundy (2001) followed Born (1920) and treated the defect as if it resides in a medium of fixed dielectric constant ε. Consider a diopside crystal in which we replace one calcium ion on an M2-site by a thorium ion of radius r. Let us initially give the thorium a charge of +2 so that the defect has zero charge. The electrostatic potential at the surface of the sphere of radius r is given by

$$\psi_r = \frac{q}{\varepsilon r}$$

where q is the charge on the surface of the sphere. If we increase the charge on the sphere by an amount dq, then the work done is $dw = \psi_r \, dq$. We now charge the sphere to its final charge which is $Z_c - Z_0$ (4–2 in this case) and obtain the electrostatic work as follows, where e_0 is the charge of the electron:

$$W = \int dw = \int_0^{(Z_c - Z_0)e_0} \psi_r \, dq$$

$$W = \int_0^{(Z_c - Z_0)e_0} \frac{q}{\varepsilon r} dq = \left[\frac{q^2}{2\varepsilon r} \right]_0^{(Z_c - Z_0)e_0}$$

$$W = \frac{(Z_c - Z_0)^2 e_0^2}{2\varepsilon r}$$

or, on a molar basis, incorporating Avogadro's number (N_A):

$$W = \frac{N_A (Z_c - Z_0)^2 e_0^2}{2\varepsilon r} = \frac{6.95 \times 10^5 (Z_c - Z_0)^2}{\varepsilon r} \text{ J mol}^{-1} \quad (27)$$

The form of Equation (27) indicates that, if the charge on the trace element is either smaller or larger than the charge on the major ion which it is replacing, then electrostatic work is done in placing it in the structure. Of particular importance is that Equation (27) indicates a parabolic dependence of the free energy of substitution (work done) on the charge difference between the substituent and the major cation. This means, e.g., that the electrostatic work done in replacing Ca^{2+} by Th^{4+} is approximately the same as that done in replacing Ca^{2+} by an uncharged atom such as argon.

Because this treatment makes no explicit provision for atomic displacements and electrostatic forces exerted by nearest neighbor atoms, it cannot be properly quantified. It is of interest, however, to see if the energies derived from it bear any relationship to those observed. Dielectric constants of silicates at room temperature are generally in the range 5–12 (Shannon, 1993), while the radius of the defect should be larger than the ionic radius of Th^{4+} (Figure 13), because the

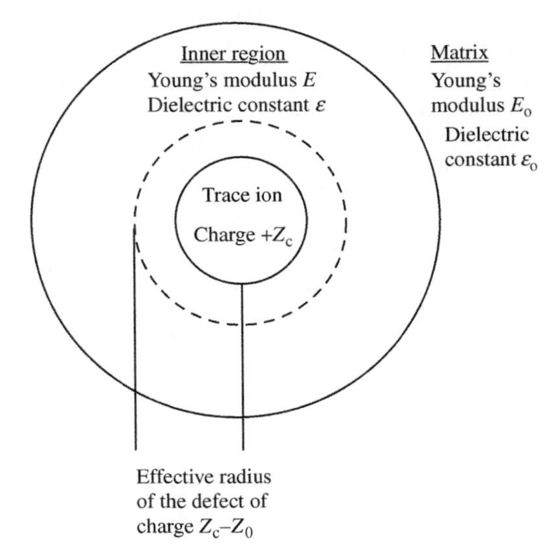

Figure 13 Schematic illustrating two main principles of trace element substitution (1) that the region around the trace defect will have different elastic (E) and electrostatic (ε) properties from the matrix and (2) that the electrostatic radius of the defect should be greater than that of the central trace ion.

sphere placed in the structure has no net charge when the central ion is Ca^{2+}. This means that the sphere of net charge +2 encompasses Th^{4+} and its surrounding oxygen atoms, and should have a radius approximately equal to that of Th^{4+} (1.05 Å) plus one oxygen diameter (2.76Å; Shannon (1976); see Figure 13). Given this defect radius of ~3.8 Å, the denominator in Equation (27) is in the range 19–46 Å, meaning that a charge difference of 1 between trace cation and major cation leads to electrostatic work of between 15 kJ mol^{-1} and 37 kJ mol^{-1}. We now consider whether this, crudely calculated, effect bears any relationship to observation.

To translate electrostatic work into an effect on the partition coefficient, we take the strain-free partition coefficient (D_o^{4+}). for the fictive ion (o^{4+})of the same charge as Th^{4+} which, unlike Th^{4+}, fits into the site without strain. We then assume that, for the reaction

$$o^{4+}_{\text{melt}} = o^{4+}_{\text{crystal}}$$

the electrostatic work dominates, i.e.,

$$\Delta G_{\text{elec}} = W_{\text{xtl}} - W_{\text{liq}}$$
$$D_o^{4+} \approx e^{[(-W_{\text{xtl}} + W_{\text{liq}})/RT]} \quad (28)$$

where W_{xtl} and W_{liq} refer, respectively, to the work done in placing the 4+ ion in the crystal and in the liquid. Although the dielectric properties of silicate melts are poorly known, it seems likely that W_{liq} is substantially smaller than W_{xtl}. This is because the melt is much more disordered than the crystal and normally contains substantially higher concentrations of potential charge-balancing atoms such as aluminum and sodium.

Figure 14 shows a plot of values of D_o as a function of charge for elements substituting into the clinopyroxene M2-site. Also shown are fitted curves based on Equations (27) and (28). As can be seen, the latter fit the data within experimental uncertainty, implying that the electrostatic work involved in transferring ions from melt to crystal is an important (but not the sole) component of the exchange free energy. Electrostatic energies in the melt must modify the form of the parabola to some extent. It appears, however, as argued above, that W_{liq} is much smaller than W_{xtl}. If the inverse were the case, then the parabolas would be inverted relative to what is observed, because highly charged cations would be rejected by the melt more strongly than the crystal and D_o^{4+} would be greater than D_o^{2+}. If one were to assume that everything except crystal effects can be ignored, then fit parameters imply that the free energy of substitution (ΔG_{elec}) is ~19 kJ mol^{-1} for unit charge entering the aluminous ($Al^{iv} = 0.195$) clinopyroxene and 28 kJ mol^{-1} for the low-Al_2O_3 pyroxene. These values agree, perhaps fortuitously, with the range estimated above from the approximate radius of the defect and the dielectric properties of silicates.

Since the electrostatic work done in replacing Ca^{2+} by Th^{4+} is the same as that done when Ar^0 replaces Ca^{2+}, it is to be anticipated that D_o^{4+} and D_o^{0+} are related to one another, possibly being of similar magnitude. Figure 14 shows that recent measurements of D_{Ar} for clinopyroxene (Chamorro *et al.*, 2002) support this suggestion,

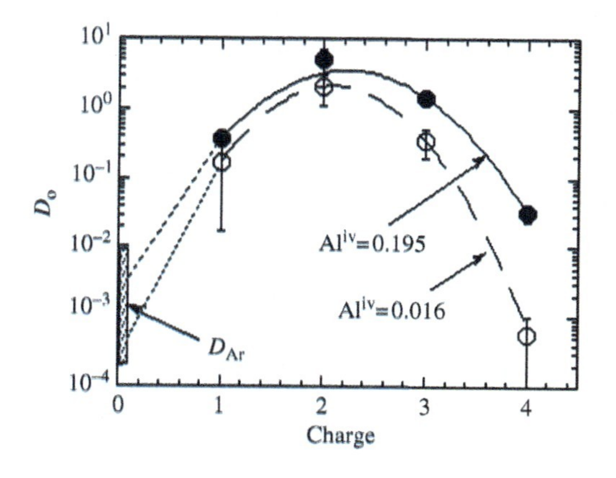

Figure 14 Plot showing the strain-free partition coefficient D_o as a function of charge for ions entering the clinopyroxene M2 site. Line labeled "$Al^{iv} = 0.195$" corresponds to 15 kb experiment of Wood *et al.* (1999); line labeled "$Al^{iv} = 0.016$" corresponds to experiment 114c of Beattie (1993a); and D_{Ar} from Chamorro *et al.* (2002).

D_{Ar} being in the range observed for partitioning of 4+ ions into the clinopyroxene M2-site. Furthermore, the experimental data of Chamorro *et al.* (2002) indicate little difference in partition coefficient between argon (atomic radius = 1.54 Å) and krypton (1.69 Å), which is in agreement with the observed linear dependence of Young's modulus on charge.

The agreement of the data in Figure 14 with predictions based on Equations (27) and (28) suggests that, for any given site in a specific mineral, the values of D_o^{4+} for ions of charge Z^+ can be parametrized in terms of D_{oo}, the partition coefficient for ions of charge Z_0, and the difference in charge between Z and Z_0 :

$$D_o^{Z+} = D_{oo} \times \exp\left\{\frac{-N_A e_0^2 (Z - Z_{0(M)})^2}{(2\varepsilon r)RT}\right\} = D_{oo} \times \exp\left\{\frac{-A(Z - Z_0)^2}{RT}\right\} \tag{29}$$

where A is a factor on the order of 25 kJ.

Equation (29) can be used in conjunction with the lattice strain model to parametrize partition coefficients for ions of different charge and radius entering sites for which data are sparse.

2.09.9 MINERAL–MELT PARTITION COEFFICIENTS

In the final section we review the available experimental data for some of the more important natural phases and place, wherever possible, the data in the context of the models discussed above. Unless otherwise stated, the data discussed refer to partitioning of trace elements between minerals and *anhydrous* silicate melts.

2.09.9.1 Clinopyroxene M2-site

REE and Y. Based on the approach of Wood and Blundy (1997), the partitioning of the rare earths between clinopyroxene and silicate melt depends on chemical compositions of the melt and pyroxene, pressure, and temperature:

$$D_o^{3+} = \left\{\frac{(\mathrm{Mg}/(\mathrm{Mg+Fe}))_{melt}}{X_{Mg}^{M1}}\right\} \times \exp\left(\frac{88{,}750 - 65.644T + 7{,}050P - 770P^2}{RT}\right) \tag{30}$$

where P is in GPa, T in K and X_{M1}^{Mg} is the atomic fraction of magnesium on the M1-site, calculated using the method of Wood and Banno (1973).

In order to calculate the cation radius at which the partition coefficient is D_o^{3+} we use the expression:

$$r_o^{3+} = 0.974 + 0.067X_{Ca}^{M2} - 0.051X_{Al}^{M1}\ \text{Å} \quad (31)$$

Equation (30) applies, of course, only to anhydrous melts. Wood and Blundy (2002) showed that there are two effects of adding water to the system, as shown schematically in Figure 15. Partition coefficients increase due to the lower temperature of crystallization and decrease due to the low activity and activity coefficients of the trace components in the melt. The net effect may be either to increase or decrease D_o (or cause no change) relative to dry, hot conditions. After calculating D_o^{3+} using Equation (30), adjustments are made for the effects of H_2O in the melt as follows:

$$D_o^{3+}\gamma_{REEMgAlSiO_6}^{melt}(1 - X_{H_2O})\frac{100}{(100 - \text{wt.\%}H_2O^{melt})} = D_o^{hydrous} \quad (32)$$

where $D_o^{hydrous}$ is the water-corrected D_o value and the activity coefficient for the melt, $\gamma_{REEMgAlSiO_6}^{melt}$, is given by

$$\gamma_{REEMgAlSiO_6}^{melt} = -0.3208 + \frac{0.0563}{(1 - X_{H_2O}^{melt})} + 1.8452(1 - X_{H_2O}^{melt}) - 0.5807(1 - X_{H_2O}^{melt})^2 \quad (33)$$

In Equations (32) and (33), $X_{H_2O}^{melt}$ is the mole fraction of H_2O in the melt. This is calculated from the weight concentration by recalculating the molecular weight of the anhydrous melt on a six-oxygen basis and then mixing the melt with H_2O, of molecular weight 18.

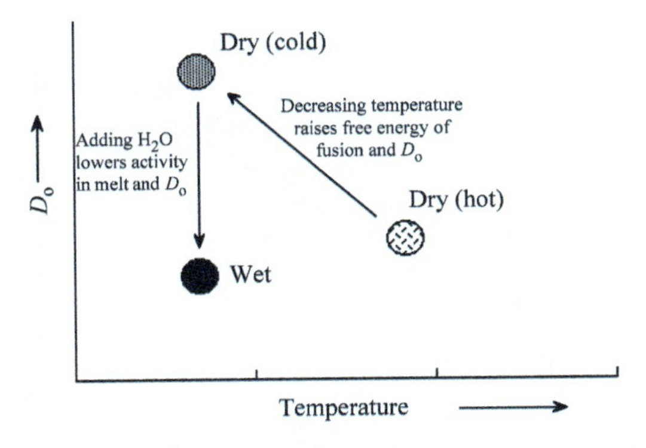

Figure 15 Schematic illustration of the combined effects of increasing H_2O content of the melt and decreasing temperature on partition coefficients. Decreasing temperature raises D_o because the entropy of fusion (ΔS_f) is positive, while adding H_2O decreases D_o because it lowers the activities of all other components in the melt.

Given D_o on either anhydrous or hydrous basis, partition coefficients of individual REE and Y can then be calculated from the Brice equation:

$$D_i = D_o \times \exp\left(\frac{-910.17E_{M2}^{3+}((r_o/2)(r_o - r_i)^2 - (1/3)(r_o - r_i)^3)}{T}\right) \quad (34)$$

In Equation (34) several of the factors of Equation (10) have been collected together, r_i is the ionic radius of the ion of interest (Å) and E_{M2}^{3+} is given by

$$E_{M2}^{3+} = 318.6 + 6.9P - 0.036T\ \text{GPa} \quad (35)$$

2+ ions. For 2+ ions we adopt the Hazen–Finger relationship for E_{M2}^{2+}:

$$E_{M2}^{2+} = \tfrac{2}{3}E_{M2}^{3+} \quad (36)$$

Using data from experiments which report both D_{Sr} and D_{Ba} (i.e., Beattie, 1993a; Hart and Dunn, 1993; Hauri *et al.*, 1994; Lundstrom *et al.*, 1994, 1998; Klemme *et al.*, 2002; Blundy and Dalton, 2000; Green *et al.*, 2000; Blundy and Brooker, 2003; Blundy, unpublished data; Chamorro, unpublished data, reported in Brooker *et al.*, 2003; Wood and Trigila, 2001) it is possible to estimate r_o^{2+}. Our best fit, consistent with trends in r_o discussed above leads to

$$r_o^{2+} = r_o^{3+} + 0.06\ \text{Å} \quad (37)$$

In general, the partition coefficient of Ca, D_{Ca}, between clinopyroxene and melt is in the range 1–4. Given a reasonable estimate of D_{Ca}, partition coefficients of other 2+ ions, particularly strontium, barium, and radium, into the M2-site should be calculated from:

$$D_i = D_{Ca} \times \exp\left(\frac{-910.17E_{M2}^{2+}[(r_o/2)(r_{Ca}^2 - r_i^2) + (1/3)(r_i^3 - r_{Ca}^3)]}{T}\right) \quad (38)$$

1+ ions. For 1+ ions we adopt the partitioning relationships for Na discussed by Blundy *et al.* (1995).

$$D_{Na} = \exp\left(\frac{10{,}367 + 2{,}100P - 165P^2}{T} - 10.27 + 0.358P - 0.0184P^2\right) \quad (39)$$

where P is in GPa and T in K. Partition coefficients for other 1+ ions can then be obtained using the Hazen–Finger relationship for E_{M2}^{1+}:

$$E_{M2}^{1+} = \tfrac{1}{3}E_{M2}^{3+} \quad (40)$$

with an r_o which should be close to the value for 2+ ions. Fortunately, results for ions larger than Na are not very sensitive to r_o which means that we can make any reasonable approximation. If we take advantage of the observed decrease in r_o with increasing charge (Figure 10), then r_o should be approximately given by

$$r_o^{1+} = r_o^{3+} + 0.12\,\text{Å} \quad (41)$$

We then obtain partition coefficients for potassium, rubidium, and caesium from D_{Na} as follows:

$$D_i = D_{Na} \times \exp\left(\frac{-910.17E_{M2}^{1+}[(r_o/2)(r_{Na}^2 - r_i^2) + \frac{1}{3}(r_i^3 - r_{Na}^3)]}{T}\right)$$

The smallest alkali, lithium, almost certainly follows magnesium into the smaller M1 site.

4+ ions. For 4+ ions entering M2, we use the parametrization of D_{Th} given by Landwehr *et al.* (2001):

$$RT \ln\left(\frac{D_{Th}\gamma_{Th}^{M2}X_{Mg}^{M1}\gamma_{Mg}^{M1}}{X_{Mg}^{melt}}\right) = 214{,}790 - 175.7T + 16{,}420P - 1{,}500P^2\ \text{J} \quad (42)$$

where X_{Mg}^{X1} and X_{Mg}^{melt} are, respectively, the atomic fractions of magnesium on the clinopyroxene M1 position, calculated on a six-oxygen basis and total magnesium atoms per six oxygens in the liquid. The γ_{Th}^{M2} and γ_{Mg}^{M1} terms are activity coefficients for the thorium and magnesium ions entering the clinopyroxene crystal sites, given by

$$\gamma_{Th}^{M2} = \exp\left(\frac{\Delta G_{strain}^{Th-M2}}{RT}\right) = \exp\left[\frac{4\pi E_{M2}^{4+}N_A}{RT}\left[\frac{r_o}{2}(r_{Th} - r_o)^2 + \frac{1}{3}(r_{Th} - r_o)^3\right]\right]$$

where r_{Th} is the eightfold ionic radius for Th^{4+} (1.041 Å) obtained by Wood *et al.* (1999) and E_{M2}^{4+} and r_o^{4+} are defined as follows:

$$E_{M2}^{4+} = \frac{4}{3}E_{M2}^{3+}$$
$$r_o^{4+} = r_o^{3+}$$

and

$$\gamma_{Mg}^{M1} = \exp\left(\frac{902(1 - X_{Mg}^{M1})^2}{T}\right)$$

where X_{Mg}^{M1} is the atomic fraction of Mg atoms on the M1-site.

The fit is best at the high pressures relevant to mantle melting. D_U can be calculated from D_{Th} using the eightfold radius of U^{4+} (0.983 Å) (Wood *et al.*, 1999) together with the values of E_{M2}^{4+} and r_o^{4+} given above and the expression:

$$D_U = D_{Th} \times \exp\left(\frac{-910.17E_{M2}^{4+}[(r_o/2)(r_{Th}^2 - r_U^2) + (1/3)(r_U^3 - r_{Th}^3)]}{T}\right) \quad (43)$$

0+ atoms. Chamorro *et al.* (2002) found D_{Ar} values (clinopyroxene–liquid) of ~3×10^{-4} for pressures to 8 GPa. As discussed above, the close similarity of D_{Kr} to D_{Ar} supports the implication of the Hazen–Finger equation that E^{0+} is close to 0 and that noble-gas partitioning is independent of atomic radius.

2.09.9.2 Clinopyroxene M1-site

4+ ions. As shown by Hill *et al.* (2000), the clinopyroxene M1-site is considerably stiffer than predicted from the Hazen–Finger relationship. For this reason, partition coefficients are very sensitive to small changes in clinopyroxene composition and we have not, as yet, been been able to attempt a full parametrization. It has been noted before, however, that the partition coefficients of elements such as zirconium, hafnium, niobium, and tantalum are strongly dependent on the aluminum content of the clinopyroxene, suggesting a strong dependence on cation charge. We take advantage of these observations to present some empirical relationships which are of practical utility.

Figure 16 is a plot of D_{Ti} versus Al^{iv}, the calculated tetrahedral aluminum content of the pyroxene per formula unit, based on microprobe

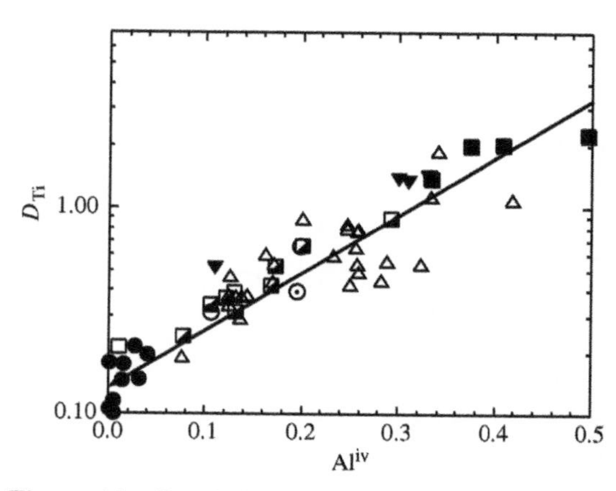

Figure 16 Correlation between clinopyroxene–melt Ti partition coefficient D_{Ti} and the tetrahedral Al content (Al^{iv}) of the pyroxene. Equation of best-fit line in the text (●: Ray *et al.* (1983); ⊙: Bristol data; □: Lundstrom *et al.* (1998); ■: Wood and Trigila (2001); ▼: Hill *et al.* (2000); △: Forsythe *et al.* (1994); Skulsky *et al.* (1994)).

analysis. As can be seen, the correlation is extremely good and can be represented by

$$\log D_{Ti} = -0.838 + 2.71\mathrm{Al}^{iv} \quad (r^2 = 0.863) \tag{44}$$

The melt and crystal compositions shown in Figure 16 cover extremely wide ranges and refer to pressures and temperatures of 1 atm to 3 GPa and 1,040–1,510 °C, respectively. We, therefore, consider that Equation (44) should work reasonably well for most purposes.

Figure 17 shows the relationship between D_{Zr} and D_{Ti}. As might be expected, a good correlation is observed and, apart from one outlier, a data point for an extremely aluminous pyroxene from Hill *et al.* (2000), the points cluster about the best-fit line:

$$D_{Zr} = -0.206 + 0.868 D_{Ti} \quad (r^2 = 0.731) \tag{45}$$

Partition coefficients for zirconium and hafnium are extremely well correlated. Fitting data from the studies shown in Figures 16 and 17 leads to

$$D_{Hf} = -0.131 + 2.345 D_{Zr} \quad (r^2 = 0.965) \tag{46}$$

5+ ions. As with 4+ ions, the lack of a suite of small 5+ cations makes it impossible to constrain r_o^{5+} and E_{M1}^{5+} by fitting of partitioning parabolas, but one would again anticipate that local charge, as represented by Al^{iv}, should be a major factor controlling partitioning. Figure 18 shows the correlation observed in experiments aimed at determining high field strength element partitioning. The best-fit line yields

$$\log D_{Ta} = -2.127 + 3.769\mathrm{Al}^{iv} \quad (r^2 = 0.686) \tag{47}$$

A good correlation between niobium and tantalum partition coefficients leads to

$$D_{Nb} = 0.003 + 0.292 D_{Ta} \quad (r^2 = 0.773) \tag{48}$$

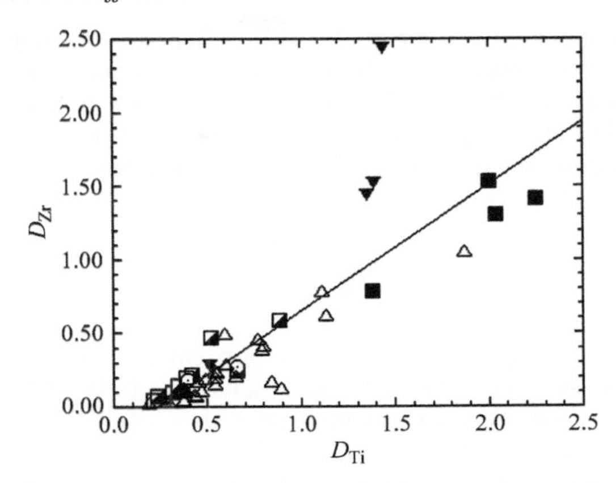

Figure 17 Correlation between clinopyroxene–melt D_{Zr} and D_{Ti}. Symbols as for Figure 16. Equation of line in the text.

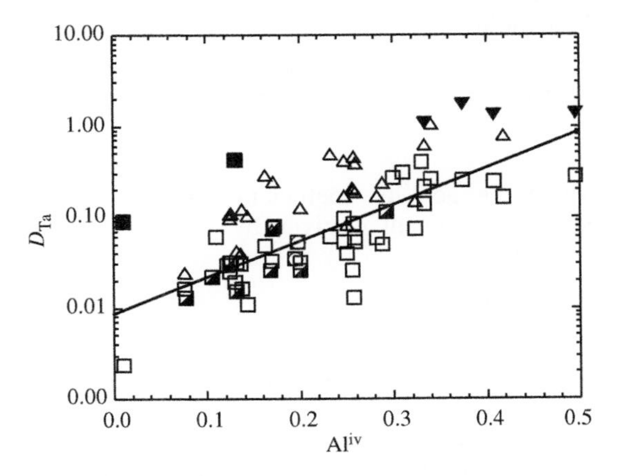

Figure 18 Relationship between clinopyroxene–melt Ta partition coefficient D_{Ta} and tetrahedral Al content (Al^{iv}) of the pyroxene. Symbols as for Figure 16. Equation of line in the text.

2.09.9.3 Garnet X-site

3+ ions. The dodecahedral (eightfold) X-site is occupied principally by calcium, magnesium, and Fe^{2+} but also hosts the lanthanides, U^{4+}, and Th^{4+}. Van Westrenen *et al.* (2001a) present a model of lanthanide and scandium partitioning between the garnet X-site and melt. These authors fitted the lattice strain model to a large number of data and derived the following expressions for r_o^{3+} and E^{3+} :

$$r_o^{3+}(\text{Å}) = 0.930X_{Py} + 0.993X_{Gr} + 0.916X_{Alm} + 0.946X_{Sp} + 1.05(X_{And} + X_{Uv}) - 0.005(P-3) \tag{49}$$

$$E^{3+}(\text{GPa}) = 3.5 \times 10^{12}(1.38 + r_o^{3+})^{-26.7} \tag{50}$$

where the X terms denote the molar fractions of the garnet components grossular (Gr), pyrope (Py), almandine (Alm), spessartine (Sp), andradite, (And), and uvarovite (Uv), and P is pressure in GPa. D_o^{3+} is described in terms of pressure and temperature via the fusion equilibrium of a fictive REE-pyrope, $\mathrm{REEMg_2Al_3Si_2O_{12}}$, with analogous activity-composition relationships to those adopted for clinopyroxene by Wood and Blundy (1997). The full expression for D_o^{3+} (as modified by Wood and Blundy, 2002) is

$$D_o^{3+} = \exp\frac{((318{,}390 + 34{,}320P - 198.2T - 2{,}470P^2)/RT)}{(\gamma_{Mg}^{gt} D_{Mg})^2} \tag{51}$$

where P is in GPa, T in K, and D_{Mg} is the magnesium partition coefficient under the conditions of interest. This can either be estimated from garnet and melt compositions or computed, for anhydrous melts from Equation (65) below. The term γ_{Mg}^{gt} accounts for nonideal mixing between Ca^{2+} and Mg^{2+} on the garnet X-site:

$$\gamma_{Mg}^{gt} = \exp\left[\frac{19{,}000(X_{Ca})^2}{RT}\right] \quad (52)$$

where X_{Ca} is the molar fraction calcium on the X-site (i.e., $X_{Gr} + X_{And}$). Equations (49)–(52) can be used to calculate partition coefficients for REE, yttrium, and scandium at any desired conditions of pressure, temperature, and garnet composition so long as the melt is anhydrous. Corrections for hydrous conditions were established by Wood and Blundy (2002) following a similar procedure to that for clinopyroxene. This yields

$$D_o^{3+}D_{Mg}^2 = \frac{D_o^{hydrous}(D_{Mg}^{hydrous})^2}{\gamma_{REEMg_2Al_3Si_2O_{12}}^{melt}(1 - X_{H_2O})(100^3/(100 - wt.\%H_2O^{melt})^3)} \quad (53)$$

where the activity coefficient for $REEMg_2Al_3Si_2O_{12}$ component is given by

$$\gamma_{REEMg_2Al_3Si_2O_{12}}^{melt} = [0.0007 + 1.433(1 - X_{H_2O}) - 1.7716(1 - X_{H_2O})^2 + 1.3379(1 - X_{H_2O})^3]^3$$

In these equations X_{H_2O} is the mole fraction of H_2O in the melt and is computed from the weight percent by calculating the molecular weight of the anhydrous melt on a 12-oxygen basis and then mixing this with H_2O of molecular weight 18 (one-oxygen basis). The left-hand side of Equation (53) is obtained by rearranging Equation (51). Provided the magnesium partition coefficient under hydrous conditions $D_{Mg}^{hydrous}$ is known or can be estimated, (53) can then be rearranged to get the D_o value for 3+ ions under hydrous conditions. Under either hydrous or anhydrous conditions, D_o is used to calculate the partition coefficients of REE, yttrium, and scandium from the Brice equation:

$$D_i = D_o \times \exp\left(\frac{-910.17E^{3+}((r_o/2)(r_o - r_i)^2 - (1/3)(r_o - r_i)^3)}{T}\right)$$

4+ ions. Uranium and thorium partitioning between garnet and melt has been studied by Beattie (1993b), La Tourrette *et al.* (1993), Salters and Longhi (1999), Van Westrenen *et al.* (1999, 2001b), Salters *et al.* (2002), Klemme *et al.* (2002), and McDade *et al.* (2003a). These data show that not only are D_U and D_{Th} highly variable ($0.001 < D_{Th} < 0.3$), but so too is the D_U/D_{Th} ratio. This variability in D_U/D_{Th} is in large part due to the large variation in the size of the X-site. If we use Equation (49) to estimate r_o^{3+} and assume $r_o^{4+} = r_0^{3+}$, we see that the variation in D_U/D_{Th} with r_o^{4+} (Figure 19) is remarkably similar to that for clinopyroxene (Figure 3). The fact that the plotted garnet partitioning data derive from a wide range of pressures, temperatures, and compositions is particularly encouraging.

The variation in D_U/D_{Th} is consistent with $E^{4+} = (4/3)E^{3+}$ and the revised eightfold ionic radii for U^{4+} and Th^{4+} of 0.983 and 1.041 Å, respectively. For those pyrope-rich garnets, characteristic of the mantle solidus garnet D_U/D_{Th} is in the range 2.4–7.5. For the chromium-rich mantle solidus garnets of Salters and Longhi (1999) and Salters *et al.* (2002), the presence of 2–4 mol.% uvarovite has a significant effect on r_o^{3+}, increasing the size of the X-site and so reducing U/Th fractionation (Figure 19). Evidently the chromium content of mantle garnets must be accounted for when estimating U–Th fractionation. For the more grossular-rich garnets characteristic of eclogites, D_U/D_{Th} has an experimental value of ~3.2 (Klemme *et al.*, 2002).

D_{Th} (and D_U) vary inversely with temperature (Figure 20). For mantle solidus garnets, the correlation is reasonably good and can be used to make a first-order estimate of D_{Th}. A more comprehensive model for D_U and D_{Th}, as a function of pressure, temperature, and melt composition, is provided by Salters *et al.* (2002).

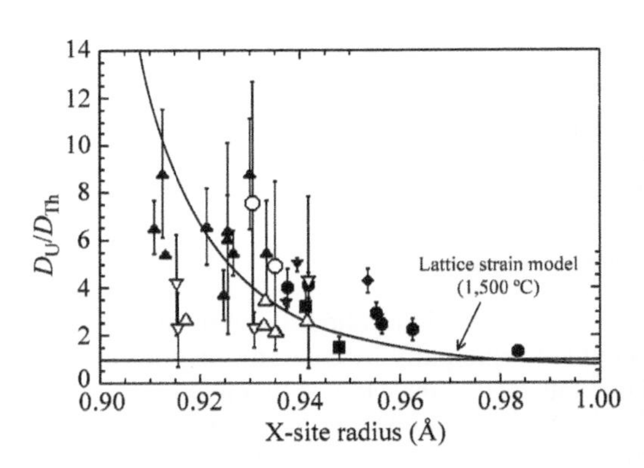

Figure 19 Observed and predicted dependence of garnet–melt D_u/D_{Th} on the radius of the X-site for 3 + ions r_o^{3+}. Solid curve was calculated from lattice strain theory and r_o^{3+} was obtained from Equation (60) (●: VanWestrenen *et al.* (1999); ■: Van Westrenen *et al.* (2000b); ▲: La Tourrette *et al.* (1993); ○: McDade *et al.* (2003a); ▽: Salters *et al.* (2002); ◆: Hauri *et al.* (1994); ▲: Beattie (1993b); △: Salters and Longhi (1999); ▼: Landwehr *et al.* (2001); and ●: Klemme *et al.* (2002)).

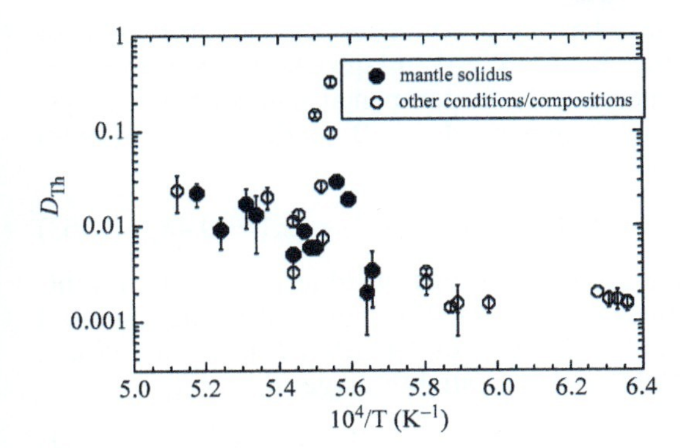

Figure 20 Plot of D_{Th} for garnet–melt partitioning versus reciprocal temperature (see text).

Their full expressions (for the molar partition coefficients, D^*) are

$$\frac{1}{2}\ln\left(\frac{D^*_U}{(X^{melt}_{FM})^4(X^{melt}_{Si})^2}\right) = 7.85 - \frac{24{,}300}{T} + 4.01(1 - X^{melt}_{Al})^2 + 10.84(1 - X^{melt}_{FM})^2 - 1.88(1 - X^{gt}_{Gr})^2 \quad (54a)$$

and

$$\frac{1}{2}\ln\left(\frac{D^*_{Th}}{(X^{melt}_{FM})^4(X^{melt}_{Si})^2}\right) = 11.46 - \frac{24{,}200}{T} + 8.6(1 - X^{melt}_{FM})^2 - 2.08(1 - X^{gt}_{Gr})^2 \quad (54b)$$

where X_{FM} is the molar cation fraction of Fe + Mg in the melt, X_{Si} is the molar cation fraction silicon in the melt, and X_{Gr} is the molar fraction grossular in the garnet.

Zr^{4+} and Hf^{4+} are rather more difficult to deal with than U^{4+} and Th^{4+}, because in compositions containing between 19% and 40% grossular they enter both X- and the octahedral Y-sites (VanWestrenen *et al.* 2001b). In compositions lower or higher in calcium content, however, they enter only the Y-sites. The reader is referred to VanWestrenen *et al.* (2001b) for discussion of the implications of these observations.

2+ ions. Partitioning of 2+ ions is based on the weight partition coefficient for magnesium, D_{Mg}. VanWestrenen *et al.* (2001a) gave an expression for calculating D_{Mg} based on the thermodynamic properties of the pyrope melting reaction. We simplified this expression by omitting heat capacity terms and refitting the anhydrous data used by VanWestrenen *et al.* to yield:

$$D_{Mg} = \frac{\exp((258{,}210 - 141.5T + 5{,}418P)/3RT)}{\exp(19{,}000X^2_{Ca}/RT)} \quad (55)$$

where P is in GPa. Note that this expression strictly applies only to anhydrous data, but that hydrous experiments are generally also reproduced to better than a factor of 1.8. In order to determine partition coefficients of other 2+ ions, we need estimates of r_o^{2+} and E^{2+}. For the limited experimental data sets where it is possible to estimate both r_o^{3+} and r_o^{2+} (VanWestrenen *et al.*, 2000b; Klemme *et al.*, 2002; McDade *et al.*, 2003a) we find that r_o^{2+} is 0.053 ± 0.004 Å larger than r_o^{3+}. This is within the range observed previously for pyroxenes and olivine. We therefore use

$$r_o^{2+} = 0.053 + r_o^{3+}\ \text{Å} \quad (56)$$

and, from the Hazen–Finger relationship,

$$E^{2+} = \tfrac{2}{3}E^{3+} \quad (57)$$

Partitioning of other 2+ cations into the eightfold coordinated X-site can then be calculated from

$$D_i = D_{Mg} \times \exp\left[\frac{-910.17E^{2+}((r_o^{2+}/2)(r^2_{Mg} - r_i^2) + (1/3)(r_i^3 - r^3_{Mg}))}{T}\right] \quad (58)$$

2.09.9.4 Plagioclase

Plagioclase has a single large cation site (M) into which most trace elements of interest partition. This site is normally occupied by calcium and sodium, with coordination increasing with increasing sodium content. For simplicity we assume VIII coordination across the entire solid solution since we have found that assumed coordination number has little effect on calculated partition coefficients. Experimental studies of plagioclase-melt partitioning have been carried out by Blundy and Wood (1994), Bindeman *et al.* (1998), Bindeman and Davis (2000), and Blundy (unpublished data). All of these studies confirm the findings of Blundy and Wood (1991) that partition coefficients are strongly dependent on plagioclase molar anorthite content (X_{An}).

2+ ions. Blundy and Wood (1991) derived linear relationships between $RT \ln D_{Sr}$ and $RT \times \ln D_{Ba}$ and X_{An} :

$$RT \ln D_{Ba}(\text{kJ mol}^{-1}) = -38.2(\pm 3.2)X_{An} + 10.2(\pm 1.8) \quad (59)$$

$$RT \ln D_{Sr}(\text{kJ mol}^{-1}) = -26.7(\pm 1.9)X_{An} + 26.8(\pm 1.2) \quad (60)$$

Equation (59) can be used to estimate D_{Ra}, using the lattice strain approach. $E^{2+}_{(M)}$ is taken as 116 GPa (Figure 5; Blundy and Wood, 1994), while $r^{2+}_{o(M)}$ is taken from Equation (12):

$$r^{2+}_{o(M)}(\text{Å}) = 1.258 - 0.057X_{An}$$

This gives D_{Ra}/D_{Ba} of 0.043 at An_{80} (1,200 °C) and 0.19 at An_{40} (900 °C). Note that as plagioclase becomes more sodic D_{Ba} and D_{Ra}/D_{Ba} increase, although Ra never becomes compatible ($D_{Ra} > 1$) in plagioclase. The maximum D_{Ra}/D_{Ba} occurs for sodic plagioclases at high temperatures.

In general, the behavior of cations with lone pairs of electrons is not well reproduced by the lattice strain model. For example, if we consider lead, then partition coefficients reported by Bindeman *et al.* (1998) may be represented as follows:

$$RT \ln D_{Pb}(\text{kJ mol}^{-1}) = -60.5(\pm 11.8)X_{An} + 25.3(\pm 7.8)$$

The expression is relatively imprecise because of the scarcity of data.

3+ ions: Bindeman *et al.* (1998) and Bindeman and Davis (2000) present SIMS analyses of Drake's (1972) experimental run products. The experiments crystallized plagioclase in the composition range An_{80}–An_{40}, which covers most terrestrial magmatic plagioclases. In the case of lanthanides, Bindeman *et al.* (1998) find a positive correlation between $RT \ln D_{Ln}$ and X_{An}. For lanthanum the relationship is

$$RT \ln D_{La}(\text{kJ mol}^{-1}) = -10.8(\pm 2.6)X_{An} - 12.4(\pm 1.8) \quad (61)$$

Lattice strain parameters for 3+ cations entering plagioclase are difficult to derive, because $r^{3+}_{o(M)}$ is clearly larger than La^{3+}, meaning that one limb of the partitioning parabola is not defined. The data are, however, are consistent with a value of $E^{3+}_{(M)}$ of 210 GPa and

$$r^{3+}_{o(M)} = r^{2+}_{o(M)} - 0.03 \text{ Å}$$

Partition coefficients for other 3+ ions may now be calculated from

$$D_i = D_{La} \times \exp\left[\frac{-191{,}136((r_o^{3+}/2)(r_{La}^2 - r_i^2) + (1/3)(r_i^3 - r_{La}^3))}{T}\right] \quad (62)$$

1+ ions. We base our model for 1+ ions on the partition coefficient for sodium, D_{Na}. Weight plagioclase-melt partition coefficients for sodium may be obtained from the modification of the Kudo–Weill plagioclase thermometer. Alternatively, experimental data in basaltic systems summarized by Bindeman *et al.* (1998) approximate

$$RT \ln D_{Na}(\text{kJ mol}^{-1}) = 2.1 - 9.4X_{An} \quad (63)$$

Partition coefficients of lithium, potassium, rubidium, and caesium are obtained (using eightfold radii) with E^{1+} of 64 GPa (Blundy and Wood, 1994) and $r^{1+}_{o(M)}$ slightly larger than $r^{2+}_{o(M)}$:

$$r^{1+}_{o(M)} = r^{2+}_{o(M)} + 0.03 \text{ Å} \quad (64)$$

This results in

$$D_i = D_{Na} \times \exp\left[\frac{-58{,}250((r_o^{1+}/2)(r_{Na}^2 - r_i^2) + (1/3)(r_i^3 - r_{Na}^3))}{T}\right] \quad (65)$$

4+ ions. Bindeman *et al.* (1998) report values for plagioclase-basaltic melt systems parametrized in the same manner as that used by Blundy and Wood (1991):

$$RT \ln D_{Ti}(\text{kJ mol}^{-1}) = -15.4 - 28.9X_{An} \quad (66)$$

$$RT \ln D_{Zr}(\text{kJ mol}^{-1}) = -15.3 - 90.4X_{An}$$

Bindeman and Davis (2000) analyzed uranium in Drake's (1972) plagioclases and found a correlation between D_U and X_{An}:

$$RT \ln D_U(\text{kJ mol}^{-1}) = -7.45 - 48.4X_{An} \quad (67)$$

The oxidation state of uranium in the experimental charges is uncertain, because some experiments were performed in air. We therefore regard Equation (67) as only indicative for U^{4+} entering plagioclase.

2.09.9.5 Olivine

3+ ions. Kennedy *et al.* (1993) and Beattie (1994) have experimentally determined lanthanide partitioning between olivine and melt at atmospheric pressure. Salters *et al.* (2002) and McDade *et al.* (2003b) present a single set of olivine-melt lanthanide partition coefficients in multiply saturated mantle melts at 1 GPa and 1.5 GPa, respectively. Taura *et al.* (1998) studied the partitioning of a large number of elements in addition to the lanthanides at pressures of 3–14 GPa. Data from all four studies can be fitted to the lattice strain model to derive best-fit values of E^{3+}_{M} and $r^{3+}_{o(M)}$. Although olivine has two octahedral (sixfold) sites (M1 and M2), they are sufficiently close in size and geometry that only one site was used for fitting (cf. Beattie, 1994). We included both scandium and octahedral aluminum in the fitting, to better resolve the parabolas at low ionic radii. A single value of E^{3+}_{M} of 360 GPa fits all data. Fitted $r^{3+}_{o(M)}$ values lie in the range 0.70–0.73 Å. There may be a slight increase in $r^{3+}_{o(M)}$ with increasing iron content, consistent with the larger site size in fayalite than forsterite. There are, however, insufficient data to quantify this relationship. For forsterite-rich olivines (>90 mol.%), a single value of $r^{3+}_{o(M)} = 0.710$ Å is adequate.

With these parameters, D^{3+}_{o} is 10% greater than D_{Sc}. Although values of the scandium partition coefficient of up to 0.5 have been reported, values in the range 0.1–0.2 are appropriate for the mantle solidus (Taura *et al.*, 1998; McDade *et al.*, 2003b). Given D_{Sc}, partition coefficients for the other 3+ ions can be obtained using their sixfold radii and the expression

$$D_i = D_{Sc} \times \exp\left[\frac{-327{,}660(0.355(r^2_{Sc} - r^2_i) + (1/3)(r^3_i - r^3_{Sc}))}{T}\right] \tag{68}$$

2+ ions. Lattice strain parameters for 2+ cations were derived from the partitioning data for Fe^{2+}, Mn, Ca, Sr, and Ba obtained by Beattie (1993a) and Kennedy *et al.* (1993). Assuming, from the Hazen–Finger relationship,

$$E^{2+} = \tfrac{2}{3}E^{3+} = 240\,\text{GPa}$$

then $r^{2+}_{o(M)}$ is close to the sixfold radius of magnesium, 0.72 Å. The partitioning of Mg^{2+} between olivine and a wide range of basaltic melts has been determined to 3 GPa by Ulmer (1989), who gives the following expression for Kd^{ol-liq}_{Mg}:

$$\log Kd^{ol-liq}_{Mg} = \log\left(\frac{Mg^{ol}}{Mg^{liq}}\right) = 1.151 - 4.82Mg + 7.3Mg^2 + 0.0028P \tag{69}$$

In Equation (69) Mg^{ol} and Mg^{liq} refer to the cation fractions of magnesium in olivine and liquid respectively and P is pressure in GPa. By cation fractions we mean

$$Mg^{ol} = \frac{Mg}{Mg + Fe + Si + \cdots}$$

$$Mg^{liq} = \frac{Mg}{Mg + Fe + Al + Ca + Si + \cdots}$$

Given liquid composition and the (perfectly adequate) assumption that olivine contains only magnesium, iron, and silicon in stoichiometric proportions, Kd is easily converted to weight partition coefficient D_{Mg}. Partitioning of other 2+ ions are then calculated from

$$D_i = D_{Mg} \times \exp\left[\frac{-218{,}440(0.36(r^2_{Mg} - r^2_i) + (1/3)(r^3_i - r^3_{Mg}))}{T}\right] \tag{70}$$

4+ ions. Partition coefficients for titanium entering olivine are generally ~0.01 (Kennedy *et al.*, 1993), while values for zirconium and hafnium are $\sim 10^{-3}$ (Kennedy *et al.*, 1993) to 10^{-4} (McDade *et al.*, 2003b) with the hafnium value slightly higher than D_{Zr}.

There are three published studies of olivine-melt partitioning of uranium and thorium. Beattie (1993a) finds $D_U = 6(\pm 1) \times 10^{-6}$ at 2.9 GPa, Salters *et al.* (2002) find $D_U = 5(\pm 7) \times 10^{-4}$ at 1 GPa, while McDade *et al.* (2003b) find $D_U = 5.9 \times 10^{-5}$ at 1.5 GPa. The data are too sparse to establish whether these differences are real. However, in all three cases $D_U > D_{Th}$, with D_U/D_{Th} values are of 3.3 ± 2.7 at 1 atm, 12 ± 5 at 1 GPa, and 4.3 at 1.5 GPa. The lattice strain model yields D_U/D_{Th} of 6.3, in broad agreement with the experimental values.

5+ ions. McDade *et al.* (2003b) report experimental data on the partitioning of niobium and tantalum between olivine and melt. At 1.5 GPa they find $D_{Nb} = 1 \times 10^{-4}$ and $D_{Ta} = 6 \times 10^{-4}$, which indicates that $r^{5+}_{o(M)}$ is smaller than r_{Nb} (0.660 Å).

1+ ions. Taura *et al.* (1998) determined values of 0.3–0.5 for D_{Li} (olivine-liquid) under upper mantle conditions at pressures of 3–15 GPa. McDade *et al.* (2003b) obtained a value of 0.29 at 1.5 GPa and 1,315 °C. Given these, values using the Hazen–Finger relationship for E^{1+} and assuming $r^{1+}_{o(M)} = r^{2+}_{o(M)} = 0.720$ Å, we obtain for the larger 1+ ions:

$$D_i = D_{Li} \times \exp\left[\frac{-109{,}220(0.36(r^2_{Li} - r^2_i) + (1/3)(r^3_i - r^3_{Li}))}{T}\right] \tag{71}$$

0+ atoms. Argon partition coefficients for olivine have recently been determined by Brooker *et al.* (1998). Their measured D_{Ar} values are $\sim 10^{-2}$, which is higher than those measured for clinopyroxene. Assuming that this value is correct, partition coefficients for the other noble gases should also be $\sim 10^{-2}$.

2.09.9.6 Orthopyroxene

Orthopyroxene has a large sixfold site, M2 and the smaller sixfold M1 site, which is similar to the clinopyroxene M1 site. We will confine our discussion to the octahedral M2, which is smaller than the equivalent (eightfold) clinopyroxene site, even after allowing for the different coordination number.

3+ ions. Lanthanides are more incompatible in orthopyroxene than in clinopyroxene, with D_o^{3+} typically lower by a factor of ~5 on the mantle solidus (Blundy and Wood, 2003; Salters *et al.*, 2002; McDade *et al.*, 2003a,b). A total of 16 runs from the studies of Kennedy *et al.* (1993), Wood *et al.* (1999), Blundy and Wood (2003), Blundy and Brooker (2003), Salters and Longhi (1999), Green *et al.* (2000) and Salters *et al.* (2002) can be fitted to the lattice strain equation. Preliminary results for lanthanides indicated that a single value of E_{M2}^{3+} (360 GPa) suffices for all data. Using this E value, the derived values of $r_{o(M2)}^{3+}$ are in the range 0.758–0.819 Å, and correlate with aluminum and calcium contents of the orthopyroxene. The following reproduces $r_{o(M2)}^{3+}$ with an average absolute deviation of 0.012 Å for the experiments fitted:

$$r_{o(M2)}^{3+} = 0.753 + 0.118\mathrm{Al}^{tot} + 0.114\mathrm{Ca} \quad (72)$$

where Al^{tot} and Ca denote atoms per six-oxygen formula unit. If we calculate D_o^{3+} for clinopyroxene under any P, T conditions of interest using Equation (30), then lower it by a factor of 5 to convert to D_o^{3+} for orthopyroxene, partition coefficients of 3+ ions are then obtained from

$$D_i = D_o^{3+} \times \exp\left[\frac{-327{,}660((r_o/2)(r_o^2 - r_i^2) + (1/3)(r_i^3 - r_o^3))}{T}\right] \quad (73)$$

4+ ions. Uranium and thorium partitioning into orthopyroxene has been studied experimentally by Beattie (1993a), Kennedy *et al.* (1993), Salters and Longhi (1999), Wood *et al.* (1999), Salters *et al.* (2002), and McDade *et al.* (2003a,b). All studies involved broadly basaltic liquid compositions. D_{Th} on the mantle solidus is ~0.001. In all cases $D_U \geq D_{Th}$ as befits the small M2 site radius. The most precise estimate of D_U/D_{Th} available is 2.52 ± 0.24 (Wood *et al.*, 1999).

2+ ions. At atmospheric pressure D_{Sr} is in the range $(9-14) \times 10^{-4}$ (Beattie, 1993a; Kennedy *et al.*, 1993). On the mantle solidus D_{Sr} is 3.7×10^{-3} at 3 GPa, 1,500 °C (McDade *et al.*, 2003a), and 0.074 at 1.5 GPa, 1,315 °C (McDade *et al.*, 2003b), suggesting a dependence on both temperature and pressure. Green *et al.* (2000) report $D_{Sr} = 0.012$ under hydrous conditions at 2 GPa. The most precise studies of D_{Ba} in orthopyroxene (Beattie, 1993a) suggest that it is 50–100 times lower than D_{Sr}. We used D_{Ca}, D_{Sr}, and D_{Ba} from the experiments of Beattie (1993a) and Kennedy *et al.* (1993) to derive $r_{o(M2)}^{2+}$, assuming that $E_{M2}^{2+} = (2/3)E_{M2}^{3+} = 240$ GPa. We obtain values which are larger than $r_{o(M2)}^{3+}$ by ~0.08 Å:

$$r_{o(M2)}^{2+} \cong r_{o(M2)}^{3+} + 0.08\ \text{Å} \quad (74)$$

Noting that magnesium is partitioned into both sites in orthopyroxene, the M2-melt partition coefficient D_{Mg}^{M2} for basaltic liquids is ~1.0. Adopting this value implies, for the larger alkaline earth cations,

$$D_i = \exp\left[\frac{-218{,}440((r_o/2)(r_{Mg}^2 - r_i^2) + (1/3)(r_i^3 - r_{Mg}^3))}{T}\right] \quad (75)$$

There is only one determination of D_{Pb} (0.009 ± 0.006) for orthopyroxene, that of Salters *et al.* (2002) at the mantle solidus at 2.8 GPa.

2.09.9.7 Amphibole

Partitioning into amphibole is complex because of the multiplicity of cation sites at which substitution may occur. There are three structurally distinct octahedral (VI) sites: M1, M2, and M3; a larger eightfold M4 site, which in clinoamphiboles is occupied by calcium and sodium and a 12-fold A-site, which may be vacant, or occupied by sodium and potassium.

3+ ions. The partitioning of lanthanum, samarium, and ytterbium was found by Tiepolo *et al.* (2000a) to be a linear function of melt silica content:

$$\ln D_{La}^{amph} = -7.8(\pm 0.6) + 10(1)X_{nf}^{melt}/X_{total}^{melt}$$

$$\ln D_{Sm}^{amph} = -7.0(\pm 0.5) + 12(1)X_{nf}^{melt}/X_{total}^{melt} \quad (76)$$

$$\ln D_{Yb}^{amph} = -5.8(\pm 0.6) + 9.5(1.0)X_{nf}^{melt}/X_{total}^{melt}$$

where X_{nf} and X_{total} are the molar fraction of network-forming cations and total cations, respectively, in the melt. Tiepolo *et al.* (2000a) take network-forming cations to be silicon plus all aluminum that is charge balanced by alkalis. We note that equally strong correlations exist between D_{Ln} and D_{Ca} for the same dataset, suggesting that calcium exchange between amphibole and melt may be equally as important as melt SiO_2 content. However, for simplicity here we adopt Tiepolo *et al.* (2000a) empirical expressions.

4+ ions. Uranium and thorium partitioning into amphibole were also studied experimentally by Tiepolo *et al.* (2000a). Under the redox conditions of their experiments (FMQ-2 log units) uranium was dominantly tetravalent. They find no correlation between D_U, D_{Th}, and crystal composition, but again find a linear correlation with silica content:

$$\ln D_{Th}^{amph} = \ln D_{U}^{amph} = -11(\pm 1) + 11(2)X_{nf}^{melt}/X_{total}^{melt} \quad (77)$$

It appears that uranium and thorium are not fractionated from each other by amphibole.

The partitioning of zirconium and hafnium between titanian amphiboles (pargasites and kaersutites) and a range of melt compositions at 1.4 GPa and 850–1,070 °C was determined by Tiepolo *et al.* (2001). Over a limited range of X_{nf}/X_{tot} of 0.50–0.68 the zirconium data fit reasonably well to the expression:

$$D_{Zr}^{amph} = -1.44 + 3.09\frac{X_{nf}}{X_{tot}} \quad (r = 0.81) \quad (78)$$

and there is an extremely good correlation between D_{Zr} and D_{Hf} :

$$D_{Hf}^{amph} = -0.0123 + 1.699 D_{Zr}^{amph} \quad (r = 0.989) \quad (79)$$

5+ ions. Amphiboles appear to fractionate niobium and tantalum from zirconium and hafnium. Tiepolo *et al.* (2001) argue that this is because zirconium and hafnium enter the M2-site, while niobium and tantalum are incorporated in the smaller M1-site in amphibole. These authors find that the ratio of niobium to zirconium partition coefficients depends on the mg-number of the amphibole, representing the changing sizes of the host sites with magnesium content:

$$\ln\frac{D_{Nb}}{D_{Zr}} = -2.76\,\text{mg\#} + 1.57 \quad (r^2 = 0.86) \quad (80)$$

where

$$\text{mg\#} = \left(\frac{\text{Mg}}{\text{Mg} + \text{Fe}}\right)_{amph}$$

Similarly, Nb/Ta fractionation was demonstrated by Tiepolo *et al.* (2000b) and shown, similarly, to depend on the size of the amphibole M1-site, as represented by mg-number and titanium content:

$$\frac{D_{Nb}}{D_{Ta}} = 2.45 - 1.26\text{mg\#} - 0.84\text{Ti}_{tot} \quad (81)$$

where Ti_{tot} is the total titanium content of the amphibole per formula unit.

2+ ions. Radium and barium both enter the large (12-fold) A-site in amphibole. Experimental values of D_{Ba} vary from 0.10 to 0.72 (Brenan *et al.*, 1995; La Tourrette *et al.*, 1995; Dalpé and Baker, 2000). Slightly lower D_{Ba} (0.05) was obtained by Andreeßen *et al.* (1996) for andesitic and basaltic andesite melts. For amphibole phenocrysts in acid volcanic rocks, Ewart and Griffin (1994) find $D_{Ba} = 0.16–0.30$. From the lattice strain model, we calculate consistent $D_{Ra}/D_{Ba} = 0.080 \pm 0.007$ for all of the amphiboles studied. There is no obvious correlation with pressure, temperature, or composition.

Tiepolo *et al.* (2000a) also studied lead partitioning between amphibole and melt, again finding a linear correlation with melt composition:

$$\ln D_{Pb(A)}^{amph} = -7.6(\pm 0.7) + 8.4(\pm 1.2)X_{nf}^{melt}/X_{total}^{melt} \quad (82)$$

2.09.9.8 Alkali-feldspar

2+ ions. Barium and strontium are well known to be compatible in alkali-feldspar. Several experimental studies demonstrate this over a wide range of pressures and temperatures (Long, 1978; Guo and Green, 1989; Icenhower and London, 1996). The Icenhower and London (1996) study provides data for D_{Mg}, D_{Ca}, D_{Sr}, and D_{Ba}, such that lattice strain parameters for 2+ cations can be derived. The partition coefficient for barium, D_{Ba}, is a strong function of the orthoclase ($KAlSi_3O_8$) content of the feldspar, increasing from values near 1 at X_{Or} of 0.1 to 18 at X_{Or} of 0.8:

$$D_{Ba}^{fsp} = 0.07 + 25X_{Or} \quad (r^2 = 0.984) \quad (83)$$

Lattice-strain model fits (using 10-fold ionic radii) indicate that $E_{(M)}^{2+}$ is in the range 55–150 GPa and a fixed $E_{(M)}^{2+}$ of 91 GPa provides an adequate fit to all data. Using $E_{(M)}^{2+} = 91$ GPa, the experimental data yield fitted $r_{o(M)}^{2+}$ values which also increase linearly with the molar fraction of orthoclase component (X_{Or}). A weighted fit gives

$$r_{o(M)}^{2+}(\text{Å}) = 1.341(\pm 0.002) + 0.207(\pm 0.005)X_{Or} \quad (84)$$

From these values it is straightforward to calculate partition coefficients for other 2+ ions as follows:

$$D_i = D_{Ba} \times \exp\left[\frac{-82{,}825((r_o^{2+}/2)(r_{Ba}^2 - r_i^2) + (1/3)(r_i^3 - r_{Ba}^3))}{T}\right] \quad (85)$$

From Equation (85) we find, for example, that Ra^{2+} becomes compatible ($D_{Ra} \geq 1$) at $X_{Or} \geq 0.13$. Evidently alkali-feldspar has a dominant influence over the behavior of radium in evolved silicic systems.

The above 2+ fit parameters can also be used to derive D_{Pb} from D_{Sr}, the nearest (in terms of ionic radius) alkaline earth cation. For the Icenhower and London (1996) experiments we again obtain a linear correlation with X_{Or} :

$$D_{Pb}/D_{Sr} = 0.801 + 1.124X_{Or} \quad (86)$$

1+ ions. Icenhower and London (1996) report the following relationship between D_{Rb} and alkali feldspar composition in peraluminous systems:

$$D_{Rb} = 0.03 + X_{Or} \qquad (r^2 = 0.927) \quad (87)$$

White (2003) has demonstrated that this relationship is inadequate for metaluminous to peralkaline systems and proposes an equation for the latter based on the agpaitic index and the SiO_2 content of the rock:

$$\ln D_{Rb} = -2.2122 + 0.0273(SiO_2) - 0.6396(AI) \quad (88)$$

where (SiO_2) refers to the weight percent concentration of SiO_2 in the whole rock and the agpaitic index is defined as the molar ratio:

$$AI = \frac{Na + K}{Al}$$

There are insufficient data for us to determine exactly the dependence of r_o^{1+} on feldspar composition, but we find that the assumption:

$$r_o^{1+} = r_o^{2+}$$

combined with $E_{(M)}^{1+}$ of 50 GPa gives an adequate fit to the available data. Partition coefficients of other 1+ ions can then be estimated from D_{Rb} as follows:

$$D_i = D_{Rb} \times \exp\left[\frac{-45{,}508((r_o^{1+}/2)(r_{Rb}^2 - r_i^2) + (1/3)(r_i^3 - r_{Rb}^3))}{T}\right] \quad (89)$$

3+, 4+, and 5+ ions. White (2003) presents a series of empirical equations for the estimation of alkali feldspar-melt partition coefficients for yttrium, the REEs, zirconium and niobium. These are of the form:

$$\ln D_Y = -7.4011 + 0.0833(SiO_2) - 0.8382(AI) - 1.3315(CaO)$$

where parameters are defined as in Equation (88) above and (CaO) is the whole rock concentration in weight percent.

2.09.9.9 Phlogopite (Biotite)

Trioctahedral micas, such as phlogopite or biotite, are characterized by four distinct cation lattice sites: tetrahedral (Z) sites occupied by silicon and aluminum; octahedral (Y) sites, denoted M1 and M2, occupied by Al, Cr, Fe^{3+}, Ti, Fe^{2+}, Mg, and Mn; and a large 12-fold coordinated interlayer X-site occupied by potassium, sodium, calcium, and other large cations. The low partition coefficients for lanthanides, uranium and thorium in phlogopite ($<2 \times 10^{-4}$) indicate, however, that large highly charged cations cannot be incorporated at the X-site, presumably because of a lack of suitable charge-balancing mechanism.

1+ ions. The data of Icenhower and London (1995) are ideal for constraining $r_{o(X)}^{1+}$ and $E_{(X)}^{1+}$, since they report partition coefficients for sodium, potassium, rubidium, and caesium, ions which span the size of the X site. For 15 experiments at 650–750 °C and 0.2 GPa, we obtain a very tight cluster of $r_{o(X)}^{1+}$ of 1.665 ± 0.007 Å and $E_{(X)}^{1+}$ of 50 ± 2 GPa. Under these conditions, D_{Rb} is ~2.0. We fixed $E_{(X)}^{1+} = 50$ GPa in order to fit the 1.5 GPa experiments of Schmidt *et al.* (1999). Significantly we derive much larger values of $r_{o(X)}^{1+}$ in the range 1.71–1.73 Å. The Schmidt *et al.* (1999) phlogopites are considerably less aluminous than those of Icenhower and London (1995) and it is likely that this change in $r_{o(X)}^{1+}$ is related to the extent of eastonite solid solution. D_{Rb} in the latter experiments is generally ~1.4, the lower value being partly due to the change in site size and partly to the higher temperature, 1,040–1,175 °C of the Schmidt *et al.* study. Given the estimates of D_{Rb}, partition coefficients of other ions may be calculated from Equation (89).

2+ ions. 2+ cations, such as strontium and barium, readily enter the X-site, with the excess positive charge balanced by lithium substitution on an M-site or aluminum on a Z-site. Phlogopite D_{Ba} is consistently larger than D_{Sr}, indicating large $r_{o(X)}^{2+}$. Unfortunately, $r_{o(X)}^{2+}$ is larger than the radius of Ba^{2+}, which makes one limb of the parabola poorly constrained. There are only five experiments that report D_{Ca}, D_{Sr}, and D_{Ba} : three from Schmidt *et al.* (1999) and two from Icenhower and London (1995). Fitting these using $E_{(X)}^{2+} = 2E_{(X)}^{1+} = 100$ GPa, we find that the calculated $r_{o(X)}^{2+}$ is 0.01–0.06 Å smaller than $r_{o(X)}^{1+}$, consistent with our observations for other phases. The difference in $r_{o(X)}^{2+}$ between the low-aluminum Schmidt *et al.* (1999) biotites and the high-aluminum Icenhower and London (1995) biotites persists.

D_{Ba} can itself be parametrized from the above experimental data, supplemented by the data of Guo and Green (1990). Figure 21 shows that $\log D_{Ba}$ is inversely correlated with temperature. There is a discrepancy, however, between D_{Ba} determined by La Tourrette *et al.* (1995) and that determined by Guo and Green (1990) at almost identical pressure and temperature. This difference must be compositional in origin, and may relate to the higher titanium content of the Guo and Green (1990) phlogopites. Until $r_{o(X)}^{2+}$ can be better parametrized in terms of biotite composition, the relationship between D_{Ba} and $D_{o(X)}^{2+}$, and hence with pressure and temperature, will remain unsure. In the meantime we suggest using Figure 21 to estimate D_{Ba}.

Schmidt *et al.* (1999) report D_{Pb} of 0.034–0.045 for two experiments with leucite lamproite

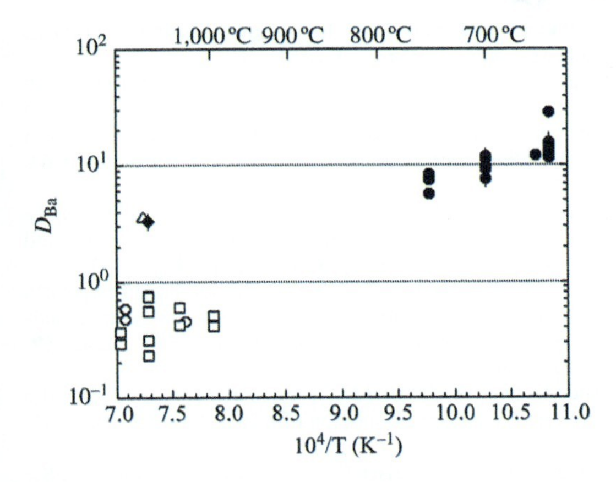

Figure 21 Plot of D_{Ba} for phlogopite-melt partitioning versus reciprocal temperature (see text) (△: La Tourrette *et al.* (1995); ◆: Green *et al.* (2000); ○: Schmidt *et al.* (1999); ●: Icenhower and London (1995); and □: Guo and Green (1990)).

melt composition; for a basanitic melt composition, La Tourrette *et al.* (1995) give $D_{Pb} = 0.10$. In all three cases D_{Pb} consistently falls below, by a factor of ~3, the parabola defined by the other 2+ cations, as previously noted for several other minerals.

2.09.9.10 Spinel

Principal experimental data on spinel-melt partitioning are those of Nielsen *et al.* (1994) for scandium, nickel, vanadium, zirconium, hafnium, niobium, tantalum, uranium, and thorium; and Horn *et al.* (1994) for scandium, vanadium, gallium, zinc, cobalt, zirconium, hafnium, niobium, and tantalum. These are supplemented by data for hydrous melts by Nielsen and Beard (2000).

2.09.9.11 Rutile

In three experiments on a doped natural tonalite under hydrous conditions at 1.8–2.5 GPa, Foley *et al.* (2000) measured D_{Zr} and D_{Hf}. We can estimate the TiO_2 content of the melt from the glass analysis in Jenner *et al.* (1994) for the same tonalite starting material at similar pressure and temprature, which gives $D_{Ti} = 84$. The data in Foley *et al.* (2000) can then be fitted (using sixfold ionic radii) with $r_0^{4+} = 0.585$ Å and $E^{4+} = 700$ GPa. From these parameters we would calculate D_U of $\sim 4 \times 10^{-5}$ and D_{Th} some three orders of magnitude smaller.

Experimental D_{Nb} and D_{Ta} are provided by Jenner *et al.* (1994) and Bennett *et al.* (2003). Foley *et al.* (2000) determine D_{Nb}, but place only lower bounds on D_{Ta}. The first two studies give D_{Nb}/D_{Ta} of 0.53 and 0.78 ± 0.14, respectively. D_{Nb} appears to be very sensitive to both temperature and, possibly, melt composition. For example, in the 3 GPa Ti-CMAS experiments of Bennett *et al.* (2003), D_{Nb} is 28 ± 3, while it is 53 in the Jenner *et al.* (1993) experiments and >100 in the Foley *et al.* (2000) experiments, both on hydrous tonalite.

ACKNOWLEDGMENTS

This research has been supported by the Natural Environment Research Council (UK) by the award of a number of research grants and studentships. The authors acknowledge the research input of a number of Bristol students and post-docs, notably Wim VanWestrenen, Richard Brooker, Karianne Law, Andy Robinson, Eddy Hill, Eva Chamorro, and John Dalton. The authors also thank their colleagues in the workshop, Fred Wheeler and Mike Dury for their unstinting support throughout a number of years of experimental research on trace-element partitioning.

REFERENCES

Anderson O. L. (1972) Patterns in elastic constants of minerals important to geophysics. In *The Nature of the Solid Earth* (ed. E. C. Robertson). McGraw-Hill, pp. 575–613.

Anderson D. L. and Anderson O. L. (1970) The bulk modulus-volume relationship for oxides. *J. Geophys. Res.* **75**, 3494–3500.

Andreeßen T., Bottazzi P., Vannucci R., Mengel K., and Johannes W. (1996) Experimental determination of trace element partitioning between amphibole and melt. In *6th Annual Goldschmidt Conference, Heidelberg, Germany*. J. Conf. Abstr. **1**, 17.

Baker M. B. and Stolper E. M. (1994) Determining the composition of high-pressure melts using diamond aggregates. *Geochim. Cosmochim. Acta* **58**, 2811–2827.

Beattie P. (1993a) The generation of uranium series disequilibria by partial melting of spinel peridotite: constraints from partitioning studies. *Earth Planet. Sci. Lett.* **117**, 379–391.

Beattie P. (1993b) Uranium-thorium disequilibria and partitioning on melting of garnet peridotite. *Nature* **363**, 63–65.

Beattie P. (1994) Systematics and energetics of trace-element partitioning between olivine and silicate melts: implications for the nature of mineral/melt partitioning. *Chem. Geol.* **117**, 57–71.

Bennett S., Blundy J., and Elliott T. (2003) The effect of sodium and titanium on trace element partitioning. *Geochim. Cosmochim. Acta* (submitted).

Bindeman I. N. and Davis A. M. (2000) Trace element partitioning between plagioclase and melt: investigation of dopant influence on partition behaviour. *Geochim. Cosmochim. Acta* **64**, 2863–2878.

Bindeman I. N., Davis A. M., and Drake M. J. (1998) Ion microprobe study of plagioclase-basalt partition experiments at natural concentration levels of trace elements. *Geochim. Cosmochim. Acta* **62**, 1175–1193.

Blundy J. D. and Brooker R. A. (2003) Trace element partitioning during melting and crystallization of mafic rocks in the lower crust. *Contrib. Mineral. Petrol.* (submitted).

Blundy J. D. and Dalton J. A. (2000) Experimental comparison of trace element partitioning between clinopyroxene and melt in carbonate and silicate systems and implications for mantle metasomatism. *Contrib. Mineral. Petrol.* **139**, 356–371.

Blundy J. D. and Wood B. J. (1991) Crystal-chemical controls on the partitioning of Sr and Ba between plagioclase feldspar silicate melts and hydrothermal solutions. *Geochim. Cosmochim. Acta* **55**, 193–209.

Blundy J. D. and Wood B. J. (1994) Prediction of crystal–melt partition coefficients from elastic moduli. *Nature* **372**, 452–454.

Blundy J. and Wood B. (2003) Trace element partitioning between minerals and melts. *Earth Planet. Sci. Lett.* **210**, 383–397.

Blundy J. D., Falloon T. J., Wood B. J., and Dalton J. A. (1995) Sodium partitioning between clinopyroxene and silicate melts. *J Geophys. Res.* **100**, 15501–15515.

Blundy J. D., Wood B. J., and Davies A. (1996) Thermodynamics of rare earth element partitioning between clinopyroxene and melt in the system $CaO–MgO–Al_2O_3–SiO_2$. *Geochim. Cosmochim. Acta* **60**, 359–364.

Born M. (1920) Volumen und hydrationswärme der ionen. *Zeitschr. Physik* **1**, 45–48.

Bourdon B., Zindler A., Elliott T., and Langmuir C. H. (1996) Constraints on mantle melting at mid-ocean ridges from global $^{238}U–^{230}Th$ disequilibrium data. *Nature* **384**, 231–235.

Brenan J. M., Shaw H. F., Ryerson F. J., and Phinney D. L. (1995) Experimental determination of trace element partitioning between pargasitic amphibole and synthetic hydrous melt. *Earth Planet. Sci. Lett.* **135**, 1–11.

Brice J. C. (1975) Some thermodynamic aspects of the growth of strained crystals. *J. Cryst. Growth* **28**, 249–253.

Brooker R. A., Wartho J.-A., Carroll M. R., Kelley S. P., and Draper D. S. (1998) Preliminary UVLAMP determinations of argon partition coefficients for olivine and clinopyroxene grown from silicate melts. *Chem. Geol.* **147**, 185–200.

Brooker R. A., Du Z., Blundy J. D., Kelley S. P., Allan N. L., Wood B. J., Chamorro E. M., Wartho J.-A., and Purton J. A. (2003) The "zero charge" partitioning behaviour of noble gases during mantle melting. *Nature* **423**, 738–741.

Burnham C. W. (1981) The Nature of multicomponent aluminosilicate melts. *Phys. Chem. Earth* **13**, 197–229.

Burns R. G. (1970) *Mineralogical Applications of Crystal Field Theory*. Cambridge University Press, Cambridge, 224pp.

Chamorro E. M., Wartho J. A., Brooker R. A., Wood B. J., Kelley S. P., and Blundy J. D. (2002) Ar and K partitioning between clinopyroxene and silicate melt to 8 GPa. *Geochim. Cosmochim. Acta* **66**, 507–519.

Condomines M., Morand P., and Allègre C. J. (1981) $^{230}Th–^{238}U$ disequilibrium in historical lavas from the FAMOUS zone (Mid-Atlantic Ridge 36° 50′N): Th and Sr isotopic geochemistry. *Earth Planet. Sci. Lett.* **55**, 247–256.

Dalpé C. and Baker D. R. (2000) Experimental investigation of large-ion lithophile-element, high-field-strength-element, and rare-earth-element partitioning between calcic amphibole and basaltic melt: the effects of pressure and oxygen fugacity. *Contrib. Mineral. Petrol.* **140**, 233–250.

Drake M. J. (1972) The distribution of major and trace elements between plagioclase feldspar and magmatic silicate liquid: an experimental study. PhD Thesis, University of Oregon Eugene, Oregon, USA (unpublished).

Dunn T. (1987) Partitioning of Hf, Lu, Ti, and Mn between olivine, clinopyroxene, and basaltic liquid. *Contrib. Mineral. Petrol.* **96**, 476–484.

Ewart A. and Griffin W. L. (1994) Application of proton-microprobe data to trace-element partitioning in volcanic rocks. *Chem. Geol.* **117**, 251–284.

Falloon T. J. and Green D. H. (1987) Anhydrous partial melting of MORB pyrolite and other peridotite compositions at 10 kbar implications for the origin of primitive MORB glasses. *Mineral. Petrol.* **37**, 181–219.

Foley S. F., Barth M. G., and Jenner G. A. (2000) Rutile/melt partition coefficients for trace elements and an assessment of the influence of rutile on the trace element characteristics of subduction zone magma. *Geochim. Cosmochim. Acta* **64**, 933–938.

Forsythe L. M., Nielsen R. L., and Fisk M. R. (1994) High-field-strength element partitioning between pyroxene and basaltic to dacitic magmas. *Chem. Geol.* **117**, 107–125.

Gaetani G. A. and Grove T. L. (1995) Partitioning of rare-earth elements between clinopyroxene and silicate melt: crystal-chemical controls. *Geochim. Cosmochim. Acta* **59**, 1951–1962.

Gale J. D. (1997) GULP—a computer program for the symmetry adapted simulation of solids. *Faraday Trans.* **93**, 629.

Gallahan W. E. and Nielsen R. L. (1992) The partitioning of Sc, Y, and the rare-earth elements between high-Ca pyroxene and natural mafic to intermediate lavas at 1 atmosphere. *Geochim. Cosmochim. Acta* **56**, 2387–2404.

Gast P. W. (1968) Trace element fractionation and the origin of tholeiitic and alkaline magma types. *Geochim. Cosmochim. Acta* **32**, 1057–1086.

Goldschmidt V. M. (1937) The principles of the distribution of chemical elements in minerals and rocks. *J. Chem. Soc. London* **140**, 655–673.

Green T. H., Blundy J. D., Adam J., and Yaxley G. M. (2000) SIMS determination of trace element partition coefficients between garnet clinopyroxene and hydrous basaltic liquids at 2–7.5 GPa and 1,080–1,200°C. *Lithos* **53**, 165–187.

Grutzeck M., Kridelbaugh S., and Weill D. (1974) The distribution of Sr and REE between diopside and silicate liquid. *Geophys. Res. Lett.* **1**, 273–275.

Guo J. and Green T. H. (1989) Barium partitioning between alkali feldspar and silicate melts at high temperature and pressure. *Contrib. Mineral. Petrol.* **102**, 328–335.

Guo J. and Green T. H. (1990) Experimental study of barium partitioning between phlogopite and silicate liquid at upper-mantle pressure and temperature. *Lithos* **24**, 83–96.

Hack P. J., Nielsen R. L. and Johnston A. D. (1994) Experimentally-determined rare-earth element and Y partitioning behaviour between clinopyroxene and basaltic liquids at pressures up to 20 kbar. *Chem. Geol.* **117**, 89–105.

Hart S. R. and Dunn T. (1993) Experimental cpx/melt partitioning of 24 trace elements. *Contrib. Mineral. Petrol.* **113**, 1–8.

Hauri E. H., Wagner T. P., and Grove T. L. (1994) Experimental and natural partitioning of Th–U–Pb and other trace elements between garnet clinopyroxene and basaltic melts. *Chem. Geol.* **117**, 149–166.

Hazen R. M. and Finger L. W. (1979) Bulk Modulus-volume relationship for cation-anion polyhedra. *J. Geophys. Res.* **84**, 6723–6728.

Hill E., Wood B. J., and Blundy J. D. (2000) The effect of Ca-Tschermaks component on trace element partitioning between clinopyroxene and silicate melt. *Lithos* **53**, 205–217.

Hill P. G. (1974) The petrology of Aden volcano, Peoples' Democratic Republic of Yemen. PhD Thesis, University of Edinburgh.

Hirose K. and Kushiro I. (1998) The effect of melt segregation on polybaric mantle melting: estimation from incremental melting experiments. *Phys. Earth Planet. Int.* **107**, 111–118.

Holland T. J. B. (1990) Activities of components in omphacite solid solutions. *Contrib. Mineral. Petrol.* **105**, 446–553.

Horn I., Foley S. F., Jackson S. E., and Jenner G. A. (1994) Experimentally determined partitioning of high field strength and selected transition elements between spinel and basaltic melt. *Chem. Geol.* **117**, 193–218.

Icenhower J. and London D. (1995) An experimental study of element partitioning among biotite muscovite and coexisting peraluminous silicic melt at 200 MPa (H_2O). *Am. Mineral.* **80**, 1229–1251.

Icenhower J. and London D. (1996) Experimental partitioning of Rb, Cs, Sr, and Ba between alkali feldspar and peraluminous melt. *Am. Mineral.* **81**, 719–734.

Jenner G. A., Foley S. F., Jackson S. E., Green T. H., Fryer B. J., and Longerich H. P. (1994) Determination of partition coefficients for trace elements in high pressure-temperature experimental run products by laser ablation microprobe-inductively coupled plasma-mass spectrometry (LAM–ICP–MS). *Geochim. Cosmochim. Acta* **58**, 5099–5103.

Jones J. H. (1995) Experimental trace element partitioning. In *Rock Physics and Phase Relations: A Handbook of Physical Constants*, Reference Shelf 3 (ed. T. J. Ahrens). American Geophysical Union, pp. 73–104.

Jones J. H. and Burnett D. S. (1987) Experimental geochemistry of Pu and Sm and the thermodynamics of trace-element partitioning. *Geochim. Cosmochim. Acta* **51**, 769–782.

Kennedy A. K., Lofgren G. E., and Wasserburg G. J. (1993) An experimental study of trace element partitioning between olivine orthopyroxene and melt in chondrules: equilibrium values and kinetic effects. *Earth Planet. Sci. Lett.* **115**, 177–195.

Klemme S., Blundy J. D., and Wood B. J. (2002) Experimental constraints on major and trace element partitioning during partial melting of eclogite. *Geochim. Comsochim. Acta* **66**, 3109–3123.

Landwehr D., Blundy J., Chamorro-Perez E. M., Hill E., and Wood B. J. (2001) U-series disequilibria generated by partial melting of spinel lherzolite. *Earth Planet. Sci. Lett.* **188**, 329–348.

La Tourrette T. Z., Kennedy A. K., and Wasserburg G. J. (1993) Thorium–uranium fractionation by garnet: evidence for a deep source and rapid rise of oceanic basalts. *Science* **261**, 739–742.

La Tourrette T., Hervig R. L., and Holloway J. R. (1995) Trace element partitioning between amphibole phlogopite and basanite melt. *Earth Planet. Sci. Lett.* **135**, 13–30.

Law K. M., Blundy J. D., Wood B. J., and Ragnarsdottir K. V. (2000) Trace element partitioning between wollastonite and carbonate–silicate melt. *Min. Mag.* **64**, 155–165.

Long P. E. (1978) Experimental determination pf partition coefficients for Rb, Sr, and Ba between alkali feldspar and silicate liquid. *Geochim. Cosmochim. Acta* **42**, 833–846.

Lundstrom C. C., Shaw H. F., Ryerson F. J., Phinney D. L., Gill J. B., and Williams Q. (1994) Compositional controls on the partitioning of U, Th, Ba, Pb, Sr, and Zr between clinopyroxene and haplobasaltic melts: implications for uranium series disequilibria in basalts. *Earth Planet. Sci. Lett.* **128**, 407–423.

Lundstrom C. C., Shaw H. F., Ryerson F. J., Williams Q., and Gill J. (1998) Crystal chemical control of clinopyroxene–melt partitioning in the Di–Ab–An system: implications for elemental fractionations in the depleted mantle. *Geochim. Cosmochim. Acta* **62**, 2849–2862.

McDade P., Wood B. J., Blundy J. D., and Dalton J. A. (2003a) Trace element partitioning at 3 GPa on the anhydrous garnet peridotite solidus. *J. Petrol.* (in press).

McDade P., Blundy J., and Wood B. (2003b) Trace element partitioning on the Tinaquillo Lherzolite solidus at 1.5 GPa. *Phys. Planet. Earth Int.* (in press).

McKay G. A., Wagstaff J., and Yang S. R. (1986) Clinopyroxene REE distribution coefficients for shergottites: the REE content of the Shergotty melt. *Geochim. Cosmochim. Acta* **50**, 927–937.

McKay G., Le L., Wagstaff J., and Crozaz G. (1994) Experimental partitioning of rare earth element and strontium: constraints on the petrogenesis and redox conditions during crystallisation of Antarctic angrite Lewis Cliff 86010. *Geochim. Cosmochim. Acta* **58**, 2911–2919.

McKenzie D. (1985) ^{230}Th–^{238}U disequilibrium and the melting processes beneath ridge axes. *Earth Planet. Sci. Lett.* **72**, 149–157.

McKenzie D. (1989) Some remarks on the movement of small melt fractions in the mantle. *Earth Planet. Sci. Lett.* **95**, 53–72.

McKenzie D. and O'Nions R. K. (1991) Partial melt distributions from inversion of rare earth element concentrations. *J. Petrol.* **32**, 1021–1091.

Mathez E. A. (1973) Refinement of the Kudo-Weill plagioclase thermometer and its application to basaltic rocks. *Contrib. Mineral. Petrol.* **44**, 61–72.

Mott N. F. and Littleton M. J. (1938) Conduction in polar crystals: I. Electrolytic conduction in solid salts. *Trans. Farad. Soc.* **34**, 485–499.

Nagasawa H. (1966) Trace element partition coefficient in ionic crystals. *Science* **152**, 767–769.

Nicholls I. A. and Harris K. L. (1980) Experimental rare earth element partition coefficients for garnet, clinopyroxene and amphibole coexisting with andesitic and basaltic liquids. *Geochim. Cosmochim. Acta* **34**, 331–340.

Nielsen R. L. and Beard J. S. (2000) Magnetite-melt HFSE partitioning. *Chem. Geol.* **164**, 21–34.

Nielsen R. L., Forsythe L. M., Gallahan W. E., and Fisk M. R. (1994) Major and trace element magnetite-melt equilibria. *Chem. Geol.* **117**, 167–191.

Onuma N., Higuchi H., Wakita H., and Nagasawa H. (1968) Trace element partition between two pyroxenes and the host lava. *Earth Planet. Sci. Lett.* **5**, 47–51.

Purton J. A., Allan N. L., Blundy J. D., and Wasserman E. A. (1996) Isovalent trace element partitioning between minerals and melts—a computer simulation model. *Geochim. Cosmochim. Acta* **60**, 4977–4987.

Purton J. A., Allan N. L., and Blundy J. D. (1997) Calculated solution energies of heterovalent cations in forsterite and diopside: implications for trace element partitioning. *Geochim. Cosmochim. Acta* **61**, 3927–3936.

Ray G. L., Shimizu N., and Hart S. R. (1983) An ion microprobe study of the partitioning of trace elements between clinopyroxene and liquid in the system diopside-albite-anorthite. *Geochim. Cosmochim. Acta* **47**, 2131–2140.

Robinson J. A. C., Wood B. J., and Blundy J. D. (1998) The beginning of melting of fertile and depleted peridotite at 1.5 GPa. *Earth Planet. Sci. Lett.* **155**, 97–111.

Salters V. J. M. and Longhi J. (1999) Trace element partitioning during the initial stages of melting beneath mid-ocean ridges. *Earth Planet. Sci. Lett.* **166**, 15–30.

Salters V. J. M., Longhi J. E., and Bizimis M. (2002) Near mantle solidus trace element partitioning at pressures up to 3.4 GPa. *Geochem. Geophys. Geosys.* **3**, 2001GC000148.

Schilling J.-G. and Winchester J. W. (1967) Rare-earth fractionation and magmatic processes. In *Mantles of the Earth and Terrestrial Planets* (ed. S. K. Runcorn). Interscience, New York, pp. 267–283.

Schmidt K. H., Bottazzi P., Vannucci R., and Mengel K. (1999) Trace element partitioning between phlogopite, clinopyroxene, and leucite lamproite melt. *Earth Planet. Sci. Lett.* **168**, 287–299.

Shannon R. D. (1976) Revised effective ionic radii and systematic studies of interatomic distances in halides and chalcogenides. *Acta Cryst.* **A32**, 751–767.

Shannon R. D. (1993) Dielectric polarizabilities of ions in oxides and fluorides. *J. Appl. Phys.* **73**, 348–366.

Shaw D. M. (1970) Trace element fractionation during anatexis. *Geochim. Cosmochim. Acta* **34**, 237–243.

Skulsky T., Minarik W., and Watson E. B. (1994) High-pressure experimental trace-element partitioning between clinopyroxene and basaltic melts. *Chem. Geol.* **117**, 127–147.

Smyth J. R. and Bish D. L. (1988) *Crystal Structures and Cation Sites of the Rock Forming Minerals.* Allen and Unwin, Boston, 332pp.

Takahashi E. and Kushiro I. (1983) Melting of a dry peridotite at high pressures and temperatures and basalt magma genesis. *Am. Mineral.* **68**, 859–879.

Taura H., Yurimoto H., Kurita K., and Sueno S. (1998) Pressure dependence on partition coefficients for trace elements between olivine and the coexisting melts. *Phys. Chem. Min.* **25**, 469–484.

Tiepolo M., Vannucci R., Oberti R., Foley S. F., Botazzi P., and Zanetti A. (2000a) Nb and Ta incorporation and fractionation in titanian pargasite and kaersutite: crystal-chemical constraints and implications for natural systems. *Earth Planet. Sci. Lett.* **176**, 185–201.

Tiepolo M., Vannucci R., Bottazzi P., Oberti R., Zanetti A., and Foley S. (2000b) Partitioning of rare earth elements, Y, Th, U, and Pb between pargasite, kaersutite, and basanite to trachyte melts: implications for percolated and veined mantle. *Geochem. Geophys. Geosys.* **1**, 2000GC000064.

Tiepolo M., Bottazzi P., Foley S. F., Oberti R., Vannucci R., and Zanetti A. (2001) Fractionation of Nb and Ta from Zr and Hf at mantle depths: the role of titanian pargasite and kaersutite. *J. Petrol.* **42**, 221–232.

Ulmer P. (1989) The dependence of the Fe^{2+}–Mg cation partitioning between olivine and basaltic liquid on pressure, temperature, and composition—an experimental study to 30 kbars. *Contrib. Mineral. Petrol.* **101**, 261–273.

VanWestrenen W., Blundy J. D., and Wood B. J. (1999) Crystal-chemical controls on trace element partitioning between garnet and anhydrous silicate melt. *Am. Mineral.* **84**, 838–847.

VanWestrenen W., Allan N. L., Blundy J. D., Purton J. A., and Wood B. J. (2000a) Atomistic simulation of trace element incorporation into garnets-comparison with experimental garnet–melt partitioning data. *Geochim. Cosmochim. Acta* **64**, 1629–1639.

VanWestrenen W., Blundy J. D., and Wood B. J. (2000b) Effect of Fe^{2+} on garnet–melt trace element partitioning: experiments in FCMAS and quantification of crystal-chemical controls in natural systems. *Lithos* **53**, 191–203.

Van Westrenen W., Wood B. J., and Blundy J. D. (2001a) A predictive thermodynamic model of garnet–melt trace element partitioning. *Contrib. Mineral. Petrol.* **142**, 219–234.

Van Westrenen W., Blundy J. D., and Wood B. J. (2001b) HFSE/REE fractionation during partial melting in the presence of garnet: implications for identification of mantle heterogeneities. *Geochem. Geophys. Geosys.* **2**, 2000GC000133.

Walter M. J. and Presnall D. C. (1994) Melting behavior of simplified lherzolite in the system $CaO–MgO–Al_2O_3–SiO_2–Na_2O$ from 7 to 35 kbar. *J. Petrol.* **35**, 329–359.

Watson E. B., Ben Othman D., Luck J. M., and Hofmann A. W. (1987) Partitioning of U, Pb, Cs, Yb, Hf, Re, and Os between chromian diopsidie pyroxene and haploba saltic liquid. *Chem. Geol.* **62**, 191–208.

White J. C. (2003) Trace element partitioning between alkali feldspar and peralkalic quartz trachyte to rhyolite magma: Part II. Empirical equations for calculating trace element partition coefficients of large-ion lithophile, high field strength and rare-earth elements. *Am. Mineral.* **88**, 330–337.

Wilson, M. I. (1988) *Petrogenesis*. Unwin Hyman, New York, 466pp.

Wood B. J. and Banno S. (1973) Garnet-orthopyroxene and orthopyroxene-clinopyroxene relationships in simple and complex systems. *Contrib. Mineral. Petrol.* **42**, 109–124.

Wood B. J. and Blundy J. D. (1997) A predictive model for rare earth element partitioning between clinopyroxene and anhydrous silicate melt. *Contrib. Mineral. Petrol.* **129**, 166–181.

Wood B. J. and Blundy J. D. (2001) The effect of cation charge on crystal–melt partitioning of trace elements. *Earth Planet. Sci. Lett.* **188**, 59–71.

Wood B. J. and Blundy J. D. (2002) The effect of H_2O on crystal–melt partitioning of trace elements. *Geochim. Comsochim. Acta* **66**, 3647–3656.

Wood B. J. and Fraser D. G. (1976) *Elementary Thermodynamics for Geologists*. Oxford University Press, Oxford, 303pp.

Wood B. J. and Trigila R. (2001) Experimental determination of aluminous clinopyroxene–melt partition coefficients for potassic liquids with applications to the evolution of the roman province potassic magmas. *Chem. Geol.* **172**, 213–223.

Wood B. J., Holland T. J. B., Newton R. C., and Kleppa O. J. (1980) Thermochemistry of jadeite–diopside clinopyroxenes. *Geochim. Cosmochim. Acta* **44**, 1363–1371.

Wood B. J., Blundy J. D., and Robinson J. A. C. (1999) The role of clinopyroxene in generating U-series disequilibrium during mantle melting. *Geochim. Cosmochim. Acta* **63**, 1613–1620.

2.10
Partition Coefficients at High Pressure and Temperature

K. Righter
NASA Johnson Space Center, Houston, TX, USA
and
M. J. Drake
University of Arizona, Tucson, AZ, USA

2.10.1 PLANETARY DIFFERENTIATION

Differentiation of terrestrial planets includes separation of a metallic core and possible later fractionation of mineral phases within either a solid or molten mantle (Figure 1). Lithophile and siderophile elements can be used to understand these two different physical processes, and ascertain whether they operated in the early Earth. The distribution of elements in planets can be understood by measuring the partition coefficient, D (ratio of concentrations of an element in different phases (minerals, metals, or melts)).

The siderophile elements (iron-loving) encompass over 30 elements and are defined as those elements for which D(metal/silicate) > 1, and are useful for deciphering the details of core formation. This group of elements is commonly broken up into several subclasses, including the slightly siderophile elements ($1 < D < 10$), moderately siderophile elements (MSEs; $10 < D < 10^4$), and highly siderophile elements (HSEs; $D > 10^4$). Because these three groups encompass a wide range of partition coefficient values, they can be very useful in trying to determine the conditions under which metal may have equilibrated with the mantle (or a magma ocean). Because metal and silicate may equilibrate by several different mechanisms, such as at the base of a deep magma ocean, or as metal droplets descend through a molten mantle, partition coefficients can potentially shed light on which mechanism may be most important, thus linking the physics and chemistry of core formation. In this chapter, we summarize metal/silicate partitioning of siderophile elements and show how they may be used to understand planetary core formation.

Once a planet is differentiated into core and mantle, a mantle will cool during convection, and can start in either a molten or solid state, depending upon the initial thermal conditions. If hot enough, minerals will crystallize from a molten mantle, and become entrained in the convecting melt, or eventually settle out at the bottom. The entrainment and settling process has been studied in detail (e.g., Tonks and Melosh, 1990), and is a potential mechanism for differentiation between the deep and shallow parts of Earth's mantle. The lithophile elements, those elements that have D(metal/silicate) < 1, fall into many different subclasses and all hold information about the deep mineral structure of the mantle. Rare-earth elements (REEs) have proven to be useful: europium anomalies have helped elucidate the role of plagioclase in lunar crust formation (e.g., Schnetzler and Philpotts, 1971; Weill *et al.*, 1974), and LREE/HREE depletion and enrichment are indicators of partial melting in the presence of garnet in the mantle. High-field-strength elements (HFSEs)—niobium, zirconium, tantalum, and hafnium—are all refractory and hence more resilient to fractionation processes such as volatility or condensation. They also have an affinity for ilmenite and rutile, and can explain differences between lunar and martian samples as well as features of Earth's continental crust (Taylor and McLennan, 1985). Alkaline-earth and alkaline elements include rubidium, strontium, barium, potassium, caesium, and calcium, some of which are involved in radioactive decay couples, e.g., Rb–Sr and K–Ar. The latter is important in understanding the contribution of radioactive decay to planetary heat production, and potential deep sources of radiogenic

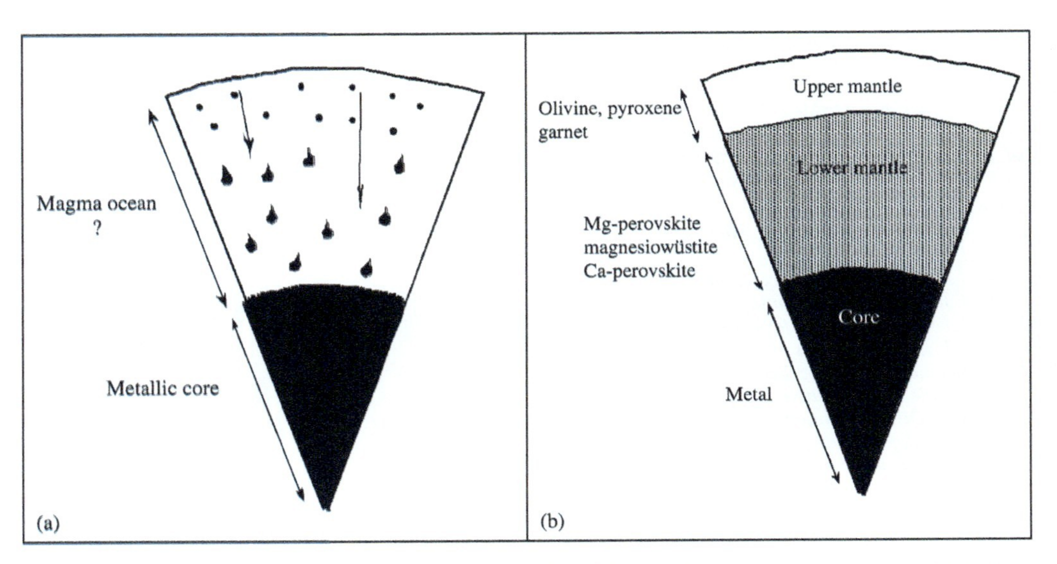

Figure 1 Schematic cross-section through the Earth, showing: (a) an early magma ocean stage and (b) a later cool and differentiated stage.

argon (see Chapter 2.06). Rubidium and potassium are further useful as tracers of hydrous phases such as mica and amphibole. Possible fractionation of any of these elements from chondritic abundances (see Chapter 2.01) can be assessed with the knowledge of partition coefficients. In this chapter we summarize our understanding of mineral/melt fractionation of minor and trace elements at high pressures and temperatures and discuss the implications for mantle differentiation.

2.10.2 EXPERIMENTAL APPROACHES

A full review of experimental techniques used to study metal/silicate and mineral/melt equilibria is beyond the scope of this effort, but a brief summary, such that the reader may appreciate the materials, scales, and techniques utilized, is in order.

2.10.2.1 Apparatuses

The pressure regime of the lower crust and upper mantle (up to 7 GPa) is attainable with the piston cylinder apparatus. Many in use today achieve pressures of 0.7–3.0 GPa, although higher pressures (up to 7 GPa) are possible (Boyd and England, 1960; Bohlen, 1984). Piston cylinder techniques are reviewed by Dunn (1993), and are based on the simple idea of squeezing on a cylindrical (uniaxial) medium within a tungsten carbide (WC) core, using a hydraulic ram. Pressure is generated with a hydraulic ram, and hydrostatically transferred to the sample through the use of pressure media such as talc + pyrex, NaCl, $BaCO_3$, or CaF_2 (Dunn, 1993). High temperatures are attained most commonly by using a graphite furnace with an internal electrical resistance.

Pressures and temperatures equivalent to that at the base of the upper mantle and top of the lower mantle can be achieved with multi-anvil devices. Hall (1958) described a tetrahedral multi-anvil device used to generate pressures high enough to synthesize diamond. Further developments led to the octahedral multi-anvil device (Kawai and Endo, 1970), refined by many Japanese scientists (e.g., Akaogi and Akimoto,1977; Onodera, 1987; Ohtani, 1987). Walker *et al.* (1990) and Walker (1991) introduced a simplified design that has made more widespread the application to the geological community (e.g., Rubie, 1999). The most common design of a multi-anvil utilizes eight WC cubes with truncated corners to generate pressure by compressing an octahedral pressure medium. Greater pressures are attained by using smaller truncations. It is possible to generate pressure up to ~30 GPa using these techniques. Heating capabilities are similar to piston cylinders, except that rhenium and $LaCrO_3$ are used to generate temperatures as high as 2,800 °C, at pressures greater than the graphite–diamond transition.

Experimentation at even higher pressures and temperatures has been limited, primarily due to the difficulty of heating samples in a diamond anvil cell (DAC). DACs (Bassett, 1979) are compact and capable of achieving pressures equivalent to that at the core–mantle boundary of the Earth and deep within Jupiter. Samples within a DAC can be heated with a laser (e.g., Boehler and Chopelas, 1991), or with resistance heaters, and there have been several applications to element partitioning and planetary differentiation (Zerr *et al.*, 1998; Tschauner *et al.*, 1999). The small size of the run products from these experiments has made characterization a challenge. However, with sub-micron spot analysis capabilities (e.g., Stadermann *et al.*, 1999) becoming more common, this may change.

2.10.2.2 The Challenge of Encapsulation

Study of geological materials at high pressures and temperatures is challenging. A further problem is determining in what the sample should be encapsulated. Metal/silicate equilibria commonly involve a metallic liquid containing iron. Noble metal capsules such as platinum and palladium are unsuitable for iron–metal/silicate partitioning experiments because they alloy with iron and form eutectic melts at relatively low temperatures. As a result, studies involving iron metal have employed refractory materials for capsules such as MgO, Al_2O_3, and graphite. Graphite capsules work well even at pressures above the diamond transition, but the run products must be ground and polished with techniques appropriate for harder materials. Rhenium, molybdenum, and tantalum, though highly reactive with iron, have high thermal stability and low solubility in silicates, and therefore have some useful applications in element partitioning experiments. Additional precaution must be taken to avoid reaction between furnace and capsule material and this can be a severe limitation as well.

2.10.2.3 Oxygen Fugacity

The pressure of oxygen at the surface and deep within planets has a fundamental control on chemical equilibria, especially those involving iron (Arculus, 1985). For instance, there are two general equilibria (Equations (1) and (2)) that are dependent upon oxygen:

$$2Fe + O_2 = 2FeO \quad (1)$$

and

$$2FeO + \tfrac{1}{2}O_2 = Fe_2O_3 \quad (2)$$

The latter is related to equilibria at the Earth's surface and in the mantle, whereas the former is relevant to equilibria involving the mantle and core. We know the oxygen fugacity in Earth's crust and mantle from studies of basalts and mantle peridotite (xenoliths and massifs). The measured Fe_2O_3/FeO ratio of a basalt of known composition can be used to infer its oxygen fugacity (Carmichael, 1991). Olivine, pyroxene, and spinel, or olivine, pyroxene, and garnet are stable in spinel and garnet lherzolite, respectively, and define (Equations (3) and (4)) oxygen fugacity:

$$\underset{\text{(olivine)}}{6Fe_2SiO_4} + \underset{\text{(gas)}}{O_2} = \underset{\text{(opx)}}{6FeSiO_3} + \underset{\text{(spinel)}}{2Fe_3O_4} \quad (3)$$

or

$$\underset{\text{(olivine)}}{8Mg_2SiO_4} + \underset{\text{(garnet)}}{5Fe_4Si_4O_{12}} + \underset{\text{(gas)}}{2O_2} = \underset{\text{(garnet)}}{4Fe_3Fe_2Si_3O_{12}} + \underset{\text{(opx)}}{16MgSiO_3} \quad (4)$$

These equilibria have helped to define the oxygen fugacity in Earth's upper mantle (Wood, 1991; O'Neill *et al.*, 1993). Finally, electrochemical measurements of oxygen potential (e.m.f.) in various mantle and magmatic samples, pioneered by M. Sato and colleagues at the USGS (Sato, 1984), have also defined oxygen fugacity in the Earth, as well as the Moon and meteorites.

As a result of these three approaches, we know that Earth's mantle records a large range of oxygen fugacities, from just above the iron-wüstite (IW) buffer, to close to the hematite-magnetite (HM) buffer (Carmichael, 1991).

Many minor and trace elements in the Earth's mantle have several valence states, such as V (2+, 3+, 4+, 5+), Cr (3+, 2+), Eu (3+, 2+), or Ru (4+, 3+). The stability of each species is dependent upon oxygen fugacity, which must be known in any partitioning study involving multivalent elements. For instance, the Eu^{3+}/Eu^{2+} ratio can be used to infer oxygen fugacities (Philpotts, 1970; Drake, 1975). Similarly, D(Cr) for olivine will be dependent upon oxygen fugacity (Hanson and Jones, 1998), due to the change in Cr^{2+}/Cr^{3+}.

Control or monitoring of oxygen fugacity in high-pressure experimental systems is challenging, but can be done in several ways. First, the capsule can be used to control oxygen fugacity. For instance, a nickel capsule coupled with NiO within can be used to buffer oxygen fugacity. Similarly, high oxygen fugacity imposed by rhenium capsules has been inferred by some experimentalists (due to the high f_{O_2} defined by the $Re-ReO_2$ buffer; Pownceby and O'Neill (1994)), although it is not clear what oxide is involved in the buffering. Second, a graphite lining can be used to buffer oxygen fugacity at the $C-CO-O_2$ surface (Holloway *et al.*, 1992). This approach can be used at pressures below the graphite–diamond transition. However, the solubility of CO_2 in silicate melt is dependent upon composition, pressure, and temperature, and the oxygen fugacity will be slightly different in each system; as a result, f_{O_2} must be monitored independently using melt Fe^{3+}/Fe^{2+}, or a direct sensor such as that outlined by Taylor *et al.* (1992). Third, in multi-anvil experiments, it is possible to include a buffer (such as Ni–NiO or $Re-ReO_2$) in the octahedral pressure medium. This will impose a specific oxygen fugacity on the system and the buffer does not affect the composition of the sample (Dobson and Brodholt, 1999).

2.10.3 METAL–SILICATE EQUILIBRIA

2.10.3.1 Partitioning Behavior of Siderophile Elements: Effects of P, T, f_{O_2}, and Composition

Partitioning of siderophile elements between metal and silicate liquid can be understood in terms of simple equilibria such as

$$\underset{\text{(metal)}}{M} + \underset{\text{(gas)}}{(n/4)O_2} = \underset{\text{(silicate liquid)}}{M^{n+}O_{n/2}} \quad (5)$$

and at equilibrium

$$-\Delta G^0/RT = \ln[aM^{n+}O_{n/2}/(aM) * (f_{O_2})^{n/4}] \quad (6)$$

where ΔG^0 is the free energy of the reaction, T is temperature, R is the gas constant, a is the thermodynamic activity of the chemical species, M is the metallic element of interest, and n is the valence in the silicate liquid (see, e.g., Capobianco *et al.*, 1993). Such equilibria would clearly be a function of temperature, pressure, and oxygen fugacity. Additional work has demonstrated that activities of siderophile elements in metal and silicate, $aM^{n+}O_{n/2}$ and aM, are dependent upon nonmetal content of metallic liquid, and silicate liquid composition. The nature of these effects will be discussed separately, and a framework for a comprehensive understanding guided by chemical thermodynamics will be presented.

2.10.3.1.1 Oxygen fugacity

The effect of oxygen pressure on the solubilities of siderophile elements in silicate melt has been understood for many years due to metallurgical interest, and has been known to the geological community since early work of Newsom and Drake (1982, 1983) on tungsten and phosphorus. Although there was a brief exploration of the possibility of zero valence dissolution of nickel and possibly cobalt at low f_{O_2} (Colson, 1992;

Ehlers *et al.*, 1992), there have been numerous studies now across a wide range of oxygen fugacities in both iron-bearing and iron-free melts for nickel, cobalt, molybdenum, tungsten, and phosphorus, demonstrating the dependence of *D*(metal/silicate) on f_{O_2} (Figure 2) (e.g., Newsom and Drake, 1982, 1983; Schmitt *et al.*, 1989; Ertel *et al.*, 1996; Holzheid *et al.*, 1994, 1997; Holzheid and Palme, 1996; Capobianco and Amelin, 1994).

The concept of relative f_{O_2} was proposed by Sato (1978). Because most solid–solid oxygen buffers are parallel to equilibria involving mineral solid solutions in T– f_{O_2} space, reporting oxygen fugacity relative to a buffer would make presentation much easier and use of absolute values of f_{O_2} would not be necessary. In addition, across the temperature range of a magmatic system (~300 °C), relative oxygen fugacities are nearly temperature independent (Figure 3). So this concept grew out of terrestrial magmatic applications. When experimental petrologists working in metal/silicate systems at high pressures and temperatures faced the problem of how to calculate oxygen fugacity in their experimental systems, the relative method was enticing for its simplicity (e.g., Hillgren *et al.*, 1994). However, it may not convey all pertinent information, as will be discussed later.

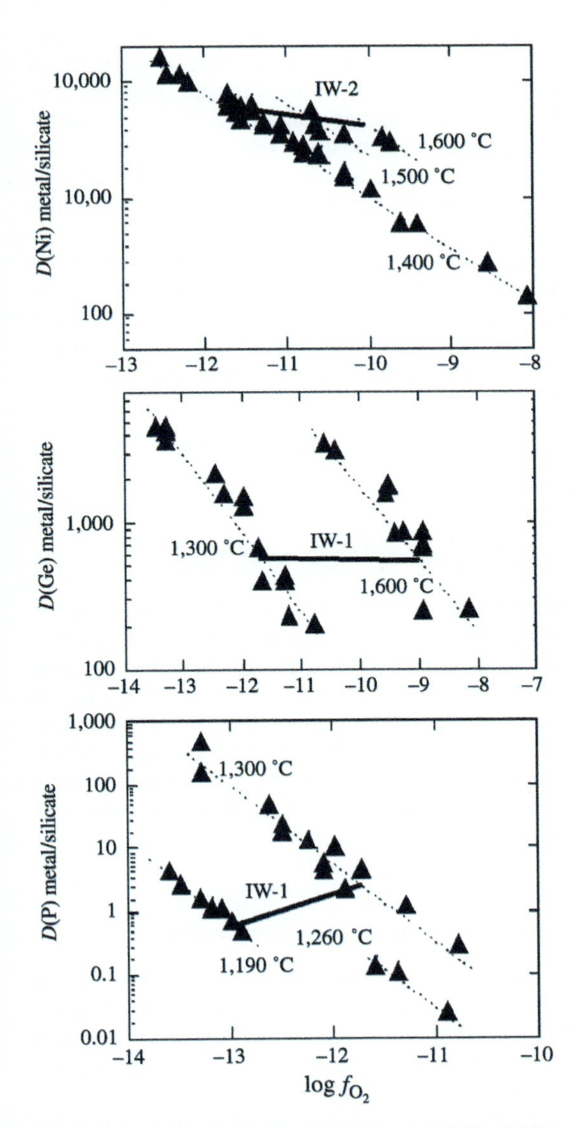

Figure 2 Effect of oxygen fugacity on *D*(Ni), *D*(Ge), and *D*(P). Note that the partition coefficients for nickel decrease, for phosphorus increase and for germanium are constant as *T* increases parallel to a buffer. "IW-1" refers to an oxygen fugacity 1 logf_{O_2} unit below the IW oxygen buffer (sources Holzheid *et al.*, 1997; Schmitt *et al.*, 1989; Hillgren, 1993; Newsom and Drake, 1983).

2.10.3.1.2 *Temperature*

A strong effect of temperature on metal/silicate partition coefficients is expected on thermodynamic grounds for those metal/silicate equilibria possessing a large entropy change (e.g., Capobianco *et al.*, 1993). Although Murthy (1991) proposed a temperature effect on metal/silicate partition coefficients, it was based on the assumption that the chemical potential is temperature independent. Thus, his work was criticized on thermodynamic grounds by O'Neill (1992) and Jones *et al.* (1992). Capobianco *et al.* (1993) showed that application of the van't Hoff equation resulted in increases in siderophility at high temperatures, the opposite of that proposed by Murthy (1991). Because Murthy's results were based on extrapolation of low-temperature data, Capobianco *et al.* (1993) cautioned against extrapolation to high temperatures and

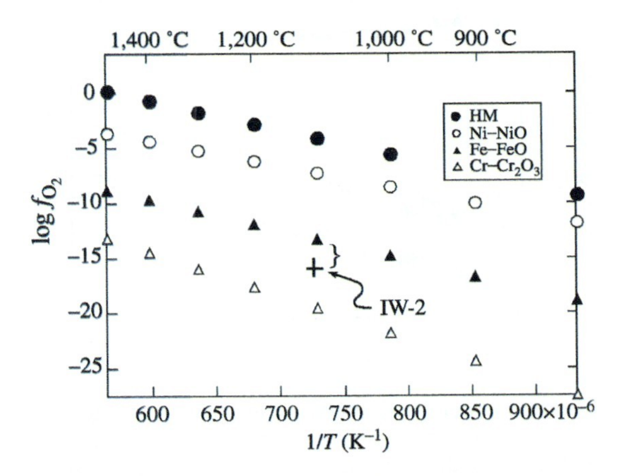

Figure 3 Temperature versus log f_{O_2} for the hematite–magnetite (HM), Ni–NiO, iron–wüstite (IW), and Cr–Cr_2O_3 buffers. Large cross illustrates the concept of relative oxygen fugacity. The cross represents an oxygen fugacity that is 2 log f_{O_2} units below the IW buffer, and is referred to as "IW-2" (sources Myers and Eugster, 1983; O'Neill, 1987; Holzheid and O'Neill, 1999).

emphasized that the best approach will be to actually make the measurements experimentally. Several studies have shown that there is a weak temperature effect, resulting in a decrease in *D*(metal/silicate) at higher temperatures for nickel and cobalt (e.g., Thibault and Walter, 1995; Walker *et al.*, 1993). This effect is predictable, using the thermodynamic approach outlined below. Although both *D*(Ni) and *D*(Co) decrease with increasing temperature, *D*(W) increases, thus verifying the caution of Capobianco *et al.* (1993) that one cannot generalize about siderophile elements—many exhibit distinct behavior. Righter *et al.* (1997) and Righter and Drake (1997, 1999, 2000) parametrized published partition coefficients for nickel and cobalt from many authors' studies, including their own. However, although there have been systematic temperature studies of metal/oxide partitioning for several siderophile elements (Figure 4), there had been no single systematic study of *D*(Ni) or *D*(Co) for metal/silicate systems across a range of temperatures, prompting a recent study by Chabot and Agee (2002).

2.10.3.1.3 Pressure

Any effect of pressure on the magnitude of the metal/silicate partition coefficient can be related to the volumetric properties of the equilibrium

$$\underset{\text{(metal)}}{\text{M}} + \underset{\text{(gas)}}{(n/4)\text{O}_2} = \underset{\text{(silicate liquid)}}{\text{MO}_{n/2}} \tag{7}$$

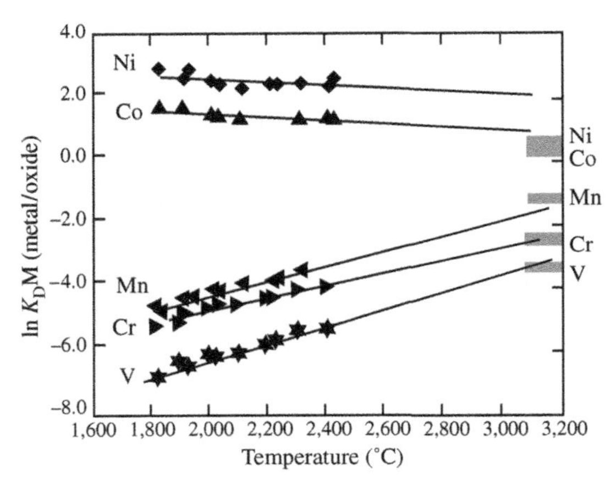

Figure 4 Metal/magnesiowüstite partition coefficients for nickel, cobalt, manganese, chromium, and vanadium at 9 GPa, and the effect of temperature (pressure 9 GPa). Partition coefficients are calculated relative to iron, according to the exchange equilibrium, M + FeO = Fe + MO. Horizontal lines at right side of the diagram indicate the values of K_D that would be required for an equilibrium explanation for these five elements in the terrestrial mantle (source Gessmann and Rubie (1998); these authors favor a high-temperature scenario to attain these concentrations in the mantle).

The volume changes of this reaction and others like it are positive, thus indicating that siderophile elements should become more siderophile with pressure. This can be seen from the results of Righter *et al.* (1997) and Righter and Drake (1997, 1999, 2000) in which the coefficients for the predictive term for pressure are positive. A common assertion, however, is that pressure reduces the metal/silicate partition coefficients (e.g., Li and Agee, 1996), but this is only true at constant relative oxygen fugacity. So it is very important to make a clear distinction between relative oxygen fugacity and absolute oxygen fugacity. The reason partition coefficients can sometimes decrease with pressure (at constant relative f_{O_2}) is that relative oxygen fugacity is pressure dependent—when pressure increases, so does oxygen fugacity and this effect can overwhelm the small, positive pressure effect. Understanding subtle effects like these make the thermochemical approach outlined below the most powerful.

One way to circumvent this problem is to normalize partitioning relations to the Fe + O_2 = FeO equilibrium, and cast the relation in the form of an exchange equilibrium. A two-element distribution coefficient, K_D, based on the equilibrium

$$\underset{\text{(metal)}}{\tfrac{1}{2}\text{Fe}} + \underset{\text{(silicate)}}{\text{MO}_{n/2}} = \underset{\text{(silicate)}}{\tfrac{n}{2}\text{FeO}} + \underset{\text{(metal)}}{\text{M}} \tag{8}$$

where

$$K_{D\,\text{M/Fe}}^{\text{met/sil}} = (X_{\text{M}}^{\text{met}}/X_{\text{MO}_{x/2}}^{\text{sil}})/(X_{\text{Fe}}^{\text{met}}/X_{\text{FeO}}^{\text{sil}})^{x/2} \tag{9}$$

(where *n* is the valence of M), is independent of oxygen fugacity, thus eliminating f_{O_2} as a variable as well. The K_Ds for these experiments then need only be corrected for the effects of temperature and variable silicate melt composition to determine the effect of pressure. When this is done for studies at high pressures, a clear decrease in the K_D with pressure is observed (Figure 5).

2.10.3.1.4 Silicate melt composition

Until the mid-1980s all experiments done on metal/silicate systems were done with basaltic liquids. Although effects of basicity or melt composition have been known in metallurgical slag systems since the 1960s (Elliot *et al.*, 1963), this concept was not recognized or applied to core formation issues until the early 1990s when silicate melt effects were observed for tungsten, molybdenum, and phosphorus (among others) (e.g., Walter and Thibault, 1995; Hillgren *et al.*, 1996; Jaeger and Drake, 2000). This is a poorly understood effect, and must relate to changing activity coefficients of siderophile elements in

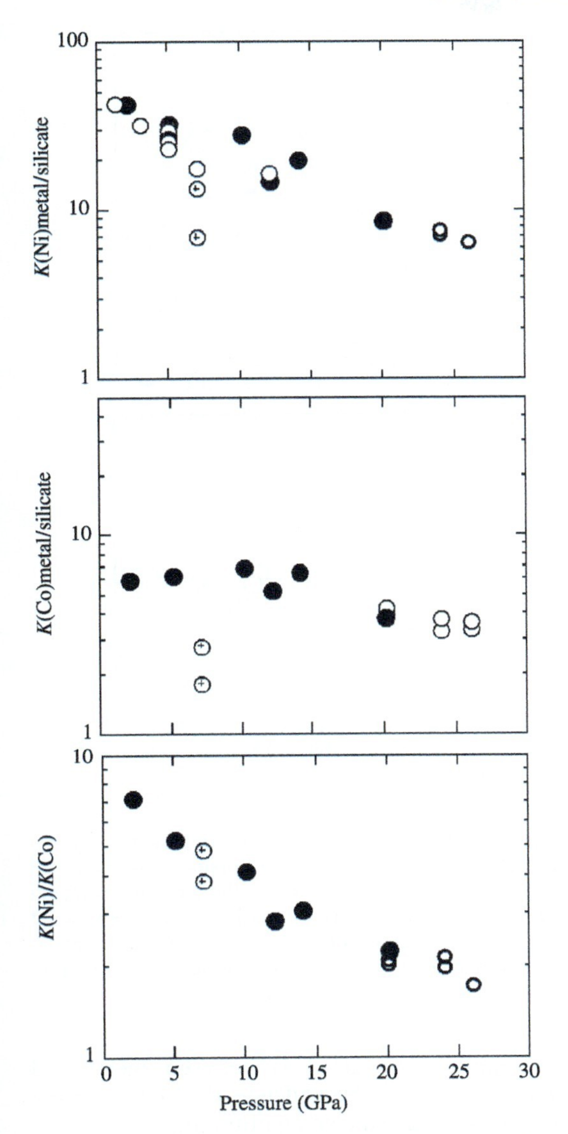

Figure 5 Effect of pressure on the K_D(Ni) and K_D(Co) for metal/silicate equilibrium. The larger effect on nickel causes a convergence of K_Ds to values that are consistent with the near chondritic Ni/Co in the primitive upper mantle (sources (**O**) Ito *et al.*, 1998; (●) Li and Agee, 1996; (○) Thibault and Walter, 1995; (⊕) Righter *et al.*, 1997).

silicate melts of variable composition (see below). The melt structural and compositional parameter, NBO/t, has been used in initial attempts to model systematic change of solubility with silicate melt composition. NBO/t (Mysen, 1991) is the ratio of nonbridging oxygens to tetrahedrally coordinated cations in a silicate melt, and is a way of estimating the degree of melt polymerization. For instance, a highly polymerized melt such as a rhyolite would have a low NBO/t value, close to 0.2, a basalt would have a value near 1, and a depolymerized melt such as a komatiite would have a value near 2 (e.g., Figure 6). So far,

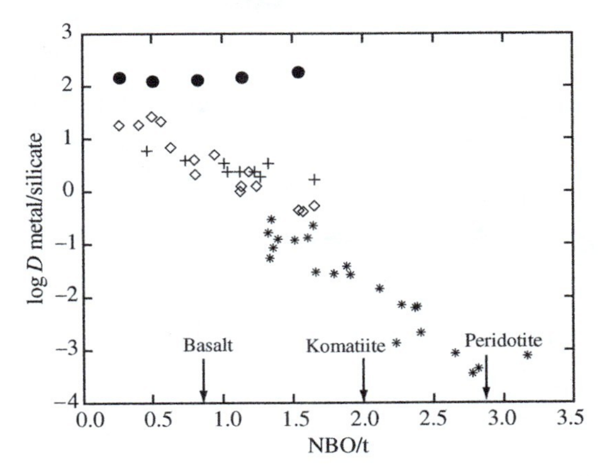

Figure 6 Effect of silicate melt composition on metal/silicate partition coefficients for cobalt (●), gallium (+), tungsten (◇), and phosphorus (∗) (Jaeger and Drake, 2000; Pak and Fruehan, 1986). NBO/t is calculated according to Mysen (1991) and corresponds to basalt values of ~1, komatiite ~1.7, and peridotite ~2.8. In general, high-valence elements such as tungsten and phosphorus are affected more strongly than lower valence elements such as cobalt (or nickel).

high-valence elements such as molybdenum, tungsten, and phosphorus are affected most significantly by melt compositional effects, with the difference between a basalt melt and a peridotite melt corresponding to nearly three orders of magnitude difference in the metal/silicate D (Figure 6). Low-valence elements such as nickel and cobalt are not affected strongly by changing melt composition (Figure 6), but a systematic understanding of why solubilities change so drastically for high- versus low-valence elements does not exist.

2.10.3.1.5 Metallic liquid composition

The effect of dissolution of nonmetals such as sulfur and carbon on the partitioning of siderophile elements between liquid and solid metal is well characterized and understood (e.g., Willis and Goldstein, 1982; Jones and Drake, 1983). Chalcophile elements typically prefer a sulfur-bearing liquid over metal, and thus have very low D(SM/LM) (where SM is solid metal and LM is liquid metal) partition coefficients (e.g., copper and silver; Figure 7). Alternatively, siderophile elements prefer iron-rich metal over a sulfur-bearing metallic liquid and have $D(\mathrm{SM/LM}) > 1$ (e.g., gallium, tungsten, or rhenium; Figure 7).

Extension of this understanding to metal/silicate systems has come through studies of Jana and Walker (1997a,b) in which it was observed that carbon and sulfur both have large effects on the magnitude of liquid metal–liquid silicate Ds. For example, for tungsten, dissolved carbon increases metal/silicate Ds, whereas sulfur

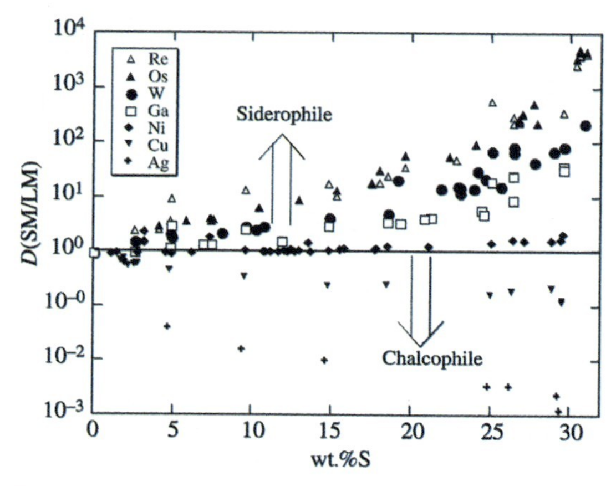

Figure 7 The effect of sulfur content of metallic liquid on the magnitude of the solid metal/liquid metal (SM/LM) partition coefficient. Note that copper and silver have an affinity for S-bearing liquid, whereas nickel, gallium, tungsten, osmium, and rhenium all prefer the solid metal. The connection to core formation is that the latter group of elements will have a lower metal/silicate partition coefficient if the metal is liquid and contains sulfur. Similar effects have been documented for carbon (Willis and Goldstein, 1982) (sources Chabot *et al.*, 2003; Malvin *et al.*, 1986; Jones and Drake, 1983; Liu and Fleet, 2001; Fleet *et al.*, 1999).

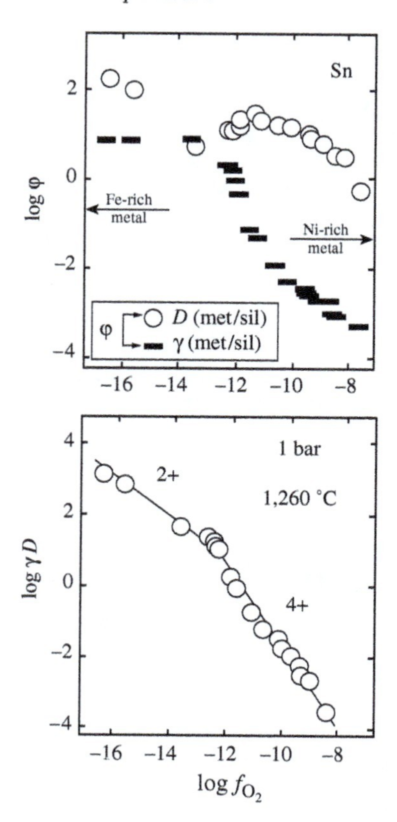

Figure 8 Variation of the activity coefficient (γ) for Sn in nickel-rich metal (high f_{O_2}) versus iron-rich metal (low f_{O_2}). When the partition coefficients are corrected for this unusual behavior, anticipated valences of 2+ and 4+ become clear (source Capobianco *et al.* (1999), who also report activity coefficients for gallium and germanium).

decreases metal/silicate *D*s. The effect of sulfur and carbon is relatively small for elements such as nickel and cobalt, but still must be accounted for in any complete understanding of partitioning behavior.

The behavior of some siderophile elements is controlled by the composition of the metal, and, in particular, the Fe/Ni ratio. Because the activity coefficient of a siderophile element, M, may be different in a nickel-rich metal than an iron-rich metal, an understanding of this effect is necessary before partition coefficients can be successfully applied to a natural system. A good example is tin, which has a low activity coefficient in iron-rich metal, and a high activity coefficient in nickel-rich metal (Figure 8; Capobianco *et al.*, 1999). Metal composition can change as a function of f_{O_2}, and this can then change the solubility in the silicate melt. In addition, the Fe/Ni ratio in metal is sometimes used to impose a specific f_{O_2} on a system. These effects are linked and must be unraveled first in order to understand tin in planetary mantles (Righter and Drake, 2000).

2.10.3.2 Quantification of *P*, *T*, f_{O_2}, and Silicate and Metallic Melt Composition Effects on *D*(metal/silicate)

Chemical thermodynamics can be used as a guide in assessing the relative importance of the factors identified above. Partitioning of siderophile elements (M) between metal and silicate liquid can be understood in terms of the following metal/silicate equilibrium:

$$\underset{\text{(metal)}}{\mathrm{M}} + \underset{\text{(gas)}}{(n/4)\mathrm{O}_2} = \underset{\text{(silicate liquid)}}{\mathrm{MO}_{n/2}} \qquad (10)$$

where $-\Delta G^0/RT = \ln K$ $(K = a(\mathrm{MO})^{n/2}/(a\mathrm{M})(f_{O_2})^{x/4}$ (where a is the thermodynamic activity of the chemical species, f_{O_2} is the oxygen fugacity, and n is the valence in the silicate liquid—e.g., Capobianco *et al.*, 1993). The partition coefficient, *D*(M), can be related to the free energy of the reaction. The expression, ln *D*(M), can be related to temperature, pressure, and oxygen fugacity by using the following equation:

$$\ln D(\mathrm{M}) = a \ln f_{O_2} + b/T + cP/T + g \qquad (11)$$

where the a is related to the valence of M in the silicate liquid, and the terms b, c, and g result from the expansion of the free energy term $(\Delta G^0 = \Delta H^0 - T\Delta S^0 + P\Delta V^0)$, and are related to $\Delta H^0/R$, $\Delta V^0/R$ and $\Delta S^0/R$, respectively, with g also capturing the conversion from molar to weight percent units and any changes in the ratios of the activity coefficients. Addition of three terms accounting for metal and silicate composition results in

$$\ln D(\mathrm{M}) = a \ln f_{\mathrm{O}_2} + b/T + cP/T + d \ln(1 - Xs) + e \ln(1 - Xc) + f(\mathrm{NBO/t}) + g \quad (12)$$

This approach has been applied to existing partitioning data for which temperature, pressure, oxygen fugacity, and silicate and metallic compositions are all well characterized (e.g., Righter and Drake, 1999, 2000), and coefficients a to g may be determined by multiple linear regression. A comparison of the calculated to the measured values shows that these expressions can predict partition coefficients with some accuracy (Figure 9). The extent to which they can be used effectively is dependent upon the input data for the regressions, and this is discussed further in the following.

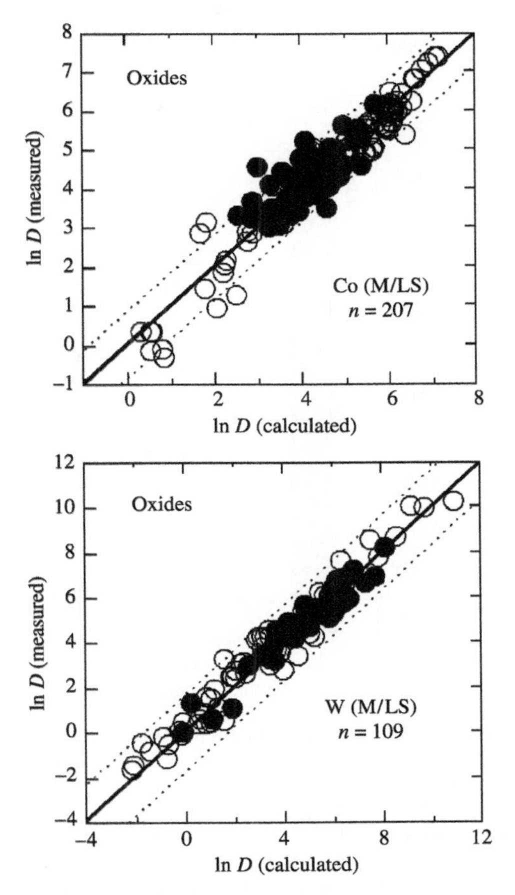

Figure 9 Comparison of ln D (calculated) versus ln D (measured) with data for cobalt ($n = 207$) and tungsten ($n = 109$), and using oxide-component-based compositional terms. Open symbols are 0.1 MPa experiments, and solid symbols are higher-pressure experiments. Dashed lines are $\pm 2\sigma$ errors on the regressions. The number of experiments used in the regression is indicated as n. M/LS refers to metal/liquid silicate for the partition coefficients, and includes experiments that have solid metal or liquid metal (source Righter and Drake, 1999).

2.10.4 MINERAL–MELT EQUILIBRIA

Partitioning of trace elements between minerals and melts has been reviewed by Irving (1978), Green (1994), and Jones (1995). Jones (1995) does include some high-pressure mineral–melt equilibria, but the emphasis in all of this previous work was on low-pressure partitioning related to mantle melting and magma chambers. The reviews form a solid foundation upon which we can build a better understanding of partitioning at high pressures. Similarly, a summary of more recent developments is provided by Chapter 2.09 in which the emphasis is on application of elastic strain theory to trace element partitioning in the mantle and magmas.

Any understanding of mantle differentiation is based on the current mineralogy of Earth's deep mantle. The phase equilibria of the deep mantle of the Earth are known to first order and have been summarized by Bina (1998) (Figure 10). Shallow-phase equilibria are dominated by olivine, pyroxenes, and garnet or spinel. The mid-mantle is dominated by garnet and beta- or gamma-$(\mathrm{Mg,Fe})_2\mathrm{SiO}_4$. The lower mantle is dominated by magnesiowüstite, magnesium perovskite ($(\mathrm{Mg,Fe})\mathrm{SiO}_3$), and a small amount of calcium perovskite. There is some controversy over whether $(\mathrm{Mg,Fe})\mathrm{SiO}_3$ breaks down in the lower mantle to form (Mg,Fe)O and SiO_2 (e.g., Boehler, 2000; Dubrovinsky *et al.*, 1998). Because we have no partitioning data for high-pressure SiO_2, this phase will not be included in the discussion below. The capacity of each of these minerals to cause elemental fractionation within the mantle during differentiation of a magma ocean will be discussed below.

2.10.4.1 Olivine/Melt and Beta-spinel/Melt Partitioning

The superchondritic Mg/Si ratio of Earth's upper mantle might arise from olivine flotation in an early magma ocean (Agee and Walker, 1988). Such an explanation would have to be consistent with other minor and trace elements, especially those known to be compatible in olivine at low pressure such as nickel, cobalt, and manganese. Olivine is a stable liquidus phase in chondritic and peridotitic compositions to ~15 GPa (Agee *et al.*, 1995; Zhang and Herzberg, 1996), and thus olivine/melt partition coefficients must be understood across a large pressure and temperature range. Furthermore, olivine undergoes a transition to spinel structure at pressures of

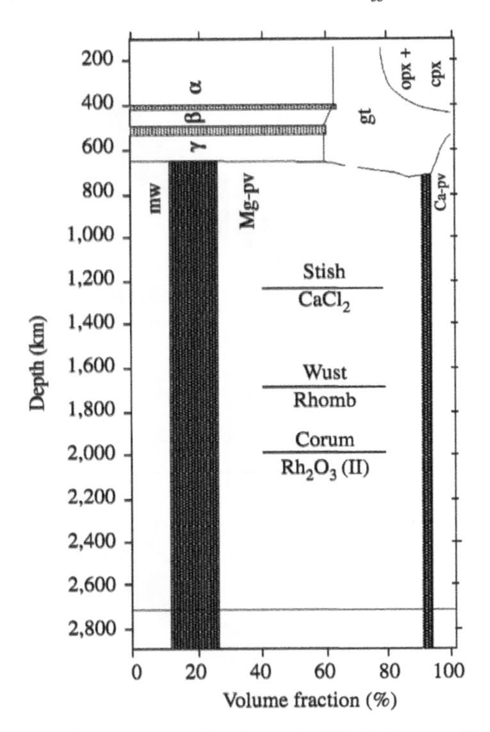

Figure 10 Deep mantle phase equilibria (source Bina, 1998). α, β, and γ refer to the different forms of Mg_2SiO_4, opx = orthopyroxene, cpx = clinopyroxene, gt = garnet, mw = magnesiowüstite, Mg-pv = magnesium perovskite, Ca-pv = calcium perovskite, and horizontal lines show the approximate locations of oxide transitions in SiO_2, FeO, and Al_2O_3.

14–15 GPa (depending upon temperature), and the structural change will undoubtedly affect the magnitude of partition coefficients.

Although hundreds of low-pressure (<1 GPa) experiments have examined olivine/melt equilibria, there has been only a small number of high-pressure trace element partitioning studies for olivine/melt pairs. McFarlane and Drake (1990) and McFarlane (1994) studied manganese, chromium, vanadium, nickel, cobalt, gallium, germanium, and scandium in peridotitic bulk compositions, to 16 GPa. Suzuki and Akaogi (1995) studied manganese, vanadium, chromium, nickel, thulium, scandium, and indium partitioning up to 14 GPa, and also investigated the effect of olivine composition on the partition coefficients for these elements. These studies show that most partition coefficients decrease with increasing pressure. Nickel and manganese have higher partition coefficients in fayalitic olivine, and nickel approaches a D of unity at close to 15 GPa. Partition coefficients for cobalt are always lower than those for nickel, by a factor of up to 2 (McFarlane and Drake, 1990). Manganese, vanadium, chromium, and scandium are all incompatible in olivine at high pressure. One experiment with beta-spinel Mg_2SiO_4 shows that nickel, manganese, chromium, and scandium all have higher Ds than in forsteritic olivine (Figure 11). The effect of olivine flotation on the upper mantle Ni/Co ratio will be discussed later in this chapter.

2.10.4.2 Garnet/Melt Partitioning

Mantle garnets are pyrope rich in the shallow mantle, but become increasingly enriched in the majorite component at higher pressures and temperatures (e.g., Akaogi and Akimoto, 1977; Putirka, 1998). As a result, garnet/melt partitioning of trace elements is likely to be sensitive to pressure and temperature changes due to crystal chemistry. The pioneering work of Kato *et al.* (1988a) reports results for a number of lithophile elements. Subsequent work by Ohtani *et al.* (1989), Yurimoto and Ohtani (1992), McFarlane (1994), Drake *et al.* (1993), Tronnes *et al.* (1992), Righter and Drake (2000), and Draper *et al.* (2003) report partition coefficients across a range of pressures and for siderophile and lithophile elements. As with low-pressure garnets, such as those relevant to melting in the shallow mantle, there are a number of trace elements that are also compatible in high-pressure garnet such as scandium, yttrium, ytterbium, and chromium. However, the partition coefficients for scandium, ytterbium, and chromium all decrease with increasing pressure, approaching unity (Figure 12). Cobalt and beryllium are compatible at high pressures, just above unity, whereas the LREE, boron, lithium, zirconium, gallium, and nickel are mildly incompatible at high pressures.

2.10.4.3 Magnesiowüstite/Melt Partitioning

The lower mantle contains ~20% (MgFe)O (Figure 10). This simple oxide is a significant host of minor and trace elements such as nickel, cobalt, manganese, and chromium. In a peridotite bulk composition, magnesiowüstite co-crystallizes with magnesium perovskite at pressure greater than 23 GPa, but it is also stable at lower pressures. Initial studies of Agee (1993), McFarlane *et al.* (1994), and McFarlane (1994) indicated compatibility of these elements in magnesiowüstite. Addition of later data sets of Gessmann and Rubie (1998) and Ohtani and Yurimoto (1996) have revealed more systematic behavior. In particular, manganese, vanadium, and chromium all show decreasing Ds with temperature, whereas D(Ni) increases (Figure 13). There are no systematic variations with pressure. Scatter in the temperature correlation may be due to small differences in oxygen fugacity (since vanadium and chromium are multivalent) or Mg# (molar $Mg/(Mg+Fe^{2+})$) of the magnesiowüstite. Future efforts should resolve these possibilities.

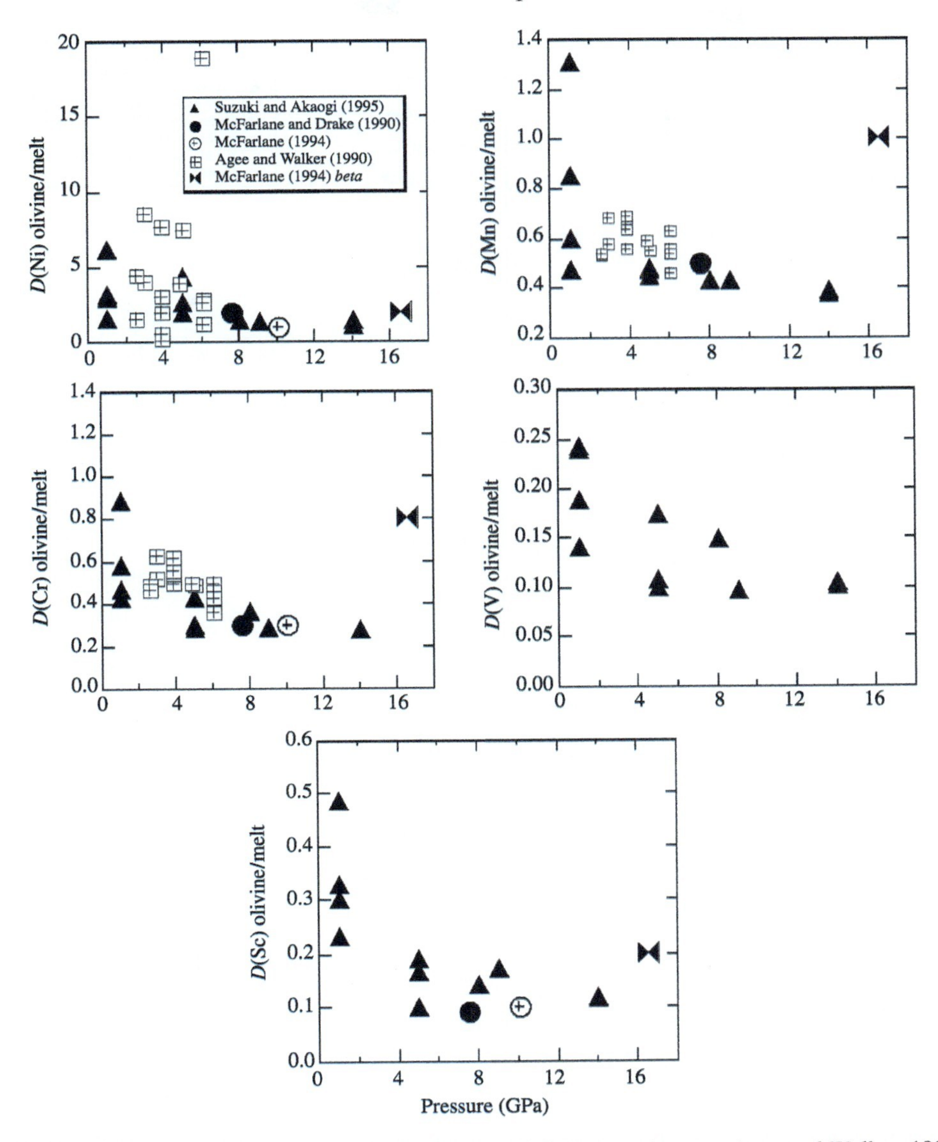

Figure 11 Summary of olivine/melt partition coefficients at high pressure (sources Agee and Walker, 1990; Suzuki and Akaogi, 1995; McFarlane and Drake, 1990; McFarlane, 1994).

However, it is clear that magnesiowüstite is a significant host phase for manganese, chromium, and vanadium and may contribute to the depletions of these elements in the Earth's upper mantle, rather than metal/silicate equilibrium.

2.10.4.4 $MgSiO_3$ Perovskite/Melt Partitioning

The lower mantle is made predominantly of magnesium perovskite, and thus a solid understanding of element partitioning between perovskite and silicate melt is relevant to the early history of the Earth. Much of the early work with perovskite was done on aluminum-free or low-aluminum materials. However, magnesium perovskite can accommodate Al_2O_3 into its structure (Wood, 2000; Stebbins *et al.*, 2001), and this will affect the partitioning relations of both major (e.g., Frost and Langenhorst, 2002) and trace elements. Not only will the presence of aluminum in the structure change the crystal chemistry, but it may also allow other minor elements such as scandium or Fe^{3+} to become more soluble (McCammon, 1997; Lauterbach *et al.*, 2000; Gasparik and Drake, 1995; Figure 14). There have been only seven published magnesium perovskite/melt trace-element partitioning experiments, all at very similar pressure and temperatures; as a result, some of the measured *D*s are presented below against the Al_2O_3 content of the perovskite, which is more variable than the pressure and temperature conditions and is as high as 6 wt.% in some run products (Figure 14).

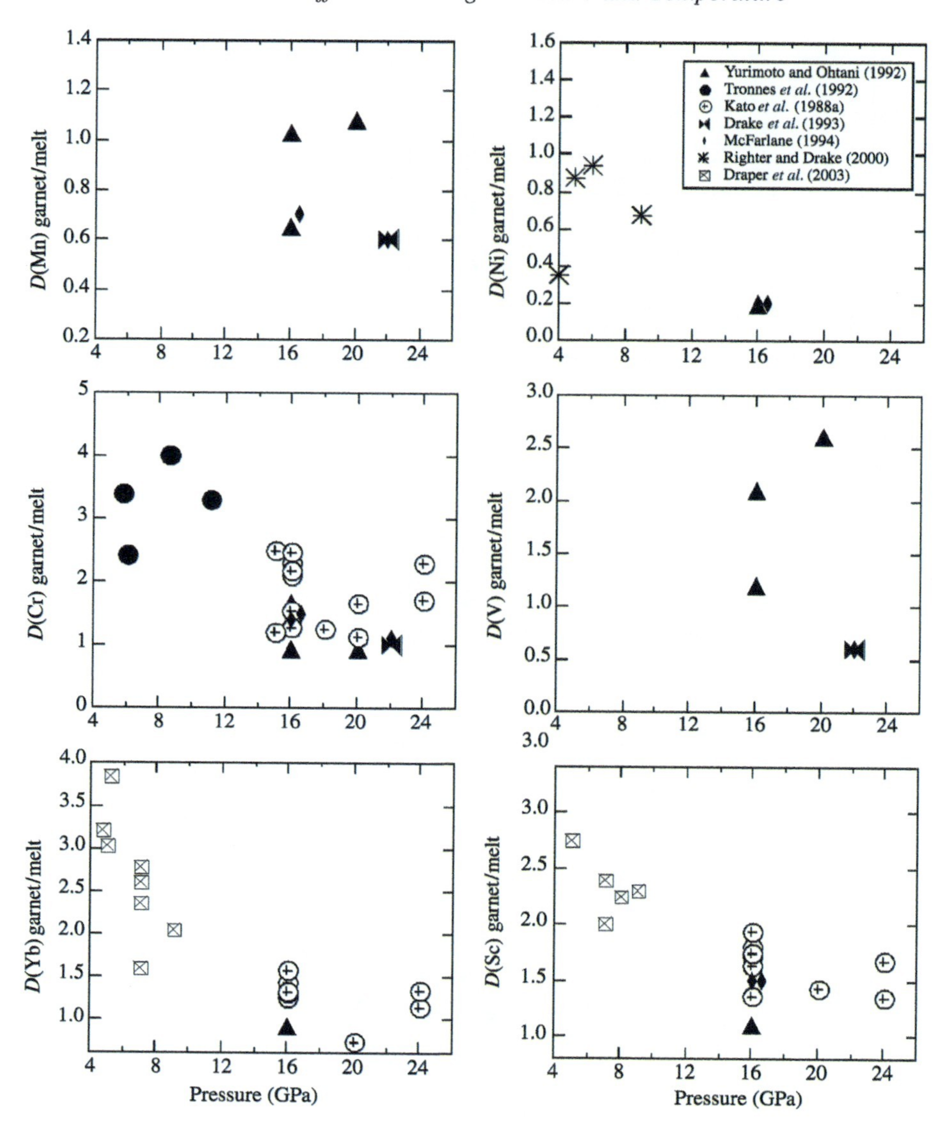

Figure 12 Summary of garnet/melt partition coefficients at high pressure (sources Kato *et al.*, 1988a; Tronnes *et al.*, 1992; Righter and Drake, 2000; McFarlane, 1994; Drake *et al.*, 1993; Ohtani *et al.*, 1989; Draper *et al.*, 2003; Yurimoto and Ohtani, 1992).

Despite the small number of experiments involving $MgSiO_3$ perovskite, there are some general observations that can be made. First, manganese, vanadium, and chromium all have *D*s approaching unity and thus are not going to be fractionated significantly by magnesium perovskite. Nickel and cobalt are either mildly incompatible or mildly compatible, but there are very few data to make any solid evaluations. Scandium, zirconium, and hafnium are compatible in magnesium perovskite, and thus may be slightly depleted in the upper mantle as a result. The magnitude of *D*(Sc) seems to be dependent upon either the aluminum content of the magnesium perovskite, or temperature. The volatile-bearing experiments of Gasparik and Drake (1995) stabilized magnesium perovskite with >4 wt.% Al_2O_3, but at lower temperatures than all previous work, due to the presence of water. Future experimentation might focus on the relative importance of these two variables. Lanthanum and ytterbium are both incompatible in magnesium perovskite, but the HREE generally have higher *D*s than the LREE (Figure 14). Hauri *et al.* (1998) and Minarik (2002) have investigated many more lithophile elements and several isotope pairs such as Sm–Nd and Lu–Hf.

2.10.4.5 $CaSiO_3$ Perovskite/Melt Partitioning

Although there is only a small amount of calcium perovskite in the lower mantle, it is a

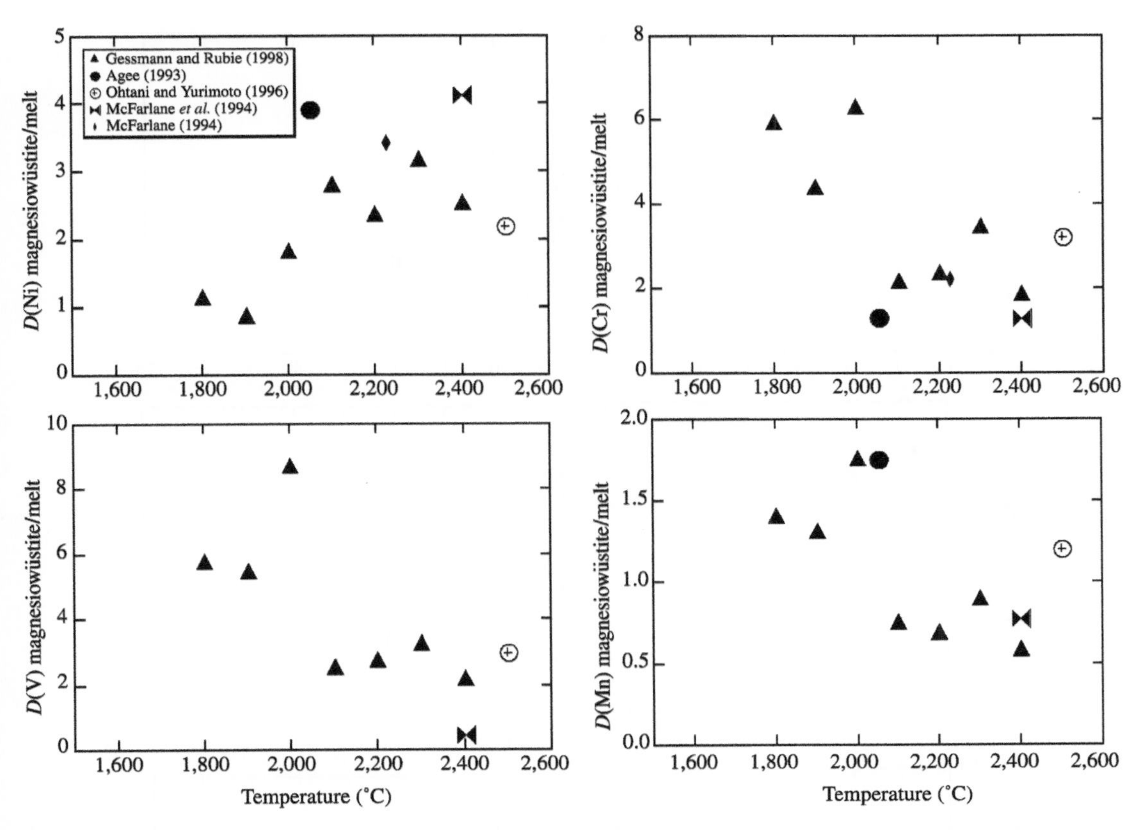

Figure 13 Summary of magnesiowüstite/melt partition coefficients at high pressure (sources Gessmann and Rubie, 1998; Ohtani and Yurimoto, 1996; Agee, 1993; McFarlane *et al.*, 1994; McFarlane, 1994).

significant host phase for a number of lithophile elements, and thus is important to understand elemental fractionation in the early Earth.

Very few studies have been undertaken on calcium perovskite, even fewer than for magnesium perovskite, and thus additional work will be necessary before we have a complete understanding of calcium-perovskite–melt equilibria. The capacity of calcium perovskite to host REE and other lithophile elements is perhaps best portrayed relative to magnesium perovskite (Taura *et al.*, 2001; Kato *et al.*, 1996; Gasparik and Drake, 1995; Corgne and Wood, 2002; Figure 15). Calcium-perovskite/melt partition coefficients for the REE, strontium, yttrium, uranium, and thorium are much higher than those for magnesium-perovskite/melt, some by a factor of ~1,000 (Figure 15). However, calcium-perovskite/melt partition coefficients for several lithophile elements including scandium, titanium, and the HSFEs zirconium, hafnium, and niobium are very similar to partition coefficients for Mg-perovskite. It is clear that any assessment of deep-mantle trace-element reservoirs must take calcium-perovskite into account—its capacity to host uranium and thorium also has implications for terrestrial heat production.

2.10.4.6 Superphase B–Melt Partitioning

High-pressure hydrous phases were unknown until the advent of multi-anvil experiments. Little is known about the abundance of such phases in the mantle, but they could be a significant repository for water if the bulk Earth water abundance is significantly greater than one Earth ocean (Abe *et al.*, 2000). If they are abundant phases, they could have a significant effect on mantle evolution.

Gasparik and Drake (1995) report Superphase B ($Mg_{10}Si_3O_{14}[OH,F]_4$)/melt partitioning data for 23 GPa and 1,500–1,600 °C. Calcium, scandium, and samarium are highly incompatible in Superphase B, while titanium is indifferent.

2.10.5 MODELS

2.10.5.1 Core Formation in the Earth

2.10.5.1.1 Basic data

Models for core formation in the Earth and other terrestrial planets are based on the distribution of siderophile elements between core and mantle. Interpretations of these data have focused on several characteristics of siderophile elements in

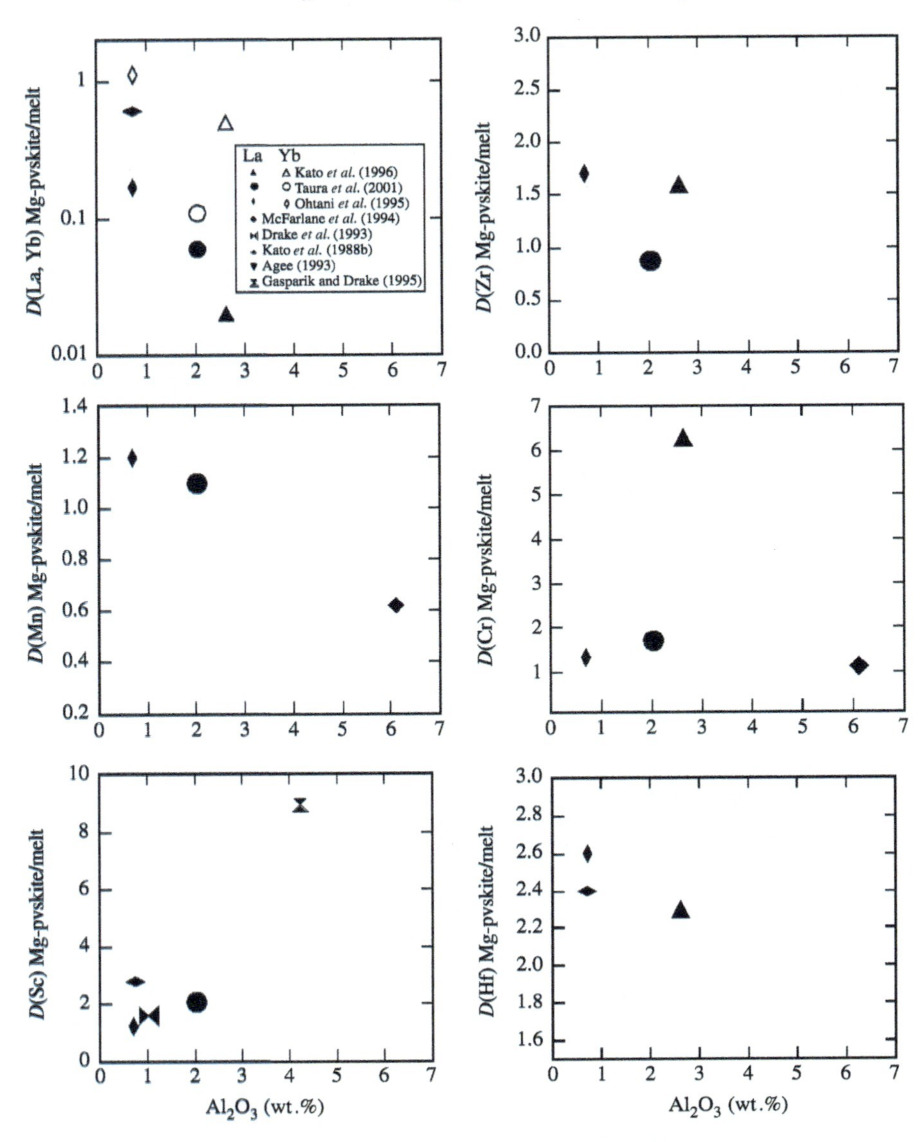

Figure 14 Summary of Mg-perovskite/melt (La, Yb, Zr, Mn, Cr, Sc, and Hf) partition coefficients at high pressure (sources Kato *et al.*, 1996; Taura *et al.*, 2001; Drake *et al.*, 1993; McFarlane *et al.*, 1994; Ohtani *et al.*, 1995; Kato *et al.*, 1988b; Gasparik and Drake, 1995; Agee, 1993).

the terrestrial mantle. First, the relative concentrations of MSEs such as nickel and cobalt in the upper mantle are close to chondritic, but both are depleted by a factor of ~10 relative to chondritic abundances. Second, the HSEs are also in chondritic ratios in primitive upper mantle, but are depleted in absolute abundance by a factor of $\sim 10^{-3}$ relative to chondritic values.

2.10.5.1.2 Heterogeneous accretion

Until the advent of multi-anvil experiments, only low-pressure metal/silicate partition coefficients were available. It was known that D(Ni) (metal/silicate) > D(Co) (metal/silicate) at low pressures, so it was concluded that the chondritic Ni/Co ratio indicated a lack of equilibration between the core and mantle. If the Earth accreted in stages, where an early, reduced stage was followed by a later, oxidized stage, the Ni/Co ratio would be chondritic (e.g., Wanke, 1981). Heterogeneous accretion was later modified to include the HSE concentrations, set by a late (postcore formation) addition of chondritic material to the mantle (e.g., Chou *et al.*, 1983). The problem with this model is that the Earth appears to have gone through a magma ocean phase (Righter *et al.*, 1997; Righter and Drake, 1999, 2000; Li and Agee, 1996), so it is unclear how primitive material accreting to the Earth could preserve its original characteristics in an environment conducive to homogenization and metal–silicate equilibrium.

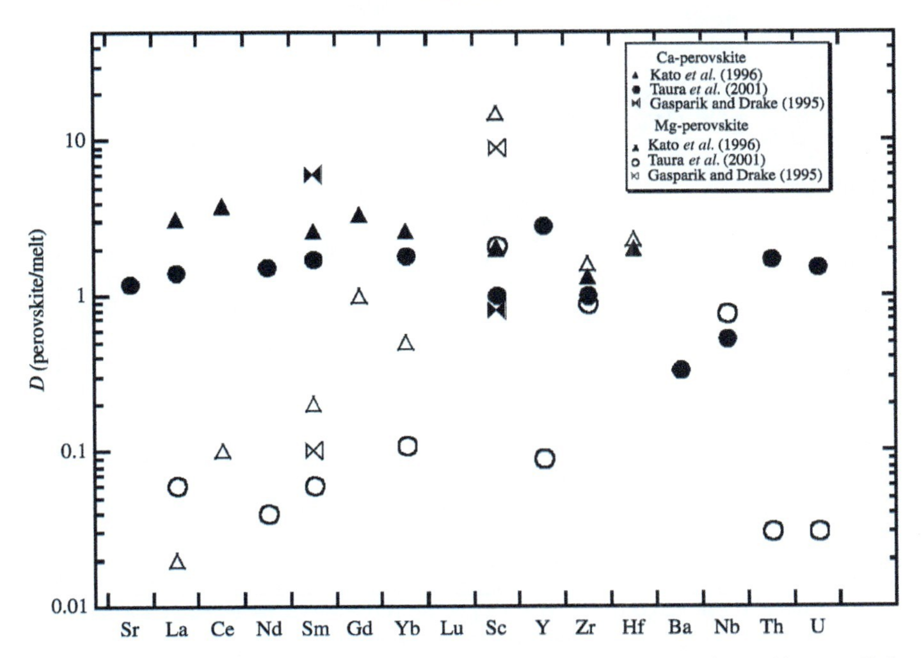

Figure 15 Summary of Ca-perovskite/melt in comparison to Mg-perovskite/melt partition coefficients (sources Gasparik and Drake, 1995; Kato *et al.*, 1996; Taura *et al.*, 2001).

2.10.5.1.3 Sulfide melt–mantle equilibrium

A subset of siderophile elements that are also chalcophile have upper mantle depletions that suggest a sulfide signature (copper, tin, and antimony; Arculus and Delano, 1981). This compositional characteristic has been interpreted as due to olivine–sulfide liquid equilibrium in the Earth (Brett, 1984), although few partition coefficients were available at the time and the model fit to mantle siderophile and chalcophile elements was quite poor. Further, it is not clear if there is enough sulfur available for the core forming metal (Dreibus and Palme, 1996), or how such a metallic liquid would move through a solid matrix (e.g., Rushmer *et al.*, 2000). Finally, because it required a mantle near its solidus, it does not appear plausible in view of the compelling evidence for a terrestrial magma ocean.

2.10.5.1.4 Inefficient core formation

If a small amount of core-forming metal was left behind in the mantle and then later oxidized, the mantle could potentially acquire a chondritic Ni/Co and even near chondritic HSE concentrations. This idea, called inefficient core formation, was advocated by Jones and Drake (1986) and provides an explanation by combining low-pressure and low-temperature metal/silicate partition coefficients and metal mobility constraints. This model awaits integration with more realistic mobility data acquired in a dynamic environment (e.g., Bruhn *et al.*, 2000) such as is appropriate for a convecting, subsolidus mantle. Again, because it required a mantle barely above its solidus, it does not appear plausible in view of the compelling evidence for a terrestrial magma ocean.

2.10.5.1.5 Magma oceans

Application of new high-pressure and high-temperature partition coefficients to the problem of Earth's excess siderophile elements has led to a resurgence of magma ocean models for the early Earth. A number of investigators (Walker *et al.*, 1993; Hillgren *et al.*, 1994; Thibault and Walter, 1995; Righter and Drake, 1995) showed that the effects of temperature and pressure on *D*(Ni) and *D*(Co) are significant. Li and Agee (1996) demonstrated that *D*(Ni) and *D*(Co) converge at higher pressures (28 GPa) by carrying out a series of experiments from 2 GPa to 20 GPa. Righter and Drake (1995, 1997) also showed that *D*(Ni) and *D*(Co) converge at higher pressures using a thermodynamics-based extrapolation of published partition coefficients. Ito *et al.* (1998) published experiments demonstrating the convergence at 28 GPa with experiments at these pressures (Figure 5). A systematic assessment of metal/silicate partition coefficients for the MSEs nickel, cobalt, molybdenum, tungsten, and phosphorus led Righter *et al.* (1997) to conclude that the excess siderophile element problem for these five elements could be resolved if their concentrations were established in a deep hydrous peridotite

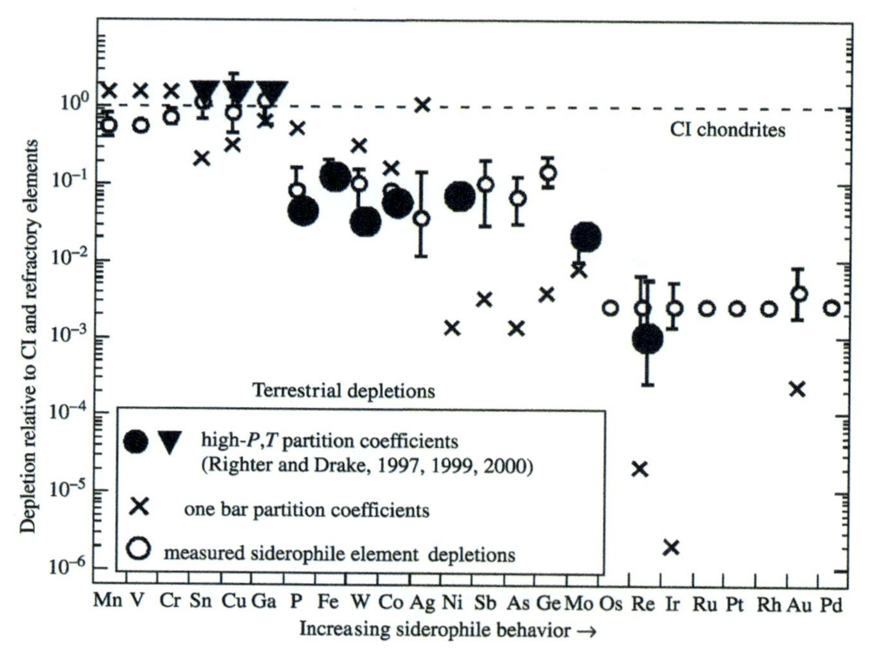

Figure 16 Comparison of observed (open) and calculated (solid) depletions of phosphorus, tungsten, cobalt, nickel, molybdenum, and rhenium (circles) together with those for gallium, tin, and copper (inverted triangles) (sources Righter and Drake, 1997, 1999, 2000). The calculated depletions utilize the partitioning expressions of Righter and Drake (1999) for conditions of 2,250 (± 300) K (1,973 °C), 27 (± 6) GPa, ΔIW = − 0.4 (± 0.3) between a hydrous peridotite (NBO/t = 2.65) magma ocean and metallic liquid. The observed depletions are those of McDonough and Sun (1995), but volatility corrected as described by Newsom and Sims (1991), where the correction is made based on comparisons to trends of lithophile volatile element depletions.

magma ocean at 25–30 GPa and 2,000–2,100 K (Figure 16). Addition of more data for nickel, cobalt, molybdenum, tungsten, and phosphorus as well as new data for rhenium, gallium, tin, and copper to this modeling approach, i.e., modeling slightly, moderately, and HSEs, confirmed this result (Righter and Drake, 1997, 1999, 2000). Finally, it was shown by Righter and Drake (1999) and Jana and Walker (1999) that addition of water to silicate melts has no additional or special effect on metal/silicate partition coefficient values (outside of already known effects of f_{O_2} and melt composition), thus showing that siderophile partitioning studies done at anhydrous conditions can be applied to both hydrous and anhydrous natural systems.

A slightly different, but still a high pressure and temperature scenario, has been proposed recently by several groups, based on a subset of the above elements. With additional data for nickel and cobalt, Li and Agee (2001) proposed that the concentrations of these elements in Earth's upper mantle could be explained by metal–silicate equilibrium in the deep mantle (40–50 GPa and ~3,000 °C). Studying nickel, cobalt, manganese, vanadium, and chromium in (Mg, Fe)O–metal systems, Gessmann and Rubie (2000) concluded that the depletions of these five elements could be satisfied in a comparably deep magma ocean. Finally, systematic studies of the effect of temperature, pressure, metal, and silicate melt composition on *D*(V), *D*(Mn), and *D*(Cr) by Chabot and Agee (2003) showed that the depletions of all three of these elements could be explained by metal–silicate equilibration only under very high-temperature conditions (3,600 °C). This is an intriguing alternative to the shallower conditions proposed by Li and Agee (1996), Righter *et al.* (1997), and Ito *et al.* (1998) (Figure 17).

More important than the variable condition of temperature and pressure proposed by different authors is the remarkable consensus that has emerged that the siderophile element fingerprint for the Earth was established in a deep magma ocean environment. It is unclear what a specific proposed temperature and pressure really means physically. Once Earth achieved the mass of Mars, it probably underwent numerous magma ocean episodes as it accreted, possibly solidifying before the next giant impact created a new magma ocean. The calculated temperature and pressure could represent the last, largest magma ocean event. Alternatively, it could represent some sort of ensemble average memory of a number of magma ocean events. The consensus on a deep magma

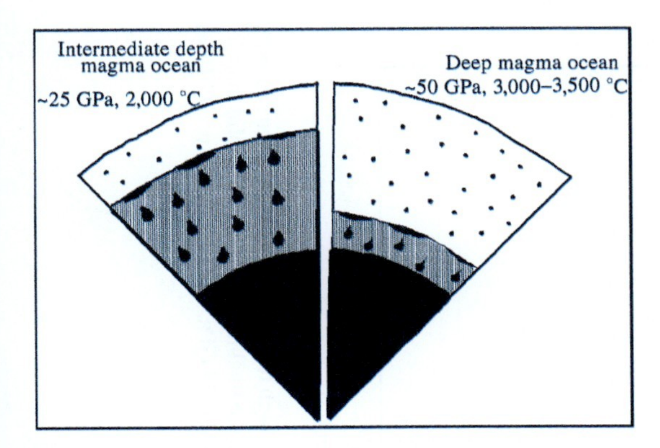

Figure 17 Schematic illustration of the two different high *P*, *T* differentiation scenarios: an intermediate depth magma ocean (proposed by Li and Agee (1996), Righter *et al.* (1997), and Ohtani *et al.* (1997)), and a deeper magma ocean (proposed by Gessmann and Rubie (2000), Li and Agee (2001) and Chabot and Agee (2002)).

ocean environment being responsible for at least the MSE fingerprint in the Earth's mantle remains robust.

2.10.5.2 Kinetics of Metal–Silicate Equilibrium

Chemical considerations of metal–silicate interaction must be linked with physical constraints, such as the likelihood and time frame of separation of metal from silicate. Segregation of metallic liquid from an olivine-rich mantle is difficult unless there is a large amount of anion (sulfur, carbon, and oxygen) dissolved in the metallic liquid to lower the dihedral angle to values <60 (e.g., Rushmer *et al.*, 2000; Figure 18). Perhaps high pressures allow a low enough dihedral angle for metal to interconnect: Shannon and Agee (1998) have shown that in a perovskite matrix, metallic liquid achieves a dihedral angle close to 60°. In addition, a dynamic environment may also allow interconnectivity of metallic liquid in a solid matrix (Rushmer *et al.*, 2000; Bruhn *et al.*, 2000), but this cannot be evaluated for natural systems until more data are available. These difficulties, coupled with the success of the magma ocean model in accounting for siderophile element abundances in Earth's mantle, have led most scholars to abandon core formation in near-mantle solidus conditions.

It is clear, however, that metal mobility through liquid silicate is rapid (Stevenson, 1990; Arculus *et al.*, 1990) and equilibration mechanisms involving liquid metal and liquid silicate have been evaluated most extensively. Since the 1990s, many new experiments on metal–silicate systems have clearly increased our knowledge and understanding of chemical equilibrium in differentiated planets. Models for a deep magma ocean have gained acceptance despite some disagreement on the specific conditions at which this may have taken place. But these models are based entirely on chemical data and have not yet been thoroughly evaluated from a physical perspective. Two classes of metal–silicate equilibrium have been proposed. First, an appealing physical aspect of the 25–30 GPa deep magma ocean noted by Walter and Thibault (1995) and Walter *et al.* (2000) is that the cusp in the peridotite liquidus at this pressure could serve as a natural floor to a magma ocean during cooling, and ponded metal at this depth could equilibrate with the overlying mantle. Second, Karato and Murthy (1997) explored the idea that metal–silicate equilibrium is attained by equilibration between metal droplets and silicate melt. Building on this idea, and results of recent partitioning studies, Rubie *et al.* (2003) compare these two modes of metal–silicate separation and equilibration. Rubie *et al.* show that metal–silicate equilibrium can only occur in a metal droplet scenario because only that mechanism is rapid enough to allow equilibration before solidification of the magma ocean. A magma ocean with a thick, hot steam atmosphere overtop will lose heat rapidly (due to thermal radiation), until the temperature is well below the peridotite liquidus (e.g., Kasting, 1988; Zahnle *et al.*, 1988). As a result, a terrestrial magma ocean will crystallize on the timescale of 1,000 yr (e.g., Hofmeister, 1983; Davies, 1990; Solomotov, 2000). Equilibration of metal ponded at the base of a deep magma ocean would not occur on this timescale. Thus, the metal droplet model seems essential for equilibrium models. Further work in this area, coupled with metal/silicate partition coefficients, should lead to a more comprehensive understanding of this equilibrium process.

2.10.5.3 Mantle Differentiation

Once Earth's core had formed, the mantle differentiated into a magnesium-perovskite- and (Mg, Fe)O-bearing lower mantle and garnet-, olivine-, and pyroxene-bearing upper mantle. If the Earth accreted cool from chondritic material, it is likely that the trace element composition of the upper mantle would resemble the chondritic material out of which the Earth was made. Clearly, the mantle is not chondritic in absolute concentrations. If the Earth accreted hot, and the mantle started in a molten state, some elements may have been fractionated during crystallization of the magma ocean. Mineral/melt partitioning data allow evaluation of element fractionation during magma ocean solidification. However, turbulent convection in a magma ocean will prevent crystal settling (and fractionation) until the crystallinity reaches values as high as 60%

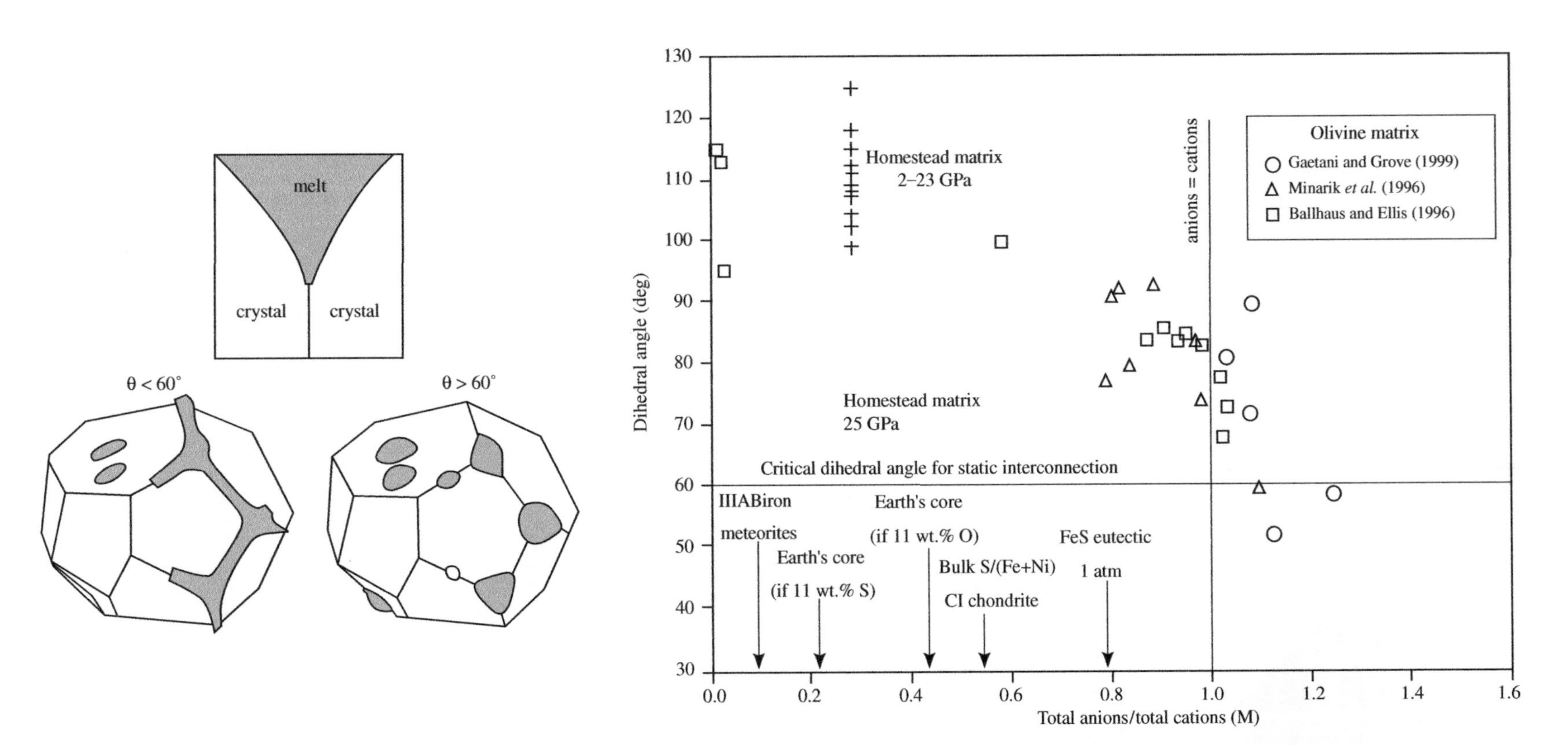

Figure 18 Metallic melt interconnectivity occurs when the dihedral angle between grains becomes <60°. At low pressures, this only occurs at high total anion content (sulfur, carbon, and oxygen). As a result, most metallic liquids relevant to core formation in the Earth and terrestrial planets have dihedral angles >60°, and thus are unable to connect. In a solid mantle, metallic liquids will be trapped and unable to percolate (after Rushmer *et al.*, 2000; homestead matrix data (2–23 GPa: Shannon and Agee (1996); 25 GPa: Shannon and Agee (1998)) and olivine data (Gaetani and Grove, 1999; Minarik *et al.*, 1996; Ballhaus and Ellis, 1996).

(Tonks and Melosh, 1990; Abe, 1993; Solomotov, 2000), thus suppressing any fractionation signature. Above 60% crystallinity, fractionation may occur. In the following, we discuss several different scenarios that have been proposed to explain compositional features in Earth's mantle.

2.10.5.3.1 *Olivine flotation (Ni/Co)*

Flotation of olivine in the mid-mantle during an early magma ocean would perhaps satisfy the superchondritic Mg/Si ratio (Agee and Walker, 1988). However, McFarlane and Drake (1990) have shown that the Ni/Co ratio of such a layer would be 25% higher than that in the observed upper mantle. In addition, the incompatible nature of platinum, palladium, gold, and rhenium in olivine compared to iridium, ruthenium, and rhodium would undoubtedly lead to nonchondritic HSE ratios such as Pd/Ir, Au/Ru, etc. in the upper mantle, and this is not observed (e.g., Righter *et al.*, 2000). As a result, early olivine flotation seems unlikely.

2.10.5.3.2 *Majorite fractionation*

Majorite-rich garnet is capable of hosting scandium, zirconium, and ytterbium and thus could have lowered the concentrations of these elements in the upper mantle if it settled during solidification of a magma ocean. Ohtani *et al.* (1989) and Yurimoto and Ohtani (1992) have shown that aluminum-depleted komatiites have trace element characteristics that are consistent with derivation from a deep, garnet-bearing source. The depletions of vanadium and chromium in the upper mantle (Figure 16) could also be attributed in part to majorite fractionation, because both of these elements are compatible in majorite-rich garnet across a range of pressures (Figure 12): majorite/melt partition coefficients for chromium and vanadium are as high as 3.5–4, much higher than metal/silicate partition coefficients for these elements at mantle oxygen fugacities. Finally, Lu/Hf and Sm/Nd ratios inferred from terrestrial basalts indicate the presence of garnet in the mantle residue (Blichert-Toft and Albarede, 1997). Further work on majorite/melt partitioning by Draper *et al.* (2002a, 2002b, 2003) should help to reveal any systematic behavior for the REE. However, it is important to note that major element trends in mantle derived liquids and residua expected from garnet fractionation are not observed (Agee and Walker, 1988).

2.10.5.3.3 *Magnesiowüstite fractionation*

Magnesiowüstite is a significant host phase for transition elements. Although nickel is compatible in magnesiowüstite, its concentration in the mantle is controlled by metal/silicate partitioning due to the much higher metal/silicate partition coefficients ($D > 30$).

Metal/silicate partition coefficients for manganese, vanadium, and chromium are all <1 at conditions of the deep mantle. Because these elements are compatible in magnesiowüstite, it is possible that their small mantle depletions (Figure 16) are due in part to magnesiowüstite in the lower mantle. The magnitude of compatibility is temperature dependent, so lower temperatures would favor greater partition coefficients for all three elements. The depletion of manganese is greatest of all three elements, yet it has the lowest garnet/melt and magnesiowüstite/melt partition coefficients. The greater depletion of manganese may thus be due to an additional effect such as volatility (Drake *et al.*, 1989).

2.10.5.3.4 *Silicate perovskite fractionation*

Because a large majority of the mantle (~70%) is made of magnesium perovskite, it has great potential for fractionating elements. Although manganese, vanadium, and chromium are all slightly compatible in perovskite, a systematic understanding is not yet possible. Given the small partition coefficients, however, it is unlikely that magnesium perovskite could account for the depletions of these three elements in the upper mantle. Hafnium is compatible in magnesium perovskite and lutetium is incompatible, making magnesium-perovskite fractionation capable of increasing the Lu/Hf ratio of the mantle. Similarly, magnesium-perovskite/melt partition coefficients for samarium and neodymium are slightly different and could be fractionated from one another, thus affecting the evolution of neodymium isotopes in the mantle. Even though melting of mantle leaving magnesium-perovskite residue would produce Lu/Hf ratios similar to that measured in some terrestrial basalt, the Sm/Nd ratios are not consistent with such a scenario (Blichert-Toft and Albarede, 1997). Garnet- or majorite-bearing mantle, however, can satisfy both ratios, as discussed above.

$CaSiO_3$ perovskite has enormous storage potential for lithophile elements. Its ability to host uranium and thorium make it an especially important phase to understand with respect to long-term storage of these heat-producing elements in the deep Earth.

2.10.6 SUMMARY AND FUTURE

Resolution of details of magma ocean scenarios awaits additional experimental work and calculations. There are a number of outstanding issues that need to be addressed before we have a full understanding that will allow a distinction between various hypotheses.

(i) There are many siderophile elements for which we have geochemical data on natural samples, but no experimental data (or at least not enough to make a prediction at high-pressure and high-temperature conditions). For example, we lack systematic high-pressure and high-temperature experimental data for most HSE necessary to make a critical assessment of their consistency with high-pressure and high-temperature magma ocean scenarios. Also, we have limited experimental data for several key MSEs such as germanium, silver, antimony, and arsenic, whose mantle depletions are well defined (e.g., Jochum and Hofmann, 1997). Finally, the volatile chalcophile elements such as indium, tellurium, selenium, bismuth, lead, and cadmium are very poorly integrated with any accretion or core-formation models.

(ii) Despite publication of the results of hundreds of metal–silicate experiments, there are gaps in the experimental database on which predictive expressions are based. Suspicions that gaps in the experimental database were bridged by fitting a linear regression form to potentially nonlinear behavior (e.g., Jones, 1998) are unfounded (Lauer and Jones, 1999; Chabot and Agee, 2002), for instance in carbon-bearing metals. However, there are very few data for molybdenum, tungsten, phosphorus, gallium, or rhenium at the high temperatures and pressures at which an equilibrium solution has been proposed (see, e.g., Righter, 2003). Moreover, an equilibrium scenario only works if the magma ocean is peridotitic, yet many of the experiments in the database have been carried out with basaltic melts. Finally, Righter and Drake (2001) show that our current knowledge of molybdenum, tungsten, and phosphorus are not consistent with the very high pressure and high temperature equilibration suggested by Gessman and Rubie (2000), Li and Agee (2001), and Chabot and Agee (2002).

(iii) We lack a systematic understanding of mineral/melt trace-element partitioning at high pressures and temperatures. Crystal chemical effects must be important, yet there are not enough data to make an evaluation at this point. Application of elastic strain theory, for instance, may increase our understanding of high-pressure and high-temperature partitioning, much the same as it has for low-pressure mineral–melt equilibria.

(iv) Laws of thermodynamics should be applied where prudent and appropriate. For example, the dependence of metal/silicate partition coefficients on silicate/melt composition has been modeled very crudely as a function of NBO/t. However, NBO/t cannot distinguish between competing or different effects of silicate–melt network modifiers such as magnesium or calcium, for example. This is really an activity problem ($a_M = f(X)$) that should be calibrated with careful experiments. In lieu of such experiments, a solid approach is to regress the effect of melt composition against a proxy for activity such as melt oxide mole fractions (Righter and Drake, 1999). An additional example is the importance of activity coefficients for some species in the metallic phase. Regression of partition coefficient data against oxygen fugacity can yield incorrect valences in systems where the activity coefficients (γ) of a siderophile element change as a function of metal composition. This effect can be accounted for by regressing against γD, instead of D, and then making a correction for the metal composition of Earth's core. It has been proposed by Capobianco *et al.* (1999) that this effect could be relevant to germanium, antimony, and arsenic. Any assessment of these elements should explore this possibility.

The drawbacks of a simplified relative f_{O_2} approach become apparent in the case of the relative volumetric properties of FeO and Fe_2O_3 in silicate melts even at relatively low pressures (Kress and Carmichael, 1991). There are two problems: first, the pressure dependence of FeO–Fe_2O_3 equilibria is different from the pressure dependence of the nickel–nickel oxide (NNO) buffer such that use of NNO as a normalization for relative f_{O_2} can be misleading. The volume change of the FMQ buffer (0.17 log f_{O_2} units per GPa) is much closer to the redox state of a silicate melt than is that of the NNO buffer (0.51 log f_{O_2} units per GPa). Second, the compressibilities of FeO and Fe_2O_3 are much different, so that even pressures of 1–2 GPa will have a large effect on their partial molar volumes. Any calculation of relative f_{O_2} should include an assessment of the volumetric properties of both the buffer phases, and the phases involved in the redox reaction.

Using previous work on magmas as an analogy to metal–silicate systems, it is clear that normalization of f_{O_2} to the IW buffer has similar drawbacks. The relative f_{O_2} calculated in the absence of information about, for example, the compressibility of iron in a core-forming liquid, or FeO in a peridotite melt, is likely to be inaccurate, especially if extrapolated across tens of GPa. For example, if the relative oxygen fugacity of an experimental run of the same composition (ΔIW value) at 20 GPa, 2,000 °C is different from that at 5 GPa, 1,700 °C, then there is a problem. The best approach to calculation of relative f_{O_2} for metal–silicate systems involves knowing not only the volumetric properties of iron and FeO for the IW buffer, but also for iron in a core-forming liquid, and FeO in a peridotite melt.

(v) Most of the partitioning data for magnesium perovskite have been obtained at similar conditions, and the effects of pressure and temperature are unknown. Because the early Earth may have been hot enough to melt well into the lower mantle, an understanding of element partitioning at higher-pressure and higher-temperature conditions would aid in evaluating the implications for mantle composition. Measurements at higher pressures (>30 GPa) would have to be done in a DAC or a cubic anvil (Mao and Hemley, 1998). Experimental advances such as these will undoubtedly provide an enhanced understanding of core and mantle evolution in the Earth.

(vi) The results of a reduced, high-pressure and high-temperature magma ocean on the early Earth present a conundrum. If the young Earth allowed metallic liquid to pass through its mantle to the core, yet the upper mantle is not reduced enough for iron metal stability, how did Earth's mantle become oxidized. The answer to this may simply be that it has become oxidized over time due to the effects of recycling and plate tectonics. However, no studies have yet revealed a secular trend of oxygen fugacity (e.g., Eggler and Lorand, 1995; Canil, 2002). The solution is linked to important developments in Earth history such as the oxygen content of the Early Archean atmosphere (Kasting *et al.*, 1993; Kump *et al.*, 2001), the origin of water, and the origin of life.

ACKNOWLEDGMENTS

The authors thank C. Agee, R. Carlson, and E. McFarlane for helpful reviews and comments on the manuscript. The work was supported in part by NASA grants NAG5-9435 and NAG5-12795 to Michael J. Drake, and NSF grant EAR-0074036 to MJD and KR.

REFERENCES

Abe Y. (1993) Thermal evolution and chemical differentiation of the terrestrial magma ocean. In *Evolution of the Earth and Planets*, Geophysical Monograph 74 (eds. E. Takahashi, R. Jeanloz, and D. Rubie). IUGG, American Geophysical Union, Washington, vol. 14, pp. 41–54.

Abe Y., Ohtani E., Okuchi T., Righter K., and Drake M. J. (2000) Water in the early Earth. In *Origin of the Earth and Moon* (eds. R. Canup and K. Righter). University of Arizona Press, Tucson, pp. 413–434.

Agee C. B. (1993) High pressure melting of carbonaceous chondrite. *J. Geophys. Res.* **98**, 5419–5426.

Agee C. B. and Walker D. (1988) Mass balance and phase density constraints on early differentiation of chondritic mantle. *Earth Planet. Sci. Lett.* **90**, 144–156.

Agee C. B. and Walker D. (1990) Aluminum partitioning between olivine and ultrabasic silicate liquid to 6 GPa. *Contrib. Mineral. Petrol.* **105**, 243–254.

Agee C. B., Li J., Shannon M. C., and Circone S. (1995) Pressure–temperature diagram for the Allende meteorite. *J. Geophys. Res.* **100**, 17725–17740.

Akaogi M. and Akimoto S. (1977) Pyroxene–garnet solid–solution equilibria in the systems $Mg_4Si_4O_{12}$–$Mg_3Al_2Si_3O_{12}$ and $Fe_4Si_4O_{12}$–$Fe_3Al_2Si_3O_{12}$ at high pressures and temperatures. *Phys. Earth Planet. Inter.* **15**, 90–106.

Arculus R. J. (1985) Oxidation status of the mantle: past and present. *Ann. Rev. Earth Planet. Sci.* **13**, 75–95.

Arculus R. J. and Delano J. W. (1981) Siderophile element abundances in the upper mantle: evidence for a sulfide signature and equilibrium with the core. *Geochim. Cosmochim. Acta* **45**, 1331–1343.

Arculus R. J., Holmes R. D., Powell R., and Righter K. (1990) Metal/silicate equilibria and core formation. In *The Origin of the Earth* (eds. H. Newsom and J. H. Jones). Oxford University Press, London, pp. 251–271.

Ballhaus C. and Ellis D. J. (1996) Mobility of core melts during Earth's accretion. *Earth Planet. Sci. Lett.* **143**, 137–145.

Bassett W. A. (1979) The diamond cell and the nature of the Earth's mantle. *Ann. Rev. Earth Planet. Sci.* **7**, 357–380.

Bina C. (1998) Lower mantle mineralogy and the geophysical perspective. In *Ultrahigh-pressure Mineralogy* (ed. R. J. Hemley). *Rev. Mineral.* Mineralogical Society of America, Washington, DC, vol. 37, pp. 205–239.

Blichert-Toft J. and Albarede F. (1997) The Lu–Hf isotope geochemistry of chondrites and the evolution of the mantle–crust system. *Earth Planet. Sci. Lett.* **148**, 243–258.

Boehler R. (2000) High pressure experiments and the phase diagram of lower mantle and core materials. *Rev. Geophys.* **38**, 221–245.

Boehler R. and Chopelas A. (1991) A new approach to laser heating in high pressure mineral physics. *Geophys. Res. Lett.* **18**, 1147–1150.

Bohlen S. R. (1984) Equilibria for precise pressure calibration and a frictionless furnace assembly for the piston-cylinder apparatus. *N. Jb. Mineral. Mh.* **9**, 404–412.

Boyd F. R. and England J. L. (1960) Apparatus for phase equilibrium measurements at pressures up to 50 kilobars and temperatures up to 1,750 °C. *J. Geophys. Res.* **65**, 741–748.

Brett R. (1984) Chemical equilibration of the Earth's core and upper mantle. *Geochim. Cosmochim. Acta* **48**, 1183–1188.

Bruhn D., Groebner N., and Kohlstedt D. L. (2000) An interconnected network of core-forming melts produced by shear deformation. *Nature* **403**, 883–886.

Canil D. (2002) Vanadium in peridotites, mantle redox and tectonic environments: Archean to present. *Earth Planet. Sci. Lett.* **195**, 75–90.

Capobianco C. J. and Amelin A. (1994) Metal/silicate partitioning of nickel and cobalt: the influence of temperature and oxygen fugacity. *Geochim. Cosmochim. Acta* **58**, 125–140.

Capobianco C. J., Jones J. H., and Drake M. J. (1993) Metal/silicate thermochemistry at high temperature: magma oceans and the "excess siderophile element" problem of the Earth's Upper Mantle. *J. Geophys. Res.* **98**, 5433–5443.

Capobianco C. J., Drake M. J., and DeAro J. A. (1999) Siderophile geochemistry of Ga, Ge and Sn: cationic oxidation states in silicate melts and the effect of composition in iron–nickel alloys. *Geochim. Cosmochim. Acta* **63**, 2667–2677.

Carmichael I. S. E. (1991) The redox state of basic and silicic magmas: a reflection of their source regions? *Contrib. Mineral. Petrol.* **106**, 129–141.

Chabot N. L. and Agee C. B. (2002) The behavior of Ni and Co during core formation. *Lunar Planet. Sci.* **XXXIII**, #1009.

Chabot N. L. and Agee C. B. (2003) Core formation in the Earth and Moon: new experimental constraints from V, Cr, and Mn. *Geochim. Cosmochim. Acta.* **67**, 2077–2091.

Chabot N. L., Campbell A. J., Jones J. H., Humayun M., and Agee C. B. (2003) Applying experimental partitioning results to iron meteorites: a test of Henry's law. *Meteorit. Planet. Sci.* **38**, 181–196.

Chou C.-L., Shaw D. M., and Crocket J. H. (1983) Siderophile trace elements in the Earth's oceanic crust and upper mantle. *J. Geophys. Res.* **88**(suppl. 2), A507–A518.

Colson R. O. (1992) Solubility of neutral nickel in silicate melts and implications for Earth's siderophile element budget. *Nature* **357**, 65–68.

Corgne A. and Wood B. J. (2002) $CaSiO_3$ and $CaTiO_3$ perovskite/melt partitioning of trace elements: implications for gross mantle differentiation. *Geophys. Res. Lett.* **29**, 39-1–39-4.

Davies G. F. (1990) Heat and mass transport in the early Earth. In *Origin of the Earth* (eds. H. Newsom and J. H. Jones). Oxford University Press, New York, pp. 175–194.

Dobson D. and Brodholt J. P. (1999) The pressure medium as a solid-state oxygen buffer. *Geophys. Res. Lett.* **26**, 259–262.

Drake M. J. (1975) The oxidation state of europium as an indicator of oxygen fugacity. *Geochim. Cosmochim. Acta* **39**, 55–64.

Drake M. J., Newsom H. E., and Capobianco C. J. (1989) V, Cr, and Mn in the Earth, Moon, EPB, and SPB and the origin of the Moon: experimental studies. *Geochim. Cosmochim. Acta* **53**, 2101–2111.

Drake M. J., McFarlane E. A., Gasparik T., and Rubie D. C. (1993) Mg-perovskite/silicate melt and majorite garnet/silicate melt partition coefficients in the system CaO–MgO–SiO_2 at high temperatures and pressures. *J. Geophys. Res.* **98**, 5427–5431.

Draper D. S., Xirouchakis D., and Agee C. B. (2002a) Effect of majorite transformation on garnet–melt trace element partitioning. *Lunar Planet. Sci.* **XXXIII**, abstract 1306.

Draper D. S., Xirouchakis D., and Agee C. B. (2002b) Garnet-melt Trace Element Partitioning from 5 to 9 GPa: Onset of Garnet to Majorite Transformation. AGU Spring Meeting, abstract #V52B-07.

Draper D. S., Xirouchakis D., and Agee C. B. (2003) Trace element partitioning between garnet and chondritic melt from 5 to 9 GPa: implications for the onset of the majorite transformation in the martian mantle. *Phys. Earth Planet. Inter.* (in press).

Dreibus G. and Palme H. (1996) Cosmochemical constraints on the sulfur content in the Earth's core. *Geochim. Cosmochim. Acta* **60**, 1125–1130.

Dubrovinsky L., Saxena S. K., Ahuja R., and Johansson B. (1998) Theoretical study of the stability of $MgSiO_3$ perovskite in the deep mantle. *Geophys. Res. Lett.* **25**, 4253–4256.

Dunn T. (1993) The piston–cylinder appartus. In *Experiments at High Pressure and Applications to the Earth's Mantle: Mineral. Assoc. Canada Short Course Handbook*. vol. 21, pp. 39–94.

Eggler D. H. and Lorand J. P. (1995) Sulfides, diamonds and mantle f_{O_2}. *Proc. 5th Int. Kimberlite Conf.* **5**, 160–169.

Ehlers K., Grove T. L., Sisson T. W., Recca S. L., and Zervas D. A. (1992) The effect of oxygen fugacity on the partitioning of nickel and cobalt between olivine, silicate melt, and metal. *Geochim. Cosmochim. Acta* **56**, 3733–3743.

Elliot J. F., Gleiser M., and Ramakrishna V. (1963) *Thermochemistry for Steelmaking*. Addison Wesley, Reading, MA, 512p.

Ertel W., O'Neill H. St. C., Dingwell D. B., and Spettel B. (1996) Solubility of tungsten in a haplobasaltic melt as a function of temperature and oxygen fugacity. *Geochim. Cosmochim. Acta* **60**, 1171–1180.

Fleet M. E., Liu M., and Crocket J. H. (1999) Partitioning of trace amounts of highly siderophile elements in the Fe–Ni–S system and their fractionation in nature. *Geochim. Cosmochim. Acta* **63**, 2611–2622.

Frost D. J. and Langenhorst F. (2002) The effect of Al_2O_3 on Fe–Mg partitioning between magnesiowüstite and magnesium silicate perovskite. *Earth Planet. Sci. Lett.* **199**, 227–241.

Gaetani G. A. and Grove T. L. (1999) Wetting of mantle olivine by sulfide melt: implications for Re/Os ratios in mantle peridotite and late-stage core formation. *Earth Planet. Sci. Lett.* **169**, 147–163.

Gasparik T. and Drake M. J. (1995) Partitioning of elements among two silicate perovskites, superphase B, and volatile-bearing melt at 23 GPa and 1,500–1,600 °C. *Earth Planet. Sci. Lett.* **134**, 307–318.

Gessmann C. K. and Rubie D. C. (1998) The effect of temperature on the partitioning of nickel, cobalt, manganese, chromium and vanadium at 9 GPa and constraints on formation of the Earth's core. *Geochim. Cosmochim. Acta* **62**, 867–882.

Gessmann C. K. and Rubie D. C. (2000) Experimental evidence for the origin of the Cr, V and Mn depletions in the mantle of the Earth and implications for the origin of the Moon. *Earth Planet. Sci. Lett.* **184**, 95–107.

Green T. H. (1994) Experimental studies of trace element partitioning applicable to igneous petrogenesis—Sedone 16 years later. *Chem. Geol.* **117**, 1–36.

Hall H. T. (1958) Some high pressure, high temperature apparatus design considerations, equipment for use at 200, 000 atmospheres and 3,000 deg. *Rev. Sci. Instru.* **29**, 267–275.

Hanson B. and Jones J. H. (1998) The systematics of Cr^{3+} and Cr^{2+} partitioning between olivine and liquid in the presence of spinel. *Am. Mineral.* **83**, 669–684.

Hauri E. H., Minarik W. G., and Fei Y. (1998) Isotopic constraints on the preservation of primordial heterogeneity in the Earth's mantle. In *Origin of the Earth and Moon Conference, Monterrey, CA*, LPI Contribution No. 957, Houston, Texas, pp. 12–13 (abstract).

Hillgren V. J. (1993) Partitioning behavior of moderately siderophile elements in Ni-rich systems: implications for the Earth and Moon. PhD Thesis, University of Arizona, 119p.

Hillgren V. J., Drake M. J., and Rubie D. C. (1994) High pressure and temperature experiments on core–mantle segregation in the accreting Earth. *Science* **264**, 1442–1445.

Hillgren V. J., Drake M. J., and Rubie D. C. (1996) High pressure and high temperature metal/silicate partitioning of siderophile elements: the importance of silicate liquid composition. *Geochim. Cosmochim. Acta* **60**, 2257–2263.

Hofmeister A. M. (1983) Effect of a Hadean terrestrial magma ocean on crust and mantle evolution. *J. Geophys. Res.* **88**, 4963–4983.

Holloway J. R., Pan V., and Gudmundsson G. (1992) High-pressure fluid-absent melting experiments in the presence of graphite: oxygen fugacity, ferric/ferrous ratio and dissolved CO_2. *Euro. J. Mineral.* **4**, 105–114.

Holzheid A., O'Neill H. St. C. (1999) The Cr–Cr_2O_3 oxygen buffer and the free energy of formation of Cr_2O_3 from high temperature electrochemical measurements. *Geochim. Cosmochim. Acta* **59**, 475–479.

Holzheid A. and Palme H. (1996) The influence of FeO in the solubilities of Co and Ni in silicate melts. *Geochim. Cosmochim. Acta* **60**, 1181–1193.

Holzheid A., Borisov A., and Palme H. (1994) The effect of oxygen fugacity and temperature on solubilities of nickel, cobalt and molybdenum in silicate melts. *Geochim. Cosmochim. Acta* **58**, 1975–1981.

Holzheid A., Palme H., and Chakraborty S. (1997) The activities of NiO, CoO and FeO in silicate melts. *Chem. Geol.* **139**, 21–38.

Irving A. J. (1978) A review of experimental studies of crystal/liquid trace element partitioning. *Geochim. Cosmochim. Acta* **42**, 743–770.

Ito E., Katsura T., and Suzuki T. (1998) Metal/silicate partitioning of Mn, Co, and Ni at high pressures and high temperatures and implications for core formation in a deep magma ocean. In *Properties of Earth and Planetary Materials at High Pressure and Temperature*, Geophysical

Monograph 101 (ed. M. H. Manghnani). AGU, Washington, DC, pp. 215–225.

Jaeger W. L. and Drake M. J. (2000) Solubilities of W, Ga and Co in silicate melts as a function of melt composition. *Geochim. Cosmochim. Acta* **64**, 3887–3895.

Jana D. and Walker D. (1997a) The impact of carbon on element distribution during core formation. *Geochim. Cosmochim. Acta* **61**, 2759–2763.

Jana D. and Walker D. (1997b) The influence of sulfur on partitioning of siderophile elements. *Geochim. Cosmochim. Acta* **61**, 5255–5277.

Jana D. and Walker D. (1999) Core formation in the presence of various C–H–O volatiles. *Geochim. Cosmochim. Acta* **63**, 2299–2310.

Jochum K. P. and Hofmann A. W. (1997) Antimony in mantle derived rocks. *Chem. Geol.* **139**, 39–49.

Jones J. H. (1995) Experimental trace element partitioning. In *Rock Physics and Phase Relations: A Handbook of Physical Constants: AGU Reference Shelf Volume 3* (ed. T. J. Ahrens). American Geophysical Union, Washington, DC, pp. 73–104.

Jones J. H. (1998) Uncertainties in modeling core formation. *Meteorit. Planet. Sci.* **33**, A79–A80.

Jones J. H. and Drake M. J. (1983) Experimental investigations of trace element fractionation in iron meteorites: II. The influence of sulfur. *Geochim. Cosmochim. Acta* **47**, 1199–1209.

Jones J. H. and Drake M. J. (1986) Geochemical constraints on core formation. *Nature* **322**, 221–228.

Jones J. H., Capobianco C. J., and Drake M. J. (1992) Siderophile elements and the Earth's formation. *Science* **257**, 1281–1282.

Karato S.-I. and Murthy V. R. (1997) Core formation and chemical equilibrium in the Earth: I. Physical considerations. *Phys. Earth Planet. Inter.* **100**, 61–79.

Kasting J. F. (1988) Runaway and moist greenhouse atmospheres and the evolution of Earth and Venus. *Icarus* **74**, 472–494.

Kasting J. F., Eggler D. H., and Raeburn S. P. (1993) Mantle redox evolution and the oxidation state of the Archean atmosphere. *J. Geol.* **101**, 245–257.

Kato T., Ringwood A. E., and Irifune T. (1988a) Experimental determination of element partitioning between silicate perovskites, garnets and liquids: constraints on early differentiation of the mantle. *Earth Planet. Sci. Lett.* **89**, 123–145.

Kato T., Ringwood A. E., and Irifune T. (1988b) Constraints on element partition coefficients between $MgSiO_3$ perovskite and liquid determined by direct measurements. *Earth Planet. Sci. Lett.* **90**, 65–68.

Kato T., Ohtani E., Ito Y., and Onuma K. (1996) Element partitioning between silicate perovskites and calcic ultrabasic melt. *Phys. Earth Planet. Inter.* **96**, 201–207.

Kawai N. and Endo S. (1970) Generation of ultrahigh hydrostatic pressures by a split sphere apparatus. *Rev. Sci. Instr.* **41**, 1178–1181.

Kress V. C. and Carmichael I. S. E. (1991) The compressibility of silicate liquids containing Fe_2O_3 and the effect of composition, temperature, oxygen fugacity and pressure on their redox states. *Contrib. Mineral. Petrol.* **108**, 82–92.

Kump L. R., Kasting J. F., and Barley M. E. (2001) Rise of atmospheric oxygen and the "upside-down" Archean mantle. *Geochem. Geophys. Geosys.* **2**, 1–10 paper no. 2000GC000114.

Lauer H. V. and Jones J. H. (1999) Tungsten and nickel partitioning between solid and liquid metal: implications for high pressure metal/silicate experiments. *Lunar Planet. Sci.* **XXX**, #1617.

Lauterbach S., McCammon C. A., van Aken P., Langenhorst F., and Seifert F. (2000) Mossbauer and ELNES spectroscopy of (Mg, Fe)(Si, Al)O_3 perovskite: a highly oxidized component of the lower mantle. *Contrib. Mineral. Petrol.* **138**, 17–26.

Li J. and Agee C. B. (1996) Geochemistry of mantle–core formation at high pressure. *Nature* **381**, 686–689.

Li J. and Agee C. B. (2001) The effect of pressure, temperature, oxygen fugacity and composition on partitioning of nickel and cobalt between liquid Fe–Ni–S alloy and liquid silicate: implications for the Earth's core formation. *Geochim. Cosmochim. Acta* **65**, 1821–1832.

Liu M. and Fleet M. E. (2001) Partitioning of siderophile elements (W, Mo, As, Ag, Ge, Ga, and Sn) and Si in the Fe–S system and their fractionation in iron meteorites. *Geochim. Cosmochim. Acta* **65**, 671–682.

Malvin D. J., Jones J. H., and Drake M. J. (1986) Experimental investigations of trace element fractionation in iron meteorites: III. Elemental partitioning in the system Fe–Ni–S–P. *Geochim. Cosmochim. Acta* **50**, 1221–1231.

Mao H. K. and Hemley R. J. (1998) New windows on the Earth's deep interior. In *Ultrahigh-pressure Mineralogy* (ed. R. J. Hemley). *Rev. Mineral.* The Mineralogical Society of America, Washington, DC, vol. 37, pp. 1–32.

McCammon C. (1997) Perovskite as a possible sink for ferric iron in the lower mantle. *Nature* **387**, 694–696.

McDonough W. F. and Sun S.-S. (1995) The composition of the Earth. *Chem. Geol.* **120**, 223–253.

McFarlane E. A. (1994) Differentiation in the early Earth: an experimental investigation. PhD Thesis. University of Arizona, 153pp.

McFarlane E. A. and Drake M. J. (1990) Element partitioning and the early thermal history of the Earth. In *Origin of the Earth* (eds. H. E. Newsom and J. H. Jones). Oxford University Press, New York, pp. 135–150.

McFarlane E. A., Drake M. J., and Rubie D. C. (1994) Element partitioning between Mg-perovskite, magnesiowüstite, and silicate melt at conditions of the Earth's mantle. *Geochim. Cosmochim. Acta* **58**, 5161–5172.

Minarik W. G., Ryerson F. J., and Watson E. B. (1996) Textural entrapment of core-forming melts. *Science* **272**, 530–532.

Minarik W. G. (2002) Trace-element perovskite/melt partitioning at the top of the upper mantle or the bottom of the magma ocean. AGU Spring Meeting, abstract #V52B-08.

Murthy V. R. (1991) Early differentiation of the Earth and the problem of mantle siderophile elements: a new approach. *Science* **253**, 303–306.

Myers J. and Eugster H. (1983) The system Fe–Si–O: oxygen buffer calibrations to 1,500 K. *Contrib. Mineral. Petrol.* **82**, 75–85.

Mysen B. O. (1991) Relations between structure, redox equilibria of iron, and properties of magmatic liquids. In *Physical Chemistry of Magmas* (eds. L. L. Perchuk and I. Kushiro). Springer, New York, pp. 41–98.

Newsom H. and Drake M. J. (1982) The metal content of the eucrite parent body: constraints from the partitioning behavior of tungsten. *Geochim. Cosmochim. Acta* **46**, 2483–2489.

Newsom H. and Drake M. J. (1983) Experimental investigations of the partitioning of phosphorus between metal and silicate phases: implications for the Earth, Moon and eucrite parent body. *Geochim. Cosmochim. Acta* **47**, 93–100.

Newsom H. and Sims K. W. W. (1991) Core formation during early accretion of the Earth. *Science* **252**, 926–933.

Ohtani E. (1987) Ultrahigh-pressure melting of a model chondritic mantle and pyrolite compositions. In *High-pressure Research in Mineral Physics*, Geophysical Monograph 39 (eds. M. H. Manghnani and Y. Syono). AGU, Washington, DC, pp. 87–93, .

Ohtani E. and Yurimoto H. (1996) Element partitioning between metallic liquid, magnesiowüstite and silicate liquid at 20 GPa and 2,500 °C: a SIMS study. *Geophys. Res. Lett.* **23**, 1993–1996.

Ohtani E., Kawabe I., Moriyama J., and Nagata Y. (1989) Partitioning of elements between majorite garnet and melt and implications for the petrogenesis of komatiite. *Contrib. Mineral. Petrol.* **103**, 263–269.

Ohtani E., Yurmoto H., Segawa T., and Kato T. (1995) Element partitioning between $MgSiO_3$ perovskite, magma, and molten iron: constraints for the earliest processes of the Earth–Moon system. In *The Earth's Central Part: Its Structure and Dynamics* (ed. T. Yukutake). Terra Scientific, Tokyo, Japan, pp. 287–300.

Ohtani E., Yurimoto H., and Seto S. (1997) Element partitioning between metallic liquid, silicate liquid, and lower-mantle minerals: implications for core formation of the Earth. *Phys. Earth Planet. Inter.* **100**, 97–114.

O'Neill H. St. C. (1987) Free energies of formation of NiO, CoO, Ni_2SiO_4, Co_2SiO_4. *Am. Mineral.* **72**, 280–291.

O'Neill H. St. C. (1992) Siderophile elements and the Earth's formation. *Science* **257**, 1282–1285.

O'Neill H. St. C., Rubie D. C., Canil D., Geiger C. A., Ross C. R., II, Seifert F., and Woodland A. B. (1993) Ferric iron in the upper mantle and in transition zone assemblages: implications for relative oxygen fugacities in the mantle. In *Evolution of the Earth and Planets*, Geophysical Monograph 74 (eds. E. Takahashi, R. Jeanloz, and D. Rubie). AGU, Washington, DC, pp. 73–88.

Onodera A. (1987) Octahedral-anvil high pressure devices. *High Temp.–High Press.* **19**, 579–609.

Pak J. J. and Fruehan R. J. (1986) Soda slag system for hot metal dephosphorization. *Met. Trans.* **17B**, 797–804.

Philpotts J. A. (1970) Redox estimation from a calculation of Eu^{2+} and Eu^{3+} concentrations in natural phases. *Earth Planet. Sci. Lett.* **9**, 257–268.

Pownceby M. I. and O'Neill H. St. C. (1994) Thermodynamic data from redox reactions at high temperatures: IV. Calibration of the $Re-ReO_2$ oxygen buffer from EMF and $NiO + Ni - Pd$ redox sensor measurements. *Contrib. Mineral. Petrol.* **118**, 130–137.

Putirka K. (1998) Garnet + liquid equilibrium. *Contrib. Mineral. Petrol.* **131**, 273–288.

Righter K. (2003) Metal/silicate partitioning of siderophile elements and core formation in the early Earth. *Ann. Rev. Earth Planet. Sci.* **31**, 135–174.

Righter K. and Drake M. J. (1995) The effect of pressure on siderophile element (Ni, Co, Mo, W and P) metal/silicate partition coefficients. *Meteoritics* **30**, 565–566.

Righter K. and Drake M. J. (1997) Metal/silicate equilibrium in a homogeneously accreting Earth: new results for Re. *Earth Planet. Sci. Lett.* **146**, 541–553.

Righter K. and Drake M. J. (1999) Effect of water on metal/silicate partitioning of siderophile elements: a high pressure and temperature terrestrial magma ocean and core formation. *Earth Planet. Sci. Lett.* **171**, 383–399.

Righter K. and Drake M. J. (2000) Metal/silicate equilibrium in the early Earth: new constraints from volatile moderately siderophile elements Ga, Sn, Cu and P. *Geochim. Cosmochim. Acta* **64**, 3581–3597.

Righter K. and Drake M. J. (2001) Constraints on the depth of an early terrestrial magma ocean. *Meteorit. Planet. Sci.* **36**, A173.

Righter K., Drake M. J., and Yaxley G. (1997) Prediction of siderophile element metal–silicate partition coefficients to 20 GPa and 2,800 °C: the effect of pressure, temperature, f_{O_2} and silicate and metallic melt composition. *Phys. Earth Planet. Inter.* **100**, 115–134.

Righter K., Walker R. J., and Warren P. W. (2000) The origin and significance of highly siderophile elements in the lunar and terrestrial mantles. In *Origin of the Earth and Moon* (eds. R. M. Canup and K. Righter). University of Arizona Press, Tucson, pp. 291–322.

Rubie D. C. (1999) Characterising the sample environment in multianvil high-pressure experiments. *Phase Trans.* **68**, 431–451.

Rubie D. C., Melosh H. J., Reid J. E., Liebske C., and Righter K. (2003) Mechanisms of metal/silicate equilibration in the terrestrial magma ocean. *Earth Planet. Sci. Lett.* **205**, 239–255.

Rushmer T., Minarik W. G., and Taylor G. J. (2000) Physical processes of core formation. In *Origin of the Earth and Moon* (eds. R. M. Canup and K. Righter). University of Arizona Press, Tucson, pp. 227–243.

Sato M. (1978) A possible role of carbon in characterizing the oxidation state of a planetary interior and originating a metallic core. *Lunar Planet. Sci.* **IX**, 990–992.

Sato M. (1984) The oxidation state of the upper mantle: thermochemical modeling and experimental evidence. *Proc. 27th Int. Geol. Congress* **11**, 405–433.

Schmitt W., Palme H., and Wanke H. (1989) Experimental determination of metal/silicate partition coefficients for P, Co, Ni, Cu, Ga, Ge, Mo, W and some implications for the early evolution of the Earth. *Geochim. Cosmochim. Acta* **53**, 173–186.

Schnetzler C. C. and Philpotts J. A. (1971) Alkali, alkaline earth, and rare earth element concentrations in some Apollo 12 soils, rocks, and separated phases. *Proc. 2nd Lunar Sci. Conf.*, 1101–1122.

Shannon M. C. and Agee C. B. (1996) High pressure constraints on percolative core formation. *Geophys. Res. Lett.* **23**, 2717–2720.

Shannon M. C. and Agee C. B. (1998) Percolation of core melts at lower mantle conditions. *Science* **280**, 1059–1061.

Solomotov V. S. (2000) Fluid dynamics of a terrestrial magma ocean. In *Origin of the Earth and Moon* (eds. R. M. Canup and K. Righter). University of Arizona Press, Tucson, pp. 323–338.

Stadermann F. J., Walker R. M., and Zinner E. (1999) Sub-micron isotopic measurements with the CAMECA Nano-SIMS. *Lunar Planet. Sci.* **XXX**, abstract #1407.

Stebbins J. F., Kroeker S., and Andrault D. (2001) The mechanism of solution of aluminum oxide in $MgSiO_3$ perovskite. *Geophys. Res. Lett.* **28**, 615–618.

Stevenson D. J. (1990) Fluid dynamics of core formation. In *The Origin of the Earth* (eds. H. Newsom and J. H. Jones). Oxford University Press, London, pp. 231–249.

Suzuki T. and Akaogi M. (1995) Element partitioning between olivine and silicate melt under high pressure. *Phys. Chem. Mineral.* **22**, 411–418.

Taura H., Yurimoto H., Kato T., and Sueno S. (2001) Trace element partitioning between silicate perovskites and ultracalcic melt. *Phys. Earth Planet. Inter.* **124**, 25–32.

Taylor J. R., Wall V. J., and Pownceby M. I. (1992) The calibration and application of accurate redox sensors. *Am. Mineral.* **77**, 284–295.

Taylor S. R. and McLennan S. M. (1985) *The Continental Crust, its Composition and Evolution: An Examination of the Geochemical Record Preserved in Sedimentary Rocks.* Blackwell, Palo Alto, CA, 312pp.

Thibault Y. and Walter M. J. (1995) The influence of pressure and temperature on the metal/silicate partition coefficients of nickel and cobalt in a model C1 chondrite and implications for metal segregation in a deep magma ocean. *Geochim. Cosmochim. Acta* **59**, 991–1002.

Tonks B. and Melosh H. J. (1990) The physics of crystal settling and suspension in a turbulent magma ocean. In *The Origin of the Earth* (eds. H. Newsom and J. H. Jones). Oxford University Press, London, pp. 151–174.

Tronnes R. G., Canil D., and Wei K. (1992) Element partitioning between silicate minerals and coexisting melts at pressures of 1–27 GPa, and implications for mantle evolution. *Earth Planet. Sci. Lett.* **111**, 241–255.

Tschauner O., Zerr A., Specht S., Rocholl A., Boehler R., and Palme H. (1999) Partitioning of nickel and cobalt between silicate perovskite and metal at pressures up to 80 GPa. *Nature* **398**, 604–607.

Walker D. (1991) Lubrication, gasketing, and precision in multianvil experiments. *Am. Mineral.* **76**, 1092–1100.

Walker D., Carpenter M. A., and Hitch C. M. (1990) Some simplifications to multianvil devices for high pressure experiments. *Am. Mineral.* **75**, 1020–1028.

Walker D., Norby L., and Jones J. H. (1993) Superheating effects on metal/silicate partitioning of siderophile elements. *Science* **262**, 1858–1861.

Walter M. J. and Thibault Y. (1995) Partitioning of tungsten and molybdenum between metallic liquid and silicate melt. *Science* **270**, 1186–1189.

Walter M. J., Newsom H., Ertel W., and Holzheid A. (2000) Siderophile elements in the Earth and Moon: metal/silicate partitioning and implications for core formation. In *Origin of the Earth and Moon* (eds. R. M. Canup and K. Righter). University of Arizona Press, Tucson, pp. 265–290.

Wanke H. (1981) Constitution of terrestrial planets. *Phil. Trans. Roy. Soc. London A* **393**, 287–302.

Weill D. F., McKay G. A., Kridelbaugh S. J., and Grutzeck M. (1974) Modeling the evolution of Sm and Eu abundances during lunar igneous differentiation. *Proc. 5th Lunar Sci. Conf.* **5**, 1337–1352.

Willis J. and Goldstein J. I. (1982) The effects of C, P and S on trace element partitioning during solidification of Fe–Ni alloys. *Proc. 13th Lunar Planet. Sci Conf., J. Geophys. Res.* **87** (suppl.), A435–A445.

Wood B. J. (1991) Oxygen barometry of spinel peridotites. In *Oxide Minerals: Petrologic and Magnetic Significance* (ed. D. H. Lindsley). *Rev. Mineral.*, The Mineralogical Society of American, Washington, DC, vol. 25, pp. 417–431.

Wood B. J. (2000) Phase transformations and partitioning relations in peridotite under lower mantle conditions. *Earth Planet. Sci. Lett.* **174**, 341–354.

Yurimoto H. and Ohtani E. (1992) Element partitioning between majorite and liquid: a SIMS study. *Geophys. Res. Lett.* **19**, 17–20.

Zahnle K., Kasting J. F., and Pollack J. B. (1988) Evolution of a steam atmosphere during Earth's accretion. *Icarus* **74**, 62–97.

Zerr A., Diegeler A., and Boehler R. (1998) Solidus of Earth's deep mantle. *Science* **281**, 243–246.

Zhang J. and Herzberg C. (1996) Melting experiments on anhydrous peridotite KLB-1: composition of magmas in the upper mantle and transition zone. *J. Geophys. Res.* **101**, 8271–8295.

2.11
Subduction Zone Processes and Implications for Changing Composition of the Upper and Lower Mantle

J. D. Morris

Washington University, St. Louis, MO, USA

and

J. G. Ryan

University of South Florida, Tampa, FL, USA

2.11.1 INTRODUCTION

With ca. forty thousand kilometers of subduction zones and convergence rates from 30 km Ma^{-1} to 180 km Ma^{-1}, subduction carries massive amounts of material into seafloor trenches, and beyond. Most of the subducting plate is made of mantle material returning to the depths from which it originated. The hydrated and altered upper oceanic section and the overlying sediments, however, carry a record of low-temperature interaction with the ocean, atmosphere, and continents. Subduction and recycling of these components into the mantle has the potential to change mantle composition in terms of volatile contents, heat-producing elements, radiogenic isotope systematics, and trace element abundances. Enrichments in volatile and potassium, uranium, and thorium contents could change the rheological, thermal, and geodynamical behavior of portions of the mantle. Changing isotope and trace-element systematics provide a means for tracking mantle mixing and the possible subduction modification of the deep mantle. A large number of studies point to possible contributions of subducted sediments and altered oceanic crust (AOC) to the mantle-source region for enriched mantle II (EMII) and high mu (HiMU) enriched oceanic island basalts. Transit through the subduction zone, however, changes the composition of the subducting sediment and AOC from that measured outboard of trenches.

This chapter focuses on subduction zone processes and their implications for mantle composition. It examines subduction contributions to the shallow mantle that may be left behind in the wedge following arc magma genesis, as well as the changing composition of the slab as it is processed beneath the fore-arc, volcanic front and rear arc on its way to the deep mantle. Much of this chapter uses boron and the beryllium isotopes as index tracers: boron, because it appears to be completely recycled in volcanic arcs with little to none subducted into the deep mantle, and cosmogenic ^{10}Be, with a 1.5 Ma half-life, because it uniquely tracks the contribution from the subducted sediments.

The focus here is on subduction processes from trench to rear arc. This chapter starts with a brief discussion of recent thermal models for the downgoing plate and the prograde metamorphic mineralogy of the oceanic crust and sedimentary veneer; the reader is referred to Schmidt and Poli (Chapter 3.17), for an extensive discussion. In the next step it uses ^{10}Be to estimate the absolute mass of sediments subducted to the volcanic arc, in comparison to that supplied to the subduction trenches. Flux balances for ^{10}Be subducted in the sediments versus that erupted in the volcanic arc provide estimates of the fraction of ^{10}Be extracted from the downgoing plate, which can be extrapolated to other elements (cf. Plank and Langmuir, 1993). It subsequently looks at chemical changes for selected elements across the subduction zone, using data from fore-arc serpentinite mud volcanoes, subduction-assemblage metamorphic rocks, high-pressure eclogites, and volcanic lavas from Kurile cross-arc transects, and examines boron-isotope systematics across the convergent margin. Lithium-isotope systematics and comparison of ^{10}Be with uranium-series systematics sometimes delineate multiple stages of subduction modification of the mantle and pinpoint the compositional effects of prior subduction modification on the upper mantle. This contribution ends with estimates of the efficiency of arsenic, antimony, potassium, caesium, rubidium, barium, strontium, uranium, thorium, lead, cerium, samarium, neodymium, lutetium, and hafnium recycling from trench to rear arc, relative to that of boron and beryllium.

2.11.2 THERMAL STRUCTURE AND MINERALOGY OF THE SUBDUCTING PLATE

Central to understanding the recycling of subducted elements in the arc or their subduction to the deep mantle is the temperature variation in the subducting slab, and the prograde mineral assemblages in the sediment, oceanic crust, and lithospheric mantle. Together, they determine where dehydration of the sediments, crust, and deeper subducting mantle occurs, and whether or not sediments or altered basaltic crust can melt. These, in turn, govern element recycling through the arc region and beyond. Current thinking about prograde metamorphism and thermal structure of the slab is very much in flux; see Schmidt and Poli (Chapter 3.17) for a more complete discussion.

2.11.2.1 Subduction Zone Thermal Models

Recent thermal models have begun to use more realistic rheologies for the slab and mantle wedge, including temperature-dependent viscosity and non-Newtonian behavior (van Keken *et al.*, 2002). In these models, the slab-surface temperature beneath the volcanic front at Honshu (with a 130 Ma old, cold plate with a convergence rate of 6 cm a^{-1}) is ~800 °C. This temperature contrasts with those of earlier iso-viscous models for the same margin, in which the slab-surface temperature at 3 GPa is 500–600 °C, depending on the specifics of the model (Peacock and Wang, 1999). A number of other thermal models for other margins indicated slab-surface temperatures beneath the volcanic front often in the range of 350–500 °C (e.g., Kincaid and Sacks, 1997; Peacock, 1996). The temperature range for the surface of the slab is quite important, in that the low end is too cool to allow sediments to melt beneath the volcanic front, while temperatures in excess of ~650–700 °C do permit water saturated sediment melting (Poli and Schmidt, 2002; see Chapter 3.17, and references therein). A slab temperature of 800 °C would allow water-saturated melting of the altered basaltic crust. Regardless of model, the temperature at the base of the subducting oceanic crust (taken as 7 km deep into the slab) beneath the volcanic front is significantly cooler than the slab surface with lesser variation as a function of rheology (~300–375 °C). There is thus a strong temperature gradient within the slab, as well as between the slab and interior of the mantle wedge, at any depth in its trajectory beneath the arc.

2.11.2.2 Prograde Metamorphism and Mineralogy of the Subducting Plate

Early studies of hydrous minerals in subduction zones tended to focus on just a few minerals, with an emphasis on discontinuous dehydration reactions. Key phases were thought to be white mica (phengite) in the metamorphosed sediments, amphibole in the altered basaltic crust, and amphibole and phlogopite in the hydrated mantle wedge (e.g., Tatsumi *et al.*, 1983; Tatsumi and Eggins, 1997; Bebout *et al.*, 1993, 1999; Moran *et al.*, 1992). Hydration of the residual mantle of the subducting plate was little discussed. In simple experimental systems, discontinuous reactions released fluids at a characteristic temperature or pressure. These reactions were envisioned as creating a flush of fluids to the mantle (e.g., a hydrous curtain). Pressure-dependent dehydration reactions (e.g., amphibole in the mantle) provided a means of localizing the volcanic front ~110 km above the downgoing slab.

More recent work shows that a wide variety of hydrous minerals can be stable and emphasizes the importance of continuous reactions in natural systems with extensive solid solution (Poli and Schmidt, 2002; see Chapter 3.17). In the metasediments, phengite (potassium-rich white mica) is an important phase; together with topaz-OH and MgAl-pumpellyite; these phases may be stable up to 900 °C and 7 GPa (Domanik and Holloway, 1996; Ono, 1998). If so, they would have the potential to carry H_2O and many elements beyond the volcanic arc. Hydrous phases in the basaltic crust that may break down in the vicinity of the volcanic arc include amphibole, lawsonite, zoisite, chloritoid, talc, and possibly phengite. Serpentine and chlorite may be important in hydrated mantle peridotite (Poli and Schmidt, 1995). In contrast to amphibole and phlogopite, dehydration reactions for many of these minerals have a strong temperature dependence, meaning that their location will depend on the specific P–T path chosen for the slab. In addition to their dehydration behavior, the importance of these minerals will depend on their water content and modal abundance in the altered basalt. For example, lawsonite carries 11.3 wt.% water and is typically 7–15% of the altered crust while amphibole constitutes a larger fraction of the assemblage (20–60%) but contains only ~2 wt.% water. Recent work has also begun to speculate on the possibility of extensive hydration of the mantle of the subducting plate, with serpentine (containing 12 wt.% water) as a dominant phase (Peacock, 2001; Ulmer and Trommsdorff, 1999; Phipps-Morgan *et al.*, 2002). In the cooler interior of the downgoing plate, serpentine, with a maximum temperature stability of 600–700 °C, could subduct to its maximum pressure stability of ~7 GPa (~220 km depth), beyond which phase A is stable.

Synthesizing a large amount of experimental data, Poli and Schmidt (2002) reach the provocative conclusion that the subsolidus assemblage of both metasediments and metabasalts of the subducting slab at pressures >2.5 GPa would be phengite + jadeitic clinopyroxene + garnet + coesite ± kyanite. Given water loss through dehydration reactions up-dip of the volcanic front, these authors envision water-undersaturated melting, which would occur at >750–800 °C at 3 GPa. These temperatures are at the highest end of the range for thermal models of slab-surface temperatures beneath the volcanic front, but may be more achievable beneath the rear arc.

Arc geochemists attempt to unravel the subduction signature recorded in the arc lavas to distinguish sediment melts from sediment or basalt-derived fluids, which could provide constraints on slab temperatures beneath the arc. The inference of sediment melting largely hinges on lava enrichment of thorium, which is a very incompatible element during melting but appears to be immobile in any aqueous fluid except those rich in fluorine (Brenan *et al.*, 1995). In many cases, a sediment melt at the volcanic front is invoked, but without evidence for a melt of the basaltic crust, as in the Mariana, Aleutian, and Central American arcs (e.g., Elliott *et al.*, 1997; Class *et al.*, 2000; George *et al.*, 2003; Patino *et al.*, 2000). In other cases, the sediment component has been proposed to be a fluid beneath the volcanic front, as in South Chile, Izu, and the Kuriles, becoming a sediment melt beneath the rear arc (Sigmarsson *et al.*, 2002; Ryan *et al.*, 1995; Plank and Kelley, 2001; Morris *et al.*, 2002b).

The contradictory interpretations just discussed are a corollary to the very active research efforts in this general area of slab temperatures and mineralogy. They also highlight the care needed in interpreting arc geochemistry. In particular, interpretations of element fluxes to the shallow or deep mantle that depend on precise dehydration or melting models must be regarded as work in progress.

2.11.3 BERYLLIUM-10 SYSTEMATICS AND SEDIMENT SUBDUCTION

Sediment subducted to the mantle represents removal of continental crust (i.e., negative crustal growth), an important parameter in models of continental growth through time. This is true because many of the particles in the sediments and much of the element budget they host are derived from the continents. To the extent that sediment on the incoming plate is either accreted or joined by the products of subduction erosion (e.g., von Huene and Scholl, 1991), the mass of sediments and the flux of sediment-hosted elements to the arc may be different from that determined outboard of the trench. The cosmogenic isotope ^{10}Be, formed in the atmosphere and decaying with a 1.5 Myr half-life (Yiou and Raisbeck, 1972), uniquely tags the youngest sediments (<10 Ma) of the incoming plate and does not build up in the mantle wedge through time (see Morris *et al.*, 2002b for a recent review of the systematics). It can thus be used to constrain the absolute volumes of sediments that bypass the accretionary prism and are subducted to the depths of magma generation. It can also be used to constrain the efficiency with which beryllium is extracted from the downgoing plate beneath the arc.

2.11.3.1 Beryllium-10 Recycling

The optimal situation for quantifying sediment subduction is where $^{10}Be/^{9}Be$ ratios in arc lavas correlate with other tracers that are also dominated by subducted contributions. Good correlations between $^{10}Be/^{9}Be$ and $(^{230}Th)/(^{232}Th)$, Ba/La, B/Be, and strontium isotopes (Nicaragua lavas; Reagan *et al.* (1994)) and with $^{143}Nd/^{144}Nd$ (oceanic Aleutian lavas; George *et al.* (2003)) point to the isotopic composition of the sediment component added to the mantle. This, in turn, may be compared with the composition of the sediment column outboard of the trench, with ^{10}Be decay during subduction taken into account. Figure 1 shows that ~100% of the incoming sediments off Central America must be subducted to provide the sediment isotope composition sampled by the Nicaraguan lavas. For the Aleutians, the oceanic lavas appear to sample a sediment component that has a lower $^{10}Be/^{9}Be$ ratio than predicted for complete sediment subduction. To generate a subducted sediment that sits on the volcano trend, either the upper 15% of the incoming sediment section could be accreted or a ~1.5–2 Ma increase in subduction/mantle storage time is required. Because this lower ^{10}Be sediment component is sampled everywhere in the oceanic arc, the lavas show uranium excess, and the magnitude of the uranium excess correlates with ^{10}Be enrichment, the interpretation of sediment accretion rather than increased mantle storage time appears more likely.

Wherever multi-isotope data sets are not available or multi-element systematics are not simple (commonly the case), flux balances for ^{10}Be are an important tool. For the Aleutian, Central American, Japan, Izu, Mariana, and Tonga margins, ^{10}Be profiles in the incoming sediment column and the arc lavas have been measured and may be compared to estimate the fraction of sediments subducted to depth (Morris *et al.*, 2002b; Valentine *et al.*, 2002). For ^{10}Be, the volcanic flux out of an arc will be a function of the efficiency with which ^{10}Be is extracted from the sediments at depth

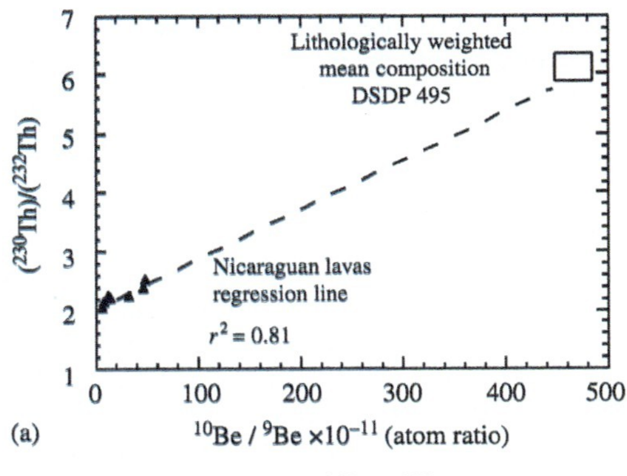

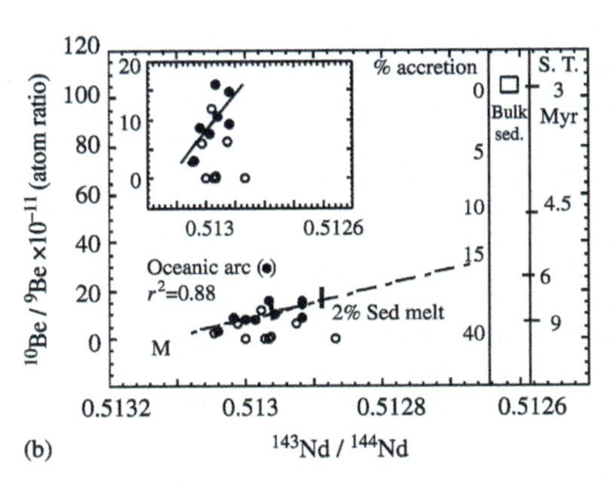

Figure 1 (a) $^{10}Be/^{9}Be$ versus $^{230}Th/^{232}Th$ for Nicaraguan arc lavas (after Reagan *et al.*, 1994). The lithologically weighted mean composition for the incoming sediment column, as measured at DSDP Site 495 off Guatemala, plots on the extension of the mixing line formed by the arc volcanoes. For Be, this composition corresponds to that measured at the trench and corrected for ^{10}Be decay during subduction. Little if any accretion of young ^{10}Be rich sediments can be tolerated before the model subducted sediment falls off the arc trend. (b) $^{10}Be/^{9}Be$ versus $^{143}Nd/^{144}Nd$ for Aleutian lavas (after George *et al.*, 2003). Isotope co-variations ($r^2 = 0.88$) for lavas from the oceanic arc point to a subducted component that has a lower $^{10}Be/^{9}Be$ than modeled for complete sediment subduction with subduction time derived from present day convergent rates and geometry. Accretion of the upper 15% of the incoming sediment column can explain the observed systematics for the oceanic lavas. For continental lavas (open circles), long subduction times due to a widening arc-trench gap to the east mean that ^{10}Be decays in transit.

as well as the amount of sediment subducted, once ^{10}Be decay during subduction has been calculated. An important aspect of any element flux-balance calculation is that it requires assumptions of steady state and element retention in the slab until the depths of magma generation; the next section provides evidence for beryllium retention to depths. Element fluxes also depend linearly on estimates of magma production rates, which are often very poorly known. Recent seismic refraction and gravity surveys, and drilling results used to date onset of arc magmatism, have led to higher magma production rates for a number of arcs than previously used (cf. Dimalanta *et al.*, 2002; Reymer and Schubert, 1984). Figure 2, however, shows why ^{10}Be calculations can be reasonably robust. Because ^{10}Be is so concentrated in the uppermost sediments, where high ^{10}Be is observed in lavas above slabs subducting at $>\sim 5$ cm a^{-1} (i.e., fast enough so that ^{10}Be does not decay during subduction), substantial sediment subduction to depth is required, whatever may be the details of the model.

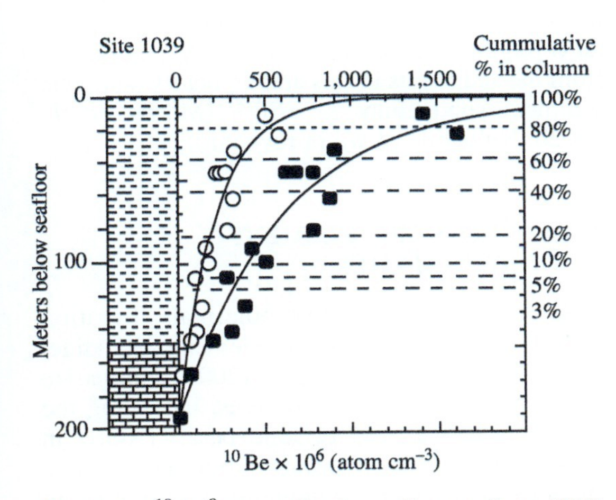

Figure 2 ^{10}Be/^{9}Be profile for sediments from ODP Site 1039, outboard of the Middle America Trench off Costa Rica. Filled circles correspond to measured values while open circles indicate the value expected beneath the volcanic arc, corrected for ^{10}Be decay during subduction assuming modern day geometry and convergence rates (Valentine *et al.*, in revision). Dashed lines and labels to the right indicate the cumulative amount of ^{10}Be present in the sediment column below each depth. Because of the exponential decay of ^{10}Be (half-life = 1.5 Ma), high ^{10}Be in arc lavas (Aleutians, mid-Middle America, Peru, S. Chile, Kuriles, Izu, Marianas, Bismarck, Tonga, Kermadec and Scotia) requires that the bulk of the sediment column subduct to depth (Morris *et al.*, 2002b). Conversely, relatively small amounts of sediment accretion would remove the ^{10}Be-rich sediments, particularly where the slow recent sedimentation rates mean ^{10}Be is in a thin layer (e.g., Marianas and Tonga); elevated ^{10}Be in lavas from these arcs precludes even minor sediment accretion.

2.11.3.2 Beryllium-10 and Sediment Dynamics

Beryllium-10 estimates of sediment subduction to depth are generally in good agreement with estimates of sediment subduction beneath the fore arc (von Huene and Scholl, 1991; Rea and Ruff, 1996). The ^{10}Be flux models (Morris *et al.*, 2002b; George *et al.*, 2003; Valentine *et al.*, 2002) and geophysical interpretations of complete sediment subduction are in agreement for Nicaragua (but not Costa Rica), Izu, Tonga, and the Marianas and for nearly complete subduction of the incoming Aleutian section. The percentages of subducted ^{10}Be recycled in the various arcs (Morris *et al.*, 2002b) are: Aleutians (oceanic section only) (10–20%), Guatemala–El Salvador–Nicaragua (31%), Costa Rica (2%), Japan (<2%), and Marianas (8–80%); where two numbers are reported, higher values use more recent, and higher, magma production rates. Similar interpretations of substantial sediment subduction to depths of magma generation would also apply to the Peru, Kurile, Izu, Bismarck, S Chile, and Kermadec arcs, which also have elevated ^{10}Be.

Since the late 1980s focused work on the magnitude of subduction erosion, which is the removal of material from the base of an old accretionary prism or from the fore-arc basement, and its subduction, has been done. Estimates of eroded material can be very large, as much as 35 km^3 per km of arc length per million years (km^3 km^{-1} Myr^{-1}) in Japan and Nicaragua–Costa Rica (von Huene and Culotta, 1989; von Huene and Lallemand, 1990; Ranero *et al.*, 2000; Ranero and von Huene, 2000; Walther *et al.*, 2000; Meschede *et al.*, 1999; Vannucchi *et al.*, 2001). The ^{10}Be data set speaks to allowable magnitudes of subduction erosion, given that old eroded material dilutes the ^{10}Be of the incoming sediment column. In Japan, arc lavas have no ^{10}Be relative to the detection limit at the time of measurement, ~1 million atom g^{-1} (Morris *et al.*, 2002a,b). A thick incoming sediment column with high recent sedimentation rates provides a large amount of ^{10}Be to the margin. ^{10}Be profiles from fore-arc sediments (Morris *et al.*, 2002a) show some imbricate thrusting and frontal accretion in Japan, also seen from drilling results, which will somewhat reduce the ^{10}Be subducted to depth. A larger effect is likely due to large amounts of subduction erosion of the Cretaceous prism proposed from fore-arc subsidence (von Huene and Culotta, 1989). Dilution of the ^{10}Be in the incoming sediment by the proposed eroded material would dilute sediment to a low enough level such that no ^{10}Be would be seen in the arc.

The situation is somewhat different in Central America. Strong evidence exists for subduction erosion along Nicaragua and Costa Rica.

Subsidence curves for the Middle-America margin suggest that erosion has pertained for longer than the 2–3 Ma subduction interval sampled by the ^{10}Be comparison between incoming sediments and arc volcanoes (Ranero *et al.*, 2000; Meschede *et al.*, 1999; Vannucchi *et al.*, 2001). And yet, the highest ^{10}Be measured in arc lavas is seen in southeastern Nicaragua (Tera *et al.*, 1986), and any significant subduction erosion or accretion proposed would pull the subducted sediment off the mantle-sediment mixing line formed by the Nicaraguan lavas (Figure 1). In Costa Rica, recent data (Morris *et al.*, 2002b; Valentine *et al.*, in revision; Kelly *et al.*, in review) show that the lavas have trace ^{10}Be enrichment. This limits the amount of dilution via subduction erosion to values significantly less than that proposed. It is likely that the precise magnitude of sediment subduction and subduction erosion along a particular margin are not constant through time. Rather, it may vary as plate age and surface topography (e.g., seamount chains, subducting ridges) affect slab buoyancy and roughness, and thus subduction erosion.

2.11.3.3 Implications for Crustal Growth Models

When looked at close up, sediment dynamics of modern arcs are complicated, and this was likely so in the past. Even so, there are some general conclusions. In many margins, the geophysical and ^{10}Be data sets indicating nearly complete sediment subduction are in agreement. Where they are not (and convergence rates are fast enough such that ^{10}Be has not completely decayed during subduction), the ^{10}Be data often can be reconciled by either small volumes of sediment underplating or large volumes of subduction erosion. This implies that the mass of material removed from the crust and carried to the mantle is likely to be comparable to, or perhaps even greater than, that being fed to the trench: an interpretation also reached by Plank and Langmuir (1998). Using the global estimates from von Huene and Scholl (1991) (which are greater than those of Rea and Ruff (1996)), the rate of contemporary sediment subduction is comparable to, or somewhat less than, the rate of magma extraction from the mantle at convergent margins, depending on the model chosen.

These rates correspond to a model predicting very low growth rates for the continents in recent geologic time (Armstrong, 1981; Reymer and Schubert, 1984; Veizer and Jansen, 1985; McLennan, 1988). Increased rates of subduction erosion may bring sediment subduction to a level comparable to modern crustal growth rates. Note, however, that improved estimates of magma production at convergent margins (Dimalanta *et al.*, 2002) are, on an average, nearly twice that of the previous values and will increase modern arc contributions to the crustal growth.

2.11.3.4 Beryllium Extraction from the Slab

The calculated percent of ^{10}Be recycled in arc volcanism (Morris *et al.*, 2002b) carries additional information, particularly for Nicaragua and the Aleutians. In those cases, the $^{10}Be/^{9}Be$ ratio and ^{10}Be concentration of the sediment component subducted to depth and incorporated in the mantle is constrained by mixing line relationships. The calculated recycling percent is thus a measure of the efficiency with which ^{10}Be is extracted from the subducted sediment beneath the volcanic front. These calculations depend on magma production rates, but are independent of assumptions about sediment dehydration or melting. Recycling for the Aleutians is estimated at ~10%; more recent magma production rates from Dimalanta *et al.* (2002) are about twice of those used previously (Morris *et al.*, 2002b; George *et al.*, 2003) and would suggest that ~20% of the subducted beryllium was extracted from the sediments. An estimate of beryllium recycling for Nicaragua, reflecting ^{10}Be extraction from the sediments in this area of nearly complete sediment subduction, is ~30% (Morris *et al.*, 2002b; Valentine *et al.*, in revision).

2.11.4 CROSS-ARC ELEMENT SYSTEMATICS

Sedimentation and hydrothermal alteration enrich the slab in a number of elements. Staudigel *et al.* (1996) and Kelley *et al.* (2003) summarize results from a number of sources, indicating the enrichment of altered oceanic basalt (AOB) in potassium, rubidium, caesium, strontium, and to a lesser extent barium and uranium, with significant redistribution of lead. AOB is also enriched in elements such as boron and lithium (Moran *et al.*, 1992; Smith *et al.*, 1995; Chan *et al.*, 1993). Relative to mid-ocean ridge basalt (MORB), subducting sediments are strongly enriched in a number of elements of interest, including potassium, rubidium, caesium, barium, strontium, uranium, thorium, lead, boron, and lithium; beryllium, samarium, neodymium, lutetium, and hafnium are moderately enriched (e.g., Plank and Langmuir, 1998).

2.11.4.1 Fore-arc Indicators of Element Distillation from the Slab

The subduction interface can be viewed as a semipermeable membrane whose properties change with depth, allowing element distillation from the slab at a wide range of pressures and temperatures. Many elements of interest to arc geochemists (such as boron, lithium, barium, strontium) are enriched in pore fluids (recovered by drilling across the decollement zone of sedimentary fore-arc prisms) to levels well above seawater compositions (Kopf and Deyhle, 2002; Kastner *et al.*, 2000; You *et al.*, 1994, 1996; Chan and Kastner, 2000; Deyhle and Kopf, 2002). Investigations of elemental releases via shallow subduction fluids are in a preliminary stage, and little is known about their net flux. It is possible that fore-arc losses of such elements from down-going slabs comprise a small part of the total budget, but they do testify to the leaky nature of the subduction interface.

Same is the case with the data from exhumed subduction assemblages. A series of articles (Moran *et al.*, 1992; Bebout, 1996; Bebout *et al.*, 1993, 1999; Ryan *et al.*, 1995, 1996a,b; Leeman *et al.*, 1990; Leeman, 1996; Noll *et al.*, 1996) show that elements such as boron, caesium, arsenic, antimony, and lead decrease sharply as water and nitrogen contents decrease with increasing metamorphic grade in metasediments from subduction assemblages. Exhumed subduction-assemblage rocks are unusual; the Catalina schist, from which much of the data come, reaches high temperatures at relatively low pressures (~700 °C at ~1.2 GPa); and higher metamorphic grade rocks need not have had the same protolith or $P-T-t$ path as the lower grade rocks. Even so, the decreasing concentrations are so large and systematic that they are widely interpreted to be reflecting element distillation from the slab in an aqueous fluid, i.e., higher-grade rocks approximately represent the residual composition of the metasedimentary section of the subducting plate following progressive fluid loss. Similar work has been carried out on eclogites from high-pressure terranes (Becker *et al.*, 2000; Becker, 2000), thought to represent subducted MORB residue. Models of eclogite composition at 500–600 °C and >900 °C show preferential loss of elements such as potassium, rubidium, caesium, barium, and lead, relative to average AOB (Staudigel *et al.*, 1996; Kelley *et al.*, 2003).

An interesting complement to subduction assemblage data sets are studies of upwelling serpentinites and associated porefluids from fore-arc seamounts in the Mariana subduction system. Two of these seamounts have now been sampled: the Conical Seamount on ODP Leg 125, and the South Chamorro seamount during ODP Leg 195 (P. Fryer and G. J. Fryer, 1987; Salisbury *et al.*, 2002). The seamounts are built on fore-arc crust 20–35 km above the downgoing Pacific plate, and are essentially mud volcanoes erupting fine-grained serpentine with entrained, highly serpentinized blocks of harzburgite and dunite, and blueschist-facies mafic crustal materials (Fryer and Mottl, 1992; Saboda *et al.*, 1992; Fryer *et al.*, 1999; Salisbury *et al.*, 2002; Guggino *et al.*, 2002). Pore fluids associated with the serpentinites are H_2O dominated, low in chloride, and have extremely high pH (>12), indicating that these fluids did not originate as seawater, but are more likely to be released during the metamorphic decomposition of slab minerals at ≤30 km depths (Mottl 1992; Haggerty, 1991). These serpentinites appear to be fragments of the uppermost mantle that has been hydrated by these fluids, and as such, they may preserve the geochemical "fingerprint" of fore-arc fluid releases.

Figure 3 includes data for those serpentinized ultramafic clasts from Conical Seamount that have been the least affected by near-surface interactions with seawater. Boron shows the largest degree of enrichment in the conical serpentinites. Lithium shows a moderate degree of enrichment, while strontium, barium, and potassium abundances cannot be easily distinguished from those of unmodified mantle. Data for associated pore fluids are largely complementary, showing substantially elevated boron, but low strontium, barium, potassium, and (interestingly) lithium abundances. New results for Leg 125 and 195 samples indicate that arsenic and caesium are also strongly enriched in fore-arc serpentines and fluids, antimony and rubidium are modestly enriched, and uranium, lead, and thorium show no elevation (Savov *et al.*, 2002). The Mariana serpentinites reflect pressure–temperature ($P-T$) conditions equivalent to some of the lowest-grade subduction association samples (<350 °C; ~1 GPa; Saboda *et al.*, 1992), and indicate that even at shallow depths of subduction, selective elemental mobility occurs: species with high solubility in the H_2O-dominated fluid phases generated at these depths are released. Interestingly, many elements which are enriched in arc lavas, such as barium, lead, and uranium, are not mobilized beneath the shallow fore-arc, indicating that these species are stably retained in slab mineral phases at shallow depths, and/or that a compositionally different fluid phase may be required to extract them from slab materials.

2.11.4.2 Cross-arc Synthesis

Changing elemental ratios (B/Be, Ce/Pb, U/Zr, Th/U, Cs/Th, Ba/Th, and Be/Th) observed as a function of increasing pressure and temperature of the subducting slab can be used to continue the investigation of element distillation, and are compiled in Figure 4. The left column shows variation with increasing metamorphic grade for largely metasedimentary subduction assemblages, with Plank and Langmuir's (1998) composition

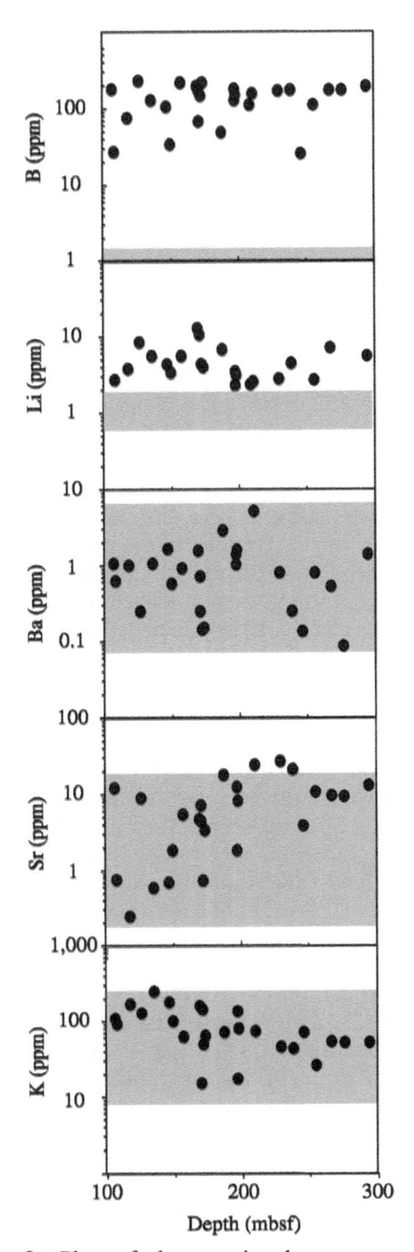

Figure 3 Plots of element abundances versus depth below seafloor (meters) for serpentinized ultramafic clasts from hole 779A in Conical Seamount, Mariana fore-arc, collected during ODP leg 125 (Ryan *et al.* (1996a)). Shaded fields represent the range of enriched and depleted mantle compositions (source McDonough and Sun, 1995).

for globally subducting sediment (GLOSS) shown for comparison. Also shown are model results for eclogite (metabasaltic) assemblages with increasing temperature, with N-MORB and "average AOB" for comparison. These data examine changes in elemental ratios in the residual subducting plate following shallower fluid loss. The central column shows variations in these ratios across the width of the Kurile volcanic arc, indicating the impact of slab-derived fluids and their elemental fractionations on the mantle wedge, which melts to produce the arc lavas. The right column shows the changing concentrations of boron, lead, uranium, thorium, caesium, barium, and beryllium across the width of the Kurile arc, to aid in understanding the recycling behavior of these elements from trench to rear arc.

A number of inferences may be drawn from Figure 4. The right-hand panel shows that elements such as boron and lead (and arsenic and antimony, not plotted) show decreasing concentrations across the width of the arc (Ryan *et al.*, 1995, 1996a,b; Noll *et al.*, 1996, 1993; Leeman, 1996). The smaller degrees of mantle melting in the rear arc mean that incompatible elements will generally show higher concentrations in rear-arc lavas (e.g., caesium, uranium, barium, beryllium, thorium, potassium, and rubidium (the latter two not shown here; see Ryan *et al.*, 1995). The exception is when slab contributions of an element to the volcanic front greatly exceed those to the rear arc (Ryan *et al.*, 1995), in which case relative source depletion of highly fluid mobile elements in the rear arc outweighs the lower degrees of melting.

The decreasing B/Be ratio with increasing metamorphic grade (left panel of Figure 4) indicates that boron begins to leave the slab at relatively low temperatures, and it continues to be distilled out of the slab as it traverses beneath the arc. Rear-arc lavas have B/Be ratios indistinguishable from MORB and OIB and there is no evidence for boron recycling into HiMU- or EMII-type OIBs in boron-abundance or boron-isotope data (e.g., Chaussidon and Jambon, 1994; Ryan *et al.*, 1996b). As noted in the literature cited, these results suggest that boron is completely recycled in the arc.

Boron-isotope data for arc lavas (Ishikawa and Nakamura, 1994; Ishikawa and Tera, 1997) have been used to argue that boron enrichment in arcs is largely derived from the AOC, with the sediment boron having largely been distilled out of the slab up-dip of the volcanic front. This argument is based in part on the fact that the boron-isotopic ratios of altered crust are greater than those of marine sediments, and were comparable to the higher boron-isotopic values encountered in arcs (e.g., Spivack and Edmond, 1987; Spivack *et al.*, 1987; Ishikawa and Nakamura, 1992, 1993). However, larger data sets for oceanic crust point to lower overall boron-isotopic ratios (Smith *et al.*, 1995) and newer data for arcs show that boron-isotopic ratios significantly higher than those of any slab constituent can occur (Straub and Layne, 2002). Boron-isotope results for the Mariana fore-arc serpentinites indicate even higher δ^{11}B values for boron released in shallow fluids

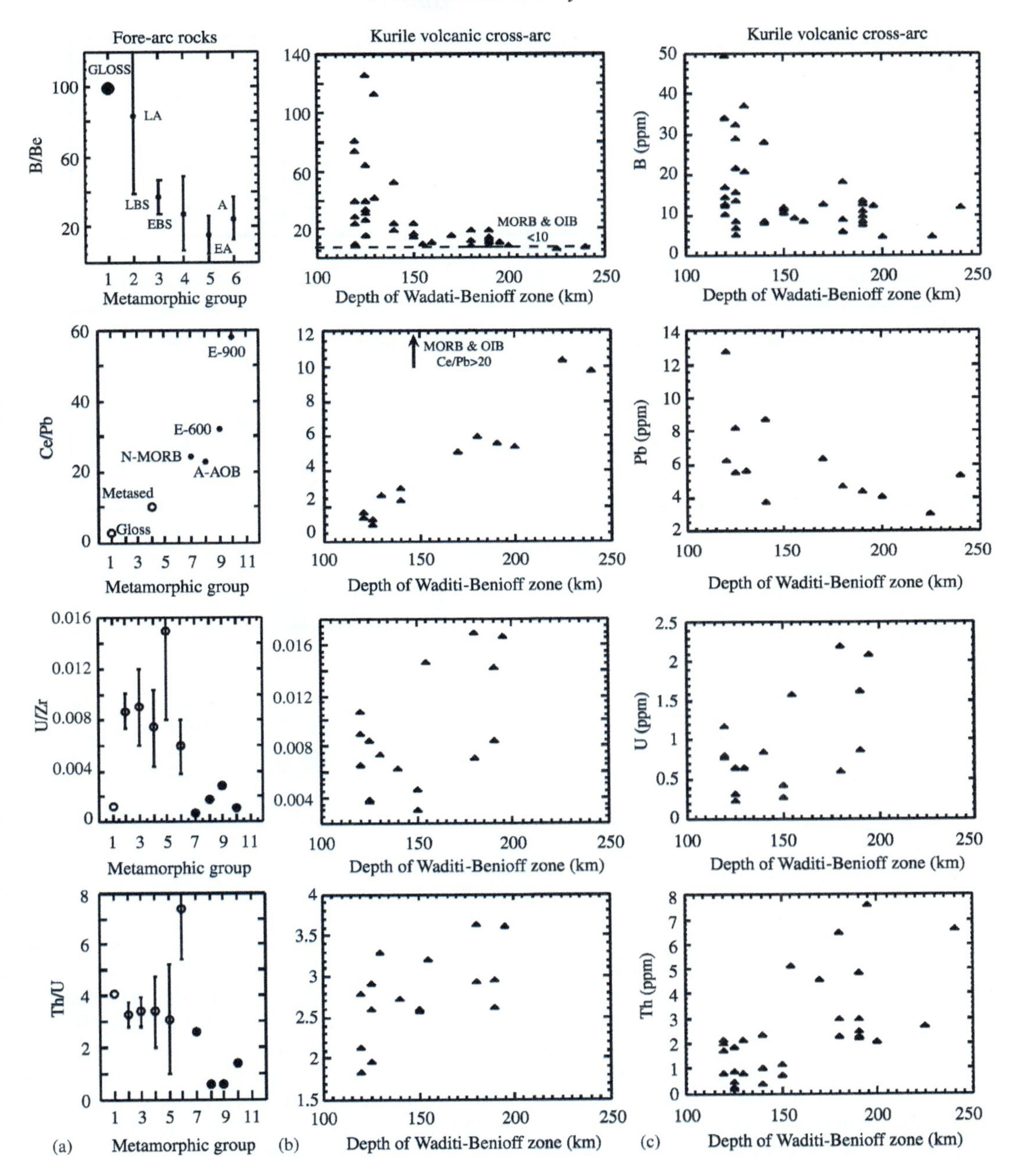

Figure 4 Compilation of cross-arc systematics for selected elements and element ratios. The left column (a) summarizes ratio data from sediments, metasedimentary and metabasaltic rocks from the fore-arcs of subduction zones. The open symbols show metasedimentary data, with metamorphic grade increasing to the right as indicated by increasing number on the x-axis. 1 = GLOSS is an estimate of globally subducting sediment (Plank and Langmuir, 1998); 2 = LA is Lawsonite–Albite grade (~0.4–0.7 GPa, 150–250 °C); 3 = LBS is Lawsonite blueschist grade (~0.7–1.1 GPa, 225-350 °C); 4 = EBS is Epidote blueschsist grade (~0.9-1.3 GPa, 375–500 °C); 5 = EA is Epidote Amphibolite grade (~0.9–1.3 GPa, 550-580 °C) and 6 = A is amphibolite grade (~0.9–1.3 GPa, 600–750 °C). Note that amphibolite grade rocks show evidence for minor melting. All sediment metamorphic data from Bebout *et al.*, 1999. Filled symbols to the right in each figure are for metabasaltic data. Point 7 = refers to unaltered N-MORB (Hofmann, 1998); 8 indicates average altered oceanic basalt from Staudigel *et al.*, 1996 and Kelley *et al.*, 2003; 9 = model altered N-MORB eclogite at 500–600 °C and 10 is for model altered N-MORB at >900 °C (from Becker *et al.*, 2000). The center column (b) shows cross-arc variations for the same ratio, measured in the Kurile volcanic arc lavas (data from Tera *et al.*, 1993; Ryan *et al.*, 1995; Noll *et al.*, 1996; and Morris, Ryan and Tera, unpublished ICP-OES and INAA data). The right column (c) shows elemental variations in the Kurile cross-arc lavas.

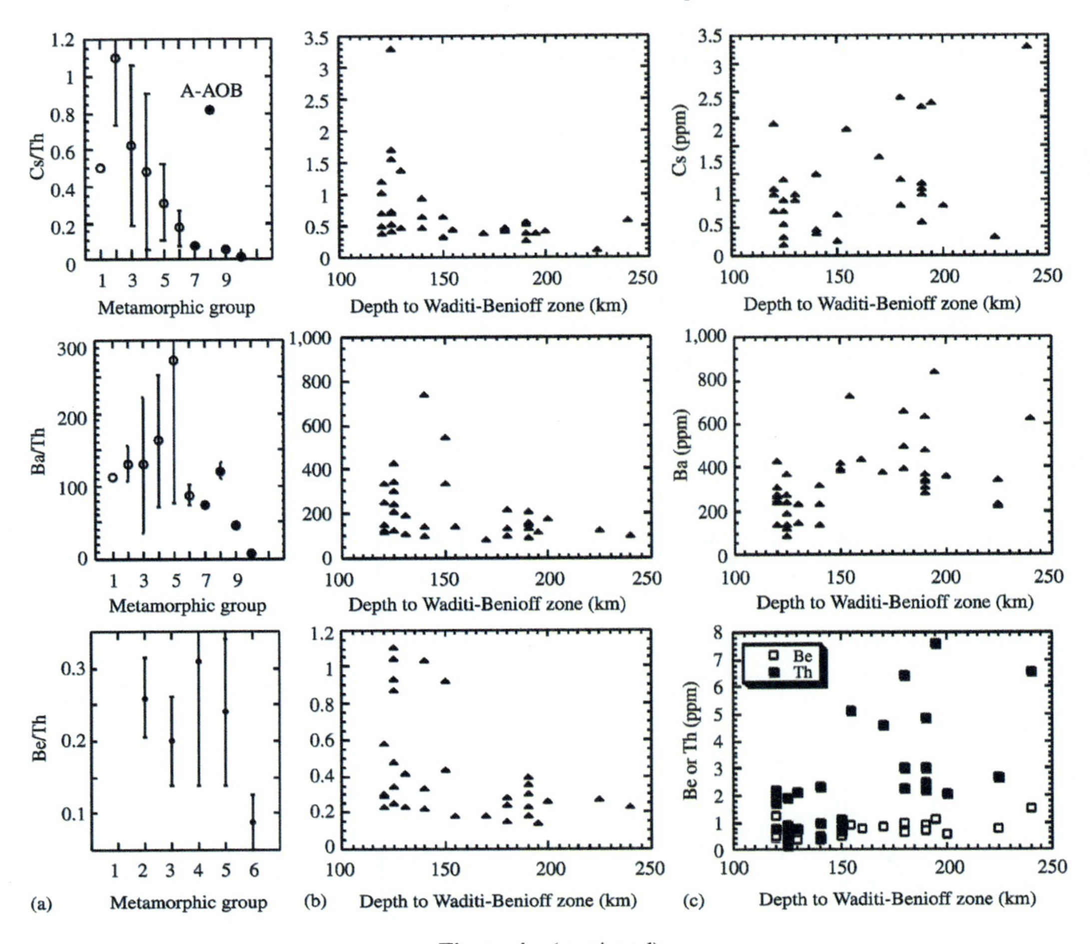

Figure 4 (continued).

(Benton *et al.*, 2001; Figure 5). It thus appears that boron is isotopically fractionated during slab-fluid releases, with the generated fluids becoming enriched in ^{11}B. Mineral analyses indicate that boron (and rubdium, caesium, and barium) are largely hosted in the white mica phengite (Bebout *et al.*, 1993, 1999; Moran *et al.*, 1992; Domanik *et al.*, 1993), which can be stable in both metasedimentary and metamorphosed altered basalt compositions (Poli and Schmidt, 2002; Schmidt and Poli, 1998). These observations suggest that boron from the altered ocean crust as well as the sediments is extracted beneath the volcanic arc.

Boron-isotopic data also show the strong decrease in slab-derived boron across the width of the arc. Figure 5(a) documents the decline in $\delta^{11}B$ across the Kurile and Izu-Bonin arcs, which follow cross-arc declines in boron abundances (Ishikawa and Nakamura, 1994; Ryan *et al.*, 1995; Ishikawa and Tera, 1997). Ishikawa and Tera (1997) interpreted the cross-arc decline of boron isotopes in the Kuriles as a simple mixture of AOB-dominated slab boron with the mantle, which should be both boron depleted, and have a low boron-isotope ratio (Figure 5(b)). As boron-isotopic data sets for MORB and OIB suites have developed, this model has become problematic because none of the MORB and OIB samples thus far examined have low enough boron-isotope ratios to be a satisfactory mixing end-member. However, if isotopic fractionation of boron is occurring during shallow slab fluid releases, and a high $\delta^{11}B$ component is removed, then the residual slab should move toward very low $\delta^{11}B$. Melts or supercritical siliceous fluid phases released at greater depths of subduction should have both very low boron abundances and very light boron-isotope signatures. It is thus possible that the low $\delta^{11}B$ mixing end-member suggested from cross-arc studies is actually the outflux from a hotter and deeper slab, mixing with an H_2O-mediated component added to the mantle at shallower depths. The mantle wedge would play a negligible role in defining the boron-isotopic signatures of arc lavas in this model, which is consistent with the observed B/Be–$^{10}Be/^{9}Be$ systematics for arcs globally (Morris *et al.*, 1990).

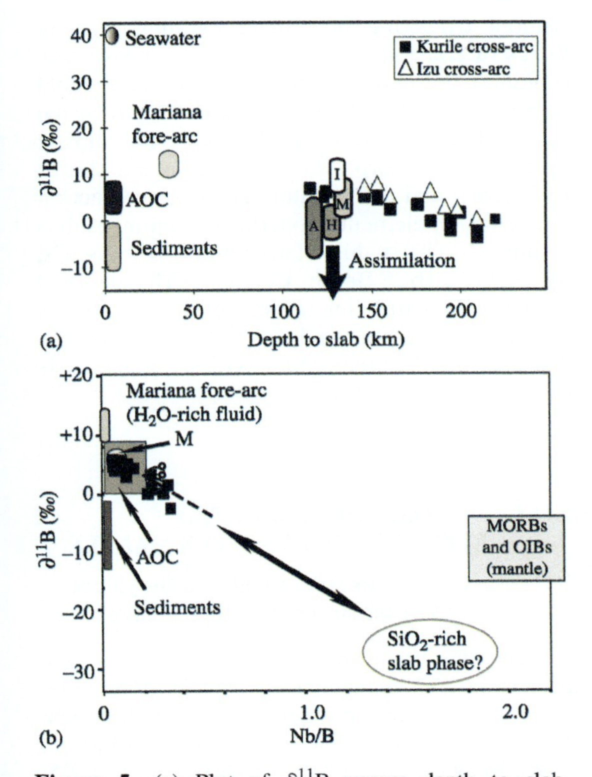

Figure 5 (a) Plot of $\delta^{11}B$ versus depth to slab, including data from the Izu-Bonin arc, field labeled I (Ishikawa and Nakamura, 1994; Straub and Layne, 2002), the Kurile arc (Ishikawa and Tera, 1997), the Mariana arc (M) (Ishikawa and Tera, 1999), Halmahera (H) (Palmer, 1991), Martinique (A) (Smith *et al.*, 1997), and the Mariana fore-arc (Benton *et al.*, 2001). Fields for seawater, altered oceanic crust (AOC), and sediments based on data from Spivack and Edmond (1987), Spivack *et al.* (1987), Ishikawa and Nakamura (1992, 1993), and Smith *et al.* (1995). B isotopes show a progressive decline in subduction zone outputs (fluids and/or melts) with increasing depth of subduction. (b) Plot of $\delta^{11}B$ versus Nb/B ratios. Data sources are the same as in (a); data for mid-ocean ridge and ocean island basalts are from Chaussidon and Jambon (1994), Chaussidon and Marty (1995) and Gurenko and Chaussidon (1997). Cross-arc data arrays form mixing trends between a high $\delta^{11}B$ slab fluid source, and materials with very low $\delta^{11}B$ and high Nb/B (=low boron). Ocean ridge and intraplate mantle sources are not an end-member in this array. A sediment melt, or high-temperature siliceous fluid phase released from the slab at greater depths would be strongly boron depleted, and should have very low $\delta^{11}B$ due to progressive devolatilization. As the mantle contains negligible amounts of boron relative to the slab, it plays only a minimal role in defining arc B abundance and B-isotopic signatures.

Returning to Figure 4, lead, like arsenic and antimony, shows a behavior similar to that of boron (Noll *et al.*, 1993) indicating that it is extracted effectively from the slab through subsolidus metamorphic reactions and also beneath the volcanic front, presumably in a fluid However, the Ce/Pb ratios for rear-arc lavas do not reach values typical of MORB or OIB, suggesting that lead is still being extracted from the slab at depths of order 200 km. Lead-isotopic data from arc lavas indicate that it is being extracted from both subducted sediments and AOC (e.g., Kay, 1980; Class *et al.*, 2000; Patino *et al.*, 2000; Miller *et al.*, 1994).

Other elements of interest (e.g., potassium, rubidium, barium, strontium, caesium, uranium, and thorium) do not show decreasing concentrations across the volcanic arc (Figure 4; Ryan *et al.*, 1995; Noll *et al.*, 1996). When the effect of decreasing degree of mantle melting in the rear arc is minimized by ratioing the element concentration to that of another incompatible element, caesium, barium and beryllium enrichment of the arc mantle decreases across the arc. Note that Cs/Th and Ba/Th ratios of lavas from the volcanic front are much higher than those of the subducting slab, particularly when barium and extensive caesium losses during prograde metamorphism are considered. The large fractionations of caesium and barium from thorium suggest a fluid component at the volcanic front, rather than a melt. However, the rear-arc lavas do not reach ratios typical of MORB or OIB mantle, indicating that slab-derived caesium, barium, and beryllium are still being added to the rear-arc mantle wedge. The corollary is that these elements, enriched in the slab by alteration or sedimentation, are not completely extracted from the slab beneath the arc. High ^{10}Be concentrations in the rear-arc lavas of the Kuriles, Bismarck, and Aleutian arcs (Tera *et al.*, 1993; Morris *et al.*, 2002b) show that beryllium is retained in the sedimentary veneer beyond the volcanic front.

The data in Figure 4 (left panel) suggest that uranium in prograde metamorphic rocks, both metasedimentary and metabasaltic, is not particularly mobile until temperatures of ~700 °C or higher. The uranium excess in arc lavas (e.g., Gill and Williams, 1990; Condomines and Sigmarsson, 1993; Turner *et al.*, 1997, 2000a,b), indicates that slab-derived uranium, thought to be transported in a fluid, is added to the mantle wedge. The increasing U/Zr ratio across the arc suggests that uranium is also added to the rear arc, although in lesser amounts than at the volcanic front. Detailed models of the combined effects of fluid and sediment addition and mantle melting are necessary to fully resolve the effects.

The concentration of elements such as potassium (Ryan *et al.*, 1995, 1996a,b) and thorium increases across the volcanic arc, even when effects of partial melting are minimized by ratioing to other incompatible elements. A suggestion that the slab component is a fluid beneath the volcanic front and a sediment melt beneath the rear arc is consistent with lead and thorium systematics in the Izu arc (Plank and Kelley, 2001) and with high ^{10}Be concentrations in rear-arc lavas from the Kurile arc (Tera *et al.*, 1993; Ryan *et al.*, 1995; Morris *et al.*, 2002b), and could explain these observations.

Potassium (Bebout *et al.*, 1999), barium, beryllium, uranium, and thorium behave similarly in the metamorphic rocks, in that their concentrations change little in metasedimentary subduction assemblages until amphibolite grade (>600 °C, with textures indicative of some melting) is reached. In contrast, potassium and barium in metabasaltic rocks appear to decrease rapidly at temperatures <500–600 °C, behaving much like rubidium and caesium (Becker *et al.*, 2000). Barium enrichment in arc lavas well above those possible from sediment addition, even through sediment melting (e.g., Kay, 1980; Elliott *et al.*, 1997; Reagan *et al.*, 1994; Lin, 1992), is indicative of barium mobility in a fluid. Becker *et al.* (2000) conclude that much of the barium enriched in arc lavas must be derived from the sediments, given the rapid decrease in barium in metamorphosed eclogite in their model. The transition in potassium, barium, and beryllium from stable in the residual mineralogy of the metasediments at <600 °C to mobile in fluids beneath the arc requires a change in fluid composition or residual mineralogy, as discussed earlier.

Several general inferences can be drawn from the compilation in Figure 4. The prograde metamorphism of sediments and basalt between trench and volcanic front mean that the slab composition delivered to the region of magma generation may be quite different from that entering the trench, and that GLOSS and AOB compositions must be used carefully in evaluating element recycling (Plank and Langmuir, 1998; Staudigel *et al.*, 1996; Becker *et al.*, 2000; Becker, 2000). For elements such as boron, lead, arsenic, and antimony, their partitioning into fluids is so effective beneath the volcanic front, and up-dip therefrom, that not even the addition of a sediment melt to the rear-arc can increase their concentration in rear-arc lavas. This implies that at least the sediment-hosted boron and lead are completely extracted from the downgoing plate beneath the arc region, and are not available to subduct to the deeper mantle. Most other slab-derived elements discussed here are enriched at the volcanic front and to comparable or lesser degrees in the rear arc. Rear-arc lavas are generally small in total volume relative to the volcanic front, and are formed by smaller degrees of partial melting; it is thus possible that element fluxes through the rear arc are a small part of the total budget in calculating element fluxes into and out of subduction zones. Alternatively, if slab components can be added to rear-arc mantle without triggering immediate melting, rear-arc fluxes of elements such as beryllium, potassium, rubidium, barium, thorium, and uranium could be an important and largely unknown part of the flux balance. Based on the behavior during prograde metamorphism and magmatism shown in Figure 4 and the literature cited, the extraction efficiency for elements from the downgoing slab is estimated to be in this relative order: B > As, Cs, Sb > Pb > Rb > Ba, Sr, Be ~U, > Th. As noted earlier, the extraction efficiency for boron is estimated at 100% and for ^{10}Be beneath the Aleutian and Nicaraguan volcanic front is estimated at 20–30%.

2.11.5 STORAGE OF SUBDUCTED ELEMENTS IN THE UPPER MANTLE

2.11.5.1 Constraints from Combined Beryllium and Uranium-series Isotope Studies

^{10}Be and uranium-series radionuclides have been measured on the same samples from a number of arcs, including Bismarck, Central America, Marianas, Aleutians, and South Chile (Gill *et al.*, 1993; Reagan *et al.*, 1994; Herrstrom *et al.*, 1995; Sigmarsson *et al.*, 1990, 2002; Elliott *et al.*, 1997; Valentine *et al.*, 2002; George *et al.*, 2003), with work in progress for Tonga, Scotia, and the Kuriles. The uranium excess, observed in many arc lavas but not in OIB or MORB suites, is often interpreted as a subduction signature. Given the plethora of subduction components (fluids or melts from sediment or AOB) and the impact of dynamic melting on the uranium-series isotopes (e.g., Elliott, 1997; George *et al.*, 2003), interpretation can be difficult. Although there is no strong correlation in all arcs studied, there is a general tendency for highest ^{10}Be and ^{10}Be/^{9}Be ratios to be seen in those members of an arc suite that show the strongest U–Th or Ra–Th disequilibria, plotting furthest from the equiline. Figure 6 illustrates this point for Nicaragua and the southern volcanic zone (SVZ) of Chile and supports the contention that the disequilibria signatures are initiated by element transport out of the downgoing plate. Equally interesting are the implications of these diagrams for multiple timescales of subduction modification of the mantle and the buildup of subduction signatures in the upper mantle over time.

The temporal resolution of the multiple parent–daughter pairs within the uranium decay chain, ^{238}U–^{230}Th (approaching secular equilibrium with a 75 ka half-life), ^{230}Th–^{226}Ra (1,500 a) and ^{235}U–^{231}Pa (~32 ka) is a powerful tool. It has been used to estimate the elapsed time between subduction modification of the mantle and lava eruption to the surface, and to identify multiple

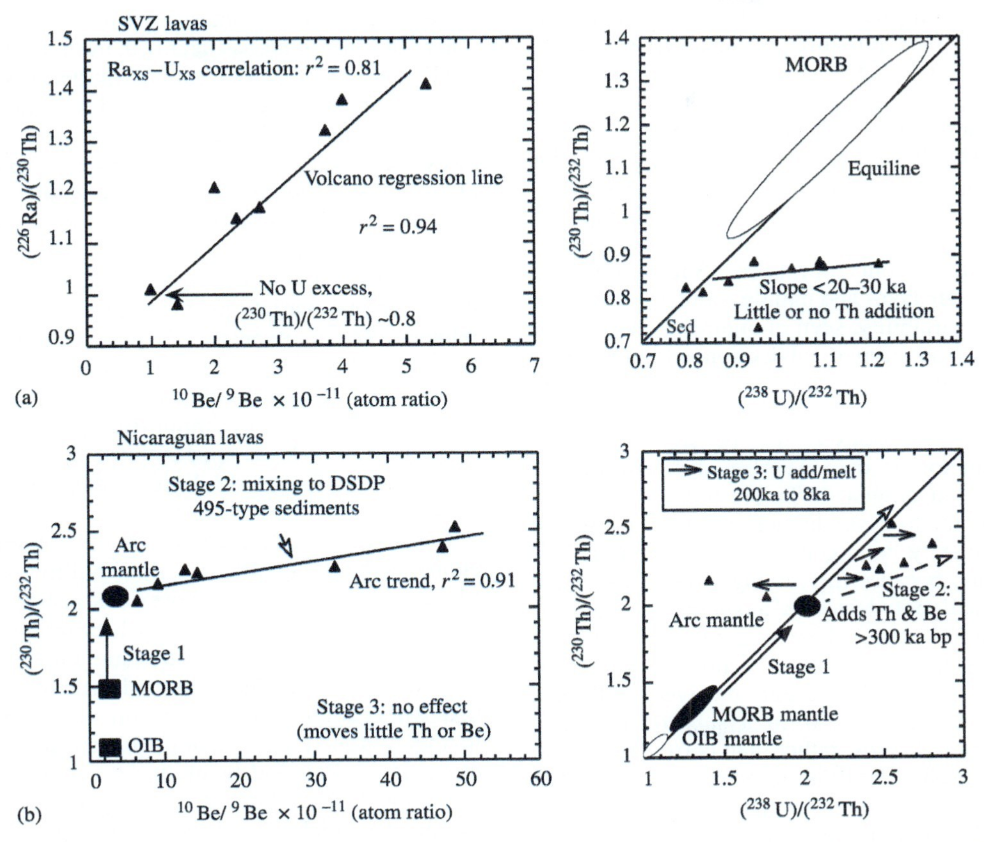

Figure 6 (a) U–Th–Ra–Be systematics for lavas from the SVZ of S. Chile, modified from Sigmarsson *et al.* (1990, 2002). Note that the well-correlated U–Th disequilibria chord intersects the equiline at a (^{230}Th)/(^{232}Th) activity ratio of about 0.8, suggesting sediment addition to MORB mantle prior to more recent addition of U, Ra, and ^{10}Be (but not Th) in the very recent past. See text for further explanation. (b) U–Th–Be systematics for Nicaraguan lavas, modified from Reagan *et al.*, 1994. At zero ^{10}Be (a possible mantle composition prior to recent sediment addition), the mixing line has a (^{230}Th)/(^{232}Th) ratio of 2, Ba/La of 24 and ^{87}Sr/^{86}Sr of 0.704, all suggesting sediment addition to the mantle prior to the event that added ^{10}Be. Three stages are envisioned for the subduction modification of the mantle. Stage 1 shifts the mantle from N-MORB values to modified mantle (arc mantle composition). Stage 2 adds ^{10}Be, Ba, Th, Sr, and B to the mantle. U may also be added (indicated by the dashed line and lines with open arrow heads), but the stage II event is old enough that any U excess will have decayed away, producing samples that plot on the equiline. Stage III adds U and Ra to the mantle in the last 200 ka to <8 ka; if Be and Th were added in stage III, the amount is small relative to that from stage II. If ^{10}Be were added in stage I (as in stage II) then stage 1 occurred prior to ca. 3–4 Ma. Note the homogeneity of the subduction-modified arc mantle, relative to the heterogeneity created during stage II.

stages in the process (e.g., Reagan *et al.*, 1994; Elliott *et al.*, 1997; Turner *et al.*, 2000a,b; Bourdon *et al.*, 1999; George *et al.*, 2003).

^{10}Be, with its 1.5 Ma half-life, adds a longer-lived subduction tracer to the arsenal, one that will decay away in the mantle on a time frame of several million years. The data for the SVZ of S Chile (Figure 6(a)) illustrate the power of the combined approach. The very well correlated U–Th, Ra–Th, and ^{10}Be/^{9}Be data indicate that uranium, radium, and ^{10}Be, but not thorium, were transported from slab to mantle to produce the nearly horizontal arrays on the disequilibria diagrams (right panel) and the strong correlations between ^{10}Be addition and uranium and radium excesses (left panel). Taken at face value, these results suggest that a slab/sediment-derived fluid was added to the mantle less than 8 ka before the lavas erupted to the surface (Turner *et al.*, 2001; Sigmarsson *et al.*, 2002). Also striking is that the intersection of the arc trend with the equiline is at a thorium activity ratio of ~0.8. This value is significantly below those for MORB and OIB, in a direction typical of that expected for many pelagic sediments, and suggests the addition also of a sediment melt to the mantle (e.g., Elliott *et al.*, 1997). The limited partitioning data for beryllium between sediments, fluids, and melts (Johnson and Plank, 1999) suggests that beryllium partitions more effectively into melts than fluids. Therefore, if the sediment melt and slab fluid were generated at about the same time, the sediment melt should also contain ^{10}Be. In this case, the arc trend near its intersection with the equiline (lowest uranium and radium excess) should have high ^{10}Be, which it clearly does not. One possible explanation is that the sediment melt signature inferred from the low (^{230}Th)/(^{232}Th) ratio reflects sediment addition long enough ago that ^{10}Be had decayed away, i.e., >2–4 Ma. It is always possible, however, that ^{10}Be simply was not added in an earlier event.

Taking a different approach, the mixing systematics for arc lavas themselves also contain information about timescales. Subduction processes create lavas and mantle sources with a wide range of variability, seen in Figures 1 and 6. If the effects of all earlier subduction additions were stripped out by melting, then recent arc lavas could show well-correlated trends terminating in mantle compositions typical of MORB or OIB. The situation is demonstrably different for the SVZ and Nicaragua. In Figure 6, well-correlated trends point to a subduction-modified mantle of approximately homogenous composition. This is intriguing, because the mixing lines in Figures 1 and 6 will not decay to a single homogenous composition, but rather to a range of compositions. If this spectrum of mantle compositions, formed in an earlier event, received additional subducted contributions in a subsequent event, then scattered fields, rather than correlated trends, would result. Creating a homogenous mantle between one episode of subduction modification and another requires some process that mixes out heterogeneity within the mantle, perhaps convective stirring over time.

The Nicaraguan lava data may be interpreted similarly, using the good correlations between ^{10}Be/^{9}Be and thorium isotopes (Figure 6(b)), strontium isotopes, Ba/La, and B/Be ratios but not with the magnitude of the uranium excess (Reagan *et al.*, 1994). As noted in the caption, three separate stages of subduction modification of the mantle are indicated. The earliest stage leaves behind an upper mantle enriched in subducted thorium, strontium, and barium.

2.11.5.2 Lithium-isotope Evidence for Mantle Storage

These results indicate that subducted elements may build up in the upper mantle and change its composition, at least over a time frame of several million years (see also Regelous *et al.* (1997) and Turner *et al.* (1997)). New evidence from lithium isotopes points in a similar direction: the surprisingly uniform δ^7Li values encountered in the Sunda, Kurile, Aleutian, and northern Central American arcs strongly suggest a mantle "buffering" effect for lithium, wherein high δ^7Li inputs from the slab are retained in mantle minerals, ultimately producing the barely elevated lithium-isotopic signature observed in most arc lavas (Chan *et al.*, 2001; Tomascak *et al.*, 2002). Only where the slab input has declined, as in the case of the Panama segment of Central America, does one see lithium-isotopic heterogeneities, and these appear to reflect diverse lithium-isotopic signatures preserved in the mantle wedge (Tomascak *et al.*, 2000; Figure 7). Edwards *et al.* (1993) contrasted B/Be and radiogenic isotope systematics in Sunda to see through slab effects to underlying mantle heterogeneities; whereas in this case the variations were presumed to be related to the involvement of a plume component, the large lithium-isotope variations in the Panama subarc mantle are more likely to have been subduction induced, as lithium-isotope variability in plume-source lavas appears limited (Tomascak *et al.*, 1999; Ryan and Kyle, 2000; Chan and Frey, 2003).

Can the timescale for upper mantle storage of elements be longer than a few million years? Much current thinking proposes that mantle melts form (and the volcanic front is localized) where subduction fluids interact with mantle hot enough to melt under the ambient thermal structure, rather than where amphibole dehydration melting occurs in the wedge (e.g., Iwamori, 1998; van Keken *et al.*, 2002; Poli and Schmidt, 2002). Melting is viewed as being triggered by a last fluid addition to a mantle which has been subduction modified in earlier events. After the last fluid addition, melting

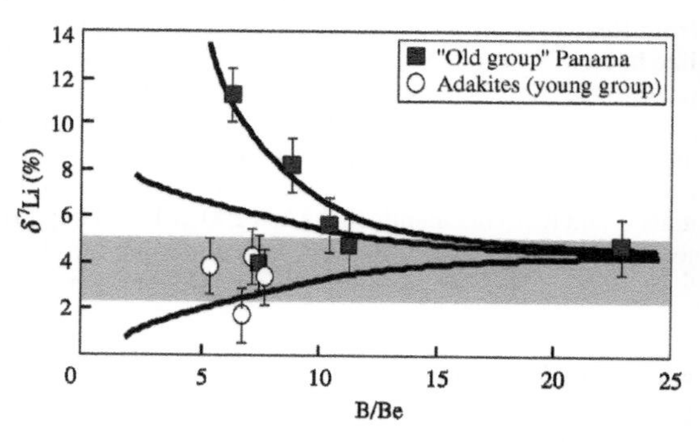

Figure 7 Plot of δ^7Li versus B/Be. Data points are from the Panama arc segment (Tomascak *et al.*, 2000); the shaded field represents data from the Sunda, Kurile, and Aleutian arcs (Tomascak *et al.*, 2002), and the northern Central American arc (Chan *et al.*, 2001). Curves represent schematic mixing arrays between a high B/Be slab component and a component (presumed to be the mantle) with variable Li-isotopic ratio. Li-isotopic ratios in most arcs span a very limited range (δ^7Li = +2 to +5‰), essentially indistinguishable from MORBs; the slab flux in Panama appears to have δ^7Li ≈ +5‰. Only where the slab influx is diminished do Li-isotopic variations appear; these may reflect heterogeneities in the mantle wedge, possibly accumulated through the time-integrated effects of slab devolatilization.

may occur very rapidly (Reagan *et al.*, 1994; Elliott *et al.*, 1997; Turner *et al.*, 2000a,b; Bourdon *et al.*, 1999; George *et al.*, 2003). This mechanism allows addition of subduction components to the upper mantle prior to the final addition and melting event. Upper mantle that had been subduction modified but did not receive a last gasp of fluid while in a hot part of the wedge may escape melting and could retain an extensive subduction signature indefinitely. Ultimately, that modified arc wedge within the upper mantle will be available to be tapped in other tectonic settings.

2.11.6 SUBDUCTION FLUXES AND MANTLE COMPOSITION

As noted earlier, quantitative element fluxes in subduction zones are tricky. They depend on magma production rates, which are often poorly known. They also depend on specific fluid-fluxing or sediment melting models, which implicitly or explicitly invoke a specific thermal structure and mineralogy for the slab, a subject of much debate.

This chapter uses the systematics of element behavior from trench to rear-arc in Section 2.11.4 to propose a relative sequence of efficiency with which elements are extracted from the downgoing plate: B > As, Sb, Cs > Pb > Rb > Ba, Sr, Be ~U, >Th. Based on cross-arc systematics and boron-isotope characteristics of MORB and OIB lavas, boron recycling appears to be approximately 100% between trench and rear arc. All subducted boron apparently winds up returned either to the oceans, or in the arc magmas that build continental crust. Based on data from the Aleutians and Nicaragua, beryllium recycling appears to be ~20–30%, beneath the volcanic front; this calculation depends on magma production rates but is independent of any specific transport models. If the ^{10}Be flux to rear-arc localities is significant, or if ^{10}Be spends appreciable time in the mantle prior to lava formation and eruption, then beryllium recycling will be higher than the rough estimate of 20–30%.

The extent to which elements such as potassium, uranium, thorium, and lead are either subducted to the deep mantle or extracted from the slab in the arc region is of major interest, having implications for the extent of internal heating possible in old subducted materials and for tracing their fate in the mantle. The Kurile cross-arc systematics strongly suggest that sediment lead is extracted completely beneath the volcanic front. In the altered basaltic crust, the lead budget is likely dominated by sulfide and oxide phases formed during hydrothermal redistribution of lead and other metals (e.g., Muhe *et al.*, 1997; Newsom *et al.*, 1986; Chauvel *et al.*, 1995; Peucker-Ehrenbrink *et al.*, 1994). Arsenic and antimony, also thought to be hosted primarily by oxide and sulfide phases (Noll *et al.*, 1996), are extracted with high efficiency, suggesting that lead too should be effectively extracted from the altered crust as well as the sediments. This interpretation is supported by lead-isotope studies of arc lavas, which require that some large fraction of lead be from the oceanic crust (e.g., Miller *et al.*, 1994; Class *et al.*, 2000). Rear-arc lavas have Ce/Pb ratios that are still distinct from those of MORB and OIB, suggesting that the slab at ~200 km depth still has lead to contribute to the mantle wedge. Taken together, these results suggest that lead recycling efficiency is closer to that of boron than beryllium, perhaps in the range of 50–75%.

Thorium appears to have recycling behavior broadly similar to beryllium (Elliott *et al.*, 1997; Johnson and Plank, 1999). Neither element shows significant depletion with increasing metamorphic grade in metasedimentary or metabasalt assemblages (Bebout *et al.*, 1999; Becker *et al.*, 2000). Their concentrations both increase across the width of the arc, but beryllium addition is less than thorium in the rear arc, relative to their concentrations at the front. Beryllium may sometimes be mobilized in a fluid beneath the volcanic front, but it is more effectively transported in sediment melts, including those feeding rear-arc volcanoes. ^{10}Be enrichment in the rear arc means that beryllium and sediment hosted elements that are equally or less incompatible during sediment melting will not be completely stripped from the slab beneath the front. Bulk partition coefficients for water-saturated sediment melting (Johnson and Plank, 1999) indicate that thorium (and cerium, samarium, neodymium, lutetium) is less incompatible than ^{10}Be during melting and does not partition effectively in fluids. Therefore, it is likely that less than 20–30% of these elements are extracted from the sediment. Elements such as caesium, rubidium, strontium, and uranium have partition coefficients for sediment melting that show incompatibility similar to beryllium, particularly at temperatures >800 °C. Rubidium and caesium show significant decreases in prograde metamorphic assemblages indicating fluid mobility up-dip of the volcanic front, which is particularly dramatic for caesium. Their combined solubility in fluids and partitioning into sediment melts suggests their total recycling is greater than that of beryllium. Uranium recycling is difficult to assess, but uranium excesses in arc magmas suggest fluid mobility beneath the arc, although Figure 4 suggests little mobility more shallowly. Together with sediment–melt partitioning similar to beryllium, uranium recycling may be comparable to that of beryllium, estimated at 20–30%.

This line of thought leads to the following estimates of the percentage of slab-hosted elements that are extracted from the downgoing plate from trench to rear arc: B (100%) > As, Sb, Cs > Pb (50–75%?) > Rb > Ba, Sr, Be (20–30%) ~ U > Th, Ce, Sm, Nd, Lu, Hf. As with any attempt to estimate subduction fluxes out of the slab, these estimates come with large, and perhaps unquantifiable uncertainties, given our limited knowledge about the temperature, mineralogy, mineral hosts for selected elements at depth, relative importance of sediment versus altered crust for each element budget, and fluid and melt partitioning behavior along the subducting slab. Even so, these results do suggest that elements that are very fluid mobile or both fluid mobile and incompatible during sediment melting could be stripped quite effectively from the slab, leaving little to subduct to the deeper mantle.

Conversely, elements that partition effectively only into sediment melts may experience less than 20–30% extraction from the slab in the arc region, with the rest subducting out of the arc window. Combined Be–U–Th systematics suggest that some fraction of slab-derived elements such as thorium, barium, and strontium could build up in the upper mantle wedge over time, being added in sediment melts to mantle that does not receive an additional increment of fluid while in a portion of the mantle wedge that is hot enough to melt. Poli and Schmidt (2002) propose that phengite is the dominant alteration phase in both the subducting sediment and AOC at pressures >2.5 GPa. With its large potassium site, this white mica is likely to be a major host of any LIL elements that subduct past the arc region. In the experiments cited, phengite plus topaz-OH and pumpellyite can be stable to relatively high temperatures and pressures (~900 °C and 7 GPa). If so, it is possible that many subducted elements bypass the volcanic front, only to be released to the upper mantle at depths ~200 km. Some fraction of such elements could be entrained by convection within the mantle wedge driven by coupling of the wedge to the downgoing slab (e.g., Kincaid and Sacks, 1997). Any incompatible elements that bypass the arc might wind up fertilizing the upper mantle beyond the arc. If so, their budget subducted to the deeper mantle would be significantly less than estimated above.

2.11.7 SUMMARY

This chapter attempts to look at the recycling behavior of subducted elements from the trench to the rear arc in the context of recent investigations of the thermal structure and prograde metamorphism of the arc, and the chemical signatures of lavas erupted behind the front above deeper subducting slabs. The goal of assessing the extent to which subducted elements are added to, and change the composition of, the upper mantle or are available for subduction to the deep mantle is very difficult via any approach. Difficulties include uncertainty about mineral hosts in subducting sediment and basalt for key elements, limited fluid and melt partitioning data, uncertainties about the P–T path of the downgoing slab and its prograde metamorphism, severely inadequate studies of magma production rates at convergent margins, limited data sets, and sometimes contradictory inferences drawn from studies of different margins. Even so, some general conclusions have been drawn throughout this chapter, using published and new results.

(i) Newer models for melt generation and localization of the volcanic front invoke fluid addition to a portion of the mantle located within a part of the convecting wedge that is hot enough to melt, rather than appealing to control by a single pressure-dependent dehydration reaction. A number of uranium-series studies indicate multiple stages of subduction modification of the mantle, with a last fluid addition rapidly triggering melting and lava eruption to the surface (perhaps within 8 ka).

(ii) For arcs with somewhat to greatly elevated ^{10}Be, such as the Kuriles, oceanic Aleutians, mid-Middle America (Guatemala, El Salvador and Nicaragua), Peru, Izu, Marianas, Bismarck, Tonga, and Kermadec arcs, nearly complete subduction of sediments to depth is required to provide sufficient ^{10}Be to the arcs. In several cases, lower ^{10}Be in the arc can be explained by either small amounts of sediment underplating or large amounts of subduction erosion. Globally, the mass of sediments subducted to the mantle currently is thus likely to be comparable to, or perhaps greater than, that carried to the trench. These values are large enough to agree with continental growth models that infer only a little or no net growth of the continents at present. Recent work, estimating arc magma production rates at nearly twice those previously used and highlighting potentially large amounts of subduction erosion, however, may change this balance considerably. In modeling sediment contributions to modern arcs, assuming complete sediment subduction to the mantle is a reasonable approximation, in the absence of large young accretionary prisms or specific ^{10}Be studies to the contrary.

(iii) Element distillation out of the subducting sediment column and the underlying altered oceanic crust can begin at relatively shallow levels in the subduction zone. For elements such as boron, arsenic, antimony, lead, caesium, potassium, barium, and rubidium, the composition of the slab delivered to the depths of magma generation may be quite different from that analyzed outboard of the trench.

(iv) Nearly 100% of the incoming boron appears to be extracted from the slab and returned to the ocean or crust. Where sediment subduction to depth (and the $^{10}Be/^{9}Be$ ratio of the sediment) is constrained by isotopic mixing relationships seen in arc lavas, the fraction of subducted ^{10}Be returned to the surface in arc lavas is ~20–30%. This estimate of the fraction of ^{10}Be extracted from the slab is dependent on a steady-state model, magma production rates and correction for ^{10}Be decay during subduction, but is independent of assumptions about melt versus fluid mobility, specific melting models or element partition coefficients.

(v) Using boron and beryllium as index elements, a relative order for the fraction of subducted elements extracted from the downgoing slab may be developed, constrained by cross-arc variation from trench to rear arc, element co-variations and sediment–melt partitioning data: B (100%) > Cs > As, Sb > Pb (50–75%?) > Rb > Ba, Sr, Be (20–30%) ~ U > Th, Ce, Sm, Nd, Lu, Hf. Any elements that remain in the slab will subduct beyond the arc window.

(vi) Lithium isotope, uranium-series and uranium-series–^{10}Be studies of arc lavas point to multiple stages of subduction modification of the mantle and indicate that elements such as thorium, strontium, and barium can build up in the upper mantle wedge. If mantle modified in an earlier event does not convect into a hot part of the mantle wedge, the subduction signature could remain indefinitely, available to be tapped by upper mantle melting in other tectonic settings.

(vii) If the potassium-muscovite phengite is the dominant host for LIL elements in both the metasedimentary and metabasaltic portions of the slab (Poli and Schmidt, 2002), then its stability to ~900 °C and 7 GPa and its breakdown behavior may mean that many subducted elements that bypass the arc are released to the upper mantle at depths ~225 km. If this is the case, the effects of subduction may be seen more profoundly in the composition of the upper mantle than the lower.

ACKNOWLEDGMENTS

The authors gratefully acknowledge collaborations and discussions with a number of colleagues, which have led to the contribution. These include Fouad Tera, Gray Bebout, Laurie Benton, Louis Brown, Roy Middleton, Jeff Klein, Marc Cafee, Bill Leeman, Mark Reagan, Jim Gill, Olegeir Sigmarsson, Rhiannon George, Simon Turner, Chris Hawkesworth, Terry Plank, Robby Valentine, Robyn Kelly, Chris Kincaid, Selwyn Sacks, Ivan Savov, Paul Tomascak, and Doug Wiens. The research efforts of USF undergraduate students Patrick Mattie, Suzie Norrell, and Steve Guggino provided key data for aspects of what has been presented here. This work was supported in part by grants to Morris from NSF-OCE and from JOI-USSAC, and by grants to Ryan from NSF (EAR, OCE, and REU) and JOI-USSAC. Review and editorial handling by Rick Carlson improved the quality of the manuscript.

REFERENCES

Armstrong R. L. (1981) Radiogenic isotopes: the case for crustal recycling on a near-steady-state no-continental-growth Earth. *Phil. Trans. Roy. Soc. London A* **301**, 443–472.

Bebout G. (1996) Volatile transfer and recycling at convergent margins: mass balance and insights from metamorphic rocks. In *Subduction: Top to Bottom*, Am. Geophys. Union Monogr. 96 (eds. G. E. Bebout, D. W. Scholl, S. H. Kirby, and J. P. Platt). American Geophysical Union, Washington, DC, pp. 179–193.

Bebout G. E., Ryan J. G., and Leeman W. P. (1993) B–Be systematics in subduction-related metamorphic rocks: characterization of the subducted component. *Geochim. Cosmochim. Acta* **57**, 2227–2237.

Bebout G. E., Ryan J. G., Leeman W. P., and Bebout A. E. (1999) Fractionation of trace elements by subduction-zone metamorphism: effect of convergent-margin thermal evolution. *Earth Planet. Sci. Lett.* **171**, 63–81.

Becker H. (2000) Re–Os fractionation in eclogites and blueschists and the implications for recycling of oceanic crust into the mantle. *Earth Planet. Sci. Lett.* **177**, 287–300.

Becker H., Jochum K. P., and Carlson R. W. (2000) Trace element fractionation during dehydration of eclogites from high-pressure terranes and the implications for element fluxes in subduction zones. *Chem. Geol.* **163**, 65–99.

Benton L. D., Ryan J. G., and Tera F. (2001) Boron isotope systematics of slab fluids as inferred from a serpentine seamount, Mariana forearc. *Earth Planet. Sci. Lett.* **187**, 273–282.

Bourdon B., Turner S., and Allègre C. (1999) Melting dynamics beneath the Tonga–Kermadec island arc inferred from $^{231}Pa–^{235}U$ systematics. *Science* **286**, 2491–2493.

Brenan J. M., Shaw H. F., Ryerson F. J., and Phinney D. L. (1995) Mineral-aqueous fluid partitioning of trace elements at 900 °C and 2.0 GPa: constraints on the trace element geochemistry of mantle and deep crustal fluids. *Geochim. Cosmochim. Acta* **59**, 3331–3350.

Chan L. H. and Frey F. A. (2003) Lithium isotope geochemistry of the Hawaiian plume: results from the Hawaii Scientific Drilling Project and Koolau Volcano. *Geochem. Geophys. Geosyst.* **4**, article 2002GC000365.

Chan L. H. and Kastner M. K. (2000) Lithium isotopic compostions of pore fluids and sediments in the Costa Rica subduction zone: implications for fluid processes and sediment contribution to the arc volcanoes. *Earth Planet. Sci. Lett.* **18**, 275–290.

Chan L.-H., Edmond J. M., and Thompson G. (1993) A lithium isotope study of hot springs and metabasalts from mid-ocean ridge hydrothermal systems. *J. Geophys. Res.* **98**, 9653–9659.

Chan L. H., Leeman W. P., and You C. F. (2001) Lithium isotopic composition of central American volcanic arc lavas: implications for modification of the subarc mantle by slab-derived fluids: correction. *Chem. Geol.* **182**, 293–300.

Chaussidon M. and Jambon A. (1994) Boron content and isotopic composition of oceanic basalts: geochemical and cosmochemical implications. *Earth Planet. Sci. Lett.* **121**, 277–291.

Chaussidon M. and Marty B. (1995) Primitive boron isotope composition of the mantle. *Science* **269**, 383–386.

Chauvel C., Goldstein S. L., and Hofmann A. W. (1995) Hydration and dehydration of oceanic crust controls Pb evolution in the mantle. *Chem. Geol.* **126**, 65–75.

Class C., Miller D. M., Goldstein S. L., and Langmuir C. L. (2000) Distinguishing melt and fluid subduction components in Umnak volcanics, Aleutian arc. *Geochem. Geophys. Geosyst.* **1**, paper number 1999GC000010.

Condomines M. and Sigmarsson O. (1993) Why are so many arc magmas close to $^{238}U-^{230}Th$ radioactive equilibrium? *Geochim. Cosmochim. Acta* **57**, 4491–4497.

Deyhle A. and Kopf A. (2002) Strong B enrichment and anomalous $\delta^{11}B$ in pore fluids from the Japan Trench forearc. *Mar. Geol.* **183**, 1–15.

Dimalanta C., Taira A., Yumul G. P., Jr., Tokuyama H., and Mochizuki K. (2002) New rates of western Pacific island arc magmatism from seismic and gravity data. *Earth Planet. Sci. Lett.* **202**, 105–115.

Domanik K. J. and Holloway J. R. (1996) The stability and composition of phengitic muscovite and associated phases from 5.5 to 11 GPa: implications for deeply subducted sediments. *Geochim. Cosmochim. Acta* **60**, 4133–4150.

Domanik K. J., Hervig R. L., and Peacock S. M. (1993) Beryllium and boron in subduction zone minerals: an ion microprobe study. *Geochim. Cosmochim. Acta* **57**, 4997–5010.

Edwards C. M. H., Morris J. D., and Thirwall M. F. (1993) Separating mantle from slab signatures in arc lavas using B/Be and radiogenic isotope systematics. *Nature* **362**, 530–533.

Elliott T. (1997) Fractionation of U and Th during melting: a reprise. *Chem. Geol.* **139**, 165–183.

Elliott T., Plank T., Zindler A., White W., and Bourdon B. (1997) Element transport from slab to volcanic front at the Mariana arc. *J. Geophys. Res.* **102**, 14991–15019.

Fryer P. and Fryer G. J. (1987) Origins of nonvolcanic seamounts in a forearc environment. In *Seamounts, Islands and Atolls*, Geophysical Monograph Series 43 (eds. B. H. Keating, P. Fryer, R. Batiza, and G. W. Boelert). American Geophysical Union, Washington, DC, pp. 61–69.

Fryer P. and Mottl M. (1992) Lithology, mineralogy, and origin of serpentine muds recovered from Conical and Torishima forearc seamounts: results from Leg 125 drilling. *Proc. ODP Sci. Results* **125**, 343–362.

Fryer P., Wheat C. G., and Mottl M. J. (1999) Mariana blueschist mud volcanism: implications for conditions within the subduction zone. *Geology* **27**, 103–106.

George R., Turner S., Hawkesworth C. J., Morris J., Nye C., Ryan J. G., and Zheng S. H. (2003) Melting processes and fluid and sediment transport rates along the Alaska–Aleutian arc from an integrated U–Th–Ra–Be isotope study. *J. Geophys. Res.* **108** (B5), 2252, 10.1029/2002JB001916.

Gill J. B. and Williams R. W. (1990) Th isotope and U-series studies of subduction-related volcanic rocks. *Geochim. Cosmochim. Acta* **54**, 1427–1442.

Gill J. B., Morris J. D., and Johnson R. W. (1993) Timescale for producing and removing the geochemical signature of island arcs: U–Th–Po and B–Be systematics in recent Papua New Guinea magmas. *Geochim. Cosmochim. Acta* **57**, 4269–4283.

Guggino S., Savov I. P., and Ryan J. G. (2002) Light element systematics of metamorphic clasts from ODP Legs 125 and 195, South Chamorro and Conical Seamounts, Mariana forearc. *EOS, Trans., AGU* **83**(19), abstract V51A-08.

Gurenko A. A. and Chaussidon M. (1997) Boron concentrations and isotopic compositions of the Icelandic mantle: evidence from glass inclusions in olivine. *Chem. Geol.* **135**, 21–34.

Haggerty J. A. (1991) Evidence from fluid seeps atop serpentine seamounts in the Mariana forearc: clues for emplacement of the seamounts and their relationship to forearc tectonics. *Mar. Geol.* **102**, 293–309.

Herrstrom E. A., Reagan M. K., and Morris J. D. (1995) Variations in lava composition associated with flow of asthenosphere beneath southern Central America. *Geology* **23**, 617–620.

Hofmann A. W. (1998) Chemical differentiation of the earth: the relationship between mantle, continental crust and oceanic crust. *Earth Planet. Sci. Lett.* **90**, 297–314.

Ishikawa T. and Nakamura E. (1992) Boron-isotope geochemistry of the oceanic crust from DSDP/ODP Hole 504B. *Geochim. Cosmochim. Acta* **56**, 1633–1639.

Ishikawa T. and Nakamura E. (1993) Boron-isotope systematics of marine sediments. *Earth Planet. Sci. Lett.* **117**, 567–580.

Ishikawa T. and Nakamura E. (1994) Origin of the slab component inferred in arc lavas from across-arc variation of B and Pb isotopes. *Nature* **370**, 205–208.

Ishikawa T. and Tera F. (1997) Source composition and distribution of the fluid in the Kurile mantle wedge: constraints from across-arc variations of B/Nb and B isotopes. *Earth Planet. Sci. Lett.* **152**, 123–138.

Ishikawa T. and Tera F. (1999) Two isotopically distinct fluid components involved in the Mariana arc: evidence from Nb/B ratios and B, Sr, Nd and Pb isotope systematics. *Geology* **27**, 83–86.

Iwamori H. (1998) Transportation of H_2O and melting in subduction zones. *Earth Planet. Sci. Lett.* **160**, 65–80.

Johnson M. C. and Plank T. (1999) Dehydration and melting experiments constrain the fate of subducted sediments. *Geochem. Geophys. Geosyst.* **1**, paper number 1999GC000014.

Kastner M., Morris J., Chan L. H., Saether O., and Luckge A. (2000) Three distinct fluid systems at the Costa Rica subduction zone: chemistry, hydrology, and fluxes, Goldschmidt 2000. *J. Conf. Abstr.* **5**, 572.

Kay R. W. (1980) Volcanic arc magmas: implications of a melting-mixing model for element recycling in the crust–upper mantle system. *J. Geol.* **88**, 497–522.

Kelly K. A., Plank T., Ludden J., and Staudigel H. (2003) Composition of altered oceanic crust at ODP Sites 801 and 1149. *Geochem. Geophys. Geosys.* **4**, no. 6, 8910. doi: 10.1029/2002GC000435.

Kelly R., Morris J., Driscoll N., McIntosh K., and Silver E. Graben development and sediment subduction in Nicaragua and Costa Rica. *Geology* (in review).

Kincaid C. and Sacks I. S. S. (1997) Thermal and dynamical evolution of the upper mantle in subduction zones. *J. Geophys. Res.* **102**, 12295–12315.

Kopf A. and Deyhle A. (2002) Back to the roots: boron geochemistry of mud volcanoes and its implications for mobilization depth and global B cycling. *Chem. Geol.* **192**, 195–210.

Leeman W. P. (1996) Boron and other fluid-mobile elements in volcanic arc lavas: implications for subduction processses. In *Subduction: Top to Bottom*. Am. Geophys. Union Monogr. 96 (eds. G. E. Bebout, D. W. Scholl, S. H. Kirby, and J. P. Platt). American Geophysical Union, Washington, DC, pp. 269–276.

Leeman W. P., Smith D. R., Hildreth W., Palacz Z., and Rogers N. W. (1990) Compositional diversity in Late Cenozoic basalts in a transect across the southern Washington Cascades. *J. Geophys. Res.* **95**, 19561–19582.

Lin P. N. (1992) Trace element and isotopic characteristics of western Pacific pelagic sediments: implications for the petrogenesis of Mariana arc magmas. *Geochim. Cosmochim. Acta* **56**, 1641–1654.

McDonough W. F. and Sun S. S. (1995) The composition of the Earth. *Chem. Geol.* **120**, 223–253.

McLennan S. M. (1988) Recycling of the continental crust. *Pure Appl. Geophys.* **128**, 683–724.

Meschede M., Zweigal P., and Kiefer E. (1999) Subsidence and extension at a convergent plate margin: evidence for subduction erosion off Costa Rica. *Terra Nova* **11**, 112–117.

Miller D. M., Goldstein S. L., and Langmuir C. H. (1994) Ce/Pb and Pb isotope ratios in arc magmas and the enrichment of Pb in the continents. *Nature* **368**, 514–520.

Moran A. E., Sisson V. B., and Leeman W. P. (1992) Boron depletion during progressive metamorphism: implications for subduction processes. *Earth Planet. Sci. Lett.* **111**, 331–349.

Morris J. D., Leeman W. P., and Tera F. (1990) The subducted component in island arc lavas: constraints from Be isotopes and B–Be systematics. *Nature* **344**, 31–36.

Morris J., Valentine R., and Harrison T. (2002a) ^{10}Be imaging of sediment accretion, subduction and erosion, NE Japan and Costa Rica. *Geology* **30**, 59–62.

Morris J., Gosse J., Brachfeld S., and Tera F. (2002b) Cosmogenic ^{10}Be and the solid earth: studies in active tectonics, geomagnetism and subduction zone processes. In *Reviews in Mineralogy* (ed. E. Grew). Mineralogical Society of America, Washington, DC, vol. 50, pp. 207–270.

Mottl M. J. (1992) Pore waters from serpentine seamounts in the Mariana and Izu-Bonin Forearcs, Leg 125: evidence for volatiles from the subducting slab. *Proc. ODP Sci. Results* **125**, 373–385.

Muhe R., Peucker-Ehrenbrink B., Devey C., and Garbe-Schonberg D. (1997) On the redistribution of Pb in the oceanic crust during hydrothermal alteration. *Chem. Geol.* **137**, 67–77.

Newsom H. E., White W. M., Jochum K. P., and Hofmann A. W. (1986) Siderophile and chalcophile element abundances in oceanic basalts: Pb isotope evolution and growth of the Earth's core. *Earth Planet. Sci. Lett.* **80**, 299–313.

Noll P. D., Jr., Newsom H. E., and Leeman W. P. (1993) Lead and tin in subduction related lavas: fluid mobility, cross-arc variations and the origin of Pb enriments in the continental crust. *EOS, Trans., AGU Suppl.* **74**, 674.

Noll P. D., Jr., Newsom H. E., Leeman W. P., and Ryan J. G. (1996) The role of hydrothermal fluids in the production of subduction zone magmas: evidence from siderophile and chalcophile trace elements and boron. *Geochim. Cosmochim. Acta* **60**, 587–611.

Ono S. (1998) Stability limits of hydrous minerals in sediment and mid-ocean ridge basalt compositions: implications for water transport in subduction zones. *J. Geophys. Res.* **103**, 18253–18267.

Palmer M. R. (1991) Boron-isotope systematics of Halmahera arc (Indonesia) lavas: evidence for involvement of the subducted slab. *Geology* **19**, 215–217.

Patino L. C., Carr M. J., and Feigenson M. D. (2000) Local and regional variations in Central American arc lavas controlled by variations in subducted sediment input. *Contrib. Mineral. Petrol.* **138**, 265–283.

Peacock S. M. (1996) Thermal and petrologic structure of subduction zones. In *Subduction: Top to Bottom*, Am. Geophys. Union Monogr. 96 (eds. G. E. Bebout, D. W. Scholl, S. H. Kirby, and J. P. Platt). American Geophysical Union, Washington, DC, pp. 119–133.

Peacock S. M. (2001) Are the lower planes of double seismic zones caused by serpentine dehydration in subducting oceanic mantle? *Geology* **29**, 299–302.

Peacock S. M. and Wang K. (1999) Seismic consequences of warm versus cool subduction metamorphism: examples from Southwest and Northeast Japan. *Science* **286**, 937–939.

Peucker-Ehrenbrink B., Hormann A. W., and Hart S. R. (1994) Hydrothermal lead transfer from mantle to continental crust: the role of metalliferous sediments. *Earth Planet. Sci. Lett.* **125**, 129–142.

Phipps-Morgan J., Ruepke L. H., Ranero C., and Hort M. (2002) Some geophysical and geochemical consequences of slab serpentinization at subduction zones. *EOS, Trans., AGU Suppl.* **83**, abstract T51-E-13.

Plank T. and Kelley K. (2001) Contrasting sediment input and output at the Izu and Mariana subduction factories. *EOS, Trans., AGU* **82** Fall Meet. Suppl. Abstract T22D-10.

Plank T. and Langmuir C. H. (1993) Tracing trace elements from sediment input to volcanic output at subduction zones. *Nature* **362**, 739–743.

Plank T. and Langmuir C. H. (1998) The chemical composition of subducting sediment and its consequences for the crust and mantle. *Chem. Geol.* **145**, 325–394.

Poli S. and Schmidt M. W. (1995) H_2O Transport and release in subduction zones: experimental constraints on basaltic and andesitic systems. *J. Geophys. Res.* **100**, 22299–22314.

Poli S. and Schmidt M. W. (2002) Petrology of subducted slabs. *Ann. Rev. Earth Planet. Sci.* **30**, 307–335.

Ranero C. R. and von Huene R. (2000) Subduction erosion along the Middle America convergent margin. *Nature* **404**, 748–752.

Ranero C. R., von Huene R., Flueh E., Duarte M., Baca D., and McIntosh K. D. (2000) A cross section of the convergent Pacific margin of Nicaragua. *Tectonics* **19**, 335–357.

Rea D. K. and Ruff L. J. (1996) Composition and mass flux of sediment entering the worlds subduction zones: implications for global sediment budgets, great earthquakes and volcanism. *Earth Planet. Sci. Lett.* **140**, 1–12.

Reagan M. K., Morris J. D., Herrstrom E. A., and Murrell M. T. (1994) Uranium series and beryllium isotope evidence for an extended history of subduction modification of the mantle below Nicaragua. *Geochim. Cosmochim. Acta* **58**, 4199–4212.

Regelous M., Collerson K. D., Ewart A., and Wendt J. I. (1997) Trace element transport rates in subduction zones: evidence from Th, Sr and Pb isotope data for Tonga–Kermadec arc lavas. *Earth Planet. Sci. Lett.* **150**, 291–302.

Reymer A. and Schubert G. (1984) Phanerozoic addition rates to the continental crust and continental growth. *Tectonics* **3**, 63–77.

Ryan J. G. and Kyle P. R. (2000) Lithium isotope systematics of McMurdo Volcanic Group lavas, and other intraplate sites. *EOS, Trans., AGU Suppl.* **81**, F1371.

Ryan J. G., Morris J., Tera F., Leeman W. P., and Tsvetkov A. (1995) Cross-arc geochemical variations in the Kurile arc as a function of slab depth. *Science* **270**, 625–627.

Ryan J. G., Morris J., Bebout G., and Leeman W. (1996a) Describing chemical fluxes in subduction zones: insight from "depth profiling" studies of arc and forearc rocks. In *Subduction: Top to Bottom*, Am. Geophys. Union Monogr. 96 (eds. G. E. Bebout, D. W. Scholl, S. H. Kirby, and J. P. Platt). American Geophysical Union, Washington, DC, pp. 263–268.

Ryan J. G., Leeman W. P., Morris J. D., and Langmuir C. H. (1996b). The boron systematics of intraplate lavas: implications for crust and mantle evolution. *Geochim. Cosmochim. Acta* **60**, 415–422.

Saboda K. L., Fryer P., and Maekawa H. (1992) Metamorphism of ultramafic clasts from Conical Seamount, sites 778, 778, and 780. *Proc. ODP Sci. Results* **125**, 431–443.

Salisbury M. H., Shinohara M., Richter C., *et al.* (2002) *Proc. ODP, Init. Repts.*, 195 [CD-ROM]. Available from: Ocean Drilling Program, Texas A&M University, College Station TX 77845-9547, USA.

Savov I. P., Ryan J. G., Chan L.-H., D'Antonio M., Mottl M., and Fryer P. (2002) Geochemistry of Serpentinites from the S. Chamorro Seamount, ODP Leg 195, Site 1200, Mariana Forearc-Iimplications For Recycling at Subduction Zones. In *Proc. V.M. Goldschmidt Conference, Davos, Switzerland.* Geochemical Society.

Schmidt M. W. and Poli S. (1998) Experimentally based water budgets for dehydrating slabs and consequences for arc magma generation. *Earth Planet. Sci. Lett.* **163**, 361–379.

Sigmarsson O., Condomines M., Morris J. D., and Harmon R. S. (1990) Uranium and ^{10}Be enrichments by fluids in Andean arc magmas. *Nature* **346**, 163–165.

Sigmarsson O., Chmeleff J., Morris J., and Lopez-Escobar L. (2002) Origin of $^{226}Ra-^{230}Th$ disequilibrium in arc lavas from southern Chile and implications for magma transfer time. *Earth Planet. Sci. Lett.* **196**, 189–196.

Smith H. J., Spivack A. J., and Hart S. R. (1995) The boron isotopic composition of altered oceanic crust. *Chem. Geol.* **126**, 119–135.

Smith H. J., Leeman W. P., Davidson J., and Spivack A. J. (1997) The B isotopic composition of arc lavas from Martinique, Lesser Antilles. *Earth Planet. Sci. Lett.* **146**, 303–314.

Spivack A. J. and Edmond J. M. (1987) Boron isotope exchange between seawater and the oceanic crust. *Geochim. Cosmochim. Acta* **51**, 1033–1043.

Spivack A. J., Palmer M. R., and Edmond J. M. (1987) The sedimentary cycle of the boron isotopes. *Geochim. Cosmochim. Acta* **51**, 1939–1949.

Staudigel H., Plank T., White W., and Schmincke H.-U. (1996) Geochemical fluxes during seafloor alteration of the basaltic upper oceanic crust: DSDP Sites 417 and 418. In *Subduction: Top to Bottom*, Am. Geophys. Union Monogr. 96 (eds. G. E. Bebout, D. W. Scholl, S. H. Kirby, and J. P. Platt). American Geophysical Union, Washington, DC, pp. 19–38.

Straub S. M. and Layne G. D. (2002) The systematics of boron isotopes in Izu arc front volcanic rocks. *Earth Planet Sci. Lett.* **198**, 25–39.

Tatsumi Y. and Eggins S. (1997) *Subduction Zone Magmatism*. Blackwell, London, 211pp.

Tatsumi Y., Sakuyama M., Fukuyama H., and Kushiro I. (1983) Generation of arc basalt magmas and thermal structure of the mantle wedge in subduction zones. *J. Geophys. Res.* **88**, 5815–5825.

Tera F., Brown L., Morris J., Sacks I. S., Klein J., and Middleton R. (1986) Sediment incorporation in island arc magmas: inferences from Be-10. *Geochim. Cosmochim. Acta* **50**, 535–550.

Tera F., Morris J. D., Ryan J., Leeman W. P., and Tsvetkov A. (1993) Significance of $^{10}Be/^{9}Be$–B correlations in lavas of the Kurile–Kamchatka arc. *EOS, Trans., AGU* **74**, 674.

Tomascak P. B., Tera F., Helz R. T., and Walker R. J. (1999) The absence of lithium isotope fractionation during basalt differentiation: new measurements by multi-collector sector ICP-MS. *Geochim. Cosmochim. Acta* **63**, 907–910.

Tomascak P. B., Ryan J. G., and Defant M. J. (2000) Lithium isotopes and light elements depict incremental slab contributions to the subarc mantle in Panama. *Geology* **28**, 507–510.

Tomascak P. B., Widom E., Benton L. J., Goldstein S. J., and Ryan J. G. (2002) The control of lithium budgets in island arcs. *Earth Planet. Sci. Lett.* **196**, 227–238.

Turner S., Hawkesworth C., Rogers N., Bartlett J., Worthington T., Hergt J., Pearce J., and Smith I. (1997) $^{238}U-^{230}Th$ disequilibria, magma petrogenesis and flux rates beneath the depleted Tonga–Kermadec island arc. *Geochim. Cosmochim. Acta* **61**, 4855–4884.

Turner S., Bourdon B., Hawkesworth C., and Evans P. (2000a) $^{226}Ra-^{230}Th$ evidence for multiple dehydration events, rapid melt ascent and the timescales of differentiation beneath the Tonga–Kermadec island arc. *Earth Planet. Sci. Lett.* **179**, 581–593.

Turner S. P., George R. M. M., Evans P. J., Hawkesworth C. J., and Zellmer G. F. (2000b) Time-scales of magma formation, ascent and storage beneath subduction-zone volcanoes. *Phil. Trans. Roy. Soc. London A* **358**, 1443–1464.

Turner S., Evans P., and Hawkesworth C. (2001) Ultra-fast source-to-surface movement of melt at island arcs from $^{226}Ra-^{230}Th$ systematics. *Science* **292**, 1363–1366.

Ulmer P. and Trommsdorff V. (1999) Phase relations of hydrous mantle subducting to 300 km. In *Mantle Petrology: Field Observations and High Pressure Experimentation: A Tribute to Francis R. (Joe) Boyd*, Spec. Publ. 6 (eds. Y. Fei, C. Bertka, and B. O. Mysen). Geochem. Soc, Houston.

Valentine R. B., Morris J. D., Ryan J. G., Kelley K. A., and Zheng S. H. (2002) ^{10}Be systematics of incoming sediments and arc lavas at the Izu and Mariana convergent margins. *EOS, Trans., AGU Suppl.* **83**.

Valentine R., Morris J., Zheng S. H., and Cardace D. Sediment accretion, erosion and subduction along the Costa Rica convergent margin: constraints provided by cosmogenic ^{10}Be. *J. Geophys. Res.* (in revision).

van Keken P. E., Kiefer B., and Peacock S. M. (2002) High-resolution models of subduction zones: implications for mineral dehydration reactions and the transport of water into the deep mantle. *Geochem. Geophys. Geosyst.* **1** 2001GC000256.

Vannucchi P., Scholl D. W., Meschede M., and McDougall-Reid K. (2001) Tectonic erosion and consequent collapse of the Pacific margin of Costa Rica: combined implications from ODP Leg 170, seismic offshore data, and regional geology of the Nicoya Peninsula. *Tectonics* **20**, 649–668.

Veizer J. and Jansen S. L. (1985) Basement and sedimentary recycling: 2. Time dimension to global tectonics. *J. Geol.* **93**, 625–643.

von Huene R. and Culotta R. (1989) Tectonic erosion at the front of the Japan Trench convergent margin. *Tectonophysics* **160**, 75–90.

von Huene R. and Lallemand S. (1990) Tectonic erosion along the Japan and Peru convergent margins. *Geol. Soc. Am. Bull.* **102**, 704–720.

von Huene R. and Scholl D. (1991) Observations at convergent margins concerning sediment subduction, subduction erosion and the growth of the continental crust. *Rev. Geophys.* **29**, 279–316.

Walther C. H., Flueh E., Ranero C. R., von Huene R., and Strauch W. (2000) An unusual crustal structure across the Pacific Margin of Nicaragua. *Geophys. J. Inter.* **141**, 759–777.

Yiou F. and Raisbeck G. M. (1972) Half-life of ^{10}Be. *Phys. Rev. Lett.* **29**, 372–375.

You C.-F., Morris J. D., Geiskes J. M., Rosenbauer R., Zheng S. H., Xu X., Ku T. L., and Bischoff J. L. (1994) Mobilization of beryllium in the sedimentary column at convergent margins. *Geochim. Cosmochim. Acta* **58**, 4887–4897.

You C.-F., Castillo P., Geiskes J. M., Chan L. C., and Spivack A. F. (1996) Trace element behavior in hydrothermal experiments: implications for fluid processes at shallow depths in subduction zones. *Earth Planet. Sci. Lett.* **140**, 41–52.

2.12
Convective Mixing in the Earth's Mantle

P. E. Van Keken

University of Michigan, Ann Arbor, MI, USA

C. J. Ballentine

University of Manchester, UK

and

E. H. Hauri

Carnegie Institution of Washington, DC, USA

NOMENCLATURE

c_p	heat capacity
dx	distance between two points after time t
dX	original infinitesimal distance between two points
$\boldsymbol{D}$	stretching tensor
Di	dissipation number
$\hat{\boldsymbol{g}}$	gravity vector
h	depth of convecting layer
$H(r)$	two-particle correlation function
P	dynamic pressure
Ra	thermal Rayleigh number
t	time
T	temperature
u_i	components of the velocity tensor
$\boldsymbol{v}$	velocity
w	upward velocity
α	thermal expansivity
$\dot{\boldsymbol{\epsilon}}$	strain-rate tensor
ϵ_μ	mixing efficiency
η	dynamic viscosity
κ	thermal diffusivity
λ	Luyaponov exponent
μ	infinitesimal stretching length

ρ_0 reference density
σ_{ij} components of the stress tensor
Φ viscous dissipation
ΔT temperature contrast across convecting layer
∇v velocity gradient
$\mathbf{\Omega}$ vorticity tensor

2.12.1 INTRODUCTION

The observed geochemical diversity of mid-oceanic ridge basalts (MORBs) and ocean island basalts (OIBs) is generally attributed to the existence of large-scale and long-term chemical heterogeneity in the Earth's silicate mantle (see Chapter 2.03). Yet, it is far from clear how this heterogeneity is generated and how it is maintained in the convecting mantle. Quantitative mixing studies indicate that the present-day Earth is convecting vigorously enough to erase significant initial heterogeneity well within the age of the Earth. This suggests that some form of layering, or barrier against convective mixing is required to explain the geochemical observations. The most basic form of layering is that of the "classical" two-layer mantle, where a depleted and well-mixed upper mantle is separated by the seismically distinct 670 km boundary from a poorly mixed and enriched lower mantle. The MORBs originate from melting of the upper mantle, whereas OIBs derive from melting of material that is brought up from the lower mantle by plumes. This model has worked well to explain a large number of observations regarding noble gas and trace element concentrations and the distribution of mantle heat sources. However, in recent years this model has come under siege. Geophysical and geodynamical observations indicate that significant mass exchange occurs through the Earth's transition zone. In addition, various geochemical observations suggest an important role for oceanic crust recycling in the plume source and demonstrate the lack of preservation of primitive mantle. The recent widespread acceptance of these fundamental problems of the classical layered mantle model has led the proposition of various alternatives such as the presence of layering below 670 km depth, or the preservation of heterogeneity in highly viscous regions in the Earth's mantle. These models appear to be able to explain one or more features better than the classical model, but often cause new conflicts with existing geochemical or geophysical observations. In addition, it is not always clear that these new conceptual models are physically realistic. With the advent of large-scale computing, geodynamical modeling has become a particularly useful tool in this arena. Using the fundamental laws of the conservation of mass, energy, and momentum, models of mantle convection can be created that allow for quantitative tests of these conceptual models. Modeling also allows for a better understanding of the physics that governs mantle flow, mantle mixing, and the distribution of chemical heterogeneity in planetary interiors.

This chapter will focus on the use of mantle convection modeling in the development of our understanding of the chemical evolution of the Earth by providing a short review of the main observations, a discussion of the physical approaches to characterize mantle mixing, and an overview of the historical and current modeling approaches to the formation, preservation, and destruction of chemical heterogeneity. Detailed reviews of the geochemical data and interpretations can be found in Zindler and Hart (1986), Silver *et al.* (1988), Carlson (1994), Hofmann (1997), Van Keken *et al.* (2002), Porcelli and Ballentine (2002), and Hauri (2002), and see Chapter 2.03.

2.12.2 GEOCHEMICAL AND GEOPHYSICAL OBSERVATIONS OF MANTLE HETEROGENEITY

2.12.2.1 Geochemical Observations Suggesting Mantle Layering

The formation of basalts by partial melting of the upper mantle at mid-oceanic ridges and hot spots provides the opportunity to determine mantle composition. Early studies of radiogenic isotopes in oceanic basalts (e.g., Faure and Hurley, 1963; Hart *et al.*, 1973; Schilling, 1973) showed fundamental chemical differences between OIBs and MORBs (see Chapter 2.03). This led to the development of the layered mantle model, which consists essentially of three different reservoirs: the lower mantle, upper mantle, and continental crust. The lower mantle is assumed primitive and identical to the bulk silicate earth (BSE), which is the bulk earth composition minus the core (see also Chapters 2.01 and 2.03). The continental crust is formed by extraction of melt from the primitive upper mantle, which leaves the depleted upper mantle as third reservoir. In this model, MORB is derived from the depleted upper mantle, whereas OIB is formed from reservoirs derived by mixing of the MORB source with primitive mantle (e.g., DePaolo and Wasserburg, 1976; O'Nions *et al.*, 1979; Allègre *et al.*, 1979).

In particular, the study of noble gas isotopes has provided a compelling case for mantle layering and the preservation of primitive mantle (see Chapter 2.06). For example, $^3He/^4He$ values for MORBs are nearly uniform, but large departures are seen for OIBs. Most oceanic hot spots have elevated $^3He/^4He$ values compared to MORBs. Helium-4 is generated by the decay of uranium and thorium. Helium-3 is not generated inside the Earth and is lost from the atmosphere upon degassing. The presence of high $^3He/^4He$ values requires, therefore, the isolation of the plume reservoir from surface melting. It has been common to equate the high $^3He/^4He$ "reservoir" with the primitive mantle. Mass flux considerations add weight to the argument for mantle layering. For example, the inferred mass flux of ^4He from the mantle is far lower than that predicted from the decay of uranium and thorium in the whole mantle (O'Nions and Oxburgh, 1983), which suggests the existence of a boundary layer that decouples the flow of helium and heat from the lower mantle. Steady-state models for lead and the noble gases similarly limit the mass exchange between upper and lower mantle (Galer and O'Nions, 1985; Kellogg and Wasserburg, 1990; O'Nions and Tolstikhin, 1994; Porcelli and Wasserburg, 1995a,b).

Further suggestions for compositional stratification come from the consideration of the Earth's heat budget. The combined heat production of the upper mantle and continental crust, based on estimates of radiogenic heat production of the crust and MORB source (Rudnick and Fountain, 1995; Van Schmus, 1995), strongly suggests that radiogenic element concentrations in the deep mantle are significantly higher than those of the upper mantle.

2.12.2.2 Problems with the Classical Layered Model

Several lines of evidence indicate problems with strict application of the layered model. It is particularly difficult to satisfy the requirement for primitive composition of the lower mantle. For example, Zindler and Hart (1986) showed that the isotopic database for the oceanic mantle could be described by mixing between depleted MORB mantle and enriched components that have compositions quite unlike primitive mantle. Similarly, the constancy of Nb/U and Ce/Pb (Hofmann, 1986; Newsom *et al.*, 1986) suggests that there is no current reservoir with primitive mantle ratios. Interestingly, there is a strong correlation between a depleted component, termed FOZO by Hart *et al.* (1992), and the high ^3He source observed in many hot spots (Hart *et al.*, 1992; Farley *et al.*, 1992; Hanan and Graham, 1996).

In recent years, we have seen the use of new isotopic systems, including those of osmium and oxygen, in the description of oceanic basalts. Osmium isotopes can trace the addition of mafic crust or melt to the mantle source (Hauri and Hart, 1993) and have been used to highlight the presence of recycled mafic crust in sources of hot spots (Shirey and Walker, 1998). In a complimentary fashion, oxygen isotope variations trace directly portions of the mantle that have interacted with water at or near the Earth's surface, which allows for identifying recycled oceanic crust in the sources of both OIBs and MORBs (Eiler *et al.*, 1997, 2000; Lassiter and Hauri, 1998; Hauri, 2002). It has become apparent that nearly all hot spots display elevated $^{187}Os/^{188}Os$, which requires involvement of mafic sources, most likely through the recycling of oceanic crust into the deep mantle. Mass balance calculations suggest the presence of a recycled oceanic crust reservoir of up to 10% of the mantle mass (Hauri and Hart, 1997). The strong evidence for the recycling of oceanic crust in the sources of hot spots is difficult to reconcile with the layered mantle unless most or all of the transition zone (which constitutes ~10% of the mass of the mantle) is composed of recycled oceanic crust and is somehow preferentially sampled by mantle plumes.

Geophysical observations are now increasingly used to argue against a form of layering at 670 km depth, in particular since it has been found that the sharp characteristics of the 670 km discontinuity (Paulssen, 1988) can be explained by a phase change in an otherwise isochemical mantle (Ito and Takahashi, 1989; see Chapter 2.02). Early reports of aseismic extensions of the slabs in the lower mantle based on travel time perturbations (Jordan, 1977; Creager and Jordan, 1984, 1986) and waveform complexity (Silver and Chan, 1986) suggested a dynamic connection between upper and lower mantle. Some of the first global tomographic models showed a pattern of high velocity in the deep mantle that correlated well with the inferred pattern of paleosubduction (Dziewonski, 1984; Dziewonski and Woodhouse, 1987). This was strengthened by a study by Lithgow-Bertelloni and Richards (1998), who suggested that tomographic models are explained quite well if the lithosphere and lower mantle are one or two orders of magnitude more viscous than the upper mantle. This is in line with earlier estimates for the viscosity of the lower mantle based on geoid and gravity anomalies above subduction zones (Hager and Richards, 1989) and geodynamic inversions (Mitrovica and Forte, 1995). More indirect, but perhaps just as convincing, are arguments provided by Davies (1998), who shows that the lack of significant plume-related topography at the Earth surface makes it inconceivable that a large thermal boundary layer exists at 670 km depth, as is required by the layered mantle hypothesis. Studies of the convective flow predicted by *S*-wave tomography also demonstrate that there is significant mass flux from the upper mantle into the lower mantle (Puster and Jordan, 1997). The quality of mantle tomography has increased dramatically in recent years, which has led to convincing images of slabs that are subducting into the lower mantle

(Fukao *et al.*, 1992; Grand, 1994; Van der Hilst *et al.*, 1997; Bijwaard *et al.*, 1998; Ritsema *et al.*, 1999; Grand, 2002). The correlation of seismically low velocities and subducted slabs is strengthened by the good agreement with paleographic reconstructions in several locations (Van der Voo *et al.*, 1999).

In summary, we have recently witnessed a shift away from the classically layered mantle model in favor of whole mantle convection models, where the buoyancy of sinking slabs is the dominant driving force. Slabs can penetrate deep into the lower mantle and with the induced return flow we would expect the mantle to mix efficiently. This leaves us with an interesting dilemma. If the mantle convects as a whole, how can it preserve the large-scale and long-lived heterogeneity seen in the geochemistry of oceanic basalts?

2.12.3 CHARACTERIZATION OF MIXING

Quantitative evaluation of the efficiency of mantle mixing provides a fundamental tool to test conceptual models for the generation and destruction of heterogeneity. In this section, we will discuss and illustrate a number of ways in which we can quantify mantle mixing and provide a short review of the major findings from the literature. The physical processes of mixing in fluids are of great interest in a large number of fields, with important implications for industrial and engineering processes and a correspondingly large body of literature. For a comprehensive review of the mathematical description of mixing and examples of fluid dynamical and engineering applications, the reader is referred to Ottino (1989).

In this section we will provide some illustrations that are based on a simple two-dimensional convection model in Cartesian geometry. Full detail on the model setup, governing equations and numerical solution, can be found in the appendix. This model is strictly for illustrative purposes: it has a number of characteristics that we think are representative of convection in the mantle (infinite Prandtl number, finite dissipation number, temperature- and pressure-dependent viscosity) but falls short in a number of others (moderately low convective vigor, two-dimensional Cartesian geometry, limited representation of rheology, no internal heating). A movie showing the convective motion is available at http://www.geo.lsa.umich.edu/~keken/treatise. The initial temperature field is shown in Figure 1(a). The flow is characterized by moderate time dependence of the boundary layers. The down- and upwelling that are near the center of the box remain at nearly stationary locations, but traveling instabilities along the top and bottom boundary layers cause significant time dependence in the center up- and downwelling. We will track two heterogeneities (numbered I and II) through this flow. The heterogeneities are identical in size and number of particles but differ in their initial position. Heterogeneity I is placed in the center upwelling, whereas heterogeneity II is placed just outside this upwelling (compare Figure 1(a) with Figure 1(e)). As we will show, the mixing behavior of these particle sets differs greatly: the heterogeneity that is started in the boundary layer encounters strong shear and reorientation and mixes with the fluid quite rapidly. Heterogeneity II remains inside a convective cell with only moderate deformation for a significant period of time. Similar extensive differences in mixing behavior within the same flow have been suggested in three-dimensional models by Ferrachat and Ricard (1998) and Van Keken and Zhong (1999). Although we may predict that this "core-in-cell" behavior (Spohn and Schubert, 1982) is not necessarily a fundamental explanation for the observed mantle heterogeneity, since mantle convection is significantly more time dependent than this study, it may well form an explanation for observed regional differences, such as the DUPAL anomaly (Hart, 1984; Castillo, 1988).

2.12.3.1 Physics of Mixing

Convective flows provide a mechanism for stirring and mixing of heterogeneities. The principal components of this process are stretching, folding, breakup, and diffusion (Ottino, 1989). Stretching and folding are illustrated in Figure 1. *Stretching* occurs in the presence of velocity gradients (such as a simple shear flow). *Folding* occurs when the material flow reverses onto itself or when the material encounters a fixed boundary (such as the core–mantle boundary and the free surface of the Earth's mantle).

Other possible mixing mechanisms include diffusion and breakup. *Breakup* is negligible in the Earth's mantle due to the lack of surface tension in solid rock. *Diffusion* in the solid phase is very slow due to the low compositional diffusion rates in mantle rocks, and is only important at very small length scales (on the order of centimeters) and perhaps at very high temperatures like those at the core–mantle boundary. The lack of interfacial tension and the similarity of properties between chemically distinct layers in the viscously deforming Earth allow us to consider the Earth's mantle as a miscible fluid. We can, therefore, focus on stretching and folding as the primary mechanisms of mantle mixing. Nevertheless, the spatial and temporal discretization that is necessary in the numerical solution of the governing equations introduces breakup- and diffusion-like processes.

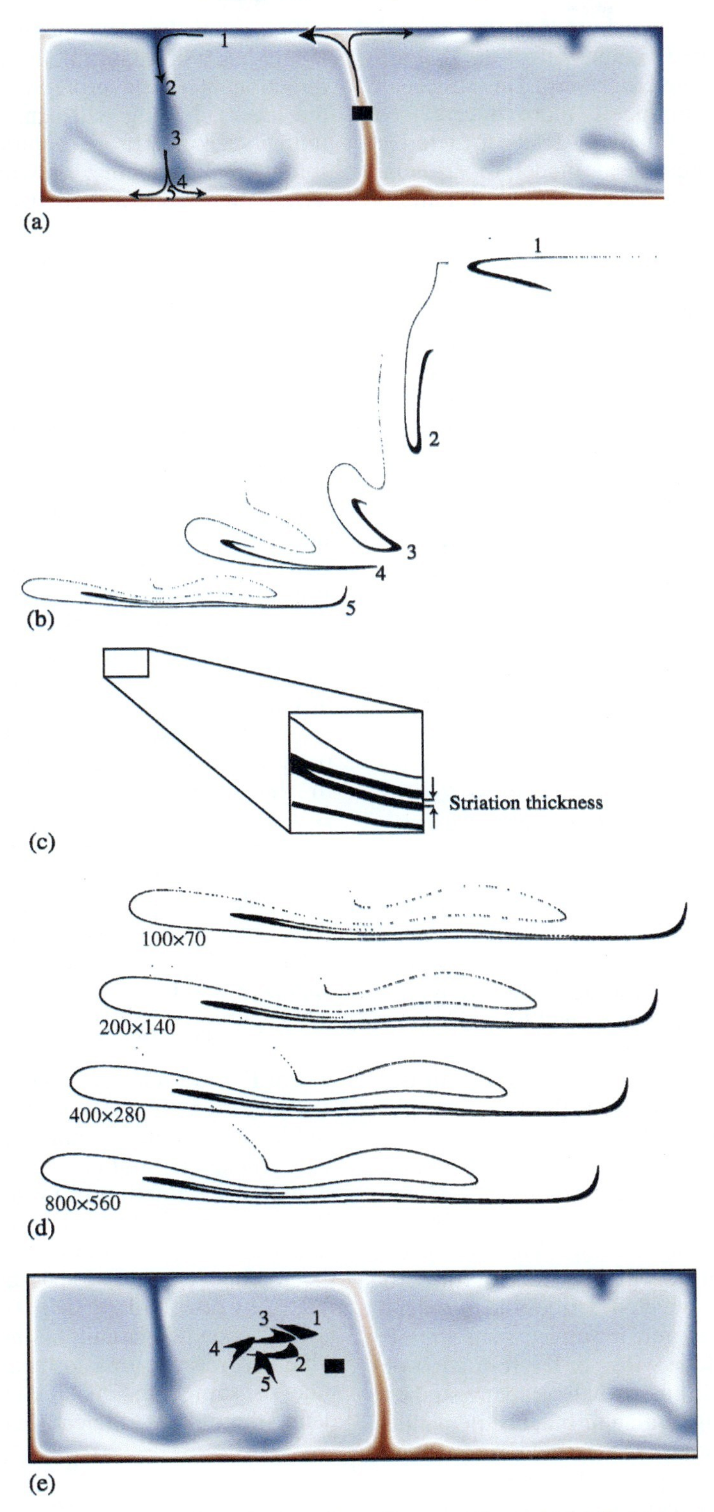

Figure 1 (a) Temperature field of a moderately time-dependent two-dimensional Cartesian convection calculation (see Appendix for details). An initial heterogeneity (number 1), discretized with 100×70 tracers, is placed in the upwelling near the center of the model. The heterogeneity is brought up to the surface and split in two sections. Arrows indicate the flow pattern that this heterogeneity will see. The numbers indicate the positions of the snapshots shown in the next frame. 1: $t = 0.001$, 2: $t = 0.002$; 3: $t = 0.0024$; 4: $t = 0.0028$; 5: $t = 0.0032$. (b) Deformation of heterogeneity 1 (left portion only) corresponding to the time intervals shown in frame (a). (c) Illustration of the definition of striation thickness. (d) Illustration of the effects of increasing the number of tracers from 100×70 to 200×140, 400×280, and 800×560. (e) Deformation of heterogeneity II that is slightly displaced from 1.

Figure 1(c) shows the breakup of the heterogeneity due to the limited number of tracers that are used to represent the heterogeneity. This effect can be mitigated by introducing more tracers (see Figure 1(d), which effectively leads to a lower sensitivity of the numerical model to artificial breakup. Similarly, the finite accuracy of numerical particle tracing leads to an effective numerical diffusivity (not illustrated).

Efficient mixing requires strong deformation. Since this occurs particularly in boundary layers, the heterogeneity of Figure 1(a) was stretched and folded quite efficiently. It is interesting to illustrate the deformation of a heterogeneity that was initially positioned just outside the boundary layer (Figure 1(e)). The snapshots correspond to the same time intervals as in Figure 1(b). Due to the lack of strong velocity gradients in the core of the convection cell, this heterogeneity is barely deformed at all. The dramatic difference in mixing efficiency between two different heterogeneities with slightly different initial condition is maintained over long computational time in this convection model, as is illustrated in Figure 2.

The processes of mixing may be conceptually simple, but the mathematical description is far from complete. The high viscosity of the Earth's mantle compared to its thermal diffusivity implies that the inertial terms in the equations of conservation of momentum (the Navier–Stokes equations for general fluids) are negligible. The resulting Stokes equations predict laminar (non-turbulent) flow, which nevertheless can be quite complicated due to nonlinear relationships between the equations of mass and heat transport and the highly temperature- and stress-dependent rheology of mantle rocks. The temporal distribution of heterogeneity in a fluid is strongly dependent on the mixing history. Small changes in the initial distribution can lead to exponentially growing differences, which means that even for simple flows we can witness chaotic or turbulent effects when observing mixing. This has been named Lagrangian turbulence in contrast to Eulerian turbulence, which describes the turbulent flow of material as, for example, seen in the airflow around airplane wings or the formation of thunderclouds. As discussed below, many studies find evidence for Lagrangian turbulence in mantle convection simulations. The turbulent nature of mixing makes a complete characterization of mixing impossible except for the simplest systems. For this reason it is essential to ask the question how we can best characterize mixing for any given situation.

In the remainder of this section we will discuss a variety of classical mixing indicators and their application to mixing in the Earth's mantle. To introduce mixing concepts it is useful to start with a simple layered situation, where each of the layers has a constant thickness. Stretching in the direction of the layering will cause thinning of the layers; folding will cause duplication of the initial stacking. If we continue this process of folding and stretching, we will witness an increasingly more complicated stacking of layers with decreasing thickness. Following the description by Ottino (1989), we can define simple measures of the efficiency of mixing such as the *striation thickness*, which can be defined as the average thickness of two neighboring layers (Figure 1(c)), or the *intermaterial density area*, which is the area of the interface per unit volume. Mixing will tend to reduce length scales by thinning and consequently reduce the striation thickness and increase the area of interfaces. In many cases, we see that length scales are reduced exponentially and that the interfacial area grows exponentially. It becomes impractical to describe the interface itself, but the exponential nature of mixing allows us to define typical timescales of mixing as expressed, for example, by the increase in interfacial area. At a local scale we can predict that the distance dx between two particles, which were originally at distance dX, will increase exponentially in efficiently mixing flows. It is convenient to define the infinitesimal stretching length $\mu = dx/dX$ in the limit of $dX \to 0$. If the mixing is efficient, the stretching length will increase exponentially, and $d(\ln \mu)/dt > 0$.

In order to describe the mechanisms of mixing, it is necessary to understand the local flow behavior. The motion around a point P can be approximated by

$$\boldsymbol{V} = \boldsymbol{v}_\mathrm{p} + d\boldsymbol{x}(\nabla \boldsymbol{v})_\mathrm{p} + \text{higher-order terms}$$

where $\boldsymbol{V}$ is the velocity in a point that is a distance $d\boldsymbol{x}$ away from point P, $\boldsymbol{v}_\mathrm{p}$ is the velocity in point P, and $\nabla \boldsymbol{v}$ is the velocity gradient tensor at point P. The change in motion can be described by a stretch and a rotation using the stretching tensor $\boldsymbol{D} = \frac{1}{2}(\nabla \boldsymbol{v} + (\nabla \boldsymbol{v})^\mathrm{T})$ and the spin or vorticity tensor $\boldsymbol{\Omega} = \frac{1}{2}(\nabla \boldsymbol{v} - (\nabla \boldsymbol{v})^\mathrm{T})$. It can be shown that the mixing length and the stretching tensor are related by $d(\ln \mu)/dt \leq (\boldsymbol{D}:\boldsymbol{D})^{1/2}$ where $(\boldsymbol{D}:\boldsymbol{D})^{1/2}$ is the second invariant of the stretching tensor which represents a basic measure of the magnitude of stretching. As a consequence, we can define the mixing efficiency as

$$\varepsilon_\mu = \frac{d(\ln \mu)/dt}{(\boldsymbol{D}:\boldsymbol{D})^{1/2}}$$

For incompressible flows it can be shown that the mixing efficiency has an upper bound of $((n-1)/n)^{1/2}$ where n is the dimension of the space. It is instructive to display the mixing behavior of simple shear flow (Figure 3(a)). After an initial rapid increase of the efficiency, due to the stretching in simple shear flow, the theoretical

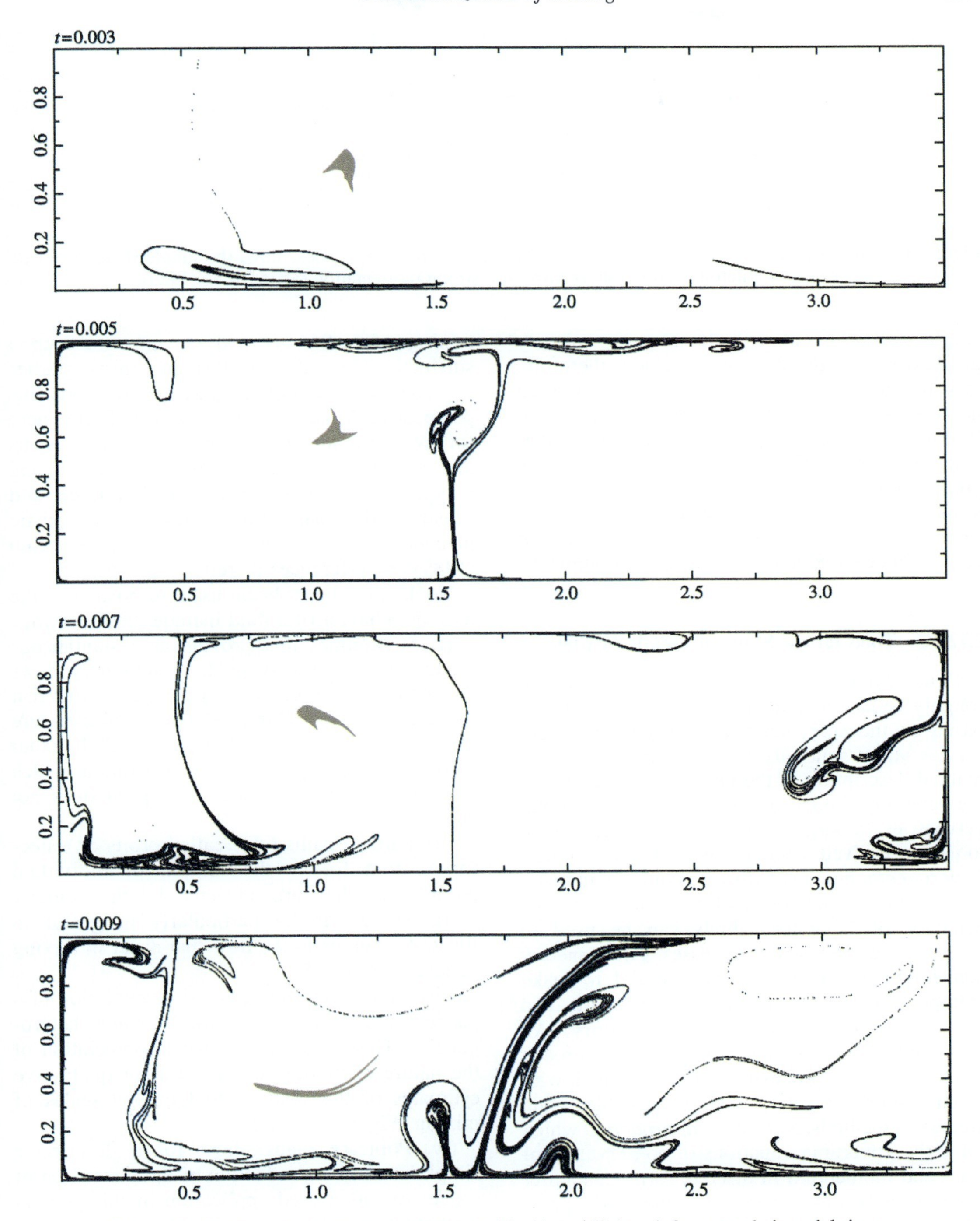

Figure 2 Deformation of heterogeneities I (black) and II (gray) for extended model time.

two-dimensional maximum of $\frac{1}{2}\sqrt{2}$ is reached. After this the efficiency decays as $1/t$, since the fluid filaments become fully aligned with the flow. An efficient way to improve on the efficiency is, therefore, to regularly reorient the filaments so that they benefit from the early high efficiency. Such reorientation can take place at stagnation points in cellular convection or by time-dependent processes. Figure 3(b) shows the time evolution of mixing efficiency for a driven cavity flow, where deformation is dominated by shearing with occasional ("weak") redirection. This is quite similar to the case of two-dimensional steady-state convection. In theory, a mixing process that combines simple shear flow with strong reorientation can retain high mixing efficiencies (Figure 3(c)). It is also interesting to note that near fixed points (e.g., the corner regions of a

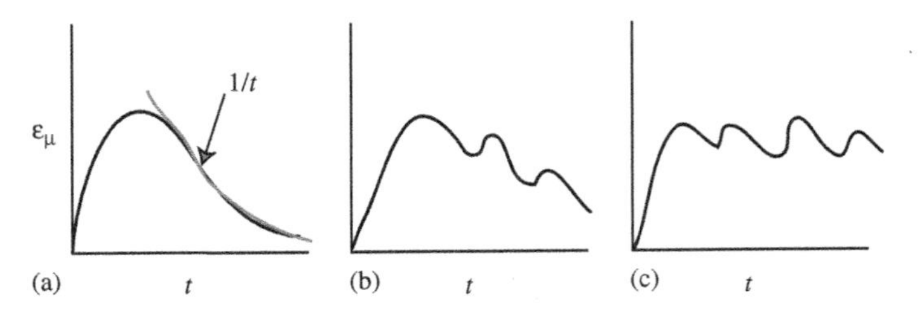

Figure 3 Illustration of the evolution of mixing efficiency for: (a) simple shear flow; (b) simple shear flow with weak reorientation; and (c) shear flow with strong reorientation (after Ottino, 1989).

single cell convection pattern), the velocity gradients are high. We can conclude, therefore, that the mixing efficiency is particularly high near fixed points. Although the description of flow around these fixed points is based on a mathematical simplification, we can expect that the predicted mixing behavior is similar near areas of strong divergence (mid-oceanic ridges, foundering of slabs at the core–mantle boundary) or convergence (subduction zones, base of plumes).

2.12.3.2 Tools for the Study of Mantle Mixing

Fluid dynamical studies of mixing can be roughly separated into three different categories based on the approach that is taken. *Analytical studies* allow generally for a continuous mathematical description of the physics. Although they are generally limited to only the simplest mixing geometries, a large number of our descriptive tools are derived from these analytical studies. *Experimental mixing studies* are common in many engineering applications and in studies where the behavior of laboratory fluids can be used to approximate large-scale processes. For planetary convection applications it is essential to make extrapolations to the much different length- and timescales than those of the laboratory. An important benefit of the experimental approach is that physical processes can be studied with much higher resolution than is possible using numerical methods. *Numerical approaches*, which solve the governing equations of conservation of mass and momentum in discrete form, can suffer from discretization errors in time and space, although convergence tests can be used to determine whether these errors have a great impact on the solution to the problem that is being investigated. An important benefit of numerical approaches is that the equations can be solved with the correct spatial and temporal dimensions without the need to extrapolate from laboratory conditions, and that data on the properties of the entire fluid can be recorded throughout the convection experiment.

Of particular interest for mantle mixing is the dispersion of heterogeneity (e.g., the efficiency of mixing of sediments and oceanic crust upon subduction into the mantle) and many studies approach mantle mixing particularly from this perspective. Most approaches we will discuss here follow from numerical convection experiments that use concepts derived from analytical and experimental studies. We will introduce and illustrate the main approaches in a heuristic manner. For a more rigorous approach, see Ottino (1989) and references therein.

Tools that have been used to visualize the mixing behavior of a fluid include the following.

(i) Streamlines that provide an instantaneous picture of the velocity field. In stationary flows the streamlines represent particle paths. Note that the term "stationary" in fluid dynamical literature means "steady state" and not "static." In other words, a stationary velocity field is one in which fluid moves, but the pattern or speed does not change.

(ii) Particle paths (also called orbits or trajectories) that follow the motion of individual particles as they are advected by the velocity field. These can be particularly insightful to illustrate regions in the flow that differ in mixing behavior (Figure 5).

(iii) Plots representing the time evolution of a particle cloud with an initially small heterogeneity. This allows for a visual appreciation of the nature of mixing and at least for qualitative estimates of the spatial and temporal scales of mixing (Figures 1 and 2).

(iv) Poincaré maps allow for a descriptive simplification by a reduction of the dimension of space. An often-used application of this general mathematical mapping is to plot the intersection of particle paths with a fixed surface. These maps are useful to identify regions of different mixing properties (e.g., laminar versus chaotic) in steady-state flows and identify temporal changes in periodic or time-dependent flows (Figure 6).

More quantitative methods that have been used to describe mantle mixing include the following.

Box counting methods. These are simple statistical approaches to describe the mixing of a small

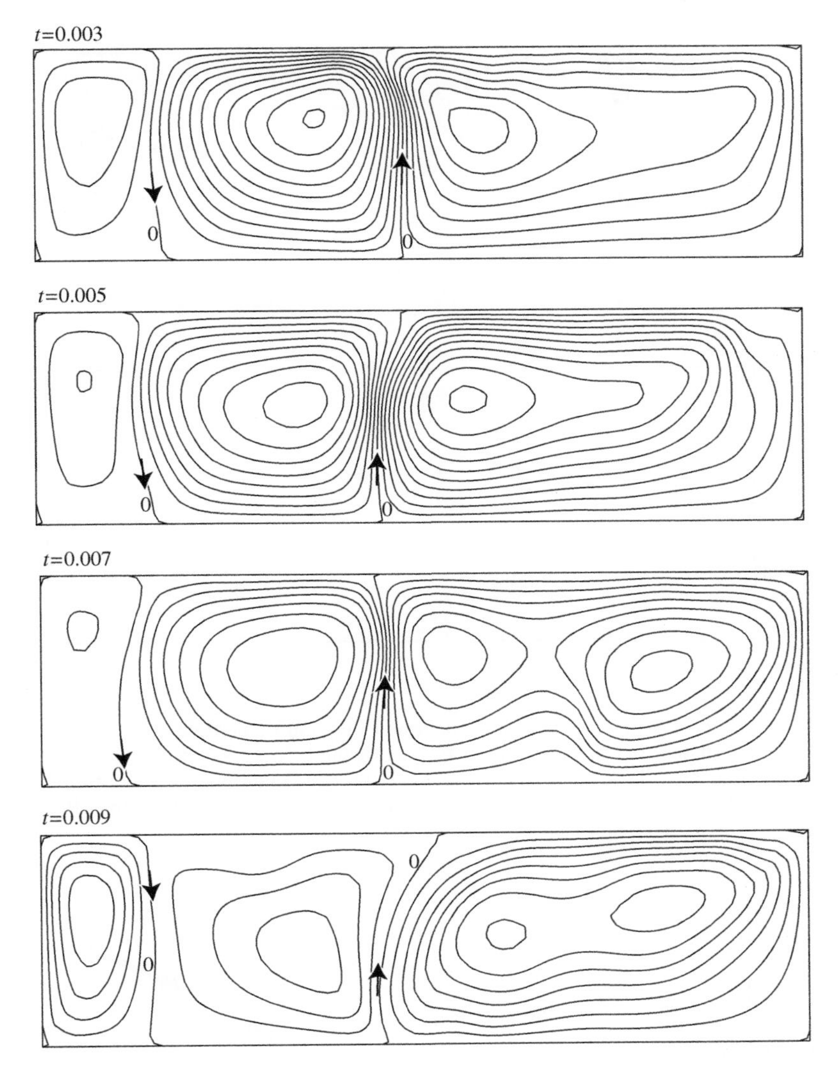

Figure 4 Streamlines corresponding to the snapshots in Figure 2. Contour interval is 50 (nondimensional) units. Note the focusing of the flow in the center upwelling due to the temperature dependence of the viscosity.

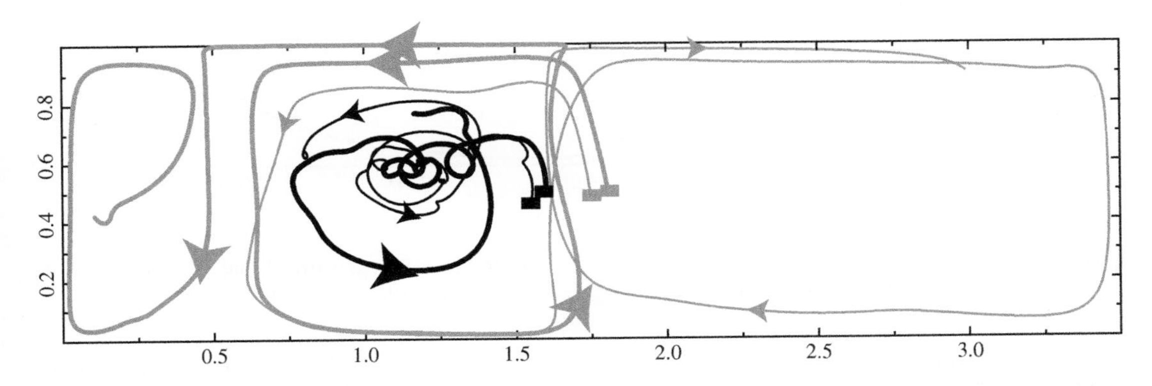

Figure 5 Particle paths corresponding to the convective mixing of the model shown in Figures 1 and 2. Each particle is followed from $t = 0$ to $t = 0.0109$. The initial position is indicated with the solid rectangles. Arrows illustrate the direction of the particle motion. The gray particle paths correspond to the efficiently mixed heterogeneity I (Figure 1(d); the black particle paths correspond to the poorly mixing heterogeneity II (Figure 1(e). Note that the gray particles traverse only about two convection circulations, but build up significant deformation in this time period (compare Figure 2).

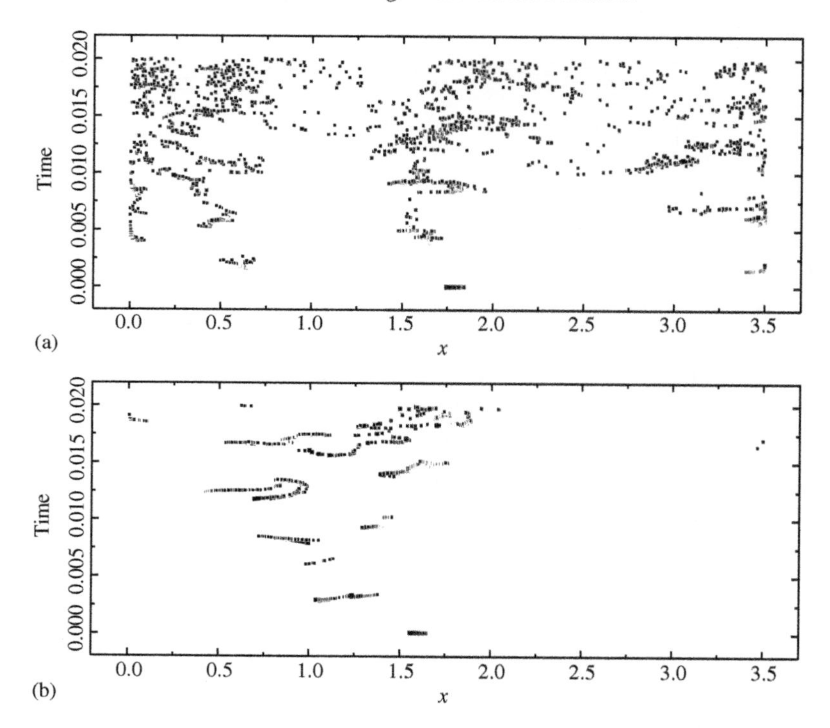

Figure 6 Poincaré plots representing the time and horizontal position of particles traveling through the surface $y = 0.5$: (a) for 21×15 particles that are distributed in heterogeneity I. b) same, but now for heterogeneity II. The general structure of the flow, the reasonably efficient mixing of (a) compared to the more periodic and poor mixing of (b) are clearly illustrated. Note the transition to apparently more efficient mixing after $t = 0.015$ in (b).

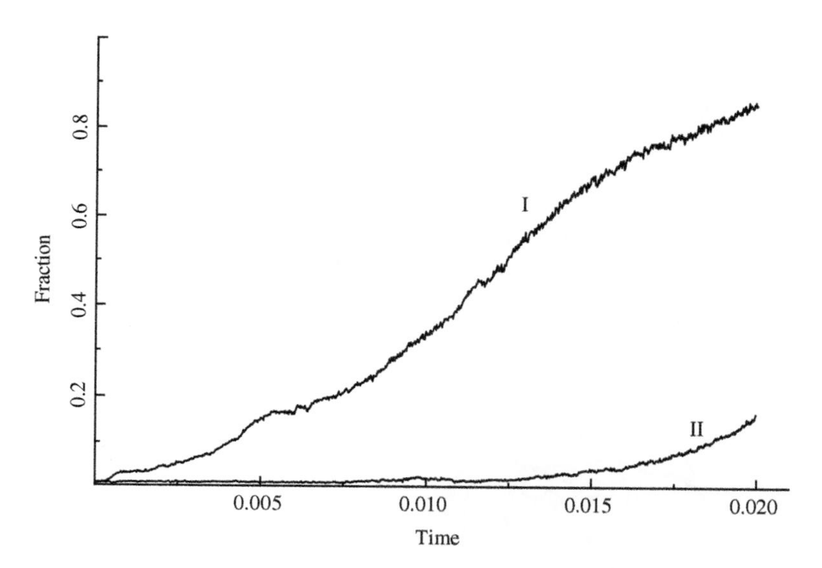

Figure 7 Time evolution of the fraction of boxes that are affected by heterogeneities I and II. In this case we used 7,000 particles to discretize the heterogeneities and a uniform grid of 70×20 boxes.

heterogeneity that use a subdivision of the computational domain into a finite number of uniform boxes. By keeping track of the concentration of the heterogeneity in each box and the related statistical distribution, we can make quantitative statements about departures of uniformity and mixing times. We can illustrate this by draping a uniform 70×20 cell grid over the convection geometry of Figure 1 and evaluating over time the number of particles in each of these boxes. Figure 7 displays the fraction of boxes that contain at least one particle of both heterogeneities (I and II). By this criterion, the first heterogeneity affects an exponentially growing number of boxes early on and becomes nearly box filling (>80% of boxes have at least one particle in them) at the end of the computation. In contrast, the second heterogeneity remains in only a limited number of boxes until $\sim t = 0.015$, when it also shows an exponential growth (see Schmalzl *et al.* (1996) for further illustrations of this technique in mantle convection models).

Strain markers. The exponential increase of interfacial area in efficient mixing suggests an approach in which one keeps track of length scales by following the stretching and folding of small line segments, which effectively work as strain markers. In practice, this is done by following two particles x_1 and x_2 that were initially close together, say at spacing dX. If the mixing is chaotic, the length of the segment between the two particles will grow exponentially: $dx = |\mathbf{x}_1 - \mathbf{x}_2| = dX \exp(\lambda t)$. After a certain time $t = \tau$, we can evaluate the *finite-time Luyaponov* exponent λ. If the mixing is chaotic, this exponent is positive and the magnitude indicates the mixing timescale. In an ideal situation, one would model the distance between the two particles by allowing for the full stretching and folding of the initial segment, which can be done, for example, by modeling the line segment with a high-resolution marker chain. For computational reasons it is, nevertheless, common to re-initialize the strain marker after significant deformation or even to ignore any folding by using the distance between the two particles as a proxy for the Luyaponov exponent. Figure 8 illustrates the potential for this approach (see the caption for details). For other examples of the use of approximations to the finite-time Luyaponov exponent in mantle

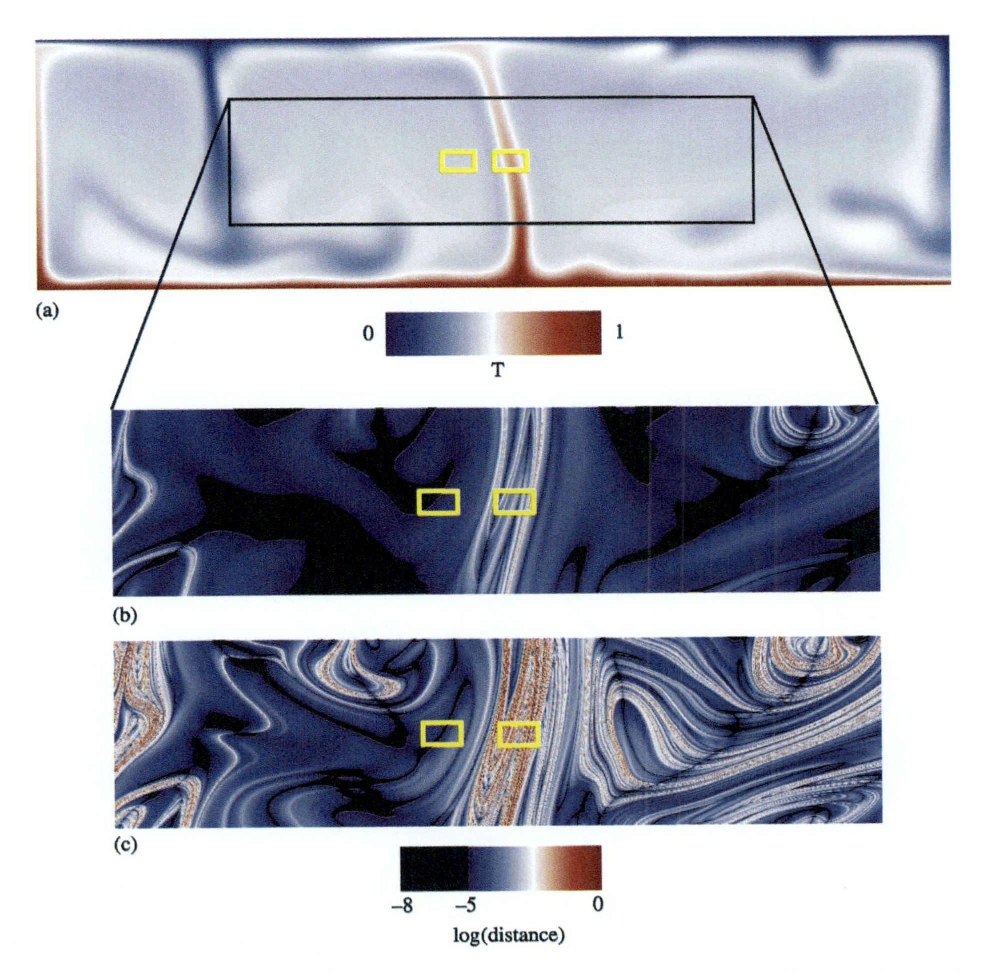

Figure 8 Illustration of the use of Luyaponov exponents to describe the efficiency of mantle mixing. (a) Temperature at $t = 0$. $1{,}000 \times 250$ particle pairs distributed uniformly in the area indicated by the black box. The particles in each pair are offset by a horizontal distance of 10^{-5}. The small yellow boxes indicate the source regions for heterogeneities I and II. We trace the distance between the particle pairs as they are advected with the flow and plot their distance at (b) $t = 0.004$ and (c) $t = 0.02$ in their original location. The black regions correspond to areas where the distance between particle pairs became smaller; the colors indicate minor stretching (dark blue) to extensive stretching (dark red).

mixing studies, see Ferrachat and Ricard (1998), Van Keken and Zhong (1999), and Farnetani and Samuel (2003).

Two-particle correlation functions. A statistical approach to quantify the mixing of a heterogeneity that is represented by many particles is to calculate the distances between each pair of particles and compute the cumulative histogram $H(r)$ which is the number of particle pairs that have a distance of less than r. The slope of $\log(H)$ versus $\log(r)$ within a particular range of r indicates the spatial dimension of the particle distribution. For example, if in a two-dimensional box calculation the slope is constant and equal to 2 for r approaching the length of the box, it is reasonable to assume that the heterogeneity has been mixed in completely. Illustrations of this method are given in Figure 9 and Schmalzl and Hansen (1994).

Viscous dissipation. The mixing efficiency of a viscous fluid is related to the viscous dissipation Φ since $\Phi = \boldsymbol{\tau}{:}\boldsymbol{D} = 2\eta\, \boldsymbol{D}{:}\boldsymbol{D}$ where η is the dynamic viscosity. Maps of viscous dissipation can, therefore, be used to qualitatively predict the differences in mixing efficiency between different regions of the mantle, or between different models of mantle convection (Figure 10).

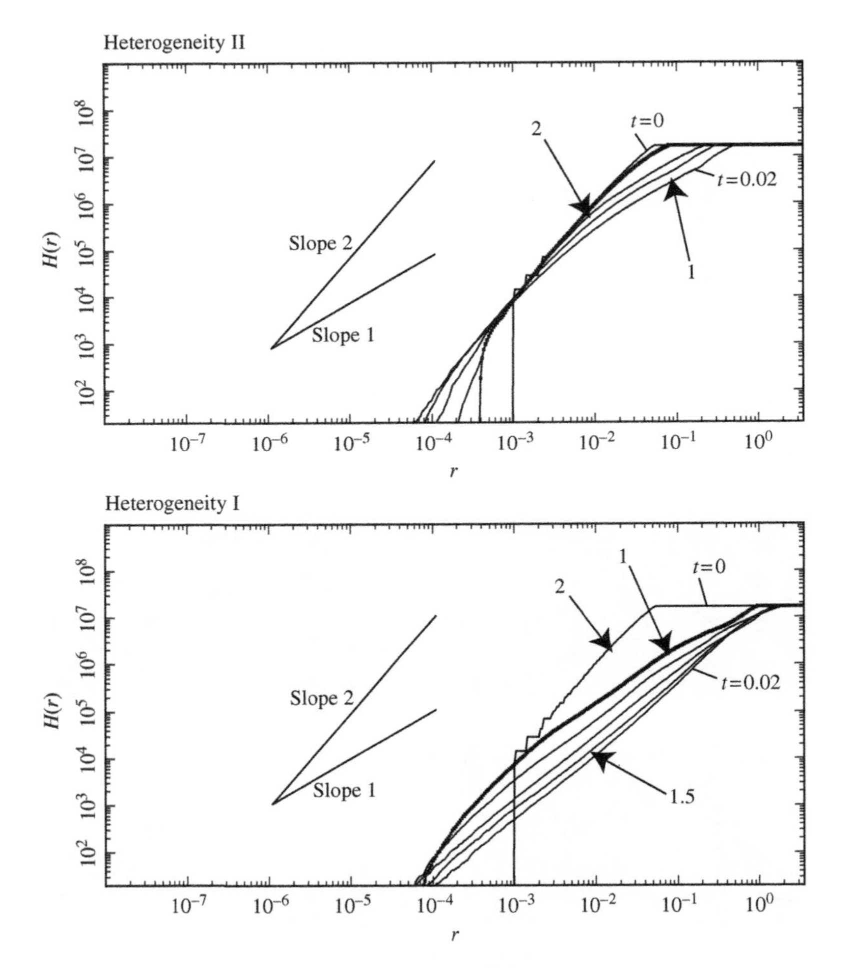

Figure 9 Two-particle correlation function $H(r)$ for heterogeneities I and II at time $t = 0$, $t = 0.004$, ... $t = 0.02$. $H(r)$ is defined as the number of particle pairs that have a distance less than r. Each heterogeneity is traced with 100×70 tracers that are at a distance 10^{-3} from each other at time $t = 0$. The graph for $t = 0.004$ is highlighted with black symbols. On the left lines of slopes 1 and 2 are indicated. The first graph is the same for both heterogeneities: at distances r less than the size of the heterogeneity the slope is 2, indicated that the heterogeneity is space filling at that scale. The position of the change in slope to horizontal indicates the spatial extent of the heterogeneity. In the early stages of mixing the slope of the correlation function becomes 1 at intermediate scales, indicating linear features. At the end of the mixing calculation the correlation function has slope 1.5 at most scales, indicating that the linear features are becoming less dominant, but that the heterogeneity has not yet become space filling at these scales. Note that the change in mixing style after $t = 0.015$ for heterogeneity II is reflected in the change in slope and change of position of the kink in the correlation function.

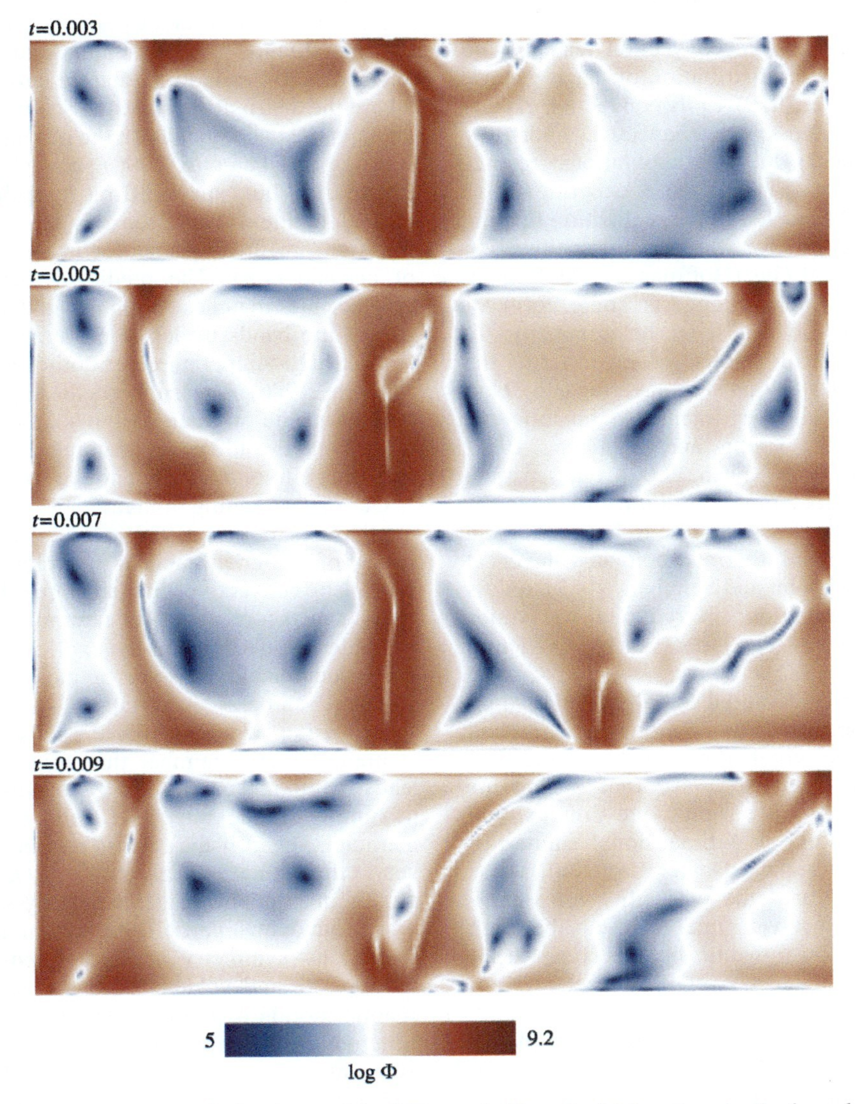

Figure 10 Viscous dissipation Φ for the model of Figure 1. Note the high values in the boundary layers and specifically in the "fixed" points of the flow.

2.12.3.3 Mantle Mixing Studies

The fluid dynamical approach to mixing provides fundamental tools, but we have to keep in mind that in our effort to understand mantle heterogeneity we have some unique circumstances to account for. First, we seem to require mechanisms to prohibit efficient mixing between "reservoirs." This contrasts strongly with the need to develop efficient mixers in industrial application to guarantee consistency and provide low-cost manufacturing techniques. Second, we have an incomplete knowledge of the distribution of heterogeneity. We only observe the chemical heterogeneity through volcanism at the Earth's surface and only at a few locations. The dynamical origin of these "leaks" of the mantle may bear directly on the observed mixing properties. Third, the isotopic chemistry that we observe at the Earth's surface is not only dependent on the distribution of heterogeneity by mantle mixing, but also on the integrated history of radioactive decay and fractionation events. Finally, we observe mantle heterogeneity through the filter of melting. The accompanying chemical and physical differentiation mechanisms may obscure the relationships between chemical differences in the melt and the mantle source heterogeneity.

Quantitative studies of mantle mixing evolved from early laboratory and numerical experiments that linked plate tectonics with convection in the planetary mantle (McKenzie, 1969; Richter, 1973; McKenzie *et al.*, 1974; Richter and Parsons, 1975). The finite strain theory for the development of seismic anisotropy can be used to describe the deformation and subsequent mixing of mantle heterogeneity as well (McKenzie, 1979). It was also shown that mantle convection leads to the relocalization and redistribution of heterogeneity (Richter and Ribe, 1979). Early convection experiments were limited to low convective vigor (low Rayleigh number) due to computational restrictions. Extrapolation from these experiments to Earthlike conditions led to the realization that the typical mixing time for whole mantle convection should be ~0.5–1 Gyr (Hoffman and McKenzie, 1985) which strongly suggests some form of layering to retain long-lived heterogeneity. In contrast, dynamical studies of layering in the mantle demonstrated that strict layering at 670 km depth causes the upper mantle to be too cold (Kopitzke, 1979; Spohn and Schubert, 1982; McNamara and Van Keken, 2000). Furthermore, the required radioelement enrichment of the lower mantle causes the lower mantle to be too hot and forms a substantial thermal boundary layer at 670 km depth (Spohn and Schubert, 1982), unless the mantle heat flow is reduced to values significantly below the global average (Richter and McKenzie, 1981).

Several studies explored the ability of a high-viscosity lower mantle to slow down mixing and increase the survival time of heterogeneity, ranging from detailed conceptual arguments (Davies, 1984) to simple one-dimensional (Loper, 1985) and two-dimensional numerical simulations (Gurnis and Davies, 1986a,b). These models demonstrated the possibility for significantly longer survival times due to the high viscosity of the lower mantle, but this comes at the cost of a much reduced overall convective vigor and too low surface heat flow.

Improved computational ability and better understanding of nonlinear systems allowed for fundamental improvements in mantle mixing studies. A detailed study by Olson *et al.* (1984) used Fourier analysis to demonstrate that the effects of mixing cascade from low to high wave numbers and demonstrated the importance of stagnation points in providing more efficient deformation. Olson *et al.* (1984) showed that mixing in a stationary cell is best described by laminar mixing. The inclusion of time dependence (e.g., Hofmann and McKenzie, 1985) allows for more efficient mixing. Christensen (1989) demonstrated a further complication: flows that are driven by time-dependent kinematic surface boundary conditions are more likely to maintain poorly mixed islands or to be characterized by laminar mixing than time-dependent dynamic models, where heterogeneities are rapidly strained and follow an exponential, or turbulent mixing law. Kellogg and Stewart (1991) confirmed the exponential stretching behavior in a 12-Fourier-mode approximation to the thermal convection equations.

As was demonstrated by several of these early mixing studies, the stretching and folding of heterogeneity leads to a "marble cake" structure, where the distribution of length scales is generally exponential and depends on the mixing efficiency. Kellogg and Turcotte (1990) showed this using a kinematic, but space-filling two-dimensional time-dependent flow based on the Lorenz equations, with a special focus on the thinning of oceanic crust. The predicted distribution of length scales matches the observed distribution of pyroxenite bands in depleted lherzolite found in several high-temperature peridotites (Allègre and Turcotte, 1986). This supports the suggestion that the pyroxenite originated from subducted oceanic crust, which would provide an intriguing geological constraint on the efficiency of mantle mixing.

Our understanding of mantle mixing has further improved in the early 1990s, particularly due to the ability of numerical models to explore the effects of realistic convective vigor and three-dimensional flows, the incorporation of new methodology to characterize mixing, and the explicit modeling of the geochemical inputs and outputs to the mantle system. Schmalzl and Hansen (1994) demonstrated using two-dimensional isoviscous models at high Rayleigh number that advective stirring leads to mixing on the large scale first and that mixing between convective cells can be quite inefficient, even if the mixing within each cell is very efficient. Schmalzl *et al.* (1995) explored mixing in a three-dimensional steady-state model. They found that particle motion in the flow is limited by symmetry surfaces imposed by the temperature and velocity field and regular, non-chaotic motion of the particles, from which they concluded that time dependence is probably essential in generating chaotic particle behavior and efficient mixing. The addition of time dependence in a three-dimensional model at moderate Rayleigh number (Schmalzl *et al.*, 1996) demonstrated that cross-cell mixing is enhanced, but the efficiency appeared to be less than in case of two-dimensional convection, probably due to the greater stability of the three-dimensional convective pattern.

Three-dimensional convective patterns can be described by a combination of poloidal (divergent and convergent) and toroidal (strike-slip) components. Ferrachat and Ricard (1998) used a kinematic model of convection driven by a surface boundary condition that mimicked a mid-oceanic ridge with a variable length transform fault. The additional toroidal energy caused a significant portion of the model to undergo chaotic mixing, although islands of laminar mixing could persist. These results suggest that it is essential to incorporate the influence of plates on mantle convection in mixing studies. Van Keken and Zhong (1999) further demonstrated the importance of toroidal motion, using a dynamic model of present-day mantle convection. In this model, toroidal motion is generated by the influence of the weak zones corresponding to the present-day plate boundaries. Several regions of laminar mixing are present in this model, but in certain areas such as near Papua New Guinea and the western US, significant toroidal motion caused cork-screw-like particle paths. This allows for large-scale transport between the convective cells and for efficient, chaotic mixing.

The increasing spatial and temporal resolution of numerical models allows for direct testing of geochemical hypotheses by tracking the evolution of particles that carry isotopic and elemental information. Christensen and Hofmann (1994) provided one of the first quantitative models that detailed the recycling of oceanic crust and its influence on the chemical evolution of the mantle. Their models included the formation of basaltic crust and depleted harzburgite upon melting of the peridotite mantle. Upon transformation to eclogite, the oceanic crust ponds at the base of the mantle and is only slowly entrained by thermal convection. The modeled lead and neodymium isotopic evolution after 3.6 Gyr provided an adequate fit to the observed MORB and HIMU ranges, which provides a strong quantitative support to the suggestion that oceanic crust recycling is of fundamental importance to the chemistry of mantle plumes (Hofmann and White, 1980; White and Hofmann, 1982). Van Keken and Ballentine (1998, 1999) developed two-dimensional cylindrical models of mantle convection to study the evolution of helium isotopes. When comparing models of similar convective vigor as measured by average surface heat flow and plate velocities, it was shown that a higher-viscosity lower mantle was not capable of preserving heterogeneity over timescales longer than ~1 Gyr. Incorporation of phase changes and moderately temperature-dependent rheology similarly could not prevent large-scale mixing, except in the case of an extreme, and probably unrealistic, dynamical influence of the endothermic phase change at 670 km depth. Similar results were obtained by Ferrachat and Ricard (2001), Davies (2002), and Stegman *et al.* (2002).

A distinct class of models that describe the chemical evolution of the Earth are the so-called "box models," in which assumptions are made about the geometry of distinct reservoirs and their interactions. For example, one can assign four distinct reservoirs in the upper mantle, lower mantle, continental crust, and atmosphere and develop differential equations that incorporate radiogenic ingrowth, chemical fractionation effects, and assumptions about mass transfer between the reservoirs. Successful models reproduce the observed isotopic ratios and/or concentrations. Applications include studies of noble gas evolution (Kellogg and Wasserburg, 1990; O'Nions and Tolstikhin, 1994; Porcelli and Wasserburg, 1995a,b; Albarède, 1998), crustal evolution (Jacobsen and Wasserburg, 1979; O'Nions *et al.*, 1979; Jacobson, 1988), mean stirring times of the mantle (Allègre and Lewin, 1995), and lead-isotope systematics (Galer and O'Nions, 1985; Kramers and Tolstikhin, 1997; Paul *et al.*, 2002).

An important advantage of this type of modeling is that the computational cost is much lower than that of full dynamical models, which allows for the efficient exploration of large parameter spaces and for data inversion to obtain optimal parameter choices (e.g., Allègre and Lewin, 1989; Coltice and Ricard, 1999). More complex physics can be integrated into the box model approach. For example, the internal mixing rate in individual reservoirs can be described in parametrized from, as was demonstrated by Kellogg *et al.* (2002). It is expected that we will see a further integration between the box and fluid dynamical modeling approaches, such as the comparison made by Coltice *et al.* (2000).

2.12.4 OUTLOOK

2.12.4.1 Conceptual Model Development

The improved geochemical and geophysical observations, coupled with enhanced computational ability, have led to an expansion of research on the chemical evolution of the Earth's mantle. The apparent contradictions between geophysical indications for whole mantle flow and the geochemical requirements for substantial and long-lived heterogeneity have led to a large set of new or revived conceptual models. Proposals include mantle layering below 670 km depth (Kellogg *et al.*, 1999; Anderson, 2002), "zoning" of the mantle due to the variable thermal and chemical properties of subducting slabs (Albarède and Van der Hilst, 2002), chemical variations hidden in small and highly viscous blobs (Becker *et al.*, 1999), and recent breakup of a layered mantle system (Allègre, 2002). For a review of the pros and cons of these various models, see, e.g., Van Keken *et al.* (2002).

An alternative model that is gaining rapid popularity incorporates the recycling of oceanic crust in whole mantle convection (e.g., Coltice and Ricard, 1999; Ferrachat and Ricard, 2001; Helffrich and Wood, 2001; Van Keken *et al.*, 2002; Davies, 2002). This idea has several attractive aspects. The formation of oceanic crust at mid-oceanic ridges and the subsequent recycling at subduction zones is the foremost modern-day method of introducing heterogeneity into the mantle. The signal of recycled oceanic crust in ocean island lavas has been clearly shown in a number of isotope systems (e.g., Hofmann and White, 1980; White and Hofmann, 1982; Hauri and Hart, 1997; Lassiter and Hauri, 1998; Eiler *et al.*, 2000; Hauri, 2002). The higher density of basalt-derived components in the mantle allows for a natural explanation for long residence times.

Commonly cited problems with this model include the difficulty in explaining the noble gas observations and the distribution of heat-producing elements (see Chapter 2.06). In particular, the observation of high $^3He/^4He$ in hot spots has been interpreted as the consequence of preservation of a "primitive" or "primordial" component, since ^3He is not produced in the Earth's interior. Yet, the oceanic basalts with high $^3He/^4He$ are nonprimitive by any other geochemical measure (e.g., Hart *et al.*, 1992; Hanan and Graham, 1996) and this would suggest that low concentrations of uranium and thorium, and consequently low concentrations of ^4He compared to ^3He are responsible for the high $^3He/^4He$ ratio. This, in turn, enables a small ^3He residue to account for the high OIB $^3He/^4He$ (Anderson, 1998b; Albarède, 1998; Coltice and Ricard, 1999). Quantitative support for this nonprimitive source of high $^3He/^4He$ is provided by a number of models (Ferrachat and Ricard, 2001; Coltice and Ricard, 2002). Other calculations show that because of the residence time of recycled material (1–1.5 Gyr), U + Th depletion of the protolith is not required if this is mixed with a small volume of ^3He rich material (Ballentine *et al.*, 2002). The latter calculations are consistent with radioelement and noble gas concentrations inferred from the Iceland plume (e.g., Hilton *et al.*, 2000).

The concentration of rare gases in the upper mantle can provide additional observational evidence. The ^3He concentration in the oceans has been used to argue for a low helium concentration in the upper mantle and a requirement for the deep burial and isolation of helium (e.g., O'Nions and Oxburgh, 1983). The related low estimates for the ^{40}Ar concentration in the upper mantle provide a similar argument for argon and potassium storage in the deep mantle. It is interesting to note that these arguments are all dependent on the assumption that the present-day concentration of helium in the oceans, which has a residence time of ~1,000 yr, allows for accurate estimate of upper mantle concentrations. It is not unreasonable to expect that due to variable volcanic activity, the present-day ^3He concentration in the oceans may not be representative of the time-averaged concentrations in the mantle (e.g., Van Keken *et al.*, 2001). In fact, an increase in the noble gas concentration of the upper mantle by a factor of 3.5 removes all requirements for layering and hidden noble gas reservoirs (Ballentine *et al.*, 2002).

An additional argument for mantle layering is based on the mantle concentration of heat-producing elements. The extremely depleted nature of N-MORB (Hofmann, 1988) suggests quite low uranium, thorium, and potassium concentrations in the upper mantle, which would require substantial deeper enrichment (e.g., Kellogg *et al.*, 1999). A new compilation of MORB analysis (Su and Langmuir, 2002) from the PetDB database (Lehnert *et al.*, 2000) suggests that the average oceanic crust is considerably less depleted than N-MORB and that consequently the upper mantle composition is far less depleted than previously assumed. This new analysis effectively removes the requirement for storage of high heat-producing element concentrations in the deep mantle.

2.12.4.2 Quantitative Modeling

Any conceptual model requires quantitative support, ideally through an integrated physical and chemical modeling approach. Even if we consider that we have come a long way in developing better geodynamical models, we are still faced with significant hurdles in the development of consistent dynamical models that would allow a full understanding of the dynamical and chemical evolution of the Earth. A number of important concepts that need better resolution include the following.

- The integration of plate tectonics and mantle convection (e.g., Tackley, 2000a). It is evident that the large-scale plate structure has an important influence on the mantle convection pattern (e.g., Anderson, 1998a). Thus far, it has been very difficult to develop models that satisfactorily produce self-consistent, and persistent, plate-tectonic style deformation on top of mantle convection, although encouraging progress has been in recent years (e.g., Trompert and Hansen, 1998; Tackley, 1998, 2000b,c; Tackley and Xie, 2002; Bercovici, 2003). This is of particular importance for the testing of conceptual mixing since otherwise it is difficult, if not impossible to use realistic temperature-dependent rheology or to accurately differentiate between mid-oceanic ridge and hot spot sampling of the mantle.
- The cause and source regions of hot spots. Since plumes are considered to deliver the signals from the deep mantle, it is essential to understand how plumes influence mantle mixing. Of particular concern are the low internal viscosity of mantle plumes and the interaction with the transition zone and lithosphere (e.g., Hauri *et al.*, 1994; Marquart and Schmeling, 2000; Farnetani *et al.*, 2002; Kumagai, 2002).
- The influence of rheological variations on mantle mixing, such as proposed for the survival of

primitive blobs (Manga, 1996; Becker *et al.*, 1999) or the mixing of oceanic crust (Spence *et al.*, 1988; Van Keken *et al.*, 1996; Merveilleux du Vignaux and Fleitout, 2001). The influence of non-Newtonian rheology on mantle mixing also requires much more detailed investigation (e.g., Ten *et al.*, 1997).

- Influence of chemical buoyancy due to melt extraction (Christensen and Hofmann, 1994; Dupeyrat *et al.*, 1995; Davies, 2002). Bulk chemistry changes upon melting are likely of fundamental importance in maintaining heterogeneity since the associated chemical buoyancy changes can effectively counteract thermally driven mixing.
- The role of the mantle transition zone. Seismological observations and petrological predictions strongly suggest an important dynamic role for the transition zone (e.g., Ringwood, 1982). This includes the influence of the phase changes, possible variations in minor and major element chemistry, and rheological variations. In particular, the 670 km discontinuity has been suggested to function as a partial chemical filter (e.g., Christensen and Yuen, 1984; Weinstein, 1992; Van Keken *et al.*, 1996).
- Structure and dynamics of the lowermost mantle. This region includes the D$''$ layer, which is characterized by major chemical and thermal variations. It is likely of fundamental importance to the chemical evolution of the mantle and may function as a (temporary) resting place for subducted slabs. It is also expected to influence the stability of mantle plumes (Davaille *et al.*, 2002; Jellinek and Manga, 2002), the entrainment and residence times of chemical heterogeneity (Olson and Kincaid, 1991; Schott *et al.*, 2002, and the thermal, chemical, and seismological characteristics compositional variations (Kellogg *et al.*, 1999; Tackley, 2002).
- Thermal evolution of the Earth. In particular, we need to understand how we can construct chemical models that provide sufficiently realistic cooling histories. It has been pointed out that the relatively low present-day heat production in the BSE as compared to the present-day heat loss requires some form of buffering or layering (e.g., Butler and Peltier, 2002). We also need to understand the tectonic regime during the Hadean and Archean. The projected higher mantle temperatures in the Archean lead to conditions that are not necessarily conductive to faster plate tectonics. The deeper melting that occurs in a hotter Earth causes more extensive melting, and the thicker basalt and depleted peridotite layers provide a strong compositional stabilizing force to plate tectonics. The tectonics of the Archean Earth may well have been dominated by thermal plumes or eclogite-driven delamination (e.g., Anderson, 1979; Vlaar *et al.*, 1994; Zegers and Van Keken, 2001).

In conclusion, the renewed interest in the integration of geophysical and geochemical approaches to develop a better understanding of the chemical evolution of the mantle has led to a significant number of new or revived conceptual models. Models that combine geodynamics and geochemistry can provide quantitative tests to these ideas. Although several hurdles still exist, we can expect that the growth of computational resources—combined with better insights into the role of chemical mixing, improved geochemical and seismological observations, and a more fundamental understanding of the interpretation of Earth structure from these observations—will ultimately allow for the development of a consistent description of the Earth's mantle evolution as it is influenced by the generation and destruction of chemical heterogeneity.

ACKNOWLEDGMENTS

The authors thank Francis Albarède, Don Anderson, Geoff Davies, Jamie Kellogg, Louise Kellogg, Dan McKenzie, Don Porcelli, and Paul Tackley for stimulating discussions and Rick Carlson for encouragement.

APPENDIX

The numerical convection model that is used to illustrate the visualization and quantification of mixing (Figures 1–10) is based on the solution of the equations governing convection in the Earth's mantle, assuming that the mantle can be described as an anelastic and weakly compressible fluid at infinite Prandtl number. Under the extended Boussinesq approximation, we can write the equation of motion as

$$-\nabla P + \nabla \cdot (\eta \dot{\boldsymbol{\epsilon}}) = RaT\hat{\boldsymbol{g}} \qquad (1)$$

and the mass conservation equation as

$$\nabla \cdot \boldsymbol{v} = 0 \qquad (2)$$

where P is dynamic pressure, η the viscosity, $\dot{\boldsymbol{\epsilon}}$ strain-rate tensor, Ra the Rayleigh number, T the temperature, and $\hat{\boldsymbol{g}}$ the gravity vector. The heat equation incorporates terms that describe viscous heating and adiabatic cooling and heating, and can be written as

$$\frac{\partial T}{\partial t} + (\boldsymbol{v} \cdot \nabla)T + \alpha DiwT = \nabla^2 T + \frac{Di}{Ra}\sigma_{ij}\frac{\partial u_i}{\partial x_j} - DiwT_0 \qquad (3)$$

where t is time, Di the dissipation number, w the vertical component of velocity, σ_{ij} the components of the stress tensor, u_i the velocity components, and T_0 the reference (surface) temperature (e.g., Jarvis and McKenzie, 1980). The equations above are nondimensional.

The Rayleigh number Ra and dissipation number Di are given by

$$Ra = \frac{\rho_0 g \alpha \Delta T h^3}{\eta_0 \kappa} \tag{4}$$

and

$$Di = \frac{\alpha_0 g h}{c_p} \tag{5}$$

where ρ is density, α the expansivity, ΔT the temperature contrast across the mantle, h the depth of the mantle, η_0 the reference viscosity, and κ the thermal diffusivity. The parameters on the right-hand side of these equations are dimensional.

We use a moderately temperature- and pressure-dependent viscosity which is written in nondimensional form as

$$\eta(T, p) = \exp(-aT + bz) \tag{6}$$

where z is depth and a and b nondimensional coefficients.

The Equations (1)–(3) are solved using a finite-element based on the general tool box Sepran (Cuvelier *et al.* (1986); http://dutita0.twi.tudelft.nl./sepran/sepran.html). The Stokes Equation (1) and incompressibility constraint (2) are solved using a penalty function approach. The heat equation is solved after Galerkin discretization. The time-dependent equations are solved using a second-order predictor–corrector method (see Van den Berg *et al.* (1993) and Van Keken and Ballentine (1999) for more details). The tracers are advanced using fourth-order Runge–Kutta method. The time step is limited to 50% of the Courant–Friedrichs–Levy criterion. The numerical implementation has been extensively tested against standard mantle convection benchmarks (Blankenbach *et al.*, 1989; Van Keken *et al.*, 1997) as well as published results by other workers.

For the model used in this chapter, we use a two-dimensional Cartesian grid with aspect ratio 3.5. The bottom and top boundary are free slip and the side boundaries are reflective. Temperatures are fixed at top ($T = 0$) and bottom ($T = 1$). We assume whole mantle convection and choose $Ra = 10^6$, $Di = 0.2$, $a = -4.60517$, and $b = 3$. There is no internal heating. The numerical grid consists of $2 \times 140 \times 40$ quadratic triangles (corresponding to 281×81 nodal points) which provides significantly higher resolution than that needed for the accurate solution of the heat equation. The higher resolution allows for more accurate particle tracking, which is more sensitive to small errors. The initial condition (Figure 1(a)) was obtained from a previous calculation that was run for sufficiently long time to reach a statistical equilibrium. The mixing calculations are for a time period from 0 to 0.02, which corresponds to a dimensional period of 5 Gyr. The convection is moderately time dependent and the overall convective vigor is low compared to that of the present-day Earth. For example, the average surface heat flow is 40 mW m^{-2} and the average velocity 1 cm yr^{-1}. The transit time (time for a particle to travel from top to bottom at average velocity) is 280 Myr, although it should be noted that the velocity in the up- and downwellings is substantially higher.

REFERENCES

Anderson D. L. (1979) Upper mantle transition zone—Eclogite. *Geophys. Res. Lett.* **6**, 433–436.

Anderson D. L. (1998a) The scales of mantle convection. *Tectonophysics* **284**, 1–17.

Anderson D. L. (1998b) The helium paradoxes. *Proc. Natl. Acad. Sci. USA* **95**, 4822–4827.

Anderson D. L. (2002) The case for irreversible chemical stratification of the mantle. *Int. Geol. Rev.* **44**, 97–116.

Albarède F. (1998) Time-dependent models of U–Th–He and K–Ar evolution and the layering of mantle convection. *Chem. Geol.* **145**, 413–429.

Albarède F. and van der Hilst R. D. (2002) Zoned mantle convection. *Phil. Trans. Roy. Soc. London A* **360**, 2569–2592, doi: 10.1098/rsta.2002.1081.

Allègre C. J. (2002) The evolution of mantle mixing. *Phil. Trans. Roy. Soc. London A* **360**, 2411–2431, doi: 10.1098/rsta.2002.1075.

Allègre C. J. and Lewin E. (1995) Isotopic systems and stirring times of the Earth's mantle. *Earth Planet. Sci. Lett.* **136**, 629–646.

Allègre C. J. and Lewin E. (1989) Chemical structure and history of the Earth: evidence from global non-linear inversion of isotopic data in a three-box model. *Earth Planet. Sci. Lett.* **96**, 61–88.

Allègre C. J. and Turcotte D. L. (1986) Implications of a two-component marble cake mantle. *Nature* **323**, 123–127.

Allègre C. J., Othman D. B., Polve M., and Richard P. (1979) The Nd–Sr isotopic correlation in mantle materials and geodynamic consequences. *Phys. Earth Planet. Inter.* **19**, 293–306.

Ballentine C. J., van Keken P. E., Porcelli D., and Hauri E. (2002) Numerical models, geochemistry, and the zero paradox noble gas mantle. *Phil. Trans. Roy. Soc. London A* **360**, 2611–2631, doi: 10.1098/rsta.2002.1083.

Becker T. W., Kellogg J. B., and O'Connell R. J. (1999) Thermal constraints on the survival of primitive blobs in the lower mantle. *Earth Planet. Sci. Lett.* **171**, 351–365.

Bercovici D. (2003) The generation of plate tectonics from mantle convection. *Earth Planet. Sci. Lett.* **205**, 107–121.

Bijwaard H., Spakman W., and Engdahl E. R. (1998) Closing the gap between regional and global travel time tomography. *J. Geophys. Res.* **103**, 30055–30078.

Blankenbach B., Busse F., Christensen U., Cserepes L., Gunkel D., Hansen U., Harder H., Jarvis G., Koch M., Marquart G., Moore D., Olson P., Schmeling H., and Schnaubelt T. (1989) A benchmark comparison for mantle convection codes. *Geophys. J. Int.* **98**, 23–38.

Butler S. L. and Peltier W. R. (2002) Thermal evolution of the Earth: models with time-dependent layering of mantle convection which satisfy the Urey ratio constraint. *J. Geophys. Res.* **107**, doi: 10.1029/2000JB0000018.

Carlson R. W. (1994) Mechanisms for Earth differentiation: consequences for the chemical structure of the mantle. *Rev. Geophys.* **32**, 337–361.

Castillo P. (1988) The DUPAL anomaly as a trace of the upwelling lower mantle. *Nature* **366**, 667–670.

Christensen U. (1989) Mixing by time-dependent convection. *Earth Planet. Sci. Lett.* **95**, 382–394.

Christensen U. R. and Hofmann A. W. (1994) Segregation of subducted oceanic crust in the convecting mantle. *J. Geophys. Res.* **99**, 19867–19884.

Christensen U. R. and Yuen D. A. (1984) The interaction of a subducting lithospheric slab with a chemical or phase boundary. *J. Geophys. Res.* **89**, 4389–4402.

Coltice N. and Ricard Y. (1999) Geochemical observations and one layer mantle convection. *Earth Planet. Sci. Lett.* **174**, 125–137.

Coltice N., Ferrachat S., and Ricard Y. (2000) Box modeling the chemical evolution of geophysical systems: case study of the Earth's mantle. *Geophys. Res. Lett.* **27**, 1579–1582.

Coltice N. and Ricard Y. (2002) On the origin of noble gases in mantle plumes. *Phil. Trans. Roy. Soc. A* **360**, 2633–2648.

Creager K. C. and Jordan T. H. (1984) Slab penetration into the lower mantle. *J. Geophys. Res.* **89**, 3031–3049.

Creager K. C. and Jordan T. H. (1986) Slab penetration into the lower mantle beneath the Mariana and other island arcs of the northwest Pacific. *J. Geophys. Res.* **91**, 3573–3589.

Cuvelier C., Segal A., and van Steenhoven A. A. (1986) *Finite Element Models and the Navier–Stokes Equations*. Reidel, Dordrecht, The Netherlands, 483pp.

Davaille A., Girard F., and Le Bars M. (2002) How to anchor hotspots in a convecting mantle? *Earth Planet. Sci. Lett.* **203**, 621–634.

Davies G. F. (1984) Geophysical and isotopic constraints on mantle convection: an interim synthesis. *J. Geophys. Res.* **89**, 6017–6040.

Davies G. F. (1998) Topography: a robust constraint on mantle fluxes. *Chem. Geol.* **145**, 479–489.

Davies G. F. (2002) Stirring geochemistry in mantle convection models with stiff plates and slabs. *Geochim. Cosmochim. Acta* **66**, 3125–3142.

DePaolo D. and Wasserburg G. (1976) Inferences about magma sources and mantle structure from variations of $^{143}Nd/^{144}Nd$. *Geophys. Res. Lett.* **3**, 743–746.

Dupeyrat L., Sotin C., and Parmentier E. M. (1995) Thermal and chemical convection in planetary mantles. *J. Geophys. Res.* **100**, 497–520.

Dziewonski A. M. (1984) Mapping the lower mantle: determination of lateral heterogeneity in P velocity up to degree and order 6. *J. Geophys. Res.* **89**, 5929–5952.

Dziewonski A. M. and Woodhouse J. H. (1987) Global images of the Earth's interior. *Science* **236**, 37–48.

Eiler J. M., Farley K. A., Valley J. W., Hauri E. H., Craig H., Hart S. R., and Stolper E. M. (1997) Oxygen isotope variations in ocean island basalt phenocrysts. *Geochim. Cosmochim. Acta* **61**, 2281–2293.

Eiler J. M., Schiano P., Kitchen N., and Stolper E. M. (2000) Oxygen isotope evidence for recycled crust in the sources of mid-ocean ridge basalts. *Nature* **403**, 530–534.

Farley K. A., Natland J. H., and Craig H. (1992) Binary mixing of enriched and undegassed (primitive?) mantle components (He, Sr, Nd, Pb) in Samoan lavas. *Earth Planet. Sci. Lett.* **111**, 183–199.

Farnetani C. G. and Samuel H. (2003) Lagrangian structures and stirring in the Earth's mantle. *Earth Planet. Sci. Lett.* **206**, 335–348.

Farnetani C. G., Legras B., and Tackley P. J. (2002) Mixing and deformation in mantle plumes. *Earth Planet. Sci. Lett.* **196**, 1–15.

Faure G. and Hurley P. M. (1963) The isotopic composition of strontium in oceanic and continental basalt: application to the origin of igneous rocks. *J. Petrol.* **4**, 31–50.

Ferrachat S. and Ricard Y. (1998) Regular vs. chaotic mantle mixing. *Earth Planet. Sci. Lett.* **155**, 75–86.

Ferrachat S. and Ricard Y. (2001) Mixing properties in the Earth's mantle: effects of the viscosity stratification and of oceanic crust segregation. *Geochem. Geophys. Geosys.* **2**.

Fukao Y., Obayashi M., Inoue H., and Nenbai M. (1992) Subducting slabs stagnant in the mantle transition zone. *J. Geophys. Res.* **97**, 4809–4822.

Galer S. J. G. and O'Nions R. K. (1985) Residence time of thorium, uranium and lead in the mantle with implications for mantle convection. *Nature* **316**, 778–782.

Grand S. P. (1994) Mantle shear structure beneath the Americas and surrounding oceans. *J. Geophys. Res.* **99**, 11591–11621.

Grand S. P. (2002) Mantle shear-wave tomography and the fate of subducted slabs. *Phil. Trans. Roy. Soc. London A* **360**, 2475–2491, doi: 10.1098/rsta.2002.1077.

Gurnis M. and Davies G. F. (1986a) Mixing in numerical models of mantle convection incorporating plate kinematics. *J. Geophys. Res.* **91**, 6375–6395.

Gurnis M. and Davies G. F. (1986b) The effect of depth-dependent viscosity on convective mixing in the mantle and the possible survival of primitive mantle. *Geophys. Res. Lett.* **13**, 541–544.

Hager B. H. and Richards M. A. (1989) Long-wavelength variations in Earth's geoid: physical models and dynamical implications. *Phil. Trans. Roy. Soc. London A* **328**, 309–327.

Hanan B. B. and Graham D. W. (1996) Lead and helium isotope evidence from oceanic basalts for a common deep source of mantle plumes. *Science* **272**, 991–995.

Hart S. R. (1984) A large-scale isotope anomaly in the Southern Hemisphere mantle. *Nature* **309**, 753–757.

Hart S. R., Schilling J. G., and Powell J. L. (1973) Basalts from Iceland and along the Reykjanes ridge: Sr isotopic geochemistry. *Nature* **246**, 104–107.

Hart S. R., Hauri E. H., Oschmann L. A., and Whitehead J. A. (1992) Mantle plumes and entrainment: isotopic evidence. *Science* **26**, 517–520.

Hauri E. H. (2002) Osmium isotopes and mantle convection. *Phil. Trans. Roy. Soc. London A* **360**, 2371–2382.

Hauri E. H. and Hart S. R. (1993) Re–Os systematics of EMII and HIMU oceanic island basalts from the south Pacific Ocean. *Earth Planet. Sci. Lett.* **114**, 353–371.

Hauri E. H. and Hart S. R. (1997) Rhenium abundances and systematics in oceanic basalts. *Chem. Geol.* **139**, 185–205.

Hauri E. H., Whitehead J. A., and Hart S. R. (1994) Geochemical and fluid dynamic aspects of entrainment in mantle plumes. *J. Geophys. Res.* **99**, 24275–24300.

Helffrich G. R. and Wood B. J. (2001) The Earth's mantle. *Nature* **412**, 501–508.

Hilton D. R., Thirlwall M. F., Taylor R. N., Murton B. J., and Nicols A. (2000) Controls on the magmatic degassing along the Reykjanes Ridge with implications for the helium paradox. *Earth Planet. Sci. Lett.* **183**, 43–50.

Hoffman N. R. A. and McKenzie D. P. (1985) The destruction of geochemical heterogeneities by differential fluid motions during mantle convection. *Geophys. J. Roy. Astron. Soc.* **82**, 163–206.

Hofmann A. W. (1986) Nb in Hawaiian magmas: constraints on source composition and evolution. *Chem. Geol.* **57**, 17–30.

Hofmann A. W. (1988) Chemical differentiation of the Earth: the relationship between mantle, continental crust, and oceanic crust. *Earth Planet. Sci. Lett.* **90**, 297–314.

Hofmann A. W. (1997) Mantle geochemistry: the message from oceanic volcanism. *Nature* **385**, 219–229.

Hofmann A. W. and White W. M. (1980) The role of subducted oceanic crust in mantle evolution. *Carnegie Yearbook* **79**, 477–483.

Ito E. and Takahashi E. (1989) Post-spinel transformations in the system Mg_2SiO_4–Fe_2SiO_4 and some geophysical implications. *J. Geophys. Res.* **94**, 10637–10646.

Jacobsen S. B. (1988) Isotopic constraints on crustal growth and recycling. *Earth Planet. Sci. Lett.* **90**, 315–329.

Jacobsen S. B. and Wasserburg G. J. (1979) The mean age of mantle and crustal reservoirs. *J. Geophys. Res.* **84**, 7411–7427.

Jarvis G. T. and McKenzie D. P. (1980) Convection in a compressible fluid with infinite Prandtl number. *J. Fluid Mech.* **96**, 515–583.

Jellinek A. M. and Manga M. (2002) The influence of a chemical boundary layer on the fixity, spacing and lifetime of mantle plumes. *Nature* **418**, 760–763.

Jordan T. H. (1977) Lithospheric slab penetration into the lower mantle beneath the sea of Okhotsk. *J. Geophys. Res.* **43**, 473–496.

Kellogg L. H. and Stewart C. A. (1991) Mixing by chaotic convection in an infinite Prandtl number fluid and implications for mantle convection. *Phys. Fluids A* **3**, 1374–1378.

Kellogg L. H. and Turcotte D. L. (1990) Mixing and the distribution of heterogeneities in a chaotically convecting mantle. *J. Geophys. Res.* **95**, 421–432.

Kellogg L. H. and Wasserburg G. J. (1990) The role of plumes in mantle helium fluxes. *Earth Planet. Sci. Lett.* **99**, 276–389.

Kellogg L. H., Hager B. H., and van der Hilst R. D. (1999) Compositional stratification in the deep mantle. *Science* **283**, 1881–1884.

Kellogg J. B., Jacobsen S. B., and O'Connell R. J. (2002) Modeling the distribution of isotopic ratios in geochemical reservoirs. *Earth Plan. Sci. Lett.* **204**, 183–202.

Kopitzke U. (1979) Finite element convection models: comparison of shallow and deep mantle convection and temperatures in the mantle. *J. Geophys.* **46**, 97–121.

Kramers J. D. and Tolstikhin I. N. (1997) Two terrestrial lead isotope paradoxes, forward transport modeling, core formation and the history of the continental crust. *Chem. Geol.* **139**, 75–110.

Kumagai I. (2002) On the anatomy of mantle plumes: effect of the viscosity ratio on entrainment and stirring. *Earth Planet. Sci. Lett.* **198**, 211–224.

Lassiter J. C. and Hauri E. H. (1998) Osmium isotope variations in Hawaiian lavas: evidence for recycled oceanic lithosphere in the Hawaiian plume. *Earth Planet. Sci. Lett.* **164**, 483–496.

Lehnert K., Su Y., Langmuir C. H., Sarbas B., and Nohl U. (2000) A global geochemical database structure for rocks. *Geochem. Geophys. Geosys.* **1**.

Lithgow-Bertelloni C. and Richards M. A. (1998) The dynamics of Cenozoic and Mesozoic plate motions. *Rev. Geophys.* **36**, 27–78.

Loper D. E. (1985) A simple model of whole-mantle convection. *J. Geophys. Res.* **90**, 1809–1836.

Manga M. (1996) Mixing of heterogeneities in the mantle: effect of viscosity differences. *Geophys. Res. Lett.* **22**, 1949–1952.

Marquart G. and Schmeling H. (2000) Interaction of small mantle plumes with the spinel-perovskite phase boundary: implications for chemical mixing. *Earth Planet. Sci. Lett.* **177**, 241–254.

Merveilleux du Vignaux N. and Fleitout L. (2001) Stretching and mixing of viscous blobs in the Earth's mantle. *J. Geophys. Res.* **106**, 30893–30908.

McKenzie D. P. (1969) Speculation on the consequences and causes of plate motions. *Geophys. J. Roy. Astron. Soc.* **18**, 1–32.

McKenzie D. P. (1979) Finite deformation during fluid flow. *Geophys. J. Roy. Astron. Soc.* **58**, 689–715.

McKenzie D. P., Roberts J. M., and Weiss N. O. (1974) Convection in the Earth's mantle: towards a numerical simulation. *J. Fluid Mech.* **62**, 465–538.

McNamara A. K. and van Keken P. E. (2000) Cooling of the Earth: a parameterized convection study of whole vs. layered models. *Geochem. Geophys. Geosys.* **1**, 15 November.

Mitrovica J. X. and Forte A. M. (1995) Pleistocene glaciation and the Earth's precession constant. *Geophys. J. Int.* **121**, 21–32.

Newsom H. E., White W. M., Jochum K. P., and Hofmann A. W. (1986) Siderophile and chalcophile element abundances in oceanic basalts, Pb isotope evolution and growth of the Earth's core. *Earth Planet. Sci. Lett.* **80**, 299–313.

Olson P. and Kincaid C. (1991) Experiments on the interaction of thermal-convection and compositional layering at the base of the mantle. *J. Geophys. Res.* **96**, 4347–4354.

Olson P., Yuen D. A., and Balsiger D. (1984) Mixing of passive heterogeneities by mantle convection. *J. Geophys. Res.* **89**, 425–436.

O'Nions R. K. and Oxburgh E. R. (1983) Heat and helium in the Earth. *Nature* **306**, 429–431.

O'Nions R. K. and Tolstikhin I. N. (1994) Behaviour and residence times of lithophile and rare gas tracers in the upper mantle. *Earth Planet. Sci. Lett.* **124**, 131–138.

O'Nions R. K., Evensen N. M., and Hamilton P. J. (1979) Geochemical modeling of mantle differentiation and crustal growth. *J. Geophys. Res.* **84**, 6091–6101.

Ottino J. (1989) *The Kinematics of Mixing: Stretching, Chaos, and Transport*. Cambridge University Press, Cambridge, 364pp.

Paul D., White W. M., and Turcotte D. L. (2002) Modelling the isotopic evolution of the Earth. *Phil. Trans. Roy. Soc. London A* **360**, 2433–2474, doi: 10.1098/rsta.2002.1076.

Paulssen H. (1988) Evidence for a sharp 670-km discontinuity as inferred from *P*-to-*S* converted waves. *J. Geophys. Res.* **93**, 10489–10500.

Porcelli D. and Ballentine C. J. (2002) Models for the distribution of Terrestrial noble gases and the evolution of the atmosphere. *Rev. Min. Geochem.* **47**, 411–480.

Porcelli D. and Wasserburg G. J. (1995a) Mass transfer of xenon through a steady-state upper mantle. *Geochim. Cosmochim. Acta* **59**, 1991–2007.

Porcelli D. and Wasserburg G. J. (1995b) Mass transfer of helium, neon, argon and xenon through a steady-state upper mantle. *Geochim. Cosmochim. Acta* **59**, 4921–4937.

Puster P. and Jordan T. H. (1997) How stratified is mantle convection? *J. Geophys. Res.* **102**, 7625–7646.

Richter F. M. (1973) Convection and the large-scale circulation of the mantle. *J. Geophys. Res.* **78**, 8735–8745.

Richter F. M. and McKenzie D. P. (1981) On some consequences and possible causes of layered mantle convection. *J. Geophys. Res.* **86**, 6133–6142.

Richter F. M. and Parsons B. (1975) On the interaction of two scales of convection in the mantle. *J. Geophys. Res.* **80**, 2529–2541.

Richter F. M. and Ribe N. M. (1979) On the importance of advection in determining the local isotopic composition of the mantle. *Earth Planet. Sci. Lett.* **43**, 212–222.

Ringwood A. E. (1982) Phase transformations and differentiation in subducted lithosphere: implications for mantle dynamics, basalt petrogenesis and crustal evolution. *J. Geol.* **90**, 611–643.

Ritsema J., van Heijst H. J., and Woodhouse J. H. (1999) Complex shear wave velocity structure imaged beneath Africa and Iceland. *Science* **286**, 1925–1928.

Rudnick R. and Fountain D. M. (1995) Nature and composition of the continental crust: a lower crustal perspective. *Rev. Geophys.* **33**, 267–309.

Schilling J. G. (1973) Icelandic mantle plume: geochemical evidence along the Reykjanes Ridge. *Nature* **242**, 565–571.

Schmalzl J. and Hansen U. (1994) Mixing the Earth's mantle by thermal convection: a scale dependent phenomenon. *Geophys. Res. Lett.* **21**, 987–990.

Schmalzl J., Houseman G. A., and Hansen U. (1995) Mixing properties of three-dimensional (3D) stationary convection. *Phys. Fluids* **7**, 1027–1033.

Schmalzl J., Houseman G. A., and Hansen U. (1996) Mixing in vigorous, time-dependent three-dimensional convection and application to Earth's mantle. *J. Geophys. Res.* **101**, 21847–21858.

Schott B., Yuen D. A., and Braun A. (2002) The influences of composition and temperature dependent rheology in thermal–chemical convection on entrainment of the D″-layer. *Phys. Earth Planet. Inter.* **129**, 43–65.

Silver P. G. and Chan W. W. (1986) Observations of body wave multipathing from broadband seismograms: evidence for lower mantle slab penetration beneath the Sea of Okhotsk. *J. Geophys. Res.* **91**, 13787–13802.

Silver P. G., Carlson R. W., and Olson P. (1988) Deep slabs, geochemical heterogeneity, and the large-scale structure of mantle convection: investigation of an enduring paradox. *Ann. Rev. Earth Planet. Sci.* **16**, 477–541.

Shirey S. B. and Walker R. J. (1998) The Re–Os isotope system in cosmochemistry and high-temperature geochemistry. *Ann. Rev. Earth Planet. Sci.* **26**, 423–500.

Spence D. A., Ockendon J. R., Wilmott P., Turcotte D. L., and Kellogg L. (1988) Convective mixing in the mantle: the role of viscosity differences. *Geophys. J.* **95**, 79–86.

Spohn T. and Schubert G. (1982) Modes of mantle convection and the removal of heat from the Earth's interior. *J. Geophys. Res.* **87**, 4682–4696.

Stegman D. R., Richards M. A., and Baumgardner J. R. (2002) Effects of depth-dependent viscosity and plate motions on maintaining a relatively uniform mid-ocean ridge basalt reservoir in whole mantle flow. *J. Geophys. Res.* **107**, 10.1029/2001JB000192.

Su Y. and Langmuir C. H. (2002) Global MORB chemistry compilation at the segment scale, http://petdb.ldeo.columbia.edu/documentation/MORBcompilation/

Tackley P. J. (1998) Self-consistent generation of tectonic plates in three-dimensional mantle convection. *Earth Planet. Sci. Lett.* **157**, 9–22.

Tackley P. J. (2000a) Mantle convection and plate tectonics: toward an integrated physical and chemical theory. *Science* **288**, 2002–2007.

Tackley P. J. (2000b) Self-consistent generation of tectonic plates in time-dependent, three-dimensional mantle convection simulations: 1. Pseudoplastic yielding. *Geochem. Geophys. Geosys.* **1**.

Tackley P. J. (2000c) Self-consistent generation of tectonic plates in time-dependent, three-dimensional mantle convection simulations: 2. Strain weakening and asthenosphere. *Geochem. Geophys. Geosys.* **1**.

Tackley P. J. (2002) Strong heterogeneity caused by deep mantle layering. *Geochem. Geophys. Geosys.* **3**.

Tackley P. J. and Xie S. (2002) The thermochemical structure and evolution of Earth's mantle: constraints and numerical models. *Phil. Trans. Roy. Soc. London A* **360**, 2593–2609, doi: 10.1098/rsta.2002.1082.

Ten A. A., Yuen D. A., Podladchikov Y. Y., Larsen T. B., Pachepsky E., and Malevsky A. (1997) Fractal features in mixing of non-Newtonian and Newtonian mantle convection. *Earth Planet. Sci. Lett.* **146**, 401–414.

Trompert R. and Hansen U. (1998) Mantle convection simulations with rheologies that generate plate-like behaviour. *Nature* **395**, 686–689.

Van den Berg A. P., van Keken P. E., and Yuen D. A. (1993) The effects of a composite non-Newtonian and Newtonian rheology on mantle convection. *Geophys. J. Int.* **115**, 62–78.

Van der Hilst R. D., Widiyantoro S., and Engdahl E. R. (1997) Evidence for deep mantle circulation from global tomography. *Nature* **386**, 578–584.

Van der Voo R., Spakman W., and Bijwaard H. (1999) Mesozoic subducted slabs under Siberia. *Nature* **397**, 246–249.

Van Keken P. E. and Ballentine C. J. (1998) Whole-mantle versus layered mantle convection and the role of a high-viscosity lower mantle in terrestrial volatile evolution. *Earth Planet. Sci. Lett.* **156**, 19–32.

Van Keken P. E. and Ballentine C. J. (1999) Dynamical models of mantle volatile evolution and the role of phase changes and temperature-dependent rheology. *J. Geophys. Res.* **104**, 7137–7169.

Van Keken P. E. and Zhong S. (1999) Mixing in a 3D spherical model of present day mantle convection. *Earth Planet. Sci. Lett.* **171**, 533–547.

Van Keken P. E., Karato S., and Yuen D. A. (1996) Rheological control of oceanic crust separation in the transition zone. *Geophys. Res. Lett.* **23**, 1821–1824.

Van Keken P. E., King S. D., Schmeling H., Christensen U. R., Neumeister D., and Doin M.-P. (1997) A comparison of methods for the modeling of thermochemical convection. *J. Geophys. Res.* **102**, 22295–22477.

Van Keken P. E., Ballentine C. J., and Porcelli D. (2001) A dynamical investigation of the heat and helium imbalance. *Earth Planet. Sci. Lett.* **188**, 421–434.

Van Keken P. E., Hauri E. H., and Ballentine C. J. (2002) Mantle mixing: the generation, preservation, and destruction of chemical heterogeneity. *Ann. Rev. Earth Planet. Sci.* **30**, 493–525.

Van Schmus W. R. (1995) Natural radioactivity of the crust and mantle. In *A Handbook of Physical Constants*, AGU References Shelf 1 (ed. T. J. Ahrens). American Geophysical Union, Washington, DC, pp. 283–291.

Vlaar N. J., van Keken P. E., and van den Berg A. P. (1994) Cooling of the Earth in the Archaean. *Earth Planet. Sci. Lett.* **121**, 1–18.

Weinstein S. A. (1992) Induced compositional layering in a convecting fluid layer by an endothermic phase transition. *Earth Planet. Sci. Lett.* **113**, 23–39.

White W. M. and Hofmann A. (1982) Sr and Nd isotope geochemistry of oceanic basalts and mantle evolution. *Nature* **296**, 821–826.

Zegers T. E. and van Keken P. E. (2001) Mid-Archaean continent formation by crustal delamination. *Geology* **29**, 1083–1086.

Zindler A. and Hart S. (1986) Chemical geodynamics. *Ann. Rev. Earth. Planet. Sci.* **14**, 493–571.

2.13
Compositional Evolution of the Mantle

V. C. Bennett
The Australian National University, Canberra, Australia

2.13.1 INTRODUCTION

The mantle is the Earth's largest chemical reservoir comprising 82% of its total volume and 65% of its mass. The mantle constitutes almost all of the silicate Earth, extending from the base of the crust (which comprises only 0.6% of the silicate mass) to the top of the metallic core at 2,900 km depth. The chemical compositions of direct mantle samples such as abyssal peridotites (Chapter 2.04) and peridotite xenoliths (Chapter 2.05), and of indirect probes of the mantle such as basalts from mid-ocean ridge basalts (MORBs) and ocean island basalts (OIBs) (Chapter 2.03), and some types of primitive granites, tell us about the compositional state of the modern mantle, with ever increasingly detailed information providing strong evidence for chemical complexity and heterogeneity at all scales (Chapter 2.03). This chemical heterogeneity must reflect the complex physical interplay of a number of distinct long-lived geochemical reservoirs that are identified primarily by their radiogenic isotopic compositions.

Many of the chapters in this volume provide detailed images of the current chemical and physical state of the Earth's mantle, whereas other contributions examine the starting composition for the Earth (Chapter 2.01). This chapter attempts to link these two areas by tracking the composition of the mantle through time. The first part of this chapter is a summary of the empirical evidence for secular change in the chemical composition of the mantle from the formation of the Earth at 4.56 Ga through to the present day. The emphasis is on results from the long-lived radiogenic isotopic systems, in particular $^{147}Sm-^{143}Nd$, $^{176}Lu-^{176}Hf$, $^{87}Rb-^{87}Sr$, and $^{187}Re-^{187}Os$ systems as these isotopic data provide some of the best constraints on the composition of the mantle in the first half of Earth history, and the timing and extent of chemical differentiation that has affected the mantle over geologic time. Selected trace element data and the "short-lived" $^{146}Sm-^{142}Nd$ isotopic systems are also considered. Understanding the origin of chemical heterogeneity in the Earth's mantle remains a fundamental focus in Earth science. Thus, the second part of this chapter is devoted to the implications of these observations for some of the key questions in mantle geochemistry such as, what are the major chemical reservoirs, when did they form, and how do they interact with each other? Was the Archean mantle substantially different from the modern mantle? How much, if any, of what we see in the modern mantle is a result of early planetary differentiation processes such as those inferred for the Moon and Mars, and how much is a result of the prolonged effects of plate tectonic processes? Has the mantle become more or less chemically heterogeneous with time? All of these questions remain active areas of research, and the intention of this chapter is primarily to present an overview of the current "state of play," which will undoubtedly evolve rapidly with further study.

2.13.1.1 The Mantle Sample

Ideally, tracing the chemical evolution of the convecting mantle would be accomplished by measuring the compositions of coherent, pristine suites of direct mantle samples, lacking metamorphic or metasomatic overprints, and with a well-determined age and geological context. Such samples would provide a range of isotopic and chemical data, which could be used to constrain the chemical and dynamical history of the mantle. Suites approaching this ideal can be found throughout the Phanerozoic, but as we move to earlier time periods in Earth's history, the mantle samples are more problematic. Abyssal peridotites are rarely preserved in the rock record and, if present, are often pervasively altered. Basalts are more commonly preserved, but it is often difficult to establish precise age constraints, tectonic settings, and source characteristics of these magmas. In the middle-to-late Archean, komatiites (high-MgO lavas representing large degree partial melts of the mantle) are often used as primary records of mantle chemistry, but questions remain regarding the comparability of komatiites to modern basalts. For example, komatiites are often argued to be the products of ancient mantle plumes (e.g., Campbell *et al.*, 1989) and as such may represent deep mantle rather than upper mantle reservoirs. Alternatively, at least some komatiites may originate in ancient subduction zone settings (e.g., Parman *et al.*, 2001) and thus be derived from the upper mantle, though potentially contaminated with subducted material and hence not representative of upper mantle compositions. Complexities in the interpretation of chemical compositions of both basalts and komatiites include not only changes in composition due to diagenetic and metamorphic alteration, but also modification of primary melt compositions due to interaction with crustal materials during their generation and ascent. For this reason, the most primitive basalts, usually those with the highest-MgO contents, are taken to be the least affected by crustal interaction and therefore the best record of mantle compositions.

Somewhat ironically the compositions of "juvenile" granites (*sensu lato*) of the continental crust that may have formed from relatively mafic sources with a limited crustal residence and with little involvement of pre-existing crustal materials are often used to trace mantle isotopic compositions throughout much of the Precambrian. Although providing a once-removed record of mantle compositions, granitoids have the benefit of enhanced preservation in the rock record. For example, all >3.6 Ga crustal remnants identified on the Earth consist largely of high-grade granitic gneisses (e.g., see the review of Early Archean terranes by Nutman *et al.* (2001)). Granitoids also offer advantages in that their ages can be determined by relatively robust methods, i.e., U–Pb isotopic compositions of zircons, and their high concentrations of lithophile elements such as rubidium, strontium, samarium, and neodymium are less prone to alteration than those of mafic rocks. Thus, much of our knowledge of the chemistry of the Earth's interior, particularly for the first half of its history, is from the study of indirect, second generation samples of the mantle.

2.13.2 RADIOGENIC ISOTOPIC COMPOSITIONS

2.13.2.1 The Use of Radiogenic Isotopic Compositions to Infer Mantle Chemistry

The signatures of long-lived radioactive isotope decay schemes provide some of the most unequivocal evidence for chemical evolution of the mantle. Excellent reviews of the application of radiogenic isotopic systems to geologic problems are provided by Faure (1986) and Dickin (1995) and only a few points are reviewed here. The isotopic composition of a rock in a given parent–daughter system at the time of crystallization is known as the *initial* isotopic composition. Initial isotopic compositions reflect the time-averaged parent/daughter ratio of the source(s) of the rock. For mantle-derived samples, the measured parent/daughter ratio preserved in the rock will, in most cases, be different from that of the mantle source owing to elemental fractionation during melting and fractional crystallization. Through time, the isotopic compositions of the sample will change in the rock owing to *in situ* decay of the parent isotopes. For young samples (less than 10 Ma), the isotopic changes are generally insignificant for most decay schemes, but for Precambrian rocks the differences in isotopic compositions between the *measured* and *initial* compositions can be quite large. The determination of accurate and precise initial isotopic compositions of old rocks requires the precise determination of concentrations of parent and daughter elements, as well as their isotopic composition, and an independent knowledge of the age of the sample, for example, through U–Pb ages of magmatic zircons. With increasing age, corrections for *in situ* decay become larger, producing potentially greater errors on the calculated initial isotopic composition. This approach is critically dependent on correct age assignment and on the geological integrity of the sample. If secondary modification of the parent/daughter ratio and/or isotopic composition has occurred, for example, through metamorphism or migration of fluids, the calculated initial isotopic composition will be in error. The degree of error depends on the age of the sample, the severity of chemical alteration, and the time at which the alteration occurred relative to formation of the original sample. Alternatively, if suites of cogenetic samples of varying parent/daughter ratios can be identified, for example, differentiated lava flows, then both the age and the initial isotopic composition of the sample suite can be determined by defining an isochron. In practice, there are often many difficulties with the assumption that all samples started with the same isotopic composition, i.e., were derived from a single source, particularly if they are from a wide geographic area, and with the assumption that all samples have maintained their chemical integrity.

Variations in the parent/daughter ratios in different mantle reservoirs arise from the partial removal of material through processes such as melt extraction, and by addition of material through crustal recycling or mixing between different reservoirs. In the case of partial melting, incompatible lithophile elements (which include large-ion lithophile elements such as potassium, rubidium, strontium, uranium, and light rare earth elements (LREEs) such as lanthanum, cerium, and neodymium) preferentially partition into the melt (Chapter 2.09). The resultant change in parent/daughter compositions such as Rb/Sr and Sm/Nd will, over time, produce isotopic differences in the residual mantle. Isotopic compositions can thus be used to identify long-lived chemical variability in the mantle, which can be related to processes of differentiation.

Since the formation of the Earth at ~4.56 Ga, the isotopic compositions of the bulk Earth have been evolving due to radioactive decay of parent isotopes. The mantle has a large chemical inertia such that small isotopic deviations in a chemical reservoir as massive as the mantle, or even portions of the mantle, must be linked to large-scale fractionation events. Because great importance is placed on small isotopic differences, normalizing schemes such as epsilon or gamma notation (described below) have been adopted for many of the commonly used long-half-life systems (e.g., Sm–Nd, Lu–Hf, and Re–Os) where sample compositions are referenced to estimated primitive mantle abundances. This has the advantage of providing an immediate and intuitive basis for comparison of isotopic compositions of a given sample relative to the primitive mantle, as well as minimizing the effects of measurement bias when comparing results from different laboratories.

2.13.2.2 ^{143}Nd Isotopic Evolution

Some of the most powerful constraints on the differentiation history of the Earth are derived from the long-lived (half-life ~106 Gyr) ^{147}Sm–^{143}Nd isotopic decay scheme. This isotopic system has developed since the early 1970s into one of the most widely applied and petrologically useful of all the decay schemes. The utility stems from several factors. The rare earth elements (REEs)—samarium and neodymium—are both refractory and lithophile, and therefore their relative abundances were not affected by core formation or modified during Earth's accretion (see Chapters 2.01 and 2.15). For this reason, the bulk silicate portion of the Earth (mantle + crust) is believed to have chondritic relative abundances of samarium and neodymium as well as the other REEs. This provides a powerful advantage compared to other schemes, for example, Rb–Sr and U–Th–Pb isotopic systems, which rely on elements of different volatilities or which may have been fractionated during core formation, creating significant uncertainties in the parent/daughter ratio and isotopic composition of the bulk silicate Earth. In contrast, the Sm/Nd and neodymium isotopic composition of chondritic meteorites provide a well-defined model for the primitive mantle of the Earth (Jacobsen and Wasserburg, 1980).

In the Earth, fractionation of the LREEs occurs primarily by magmatic processes such as partial melting and fractional crystallization. Extraction of LREE-enriched (low Sm/Nd) continental crust from the primitive mantle has resulted in an LREE-depleted (high Sm/Nd) upper mantle with radiogenic $^{143}Nd/^{144}Nd$ compositions, and complimentary unradiogenic or low $^{143}Nd/^{144}Nd$ ratios in the average felsic crust, as compared to the bulk Earth. Owing to the genetic importance of small deviations from chondritic or primitive mantle compositions in the $^{147}Sm-^{143}Nd$ isotopic system, epsilon (ε_{Nd}) notation is used to reference the $^{143}Nd/^{144}Nd$ isotopic composition of a sample to the primitive mantle, where $\varepsilon_{Nd} = [(^{143}Nd/^{144}Nd_{(t)\text{sample}}/^{143}Nd/^{144}Nd_{(t)\text{CHUR}} - 1) \times 10^4]$ with $t =$ the crystallization age of the rock, and CHUR the $^{143}Nd/^{144}Nd$ isotopic composition of a chondritic uniform reservoir (= primitive mantle) at time t, where $^{147}Sm/^{144}Nd_{CHUR} = 0.1966$, and present-day $^{143}Nd/^{144}Nd_{CHUR} = 0.512638$ for the analytical protocols used in most laboratories.

Classic studies of oceanic basalts have demonstrated that "normal" or LREE-depleted mid-ocean ridge basalts (N-MORB) typically exhibit a relatively narrow range of neodymium compositions and define a modern depleted mantle $\varepsilon_{Nd} = +10 \pm 2$ (e.g., Dickin, 1995; Chapter 2.03). This key observation implies that the convecting upper mantle which is the source of N-MORB is characterized by long-term depletion in the LREEs and, by inference, other incompatible lithophile elements, compared with the bulk silicate Earth. In contrast, OIBs have a much greater range of compositions, extending from values similar to those of N-MORB, to much less radiogenic values (e.g., Hofmann, 1997). The neodymium isotopic compositions of MORBs and OIBs appear to be reflecting complex processes of melt depletion and crustal recycling, which have operated throughout much, if not all, of Earth history, but how the mantle achieved its present-day composition remains a matter of active research. A number of studies have compiled initial neodymium isotopic compositions of both mafic and felsic samples using various selection criteria in order to define the evolution of the mantle through time. Among the most commonly used reference curves are those of DePaolo (1981), DePaolo *et al.* (1991), and Goldstein *et al.* (1984). In principle, knowing the isotopic evolution of the mantle through time should provide first-order constraints on problems such as the growth rate of continental crust and development of enriched reservoirs in the mantle; however, a number of issues complicate a straightforward interpretation

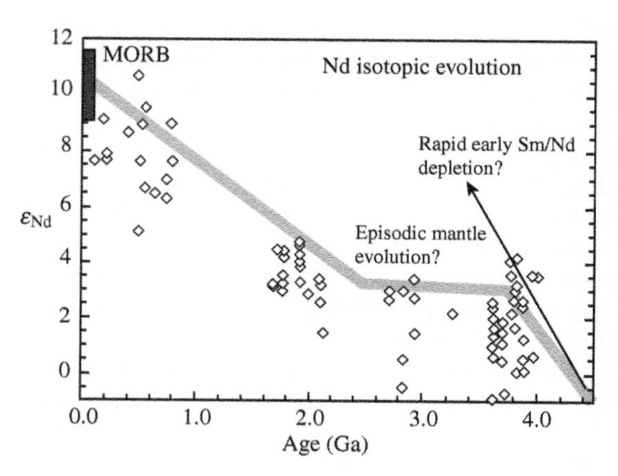

Figure 1 ε_{Nd} evolution of the depleted mantle as defined by juvenile granites and Phanerozoic ophiolites. The inflections in the evolution curve may represent rapid early depletion of the mantle, followed by mixing of more highly depleted mantle reservoirs with either less depleted mantle, or enriched crustal components, between the early and late Archean (sources Jacobsen and Dymek, 1988; Bowring and Housh, 1995; Bennett *et al.*, 1993; Collerson *et al.*, 1991; Baadsgaard *et al.*, 1986; Moorbath *et al.*, 1997; Vervoort and Blichert-Toft, 1999; and the compilation of Shirey, 1991).

of the data. For example, at any given time, the mantle appears to be characterized by a range of initial neodymium compositions. As incorporation of older continental crust typically lowers the apparent initial ε_{Nd}, the most positive ε_{Nd} composition at any time is usually taken as most representative of the depleted mantle composition. A feature common to all depleted mantle curves is the clear, monotonic increase in ε_{Nd} compositions from at least 2.7 Ga to the present day (Figure 1), which is a reasonably well-established feature of the mantle.

2.13.2.3 ^{143}Nd Isotopic Evolution of the Archean and Hadean Mantle

Mantle evolution curves for neodymium isotopic compositions in the Earth prior to 2.7 Ga are much less well defined than for younger time periods. Many studies have elected simply to extrapolate a linear evolution from 2.7 Ga to the composition of the primitive mantle ($\varepsilon_{Nd} = 0$) at 4.56 Ga (e.g., Goldstein *et al.*, 1984). In contrast, other studies have attempted a more precise definition of the neodymium mantle evolution curve for the Archean based on observed compositions (e.g., Bennett *et al.*, 1993; Bowring and Housh, 1995; Vervoort and Blichert-Toft, 1999). The justification of these efforts is that Archean mantle chemistry, particularly for the long-lived isotopic systems, provides a window into the first 700 Myr of Earth's history, prior to establishment of an abundant rock record. This key period in Earth history includes the transition from differentiation processes operative on Earth in the period immediately after accretion, to plate tectonic processes similar to those observed today. However, determining precise isotopic compositions for the ancient mantle, particularly for the first 2 Gyr of Earth history, is a much more complex proposition than defining the composition of the modern mantle. Careful integration of geochemical, field, and geochronologic observations is required (cf. Nutman *et al.*, 2000), and even so, the results from Archean samples are controversial, with much of the discussion centered on possible roles of alteration and open system chemical behavior in modifying their isotopic signatures (Gruau *et al.*, 1996).

Beginning with the first studies of Archean samples (Hamilton *et al.*, 1979; McCulloch and Compston, 1981), neodymium isotopic investigations of ancient rocks have established and continued to reveal the depleted nature of the mantle in the early Earth. The oldest crustal remnants, the ~4 Ga Acasta gneisses, are characterized by a range of initial ε_{Nd} values from +4 to −4 (Bowring and Housh, 1995; Bowring and Williams, 1999). In fact, almost all >3.6 Ga gneiss suites have initial values greater than $\varepsilon_{Nd} = 0$, and some appear to have values as positive as $\varepsilon_{Nd} = +4$ (Figure 1). These include samples of the Acasta gneiss complex, Canada (Bowring and Housh, 1995), the Uivak gneisses of Labrador (Collerson *et al.*, 1991), and the Itsaq complex (containing the Amîtsoq gneisses, the Isua Supracrustal Belt, and Akilia association supracrustal rocks; Nutman *et al.* (1996)) of southwest Greenland (Bennett *et al.*, 1993; Jacobsen and Dymek, 1988; Moorbath *et al.*, 1997). The veracity of these unexpectedly positive ε_{Nd} isotopic compositions in Early Archean rocks has, however, been strenuously questioned (e.g., Moorbath *et al.*, 1997). Early Archean rocks are invariably preserved in complex granite–gneiss terrains, which have undergone a prolonged history of metamorphism and tectonism. The question, therefore, arises as to whether these rocks do, in fact, preserve their original initial isotopic ratios, or are these highly positive ε_{Nd} compositions artifacts of subsequent, perhaps subtle, alteration. Alteration is not an uncommon problem, for example, disturbance of initial strontium ratios of Archean rocks is generally recognized as being pervasive due to the mobility of rubidium in crustal fluids. For the Sm–Nd system, the geochemical similarity and generally immobile nature of these elements in most geologic environments has led to the expectation that they are generally robust against secondary disturbance. This confidence has been justified for many Precambrian terranes as, for example, shown by the regionally coherent data sets compiled in DePaolo *et al.* (1991), but exceptions do exist, as demonstrated by McCulloch and Black (1984), in high-grade metamorphic rocks from Enderby Land in the Antarctic, where resetting of the Sm–Nd system due to younger granulite grade metamorphism has been documented.

Disturbance of the Sm–Nd systematics of ancient gneisses cannot be known *a priori*; however, various criteria can be used to evaluate the integrity of data sets. For example, Bennett *et al.* (1993) reported internally consistent compositions with relatively high initial ε_{Nd} for a suite of 3.76–3.87 Ga rocks from the Itsaq complex of southwest Greenland, with 12 out of 14 analyzed samples falling within the range of $\varepsilon_{Nd} = 0$ to +4.5. This area of Greenland contains the most extensive exposures of well-preserved, locally homogeneous ancient gneisses in the world, and the crystallization ages for all of the samples in that study were determined by high precision U–Pb analysis of individual zircons. Only those samples with simple concordant zircon populations yielding well-defined crystallization ages and showing the least evidence for either metamorphic overprinting or lead mobility were used in the neodymium isotopic studies. Although unmodified zircon U–Pb characteristics do not prove that the whole-rock Sm–Nd system has also been undisturbed, it is evidence that large amounts of fluids have not interacted with these samples. An alternative approach to determining the initial neodymium isotopic compositions of gneisses was used by Moorbath *et al.* (1997). By grouping together 24 Early Archean gneiss samples from various Archean terranes in southwest Greenland, they were able to generate a data array, whose slope indicated an age of 3,640 ± 120 Ma with an initial $\varepsilon_{Nd} = 0.9 \pm 1.4$. In the same study a data array for 58 Isua metasediments (felsites and mica schists) suggested an age of 3,776 ± 52 Ma with initial $\varepsilon_{Nd} = 2.0 \pm 0.6$. The gneiss "isochron" serves to illustrate the problems with the regional whole-rock isochron approach. As discussed in detail by McGregor (2000), this data set incorporated a wide range of lithologies, with a ~300 Myr range of Early Archean ages such that the samples could not have been derived from the same source; many of these samples were identified as being from evolved granite suites interpreted to form with significant pre-existing, low-ε_{Nd} crustal components (e.g., Nutman *et al.*, 1984). Therefore, it is unclear how stringent a constraint these data place on mantle evolution.

Although there is some debate as to whether initial ε_{Nd} values were as positive as +3 to +4, in the early Earth, it is quite clear that ε_{Nd} values of *at least* +2 at 3.8 Ga were prevalent in the Early Archean upper mantle. Confirmation of very positive ($\varepsilon_{Nd} > 4$) values in rocks older than

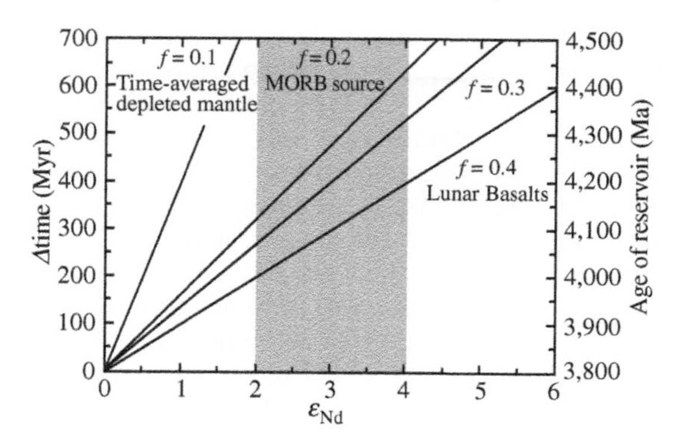

Figure 2 The relationship between f, the chondrite normalized $^{147}Sm/^{144}Nd$, ε_{Nd} evolution, and time. The curves represent the rate of change of ε_{Nd} for different f values in the Early Archean mantle. Δtime is the difference between the age of the Earth (~4.5 Ga) and the sample age. The shaded area shows the range of initial ε_{Nd} compositions that have been proposed for ~3.8 Ga mantle based on data from Early Archean gneisses. Depending on the age of the mantle fractionation event, the Early Archean upper mantle may have been equally, or even more LREE depleted than the modern N-MORB source mantle.

3.7 Ga would be extremely important, as this would require extreme early differentiation of the Earth's mantle within the first 100–400 Myr of Earth history. As shown in Figure 2, to generate ε_{Nd} of +2 to +4 at 3.8 Ga requires $^{147}Sm/^{144}Nd$ ratios in the pre-3.8 Ga upper mantle similar to those in the present-day depleted mantle, yet the modern mantle records the effects of extraction of the whole of the continental crust. Thus, neodymium isotopic compositions of the early preserved continental crust provide strong evidence that portions of the Earth's upper mantle in the Early Archean were significantly lithophile-element-depleted requiring very early (>4.0 Ga) differentiation.

2.13.2.4 ^{176}Hf Isotopic Evolution

The $^{176}Lu-^{176}Hf$ isotopic system (half-life ~37 Gyr) is, in many ways, chemically similar to $^{147}Sm-^{143}Nd$. In both isotopic schemes the parent and daughter elements are refractory lithophile elements, such that their relative abundances in the Earth were probably not modified during accretion, nor did they participate in core formation. Thus, as for the Sm–Nd system, the compositions of chondritic meteorites can, in principle, be used to establish bulk silicate Earth isotopic compositions and Lu/Hf ratios directly. The potential, therefore, exists for establishing a precise isotopic baseline to use for recognizing fine-scale deviations in isotopic compositions, which can then be used to reveal the differentiation processes that have affected mantle chemical evolution. In practice, the most recent data for chondritic meteorites (e.g., Bizzarro *et al.*, 2003) show a range of Lu/Hf and measured $^{176}Hf/^{177}Hf$ compositions making the selection of representative bulk silicate Earth compositions ambiguous.

As for neodymium, ε notation is used in the Lu–Hf system where $\varepsilon_{Hf} = [(^{176}Hf/^{177}Hf_{(t)sample}/^{176}Hf/^{177}Hf_{(t)CHUR} - 1) \times 10^4]$, and t is the crystallization age of the sample. The commonly used chondritic reference compositions (CHUR) are $^{176}Lu/^{177}Hf = 0.0332$ and present-day $^{176}Hf/^{177}Hf = 0.282772$ (Blichert-Toft and Albarède, 1997). Partial melting of the mantle typically results in a melt with lower Lu/Hf, and a complementary residual mantle with higher Lu/Hf evolving over time to positive ε_{Hf} values. The modern depleted upper mantle, as reflected in N-MORB compositions, is characterized by $\varepsilon_{Hf} \sim +16 \pm 4$ (e.g., Salters and White, 1998), as compared with $\varepsilon_{Nd} \sim +10 \pm 2$. Early efforts at determining hafnium isotopic compositions using thermal ionization mass spectrometry (TIMS) (Patchett and Tatsumoto, 1980; Patchett, 1983) served to demonstrate the generally parallel behavior of the $^{176}Lu-^{176}Hf$ and $^{147}Sm-^{143}Nd$ systems in almost all geochemical environments. These initial studies also outlined the strong correlations between ε_{Nd} and ε_{Hf} isotopic compositions in oceanic basalts, with ε_{Hf} values being about twice that of ε_{Nd}. Owing to the analytical difficulties of making precise hafnium isotopic measurements by TIMS, as well as its seeming redundancy with the $^{147}Sm-^{143}Nd$ system, the widespread application of the $^{176}Lu-^{176}Hf$ isotopic system was delayed until the advent of plasma source multicollector mass spectrometers in the mid-1990s. This new technology resulted in significantly improved ionization efficiency for hafnium and enabled simpler chemical separation schemes (cf. review of Blichert-Toft (2001)). More recent studies based on larger numbers of samples have served to confirm the generally well-correlated behavior of Lu/Hf and Sm/Nd ratios during mantle melting throughout much of Earth history, with initial isotopic compositions of samples of juvenile crust defining the relationship $\varepsilon_{Hf} = 1.4\varepsilon_{Nd} + 2.1$ ($r^2 = 0.7$) for at least the last 3 Gyr (Vervoort and Blichert-Toft, 1999). The general conclusions regarding mantle evolution as determined from hafnium isotopic compositions are, therefore, similar to those from neodymium isotopic compositions, which is that the mantle has become increasingly depleted through time with a near linear evolution throughout much of the Proterozoic and Phanerozoic (Figure 3). This is largely due to extraction of continental crust coupled with some degree of crustal recycling

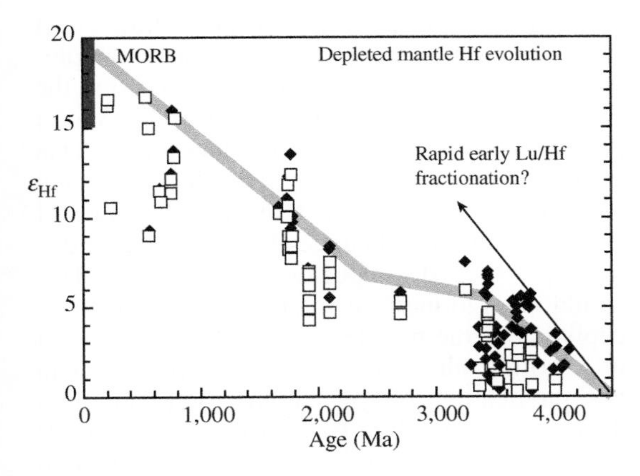

Figure 3 Mantle evolution defined by the initial ε_{Hf} of juvenile granites and their zircon separates. The ε_{Hf} compositions (closed symbols) have been recalculated using the most recently proposed ^{176}Lu decay constant of Bizzarro *et al.* (2003). For comparison are shown the initial values (open symbols) calculated using the Blichert-Toft and Albarède (1997) decay constant. Note that the change of decay constant results in an increase in calculated initial ε_{Hf} compositions of some Early Archean samples. The current database suggests rapid Lu/Hf fractionation in the early Earth and the possibility of episodic changes in mantle evolution similar to that seen in ε_{Nd} evolution (sources Vervoort and Blichert-Toft, 1999; Amelin *et al.*, 1999, 2000; Salters, 1996).

back into the mantle. In detail, however, combined Hf–Nd isotopic studies, particularly for the Early Archean, have engendered much interest and controversy. As discussed in the following section, a particularly vexing problem for a precise comparison of Nd and Hf isotopic compositions of early crustal rocks is the current controversy over uncertainties in the decay constant for ^{176}Lu. Fewer concerns have been expressed over the integrity of the Lu–Hf isotopic system during alteration, despite the fact that these elements show a greater contrast in compatibility than Sm and Nd, and can be hosted in different phases (e.g., zircon versus garnet).

2.13.2.5 ^{176}Hf Isotopic Constraints on the Evolution of the Archean and Hadean Mantle

As described in Section 2.13.2.4, there has been, and continues to be, much debate over the veracity of the highly positive ε_{Nd} isotopic compositions observed in >3.6 Ga rocks. Lutetium and hafnium are sited in different phases than are samarium and neodymium, such that metamorphism and alteration should affect each isotopic system differently. One motivation for much of the hafnium isotopic work on Archean samples owes to its abundance in the mineral zircon ($ZrSiO_4$), which is a common accessory phase in most granitoids, including ancient gneisses. Zircon typically contains 1–3 wt.% hafnium with only limited amounts (<1 ppm) of lutetium, resulting in extremely low Lu/Hf ratios. These very low Lu/Hf ratios require almost no correction for *in situ* decay even for very ancient (>3.6 Ga) zircons. In addition, zircon is extremely resistant to metamorphism, and is often the only relict igneous phase remaining through the high-grade metamorphic events that affect most Early Archean samples. Zircon is also the mineral of choice for direct U–Pb age determinations using either *in situ* techniques such as ion microprobe and laser-ablation ICPMS, or conventional solution methods and thermal ionization mass spectrometry. This means that both ages and initial hafnium isotopic compositions can be determined on the same phase, largely eliminating problems of age correlations, correction for *in situ* decay, and secondary disturbance of the isotopic system. Combining the integrity of the hafnium system in zircons, with the expected $\varepsilon_{Hf}-\varepsilon_{Nd}$ correlations derived from younger rocks, would seem to be an ideal basis for testing the validity of the positive ε_{Nd} values in Early Archean samples, by comparing hafnium isotopic compositions in zircons with whole-rock initial neodymium isotopic compositions of the same sample. With this objective, data were obtained for a suite of >3.6 Ga gneisses from Southwest Greenland (Vervoort *et al.*, 1996), many of them the same samples used to propose rapid ^{143}Nd evolution in the early Earth (Bennett *et al.*, 1993). Vervoort *et al.* (1996) found ubiquitous positive initial ε_{Hf} values in the zircons, requiring derivation from a previously depleted mantle. However, the ε_{Hf} values were not as high as expected on the basis of the assumed neodymium correlations, with correlation closer to 1 : 1 $\varepsilon_{Nd}\varepsilon_{Hf}$ instead of 1 : 1.4 as found for younger rocks see Section 2.13.2.4. Their conclusions were that the Earth's mantle was depleted before 3.8 Ga, but not to the extent indicated by the ε_{Nd} values, and that there was no evidence for large amounts of early crust, either felsic or basaltic (Vervoort *et al.*, 1996). The ^{147}Sm–^{143}Nd data were considered to be in error as a consequence of open system disturbance (e.g., Gruau *et al.*, 1996). Subsequent higher precision work on a more extensive suite of Early Archean samples also produced samples with initial ε_{Hf} values from +0 to +4.6 and these were used to reassert these interpretations (Vervoort and Blichert-Toft, 1999). Even though the ε_{Hf} values found by these studies were not as extreme as expected based on the assumed correlation with ε_{Nd}, the new hafnium data did confirm a major conclusion of the neodymium results, that of nonlinear isotopic mantle evolution. The hafnium evolution curve obtained from the zircons parallels the neodymium evolution curve obtained from whole-rock compositions, and describes rapid change in isotopic compositions from 4.56 Ga to ~3.6 Ga. This is followed by near-constant or even decreasing hafnium and neodymium isotopic compositions from 3.8 Ga to 2.7 Ga, and nearly linear evolution from the Proterozoic onward (Figure 3). The apparent buffering of mantle compositions between ~3.8 Ga and 2.7 Ga provides strong evidence for a major reorganization of mantle and/or crustal reservoirs during this time (Section 2.13.4).

An additional consideration in the interpretation of the Hf –Nd isotopic data is the validity of the assumption of well-defined $\varepsilon_{Hf}:\varepsilon_{Nd}$ correlations for individual, limited data sets. There are many exceptions to this rule resulting from variable petrogenetic histories of sample suites. Despite the general global $\varepsilon_{Hf}-\varepsilon_{Nd}$ correlations, in detail any given suite of rocks including MORBs can yield a wide range of $\varepsilon_{Hf}/\varepsilon_{Nd}$ ratios from less than 1 to greater than 2. This dispersion most likely results from variable amounts of residual garnet, and the timing of the last melt extraction event from the depleted mantle. The amount of residual garnet exerts a strong control over both the Lu/Hf ratios in the derived magmas (e.g., Salters, 1996) and over the Lu/Hf of the residual mantle which controls its hafnium isotopic evolution. For example, basalts from the Rio Grande Rift that are derived from shallow melting of previously depleted spinel peridotite (Johnson and Beard, 1993) all have $\varepsilon_{Hf}/\varepsilon_{Nd}\sim1$, which lies below the canonical mantle correlation. The $\varepsilon_{Hf}/\varepsilon_{Nd}$ compositions of these lavas must reflect specific petrogenetic processes rather than open system isotopic disturbances, and highlights the uncertainties involved in assuming specific $\varepsilon_{Hf}:\varepsilon_{Nd}$ correlations for the Early Archean mantle.

Disturbing uncertainties in the interpretation of hafnium isotopic data from Early Archean samples have been introduced by recent proposed revisions of the ^{176}Lu decay constant. The most widely adopted decay constant for ^{176}Lu ($(1.94 \pm 0.7) \times 10^{-11}$ yr^{-1}) was derived from a slope of eucrite meteorites of known age (Patchett and Tatsumoto, 1980). This value was used until 1997 when a more precise estimate ($\lambda\,^{176}\text{Lu} = (1.93 \pm 0.03) \times 10^{-11}$ yr^{-1}) based on counting experiments (Sguigna *et al.*, 1982) was adopted (Blichert-Toft and Albarède, 1997). Scherer *et al.* (2001) combined results from direct counting experiments with isochron determinations on minerals of known U–Pb ages and high Lu/Hf ratios to suggest a new decay constant, which is significantly different from previously determined values ($\lambda\,^{176}\text{Lu} = (1.865 \pm 0.015) \times 10^{-11}$ yr^{-1}). The ~4% difference between the values of Scherer *et al.* (2001) and Blichert-Toft and Albarède (1997) has little effect (less than an epsilon unit) on most samples younger than 2.7 Ga, but large effects on the calculated ε_{Hf} values of low Lu/Hf, >3.6 Ga zircons and whole rocks. If ε_{Hf} values for >3.6 Ga gneisses are recalculated using the decay constant of Scherer *et al.* (2001) and the CHUR values of Blichert-Toft and Albarède (1997), near chondritic ($\varepsilon_{Hf} = 0 \pm 3$) compositions are obtained. This would seem to indicate no significant amounts of depletion of the mantle at 3.6 Ga. At the time of this writing, the situation remains in flux with the publication of yet another new lutetium decay constant by Bizzarro *et al.* (2003) ($\lambda\,^{176}\text{Lu} = (1.983 \pm 0.033) \times 10^{-11}$ yr^{-1}) based on a precise isochron determined from chondritic meteorites. The present-day chondritic (and therefore bulk silicate Earth) hafnium isotopic compositions and Lu/Hf ratio determined from this meteorite isochron agree with the compositions determined by Blichert-Toft and Albarède (1997), and would indicate a much more positive initial ε_{Hf} for the Early Archean gneisses ($\varepsilon_{Hf} = +5 \pm 1$; Figure 3). Interestingly, the newly revised compositions would be consistent with an $\varepsilon_{Nd}-\varepsilon_{Hf}$ correlation of ~1 : 1.4 for many of the Early Archean samples. The initial hafnium isotopic compositions of the low Lu/Hf zircon separates are also not affected by the change in decay constants; however, their calculated initial ε_{Hf} values change dramatically owing to the large age corrections necessary for the present-day chondritic reference compositions.

Uncertainties in the interpretation of hafnium isotopic compositions in the early Earth are highlighted by consideration of the most ancient terrestrial materials, the ≥4.0 Ga zircons from the Jack Hills of Western Australia (Froude *et al.*, 1983; Wilde *et al.*, 2001). Early researchers recognized that determination of the hafnium isotopic compositions of these grains could provide the oldest direct constraints on early mantle evolution, but were unable to obtain sufficiently precise measurements by ion microprobe (Kinny *et al.*, 1991). Amelin *et al.* (1999), using high precision solution methods, determined U–Pb ages and hafnium isotopic compositions from the same single zircon grains with the results from the oldest (>3.9 Ga) zircons giving negative initial ε_{Hf} compositions using the half-life reported by Blichert-Toft and Albarède, 1997. This was interpreted as indicating derivation of the zircons from a very old (>4.3 Ga) enriched crustal reservoir. Use of the half-life of Scherer *et al.* (2001) changes the calculated initial ε_{Hf} of these grains to even more negative values, as low as −8. A strict interpretation of these results would be that some enriched (i.e., granitic) crust formed at >4.4 Ga survived until ~4.0 Ga, and was then reworked to produce the ~4.0 zircons with very negative initial ε_{Hf} isotopic compositions (Amelin *et al.*, 1999). Despite the evidence that this would provide for early enriched crust, there is, paradoxically, no evidence from zircons from the ~3.6 Ga to 3.8 Ga Greenland gneisses to suggest the presence of either strongly enriched or of depleted reservoirs (Figure 4).

Adoption of the more recently proposed decay constant for ^{176}Lu (Bizzarro *et al.*, 2003) would change the hafnium picture for the Early Archean dramatically. Initial ε_{Hf} values calculated using this half-life result in strongly positive, rather than negative, ε_{Hf} values for the >3.9 Ga Jack Hills zircons (ε_{Hf} up to +4) indicating crystallization of the zircons from magmas derived from a depleted mantle source, rather than enriched crust, with depletion starting at >4.3 Ga (Figure 4). Although the initial ε_{Hf} values determined for >3.9 Ga zircons using *any* of the proposed lutetium half-lives require the early (>4.3 Ga) formation of differentiated chemical reservoirs, the most recently determined half-life apparently provides evidence for the early differentiation of the Earth into enriched crust and depleted mantle reservoirs, and results in a coherent evolutionary picture for the ε_{Hf} and ε_{Nd} isotopic compositions of Early Archean gneisses.

The large potential of the Lu–Hf system for resolving questions of the amount of mantle depletion in the early Earth will remain unfulfilled until both the discrepancies in the ^{176}Lu decay constant as determined from terrestrial (e.g. Scherer *et al.*, 2001) and meteoritic materials

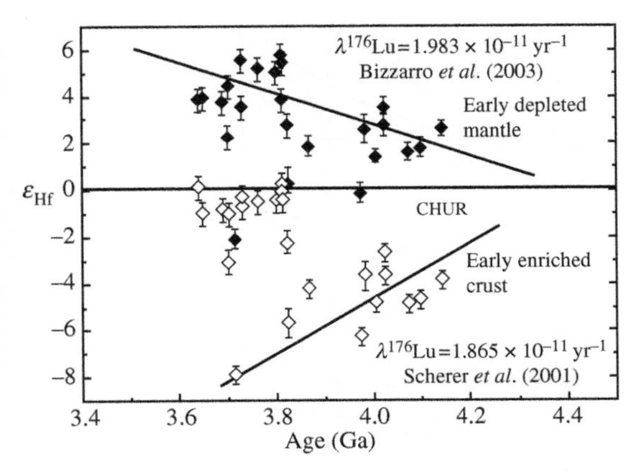

Figure 4 Initial ε_{Hf} values of >3.9 Ga detrital zircons from the Jack Hills, Western Australia (Amelin *et al.*, 1999) and Early Archean Itsaq complex gneisses (Vervoort and Blichert-Toft, 1999). The same data have been recalculated using two recently proposed revisions to the ^{176}Lu decay constant: the closed symbols use the decay constant proposed by Bizzarro *et al.* (2003). The open samples are calculated using the ~6% lower decay constant proposed by Scherer *et al.* (2001). The present-day chondrite parameters used are ^{176}Hf/^{177}Hf = 0.282772 and ^{176}Lu/^{177}Hf = 0.0332 (Blichert-Toft and Albarède, 1997).

(e.g., Patchett and Tatsumoto, 1980; Blichert-Toft and Albarède, 1997; Bizzarro *et al.*, 2003) are understood, and accurate bulk silicate Earth parameters for Lu–Hf are defined. Given early assumptions of bulk-earth ^{176}Hf/^{177}Hf, it appeared that the well-defined Nd–Hf isotopic correlation in modern oceanic basalts did not pass through bulk Earth compositions for hafnium as they did for neodymium, but instead were 2–3 epsilon units above chondritic values. This resulted in proposals for a "missing" low Lu/Hf reservoir in the deep Earth (e.g., Blichert-Toft and Albarède, 1997) or for early Hf fractionation perhaps by crystallization and isolation of small amounts (~1%) of high hafnium material (magnesium perovskite) from a >3.8 Ga magma ocean (Salters and White, 1998). With the large range in chondritic ^{176}Hf/^{177}Hf compositions recently presented (Bizzarro *et al.*, 2003), some of which include lower hafnium isotopic compositions, the conclusion of a nonchondritic mantle with implied missing complimentary reservoirs based on this observation is no longer as firm. However, regardless of the choice of hafnium parameters, the Hf–Nd isotopic arrays of oceanic basalts seem to require a third component, in addition to continental crust and depleted mantle to explain their distribution of compositions, with the most plausible component suggested to be ancient subducted basalts (Blichert-Toft and Albarède, 1997). Other evidence for a missing hafnium component in the Earth comes from unradiogenic hafnium isotopic compositions measured in carbonatites and kimberlites (e.g., Nowell *et al.*, 1999). These melts are characterized by negative initial ε_{Hf} values with one possible explanation being derivation from an ancient reservoir having low Lu/Hf stored in the deep Earth (Bizzarro *et al.*, 2002).

2.13.2.6 ^{142}Nd Isotopic Signatures

In addition to the long-lived ^{147}Sm–^{144}Nd pair, the Sm–Nd isotopic system also contains the short half-life (103 Myr) ^{146}Sm–^{142}Nd decay scheme. All ^{146}Sm has now decayed such that it is no longer possible to generate variations in ^{142}Nd/^{144}Nd compositions. Any deviation from modern ^{142}Nd/^{144}Nd compositions must, therefore, reflect preservation of events occurring while ^{146}Sm was still "alive." With the precision of current measurement techniques (~5–10 ppm) ^{142}Nd/^{144}Nd compositions effectively closed at ~4.2 Ga. Thus, variations in ε_{142} (defined as deviations in parts of 10^4 of ^{142}Nd/^{144}Nd in a sample compared to present-day ^{142}Nd/^{144}Nd) must reflect the effects of differentiation processes occurring during the first 350 Myr of Earth history, i.e., >4.20 Ga. Either positive or negative ε_{142} anomalies could result if reservoirs with fractionated REE abundances formed early in Earth history and survived long enough to be sampled and preserved within the oldest terrestrial rocks. As elaborated in Harper and Jacobsen (1992), the size of any ε_{142} effect is a function of both the time of formation of a fractionated reservoir (T) and the degree of fractionation f (f for isotope ratios such as for Sm–Nd system is defined as $f_{Sm/Nd} = [(^{147}Sm/^{144}Nd)_{sample}/(^{147}Sm/^{144}Nd)_{CHUR} - 1])$ such that

$$\varepsilon_{142}(t) = f\,Q_{142}[^{146}Sm/^{144}Sm] \times [e^{-\lambda 146(T_0-T)} - e^{-\lambda 146(T_0-t)}]$$

where Q_{142} ($354 \times 10^{-9}\ yr^{-1}$) is a constant, $T_0 = 4.56$ Ga, T is the age of the mantle differentiation event resulting in a change of Sm/Nd, and t is the crystallization age of the rock.

ε_{142} effects are well documented in basalts from both the Moon (e.g., Nyquist *et al.*, 1995) and Mars (e.g., Harper *et al.*, 1995). For example, Apollo 17 mare basalts are characterized by $\varepsilon_{142} = +0.25$ and with very positive $\varepsilon_{143} = +7$. These compositions are consistent with derivation of the Apollo 17 basalts from mantle cumulates having extreme Sm/Nd fractionation ($f_{Sm/Nd} = 0.25$–0.4) that formed during crystallization of a lunar magma ocean at ~4.3 Ga. In contrast to the Moon, which has been tectonically dead for billions of years and maintains extensive areas of >3.9 Ga rocks, the chance of finding preserved ε_{142} effects in terrestrial samples is substantially less likely. Harper and Jacobsen (1992) reported the first terrestrial ^{142}Nd anomaly ($\varepsilon_{142} = +0.33$), which was measured in a ~3.7 Ga metasediment from the Isua supracrustal belt, southwest Greenland. This measurement was analytically challenging and the results controversial; attempts to repeat the measurement using an aliquot of the same sample in a different laboratory yielded equivocal results, but with "strong hints of a ^{142}Nd excess" (Sharma *et al.*, 1996). Other labs working on a range of Early Archean rocks did not unambiguously identify any other ^{142}Nd excesses (e.g., Goldstein and Galer, 1992; Regelous and Collerson, 1996; McCulloch and Bennett, 1993). The conclusion in the mid-1990s was that ^{142}Nd variations were not present in >3.8 Ga rocks, which would suggest that the Sm/Nd fractionation event responsible for widespread positive ε_{143} values in Early Archean rocks must have occurred after ~4.3 Ga (Figure 5), with no preserved evidence of very early differentiation. Recently, the situation has changed dramatically with three different laboratories using improved analytical techniques, in one case employing a multicollector ICP-MS and in

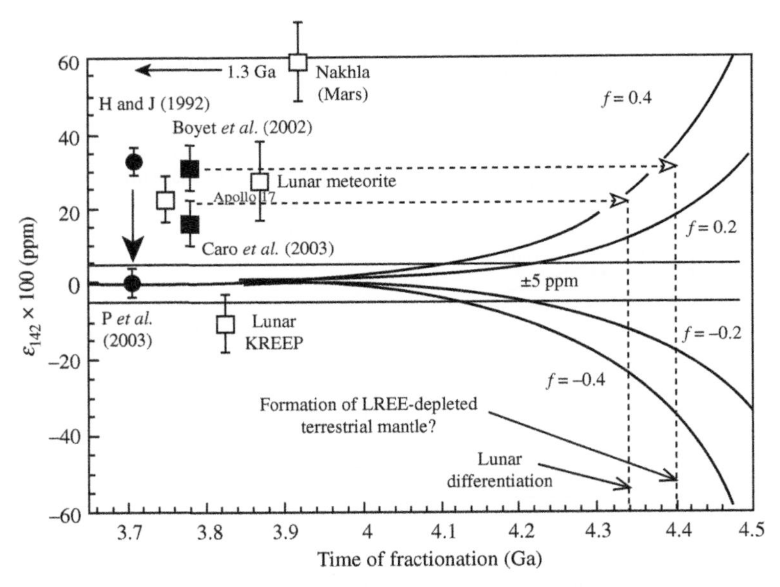

Figure 5 Plot of variations in ^{142}Nd (which arise from the decay of the extinct nuclide ^{146}Sm) versus time expressed as ppm differences from present-day compositions. An initial $^{146}Sm/^{144}Sm = 0.007$ is assumed (Prinzhofer *et al.*, 1992). If an LREE-depleted mantle (high Sm/Nd) formed within the first ~200–300 Ma of Earth history, then with the current 5–10 ppm analytical precision, a positive ^{142}Nd deviation would be expected to be observed in any sample derived from this source. Lunar samples (Nyquist *et al.*, 1995) and martian meteorites (Harper *et al.*, 1995) show clear evidence of ^{142}Nd variations reflecting very early differentiation on these bodies. ^{142}Nd anomalies have now been suggested for some 3.7–3.8 Ga terrestrial samples from southwest Greenland (Harper and Jacobsen, 1992; Boyet *et al.*, 2002; Caro *et al.*, 2003). Most of the ~3.8 Ga gneisses that have been measured do not show these effects and recent re-measurement of the original Harper and Jacobsen sample by Papanastassiou *et al.* (2003) has not confirmed the original result (see text). The different f curves are for different degrees of depletion (20% and 40%) of Sm/Nd in the Early Archean mantle.

the other two a new generation thermal ionization mass spectrometer, now continuing the search for ^{142}Nd excesses in ancient terrestrial rocks. Two labs have reported measuring ε_{142} anomalies from >3.7 Ga rocks from the Isua area (Boyet *et al.*, 2002; Caro *et al.*, 2003). Surprisingly, re-analysis of the original sample that was purported to have yielded the first terrestrial positive ε_{142} (Harper and Jacobsen, 1992) by Papanastassiou *et al.* (2003) using more modern instrumentation has failed to detect a resolvable excess of ^{142}Nd. Thus, at present the situation with respect to terrestrial ^{142}Nd effects is unclear, with several laboratories actively pursuing the problem. If ^{142}Nd variations are confirmed, a very different picture for the timing of early terrestrial differentiation emerges. Interpretations of these signatures are less equivocal than many other types of chemical indicators, as they can only result from very ancient events; any subsequent alteration or mixing with younger materials can only dilute or erase these signatures of early differentiation events. Verification of widespread ε_{142} effects in ancient terrestrial rocks would suggest the formation of major silicate reservoirs with highly fractionated LREE before ~4.3 Ga (Figure 5). This would implicate early planetary scale differentiation processes associated with Earth's formation as a cause of depletion of the Hadean upper mantle. Measurement of additional ancient terrestrial materials, including the 4.0–4.3 Ga zircons from Western Australia, has the potential to provide the least equivocal and tightest constraints on the timing and processes of early mantle depletion.

2.13.2.7 ^{187}Os Isotopic Evolution

Advances in analytical methods (e.g., Shirey and Walker, 1995; Creaser *et al.*, 1991) since the early 1990s have allowed the previously somewhat obscure ^{187}Re–^{187}Os decay scheme to come into its own. Owing largely to the chalcophile or siderophile characteristics of the parent and daughter elements, and thus their contrasting behavior during melting processes as compared with the lithophile-element-based radiogenic isotopic schemes, for example, Rb–Sr and Sm–Nd, ^{187}Os isotopic compositions have provided a significant new perspective on mantle evolution processes. Shirey and Walker (1998) provide an excellent overview of the Re–Os system and its application to geochemistry. The fundamentals of this system are that osmium is compatible during mantle melting, whereas rhenium is moderately compatible, resulting in mantle materials, including abyssal peridotites, xenoliths, and ultramafic portions of ophiolites being highly enriched in osmium (~3 ppb) relative to crustal rocks (~0.03 ppb). Thus, for this system the best recorders of mantle compositions are direct mantle samples and their high-osmium, low-Re/Os mineral phases such as chrome spinel and sulfides, rather than mantle-derived basalts or juvenile granitoids. In contrast to all the other isotopic decay schemes that have been discussed here, rhenium and osmium are both refractory, and highly siderophile elements when metal is present. Thus, they probably accreted to the Earth in their original chondritic abundances, but were quantitatively stripped from the silicate Earth and sequestered in the core (cf. Chapter 2.01 and Chapter 2.15) during early planetary differentiation. The mantle's current compliment of highly siderophile elements, including rhenium and osmium, likely results from a late meteoritic veneer added postcore formation (Chou, 1978), although metal–silicate equilibrium in a deep magma ocean may also have played a role (Righter and Drake, 1997; Chapter 2.10). The result was an upper mantle (there are few constraints on the highly siderophile element composition of the lower mantle) with highly siderophile elements present in chondritic relative proportions, but at greatly diminished (0.8%) concentrations compared to chondritic precursor materials. Although it is not possible to know *a priori* the osmium isotopic composition and Re/Os for the silicate Earth, it is apparent from the osmium systematics of fertile peridotite xenoliths that the ^{187}Os isotopic composition of the modern primitive upper mantle is within the range of chondritic meteorites, and in detail is more similar to enstatite and ordinary chondrite compositions rather than carbonaceous chondrites (Meisel *et al.*, 1996, 2001). Isochrons determined from iron meteorites, which have high concentrations of osmium (several ppm) and a range of Re/Os ratios, have provided precise constraints on the initial ^{187}Os composition of solar-system materials and on the ^{187}Re decay constant (1.666×10^{-11} yr^{-1}; e.g., Smoliar *et al.* (1996) and Shen *et al.* (1996)). Owing to the large ^{187}Os isotopic variations in terrestrial materials resulting from more extreme parent–daughter fractionations during melt formation as compared to, for example, the Sm–Nd system, osmium isotopic data are often reported in percent deviations from reference compositions. This is referred to as "γ-osmium," where $\gamma_{Os(t)} = [(^{187}Os/^{188}Os_{sample(t)})/(^{187}Os/^{188}O_{chon(t)}) - 1] \times 100$ (Walker *et al.*, 1989). By convention, a present-day average chondritic reference value of $^{187}Os/^{188}Os = 0.1270$ is used, although chondrites can have a range of osmium isotopic compositions. This chondritic reference value is slightly lower than the best estimate for primitive upper mantle (PUM) of $^{187}Os/^{188}Os = 0.1296 \pm 8$ (Meisel *et al.*, 2001), such that present-day PUM would have $\gamma_{Os} \sim +2$.

Another significant difference between the Re–Os system and other commonly used isotopic systems (e.g., Sm–Nd, Lu–Hf) is that the crust is not likely to be the primary compliment to any depletions in the upper mantle. Because the volume of the crust is small compared with the mantle and with very low osmium concentrations, the formation of the continental crust has not significantly affected the osmium isotopic evolution of the upper mantle (e.g., Martin *et al.*, 1991). The two most widely cited estimates of crustal compositions both suggest average concentrations for osmium of ~0.03 ppb, and for rhenium of 0.4–0.5 ppb (Esser and Turekian, 1993; Peucker-Ehrenbrink and Jahn, 2001). A higher rhenium concentration (~2 ppb) for the continental crust has been suggested on the basis of measurements of primitive-arc lavas from the South Pacific (Sun *et al.*, 2003). If the higher rhenium concentrations in arcs are shown to characterize average continental crust, then the crust will play a more significant, but still minor role as the compliment to long-term mantle Re/Os depletions. The most probable mechanism to balance any mantle depletion in ^{187}Os is through long-term (>10^9 yr) storage of significant amounts of high-Re/Os material in the deep mantle, with the most likely reservoir being subducted oceanic crust. The existence of a reservoir of ancient basalt in the deep mantle is also attractive in that it can provide a source for the high-^{187}Os component of OIBs (e.g., Hauri and Hart, 1997). Thus, the high potential of the Re–Os system to reveal previously unknown reservoirs within the mantle is of great interest for the study of mantle evolution.

The modern MORB source mantle is characterized by a range of ^{187}Os isotopic compositions, extending from chondritic to subchondritic compositions. The commonly cited average ^{187}Os/^{188}Os for abyssal peridotite compositions is 0.1247 (Snow and Reisberg, 1995), which is 3% below the nominal PUM composition. Other studies present both higher and lower ^{187}Os/^{188}Os compositions for the upper mantle (e.g., Martin *et al.*, 1991; Hattori and Hart, 1991; Roy-Barman and Allègre, 1994; Schiano *et al.*, 1997). At least some of the variability may be due to alteration by high ^{187}Os seawater, but much of the range measured in modern MORB likely represents true isotopic variability within the mantle. Unfortunately, it is not possible to specify the prevalent, or average, ^{187}Os/^{188}Os ratio of the upper mantle. If the majority of the upper mantle has close to chondritic compositions, then it is not necessary to call upon large amounts of storage of mafic crust to balance mantle depletions. Alternatively, if much of the mantle is characterized by unradiogenic compositions, perhaps as low as ^{187}Os/^{188}Os = 0.1247 (Snow and Reisberg, 1995), then substantial amounts of basalt are needed, with the exact amount depending on the average storage age of the mafic material (Figure 6). For example, assuming a 2×10^9 yr average storage age, and assuming the depleted mantle represents 50% of the total mantle, ~10% of the mantle by mass would have to consist of mafic slabs to account for this level of depletion (Walker *et al.*, 2002; Bennett *et al.*, 2002). If longer storage times or smaller degrees of mantle depletion are assumed, then less stored mafic crust is required. If the mantle is less depleted, for example, with a ^{187}Os/^{188}Os of 0.1281 as some studies suggest (Walker *et al.*, 2002), then only 2–3% of the mantle, by mass, would need to consist of isolated 1.5–2.0 Ga mafic crust. This is still a significant amount of stored material considering that the entire continental crust represents ~0.6% by mass of the mantle.

Attempts have been made to define the ^{187}Os composition of the Phanerozoic mantle through the analysis of ophiolites representing upper

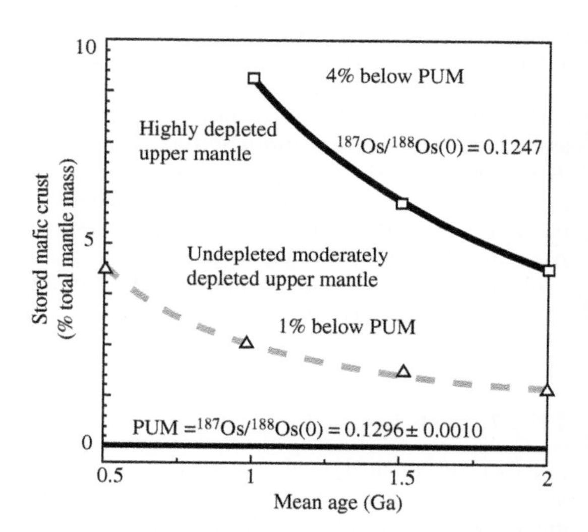

Figure 6 Effect of basalt storage on the ^{187}Os/^{188}Os isotopic composition of the upper mantle (representing 50% of total mantle, by mass) as a function of reservoir mass, age, and ^{187}Os depletion of the upper mantle. The results are presented in terms of mass of stored mafic crust, as a percentage of total mass of the mantle. Mass balance calculations assume that the "missing" Re, complementing depletion in the upper mantle, resides in subducted basaltic crust existing as a stored reservoir within the mantle. (Basaltic crust assumed to have 0.900 ng g^{-1} Re and 0.005 ng g^{-1} Os; upper mantle has 0.280 ng g^{-1} Re and 3.1 ng g^{-1} Os). Depletion of ^{187}Os is with reference to primitive upper mantle composition of Meisel *et al.* (2001) of ^{187}Os/^{188}Os = 0.1296 ± 0.0010. The upper curve assumes that the entire upper mantle is depleted to the extent of average abyssal peridotites, with ^{187}Os/^{188}Os = 0.1247 (Snow and Reisberg, 1995). The other curves are for more moderate amounts of mantle depletion. PUM indicates primitive upper mantle estimates. (after Walker *et al.*, 2002; Bennett *et al.*, 2002).

mantle compositions (e.g., Walker *et al.*, 1996, 2002; Hattori and Hart, 1991). The mineral chromite, a common phase in ultramafic samples, is particularly useful in this regard owing to its high osmium concentrations (ppb levels), very low Re/Os ratios, and resistance to alteration. Thus, initial osmium isotopic compositions determined from chromites are considered to be accurate records of initial mantle ^{187}Os. The compositions determined from a range of ophiolites show a scatter about the PUM evolution curve, with the majority (~68%) of samples falling within ±1% (1 γ_{Os} unit) of the reference curve (Figure 7). In contrast to ophiolites, which are taken to represent MORB source mantle, OIBs, which are generally considered to be derived from mantle plumes, have a wide range of compositions from chondritic, overlapping with MORB mantle, to extremely radiogenic compositions (e.g., Reisberg *et al.*, 1993; Hauri and Hart, 1993; Marcantonio *et al.*, 1995).

Fewer reliable osmium isotopic data exist for Precambrian samples owing to the problems of disturbance of rhenium and osmium during alteration and metamorphism. Key points used to define mantle evolution are indicated in Figure 7. The osmium isotopic constraints on the composition of the Archean mantle are, in many cases, derived from Late Archean komatiite suites. Komatiites are high-MgO lavas that often form differentiated flow sequences, such that Re–Os isochrons can be determined; chromite-rich layers are commonly present and provide a robust estimate of the initial osmium isotopic

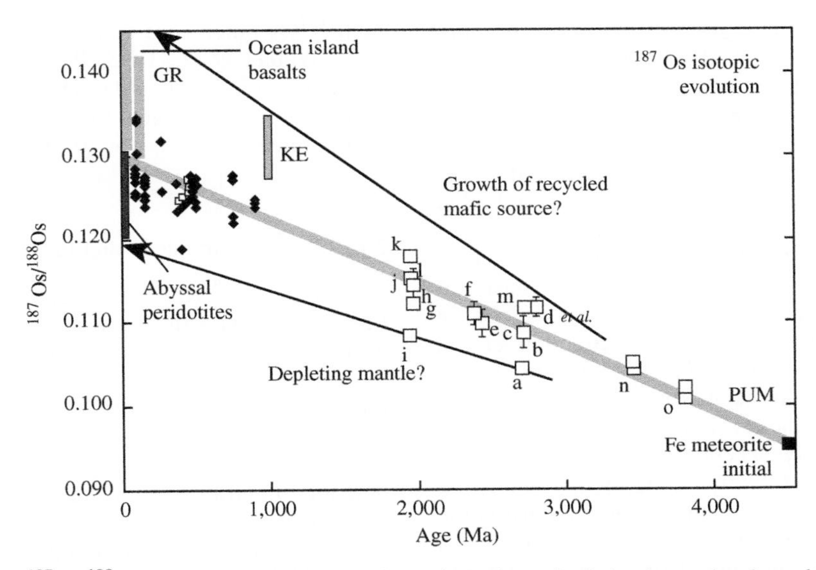

Figure 7 Initial ^{187}Os/^{188}Os isotopic compositions of mantle and mantle-derived samples through time. The bold line represents terrestrial ^{187}Os/^{188}Os isotopic evolution as defined by a solar-system initial composition from a IIA iron meteorite isochron (Smoliar *et al.*, 1996) and a ^{187}Os/^{188}Os composition for modern upper mantle of 0.1296 ± 0.0008 from the Re–Os systematics of peridotite xenoliths (Meisel *et al.*, 1996, 2001). These compositions are within the ranges determined for ordinary and enstatite chondrites (Meisel *et al.*, 2001). Also shown are the range of compositions measured in abyssal peridotites presenting modern upper mantle compositions and OIBs. The small black diamonds are initial compositions determined from Phanerozoic ophiolites (Walker *et al.*, 2002) which represent upper mantle compositions. Gorgona komatiites GR (Brandon *et al.*, 2003) and Keweenawan flood basalts KE (Shirey, 1997) represent likely plume-derived materials. Other data sources: a—isochron initial from Late Archean komatiites from the Abitibi Greenstone belt, Ontario (Walker and Stone, 2001); b—isochron initial from Late Archean komatiites from Munro Township, Ontario (Shirey, 1997); c—isochron initial from Late Archean komatiites from the Kambalda Greenstone belt, Western Australia (Foster *et al.*, 1996); d—isochron initial from Late Archean komatiites of the Kostomuksha greenstone belt, NW Baltic Shield (Puchtel *et al.*, 2001a) e, f—initial compositions from two isochrons from Early Proterozoic komatiitic basalts of the Vetreny belt, Baltic Shield (Puchtel *et al.*, 2001b); g—initial composition determined from high-Os, low-Re/Os laurite grain from the Early Proterozoic Outokumpu ophiolite, Finland (Walker *et al.*, 1996); h—initial composition from low-Re/Os, high-Os phases of Early Proterozoic mafic–ultramafic sill in the Onega plateau flood basalt province, Baltic Shield (Puchtel *et al.*, 1999); i, j, k—three initial compositions from disseminated oxides and chromitite boulders within the Early Proterozoic Jormua Ophiolite Complex, northeastern Finland (Tsuru *et al.*, 2000); l—isochron initial from Early Proterozoic spinel peridotite xenoliths, southeastern Australia (Handler *et al.*, 1997); m—initial compositions from chromites separated from Belingwe komatiites, Zimbabwe (Walker and Nisbet, 2002); n—initial compositions from chromites from the 3.46 Ga komatiites, Western Australia (Bennett *et al.*, 2002); and o—initial compositions from low Re/Os whole-rock and olivine and spinel mineral separates from 3.8 Ga peridotites SW Greenland (Bennett *et al.*, 2002).

composition of the lavas at the time of eruption. A disadvantage to the use of komatiites in defining mantle evolution curves is that they are generally considered to be derived either from deep mantle sources analogous to modern plumes (e.g., Campbell *et al.*, 1989; Puchtel *et al.*, 2001a), or from convergent margin settings (e.g., Parman *et al.*, 2001) rather than depleted upper mantle analogous to the MORB source. Samples from several Late Archean komatiites lie on a chondrite evolution curve (Figure 7). However, in one case, initial osmium compositions determined from komatiites suites define subchondritic values (Abitibi Greenstone Belt; Walker and Stone, 2001), and in two suites, superchondritic initial ^{187}Os were determined (Kostomuksha; Puchtel *et al.* (2001a) and Belingwe; Walker and Nisbet (2002)). Three plausible origins for the radiogenic composition of some komatiite suites include contamination with high-^{187}Os crust, the presence of a long-lived enriched source with high Re/Os stored in the deep mantle, or incorporation of high-^{187}Os outer core materials (cf. Walker and Nisbet, 2002; Puchtel *et al.*, 2001a). Crustal contamination can largely be eliminated as an option on the basis of the trace element and other isotopic chemistry of the studied komatiites suites. Addition of outer core material to some mantle plumes has been suggested on the basis of correlated enrichments of ^{187}Os and ^{186}Os in some OIBs and flood basalts (Brandon *et al.*, 1998, 1999; Walker *et al.*, 1995) and remains a possibility for Archean komatiites, although it would require earlier inner core crystallization than is generally accepted (Walker and Nisbet, 2002). The most likely scenario appears to be some form of ancient subducted mafic crust in the mantle source regions of komatiites, which has also been suggested from trace element considerations of various komatiite suites (e.g., Campbell, 2002).

The oldest samples that provide direct constraints on the osmium isotopic composition of the upper mantle are rare spinel peridotites contained within the Early Archean Itsaq gneiss complex of southwest Greenland that are interpreted to be ~3.81 Ga abyssal peridotites (Friend *et al.*, 2002). The measured and initial compositions determined from low-Re/Os spinel and olivine mineral separates from these peridotites are the most "primitive," in the sense of closest to solar-system initial compositions, ^{187}Os isotopic compositions yet obtained on any terrestrial material (Bennett *et al.*, 2002). This shows that at least some, if not all, of the Early Archean upper mantle was characterized by chondritic ^{187}Os/^{188}Os isotopic compositions. Osmium isotopic constraints from this time period (~3.8–3.9 Ga) are of particular interest as they provide a rough constraint on the timing of the addition of the "late veneer" of highly siderophile elements to the Earth as any meteoritic component must have been added to the Earth and incorporated into the upper mantle by 3.81 Ga, the age of the peridotite.

The emerging picture for the ^{187}Os isotopic evolution of the mantle is that the upper mantle was characterized by chondritic ^{187}Os/^{188}Os isotopic compositions since at least 3.8 Ga. In addition, data from Archean komatiites require both enriched (high Re/Os) and depleted (low Re/Os) reservoirs, leading to both radiogenic and unradiogenic ^{187}Os compositions compared to PUM, to have formed prior to 3.0 Ga. The enriched reservoirs may have been early manifestations of plume sources formed by storage of subducted mafic material. A speculative alternative possibility for high ^{187}Os in some komatiites and modern OIBs is interaction with, or incorporation of, small amounts of osmium-rich, high-^{187}Os outer core at the base of the mantle (e.g., Walker *et al.*, 1995; Puchtel *et al.*, 2001a; Brandon *et al.*, 1999), although it appears unlikely that the high-^{187}Os signature characteristic of most plume-related lavas is due to this mechanism.

2.13.2.8 ^{87}Sr Isotopic Evolution

Despite the ^{87}Rb–^{87}Sr isotopic system (half-life ~48.8 Gyr) being in routine use for more than 40 years, there are still few reliable constraints on the strontium isotopic evolution of the mantle for much of Earth's history. The determination of accurate initial strontium isotopic compositions in ancient samples is extremely difficult owing to the potential mobility and exchange of both rubidium and strontium by weathering and metamorphism, leading to erroneous corrections for *in situ* ^{87}Rb decay. Unlike refractory lithophile elements, for example, samarium and neodymium, rubidium was volatile during accretion such that the Earth formed with a much lower Rb/Sr (estimated to be 0.025–0.03; e.g., Hurley (1968) and DePaolo (1979)) than either the Sun (0.5) or primitive meteorites (~0.3). Thus, analysis of meteorites cannot provide information on bulk silicate Earth compositions, although they do provide the starting isotopic composition for planetary evolution. This benchmark is generally known as "basaltic achondrite best initial" (BABI; Papanastassiou and Wasserburg, 1969) with a composition of ^{87}Sr/^{86}Sr = 0.69898 ± 5 at 4.5 Ga.

The modern distribution of strontium isotopic ratios in basalts from both oceanic and continental regions is well documented (e.g., Hofmann, 1997; Chapter 2.03; Dickin, 1995), with the composition of the upper mantle today defined from analyses of N-MORBs (^{87}Sr/^{86}Sr = 0.7025 ± 0.0005). Rubidium is one of the most incompatible elements during melting, and it is evident that there has been a large transfer of rubidium from the mantle to the crust, with estimates that up to 58% of the total mantle rubidium now resides in the crust (Rudnick and Fountain, 1995). Crustal extraction has left the depleted mantle with a low Rb/Sr and evolving to less radiogenic compositions than characterize the bulk Earth. The Rb/Sr and ^{87}Sr isotopic composition of the bulk silicate Earth are not known directly but are estimated on the basis of well-established correlations with neodymium isotopes in modern oceanic basalts (the mantle array) with high ε_{Nd} corresponding to low $^{87}Sr/^{86}Sr$. The strontium composition at $\varepsilon_{Nd} = 0$ is generally taken to be bulk silicate Earth with $^{87}Sr/^{86}Sr = 0.7045 \pm 0.0005$ (DePaolo and Wasserburg, 1976). Modern OIBs typically have very high $^{87}Sr/^{86}Sr$ (e.g., Hofmann, 1997) compared to N-MORBs, often taken as evidence of recycled crustal components in their sources.

There is a substantial range in observed initial strontium isotopic compositions in mantle-derived samples at any given time period (e.g., Jahn *et al.*, 1980), likely resulting from post-crystallization rubidium mobility, the possibility of contamination of magmas with very radiogenic, strontium-rich, older crustal rocks during generation and emplacement, as well as true source heterogeneity. Owing to the uncertainties that exist in determining ^{87}Sr in Precambrian rocks, there have been few attempts to establish strontium isotopic evolution curves for the mantle analogous to neodymium or hafnium isotopic evolution. Since the early days of modern isotope geochemistry however, it has been recognized that establishing a mantle evolution curve for strontium (e.g., Hurley, 1968; Machado *et al.*, 1986) could provide potentially unique information on crust–mantle dynamics. The system is in many ways chemically similar to hafnium and neodymium in that the main parent–daughter fractionation is controlled by melt extraction from the mantle. However, there are important differences with strontium being more soluble in subduction zone fluids and in being readily exchanged between the ocean and crust at mid-ocean ridges and in extended weathering of the oceanic crust (e.g., Staudigel *et al.*, 1995). Thus, differences between ^{87}Sr compared with ^{143}Nd and ^{176}Hf evolution curves for the mantle are likely to be controlled by fluid-dominated processes as well as crust extraction and recycling. The oldest strontium isotopic compositions for the Earth are from 3.45 Ga barites ($BaSO_4$ is a very high strontium, low rubidium concentration mineral) from the North Pole dome in the Pilbara Block of Western Australia (McCulloch, 1994) and from relict igneous clinopyroxenes isolated from 3.45 Ga komatiites from the Onverwacht Group, South Africa (Jahn and Shih, 1974). Both of these yield similar compositions (0.70050 ± 1), yet neither is a particularly satisfying constraint on early mantle compositions. Barites form as either hydrothermal or evaporite minerals with their strontium isotopic composition reflecting that of the host fluids—they are not direct mantle samples. Komatiites are large-degree partial melts from the mantle; however, they are readily contaminated by interaction with the crust. All Archean compositions including those from well-preserved Late Archean Greenstone belts (e.g., Machado *et al.*, 1986) record higher than expected initial $^{87}Sr/^{86}Sr$, which may reflect crustal extraction and recycling processes, timescales of Earth accretion, contamination, or all of these effects. For example, the barite (3.45 Ga) strontium isotopic composition projects to a much higher than expected initial composition for the Earth based on the solar-system (BABI) value and reasonably estimated terrestrial Rb/Sr ratios. That is, if the Earth accreted from high-Rb/Sr materials similar to chondrites and lost rubidium during the accretion process, then the Earth's initial $^{87}Sr/^{86}Sr$ should be close to BABI (~0.699), if accretion occurred within the first 10 Myr of Earth history. The difference between the calculated and expected initial $^{87}Sr/^{86}Sr$ compositions for the Earth has been used to estimate an accretion interval, or more precisely, the timing of volatility-related fractionation of rubidium from strontium of ~80 Ma (McCulloch, 1994).

Although $^{87}Sr/^{86}Sr$ isotopic evolution for the mantle is only poorly constrained at present, this system has tremendous potential for revealing details of crust-formation processes through time as well as for providing precise timescales on Earth formation. The test of this awaits determination of accurate initial strontium isotopic compositions from Precambrian samples. The potential for acquisition of this type of data is high, as advances in techniques including both *in situ* measurements (e.g., Christiensen *et al.*, 1996) and low-level solution work (e.g., Mueller *et al.*, 2000) now allow precise determination of isotopic compositions from extremely small amounts ($<10^{-9}$ g) of strontium. This ability makes feasible analyses of rare, relict high-strontium, low-rubidium phases which may be preserved within ancient rocks and minerals and may accurately record mantle compositions.

2.13.3 TRACE-ELEMENT VARIATIONS: Nb/Th AND Nb/U IN THE MANTLE THROUGH TIME

The evidence of secular variation in the mantle presented to this point has been based on the use of radiogenic isotopic compositions. Variations in radiogenic isotopic compositions evolve from changes in trace-element ratios, in this case parent–daughter pairs, as a result of chemical fractionation of mantle sources. Another approach to investigating how the mantle has changed through time is through measurement of trace element concentrations directly, in materials whose compositions are representative of their mantle source regions. Primitive basalts, i.e., high-MgO picrites and komatiites, are often used for this purpose as they are large-degree (>10%) melts of their mantle sources, which results in minimal fractionation during melt extraction for many element pairs. In contrast to isotopic compositions that provide time-averaged trace element information, measured trace element ratios provide an instantaneous image of a mantle source. If the trace element signatures can be linked to specific sources, or processes, they can be used in mass balances between these sources. An example of this approach is through the use of Nb/U/Th ratios to trace crustal growth from depleting mantle.

Modern MORBs have nearly constant Nb/U and Nb/Th that are higher than primitive mantle values. This appears to reflect removal of continental crust, which has ratios lower than the primitive mantle (Hofmann *et al.*, 1986). Because niobium, thorium, and uranium are all highly incompatible during mantle melting, in principle it should be possible to trace the evolution of depleted sources through time from the trace element compositions of mafic lavas. In detail, this is not entirely straightforward because of the possibility of contamination from older continental crust, the ready mobility of uranium in crustal fluids, and uncertainties over the tectonic environment of ancient lavas. However, several examples of oceanic basalts appear to preserve their primary magmatic characteristics sufficiently to allow a preliminary examination of the record of mantle evolution recorded in the diagnostic trace element ratios of these lavas (Jochum *et al.*, 1991; Sylvester *et al.*, 1997; Collerson and Kamber, 1999; Kerrich *et al.*, 1999; Polat and Kerrich, 2000; Campbell, 2002; Condie, 2003). The tectonic setting, i.e., plume environments versus mid-ocean ridge versus convergent margin setting, of many of the mafic lava successions in the ancient rock record is uncertain. The geochemistry of these lavas may, therefore, provide a perspective that is biased more toward the evolution of particular mantle sources, for example, plume source, rather than depleted mantle analogous to the source of modern N-MORB.

Nb/U ratios do not show a clear pattern of temporal evolution with, for example, Late Archean basalts having a range of ratios from primitive mantle (~30) to modern basalt values (~47) (Figure 8(a)). The high Nb/U in Late Archean basalts has been interpreted by Sylvester *et al.* (1997) as indicating that the source of these basalts underwent the same level of melt extraction as the modern depleted mantle, requiring equivalent amounts of crustal extraction. Thus, either the mass of continental crust was the same as at present, or the volume of depleting mantle was proportionally smaller. In contrast, Collerson and Kamber (1999) argue that high Nb/U ratios reflect preferential recycling of uranium once the atmosphere became globally oxidizing, rather than large amounts of crustal extraction. The case for systematic variation of Nb/Th in mafic lavas through time is much clearer. Archean basalts and komatiites inferred to represent equivalents of modern oceanic basalts show a relatively restricted range of Nb/Th that cluster near the primitive mantle values or extend to values that are considerably less than those of modern N-MORB (Figure 8(b)). Proterozoic mafic lavas have, on average, Nb/Th

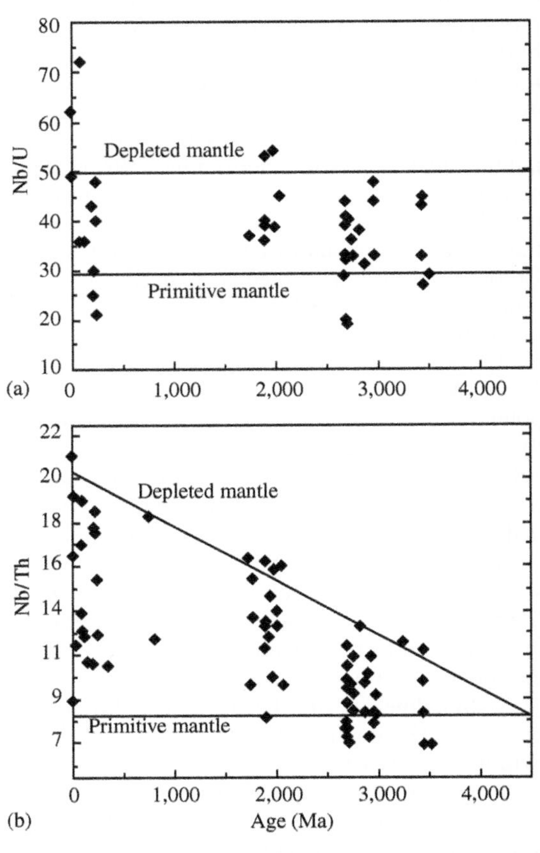

Figure 8 (a) Nb/U compositions of oceanic basalts through time showing little variation from Archean to the present. (b) Nb/Th compositions of oceanic basalts. Although a range of compositions between primitive mantle and depleted mantle compositions are apparent at all times periods, the data suggest a progressive increase in Nb/Th that has been linked to extraction of the continental crust (see text). (source Condie, 2003).

values that are intermediate between those of Archean lavas and the modern depleted mantle as represented by N-MORB. Nb/Th ratios of Phanerozoic lavas are typically higher than those of Precambrian basalts, and range from moderately depleted compositions to values typical of modern MORB. Both Collerson and Kamber (1999) and Condie (2003) interpret the apparent increase in maximum Nb/Th through time as representing progressive extraction of continental crust with Condie linking apparent inflections in the trends, particularly at 2.7 Ga, to either a period of rapid growth of continental crust or a reorganization of mantle reservoirs. By examining the systematics of Sm/Nd combined with U/Th/Nb data in basalts through time, Campbell (2002) argues that extraction of both continental and oceanic crust from the mantle sources of the basalts is required to account for their signatures.

2.13.4 ORIGIN OF CHEMICAL VARIATIONS IN THE EARTH'S MANTLE

In the first part of this chapter, some of the primary lines of evidence for long-term chemical changes in the Earth's mantle were summarized. In this section, the processes that can modify mantle chemical compositions are considered. The focus here is on consideration of potential mechanisms that could have created the chemical signatures preserved in the oldest rocks, and the subsequent processes that result in the patterns of chemical evolution present in the mantle.

2.13.4.1 Continental Crust Extraction

The most significant control on secular evolution of the upper mantle for at least the last 3.6 Ga, and perhaps before that as well, is extraction of the continental crust and return of crustal materials via recycling at subduction zones. That the continental crust is the compliment to the depleted upper mantle for many trace elements is well demonstrated by comparisons of element variation diagrams (Hofmann, 1988; Chapter 2.03) in which the depletions in incompatible elements in MORB are mirrored by enrichments in the continental crust. Continental crust formation is clearly a multistage process that, in the simplest description, involves production and differentiation of mafic magmas through a series of complex stages to form more evolved (felsic) magmas. Implicit in this scenario is that the mafic residuum complementary to the felsic crust must be returned to the convecting mantle. Various studies (e.g., Reymer and Schubert, 1984; Taylor and McLennan, 1985) have estimated continental crust production rates based on compilation of ages of juvenile granites. Most crustal production curves have a sigmoidal shape, with slow growth in the Archean, rapid crust production from the Late Archean throughout the Proterozoic, and declining additions of new crust from the mantle in the Phanerozoic. In fact, these crustal growth curves are actually crustal survival curves that reflect the balance between recycling and new additions. The mechanisms that produced the crust through time must have evolved, if not changed qualitatively, owing to changes in mantle thermal regimes.

There is now a large and detailed literature describing quantitative models of crustal growth scenarios and associated mantle depletion using a variety of isotopic and trace element constraints. These models approximate the Earth using a variety of homogeneous reservoirs, with mass transport between them and each subject to chemical fractionation effects. Some of the fundamental papers include, but are not limited to, O'Nions *et al.* (1979), Jacobsen and Wasserburg (1979), Allègre *et al.* (1983), Jacobsen (1988), DePaolo (1983), Chase and Patchett (1988), Smith and Ludden (1989), Galer *et al.* (1989), McCulloch and Bennett (1994), and Albarède (2001). Most of these models use simple assumptions about melt extraction processes, and assume compositional homogeneity for given reservoirs. Recent models (Kellogg *et al.*, 2002) have attempted to incorporate heterogeneity in reservoirs caused by mass transport and then examine the effects of sampling on different length scales.

Common to all crustal growth models is that simple crustal extraction using the known amount of Early Archean crust cannot readily account for the degree of depletion evident in some interpretations of the initial ε_{Nd} and ε_{Hf} compositions of the Archean mantle (Section 2.13.2). In the next sections, the effects of processes other than progressive growth of the continental crust from a depleting mantle are considered.

2.13.4.2 Early Differentiation

The question is still open as to whether large-scale processes associated with the formation of the Earth had a significant long-term effect on the chemical evolution of the mantle. Whether or not the Earth was largely or totally molten shortly after accretion is a topic of considerable interest and conjecture (e.g., Abe, 1993). The isotopic compositions of martian meteorites (Harper *et al.*, 1995; Borg *et al.*, 1997; Bogard *et al.*, 2001; Nyquist *et al.*, 2001; Brandon *et al.*, 2000) and lunar samples (Tera *et al.*, 1974; Nyquist *et al.*, 1979, 1995; Carlson and Lugmair, 1979, 1988; Norman *et al.*, 2003) provide compelling evidence for rapid planetary differentiation and formation of incompatible element enriched and depleted reservoirs on Mars and the Moon within the first 200 Myr of solar-system history. On the Moon, this early differentiation is generally attributed to crystallization of a global magma ocean, with flotation of a plagioclase-rich crust and formation of a complementary ultramafic (olivine + pyroxene) cumulate mantle (Longhi, 1980). Various calculations suggest that the Earth should also have been extensively molten early in its history (Stevenson, 1987; Tonks and Melosh, 1993). The observation that the apparently rapid ε_{Nd} evolution on the early Earth (Figure 1) may be similar in magnitude to that inferred for the lunar mantle raises the possibility that grossly similar fractionation events could have occurred on both bodies, although there remains little firm evidence for early global differentiation of the silicate Earth (see Chapter 2.10).

With the exception of rare 4.0–4.4 Ga zircons from Western Australia (Froude *et al.*, 1983; Wilde *et al.*, 2001), and two high-grade gneisses (Bowring and Williams, 1999), no terrestrial materials older than 4.0 Ga are preserved, although these materials do argue for at least limited amounts of early crust formation (Wilde *et al.*, 2001). The evidence for severe early fractionation derives not from the rock record directly, but largely from interpretations of the ^{176}Hf and ^{143}Nd isotopic database, which require extreme Lu/Hf and Sm/Nd fractionation of at least parts of the mantle prior to 4.3 Ga (e.g., Albarède *et al.*, 2000; Bennett *et al.*, 1993; Bowring and Housh, 1995; Bizzarro *et al.*, 2003; Amelin *et al.*, 1999, 2000; Collerson *et al.*, 1991). Additional evidence may include recently determined ^{142}Nd compositions in ~3.8 Ga samples from Isua (Boyet *et al.*, 2002; Caro *et al.*, 2003), indicating early (>4.3 Ga) fractionation, and possibly the need to create a "lost" reservoir with low Lu/Hf early in Earth history (Blichert-Toft and Albarède, 1997; Bizzarro *et al.*, 2002; Salters and White, 1998). Regardless of the eventual resolution of the current uncertainties over the ^{176}Lu decay constant (Section 2.13.2), ^{176}Hf isotopic evidence from the oldest zircons also seems to point to very early (>4.3 Ga) mantle differentiation (Amelin *et al.*, 1999; Scherer *et al.*, 2001; Bizzarro *et al.*, 2003), although it is, at present, not certain if they are derived from enriched or depleted mantle sources.

Although not strictly related to mantle depletion and crust formation, ^{182}Hf–^{182}W isotopic compositions do provide clear evidence for early planetary differentiation of the Earth, Moon, and Mars related to core formation. The results (Kleine *et al.*, 2002; Yin *et al.*, 2002; Schoenberg *et al.*, 2002) from this short half-life ($t^{1/2} \sim 8$ Myr) system provide convincing evidence that metal last equilibrated with silicate within the first 30 Myr of solar-system history. Tungsten-182 forms from the decay of ^{182}Hf; tungsten is siderophile and hafnium is lithophile, such that core extraction leaves the remaining silicate mantle with a high Hf/W. The ^{182}W isotopic composition of samples from the silicate Earth, Moon, and Mars differs from that of chondritic meteorites (Lee and Halliday, 1995; Kleine *et al.*, 2002; Yin *et al.*, 2002; Schoenberg *et al.*, 2002), requiring that metal–silicate fraction occurred while ^{182}Hf was still "alive," i.e., within ~30 Myr of injection of short-lived isotopes into the solar nebula (Cameron, 1995). On both the Moon and Mars, early core segregation was accompanied by differentiation of major silicate reservoirs. The question remains of whether this occurred on the Earth and if so, if any evidence is preserved.

2.13.4.3 Massive Early Crust Formation

An intriguing alternative view of crust formation championed by Armstrong (1981, 1991) and advocated by Bowring and Housh (1995) is that a continental crustal reservoir of approximately present-day mass formed very early in Earth's history (pre-4 Ga), and since then, has remained constant in mass. As we know from geochronological studies that the continents are composed of rocks of many different ages, and new crust is seen forming today in island arcs, however, clearly there have been ongoing additions to the continents. In the "no-growth" or "steady-state" recycling model, the "new" additions to the continental crust must, therefore, be matched by loss via return or recycling to the mantle of an equivalent mass of continental crust. Isotopic evolution curves alone cannot discriminate between these two very different scenarios of crustal evolution (i.e., progressive growth versus no-growth), because the continental mass and depleted mantle mass variables are interrelated and do not allow a unique solution. Production of large amounts of continental crust early in the Earth's history would result in a complementary depletion in the Early Archean mantle, and this depletion could subsequently be buffered to produce inflections in mantle evolution curves, by recycling of enriched continental crust back into the mantle. In terms of hafnium or neodymium isotopic evolution modeling, this is mathematically equivalent to progressively growing the continental crust and complementary depleted mantle. The distinction between these two very different modes for the production and preservation of continental crust is, however, important not only for the understanding of the chemical evolution of continental crust and the depleted mantle, but also in understanding how the styles of mantle convection may have changed through time.

The main criticisms against this style of early differentiation are the lack of any physical evidence for extensive amounts of early formed crust. Owing to its low density relative to the mantle, continental crust is difficult to subduct completely. At present the recognized Early Archean terranes occupy less than 5% of the preserved continental crust (Nutman *et al.*, 2001), with only one occurrence of $\geq$4.0 Ga rocks known (Bowring and Williams, 1999). Ion-probe investigations of the U–Pb ages of individual zircon grains from detrital populations (zircons are a common component in felsic rocks and are preserved in the sedimentary record through multiple generations of crustal reworking) in Archean sediments from all Early Archean terranes provide little evidence for continental crust older than 3.9 Ga (Nutman, 2002). Similarly, hafnium isotopic studies of detrital zircons preserved in Archean sediments provide no evidence for abundant continental crust older than 4.0 Ga (Stevenson and Patchett, 1990).

2.13.4.4 Early Basaltic Crusts

An alternative to massive amounts of felsic continental crust to account for early depletion in the mantle is the production and storage of an early basaltic crust. Models invoking mafic–ultramafic crust (e.g., Chase and Patchett, 1988) and alkalic crust (Galer and Goldstein, 1991) have been proposed. Although extraction of mafic crust is less effective at depleting the mantle in lithophile elements than production of continental crust, substantially greater volumes of oceanic crust have been produced at oceanic spreading centers throughout geologic history. It is also expected that owing to higher heat flow in the early Earth, basaltic crust production rates would have been higher than present-day rates. Quantitative models demonstrate that it is theoretically possible to create the ^{143}Nd isotopic signatures of Early Archean rocks by extraction and long-term storage ($>10^9$ yr) of an LREE-enriched basaltic crust (e.g., Chase and Patchett, 1988). A difficulty with this model includes the problem of explaining enhanced storage of basaltic crust in the Archean as opposed to the post-Archean. In another version of the basalt storage model, Galer and Goldstein (1991) proposed the formation of a global mafic crust resulting from small-degree, alkalic partial melts that formed soon after Earth accretion. In principle, this model can also account for the observed isotopic and trace element characteristics of Early Archean rocks, but there is little direct evidence to support this type of process. No samples matching the expected compositions and ages of an early alkalic crust have been identified, but this may be a matter of lack of preservation. As with all of the proposed processes of early mantle differentiation, the absence of strong evidence supporting any specific model does not necessarily imply that the process did not occur in the early history of the Earth; rather it may be an indicator of the scale of rehomogenization and vigor of mantle convection in the first 500 Myr of Earth history (see Chapter 2.12).

2.13.4.5 Implications of Neodymium and Hafnium Isotopic Evolution Curves: Episodic Mantle Evolution?

Evolution curves for ^{143}Nd and ^{176}Hf isotopic compositions from 3.8 Ga to the present have been presented by a number of studies (DePaolo, 1981; DePaolo *et al.*, 1991; Shirey, 1991; Bowring and Housh, 1995; Smith and Ludden, 1989; Bennett *et al.*, 1993; Vervoort and Blichert-Toft, 1999; Blichert-Toft *et al.*, 1999). Although the general outline of mantle evolution for the hafnium and neodymium system is known, the details, particularly for the Archean, are still uncertain. The shapes of these curves provide important clues to mantle dynamics through time that cannot come simply from examination of the distribution of isotopic compositions in the modern Earth. Comprehensive analyses of mantle evolution curves reveal that, in general, linear increases in ε_{Nd} values through time are not consistent with progressive growth by depletion of a fixed volume of the mantle (Jacobsen, 1988; DePaolo, 1983). Rather an apparent linear evolution requires either a buffering of mantle Sm/Nd compositions by crustal recycling (DePaolo, 1980, 1983), an increase in the mass of depleting mantle with time (McCulloch and Bennett, 1994), or continuous replenishment of the upper mantle from a less depleted mantle reservoir, possibly the lower mantle (Patchett and Chauvel, 1984; Coltice *et al.*, 2000). That the present-day mantle compositions cannot represent a simple continuous process of crustal extraction through time can be discerned from a comparison of the Sm/Nd of the present-day upper mantle with the time-averaged upper mantle Sm/Nd based on ^{143}Nd compositions. From measurements of MORB, we know that the depleted upper mantle today is characterized by ^{147}Sm/^{144}Nd ratios ~25% higher than primitive mantle, yet the time-averaged ^{147}Sm/^{144}Nd over 4.5 Ga, based on an $\varepsilon_{Nd} = +10$ today in the upper mantle, requires an ^{147}Sm/^{144}Nd only 10% higher than primitive mantle compositions. If the mantle had been characterized by its present degree of LREE depletion for a substantial period of time, it would have a higher present-day ε_{Nd}. The difference between the current and integrated

Sm/Nd requires mixing with less depleted material (lower Sm/Nd) at different time periods, which could be either recycled continental crust or more primitive mantle.

An additional feature of the mantle evolution curves for both ^{143}Nd and ^{176}Hf for the Proterozoic and Archean (Figures 1 and 3) suggests periods of relatively constant isotopic compositions compared with periods of more rapid change. The apparent inflections in the evolution curves could be the artifact of a limited or compromised database; however, the presence of the same patterns in two different systems suggests that they may accurately reflect mantle evolution. Owing to the isotopic inertia of the mantle for long half-life systems, changes of this sort must reflect large-scale and persistent changes in the degree of depletion of the mantle. The correspondence between the timing of "kinks" in isotopic evolution curves particularly at ~2.7 Ga, and apparent periods of enhanced continental crust production as seen from crustal growth curves, has been noted in various studies with attempts at a genetic linkage between these two sets of observations (Stein and Hofmann, 1994; McCulloch and Bennett, 1994; Condie, 1997, 2003; Davies, 1995). The concept of "episodic mantle" evolution or transient mantle chemical regimes has been incorporated into a number of mantle models. These types of models postulate that the size of the depleting mantle has increased periodically through time (e.g., McCulloch and Bennett, 1994) and adjust the size of depleting mantle and crustal growth to match known isotopic and chemical secular trends, including rapid early depletions of the mantle. Changes in degree of mantle depletion occur when more primitive mantle is incorporated into depleted mantle regions. Alternatively, other models postulate intermittent periods of increased communication between the upper and lower mantles (mantle avalanches) during which the mantle convection regime may alternate between whole mantle and layered mantle convection (Stein and Hofmann, 1994; Davies, 1995; Condie 2002). In this type of scenario, mixing between less depleted and more depleted mantle reservoirs replenishes the upper mantle in trace elements and may contribute to enhanced crustal growth rates.

2.13.4.6 Mafic Reservoirs in the Deep Earth

Regardless of the role of basaltic crust in creating early mantle depletions, there is increasing chemical and seismic evidence for the presence of long-lived chemical reservoirs in the deep mantle, perhaps at the D″ layer at the base of the mantle. Some of the strongest chemical evidence includes the range of trace element and isotopic compositions of modern OIBs (cf. Chapter 2.03), and their older equivalents, including continental flood basalts, and komatiites. OIBs require at least four (five if primitive mantle is considered) long-lived, likely deep mantle compositional end-members to explain the range of isotopic (lead, hafnium, neodymium, osmium, and strontium) and chemical variations (Hofmann, 1997). Additional evidence for deep mantle reservoirs derives from ^{187}Os isotopic mass balance considerations in both the modern (e.g. Hauri and Hart, 1997) and 2.7 Ga mantle (Walker and Nisbet, 2002) that seem to require a cryptic mafic reservoir to balance depletions in the upper mantle and to provide a high-^{187}Os component to both modern and ancient plume magmas. The most suitable place to hide such a reservoir appears to be deep in the mantle.

Although the continental crust clearly provides a good geochemical complement to lithophile element depletions in the mantle, there are some discrepancies. For example, the high-field-strength elements niobium, tantalum, and titanium are anomalously depleted in both the depleted mantle source of MORB and continental crust compared to primitive mantle abundances (McDonough, 1991). Melt–fluid processes at subduction zones likely result in relative enrichments in these elements in subducted basaltic crust, such that storage of fluid-modified oceanic basalt equal to 2% of the mass of the mantle could account for these elemental characteristics. Similarly, Rudnick *et al.* (2000) considered the mass balance of tantalum, niobium, and aluminum in the silicate Earth, and proposed that storage of basaltic crust in the form of eclogite equal to 6% of the mass of the mantle could be the source of the missing low Nb/Ta reservoir complementary to the continental crust and depleted mantle. A substantial basaltic reservoir stored in the deep mantle has also been proposed to account for the combined Sm–Nd and U–Th–Nb characteristics of late Archean to modern oceanic basalts (Campbell, 2002). Although any individual line of evidence may be disputed, taken together the various isotopic and trace element observations provide compelling evidence for the existence of mafic reservoirs in the deep mantle.

2.13.4.7 Core Interactions?

It is increasingly apparent that there are exchanges and interactions between all the major chemical domains within the silicate earth. Communication between the upper mantle and crust occurs at ridges, above mantle plumes, and at volcanic arcs. Geochemical studies of OIBs, as well as seismic tomography of mantle structure beneath subduction zones demonstrate that there is communication between the upper and lower mantles across the transition zone. The degree to which there is chemical communication across the core–mantle boundary is currently unknown. Increased impetus for considering the possible role of core interactions in mantle chemistry stems from consideration of plume dynamics, which suggest that some OIBs originate at the core–mantle boundary, as well as from tomographic images that suggest that slabs may sink to the base of the mantle to form an enriched mafic reservoir at the D″ layer (Kellogg *et al.*, 1999). Various studies have looked at lithophile/siderophile element ratios in basalts through time as potential tests of secular evolution of the mantle due to core interaction. As the core must be highly enriched in siderophile elements compared to the mantle, basalts generated from the core–mantle boundary layer might be expected to carry a core signature in the form of higher siderophile element contents or fractionated siderophile element patterns. To date, none of these trace element approaches (e.g. Newsom *et al.*, 1986; Bennett *et al.*, 2000) have provided conclusive evidence of a core signature or changes in the siderophile element composition of the mantle through time. Part of the difficulty is that the petrogenetic processes that lead to generation of basalts can also result in modification of highly siderophile element patterns, masking any core signature. The high-$^{187}Os/^{188}Os$ isotopic compositions of some OIBs (e.g. Widom and Shirey, 1996) and some komatiites (Walker and Nisbet, 2002; Puchtel *et al.*, 2001a) relative to upper mantle compositions have also been suggested to result from incorporation of small amounts of core material; however, these compositions may also be modeled by incorporation of aged, recycled mafic slabs. Perhaps the strongest indicator of possible core interaction is from the measurement of ^{186}Os excesses in Hawaiian basalts and the Nor'ilsk flood basalts (Brandon *et al.*, 1999; Walker *et al.*, 1995). Osmium-186 forms from the decay of ^{190}Pt; owing to the long half-life and small natural abundance of the parent isotope, measurable deviations from normal ^{186}Os compositions can only be generated over long time periods ($>10^9$ yr) in environments with high Pt/Os ratios. One of the few places where such an environment may occur in the Earth is in the outer core, where liquid metal/solid metal fractionation during core crystallization is expected to have produced an outer core with high Pt/Os and high Re/Os that will evolve to compositions with high ^{186}Os and ^{187}Os (cf. Chapter 2.15). The existence of coupled ^{186}Os–^{187}Os enrichments in the Hawaiian basalts and Nor'ilsk flood basalts provides a strong argument in support of some form of geochemical communication between the outer core and lower mantle.

2.13.5 NATURE OF CHEMICAL LAYERING IN THE MANTLE

A continuing quest in Earth sciences is the integration and in some cases reconciliation of geochemical and geophysical observations to provide an accurate picture of mantle structure. Since the early 1970s, it has been apparent from mass balance considerations of continental crust and MORB mantle compositions that the entire mantle cannot be depleted to the same degree as the MORB source mantle. Most estimates based on continental crust–N-MORB source mantle mass balances, suggest that between 30% and 60% of the mantle is depleted. Therefore, models of whole mantle convection that imply that the mantle is well mixed and with a uniform composition (Figure 9(a)) are not supported by the chemical data. (If there is a range in degree of depletion of the mantle with only very limited portions as depleted as the N-MORB source, then this style of model might be applicable.) Simple three-box models consisting of three components—crust, depleted mantle, and primitive mantle (e.g. DePaolo, 1980; Jacobsen, 1988)—provided a conceptually simple image of mantle structure whereby a more depleted upper mantle lies above a less depleted, or even primitive mantle. In these models, which can account for many of the chemical features of the crust and upper mantle, the amount of continental crust progressively increases with time, with the rate of exchange (i.e., rate of recycling) between the crust and upper mantle being a free parameter (Jacobsen, 1988). A two-layer mantle structure (Figure 9(b)) in which ~30% of the mantle is depleted is compatible with these types of models. In these models, the depleted mantle lay above the transition zone at 660 km, which was in accord with proposals based on mantle dynamics (cf. Chapter 2.12). The transition zone was considered an effective barrier to chemical exchange, with the lower mantle below the transition zone retaining its primitive composition. The current recognition that the depleted mantle must occupy more than the volume above the transition zone (~1/3 of the mantle), as well as the need for multiple deep mantle reservoirs based on plume-related magma chemistry and the emerging view of mantle thermal and density complexity that is apparent from seismic tomography, now drives the consideration of other types of mantle structure.

Chemical mass balance models can actually only provide information on the relative amounts of different reservoirs, not on their spatial distribution. A mantle structure in which blobs of relict primitive mantle are maintained throughout a variously depleted and enriched lower mantle may be possible (e.g., Becker *et al.* (1999); Figure 9(c)), or true primitive mantle

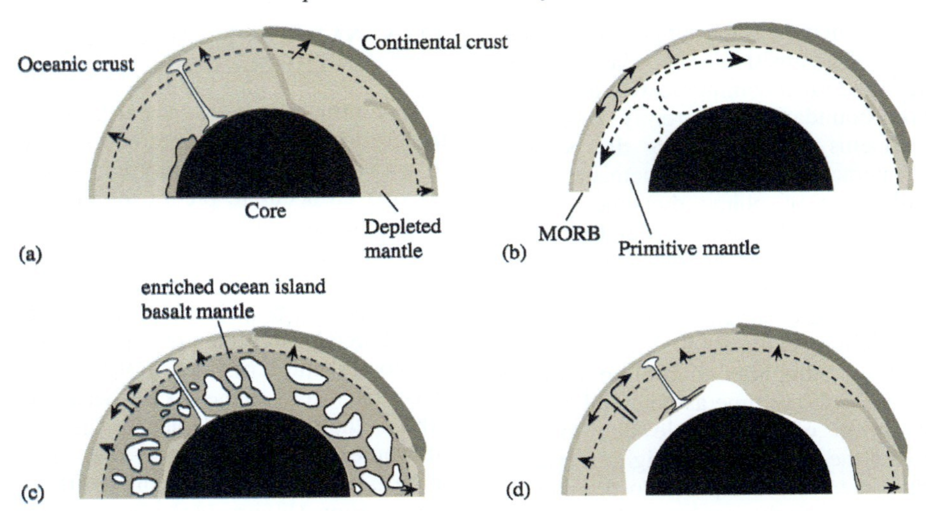

Figure 9 Examples of models proposed for the chemical structure of the terrestrial mantle. (a) Whole mantle convection with depletion of the entire mantle. Some subducted slabs pass through the transition zone to the core–mantle boundary. Plumes arise from both the core–mantle boundary and the transition zone. This model is not in agreement with isotopic and chemical mass balances. (b) Two-layer mantle convection, with the depleted mantle above the 660 km transition zone and the lower mantle retaining a primitive composition. (c) "Blob" model mantle where regions of more primitive mantle are preserved within a variously depleted and enriched lower mantle. (d) Chemically layered mantle with lower third above the core comprising a heterogeneous mixture of enriched (mafic slabs) and more primitive mantle components, and the upper two-thirds of the mantle is depleted in incompatible elements (see text) (after Albarède and van der Hilst, 1999).

might not exist within the Earth at all. Despite the range of trace element compositions present in OIBs, none of the likely mixing end-members appears to have primitive compositions (Hofmann, 1997; Chapter 2.03). The strongest arguments for the continued existence of a primitive reservoir are from noble gas compositions, in particular $^3He/^4He$ (Chapter 2.06). Consideration of rates of basalt production at ridges through time reveals that the entire mantle could have been processed at least once. Additionally, estimates that place the amount of depleted mantle at ~50% of the total mantle are based on the assumption that the degree of depletion is equivalent to that of highly depleted N-MORB source mantle. If enriched reservoirs other than the crust exist within the mantle, or extensive areas of mantle with more moderate levels of depletions are present, for example, as advocated for the FOZO composition (FOcus ZOne, taken as the midrange compositions of lead, strontium, neodymium isotopic arrays for MORBs and OIBs; Hart *et al.* (1992)) then only small (10–30%) amounts, if any, of truly primitive mantle may remain.

A significant new constraint on mantle chemical structure is provided by global seismic tomographic images. These images strongly suggest slab penetration into the lower mantle, with several studies showing narrow features of high seismic velocity extending from the sites of present-day subduction into the lower mantle (van der Hilst *et al.*, 1991; Bijwaard *et al.*, 1998). In addition, these models reveal complex lower mantle structures that have been interpreted as sizable areas of heterogeneous materials. These direct observations of mantle structures provide firm evidence that the transition zone, at least at the present time, is not an impermeable chemical boundary layer, and have led to the development of a radical new class of models of mantle chemical structure (Kellogg *et al.*, 1999; van der Hilst and Karason, 1999; Albarède and van der Hilst, 1999) where the bottom 1,000 km of the mantle above the core, i.e., ~33% of the mantle, is composed of a seismically distinct chemical layer (Figure 9(d)). This region would contain undegassed material formed early in Earth history, and would also be the site of stored subducted slabs. The remaining 67% of the mantle would comprise depleted mantle similar to the MORB source (Albarède and van der Hilst, 1999).

Tomographic images, however, only provide a snapshot of present-day mantle structure. Whether or not the picture that is emerging, in which an apparently high level of communication exists between upper and lower mantle, has been the case throughout Earth's history is not clear. Various lines of geophysical argument have suggested that owing to a differing thermal structure, the transition zone may, at least at times, have been a more significant chemical barrier, and that there may have been greater tendency for higher degrees of mantle layering earlier in Earth's history (e.g., Machtel and Weber, 1991; Davies, 1995; Tackley *et al.*, 1993).

2.13.6 FUTURE PROSPECTS

The study of the chemical evolution of the mantle is a highly active growth field, fueled by analytical advances enabling the measurement of a wider range of elements and isotopes at unprecedented precisions on ever smaller samples. The refinement of seismic imaging techniques is providing new constraints on mantle structure, and greater emphasis is being placed on understanding Earth in a planetary context. Considerable potential exists in the near term for significant advances in our understanding of mantle structure and evolution of the chemical and physical structure of the Earth since its formation. The mantle is increasingly appreciated as an immense and key part of the Earth system; it is impossible to have a true understanding of any part of the planet or its evolution without consideration of this vast and largely unknown domain. The mantle is the ultimate source and sink of elements to the crust, with complex interactions between the hydrosphere, atmosphere, lithosphere, and perhaps the core, all playing a role in establishing its composition.

This chapter has highlighted the results of only a few chemical observations on mantle evolution. Whole new types of analytical approaches including analyses of the stable isotopes of elements such as lithium, boron, copper, iron, and magnesium, and isotopic variations of the decay products of extinct or rare nuclides including ^{142}Nd, ^{182}W, ^{186}Os, ^{107}Ag and trace element constraints, as, for example, from platinum group element chemistry that are yet to be fully applied to understanding mantle chemistry. Emerging with these new types of analytical approaches is long-needed improved sophistication of the models applied to quantitatively understanding mantle chemistry, with incorporation of complex melting processes and consideration of mixing timescales. Intensive geological investigations of the Earth's oldest terranes are producing new finds of ancient materials to which the new high precision techniques can be applied. Thus, the answers to key questions of mantle chemical evolution will be found, and will result from a synthesis of geochemical, geophysical, theoretical, and experimental observations.

ACKNOWLEDGMENTS

M. Norman is thanked for detailed comments and suggestions and R. Carlson for his polite patience and superb editorial skills.

REFERENCES

Abe Y. (1993) Physical state of the very early Earth. *Chem. Geol.* **30**, 223–235.

Albarède F. (2001) Radiogenic ingrowth in systems with multiple reservoirs: applications to the differentiation of the mantle-crust system. *Earth Planet. Sci. Lett.* **189**, 59–73.

Albarède F. and van der Hilst R. D. (1999) New mantle convection model may reconcile conflicting evidence. *Am. Geophys. Union, EOS* **80**, 535–539.

Albarède F., Blichert-Toft J., Vervoort J. D., Gleason J. D., and Rosing M. (2000) Hf–Nd isotope evidence for a transient dynamic regime in the early terrestrial mantle. *Nature* **404**, 488–490.

Allègre C. J., Hart S. R., and Minster J.-F. (1983) Chemical structure and evolution of the mantle and continents determined by inversion of Nd and Sr data. *Earth Planet. Sci. Lett.* **66**, 177–190.

Amelin Y., Lee D. C., and Halliday A. N. (1999) Nature of the Earth's earliest crust from hafnium isotopes in single detrital zircons. *Nature* **399**, 252–255.

Amelin Y., Lee D. C., and Halliday A. N. (2000) Early-middle Archaean crustal evolution from Lu–Hf and U–Pb isotopic studies of single zircon grains. *Geochim. Cosmochim. Acta* **64**, 4205–4225.

Armstrong R. L. (1981) Radiogenic isotopes: the case for recycling on a near-steady-state no-continental-growth, Earth. *Phil. Trans. Roy. Soc. London* **A301**, 443–472.

Armstrong R. L. (1991) The persistent myth of crustal growth. *Austral. J. Earth Sci.* **38**, 613–640.

Baadsgaard H., Nutman A. P., and Bridgwater D. (1986) Geochronology and isotopic variation of the early Archaean Amitsoq Gneisses of the Isukasia area, southern West Greenland. *Geochim. Cosmochim. Acta* **50**, 2173–2183.

Becker T. W., Kellogg J. B., and O'Connell R. J. (1999) Thermal constraints on the survival of primitive blobs in the lower mantle. *Earth Planet. Sci. Lett.* **171**, 351–365.

Bennett V. C., Nutman A. P., and McCulloch M. T. (1993) Nd isotopic evidence for transient highly-depleted mantle reservoirs in the early history of the Earth. *Earth Planet. Sci. Lett.* **119**, 299–317.

Bennett V. C., Norman M. N., and Garcia M. O. (2000) Correlated platinum group element and Re abundances with isotopic compositions in Hawaiian picrites: the role of sulphides. *Earth Planet. Sci. Lett.* **183**, 513–526.

Bennett V. C., Nutman A. P., and Esat T. (2002) The osmium mantle evolution curve and limits on accretion and differentiation of the Earth from Archean (3.4–3.8 Ga) Ultramafic Rocks. *Geochim. Cosmochim. Acta* **66**, 2615–2630.

Bijwaard H., Spakman W., and Engdahl E. R. (1998) Closing the gap between regional and global travel time tomography. *J. Geophys. Res.* **103**, 30055–30078.

Bizzarro M., Simonetti A., Stevenson R., and David J. (2002) Hf isotope evidence for a hidden mantle reservoir. *Geology* **30**, 771–774.

Bizzarro M., Baker J., Haack H., Ulfbeck D., and Rosing M. (2003) Early history of the Earth's crust–mantle system inferred from hafnium isotopes in chondrites. *Nature* **421**, 931–933.

Blichert-Toft J. (2001) On the Lu–Hf isotope geochemistry of silicate rocks. *Geostand. Newslett.* **25**, 41–56.

Blichert-Toft J. and Albarède F. (1997) The Lu–Hf isotope geochemistry of chondrites and the evolution of the mantle–crust system. *Earth Planet. Sci. Lett.* **148**, 243–258.

Blichert-Toft J., Albarède F., Rosing M., Frei R., and Bridgwater D. (1999) The Nd and Hf isotope evolution of the mantle through the Archean: results from the Isua supracrustals, West Greenland, and from the Birimian terranes of West Africa. *Geochim. Cosmochim. Acta* **63**, 3901–3914.

Bogard D. D., Clayton R. N., Marti K., Owen T., and Turner G. (2001) Chronology and evolution of Mars. *Space Sci. Rev.* **96**, 425–458.

Borg L. E., Nyquist L. E., Taylor L. A., Wiesmann H., and Shih C.-Y. (1997) Constraints on martian differentiation processes from Rb–Sr and Sm–Nd isotopic analyses of the basaltic shergottite QUE94201. *Geochim. Cosmochim. Acta* **61**, 4915–4931.

Brandon A. D., Walker R. J., Morgan J. W., Norman M. D., and Prichard H. (1998) Coupled ^{186}Os and ^{187}Os evidence for core-mantle interaction. *Science* **280**, 1570–1573.

Brandon A. D., Norman M. D., Walker R. J., and Morgan J. W. (1999) ^{186}Os and ^{187}Os Systematics of Hawaiian picrites. *Earth Planet. Sci. Lett.* **174**, 24–42.

Brandon A. D., Walker R. J., Morgan J. W., and Goles G. G. (2000) Re–Os isotopic evidence for early differentiation of the martian mantle. *Geochim. Cosmochim. Acta* **64**, 4083–4095.

Brandon A. D., Walker R. J., Puchtel I. S., Becker H., Humayun M., and Revillon S. (2003) ^{186}Os–^{187}Os systematics of Gorgona Island Komatiites: implications for early growth of the inner core. *Earth Planet. Sci. Lett.* **206**, 411–426.

Bowring S. A. and Housh T. (1995) The Earth's early evolution. *Science* **269**, 1535–1540.

Bowring S. A. and Williams I. S. (1999) Priscoan (4.00–4.03 Ga) orthogneisses from northwestern Canada. *Contrib. Mineral. Petrol.* **134**, 3–16.

Boyet M., Albarède F., Rosing M., Garcia M. O., and Pik R. (2002) Vestige of a beginning: a quest for ^{142}Nd anomalies in the Earth. *EOS, Trans., AGU* **83**(47) V41C-04, 1446, Fall Meet. Suppl. Abstr.

Cameron A. G. W. (1995) The first ten million years in the solar nebula. *Meteoritics* **30**, 133–161.

Campbell I. (2002) Implications of Nb/U, Th/U and Sm/Nd in plume magmas for the relationship between continental and oceanic crust formation and the development of the depleting mantle. *Geochim. Cosmochim. Acta* **66**, 1651–1661.

Campbell I., Griffiths R. W., and Hill R. E. T. (1989) Melting in an Archean mantle plume: heads its basalts, tails its komatiites. *Nature* **339**, 697–699.

Carlson R. W. and Lugmair G. W. (1979) Sm–Nd constraints on early differentiation and the evolution of KREEP. *Earth Planet. Sci. Lett.* **45**, 123–132.

Carlson R. W. and Lugmair G. W. (1988) The age of ferroan anorthosite 60025: oldest crust on a young Moon? *Earth Planet. Sci. Lett.* **90**, 119–130.

Caro G., Bourdon B., Birck J. L., and Moorbath S. (2003) ^{146}Sm–^{142}Nd evidence from Isua metamorphosed sediments for early differentiation of the Earth's mantle. *Nature* **243**, 428–432.

Chase C. G. and Patchett P. J. (1988) Stored mafic/ultramafic crust and early Archean mantle depletion. *Earth Planet. Sci. Lett.* **91**, 66–72.

Christensen J. N., Halliday A. N., Lee D.-C., and Hall C. M. (1996) *In situ* Sr isotopic analysis by laser ablation. *Earth Planet. Sci. Lett.* **136**, 79–85.

Chou C. L. (1978) Fractionation of siderophile elements in the Earth's upper mantle. *Proc. Lunar Planet. Sci. Conf.* **9**, 219–230.

Collerson K. D., Campbell L. M., Weaver B. L., and Palacz Z. A. (1991) Evidence of extreme mantle fractionation in early Archaean ultramafic rocks from northern Labrador. *Nature* **349**, 209–214.

Collerson K. D. and Kamber B. S. (1999) Evolution of the continents and the atmosphere inferred from Th–U–Nb systematics of the depleted mantle. *Science* **283**, 1519–1522.

Coltice N., Albarède F., and Gillet P. (2000) ^{40}K–^{40}Ar constraints on recycling continental crust into the mantle. *Science* **288**, 845–847.

Condie K. C. (1997) Episodic continental growth and super continents: a mantle avalanche connection? *Earth Planet. Sci. Lett.* **163**, 97–108.

Condie K. C. (2003) Incompatible element ratios in oceanic basalts and komatiites: tracking deep mantle sources and continental growth rates with time. *Geochem. Geophys. Geosyst.* **4**(1), 1005 doi:10.1029/2002GC000333.

Creaser R. A., Papanastassiou D. A., and Wasserburg G. J. (1991) Negative thermal in mass spectrometery of osmium, rhenium, and iridium. *Geochim. Cosmochim. Acta* **55**, 397–401.

Davies G. F. (1995) Punctuated tectonic evolution of the Earth. *Earth Planet. Sci. Lett.* **136**, 363–379.

DePaolo D. J. (1979) Inferences from correlated Nd and Sr isotopic variations for the chemical evolution of the crust and mantle. *Earth Planet. Sci. Lett.* **43**, 201–211.

DePaolo D. J. (1980) Crustal growth and mantle evolution: inferences from models of element transport and Nd and Sr isotopes. *Geochim. Cosmochim. Acta* **44**, 1185–1196.

DePaolo D. J. (1981) Neodymium isotopes in the Colorado Front Range and crust–mantle evolution in the Proterozoic. *Nature* **291**, 193–196.

DePaolo D. J. (1983) The mean life of continents: estimates of continent recycling rates from Nd and Hf isotopic data and implications for mantle structure. *Geophys. Res. Lett.* **10**, 705–708.

DePaolo D. J. and Wasserburg G. J. (1976) Inferences about magma sources and mantle structure from variations of Nd-143/ Nd-144. *Geophys. Res. Lett.* **3**, 743–746.

DePaolo D. J., Linn A. M., and Schubert G. (1991) The continental crustal age distribution; methods of determining mantle separation ages from Sm–Nd isotopic data and application to the southwestern United States. *J. Geophys. Res.* **96**, 2071–2088.

Dicken A. P. (1995) *Radiogenic Isotope Geology*. Cambridge University Press, Cambridge, UK, 490pp.

Esser B. K. and Turekian K. K. (1993) The osmium isotopic composition of the continental crust. *Geochim. Cosmochim. Acta* **57**, 3093–3104.

Faure G. (1986) *Principles of Isotope Geology*, 2nd edn. Wiley, New York, 460pp.

Foster J. G., Lambert D. D., Frick L. R., and Maas R. (1996) Re–Os isotopic evidence for genesis of Archaean nickel ores in uncontaminated komatiites. *Nature* **382**, 703–706.

Friend C. R. L. F., Bennett V. C., and Nutman A. P. (2002) Abyssal peridotites >3,800 Ma for southern West Greenland: field relationships, petrography, geochronology, whole-rock and mineral chemistry of dunite and harzburgite inclusions in the Itsaq Gneiss Complex. *Contrib. Mineral. Petrol.* **143**, 71–92.

Froude D., Ireland T. R., Kinny P. D., Williams I. S., Compston W., Williams I. R., and Meyers J. S. (1983) Ion microprobe identification of 4100–4200 terrestrial zircons. *Nature* **304**, 616–618.

Galer S. J. G. and Goldstein S. L. (1991) Early mantle differentiation and its thermal consequences. *Geochim. Cosmochim. Acta* **55**, 39–227.

Galer S. J. G., Goldstein S. L., and O'Nions R. K. (1989) Limits on chemical and convective isolation in the Earth's interior. *Chem. Geol.* **75**, 257–290.

Goldstein S. L. and Galer S. J. G. (1992) On the trail of early mantle differentiation: ^{142}Nd/^{144}Nd ratios of early Archean rocks. *EOS, Trans., AGU* **73**(suppl.), 323.

Goldstein S. L., O'Nions R. K., and Hamilton P. J. (1984) A Sm–Nd isotopic study of atmospheric dusts and particulates from major river systems. *Earth Planet. Sci. Lett.* **70**, 221–236.

Gruau G. M., Rosing M., Bridgwater D., and Gill R. C. O. (1996) Resetting of Sm–Nd systematics during metamorphism of >3.7-Ga rocks: implications for isotopic models of early Earth differentiation. *Chem. Geol.* **133**(1–4), 225–240.

Hamilton P. J., Evensen N. M., and O'Nions R. K. (1979) Sm-Nd systematics of Lewisian gneisses: implications for the origin of granulites. *Nature* **277**, 25–28.

Handler M. R., Bennett V. C., and East T. M. (1997) The persistence of off-cratonic lithospheric mantle. Os-isotopic systematics of variably metasomatised southeast Australian xenoliths. *Earth Planet. Sci. Lett.* **151**, 61–75.

Harper C. L., Jr. and Jacobsen S. B. (1992) Evidence from coupled $^{147}Sm-^{143}Nd$ and $^{146}Sm-^{142}Nd$ for very early (4.5 Ga) differentiation of the Earth's mantle. *Nature* **360**, 728–732.

Harper C. L., Jr. Nyquist L. E., Bansal B. M., Wiesmann H., and Shih C.-Y. (1995) Rapid accretion and early differentiation of Mars indicated by $^{142}Nd/^{144}Nd$ in SNC meteorites. *Science* **267**, 213–216.

Hart S. R., Hauri E. H., Oschmann L. A., and Whitehead J. A. (1992) Mantle plumes and entrainment: isotopic evidence. *Science* **256**, 517–520.

Hattori H. and Hart S. R. (1991) Osmium isotope ratios of platinum group elements associated with ultramafic intrusions. *Earth. Planet. Sci. Lett.* **107**, 499–514.

Hauri E. H. and Hart S. R. (1993) Re–Os isotope systematics of EMII and HIMU oceanic island basalts from the south Pacific Ocean. *Earth Planet. Sci. Lett.* **114**, 353–371.

Hauri E. H. and Hart S. R. (1997) Rhenium abundances and systematics in oceanic basalts. *Chem. Geol.* **139**, 185–207.

Hofmann A. W. (1988) Chemical differentiation of the Earth: the relationship between mantle, continental crust and oceanic crust. *Earth Planet. Sci. Lett.* **90**, 297–314.

Hofmann A. W. (1997) Mantle geochemistry, the message from oceanic volcanism. *Nature* **385**, 219–229.

Hofmann A. W., Jochum K. P., Seufert M., and White W. M. (1986) Nb and Pb in oceanic basalts; constraints on mantle evolution. *Earth Planet. Sci. Lett.* **79**, 33–45.

Hurley P. M. (1968) Absolute abundance and distribution of Rb, K, and Sr in the Earth. *Geochim. Cosmochim. Acta* **32**, 273–283.

Jacobsen S. B. (1988) Isotopic constraints on crustal growth and recycling. *Earth. Planet. Sci. Lett.* **90**, 315–329.

Jacobsen S. B. and Dymek R. F. (1988) Nd and Sr isotope systematics of clastic metasediments from Isua, West Greenland: identification of pre-3.8 Ga differentiated crustal components. *J. Geophys. Res.* **93**, 338–354.

Jacobsen S. B. and Wasserburg G. J. (1979) The mean age of mantle and crustal reseviors. *J. Geophys. Res.* **84**, 7411–7427.

Jacobsen S. B. and Wasserburg G. J. (1980) Sm–Nd isotopic evolution of chondrites. *Earth Planet. Sci. Lett.* **50**, 139–155.

Jahn B.-M. and Shih C. (1974) On the age of the Onverwacht Group, Swaziland Group, South Africa. *Geochim. Cosmochim. Acta* **38**, 873–885.

Jahn B.-M., Vidal P., and Tilton G. R. (1980) Archaean mantle heterogeneity: evidence from chemical and isotopic abundances in Archaean igneous rocks. *Phil. Trans. Roy. Soc. London* **A297**, 353–364.

Jochum K. P., Arndt N. T., and Hofmann A. W. (1991) Nb–Th–La in komatiites and basalts: constraints on komatiite petrogenesis. *Earth Planet. Sci. Lett.* **107**, 272–289.

Johnson C. M. and Beard B. L. (1993) Evidence from hafnium isotopes for ancient sub-oceanic beneath the Rio Grande Rift. *Nature* **362**, 441–444.

Kellogg J. B., Jacobsen S. B., and O'Connell R. J. (2002) Modeling the distribution of isotopic ratios in geochemical reservoirs. *Earth Planet Sci. Lett.* **204**, 183–202.

Kellogg L. H., Hager B. H., and van der Hilst R. D. (1999) Compositional stratification in the deep mantle. *Science* **283**, 1881–1884.

Kerrich R., Wyman D., Holings P., and Polat A. (1999) Variability of Nb/U and Th/La in the 3.0–2.7 Ga superior province ocean plateau basalts: implications for the timing of continental growth and lithosphere recycling, *Earth Planet. Sci. Lett.* **168**, 101–115.

Kinny P. D., Compston W., and Williams I. S. (1991) A reconnaissance ion-probe study of hafnium isotopes in zircons. *Geochim. Cosmochim. Acta* **55**, 849–859.

Kleine T., Muenker C., Mezger K., and Palme H. (2002) Rapid accretion and early core formation on asteroids and the terrestrial planets from Hf–W chronometry. *Nature* **418**, 952–955.

Lee D.-C. and Halliday A. N. (1995) Hafnium–tungsten chronometry and the timing of terrestrial core formation. *Nature* **378**, 771–774.

Longhi J. (1980) A model of early lunar differentiation. *Proc. Lunar Planet. Sci. Conf.* **11**, 289–315.

Machado N., Brooks C., and Hart S. R. (1986) Determination of initial $^{87}Sr/^{86}Sr$ and $^{143}Nd/^{144}Nd$ in primary minerals from mafic and ultramafic rocks: experimental procedure and implications for isotopic characteristics of the Archean mantle under the Abitibi greeenstone belt, Canada. *Geochim. Cosmochim. Acta* **50**, 2335–2338.

Machetel P. and Weber P. (1991) Intermittant layered convection in a model mantle with an endothermic phase change at 670 km. *Nature* **350**, 55–57.

Marcantonio F., Zindler A., Staudigel H., and Schminke H. (1995) Os isotope systematics of La Palma, Canary Islands: evidence for recycled crust in the mantle source of HIMU ocean islands. *Earth Planet. Sci. Lett.* **133**, 397–410.

Martin C. E., Esser B. K., and Turekian K. K. (1991) Re–Os isotopic constraints on the formation of mantle and crustal reservoirs. *Austral. J. Earth Sci.* **38**, 569–576.

McCulloch M. T. (1994) Primitive $^{87}Sr/^{86}Sr$ from an Archean barite and conjecture on the Earth's age and origin. *Earth Planet. Sci. Lett.* **126**, 1–13.

McCulloch M. T. and Bennett V. C. (1993) Evolution of the early Earth: constraints from $^{143}Nd-^{142}Nd$ isotopic systematics. *Chem. Geol.* **30**, 237–255.

McCulloch M. T. and Bennett V. C. (1994) Progressive Growth of the Earth's continental crust and depleted mantle. geochemical constraints. *Geochim. Cosmochim. Acta* **58**, 4717–4738.

McCulloch M. T. and Black L. P. (1984) Sm–Nd isotopic systematics of Enderby Land granulites and evidence for the redistribution of Sm and Nd during metamorphism. *Earth Planet. Sci. Lett.* **71**, 46–58.

McCulloch M. T. and Compston W. (1981) Sm–Nd age of Kambalda and Kanowna greenstones and heterogeneity in the Archaean mantle. *Nature* **294**, 322–327.

McDonough W. F. (1991) Partial melting of subducted oceanic crust and isolation of its residual eclogitic ecology. *Phil. Trans. Roy. Soc. London Ser. A* **335**, 407–418.

McGregor V. M. (2000) Initial Pb of the Amîtsoq gneiss revisited: implications for the timing of early Archaean crustal evolution in West Greenland. *Chem. Geol.* **166**, 301–308.

Meisel T., Walker R. J., and Morgan J. W. (1996) The osmium isotopic composition of the primitive upper mantle. *Nature* **383**, 517–520.

Meisel T., Walker R. J., Irving A. J., and Lorand J.-P. (2001) Osmium isotopic compositions of mantle xenoliths: a global perspective. *Geochim. Cosmochim. Acta* **65**, 1311–1323.

Moorbath S., Whitehouse M. J., and Kamber B. S. (1997) Extreme Nd-isotope heterogeneity in the early Archaean-fact or fiction? Case histories from northern Canada and West Greenland. *Chem. Geol.* **135**, 213–231.

Mueller W., Mancktelow N., and Meier M. (2000) Rb–Sr microchrons of synkinematic mica in mylonites: an example from the DAV Fault of the Eastern Alps. *Earth Planet. Sci. Lett.* **180**, 385–397.

Newsom H. E., White W. M., Jochum K. P., and Hofmann A. W. (1986) Siderophile and chalcophile element abundances in oceanic basalts, Pb isotope evolution and growth of the Earth's core. *Earth Planet. Sci. Lett.* **80**, 299–313.

Norman M., Borg L., Nyquist L., and Bogard D. (2003) Chronology, geochemistry, and petrology of a ferroan noritic anorthosite clast from Descartes breccia 67215: clues to the age, origin, structure, and impact history of the lunar crust. *Meteorit. Planet. Sci.* **38**, 645–661.

Nowell G. M., Pearson D. G., Kempton P. D., Noble S. R., and Smith C. B. (1999) Origins of kimberlites: a Hf isotope perspective. *Proc. Int. Kimberlite Conf.* **7**, 616–624.

Nutman A. P. (2002) On the scarcity of >3900 Ma detrital zircons in greater-than or equal to 3500 Ma metasediments. *Precamb. Res.* **105**, 93–114.

Nutman A. P., Bridgwater D., and Fryer B. (1984) The iron rich suite from the Amitsoq gneisses of southern West Greenland: early Archaean plutonic rocks of mixed crustal and mantle origin. *Contrib. Mineral. Petrol.* **87**, 24–34.

Nutman A. P., McGregor V. R., Friend C. R. L., Bennett V. C., and Kinny P. D. (1996) The Itsaq Gneiss Complex of southern West Greenland: the world's most extensive record of early crustal evolution (3,900–3,600 Ma). *Precamb. Res.* **78**, 1–39.

Nutman A. P., Bennett V. C., Friend C. R. L., and McGregor V. R. (2000) The Early Archaean Itsaq Gneiss complex of southern West Greenland: the importance of geology in interpreting geochronological and isotopic data on early terrestrial evolution. *Geochim. Cosmochim. Acta* **64**, 3035–3060.

Nutman A. P., Friend C. R. L., and Bennett V. C. (2001) Review of the oldest (4400–3600 Ma) geological and mineralogical record: glimpses of the beginning. *Episodes* **24**, 1–9.

Nyquist L. E., Shih C. Y., Wooden J. L., Bansal B. M., and Wiesmann H. (1979) The Sr and Nd isotopic record of Apollo 12 basalts: implications for lunar geochemical evolution. *Proc. Lunar Planet. Sci. Conf.* **10**, 77–114.

Nyquist L. E., Wiesmann H., Shih C.-Y., Keith J. E., and Harper C. L. (1995) ^{146}Sm–^{142}Nd formation interval for the lunar mantle. *Geochim. Cosmochim. Acta* **59**, 2817–2837.

Nyquist L. E., Bogard D. D., Shih C.-Y., Greshake A., Stoffler D., and Eugster O. (2001) Ages and geological histories of martian meteorites. *Space Sci. Rev.* **96**, 105–164.

O'Nions R. K., Evensen N. M, and Hamilton P. J. (1979) Geochemical modelling of mantle differentiation and crustal growth. *J. Geophys. Res.* **84**, 6091–6101.

Papanastassiou D. A. and Wasserburg G. J. (1969) Initial strontium isotopic abundances and the resolution of small time differences in the formation of planetary objects. *Earth Planet. Sci. Lett.* **5**, 361–367.

Papanastassiou D. A., Sharma M., Ngo H. H., Wasserburg G. J., and Dymek R. F. (2003) No ^{142}Nd excess in the early Archean Isua gneiss IE 715-28. *Lunar Planet. Sci. Conf.* **XXXIV**, 1851.

Parman S. W., Grove T. L., and Dann J. C. (2001) The production of Barberton Komatiites in an Archean subduction zone. *Geophys. Res. Lett.* **28**, 2513–2516.

Patchett P. J. (1983) Importance of the Lu–Hf isotopic system in studies of planetary chronology and chemical evolution. *Geochim. Cosmochim. Acta* **47**, 81–91.

Patchett P. J. and Chauvel C. (1984) The mean life of continents is not currently constrained by Nd and Hf isotopes. *Geophys. Res. Lett.* **11**, 151–153.

Patchett P. J. and Tatsumoto M. (1980) Hafnium isotope variations in oceanic basalts. *Geophys. Res. Lett.* **7**, 1077–1080.

Peucker-Ehrenbrink B. and Jahn B.-M. (2001) Rhenium–osmium isotope systematics and platinum group element concentrations: loess and the upper continental crust. *Geochem. Geophys. Geosys. G* **3** 2001GC000172.

Polat A. and Kerrich R. (2000) Archean greenstone belt magmatism and the continental growth-mantle evolution connection: constraints from Th–U–Nb-LREE systematics of the 2.7 Ga Wawa subprovince, Superior Province, Canada. *Earth. Planet. Sci. Lett.* **175**, 41–54.

Prinzhofer A., Papanastassiou D. A., and Wasserburg G. J. (1992) Samarium-neodymium evolution of meteorites. *Geochim. Cosmochim. Acta* **56**, 797–815.

Puchtel I. S., Brügmann G. E., and Hofmann A. W. (1999) Precise Re–Os mineral isochron and Pb–Nd–Os isotope systematics of a mafic-ultramafic sill in the 2.0 Ga Onega plateau (Baltic Shield). *Earth Planet. Sci. Lett.* **170**, 447–461.

Puchtel I. S., Brügmann G. E., and Hofmann A. W. (2001a) ^{187}Os-enriched domain in an Archean mantle plume: evidence from 2.8 Ga komatiites of the Kostomuksha greenstone belt, NW Baltic Shield. *Earth Planet. Sci. Lett.* **186**, 513–526.

Puchtel I. S., Brügmann G. E., Hofmann A. W., Kukikov V. S., and Kulikova V. V. (2001b) Os isotope systematics of komatiitic basalts from the Vetreny belt, Baltic Shield: evidence for a chondritic source of the 2.45 Ga plume. *Contrib. Mineral. Petrol.* **140**, 588–599.

Regelous M. and Collerson K. D. (1996) ^{147}Sm–^{143}Nd, ^{146}Sm–^{142}Nd systematics of early Archaean rocks and implications for crust–mantle evolution. *Geochim. Cosmochim. Acta* **60**, 3513–3520.

Reisberg L., Zindler A., Marcantonio F., White W., and Wyman D. (1993) Os isotope systematics in ocean island basalts. *Earth Planet. Sci. Lett.* **120**, 149–167.

Reymer A. and Schubert G. (1984) Phanerozoic addition rates to the continental crust and crustal growth. *Tectonics* **3**, 63–77.

Righter K. and Drake M. J. (1997) Metal-silicate equilibrium in a homogeneously accreting earth: new results for Re. *Earth Planet. Sci. Lett.* **146**, 541–553.

Roy-Barman M. and Allègre C. J. (1994) ^{187}Os/^{186}Os ratios of mid-ocean ridge basalts and abyssal peridotites. *Geochim. Cosmochim. Acta* **58**, 5043–5054.

Rudnick R. L. and Fountain D. M. (1995) Nature and composition of the continental crust: a lower crust perspective. *Rev. Geophys.* **33**, 267–309.

Rudnick R. L., Barth M., Horn I., and McDonough W. F. (2000) Rutile bearing refractory eclogites: missing link between continents and depleted mantle. *Science* **287**, 278–281.

Salters V. J. M. (1996) The generation of mid-ocean ridge basalts from the Hf and Nd isotope perspective. *Earth Planet. Sci. Lett.* **141**, 109–123.

Salters V. J. M. and White W. M. (1998) Hf isotope constraints on mantle evolution. *Chem. Geol.* **145**, 447–460.

Scherer E., Muenker C., and Mezger K. (2001) Calibration of the lutetium–hafnium clock. *Science* **293**, 683–687.

Schiano P., Birck J.-L., and Allegere C.-J. (1997) Osmium–strontium–neodymium-lead isotopic covariations in mid ocean ridge basalt glasses and the heterogeneity of the upper mantle. *Earth Planet. Sci. Lett.* **150**, 363–379.

Schoenberg B. S., Kamber B. S., Collerson K. D., and Eugster O. (2002) New W-isotope evidence for rapid terrestrial accretion and very early core formation. *Geochim. Cosmochim. Acta* **66**, 3151–3160.

Sguigna A. P., Larabee A. J., and Waddington J. C. (1982) The half-life of ^{176}Lu by a γ–γ measurement. *Can. J. Phys.* **60**, 361–364.

Sharma M., Papanastassiou D. A., Wasserburg G. J., and Dymek R. F. (1996) The issue of the terrestrial record of ^{146}Sm. *Geochim. Cosmochim. Acta* **60**, 2037–2047.

Shen J. J., Papanastassiou D., and Wasserburg G. J. (1996) Precise Re–Os determinations and systematics of iron meteorites. *Geochim. Cosmochim. Acta* **60**, 2887–2900.

Shirey S. B. (1991) The Rb–Sr, Sm–Nd and Re–Os isotopic systems: a summary and comparison of their applications to the cosmochronology and geochronology of igneous rocks. In *Applications of Radiogenic Isotope Systems to Problems in Geology*, Short Course Handbook (eds. L. Heaman and J. Ludden). Mineral. Assoc. Can., Nepean, vol. 19, pp. 109–166.

Shirey S. B. (1997) Initial Os isotopic composition of Munro Township, Ontario komatiites revisited: additonal evidence for near chondritic, Late-Archean convecting mantle beneath the Superior Province. *7th Annual Goldschmidt Conf. LPI Contrib.* **921**, 193 (Abstr.).

Shirey S. B. and Walker R. J. (1995) Carius tube digestion for low blank Re–Os analysis. *Anal. Chem.* **67**, 2136–2141.

Shirey S. B. and Walker R. J. (1998) The Re–Os isotope system in cosmochemistry and high temperature geochemistry. *Ann. Rev. Earth Planet. Sci.* **26**, 423–500.

Snow J. E. and Reisberg L. (1995) Os isotopic systematics of the MORB mantle: results from altered abyssal peridotites. *Earth Planet. Sci. Lett.* **133**, 411–421.

Smith A. D. and Ludden J. N. (1989) Nd isotopic evolution of the Precambrian mantle. *Earth Planet. Sci. Lett.* **93**, 14–22.

Smoliar M. I., Walker R. J., and Morgan J. W. (1996) Re–Os ages of group IIA, IIIA, IVA, and IVB iron meteorites. *Science* **271**, 1099–1102.

Staudigel H., Davies G. R., Hart S. R., Marchant K. M., and Smith B. M. (1995) Large scale isotopic Sr, Nd, and O isotopic anatomy of altered oceanic crust: DSDP/ ODP sites 417/418. *Earth Planet. Sci. Lett.* **130**, 169–185.

Stein M. and Hofmann A. W. (1994) Mantle plumes and episodic crustal growth. *Nature* **372**, 63–68.

Stevenson D. J. (1987) Origin of the Moon: the collision hypothesis. *Ann. Rev. Earth Planet. Sci.* **15**, 271–315.

Stevenson R. K. and Patchett P. J. (1990) Implications for the evolution of continental crust from Hf isotope systematics of Archean detrital zircons. *Geochim. Cosmochim. Acta* **54**, 1683–1697.

Sun W., Bennett V. C., Eggins S. M., Kamenetsky V. S., and Arculus R. J. (2003) Enhanced mantle-to-crust rhenium transfer in undegassed arc magmas. *Nature* **422**, 294–297.

Sylvester P. J., Campbell I. H., and Bowyer D. A. (1997) Niobium/Uranium evidence for early formation of the continental crust. *Science* **275**, 521–523.

Tackley P., Stevenson D. J., Glatzmaier G. A., and Schubert G. (1993) Effects of endothermic phase transition at 670 km depth in a spherical model of convection in the Earth's mantle. *Nature* **361**, 699–704.

Taylor S. R. and McLennan S. M. (1985) *The Continental Crust: Its Composition and Evolution.* Blackwell, Oxford, England, 312p.

Tera F., Papanastassiou D. A., and Wasserburg G. J. (1974) Isotopic evidence for a terminal lunar cataclysm. *Earth Planet. Sci. Lett.* **22**, 1–21.

Tonks W. B. and Melosh H. J. (1993) Magma ocean formation due to giant impacts. *J. Geophys. Res.* **98**, 5319–5333.

Tsuru A., Walker R. J., Kontinen A., Peltonen P., and Hanski E. (2000) Re–Os systematics of the 1.95 Ga Jormua Ophiolite complex, northeastern Finland. *Chem. Geol.* **164**, 123–141.

van der Hilst R. D. and Karason H. (1999) Compositonal heterogeneity in the bottom 1000 km of Earth's mantle: towards a hybrid convection model. *Science* **283**, 1885–1888.

van der Hilst R. D., Engdahl R., Spakman W., and Nolet G. (1991) Tomographic imaging of subducted lithosphere below Northwest Pacific island arcs. *Nature* **353**, 37–43.

Vervoort J. D., Patchett P. J., Gehrels G. E., and Nutman A. P. (1996) Constraints on early Earth differentiation from hafnium and neodymium isotopes. *Nature* **379**, 624–627.

Vervoort J. D. and Blichert-Toft J. (1999) Evolution of the depleted mantle: Hf isotope evidence from juvenile rocks through time. *Geochim. Cosmochim. Acta* **63**, 533–556.

Walker R. J. and Nisbet E. (2002) ^{187}Os isotopic constraints on Archean mantle dynamics. *Geochim. Cosmochim. Acta* **66**, 3317–3325.

Walker R. J. and Stone W. E. (2001) Os isotope constraints on the origin of the 2.7 Ga Boston Creek Flow, Ontario, Canada. *Chem. Geol.* **175**, 567–579.

Walker R. J., Carlson R. W., Shirey S. B., and Boyd F. R. (1989) Os, Sr, Nd, and Pb isotope systematics of Southern African peridotite xenoliths: implications for the chemical evolution of the subcontinental mantle. *Geochim. Cosmochim. Acta* **53**, 1583–1595.

Walker R. J., Morgan J. W., and Horan M. (1995) Osmium-187 enrichments in some plumes: a consequence of for core-mantle interaction? *Science* **269**, 819–822.

Walker R. J., Hanski E., Vuollo J., and Liipo J. (1996) The Os isotopic composition of Proterozoic upper mantle: evidence for chondritic upper mantle from the Outokumpo ophiolite, Finland. *Earth Planet. Sci. Lett.* **141**, 161–173.

Walker R. J., Prichard H. M., Ishiwatari A., and Pimentel M. (2002) The osmium isotopic composition of convecting upper mantle deduced from ophiolite chromites. *Geochim. Cosmochim. Acta* **66**, 329–345.

Widom E. and Shirey S. B. (1996) Os isotope systematics in the Azores: implications for mantle plume sources. *Earth Planet. Sci. Lett.* **142**, 451–465.

Wilde S. A., Valley J. W., Peck W. H., and Graham C. M. (2001) Evidence from detrital zircons for the existence of continental crust and oceans on the Earth 4.4 Gyr ago. *Nature* **409**, 175–178.

Yin Q., Jacobsen S. B., Yamashita K., Blichert-Toft J., Telouk P., and Albarède F. (2002) A short timescale for terrestrial planet formation from Hf–W chronometry of meteorites. *Nature* **418**, 949–952.

2.14
Experimental Constraints on Core Composition

J. Li

University of Illinois at Urbana Champaign, IL, USA

and

Y. Fei

Carnegie Institution of Washington, DC, USA

2.14.1 INTRODUCTION

The Earth's core was discovered in 1906, when Oldham inferred the existence of a low-velocity region inside the Earth from changes in the amplitude of compressional waves traveling through the Earth's interior (Oldham, 1906). Over the last century, a wealth of knowledge has been obtained on the nature and dynamics of the core (Figure 1; Dziewonski and Anderson, 1981; Stacey, 1992). Residing in the center of the planet, the core has a radius of 3,480 km, more than half of the Earth's radius. It occupies roughly one-eighth of the Earth's volume, and accounts for nearly one-third of its mass. The mass fraction in the core

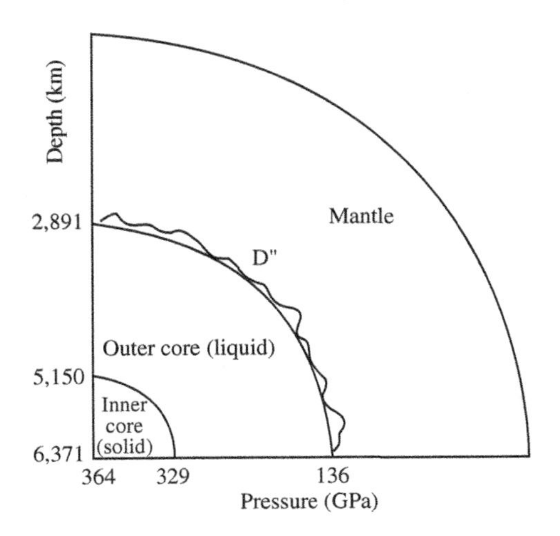

Figure 1 Cross-section of the Earth showing its layered structure (source Dziewonski and Anderson, 1981).

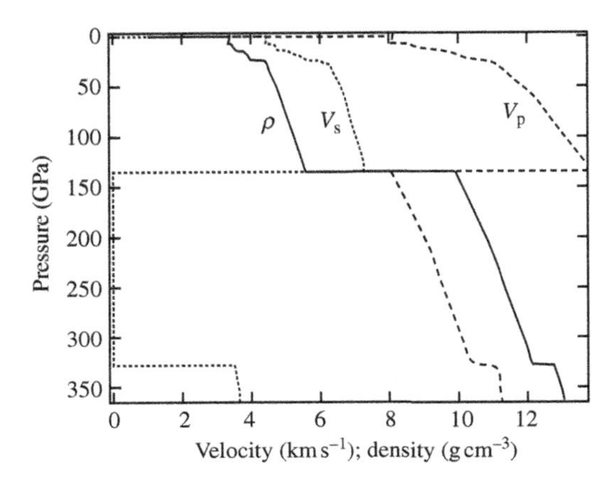

Figure 2 Preliminary reference Earth model (PREM) (source Dziewonski and Anderson, 1981).

is much higher than its volume fraction, because the density in the core is much higher than that of the mantle (Figure 2). With the density jumping from 5.5 g cm^{-3} to 9.9 g cm^{-3}, the density contrast at the core–mantle boundary (CMB) is the largest in the whole planet. Based on analyses of the velocity and attenuation of seismic waves, the core has been established to have a layered structure. The central part comprising less than 5% of the core's mass or volume is solid, while the rest is largely molten. The pressure in the core ranges from 136 GPa (1,380 kbar) at the CMB to 360 GPa at the very center. In order to stay in the liquid state under such high pressures, the temperature in the core must be high as well. The temperature of the liquid–solid interface at the inner–outer core boundary (ICB) is estimated at $5{,}400 \pm 400$ K (e.g., Brown and McQueen, 1986; Boehler, 2000; Hemley and Mao, 2001).

Combining geochemical and seismological observations on the Earth with laboratory measurements on relevant materials, more than 80 wt.% of the core has been deduced to consist of iron. Other elements with significant concentrations in the core include nickel (~5 wt.%) and one or more elements that are lighter than iron (e.g., Stevenson, 1981; Jacobs, 1987; Jeanloz, 1990). The age of the core has been determined using isotope geochronometers. According to the most recent measurements based on tungsten–hafnium systematics, most of the core–mantle segregation took place in less than 30 Myr (Yin *et al.*, 2002; Kleine *et al.*, 2002). In other words, the core is almost as old as the Earth itself. Core formation occurred as soon as the Earth accreted, or simultaneously with accretion.

Despite its old age, the core is dynamically active. The geomagnetic field observed on the Earth's surface is believed to originate in the outer core through the convection of the liquid conductive metal (Jacobs, 1987). A number of observations have been made recently that are indicative of chemical reaction and dynamical coupling between the core and the mantle (e.g., Lay *et al.*, 1998). Moreover, the core is widely accepted to be a major energy source for our planet. Not only did it acquire a large amount of energy early on during core–mantle segregation from radioactive decay of short-life isotopes, but it is also capable of producing more heat through solidification of the liquid outer core. Speculation suggests that the core contains a significant amount of potassium, which could have been generating heat over the history of the Earth (e.g., Jacobs, 1987; see Chapter 2.15).

The surface of the core is only 2,900 km in depth. However, drilling into the Earth's interior turns out to be much more challenging than flying into outer space. Spacecrafts have reached Jupiter, more than 600 million km away from the Earth, but the deepest hole we have successfully drilled has reached less than 14 km below the Earth's surface. Volcanic eruptions are unlikely to bring up pristine samples of the core to the surface of the Earth. To date, the most direct observations of the core have come from seismological studies using remote-sensing techniques. Due to the complex structure of the Earth's interior, seismic investigations require extensive data coverage, sophisticated modeling and efficient data analytical methods. Deciphering geochemical core signatures carried by mantle plumes faces similar challenges. In addition, experimental and computational simulations have been hindered by the necessity to approach extreme pressure and temperature conditions prevalent in the core. For these reasons, many fundamental issues concerning the Earth's core remain controversial or poorly understood.

Along with steady improvement in observational, experimental, and computational techniques, research interest in the Earth's core has been growing over the last few decades. Issues of current interest include the timing, duration, and mechanism of core formation, the identity and abundance of light elements in the core, the thermal and chemical evolution of the core, the possibility of ongoing radioactive decay in the core, evidence for continuing core–mantle interaction, structure and dynamics of the outer and inner core, and the origin, structure, and evolution of the geomagnetic field. Progress in the study of the Earth's core has been reviewed by a number of researchers with different perspectives (e.g., Stevenson, 1981; Jacobs, 1987; Jeanloz, 1990; Poirier, 1994; Hillgren *et al.*, 2000). The aim of this chapter is to provide an updated summary of our understanding of the composition of the Earth's core, with an emphasis on experimental constraints. Another chapter focusing on cosmochemical and geochemical observations can be found in Chapter 2.15.

We will start with a description of experimental and analytical techniques used in the studies of the Earth's core. Following a review of geophysical and geochemical evidence for iron being the most abundant element in the core, we will provide a summary of experimental data on the phase diagram, equation-of-state (EOS), and physical properties of iron and discuss their implications for the core composition. The discussion of the role of nickel will be brief, as the number of experimental studies of nickel is limited. A large portion of the chapter will be devoted to constraints on the light element composition of the core, a highly controversial subject, with direct implications for diverse topics concerned with the origin, evolution, and current state of the Earth. Finally, we summarize the latest experimental results on potassium, niobium, rhenium, and osmium, some of the interesting minor and trace elements in the core.

2.14.2 GENERAL APPROACHES

2.14.2.1 Experimental Methods

Various experimental apparatus have been invented and developed to study the chemistry and physics of materials under the high pressures and temperatures prevalent in the Earth's interior. Most commonly used apparatus for experiments under static high pressure include the piston-cylinder apparatus, multi-anvil apparatus, and diamond-anvil cell (Bertka and Fei, 1997; Mao and Hemley, 1998). According to a basic principle of mechanics, when a given force is applied to a surface, the pressure on the surface is inversely proportional to the area of the surface. Applying the same force to a smaller surface can generate higher pressure. In the piston-cylinder or multi-anvil apparatus, a hydraulic press drives a ram with an area of $\sim$0.1–1 m^2. The force on the ram is transmitted to the flat surface of a tungsten carbide piston (piston-cylinder), or the truncated corners of several tungsten carbide or sintered diamond cubes (multi-anvil), with a total area of $\sim 10^{-4}-10^{-5}$ m^2. As the force is concentrated on the smaller area, high pressures ($\sim$1–10 GPa) can be generated. Due to large deviatoric stress, the maximum pressure in the piston-cylinder apparatus is normally less than 6 GPa. In the multi-anvil apparatus, $\sim$30 GPa can be achieved using tungsten carbide cubes, and $\sim$50 GPa using sintered diamond cubes. Compared with piston-cylinder and multi-anvil apparatuses, the diamond-anvil cell is a more compact device. Applying force through wrenches and screws, the pressure produced on the culets of diamond anvils can be as high as 300 GPa. Higher pressure usually corresponds to smaller sample volume. As a rule of thumb, the ratio of sample volumes for the piston-cylinder apparatus, multi-anvil apparatus, and diamond-anvil cell is approximately $10^6 : 10^4 : 1$.

High temperature in the piston-cylinder and multi-anvil apparatuses is produced by resistive heating. Commonly used heater materials include graphite, rhenium foil, and $LaCrO_3$. Stable heating for durations of days can be achieved at temperatures below 2,000 K. At higher temperatures, due to hotspots in the heater and possible reactions between the heater and surrounding materials, stable heating is usually limited to hours or less. Multi-anvil experiments have been performed at temperatures above 2,800 K (e.g., Bertka and Fei, 1997). Resistive heating is also used in the diamond-anvil cell. External resistive heating in the diamond-anvil cell has been limited to relatively low temperatures (<1,100 K; Mao *et al.*, 1991; Fei *et al.*, 1992). Internal resistive heating in diamond-anvil cell can reach higher temperatures, as long as the heater and the sample do not react. The most widely used heating method in the diamond-anvil cell is by absorption of laser radiation. Using a Nd : YAG (yttrium–aluminum–garnet) or a CO_2 laser, temperatures as high as 6,000 K can be reached. The introduction of double-sided laser heating and an electronic feedback system has helped to reduce temperature gradients and has improved temperature stability significantly (Mao and Hemley, 1998; Boehler, 2000).

Another important variable in the study of core composition is oxygen fugacity. Accurately controlled and measured oxygen fugacity can be obtained over a wide range using a gas-mixing furnace, but only at ambient pressure. In the piston-cylinder apparatus, multi-anvil apparatus, and diamond-anvil cell, limited control of oxygen fugacity can be achieved by using different capsule and sample materials (e.g., Rubie, 1999).

The use of the gas-mixing furnace, piston-cylinder apparatus, multi-anvil apparatus, and diamond-anvil cell allows material properties to be investigated over a large pressure–temperature–composition space. Until recently, sample characterization in piston-cylinder and multi-anvil experiments was carried out mainly after quenching and decompression. Products of chemical reactions are analyzed using optical microscope, X-ray diffractometer, electron microprobe, etc. The results are used to construct phase diagrams, detect chemical reactions, determine element partitioning, characterize wetting behavior, and measure densities. The transparency of diamonds to light allows *in situ* optical observations and laser-excited spectroscopy to be performed, in addition to X-ray diffraction, electrical and magnetic measurements, and acoustic interferometry (Mao and Hemley, 1998; Hemley and Mao, 2001). *In situ* measurements provide data on phase transitions, EOS, phase segregation, elastic properties, plastic deformation, transport properties, etc. Due to the small sample size under high pressure, some *in situ* measurements are only possible with the application of synchrotron facilities. X-ray tomography and rheology studies with radial diffraction on stressed samples are among the most recent *in situ* analytical techniques facilitated by synchrotron radiation (e.g., Merkel *et al.*, 2002; Mao *et al.*, 1999, 2001). Synchrotron facilities have also been applied to the multi-anvil apparatus (e.g., Irifune *et al.*, 1998; Hirose *et al.*, 2000).

Phase transition and EOS measurements have been carried out under dynamic pressure as well, using shock-wave facilities (Ahrens, 1980). Upon explosion of gunpowder, light gas in a large-bore tube is compressed between a piston and a rupture disc. When the disc breaks apart, the gas expands into a small-bore barrel, accelerating a projectile in front of it. The projectile flies through the barrel and impacts the sample, generating a high-pressure shock wave and adiabatic heating. By measuring temperature and velocities of the projectile and the shock wave, one can calculate the density of the sample along the Hugoniot.

2.14.2.2 Theoretical Studies

The impact of computers on high-pressure theoretical studies may be compared with that of synchrotron facilities on high-pressure experimental studies. Steady improvements in computational methods have enabled calculations of structure, stability, and elastic properties of simple systems under pressures and temperatures of the Earth's interior (e.g., Stixrude and Brown, 1998; Boness and Brown, 1990; Sherman, 1995, 1997; Steinle-Neumann and Stixrude, 1999; Steinle-Neumann *et al.*, 2001; Vocadlo and Dobson, 1999; Vocadlo *et al.*, 2000; Alfé *et al.*, 1999a, 2000a, 2001, 2002a). Simulation techniques can be divided into two categories: the calculation of the electronic structure at a given arrangement of nuclei and the calculation of the motion of the nuclei. To calculate the electronic structure, density functional theory (DFT) is the most commonly used technique. These calculations have no free parameters to adjust and are therefore also termed *ab initio*. Temperature can be included in the form of thermal motion by moving nuclei according to the forces calculated from the computational intensive DFT or from simplified pseudo-potential models, where interatomic potentials are approximated by simple functions. Monitoring the motion of the atomic nuclei and extracting the dynamical properties is referred to as molecular dynamics. Because of the high computational requirements, simulations are normally restricted to simple systems with only a small number of particles compared to the vast number of atoms in even the tiniest crystal.

2.14.3 MAJOR ELEMENTS IN THE CORE

2.14.3.1 Iron

2.14.3.1.1 Evidence for iron being the dominant component of the core

At the end of the eighteenth century, the average density of the Earth as a whole was already known to be twice that of the common crust rocks. The original concept of an iron core was purely based on meteoritic observations. Stony meteorites provide the clue that iron is more abundant in accretion materials than in the Earth's mantle (Cox, 1989). A serious proposal of an iron core was not presented until the recognition of the two main classes of meteorites, stony and iron meteorites, in the mid-nineteenth century. The iron meteorites had been assumed to come from the core of a fragmented planet. The idea of an iron core was generally accepted once Oldham (1906) confirmed the presence of a core.

Since the discovery of the Earth's core about a century ago, the idea of iron being the dominant component of the core has gained firm supporting evidence from cosmochemical observations, refined seismic data, high-pressure experimentation, and theories of geomagnetism. Strong support for the idea of an iron core comes from the reasonably close match between the seismologically inferred sound velocity and density of the core and the measured experimental values for iron by shock compression (Al'tshuler, 1962; Birch, 1964; McQueen and Marsh, 1966; Press, 1968; Jeanloz, 1979; Brown and McQueen, 1986). This idea was further supported by the

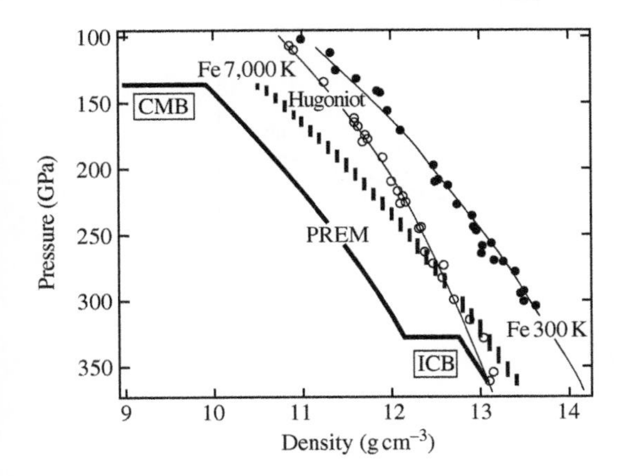

Figure 3 Comparison of the measured density of iron with the density of the core from PREM (Dziewonski and Anderson, 1981). The solid circles are the measured densities of ε-Fe (Brown *et al.*, 2000) from static experiments at 300 K (Mao *et al.*, 1990). The open circles represent densities of iron on the Hugoniot from dynamic experiments (Brown *et al.*, 2001). The dashed lines are the calculated isothermal compression curve at 7,000 K (Dubrovinsky *et al.*, 2000).

density measurements of iron at core pressures by static compression (Mao *et al.*, 1990; Dubrovinsky *et al.*, 2000). Figure 3 shows a comparison of the densities of iron determined from shock compression (Brown and McQueen, 1986) and from static compression (Mao *et al.*, 1990; Dubrovinsky *et al.*, 2000), together with the density profile derived from seismic data (PREM model; Dziewonski and Anderson, 1981). As reviewed by Jeanloz (1990), additional evidence for an iron core comes from our current understanding of the generation of the Earth's magnetic field for which a metallic liquid outer core is required (e.g., Merrill and McElhinny, 1983; Jacobs, 1987; Merrill *et al.*, 1996).

2.14.3.1.2 *Phase diagram of iron at high pressure and temperature*

Geophysicists have spent considerable efforts to develop experimental techniques to determine the phase diagram and physical properties of iron at high pressure and temperature since the 1950s. Much of our initial knowledge on the density and phase transformation of iron at high pressures is from dynamic shock wave experiments (e.g., Bancroft *et al.*, 1956; McQueen and Marsh, 1966; Barker and Hollenbach, 1974; Brown and McQueen, 1986). However, structural information for high-pressure polymorphs of iron was obtained from static compression experiments combined with *in situ* X-ray diffraction measurements (e.g., Jamieson and Lawson, 1962; Takahashi and Bassett, 1964; Clendenen and Drickamer, 1964).

Figure 4(a) shows the phase diagram of iron at pressures below 200 GPa, based on static experiments. Upon heating at room pressure, the structure of metallic iron changes from the body-centered cubic (b.c.c.) α-phase to the face-centered cubic (f.c.c.) γ-phase. The f.c.c. γ-phase transforms to the b.c.c. δ-phase before melting at ~1,700 K (e.g., Strong, 1959; Claussen, 1960; Omel'chenko *et al.*, 1969; Strong *et al.*, 1973; Mirwald and Kennedy, 1979). Upon compression at room temperature, the α-phase transforms to the hexagonal close-packed (h.c.p.) ε-phase at ~10 GPa. This transition is rather sluggish, as indicated by the large hysteresis in the reported data. Precise determination of this transition boundary was attempted by *in situ* measurements under simultaneous high pressure and temperature conditions, using a cubic-type multi-anvil apparatus and synchrotron X-ray diffraction (Akimoto, 1987) and an externally heated diamond-anvil cell at synchrotron facilities (Huang *et al.*, 1987; Manghnani and Syono, 1987). The results of Akimoto (1987) and Manghnani and Syono (1987) are in general agreement. The $\alpha-\varepsilon-\gamma$ triple point was located at 8.3 GPa and 713 K (Akimoto, 1987; Table 1).

The $\varepsilon-\gamma$ transition boundary was determined by measuring the resistance changes during the transition in a "high-compression belt" apparatus (Bundy, 1965) and in an internally heated diamond-anvil cell (Boehler, 1986; Mao *et al.*, 1987). The boundary was also determined by *in situ* X-ray diffraction measurements in an internally heated diamond-anvil cell (Boehler, 1986; Dubrovinsky *et al.*, 1998), in a laser-heated diamond-anvil cell (Shen *et al.*, 1998), and in a multi-anvil apparatus (Funamori *et al.*, 1996; Uchida *et al.*, 2001). The boundaries determined by Mao *et al.* (1987), Shen *et al.* (1998), and Uchida *et al.* (2001) are in good agreement, but are all at ~75 K higher temperature (or ~2 GPa lower pressure) than the boundary determined by Funamori *et al.* (1996), Boehler (1986), and Bundy (1965).

Experiments have established that there are four polymorphs in solid iron (α-, γ-, δ-, and ε-phases). Saxena *et al.* (1993) proposed a fifth iron phase (β-phase) based on changes in thermal emission while laser heating the sample in a diamond-anvil cell. Boehler (1993) also observed similar changes in optical properties of iron in the same $P-T$ range. Subsequent *in situ* X-ray diffraction measurements in the laser-heated diamond-anvil cell supported the occurrence of this new iron phase, although the structure of this phase is still under debate (Saxena *et al.*, 1996; Yoo *et al.*, 1996; Andrault *et al.*, 1997, 2000; Saxena and Dubrovinsky, 2000). However, this phase was not observed in

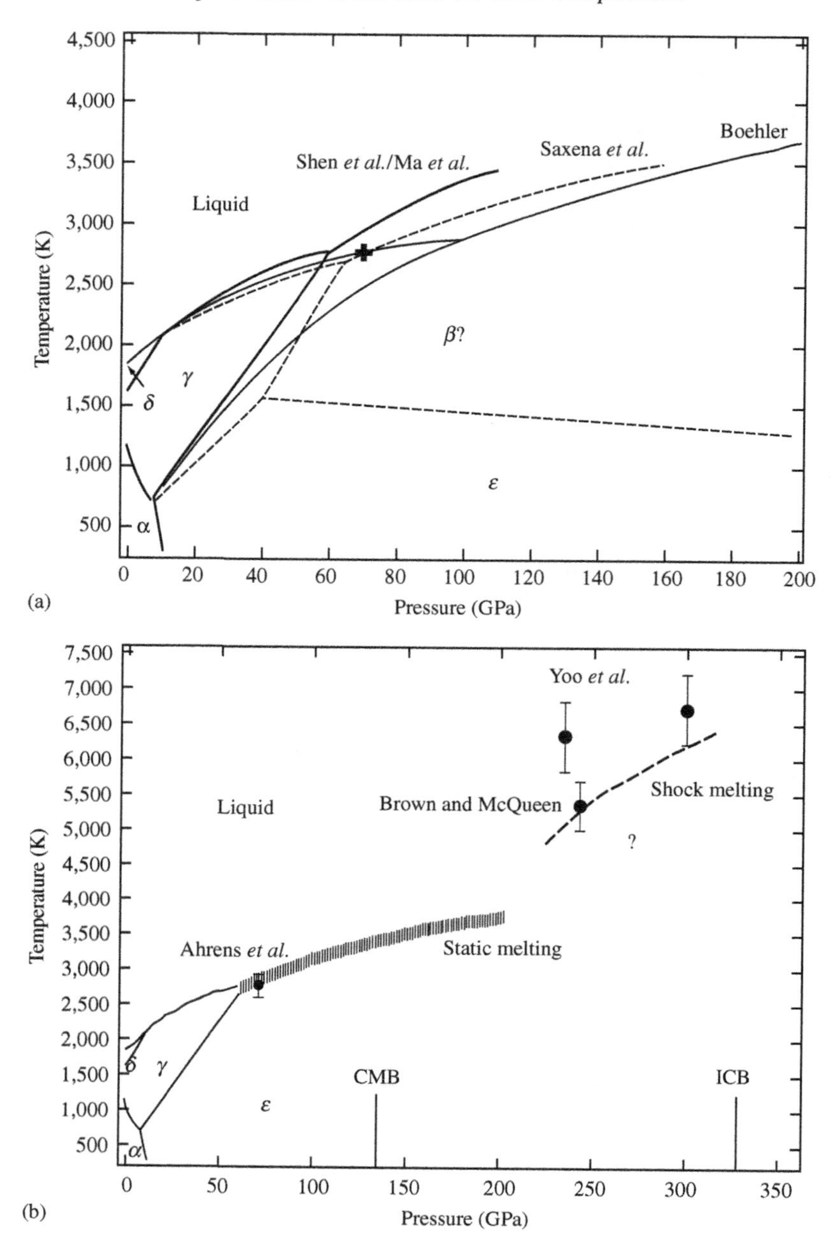

Figure 4 (a) Phase diagram of iron under 200 GPa, determined by static experiments (data from heavy solid lines: Shen *et al.* (1998) and Ma *et al.* (2001); thin solid lines: Boehler (1993); the dashed lines: Saxena *et al.* (1994); the cross represents the melting temperature at 71 GPa determined by recent shock wave experiments (source Ahrens *et al.*, 2002). (b) Melting temperatures of iron at core pressures, determined by static experiments and by dynamic experiments. The extrapolation of the static melting curve cannot match the melting temperatures inferred from shock-wave experiments at core pressures (sources Brown and McQueen, 1986; Yoo *et al.*, 1993).

other experiments (Shen *et al.*, 1998; Ma *et al.*, 2001; also see Hemley and Mao, 2001). Further experimental investigations are needed to resolve the discrepancies.

Experimental results show that the γ-phase is stable up to pressure between 60 GPa and 100 GPa along the melting curve of iron (Figure 4(a)). By re-analyzing the existing Hugoniot data (Brown and McQueen, 1986; Brown *et al.*, 2000), Brown (2001) concluded that the discontinuity in the sound velocity and density at 200 GPa is caused by a solid–solid phase transition with a density difference of $\sim 0.7\%$. The newly found β-phase may or may not be the subsolidus phase inferred from the shock-wave data (Andrault *et al.*, 2000).

Table 1 Transition points in iron at high pressure and temperature.

P (GPa)	*T* (K)	*Techniques*	*Method*	*References*
α–ε–γ triple point				
11	763	Resistance	Multi-anvil	Bundy (1965)
8.3	713	X-ray	Multi-anvil	Akimoto (1987)
8	680	X-ray	Multi-anvil	Uchida *et al.* (2001)
ε–γ–l triple point				
100	2,900	Resistance	Laser	Boehler (1993)
75	2,500	Resistance	Resistance	Boehler (1986)
65	2,700	Laser power	Laser	Saxena *et al.* (1993)
50	2,500	X-ray	Laser	Yoo *et al.* (1995)
60	2,800	X-ray	Laser	Shen *et al.* (1998)
Estimated melting temperature at ICB				
330	5,800		Shock wave	Brown and McQueen (1986)
330	6,830		Shock wave	Yoo *et al.* (1993)
330	5,300		Shock wave	Ahrens *et al.* (2002)
330	6,130		Static	Saxena *et al.* (1994)
330	5,000		Static	Boehler (2000)
330	5,700		Theory	Steinle-Neumann *et al.* (2001)

The melting curve of iron at high pressures has been studied extensively because of its importance for inferring the temperature of the Earth's core (Sterrett *et al.*, 1965; Strong *et al.*, 1973; Liu and Bassett, 1975; Boehler, 1986, 1993, 2000; Williams *et al.*, 1987; Williams and Jeanloz, 1990; Saxena *et al.*, 1994; Yoo *et al.*, 1995; Shen *et al.*, 1998; Ma *et al.*, 2001). The melting data from static experiments are in general agreement at pressures below 60 GPa, except for those of Williams and co-workers (1987, 1990). However, the location of the ε–γ–liquid triple point varies from 50 GPa and 2,500 K (Yoo *et al.*, 1995) to 100 GPa and 2,900 K (Boehler, 1993) (Table 1). The melting temperatures determined at pressures above 60 GPa vary considerably among different static experiments. At 100 GPa they range from 2,900 K (Boehler, 1993; Saxena *et al.*, 1994) to 4,100 K (Williams *et al.*, 1987). More recent *in situ* X-ray diffraction measurements in a double-sided laser heating diamond-anvil cell showed a melting temperature of 3,400 K at 100 GPa, between the two extreme melting temperatures (Ma *et al.*, 2001; also see Hemley and Mao, 2001).

In Figure 4(b) we compare melting temperatures from the static and dynamic experiments up to the core pressures. Recent shock-wave experiments on preheated iron yield a melting temperature of 2,775 K at 71 GPa (Ahrens *et al.*, 2002). This result is in good agreement with the measured melting temperatures under static pressure conditions by Boehler (1993) and Saxena *et al.* (1994). Shock compression data at pressures between 77 GPa and 400 GPa revealed two discontinuities in sound velocities at 200 GPa and 243 GPa along the Hugoniot (Brown and McQueen, 1986). The discontinuity at 243 GPa was believed to indicate the onset of melting. The estimated melting temperature at 243 GPa is located between 5,000 K and 5,700 K, based on the calculated temperatures along the Hugoniot. This shock melting temperature is over 1,000 K higher than that extrapolated from the static experiments (Boehler, 1993; Saxena *et al.*, 1994; Boehler, 2000). Measured shock temperatures gave an even higher melting temperature, which is 6,350 K at 235 GPa (Yoo *et al.*, 1993). The occurrence of another new subsolidus phase along the Hugoniot at ~200 GPa would reconcile the dynamic and static data on the melting temperatures of iron (Brown, 2001; Ahrens *et al.*, 2002).

2.14.3.1.3 *Physical properties of ε-Fe and liquid Fe at high pressure and temperature*

As reviewed in the previous section, the Earth's core consists of a liquid outer core and a solid inner core. The ICB pressure is ~330 GPa. The estimated temperature at this boundary ranges from 5,000 K (e.g., Boehler, 2000) to 5,800 K (e.g., Brown and McQueen, 1986), based on estimates of the melting temperature of iron at high pressure. So far, no direct static or dynamic measurements of iron under inner-core conditions have been made. The maximum pressure achieved in static experiments is ~200 GPa at temperatures relevant to the Earth's core (Boehler, 1993). Although shock-wave experiments have achieved a maximum pressure of 442 GPa (Brown *et al.*, 2000), the calculated Hugoniot temperature at the ICB pressure is over 8,000 K, which is far above the melting temperature of iron. Our current understanding of the phase diagram of iron favors h.c.p. ε-Fe phase as the most likely candidate for the solid inner core. All models of the inner core have been based on the h.c.p. structure of ε-Fe.

The volume–pressure relationship for ε-Fe has been determined up to 300 GPa at room temperature (Figure 3; Mao *et al.*, 1990). Determining the temperature dependence of the bulk modulus of iron is crucial for accurate comparison between the density of iron and that of the inner core. Several *in situ* X-ray diffraction studies in the diamond-anvil cell (Huang *et al.*, 1987; Dubrovinsky *et al.*, 1998, 2000) and in the multi-anvil apparatus (Funamori *et al.*, 1996; Uchida *et al.*, 2001) have provided limited data on the thermal expansion of ε-Fe at high pressures. The data of Dubrovinsky *et al.* (2000) covered the largest pressure–temperature range in the static experiments (up to 300 GPa and 1,300 K). Their calculated 7,000 K isotherm for ε-Fe is plotted in Figure 3. These data suggest that the inner core is ~2.5% lighter than ε-Fe under the same conditions.

The shear and compressional acoustic wave velocities for the inner core are the direct output parameters from seismological observations. In order to make a direct comparison between the seismic data and measured physical properties, measurements of the acoustic velocities for iron at core pressures are required. Only very recently has it become possible to measure the elastic constants of ε-Fe at high pressures and room temperature (Mao *et al.*, 1999; Lübbers *et al.*, 2000; Fiquet *et al.*, 2001; Anderson *et al.*, 2001). Recent advances in theory and computational methods have also provided new tools for computing the elastic constants of ε-Fe at core pressures (Stixrude and Cohen, 1995; Söderlind *et al.*, 1996; Cohen *et al.*, 1997; Steinle-Neumann and Stixrude, 1999) and core conditions (Laio *et al.*, 2000; Steinle-Neumann *et al.*, 2001; Alfé *et al.*, 2001). There is considerable disagreement on the elastic constants of ε-Fe between experimental results and theoretical calculations. The differences in the aggregate shear (V_s) and compressional (V_p) wave velocities are smaller (Hemley and Mao, 2001; Steinle-Neumann *et al.*, 2001). Further improvement of theory and experiment is required to resolve the discrepancies.

Seismological observations revealed inner core anisotropy, i.e., seismic waves travel faster along the Earth's polar axis than in the equatorial direction (Morelli *et al.*, 1986; Woodhouse *et al.*, 1986; Creager, 1992; Song and Helmberger, 1993; Vinnik *et al.*, 1994). There is also seismological evidence for differential rotation between the inner core and the rest of the Earth (Song and Richards, 1996; Su *et al.*, 1996). Full interpretation of these seismic observations requires quantitative determination of the elastic anisotropy and knowledge of the elastic constant tensor of iron under inner-core conditions. Recent experimental and theoretical investigations of the elasticity of iron at high pressures provide several explanations for the anisotropy of the inner core (Stixrude and Cohen, 1995; Mao *et al.*, 1999; Steinle-Neumann *et al.*, 2001).

The outer core is in a liquid state. Unfortunately, there are very limited data on the physical properties of liquid iron at high pressures and temperatures. Shock-wave studies provide data on the density of liquid iron at very high pressures (>243 GPa) along the Hugoniot (Brown and McQueen, 1986). Static data on the structure and density of liquid iron are limited to pressures less than 5 GPa (e.g., Sanloup *et al.*, 2000a,b; Balog *et al.*, 2001). Anderson and Ahrens (1994) derived an EOS for liquid iron based on available experimental data. There is no immediate solution to fill the pressure gap between static and dynamic experiments. For the moment, we have to rely on theoretical calculations of the structure of liquid iron under pressure (Stixrude and Brown, 1998; Alfé *et al.*, 2000c).

2.14.3.2 Nickel

According to chondritic bulk Earth models, the Earth's core contains ~5 wt.% nickel (McDonough and Sun, 1995; see Chapter 2.15). Static compression of iron and an Fe–Ni alloy with 20 wt.% nickel at room temperature to megabar pressure demonstrates that within the experimental uncertainties, the EOS of the Fe–Ni alloy is indistinguishable from that of pure iron at core pressures (Mao *et al.*, 1990). A recent study on the phase relations of an Fe–Ni alloy with 10 wt.% nickel to 86 GPa and 2,382 K has reached a similar conclusion (Lin *et al.*, 2002a). Given the uncertainty in the seismic velocity models and considering the presence of light elements, a core containing 5 wt.% nickel is consistent with seismic observations.

Differences in the structure parameters and phase stability fields between Fe–Ni alloys and pure iron have been observed at high temperature (Lin *et al.*, 2002a). The Ni–S binary phase diagram at ambient pressure contains more intermediate compounds than the Fe–S counterpart (Baker, 1992). Certain Ni–S compounds that are stable at ambient pressure are only observed for Fe–S under high pressure (Fei *et al.*, 1997). The chemistry of nickel is also considerably different from that of iron. At ambient conditions, the partition coefficient of nickel between liquid alloy and liquid silicate is nearly three orders of magnitude larger than that of iron. The difference becomes smaller at higher pressures and temperatures (e.g., Holzheid *et al.*, 1994; Li and Agee, 1996).

Compared to iron and other possible components of the core, experimental studies on nickel are limited. More data are needed to evaluate the effects of nickel on the physical properties and chemistry of the core.

2.14.4 LIGHT ELEMENTS IN THE CORE

2.14.4.1 Indications for the Presence of Light Elements in the Core

The presence of light elements in the core was first proposed by Francis Birch in the early 1950s (Birch, 1952). Recognizing that the density of the liquid core is ~20% lower than the calculated density of iron at core pressures, Birch suggested that the core is made of alloys of iron and elements lighter than iron (Birch, 1964). As shown in the previous section, over the years, the density and sound velocity profiles of the Earth's interior have been determined to increasingly higher precision, and the phase diagram and EOS of iron have been measured to core pressures and to high temperatures. Progress in seismology and mineral physics not only provides tighter constraints on the density deficit in the liquid outer core, but has also revealed a small but non-negligible amount of density deficit in the solid inner core. Estimates for the density deficit relative to solid iron varies between 6% and 10% for the outer core (Stevenson, 1981; Anderson and Isaak, 2002). The density deficit in the inner core is estimated at $2 \pm 1\%$ (Jephcoat and Olson, 1987; Shearer and Masters, 1990; Stixrude *et al.*, 1997; Dubrovinsky *et al.*, 2000; Anderson and Ahrens, 1994; Figure 3). The uncertainties in the estimated density deficits are mainly due to the uncertainties in the core temperature, and due to the limited experimental data on the EOS of iron and Fe–Ni alloys at high temperatures.

In addition to the observed density deficits, there are indications for the presence of light elements in the core from cosmochemical and geochemical studies as summarized in Chapter 2.15. Based on these arguments, the most likely candidates for the light elements in the core include silicon, sulfur, and carbon.

Yet another indication for the presence of light elements in the core comes from the study of geomagnetism. The operation of the geodynamo has been proposed to require compositional buoyancy in the core to drive convection. Compositional buoyancy is produced at the ICB if the light component of the outer-core partitions preferably into the liquid phase upon solidification.

2.14.4.2 Constraints on and Significance of Light Elements in the Core

In the last few decades, the number of studies related to the issue of light elements in the core has been increasing steadily. Various constraints were explored in an attempt to identify the principal light elements in the core. These constraints can be grouped into three categories. The first category is based on compositional studies of the Earth's mantle. The bulk Earth composition has been estimated from cosmochemical studies on the solar photosphere and meteorites, and geochemical studies on samples from the Earth's crust and mantle (e.g., McDonough and Sun, 1995; O'Neill and Palme, 1998; Kargel and Lewis, 1993; Allègre *et al.*, 1995; see Chapter 2.15). Assuming that the Earth's lower mantle has the same composition as the upper mantle, one can calculate the light element composition of the core by mass balance (see Chapter 2.15).

The second category involves the core alone. From the densities of iron–light-element alloys, we can calculate the amount of light elements needed to explain the observed density deficits in the outer and inner cores. Not only should the density and sound velocity of the alloy match that of the core, but it should also have the appropriate compressibility and thermal expansivity to reproduce the gradients of density and sound velocity in each layer. Moreover, the presence of immiscible liquids on a large scale in the outer core is generally assumed to be incompatible with seismic observations. Hence, the alloy must form one liquid under core conditions. For the solid inner core, the rheology, structure, and elastic properties of the iron–light-element alloy should be able to produce the observed anisotropy in seismic velocity. When the inner and outer cores are considered together, the dominant light element must partition preferentially into the outer core, since the density deficit in the outer core is larger than that in the inner core. Quantitatively, solid/liquid partitioning of the principal light elements must reproduce the density deficits in both the outer and inner cores. For a core made of iron and one light element, freezing of the binary liquid must produce a solid containing a smaller amount of the light element. If the relevant portion of the binary phase diagram has a eutectic point, then the core composition must lie on the iron-rich side of the eutectic. If a continuous solid solution is formed, then the iron-rich end-member must have a higher melting temperature.

The third category involves both the core and the mantle. Given sufficient abundance in the bulk Earth, elements with strong affinity for the core-forming alloys will become significant components of the core. Interaction between the Earth's core and mantle may have taken place at various stages of the Earth's history, at different depths, and under a range of chemical environments. The affinity of an element for the core depends on the conditions of core formation (see Chapter 2.10). After initial core–mantle differentiation, element distribution may be modified by chemical interactions and dynamical processes occurring at the CMB. This modification is deemed minor, as it has been argued previously that core–mantle interaction is likely to be very inefficient and sluggish (e.g., Poirier *et al.*, 1998; Gessmann *et al.*, 2001). Nevertheless, the products of chemical interaction at the current CMB must reproduce seismic observations of the D″ layer.

Geochemical studies suggest that core formation took less than 30 Myr and the mantle contains no more than 1 wt.% core-forming alloy. Such a rapid and efficient core formation may require that alloy–silicate separation occur in a magma ocean (Stevenson, 1981). If the iron–light-element alloy had to sink through the solid mantle to reach its present location, then its surface tension with respect to mantle minerals (also discussed in terms of interfacial energy, wetting angle, or dihedral angle) would have to be sufficiently low to allow formation of a network that connects the liquid alloy at shallow depths to the center of the planet (e.g., Stevenson, 1981; Minarik *et al.*, 1996; Shannon and Agee, 1998). Hence, the effect of light elements on the wetting behavior of liquid iron-alloys with respect to mantle minerals may provide additional constraints on the light element composition of the core.

The core accounts for one-third of the Earth's mass. The nature and abundance of the light elements in the core are of fundamental importance to the study of the Earth and the solar system. Once identified, the abundance and distribution of light elements in the core would place constraints on a variety of issues including core formation models, volatile element budget in the bulk Earth, thermal structure and evolution of the core, and convection in the liquid outer core. For instance, the temperature at the CMB might depend on the identity and concentration of light elements in the core. Whether or not compositional buoyancy is important in the outer core would affect the pattern of core convection, hence the structure and evolution of the geodynamo.

2.14.4.3 Review of Existing Data

2.14.4.3.1 Constraints from the mantle

Based on geochemical and cosmochemical considerations, the candidates for the dominant light element in the core include hydrogen, carbon, oxygen, silicon, and sulfur. Phosphorous is also considered sometimes. A number of recent estimates for the light element content of the core are given in Table 2. The discrepancies among these estimates are primarily due to differences in the estimated abundances of volatile elements in the bulk Earth, and due to differences in the estimated compositions of the mantle. An extensive and elaborate review of constraints on core composition from geochemical and cosmochemical observations are given in Chapter 2.15.

2.14.4.3.2 Constraints from the core

Although there is no *a priori* reason to assume that the core contains only one light element, a binary iron–light-element alloy core serves as a useful end-member model until sufficient experimental data become available to allow consideration of more complicated and realistic alternatives.

Experimental results on the phase diagrams of binary systems containing iron and one light element are shown in Figure 5. Not enough data are available to construct the relevant phase relations for the Fe–H binary system. The Fe–C binary system exhibits eutectic behavior at 1 bar (Baker, 1992; Figure 5(a)). The eutectic temperature increases slightly upon compression. The composition of the eutectic point shifts to lower carbon content with increasing pressure. Based on a thermodynamic calculations, Wood (1993) predicted that the under core pressures, the eutectic composition contains so little carbon that Fe_3C instead of iron will be the solidus phase in the inner core.

The phase diagram of the Fe–O binary system at ambient pressure exhibits a large liquid immiscibility gap (Baker, 1992; Figure 5(b)).

Table 2 Geochemical constraints on light elements in the bulk core.

References	*H* (wt.%)	*C* (wt.%)	*O* (wt.%)	*Si* (wt.%)	*S* (wt.%)
McDonough and Sun (1995)	0.06	0.2	5.8	0	1.9
Wood (1993)		2–4			
Allègre *et al.* (1995)			4.1	7.35	2.3
O'Neill *et al.* (1998)			Too volatile		
Wänke and Driebus (1997)				14	
Kargel and Lewis (1993)					1.8–4.1
Dreibus and Palme (1995)					<2

Ringwood and Hibberson (1989) reported that the solubility of oxygen in liquid iron increases with pressure. They predicted that the liquid immiscibility gap should disappear at core pressures. Based on theoretical calculations, Alfé *et al.* (2000a,c) found that up to a few wt.% oxygen may dissolve in h.c.p. iron under core conditions. However, a more recent study by O'Neill *et al.* (1998) shows that the solubility of oxygen in liquid iron decreases with increasing pressure at least up to 25 GPa. Since oxygen is a major element in the Earth's mantle, the partitioning of oxygen between iron and FeO and that between iron and mantle silicates are directly linked. As will be shown later, the effect of pressure on the solubility of oxygen in the liquid iron-alloys that are in equilibrium with mantle silicates was found to be negative or unresolved within experimental uncertainties.

Iron and silicon form continuous solid solutions at ambient pressure (Baker, 1992; Figure 5(b)). The subsolidus phase relations in the Fe–Si system have been studied to 140 GPa and >3,800 K (Zhang and Guyot, 1999a,b; Lin *et al.*, 2002b; Dubrovinsky, 2003, Figure 5(c)). Silicon alloyed with iron stabilizes the b.c.c. phase. As a result, an inner core containing more than 4 wt.% silicon would be a mixture of a silicon-rich b.c.c. phase and a silicon-poor h.c.p. phase. Dubrovinsky *et al.* (2003) found that Fe–Si alloy dissociates into iron and CsCl-structured FeSi compound at the CMB pressure. In the Fe–Si binary system, the melting interval at 1 bar is so narrow that the density difference between solid and liquid phases is too small to account for the observed density jump at the ICB, and the solidification of the liquid reaches completion over a temperature interval of less than 50°. How the melting interval changes with pressure is not known.

The Fe–S system has been studied extensively to 25 GPa (Figure 5(d)). There is a eutectic point between iron and FeS at low pressures. Due to formation of Fe_3S_2 at 14 GPa, and formation of

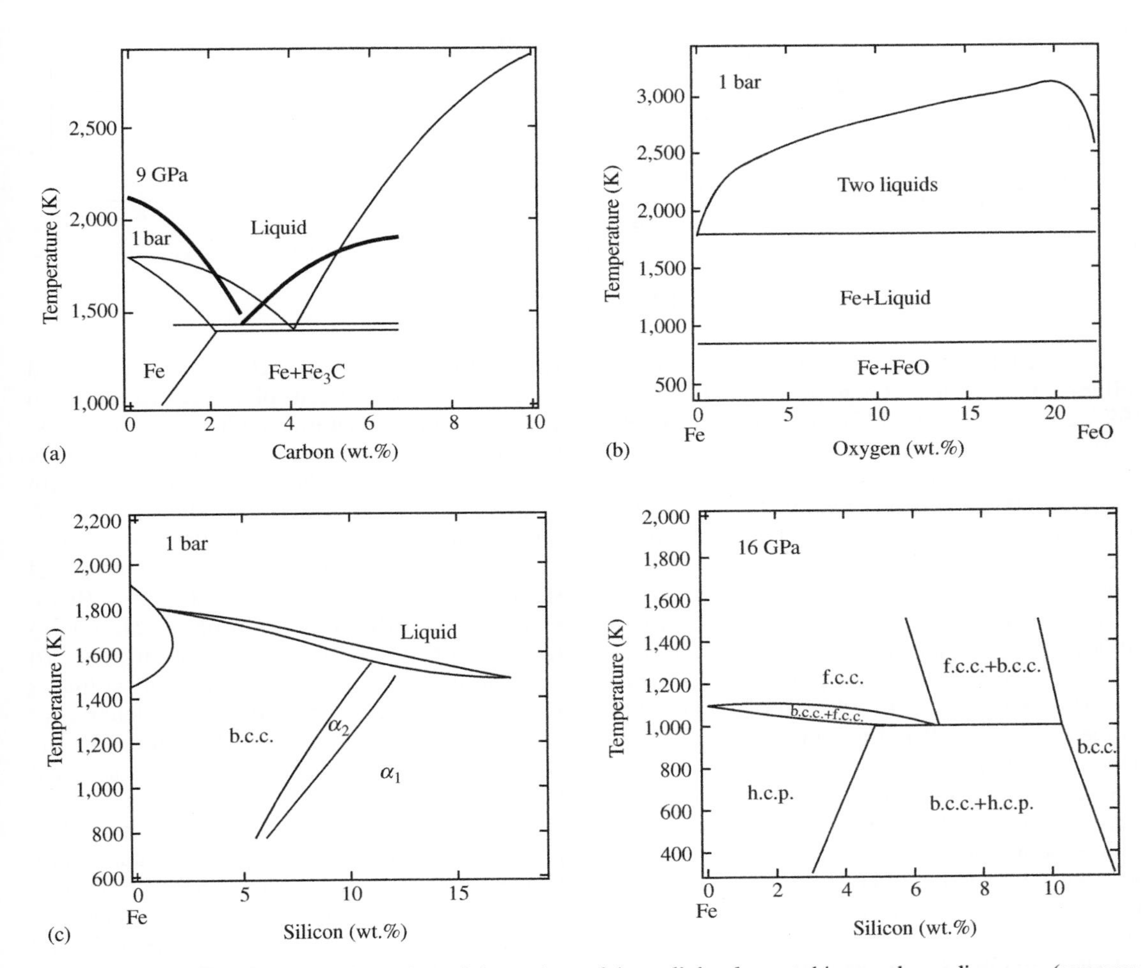

Figure 5 Simplified version of the iron-rich portion of iron–light-element binary phase diagrams (sources: (a) Fe–C, 1 bar data, Baker (1992); 9 GPa data, Shterenberg *et al.* (1975); (b) Fe–O, 1 bar data, Baker (1992); (c) Fe–Si, 1 bar data, Baker (1992); 16 GPa data, Lin *et al.* (2002b); (d) Fe–S, 1 bar data, Baker (1992); 14 GPa data, modified from Fei *et al.* (1997); 21 GPa data, Fei *et al.* (2000); 25 GPa data, Li and Agee (2001)).

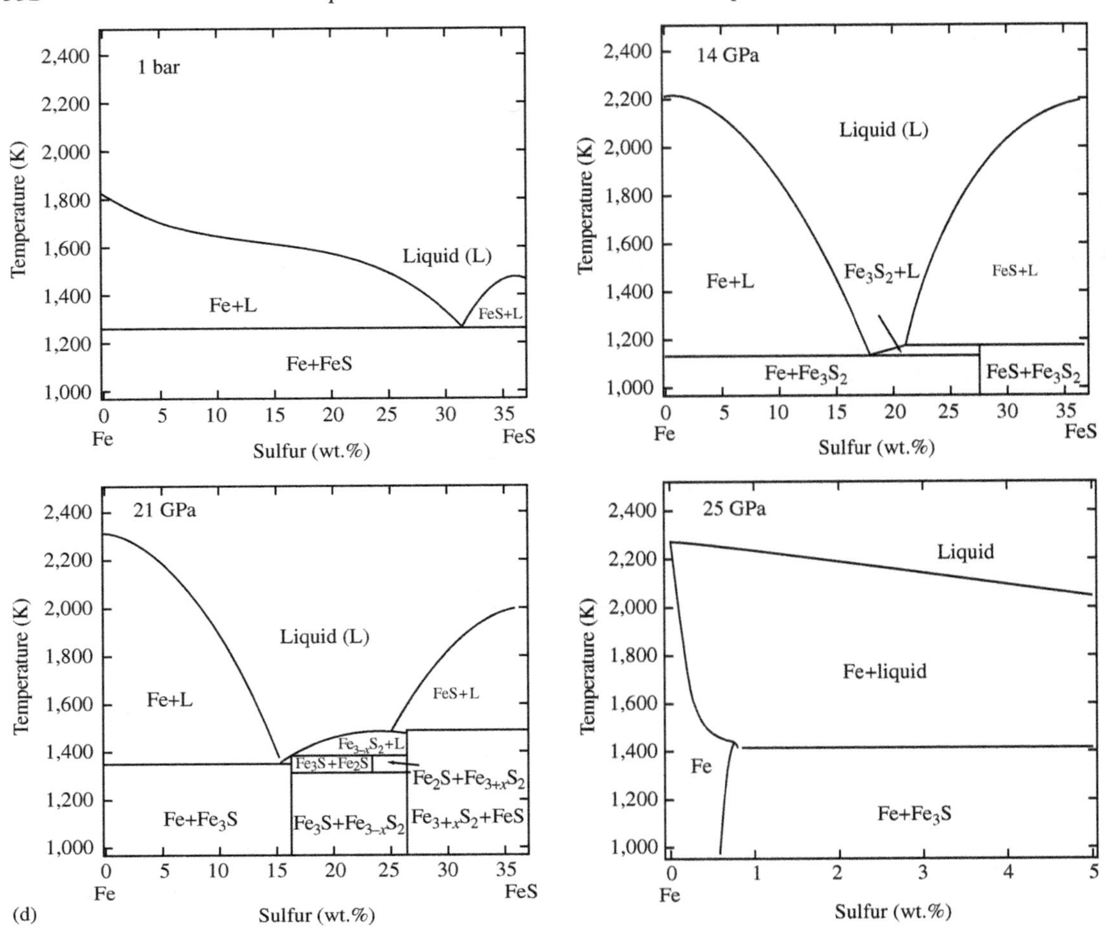

Figure 5 (continued).

Fe_3S at 21 GPa, new eutectic points with lower sulfur contents appear at high pressures (Fei *et al.*, 1997, 2000). Although sulfur is virtually insoluble in solid iron at 1 bar, a limited solid solution between iron and sulfur was observed at 25 GPa (Li *et al.*, 2001). This observation supports sulfur as the principal light element in the outer core. It also indicates that the Earth's inner core may contain a non-negligible amount of sulfur.

Theoretical results from Boness and Brown (1990) and Sherman (1995) favored sulfur over oxygen as the principal light element in the inner core. However, Alfé *et al.* (2000c, 2002b) found that all the binary models fail to reproduce the density jump at the ICB: while the partitioning of sulfur or silicon between solid and liquid iron leads to a density difference that is too small to be consistent with seismic observation, the density contrast between solid and liquid Fe–O alloys is too large. These conclusions remain to be tested by experiments.

Extensive efforts have been made to determine the phase stability fields and structures of alloys and compounds consisting of iron and one of the potential light elements at elevated pressures and temperatures. Most iron–light-element alloys and compounds have been observed to undergo complex phase transitions with increasing pressure and temperature (Figure 6). In the Fe–S binary system, new compounds and solid solutions form at high pressure and temperature (Fei *et al.*, 1997, 2000). Indications for metallization of FeO at megabar pressures have been observed in several studies (Knittle and Jeanloz, 1986, 1991; Badro *et al.*, 1999). Recently, Dobson *et al.* (2002) reported a new high-pressure phase of FeSi. To estimate the amount of each light element that is needed to explain the observed density deficits, we need to know its partial molar volume in the corresponding iron-alloy at core pressures and temperatures.

Presently EOS data are limited to crystalline phases at relative low pressures and/or low temperatures (Figure 7). These data demonstrate that all the proposed light elements are capable of reducing the density of iron as expected. The efficiency of density reduction (or the amount of light element needed to account for the core density deficits) depends on the structure and EOS of the alloys or compounds. Assuming that the

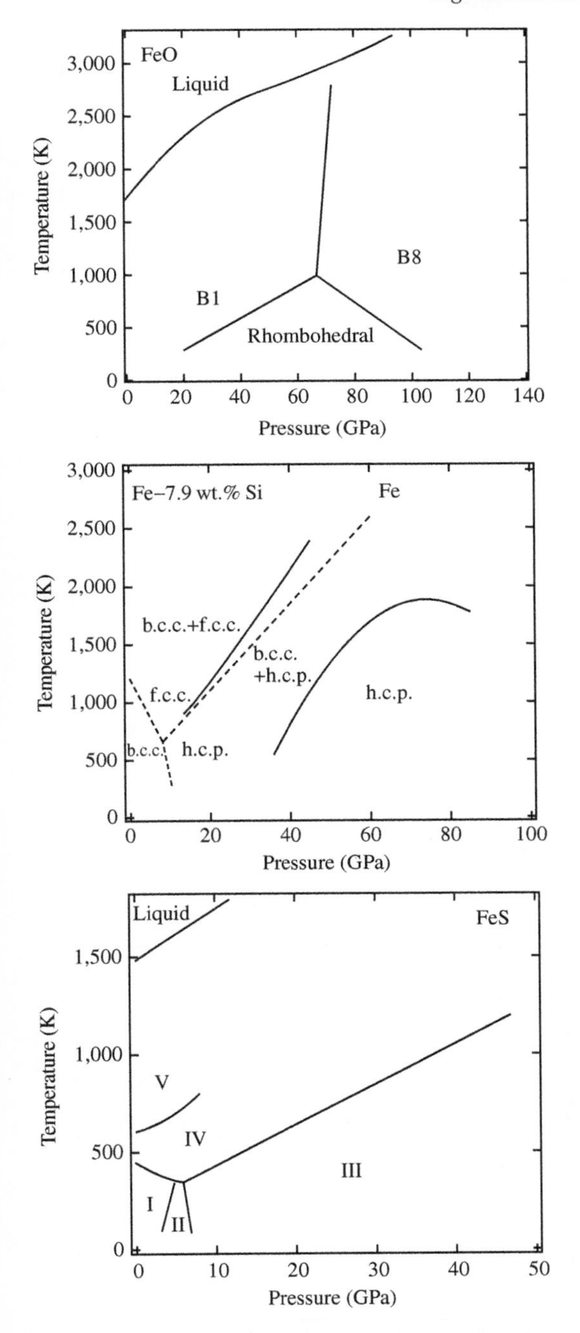

Figure 6 Phase diagram of iron–light-element alloys or compounds at high pressure and temperature (data for FeO are from Fei and Mao (1994); for Fe–Si alloy containing 7.9 wt.% silicon with a comparison to pure iron are from Lin *et al.* (2002b); for FeS are from Fei *et al.* (1995)).

thermal expansivity and compressibility of iron–light-element alloys and compounds are comparable to that of iron, we can calculate the relative density deficit with respect to crystalline iron as a function of the mass fraction of a light element in an alloy or compound, following Poirier (1994). Ignoring the difference between solid and crystalline phases, the amount of light element in a binary alloy corresponding to 10% density deficit in the core is roughly 1 wt.% hydrogen, 6 wt.% oxygen, 13 wt.% silicon, and 14 wt.% sulfur. Using the EOS of $Fe_{0.9}S$ leads to a much smaller amount of sulfur (8 wt.%). The maximum amount of density reduction with carbon in Fe_3C is less than 5%. As shown in Table 3 the compressibilities of some iron–light-element alloys and compounds differ considerably from that of pure iron. Thermal expansivity data for alloys and compounds are largely missing. Therefore, the above estimates are likely to change when new experimental data become available.

In comparison, *ab initio* calculations of the electronic structure of Fe–S and Fe–Si compounds by Sherman (1997) indicate that due to large excess volumes of mixing as little as 2–8 wt.% of sulfur is enough to account for the density deficit in the core. Alfé *et al.* (1999b) estimated that 9–11 wt.% oxygen is required to explain the density deficit.

Experimental work on the melting behavior of iron–light-element systems have been mainly focused on sulfur and oxygen, with a few data reported for hydrogen, silicon, and carbon (Figure 8). Iron hydride and Fe–Si alloy melt at substantially lower temperatures than iron (Yagi, 1995; Okuchi, 1998; Yang and Secco, 1999). The melting temperature of FeO is lower than iron at 1 bar (Baker, 1992). The situation reverses at high pressure, as the melting curve for FeO has a steeper Clapeyron slope (Boehler, 1992; Knittle and Jeanloz, 1991). Consequently, if iron and FeO form a complete solid solution under core conditions, then oxygen cannot be the principal light element in the core. The melting temperatures of Fe–FeO mixtures are lower than that of iron, indicating eutectic behavior up to 120 GPa (Boehler, 1992). Addition of sulfur to the mixture decreases the melting temperature further at low pressure, but the difference between the melting temperature of Fe–O–S mixture and that of iron gets smaller as pressure increases. Large discrepancies exist between the melting temperatures of FeO and FeS that are determined by different authors. Temperature appears to have been overestimated in the pioneering work by Williams and Jeanloz (1990) and Knittle and Jeanloz (1991).

The eutectic temperature of the Fe–S binary system has been measured in a multi-anvil apparatus to 25 GPa (Fei *et al.*, 1997, 2000; Li *et al.*, 2001), and in a laser-heated diamond-anvil cell to 62 GPa (Boehler, 1996). Within the common pressure range, the diamond-anvil cell results are higher than the multi-anvil apparatus results by as much as 400°. Despite the discrepancies, the existing data show that between

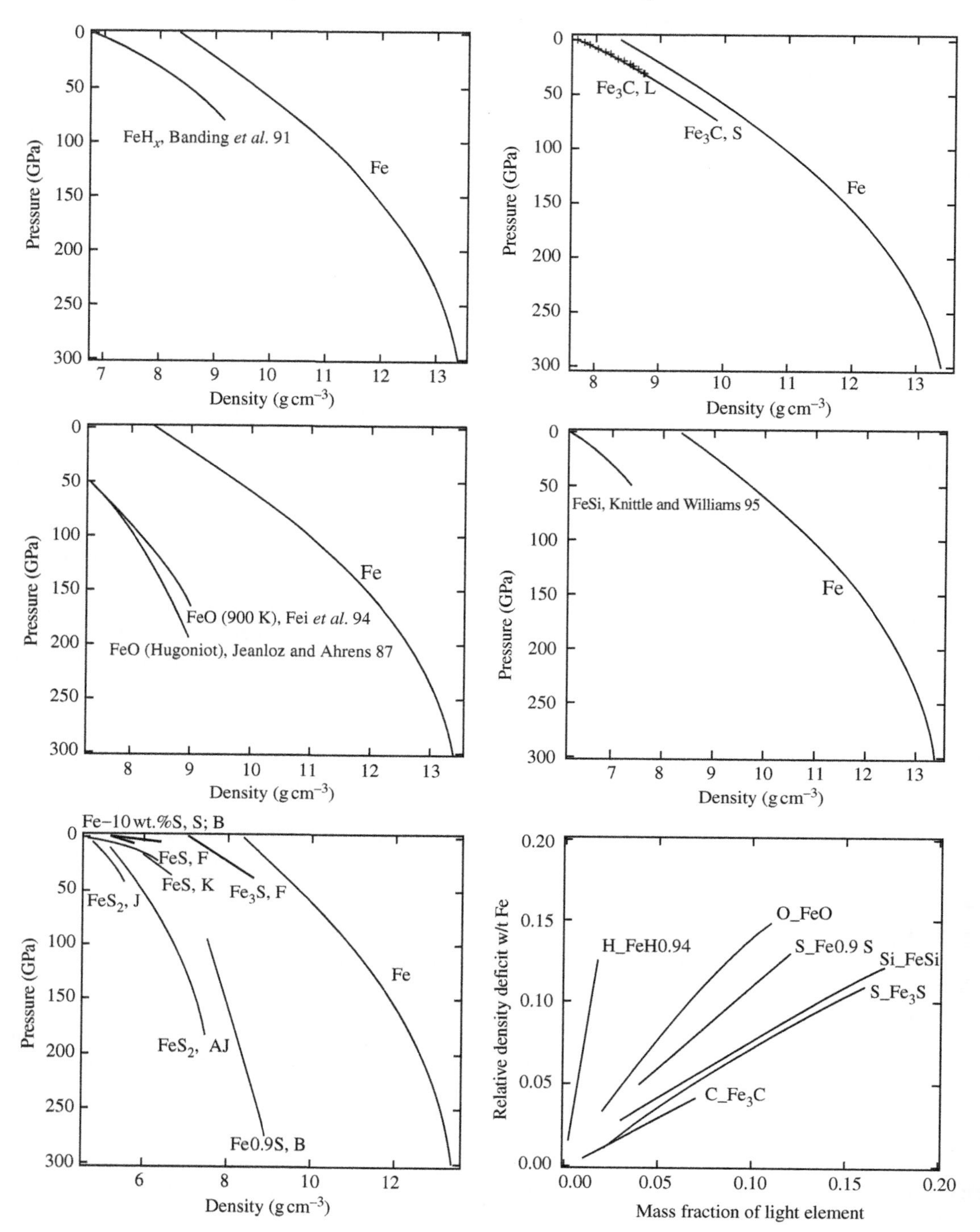

Figure 7 EOS data for iron and iron–light-element alloys and compounds. (Data for iron are from Mao *et al.* (1990); for FeH_x are from Badding *et al.* (1991); for Fe_3C are from Scott *et al.* (2001, curve, S) and Li *et al.* (2002, crosses, L); static data for FeO at 900 K are from Fei and Mao (1994); shock data for FeO are from Jeanloz and Ahrens (1980); for FeSi are from Knittle and Williams (1995); for Fe–S alloy containing 10 wt.% sulfur are from Sanloup *et al.* (2000b, S) and Balog *et al.* (2001, B); for Fe_3S are from Fei *et al.* (2000); for FeS are from Fei and Prewitt (1995, F) and Kavner *et al.* (2001, K); and for FeS_2 are from Jephcoat (1985, J) and Ahrens and Jeanloz (1987, AJ). The summary plot shows relative density deficit with respect to pure iron as a function of mass fraction of a given light element in a given iron–light-element alloy or compound, calculated from the data in the first five plots.

Table 3 Elastic properties of iron and iron–light-element alloys.

Phase	K_{T_0} (GPa)	$K_{T'}$	α_0 ($10^{-5}K-1$)	P_{max} (GPa)	T_{max} (K)	*References*
Fe	165 (±2.1)	5.33 (±0.09)		330	300	Mao *et al.* (1990)
Fe	135 (±19)	6.0 (±0.4)	5.5 (± 0.2)	20	1,800	Uchida *et al.* (2001)
Fe	164 (±3)	5.36 (±0.16)	5.8 (±0.4)	68	2,000	Dubrovinski *et al.* (1998)
Fe–10 wt.%S	42–48	4–7		6	1,700	Sanloup *et al.* (2000b)
FeS	54 (±6)	Fixed at 4		40	800	Fei *et al.* (1995)
Fe_3S	170 (±8)	2.6 (±0.5)		43	300	Fei *et al.* (2000)
FeO	172 (±14)	4.3 (±0.6)		96	900	Fei and Mao (1994)
FeSi	209 (±6)	3.5 (±0.4)		50	300	Knittle and Williams (1995)
Fe_3C	175 (±4)	5.2 (±0.3)		73	300	Scott *et al.* (2001), Li *et al.* (2002)
FeH_x	121 (±19)	5.3 (±0.9)		62	300	Badding *et al.* (1991)

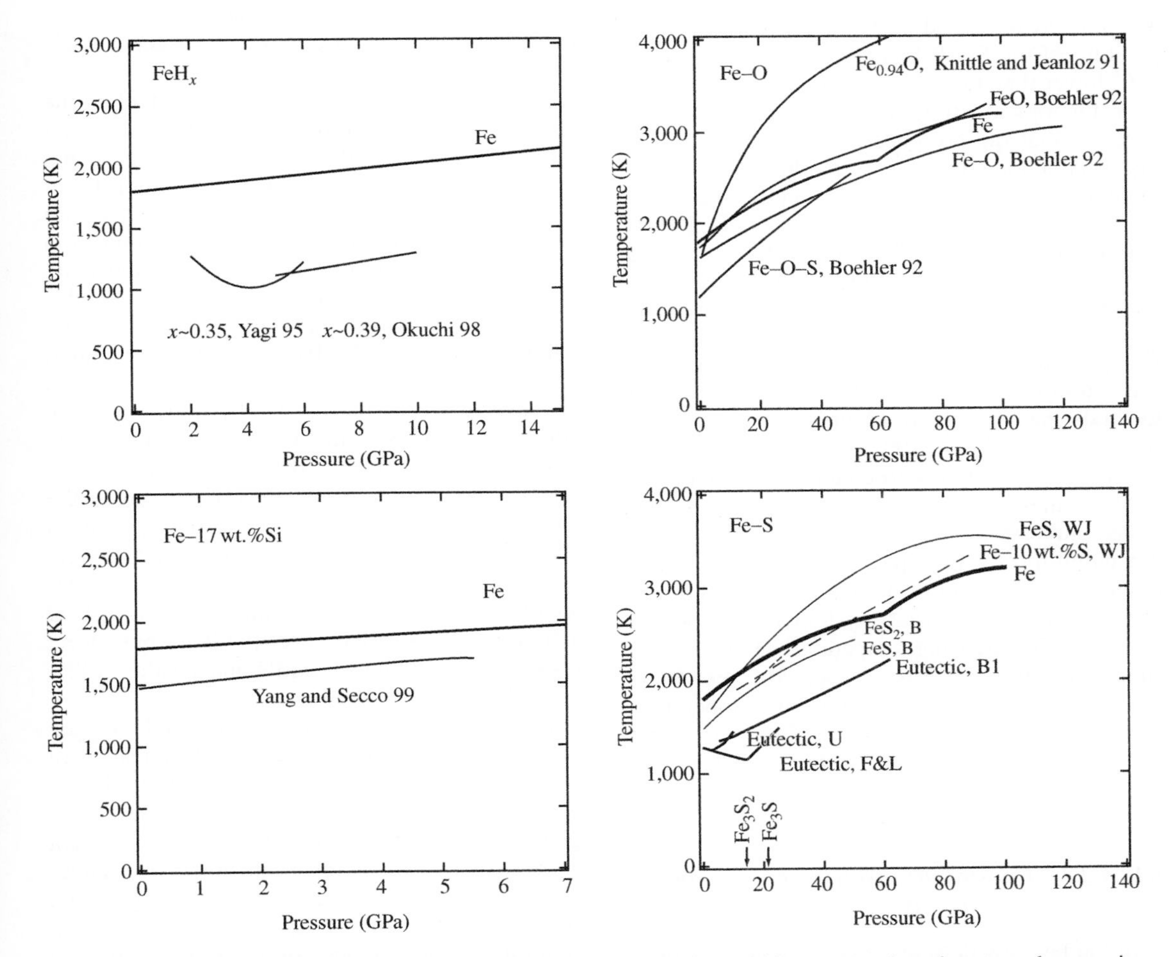

Figure 8 Melting curves of iron–light-element alloys, compounds, and binary eutectic points near the pure iron end-member (sources: melting curve of pure iron is from Shen *et al.* (1998); data for FeH_x are from Yagi (1995) and Okuchi (1998); for Fe–O eutectic and Fe–O–S eutectic are from Boehler (1992); for Fe–Si alloy containing 17 wt.% silicon are from Yang and Secco (1999); for Fe–S alloy containing 10 wt.% sulfur are from Williams and Jeanloz (1990, WJ); for Fe–S eutectic are from Usselman (1975, U), Fei *et al.* (1997), Li *et al.* (2001) (F&L), and Boehler (1996, B1); for FeS are from Williams and Jeanloz (1990, WJ), and Boehler *et al.* (1996, B); for FeS_2 are from Boehler (1996, B)). Formation of Fe_3S_2 and Fe_3S is indicated by arrows.

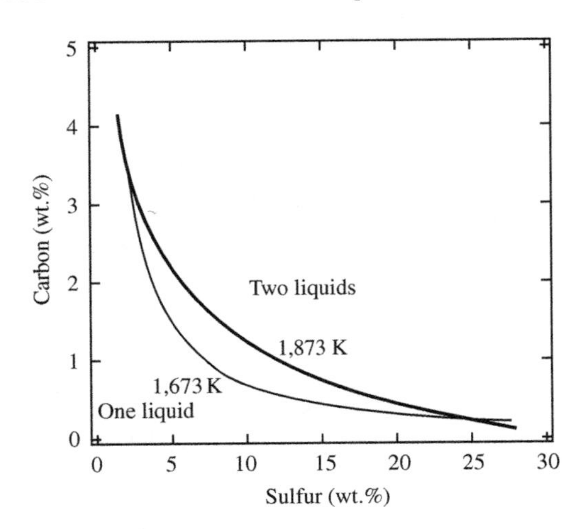

Figure 9 Boundary of miscibility in Fe–C–S liquid (source Wang *et al.* (1991)).

ambient pressure and 62 GPa, the eutectic temperature of the Fe–S binary system is more than 500° lower than the melting temperature of iron.

A theoretical study by Alfé *et al.* (1999b) suggests that the presence of oxygen lowers the viscosity of liquid iron at core conditions. Sulfur on the other hand is found to have a negligible effect on the viscosity of liquid iron (Alfé *et al.*, 2000b; Vocadlo *et al.*, 2000). If the viscosity of the core can be constrained by the study of core dynamics, it may be possible to exclude certain light elements based on their effects on the viscosity of liquid iron. However, this is unrealistic at the present time, as estimates of outer core viscosity based on different methods vary by up to 12 orders of magnitude (Poirier, 1988; Secco, 1995).

As the core is likely to contain more than one light element (e.g., Stevenson, 1981), initial efforts have been undertaken to investigate the compatibility of multiple light elements in a liquid. Wang *et al.* (1991) found a vast miscibility gap in the Fe–S–C ternary system at 1 bar, which shrinks with increasing temperature (Figure 9). A similar miscibility gap was observed in Fe–S–Si system at 1 bar, which widens as temperature increases (Raghavan, 1988). The effect of pressure on miscibility has not been studied.

2.14.4.3.3 Constraints involving core–mantle interactions

In the last few decades, a considerable number of experimental studies have been carried out to determine partitioning of the candidate light elements between core and mantle at relevant pressures, temperatures, compositions, and redox conditions. The effects of light elements on the wetting behavior of liquid iron-alloy with respect to mantle minerals have also been evaluated. These studies show that element partitioning between core alloys and mantle silicates and wetting behavior of core-forming alloys are functions of pressure, temperature, composition, and oxygen fugacity, and the effects differ from one element to another (Figures 10(a) and (b)).

Relatively weak positive temperature dependence has been observed for the alloy/silicate partitioning of hydrogen (Okuchi, 1997). The effect of temperature on the partitioning of silicon between iron-rich alloy and silicate is positive as well (Gessmann *et al.*, 2001; Li and Agee, 2001). Furthermore, Dubrovinsky *et al.* (2003) reported chemical reactions between iron and silica at high temperature and below 40 GPa. The partitioning of oxygen into the liquid alloy is significantly enhanced at high temperatures (Ohtani *et al.*, 1997; Li and Agee, 2001; Dubrovinsky *et al.*, 2001). In contrast, the partitioning of sulfur between liquid alloy and silicate decreases with increasing temperature (Li and Agee, 2001; Holzheid and Grove, 2002).

Pressure has a strong and positive effect on the partitioning of sulfur between liquid metal and liquid silicate (Li and Agee, 2001; Holzheid and Grove, 2002). As mentioned earlier, inconsistent results have been obtained on the effect of pressure on the partitioning of oxygen between liquid iron alloys and mantle oxide or silicate. Opposite pressure dependence was observed on the solubility of oxygen in liquid iron alloy (Ringwood and Hibberson, 1989; O'Neill *et al.*, 1998). Wendlandt (1982) and Holzheid and Grove (2002) found lower oxygen contents in the Fe–S alloy that are in equilibrium with silicates at higher pressure, while Hillgren and Boehler (2000) and Li and Agee (2001) did not observe any effect of pressure on the partitioning of oxygen between liquid iron alloys and silicates. Dramatic changes in the partitioning of oxygen between core and mantle due to high-spin/low-spin transition and metallization of FeO at megabar pressure have been predicted but not yet tested by experiments (Knittle and Jeanloz, 1986, 1991; Badro *et al.*, 1999).

Partitioning of silicon, sulfur, and oxygen between liquid metal and liquid silicate are sensitive to oxygen fugacity (Figure 10(c); Gessmann and Rubie, 1998; Gessmann *et al.*, 2001; Kilburn and Wood, 1997). A significant amount of silicon dissolves in liquid iron under highly reducing conditions, while oxygen and sulfur enter the core under relatively oxidizing conditions. The composition of silicate melt also has an effect on the partitioning of sulfur between liquid metal and liquid silicate. Increasing the degree of polymerization of the silicate melt lowers the solubility of sulfur in liquid silicate (Figure 10(c); Holzheid and Grove, 2002).

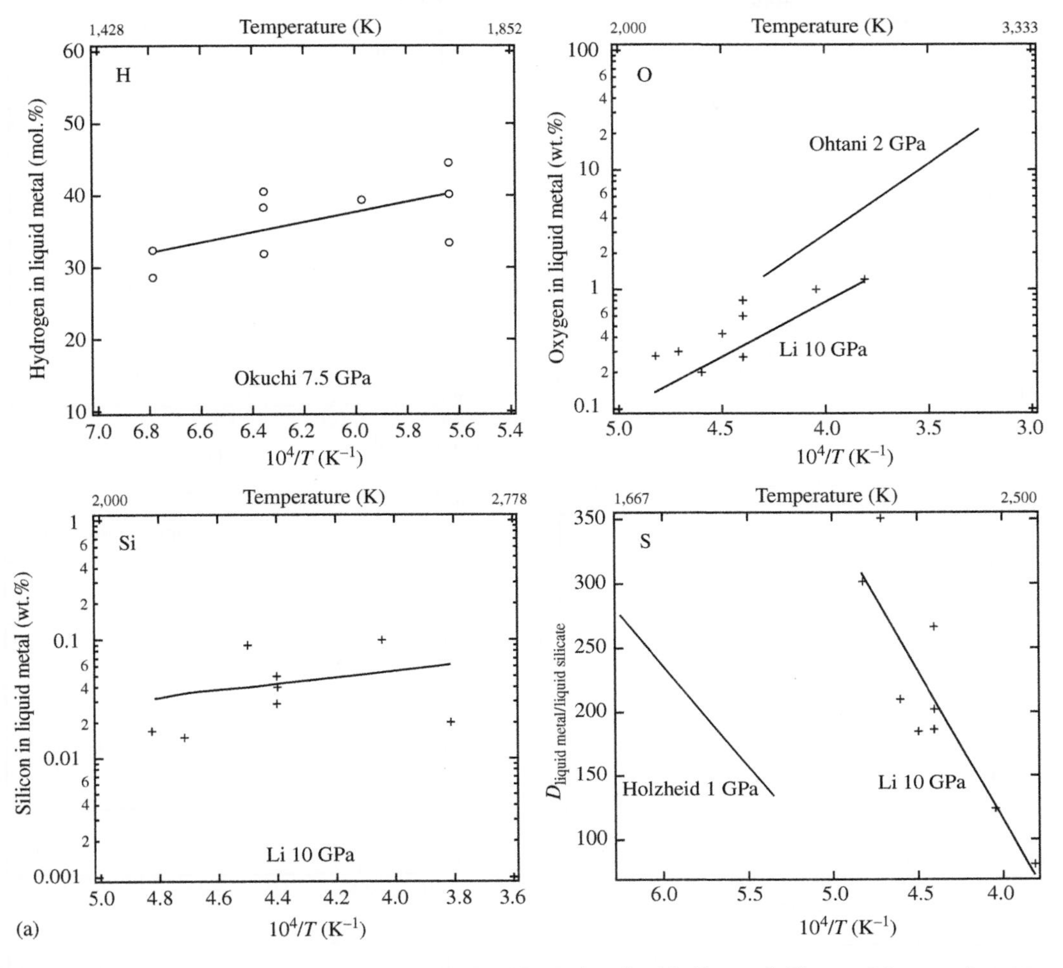

Figure 10 Partitioning of light elements between liquid metal or liquid alloy and silicates. Lines are least-squares fits to data. (a) Effect of temperature (data are from Okuchi (1997), Ohtani and Ringwood (1984), Li and Agee (2001), and Holzheid and Grove (2002)). The oxygen fugacities in Li and Agee's experiments vary between iron-wüstite buffer and two log units below it. (b) Effect of pressure (data from Hillgren and Boehler (2000), O'Neill *et al.* (1998), Li and Agee (2001), and Holzheid and Grove (2002)). (c) Effect of oxygen fugacity (f_{O_2}) and degree of polymerization of the silicate melt (described by NBO/t, the ratio of nonbridging oxygen to tetrahedrally coordinated cations) (sources: Holzheid and Grove, 2002, Gessmann and Rubie, 1998; Gessmann *et al.*, 2001; and Kilburn and Wood, 1997).

Experimental data on the interfacial energy between liquid metal and solid silicates and oxides are scarce (Figure 11). Chaklader *et al.* (1981) observed that at ambient pressure the interfacial energy between liquid iron and alumina decreases from 2.4 J m^{-2} to 0.6 J m^{-2} upon an addition of 1 at.% oxygen. Iida and Guthrie (1988) showed that 1 at.% oxygen or sulfur lowers the surface tension of liquid iron by a factor of 3–4. Carbon does not seem to affect the surface tension to any noticeable degree. Minarik *et al.* (1996) measured dihedral angle between Fe–Ni–O–S liquid and olivine at high pressures, and confirmed the reduction of interfacial energy due to the presence of oxygen and sulfur. The dihedral angle of Fe–Ni–S melt in equilibrium with mantle minerals (olivine, β-phase, and γ-phase) is barely affected by pressure or mineral structure. However, a major reduction in the dihedral angle was observed as the pressure reaches the perovskite stability field (Shannon and Agee, 1998).

2.14.4.3.4 *Summary*

Major constraints on the light element composition of the Earth's core are summarized in Tables 4 and 5. Hydrogen is the least understood element among the five top candidates. The situation is not likely to change soon, as estimating the bulk Earth budget of hydrogen and studying the physical properties and chemical affinity of hydrogen under

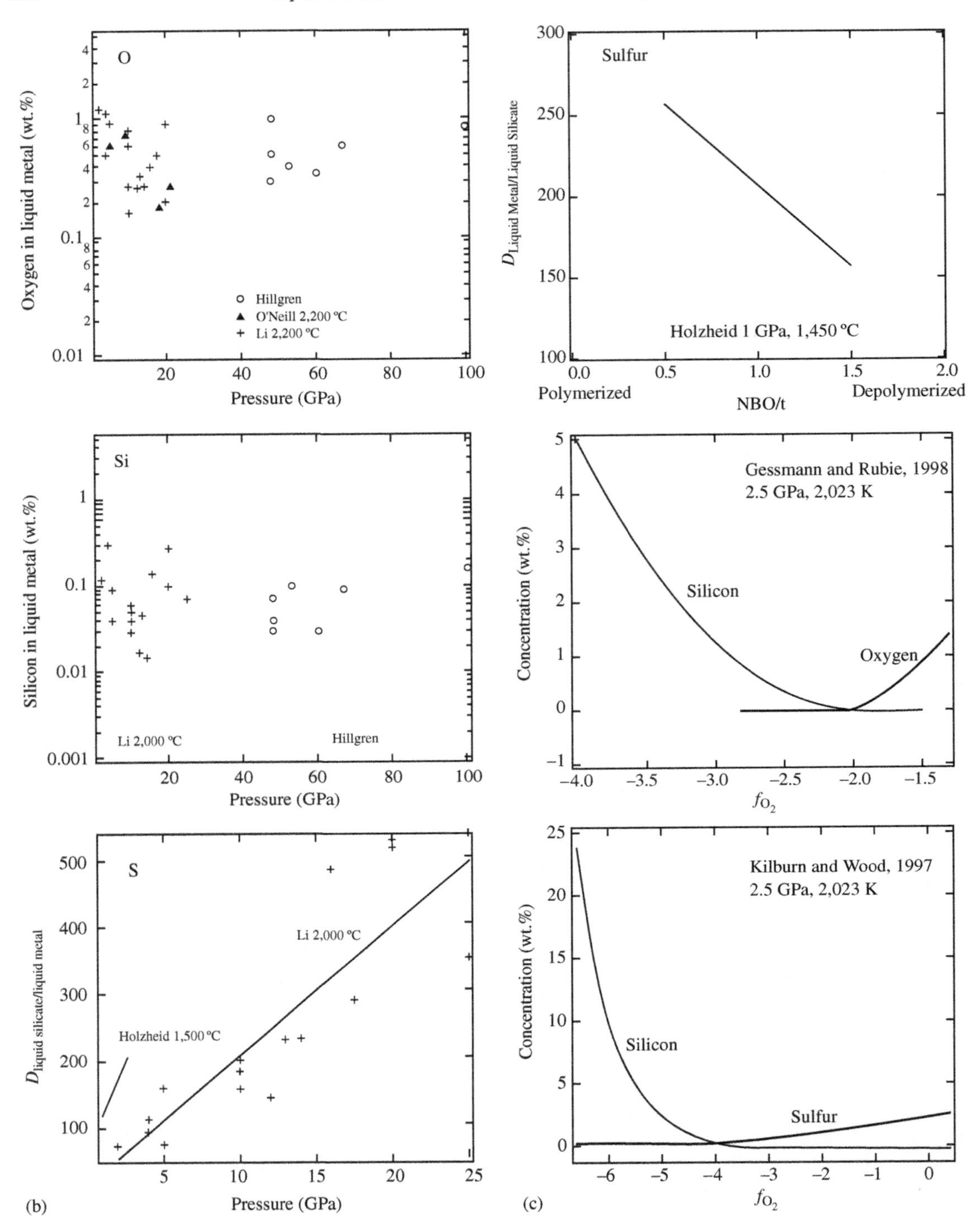

Figure 10 (continued).

high pressure and high temperature are both extremely challenging. In a comprehensive review, Williams and Hemley (2001) have discussed various issues concerning hydrogen in the deep Earth. Existing data suggest that carbon has the required chemical affinity for iron. However, the amount of carbon needed to account for the core density deficit is much higher than the amount deemed feasible from the geochemical and cosmochemical point of view (see Chapter 2.15). Furthermore, the eutectic composition at core pressures may contain too little carbon to account for the density deficits in the outer and inner cores simultaneously, and the surface tension of Fe–C alloy may be too high to allow fast metal–silicate segregation in a low temperature core-formation

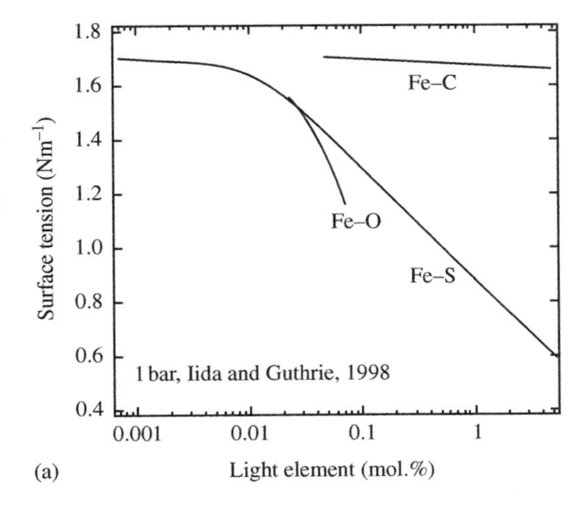

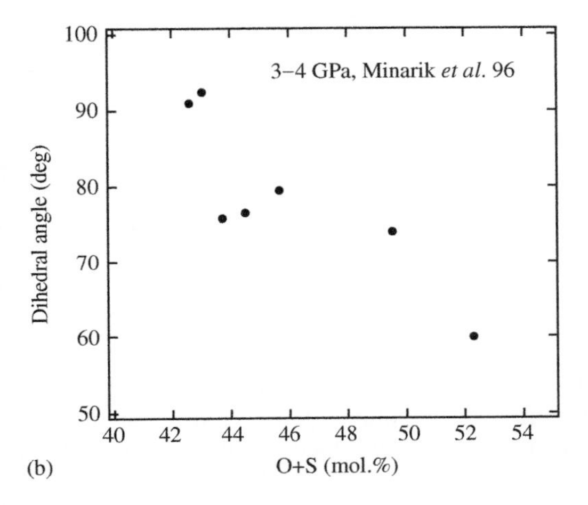

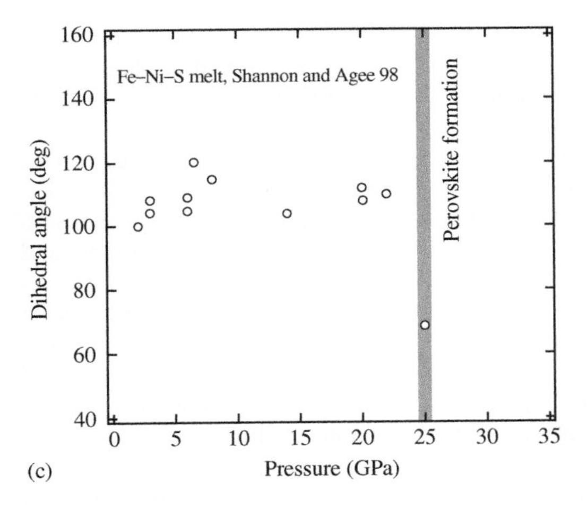

Figure 11 Wetting behavior of iron–light-element alloys. (a) Effect of light element on surface tension at 1 bar (Iida and Guthrie, 1988). (b) Effect of light elements on dihedral angle at 3–4 GPa (Minarik *et al.*, 1996). (c) Effect of pressure on dihedral angle of Fe–Ni–S melt in equilibrium with mantle minerals (olivine, β-phase, and γ-phase below 25 GPa, magnesiowüstite, perovskite, and γ-spinel at 25 GPa, Shannon and Agee (1998)).

regime, where a significant portion of the mantle remains solid. For these reasons, carbon is not considered a leading candidate.

The arguments for and against oxygen are equally strong. Oxygen is the most abundant element in the whole planet. Although the exact amount is not known, incorporating a few percent oxygen in the core is not likely to cause a problem for the bulk Earth oxygen budget. Inconsistent results on the affinity of oxygen for liquid iron-alloys have been obtained. Further studies are needed to resolve the discrepancies in the observed partitioning of oxygen between iron-alloys and silicates/oxides at high pressures and temperatures. If the melting temperature of FeO is higher than iron under ICB pressures, then oxygen can be excluded as the sole light element in the core. Further constraints on oxygen may be obtained in the near future, as more data on thermoelastic properties, phase diagram, and chemical reactions involving oxygen at core conditions are acquired.

Silicon is also abundant in the Earth. The main argument against silicon is its strong lithophile character at the present-day oxygen fugacities in the Earth's interior. However, high temperature and/or low oxygen fugacity enable significant amounts of silicon to dissolve in liquid iron. Such conditions may have occurred at the base of an early magma ocean. In the absence of oxygen, silicon alloys with iron readily. However, if the Fe–Si binary phase diagram at core pressure is similar to that at 1 bar, then the silicon contents in the solid and liquid phase are too similar to simultaneously satisfy the density deficits in both the outer and inner cores. The relatively small pressure derivative of the bulk modulus (K') of FeSi has also been used as an argument against silicon being the primary light alloying element in the core. The efficiency of silicon in density reduction is controversial. More experimental data under high pressure are needed to further evaluate the possibility of silicon as the sole light element in the core.

Sulfur is a prime candidate for the principal light element in the core. It has strong affinity for iron, reduces density and surface tension of iron, preferentially partitions into the liquid phase upon freezing, and dissolves into solid iron under high pressure and temperature. Until recently, the only strong objection for sulfur came from geochemical considerations (Dreibus and Palme, 1995; see Chapter 2.15). Theoretical studies indicate that the sulfur contents in liquid and solid iron under core pressure may be too similar to satisfy the density deficits in both reservoirs. This issue can be resolved by experimental studies in the near future.

Combinations between silicon and oxygen or silicon and sulfur are unlikely because they are

Table 4 Summary of constraints on the dominant light element in the core.

	H	*C*	*O*	*Si*	*S*
Type I (based on element depletion in the mantle)					
Available for the core, wt.%	?	2–4?	Yes	5–14	<2–5?
Type II (based on physical properties and chemistry of the core)					
Needed in the core, wt.%	1	Too much?	6	7.3–13	5–14
Partition of ICB	No?	?	?	No?	Yes?
K' of Fe–light-element alloy	?	?	?	No	?
Type III (considering core-mantle interaction)					
Partition at CMB	Yes	Yes	High pressure?	Low f_{O_2}	Yes
Surface tension	?	No?	Yes	?	Yes

Yes/No: whether a constraint supports a given element to be the dominant light element in the core.
?: more experimental data are needed.

mutually exclusive during core–mantle equilibrium. Certain combinations between silicon and carbon, sulfur and carbon, and sulfur and oxygen are permitted as long as the composition falls outside of the miscibility gap, assuming that the outer core is a uniform liquid. Ternary miscibility fields at high pressure are yet to be determined.

Table 5 Possible combination of light elements in the core.

	H	*C*	*O*	*Si*	*S*
H					
C	Unknown				
O	Unknown	Unknown			
Si	Unknown	Possible	Unlikely		
S	Unknown	Possible	Possible	Unlikely	

Blank: redundant combinations.

2.14.5 MINOR AND TRACE ELEMENTS IN THE CORE

Many elements are present in the core at low-concentration levels (<1wt.%). Some of the minor and trace elements may hold critical information on the origin, evolution, and current processes in the Earth's interior. A few examples are given below, where experimental data on minor and trace elements provide important constraints on the thermal state of the Earth, bulk Earth composition, and differentiation of the core.

2.14.5.1 Potassium

As a possible radiogenic heat source in the Earth, the presence of an appreciable amount of potassium in the core was suggested over 30 years ago (Lewis, 1971; Hall and Murthy, 1971). The estimated concentration of potassium in the mantle is 240 ppm (McDonough and Sun, 1995). If the concentration of potassium in the core is at a comparable level, the present-day heat production due to ^{40}K would be on the order of 10^{12} W, enough to drive the geodynamo. Radiogenic heat production due to ^{40}K also has direct implications for convection in the outer core, heat flux at the CMB, and the dynamics of the lower mantle. Recent studies of plume dynamics and considerations of the age of the inner core have inspired a renewed interest in the potassium content of the core (Murthy *et al.*, 2003, and references therein; see Chapter 2.15).

Potassium is generally considered lithophile, hence mainly concentrated in the Earth's crust. Thermodynamic calculations indicate that the affinity of potassium for iron-metal may be enhanced by the presence of sulfur (Lewis, 1971; Hall and Murthy, 1971). Parker *et al.* (1997) observed formation of a K–Ni compound at 31 GPa. The compressional behavior of potassium suggests higher affinity for iron with increasing pressure. Experimental studies on the partitioning of potassium between iron-metal and silicate have produced ambiguous and controversial results (Oversby and Ringwood, 1972; Goettel and Lewis, 1973; Ito and Morooka, 1993; Ohtani *et al.*, 1997; Murrell and Burnett, 1986; Chabot and Drake, 1999; Gessmann and Wood, 2002; Murthy *et al.*, 2003). The discrepancies cannot be explained by differences in experimental pressure, temperature, and/or the presence of sulfur. Murthy *et al.* (2003) found that potassium in Fe–S alloy dissolves in oil-based polishing agents, which were routinely employed in sample preparation for the earlier studies. As a result, a serious experimental artifact was introduced into the measured partition coefficients. This artifact may explain the observed inconsistencies. Using a dry polishing technique, Murthy *et al.* (2003) showed that the solubility of potassium increases with increasing temperature, and that the presence of sulfur enhances the affinity of potassium for the metal significantly.

2.14.5.2 Niobium

The issue of niobium in the core is of particular interest for the chondritic model of the bulk Earth. Niobium has always been thought to be refractory and lithophile, yet it is depleted in the upper mantle relative to other refractory and lithophile elements. Failure to locate hidden niobium-rich reservoirs in the lower mantle or the core would lead to serious problems for the well-established chondritic model. Recently, Wade and Wood (2001) studied partitioning of niobium between liquid metal and liquid silicate under high pressure and temperature. They found that niobium becomes more siderophile with increasing pressure, hence opening up the possibility of storing niobium in the core.

2.14.5.3 Platinum, Rhenium, and Osmium

Recent observations of mantle samples with extremely radiogenic $^{186}Os/^{187}Os$ isotope compositions and old Re–Os minimum ages indicate that they may be carrying the signature of inner–outer core differentiation (Meibom and Frei, 2002; Brandon *et al.*, 1998; Shirey and Walker, 1998). A critical test of this hypothesis requires data on the partitioning of the parent elements platinum and rhenium and the daughter element osmium between solid and liquid metal in the relevant pressure–temperature regime. A preliminary experimental study by Walker (2000) showed that the partition coefficients of these elements to 10 GPa are in general agreement with those required to explain the observed isotopic composition. The effects of pressure, temperature, and composition on partitioning of platinum, rhenium, and osmium between solid and liquid metal need to be studied in the future.

2.14.6 CONCLUSIONS AND OUTLOOK

The composition of the Earth's core is a fundamental issue in the study of the Earth. It is directly linked to the volatile element budget of the bulk Earth, the conditions and mechanisms of core formation, the nature and dynamics of core–mantle interaction, and the origin and evolution of the geomagnetic field.

That iron is the dominant element in the core has been well established. Recent experimental and theoretical work has substantially expanded and refined our knowledge on the physics and chemistry of iron under the pressure–temperature regime that is relevant to the core. New data on high-pressure, high-temperature phase transitions and thermal EOS of iron promise to provide a better estimate on the density deficit in the core. Convergence of estimates of the melting curve of iron places a tighter constraint on the temperature at the ICB. Experimental measurements and theoretical calculations of elastic properties and plastic deformation of iron offer new interpretations for the inner core anisotropy. Preliminary results have been obtained on the properties of liquid iron, which allow a more direct comparison between laboratory measurements and seismic observations.

Geochemical studies suggest that the core contains ~5 wt.% nickel. This is consistent with seismic observations, as experimental measurements found that the presence of nickel has little effect on the density of iron at core pressures. More work is needed to elucidate the role of nickel in the chemical evolution of the Earth and the core.

A large number of experimental and theoretical studies carried out in recent years have laid a solid foundation for further investigations of the light-element composition of the core. Several major candidates for the principal light elements in the core have been identified, including hydrogen, oxygen, carbon, sulfur, and silicon. The phase diagram and EOS of relevant alloys and compounds of iron and one or more light elements have been studied under high pressure and high temperature. These data provide rough estimates for the amounts of light elements needed to account for the density deficit in the outer core. They also demonstrate potential tests for the principal light element in the core (Table 4). Motivated by interests in core formation, an appreciable number of experimental studies have been carried out to determine the effects of pressure, temperature, oxygen fugacity, and composition on the partitioning of the light elements between core-forming alloys and mantle silicate. In systems containing more than one light element, mutually exclusive behavior has been observed. Current data suggest that hydrogen, carbon, and sulfur partition into the core readily. Incorporation of oxygen in the core requires high temperature. Silicon enters the core only under highly reducing conditions and/or at high temperature.

The pressure–temperature–oxygen fugacity–composition coverage of experimental data varies significantly, with sulfur and oxygen being the most-studied elements, and hydrogen and carbon being the least studied. Difference in data coverage is mostly due to experimental and analytical difficulties. Since most of the experimental studies are still limited to relatively low pressures and temperatures compared to those prevalent in the Earth's core, and are limited to simplified compositions, we are far from narrowing down the list of candidate light elements in the core.

Despite the uncertainties in light element composition, experiments have been conducted to study the behavior of trace elements that are of interest in the core. The solubility of potassium is enhanced by high temperature and the presence of sulfur. The question of a potassium-powered geodynamo is still open. Niobium becomes slightly siderophile under high pressure. Hence, the depletion of niobium in the Earth's mantle may not pose a challenge for the chondritic bulk Earth model. Partitioning of rhenium and osmium between solid and liquid metals under relatively low pressures is shown to support the interpretation of an osmium isotope signature from the core.

To paraphrase Brett (1984), discussions on the composition of the core still suffer from too few data and many extrapolations. However, the situation has been improving steadily. State-of-the-art experimental techniques are capable of reaching the pressure and temperature of the outer core. The phase diagram and EOS of iron need to be extended to higher temperatures. Experimental data on element partitioning, wetting behavior, phase diagrams and EOS of iron–light-element alloys at higher pressures and higher temperatures and with more realistic compositions are possible and urgently needed.

ACKNOWLEDGMENTS

The authors thank David Rubie, Richard Carlson, and Willem van Westrenen for their thoughtful, constructive, and meticulous reviews.

REFERENCES

Ahrens T. J. (1980) Dynamic compression of Earth materials. *Science* **207**, 1035–1041.

Ahrens T. J. and Jeanloz R. (1987) Pyrite: shock compression, isentropic release, and composition of the Earth's core. *J. Geophys. Res.* **92**, 10363–10375.

Ahrens T. J., Holland K. G., and Chen G. Q. (2002) Phase diagram of iron, revised core temperatures. *Geophys. Res. Lett.* **29**(7), article no. 1150, doi: 10.1029/2001GL014350.

Akimoto S. (1987) High-pressure research in geophysics: past, present, and future. In *High-pressure Research in Mineral Physics* (eds. M. H. Manghnani and Y. Syono). Terra Scientific Publishing Company, Tokyo, pp. 1–13.

Alfé D., Gillan M. J., and Price G. D. (1999a) The melting curve of iron at the pressures of the Earth's core from *ab initio* calculations. *Nature* **401**, 462–464.

Alfé D., Price G. D., and Gillan M. J. (1999b) Oxygen in the Earth's core: a first-principles study. *Phys. Earth Planet. Inter.* **110**, 191–210.

Alfé D., Gillan M. J., and Price G. D. (2000a) Constraints on the composition of the Earth's core from *ab initio* calculations. *Nature* **405**, 172–175.

Alfé D., Kresse G., and Gillan M. J. (2000b) Structure and dynamics of liquid iron under Earth's core conditions. *Phys. Rev. B* **61**, 132–142.

Alfé A., Price G. D., and Gillan M. J. (2000c) Thermodynamic stability of Fe/O solid solution at inner-core conditions. *Geophys. Res. Lett.* **27**, 2417–2420.

Alfé D., Price G. D., and Gillan M. J. (2001) Thermodynamics of hexagonal-close-packed iron under Earth's core conditions. *Phys. Rev. B* **64**, article no. 045123.

Alfé D., Price G. D., and Gillan M. J. (2002a) Iron under Earth's core conditions: liquid-state thermodynamics and high-pressure melting curve from *ab initio* calculations. *Phys. Rev. B* **65**, article no. 165118.

Alfé D., Gillan M. J., and Price G. D. (2002b) *Ab initio* chemical potentials of solid and liquid solutions and the chemistry of the Earth's core. *J. Chem. Phys.* **116**, 7127–7136.

Allègre C. J., Poirier J. P., Humler E., and Hofmann A. W. (1995) The chemical composition of the Earth. *Earth Planet. Sci. Lett.* **134**, 515–526.

Al'tshuler L. V. (1962) Composition and state of matter in the deep interior of the Earth. *Phys. Earth Planet. Inter.* **5**, 295–300.

Anderson O. L. and Isaak D. G. (2002) Another look at the core density deficit of Earth's outer core. *Phys. Earth Planet. Inter.* **131**, 19–27.

Anderson O. L., Dubrovinsky L., Saxena S. K., and Lebihan T. (2001) Experimental vibrational grüneisen ratio values for epsilon-iron up to 330 GPa at 300 K. *Geophys. Res. Lett.* **28**, 399–402.

Anderson W. W. and Ahrens T. J. (1994) An equation of state for liquid iron and implications for the Earth's core. *J. Geophys. Res.* **99**, 4273–4284.

Andrault D., Fiquet G., Kunz M., Visocekas F., and Hausermann D. (1997) The orthorhombic structure of iron: an *in situ* study at high temperature and high pressure. *Science* **278**, 831–834.

Andrault D., Fiquet G., Charpin T., and Bihan T. L. (2000) Structure analysis and stability field of beta-iron at high *P* and *T*. *Am. Mineral.* **85**, 364–371.

Badding J. V., Mao H. K., and Hemley R. J. (1991) High-pressure chemistry of hydrogen in metals: *in-situ* study of iron-hydrite. *Science* **253**, 421–424.

Badro J., Struzhkin V. V., Shu J., Hemley R. J., Mao H. K., Rueff J.-P., Kao C. C., and Sheu G. (1999) Magnetism in FeO at megabar pressures from X-ray emission spectroscopy. *Phys. Rev. Lett.* **83**, 4101–4104.

Baker H. (1992) *ASM Handbook: Alloy Phase Diagrams*. ASM International, Materials Park.

Balog P. S., Secco R. A., and Rubie D. C. (2001) Density measurements of liquids at high pressure: modifications to the sink/float method by using composite spheres, and application to Fe–10 wt.%S. *High Press. Res.* **21**, 237–261.

Bancroft D., Peterson E. L., and Minshall S. (1956) Polymorphism of iron at high pressure. *J. Appl. Phys.* **27**, 291–298.

Barker L. M. and Hollenbach R. E. (1974) Shock wave study of the alpha–epsilon phase transition in iron. *J. Appl. Phys.* **45**, 4872–4887.

Bertka C. M. and Fei Y. (1997) Mineralogy of the Martian interior up to core–mantle boundary pressures. *J. Geophys. Res.* **103**, 5251–5264.

Birch F. (1952) Elasticity and constitution of the Earth's interior. *J. Geophys. Res.* **57**, 227–286.

Birch F. (1964) Density and composition of the mantle and the core. *J. Geophys. Res.* **69**, 4377–4388.

Boehler R. (1986) The phase diagram of iron to 430 kbar. *Geophys. Res. Lett.* **13**, 1153–1156.

Boehler R. (1992) Melting of the Fe–FeO and the Fe–FeS systems at high pressure: constraints on core temperatures. *Earth Planet. Sci. Lett.* **111**, 217–227.

Boehler R. (1993) Temperatures in the Earth's core from melting-point measurements of iron at high static pressures. *Nature* **363**, 534–536.

Boehler R. (1996) Fe–FeS eutectic temperatures to 620 kbar. *Phys. Earth Planet. Inter.* **96**, 181–186.

Boehler R. (2000) High-pressure experiments and the phase diagram of lower mantle and core materials. *Rev. Geophys.* **38**, 221–245.

Boness D. A. and Brown J. M. (1990) The electronic band structures of iron, sulfur, and oxygen at high pressures and the Earth's core. *J. Geophys. Res.* **95**, 21721–21730.

Brandon A., Walker R. J., Morgan J. W., Norman M. D., and Prichard H. M. (1998) Coupled ^{186}Os and ^{187}Os evidence for core–mantle interaction. *Science* **280**, 1570–1573.

Brett R. (1984) Chemical equilibrium of the Earth's core and upper mantle. *Geochim. Cosmochim. Acta* **48**, 1183–1188.

Brown J. M. (2001) The equation of state of iron to 450 GPa: another high pressure solid phase? *Geophys. Res. Lett.* **28**, 4339–4342.

Brown J. M. and McQueen R. G. (1986) Phase transitions, Grüneisen parameter, and elasticity for shocked iron between 77 GPa and 400 GPa. *J. Geophys. Res.* **91**, 7485–7494.

Brown J. M., Fritz J. N., and Hixson R. S. (2000) Hugoniot data for iron. *J. Appl. Phys.* **88**, 5496–5498.

Bundy F. P. (1965) Pressure–temperature phase diagram of iron to 200 kbar, 900 °C. *J. Appl. Phys.* **36**, 616–620.

Chabot N. L. and Drake M. J. (1999) Potassium solubility in metal: the effects of composition at 15 kbar and 1,900 degrees C on partitioning between iron alloys and silicate melts. *Earth Planet. Sci. Lett.* **172**, 323–335.

Chaklader A. C. D., Gill W. W., and Mehrotra S. P. (1981) Predictive model for interfacial phenomena between molten metals and sapphire in varying oxygen partial pressures. In *Surfaces and Interfaces in Ceramics and Ceramic–Metal Systems* (eds. J. Pask and A. Evan). Plenum, New York, pp. 421–431.

Claussen W. F. (1960) Detection of the $\alpha-\gamma$ iron phase transformation by differential thermal analysis. *Rev. Sci. Instr.* **31**, 878–881.

Clendenen R. L. and Drickamer H. G. (1964) The effect of pressure on the volume and lattice parameters of ruthenium and iron. *J. Phys. Chem. Solids* **25**, 841–845.

Cohen R. E., Mazin I. I., and Isaak D. G. (1997) Magnetic collapse in transition metal oxides at high pressure: implication for the Earth. *Science* **275**, 654–657.

Cox P. A. (1989) *The Elements, Their Origin, Abundance and Distribution.* Oxford University Press, New York.

Creager K. C. (1992) Anisotropy of the inner core from differential travel times of the phases pkp and pkikp. *Nature* **356**, 309–313.

Dreibus G. and Palme H. (1995) Cosmochemical constraints on the sulfur content in the Earth's core. *Geochim. Cosmochim. Acta* **60**, 1125–1130.

Dobson D., Vocaldo L., and Wood I. G. (2002) A new high-pressure phase of FeSi. *Am. Mineral.* **87**, 784–787.

Dubrovinsky L., Annerstin H., Dubrovinskaia N., Westman F., Harryson H., Fabrichnaya O., Carlson S. (2001) Chemical interaction of Fe and Al_2O_3 as a source of heterogeneity at the Earth's core–mantle boundary. *Nature* **412**, 527–529.

Dubrovinsky L., Dubrovinskaia N., Langenhorst F., Dobson D., Rubie D., Gessman C., Abrikesov I. A., Johansson B., Baykov V. I., Vitos L., Le Bihan T., Grcichton W. A., Dmitniev V., and Weber H.-P. (2003) Iron–silica interaction at extreme conditions and the electrically conducting layer at the base of Earth's mantle. *Nature* **422**, 58–61.

Dubrovinsky L. S., Saxena S. K., and Lazor P. (1998) High-pressure and high-temperature *in situ* X-ray diffraction study of iron and corundum to 68 GPa using an internally heated diamond anvil cell. *Phys. Chem. Mineral.* **25**, 434–441.

Dubrovinsky L. S., Saxena S. K., Tutti F., and Rekhi S. (2000) *In situ* X-ray study of thermal expansion and phase transition of iron at multimegabar pressure. *Phys. Rev. Lett.* **84**, 1720–1723.

Dziewonski A. M. and Anderson D. L. (1981) Preliminary reference Earth model. *Phys. Earth Planet. Inter.* **25**, 297–356.

Fei Y. and Mao H.-K. (1994) *In situ* determination of the Ni–As phase of FeO at high pressure and temperature. *Science* **266**, 1678–1680.

Fei Y., Mao H.-K, Shu J., and Hu J. Z. (1992) *P–V–T* equation of state of magnesiowüstite ($Mg_{0.6}Fe_{0.4}$)O. *Phys. Chem. Mineral.* **18**, 416–422.

Fei Y., Prewitt C. T., Mao H. K., and Bertka C. M. (1995) Structure and density of FeS at high pressure and high temperature and the internal structure of Mars. *Science* **268**, 1892–1894.

Fei Y., Bertka C. M., and Finger L. W. (1997) High-pressure iron–sulfur compound Fe_3S_2, and melting relations in the system Fe–FeS. *Science* **275**, 1621–1623.

Fei Y., Li J., Bertka C. M., and Prewitt C. T. (2000) Structure type and bulk modulus of Fe_3S, a new iron–sulfur compound. *Am. Mineral.* **85**, 1830–1833.

Fiquet G., Badro J., Guyot F., Requardt H., and Krisch M. (2001) Sound velocities of iron to 110 gigapascals. *Science* **291**, 468–471.

Funamori N., Yagi T., and Uchida T. (1996) High-pressure and high-temperature *in situ* X-ray diffraction study of iron to above 30 GPa using MA8-type apparatus. *Geophys. Res. Lett.* **23**, 953–956.

Gessmann C. K. and Rubie D. C. (1998) The effect of temperature on the partitioning of nickel, cobalt, manganese, chromium, and vanadium at 9 GPa and constraints on formation of the Earth's core. *Geochim. Cosmochim. Acta* **62**, 867–882.

Gessmann C. K. and Wood B. J. (2002) Potassium in the Earth's core? *Earth Planet. Sci. Lett.* **200**, 63–78.

Gessmann C. K., Wood B. J., Rubie D. C., and Kilburn M. R. (2001) Solubility of silicon in liquid metal at high pressure: implications for the composition of the Earth's core. *Earth Planet. Sci. Lett.* **184**, 367–376.

Goettel K. A. and Lewis J. S. (1973) Comments on a paper by V. M. Oversby and A. E. Ringwood. *Earth Planet. Sci. Lett.* **18**, 148–150.

Hall H. T. and Murthy V. R. (1971) The early chemical history of the Earth: some critical elemental fractions. *Earth Planet. Sci. Lett.* **11**, 239–244.

Hemley R. J. and Mao H. K. (2001) *In situ* studies of iron under pressure: new windows on the Earth's core. *Int. Geol. Rev.* **43**, 1–30.

Hillgren V., Gessmann C. K., and Li J. (2000) An experimental perspective on the light element in Earth's core. In *Origin of the Earth and Moon* (eds. R. M. Canup and K. Righter). University of Arizona Press, Tucson, pp. 245–263.

Hillgren V. J. and Boehler R. (2000) High pressure geochemistry in the diamond cell to 100 GPa and 3,300 K. *Science and Technology of High Pressure AIRAPT-17, Proceedings of AIRAPT-17* (eds. M. H. Manghnani, W. J. Nellis and M. F. Nicol). University Press, Hyderabad, India, pp. 609–611.

Hirose K., Fei Y., Ono S., Yagi T., and Funakoshi K.-I. (2000) *In situ* measurements of the phase transition boundary in $Mg_3Al_2Si_3O_{12}$: implications for the nature of the seismic discontinuities in the Earth's mantle. *Earth Planet. Sci. Lett.* **5705**, 1–7.

Holzheid A. and Grove T. L. (2002) Sulfur saturation limits in silicate melts and their implications for core formation scenarios for terrestrial planets. *Am. Mineral.* **87**, 227–237.

Holzheid J. R., Borisov A., and Palme H. (1994) The effect of oxygen fugacity and temperature on solubilities of nickel, cobalt, and molybdenum in silicate melts. *Geochim. Cosmochim. Acta* **58**, 1975–1981.

Huang E., Bassett W. A., and Weathers M. S. (1987) Phase relationships in Fe–Ni alloys at high pressures and temperatures. *J. Geophys. Res.* **93**, 7741–7746.

Iida T. and Guthrie R. I. L. (1988) *The Physical Properties of the Liquid Metals.* Clarendon Press, Oxford.

Irifune T., Nishiyama N., Kuroda K., Inoue T., Isshiki M., Utsumi W., Funakoshi K.-I., Urakawa S., Uchida T., Katsura T., and Ohtaka O. (1998) The postspinel phase boundary in Mg_2SiO_4 determined by *in situ* X-ray diffraction. *Science* **279**, 1698–1700.

Ito E. and Morooka K. (1993) Dissolution of K in molten iron at high pressure and temperature. *Geophys. Res. Lett.* **20**, 1651–1654.

Jacobs J. A. (1987) *The Earth's Core*. Academic Press, Orlando.

Jamieson J. C. and Lawson A. W. (1962) X-ray diffraction studies to 100 kbar. *J. Appl. Phys.* **33**, 776–780.

Jeanloz R. (1979) Properties of iron at high pressures and the state of the core. *J. Geophys. Res.* **84**, 6059–6069.

Jeanloz R. (1990) The nature of the Earth's core. *Ann. Rev. Earth Planet. Sci.* **18**, 357–386.

Jeanloz R. and Ahrens T. J. (1980) Equations of state of FeO and CaO. *Geophys. J. Roy. Astro. Soc.* **62**, 505–528.

Jephcoat A. (1985) Hydrostatic compression studies on iron and pyrite to high pressures: the composition of the Earth's core and the equation of state of solid argon. PhD Thesis, John Hopkins, Baltimore, Maryland. 214pp.

Jephcoat A. and Olson P. (1987) Is the inner Core of the Earth pure iron? *Nature* **325**, 332–335.

Kargel J. S. and Lewis J. S. (1993) The composition and early evolution of Earth. *Icarus* **105**, 1–25.

Kavner A., Duffy T. S., and Shen G. Y. (2001) Phase stability and density of FeS at high pressures and temperatures: implications for the interior structure of Mars. *Earth Planet. Sci. Lett.* **185**, 25–33.

Kilburn M. R. and Wood B. J. (1997) Metal-silicate partitioning and the incompatibility of S and Si during core formation. *Earth Planet. Sci. Lett.* **152**, 139–148.

Kleine T., Münker C., Mezger K., and Palme H. (2002) Rapid accretion and early core formation on asteroid and the terrestrial planets from Hf–W chronometry. *Nature* **418**, 952–956.

Knittle E. and Jeanloz R. (1986) High-pressure metallization of FeO and implications for the Earth's core. *Geophys. Res. Lett.* **13**, 1541–1544.

Knittle E. and Jeanloz R. (1991) Earth's core–mantle boundary: results of experiments at high pressures and temperatures. *Science* **251**, 1438–1443.

Knittle E. and Williams Q. (1995) Static compression of ε-FeSi and an evaluation of reduced silicon as a deep Earth constituent. *Geophys. Res. Lett.* **22**, 445–448.

Laio A., Bernard S., Chiarotti G. L., Scandolo S., and Tosatti E. (2000) Physics of iron at Earth's core conditions. *Science* **287**, 1027–1030.

Lay T., Williams Q., and Garnero E. J. (1998) The core–mantle boundary layer and deep Earth dynamics. *Nature* **392**, 461–468.

Lewis J. S. (1971) Consequences of the presence of sulfur in the core of the Earth. *Earth Planet. Sci. Lett.* **11**, 130–134.

Li J. and Agee C. B. (1996) Geochemistry of mantle–core differentiation at high pressure. *Nature* **381**, 686–689.

Li J. and Agee C. B. (2001) Element partitioning constraints on the light element composition of the Earth's core. *Geophys. Res. Lett.* **28**, 81–84.

Li J., Fei Y., Mao H. K., Hirose K., and Shieh S. R. (2001) Sulfur in the Earth's inner core. *Earth Planet. Sci. Lett.* **193**, 509–514.

Li J., Mao H. K., Fei Y., Gregoryanz E., Eremets M., and Zha C. S. (2002) Compression of Fe_3C to 30 GPa at room temperature. *Phys. Chem. Mineral.* **19**, 166–169.

Lin J. F., Heinz D. L., Campbell A. J., Devine J. M., Mao W. L., and Shen G. (2002a) Iron–nickel alloy in the Earth's core. *Geophys. Res. Lett.* **29**, 109–111.

Lin J. F., Heinz D. L., Campbell A. J., Devine J. M., and Shen G. Y. (2002b) Iron–silicon alloy in Earth's core? *Science* **295**, 313–315.

Liu L. G. and Bassett W. A. (1975) The melting of iron up to 200 kbar. *J. Geophys. Res.* **81**, 3777–3782.

Lübbers R., Grunsteudel H. F., Chumakov A. I., and Wortmann G. (2000) Density of phonon states in iron at high pressure. *Science* **287**, 1250–1253.

Ma Y., Mao H. K., Hemley R. J., Gramsch S. A., Shen G., and Somayazulu M. (2001) Two-dimensional energy dispersive X-ray diffraction at high pressures and temperatures. *Rev. Sci. Instr.* **72**, 1302–1305.

Manghnani M. and Syono Y. (1987) *High-pressure Research in Mineral Physics*. Terra Scientific Publishing Company, Tokyo.

Mao H.-K. and Hemley R. J. (1998) New windows on the Earth's deep interior. In *Ultra-high-pressure Mineralogy: Physics and Chemistry of the Earth's Deep Interior* (ed. R. J. Hemley). Mineralogical Society of America, Washington, DC, pp. 1–32.

Mao H. K., Bell P. M., and Hadidiacos C. (1987) Experimental phase relations of iron to 360 kbar, 1,400 °C, determined in an internally heated diamond-anvil apparatus. In *High-pressure Research in Mineral Physics* (eds. M. H. Manghnani and Y. Syono). Terra Scientific Publishing Company, Tokyo, pp. 135–138.

Mao H. K., Wu Y., Chen L. C., and Shu J. F. (1990) Static compression of iron to 300 GPa and $Fe_{0.8}Ni_{0.2}$ alloy to 260 GPa: implications for compositions of the core. *J. Geophys. Res.* **95**, 21,737–21,742.

Mao H.-K., Hemley R. J., Fei Y., Shu J., Chen L., Jephcoat A. P., Wu Y., and Bassett W. A. (1991) Effect of pressure, temperature, and composition on lattice parameters and density of $(Fe,Mg)SiO_3$-perovskites to 30 GPa. *J. Geophys. Res.* **96**, 8069–8079.

Mao H. K., Shu J. F., Shen G. Y., Hemley R. J., Li B. S., and Singh A. K. (1999) Elasticity and rheology of iron above 220 GPa and the nature of the Earth's inner core. *Nature* **399**, 741–743.

Mao H. K., Xu J., Struzhkin V. V., Shu J., Hemley R. J., (2001) Phonon density of states of iron up to 153 gigapascals. *Science* **292**, 914–916.

McDonough W. F. and Sun S.-S. (1995) The composition of the Earth. *Chem. Geol.* **120**, 223–253.

McQueen R. G. and Marsh S. P. (1966) Shock-wave compression of iron–nickel alloys and the Earth's core. *J. Geophys. Res.* **71**, 1751–1756.

Meibom A. and Frei R. (2002) Evidence for an ancient osmium isotopic reservoir in Earth. *Science* **296**, 516–518.

Merkel S., Jephcoat A. P., Shu J., Mao H.-K., Gillet P., and Hemley R. J. (2002) Equation of state, elasticity, and shear strength of pyrite under high pressure. *Phys. Chem. Mineral.* **29**, 1–9.

Merrill R. T. and McElhinny M. W. (1983) *The Earth's Magnetic Field*. Academic Press, San Diego.

Merrill R. T., McElhinny M. W., and McFadden P. L. (1996) *The Magnetic Field of the Earth*. Academic Press, San Diego.

Minarik W. G., Ryerson F. J., and Watson E. B. (1996) Textural entrapment of core-forming melts. *Science* **272**, 530–533.

Mirwald P. W. and Kennedy G. C. (1979) The Curie point and the $\alpha-\gamma$ transition of iron to 53 kbar—a reexamination. *J. Geophys. Res.* **84**, 656–658.

Morelli A., Dziewonski A. M., and Woodhouse J. H. (1986) Anisotropy of the inner core inferred from pkikp travel times. *Geophys. Res. Lett.* **13**, 1545–1548.

Murrell M. T. and Burnett D. S. (1986) Partitioning of K, U, and Th between sulfide and silicate liquids: implications for radioactive heating of planetary cores. *J. Geophys. Res.* **91**, 8126–8136.

Murthy R., van Westrenen W., and Fei Y. (2003) Experimental evidence that potassium is a substantial radioactive heat source in planetary cores. *Nature.* **423**, 163–165.

O'Neill H. S. C. and Palme H. (1998) Composition of the silicate Earth: implications for accretion and core formation. In *The Earth's Mantle* (ed. I. Jackson). Cambridge University Press, Cambridge, pp. 3–126.

O'Neill H. St., C., Canil D., and Rubie D. C. (1998) Oxide–metal equilibria to 2,500 °C and 25 GPa: implications for core formation and the light component in the Earth's core. *J. Geophys. Res.* **103**, 12,239–212,260.

Ohtani E. and Ringwood A. E. (1984) Composition of the core: I. Solubility of oxygen in molten iron at high temperatures. *Earth Planet. Sci. Lett.* **71**, 85–93.

Ohtani E., Yurimoto H., and Seto S. (1997) Element partitioning between metallic liquid, silicate liquid and lower-mantle minerals: implications for core formation of the Earth. *Phys. Earth Planet. Inter.* **100**, 97–114.

Okuchi T. (1997) Hydrogen partitioning into molten iron at high pressure: implications for Earth's core. *Science* **278**, 1781–1784.

Okuchi T. (1998) The melting temperature of iron hydride at high pressures and its implications for the temperature of the Earth's core. *J. Phys.-Cond. Matt.* **10**, 11595–11598.

Oldham R. D. (1906) Constitution of the interior of the Earth as revealed by earthquakes. *Quart. J. Geol. Soc.* **62**, 456–475.

Omel'chenko A. V., Soshnikov V. I., and Estrin E. I. (1969) Effect of pressure on the Curie point and $\alpha-\gamma$ transformation of iron. *Fiz. Metal. Metalloved.* **28**, 77–83.

Oversby V. M. and Ringwood A. E. (1972) Partitioning of potassium between silicates and sulphide melts: experiments relevant to the Earth's core. *Phys. Earth Planet. Inter.* **6**, 161–166.

Parker L. J., Hasegawa M., Atou T., and Badding J. V. (1997) High-pressure synthesis of alkali metal–transition metal compounds. *Euro. J. Solid State Inorg. Chem.* **34**, 693–704.

Poirier J. P. (1988) Transport properties of liquid metals and viscosity of the Earth's core. *Geophys. J.* **92**, 99–105.

Poirier J. P. (1994) Light elements in the Earth's outer core: a critical review. *Earth Planet. Sci. Lett.* **85**, 319–337.

Poirier J. P., Malavergne V., and Mouël J. L. L. (1998) Is there a thin electrically conducting layer at the base of the mantle? In: *The Core–Mantle Boundary Region* (eds. M. Gurnis, M. E. Wysession, E. Knittle, and B. A. Buffet). American Geophysical Union, Washington, DC, pp. 131–137.

Press F. (1968) Density distribution in Earth. *Science* **160**, 1218–1221.

Raghavan V. (1988) *Phase Diagrams of Ternary Iron Alloys: Part 2. Ternary Systems Containing Iron and Sulphur.* Indian Institute of Metals, Calcutta.

Ringwood A. E. and Hibberson W. (1989) The system Fe–FeO revisited. *Phys. Chem. Mater.* **17**, 313–319.

Rubie D. C. (1999) Characterizing the sample environment in multi-anvil high-pressure experiments. *Phase Trans.* **68**, 431–451.

Sanloup C., Guyot F., Gillet P., Fiquet G., Hemley R. J., Mezouar M., and Martinez I. (2000a) Structural changes in liquid Fe at high pressures and high temperatures from synchrotron X-ray diffraction. *Europhys. Lett.* **52**, 151–157.

Sanloup C., Guyot F., Gillet P., Fiquet G., Mezouar M., and Martines I. (2000b) Density measurements of liquid Fe–S alloys at high-pressure. *Geophys. Res. Lett.* **27**, 811–814.

Saxena S. K. and Dubrovinsky L. S. (2000) Iron phases at high pressures and temperatures: phase transition and melting. *Am. Mineral.* **85**, 372–375.

Saxena S. K., Shen G., and Lazor P. (1993) Experimental evidence for a new iron phase and implications for Earth's core. *Science* **260**, 1312–1314.

Saxena S. K., Shen G., and Lazor P. (1994) Temperatures in the Earth's core based on melting and phase transformation experiments on iron. *Science* **264**, 405–407.

Saxena S. K., Dubrovinsky L. S., and Häggkvist P. (1996) X-ray evidence for the new phase β-iron at high temperature and high pressure. *Geophys. Res. Lett.* **23**, 2441–2444.

Scott H. P., Williams Q., and Knittle E. (2001) Stability and equation of state of Fe_3C to 73 GPa: implications for carbon in the Earth's core. *Geophys. Res. Lett.* **28**, 1875–1878.

Secco R. A. (1995) Viscosity of the outer core. In: *Mineral Physics and Crystallography: A Handbook of Physical Constants* (ed. T. J. Ahrens). American Geophysical Union, Washington, DC, 218pp.

Shannon M. C. and Agee C. B. (1998) Percolation of core melts at lower mantle conditions. *Science* **280**, 1059–1061.

Shearer P. and Masters G. (1990) The density and shear velocity contrast at the inner core boundary. *Geohys. J. Int.* **102**, 408–491.

Shen G., Mao H. K., Hemley R. J., Duffy T. S., and Rivers M. L. (1998) Melting and crystal structure of iron at high pressures and temperatures. *Geophys. Res. Lett.* **25**, 373–376.

Sherman D. M. (1995) Stability of possible Fe–FeS and Fe–FeO alloy phases at high pressure and the composition of the Earth's core. *Earth Planet. Sci. Lett.* **132**, 87–98.

Sherman D. M. (1997) The composition of the Earth's core: constraints on S and Si vs. temperature. *Earth Planet. Sci. Lett.* **153**, 149–155.

Shirey S. B. and Walker R. J. (1998) The Re–Os isotope system in cosmochemistry and high-temperature geochemistry. *Ann. Rev. Earth Planet. Sci.* **26**, 423–500.

Shterenberg L. E., Slesarev V. N., Korsunskaya I. A., and Kamenetskaya D. S. (1975) The experimental study of the interaction between the melt carbides and diamond in the iron–carbon system at high pressures. *High Temp.–High Press.* **7**, 517–522.

Söderlind P., Moriarty J. A., and Wills J. M. (1996) First-principles theory of iron up to Earth-core pressures: structural, vibrational, and elastic properties. *Phys. Rev. B* **53**, 14063–14072.

Song X. D. and Helmberger D. (1993) Anisotropy of the Earth's inner core. *Geophys. Res. Lett.* **20**, 2591–2594.

Song X. D. and Richards P. G. (1996) Seismological evidence for differential rotation of the Earth's inner core. *Nature* **382**, 221–224.

Stacey F. D. (1992) *Physics of the Earth.* Bookfield Press, Brisbane.

Steinle-Neumann G. and Stixrude L. (1999) First-principles elastic constants for the hcp transition metals Fe, Co and Re at high pressure. *Phys. Rev. B* **60**, 791–799.

Steinle-Neumann G., Stixrude L., Cohen R. E., and Gulseren O. (2001) Elasticity of iron at the temperature of the Earth's inner core. *Nature* **413**, 57–60.

Sterrett K. F., Jr., Klement W., Jr., and Kennedy G. C. (1965) Effect of pressure on the melting of iron. *J. Geophys. Res.* **70**, 1979–1984.

Stevenson D. J. (1981) Models of the Earth's core. *Science* **214**, 611–619.

Stixrude L. and Brown M. J. (1998) The Earth's core. In *Ultrahigh-pressure Mineralogy; Physics and Chemistry of the Earth's Deep Interior* (ed. R. J. Hemley). Mineralogical Society of America, pp. 261–282.

Stixrude L. and Cohen R. E. (1995) High-pressure elasticity of iron and anisotropy of Earth's inner core. *Science* **267**, 1972–1975.

Stixrude L., Wasserman E., and Cohen R. E. (1997) Composition and temperature of Earth's inner core. *J. Geophys. Res.* **102**, 24729–24739.

Strong H. M. (1959) Fusion curve of iron to 96,000 atmospheres. *J. Geophys. Res.* **64**, 653.

Strong H. M., Tuft R. E., and Hanneman R. E. (1973) The iron fusion curve and γ-δ-λ triple point. *Rep. 73CRD017*, Gen. Elec. Corp., Schenectady, New York.

Su W., Dziewonski A. M., and Jeanloz R. (1996) Planet within a plant: rotation of the inner core of the Earth. *Science* **274**, 1883–1887.

Takahashi T. and Bassett W. A. (1964) High-pressure polymorph of iron. *Science* **145**, 483–486.

Uchida T., Wang Y. B., Rivers M. L., and Sutton S. R. (2001) Stability field and thermal equation of state of epsilon-iron determined by synchrotron X-ray diffraction in a multianvil apparatus. *J. Geophys. Res.* **106**, 21799–21810.

Usselman T. M. (1975) Experimental approach to the state of the core: Part 1. The liquidus relations of the Fe-rich portion of the Fe–Ni–S system from 30 to 100 kb. *Am. J. Sci.* **275**, 278–290.

Vinnik L., Romanowicz B., and Breger L. (1994) Anisotropy in the center of the inner core. *Geophys. Res. Lett.* **21**, 1671–1674.

Vocadlo L. and Dobson D. (1999) The Earth's deep interior: advances in theory and experiment. *Phil. Trans. Roy. Soc. London Ser. A: Math. Phys. Eng. Sci.* **357**, 3335–3357.

Vocadlo L., Alfé D., Price G. D., and Gillan M. J. (2000) First principles calculations on the diffusivity and viscosity of liquid Fe–S at experimentally accessible conditions. *Phys. Earth Planet. Inter.* **120**, 145–152.

Wade J. and Wood B. J. (2001) The Earth's "missing" niobium may be in the core. *Nature* **409**, 75–78.

Walker D. (2000) Core participation in mantle geochemistry: Geochemical Society Ingersoll Lecture, GSA, Denver, October 1999. *Geochim. Cosmochim. Acta* **64**, 2897–2911.

Wang C., Hirama J., Nagasaka T., and Ban-Ya S. (1991) Phase equilibria of liquid Fe–S–C ternary system. *ISIJ Int.* **31**, 1292–1299.

Wänke H. and Dreibus G. (1997) New evidence for silicon as a major light element in the Earth's core. Abstracts of papers submitted to *Lunar Planet. Sci. Conf.* **28**, 1495–1496.

Wendlandt R. F. (1982) Sulfide saturation of basalt and andesite melts at high pressures and temperatures. *Am. Mineral.* **67**, 877–885.

Williams Q. and Hemley R. J. (2001) Hydrogen in the deep Earth. *Ann. Rev. Earth Planet. Sci.* **29**, 365–418.

Williams Q. and Jeanloz R. (1990) Melting relations in the iron–sulfur system at ultra-high pressures: implications for the thermal state of the Earth. *J. Geophys. Res.* **95**, 19299–19310.

Williams Q., Jeanloz R., Bass J., Svendsen B., and Ahrens T. J. (1987) The melting curve of iron to 250 gigapascals: a constraint on the temperature at Earth's center. *Science* **236**, 181–182.

Wood B. J. (1993) Carbon in the core. *Earth Planet. Sci. Lett.* **117**, 593–607.

Woodhouse J. H., Giardine D., and Li X. D. (1986) Evidence for inner core anisotropy from free oscillations. *Geophys. Res. Lett.* **13**, 1549–1552.

Yagi T. (1995) Formation of iron hydrides under the condition of the Earth's interior-implication for the core formation process. In *The Earth's Central Part: Its Structure and Dynamics* (ed. T. Yukutake). Terra Scientific Publishing Company, Tokyo, pp. 13–28.

Yang H. T. and Secco R. A. (1999) Melting boundary of Fe–17%Si up to 5.5 GPa and the timing of core formation. *Geophys. Res. Lett.* **26**, 263–266.

Yin Q., Jacobsen S. B., Yamashita K., Blichert-Toft J., Télouk P., and Albarède F. (2002) A short timescale for terrestrial planet formation from Hf–W chronometry of meteorites. *Nature* **418**, 949–952.

Yoo C. S., Holmes N. C., Ross M., Webb D. J., and Pike C. (1993) Shock temperatures and melting of iron at Earth core conditions. *Phys. Rev. Lett.* **70**, 3931–3934.

Yoo C. S., Akella J., Campbell A. J., Mao H. K., and Hemley R. J. (1995) Phase diagram of iron by *in situ* X-ray diffraction: implications for Earth's core. *Science* **270**, 1473–1475.

Yoo C. S., Söderlind P., and Campbell A. J. (1996) dhcp as possible new e phase of iron at high pressure and temperature. *Phys. Rev. Lett.* **214**, 65–68.

Zhang J. and Guyot F. (1999a) Experimental study of the bcc–fcc phase transformations in the Fe-rich system Fe–Si at high pressures. *Phys. Chem. Mineral.* **26**, 419–424.

Zhang J. and Guyot F. (1999b) Thermal equation of state of iron and $Fe_{0.91}Si_{0.09}$. *Phys. Chem. Mineral.* **26**, 206–211.

2.15
Compositional Model for the Earth's Core

W. F. McDonough
University of Maryland, College Park, USA

2.15.1 INTRODUCTION

The remote setting of the Earth's core tests our ability to assess its physical and chemical characteristics. Extending out to half an Earth radii, the metallic core constitutes a sixth of the planet's volume and a third of its mass (see Table 1 for physical properties of the Earth's core). The boundary between the silicate mantle and the core (CMB) is remarkable in that it is a zone of greatest contrast in Earth properties. The density increase across this boundary represents a greater contrast than across the crust-ocean surface. The Earth's gravitational acceleration reaches a maximum (10.7 m s^{-2}) at the CMB and this boundary is also the site of the greatest temperature gradient in the Earth. (The temperature at the base of the mantle (~2,900 °C) is not well established, and that at the top of the inner core is even less securely known (~3,500–4,500 °C).) The pressure range throughout the core (i.e., 136 GPa to >360 GPa) makes recreating environmental conditions in most experimental labs impossible, excepting a few diamond anvil facilities or those with high-powered, shock-melting guns (see Chapter 2.14). Thus, our understanding of the core is based on very few pieces of direct evidence and many fragments of indirect observations. Direct evidence comes from seismology, geodesy, geo- and paleomagnetism, and, relatively recently isotope geochemistry (see Section 2.15.6). Indirect evidence comes from geochemistry, cosmochemistry, and meteoritics; further constraints on the core system are gained from studies in experimental petrology, mineral physics, *ab initio* calculations, and evaluations of the Earth's energy

Table 1 Physical properties of the Earth's core.

		Units	*Refs.*
Mass			
Earth	5.9736E + 24	kg	1
Inner core	9.675E + 22	kg	1
Outer core	1.835E + 24	kg	1
Core	1.932E + 24	kg	1
Mantle	4.043E + 24	kg	1
Inner core to core (%)	5.0%		
Core to Earth (%)	32.3%		
Depth			
Core–mantle boundary	3,483 ± 5	km	2
Inner–outer core boundary	1,220 ± 10	km	2
Mean radius of the Earth	6,371.01 ± 0.02	km	1
Volume relative to planet			
Inner core	7.606E + 09 (0.7%)	km^3	
Inner core relative to the bulk core	4.3%		
Outer core	1.694E + 11 (15.6%)	km^3	
Bulk core	1.770E + 11 (16.3%)	km^3	
Silicate earth	9.138E + 11 (84%)	km^3	
Earth	1.083E + 12	km^3	
Moment of inertia constants			
Earth mean moment of inertia (I)	0.3299765	Ma^2	1
Earth mean moment of inertia (I)	0.3307144	MR_0^2	1
Mantle: I_m/Ma^2	0.29215	Ma^2	1
Fluid core: I_f/Ma^2	0.03757	Ma^2	1
Inner core: I_{ic}/Ma^2	2.35E−4	Ma^2	1
Core: $I_{f+ic}/M_{f+ic}a_f^2$	0.392	Ma^2	1

1—Yoder (1995), 2—Masters and Shearer (1995).
M is the Earth's mass, a is the Earth's equatorial radius, R_0 is the radius for an oblate spheroidal Earth, I_m is the moment of inertia for the mantle, I_f is the moment of inertia for the outer (fluid) core, I_{ic} is the moment of inertia for the inner core, and $I_{f+ic}/M_{f+ic}a_f^2$ is the mean moment of inertia for the core.

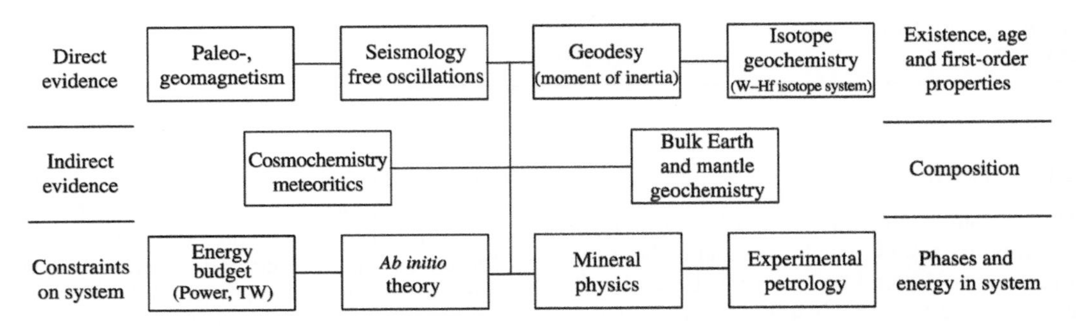

Figure 1 The relative relationship between disciplines involved in research on the Earth's core and the nature of data and information that come from these various investigations. Studies listed in the upper row yield direct evidence on properties of the core. Those in the middle row yield indirect evidence on the composition of the Earth's core, whereas findings from disciplines listed on the bottom row provide descriptions of the state conditions for the core and its formation.

budget (e.g., geodynamo calculations, core crystallization, heat flow across the core–mantle boundary). Figure 1 provides a synopsis of research on the Earth's core, and the relative relationship between disciplines. Feedback loops between all of these disciplines refine other's understanding of the Earth's core.

2.15.2 FIRST-ORDER GEOPHYSICS

The Earth's three-layer structure (the core, the silicate shell (mantle and crust), and the atmosphere–hydrosphere system) is the product of planetary differentiation and is identified as the most significant geological process to have occurred since the formation of the Earth. Each layer is distinctive in its chemical composition, the nature of its phase (i.e., solid, liquid, and gas), and physical properties. Evidence for the existence and nature of the Earth's core comes from laboratory studies coupled with studies that directly measure physical properties of the Earth's interior including its magnetic field, seismological profile, and orbital behavior, with the latter providing a coefficient of the moment of inertia and a model for the density distribution in the Earth.

There is a long history of knowing indirectly or directly of the existence of Earth's core. Our earliest thoughts about the core, albeit indirect and unwittingly, may have its roots in our understanding of the Earth's magnetic field. The magnetic compass and its antecedents appear to be ~2,000 yr old. F. Gies and J. Gies (1994) report that Chinese scholars make reference to a south-pointing spoon, and claim its invention to ca. AD 83 (Han dynasty). A more familiar form of the magnetic compass was known by the twelfth century in Europe. With the discovery of iron meteorites followed by the suggestion that these extraterrestrial specimens came from the interior of fragmented planets in the late-nineteenth century came the earliest models for planetary interiors. Thus, the stage was set for developing Earth models with a magnetic and metallic core. Later development of geophysical tools for peering into the deep Earth showed that with increasing depth the proportion of metal to rock increases with a significant central region envisaged to be wholly made up of iron.

A wonderful discussion of the history of the discovery of the Earth's core is given in the Brush (1980) paper. The concept of a core perhaps begins with understanding the Earth's magnetic field. Measurements of the Earth's magnetic field have been made since the early 1500s. By 1600 the English physician and physicist, William Gilbert, studied extensively the properties of magnets and found that their magnetic field could be removed by heating; he concluded that the Earth behaved as a large bar-magnet. In 1832, Johann Carl Friedrich Gauss, together with Wilhelm Weber, began a series of studies on the nature of Earth's magnetism, resulting in the 1839 publication of *Allgemeine Theorie des Erdmagnetismus* (*General Theory of the Earth's Magnetism*), demonstrating that the Earth's magnetic field was internally generated.

With the nineteenth-century development of the seismograph, studies of the Earth's interior and core accelerated rapidly. In 1897 Emil Wiechert subdivided the Earth's interior into two main layers: a silicate shell surrounding a metallic core, with the core beginning at ~1,400 km depth. This was the first modern model of the Earth's internal structure, which is now confirmed widely by many lines of evidence. Wiechert was a very interesting scientist; he invented a seismograph that saw widespread use in the early twentieth century, was one of the founders of the Institute of Geophysics at Göttingen, and was the PhD supervisor of Beno Gutenberg. The discoverer of the Earth's core is considered to be Richard Dixon Oldham, a British seismologist, who first distinguished *P* (compressional) and *S* (shear) waves following his studies of the Assam earthquake of 1897. In 1906 Oldham observed that *P* waves arrived later than expected at the surface antipodes of epicenters and recognized this as evidence for a dense and layered interior. Oldham placed the depth to the core–mantle boundary at 3,900 km. Later, Gutenberg (1914) established the core–mantle boundary at 2,900 km depth (cf. the modern estimate of $2,891 \pm 5$ km depth; Masters and Shearer, 1995) and suggested that the core was at least partly liquid (Gutenberg, 1914). Subsequently, Jeffreys (1926) established that the outer core is liquid, and Lehmann (1936) identified the existence of a solid inner core using seismographic records of large earthquakes, which was later confirmed by Anderson *et al.* (1971) and Dziewonski and Gilbert (1972) using Earth's free-oscillation frequencies. Finally, Washington (1925) and contemporaries reported that an iron core would have a significant nickel content, based on analogies with iron meteorites and the cosmochemical abundances of these elements.

The seismological profile of the Earth's core (Figure 2) combined with the first-order relationship between density and seismic wave speed velocity (i.e., $V_p = ((K + 4/3\mu)/\rho)^{0.5}$, $V_s = (\mu/\rho)^{0.5}$, $\mathrm{d}\rho/\mathrm{d}r = -GM_r\rho(r)/r^2\Phi$ (the latter being the Adams–Williamson equation), where V_p is

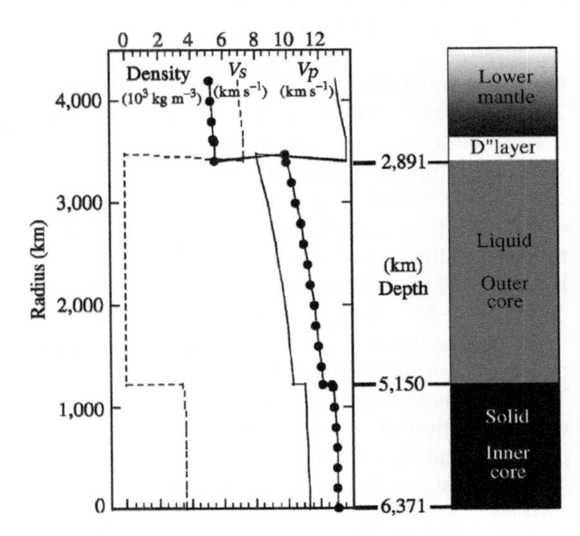

Figure 2 Depth versus *P*- and *S*-wave velocity and density for the PREM model (after Dziewonski and Anderson, 1981 and Masters and Shearer, 1995).

the P-wave velocity, V_s is the S-wave velocity, K is the bulk modulus, μ is the shear modulus, ρ is the density, $\rho(r)$ is the density of the shell within radius r, G is gravitational constant, M_r is the mass of the Earth within radius r, and $\Phi = V_p^2 - (4/3)V_s^2)$ provides a density profile for the core that, in turn, is perturbed to be consistent with free oscillation frequencies (Dziewonski and Anderson, 1981). Combining seismological data with mineral physics data (e.g., equation of state (EOS) data for materials at core appropriate conditions) from laboratory studies gives us the necessary constraints for identifying the mineralogical and chemical constituents of the core and mantle.

Birch (1952) compared seismically determined density estimates for the mantle and core with the available EOS data for candidate materials. He argued that the inner core was "a crystalline phase, mainly iron" and the liquid outer core is perhaps some 10–20% less dense than that expected for iron or iron–nickel at core conditions. Later, Birch (1964) showed that the Earth's outer core is ~10% less dense than that expected for iron at the appropriate pressures and temperatures and proposed that it contained (in addition to liquid iron and nickel) a lighter alloying element or elements such as carbon, or hydrogen (Birch, 1952) or sulfur, silicon, or oxygen (Birch, 1964).

Uncertainties in estimates of the composition of the Earth's core derive from uncertainties in the core density (or bulk modulus, or bulk sound velocity) and in that of candidate materials (including pure liquid iron) when calculated for the temperatures and pressures of the outer core. Although there is an excellent agreement between the static compression data for ε-Fe (and Fe–Ni mixtures) at core pressures (Mao *et al.*, 1990) and isothermal-based Hugoniot data for ε-Fe (Brown and McQueen, 1986), extrapolation of these data to core conditions requires knowledge of the thermal contribution to their EOS (see Chapter 2.14). Boehler (2000) calculated an outer core density deficit of ~9% using these data coupled with an assumed value for the pressure dependence of α (thermal expansion coefficient) and outer core temperatures of 4,000–4,900 K. In a review of these and other data, Anderson and Isaak (2002) concluded that the core density deficit is ~5% (with a range from 3% to 7%, given uncertainties) and argued that the density deficit is not as high as the often-cited ~10%. Their revised estimate is derived from a re-examination of EOS calculations with revised pressure and temperature derivatives for core materials at inner–outer core boundary conditions over a range of temperatures (4,800–7,500 K). This is a topic of much debate and a conservative estimate of the core density deficit is ~5–10%.

The solid inner core, which has a radius of 1,220 ± 10 km (Masters and Shearer, 1995), represents 5% of the core's mass and <5% of its volume. It is estimated to have a slightly lower density than solid iron and, thus, it too would have a small amount of a light element component (Jephcoat and Olson, 1987). Birch (1952) may have recognized this when he said that it is "a crystalline phase, mainly iron." Like the outer core, uncertainties in the amount of this light element component is a function of seismically derived density models for the inner core and identifying the appropriate temperature and pressure derivatives for the EOS of candidate materials. Hemley and Mao (2001) have provided an estimate of the density deficit of the inner core of 4–5%.

The presence of an iron core in the Earth is also reflected in the Earth's shape. The shape of the Earth is a function of its spin, mass distribution, and rotational flattening such that there is an equatorial bulge and flattening at the poles. The coefficient of the moment of inertia for the Earth is an expression that describes the distribution of mass within the planet with respect to its rotational axis. If the Earth was a compositionally homogenous planet having no density stratification, its coefficient of the moment of inertia would be $0.4Ma^2$, with M as the mass of the Earth and a as the equatorial radius. The equatorial bulge, combined with the precession of the equinoxes, fixes the coefficient of the moment of inertia for the Earth at $0.330Ma^2$ (Yoder, 1995) reflecting a marked concentration of mass at its center (see also Table 1).

Finally, studies of planets and their satellites show that internally generated magnetic fields do not require the existence of a metallic core, particularly given the diverse nature of planetary magnetic fields in the solar system (Stevenson, 2003). Alternatively, the 500+ years of global mapping of the Earth's magnetic field in time and space demonstrates the existence of the Earth's central magnetic core (Bloxham, 1995; Merrill *et al.*, 1996). The generation of this field in the core also requires the convection of a significant volume of iron (or similar electrically conducting material) as it creates a self-exciting dynamo (Buffett, 2000). In the Earth, as with the other terrestrial planets, iron is the most abundant element, by mass (Wänke and Dreibus, 1988). Its high solar abundance is the result of a highly stable nuclear configuration and processes of nucleosynthesis in stars.

2.15.3 CONSTRAINING THE COMPOSITION OF THE EARTH'S CORE

The major "core" issues in geochemistry include: (i) its composition (both inner and outer core), (ii) the nature and distribution of the light element, (iii) whether there are radioactive elements in the core, (iv) timing of core formation, and (v) what evidence exists for core–mantle exchange. The answers to some or all of these questions provide constraints on the conditions (e.g., P, T, f_{O_2}) under which the core formed.

2.15.3.1 Observations from Meteorites and Cosmochemistry

That the core is not solely an Fe–Ni alloy, but contains ~5–10% of a light mass element alloy, is about the extent of the compositional guidance that comes from geophysics. Less direct information on the makeup of the Earth is provided by studies of meteorites and samples of the silicate Earth. It is from these investigations that we develop models for the composition of the bulk Earth and primitive mantle (or the silicate Earth) and from these deduce the composition of the core.

The compositions of the planets in the solar system and those of chondritic meteorites provide a guide to the bulk Earth composition (see Chapter 2.01). However, the rich compositional diversity of these bodies presents a problem insofar as there is no single meteorite composition that can be used to characterize the Earth. The solar system is compositionally zoned; planets with lesser concentrations of volatile elements are closer to the Sun. Thus, as compared to Mercury and Jupiter, the Earth has an intermediate uncompressed density (roughly a proportional measure of metal to rock) and volatile element inventory, and is more depleted in volatile elements than CI-chondrites, the most primitive of all of the meteorites.

There is a wide range of meteorite types, which are readily divided into three main groups: the irons, the stony irons and the stones (see also Volume 1 of the Treatise). With this simple classification, we obtain our first insights into planetary differentiation. All stony irons and irons are differentiated meteorites. Most stony meteorites are chondrites, undifferentiated meteorites, although lesser amounts are achondrites, differentiated stony meteorites. The achondrites make up ~4% of all meteorites, and <5% of the stony meteorites. A planetary bulk composition is analogous to that of a chondrite, and the differentiated portions of a planet—the core, mantle, and crust—have compositional analogues in the irons, stony irons (for core–mantle boundary regions), and achondrites (for mantle and crust).

Among the chondrites there are three main classes: the carbonaceous, enstatite, and ordinary chondrites. One simple way of thinking about these three classes is in terms of their relative redox characteristics. First, the carbonaceous chondrites, some of which are rich in organic carbon, have more matrix and Ca–Al inclusions and are the most oxidized of the chondrites, with iron existing as an FeO component in silicates. Second, the enstatite chondrites are the most reduced, with most varieties containing native metals, especially iron. Finally, the ordinary chondrites, the most abundant meteorite type, have an intermediate oxidation state (see review chapters in Volume 1 and Palme (2001)). Due to chemical and isotopic similarities, some researchers have argued that the bulk Earth is analogous to enstatite chondrites (Javoy, 1995). In contrast, others believe that the formation of the Earth initially began from materials such as the enstatite chondrites with the later 20–40% of the planet's mass forming from more oxidized accreting materials like the carbonaceous chondrites (Wänke, 1987; Wänke and Dreibus, 1988). As of early 2000s, we do not have sufficient data to resolve this issue and at best we should treat the chondrites and all meteoritic materials as only a guide to understanding the Earth's composition.

A subclass of the carbonaceous chondrites that uniquely stands out among all others is the CI (or C1) carbonaceous chondrite. These chondrites possess the highest proportional abundances of the highly volatile and moderately volatile elements, are chondrule free, and they possess compositions that match that of the solar photosphere when compared on a silicon-based scale (see also Chapter 2.01). The photosphere is the top of the Sun's outer convection zone, which can be thought of as an analogue to the Sun's surface. The Sun's photospheric layer emits visible light and hence its composition can be measured spectroscopically. This, plus the fact that the Sun contains >99.9% of the solar system's mass, makes the compositional match with CI carbonaceous chondrites seem all that more significant.

For this review the Earth's composition will be considered to be more similar to carbonaceous chondrites and somewhat less like the high-iron end-members of the ordinary or enstatite chondrites, especially with regard to the most abundant elements (iron, oxygen, silicon, and magnesium) and their ratios. However, before reaching any firm conclusions about this assumption, we need to develop a compositional model for the Earth that can be compared with different chondritic compositions. To do this we need to: (i) classify the elements in terms of their properties in the nebula and the Earth and (2) establish the absolute abundances of the refractory and volatile elements in the mantle and bulk Earth.

2.15.3.2 Classification of the Elements

Elements can be classified according to their volatility in the solar nebular at a specific partial pressure (Larimer, 1988). This classification scheme identifies the major components (e.g., magnesium, iron, silicon, and nickel), which are intermediate between refractory and volatile, and then assigns the other, less abundant elements to groups based on volatility distinguishing refractory (condensation temperatures >1,250 K), moderately volatile (condensation temperatures <1,250 K and >600 K), and highly volatile (condensation temperatures <600 K) elements, depending on their sequence of condensation into mineral phases (metals, oxides, and silicates) from a cooling gas of solar composition (Larimer, 1988). In terms of accretionary models for chondrites and planetary bodies, it is often observed that a model assuming a 10^{-4} atm partial pressure best fits the available data (Larimer, 1988). Those with the highest condensation temperatures (>1,400 K) are the refractory elements (e.g., calcium, aluminum, titanium, zirconium, REE, molybdenum, and tungsten), which occur in all chondrites with similar relative abundances (i.e., chondritic ratios of Ca/Al, Al/Ti, Ti/Zr). Major component elements (aside from oxygen and the gases) are the most abundant elements in the solar system, including silicon, magnesium, and iron (as well as cobalt and nickel); these elements have condensation temperatures of ~1,250 K. Moderately volatile elements (e.g., chromium, lithium, sodium, potassium, rubidium, manganese, phosphorus, iron, tin, and zinc) have condensation temperatures of ~1,250–600 K (Palme *et al.*, 1988), whereas highly volatile elements (e.g., thallium, cadmium, bismuth, and lead) have condensation temperatures ~600–400 K. Below this temperature the gas-phase elements (carbon, hydrogen, and nitrogen) condense. The relative abundance ratios of the major components, moderately volatile and highly volatile elements all vary considerably between the different types of chondritic meteorites. Figure 3 illustrates the differing proportions of the major component elements in chondrite groups and the Earth, which together with oxygen make up some 90% of the material in the Earth and other terrestrial planets.

Elements can also be classified according to their chemical behavior based on empirical observations from meteorites and systems in the Earth; this leads to the following groups: lithophile, siderophile, chalcophile, or atmophile. The lithophile elements are ones that bond readily with oxygen and are concentrated in the silicate shell (crust and mantle) of the Earth. The siderophile elements readily bond with iron and are concentrated in the core. The chalcophile elements bond readily with sulfur and are distributed between the core and mantle, with a greater percentage of them likely to be in the core. Finally, the atmophile elements (e.g., hydrogen, carbon, nitrogen, oxygen, and noble gases) are gaseous and are concentrated in the atmosphere–hydrosphere system surrounding the planet. A combination of these two different classification schemes provides a better understanding of the relative behavior of the elements, particularly during accretion and large-scale planetary differentiation.

Developing a model for the composition of the Earth and its major reservoirs can be established in a four-step process. The first involves estimating the composition of the silicate Earth (or primitive mantle, which includes the crust plus mantle after core formation). The second step involves defining a volatility curve for the planet, based on the abundances of the moderately volatile and highly volatile lithophile elements in the silicate Earth, assuming that none have been sequestered into the core (i.e., they are truly lithophile). The third step entails calculating a bulk Earth composition using the planetary volatility curve established in step two, chemical data for chondrites, and

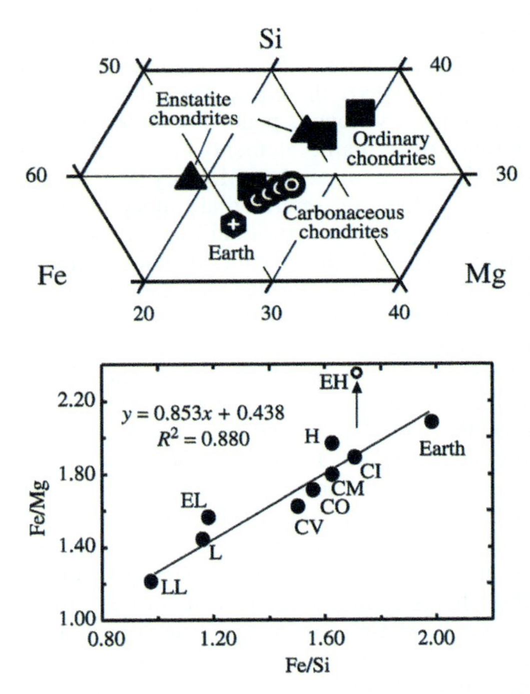

Figure 3 A ternary plot (upper) and binary ratio plot (lower) of the differing proportions (in wt.%) of Si, Fe, and Mg (three out of the four major elements) in chondrites and the Earth. These elements, together with oxygen, constitute >90% by mass of chondrites, the Earth, and other terrestrial planets. Data for the chondrites are from Wasson and Kellemeyn (1988) and for the Earth are from Table 2. The regression line is derived using only chondrites and does not include the EH data.

the first-order features of the planets in the solar system. Finally, a core composition is extracted by subtracting the mantle composition from the bulk planetary composition, revealing the abundances of the siderophile and chalcophile elements in the core. Steps three and four are transposable with different assumptions, with the base-level constraints being the compositions of meteorites and the silicate Earth and the solar system's overall trend in the volatile element abundances of planets outward from the Sun.

2.15.3.3 Compositional Model of the Primitive Mantle and the Bulk Earth

The silicate Earth describes the solid Earth minus the core. There is considerable agreement about the major, minor, and trace element abundances in the primitive mantle (Allegre *et al.*, 1995; McDonough and Sun, 1995; see Chapter 2.01). The relative abundances of the lithophile elements (e.g., calcium, aluminum, titanium, REE, lithium, sodium, rubidium, boron, fluorine, zinc, etc.) in the primitive mantle establish both the absolute abundances of refractory elements in the Earth and the planetary signature of the volatile element depletion pattern (Figure 4). The details of how these compositional models are developed can be found in Allegre *et al.* (1995); McDonough and Sun (1995), and Palme and O'Neill (Chapter 2.01). A model composition for the silicate Earth is given in Table 2, which is adapted from McDonough (2001); Palme and O'Neill (Chapter 2.01) present a similar model.

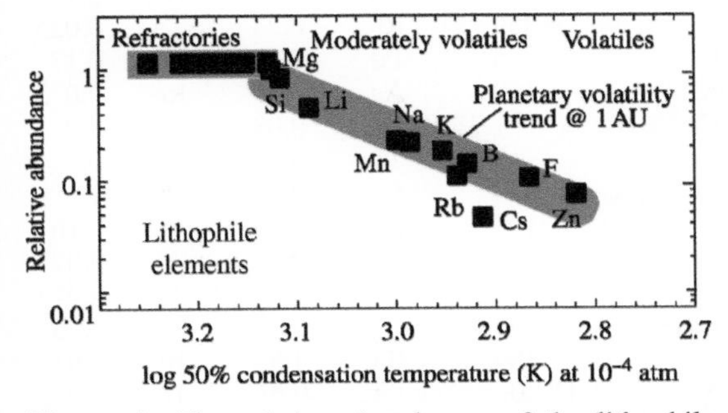

Figure 4 The relative abundances of the lithophile elements in the primitive mantle (or silicate Earth) plotted versus the log of the 50% condensation temperature (K) at 10^{-4} atm pressure. The relative abundances of the lithophile elements are reported as normalized to CI carbonaceous chondrite on an equal basis of Mg content. The planetary volatility trend (negative sloping shaded region enclosing the lower temperature elements) establishes integrated flux of volatile elements at 1 AU. Data for condensation temperatures are from Wasson (1985); chemical data for the chondrites are from Wasson and Kellemeyn (1988) and for the Earth are from Table 2.

A first-order assumption is that lithophile elements, inclusive of the refractory, moderately volatile, and highly volatile ones, are excluded from the core. The moderately volatile and highly volatile lithophiles are depleted relative to those in CI-chondrites. Together, the lithophiles describe a coherent depletion or volatility pattern. This negative correlation (Figure 4) thus establishes the planetary volatile curve at ~1 AU, which is an integrated signature of accreted nebular material in the coalescing region of the proto-Earth. By comparison, Mars has a less depleted abundance pattern (Wänke, 1981), whereas Mercury has a more depleted abundance pattern (BVSP, 1981). The most significant feature of this pattern is that potassium follows all of the other moderately volatile and highly volatile lithophiles. This observation demonstrates that the potassium budget of the silicate Earth is sufficient to describe that in the planet and argue against any sequestration of potassium into the core.

Data for the content of lithophile elements in the Earth plus knowledge of the iron content of the mantle and core together establish a bulk Earth compositional model (McDonough, 2001). This model assumes chondritic proportions of Fe/Ni in the Earth, given limited Fe/Ni variation in chondritic meteorites (see below). This approach yields

Table 2 The composition of the silicate Earth.

H	100	Zn	55	Pr	0.25
Li	1.6	Ga	4	Nd	1.25
Be	0.07	Ge	1.1	Sm	0.41
B	0.3	As	0.05	Eu	0.15
C	120	Se	0.075	Gd	0.54
N	2	Br	0.05	Tb	0.10
O (%)	44	Rb	0.6	Dy	0.67
F	15	Sr	20	Ho	0.15
Na (%)	0.27	Y	4.3	Er	0.44
Mg (%)	22.8	Zr	10.5	Tm	0.068
Al (%)	2.35	Nb	0.66	Yb	0.44
Si (%)	21	Mo	0.05	Lu	0.068
P	90	Ru	0.005	Hf	0.28
S	250	Rh	0.001	Ta	0.037
Cl	17	Pd	0.004	W	0.029
K	240	Ag	0.008	Re	0.0003
Ca (%)	2.53	Cd	0.04	Os	0.003
Sc	16	In	0.01	Ir	0.003
Ti	1,200	Sn	0.13	Pt	0.007
V	82	Sb	0.006	Au	0.001
Cr	2,625	Te	0.012	Hg	0.01
Mn	1,045	I	0.01	Tl	0.004
Fe (%)	6.26	Cs	0.021	Pb	0.15
Co	105	Ba	6.6	Bi	0.003
Ni	1,960	La	0.65	Th	0.08
Cu	30	Ce	1.68	U	0.02

Concentrations are given in $\mu g\ g^{-1}$ (ppm), unless stated as "%," which are given in wt.%.

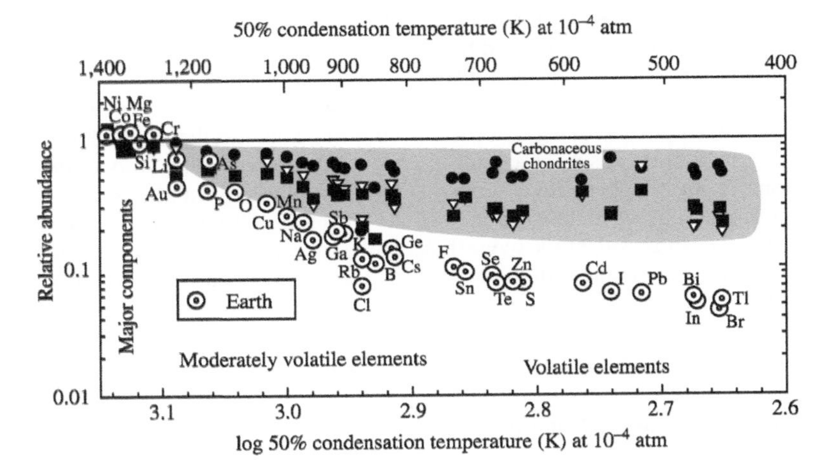

Figure 5 The relative abundances of the elements in the Earth and various carbonaceous chondrites plotted versus the log of the 50% condensation temperature (K) at 10^{-4} atm pressure. Data are normalized to CI carbonaceous chondrite on an equal basis of Mg content. The overall volatility trend for the Earth is comparable to that seen in these chondrites. The carbonaceous chondrites include CM (filled circles), CV (filled squares), and CO (open triangles) and define the shaded region. Data for condensation temperatures are from Wasson (1985); chemical data for the chondrites are from Wasson and Kellemeyn (1988) and for the Earth are from Table 3.

an Fe/Al of 20 ± 2 for the Earth. Aluminum, a refractory lithophile element, is considered the least likely of the lithophile elements (e.g., silicon, magnesium, and calcium) to be incorporated in the core. Thus, an aluminum content for the mantle translates directly into the aluminum content for the bulk Earth. This tightly constrained Fe/Al value also provides a first-order compositional estimate of the planet that requires no knowledge of light elements in the core.

Chondritic meteorites display a range of Fe/Al ratios, with many having a value close to 20 (Allegre *et al.*, 1995), although high Fe/Al values (35) are found in the iron-rich (EH) enstatite chondrites (Wasson and Kallemeyn, 1988). Combining these data and extending the depletion pattern for the abundances of nonrefractory, nonlithophile elements provides a model composition for the bulk Earth (Figure 5). A model composition for the bulk Earth is given in Table 3, which is adapted from McDonough (2001); Palme and O'Neill (Chapter 2.01) present a similar model. In terms of major elements this Earth model is iron and magnesium rich and coincident with the Fe/Mg–Fe/Si compositional trend established by chondrites (Figure 3). The Earth's volatility trend is comparable, albeit more depleted, than that of other carbonaceous chondrites (data in gray field in Figure 5).

Table 3 The composition of the bulk Earth.

H	260	Zn	40	Pr	0.17
Li	1.1	Ga	3	Nd	0.84
Be	0.05	Ge	7	Sm	0.27
B	0.2	As	1.7	Eu	0.10
C	730	Se	2.7	Gd	0.37
N	25	Br	0.3	Tb	0.067
O (%)	29.7	Rb	0.4	Dy	0.46
F	10	Sr	13	Ho	0.10
Na (%)	0.18	Y	2.9	Er	0.30
Mg (%)	15.4	Zr	7.1	Tm	0.046
Al (%)	1.59	Nb	0.44	Yb	0.30
Si (%)	16.1	Mo	1.7	Lu	0.046
P	715	Ru	1.3	Hf	0.19
S	6,350	Rh	0.24	Ta	0.025
Cl	76	Pd	1	W	0.17
K	160	Ag	0.05	Re	0.075
Ca (%)	1.71	Cd	0.08	Os	0.9
Sc	10.9	In	0.007	Ir	0.9
Ti	810	Sn	0.25	Pt	1.9
V	105	Sb	0.05	Au	0.16
Cr	4,700	Te	0.3	Hg	0.02
Mn	800	I	0.05	Tl	0.012
Fe (%)	32.0	Cs	0.035	Pb	0.23
Co	880	Ba	4.5	Bi	0.01
Ni	18,200	La	0.44	Th	0.055
Cu	60	Ce	1.13	U	0.015

Concentrations are given in $\mu g\ g^{-1}$ (ppm), unless stated as "%," which are given in wt.%.

2.15.4 A COMPOSITIONAL MODEL FOR THE CORE

As stated earlier, the Earth's core is dominantly composed of a metallic Fe–Ni mixture. This fact is well established by seismic data (*P*-wave velocity, bulk modulus, and density), geodynamo observations (the need for it to be reasonably good electrical conductor), and cosmochemical constraints. This then requires that the core, an iron- and nickel-rich reservoir, chemically balances the silicate Earth to make up a primitive, chondritic planet. Many iron meteorites, which are mixtures of iron and nickel in various proportions, are pieces of former asteroidal cores. These meteorites provide insights into the compositions of smaller body cores, given they are products of low-pressure differentiation, whereas the Earth's core likely formed under markedly different conditions (see Chapter 2.10). Thus, the Earth's core superficially resembles an iron meteorite; however, such comparisons are only first-order matches and in detail we should anticipate significant differences given contrasting processes involved in their formation.

2.15.4.1 Major and Minor Elements

A compositional model for the primitive mantle and bulk Earth is described above, which indirectly prescribes a core composition, although it does not identify the proportion of siderophile and chalcophile elements in the core and mantle. The mantle abundance pattern for the lithophile elements shown in Figure 4 provides a reference state for reviewing the abundances of the siderophile and chalcophile elements in the silicate Earth, which are shown in Figure 6. All of the siderophile (except gallium) and chalcophile elements plot below the shaded band that defines the abundance pattern for the lithophile elements. That these nonlithophile elements fall below this band (i.e., the planetary volatility trend) indicates that they are depleted in the mantle, and therefore the remaining planetary complement of these elements are in the core. The relative effects of core subtraction are illustrated in both panels with light-gray arrows, extending downward from the planetary volatility trend. The displacement length below the volatility trend (or length of the downward-pointing arrow) reflects the element's bulk distribution coefficient between core and mantle (e.g., bulk $D^{\text{metal/silicate}}$ for Mo > P ≈ Sb).

By combining the information derived from Figures 4–6, one can construct a compositional model for the Earth's core (Table 4), which is adapted from McDonough (1999). A first-order comparison of the composition of the bulk Earth, silicate Earth, and core in terms of weight percent and atomic proportion is presented in Table 5.

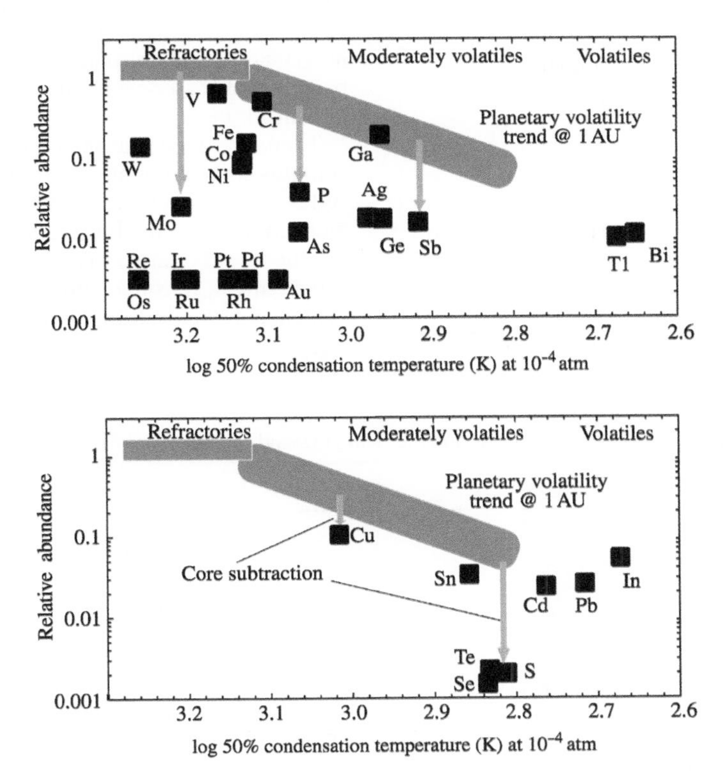

Figure 6 The relative abundances of the siderophile elements (upper panel) and chalcophile elements (lower panel) in the primitive mantle (or silicate Earth) plotted versus the log of the 50% condensation temperature (K) at 10^{-4} atm pressure. Data are normalized to CI carbonaceous chondrite on an equal basis of Mg content. The gray shaded region illustrates the relative abundances of the lithophile elements as reported in Figure 4. The light gray, downward pointing arrows reflect the element's bulk distribution coefficient between core and mantle during core formation; the longer the length of the arrow, the greater the bulk D (data sources are as in Figure 4).

The compositional model for the core has a light element composition that seeks to fit the density requirements for the outer core and is consistent with cosmochemical constraints. Significantly, along with iron and nickel the core contains most of the planet's sulfur, phosphorus, and carbon budget. Finally, this model composition is notable in that it is devoid of radioactive elements. The discussion that follows reviews the issues associated with compositional models for the core.

Table 4 The composition of the Earth's core.

H	600	Zn	0	Pr	0
Li	0	Ga	0	Nd	0
Be	0	Ge	20	Sm	0
B	0	As	5	Eu	0
C (%)	0.20	Se	8	Gd	0
N	75	Br	0.7	Tb	0
O (%)	0	Rb	0	Dy	0
F	0	Sr	0	Ho	0
Na (%)	0	Y	0	Er	0
Mg (%)	0	Zr	0	Tm	0
Al (%)	0	Nb	0	Yb	0
Si (%)	6.0	Mo	5	Lu	0
P (%)	0.20	Ru	4	Hf	0
S (%)	1.90	Rh	0.74	Ta	0
Cl	200	Pd	3.1	W	0.47
K	0	Ag	0.15	Re	0.23
Ca (%)	0	Cd	0.15	Os	2.8
Sc	0	In	0	Ir	2.6
Ti	0	Sn	0.5	Pt	5.7
V	150	Sb	0.13	Au	0.5
Cr (%)	0.90	Te	0.85	Hg	0.05
Mn	300	I	0.13	Tl	0.03
Fe (%)	85.5	Cs	0.065	Pb	0.4
Co	0.25	Ba	0	Bi	0.03
Ni (%)	5.20	La	0	Th	0
Cu	125	Ce	0	U	0

Concentrations are given in $\mu g\ g^{-1}$ (ppm), unless stated as "%," which are given in wt.%.

2.15.4.2 The Light Element in the Core

Given constraints of an outer core density deficit of 5–10% and a host of candidate elements (e.g., hydrogen, carbon, oxygen, silicon, and sulfur), we need to evaluate the relative potential of these elements to explain core density deficit. Uniformly, the bolstering of one's view for these components in the core involve metallurgical or cosmochemical arguments, coupled with the identification of candidate minerals found in meteorites, particularly iron meteorites and reduced chondrites (the classic example being the high-iron (EH) enstatite chondrite).

Washington (1925), of the Carnegie Institution of Washington, developed a model for the chemical composition of the Earth based on the Wiechert structural model, the Oldham–Gutenburg revised core radius, and the newly derived Adams–Williamson relationship (Williamson and Adams, 1923) for determining the density profile of the planet. Washington's model for the core assumed an average density for the core of $\sim 10\ g\ cm^{-3}$ (cf. $\sim 11.5\ g\ cm^{-3}$ for today's models), and a "considerable amount, up to $\sim$5% or so, of phosphides (schreibersite, $(Fe,Ni)_3P$), carbides (cohenite, Fe_3C), sulfides (troilite, FeS) and carbon (diamond and graphite)." This amazing and insightful model, which is now ~80 yr old, provides us with a good point from which to consider the light element component in the core.

There are good reasons to assume that the core contains some amount of carbon, phosphorus, and sulfur. These three elements are among the 12 most common in the Earth that account for >99% of the total mass (Table 5), as based on geochemical, cosmochemical, and meteoritical evidence. Seven out of 12 of these elements (not including carbon, phosphorus, and sulfur) are either refractory or major component elements,

Table 5 The composition of the bulk Earth, mantle, and core and atomic proportions for abundant elements.

wt.%	*Earth*	*Mantle*	*Core*	*Atomic prop.*	*Earth*	*Mantle*	*Core*
Fe	32.0	6.26	85.5	Fe	0.490	0.024	0.768
O	29.7	44	0	O	0.483	0.581	0.000
Si	16.1	21	6	Si	0.149	0.158	0.107
Mg	15.4	22.8	0	Mg	0.165	0.198	0.000
Ni	1.82	0.20	5.2	Ni	0.008	0.001	0.044
Ca	1.71	2.53	0	Ca	0.011	0.013	0.000
Al	1.59	2.35	0	Al	0.015	0.018	0.000
S	0.64	0.03	1.9	S	0.005	0.000	0.030
Cr	0.47	0.26	0.9	Cr	0.002	0.001	0.009
Na	0.18	0.27	0	Na	0.002	0.002	0.000
P	0.07	0.009	0.20	P	0.001	0.000	0.003
Mn	0.08	0.10	0.03	Mn	0.000	0.000	0.000
C	0.07	0.01	0.20	C	0.002	0.000	0.008
H	0.03	0.01	0.06	H	0.007	0.002	0.030
Total	99.88	99.83	99.97	Total	1.000	1.000	1.000

and so their abundances in the Earth are relatively fixed for all planetary models (see also Figure 2). The remaining five elements are sodium, chromium, carbon, phosphorus, and sulfur (Table 5); all of these are highly volatile to moderately volatile and estimates of their abundances in the bulk Earth and core are established from cosmochemical constraints. A significant question concerning the abundance of carbon, phosphorus, and sulfur in the core, however, is whether their incorporation into the core can account for the density discrepancy?

The planetary volatility trend illustrated in Figures 4 and 5 does not extend out to the lowest temperature components, including the ices and gases (e.g., hydrogen, carbon, nitrogen, oxygen, and the noble gases). Estimates for the Earth's content of these components (Figure 7) are from McDonough and Sun (1995) and McDonough (1999, 2001) and are based on data for the Earth's mantle and a comparison of carbonaceous chondrite data. Figure 7 provides a comparison of the Earth's estimate of these elements relative to the data for chondrites; the estimate for the Earth comes from an extrapolation of the trend shown in Figure 5. Although these extrapolations can only provide an approximate estimate, the abundance of carbon in the Earth is suggested to be of the order <0.1 wt.%. This estimate translates to a core having only ~0.2 wt.% carbon (Tables 3 and 4). By comparison Wood (1993) estimated a factor of 10–20 times more carbon in the core. Wood's estimate seems most unlikely insofar as it is inconsistent with data for meteorites, which are not markedly enriched in highly volatile elements (Figure 7).

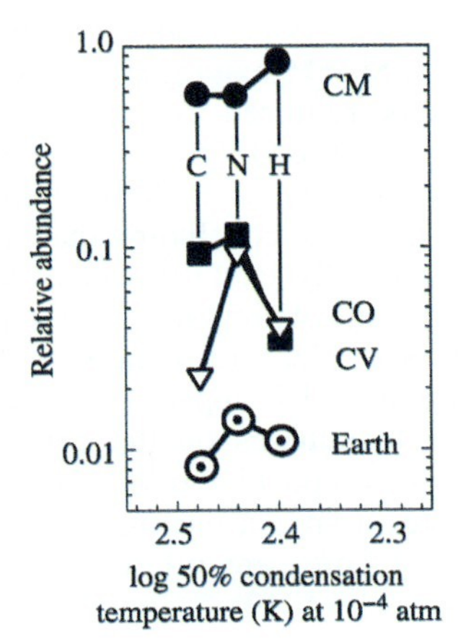

Figure 7 The relative abundances of C, N, and H in the Earth plotted versus the log of the 50% condensation temperature (K) at 10^{-4} atm pressure. The Earth's estimate is based on compositional estimates of these gases in the mantle and the Earth's surface, as well as by comparison with data for carbonaceous chondrites (data sources are as in Figure 5).

This view is untenable when compared with data trends in Figure 5 for the Earth and the carbonaceous chondrites. It is also noted that the Earth's budget for hydrogen and nitrogen are such that the core would likely contain a minor amount of these elements. The consequences of having hydrogen in the core are significant and have been reviewed by Williams and Hemley (2001).

There is ~90 ppm of phosphorus in the silicate Earth (McDonough *et al.*, 1985), and the bulk Earth is estimated to have ~0.1 wt.% phosphorus. Using the relationships in Figure 6 the core is thus estimated to have ~0.20 wt.% phosphorus (Table 4). Thus, 90% of the planet's inventory of phosphorus is in the core (Table 6) and the core's metal/silicate phosphorus enrichment factor is ~22. Similarly, the core hosts ~90% of the planet's carbon budget, and has a metal/silicate enrichment factor only slightly lower at ~17.

The sulfur content of the core is said to be ~1.5–2 wt.% (McDonough and Sun, 1995; Dreibus and Palme, 1996). This number is based on calculating the degree of sulfur depletion in the silicate Earth relative to the volatility trend (Figure 6). Figure 8 illustrates the problem with suggesting that the core contains 10% sulfur, which is commonly invoked as the light element required to compensate for the density deficit in the outer core. Accordingly, the total sulfur, carbon, and phosphorus content of the core constitute only a minor fraction (~2.5 wt%) and this mixture of light elements cannot account for the core's density discrepancy. Thus, it is likely that there is another, more abundant, light element in the core in addition to these other components.

A model core composition has been constructed using silicon as the other light element in the outer core, which is also consistent with evidence for

Table 6 The metal/silicate enrichment factor and the proportion of element in the core relative to the planet.

Elements	*Metal/silicate enrichment factor*	*% of planetary inventory in the core*
Re, PGE	>800	98
Au	~500	98
S, Se, Te, Mo, As	~100	96
N	~40	97
Ni, Co, Sb, P	~25	93
Ag, Ge, C, W	~17	91
Fe (%)	~14	87
Cl, Br, and I	10–15	85
Bi and Tl	~10	80
H and Hg	~6	70
Cu, Sn, Cd, Cr	3–4	60–65
Cs and Pb	~3	55–60
V	~2	50
Si and Mn	0.3	~10

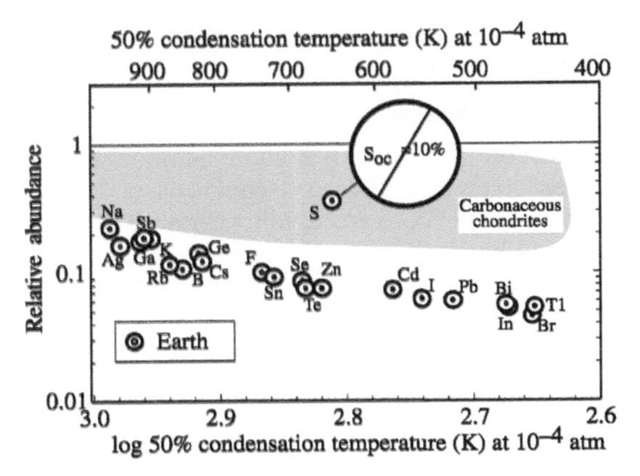

Figure 8 An illustration showing where S would plot if the core contained 10 wt.% sulfur so to account for the core's density discrepancy (see text for further discussion). The relative abundances of the elements in the Earth are plotted versus the log of the 50% condensation temperature (K) at 10^{-4} atm pressure. Data are normalized to CI carbonaceous chondrite on an equal basis of Mg content. The overall volatility trend for the Earth is comparable to that seen in these chondrites. The shaded region for the carbonaceous chondrites is the same as in Figure 5 (data sources are as in Figure 5).

core formation at high pressures (e.g., 20–30 GPa; see Chapter 2.14). This model is at best tentative, although comparisons of Mg/Si and Fe/Si in the Earth and chondrites (Figure 2) show that it is permissible. Silicon is known to have siderophilic behavior under highly reducing conditions and is found as a metal in some enstatite chondrites. A number of earlier models have suggested silicon as the dominant light element in core (Macdonald and Knopoff, 1958; Ringwood, 1959; Wänke, 1987; O'Neill, 1991b; Allegre *et al.*, 1995; O'Neill and Palme, 1997). The estimate for silicon in the core is based on the volatility curve for lithophile elements in the Earth (Figure 4).

An alternative case can be made for oxygen as the predominant light element in the core. On the grounds of availability, oxygen is a good candidate; it is the second most abundant element in the Earth and only a few percent might be needed to account for the core's density discrepancy. However, O'Neill *et al.* (1998) point out that oxygen solubility in iron liquids increases with temperature but decreases with pressure and thus showed that only ~2% or less oxygen could be dissolved into a core forming melt. The planetary volatility trend provides no guidance to the core's oxygen abundance. A plot of the log 50% condensation temperature versus element abundance (i.e., Figure 5) does not consider oxygen, because its 50% condensation temperature is not considered in systems where it is the dominant element in rocks and water ice. The core and Earth model composition, assuming oxygen as the light element in the core, is presented in Table 7, along with that for the silicon-based model. Both model compositions attempt to fit the density requirements for the outer core by assuming a mean atomic number of ~23, following Birch (1966). In terms of the light-element-alloy component in the core, this results in ~9% (by weight) for the silicon-based model and ~6% (by weight) for the oxygen-based model (Table 7).

Table 7 Compositional comparison of two models for the Earth and core.

wt.%	*Si-bearing*		*O-bearing*	
	Earth	*Core*	*Earth*	*Core*
Fe	32.0	85.5	32.9	88.3
O	29.7	0	30.7	3
Si	16.1	6	14.2	0
Ni	1.82	5.2	1.87	5.4
S	0.64	1.9	0.64	1.9
Cr	0.47	0.9	0.47	0.9
P	0.07	0.20	0.07	0.20
C	0.07	0.20	0.07	0.20
H	0.03	0.06	0.03	0.06
Mean atomic #		23.5		23.2
Atomic proportions				
Fe		0.768		0.783
O		0.000		0.093
Si		0.107		0.000
Ni		0.044		0.045
S		0.030		0.029
Cr		0.009		0.009
P		0.003		0.003
C		0.008		0.008
H		0.030		0.029
Total		1.000		1.000

Less attractive models that consider complex mixtures (e.g., Si–O mixture) are unlikely, given the conditions required for core formation. O'Neill *et al.* (1998), Hillgren *et al.* (2000), and Li and Fei (Chapter 2.14) have reviewed the literature on the topic concluding that silicon and oxygen are mutually exclusive in metallic iron liquids over a range of pressures and temperatures. Until there is a clear resolution as to which compositional model is superior, we must entertain multiple hypotheses on the core's composition. The two compositional models for the core presented here (a silicon-bearing core versus an oxygen-bearing core) are offered as competing hypotheses.

2.15.4.3 Trace Elements in the Core

The abundance of trace siderophile elements in the bulk Earth (and that for the core) may be constrained by examining their abundance ratios

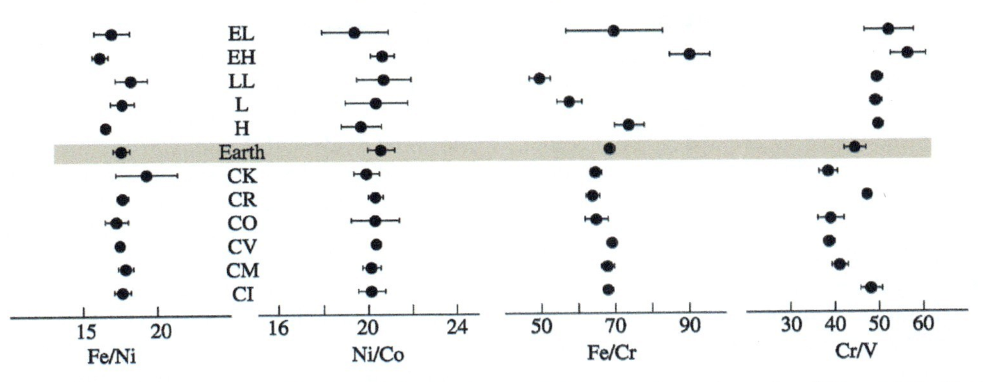

Figure 9 A plot of the variation in Fe/Ni, Ni/Co, Fe/Cr, and Cr/V values in various chondrites and the Earth. The different groups of carbonaceous chondrites include CI, CM, CV, CO, CR, and CK; the ordinary chondrites include H-, L-, and LL-types, and the enstatite chondrites include EH- and EL-types. The error bars represent the 1 SD of the data population. The Earth's composition is shown in the shaded bar and data are from Table 3 (including data are from various papers of Wasson and Kellemeyn cited in Wasson and Kellemeyn (1988)).

in chondrites. Figure 9 presents data for various groups of chondrites, which show limited variation for Fe/Ni (17.5) and Ni/Co (20), and slightly more variation for Fe/Cr (67) and Cr/V (45). Using the iron content for the core and the silicate Earth abundances for iron, nickel, cobalt, chromium, and vanadium, the Earth's core composition is established by assuming chondritic ratios of the elements for the planet. Following similar lines of reasoning for siderophile and chalcophile elements, the trace element composition of the core is also determined (Tables 4 and 6).

Based on these results, the core appears to be rich in chromium and vanadium (i.e., 50–60% of the planet's budget for these elements, Table 6), with a minor amount of manganese in the core (~10% of the planet's budget). Discussions relating to incorporation of chromium into the core usually also involve that for manganese and vanadium, because the partitioning behavior of these three elements during core formation may have been similar (Ringwood, 1966; Dreibus and Wänke, 1979; Drake *et al.*, 1989; Ringwood *et al.*, 1990; O'Neill, 1991a; Gessmann and Rubie, 2000). However the model presented here does not take into account element partitioning behavior during core formation, it is solely based on the planetary volatility trend and a model composition for the silicate Earth.

The minor amount of manganese in the core reflects the volatility model assumed for this element. O'Neill and Palme (1997) argue that manganese and sodium have similar volatilities based on the limited variation in Mn/Na ratios in chondrites (see also Chapter 2.01). However, a plot of Na/Ti versus Mn/Na in chondrites (Figure 10) shows that indeed Mn/Na varies as a function of volatility; this illustration monitors volatility by comparing titanium, a refractory lithophile element, with sodium, a moderately

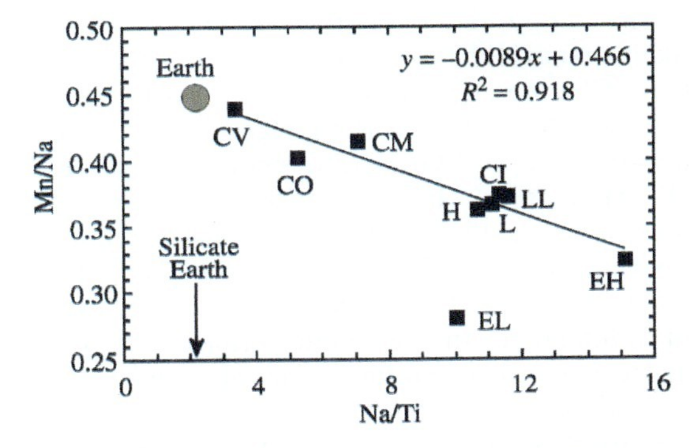

Figure 10 A plot of the variation in Na/Ti versus Mn/Na ratios in chondritic meterorites and the Earth. Data for chondrites are from Wasson and Kellemeyn (1988). The value for the Na/Ti ratio of the silicate Earth is indicated with an arrow (data from McDonough and Sun, 1995). The regression line, R^2 value, and the coefficients for the line equation are derived from the data for chondrites, not including the low-Fe enstatite chondrite. This regression and the Na/Ti ratio of the silicate Earth together provide a method to estimate the Mn/Na ratio for the Earth and indicate that the core is likely to contain a small fraction of the Earth's Mn budget.

lithophile element. Therefore, given the planetary volatility trend (Figure 4) and a reasonably well-constrained value for Na/Ti in the silicate Earth (McDonough and Sun, 1995), one estimates a planetary Mn/Na value of 0.45 for the Earth, implying that the core hosts ~10% of the planet's manganese budget.

The behavior of gallium, a widely recognized siderophile element, during core formation appears to be the most anomalous; this is most clearly illustrated by noting that gallium plots directly on the planetary volatility trend (Figure 6, top panel), indicating its undepleted character in the mantle. This result implies that there is little to no gallium in the Earth's core, which is a most unexpected result. The silicate Earth's gallium content of 4 ppm is well established and there is little uncertainty to this number (McDonough, 1990). In the silicate Earth gallium follows aluminum (these elements are above one another on the periodic table) during magma generation, as well as during the weathering of rocks, with overall limited and systematic variations in Al/Ga values in rocks. That gallium plots within the field defined by the moderately volatile and highly volatile lithophile elements (Figure 6) suggests that either the assumed temperature at 50% condensation is incorrect (unlikely given a wide spectrum of supporting meteorite data), or gallium behaves solely as a lithophile element during core formation. If the latter is true, then determining under what conditions gallium becomes wholly lithophile provides an important constraint on core formation.

The composition of the Earth's core, which was likely established at relatively high pressures (~20 GPa; see Chapters 2.10 and 2.14), can be compared with that of iron meteorites, which are low-pressure (<1 GPa) differentiates. Wasson's (1985) chemical classification of iron meteorites uses nickel, gallium, germanium, and iridium to divide them into 13 different groups. He shows that gallium is clearly a siderophile element found in abundance in the metal phases of iron meteorites. Also, gallium is highly depleted in achondrites. A comparison of the composition of the Earth's core with that of different iron meteorites is given in Figure 11. The Earth's core and some iron meteorites have comparable nickel, germanium, and iridium contents, albeit on the low end of the nickel spectrum. In contrast, the gallium content of the Earth's core (Figure 11) is substantially lower than that found in all iron meteorites, which may reflect the markedly different conditions under which core separation occurred in the Earth.

It has been suggested that there is niobium in the core (Wade and Wood, 2001). This suggestion is based on the observation that niobium is siderophile under reducing conditions (it is not uncommon to find niobium in steels) and if core extraction were sufficiently reducing, then some niobium would have been sequestered into the core. In addition, Wade and Wood (2001) observed that the partitioning data for niobium mimicked that for chromium and vanadium. Given the distribution of chromium and vanadium between the core and mantle, it is expected that a considerable portion of the Earth's niobium budget is hosted in the core. The Wade and Wood model was, in part, developed in response to the observations of McDonough (1991) and Rudnick *et al.* (2000), who reported that niobium and tantalum are depleted in the upper mantle and crust and that both reservoirs have low Nb/Ta values relative to chondrites. These observations lead to the suggestion that refractory components of subducting oceanic crust would contain the complementary niobium- and tantalum-enriched reservoir of the silicate Earth (McDonough, 1991; Rudnick *et al.*, 2000). However, Wade and Wood (2001) proposed an alternative model in which niobium, but not tantalum, is extracted into the core. To address this issue it is useful to examine the relative abundances of Nb–Ta–La in the crust–mantle system, because this triplet may characterize silicate Earth processes and reservoirs.

The range of Nb/Ta and La/Ta values in the continental crust and depleted mantle (MORB source) are given in Figure 12. This illustration

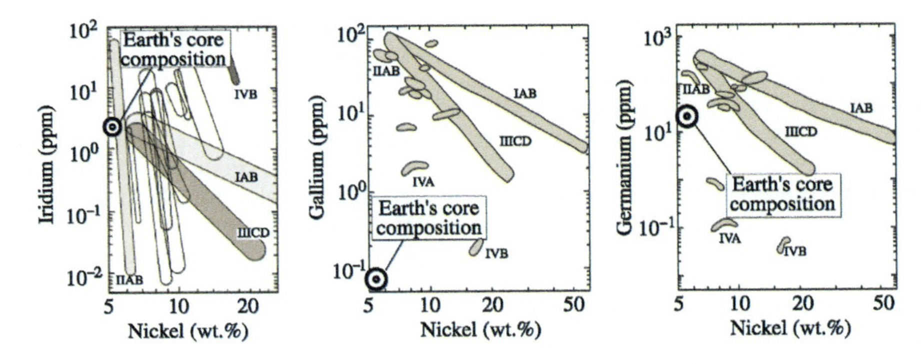

Figure 11 A plot of the variation in Ni versus Ir, Ni versus Ga, and Ni versus Ge in iron meteorites and the Earth. Data for the iron meteorites are adapted from the work of Wasson (1985). Data for the Earth's core are from Table 4. The plot of Ni–Ir shows that the composition of the Earth's core is comparable to that of various iron meteorites, whereas the Earth's core appears to have a slightly lower Ge content and a markedly lower Ga; the latter being unlike anything seen in iron meteorites. These four elements are the ones that are used to define the chemical classification of iron meteorites (reproduced by permission of W. H. Freeman from *Meteorites, Their Record of Early Solar-system History*, **1985**, p. 41, 42).

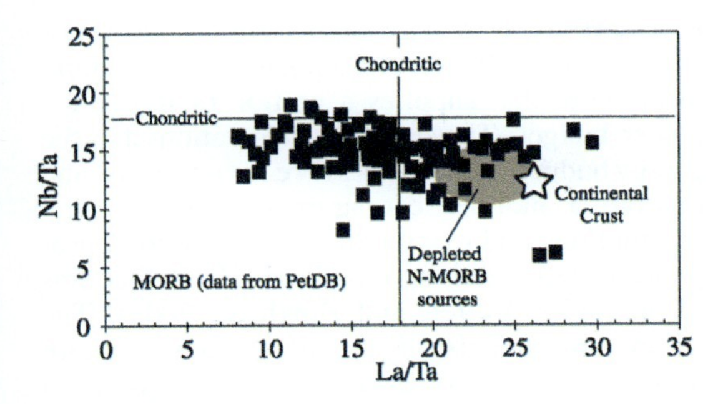

Figure 12 A plot of the Nb/Ta and La/Ta variation in MORB and the continental crust. The continental crust, MORBs, and their source regions all plot below the chondritic Nb/Ta value. Likewise, the continental crust plots and depleted MORB source regions are strongly depleted in Ta relative to La. See text for further details (data for MORB are from the PetDB resource on the web (http://petdb.ldeo.columbia.edu/petdb/); the estimate of the continental crust from Rudnick and Gao (Chapter 3.01).

shows that both of these major silicate reservoirs are clearly depleted in Nb/Ta relative to chondrites. In addition, both the continental crust and the depleted source regions of MORBs plot in the field that characterizes depletions of both niobium and tantalum relative to lanthanum, and tantalum relative to niobium. Niu and Batiza (1997) showed that during melting $D_{Nb} < D_{Ta} < D_{La}$ such that increasing melt extraction depletes the MORB source regions respectively and progressively in these elements so that they would plot in the same field as continental crust (Figure 12). This demonstrates that both the production of continental and oceanic crusts result in the production of a crustal component that, when processed through a subduction zone filter, generates residues with high Nb/Ta and low La/Ta that remain in the mantle. Is the core another niobium-enriched reservoir in the Earth? As of early 2000s, this is an unresolved issue, but crucial tests of this hypothesis will be gained by further examination of silicate Earth samples, iron meteorites, and further tests from experimental petrology.

Finally, Table 4 lists the halides—chlorine, bromium, and iodine—in the core. McDonough and Sun (1995) noted the marked depletion of these elements in the silicate Earth and suggested that this effect is due possibly to their incorporation into the Earth's core, or that the region of the nebula at 1 AU was anomalously depleted in the halides. There are iron halides, some of which are found in chondrites. However, such halides in chondrites are believed to be decompositional products created during terrestrial weathering (Rubin, 1997).

2.15.5 RADIOACTIVE ELEMENTS IN THE CORE

Those that have suggested the presence of radioactive elements in the Earth's core have usually done so in order to offer an alterative explanation for the energy needed to run the geodynamo, and/or as a way to explain Earth's volatile elements inventory. Potassium is commonly invoked as being sequestered into the Earth's core due to: (i) potassium sulfide found in some meteorites; (ii) effects of high-pressure *s–d* electronic transitions; and/or (iii) solubility of potassium in Fe–S (and Fe–S–O) liquids at high pressure. Each of these is considered below and rejected.

The cosmochemical argument for potassium in the core is based on the presence of a potassium iron sulfide (djerfisherite) and sodium chromium sulfide (carswellsilverite) in enstatite chondrites, and the plausibility of these phases in core-forming liquids (Lodders, 1995). However, this hypothesis does not consider that enstatite chondrites also contain a myriad of other (and more abundant) sulfides, including niningerite ((Mg,Fe)S), titanium-bearing troilite (FeS), ferroan alabandite ((Mn,Fe)S), and oldhamite (CaS). These common, higher-temperature sulfide phases contain substantial concentrations of REE and other refractory lithophile elements (see review in Brearley and Jones (1998)). If these were incorporated into the Earth's core, the composition of the silicate Earth would be grossly changed on both an elemental and isotopic level. However, there is no evidence, even at the isotopic level (e.g., Sm/Nd and Lu/Hf systems), for REE depletion in the silicate Earth and, thus, it is unlikely that such sulfides were incorporated into the core. The mere identification of a potassium-bearing sulfide does not demonstrate the existence of potassium in the core; it simply allows for the possibility. Plausibility arguments need to be coupled with corroborating paragenetic evidence that is also free of negating geochemical consequences.

The *s–d* electronic transitions occur at higher pressures, particularly for larger alkali metal ions (e.g., caesium, rubidium, and potassium). Under high confining pressures the outer most *s*-orbital electron transforms to a d^1-orbital configuration, resulting in transition metal-like ions. This electronic transition changes the chemical characteristics of the ion making it more siderophilic and potentially allowing it to be sequestered into the core. It has been suggested that some amount of caesium (see Figure 4) may have been sequestered into the Earth's core via this mechanism (McDonough and Sun, 1995). However, data for rubidium and potassium show that this effect is unlikely to have taken place based on the depletion pattern for the moderately volatile lithophile elements.

The third argument for the presence of radioactivity in the core usually involves finding a condition ($P-T-X-f_{O_2}$) under which potassium is soluble (Hall and Rama Murthy, 1971; Lewis, 1971; Chabot and Drake, 1999). Gessmann and Wood (2002) demonstrated that potassium is soluble in Fe–S and Fe–S–O liquids at high pressure; these authors argued that potassium was sequestered into the core (see also Chapter 2.14). However, these experimental studies suffer from either examining only simple liquid systems (e.g., synthetic Fe–Ni–S), or overlooking the consequences of other minor and trace elements (e.g., Th, U, and REE). In the case of the Gessmann and Wood (2002) study, calcium is also incorporated into the metallic liquid and the consequences of this are that it creates even more problems. For example, the silicate Earth has a calcium content that is in chondritic proportions to other refractory lithophile elements (e.g., Ca/Al, Ca/Sc, Ca/Yb), demonstrating that there was neither calcium nor potassium incorporated into the core. Similar arguments can also be made for uranium and thorium, which are also based on ratios with other refractory lithophile elements. No experimental evidence exists that shows similar solubility for uranium and thorium (i.e., two elements with significantly different siderophilic behavior) in Fe–S and Fe–S–O liquids at high pressure that does not incorporate other refractory lithophile elements.

An Earth's core containing a significant amount of radioactive elements has been proposed by Herndon (1996). This model envisages a highly reduced composition for the whole Earth and, in particular, for the core. Unfortunately, Herndon has developed a core compositional model that is inconsistent with chemical and isotopic observations of the Earth's mantle and a chondritic planetary composition. Herndon's core contains significant quantities of calcium, magnesium, uranium, and other elements typically considered lithophile. Drawing upon analogies with enstatite chondrites (highly reduced meteorites), Herndon has suggested that these elements were extracted into a metal phase as sulfides (e.g., oldhamite and niningerite). However, these phases are known to grossly fractionate many lithophile elements from one another (Crozaz and Lundberg, 1995), which would lead to a mantle with significantly non-chondritic ratios of Sm/Nd, Lu/Hf, and Th/U (element pairs constrained by isotopic evidence), as well Ca/Ti, Ca/Al, Ti/Sc, and others (element pairs whose bulk mantle properties are well constrained to be chondritic in the mantle), and this is not observed (McDonough and Sun, 1995).

Finally, there is a question of the need for radioactive heating of the core to support the necessary energy budget. Some geophysicists have speculated that there is either potassium or uranium in the core that supplies a portion of the core's power budget. These model calculations for the energy budget in the Earth's core are nontrivial and involve a number of parameters, with many assumptions and extrapolations of data to appropriate core conditions (Gubbins, 1977; Gubbins *et al.*, 1979; Buffett and Bloxham, 2002). The competing models of the geodynamo require different amounts of energy to drive convection in the outer core, and the details of the various models are vastly different (Glatzmaier, 2002). Labrosse *et al.* (2001) proposed a model for the timing and rate of inner core solidification that requires radioactive heating. Likewise, Anderson (2002) examined the energy balance at the CMB and concluded that there is a need for some amount of radioactive heating in the core. In contrast, no radioactive heating is required in other models of the core's energy budget (Stacey, 1992), which is consistent with geochemical evidence for its general absence in the core.

2.15.6 TIMING OF CORE FORMATION

Defining the age and duration of core formation depends on having an isotope system in which the parent–daughter isotope pairs are fractionated by core subtraction over a time interval within the functional period of the system's half-life. Fortunately, analytical advances in the W–Hf isotope system provide us with a tool to gauge the timing of core formation (Kleine *et al.*, 2002; Yin *et al.*, 2002).

The W–Hf isotope system involves the decay of ^{182}Hf to ^{182}W with a half-life of 9 Myr (thus the system became extinct within the first 100 Myr of Earth history). Both hafnium and tungsten are refractory elements (lithophile and siderophile, respectively) and thus their relative concentration in the Earth is set at chondritic. Some 90% of the Earth's budget of tungsten is hosted in the core (Table 6), whereas all of the planet's hafnium is hosted in the silicate Earth. Early studies found that iron meteorites have lower ^{182}W/^{184}W isotopic compositions (by about some $4\varepsilon_{^{182}W}$ units, where ε units express difference in parts per 10,000) than the Earth (Lee and Halliday, 1995; Harper and Jacobsen, 1996; Horan *et al.* (1998)); the Lee and Halliday (1995) study found no difference between the Earth and chondrites for their tungsten isotopic compositions. The findings of Lee and Halliday (1995), however, have been challenged by Yin *et al.* (2002), Kleine *et al.* (2002), and Schoenberg *et al.* (2002), who found that the Earth's tungsten isotopic composition is some $2\varepsilon_{^{182}W}$ units higher than that of chondrites. (These studies measured some of the same chondrites as reported in the Lee and Halliday (1995) study and were able to resolve the compositional differences between the Earth and chondrites.) This difference means that core separation was very early, and happened prior to the effective decay of the ^{182}Hf system such that the tungsten remaining in the silicate Earth became enriched in ^{182}W relative to that in the core. These studies demonstrate that much of the core's separation must have been completed by ~30 Ma after t_0 (4.56 Ga) in order to explain the Earth's higher $\varepsilon_{^{182}W}$ signature (Figure 13). There are possible scenarios in which one could argue for significantly shorter, but not longer time interval for core formation (Kleine *et al.*, 2002; Yin *et al.*, 2002). By implication the core must have an $\varepsilon_{^{182}W}$ of about −2.2 compared to the zero value for the silicate Earth.

The U–Pb, Tc–Ru (^{98}Tc has a half-life between 4 Ma and 10 Ma), and Pd–Ag (^{107}Pd has a half-life of 9.4 Ma) isotope systems have also been examined in terms of providing further insights into the timing of core separation. Overall, the results from these systems are definitive, but not very instructive. The extinct systems of Tc–Ru and Ag–Pd have parent and daughter isotopes that are siderophilic and so were strongly partitioned into the core during its formation. The absence of isotopic anomalies in these systems indicates that core separation left no signature on the silicate Earth. The extant U–Pb system (^{235}U with a half-life of 0.7 Ga) has also been examined with respect to the incorporation of lead into the core, with the result being that core separation must have happened within the first 100 Myr of the Earth's formation in order to reconcile the lead isotopic evolution of the silicate Earth (see review of Galer and Goldstein (1996)).

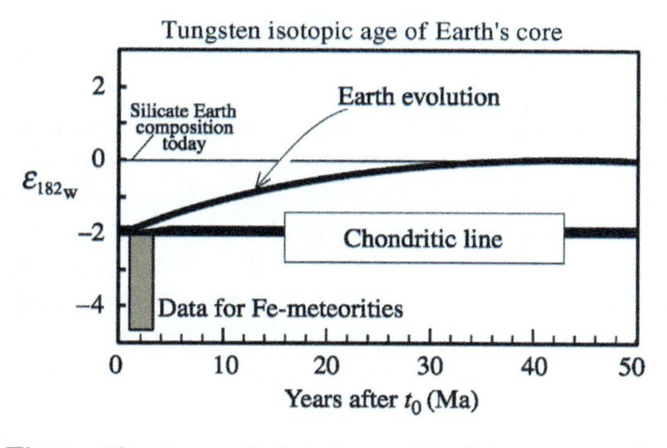

Figure 13 An evolutionary model of time versus the $\varepsilon_{^{182}W}$ composition of the silicate Earth for the first 50 of Earth's history. The higher $\varepsilon_{^{182}W}$ composition of the Earth relative to chondrites can only be balanced by a complementary lower than chondrites reservoir in the core. Extraction age models for the core are a function of the decay constant, the difference between the silicate Earth and chondrites, the proportion of W and Hf in the mantle and core and the rate of mass extraction to the core. Details of these models are given in the above citations, with the upper limit of the age curves shown here (sources Yin *et al.*, 2002; Kleine *et al.*, 2002; Schoenberg *et al.*, 2002).

2.15.7 NATURE OF CORE FORMATION

Core formation is not a well-understood process. Constraints for this process come from pinning down the timing of the event, characterizing its bulk chemical properties, and establishing a bulk Earth compositional model. The W–Hf isotope studies dictate that core formation happened early and was virtually completed within 30 Ma of solar system formation. The findings of Li and Agee (1996) and related studies (see Chapter 2.14) demonstrate that the integrated pressure and temperature of core formation was accomplished at mid- to upper-mantle conditions, not in predifferentiated planetismals. This finding, however, does not preclude the accretion of predifferentiated planetismals; it simply requires that these additions were rehomogenized back into the larger and still evolving Earth system. Finally, the nickel content of the silicate Earth places significant restrictions on oxidation potential of the mantle during core formation. These findings have led to the competing hypotheses of homogeneous and heterogeneous planetary accretion (Wänke, 1981; Jones and Drake, 1986). The former envisages the composition of accreting materials to remain constant throughout Earth's growth history, whereas heterogeneous accretion models postulate that there was a significant compositional shift during the latter stages of the Earth's growth history. These models were developed in order to account for the observed chemical features of the mantle.

The homogeneous accretion model requires a fairly restricted set of conditions to attain the silicate Earth composition observed today (Jones and Drake, 1986). Continued support for this model is wanning given its failure to reconcile a number of rigorous chemical and isotopic constraints (see reviews of O'Neill, 1991b; O'Neill and Palme, 1997; and Palme and O'Neil (Chapter 2.01)). For example, it is well established from osmium isotope studies (Meisel *et al.*, 1996; Walker *et al.*, 1997, 2002) that the mantle abundances of rhenium, osmium, and platinum are in chondritic proportions (to within 3% and 10% uncertainty, respectively, for Re/Os and Pt/Os) and as of 2003 no model of homogeneous accretion has been successful in generating such a result. In order to address these and other issues (e.g., the high nickel, sulfur, and selenium content of the mantle) many have appealed to models of heterogeneous accretion (Morgan *et al.*, 1981; Wänke, 1981; Wänke and Dreibus, 1982;

Ringwood, 1984; O'Neill, 1991b; O'Neill and Palme, 1997).

Heterogeneous accretion models for the formation of the Earth advocate the initial accretion of refractory, less-oxidized components that make up the bulk of the planet (some 50–80%), followed by the accretion of a lower-temperature, more oxidized component (e.g., perhaps comparable to carbonaceous chondrites). The overall nature of the initially refractory material is not well characterized, but it could have affinities to ordinary or enstatite chondrites. These two-component mixing models seek to reconcile the observational constraints from chemical and isotopic studies of the silicate Earth. As of early 2000s, we do not have sufficient data to identify in detail the nature of these two components of accretion if they existed.

Data for sulfur, selenium, and tellurium (the sulfonic elements, the latter two are also strongly chalcophile and sit below sulfur on the periodic table) show that these elements were sequestered into the core in equal proportions (Figure 6). The upper and lower panels of Figure 6 also show that the mantle content of the highly siderophile elements (HSEs)—rhenium and gold—and the platinum group elements (PGEs)—ruthenium, rhodium, palladium, osmium, iridium, and platinum—are depleted to approximately the same level as that for the sulfonic elements. This is consistent with all of these elements being delivered to the mantle by either the entrainment of small amounts of a core material in plumes coming off of the early CMB (McDonough, 1995) or a model that invokes the addition of a late stage veneer component added to the Earth (Kimura *et al.*, 1974; Chou, 1978; Morgan *et al.*, 1981). Quantitatively, it has been demonstrated that either an endogenous delivery mechanism (former model) or an exogenous delivery model (the latter model) is consistent with the early (ca. pre-4.0 Ga) addition of a sulfonic-HSE component to the mantle (McDonough, 1995).

The endogenous delivery mechanism is inefficient in that it requires core entrainment by plumes that arise off of a newly formed CMB. Kellogg and King (1993) and Kellogg (1997) have shown that such plumes can incorporate $\sim<1\%$ of core material and that this material can be re-entrained back into the mantle. However, such plumes would have been considerably more vigorous during the Hadean, assuming a significant temperature contrast across the core–mantle boundary (i.e., established some 10^5–10^8 yr following accretion and core formation) and a higher-temperature state of the planet resulting from accretion, Moon formation, and core separation. Therefore, it is likely that some degree of entrainment of core material into the mantle occurs in the aftermath of establishing a core–mantle boundary.

Walker *et al.* (2002) demonstrated that the primitive upper mantle has $^{187}Os/^{188}Os$ values similar to ordinary and enstatite chondrites, and that this mantle signature is distinct from that of carbonaceous chondrites. These observations translate to different Re/Os values in different chondrite classes (ordinary, enstatite, and carbonaceous), with the mantle having Re/Os a value unlike that of carbonaceous chondrites. This raises the importance of this late addition (i.e., the sulfonic-HSE signature material), given its distinctive composition. Therefore, the diagnostic sulfonic-HSE signature of the mantle reflects either the nature of the very earliest accreted material delivered to the forming Earth (the endogenous model) or that delivered at the final stages of accretion (the exogenous model).

If the sulfonic-HSE signature derives from material extracted from the core, we can use its HSE signature to characterize the nature of material delivered to the early accreting Earth. Standard heterogeneous accretion models argue that this early accreting material is reduced, with compositional characteristics comparable to ordinary chondrites. Thus, the observations of Walker *et al.* (2002) on Re/Os values of the silicate Earth are consistent with the early accretion of a reduced component. Alternatively, the exogenous delivery model (i.e., late veneer component) would contradict standard heterogeneous accretion models, which envisage accretion of an initial reduced component followed by the subsequent accretion of a more oxidized component. Thus, the exogenous model requires yet another, final shift in the oxidation state of the late accretion material.

In summary, core formation was early and fast and was accomplished at mid- to upper-mantle conditions in a hot energetic Earth. Given silicon as the dominant light element component in the core, then core–mantle equilibrium occurred under fairly reducing conditions. (If instead oxygen was the dominant light element component in the core, then core–mantle equilibrium occurred under fairly oxidizing conditions.)

2.15.8 THE INNER CORE, ITS CRYSTALLIZATION, AND CORE–MANTLE EXCHANGE

The solid inner core represents only $\sim5\%$ of the core's mass and $\sim4\%$ of its volume. Geophysical models of the inner core have identified its structure, elastic properties, and modeled its crystallization history. This, without question, is the most remote region of the planet and little is known of its properties and origins. There are no direct insights to be gained from either compositional studies of the mantle or those of meteorites. The high-pressure conditions of the inner core limit the applicability of any insights drawn from analogies with iron meteorites, which were formed at $\lhd$1 GPa conditions.

Labrosse *et al.* (1997, 2001) examined the power budget for the core and its implications for inner core crystallization. These calculations generally find that inner core crystallization began in the latter half of Earth's history (ca. 1–2 Ga) and that some amount of radioactive heating is necessary to extend the age of crystallization back in time (Labrosse *et al.*, 2001). Similarly, Brodholt and Nimmo (2002) concluded that models for inner core crystallization could perhaps be developed with long lifetimes (~2.5 Ga) for inner core crystallization with some potassium in the core producing radioactive heating. A long protracted history for inner core crystallization, however, would satisfy those who view the paleomagnetic record in 3.5 Ga old rocks as evidence for an inner core, which gave rise to the early Earth's geomagnetic field. The fundamental problem with developing an early inner core (i.e., older than 2.5 Ga) is with Earth's cooling rates, and the power needed to drive the geodynamo.

Isotopic studies have also considered ways in which to constrain the timing of inner core crystallization. Walker *et al.* (1995) argued that enrichment of $^{187}Os/^{188}Os$ in some plume-derived systems relative to the ambient mantle was a signature from the outer core delivered by CMB-originating plumes. The origin of this isotopic signature would be due to inner core crystallization, which produces an outer core relatively enriched in ^{187}Re (the parent isotope) but still overall depleted in rhenium and osmium. Following this, Brandon *et al.* (1998) found coupled enrichments in $^{187}Os/^{188}Os$ and $^{186}Os/^{188}Os$ similar to those predicted by Walker *et al.* (1995), which provided further support for Walker's model of inner core crystallization that left behind an outer core relatively enriched in ^{187}Re and ^{190}Pt (the parent isotopes) with respect to osmium. These constraints, however, argued for significant element fractionation due to inner core crystallization that was relatively early and rapid in the Earth's history (<1 Ga) in order to obtain the observed elevated isotopic compositions. Brandon *et al.* (2003) have extended these observations to include the Cretaceous komatiite suite from Gorgona Island and again re-enforced a model of early and rapid growth of the inner core, conclusions that are not mutually exclusive with findings from Labrosse *et al.* (2001) but are more difficult to reconcile.

In summary, there are two models of inner core crystallization: one involving early and rapid growth (osmium isotopic model) and one involving later, slow growth (energy balance model). These models address very different problems and concerns, are mutually independent, and reach somewhat divergent conclusions. The inner core exists and began forming after core formation (i.e., after the first 30 Ma of Earth's history). In addition, the generation and maintenance of a planetary dynamo does not require inner core growth (Stevenson, 2003). Thus, as of 2003, we are unable to resolve the issue of when inner core crystallization began.

A final observation on the amount of core–mantle exchange, albeit on a less sensitive scale, can be gained from studies of peridotites. It is recognized that by ~3.8 Ga, when we begin to have a substantial suite of crust and mantle samples, the mantle's composition is relatively fixed as far as key ratios of lithophile-to-siderophile elements in mantle samples. McDonough and Sun (1995) showed that ratios of Mg/Ni and Fe/Mn in the mantle have been fixed (total of ±15% SD for both ratios) for mantle peridotites spanning the age range 3.8 Ga to present (their Figure 7), which is inconsistent with continued core–mantle exchange. At a finer scale, there is ~20% variation in P/Nd values of Archean to modern basalts and komatiites, which because of the core's high P/Nd value (virtually infinity) and mantle's low value (~70 ± 15) restricts mass exchange between the core and mantle to <1%. Collectively, these and other ratios of lithophile (mantle)-to-siderophile (core) elements bound the potential core–mantle exchange to <1% by mass since core formation. The suggested mass fraction of core–mantle exchange based on Re–Os and Pt–Os isotopic studies is at a much smaller scale by at least two orders of magnitude, the scaling is only relative to the mass of the upwelling plume.

2.15.9 SUMMARY

An estimate of the density deficit in the core is ~5–10% (Boehler, 2000; Anderson and Isaak, 2002); the uncertainty in this estimate is dominantly a function of uncertainties in the pressure and temperature derivatives of EOS data for candidate core materials and knowledge of the temperatures conditions in the core. A tighter constraint on this number will greatly help to refine chemical and petrological models of the core. A density deficit estimate for the inner core is 4–5% (Hemley and Mao, 2001).

The Fe/Ni value of the core (16.5) is well constrained by the limited variation in chondritic meteorites (17.5 ± 0.5) and the mantle ratio (32), as well as the mass fraction of these elements in the two terrestrial reservoirs. The total content of sulfur, carbon, and phosphorus in the core represents only a minor fraction (~2.5 wt.%) of the light element component and this mixture is insufficient to account for the core's density discrepancy. A model composition of the core using silicon as the additional light element in the outer core is preferred over an alternative composition using oxygen as the other light element. Within the limits of our resolving power, either model is tenable.

The trace element content of the core can be determined by using constraints derived from the composition of the mantle and that of chondritic meteorites. This approach demonstrates that there is no geochemical evidence for including any radioactive elements in the core. Relative to the bulk Earth, the core contains about half of the Earth's vanadium and chromium budget and it is equivocal as to whether the core hosts any niobium and tantalum. As compared to iron meteorites, the core is depleted in germanium and strongly depleted in (or void of) gallium. Collectively, the core's chemical signature provides a robust set of restrictions on core formation conditions (i.e., pressure, temperature, and gas fugacity).

The W–Hf isotope system constrains the age of core–mantle differentiation to within the first 30 Ma years of Earth's history (Kleine *et al.*, 2002; Yin *et al.*, 2002). However, the age of inner core crystallization is not resolved.

ACKNOWLEDGMENTS

I thank Rick Carlson for the invitation to contribute to this volume. Also, I am very grateful to Rick Carlson, Rus Hemley, Guy Masters, Hugh O'Neill, Herbert Palme, Bill Minarik and others for review comments on this manuscript and for the many discussions relating to core and mantle phenomena that we have had over the years.

REFERENCES

Allegre C. J., Poirier J.-P., Humler E., and Hofmann A. W. (1995) The chemical composition of the Earth. *Earth Planet. Sci. Lett.* **134**, 515–526.

Anderson D. L., Sammis C., and Jordan T. (1971) Composition and evolution of mantle and core. *Science* **171**, 1103.

Anderson O. L. (2002) The power balance at the core–mantle boundary. *Phys. Earth Planet. Inter.* **131**, 1–17.

Anderson O. L. and Isaak D. G. (2002) Another look at the core density deficit of Earth's outer core. *Phys. Earth Planet. Inter.* **131**, 19–27.

Birch F. (1952) Elasticity and constitution of the Earth's interior. *J. Geophys. Res.* **57**, 227–286.

Birch F. (1964) Density and composition of mantle and core. *J. Geophys. Res.* **69**, 4377–4388.

Birch F. (1966) Evidence from high-pressure experiments bearing on density and composition of Earth. *Geophys. J. Roy. Astro. Soc.* **11**, 256.

Bloxham J. (1995) Global magnetic field. In *Global Earth Physics: A Handbook of Physical Constants*, Vol. AGU Reference Shelf (ed. T. J. Ahrens). American Geophysical Union, Washington, DC, pp. 47–65.

Boehler R. (2000) High-pressure experiments and the phase diagram of lower mantle and core materials. *Rev. Geophys.* **38**, 221–245.

Brandon A., Walker R. J., Morgan J. W., Norman M. D., and Prichard H. (1998) Coupled ^{186}Os and ^{187}Os evidence for core–mantle interaction. *Science* **280**, 1570–1573.

Brandon A. D., Walker R. J., Puchtel I. S., Becker H., Humayun M., and Revillon S. (2003) $^{186}Os/^{187}Os$ systematics of Gorgona Island komatiites: implications for early growth of the inner core. *Earth Planet. Sci. Lett.* **206**, 411–426.

Brearley A. J. and Jones R. H. (1998) Chondritic meteorites. In *Planetary Materials* (ed. J. J. Papike). Mineralogical Soc. Amer., Washington, DC, vol. 36, pp. 3-01–3-398.

Brodholt J. and Nimmo F. (2002) Earth science—core values. *Nature* **418**(6897), 489–491.

Brown J. M. and McQueen R. G. (1986) Phase transitions Güneisen parameter and elasticity for shocked iron between 77 GPa and 400 GPa. *J. Geophys. Res.* **91**, 7485–7494.

Brush S. G. (1980) Discovery of the Earth's core. *Am. J. Phys.* **48**, 705–724.

Buffett B. A. (2000) Earth's core and the geodynamo. *Science* **288**(5473), 2007–2012.

Buffett B. A. and Bloxham J. (2002) Energetics of numerical geodynamo models. *Geophys. J. Int.* **149**, 211–224.

BVSP (1981) *Basaltic Volcanism on the Terrestrial Planets*. Pergamon, New York.

Chabot N. L. and Drake M. J. (1999) Potassium solubility in metal: the effects of composition at 15 kbar and 1,900 °C on partitioning between iron alloys and silicate melts. *Earth Planet. Sci. Lett.* **172**, 323–335.

Chou C. L. (1978) Fractionation of siderophile elements in the Earth's upper mantle. *Proc. Lunar Sci. Conf.* **9**, 219–230.

Crozaz G. and Lundberg L. (1995) The origin of oldhamite in unequilibrated enstatite chondrites. *Geochim. Cosmochim. Acta* **59**, 3817–3831.

Drake M. J., Newsom H. E., and Capobianco C. J. (1989) V, Cr, and Mn in the Earth, Moon, EPB and SPB and the origin of the Moon: experimental studies. *Geochim. Cosmochim. Acta* **53**, 2101–2111.

Dreibus G. and Palme H. (1996) Cosmochemical constraints on the sulfur content in the Earth's core. *Geochim. Cosmochim. Acta* **60**(7), 1125–1130.

Dreibus G. and Wänke H. (1979) On the chemical composition of the moon and the eucrite parent body and comparison with composition of the Earth: the case of Mn, Cr and V. *Lunar Planet Sci. (Abstr.)* **10**, 315–317.

Dziewonski A. and Anderson D. L. (1981) Preliminary reference Earth model. *Phys. Earth Planet. Inter.* **25**, 297–356.

Dziewonski A. and Gilbert F. (1972) Observations of normal modes from 84 recordings of the Alsakan earthquake of 1964 March 28. *Geophys. J. Roy. Astron. Soc.* **27**, 393–446.

Galer S. J. G. and Goldstein S. L. (1996) Influence of accretion on lead in the Earth. In *Earth Processes: Reading the Isotopic Code* (eds. A. Basu and S. R. Hart). American Geophysical Union, Washington, DC, pp. 75–98.

Gessmann C. K. and Rubie D. C. (2000) The origin of the depletion of V, Cr, and Mn in the mantles of the Earth and Moon. *Earth Planet. Sci. Lett.* **184**, 95–107.

Gessmann C. K. and Wood B. J. (2002) Potassium in the Earth's core? *Earth Planet. Sci. Lett.* **200**(1–2), 63–78.

Gies F. and Gies J. (1994) *Cathedral, Forge, and Waterwheel: Technology and Invention in the Middle Ages*. Harper Collins, New York.

Glatzmaier G. A. (2002) Geodynamo simulations: How realistic are they? *Ann. Rev. Earth Planet. Sci.* **30**, 237–257.

Gubbins D. (1977) Energetics of the Earth's core. *J. Geophys.* **47**, 453–464.

Gubbins D., Masters T. G., and Jacobs J. A. (1979) Thermal evolution of the Earth's core. *Geophys. J. Roy. Astron. Soc.* **59**(1), 57–99.

Gutenberg B. (1914) Über Erdbebenwellen VIIA. Nachr. Ges. Wiss. Göttingen Math. Physik. Kl, 166. *Nachr. Ges. Wiss. Göttingen Math. Physik.* **125**, 116.

Hall H. T. and Rama Murthy V. (1971) The early chemical history of the Earth: some critical elemental fractionations. *Earth Planet. Sci. Lett.* **11**, 239–244.

Harper C. L. and Jacobsen S. B. (1996) Evidence for Hf-182 in the early solar system and constraints on the timescale of terrestrial accretion and core formation. *Geochim. Cosmochim. Acta* **60**(7), 1131–1153.

Hemley R. J. and Mao H. K. (2001) *In situ* studies of iron under pressure: new windows on the Earth's core. *Int. Geol. Rev.* **43**(1), 1–30.

Herndon J. M. (1996) Substructure of the inner core of the Earth. *Proc. Natl. Acad. Sci. USA* **93**(2), 646–648.

Hillgren V. J., Gessmann C. K., and Li J. (2000) An experimental perspective on the light element in the earth's core. In *Origin of the Earth and Moon* (eds. R. M. Canup and K. Righter). University of Arizona Press, Tucson, pp. 245–263.

Horan M. F., Smoliar M. I., and Walker R. J. (1998) ^{182}W and $^{187}Re-^{187}Os$ systematics of iron meteorites: chronology for melting, differentiation and crystallization in asteroids. *Geochim. Cosmochim. Acta* **62**, 545–554.

Javoy M. (1995) The integral enstatite chondrite model of the Earth. *Geophys. Res. Lett.* **22**, 2219–2222.

Jeffreys H. (1926) The rigidity of the Earth's central core. *Mon. Not. Roy. Astron. Soc. Geophys. Suppl.* **1**(7), 371–383.

Jephcoat A. and Olson P. (1987) Is the inner core of the Earth pure iron. *Nature* **325**(6102), 332–335.

Jones J. H. and Drake M. J. (1986) Geochemical constraints on core formation in the Earth. *Nature* **322**, 221–228.

Kellogg L. H. (1997) Growing the Earth's D" layer: effect of density variations at the core–mantle boundary. *Geophys. Res. Lett.* **24**, 2749–2752.

Kellogg L. H. and King S. D. (1993) Effect of mantle plumes on the growth of D'' by reaction between the core and mantle. *Geophys. Res. Lett.* **20**, 379–382.

Kimura K., Lewis R. S., and Anders E. (1974) Distribution of gold and rhenium between nickel–iron and silicate melts: implications for the abundances of siderophile elements on the Earth and Moon. *Geochim. Cosmochim. Acta* **38**, 683–701.

Kleine T., Münker C., Mezger K., and Palme H. (2002) Rapid accretion and early core formation on asteroids and the terrestrial planets from Hf–W chronometry. *Nature* **418**, 952–955.

Labrosse S., Poirier J. P., and LeMouel J. L. (1997) On cooling of the Earth's core. *Phys. Earth Planet. Inter.* **99**(1–2), 1–17.

Labrosse S., Poirier J. P., and LeMouel J. L. (2001) The age of the inner core. *Earth Planet. Sci. Lett.* **190**, 111–123.

Larimer J. W. (1988) The cosmochemical classification of the elements. In *Meteorites and the Early Solar System* (eds. J. F. Kerridge and M. S. Matthews). University of Arizona Press, Tucson, pp. 375–389.

Lee D.-C. and Halliday A. N. (1995) Hafnium–tungsten chronometry and the timing of terrestrial core formation. *Nature* **378**, 771–774.

Lehmann I. (1936) *P' Publ. Bur. Cent. Seism. Int. Ser. A* **14**, 87–115.

Lewis J. S. (1971) Consequences of presence of sulfur in core of Earth. *Earth Planet. Sci. Lett.* **11**, 130.

Li J. and Agee C. B. (1996) Geochemistry of mantle–core differentiation at high pressure. *Nature* **381**(6584), 686–689.

Lodders K. (1995) Alkali elements in the Earths core—evidence from enstatite meteorites. *Meteoritics* **30**(1), 93–101.

Macdonald G. J. F. and Knopoff L. (1958) On the chemical composition of the outer core. *Geophys. J. Roy. Astron. Soc.* **1**(4), 284–297.

Mao H. K., Wu Y., Chen L. C., Shu J. F., and Jephcoat A. P. (1990) Static compression of iron to 300 GPa and $Fe_{0.8}Ni_{0.2}$ to 260 GPa: implications for composition of the core. *J. Geophys. Res.* **95**, 21737–21742.

Masters T. G. and Shearer P. M. (1995) Seismic models of the Earth: elastic and anelastic. In *Global Earth Physics: A Handbook of Physical Constants*. Vol. AGU Reference Shelf (ed. T. J. Ahrens). American Geophysical Union, Washington, DC, pp. 88–103.

McDonough W. F. (1990) Comment on "Abundance and distribution of gallium in some spinel and garnet lherzolites" by D. B. McKay and R. H. Mitchell. *Geochim. Cosmochim. Acta* **54**, 471–473.

McDonough W. F. (1991) Partial melting of subducted oceanic crust and isolation of its residual eclogitic lithology. *Phil. Trans. Roy. Soc. London A* **335**, 407–418.

McDonough W. F. (1995) An explanation for the abundance enigma of the highly siderophile elements in the earth's mantle. *Lunar Planet. Sci. (Abstr.)* **26**, 927–928.

McDonough W. F. (1999) Earth's Core. In *Encyclopedia of Geochemistry* (eds. C. P. Marshall and R. W. Fairbridge). Kluwer Academic, Dordrecht, pp. 151–156.

McDonough W. F. (2001) The composition of the Earth. In *Earthquake Thermodynamics and Phase Transformations in the Earth's Interior* (eds. R. Teisssyere and E. Majewski). Academic Press, San Diego, vol. 76, pp. 3–23.

McDonough W. F. and Sun S.-S. (1995) The composition of the Earth. *Chem. Geol.* **120**, 223–253.

McDonough W. F., McCulloch M. T., and Sun S.-S. (1985) Isotopic and geochemical systematics in Tertiary–Recent basalts from southeastern Australia and implications for the evolution of the sub-continental lithosphere. *Geochim. Cosmochim. Acta* **49**, 2051–2067.

Meisel T., Walker R. J., and Morgan J. W. (1996) The osmium isotopic composition of the Earth's primitive upper mantle. *Nature* **383**, 517–520.

Merrill R. T., McElhinney M. W., and McFadden P. L. (1996) *The Magnetic Field of the Earth*. Academic Press, San Diego.

Morgan J. W., Wandless G. A., Petrie R. K., and Irving A. J. (1981) Composition of the Earth's upper mantle: 1. Siderophile trace-elements in Ultramafic Nodules. *Tectonophysics* **75**(1–2), 47–67.

Niu Y. and Batiza R. (1997) Trace element evidence from seamounts for recycled oceanic crust in eastern Pacific mantle. *Earth Planet. Sci. Lett.* **148**, 471–783.

O'Neill H. S. C. (1991a) The origin of the Moon and the early history of the Earth—a chemical model: Part 1. The Moon. *Geochim. Cosmochim. Acta* **55**, 1135–1157.

O'Neill H. S. C. (1991b) The origin of the Moon and the early history of the Earth—a chemical model: Part 2. The Earth. *Geochim. Cosmochim. Acta* **55**, 1159–1172.

O'Neill H. S. C. and Palme H. (1997) Composition of the silicate Earth: implications for accretion and core formation. In *The Earth's Mantle: Structure, Composition, and Evolution—The Ringwood Volume* (ed. I. Jackson). Cambridge University Press, Cambridge, pp. 1–127.

O'Neill H. S., Canil D., and Rubie D. C. (1998) Oxide–metal equilibria to 2,500 degrees C and 25 GPa: implications for core formation and the light component in the Earth's core. *J. Geophys. Res.-Solid Earth* **103**(B6), 12239–12260.

Palme H. (2001) Chemical and isotopic heterogeneity in protosolar matter. *Phil. Trans. Roy. Soc. London* **359**, 2061–2075.

Palme H., Larimer J. W., and Lipchutz M. E. (1988) Moderately volatile elements. In *Meteorites and the Early Solar System* (eds. J. F. Kerridge and M. S. Matthews). University of Arizona Press, Tucson, pp. 436–461.

Ringwood A. E. (1959) On the chemical evolution and density of the planets. *Geochim. Cosmochim. Acta* **15**, 257–283.

Ringwood A. E. (1966) The chemical composition and origin of the Earth. In *Advances in Earth Sciences* (ed. P. M. Hurley). MIT Press, Cambridge, pp. 287–356.

Ringwood A. E. (1984) The Bakerian Lecture, 1983—the Earth's core—its composition, formation and bearing upon the origin of the Earth. *Proc. Roy. Soc. London Ser. A: Math. Phys. Eng. Sci.* **395**(1808), 1–46.

Ringwood A. E., Kato T., Hibberson W., and Ware N. (1990) High-pressure geochemistry of Cr, V, and Mn and implications for the origin of the Moon. *Nature* **347**, 174–176.

Rubin A. E. (1997) Mineralogy of meteorite groups. *Meteorit. Planet. Sci.* **32**, 231–247.

Rudnick R. L., Barth M., Horn I., and McDonough W. F. (2000) Rutile-bearing refractory eclogites: missing link between continents and depleted mantle. *Science* **287**, 278–281.

Schoenberg R., Kamber B. S., Collerson K. D., and Eugster O. (2002) New W-isotope evidence for rapid terrestrial accretion and very early core formation. *Geochim. Cosmochim. Acta* **66**, 3151–3160.

Stacey F. D. (1992) *Physics of the Earth*. Brookfield Press, Brisbane.

Stevenson D. J. (2003) Planetary magnetic fields. *Earth Planet. Sci. Lett.* **208**, 1–11.

Wade J. and Wood B. J. (2001) The Earth's "missing" niobium may be in the core. *Nature* **409**(6816), 75–78.

Walker R. J., Morgan J. W., and Horan M. F. (1995) Osmium-187 enrichments in some plumes: evidence for core–mantle interactions. *Science* **269**, 819–822.

Walker R. J., Morgan J. W., Beary E., Smoliar M. I., Czamanske G. K., and Horan M. F. (1997) Applications of the $^{190}P-^{186}Os$ isotope system to geochemistry and cosmochemistry. *Geochim. Cosmochhim. Acta* **61**, 4799–4808.

Walker R. J., Horan M. F., Morgan J. W., Becker H., Grossman J. N., and Rubin A. E. (2002) Comparative $^{187}Re-^{187}Os$ systematics of chondrites: implications regarding early solar system processes. *Geochim. Cosmochim. Acta* **66**, 4187–4201.

Wänke H. (1981) Constitution of terrestrial planets. *Phil. Trans. Roy. Soc. London A* **303**, 287–302.

Wänke H. (1987) Chemistry and accretion of Earth and Mars. *Bull. Soc. Geol. Fr.* **3**(1), 13–19.

Wänke H. and Dreibus G. (1982) Chemical and isotopic evidence for the early history of the Earth–Moon system. In *Tidal Friction and the Earth's Rotation* (eds. P. Brosche and J. Sundermann). Springer, Berlin, pp. 322–344.

Wänke H. and Dreibus G. (1988) Chemical composition and accretion history of terrestrial planets. *Phil. Trans. Roy. Soc. London A* **325**, 545–557.

Washington H. S. (1925) The chemical composition of the Earth. *Am. J. Sci.* **9**, 351–378.

Wasson J. T. (1985) *Meteorites, Their Record of Early Solar-system History*. W. H. Freeman, New York, 267pp.

Wasson J. T. and Kallemeyn G. W. (1988) Compositions of chondrites. *Phil. Trans. Roy. Soc. London A* **325**, 535–544.

Williams Q. and Hemley R. J. (2001) Hydrogen in the deep Earth. *Ann. Rev. Earth Planet. Sci.* **29**, 365–418.

Williamson E. D. and Adams L. H. (1923) Density distribution in the Earth. *J. Washington Acad. Sci.* **13**, 413–428.

Wood B. J. (1993) Carbon in the Core. *Earth Planet. Sci. Lett.* **117**(3–4), 593–607.

Yin Q., Jacobsen S. B., Yamashita K., Blichert-Toft J., Télouk P., and Albarede F. (2002) A short timescale for terrestrial planet formation from Hf–W chronometry of meteorites. *Nature* **418**, 949–952.

Yoder C. F. (1995) Astrometric and geodetic properties of the Earth and the solar system. In *Global Earth Physics: A Handbook of Physical Constants*. Vol. AGU Reference Shelf (ed. T. J. Ahrens). American Geophysical Union, Washington, DC, pp. 1–31.

Volume Subject Index

The index is in letter-by-letter order, whereby hyphens and spaces within index headings are ignored in the alphabetization (e.g. Arabian–Nubian Shield precedes Arabian Sea). Terms in parentheses are excluded from the initial alphabetization. In line with normal materials science practice, compound names are not inverted but are filed under substituent prefixes.

The index is arranged in set-out style, with a maximum of three levels of heading. Location references refer to the page number. Major discussion of a subject is indicated by bold page numbers. Page numbers suffixed by *f* or *t* refer to figures or tables.

Printed and bound by CPI Group (UK) Ltd, Croydon, CR0 4YY
08/06/2025
01896869-0018